2020 22nd European Conference on Power Electronics and Applications (EPE'20 ECCE Europe)

Lyon, France
7-11 September 2020

Pages 1-685

IEEE Catalog Number: CFP20850-POD
ISBN: 978-1-7281-9807-1

Copyright © 2020, EPE Association
All Rights Reserved

*** *This is a print representation of what appears in the IEEE Digital Library. Some format issues inherent in the e-media version may also appear in this print version.*

IEEE Catalog Number: CFP20850-POD
ISBN (Print-On-Demand): 978-1-7281-9807-1
ISBN (Online):   978-9-0758-1536-8

Additional Copies of This Publication Are Available From:

Curran Associates, Inc
57 Morehouse Lane
Red Hook, NY 12571 USA
Phone: (845) 758-0400
Fax: (845) 758-2633
E-mail: curran@proceedings.com
Web: www.proceedings.com

2020 22nd European Conference on Power Electronics and Applications (EPE'20 ECCE Europe)

Lyon, France
7-11 September 2020

Pages 1-685

IEEE Catalog Number: CFP20850-POD
ISBN: 978-1-7281-9807-1

TABLE OF CONTENTS

VALIDATION OF THERMAL STRESS MODELING IN PV INVERTERS UNDER MISSION PROFILE OPERATION 1
Ariya Sangwongwanich, Huai Wang, Frede Blaabjerg

ON THE LIMITATIONS OF USING A LTI MODELLING APPROACH FOR CONTROL TUNING OF VSC-HVDC SYSTEMS 9
Pablo Briff, Julián Freytes, Guillaume De-Preville, Jiaqi Li, Omar Jasim

A VOLTAGE CONTROL METHOD FOR POWER DISTRIBUTION LINES UTILIZING DISPERSED CUSTOMER RESOURCES 19
Hiroki Ishihara, Kaho Nada, Miwako Tanaka, Sadayuki Inoue, Akiko Kuwata, Tomihiro Takano

PERFORMANCE COMPARISON BETWEEN SIC AND SI INVERTER MODULES IN AN ELECTRICAL VARIABLE TRANSMISSION APPLICATION 27
Mauricio Dalla Vecchia, Simon Ravyts, Florian Verbelen, Jeroen Tant, Peter Sergeant, Johan Driesen

SEAMLESS INTEGRATION OF FEEDFORWARD AND FEEDBACK CONTROL OF BALANCE OF ARM CAPACITOR VOLTAGES IN STATCOMS BASED ON CHAIN LINKS OF H BRIDGE MODULES 37
D. Basic, N. Lapassat

ASYNCHRONIZED ELECTROMECHANICAL CONVERTER IN THE ELECTRICAL SUPPLY SYSTEM OF POWERFUL ENERGY CONSUMERS 47
Aleksey G. Vorontsov, Mikhail V. Pronin, Anastasiia D. Stotckaia, Vasiliy V. Glushakov, Pavel V. Sokur

SYMMETRIC AND ASYMMETRIC OPERATING MODES OF HYBRID CASCADE FREQUENCY CONVERTERS 56
Aleksey G. Vorontsov, Vasiliy V. Glushakov, Mikhail V. Pronin, Anastasiia D. Stotckaia

SYSTEM FREQUENCY DYNAMIC RESPONSE OF A NOVEL, SELF-SYNCHRONIZING INVERTER IN A HIGH RENEWABLE PENETRATION GRID 65
Christian Perenyi, Moath Alqatamin, Thibaut Harzig, Michael McIntyre, Brandon M. Grainger

ROTOR POSITION ESTIMATION WITH HALL-EFFECT SENSORS IN BEARINGLESS DRIVES 75
Patricio Peralta, Jacopo Leo, Yves Perriard

NON-UNIT ROCOV SCHEME FOR PROTECTION OF MULTI-TERMINAL HVDC SYSTEMS 85
María José Pérez-Molina, Pablo Eguia, Marene Larruskain, Garikoitz Buigues, Esther Torres

MODELLING OF CONVERTER SYSTEMS PARALLELED VIA INTERPHASE TRANSFORMERS IN CYCLIC CASCADE TOPOLOGY AND OPTIMIZATION OF PWM CARRIER SHIFTS 95
D. Basic, H. Baërd, S. Siala

MEASUREMENT AND CALCULATION METHOD OF WIRELESS POWER TRANSFER COIL EQUIVALENT SERIES RESISTANCE UNDER THE VEHICLE 105
Norihito Kimura, Hiroaki Yuasa

DESIGN OF A CIRCUMSCRIBING POLYGON WIDE BANDGAP BASED INTEGRATED MODULAR MOTOR DRIVE TOPOLOGY WITH THERMALLY DECOUPLED WINDINGS AND POWER CONVERTERS 115

Abdalla Hussein Mohamed, Hendrik Vansompel, Peter Sergeant

LIMITS OF ENHANCED DESATURATION DETECTION METHOD WITH ADAPTIVE BLANKING FOR GAN HEMTS 124

Jan Schmitz, Markus Meißner, Steffen Bernet

CURRENT CONTROL OF A GRID-CONNECTED SINGLE-PHASE VOLTAGE-SOURCE INVERTER WITH LCL FILTER 134

Alfonso Parreño Torres, Fco. Javier López-Alcolea, Pedro Roncero-Sánchez, Javier Vázquez, Emilio J. Molina-Martínez, Felix García-Torres

FOUR SWITCH BUCK/BOOST CONVERTER FOR DC MICROGRID APPLICATIONS 143

Matthias Schulz, Nico Schleippmann, Kilian Gosses, Bernd Wunder, Martin März

STABILITY INVESTIGATION OF THREE-PHASE GRID-TIED PV INVERTERS WITH IMPEDANCE-BASED METHOD 153

Zhiqing Yang, Wanchao Gou, Xian Luo, Chirag Shah, Nurhan Rizqy Averous, Rik W. De Doncker

STABILITY INVESTIGATION OF LARGE-SCALE PV PARKS WITH EIGENVALUE-BASED METHOD 163

Zhiqing Yang, Christian Bendfeld, Jin Qiang, Benedict Mortimer, Rik W. De Doncker

COMPACT CORE LOSS MODEL BASED ON AN EFFECTIVE FREQUENCY FOR ARBITRARY CORE EXCITATIONS INCLUDING DC-BIAS 173

Erika Stenglein, Manfred Albach, Thomas Dürbaum

ASSESSMENT OF AGING AND PERFORMANCE DEGRADATION OF SUPERCAPACITORS INTEGRATED INTO A MODULAR MULTILEVEL CONVERTER 183

F. Errigo, L. Chédot, F. Morel, P. Venet, A. Sari, A. Hijazi, R. A. Peña

SEPARATION OF MAGNETIC FLUX DENSITY TRAJECTORIES INTO SUBLOOPS FOR THE PREDICTION OF HYSTERESIS LOSS 193

Erika Stenglein, Manfred Albach, Thomas Dürbaum

INFLUENCE OF GENERALIZED DISCONTINUOUS PULSE WIDTH MODULATION (GDPWM) ON THE DC-LINK CURRENT AND VOLTAGE RIPPLE IN BATTERY-FED PWM INVERTER SYSTEMS 203

Panagiotis Mantzanas, Alexander Bucher, Daniel Kuebrich, Alexander Pawellek, Christian Hasenohr, Harald Hofmann, Thomas Duerbaum

AUTOMATED DESIGN METHOD FOR SINE WAVE FILTERS IN MOTOR DRIVE APPLICATIONS WITH SIC-INVERTERS 213

Thorben Schobre, Regine Mallwitz

A SYMMETRICAL BOOST CONVERTER WITH REDUCED COMMON-MODE LEAKAGE CURRENTS FOR EV APPLICATIONS 223

Caniggia Viana, Netan Yakop, Damien Frost, Peter Lehn

MODELING AND ANALYSIS OF CONDUCTED EMI ON FLYBACK CONVERTER USING POWER MANAGEMENT IC WITH CHAOTIC SUPPRESSION EMI 231

Diao Jiaqi, Yang Ru, Liu Zuolian, Yang Hong, Jie Hai

HIGH PERFORMANCE DRIVE INVERTER FOR AN ELECTRIC TURBO COMPRESSOR IN FUEL CELL APPLICATIONS 241
  N. Langmaack, G. Tareilus, R. Mallwitz

DEVELOPMENT OF AN ALGORITHM FOR THE AUTOMATION OF THE MODELLING PROCESS OF POWER CONVERTERS 251
  Jon Anzola, Iosu Aizpuru, Asier Arruti

A NOVEL FULLY DISTRIBUTED COST OPTIMAL CONTROL METHOD FOR DC MICROGRID 260
  Qingping Xia, Hua Han, Yao Liu, Zhangjie Liu, Yao Sun, Mei Su

MEASUREMENT OF DYNAMIC ON-STATE RESISTANCE OF HIGH-VOLTAGE GAN-HEMTS UNDER REAL APPLICATION CONDITIONS 266
  Benedikt Kohlhepp, Carsten Kuring, Stefan Peller, Daniel Kübrich

ANALYSIS OF DC-SIDE FAULT RESPONSE OF MMCS WITH CONTROLLED FAULT BLOCKING CAPABILITY FOR DIFFERENT TRANSMISSION LINE TYPES 276
  Willem Leterme, Paul D. Judge, Tim C. Green

A HYBRID SERIES-PARALLEL MICROGRID AND ITS LOW-DEPENDENT COMMUNICATION CONTROL 285
  Lang Li, Yao Sun, Hua Han, Mei Su

ADAPTIVE VOLTAGE CONTROL OF ISLANDED RES-BASED RESIDENTIAL MICROGRID WITH INTEGRATED FLYWHEEL/BATTERY HYBRID ENERGY STORAGE SYSTEM 292
  Linda Barelli, Gianni Bidini, Ermanno Cardelli, Dana-Alexandra Ciupageanu, Andrea Ottaviano, Dario Pelosi, Simone Castellini, Gheorghe Lazaroiu

AN IMPROVED λ-CONSENSUS CONTROL METHOD FOR DC MICROGRIDS 302
  Siqi Fu, Yao Sun, Zhangjie Liu, Hua Han, Mei Su

DECREASE OF POWER ELECTRONIC SWITCHING LOSSES USING VARIABLE SWITCHING EVENTS 307
  Hannes Ramm, Michael Homann, Torben A. Schulze, Faical Turki, Heiko Rabba

OPTIMIZATION OF MEDIUM-FREQUENCY TRANSFORMERS WITH LARGE CAPACITY AND HIGH INSULATION REQUIREMENT 317
  Xuan Guo, Chi Li, Zedong Zheng, Yongdong Li

IMPROVED SOC BALANCING AND ACTIVE POWER SHARING CONTROL METHOD IN HIGHLY RESISTIVE LINE MICROGRID 326
  Yuanhao Zhu, Hua Han, Guangze Shi, Zhangjie Liu, Yao Sun, Mei Su

TECHNO-ECONOMIC ANALYSIS OF SECOND-LIFE LITHIUM-ION BATTERIES INTEGRATION IN MICROGRIDS 332
  Camille Birou, Xavier Roboam, Hugo Radet, Fabien Lacressonnière

DESIGN, MODELLING, AND TEST OF A SOLID-STATE MAIN BREAKER FOR HYBRID DC CIRCUIT BREAKER 342
  Jiawen Xi, Xiaoze Pei, Xianwu Zeng, Liyong Niu

MODEL PREDICTIVE CONTROL FOR THREE-PHASE SPLIT-SOURCE INVERTER 352
  Youssuf Elthokaby, Islam Mohamed, Naser Abdel-Rahim

HARDWARE IMPLEMENTATION STUDY OF VARIABLE SPEED WIND-TURBINE-DFIG IN STAND-ALONE MODE 362
Fayssal Amrane, Bruno Francois, Azeddine Chaiba

INFLUENCE OF WIRE-BONDING LAYOUT ON RELIABILITY IN IGBT MODULE 370
Lubin Han, Lin Liang, Wei Xin, Fang Luo

RAIL POTENTIAL CALCULATION MODEL FOR DC RAILWAY POWER SUPPLY EQUIPPED WITH VOLTAGE LIMITING DEVICE 377
Shota Kimura, Tsutomu Miyauchi, Kenji Oguma, Hirotaka Takahashi, Keiko Teramura

HOMOGENIZATION OF CURRENT DISTRIBUTION IN PARALLEL CONNECTION OF INTERLEAVED WINDING LAYERS OF HIGH-FREQUENCY TRANSFORMERS BY OPTIMIZING DISTANCE BETWEEN WINDING LAYERS 386
Ryo Murata, Tomohide Shirakawa, Kazuhiro Umetani, Eiji Hiraki, Hiroto Mizutani, Takaaki Takahara, Osamu Mori

REAL-TIME PARAMETERS IDENTIFICATION OF LITHIUM-ION BATTERIES MODEL TO IMPROVE THE HIERARCHICAL MODEL PREDICTIVE CONTROL OF BUILDING MICROGRIDS 396
Daniela Yassuda Yamashita, Ionel Vechiu, Jean-Paul Gaubert

IMPACT OF DC FAULT BLOCKING CAPABILITY ON THE SIZING OF THE DC-DC MODULAR MULTILEVEL CONVERTER 406
J. D. Paez, F. Morel, S. Bacha, Piotr Dworakowski, D. Frey

OPTIMIZATION OF HIGH FREQUENCY MAGNETIC DEVICES WITH CONSIDERATION OF THE EFFECTS OF THE MAGNETIC MATERIAL, THE CORE GEOMETRY AND THE SWITCHING FREQUENCY 416
Sobhi Barg, Muhammad Farhan Alam, Kent Bertilsson

REAL TIME CONTROL HARDWARE IN THE LOOP TEST OF A NOVEL MVDC SOLID-STATE BREAKER 424
Alessio Clerici, Riccardo Chiumeo, Chiara Gandolfi

IGBT LIFETIME ESTIMATION IN A MODULAR MULTILEVEL CONVERTER FOR BIDIRECTIONAL POINT-TO-POINT HVDC APPLICATION 433
Diego Velazco, Guy Clerc, Emmanuel Boutleux, François Wallart, Laurent Chédot

OPTIMIZATION DESIGN FOR SIC DRIFT STEP RECOVERY DIODE (DSRD) 443
Xiaoxue Yan, Lin Liang, Ziyue Wang, Guoqiang Tan

DISCRETE SUPER-TWISTING SLIDING MODE CURRENT CONTROLLER FOR INDUCTION MOTOR DRIVES 450
Tianqing Wang, Bo Wang, Yong Yu, Yangming Zhu, Dianguo Xu

NEW GRID-CONNECTED MULTILEVEL BOOST CONVERTER TOPOLOGY WITH INHERENT CAPACITORS VOLTAGE BALANCING USING MODEL PREDICTIVE CONTROLLER 460
Rasoul Shalchi Alishah, Kent Bertilsson, Frede Blaabjerg, Mohd. Ali Jagabar Sathik, Ali Yahya Rezaee

DCM OPERATION OF SINGLE-SWITCH HIGH STEP-UP DC-DC CONVERTER WITH THREE-WINDING COUPLED INDUCTOR 467
Masataka Minami, Genki Hase

POWER LOSSES CALCULATION FOR MEDIUM VOLTAGE DC/DC CURRENT-FED SOLID STATE TRANSFORMER FOR BATTERY GRID-CONNECTED 471
E. K. Hussain, Mohammad Abusara, S. M. Sharkh

MODELLING AND EXPERIMENTAL VALIDATION OF A POLE-TO-GROUND PROTECTION DEVICE IN LOW VOLTAGE DC MICROGRIDS 480
L. Hallemans, G. Govaerts, G. Van Den Broeck, S. Ravyts, M. M. Alam, P. Van Tichelen, J. Driesen

DESIGN OF A DUAL ACTIVE BRIDGE CONVERTER FOR ON-BOARD VEHICLE CHARGERS USING GAN AND INTO TRANSFORMER INTEGRATED SERIES INDUCTANCE 490
K. Siebke, M. Giacomazzo, R. Mallwitz

AN EXPERIMENTAL ANALYSIS OF CIRCULATING CURRENT CONTROL CIRCUIT FOR OUTPUT POWER FROM VIBRATION GENERATOR FOR VIBRATION INCLUDING THE THIRD HARMONICS 498
Masataka Minami, Akito Nakagaki, Genki Hase

IMPLEMENTATION OF CONTROL STRATEGY FOR STEP-DOWN DC-DC CONVERTER BASED ON PIEZOELECTRIC RESONATOR 503
Mustapha Touhami, Ghislain Despesse, François Costa, Benjamin Pollet

THERMAL IMPEDANCES AND TEMPERATURE SENSORS: A COMBINED APPROACH FOR A NOVEL THERMAL MODEL OF POWER SEMICONDUCTORS 512
Maria De Lauretis, Jonas Millinger, Erik Baker, Martin Karlsson, Diane -Perle Sandik

A 3A LOW VOLTAGE LASER DIODE DRIVER IC IN A CMOS TECHNOLOGY FOR AN ITOF-BASED 3D IMAGE SENSOR 522
Romain David, Bruno Allard, Xavier Branca, Charles Joubert

COMPARISON OF DECOUPLING TECHNIQUES VIA DISCRETE LUENBERGER STYLE OBSERVER FOR VOLTAGE ORIENTED CONTROL 532
Gyanendra Kumar Sah, Michael Schütt, Hans-Günter Eckel

VARIABLE SWITCHING POINT PARALLEL PREDICTIVE CURRENT CONTROL (VSP3CC) FOR INDUCTION MOTOR 542
Qing Chen, Ralph Kennel

OPERATION OF AN EXTERNALLY EXCITED SYNCHRONOUS MACHINE WITH A HYBRID MULTILEVEL INVERTER 551
C. Terbrack, J. Stöttner, C. Endisch

A FACILITY FOR MIXED FLOWING GAS TESTING OF AND EXPERIMENTATION WITH POWER ELECTRONIC COMPONENTS AND SYSTEMS 563
Juuso Rautio, Janne Jäppinen, Tommi J. Kärkkäinen, Markku Niemelä, Pertti Silventoinen, Mika Kiviniemi, Joonas Leppänen, Jonny Ingman

IMPACT OF IMPLEMENTATION OF AUXILIARY BIAS-WINDINGS ON CONTROLLABLE INDUCTORS FOR POWER ELECTRONIC CONVERTERS 571
Jonas Pfeiffer, Pierre Küster, Yeliz Erenler, Ziyad H. S. Qashlan, Peter Zacharias

APPROXIMATED SLIDING-MODE CONTROL OF PARALLEL-CONNECTED GRID INVERTERS 581
Albrecht Gensior

EQUIVALENT MODEL AND CONTROL OF A NEUTRAL POINT SUPPLY SYNRM DRIVE.............. 590
Xiaokang Zhang, Jean-Yves Gauthier, Xuefang Lin-Shi

IMPROVEMENTS ON SIGNAL-TO-NOISE RATIO IN FEEDBACK MEASUREMENT IN
DC/DC CONVERTERS.. 598
Fernando Davalos Hernandez, Federico Ibanez, Sebastian Gutierrez, Wilmar Martinez

APPROACH OF AN ACTIVE DEVICE PROTECTION FOR DRIVE INVERTERS AGAINST
SHORT CIRCUIT FAULTS IN AN OPEN INDUSTRIAL DC GRID.. 608
Simon Puls, Urs Obernolte, Martin Ehlich, Holger Borcherding

A NEW DESIGN OF AN AIR CORE TRANSFORMER FOR ELECTRIC VEHICLE ON-
BOARD CHARGER.. 618
Valentin Rigot, Tanguy Phulpin, Daniel Sadarnac, Jihen Sakly

ENABLING FOIL WINDINGS OF MEDIUM-FREQUENCY TRANSFORMERS FOR HIGH
CURRENTS 627
Thomas B. Gradinger, Uwe Drofenik, Filip Grecki

A HIGH-EFFICIENCY WIRELESS POWER TRANSFER SYSTEM FOR UNMANNED
AERIAL VEHICLE CONSIDERING CARBON FIBER BODY 637
Kai Song, Peng Zhang, Zhengxin Chen, Guang Yang, Jinhai Jiang, Chunbo Zhu

ANALYTICAL COMPUTATION OF NORMAL AND FAULT-TOLERANT ACTIVE SHORT
CIRCUIT OPERATION OF ANISOTROPIC SYNCHRONOUS DOUBLE STAR MACHINES 644
Michael Gleissner, Johannes Häring, Wolfgang Wondrak, Mark-M. Bakran

FULL-SILICON 98.7% EFFICIENT THREE-PHASE FIVE-LEVEL 3-PORT UPS
ARCHITECTURE WITH WIDE VOLTAGE RANGE BATTERY BASED ON MULTIPLEXED
TOPOLOGY 654
Kepa Odriozola, Thierry A. Meynard, Alain Lacarnoy

ON-GRID/OFF-GRID DC MICROGRID OPTIMIZATION AND DEMAND RESPONSE
MANAGEMENT 667
Wenshuai Bai, Manuela Sechilariu, Fabrice Locment

SHEDDING AND RESTORATION ALGORITHMS FOR AN EV CHARGING STATION TO
MAXIMIZE AVAILABLE POWER... 677
Dian Wang, Fabrice Locment, Manuela Sechilariu

EFFICIENCY AND COST COMPARISON OF B6 AND HYBRID ANPC CONVERTERS FOR
TRACTION DRIVES 686
Johannes Häring, Michael Gleissner, Wolfgang Wondrak, Mark-M. Bakran

DESIGN AND CONTROL OF A KE (KINETIC ENERGY) - COMPENSATED
GRAVITATIONAL ENERGY STORAGE SYSTEM 696
Alfred Rufer

A NOVEL POWER FLOW CONTROL STRATEGY FOR HETEROGENEOUS BATTERY
ENERGY STORAGE SYSTEMS BASED ON PROGNOSTIC ALGORITHMS FOR
BATTERIES 707
Markus Muehlbauer, Samantha Klier, Herbert Palm, Oliver Bohlen, Michael A. Danzer

AN IGCT-BASED MULTI-FUNCTIONAL MMC SYSTEM WITH COMMUTATION AND
SWITCHING.. 718
Chaoqun Xu, Mingzhu Guo, Biao Zhao, Bojin Tang, Zhanqing Yu, Dongling Zhai, Chunpin
Ren

COMMON-MODE NOISE MODELLING AND RESONANT ESTIMATION IN A THREE-PHASE MOTOR DRIVE SYSTEM: 9-150 KHZ FREQUENCY RANGE 726
Hansika Rathnayake, Amir Ganjavi, Firuz Zare, Dinesh Kumar, Pooya Davari

POLYNOMIAL MULTI-VARIABLE CONTROL STRATEGY FOR FLUX BALANCING IN DUAL ACTIVE BRIDGE CONVERTER 736
Pierre-Baptiste Steckler, Jean-Yves Gauthier, Xuefang Lin-Shi, François Wallart

ENHANCED POWER SYSTEM DAMPING ESTIMATION VIA OPTIMAL PROBING SIGNAL DESIGN 745
S. Boersma, X. Bombois, L. Vanfretti, V. Peric, J-C. Gonzalez-Torres, R. Segur, A. Benchaib

IMPROVED HIGH STEP-UP BOOST-BASED DC/DC CONVERTER WITH BUILT-IN TRANSFORMER AND ACTIVE CLAMP FOR DC MICROGRIDS 755
Konstantinos Zaoskoufis, Emmanuel C. Tatakis

ELIMINATION/MITIGATION OF OUTPUT VOLTAGE HARMONICS FOR MULTILEVEL CONVERTERS OPERATED AT FUNDAMENTAL SWITCHING FREQUENCY USING MATLAB'S GENETIC ALGORITHM OPTIMIZATION 765
Anton Kersten, Manuel Kuder, Arthur Singer, Weiji Han, Torbjörn Thiringer, Thomas Weyh, Richard Eckerle

EVALUATION OF DRIVE TOPOLOGIES FOR MACRO SCALE SYNCHRONOUS ELECTROSTATIC MACHINES 777
Peter Killeen, Daniel C. Ludois

DECENTRALIZED VOLTAGE REGULATION IN ISLANDED DC MICROGRIDS IN THE PRESENCE OF DISPATCHABLE AND NON-DISPATCHABLE DC SOURCES 787
Mohammadreza Nabatirad, Reza Razzaghi, Behrooz Bahrani

AN ULTRA-FAST GATE DRIVER WITH OVER CURRENT PROTECTION FOR GAN POWER TRANSISTORS 797
Qingqing Nie, Han Peng, Yong Kang

A NEW GAN HYBRID RESONANT-CLAMPING GATE DRIVER FOR HIGH FREQUENCY SIC MOSFETS 804
Ziyue Dang, Han Peng, Hao Peng, Yong Kang, Yu Chen, Xudan Liu, Maojun He

MAINTENANCE SCHEDULING IN POWER ELECTRONIC CONVERTERS CONSIDERING WEAR-OUT FAILURES 810
Saeed Peyghami, Frede Blaabjerg, Jose Rueda Torres, Peter Palensky

AC/DC DYNAMIC INTERACTIONS OF MMC-HVDC IN GRID-FORMING FOR WIND-FARM INTEGRATION IN AC SYSTEMS 820
Rayane Mourouvin, Kosei Shinoda, Jing Dai, Abdelkrim Benchaib, Seddik Bacha, Didier Georges

A DESIGN OF SOLID STATE POWER CONTROLLER FOR A BIDIRECTIONAL DC-DC CONVERTER IN AN AERONAUTIC CONTEXT 829
Hassan Cheaito, Bruno Allard, Guy Clerc, Joris Pallier, Pascal Pommier-Petit

A NEW APPROACH OF RESONANT CONVERTER USING LARGE AIR GAP TRANSFORMER 835
Michael Finkenzeller, Monika Poebl, Thomas Komma

REDUCED CAPACITOR SIZE AND ON-STATE LOSSES IN ADVANCED MMC SUBMODULE TOPOLOGIES.. 843
Christopher Dahmen, Rainer Marquardt

STABILITY AND ROBUSTNESS ANALYSIS OF FRACTIONAL PROPORTIONAL RESONANT CONTROLLERS IN CURRENT-CONTROLLED VOLTAGE-SOURCE-INVERTERS 853
Daniel Heredero-Peris, Cristian Chillón-Antón, Daniel Montesinos-Miracle

EMPLOYING VIRTUAL SYNCHRONOUS GENERATOR WITH A NEW CONTROL TECHNIQUE FOR GRID FREQUENCY STABILIZATION 863
Meysam Saeedian, Bahman Eskandari, Kumars Rouzbehi, Shamsodin Taheri, Edris Pouresmaeil

A HYBRID PULSE WIDTH MODULATION TECHNIQUE WITH TEMPERATURE CONTROL FOR MODULAR MULTILEVEL CONVERTERS 871
Ara Bissal, Waqas Ali, Rob Leedham, Mark Snook, Ibrahim Elsabrouty, Ilknur Colak

DESIGN FLOW OF A COMPACT HIGH-FREQUENCY DC/DC CONVERTER WITH OPTIMUM AVERAGE EFFICIENCY IN A WIDE OPERATION RANGE................................ 880
Maximilian Nitzsche, Matthias Zehelein, Julian Weimer, Dominik Koch, Jörg Roth-Stielow

ANALYSIS OF THE TRANSFORMER MODULARIZATION FOR HIGH FREQUENCY ISOLATED HIGH VOLTAGE GENERATOR WITH THE SILICON CARBIDE DEVICES 892
Saijun Mao, Popovic Jelena, Jan Abraham Ferreira

IMPROVED DIRECT-MODEL PREDICTIVE CONTROL WITH A SIMPLE DISTURBANCE OBSERVER FOR DFIGS 900
Mohamed Abdelrahem, Christoph Hackl, José Rodríguez, Ralph Kennel

MODELING OF SIC-MOSFET CONVERTER LEG INCLUDING PARASITICS OF PRINTED CIRCUIT BOARD LAYOUT AND DEVICE PACKAGING 909
M. Pulvirenti, L. Salvo, A. G. Sciacca, G. Scelba, M. Cacciato

PERFORMANCE ANALYSIS OF RL DAMPER IN GAN-BASED HIGH-FREQUENCY BOOST CONVERTER.. 919
A. Gutierrez, E. Marcault, C. Alonso, D. Tremouilles

RAPID IMPEDANCE ESTIMATION ALGORITHM FOR MITIGATION OF SYNCHRONIZATION INSTABILITY OF PARALLELED CONVERTERS UNDER GRID FAULTS.. 927
Mads Graungaard Taul, Robert Eric Betz, Frede Blaabjerg

ADAPTIVE THERMAL CONTROL FOR MOSFET-BASED MODULAR MULTILEVEL CONVERTER... 937
Tianxiang Yin, Lei Lin, Chen Xu

ELECTRIC IMPULSE TECHNOLOGY – BREAKING ROCK 944
Matthias Voigt, Erik Anders, Franziska Lehmann, Margarita Mezzetti, Frank Will

IMPACT OF COMBINED THERMO-MECHANICAL AND ELECTRO-CHEMICAL STRESS ON THE LIFETIME OF POWER ELECTRONIC DEVICES 954
Felix Hoffmann, Stefan Schmitt, Nando Kaminski

CURRENT CONTROL AND FPGA–BASED REAL–TIME SIMULATION OF GRID–TIED INVERTERS 962
Sabin Carpiuc, Matthias Schiesser, Carlos Villegas

IMPACT OF CONTROL LOOPS ON THE LOW-FREQUENCY PASSIVITY PROPERTIES OF GRID-FORMING CONVERTERS 969

Mebtu Beza, Massimo Bongiorno, Anant Narula

GRID IMPEDANCE ESTIMATION WITH OVERSAMPLING FOR GRID-CONNECTED CONVERTERS 979

Niklas Himker, Robin Strunk, Axel Mertens

LOW SPEED SENSORLESS CURRENT CONTROL FOR PMSM WITH SEARCH-BASED OBSERVER (SBO) 989

K. Scicluna, C. Spiteri Staines, R. Raute

INSIGHT INTO THE PECULIARITIES OF OPTIMIZED PULSE PATTERNS FOR PERMANENT-MAGNET SYNCHRONOUS MACHINES 998

Georgios Darivianakis, Ioannis Tsoumas

INVESTIGATING THE EFFECT OF DIFFERENT PARAMETERS ON HARMONICS AND EMI EMISSIONS AT THE FREQUENCY RANGE OF 0–9 KHZ 1006

Amir Ganjavi, Hansika Rathnayake, Firuz Zare, Dinesh Kumar, Amin Abbosh, Pooya Davari

FIVE-LEVEL NESTED INVERTER WITH NEUTRAL POINT CONNECTION 1016

Juhamatti Korhonen, Aleksi Mattsson, Heikki Järvisalo, Pertti Silventoinen, William Giewont, Dan Isaksson

ELECTRIC SPRING-BASED SMART WATER HEATER FOR LOW VOLTAGE MICROGRIDS 1025

Alexander Micallef, Racquel Ellul, John Licari

ENERGY-BALANCING OF A MODULAR MULTILEVEL CONVERTER USING AN ONLINE TRAJECTORY PLANNING ALGORITHM 1030

Qiuye Gui, Jan Lasse Gnärig, Hendrik Fehr, Albrecht Gensior

CAPACITOR SIZE COMPARISON ON HIGH-POWER DC-DC CONVERTERS WITH DIFFERENT TRANSFORMER WINDING CONFIGURATIONS ON THE AC-LINK 1040

Babak Khanzadeh, Torbjörn Thiringer, Yuhei Okazaki

DYNAMIC CHARACTERISTICS VERIFICATION OF LINEAR INDUCTION MOTOR BY SIMULTANEOUS PROPULSION AND LEVITATION CONTROL 1047

Shota Nakatani, Daichi Okamori, Toshimitsu Morizane, Hideki Omori

'IG,VGS' MONITORING FOR FAST AND ROBUST SIC MOSFET SHORT-CIRCUIT PROTECTION WITH HIGH INTEGRATION CAPABILITY 1057

Yazan Barazi, François Boige, Nicolas Rouger, Jean-Marc Blaquiere, Frédéric Richardeau

FAULT-TOLERANT CONTROL OF SERIES CONNECTABLE MODULAR FULL-BRIDGE INVERTER MITIGATING OPEN SWITCH FAULTS 1067

Juris Arrozy, Darian V. Retianza, Jorge L. Duarte, Henk Huisman

DESIGN AND CONTROL OF A MODULAR POWER ELECTRONIC BACK-TO-BACK CONVERTER FOR WAVE ENERGY HARVESTING APPLICATIONS 1076

Mattia Mantellini, Riccardo Morici, Marcos Blanco, Marcos Lafoz, Gustavo Navarro, Jorge Torres, Jorge Najera, Miguel Santos

INTELLIGENT HIGH CURRENT SENSOR FOR VARIOUS FREQUENCY 1086

Bohumil Skala, Vladimir Kindl, Pavel Turjanica, Ales Voborník, Libor Polacek, Josef Stengl, Vladimir Pavlicek, Jiri Fort

FAIL-SAFE SWITCHING-CELLS ARCHITECTURES BASED ON MONOLITHIC ON-CHIP FUSE 1096
Amirouche Oumaziz, Emmanuel Sarraute, Frédéric Richardeau, Abdelhakim Bourennane

HOW GOOD ARE THE DESIGN TOOLS IN POWER ELECTRONICS? 1106
Thomas Lagier, Piotr Dworakowski, Laurent Chédot, François Wallart, Bruno Lefebvre, Jose Maneiro, Juan Páez, Philippe Ladoux, Cyril Buttay

ANALYSIS OF THE IMPACT OF MANUFACTURING DISSYMMETRY ON CURRENT DISTRIBUTION FOR MAGNETICALLY COUPLED INTERLEAVED INVERTERS 1118
Rita Mattar, Mickael Petit, Eric Monmasson, Stéphane Lefebvre, Christelle Saber, Cyrille Gautier, Marwan Ali

POWER FLOW CONTROL USING A BIDIRECTIONAL Z-SOURCE INVERTER–BASED STATIC SYNCHRONOUS SERIES COMPENSATOR 1128
Xuejiao Pan, Han Huang, Li Zhang

INVESTIGATION OF HARMONICS CONTENT IN PWM NATURAL AND REGULAR SAMPLING INCLUDING DEAD TIME AND LOAD CURRENT PHASE 1138
Tonny Wederberg Rasmussen, Anushruti Vashishtha, Ankit Jotwani

USING A WEB SCRAPING ALGORITHM FOR COMPONENT MODEL GENERATION IN MULTIOBJECTIVE OPTIMIZATION OF POWER ELECTRONIC APPLICATIONS 1148
Marcel Gladen, Volker Staudt

IMPACT ON THE ELECTRICAL CHARACTERISTICS, WAVEFORMS AND LOSSES OF THE ZERO-SEQUENCE INJECTION ON THE MODULAR MULTILEVEL CONVERTER 1158
Francois Gruson, Pierre Vermeersch, Philippe Delarue, Philippe Le Moigne, Frédéric Colas, Haibo Zhang, Moez Belhaouane, Xavier Guillaud

WIDE BANDWIDTH CURRENT SENSOR FOR COMMUTATION CURRENT MEASUREMENT IN FAST SWITCHING POWER ELECTRONICS 1168
Philipp Ziegler, Nathan Tröster, Dimitri Schmidt, Johannes Ruthardt, Manuel Fischer, Jörg Roth-Stielow

A SERIES–PARALLEL-TYPE RESONANT CIRCUIT WIRELESS POWER TRANSFER SYSTEM WITH A DUAL ACTIVE BRIDGE DC–DC CONVERTER 1177
Kohei Sugiyama, Taishi Kitamura, Shuto Uwai, Takahiro Yano, Yoshitaka Kawabata

STRAY VOLTAGE CAPTURE FOR ROBUST AND ULTRA-FAST SHORT CIRCUIT DETECTION IN POWER ELECTRONICS WITH HALF-BRIDGE STRUCTURE: THE LIMITATION AND IMPLEMENTATION 1186
Darian Verdy Retianza, Jeroen Van Duivenbode, Henk Huisman

ON THE INFLUENCE OF THE STATOR WINDING TOPOLOGY ON THE ELECTROMAGNETIC EMISSIONS OF FRACTIONAL HORSEPOWER BLDC MOTORS 1196
Felix Krall, Annette Muetze

IMPACT OF SILICON CARBIDE DEVICES IN 2 MW DFIG BASED WIND ENERGY SYSTEM 1205
Antxon Arrizabalaga, Aitor Idarreta, Mikel Mazuela, Iosu Aizpuru, Unai Iraola, José Luis Rodriguez, Daniel Labiano, Ibrahim Alisar

SMALL-SIGNAL STABILITY OF HVDC SYSTEM COMPRISING DC REACTORS 1215
Kosei Shinoda, Abdelkrim Benchaib, Jing Dai

MODEL PREDICTIVE CONTROL FOR THE REDUCTION OF DC-LINK CURRENT RIPPLE IN TWO-LEVEL THREE-PHASE VOLTAGE SOURCE INVERTERS 1224

Junzhong Xu, Fei Gao, Thiago Batista Soeiro, Linglin Chen, Luca Tarisciotti, Houjun Tang, Pavol Bauer

CARRIER-BASED MODULATED MODEL PREDICTIVE CONTROL FOR VIENNA RECTIFIERS 1233

Junzhong Xu, Fei Gao, Thiago Batista Soeiro, Linglin Chen, Luca Tarisciotti, Houjun Tang, Pavol Bauer

NEW HIGH-EFFICIENCY POWER GENERATION USING POSITION SENSOR-LESS PERMANENT MAGNET SYNCHRONOUS GENERATOR 1243

Somi Takeuchi, Hiroyuki Takahashi, Shota Yamada, Yoshitaka Kawabata

ACTIVE CLAMPING METHOD FOR SIC MOSFET HIGH POWER MODULES - BENEFITS AND LIMITS 1252

Robert W. Maier, Mark-M. Bakran

PREDICTIVE TORQUE CONTROL OF INDUCTION MACHINE WITH AN ADAPTIVE OBSERVER FOR TRAJECTORY PLANNING OF SERVO PRESS 1262

Qi Li, Jianbo Gao, Qiwu Wang, Ralph Kennel

FUTURE GRID STABILITY, A COST COMPARISON OF GRID-FORMING AND SYNCHRONOUS CONDENSER BASED SOLUTIONS 1270

Thibault Prevost, Guillaume Denis, Clementine Coujard

DEMONSTRATION OF THE SHORT-CIRCUIT RUGGEDNESS OF A 10 KV SILICON CARBIDE BIPOLAR JUNCTION TRANSISTOR 1279

Besar Asllani, Hervé Morel, Pascal Bevilacqua, Dominique Planson

LOSS MINIMIZATION OF TRACTION SYSTEMS IN BATTERY ELECTRIC VEHICLES USING VARIABLE DC-LINK VOLTAGE TECHNIQUE — EXPERIMENTAL STUDY 1289

Libo Liu, Boyang Li, Gunther Götting, Yusheng Xiang, Qusay Salem, Muhammad Hamid, Jian Xie

DIRECT MULTIVARIABLE CONTROL FOR MMC: DIGITAL SIGNAL PROCESSING AND EXPERIMENTAL RESULTS 1297

Daniel Dinkel, Claus Hillermeier, Rainer Marquardt

STATE OF CHARGE CONTROL FOR A FREQUENCY-SUPPORTING STORAGE SYSTEM BASED ON AN AUTO-REGRESSIVE FREQUENCY FORECAST 1306

A. Bolzoni, R. Todd, Q. Zhu, A. J. Forsyth

DESIGN OF A WIDE INPUT VOLTAGE RANGE CURRENT-FED DC/DC CONVERTER WITHIN A REDUCED DUTY-CYCLE RANGE 1316

Michael Gerstner, Martin Maerz, Armin Dietz

AN IMPROVED CONTROL STRATEGY FOR RENEWABLE ENERGY SOURCES (RES) BASED DC MICROGRID WITH ENHANCED SYSTEM STABILITY AND CONTROL PERFORMANCE 1326

Muhammad Adnan Mumtaz, Zheng Yan

TRANSIENT VOLTAGE DIP MITIGATION SYSTEM BASED ON HYBRID MODULAR MULTILEVEL CONVERTERS 1336

Manuel Colmenero, Francisco R. Blanquez, Karsten Kahle

A LOSS-COMPENSATED CONTROL SCHEME FOR SIC-BASED DUAL ACTIVE BRIDGE CONVERTER 1346
Ishan Pendharkar, Tobias Strittmatter, Paula Diaz Reigosa, Nicola Schulz

EXPERIMENTAL HYBRID AC/DC-MICROGRID PROTOTYPE FOR LABORATORY RESEARCH 1354
Enrique Espina, Claudio Burgos-Mellado, Juan S. Gomez, Jacqueline Llanos, Erwin Rute, Alex Navas F., Manuel Martínez-Gómez, Roberto Cárdenas, Doris Sácz

EXPERIMENTAL AND NUMERICAL CHARACTERIZATION OF PCB-EMBEDDED POWER DIES USING SOLDERLESS PRESSED METAL FOAM 1363
S. Bensebaa, M. Berkani, S. Lefebvre, M. Petit, N. Schmitt

FEASIBILITY STUDY OF A SUPERCONDUCTING POWER FILTER FOR HVDC GRIDS 1373
Loïc Quéval, Olivier Despouys, Frédéric Trillaud, Bruno Douine

POWER DECOUPLING METHOD OF DC TO SINGLE-PHASE AC CONVERTER USING FLYING CAPACITOR DC/DC CONVERTER WITH BOUNDARY CURRENT MODE 1380
Hiroki Watanabe, Keisuke Kusaka, Jun-Ichi Itoh

AN ARCHITECTURE FOR LEVEL-3 EV BATTERY CHARGER STATIONS USING INTEGRATED SOLID STATE TRANSFORMER (I-SST) 1390
Erick I. Pool-Mazun, Prasad Enjeti, Gerardo Escobar, Ira Pitel

LQR AND H-INFINITY CONTROL OF VOLTAGE SOURCE INVERTERS FOR AC MICROGRIDS 1400
Tenorio Jorge, Jose Miguel Ramirez Scarpetta, Fabio Andrade

FAMILY OF SPLITTING CURRENT SINGLE-LOOP CONTROL FOR *LCL*- TYPE GRID-CONNECTED INVERTER 1410
Yuying He, Xuehua Wang, Xinbo Ruan, Guoxing Su, Fuxin Liu

ANALYSIS AND DESIGN OF HIGH-POWER SINGLE-STAGE THREE-PHASE DIFFERENTIAL-BASED FLYBACK INVERTER FOR PHOTOVOLTAIC APPLICATIONS 1417
Ahmed Ismail M. Ali, Mahmoud A. Sayed, Takaharu Takeshita

INVESTIGATION OF IMPROVEMENT OF MODELING PRECISION FOR CONDUCTED NOISE ON ISOLATED AC/DC CONVERTER USING SIC DEVICES 1425
Kazuki Kuwana, Kohei Mitani, Wataru Kitagawa, Takaharu Takeshita

PASSIVITY-BASED DESIGN FOR THE PLUG-AND-PLAY SINGLE-LOOP CONTROLLED LCL-FILTERED INVERTER 1435
Yuying He, Xuehua Wang, Xinbo Ruan, Yixiao Ma, Fuxin Liu

CHARACTERISTICS OF AN INTEGRATED MOTOR CONTROLLED INDEPENDENTLY BY MULTI-INVERTERS TO ACHIEVE HIGH EFFICIENCY AND A WIDE SPEED RANGE 1442
Kazuto Sakai, Yano Hideaki

AN ISOLATED MEDIUM-VOLTAGE AC-DC CONVERTER USING LEVEL-SHIFTED PWM CONTROL OF A MODULAR MATRIX CONVERTER 1450
Kohei Budo, Takaharu Takeshita

DETAILED SIMULATION MODEL OF AN ASYMMETRICAL HALF-BRIDGE PWM CONVERTER WITH SYNCHRONOUS RECTIFICATION INCLUDING PARASITIC ELEMENTS 1460
Benedikt Kohlhepp, Valentin Zeller, Markus Barwig, Thomas Dürbaum

ELECTRICAL PROPERTY VARIABILITY OF GAN TRANSISTORS IN PARALLEL AND THEIR IMPACT ON FAST SWITCHING OPERATIONS 1470
Thilini Wickramasinghc, Bruno Allard, Réne Escofficr, Marc Plissonnicr

A COMPARISON BETWEEN DIFFERENT MODELS OF THE MODULAR MULTILEVEL CONVERTER 1479
Rafael Coelho-Medeiros, Bogdan Džonlaga, Jean-Claude Vannier, Jing Dai, Loic Queval, Philippe Egrot

PACKAGING TECHNOLOGY FOR THE IMPROVEMENT OF POWER CYCLING CAPABILITY OF HVIGBTS 1489
Kenji Hatori, Keiichi Nakamura, Nobuhiko Tanaka, Yasuhiro Sakai, Norikazu Sakai, Kenji Ota, Takeshi Higashihata, Eckhard Thal, Nils Soltau

A BIDIRECTIONAL DAB-LLC DCX TO ACHIEVE VOLTAGE REGULATION AND WIDE ZVS RANGE CAPABILITY 1498
Yuefeng Liao, Tao Peng, Mei Su, Yao Sun, Weijing Xiong, Guo Xu

SALIENCY SELECTION FOR SEARCH-BASED AC MACHINE LOW AND ZERO SPEED ESTIMATION METHODS 1506
K. Scicluna, C. Spiteri Staines, R. Raute

GENETIC ALGORITHM BASED MULTI OBJECTIVE OPTIMIZATION FOR INDUCTOR DESIGN 1515
Thorben Schobre, Raquel González Aríztegui, Regine Mallwitz

DIGITAL SMART DRIVER FOR SIC MOSFETS 1524
Nerea Arandia, José Ignacio Garate, Jon Mabe, Ander Ordoño

FASTER SWITCHING WITH LESS OVERVOLTAGE - OPERATING A SIC-MOSFET AT ITS SPEED LIMIT 1533
Pablo Rodriguez De Mora, Mark-M. Bakran

THE ENERGY RING TO SUPPLY THE EXPOELECTRIC'18 SHOW WITH RENEWABLE ENERGY SOURCES AND ELECTRIC VEHICLES 1542
Cristian Chillón-Antón, Daniel Heredero-Peris, Francesc Girbau-Llistuella, Paula González-Fontderubinat, Marc Llonch-Masachs, Daniel Montesinos-Miracle, Oriol Gomis-Bellmunt

IMPEDANCE-BASED MODELING OF A THREE-LEVEL CONVERTER UNDER BALANCED AND UNBALANCED CONDITION FOR THE STABILITY ANALYSIS OF BIPOLAR LVDC GRIDS 1551
T. Roose, G. Van Den Broeck, M. M. Alam, J. Beerten

LCL FILTER DESIGN FOR THREE PHASE AC-DC CONVERTERS CONSIDERING SEMICONDUCTOR MODULES AND MAGNETICS COMPONENTS PERFORMANCE 1561
Marco Stecca, Thiago Batista Soeiro, Laura Ramirez Elizondo, Pavol Bauer, Peter Palensky

SWITCHING BEHAVIOR AND COMPARISON OF 600V SMD WIDE BANDGAP POWER DEVICES 1569
Markus Meißner, Jan Schmitz, Steffen Bernet

ANALYSIS OF THE COUPLING BETWEEN THE OUTER AND INNER CONTROL LOOPS OF A GRID-FORMING VOLTAGE SOURCE CONVERTER 1579
T. Qoria, F. Gruson, F. Colas, X. Kestelyn, X. Guillaud

INFLUENCE OF DIFFERENT PULSE-WIDTH MODULATION METHODS ON MAGNET LOSSES IN PERMANENT MAGNET SYNCHRONOUS MACHINES.. 1589
Narciso G. Marmolejo, Xiaohu Tang, Martin Doppelbauer

RESONANT DC/DC CONVERTER WITH CLASS ϕ_2 INVERTER AND CLASS DE RECTIFIER BASED ON GAN HEMT 1599
Cai Si-Yuan, He Jun-Ping, Li Zi-Fan

FOUR-LEVEL INVERTER WITH VARIABLE VOLTAGE LEVELS FOR HARDWARE-IN-THE-LOOP EMULATION OF THREE-PHASE MACHINES 1605
Manuel Fischer, Johannes Ruthardt, Vasken Ketchedjian, Philipp Ziegler, Maximilian Nitzsche, Jörg Roth-Stielow

POWDER INJECTION MOLDING IN THE FABRICATION OF SOFT FERRITE MATERIAL FOR POWER ELECTRONICS 1613
J-S Ngoua-Teu, U. Soupremanien, P. Sallot, G. Delette, M. Bohnke

MODULATION SCHEME WITH COMMON MODE AND DIFFERENTIAL MODE VOLTAGE ELIMINATION FOR A FIVE LEVEL INVERTER FED OPEN END WINDING INDUCTION MOTOR DRIVE 1619
Greeshma Nadh, Durga Nair S., Arun Rahul S.

A FAST AND ROBUST MODEL OF DUAL-ACTIVE BRIDGE CONVERTERS IN REAL-TIME SIMULATION 1627
Ming Jia, Philipp Joebges, Rik W. De Doncker

DUAL INTERLEAVED 3.6 KW LLC CONVERTER OPERATING IN HALF-BRIDGE, FULL-BRIDGE AND PHASE-SHIFT MODE AS A SINGLE-STAGE ARCHITECTURE OF AN AUTOMOTIVE ON-BOARD DC-DC CONVERTER 1638
Philipp Rehlaender, Sergey Tikhonov, Frank Schafmeister, Joachim Bocker

SWITCHING LOSS ESTIMATION USING A VALIDATED MODEL OF 650 V GAN HEMTS............ 1648
Joao Oliveira, Florent Loiselay, Hervè Morel, Dominique Planson

REDUCTION OF CONDUCTION LOSSES IN RESONANT CONVERTERS BY CONNECTING THREE SINGLE-PHASE INVERTERS TO A COMMON GENERATOR 1658
Sergio Tárraga, John Paul Mayorga, Esther De Jódar, José Villarejo

COMPARISON OF DIFFERENT LOW VOLTAGE MULTILEVEL CONVERTER TOPOLOGIES FOR DISTRIBUTED POWER GENERATION 1666
Ingmar Kaiser, Hans-Günter Eckel

LOSS DISTRIBUTION COMPARISON OF VARIABLE AND FIXED INDUCTOR DAB CONVERTERS.. 1675
Erik Smailus, Gerd Griepentrog, Markus Pfeifer, Marcel Lutze

DESIGN BY OPTIMIZATION OF MULTIPHASE INVERTER FOR ELECTRIC VEHICLE DRIVE.. 1685
Nasreddine Kesbia, Jean-Luc Schanen, Hadi Alawieh, Lauric Garbuio, Yvan Avenas

OPTIMAL TORQUE/SPEED CHARACTERISTICS OF A FIVE-PHASE SYNCHRONOUS MACHINE UNDER PEAK OR RMS CURRENT CONTROL STRATEGIES 1693
Tiago José Dos Santos Moraes, Hailong Wu, Eric Semail, Ngac Ky Nguyen, Duc Tan Vu

COMPARATIVE STUDY OF TWO CONTROL TECHNIQUES OF REGENERATIVE
BRAKING POWER RECOVERING INVERTER BASED DC RAILWAY SUBSTATION 1700
  Youssef Krim, Khaled Almaksour, Hervé Caron, Tony Letrouvé, Christophe Saudemont,
  Bruno Francois, Benoit Robyns

JUNCTION TEMPERATURE CONTROL STRATEGY FOR LIFETIME EXTENSION OF
POWER SEMICONDUCTOR DEVICES 1709
  Johannes Ruthardt, Hendrik Schulte, Philipp Ziegler, Manuel Fischer, Maximilian Nitzsche,
  Jörg Roth-Stielow

HIGH DYNAMIC POWER BALANCING FOR DUAL TWO-LEVEL INVERTERS DURING
HIGH-SPEED MACHINE OPERATION 1718
  Johannes Büdel, Johannes Teigelkötter, Alexander Stock, Christian Herkommer, Kai
  Kuhlmann

CHARGING HIGH VOLTAGE CAPACITORS IN PULSED POWER APPLICATIONS WITH A
CAPACITOR DIODE VOLTAGE MULTIPLIER OF REDUCED SIZE AND LOWER RIPPLE
CURRENTS 1727
  Tristan Weinert, Wolfgang Oberschelp, Günter Schröder

REVIEW OF OPTIMIZATION METHODS FOR THE DESIGN OF POWER ELECTRONICS
SYSTEMS 1737
  Mylène Delhommais

A FLEXIBLE POWER CROSSBAR-BASED ARCHITECTURE FOR SOFTWARE-DEFINED
POWER DOMAINS 1747
  Francesco Di Gregorio, Gilles Sassatelli, Abdoulaye Gamatié, Arnaud Castelltort

IMPACT OF GRID-FORMING CONTROL ON THE INTERNAL ENERGY OF A MODULAR
MULTILEVEL CONVERTER 1756
  Ebrahim Rokrok, Taoufik Qoria, Antoine Bruyere, Bruno Francois, Haibo Zhang, Moez
  Belhaouane, Xavier Guillaud

COMBINING MULTIPLE TEMPERATURE-SENSITIVE ELECTRICAL PARAMETERS
USING ARTIFICIAL NEURAL NETWORKS 1766
  Daniel Herwig, Torben Brockhage, Axel Mertens

SINGLE-PHASE MEASUREMENT OF THE OUTPUT IMPEDANCE OF THE FOUR-
QUADRANT CASCADED H-BRIDGE CONVERTER CELL USING WIDEBAND SIGNALS 1776
  Marko Petkovic, Dražen Dujic

A NOVEL THREE-PHASE PFC DIODE RECTIFIER BY LC NETWORK CIRCUITS FOR
HIGH FREQUENCY GENERATOR 1786
  Shin-Ichi Motegi, Yasuyuki Nishida

FREQUENCY-DOMAIN SIMULATION OF POWER ELECTRONIC SYSTEMS BASED ON
MULTI-TOPOLOGY EQUIVALENT SOURCES MODELLING METHOD 1793
  Stephane Vienot, Arnaud Videt, Nadir Idir, Lamine Kone, Sébastien Weiss, Frederic Lafon

MODULAR MULTILEVEL CONVERTER WITH DISTRIBUTED GALVANIC ISOLATION: A
DECENTRALIZED VOLTAGE BALANCING ALGORITHM WITH SMART GATE DRIVERS 1803
  Darbas Corentin, Ginot Nicolas, Olivier Jean-Christophe, Poitiers Frédèric

COMPARISON AND OPTIMIZATION OF MAGNETICALLY COUPLED AND NON-
COUPLED MAGNETIC DEVICES IN INTERLEAVED OPERATION 1813
  Peter Zacharias, Alejandro Aganza-Torres

EXPERIMENTAL TUNING AND DESIGN GUIDELINES OF A DYNAMICALLY
RECONFIGURED WEIGHTING FACTOR FOR THE PREDICTIVE TORQUE CONTROL OF
AN INDUCTION MOTOR 1823
 Ilker Sahin, Ozan Keysan, Eric Monmasson

COMPENSATION OF TEMPERATURE DEPENDENCE IN A MODULE PARASITIC BASED
CURRENT MEASUREMENT SYSTEM 1831
 Frank Lautner, Mark-M. Bakran

DEVELOPMENT AND IMPLEMENTATION OF A LOW-COST RESEARCH PLATFORM
FOR CONTROL APPLICATIONS FOR INVERTER-BASED GENERATORS 1841
 Jesus D. Vasquez Plaza, Juan F. Patarroyo-Montenegro, Fabio Andrade

CONTROL OF PARALLEL CONNECTED VOLTAGE SOURCE INVERTERS IN A
MICROGRID FOR EXPERIMENTAL TESTING 1850
 *Jesus D. Vasquez-Plaza, Jorge Tenorio, J. M. Ramírez-Scarpetta, Jose Alex Restrepo, Fabio
 Andrade*

OPTIMIZATION STRATEGY FOR THE SIZING OF PASSIVE MAGNETIC COMPONENTS 1858
 *Guillaume Devos, Maya Hage-Hassan, Philippe Dessante, Cyrille Gautier, Adrien Mercier,
 Eric Labouré*

EXPLOITING A MULTI-PORT TRANSFORMER FOR MINIMAL DC-LINK CAPACITANCE
FOR AN AUTOMOTIVE ONBOARD CHARGER 1866
 Franz Vollmaier, Alexander Connaughton, Thomas Langbauer, Klaus Krischan

DESIGN AND OPTIMIZATION OF HIGH-EFFICIENCY 1W 500V-12V ISOLATED LOW-
COST DC/DC CONVERTER 1874
 Etienne Foray, Christian Martin, Bruno Allard

CHALLENGES IN CALIBRATING AN UNCONVENTIONAL PARTIAL DISCHARGE
MEASUREMENT SYSTEM FOR PULSED VOLTAGES 1885
 Markus Fürst, Mark-M. Bakran

ELECTROTHERMAL MODELING OF GAN POWER TRANSISTOR FOR HIGH
FREQUENCY POWER CONVERTER DESIGN 1895
 *Loris Pace, Florian Chevalier, Arnaud Videt, Nicolas Defrance, Nadir Idir, Jean-Claude De
 Jaeger*

MODELING AND FAULT DETECTION IN PHOTOVOLTAIC SYSTEMS USING THE I-V
SIGNATURE 1905
 *Abdelhadi Benzagmout, Thierry Talbert, Olivier Fruchier, Thierry Martire, Philippe
 Alexandre, Carolina Penin*

EFFICIENCY REQUIREMENTS FOR PASSIVELY COOLED CONVERTERS WITH
THERMAL MEASUREMENT BASED 3D-FEM SIMULATION 1915
 Julian Weimer, Dominik Koch, Maximilian Nitzsche, Matthias Zehelein, Ingmar Kallfass

GENERIC CONTROL LAW FOR DC AND AC MACHINES 1923
 Pierre-Philippe Robet, Maxime Gautier, Yannick Aoustin

A HIGH PERFORMANCE 48-TO-8 V MULTI-RESONANT SWITCHED-CAPACITOR
CONVERTER FOR DATA CENTER APPLICATIONS 1934
 Rose A. Abramson, Zichao Ye, Robert C. N. Pilawa-Podgurski

SISO CONTROL STRATEGY OF RESONANT DUAL ACTIVE BRIDGE WITH A TUNED CLC NETWORK 1944
   Meiqi Wang, Bo Yang, Lie Xu, Jing Li, David Gerada, Chunyang Gu, He Zhang, Chris Gerada, Yongdong Li

IMPACT OF STEADY-STATE GRID-FREQUENCY DEVIATIONS ON THE PERFORMANCE OF GRID-FORMING CONVERTER CONTROL STRATEGIES 1952
   Anant Narula, Massimo Bongiorno, Mebtu Beza, Jan R Svensson, Xavier Guillaud, Lennart Harnefors

A GENERAL METHOD TO DAMP WIND TURBINE SSR WITH DIFFERENT TRANSMISSION SYSTEMS 1962
   Ignacio Vieto, Jian Sun

A TEST SCHEME FOR THE COMPREHENSIVE QUALIFICATION OF MMC SUBMODULE BASED ON 10 KV SIC MOSFETS UNDER HIGH DV/DT 1972
   Xingxuan Huang, Shiqi Ji, Dingrui Li, Cheng Nie, William Giewont, Leon M. Tolbert, Fred Wang

PWM GAIN LINEARIZATION ALGORITHM FOR MEDIUM VOLTAGE SOURCE INVERTER 1982
   Hamza El Jihad, Sami Siala, Elise Savarit

AUTO-COMMISSIONING OF ACOUSTIC CONTROL OF IM DRIVE USING BAYESIAN OPTIMIZATION 1992
   Michal Kroneisl, Václav Šmídl

EXPERIMENTAL EMI STUDY OF A 3-PHASE 100KW 1200V DUAL ACTIVE BRIDGE CONVERTER USING SIC MOSFETS 2000
   Hadiseh Geramirad, Florent Morel, Piotr Dworakowski, Philippe Camail, Bruno Lefebvre, Thomas Lagier, Christian Vollaire

MODELING OF A DAB UNDER PHASE-SHIFT MODULATION FOR DESIGN AND DM INPUT CURRENT FILTER OPTIMIZATION 2010
   Glauber De Freitas Lima, Yves Lembeye, Fabien Ndagijimana, Jean-Christophe Crebier

ACTIVE CURRENT AND ENERGY CONTROL FOR THE QUASI-THREE-LEVEL OPERATION MODE OF AN EXTENDED MODULAR MULTILEVEL CONVERTER TOPOLOGY 2020
   Malte Lorenz, Jakub Kucka, Axel Mertens

TORQUE RIPPLE REDUCTION TECHNIQUE FOR A SWITCHED RELUCTANCE MOTOR 2029
   Krzysztof Jackiewicz, Arkadiusz Kaszewski, Andrzej Stras, Bartlomiej Ufnalski, Tomasz Balkowiec

EXPERIMENTAL VALIDATION OF THE PERFORMANCES OF AN INVERTER SIZED WITH OPTIMIZATION METHODS 2039
   Adrien Voldoire, Jean-Luc Schanen, Jean-Paul Ferrieux, Alexis Derbey, Cyrille Gautier, Marwan Ali

INFLUENCE OF SYSTEM PARAMETERS IN VARIABLE SPEED AC-INDUCTION MOTOR DRIVES ON PARASITIC ELECTRIC BEARING CURRENTS 2049
   Martin Weicker, Guilherme Bello, Dennis Kampen, Andreas Binder

PLASMA IMPACT ON OVERVOLTAGE SHORT-CIRCUIT FAILURES IN ANPC CONVERTERS 2059
   David Hammes, Sidney Gierschner, Dietmar Krug, Hans-Günter Eckel

NOVEL SOFT-SWITCHING INTERLEAVED BOOST CONVERTERS FOR RENEWABLE ENERGY CONVERSION SYSTEMS 2068
Madhuchandra Popuri, V. V. Subrahmanya Kumar Bhajana, Pavel Drabek, Manoj Kumar Maharana

POWER DENSITY OF PLANAR TRANSFORMERS DESIGNED WITH COMMERCIAL STANDARD CORES 2078
Reda Bakri, Xavier Margueron, Jean Sylvio Ngoua Teu Magambo, Philippe Le Moigne, Nadir Idir

EFFECTS OF PV PANEL AND BATTERY DEGRADATION ON PV-BATTERY SYSTEM PERFORMANCE AND ECONOMIC PROFITABILITY 2088
Monika Sandelic, Ariya Sangwongwanich, Frede Blaabjerg

FULL SENSORLESS OPERATION OF INDUCTION MACHINES BASED ON ONLINE IDENTIFICATION OF SALIENCIES USING HARMONIC COMPENSATION LUTS IN TRACTION APPLICATIONS 2098
E. Rodriguez Montero, M. Vogelsberger, T. Wolbank

MITIGATING DRAIN SOURCE VOLTAGE OSCILLATION WITH LOW SWITCHING LOSSES FOR SIC POWER MOSFETS USING FPGA-CONTROLLED ACTIVE GATE DRIVER 2106
Zheming Li, Robert W. Maier, Mark-M. Bakran

ONLINE TRAJECTORY PLANNING DURING LOW-VOLTAGE FRT OF A MODULAR MULTILEVEL CONVERTER 2116
Hendrik Fehr, Albrecht Gensior

EVALUATING FREQUENCY STABILITY WITH CONSIDERATION OF LOAD TYPE IN DIFFERENT SHARE OF RENEWABLES AND EMULATED INERTIA IN CASE OF SYSTEM SPLIT 2126
Nastaran Fazli, Sidney Gierschner, Hans-Günter Eckel

DISCRETE-TIME DIRECT POLE PLACEMENT FOR STABILITY ENHANCEMENT OF LCL-FILTERED INVERTERS IN THE SYNCHRONOUS-REFERENCE FRAME 2135
Pei Cai, Xiaohua Wu, Yongheng Yang, Wenli Yao, Weilin Li, Frede Blaabjerg

ON THE SWITCHING-INDUCED DC-LINK VOLTAGE RIPPLE IN THREE-LEVEL CONVERTERS WITH A NEUTRAL POINT 2145
Ioannis Tsoumas, Tobias Geyer

EFFECT OF PASSIVE INVERTER OUTPUT MOTOR FILTERS ON DRIVE SYSTEMS 2153
Dennis Kampen, Martin Weicker

IMPACT OF THE NEUTRAL POINT POTENTIAL RIPPLE ON THE GRID SIDE HARMONICS OF A 3LNPC BACK-TO-BACK CONVERTER EMPLOYED IN A MEDIUM VOLTAGE WECS 2163
Ioannis Tsoumas

TWO-LAYER GENETIC ALGORITHM FOR THE CHARGE SCHEDULING OF ELECTRIC VEHICLES 2172
Nikolaos T. Milas, Dimitris A. Mourtzis, Panagiotis I. Giotakos, Emmanuel C. Tatakis

SIX-PHASE PMSM DRIVE INVERTER TESTING ON A HIGH PERFORMANCE POWER HARDWARE-IN-THE-LOOP TESTBED 2182
Yasser Rahmoun, Patrick Winzer, Alexander Schmitt, Horst Hammerer

AN IMPROVED BIDIRECTIONAL HYBRID SWITCHED INDUCTOR CONVERTER............................ 2192
  Dan Hulea, Mihaita Gireada, Danut Vitan, Octavian Cornea, Nicolae Muntean

HYBRID MULTIPLE CHOPPER CELLS OF PWM AND SQUARE-WAVE OPERATION FOR
SOLID-STATE TRANSFORMER 2200
  Naoto Kikuchi, Jun-Ichi Itoh, Keisuke Kusaka, Hoai Nam Le

A NEW ZVS ZONE IDENTIFICATION FOR DUAL ACTIVE BRIDGE WITH A GENERAL
MODULATION OBJECTIVE 2210
  Suman Maharana, Dipankar De, Alberto Castellazzi

SINGLE-STAGE BOOST MODULAR MULTILEVEL CONVERTER (BMMC) FOR ENERGY
STORAGE INTERFACE 2220
  Ahmed Abdelhakim, Frede Blaabjerg, Hans-Peter Nee

LOW VOLTAGE GAN-BASED GATE DRIVER TO INCREASE SWITCHING SPEED OF
PARALLELED 650 V E-MODE GAN HEMTS 2230
  Raffael Risch, Jürgen Biela

GATE STRESSES AND THRESHOLD VOLTAGE INSTABILITY IN NORMALLY-OFF GAN
HEMTS 2241
  Jose Ortiz Gonzalez, Burhan Etoz, Olayiwola Alatise

NEW ENERGY MANAGEMENT ALGORITHM BASED ON FILTERING FOR ELECTRICAL
LOSSES MINIMIZATION IN BATTERY-ULTRACAPACITOR ELECTRIC VEHICLES 2251
  *Bakou Traoré, Moustapha Doumiati, Cristina Morel, Jean-Christophe Olivier, Ousmane
  Soumaoro*

MECHANISTIC POWER MODULE DEGRADATION MODELLING CONCEPT WITH
FEEDBACK 2258
  Martin Bendix Fogsgaard, Paula Diaz Reigosa, Francesco Iannuzzo, Michael Hartmann

EXPERIMENTAL VALIDATION AND COMPARISON OF A SIC MOSFET BASED 100 KW
1.2 KV 20 KHZ THREE-PHASE DUAL ACTIVE BRIDGE CONVERTER USING TWO
VECTOR GROUPS 2265
  *Thomas Lagier, Piotr Dworakowski, Cyril Buttay, Philippe Ladoux, Andrzej Wilk, Philippe
  Camail, Elissa Cresenta Anak Justin*

IMPEDANCE ANALYSIS OF AN AUTOMOTIVE DC BUS.. 2274
  Michael Schlüter, Marius Gentejohann, Sibylle Dieckerhoff

A NEW DUAL-MODE MPPT ALGORITHM APPLIED TO A QUADRATIC CONVERTER IN
A SOLAR ENERGY SYSTEM 2284
  Ahmad Ghamrawi, Jean-Paul Gaubert, Driss Mehdi

THERMAL MODEL DEVELOPMENT FOR SIC MOSFETS ROBUSTNESS ANALYSIS
UNDER REPETITIVE SHORT CIRCUIT TESTS 2293
  M. Pulvirenti, D. Cavallaro, N. Bentivegna, S. Cascino, E. Zanetti, M. Saggio

COMPENSATION OF THE RADIAL AND CIRCUMFERENTIAL MODE 0 VIBRATION OF A
PERMANENT MAGNET ELECTRIC MACHINE BASED ON AN EXPERIMENTAL
CHARACTERISATION 2303
  Jan Andresen, Stephan Vip, Axel Mertens, Sebastian Paulus

MEASUREMENT BASED MODEL FOR THE CALCULATION OF CURRENT
DISTRIBUTIONS BETWEEN PARALLELED POWER SEMICONDUCTORS DURING HIGH
CURRENT OPERATION.. 2312
Julian Da Cunha

DUAL-LOOP CONTROL SCHEME WITH OPTIMIZED TYPE-III CONTROLLER BASED ON
GENETIC ALGORITHM FOR 6-PHASE INTERLEAVED CONVERTER IN ELECTRIC
VEHICLE DRIVETRAINS 2320
*Dai-Duong Tran, Sajib Chakraborty, Thomas Geury, Joeri Van Mierlo, Mohamed El
Baghdadi, Omar Hegazy*

HIGH SENSITIVITY CURRENT TRANSFORMER WITH LOW SETTLING TIME, FOR
MAGNIFIED AC CURRENT MEASUREMENTS IN PULSED APPLICATIONS................................... 2331
Georgios Tsolaridis, Pascal Seiler, Juergen Biela

LOSS SEPARATION IN HARD- AND SOFT-SWITCHING GAN HEMTS OPERATED IN A 10
KW ISOLATED DC/DC CONVERTER.. 2341
Jan Böcker, Sören Heucke, Sibylle Dieckerhoff

A SWITCHED-MODE POWER AMPLIFIER FOR ION ENERGY CONTROL IN PLASMA
ETCHING 2350
Qihao Yu, Erik Lemmen, Korneel Wijnands, Bas Vermulst

EXPLORING THE BOUNDARIES AND EFFECTS OF THE DISCONTINUOUS
CONDUCTION MODE IN H-BRIDGE INVERTER WITH DEAD-TIME.. 2358
Qihao Yu, Erik Lemmen, Korneel Wijnands, Bas Vermulst

FIGURES-OF-MERIT AND CURRENT METRIC FOR THE COMPARISON OF IGCTS AND
IGBTS IN MODULAR MULTILEVEL CONVERTERS 2366
*Arthur Boutry, Cyril Buttay, Dong Dong, Rolando Burgos, Bruno Lefebvre, Florent Morel,
Colin Davidson*

ZERO-CURRENT SWITCHING WITH LC RESONANT TANK CIRCUIT AND CAPACITOR
ISOLATION DC-DC CONVERTER.. 2376
Hideki Jonokuchi, Osamu Nakashima, Daichi Hiwatari, Hiroshi Hirayama

A FULL STATE-VARIABLE PREDICTIVE CONTROL OF BI-DIRECTIONAL BOOST
CONVERTERS WITH GUARANTEED STABILITY 2386
Yu Li, Zhenbin Zhang, Ralph Kennel

SYSTEM-LEVEL RELIABILITY ANALYSIS OF A REPAIRABLE POWER ELECTRONIC-
BASED POWER SYSTEM CONSIDERING NON-CONSTANT FAILURE RATES 2393
Amirali Davoodi, Yongheng Yang, Tomislav Dragicevic, Frede Blaabjerg

AN EFFICIENCY ANALYSIS OF A FERRITE MAGNET ASSISTED SYNCHRONOUS
RELUCTANCE MACHINE FOR LOW POWER DRIVES INCLUDING FLUX WEAKENING 2403
Matthias Hofer, Mario Nikowitz, Thomas Kirowitz, Manfred Schrödl

HIGH PERFORMANCE LQR CONTROL OF MODULAR MULTILEVEL CONVERTERS
WITH SIMPLE CONTROL STRUCTURE AND IMPLEMENTATION 2409
Min Jeong, Simon Fuchs, Jürgen Biela

FAULT DETECTION AND CLASSIFICATION BASED ON DEEP LEARNING IN LVDC OFF-
GRID SYSTEM 2419
Iurii Demidov, Antti Pinomaa, Andrey Lana, Olli Pyrhönen

AN INPUT-SERIES OUTPUT-INDEPENDENT FULL-BRIDGE DUAL ACTIVE BRIDGE
CONVERTER WITH SOFT-SWITCHING CHARACTERISTICS FOR CHARGING AND
BALANCING ELECTRIC VEHICLE BATTERY STACKS 2429
Alex V. Mirtchev, Emmanuel C. Tatakis

A METHOD TO SEARCH GLOBAL MAXIMA BY PERMANENT MONITORING OF
VOLTAGE AND CURRENT OF EACH PV PANEL 2439
Shailendra Rajput, Moshe Averbukh

SURVEY AND COMPARISON OF 1D/2D ANALYTICAL MODELS OF HF LOSSES IN LITZ
WIRE 2446
Qingchao Meng, Jürgen Biela

HIGH-FREQUENCY SIC-BASED MEDIUM VOLTAGE QUASI-2-LEVEL FLYING
CAPACITOR DC/DC CONVERTER WITH ZERO VOLTAGE SWITCHING.................................. 2457
Rafal Kopacz, Przemyslaw Trochimiuk, Grzegorz Wrona, Jacek Rabkowski

SMART FUEL CELL MODULE (6.5 KW) FOR A RANGE EXTENDER APPLICATION 2467
Pascal Bazin, Bruno Beranger, Jacques Ecrabey, Laurent Garnier, Sylvain Mercier

IMPACT OF THE INITIAL TRANSIENT INTERRUPTION VOLTAGE (ITIV) ON THE
DESIGN AND OPERATION OF HYBRID CURRENT-INJECTION DC CIRCUIT BREAKERS 2475
Andreas Jehle, Jürgen Biela

FOUR QUADRANT BUS-TIE SWITCH FOR PROTECTION OF SHIPBOARD POWER
SYSTEMS 2486
Gabriele Ulissi, Seong-Yong Lee, Drazen Dujic

ESTIMATION OF AN UNBALANCED GRID IMPEDANCE USING A THREE-PHASE
POWER CONVERTER 2495
Jarno Kukkola, Ville Pirsto, Mikko Routimo, Marko Hinkkanen

FAULT DIAGNOSIS OF HVDC TRANSMISSION SYSTEM USING WAVELET ENERGY
ENTROPY AND THE WAVELET NEURAL NETWORK 2505
Cuicui Liu, Feng Wang, Fang Zhuo, Ziqian Zhang

REDUCING THE ENERGY STORAGE REQUIREMENTS OF MODULAR MULTILEVEL
CONVERTERS WITH OPTIMAL CAPACITOR VOLTAGE TRAJECTORY SHAPING 2513
Simon Fuchs, Min Jeong, Jürgen Biela

LEAKAGE INDUCTANCE MODELLING OF TRANSFORMERS: ACCURATE AND FAST
MODELS TO SCALE THE LEAKAGE INDUCTANCE PER UNIT LENGTH 2524
Richard Schlesinger, Jürgen Biela

A GAN-BASED DC/DC CONVERTER FOR E-VEHICLES APPLICATIONS 2535
Eduardo F. De Oliveira, Sebastian Sprunck, Jonas Pfeiffer, Peter Zacharias

THEORY OF INFLUENCING THE BREATHING MODE AND TORQUE PULSATIONS OF
PERMANENT MAGNET ELECTRIC MACHINES WITH HARMONIC CURRENTS 2545
Jan Andresen, Stephan Vip, Axel Mertens, Sebastian Paulus

POWER HARDWARE IN THE LOOP SYSTEM BASED ON INTERLEAVED CONVERTER
AND FPGA - APPLICATION TO DC AND AC SIDE EMULATION FOR PHOTOVOLTAIC
INVERTER TESTING 2554
R. Kadri, R. Bakri, A. Omrane, F. Colas, F. Delpech

IMPLEMENTATION OF TAPIR SWITCHING CELLS WITH INTEGRATED DIRECT AIR-COOLING FOR SIC POWER DEVICES 2564

Wendpanga Fadel Bikinga, Kouceila Alkama, Bachir Mezrag, Jean Michel Guichon, Yvan Avenas

EFFECT OF UNIPOLAR AND BIPOLAR SPWM ON THE LIFETIME OF DC-LINK CAPACITORS IN SINGLE-PHASE VOLTAGE SOURCE INVERTERS 2573

Silpa Baburajan, Saeed Peyghami, Dinesh Kumar, Frede Blaabjerg, Pooya Davari

TRANSIENT THERMAL MODELS OF CAPACITORS AND INDUCTORS FOR SYSTEM OPTIMIZATION 2583

Vasilios Karaventzas, Juergen Biela, Felix Rodriguez Mateos

ENERGY MANAGEMENT FOR ISOLATED RENEWABLE-POWERED MICROGRIDS USING REINFORCEMENT LEARNING AND GAME THEORY 2594

Rui Hu, Alexis Kwasinski

ALL-GAN BIDIRECTIONAL ANPC-BASED RESONANT DC-DC CONVERTER 2603

Tino Kahl, Laurenz Wernicke, Sibylle Dieckerhoff, Christopher Fromme, Marvin Tannhäuser, Ag Siemens

LIFETIME ESTIMATION AND DIMENSIONING OF THE MACHINE-SIDE CONVERTER FOR PUMPING-CYCLE AIRBORNE WIND ENERGY SYSTEM 2613

Bakr Bagaber, Patrick Junge, Axel Mertens

A DESIGN OF HIGH-POWER INVERTER CIRCUIT INCLUDING GAN POWER DEVICES 2623

Takashi Sawada, Hiroshi Tadano, Koji Shiozaki

SPEED SENSORLESS COMMISSIONING OF RESONATING MECHANICAL SYSTEM IN ELECTRIC DRIVES 2630

A. Putkonen, N. Nevaranta, O. Liukkonen, M. Niemelä, O. Pyrhönen

CONTROL OF A TWO-STAGE, SINGLE-PHASE GRID-TIED, GAN BASED SOLAR MICRO-INVERTER 2638

Anthony Bier, Van Sang Nguyen, Stéphane Catellani, Jérémy Martin

A DC/DC BUCK-BOOST CONVERTER CONTROL USING SLIDING SURFACE MODE CONTROLLER AND ADAPTIVE PID CONTROLLER 2648

Bassem Saleh, Ahmed Teirelbar, Amr Wasfi

SENSORLESS NEUTRAL POINT VOLTAGE STABILIZATION IN THREE-PHASE FOUR-WIRE CONVERTERS 2656

Xinwei Xu, Gabriel Tibola, Jorge L. Duarte

BIDIRECTIONAL ISOLATED RIPPLE CANCEL TRIPLE ACTIVE BRIDGE DC-DC CONVERTER 2666

Takahiro Ohta, Pin-Yu Huang, Yuichi Kado

DESIGN OF THE SPEED SENSORLESS FIELD ORIENTED CONTROL SYSTEM FOR INDUCTION MOTORS CONSIDERING SUDDEN CHANGE OF THE ROTOR SPEED 2675

Yoshiki Sakurazawa, Osamu Yamazaki, Kazuaki Yuki, Yosuke Nakazawa, Kenji Natori, Keiichiro Kondo

EFFICIENCY POTENTIAL OF SOLID-STATE PULSE MODULATORS USING SIC DEVICES 2684

Spyridon Stathis, Michael Jaritz, Sebastian Blume, Jürgen Biela

EFFICIENT AND SCALABLE POWER CONTROL IN MULTI-PORT ACTIVE-BRIDGE CONVERTERS 2695
Soleiman Galeshi, David Frey, Yves Lembeye

COMPARISON OF PRESS-PACK AND WIRE-BONDING TECHNOLOGIES FOR SIC MOSFETS UNDER SHORT-CIRCUIT CONDITIONS 2704
Ran Yao, Francesco Iannuzzo, Amir Sajjad Bahman, Hui Li

ERROR INDUCED BY THE OPTICAL PATH OF A HIGH ACCURACY AND HIGH BANDWIDTH OPTICAL CURRENT MEASUREMENT SYSTEM 2712
Stefan Rietmann, Jürgen Biela

ANALYSIS OF THE RMS CURRENT STRESS ON THE DC LINK CAPACITORS OF THE FOUR PHASE 3-LEVEL T-TYPE VOLTAGE SOURCE CONVERTER 2723
Zoran Miletic, Werner Tremmel, Roland Bründlinger, Johannes Stöckl, Petar J. Grbovic

AN ADAPTIVE DROOP CONTROL METHOD FOR INTERLINK CONVERTER IN HYBRID AC/DC MICROGRIDS 2733
Mohammad S. Golsorkhi, Rasool Heydari, Mehdi Savaghebi

SIMPLIFIED CALCULATION OF PARASITIC ELEMENTS AND MUTUAL COUPLINGS OF WIDE-BANDGAP POWER SEMICONDUCTOR MODULES 2743
Mohammad Ali, Jens Friebe, Axel Mertens

VARIABLE-SPEED-DRIVE-BASED SENSORLESS ESTIMATION OF PUMP SYSTEM RESERVOIR FLUID LEVEL 2753
Santeri Pöyhönen, Aleksi Simola, Jero Ahola

ANALYSIS OF SWITCHING PERFORMANCE AND EMI EMISSION OF SIC INVERTERS UNDER THE INFLUENCE OF PARASITIC ELEMENTS AND MUTUAL COUPLINGS OF THE POWER MODULES 2763
Mohammad Ali, Jan-Kaspar Müller, Jens Friebe, Axel Mertens

WIRE-WOUND MULTI-PHASE STATOR BASED EMEH WITH MPPT SELF-POWERED ENERGY MANAGEMENT SYSTEM 2773
Mahmoud Shousha, Dragan Dinulovic, Talha Zafar, Michael Brooks, Martin Haug

COMPARISON OF OPTIMIZED MOTOR-INVERTER SYSTEMS USING A STACKED POLYPHASE BRIDGE CONVERTER COMBINED WITH A 3-, 6-, 9-, OR 12-PHASE PMSM 2780
Thilo Bringezu, Jürgen Biela

DESIGN OF A PULSE MODULATOR BASED ON TRANSMISSION LINES FOR GENERATING FAST CURRENT PULSES FOR PLASMA DRILLING 2791
Oliver Keel, Melissa Artiglia, Juergen Biela

ANALYSIS OF CURRENT IN PULSATING DC LINK CONVERTER WITH ZERO VOLTAGE TRANSITION 2802
Daniele Marciano, Giovanni Busatto, Carmine Abbate, Annunziata Sanseverino, Davide Tedesco, Francesco Velardi

SIGNAL INJECTION FOR SENSORLESS CURRENT SHARING WITH EXPERIMENTAL VERIFICATION ON 1 MHZ GAN PROTOTYPE 2812
N. Boškovic, J. Duarte, E. A. Lomonova

MODELLING AND ANALYSIS OF SENSORLESS CURRENT SHARING APPROACH 2820
N. Boškovic, J. Duarte

PWM-INDUCED HARMONIC POWER IN 75 KW IM DRIVE SYSTEM 2829
Lassi Aarniovuori, Hannu Kärkkäinen, Markku Niemelä, Juha Pyrhönen

PROPOSAL OF BOOST CONVERTER WITHOUT REACTOR USING OPEN-ENDED
WINDING PMSM FOR PHOTOVOLTAIC PUMP SYSTEM 2838
Akihiro Okazaki, Sari Maekawa

THE PROPOSAL OF DISCRIMINATING STABLE CONTROL BANDWIDTH USING ANN IN
SENSORLESS SPEED CONTROL SYSTEM FOR PMSM 2844
Ami Tanaka, Sari Maekawa

COST FUNCTION DESIGN FOR STABILITY ASSESSMENT OF MODULATED MODEL
PREDICTIVE CONTROL 2851
Jordan P. Zucuni, Fernanda Carnielutti, Humberto Pinheiro, Margarita Norambuena, Jose
Rodriguez

A ROBUST FUZZY-BASED CONTROL TECHNIQUE FOR WIND FARM TRANSIENT
VOLTAGE STABILITY USING SVC AND STATCOM: COMPARISON STUDY 2860
Reza Ebrahimi, Vahid Eslampanah, Hossein Madadi Kojabadi, Mohammadreza Azizian,
Naser Nourani Esfetanaj, Dao Zhou

TEMPERATURE EVOLUTION AS AN EFFECT OF WIRE-BOND FAILURES IN A MULTI-
CHIP IGBT POWER MODULE 2865
N. Degrenne, R. Delamea, S. Mollov

COST OF ENERGY ASSESSMENT OF WIND TURBINE CONFIGURATIONS 2873
Catalin Dincan, Philip Kjær, Lars Helle

ENERGY MANAGEMENT IN A MULTI-SOURCE SYSTEM USING ISOLATED DC-DC
RESONANT CONVERTERS 2881
M. Arazi, A. Payman, M. B. Camara, B. Dakyo

LONG-TERM CLIMATE IMPACT ON IGBT LIFETIME 2888
Martin Vang Kjaer, Yongheng Yang, Huai Wang, Frede Blaabjerg

COMMUNICATION-FREE SECONDARY FREQUENCY AND VOLTAGE CONTROL OF
VSC-BASED MICROGRIDS: A HIGH-BANDWIDTH APPROACH 2898
Rasool Heydari, Mohammad S. Golsorkhi, Mehdi Savaghebi, Tomislav Dragicevic, Frede
Blaabjerg

OFFSHORE WIND FARM LAYOUT OPTIMIZATION CONSIDERING WAKE EFFECTS 2907
Asma Dabbabi, Salvy Bourguet, Rodica Loisel, Mohamed Machmoum

SMALL-SIGNAL STABILITY ANALYSIS OF SMART GRIDS CONSIDERING HIGH
PENETRATION OF POWER ELECTRONICS CONVERTERS AND ENERGY MARKETS 2917
Javiera Meneses, Patricio Mendoza-Araya

COMPONENT-LEVEL RELIABILITY ASSESSMENT OF A DIRECT-DRIVE PMSG WIND
POWER CONVERTER CONSIDERING LONG-TERM AND SHORT-TERM THERMAL
CYCLES 2928
Shuaichen Ye, Dao Zhou, Frede Blaabjerg

A SUBMODULE IMPLEMENTATION FOR PARALLEL CONDUCTION OF DIODES IN
MODULAR MULTILEVEL CONVERTERS 2938
Martin Geske, Duro Basic, Christian Keller, Thomas Brückner

EVALUATION OF THE I_{MAX}-F_{SW}-DV/DT TRADE-OFF OF HIGH VOLTAGE SIC MOSFETS BASED ON AN ANALYTICAL SWITCHING LOSS MODEL 2946
Anliang Hu, Jürgen Biela

PROTECTION MEASURES FOR MODULAR MULTILEVEL CONVERTERS IN CASE OF DC SHORT-CIRCUIT FAULTS 2957
Martin Geske, Duro Basic, Roland Jakob, Christian Keller, Thomas Brückner

INVESTIGATION ON PARALLEL OPERATION OF TWO MMC-HVDC LINKS IN GRID FORMING CONNECTED TO AN EXISTING NETWORK 2967
H. Saad, P. Rault, S. Dennetière

MODELLING AND EXPERIMENTAL VALIDATION OF A LABORATORY-SCALED HVDC CABLE EMULATOR TESTED IN AN MMC-BASED PLATFORM 2977
Enric Sánchez-Sánchez, Adrià Junyent-Ferré, Eduardo Prieto-Araujo, Oriol Gomis-Bellmunt, Tim Green

DAISY CHAIN PN CELL FOR MULTILEVEL CONVERTER USING GAN FOR HIGH POWER DENSITY 2987
Faheem Ahmad, Asger Bjørn Jørgensen, Szymon Michal Beczkowski, Stig Munk-Nielsen

GRID-FREQUENCY VIENNA RECTIFIER AND ISOLATED CURRENT-SOURCE DC-DC CONVERTERS FOR EFFICIENT OFF-BOARD CHARGING OF ELECTRIC VEHICLES 2996
Jacek Rabkowski, Andrei Blinov, Denys Zinchenko, Grzegorz Wrona, Mariusz Zdanowski

UNIDIRECTIONAL THYRISTOR-BASED DC-DC CONVERTER FOR HVDC CONNECTION OF OFFSHORE WIND FARMS 3006
Pierre Le Métayer, Piotr Dworakowski, Jose Maneiro

INDUCTOR SIZE EVALUATION OF AN ELECTROMAGNETIC INTERFERENCE FILTER FOR A TWO-LEVEL POWER FACTOR CORRECTION RECTIFIER USING DIFFERENT MODULATION TECHNIQUES 3015
Mohammad Najjar, Alireza Kouchaki, Morten Nymand

EVALUATION OF MMCS FOR HIGH-POWER LOW-VOLTAGE DC-APPLICATIONS IN COMBINATION WITH THE MODULE LLC-DESIGN 3024
Roland Unruh, Frank Schafmeister, Joachim Böcker

IRON LOSS CHARACTERISTICS OF MNZN FERRITES UNDER GAN INVERTER EXCITATION IN THE MHZ ORDER 3034
Wilmar Martinez, Camilo Suarez, Federico Ibanez

VIBRATION SUPPRESSION AND CONTROL PARAMETER DESIGN OF A SENSORLESS PMSM ROTARY COMPRESSOR DRIVE 3044
Tao Li, Chaohui Liang

3D PCB PACKAGE FOR GAN INVERTER LEG WITH LOW EMC FEATURE 3054
Pawel B. Derkacz, Jean-Luc Schanen, Pierre-Olivier Jeannin, Piotr Musznicki, Piotr J. Chrzan, Mickael Petit

ESTIMATION OF THE WINDING LOSSES OF MEDIUM FREQUENCY TRANSFORMERS WITH LITZ WIRE USING AN EQUIVALENT PERMEABILITY AND CONDUCTIVITY METHOD 3064
Mohammad Kharezy, Morteza Eslamian, Torbjörn Thiringer

IMPROVEMENT OF DRIVING EFFICIENCY OF PMSM BY USING MODIFIED TRAPEZOIDAL MODULATING SIGNAL 3071
Kento Betto, Satoshi Joryo, Toshimitsu Morizane

DESIGN AND CONTROL OF A VIRTUAL DC-LINK FOR A FULL GAN-BASED SINGLE PHASE CONVERTER WITH HIGH POWER DENSITY 3081
Yugandhara H. Wankhede, Leon Fauth, Jens Friebe

USING BOTH THE CIRCULATING CURRENTS AND THE COMMON-MODE VOLTAGE FOR THE BRANCH ENERGY CONTROL OF MODULAR MULTILEVEL CONVERTERS 3091
Rebecca Dierks, Jakub Kucka, Axel Mertens

ANALYTICAL HARMONIC CURRENT MODEL FOR A PERMANENT MAGNET ASSISTED SYNCHRONOUS RELUCTANCE MOTOR (PMA-SYNRM) FED BY PWM INVERTER 3101
Jessica Neumann, Carole Hénaux, Maurice Fadel, Etienne Founier, Dany Prieto, Mathias Tientcheu Yamdeu

GENERALIZED SMALL-SIGNAL AVERAGED SWITCH MODEL ANALYSIS OF A WBG-BASED INTERLEAVED DC/DC BUCK CONVERTER FOR ELECTRIC VEHICLE DRIVETRAINS 3111
Sajib Chakraborty, Dai-Duong Tran, Joeri Van Mierlo, Omar Hegazy

ADAPTIVE PREDICTIVE-DPC FOR LCL-FILTERED GRID CONNECTED VSC WITH REDUCED NUMBER OF SENSORS 3119
Hosein Gholami-Khesht, Pooya Davari, Frede Blaabjerg

FPGA IMPLEMENTATION OF MODIFIED SPACE VECTOR MODULATION (SVM) FOR HIGH-FREQUENCY HYBRID ACTIVE NEUTRAL-POINT-CLAMPED (NPC) POWER FACTOR CORRECTION RECTIFIER 3129
Mohammad Najjar, Alireza Kouchaki, Morten Nymand

ENHANCED FLUX CONTROL INCLUDING A CLOSED LOOP VOLTAGE CONTROLLER TO OPTIMIZE THE VOLTAGE USAGE AND THE TORQUE COMPUTATION FOR A 48V IPMSM 3137
Felix Bertele, Ulrich Ammann, Christoph Cheshire, Tobias Röser

EXTENDED BOOST PV INVERTER TOPOLOGY FOR THE REDUCTION OF COMMON-MODE LEAKAGE CURRENT IN THREE-PHASE APPLICATIONS 3146
Georgios I. Orfanoudakis, Eftychios Koutroulis, Michael A. Yuratich, Suleiman M. Sharkh

A ROBUST CONTROL DESIGN TO REAL-TIME CONDITIONS AND MODELLING OF A MICROGRID 3156
Iréna Horvatic, Delphine Riu, Moataz Elsied, Sébastien Benjamin

DESIGN OF MODULAR LOW-PROFILE FREQUENCY CONVERTER FOR MULTI-MOTOR MANIPULATORS 3166
Tomas Glasberger, Zdenek Kehl, Tomas Kosan, Jan Molnar

STUDY OF THE CONTROL OF A NEW AC VOLTAGE STABILIZER USING LINEAR CONTROLLER WITH REFERENCE FRAME TRANSFORMATION 3172
Bunthern Kim, Etienne Boulaud, Emile Boisaubert, Sokchea Am, Phok Chrin

HYBRID ENERGY STORAGE SYSTEM FOR MVDC-GRIDS 3179
Florian Mahr, Johann Jaeger, Stefan Henninger, Hubert Rubenbauer

A COMBINED MODEL FOR OPTIMAL POWER FLOW APPLIED TO MT-HVDC SYSTEMS 3189
Fernando Torres, Javier Muñoz, Fredy Muñoz, Claudio Roa

CHARACTERIZATION OF LITHIUM ION SUPERCAPACITORS 3198

Zeyang Geng, Felix Mannerhagen, Torbjöm Thiringer

GREY WOLF OPTIMIZER BASED PREDICTIVE TORQUE CONTROL FOR ELECTRIC
VEHICLE APPLICATIONS 3205

*Ali Djerioui, Azeddine Houari, Mohamed Machmoum, Malek Ghanes, Tedjani Mesbahi,
Mohamed Fouad Benkhoris*

OPERATION PRINCIPLE AND PERSPECTIVE PERFORMANCES OF METAL OXIDE
VACUUM FIELD EFFECT TRANSISTOR - MOVFET 3210

Davide Patti, G. Busatto, G. Golluccio, D. Marciano, A. Sanseverino, F. Velardi

IMPROVED METHODOLOGY FOR PREDICTING CORRELATED COLOR TEMPERATURE
IN MIXED LED LIGHTING SOURCES 3217

Thais E. Bolzan, Bruno F. Almeida, Renan R. Duarte, Vitor C. Bender, Rafael A. Pinto

DC MICROGRID CONCEPT FOR MINE ENVIRONMENT 3227

Jooa Pursiainen, Jenni Rekola, Raimo Juntunen, Mikko Valtee, Pasi Peltoniemi

A COMPARISON OF TWO-STAGE INVERTER AND QUASI-Z-SOURCE INVERTER FOR
HYBRID ENERGY STORAGE APPLICATIONS 3237

V. Castiglia, R. Miceli, F. Blaabjerg, Y. Yang

STATE ESTIMATION FOR MEDIUM AND LOW VOLTAGE DISTRIBUTION GRIDS
BASED ON NEAR REAL-TIME GRID MEASUREMENTS AND DELAYED SMART
METERS DATA 3247

Mohammad Rayati, Thomas Pidancier, Mauro Carpita, Mokhtar Bozorg

GROUND FAULT ACTIVE COMPENSATION IN EMULATED DISTRIBUTION GRID OF 10
KV 3257

*Tomáš Komrska, Antonín Glac, Jakub Talla, Bohumil Skala, Jan Štepánek, Lubeš Streit,
Zdenek Peroutka*

MODELING OF A POWER TRANSFORMER INCLUDING HIGHER ORDER RESONANCES 3263

Lukas Reißenweber, Alexander Stadler

A COMPARISON OF TWO STATE-SPACE MODELS OF AN INDUCTION MACHINE
CONSIDERING DIFFERENT SETS OF WINDING DISTRIBUTION HARMONICS 3272

Julien Cordier, Stefan Klass, Ralph Kennel

PERFORMANCE IMPROVEMENT FOR PLUG-IN REVERSE CONDUCTING IGBTS
THROUGH GATE-VOLTAGE OBSERVATION 3282

Daniel Lexow, Hans-Günter Eckel

DIFFERENTIAL FLATNESS FOR SMOOTH TRANSITION BETWEEN GRID-CONNECTED
AND STANDALONE MODE OF THREE-PHASE INVERTER 3289

*Abdelhakim Saim, Azeddine Houari, Mourad Ait-Ahmed, Mohamed Machmoum, Josep. M
Guerrero*

DIFFERENTIAL MODEL EMI FILTER ANALYSIS FOR INTERLEAVED BOOST PFC
CONVERTERS CONSIDERING OPTIMAL PHASE SHIFTING 3295

Naser Nourani Esfetanaj, Yamen Saad, Omar Ahmed Sakaria, Huai Wang, Pooya Davari

MODULAR HYBRID DC BREAKER-BASED ADAPTIVE AUTO-RECLOSING METHOD
FOR MMC-HVDC SYSTEMS 3305

Hossein Iman-Eini, M. Langwasser, L. Camurca, Marco Liserre

MULTISTEP MPC OF DUAL INVERTER FOR SWITCHING LOSSES OPTIMIZATION 3314
Martin Votava, Tomas Glasberger, Zdenek Peroutka

A HIGH-EFFICIENCY CONTROL OF A DOUBLE-INPUT CONVERTER FOR RENEWABLE
ENERGIES AND HYBRID VEHICLES.. 3321
Mario Marchesoni, Massimiliano Passalacqua, Luis Vaccaro

DEAD-TIME INFLUENCE ON FAST SWITCHING PULSED POWER CONVERTERS
DESIGN - A HIGH CURRENT APPLICATION FOR ACCELERATOR'S MAGNETS 3330
*Ludovic Horrein, Jean-Marc Cravero, Philippe Delarue, Alain Bouscayrol, Davide Aguglia,
Carmen Ortega-Perez*

DYNAMIC CHARACTERIZATION OF A SIC-MOSFET HALF BRIDGE IN HARD- AND
SOFT-SWITCHING AND INVESTIGATION OF CURRENT SENSING TECHNOLOGIES 3340
Janine Ebersberger, Jan-Kaspar Müller, Axel Mertens

POWER SUPPLY DESIGN CONSIDERATIONS FOR 400HZ AIRCRAFT APPLICATIONS 3348
Bilal Ahmad, Jorma Kyyrä, Juha Mäkelä

DC CAPACITOR VOLTAGE FEEDBACK METHOD FOR A PEAK VOLTAGE
SUPPRESSION CONTROL WITH MULTIPLE LEG-SHORT-CIRCUITS USING SIC-
MOSFETS EMPLOYED IN POWER CONVERTERS 3358
Tomoyuki Mannen, Takanori Isobe, Keiji Wada

INVESTIGATION OF BOND WIRE LIFT-OFF BY ANALYZING THE CONTROLLER
OUTPUT VOLTAGE HARMONICS FOR THE PURPOSE OF CONDITION MONITORING 3366
Firat Yüce, Marc Hiller

FRUGAL INNOVATION FOR SUSTAINABLE RURAL ELECTRIFICATION 3376
Bunthern Kim, Phok Chrin, Maria Pietrzak-David, Pascal Maussion

A CURRENT-MODULUS DERIVATIVE-BASED PROTECTION METHOD IN A FLEXIBLE
DC GRID 3385
Jianquan Liao, Niancheng Zhou, Qianggang Wang

COMPARATIVE ASSESSMENT OF VOLTAGE MODULATION METHODS FOR
ASYMMETRIC SIX-PHASE MACHINES 3393
R. S. Kanchan, Omer Ikram Ul Haq, Luca Peretti

SIMULATION AND MEASUREMENT-BASED ANALYSIS OF EFFICIENCY
IMPROVEMENT OF SIC MOSFETS IN A SERIES-PRODUCTION READY 300 KW / 400 V
AUTOMOTIVE TRACTION INVERTER... 3403
*A. Nisch, M. Heller, W. Wondrak, A. Bucher, C. Hasenohr, K. Kefer, B. Lunz, A. Pawellek, A.
Smit, M. Gärtner, N. Twardon, U. Kirchenberger*

VALIDITY OF POWER CYCLING LIFETIME MODELS FOR MODULES AND EXTENSION
TO LOW TEMPERATURE SWINGS 3413
Josef Lutz, Christian Schwabe, Guang Zeng, Lukas Hein

ROADMAP FOR DC.. 3422
Pavol Bauer

THE ROLE OF COLLABORATIVE RESEARCH TO SUPPORT INNOVATION FOR CLEAN
ENERGY TRANSITION 3424
Hubert De La Grandiere

THOMAS EDISON VINDICATED — THE RESURGENCE OF DC IN MV AND HV POWER GRIDS 3425

Colin Davidson

INTEGRATION OF ELECTRIC MOBILITY IN THE FRENCH PUBLIC ELECTRICITY DISTRIBUTION NETWORK 3426

Anne-Sophie Cochelin

A CRITICAL ROLE FOR R&I FOR CLEAN ENERGY FOR THE EU GREEN AND DIGITAL RECOVERY 3427

Hélène Chraye

Author Index

Validation of Thermal Stress Modeling in PV Inverters under Mission Profile Operation

Ariya Sangwongwanich, Huai Wang, and Frede Blaabjerg
Department of Energy Technology, Aalborg University, Aalborg, Denmark
E-Mail: ars@et.aau.dk, hwa@et.aau.dk, fbl@et.aau.dk

Keywords

«Reliability», «Mission profile», «Photovoltaic», «Power semiconductor device», «Thermal stress».

Abstract

This paper quantifies the accuracy of thermal stress modeling in PV inverters under real mission profile operation. The estimated thermal stress profiles obtained from a lumped thermal network under one-day mission profiles are compared with the experimental measurement. According to the results, the average estimation error is well below 1.5 % even under highly dynamics mission profile conditions.

Introduction

Reliability is one of the key aspects of the design and development of Photovoltaic (PV) inverters. Power devices such as Insulated-Gate Bipolar Transistors (IGBTs) are one the most reliability-critical components in the PV inverters, which are subjected to high thermal stress during operation [1]. To ensure reliable and robust operation of PV inverters, thermal stress analysis of the power devices needs to be carried out during the design phase by considering the real operating conditions of the PV inverter, also referred to as mission profiles [2].

In that regard, thermal modeling of the power devices in PV inverters is essential. On one hand, the thermal model is used to ensure that the thermal stress of the power devices under the worst-case scenario (e.g., maximum loading condition) is still within its maximum limit in order to ensure safe operation and also identify a robust design margin. On the other hand, it is also employed during the reliability prediction of the power devices [3], where the dynamic loading from the mission profile needs to be translated into the thermal stress of the power device through a thermal model. The obtained thermal stress is then applied to the lifetime prediction (e.g., through cycle counting algorithm) to estimate the probability of wear-out failure of the power devices under certain mission profiles. Since the thermal stress dynamic is dictated by the loading and ambient temperature condition of the PV inverter, the mission profile is required as an input of the thermal stress analysis. In PV applications, the mission profile parameter consists of the solar irradiance and temperature, whose time-spans are in a range of days to months. Therefore, a simplified thermal model, e.g., based on a lumped thermal network, is normally preferred to be used under such a long-term simulation due to: 1) low computational burden and 2) simple parameterization (e.g., from datasheet), and it has been applied by the previous researches [4]-[9].

However, there is still a lack of validation in terms of thermal stress modeling accuracy, especially when comparing the simulation results with the real thermal stress of the PV inverter, e.g., the junction temperature of the power devices, under real-field mission profile operation. Since the lumped thermal network is, to a certain degree, a simplified representation of the thermal network, it will inevitably introduce a certain error in the thermal stress estimation. The estimation error will introduce uncertainty in the reliability and robustness analysis and thus needs to be quantified.

Accordingly, a validation of thermal stress modeling accuracy of the power device in PV inverters under mission profile operation is carried out in this paper. A test-bench of a PV inverter, which allows the experimental measurement of power device thermal stress under mission profile operation, will be discussed. Then, a step-by-step thermal stress modeling method is provided, including power loss and thermal impedance characterization. Afterwards, the mission profiles are applied to the thermal model, and the obtained thermal stress profiles from simulations are compared with the experimental

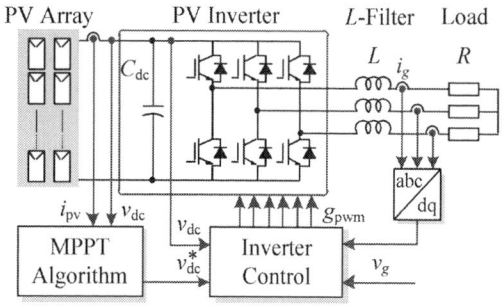

Fig. 1. System diagram of the three-phase PV inverter test-bench.

Table I. Parameters of PV Inverter Test-Bench

PV array rated power	2500 W
Output current (rated)	$i_g = 30$ A
DC-link voltage	$v_{dc} = 400\text{-}600$ V
DC-link capacitance	$C_{dc} = 340$ μF
Filter inductance	$L = 2.5$ mH
Resistive load	$R = 16.5\ \Omega$
Switching frequency	$f_{sw} = 10$ kHz
AC output frequency	$f_g = 50$ Hz
Ambient temperature	$T_a = 25\ °C$

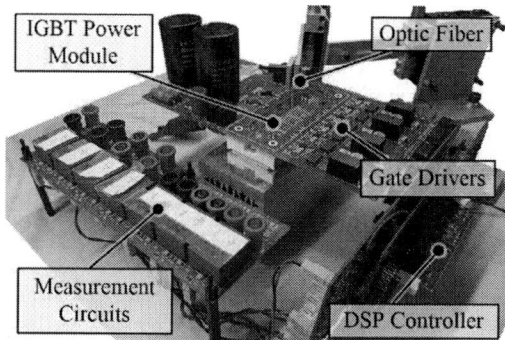

Fig. 2. Hardware prototype of the three-phase PV inverter test-bench with the IGBT junction temperature measurement using an optic fiber.

measurements from the PV inverter test-bench, where the error from the IGBT junction temperature estimation is measured. Finally, concluding remarks are given in the last section.

Real-Field Thermal Stress of PV Inverters

Test-Bench for PV Inverters

In order to validate the thermal stress modeling, a test-bench of PV inverter, which is capable of emulating the mission profile operation is required. In this work, a two-level three-phase PV inverter shown in Fig. 1 is used as the power stage of the PV inverter test-bench. A PV simulator is employed to emulate the electrical behavior of PV arrays under mission profile operations. The MPPT algorithm is implemented in the control of the PV inverter together with the dc-link voltage controller and current controller as discussed in [10]. The extracted PV power is then delivered to the load. The prototype of the PV inverter test-bench is shown in Fig. 2 and the system parameters are provided in Table I.

The power devices are realized by a three-phase IGBT power module. An opened power module is used together with a customized Printed Circuit Board (PCB), as it is shown in Fig. 2. This allows a direct measurement of the IGBT junction temperature during the operation (e.g., mission profile) by using an optic fiber as it is demonstrated in Fig. 2.

Thermal Stress under Mission Profile Operation

Two daily mission profiles are applied to the test-bench in order to obtain real-field thermal stress profiles from the experiment. The first mission profile is shown in Fig. 3(a), which represents a typical clear-day mission profile of the PV inverter, where the PV output power changes smoothly during the day. In another case, a relatively high-dynamic mission profile is selected as shown in Fig. 3(b), which

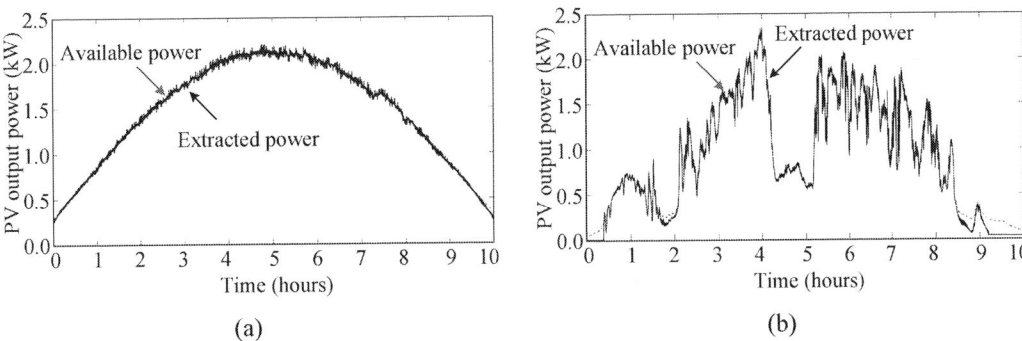

(a)   (b)

Fig. 3. PV power extraction of the PV inverter test-bench under one-day mission profile operations: (a) clear day and (b) cloudy day.

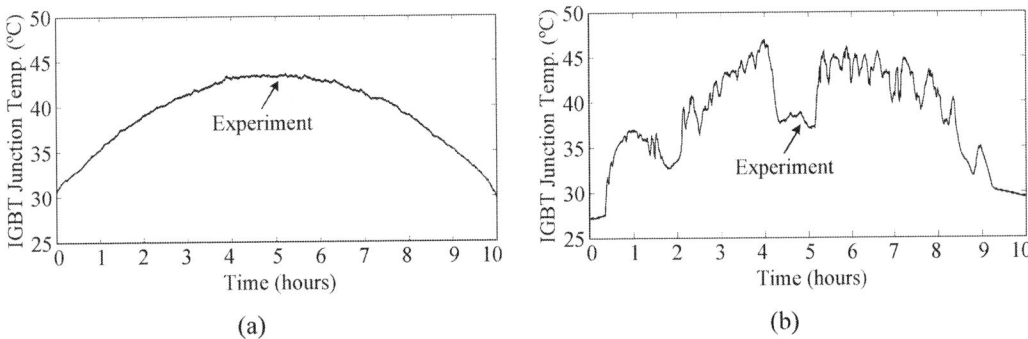

(a)   (b)

Fig. 4. The experimental measurement of thermal stress of PV inverter under one-day mission profile operations: (a) clear day and (b) cloudy day.

occurs during the cloudy-day condition. The fluctuation in the PV power profile during the cloudy-day condition will introduce the high-dynamic thermal stress of the PV inverter, which will challenge the accuracy of the thermal modeling [11]. These two mission profile conditions are considered as benchmark cases in this paper.

The experimental measurements of the obtained thermal stress profiles of the PV inverter when applying the mission profiles in Fig. 3 are shown in Fig. 4(a) and Fig. 4(b) for the clear-day and cloudy-day mission profile conditions, respectively. In this case, the sampling rate of the mission profile applied to the PV simulator is 1 minute, while the sampling rate of the temperature measurement is 1 kHz (with the data acquisition, e.g., averaging, period of 1 second). In this way, several sampling points of the temperature measurement can be obtained for each load change. It should be noted that the experimental test is carried out in real-time, where the testing time is 10 hours (corresponding to the PV inverter loading period during the day).

Thermal Stress Modeling of PV Inverters

In the prototype, a 1200V/50A three-phase IGBT module [12] is used as the power stage. The thermal stress modeling of the IGBT in the PV inverter, which includes the power losses and thermal impedance characterizations, will be discussed in the following.

Power Losses Model

The power loss of the IGBT consists of switching loss $P_{S,sw}$ and conduction loss $P_{S,con}$. In this work, a look-up table obtained from the datasheet is used from calculating the average power losses during operation (for the purpose of a long-term simulation), as it is shown in Fig. 5. The total power losses dissipated in the IGBT can be obtained as given in the following:

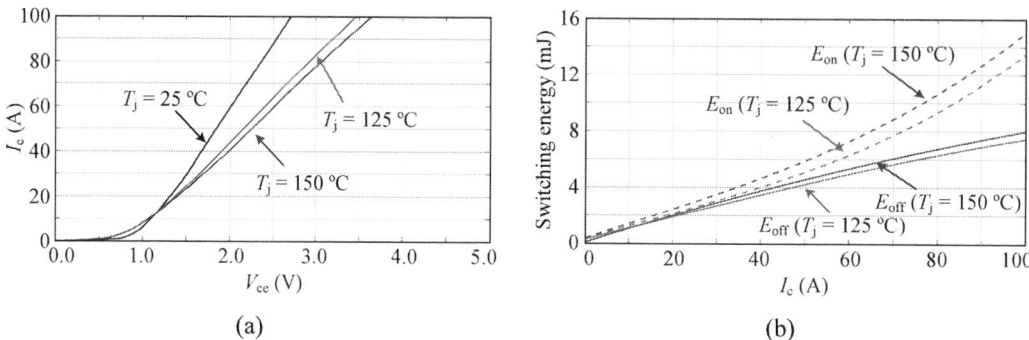

(a)   (b)

Fig. 5. Power loss characteristic of the IGBT at different collector-emitter currents I_c and junction temperature T_j: a) Output characteristic and b) Switching losses, where E_{on} and E_{off} are the energy loss during turn-on and turn-off, respectively [12].

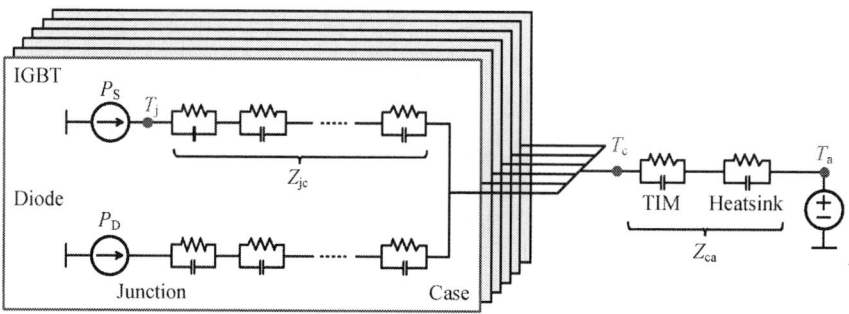

Fig. 6. Thermal model of three-phase IGBT module in PV inverter based on lumped thermal network (TIM: Thermal Interface Material).

$$P_S = P_{S,sw}(f_{sw}, E_{on}, E_{off}) + P_{S,con}(i_c, v_{ce}) \tag{1}$$

where f_{sw} is the switching frequency, E_{on} and E_{off} are the turn-on and turn-off energy, respectively, i_c and v_{ce} are the collector-emitter current and voltage of the IGBT during conduction, respectively. Notably, the power losses are affected by the junction temperature, which is taken into consideration in the look-up table. The detailed power losses calculation method can be found in [13]. A similar approach has also been applied for the power loss calculation of the diodes. Therefore, it will not be repeated here.

Thermal Model

A thermal model of the three-phase IGBT power module used in this paper is shown in Fig. 6. The thermal impedance network is based on the Foster's network, whose parameters can easily be fitted from the experimental results (and usually available in the datasheet). In general, the junction temperature of the IGBT is contributed by the temperature drop inside the power module T_{jc} (e.g., between junction and case of the IGBT power module), between the case and ambient T_{ca}, and the ambient temperature T_a following:

$$\begin{aligned} T_j(t) &= T_{jc}(t) + T_{ca}(t) + T_a(t) \\ &= P_S \cdot Z_{jc}(t) + 6 \cdot (P_S + P_D) \cdot Z_{ca}(t) + T_a(t) \end{aligned} \tag{2}$$

where P_S and P_D are the total power losses of each IGBT and diode. Z_{jc} is the thermal impedance between the junction and case of the IGBT, while Z_{ac} is the thermal impedance between the case and ambient condition, representing the Thermal Interface Material (TIM) and the heatsink (and also the cooling systems).

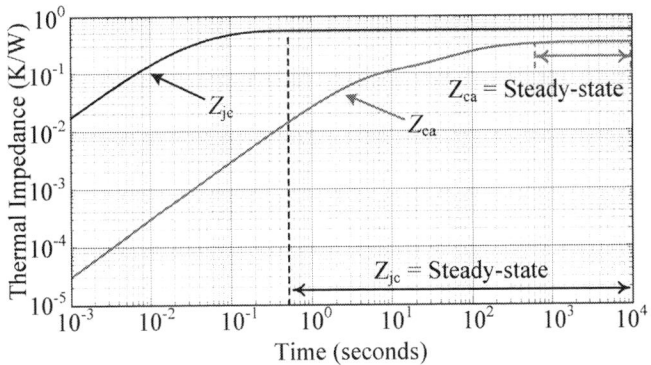

Fig. 7. Thermal impedance of the IGBT module between junction and case Z_{jc} and between the case and ambient Z_{ca}

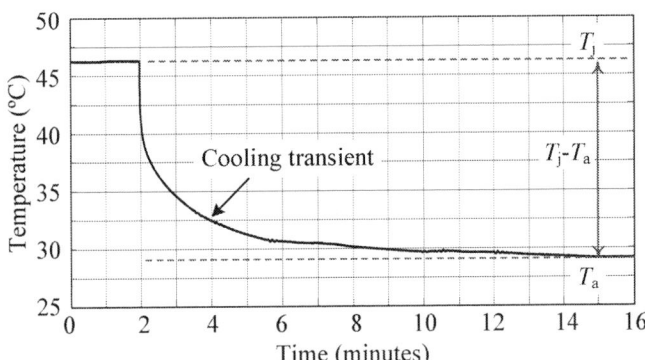

Fig. 8. Cooling curve of the IGBT module where the power losses are supplied until $t = 2$ minutes.

Table II. Parameters of thermal impedance between junction and case Z_{jc} [12].

Layer i	1	2	3	4
$R_{jc,i}$	0.0324	0.1782	0.1728	0.1566
$C_{jc,i}$	0.3086	0.1122	0.2894	0.6386

Table III. Parameters of thermal impedance between case and ambient Z_{ca}.

Layer i	1	2	3
$R_{ca,i}$	0.0670	0.1737	0.0869
$C_{ca,i}$	6,157	404.72	37.335

Extraction of Thermal Impedance

The accuracy of the lumped thermal network relies on the parameters of the *RC* circuit (i.e., thermal impedance). According to (2), there are two main thermal impedance networks that need to be parameterized: Z_{jc} and Z_{ca}. The parameter of the thermal impedance network inside the IGBT module Z_{jc} is usually provided by the manufacturer (e.g., datasheet). In this case, the thermal impedance between the junction and case Z_{jc} is obtained from the datasheet [12], as it is shown in Fig. 7 and Table II.

On the other hand, the thermal impedance between the case and ambient Z_{ca} is strongly dependent on the design of the cooling system and also on the applied TIM [14]. Thus, their parameters are usually not provided by the manufacturer and need to be characterized, e.g., via experiments. Although the thermal resistance and capacitance can be determined analytically from the geometry and material properties of the cooling system, it requires a relatively complex modeling effort. In practice, the parameters of the thermal impedance Z_{ca} can be obtained from the experimental result during the cooling phase of the IGBT module [15]. An example of the IGBT cooling curve is shown in Fig. 8, where a

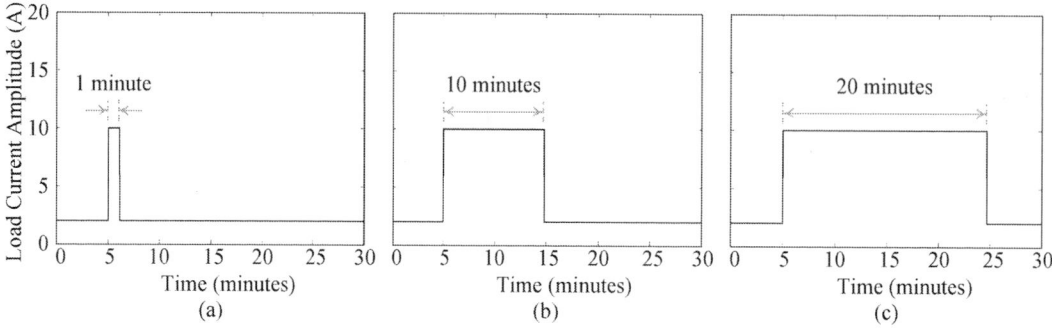

Fig.9. Load current amplitude of the PV inverter test-bench when applying different load dynamics (e.g., load duration): a) 1 minute, b) 10 minutes, and c) 20 minutes, is applied.

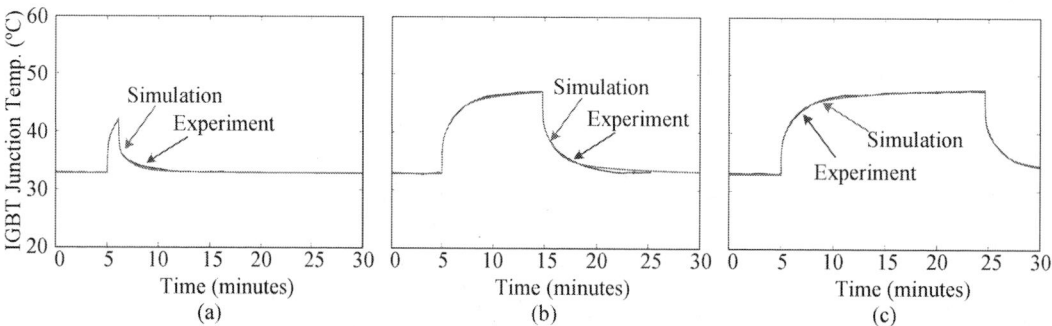

Fig.10. Experimental and simulation results of thermal stress when applying different load dynamics (e.g., load duration): a) 1 minute, b) 10 minutes, and c) 20 minutes, is applied.

certain power loss is initially applied to the IGBT and then removed after $t = 2$ minutes. In that case, the transient thermal impedance between the case and ambient Z_{ca} can be calculated by the following:

$$Z_{ca}(t) = \frac{T_j(t) - T_a(t) - P_S \cdot Z_{jc}(t)}{6 \cdot (P_S + P_D)} \qquad (3)$$

An example of the thermal impedance between the case and ambient Z_{ca} derived from the cooling curve is shown in Fig. 7 and its parameters are provided in Table III. By doing so, all the required thermal impedance parameters of the PV inverter are obtained.

Validation of Thermal Stress Modeling under Mission Profile Operation

Thermal Stress under Step-Load

A step-load is first applied to validate the accuracy of the thermal model under dynamic loading conditions. In Fig. 9, a step-change is applied to the PV inverter output current (i.e., from 2 A to 10 A and then back to 2 A) at $t = 5$ minutes with different load pulse duration, e.g., 1 minute, 10 minutes and 20 minutes, and the corresponding thermal stress profiles are shown in Fig. 10. It can be seen from the results in Fig. 10(a) that the junction temperature is still in the transient phase when the load duration is 1 minute. In contrast, the junction temperature is reaching its steady-state value when load durations of 10 minutes and 20 minutes are applied as it is shown in Fig. 10(b) and (c), which corresponds to the time-constant of the thermal impedance Z_{ca} in Fig. 6. In all cases, the junction temperature can be estimated accurately with the thermal model where the dynamics of the thermal stress can be fully captured. These results validate the accuracy of the thermal stress modeling under short-term operation.

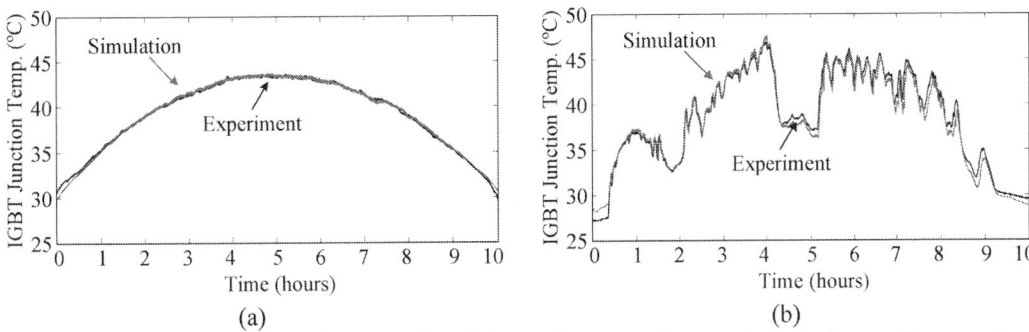

(a)   (b)

Fig. 11. Experimental and simulation results of thermal stress under one-day mission profile operation: (a) clear day and (b) cloudy day.

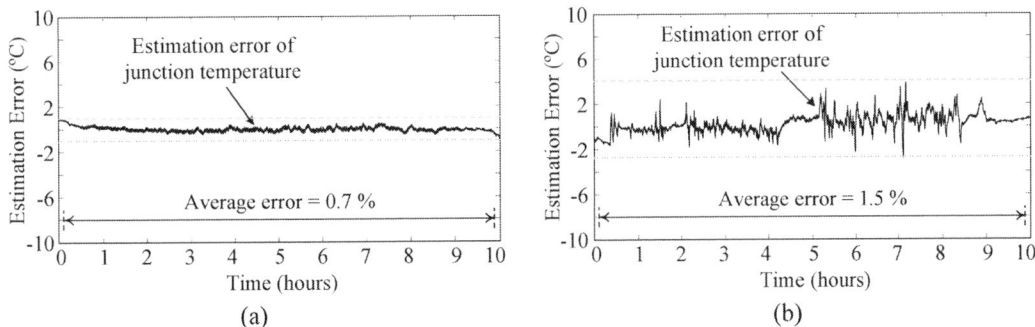

(a)   (b)

Fig. 12. Error in the thermal stress estimation (between simulations and experiments) under one-day mission profile operation: (a) clear day and (b) cloudy day.

Thermal Stress under Mission Profiles

The thermal stress estimation under long-term operation is considered by applying the one-day mission profiles in Fig. 3 to the thermal model in Fig. 6. The obtained thermal stress under the clear-day mission profile condition is shown in Fig. 11(a) together with the experimental results. According to the results, the estimated thermal stress from the thermal model is well aligned with the experimental measurement during the entire operation. The mission profile during a cloudy-day condition in Fig. 3(b), which highly challenges the accuracy of the thermal modeling under dynamics conditions, is also applied to the thermal model. It can be seen from the comparison between the estimated thermal stress profile and the experimental result in Fig. 11(b) that the dynamics of the thermal stress under mission profile operation can be well captured with the thermal model even under a highly fluctuating mission profile condition.

Model Accuracy

The accuracy of the thermal stress modeling is measured from the error between the estimated junction temperature profile (simulation) and the experimental measurement under the same mission profile condition. The error in the thermal stress modeling during clear-day and cloudy-day conditions is shown in Fig. 12. It can be seen from the results that the error between the estimated and the measured junction temperature in Fig. 12 is relatively small, especially for the case with the clear-day mission profile. The error in the thermal stress estimation is higher during the cloudy day due to the higher dynamics of the mission profile condition, which affects the thermal modeling accuracy. Nevertheless, the maximum estimation error is well below 4 °C, which only occurs during very fast load changing conditions. The average errors during the entire operation are only 0.7 % and 1.5 % for the clear-day and cloudy-day mission profile conditions, respectively. This result demonstrates the accuracy of the thermal stress estimation using the lumped thermal model for the power device in PV inverters.

Conclusion

Thermal stress modeling of the power devices plays a key role in the reliability modeling and robustness validation of PV inverters. A lumped thermal network is normally used for long-term thermal stress modeling, e.g., mission profile operation of PV inverters. However, there is still a lack of validation in terms of modeling accuracy, especially, when considering mission profile operation. In this paper, a comprehensive comparison between the thermal stress estimated from the thermal model and the real-field measurement is carried out. Two daily mission profiles under clear-day and cloudy-day conditions are applied to the PV inverter test bench and the thermal stress of the PV inverter (i.e., the junction temperature of the power device) is measured experimentally using an optic fiber. The same mission profiles are also applied to the thermal model of the PV inverter and the estimated thermal stress is compared with the experimental results. According to the results, the average estimation error is 0.7 % and 1.5 % for the clear-day and cloudy-day mission profile conditions, respectively. The maximum deviation in the junction temperature estimation is below 4 °C during the entire operation, which validates the thermal stress modeling accuracy under mission profile operation.

References

[1] P. Hacke, S. Lokanath, P. Williams, A. Vasan, P. Sochor, G. TamizhMani, H. Shinohara, and S. Kurtz, "A status review of photovoltaic power conversion equipment reliability, safety, and quality assurance protocols," *Renew. Sustain. Energy Rev.*, vol. 82, pp. 1097–1112, 2018.

[2] A. Wintrich, U. Nicolai, W. Tursky, and T. Reimann, *Application manual power semiconductors. Semikron international GmbH*, 2015

[3] M. Musallam, C. Yin, C. Bailey, and M. Johnson, "Mission profile-based reliability design and real-time life consumption estimation in power electronics," *IEEE Trans. Power Electron.*, vol. 30, no. 5, pp. 2601–2613, May 2015.

[4] N. Sintamarean, H. Wang, F. Blaabjerg, and P. P. Rimmen, "A design tool to study the impact of mission-profile on the reliability of SiC-based PV-inverter devices," *Microelectron. Reliab.*, vol. 54, no. 9, pp. 1655–1660, 2014.

[5] P. D. Reigosa, H. Wang, Y. Yang, and F. Blaabjerg, "Prediction of bond wire fatigue of IGBTs in a PV inverter under a long-term operation," *IEEE Trans. Power Electron.*, vol. 31, no. 10, pp. 7171–7182, Oct. 2016

[6] C. Felgemacher, S. Araujo, C. Noeding, P. Zacharias, A. Ehrlich, and M. Schidleja, "Evaluation of cycling stress imposed on IGBT modules in PV central inverters in sunbelt regions," in *Proc. CIPS*, pp. 1–6, Mar. 2016

[7] A. Sangwongwanich, Y. Yang, D. Sera, and F. Blaabjerg, "Lifetime evaluation of grid-connected PV inverters considering panel degradation rates and installation sites," *IEEE Trans. Power Electron.*, vol. 33, no. 2, pp. 1225–2361, Feb. 2018.

[8] J.M.S. Callegari, M.P. Silva, R.C. de Barros, E.M.S. Brito, A.F. Cupertino, and H.A. Pereira, "Lifetime evaluation of three-phase multifunctional PV inverters with reactive power compensation, " *Electr. Power Syst. Res.*, vol. 175, 2019.

[9] R. K. Gatla, W. Chen, G. Zhu, J. V. Wang, and S. S. Kshatri, "Lifetime comparison of IGBT modules in Grid-connected Multilevel PV inverters Considering Mission Profile," in *Proc. ICPE 2019 - ECCE Asia*, Busan, Korea (South), 2019, pp. 2764-2769.

[10] F. Blaabjerg, R. Teodorescu, M. Liserre, and A.V. Timbus, "Overview of control and grid synchronization for distributed power generation systems," *IEEE Trans. Ind. Electron.*, vol. 53, no. 5, pp. 1398–1409, Oct. 2006.

[11] A. Sangwongwanich, H. Wang and F. Blaabjerg, "Impact of mission profile dynamics on accuracy of thermal stres modeling in PV inverters," in *Proc. ECCE 2020*, Detroit, MI, USA, 2020.

[12] *FS50R12KT4_B15*, rev. 3.0, Infineon Technologies AG, Germany, 2009.

[13] *IGBT Power Losses Calculation Using the Data-Sheet Parameters*, Infineon Technologies AG, 2009, rev. 1.1.

[14] M. Schulz, S. Allen, and W. Pohl, "The crucial influence of thermal interface material in power electronic design," in *Proc. 29th Annu. IEEE Semicond. Therm. Meas. Manage. Symp.*, Mar. 2013, pp. 251–254

[15] *Transient thermal measurements and thermal equivalent circuit models*, Infineon Technologies AG, 2018, rev. 1.1.

On the Limitations of Using a LTI Modelling Approach for Control Tuning of VSC-HVDC Systems

Pablo Briff [**], Julián Freytes [*], Guillaume de-Preville [*], Jiaqi Li [*], Omar Jasim [**]

GE Renewable Energy

[**] The Lord Nelson Building, William Bagnall Drive, Stafford, ST16 1WS, United Kingdom

[*] 102 Avenue de Paris, Massy, 91300, France

Corresponding authors: {pablo.briff} {julian.freytes} @ge.com

URL: http://www.ge.com

Abstract

This paper investigates the tuning methods of a High Voltage Direct Current (HVDC) Voltage Sourced Converter (VSC) system using double dq synchronous rotating reference frames. This work shows, and confirms by simulation results, that commonly used Linear Time-Invariant (LTI) tuning methods may lead to undesired behaviours when the underlying system is Linear Time-Periodic (LTP). In order to find the optimal tuning parameters, LTP system has been tuned using a machine learning method called Gradient Descent. As a result of this work, new LTP-based tuning methods may be used in the field of VSC-HVDC control.

Keywords: HVDC, VSC, MMC, Linear systems, Periodic systems.

Introduction

High Voltage Direct Current (HVDC) systems are gaining prevalence in the Power Electronics (PE) industry. With an expected penetration of up to 35% of Renewable Energy Systems (RES) in the UK by 2030 and 50% in Europe by 2025 [1], the predictable behaviour of the control algorithms of PE-based converters becomes crucial.

Among all the different control strategies for PE converters, the dual vector control technique is widely used in the industry for the accurate control of the grid currents [2]. When the ac system is balanced and harmonic free, the grid positive sequence currents are defined by constant values in steady state when representing them in the dq synchronous rotating reference frame. However, when the system is unbalanced, negative sequence components appear as oscillations with double grid frequency on the currents. For this reason, the negative sequence currents may be also controlled within a synchronous reference frame rotating in opposite direction as the positive sequence.

Generally speaking, dq-based controllers rely on Proportional-plus-Integral (PI) controllers to attain zero steady-state error in the demanded quantities. Once the control structure is defined, the parameter (i.e., gains) tuning of the PI controllers is fundamental for the system reliability, dynamic performance, and stability [3, 4]. Moreover, the correct behaviour and stability of grid-feeding converters relies on accurate grid current control. Several methods are proposed in the literature for the tuning of PI controllers using simplified Linear Time-Invariant (LTI) modelling approaches as in [5, 6], where all the PI controllers of the converter are tuned all at once. In [7], and after some modelling simplifications for the Modular Multilevel Converter (MMC), the inner and outer loops are accurately tuned. In [8, 9], several methods for tuning purposes are described and analysed. While the previous references deduce very useful tuning methods, their results are only verified considering classical 2-level Voltage Sourced Converters (VSCs) with positive-sequence controllers. For vector controllers with both positive and negative sequence, more modelling efforts are needed to accurately compare the effectiveness of the tuning methods.

Even if the grid current dynamics can be expressed by a sub-set of four differential equations (two for the positive sequence, and two for the negative sequence), it is well-known that the resolution of both positive and negative sequence currents cannot be attained instantaneously. Moreover, modelling the grid current dynamics considering all the components of the controller leads to a Linear Time-Periodic (LTP) system, since the resulting model contains a time-varying periodic state matrix. This means that traditional LTI tuning methods may not be accurate or reliable when applied to LTP systems. Indeed, that is the motivation in [10], to study the complete MMC directly on the LTP modelling approach. However, the tuning techniques for LTP systems are less known in the power systems community, which translates into a lack of comprehensive methods to attain predictable LTP systems tuning.

This paper explores the limitation of using LTI methods to tune a VSC system in the dq frame. Furthermore, the application of a machine learning method, the Gradient Descent algorithm, to find the optimal tuning parameters of the VSC system is investigated. Ultimately, this papers proposes a first attempt at closing the gap between the applicability of vector control tuning methods for both LTI and LTP models for VSC-HVDC applications. A summary of the contributions of this paper is shown in Fig. 1.

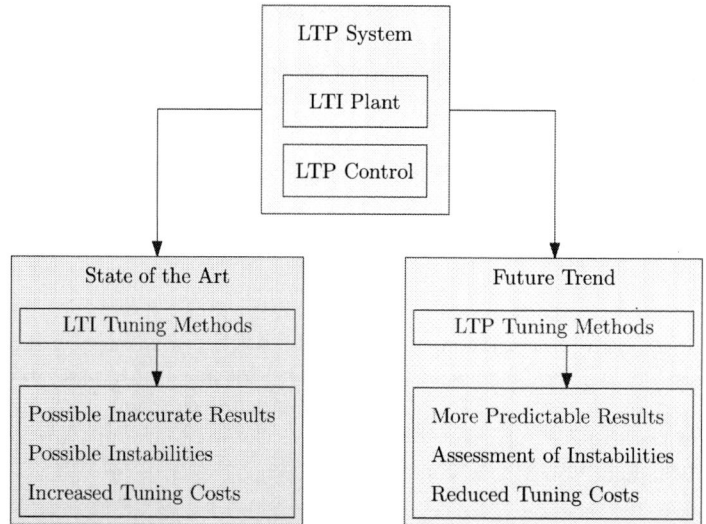

Figure 1: Linear Time-Period Tuning Methods: State of the Art vs. Expected Future Trends.

System Model

The single line diagram of a VSC-HVDC converter is shown in Fig. 2. The converter is assumed to be controlled with a grid-following strategy, imposing the power flow in the dc-link. The ac-side currents $\mathbf{i_{abc}}$ are regulated modulating the ac-side converter voltage $\mathbf{v_{mabc}}$. The inductance L_{eq}^{ac} and resistance R_{eq}^{ac} models the valve reactors and the ac transformer used to interface the converter with the ac grid.

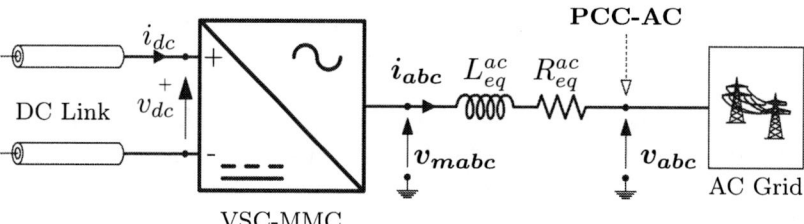

Figure 2: VSC-MMC ac equivalent circuit.

This equivalent circuit is used as a reference for converter modelling with the aim of developing tuning methodologies for the current controllers.

Linear Time-Periodic System in Double Synchronous Reference Frame

The system from Fig. 2 can be controlled with classical current controllers in dq positive-negative synchronous reference frame. The dynamic model in (1) expresses the four current equations.

$$L_{eq}^{ac}\frac{d\mathbf{i_{dqpn}}}{dt} = \mathbf{v_{mdqpn}} - \mathbf{v_{dqpn}} - R_{eq}^{ac}\mathbf{i_{dqpn}} - \omega \mathbf{L_{eq}^{ac}} \begin{bmatrix} 0 & 1 & 0 & 0 \\ -1 & 0 & 0 & 0 \\ 0 & 0 & 0 & -1 \\ 0 & 0 & 1 & 0 \end{bmatrix} \cdot \mathbf{i_{dqpn}}. \tag{1}$$

The currents $\mathbf{i_{dqpn}}$ from (1) cannot be measured instantaneously in reality since both sequences are coupled. The mathematical expression of this coupling can be found by transforming the positive (negative) sequence currents $\mathbf{i_{dqp}} = \begin{bmatrix} \mathbf{i_{dp}}, \mathbf{i_{qp}} \end{bmatrix}^{\top}$ ($\mathbf{i_{dqn}} = \begin{bmatrix} \mathbf{i_{dn}}, \mathbf{i_{qn}} \end{bmatrix}^{\top}$) by using $\mathbf{T_{dqp}^{-1}}$ ($\mathbf{T_{dqn}^{-1}}$) to transform the variables in the abc frame [see (2), with

$\theta = \omega t$]. Then, both components are added to form the total current in *abc* frame, and further transformed back to the *dq* positive (negative) frame with $\mathbf{T_{dqp}}$ ($\mathbf{T_{dqn}}$), to obtain $\tilde{\mathbf{i}}_{\mathbf{dqp}}$ ($\tilde{\mathbf{i}}_{\mathbf{dqn}}$). The mathematical formulation is expressed in (3).

$$\mathbf{T_{dqp}} = \begin{bmatrix} +\cos(\theta) & +\sin(\theta) \\ -\sin(\theta) & +\cos(\theta) \end{bmatrix}; \quad \mathbf{T_{dqn}} = \begin{bmatrix} +\cos(\theta) & -\sin(\theta) \\ +\sin(\theta) & +\cos(\theta) \end{bmatrix} \tag{2}$$

$$\tilde{\mathbf{i}}_{\mathbf{dqp}} = \mathbf{T_{dqp}}\left(\mathbf{T_{dqp}^{-1}}\mathbf{i_{dqp}} + \mathbf{T_{dqn}^{-1}}\mathbf{i_{dqn}}\right) = \mathbf{i_{dqp}} + \mathbf{T_{dqp}}\mathbf{T_{dqn}^{-1}}\mathbf{i_{dqn}} \tag{3a}$$

$$\tilde{\mathbf{i}}_{\mathbf{dqn}} = \mathbf{T_{dqn}}\left(\mathbf{T_{dqp}^{-1}}\mathbf{i_{dqp}} + \mathbf{T_{dqn}^{-1}}\mathbf{i_{dqn}}\right) = \mathbf{i_{dqn}} + \mathbf{T_{dqn}}\mathbf{T_{dqp}^{-1}}\mathbf{i_{dqp}}. \tag{3b}$$

For a given variable z, $\tilde{z}$ denotes that the variable has oscillations in steady-state. References are denoted as z^*, measured quantities are denoted as z^{meas}, estimated quantities are denoted as z^{est}, and feedforward quantities are denoted as z^{ff}.

After solving (3), the obtained results for $\tilde{\mathbf{i}}_{\mathbf{dqpn}} = \left[\tilde{\mathbf{i}}_{\mathbf{dp}}, \tilde{\mathbf{i}}_{\mathbf{qp}}, \tilde{\mathbf{i}}_{\mathbf{dn}}, \tilde{\mathbf{i}}_{\mathbf{qn}}\right]^\top$ are depicted in (4).

$$\tilde{i}_{dp} = i_{dp} + i_{dn}\cos(2\omega t) + i_{qn}\sin(2\omega t) \tag{4a}$$

$$\tilde{i}_{qp} = i_{qp} + i_{qn}\cos(2\omega t) - i_{dn}\sin(2\omega t) \tag{4b}$$

$$\tilde{i}_{dn} = i_{dn} + i_{dp}\cos(2\omega t) - i_{qn}\sin(2\omega t) \tag{4c}$$

$$\tilde{i}_{qn} = i_{qn} + i_{qp}\cos(2\omega t) + i_{dn}\sin(2\omega t). \tag{4d}$$

With (1) and (4), a suitable mathematical model can be deduced for studying the impact of the current controller methodologies on the system dynamics. The complete system model in *dq* positive-negative frame and its control is depicted in Fig. 3.

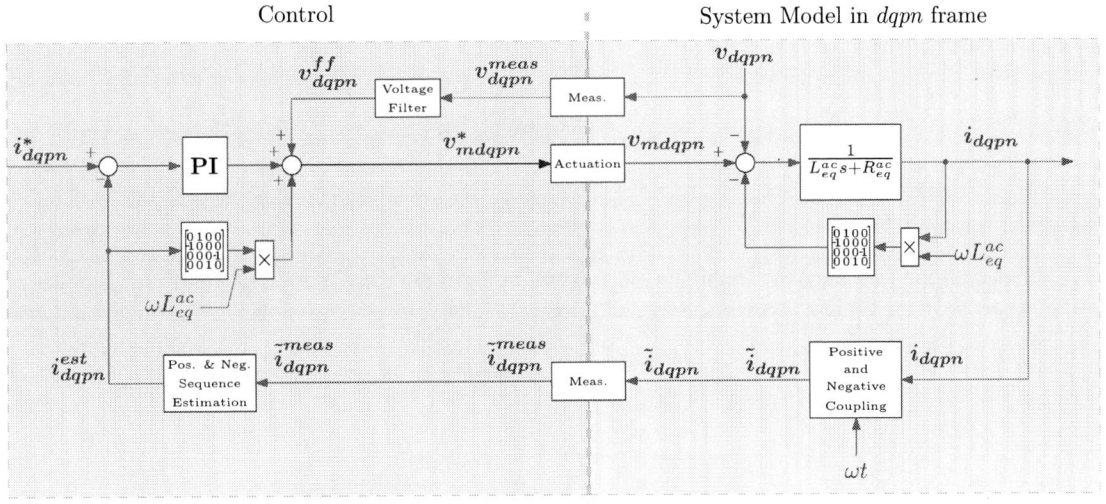

Figure 3: Control and system model block diagram.

From Fig. 3, the differential equations can be gathered to create a linear State-Space model in the following form:

$$\dot{\mathbf{x}}(t) = \mathbf{A_{LTP}}(t) \cdot \mathbf{x}(t) + \mathbf{B_{LTP}}(t) \cdot \mathbf{u}(t) \tag{5a}$$

$$\mathbf{y}(t) = \mathbf{C_{LTP}}(t) \cdot \mathbf{x}(t) + \mathbf{D_{LTP}}(t) \cdot \mathbf{u}(t) \tag{5b}$$

where $\dot{\mathbf{x}}(t) \triangleq d\mathbf{x}(t)/dt$, whereas $\mathbf{x}(t)$ and $\mathbf{u}(t)$ are expressed in (6) and (7), respectively. It is important to highlight that the final LTP model includes the system model in *dqpn* frame as well as the control strategy.

$$\mathbf{x}(t) = \left[\ \underbrace{\mathbf{i_{dpqn}}^\top}_{\text{RL circuit}}, \underbrace{\chi_{\mathbf{1dqpn}}^\top, \chi_{\mathbf{2dqpn}}^\top}_{\text{Notch Filters}}, \underbrace{\chi_{\mathbf{PIdqpn}}^\top}_{\text{PI Integrators}}, \underbrace{\chi_{\mathbf{vdqpn}}^\top}_{\text{Voltage Filters}}\ \right]^\top \in \mathfrak{R}^{20}. \tag{6}$$

Since the matrices $\mathbf{A}(t), \mathbf{B}(t)$ are periodic with period $T \triangleq 2\pi/\omega$, i.e. $\mathbf{A}(t) = \mathbf{A}(t+T)$ and $\mathbf{B}(t) = \mathbf{B}(t+T)$, this gives rise to a LTP system.

$$\mathbf{u}(t) = [i_{dp}^*, i_{qp}^*, i_{dn}^*, i_{qn}^*, v_{dp}, v_{qp}, v_{dn}, v_{qn}] \in \mathfrak{R}^8. \tag{7}$$

The equivalent Multiple-Input Multiple-Output (MIMO) block diagram is shown in Fig. 4. The Bounded-Input Bounded-Output (BIBO) stability and oscillatory response of the system shown in Fig. 4 will be studied in the remainder of the paper.

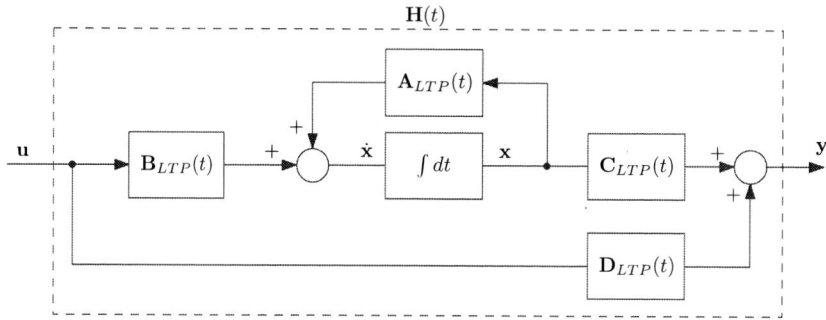

Figure 4: MIMO block diagram from **u** to **y**.

The transfer function $\mathbf{H}(t)$ relates the input and the output as

$$\mathbf{y}(t) = \mathbf{H}(t) \cdot \mathbf{u}(t). \tag{8}$$

When applying a step at the input $\mathbf{u}(t)$, the tracking error of the output is defined as

$$\mathbf{e}(t) = \mathbf{u}(t) - \mathbf{y}(t). \tag{9}$$

Next, the stability and convergence properties of the LTP system is studied by observing the output $\mathbf{y}(t)$ when a step is applied at the input $\mathbf{u}(t)$.

LTP Model Validation

The detailed model from Fig. 2 and the simplified model given in Fig. 3 are compared with time-domain simulations where a step on i_{dp}^* of 0.1 pu at $t = 0.1$ s is provided. Comparison results are shown in Fig. 5, where, on the one hand, it shows that the simplified model reproduces accurately the complete system behaviour; on the other hand, it highlights the couplings between dq and positive and negative sequence (even if the classical decoupling terms "ωL" are used in the current controller, see [11]).

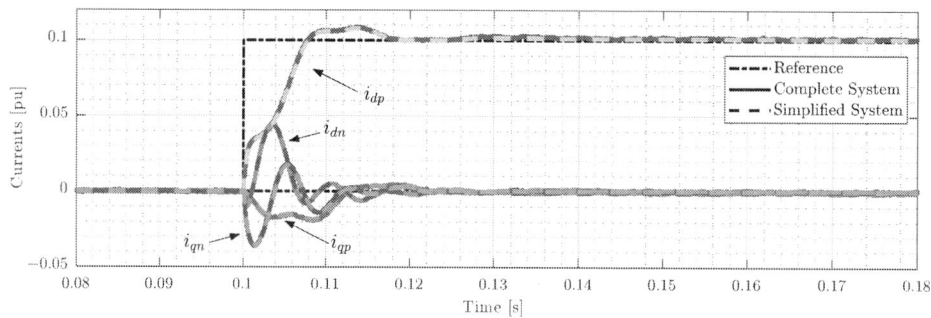

Figure 5: LTP simplified model validation - "Complete System": Fig. 2, "Simplified System": Fig. 3.

LTI Tuning Methodologies

Pole Placement

The most adopted method for tuning the vector control is the classical second-order pole placement method, which is widely disclosed in the literature. In the framework of the tuning of the grid current control, the system from Fig. 6a is first simplified to Fig. 3. The closed loop system in Fig. 6a, named $T_f(s)$, can be used to create simplified but effective tuning methodologies as long as the original system is well represented by the LTI version. From the characteristic equation from $T_f(s)$, the gains for K_p and K_i can be selected as in (10), where $\omega_{n,sys}$ is the natural frequency, and ξ_{sys} the damping factor.

$$K_p = 2\,\xi_{sys}\,\omega_{n,sys}\,L_{eq}^{ac} - R_{eq}^{ac}; \quad K_i = \omega_{n,sys}^2\,L_{eq}^{ac} \tag{10}$$

It is customary to relate $\omega_{n,sys}$ to the response time $t_{r,5\%}$, defined as the time where the signal comes into a band of 5% with respect to a reference step, as $\omega_{n,sys} = (3/t_{r,5\%})$.

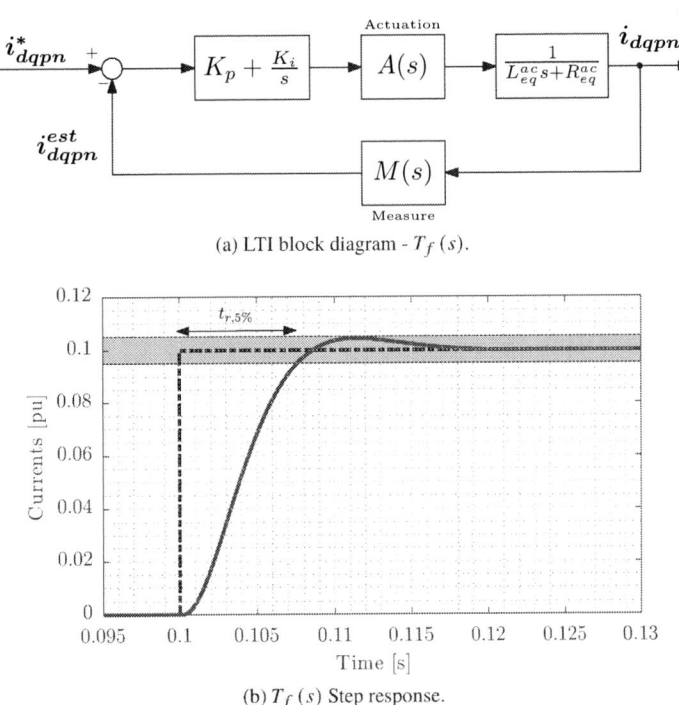

(a) LTI block diagram - $T_f(s)$.

(b) $T_f(s)$ Step response.

Figure 6: Simplified closed loop transfer function.

As an example, for a response time of $t_{r,5\%} = 8$ ms and a damping $\xi_{sys} = 0.707$, the natural frequency is $\omega_{n,sys} = 375$ rad/s. It should be noted that for obtaining the correct response time, the zero in the closed loop transfer function of the system in Fig. 6a should be compensated for by using an Integral-Proportional (IP) controller or by filtering the reference. When applying (10) to the LTI system in Fig. 6a, the response time is controlled to a prescribed value, as shown in Fig. 6b. However, when these gains are applied to the LTP model from (5), the results are highly oscillatory, see Fig. 7. This confirms that, in general, LTI-based methods are not well-suited for LTP systems.

Symmetrical Optimum

Another well-known LTI tuning method based on $T_f(s)$ is the Symmetrical Optimum (SO) [8], which is used to produce the results in Fig. 8. Notice that, while the results have improved with respect to Fig. 7, the prescribed response time of $t_{r,5\%} = 8$ ms is still not satisfied. This is due to the limitation of LTI-based methods when applied to LTP systems.

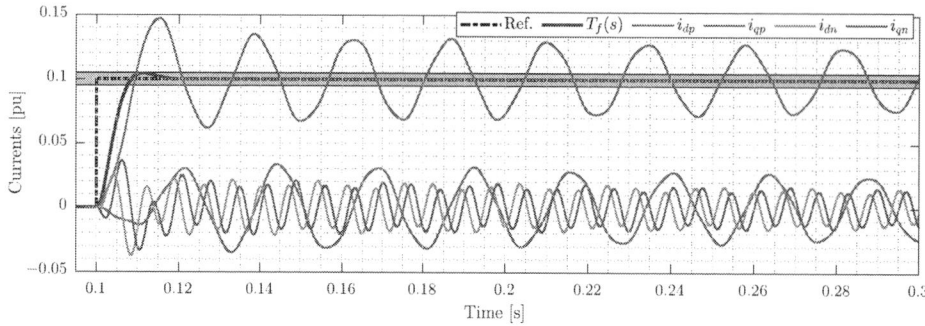

Figure 7: Oscillatory time-domain response using LTI tuning methods on a LTP system.

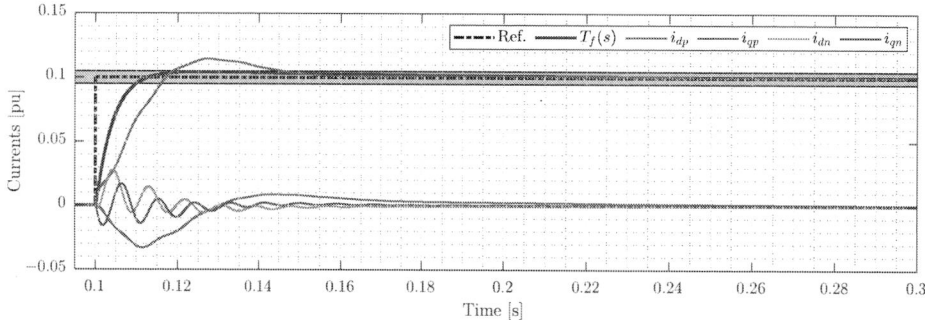

Figure 8: Enhanced system response using the symmetrical optimum method on a LTP system.

Gradient Descent Method

In this section, the Gradient Descent (GD) method, commonly used in machine learning applications, will be used to find the optimal gains K_p, K_i based on a predefined cost function.

Two cost functions based on the tracking error $\mathbf{e}(t)$ defined in (9) are explored in this paper: the integral of time multiplied by the absolute error (ITAE) [12] and the integral of the energy of the error (IEE). The cost for the ITAE metric is given by

$$J_{itae} = \int_{-\infty}^{+\infty} t|e(t)|dt \tag{11}$$

where $|e(t)| = \sqrt{\mathbf{e}^T(t) \cdot \mathbf{e}(t)}$. Meanwhile, the cost function for the IEE metric is

$$J_{iee} = \int_{-\infty}^{+\infty} \mathbf{e}^T(t) \cdot \mathbf{e}(t)dt. \tag{12}$$

The update rule for gains $\mathbf{K}$ the GD method is given by

$$\mathbf{K}(n+1) = \mathbf{K}(n) - \eta \nabla_{\mathbf{K}} J \tag{13}$$

where n is the iteration index, η is the learning rate, and $\nabla_{\mathbf{K}} J$ is the gradient of J with respect to the gains vector $\mathbf{K}$. The learning rate is initially set to a small value, e.g. $\eta = 0.01$ and increased progressively.

The algorithm starts by using the gains obtained from the SO method, $\mathbf{K_{SO}}$, in the LTP mathematical model described in (5). A cost functions J, e.g. the ones described in (11),(12), is computed to obtain the metric for the gains $\mathbf{K}$. The analytical stability of the system given the selected gains are checked in order to only choose stable solutions based on stability methods for LTP systems such as Poincaré multipliers, inverse Nyquist diagrams, eigenvalues of the linearised system model or the LTP Nyquist criterion as described in [10, 13, 14]. If the cost obtained is smaller than a previously found minimum, the gains candidate vector $\mathbf{K}$ is stored. At each iteration, the gains vector $\mathbf{K}$ is then

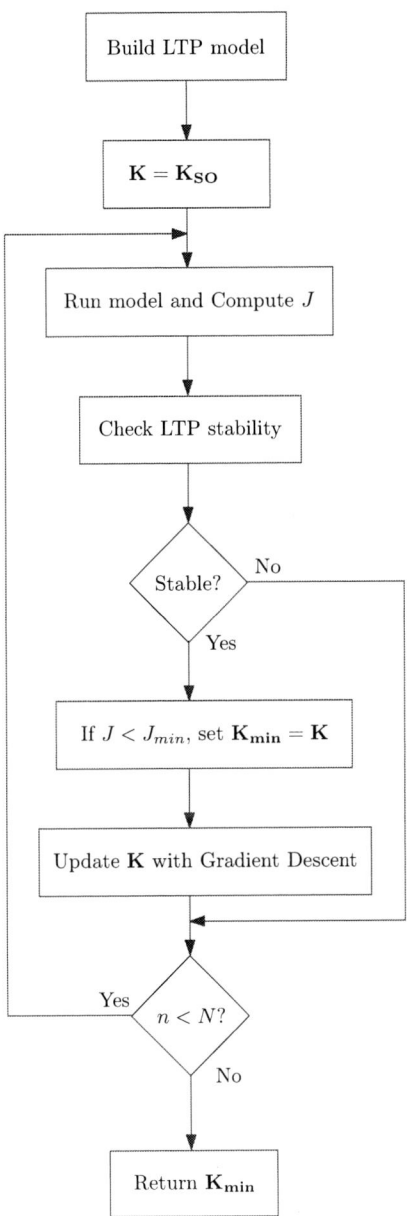

Figure 9: Optimal gains search method using Gradient Descent and different cost functions.

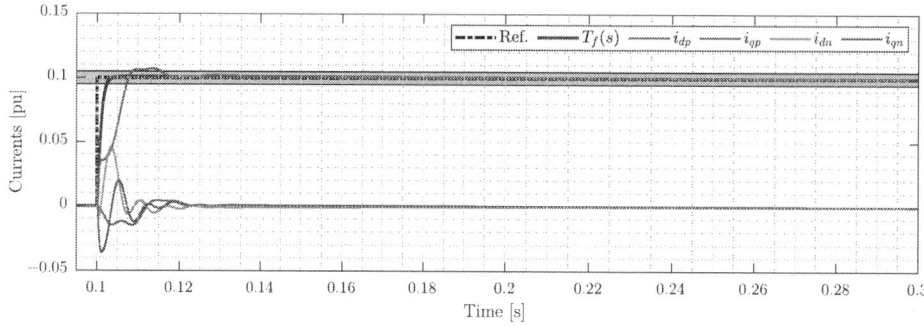

Figure 10: System response using the Gradient Descent algorithm for the ITAE metric.

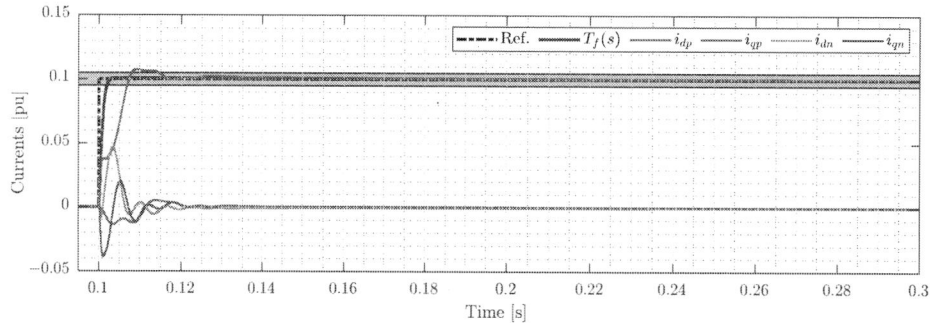

Figure 11: System response using the Gradient Descent algorithm for the IEE metric.

updated using GD as described in (13). The process continues until the number of iterations reaches a preset value, e.g. $N = 100$ has been used in this work. The gains search algorithm is summarised in Fig. 9.

Fig. 10 shows the system's step response by using the calculated K_p, K_i gains with the GD algorithm and the ITAE metric. Similarly, Fig. 11 shows the step response using GD and the IEE metric. It can be observed that the settling time with both ITAE and IEE is enhanced with respect to that of the SO method. Furthermore, by comparing Fig. 10 and 11, the gains obtained for this system with the IEE metric slightly outperform those obtained with ITAE.

Fig. 12 shows the cost J for the GD method and the IEE metric. It can be seen that the cost flattens as the number of iterations n grows. Fig. 13 shows the gains found using the GD search method. Although these gains consistently increase with n, their contribution to lowering J is less relevant as n increases. This can be confirmed by observing that the gradient of J with respect to $\mathbf{K}$ asymptotically approaches zero as n increases, as shown in Fig. 14.

In summary, these numerical methods show the power of defining an effective metric as a function of the tracking error in order to settle the response within acceptable boundaries.

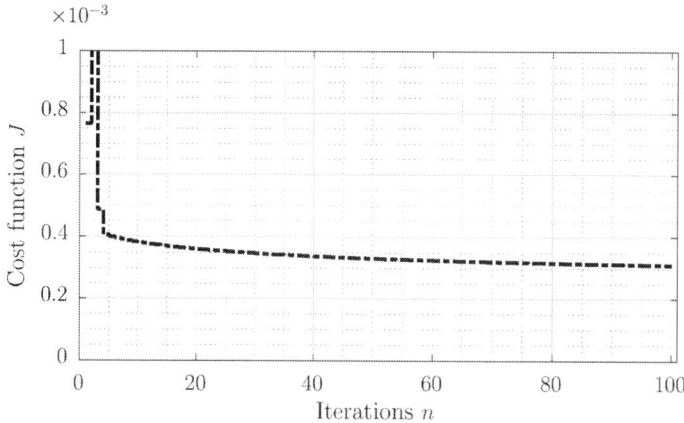

Figure 12: Cost function using Gradient Descent for the IEE metric as a function of the iteration number.

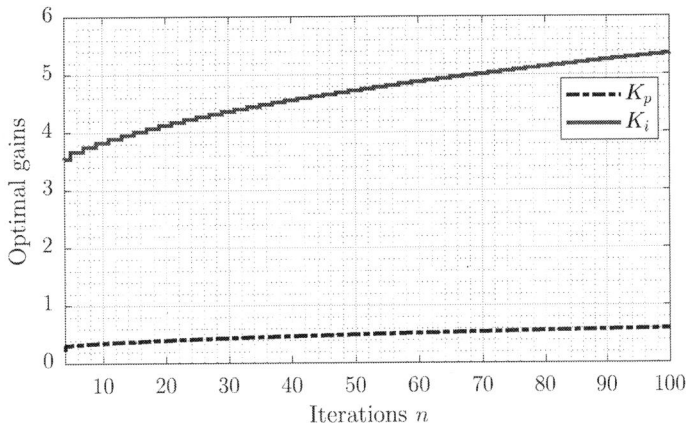

Figure 13: Optimal gains using Gradient Descent for the IEE metric as a function of the iteration number.

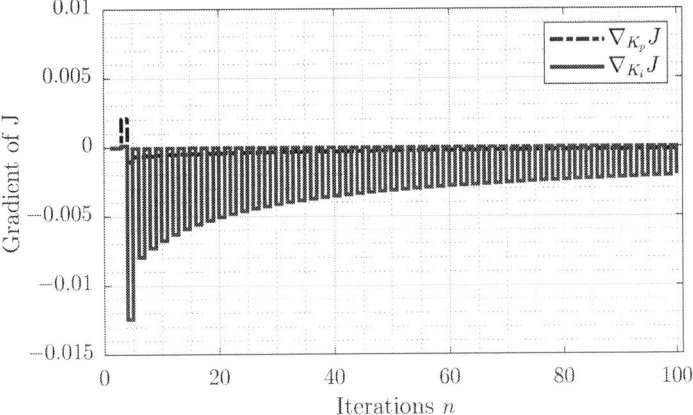

Figure 14: Gradient of the cost function using the Gradient Descent algorithm for the IEE metric.

Conclusions

This paper has presented the challenges faced by the control designer when using LTI tuning methods on an LTP VSC-HVDC control system. The periodicity of the system arises when working in the dq-frame and when resolving the positive and negative sequence components to operate under unbalanced grid voltage conditions. The problem has been demonstrated in a computer simulation environment, where an LTP control system has been tuned using the commonly-used pole placement LTI method. An alternative LTI tuning method, i.e. symmetrical optimum, applied to a LTP system has been investigated, confirming the limitation of LTI-based methods on LTP systems by simulation results. Finally, the LTP system has been optimally tuned using a machine learning search method called Gradient Descent. The results show that LTP system can be iteratively tuned by defining a cost function until a minimum is found. This numerical approach can serve as a basis to motivate the development of tuning methods for LTP systems in the field of VSC-HVDC control.

References

[1] ENTSO-E, "Scenario outlook and adequacy forecast 2014-2030," 2014. [Online]. Available: https://docs.entsoe.eu/dataset/scenario-outlook-adequacy-forecast-2014-2030

[2] R. Teodorescu, M. Liserre, and P. Rodríguez, *Grid converters for photovoltaic and wind power systems.* IEEE/John Wiley & Sons, 2011, vol. 29.

[3] N. R. Chaudhuri, R. Oliveira, and A. Yazdani, "Stability analysis of vector-controlled modular multilevel converters in linear time-periodic framework," *IEEE Transactions on Power Electronics*, vol. 31, no. 7, pp. 5255–5269, 2015.

[4] B. Porter, A. H. Jones, and C. B. McKeown, "Real-time expert tuners for pi controllers," *IEE Proceedings D - Control Theory and Applications*, vol. 134, no. 4, pp. 260–263, July 1987.

[5] S. D'Arco, J. A. Suul, and O. B. Fosso, "Automatic tuning of cascaded controllers for power converters using eigenvalue parametric sensitivities," *IEEE Transactions on Industry Applications*, vol. 51, no. 2, pp. 1743–1753, March 2015.

[6] S. Arunprasanth, U. D. Annakkage, C. Karawita, and R. Kuffel, "Generalized frequency-domain controller tuning procedure for vsc systems," *IEEE Transactions on Power Delivery*, vol. 31, no. 2, pp. 732–742, April 2016.

[7] S. Sanchez, G. Bergna, and E. Tedeschi, "Tuning of control loops for grid-connected modular multilevel converters under a simplified port representation for large system studies," in *2017 Twelfth International Conference on Ecological Vehicles and Renewable Energies (EVER)*, April 2017, pp. 1–8.

[8] J. A. Suul, M. Molinas, L. Norum, and T. Undeland, "Tuning of control loops for grid connected voltage source converters," in *2008 IEEE 2nd International Power and Energy Conference*, Dec 2008, pp. 797–802.

[9] C. Bajracharya, M. Molinas, J. A. Suul, T. M. Undeland *et al.*, "Understanding of tuning techniques of converter controllers for vsc-hvdc," in *Nordic Workshop on Power and Industrial Electronics (NORPIE/2008), June 9-11, 2008, Espoo, Finland.* Helsinki University of Technology, 2008.

[10] V. Salis, A. Costabeber, S. M. Cox, P. Zanchetta, and A. Formentini, "Stability boundary analysis in single-phase grid-connected inverters with pll by ltp theory," *IEEE Transactions on Power Electronics*, vol. 33, no. 5, pp. 4023–4036, 2017.

[11] A. G. Yepes, A. Vidal, J. Malvar, Ó. López, J. Doval-Gandoy, and F. D. Freijedo, "Ineffectiveness of orthogonal axes cross-coupling decoupling technique in dual sequence current control," in *2013 IEEE Energy Conversion Congress and Exposition.* IEEE, 2013, pp. 1047–1053.

[12] D. Maiti, A. Acharya, M. Chakraborty, A. Konar, and R. Janarthanan, "Tuning pid and pi/λ d δ controllers using the integral time absolute error criterion," in *2008 4th International Conference on Information and Automation for Sustainability.* IEEE, 2008, pp. 457–462.

[13] V. Salis, A. Costabeber, S. M. Cox, and P. Zanchetta, "Stability assessment of power-converter-based ac systems by ltp theory: Eigenvalue analysis and harmonic impedance estimation," *IEEE Journal of Emerging and Selected Topics in Power Electronics*, vol. 5, no. 4, pp. 1513–1525, 2017.

[14] S. Golestan, J. M. Guerrero, and J. C. Vasquez, "Modeling and stability assessment of single-phase grid synchronization techniques: Linear time-periodic versus linear time-invariant frameworks," *IEEE Transactions on Power Electronics*, vol. 34, no. 1, pp. 20–27, 2018.

A Voltage Control Method for Power Distribution Lines Utilizing Dispersed Customer Resources

Hiroki Ishihara, Kaho Nada, Miwako Tanaka, Sadayuki Inoue,
Akiko Kuwata and Tomihiro Takano
MITSUBISHI ELECTRIC CORPORATION
8-1-1 Tsukaguchi-Honmachi
Amagasaki City, Hyogo, Japan
Tel.: +81-6-6497-7655.
E-Mail: Ishihara.Hiroki@ab.MitsubishiElectric.co.jp

Keywords

«distribution system», «voltage control», «photovoltaic», «power conditioning system», «reactive power control».

Abstract

This paper proposes a voltage control method in cooperation with a tap changer, such as a step voltage regulator, for power distribution lines using power conditioning systems. Generally, a town which plays a role of a virtual power plant has an energy management system and consists of a lot of power conditioning systems in houses. In this paper, it is clarifies that the method is able to control the voltage at the connecting point of the distribution system properly cooperating with the step voltage regulator under several voltage fluctuation conditions. Furthermore, it is shown that a communication between the energy management system and power conditioning systems enables the system to resolve unevenness output of reactive power of the power conditioners.

Introduction

According to the increase of the amount of renewable energy (RE) such as solar energy in a power distribution system, the voltage of the system may fluctuate rapidly due to weather conditions. Consequently, controlling the voltage at a proper level will be more important issue [1,2]. In the distribution system, the voltage control is usually achieved by voltage regulators such as a step voltage regulator (SVR) using a tap changer. However, it is difficult that the SVR operates quickly against rapid voltage fluctuations due to the tap changing mechanism, and its lifetime may be shorten depending on the number of tap operation times in case of drastic and frequent changes of power generation from RE [3]. Therefore, the static var compensators (SVCs) are installed to the distribution system to control the voltage much faster than the SVR by outputting reactive power using a static power converter [4,5]. While the SVCs response to the fluctuations quickly, a number of SVCs and its cost may be larger if we take the increase of the amount of RE into account.

Power conditioning systems (PCSs) for photovoltaics (PVs) and batteries connected to the distribution system can output reactive power as well as SVCs as mentioned in [6]-[11]. [6] shows an example for voltage regulation using PV PSCs. A method which changing power factor of each PCS for voltage regulation of distribution line is shown in [7]. [8] shows that reactive power injected by distributed generators is valid for voltage control assuming whole a day scenario. Therefore, the connected PCSs can contribute to maintain the voltage if they output enough reactive power in a cooperated manner and may be a popular method in the future. [9] refers to the corporation of tap changers and voltage regulators for stabilization of grid voltage. However, the voltage controller time constant of adjusting reactive power output reference for the local power electronics devices such as PCSs is not focused deeply. Although [10, 11] propose voltage control methods with communication networks between local devices and upper control systems, the cooperated operation is difficult without an additional communication network since the PCSs are installed independently from each other in the distribution system in the usual way. Furthermore, when the PCSs are used as resources of reactive power, output should be even for each PCS in terms of equity of each customer.

In recent years, virtual power plants (VPPs) are investigated in order to obtain enough adjusting power using secondary batteries of customers [12]. The VPP includes a lot of customers with PCSs for secondary batteries and PVs, therefore sufficient resources of reactive power can be available for controlling the voltage of the distribution system. In general, the VPP is controlled by an energy management system (EMS) and home energy management systems (HEMSs) of customers, so that the communication network is already established and the communication between EMS and each PCS can be used for the cooperated operation of PCSs.

This paper proposes a voltage control method using the PCSs in the VPP to realize stabilization of the distribution system, and to cooperate with the SVR as well. The simulation results will be shown to verify the performance of the proposed method.

A method of control voltage of PCSs

In this paper, we assume that named "VPP town" plays a role of the VPP. Fig. 1 shows the power distribution system for study, the VPP town is located at 8km from the substation and the SVR is located at a middle point between the substation and the VPP town. The tap controller estimates the voltage of connected point of the VPP town (V_{VPP}) and sends operation order to the tap changer, and the SVR operates to maintain the V_{VPP} in the proper range. A configuration of the VPP town is as shown in Fig. 2. There are 40 pole transformers (6600V/210V) in the middle voltage (6600V) network in the VPP town, each pole transformer feeds power to 10 consumers. In other words, the VPP town consists of 400 consumers.

Each customer of the VPP town has a PV with a PV-PCS and a secondary battery with a battery-PCS as shown in Fig. 3. The HEMS communicates with a smart meter and PCSs in a house. The information such as active power, reactive power and the voltage of the connected point of each house is sent to the EMS, and the EMS sends some orders to the PCSs via the HEMSs in several minutes intervals. However, no information is available for all devices during the intervals.

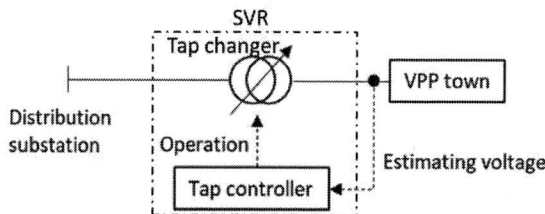

Fig.1 Power distribution system for study.

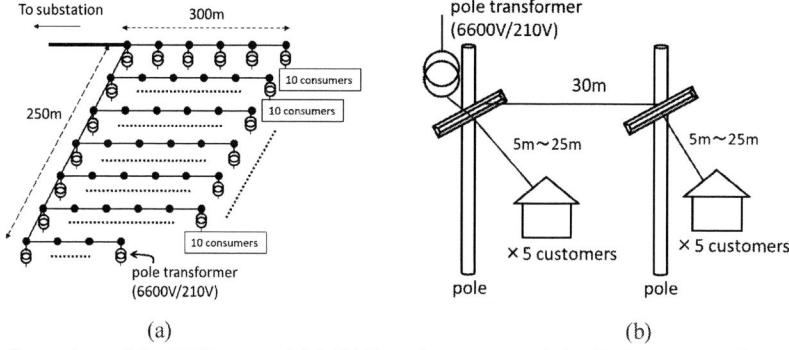

Fig.2 Configuration of the VPP town. (a) Middle voltage network in the VPP town (b) Low voltage network under a pole transformer.

Considering restrictions as mentioned above, the block diagram of the proposed voltage control of the PCS is shown in Fig. 4. The proposed control consists of a low pass filter, a dead band, a controller and a limiter. The V_{ref} is calculated from the voltage at the connected point of a house (V_{det}) through the first order low pass filter to eliminate high frequency fluctuation components. The range of the dead band is given by the EMS. The controller outputs reactive power reference Q_{ref} to control the V_{det} to the V_{ref} when the voltage difference deviates from the dead band. In this method, the controller outputs the Q_{ref} only if the V_{det} fluctuates rapidly because the V_{ref} contains low frequency components of the V_{det}. If the generation of the PV varies largely and stays (namely a long period fluctuation), the PCSs control voltage to reduce voltage fluctuations at first, and then they reduce the output of reactive power gradually by the low pass filter. Futher reduction of the reactive power will follow when the SVR changes its tap position. In the case of short period fluctuations occur by a drastic variation of the PV generations, only the PCSs control the voltage during the fluctuations without the tap change of the SVR. The controller is designed in order to output reactive power faster than the SVR operation and prevent interactions between the PCSs.

The SVR is suitable for compensating slow voltage fluctuations caused by the slow change of demands during a day because the SVR operation is relatively slow due to the tap changer. The PCSs can control voltage rapidly, so it is suitable to reduce rapid voltage fluctuations. As these reason, the proposed method is valid to stabilize the voltage of the distribution system with the SVR and the PCSs of the costumers in a community can operate as the VPP.

Generally, the voltage of the farer points fluctuates larger in radial construction. If the ranges of the dead band ($\pm\Delta V$) of the PCSs are the same value, PCSs in farer houses from substation have to output much reactive power because the error of V_{det} to the $V_{ref} \pm \Delta V$ may be larger. Therefore, the EMS gives proper ranges of the dead band for customers to reduce the uneven output of the reactive power.

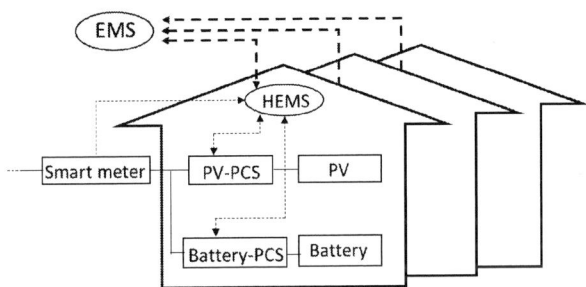

Fig.3 Equipment of the consumers in the VPP town.

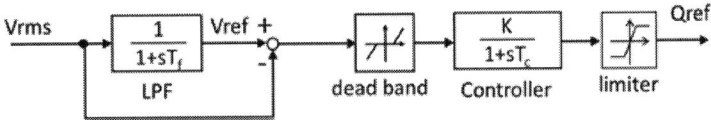

Fig.4 Block diagram of voltage control.

Simulation model for verification

The proposed method of the PCSs cooperating with the SVR will be verified by simulation results using the MATLAB/Simulink to show the stabilization of the voltage in the distribution system. The 10 houses under the pole transformer are replaced with a single customer cluster in order to decrease complexity, so that the VPP town consists of 40 consumer clusters in the simulation model. A power flow reference from/to the VPP town is not ordered and each cluster outputs active power of the PV generation to the substation in this simulation. In addition, the direction of positive active power is to the substation and positive reactive power implies that the PCS operates as an inductive mode in this simulation.

The specifications of the PV-PCS, the Battery-PCS and the SVR settings are shown in Table I. The SVR controls the secondary voltage within $\pm1.5\%$ when the V_{VPP} deviates from the SVR dead band width in 45 seconds continuously. The voltage of the node n (V_n) is approximately calculated from the voltage of one closer node to the substation (V_{n-1}) as

$$V_n \sim V_{n-1} + \frac{R_n P_n}{V_{n-1}} \tag{1}$$

where R_n is a resistance between node n and node n-1, P_n is an active power flowing from the node n to the node n-1. The dead band width of the node n (ΔV_n) is introduced as

$$\Delta V_n = \frac{V_n}{V_r} \times \Delta V_r \tag{2}$$

where V_r and ΔV_r are the voltage and the dead band width of the reference point, respectively. In this study, the reference point is the node of the end point of the VPP town. If the voltage of the substation is constant, V_r is also calculated from Eq. (1). ΔV_r is given for being smaller than the band width of the SVR because the PCSs have to control voltage faster than the SVR. However, the PCSs have to maintain larger output reactive power in a longer time if the ΔV_r is too small. Therefore, the ΔV_r is selected as $\pm0.75\ \%$ which is a half of the SVR band width. The configuration of the VPP town simulation model is shown in Fig. 5. The particular 4 nodes of the VPP town model are named as node A, B, C and D as shown in Fig. 5. The node A is the closest point to the substation, the node B and C are located at middle points, the node D is the end point of the high voltage line of the VPP town model, respectively.

Table I. Simulation parameters of the VPP town model.

The voltage of the substation	6600 V
Rated active power (PV per a customer cluster)	40 kW
Rated apparent power (PV-PCS per a customer cluster)	40 kVA
Rated apparent power (Battery-PCS per a customer cluster)	20 kVA
SVR voltage reference	6732 V (1.02 p.u.)
SVR dead band width	±1.5%
SVR time delay	45sec
Voltage variation by SVR operation	±1.5%

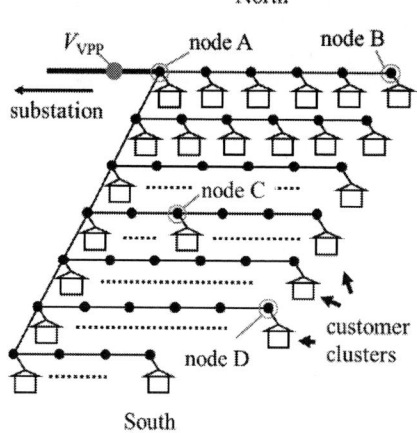

Fig.5 Configuration of the simulation model.

Simulation results

The first simulation condition is the case of a short period fluctuation. The power generation of the PV in the each cluster increases 0.3 p.u. to 0.7 p.u. in 4 seconds, then its decreases 0.7 p.u. to 0.3 p.u. in 4 seconds after the certain period of time. In addition, generation of the PV varies sequentially in 4 seconds difference from northen clusters to southern clusters, so that the sum of the PV generation of the VPP town varies as shown in Fig. 6. Three cases are investigated in terms of the short period fluctuation as below. PCSs in VPP town output zero reactive power in the case 1, whereas PCSs output reactive power to control V_{VPP} in the case 2 and the case 3. The dead band widths of all nodes are the same value ±0.75 % in the case 2, the value of each node is given by EMS based on eq. (2) in the case 3.

Fig. 7 (a) shows the waveform of the V_{VPP} and Fig. 7 (b) shows the reactive power from the node A, B, C and D in the case 1. As power generation of the PVs increase, the V_{VPP} rises as well. The tap changer of the SVR operates after the V_{VPP} exceeds the SVR's band width at 45 seconds and then the V_{VPP} steps down into the range of the SVR's dead band. Subsequently, the generation of PVs turns to decrease in a few minutes in this case of a short period fluctuation, the V_{VPP} also decreases to the limit of the SVR's band width, and then the tap changer of the SVR operates to control the V_{VPP} within the dead band again. Consequently, the SVR operates 2 times in this situation. It is not suitable for the SVR because, as mentioned above, frequent operations of the tap changer of the SVRs may reduce its lifetime. Fig. 8 and 9 show the results in the case 2 and 3 respectively. In both results, as generation of PVs increases, the V_{VPP} also rises, and the voltages of the clusters rise as well consequently. Then the PCSs output reactive power and the V_{VPP} fluctuation is reduced. When the generation of the PVs decreases, subsequently, the PCSs reduce the output of reactive power to zero. Consequently, the V_{VPP} is controlled in the range of the SVR band width, therefore the tap changer of the SVR doesn't need to operate.

The comparisons of the reactive power output between the four nodes are shown in Fig. 10 and Fig. 11. Fig. 10 shows the peak reactive power output and Fig. 11 shows the integration of the output reactive power from the node A, B, C and D, respetively, in the case 2 and 3. In the results of the case 2, the farer nodes from the substation tend to output more reactive power and the integration of the reactive power of the node D is 142 % of the node A. On the contrary, the integrateion of the reactive power of the node D is reduceed to 120 % of the node A as shown in Fig. 11. As a result, it is clarified that the proposed voltage control method is valid for SVRs in terms of the number of operation and be able to resolve uneven reactive power output from the nodes in the VPP town.

In the second case of a long period fluctuation, the generation of the PVs of the VPP town model increases 0.3 p.u. to 0.7 p.u., then it stays at 0.7 p.u. as shown in Fig.12. The other simulation conditions are the same as the case 3. As shown in Fig. 13, the V_{VPP} fluctuation is reduced at first by the PCSs, then the PCSs reduce reactive power gradually because the input of the controller of each PCS is getting smaller by the effect of the low pass filter. Finally, the SVR changes its tap when the V_{VPP} exceeds the SVR's band width in 45 seconds, then the PCSs reduce the output reactive power further.

These simulation results show that the stabilization of the voltage of the distribution system is achieved by the proposed voltage control method of the PCSs cooperating with the SVR.

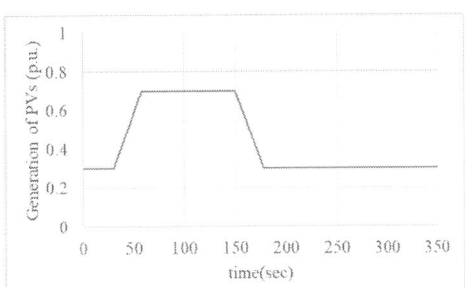

Fig. 6 The PV generation of the VPP town in a short period fluctuation.

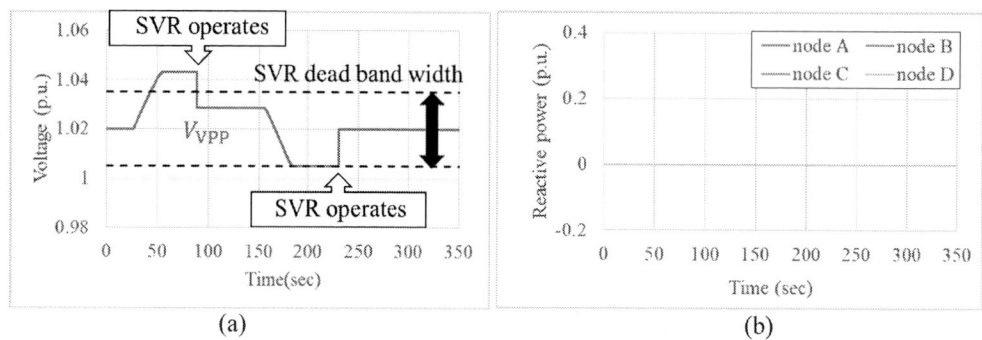

(a)   (b)

Fig. 7 Simulation results in the case 1. (a) V_{VPP}. (b) Reactive power.

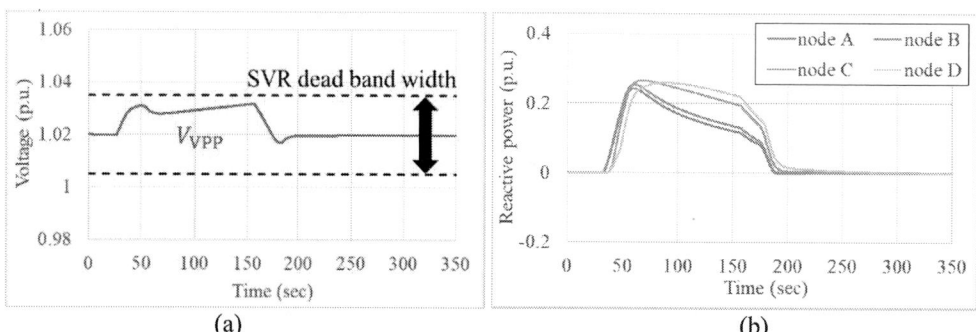

(a)   (b)

Fig. 8 Simulation results in the case 2. (a) V_{VPP}. (b) Reactive power.

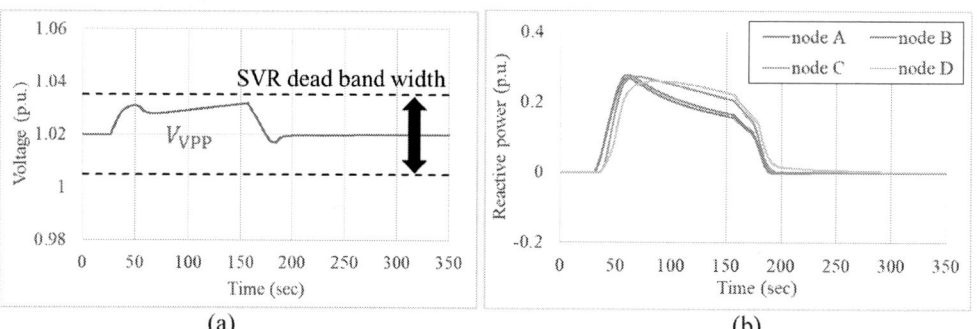

(a)   (b)

Fig. 9 Simulation results in the case 3. (a) V_{VPP}. (b) Reactive power.

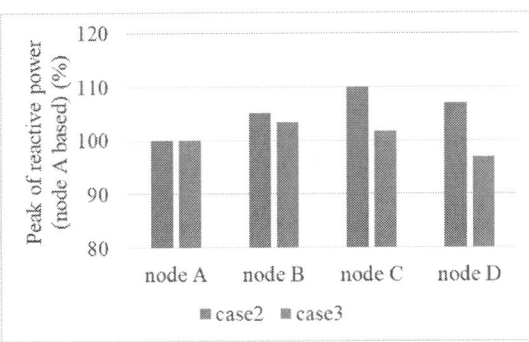

Fig. 10 The peak ratio of the reactive power output from node A, B, C and D.

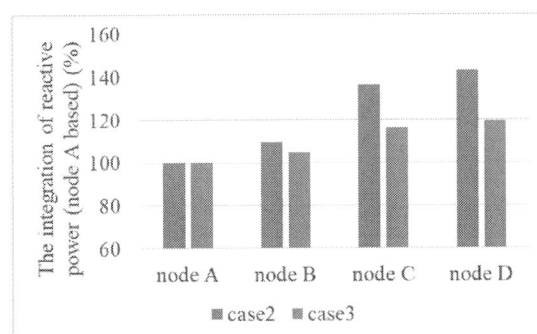

Fig. 11 The integration ratio of the reactive power output from node A, B, C and D.

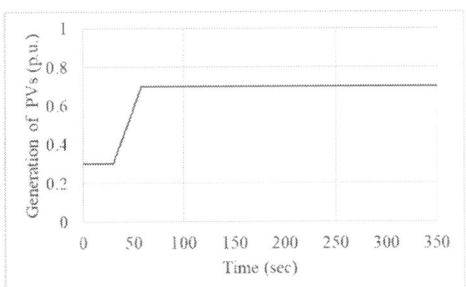

Fig. 12 Generation of the VPP town in a long period fluctuation.

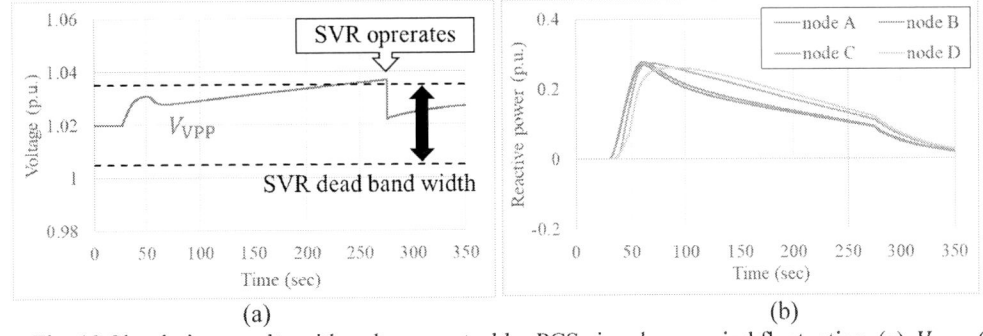

(a)   (b)

Fig. 13 Simulation results with voltage control by PCSs in a long period fluctuation. (a) V_{VPP}. (b) Reactive power.

Conclusion

In this paper, the voltage control method of the PCSs of the customers serving as VPP is proposed for stabilization of the distribution system voltage. As simulation results, it is clarifies that the proposed method of the PCSs can control distribution system voltage coordinating with the SVR. Furthermore, it is indicated that proper range of the dead band given by the EMS is valid for resolving uneven reactive power outputs of the PCSs in the VPP town. In other words, the proposed method can be realized without too much communication cost because a complicated network system or frequent communication between PCSs and EMS is not required.

If a battery-PCS and a PV-PCS are in a house, as well as in this paper, both of them may be able to output reactive power. In the situation of the town working as a VPP, the amount of active power from/to each house is ordered by higher control system, a charging/discharging control by a battery-PCS is more important. Therefore, how to determine the contribution of reactive power output considering economically optimal of each customers has to investigate as a future work.

References

[1] A. Woyte, V. V. Thong, R. Belmans, and J Nijs,"Voltage fluctuations on distribution level introduced by photovoltaic systems", *IEEE Trans. Energy Convers.*, Vol. 21, No. 1, pp.202-209, 2006.

[2]C. J. Dent, L. F. Ochoa, and G. P. Harrison, "Network distributed generation capacity analysis using OPF with voltage step constraints", *IEEE Trans. Power Syst.*, Vol. 25, No. 1, pp. 296–304, 2010.

[3] F. C. L. Trindade, K. V. do Nascimento, and J. C. M. Vieira, "Investigation on voltage sags caused by dg anti-islanding protection", *IEEE Trans. Power Deliv.*, Vol. 28, No. 2, pp. 972–980, 2013.

[4] H. Hayashi, "Trend and future view of voltage control for distribution systems with distributed generators", *IEEJ Trans. PE*, Vol. 129, No. 4, pp.491-494, 2009.

[5] H.Hatta and H. Kobayashi, "A study of centralized voltage control method for distribution system with distributed generation", Proceedings of 19th International Conference on Electricity Distribution, Paper No. 330, 2006.

[6] L. J. Borle, M. S. Dymond, and C. V. Nayer, "Development and testing of a 20-kW grid interactive photovoltaic power conditioning system in western australia", *IEEE Trans. Ind. APPL.*, Vol. 33, No. 2, pp.502-508, 1997.

[7] P. M. S. Carvalho, P. F. Correia, and L. A. F M. Ferreira, "Distributed reactive power generation control for voltage rise mitigation in distribution networks", *IEEE Trans. Power Syst.*, Vol. 23, No. 2, pp. 766-,772, 2008.

[8] H. Zhu and H. J. Liu, "Fast Local Voltage Control Under Limited Reactive Power: Optimality and Stability Analysis", *IEEE Trans. Power Syst.*, Vol. 31 , No. 5, pp. 3794-3803, 2016.

[9] W. Ren and H. Ghassempouraghamolki, "Tuning of Voltage Regulator Control in Distribution Systems with High Renewable Penetration", in Proc. *IEEE PES General Meeting*, Portland OR, 2018.

[10] A. Momeneh, M. Castilla, J. Miret, P. Marti, and M. Velsco, "Comparative study of reactive power control methods for photovoltaic inverters in low-voltage grids", *IET Renew. Power Gener.* Vol. 10, No. 3, pp.310–318, 2016.

[11] S. Y. M. Mousavi, A. Jalilian, M. Savaghebi, and J. M. Guerrero, "Coordinated control of multifunctional inverters for voltage support and harmonic compensation in a grid-connected microgrid", Electric Power Systems Research, vol. 155, pp. 254-264, 2018.

[12] E. Dall'Anese, S. S. Guggilam, A. Simonetto, Y. C. Chen, and S. V. Dhople,"Optimal regulation of virtual power plants", *IEEE Trans. Power Syst.*, Vol. 33, No. 2, pp.1868-1881, 2018.

Performance Comparison Between SiC and Si Inverter Modules in an Electrical Variable Transmission Application

Mauricio Dalla Vecchia[*], Simon Ravyts[*], Florian Verbelen[+], Jeroen Tant[*],
Peter Sergeant[+] and Johan Driesen[*]
[*]KU Leuven - EnergyVille - Dept. Electrical Engineering (ESAT), Div. ELECTA
[*]Kasteelpark Arenberg 10, bus 2445
[*]Leuven, Belgium
[+]Department of Electromechanical, Systems and Metal Engineering
[+]FlandersMake@UGent, Belgium
Email: simon.ravyts@kuleuven.be
URL: https://www.energyville.be

Keywords

≪SiC MOSFET≫, ≪DC/AC inverter≫, ≪EVT≫, ≪HEV≫

Abstract

This paper evaluates the performance of Silicon Carbide MOSFET and Silicon IGBT modules in a three-phase inverter for Electrical Variable Transmission systems. For this purpose, two practical inverter set-ups were developed and compared. An increase of several percentage points is visible over the entire operating range for the Silicon Carbide prototype. The total energy efficiency increased by 3.7% for the rotor and by 11.2% for the stator, for the same test conditions.

Introduction

Governments all over the world are forcing the automotive industry, via legislation, to enhance their drive trains in order to reduce fuel consumption and emissions. One of the solutions to increase the efficiency of the traditional Internal Combustion Engine (ICE) based vehicle is to enhance the drive train with a power split transmission. A well know example of such a vehicle, also known as a Hybrid Electrical Vehicle (HEV) is the Toyota Prius. Due to this power split, the ICE can be operated independently from the requested power at the wheels. As a consequence, the operating points of the ICE can be chosen at the optimal operating line. Important side-effects are friction and thus wear in the planetary gears that enable this power split. Moreover, valuable space is taken by the electrical machines that support the traction of the vehicle.

To eliminate the planetary gear, a component can be used that is called the Electrical Variable Transmission (EVT) [1, 2]. The device consists of an inner rotor, an outer rotor and a stator of which the inner rotor and stator are equipped with a distributed three-phase winding [3]. In a HEV, the ICE is connected to the inner rotor while the wheels are connected to the outer rotor [4]. Both shafts are electrically coupled via inverters and the DC-bus, which includes a storage device (i.e. battery). Power can thus be transferred from inner rotor to outer rotor by means of an electric and/or electromagnetic path. Because the power is converted via two separate paths, the EVT can be considered as a power split device.

Wide Band-Gap (WBG) semiconductors are on the rise nowadays and impacts the transportation industry [5]. Gallium Nitride (GaN) and Silicon Carbide (SiC) stand out among the other WBG materials for applications in power electronics. For low voltage/low power applications, GaN proves to be a competitor to the state-of-the-art Si MOSFET technology in applications of, for instance, photovoltaics, battery

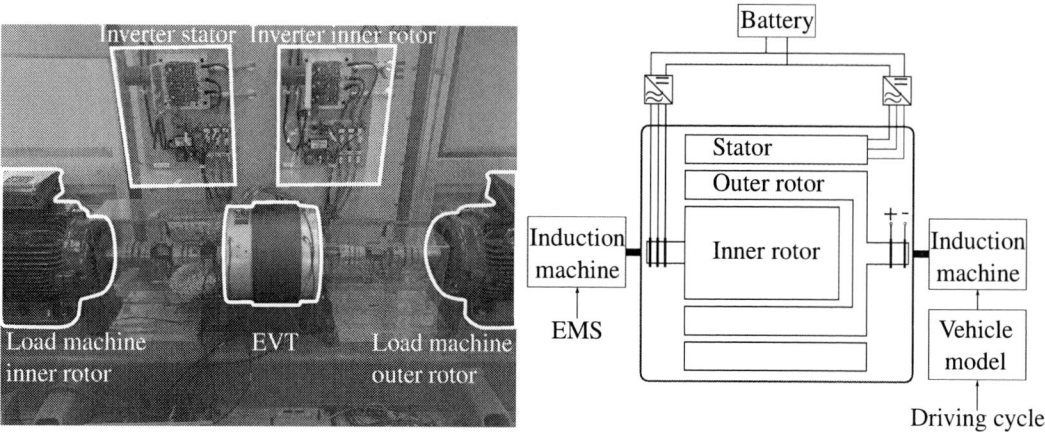

Fig. 1: Experimental setup (left hand side) as well as a schematic representation (right hand side).

energy storage systems or fuel cells [6]. SiC, on the other hand, is considered as a competitor for Si IGBTs for high voltage/ high power applications [7]. The drawbacks of SiC is mainly found in higher component costs, increased ElectroMagnetic Interference (EMI) due to the high di/dt and dv/dt that occur at the switching instants and the lower short-circuit tolerance, requiring faster gate drivers [8].

In general, WBG semiconductors present better electrical characteristics compared to state-of-the-art Si technology. Therefore, the aim of this paper is to develop and experimentally validate the performance, in terms of efficiency, of two DC-AC inverters for use in EVT. The efficiency of SiC MOSFETs will be compared with state-of-the-art Si IGBT technology. A brief discussion about the system under study is presented first, followed by a description of the inverter and experimental results that highlight the superior performance of WBG semiconductors.

System description

The EVT is an electrical machine that consists of an inner rotor, an outer rotor and a stator (see right hand side Fig. 1). Both the inner rotor as the stator are equipped with a distributed three-phase winding. The outer rotor has a single layer of permanent magnet material and a flux bridge, located underneath the DC-field winding (see Fig. 2).

Due to the flux bridge only a part of the permanent magnet flux is linked with the stator. Consequently, the stator flux linkage is low as well as the stator torque. The stator is thus inherently flux weakened as the magnetic field of the magnets links almost completely with the inner rotor. The advantage of this layout is that the induced losses in the stator by the PM field are low as well [9].

During the occasion that high stator torque is required, i.e. high acceleration or higher loads due to uphill driving, the DC-field current can be adapted to increase the stator flux linkage. Note that while the stator flux linkage can be modulated, the inner rotor flux linkage remains invariant for variations of the DC-field current [9].

As the DC-field winding has an impact on both the losses as the actual maximum stator torque, defining the optimal current setpoint is crucial for the performance of the EVT based HEV. To find this optimal current setpoint and in addition the current setpoints in stator and inner rotor for given inner rotor and stator torque, an algorithm is developed to track the optimum (in terms of losses) offline [10]. The results are stored in a look-up table and are used at the setup as setpoint generator for the currents. To control the currents in the stator and inner rotor, two three-phase inverters are used with a common DC-bus (see Fig. 1, left hand side). More details on the inverters are given in the next section. The current in the DC-field winding is controlled via a DC source.

In order to load the shafts of the EVT, two induction machines are used (see Fig. 1, left hand side). These

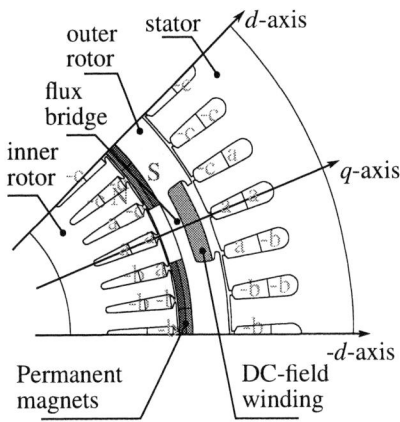

Fig. 2: Cross-sectional view of the EVT considered [11].

Table I: Electrical properties of the tested power modules.

	Si SKiM459GD12E4	SiC CAS300M12BM2
V_{DS}	1200 V	1200 V
$V_{CE,sat}$	1.85 V	/
$R_{CE}; R_{DS,on}$	3.7 mΩ	4.2 mΩ
I_D	452 A @ 70 °C	293 A @ 90 °C
C_{iss}	26.4 nF	19.3 nF
C_{oss}	1.74 nF	2.57 nF
Q_c	2.55 μC	3.2 μC
$t_{d,on}$	276 ns	76 ns
t_r	55 ns	68 ns
$t_{d,off}$	538 ns	168 ns
t_f	114 ns	43 ns

induction machines are controlled with an industrial drive. The induction machine connected to the inner rotor emulates the behavior of an ICE controlled by an Energy Management Strategy (EMS) [12], while the other induction machine emulates the load behavior coming from a vehicle. The set points for both induction machines are generated in a dSPACE platform.

Three-phase inverter

The inverter topology applied in this work is shown in Fig. 3a. A regular three-phase inverter, consisting of three half bridges, is the interface between the battery pack and the inner rotor or stator machine. The battery is emulated by a variable DC power supply. For further comparison between WBG SiC modules and the regular state-of-the-art Si modules, two inverters were tested under the same operational conditions. A space vector modulation scheme for three-phase inverters was implemented and the control signals were generated by a dSPACE module. Both prototypes are also displayed in Figs. 3b and 3c. Fig. 3b shows the Si inverter while Fig. 3c displays the SiC inverter developed for testing purposes and performance comparison. Both prototypes employ water cooling.

The main specification of the Si and SiC modules used in the inverter are shown in Table I. It can be noticed that the WBG SiC module has no forward voltage drop as it is a unipolar device. In contrast, the IGBT is a bipolar switching device with a forward voltage drop $V_{CE,sat}$. Furthermore, the SiC module has improved transient characteristics compared to its Si competitor, which reduces drastically the switching losses under normal operation. The other specifications are comparable and present similar values.

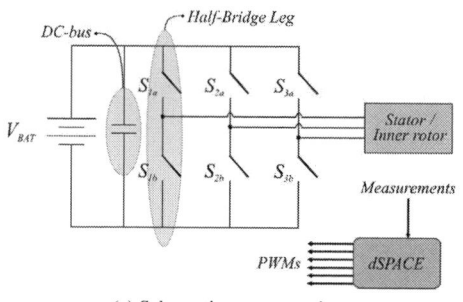

(a) Schematic representation.

(b) Si IGBT inverter.

(c) SiC MOSFET inverter.

Fig. 3: Three-phase inverter for interconnection of the battery pack and the inner rotor/stator.

To avoid EMI problems related to the high switching speed of the SiC MOSFETs, the capacitor banks were carefully designed to minimize the stray inductance between the legs and the DC bus. For the bulk capacitance, two $310\mu F$ Cornell Dubilier 947D311K132CFRSN film capacitors were used for their low Equivalent Series Resistance (ESR) and inductance (ESL), to guarantee a stable DC-link voltage. Furthermore, an extra film capacitor bank of 75μ F using Epcos M115616276 was placed on top of the SiC power modules to minimize the influence of stray inductances and improve filtering. The implemented driver is the Wolfspeed CGD15HB62P1.

Experimental results

In order to evaluate the performance of the Si and SiC inverters, both are imposed to the same set of power cycles, guaranteeing similar testing conditions. As there are two inverters, one for the stator and one for the rotor, which are typically subjected to different power cycles in a HEV, two power cycles are defined. The cycles are characterized by four operational modes that can occur during driving (see Fig.

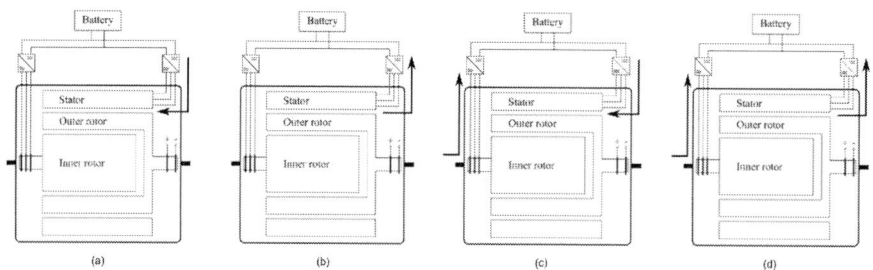

Fig. 4: Considered electrical power flows in the power cycles. The arrows specify the electrical power flow. (a) Mode a: pure electrical driving (acceleration). (b) Mode b: pure electrical driving (regenerative braking). (c) Mode c: charging via ICE while acceleration. (d) Mode (d): charging via ICE as well as due to regenerative braking.

4). The first 2 modes consider pure electric driving (Fig. 4 (a) and (b), acceleration and regenerative braking, respectively). In the third and fourth mode (Fig. 4 (c) and (d), respectively), the battery is charged via the ICE while the vehicle is being accelerated or decelerated. Please note that the authors are well aware that there are many more possible combinations and that the considered power flows do not necessarily result in the most efficient system level behavior (vehicle). However, enough variability in the power flow is considered to benchmark the SiC inverter with the Si inverter.

The operational switching frequency was defined as $f_s = 10$ kHz and a peak power of 10 kW is processed. In Figs. 5, 6, 7 and 8, the measured currents, the processed power, the losses and the instantaneous efficiency are plotted over the time duration of the implemented torque profile. The measurements were done using a Yokogawa WT1800 power analyzer. For a driving cycle of about 20 minutes, one measure per 50ms is acquired which leads to a total of about 24 thousand points measured per test (all points displayed in the measured Figures). Not only the measured datapoints but also the calculated error bars are displayed. The error was calculated using standard methods for uncertainty analysis [13] and consider the scale/range of the measurements.

The experimental results are shown four times: twice for the Si and SiC test of the rotor and twice for the Si and SiC test of the stator. A measured negative electrical power in the inverter connected to the inner rotor windings means that power is flowing from the emulated ICE to the battery (mode c and d displayed in Fig. 4). For the stator, a positive power flow means acceleration of the vehicle (mode a or c displayed in Fig. 4), while a negative power flow means regenerative braking (mode b or d displayed in Fig. 4).

Both for the rotor and stator a significant improvement is visible for the measured losses over the entire torque profile. The reduction in losses leads to a strong efficiency increase, as visible in Fig. 9. The overall energy efficiency was twice improved by implementing SiC MOSFETs and lead to an increase of 3.7% and 11.2% for respectively the rotor and the stator. The results are summarized in Table II. The performance difference between the rotor and stator are a consequence of the defined driving cycle. More reactive power circulate through the inverter connected to the stator.

Table II: Measured energy efficiency (in %) over the pre-defined torque profile.

	Si IGBT	SiC MOSFET	Difference
Rotor	92.9 ± 1.6	96.6 ± 1.7	+3.7
Stator	77.6 ± 2.2	88.8 ± 2.5	+11.2

Conclusions

This works aim at testing the overall performance of Si and SiC inverters operating in an Electrical Variable Transmission system. Under similar testing conditions, the SiC MOSFET inverters present better performance compared with its Si IGBT modules competitor for a 10kHz operational switching frequency and 10kW of processed power. An improvement of around 3.7% and 11.2% is observed for the SiC inverters connected to the inner rotor and stator machines, respectively.

Acknowledgments

This work was carried out for the EMTechno project (project ID: IWT150513) supported by VLAIO and Flanders Make, the strategic research centre for the manufacturing industry in Belgium.

Performance Comparison Between SiC and Si Inverter Modules in an Electrical
Variable Transmission Application

DALLA VECCHIA Mauricio

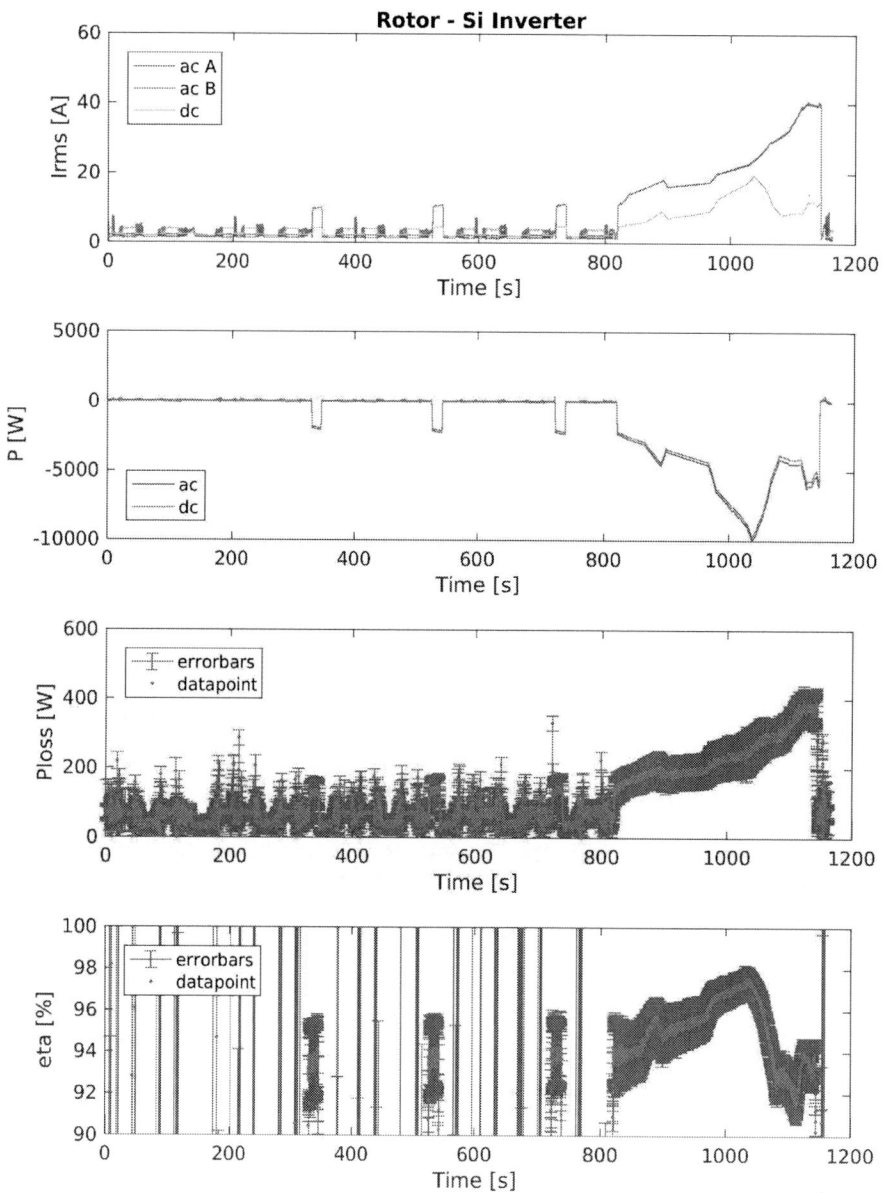

Fig. 5: Performance of the rotor when the Si IGBT inverter is used.

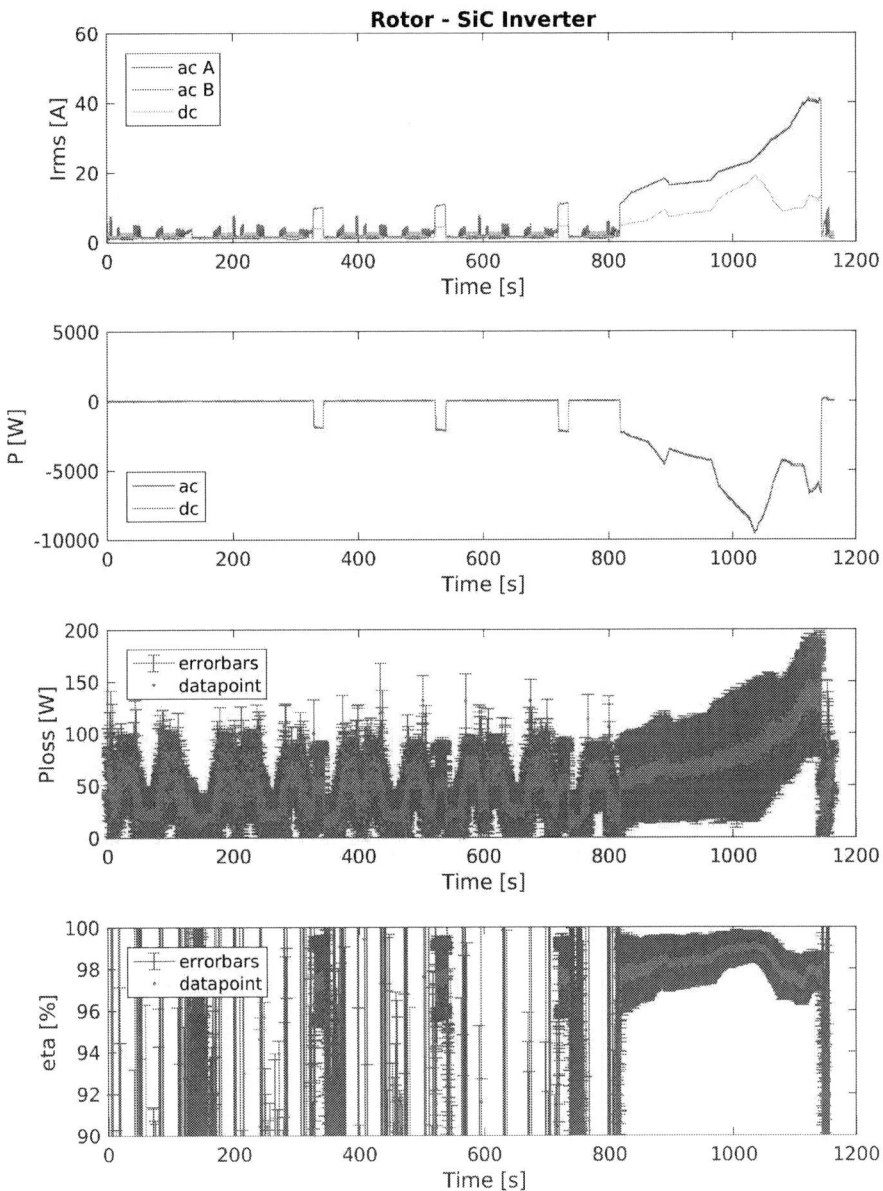

Fig. 6: Performance of the rotor when the SiC MOSFET inverter is used.

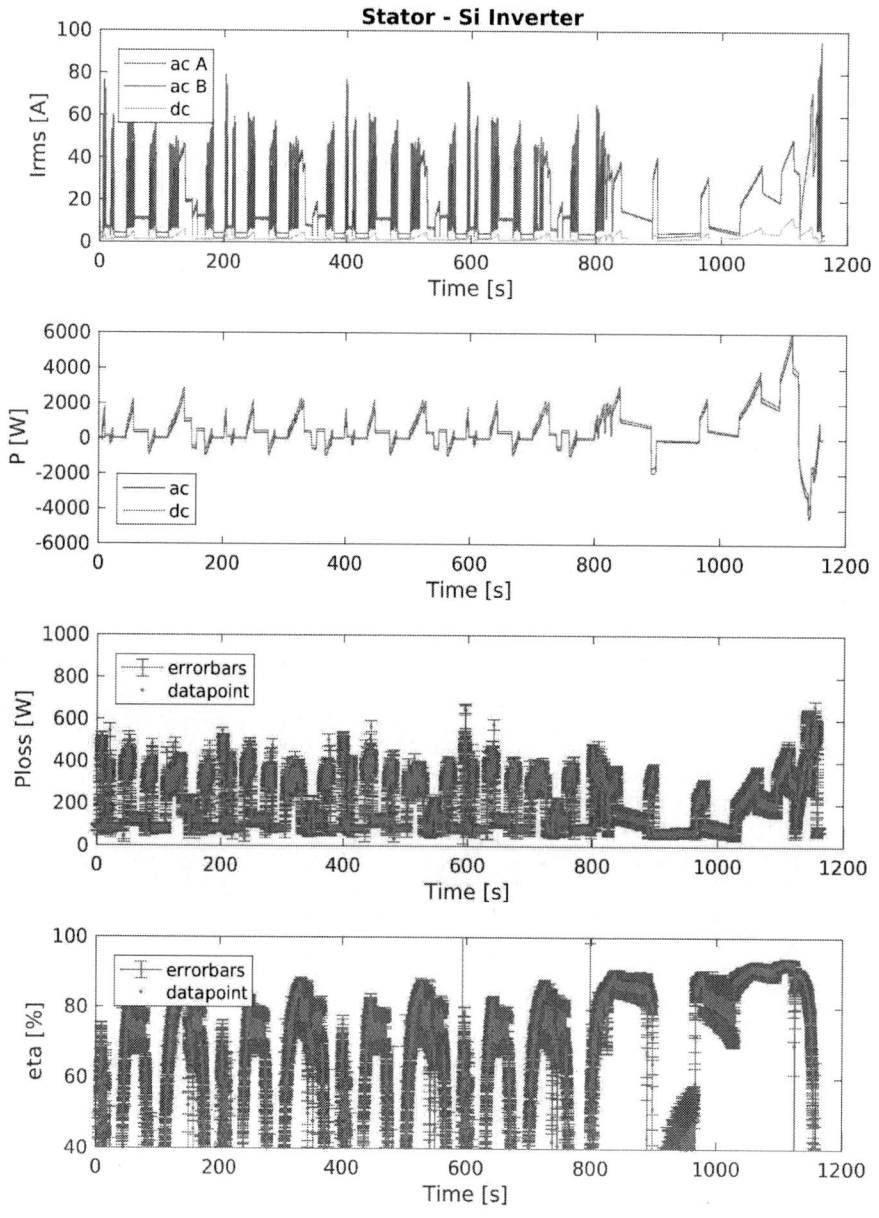

Fig. 7: Performance of the stator when the Si IGBT inverter is used.

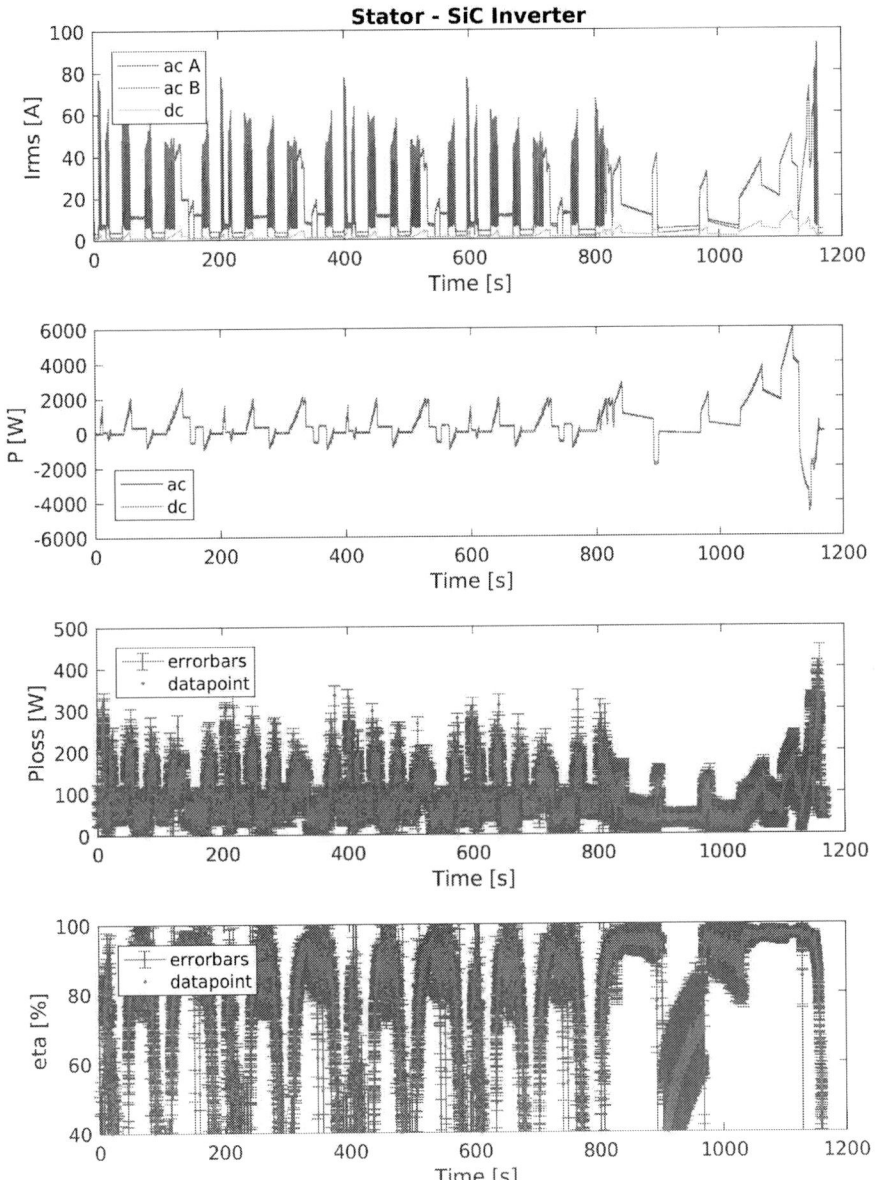

Fig. 8: Performance of the stator when the SiC MOSFET inverter is used.

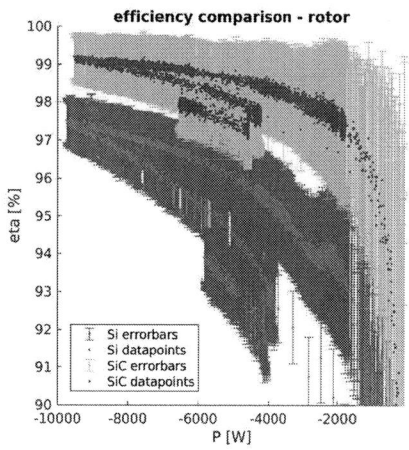 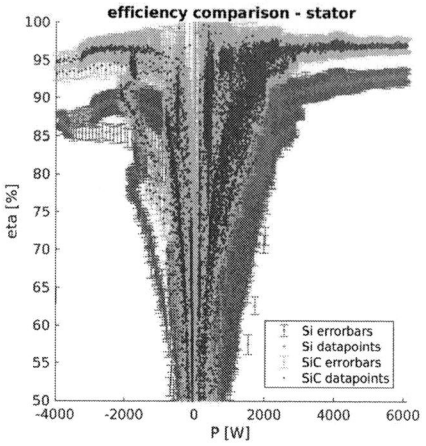

(a) Energy efficiency: 96.6% (SiC) and 92.9% (Si).   (b) Energy efficiency: 88.8% (SiC) and 77.6% (Si).

Fig. 9: Efficiency comparison of SiC vs Si inverters connected to the inner rotor/stator.

References

[1] M. Hoeijmakers and J. Ferreira, "The electric variable transmission", IEEE Transactions on Industrial Applications, 42 (4), 2006, pp. 1092-1100.

[2] F. Verbelen, A. Abdallh, H. Vansompel, K. Stockman and P. Sergeant, "Sizing Methodology based on Scaling Laws for a Permanent Magnet Electrical Variable Transmission", IEEE Transactions on Industrial Electronics, 67 (3), 2019, pp. 1739-1749.

[3] M. Hoeijmakers, "Electromechanical converter, Patent US 7 164 219", 2007.

[4] E. Vinot, R. Trigui, Y. Cheng, C. Espanet, A. Bouscayrol and V. Reinbold, "Improvement of an EVT-Based HEV Using Dynamic Programming", IEEE Transactions on Vehicular Technology, 63 (1), 2014, pp. 40-50.

[5] P. Shamsi, M. McDonough and B. Fahimi, "Wide-Bandgap Semiconductor Technology: Its impact on the electrification of the transportation industry.," in IEEE Electrification Magazine, vol. 1, no. 2, pp. 59-63, Dec. 2013, doi: 10.1109/MELE.2013.2293931.

[6] Dalla Vecchia, M.; Ravyts, S.; Van den Broeck, G.; Driesen, J. Gallium-Nitride Semiconductor Technology and Its Practical Design Challenges in Power Electronics Applications: An Overview. Energies 2019, 12, 2663.

[7] A. Anthon, Z. Zhang, M. A. E. Andersen, D. G. Holmes, B. McGrath and C. A. Teixeira, "The Benefits of SiC mosfets in a T-Type Inverter for Grid-Tie Applications," in IEEE Transactions on Power Electronics, vol. 32, no. 4, pp. 2808-2821, April 2017, doi: 10.1109/TPEL.2016.2582344.

[8] D. Sadik et al., "Short-Circuit Protection Circuits for Silicon-Carbide Power Transistors," in IEEE Transactions on Industrial Electronics, vol. 63, no. 4, pp. 1995-2004, April 2016, doi: 10.1109/TIE.2015.2506628.

[9] J. Druant, H. Vansompel, F. De Belie, J. Melkebeek and P. Sergeant, "Torque Analysis on a Double Rotor Electrical Variable Transmission With Hybrid Excitation", IEEE Transactions on Industrial Electronics, 64 (1), 2017, pp. 60-68.

[10] J. Druant, H. Vansompel, F. De Belie and P. Sergeant, "Optimal Control for a Hybrid Excited Dual Mechanical Port Electric Machine", IEEE Transactions on Energy Conversion, 32 (2), 2017, pp. 599-607.

[11] J. Druant, H. Vansompel, F. De Belie, and P. Sergeant, "Loss Identification in a Double Rotor Electrical Variable Transmission", IEEE Transactions on industrial electronics, 64 (10), 2017, pp. 7731–7740.

[12] M. Vafaeipour, M. El Baghdadi, J. Van Mierlo, O. Hegazy, F. Verbelen and P. Sergeant, "An ECMS-based approach for energy management of a HEV equipped with an electrical variable transmission", Fourteenth International Conference on Ecological Vehicles and Renewable Energies (EVER), 2019.

[13] JCGM 100:2008 "Evaluation of measurement data — Guide to the expression of uncertainty in measurement"

Seamless integration of feedforward and feedback control of balance of arm capacitor voltages in STATCOMs based on chain links of H bridge modules

D. Basic, N. Lapassat

General Electric, Power Conversion, 18 Avenue de Québec, Villebon-sur -Yvette, 91140
France
e-mail: duro.basic@ge.com, nicolas.lapassat@ge.com

URL: https://www.gepowerconversion.com/

Keywords

STATCOM, H bridge, Multilevel Converters.

Abstract

This paper presents a seamless integration of the feed-forward and feedback controls of balance of average arm capacitor dc bus voltages in the star and delta connected STATCOM converters, based on chain links of H bridge cells/modules, which are operated in general imbalanced situations.

Introduction

High power multilevel converters, suitable for STATic VAr COMpensators (STATCOMs) applications, based on chain links of full H bridge modules with floating dc bus capacitors, can be constructed if three converter chains (arm) are connected in either the star or delta topology (Fig. 1). Average power across the converter arms must be actively controlled in order to balance voltages of floating dc bus capacitors. When such multilevel converter topologies are used in balanced three phase operation, the arm level dc bus voltage balancing is not an issue. However, in imbalanced operation, even if the total three phase power exchange with the grid is kept at zero, the individual phase powers may not be zero. In such situations the arm converter dc bus voltages rapidly diverge, and an active control is required to keep them balanced. To achieve zero active powers in all arms in the star topology, the zero-sequence fundamental voltage can be injected into the arm converter phase voltages [1]. Similarly, in the delta topology, the zero-sequence fundamental current may be circulated through the arm converters [1], [4]. The zero-sequence injection needed to balance the arm powers/voltages may be synthetized using the feedback control of the arm dc bus voltages, either in the star [2] or delta [3] STATCOMs. However, relationship between the zero-sequence injection and produced balancing power is a non-linear function of the converter arm currents and voltages. To improve the dc bus control performances, the required zero-sequence injection can be found by using the feed-forward calculations. A possibility is to use the classical Steinmetz's equations to map the load phase reactive currents into the converter arm current references. This method has been employed in a delta STATCOM used for compensation of load current imbalance [4]. An alternative approach is to perform the feedforward calculations of required zero-sequence voltage/current injections using the converter arm currents and voltages as the input. The feedforward balancing of the star/delta STATCOMs presented in [5] considers the STATCOM which compensates only the positive sequence currents. In the subsequent work [6] and [7], the feedforward voltage balancing has been employed in cases when the load current imbalance is compensated. In more recent work [8]- [11] it has been shown that, in the star STACOMs, the required zero sequence voltage injection may become unbounded when the converter current imbalance approaches unity. This singularity point may be encountered when simultaneous compensations of the load positive sequence reactive and negative sequence currents is required. Duality between of star and delta STATCOM topologies and equations for the feedforward calculations of the required zero sequence injections (in case of arbitrary imbalances of the arm voltages and currents) has been discussed in [10]-[11]. According to the duality principle, singularity in the arm power balancing in the delta STATCOMs may be encountered when the voltage imbalance approaches unity[10]-[12]12. In vicinity of the singular points, relationship between the arm current or/and voltage imbalance and required zero sequence injection is highly nonlinear.

This paper presents an approach for seamless integration of the feed-forward feed-back controls of balance of the arm dc bus voltages, suitable for the star and delta STATCOMs, operated with balanced or imbalanced currents and/or voltages. The controller is built around the so-called arm power imbalance vector and the compact mapping between symmetrical components of the converter arm voltages and currents [11]. The feedforward control is based on direct synthesis of the arm power imbalance vector using the symmetrical components the converter arm voltages currents. The feedforward control is supplemented by the feedback control [13]. It trims the power imbalance vector calculated in the feedforward manner to ensure precise steady state control of average values of the arm dc bus voltages. In this manner, the feedforward and feedback controllers are seamlessly integrated. The resulting imbalance power vector is used as a reference for the inverse mappings (configurable for the star and delta topologies [11]) to deduce the zero-sequence voltage or current injection needed to compensate for the arm power imbalance. In such manner, transfer function of the balancing loop is linearized (even in vicinity of the singular points) and synthesis of parameters of the feedback PI controller is simplified. The controller performances are illustrated by several simulation results of operation of the star and delta STATCOMs, employed for compensation of imbalanced and dynamic loads in imbalanced grid voltage situations. The results presented demonstrate good dynamic disturbance rejection and steady state accuracy of the arm level dc bus voltage balancing. Such performances are achievable thanks to seamless integration of the feedforward and feedback control and linearization of the control loop transfer function.

Imbalance power vector and compensation via zero sequence injection

Converter topologies

The star and delta STATCOM converter topologies considered in this paper are shown in Fig. 1. The converters are constructed from 3 converter arms composed of chain links of N H bridge modules per converter arms. In series with each converter arm an inductor L_{arms} is connected. This inductor limits PWM current ripple and allows control of the arm currents. In order to balance the converter arm powers, the zero-sequence voltage v_0 and current i_0 are added to the converter arm voltages (v_1, v_2 and v_3) and currents (i_1, i_2 and i_3) of the star and delta STATCOM respectively.

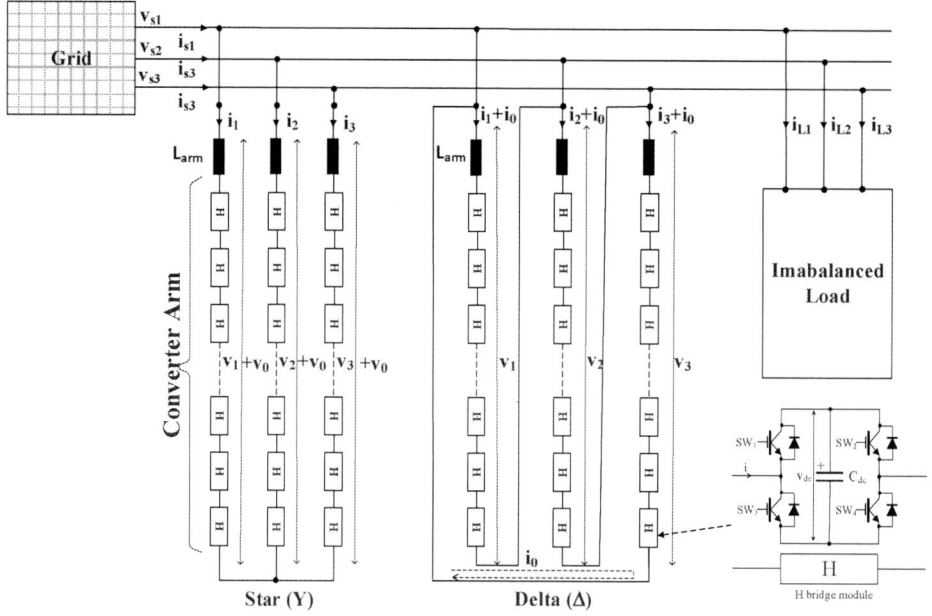

Fig. 1: The star and delta STATCOM converter topologies.

Imbalance power vector

The required zero sequence injection may be calculated from the converter imbalance arm power vector. It is defined in the following way [11] (P_1, P_2 and P_3 are average arm powers and $a=e^{j2\pi3}$):

$$\vec{P}_{imb} = \frac{2}{3}\left(P_1 + a^2 P_2 + a P_3\right) \tag{1}$$

The converter arm powers can be further expressed via the positive and negative sequence components of the converter arm voltages and currents [11] (subscripts p, n and 0 are used in the further text to denote the positive, negative and zero sequence components):

$$\vec{P}_{imb} = V_p I_n^* + V_n^* I_p \tag{2}$$

The arm power imbalance produced by imbalance of the arm converter voltages and/or currents is fully characterized by the imbalanced power vector (2). It can be compensated by producing additional power imbalance vector by injecting the zero-sequence voltage or current, without altering the STATCOM current injections into the power grid. Thus, the power imbalance vector (2) will be used in the feedforward calculation of the required zero sequence injection and as a base for integration of the feedforward and feedback controls.

Zero voltage injection (star STATCOM)

In the star connected STATCOMs, the power imbalance vector $\vec{P}_{v0}$ produced by the zero-sequence voltage injection is [11]:

$$\vec{P}_{v0} = I_n V_0^* + I_p^* V_0 \tag{3}$$

To cancel the converter imbalance power vector $\vec{P}_{imb}$, i.e. to re-balance the arm powers and capacitor voltages in an arbitrary imbalanced situation via $\vec{P}_{v0}$, the following zero sequence voltage injection is required [11]:

$$V_0 = \frac{\vec{P}_{imb}^* I_n - \vec{P}_{imb} I_p}{|I_p|^2 - |I_n|^2} \qquad V_0 = \frac{I_p^* I_n V_n - I_p^2 V_n^* + I_n^2 V_p^* - I_p I_n^* V_p}{|I_p|^2 - |I_n|^2} \tag{4}$$

Thus (4) defines the inverse map linking the power imbalance vector with the zero-sequence voltage reference to be injected into the arm voltages in the star STATCOMs.

Zero current injection (delta STATCOM)

In the delta connected STATCOMs, the zero-sequence current injection produces arm power imbalance which is characterised by the power imbalance vector $\vec{P}_{i0}$ [11]:

$$\vec{P}_{i0} = V_n I_0^* + V_p^* I_0 \tag{5}$$

It can be controlled in such a way to cancel the imbalance power vector $\vec{P}_{imb}$. For that, the following zero sequence current injection is required [11]:

$$I_0 = \frac{\vec{P}_{imb}^* V_n - \vec{P}_{imb} V_p}{|V_p|^2 - |V_n|^2} \qquad I_0 = \frac{V_p^* V_n I_n - V_p^2 I_n^* + V_n^2 I_p^* - V_p V_n^* I_p}{|V_p|^2 - |V_n|^2} \tag{6}$$

Consequently (6) represents the inverse map linking the power imbalance vector with reference of the zero-sequence current to be injected into the arm currents in the delta STATCOM topology.

STATCOM voltage-current decomposition into symmetrical components

Several methods can be employed for the decomposition of the converter three phase voltage/current systems into the positive and negative sequence components which are needed for the feed-forward calculation of the zero-sequence injections. To ensure accurate extraction of the positive or negative sequence components and full suppression of all higher order harmonics in the input (voltage/current) signals, the simplest approach is to place moving average filters (MAFs) in two Synchronous Reference Frames (SRFs), rotating at the synchronous angular frequency in the opposite directions (Fig. 2). For effective filtering by the SRF based MAFs the ratio between the fundamental period and sampling period should be an integer (otherwise the notches produced by the MAF are not aligned with the harmonic frequencies).

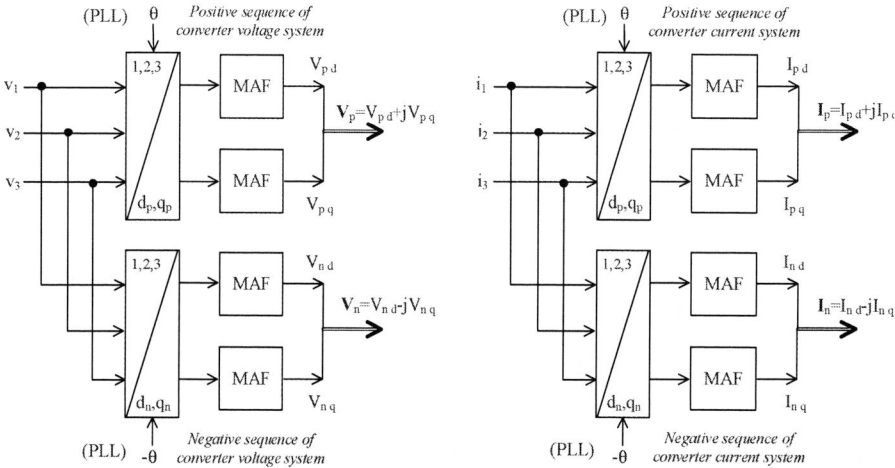

Fig. 2: Detection of positive and negative sequence components of the converter voltages and currents based on SRF MAF.

Instead of MAFs, different variants of Double SRF based Low Pass Filtering (LPF) can be used. For synchronisation with the grid, angular base (θ) is generated using either a Phase Locked Loop (PLL) or Frequency Locked Loop (FLL).

Average dc bus voltage control

Control of the average dc bus of all three arms can be performed in the classical manner, by adjusting active component of the STATCOM positive sequence fundamental current. For that a PI controller, driven by difference between the dc bus voltage reference and average dc bus voltage feedback, can be used. The average arm dc bus voltage is calculated by finding average value of arm level dc bus voltages:

$$v_{dc\,ave} = \frac{v_{dc\,arm\,1} + v_{dc\,arm\,2} + v_{dc\,arm\,3}}{3} \tag{7}$$

where the arm dc bus voltages are equal to sum of all dc bus voltages in the arm:

$$v_{dc\,arm\,i} = v_{dc\,i1} + v_{dc\,i2} + \dots + v_{dc\,i\,N} \qquad i=1, 2, 3 \tag{8}$$

In Fig. 3 the average dc bus voltage control is integrated with the balancing controller. It is performed indirectly, via controlling amount of the total stored energy in all converter dc bus capacitors.

Balancing of arm capacitor voltages

Block diagram of the arm capacitor voltage balance controller is shown in Fig. 3. The controller has the following blocks:

- feed-forward calculator of imbalance power vector,

- feed-back PI control for trimming of the imbalance power vector and

- inverse mapping for zero sequence reference calculation.

The feedforward calculator and feedback control together define the set point for the imbalance power vector which is to be eliminated via the zero-sequence signal injection. The imbalance power vector is used as an input to the inverse mapping block which calculates the required zero sequence signal injection. Then the calculated zero sequence reference is added to the converter arm phase voltage (star converter) or current (delta converter) references.

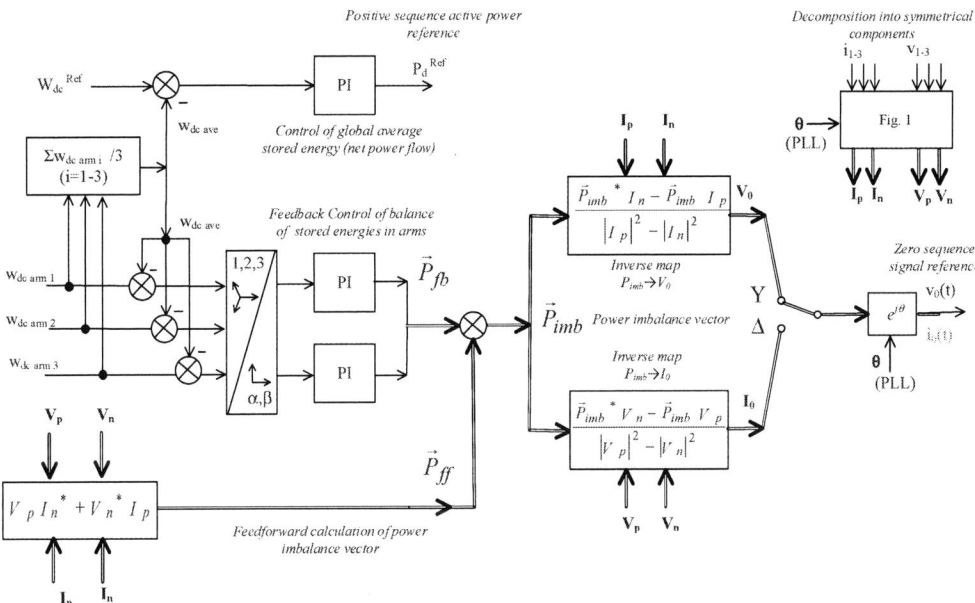

Fig. 3: Block diagram of the integrated feedforward-feedback control of the arm dc bus voltage balance (configurable for the star and delta connected STATCOMs).

The feedback control is based on the stored energy control (rather than on direct voltage control) as show in Fig. 4. In this way the transfer function between the stored energy and power is a linear function (simple integrator).

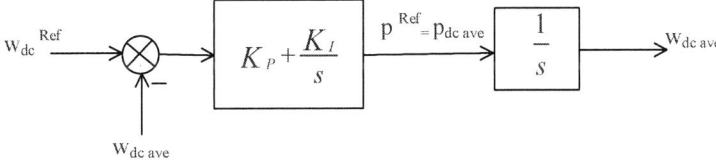

Fig. 4: Feedback control of mean value of stored energy in the arm capacitors.

For the feedback control of balance of the arm dc bus voltages, two additional PI controllers are employed in Fig. 3. These controllers directly trim the calculated power imbalance vector using the 3 arm voltage error signals transformed into a vector (according to Eq. (1), using the $1,23 \rightarrow \alpha,\beta$ transformation). The energies stored in arm capacitors are estimated from the arm capacitor voltages and equivalent arm capacitance $C_{eq}=C_{dc}/N$ (C_{dc} is dc bus capacitance of the H bridge module in Fig. 1):

$$w_{dc\,arm\,i} = \frac{1}{2} C_{eq}\, v_{dc\,arm\,i}^{\,2} \qquad i=1,2,3 \tag{9}$$

The native 2nd harmonic ripple should be removed from the arm dc bus voltage signals to prevent the ripple propagation into the converter current/voltage references. It can be accomplished by using the 2nd harmonic notch filter (NF) as shown in Fig. 5. Optionally a low pass filter can be added in series with the notch filter.

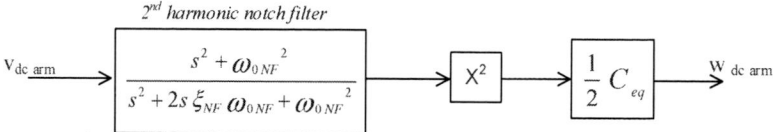

Fig. 5: Average arm capacitor stored energy estimation and feedback control.

Bandwidth of the feedback control loop can be set to a relatively low value of 5Hz. It has been found via simulations that this bandwidth value gives satisfactory results when the feedforward control is used. The converter parameters are summarized in Table I.

Table I: **Converter Parameters.**

Parameter	ω_{0NF}	ξ_{NF}	ω_{BW}	ξ	K_P	K_I	C_{eq} (arm)
Value	$2\pi100$	0.5	$2\pi5$	1	$2\,\xi\,\omega_{BW}$	$\omega_{BW}^{\,2}$	C_{dc}/N

Simulation results

In this section several simulation results illustrating the STATCOM operation in imbalanced situations are presented. The system parameters used in the simulation are listed in Table II. Nominal current of the arm converters (i.e. H bridge modules) is 1450Arms.

Table II: **System parameters.**

Parameter	$V_{s\,l\text{-}l}$	f_s	N (delta)	L_{arm} (delta)	N (star)	L_{arm} (star)	V_{dc} (module)	C_{dc} (module)
Value	11kV	50Hz	9	3.9mH	6	1.3mH	2500V	15mF

Simulation of star STATCOM in imbalanced operation

In the simulations which results are shown in Fig. 6, a star STATCOM is balancing an imbalanced load which negative sequence current is approximately 30% of the positive sequence current. The grid voltages are also imbalanced, the negative sequence voltage is 20% of the positive sequence voltage. At instant t=0.25s, the STATCOM start injecting the compensation currents which results in balanced grid side currents which are in phase with the positive sequence voltage. Two results in Fig. 6 illustrate perturbance of the arm capacitor voltages at the start of the load compensation when the feedback control only (Fig. 6, left), and the combined feedforward +feedback control (Fig. 6, right) are used. With the feedforward control, a significant reduction of perturbation of balance of the arm capacitor voltages can be observed. The arm powers and capacitor voltage balance are achieved via injection of the zero-sequence voltage. This causes a significant increase of the arm voltages when the STATCOM compensation is activated. Thus, the star connected STATCOM must have a significant built-in voltage margin when the imbalanced load compensation is required [11].

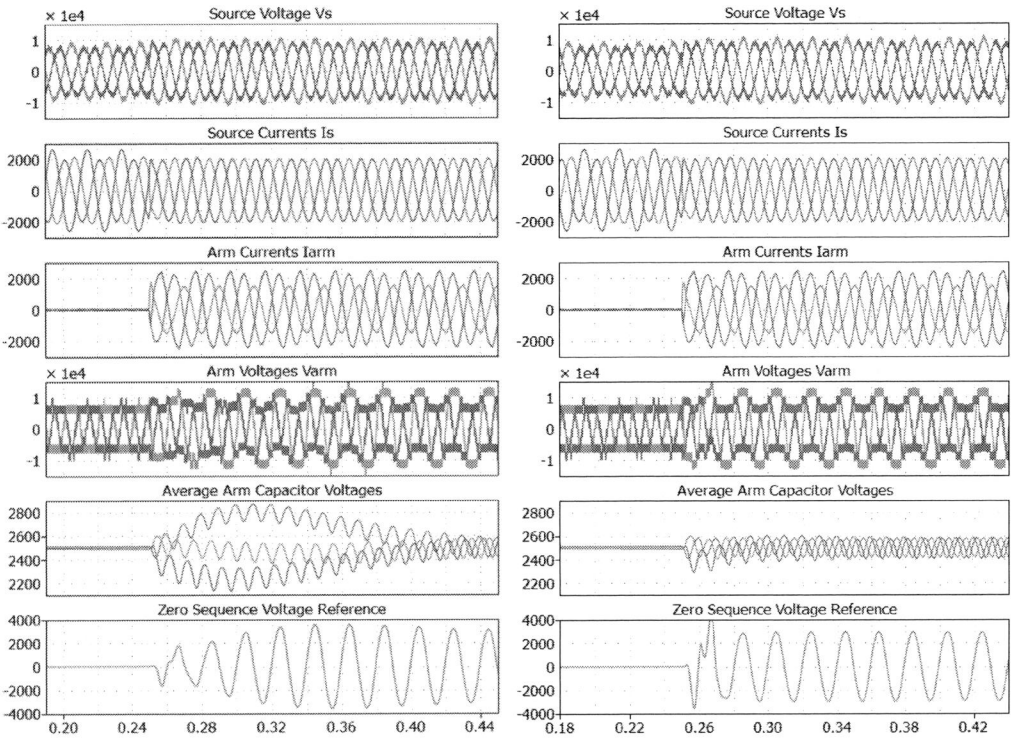

Fig. 6: Performance of the arm dc bus capacitor voltage balance control in the star STATCOM with the feedback PI control (left) and feedforward-feedback control (right).

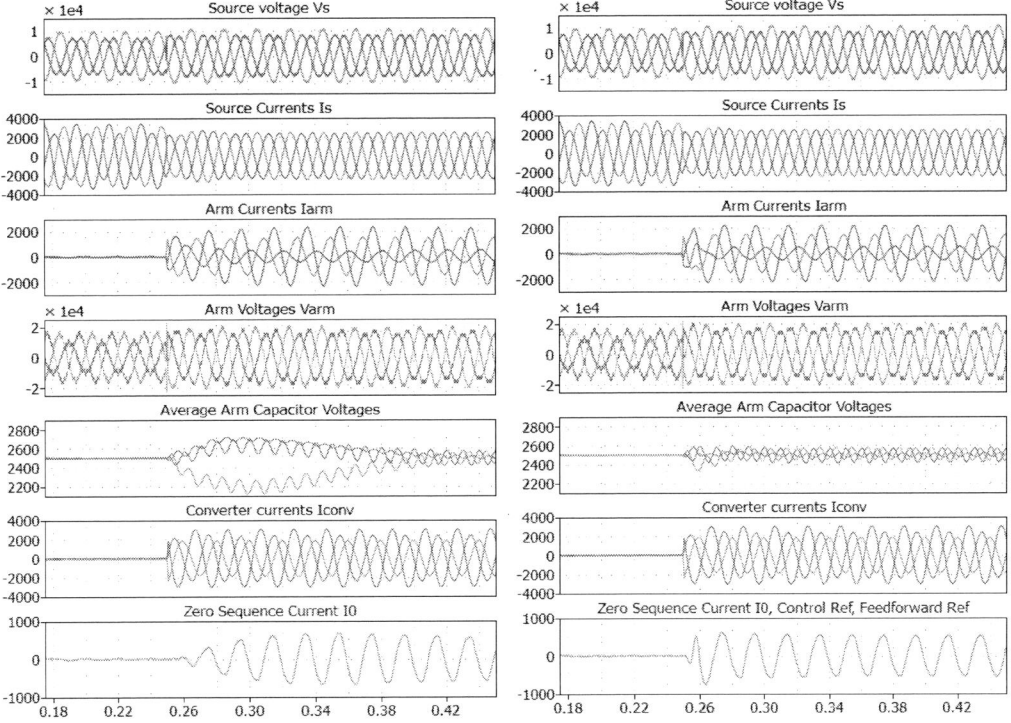

Fig. 7: Performance of the arm dc bus capacitor voltage balance control in the delta STATCOM with the feedback PI control (left) and feedforward-feedback control (right).

Simulation of delta STATCOM in imbalanced operation

In the simulation results shown in Fig. 7, a delta STATCOM is used to rebalance an imbalanced load under imbalanced grid voltages. The load current imbalance is 30%, while the grid voltage imbalance is 20%. Similarly, as in the previous case, at instant t=0.25s, the STATCOM starts injecting the compensation currents to produce balanced grid side currents which are in phase with the positive sequence voltages. Again, a significant reduction of perturbations of the arm capacitor voltages is achieved when using the feedforward control. The arm powers and capacitor voltages are balanced via the zero-sequence current injection. This causes a significant increase of the arm currents once the STATCOM compensation is activated. Thus, the delta connected STATCOM must have built in a significant current margin if load compensation in imbalanced situations is required [12].

Compensation of electrical arc furnace by delta STATCOM

The feedforward control improves balancing of the arm capacitor voltages in cases when (for example) delta STATCOM is used for compensation of highly dynamic loads, producing random current imbalances such as Electrical Arc Furnace (EAF). The simulation results shown in Fig. 8 illustrate perturbations of the arm capacitor voltages with the feedback based control (Fig. 8, left) and the feedforward + feedback control (Fig. 8, right). In the latter case the arm capacitor voltages are well balanced. The residual (balanced) arm capacitor voltage perturbations (comparable to that which would be seen in converters with common energy storages) are produced by partial compensation of dynamic fluctuations of the EAF active power.

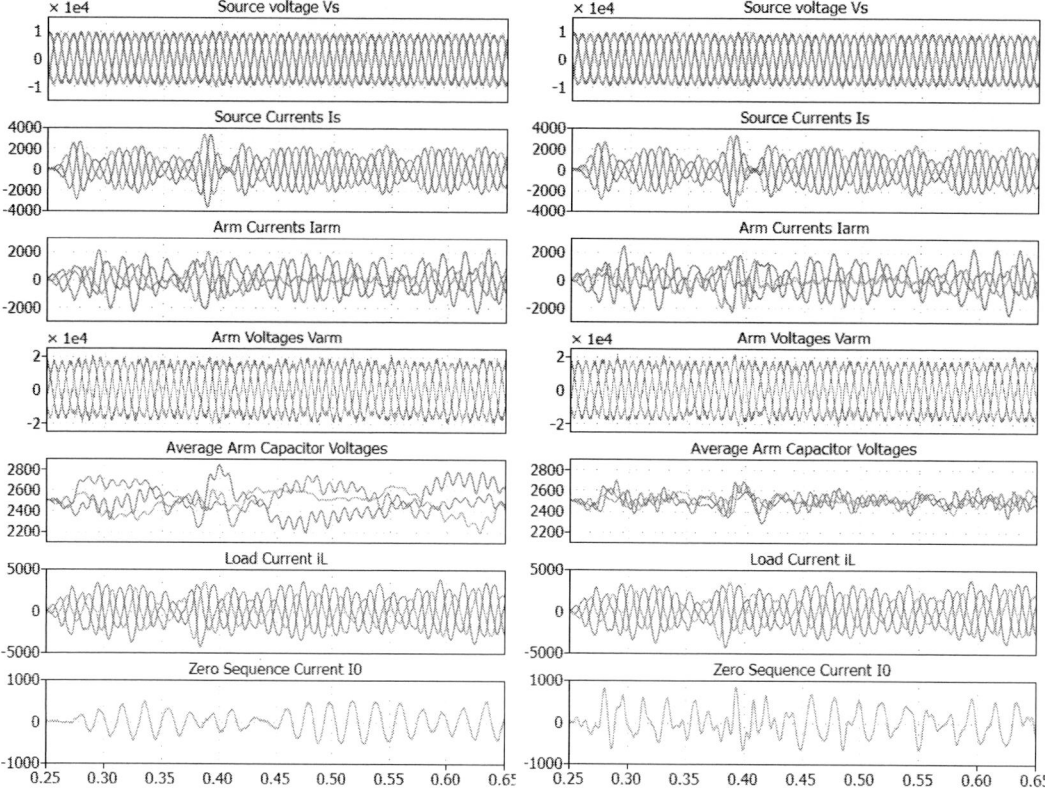

Fig. 8: Simulation of the arm dc bus capacitor voltage balance control during compensation of electrical arc furnace load with the delta STATCOM: with the feedback PI control (left) and feedforward-feedback control (right).

Conclusion

This paper presents an optimized approach for control of balance of the arm capacitor voltages in STATCOM converters based on chain links of H bridge modules which are operated with imbalanced arm voltages and/or currents. The proposed approach seamlessly integrates the feedforward, feedback PI controls and mappings between symmetrical components of the STATCOM:

- disturbing, differential mode, positive and negative sequence, and
- stabilizing, common mode, zero sequence

arm voltages and currents.

Main features of the proposed implementation can be summarized as follows:

- The arm power imbalance vector introduced in [11] is utilized for seamless integration of the feedforward and feedback control loops and linearization of the control loop transfer function.
- The feedforward control is based on direct mapping between the (differential mode) positive/negative sequence symmetrical components of the converter arm voltage/currents and the power imbalance vector. The feedforward control allows detection the arm power imbalance prior significant imbalance of the arm capacitor voltages is developed. It enables improved balancing performance in transient situations or in compensation of random EAF loads.
- The feedback control is based on two PI controllers. It provides fine trimming of the imbalance power vector defined by the feedforward controller and good steady state accuracy.
- The resulting power imbalance vector (synthetized by the feedforward and feedback controls) is used as a reference for calculation of the required balancing zero sequence injection.
- The inverse mapping (functional relationship) is employed to convert the power imbalance vector reference into the required zero sequence injection. It is configurable for the star and delta STATCOM topologies. Utilization of this inverse mapping linearizes the control loop, simplifies synthesis of the PI controller and improves performance of the feedback control.

References

[1] R.E. Betz, T. Summers, T. Furney, 'Symmetry Compensation using a H-Bridge Multilevel STATCOM with Zero Sequence Injection', Conference Record of the 2006 IEEE Industry Applications Conference, 41st Annual IAS Meeting, Vol. 4, August 2006, pp. 1724-1731.

[2] H. Akagi, S. Inoue, T. Yoshii, 'Control and Performance of a Transformer less Cascade PWM STATCOM with a Star Configuration', IEEE Transactions on Industry Applications, Vol. 43, No. 4, July/August 2007, pp. 1041-1049.

[3] M. Hagiwara, R. Maeda, H. Akagi, 'Negative Sequence Reactive Power Control by PWM STATCOM Based on Modular Multilevel Cascaded Converter', 2011 IEEE Energy Conversion Congress and Exposition (ECCE), 2011 , pp. 3728 – 3735.

[4] F. Z. Peng, J. Wang, 'A Universal STATCOM with Delta-Connected Cascade Multilevel Inverter', 35th Annual IEEE Power Electronics Specialist Conference, Aachen, Germany, 2004, pp. 3529-3533.

[5] T. J. Summers, R. E. Betz, G. Mirezaeva, 'Phase Leg Voltage Balancing of a Cascaded H-Bridge Converter Based STATCOM using Zero Sequence Injection', 13th European conference on Power Electronics and Applications, EPE'09, 2009, pp. 1-10.

[6] Q. Song, W. Liu, 'Control of a Cascade STATCOM With Star Configuration Under Unbalanced Conditions', IEEE Transactions on Power Electronics, Vol. 24. No. 1, January 2009, pp. 45-58.

[7] L. Tan, S. Wang, P. Wang, Y. Li, 'High Performance Controller with Effective Voltage Balance Regulation for a Cascade STATCOM with Star Configuration Under Unbalance Conditions', 15th European conference on Power Electronics and Applications, EPE'13, 2013, pp. 1-10.

[8] G. Postiglione, G. Borghetti, G. Torre, P. Bordignon, 'Transformerless STATCOM based on multilevel converter for grid voltage restoring', PCIM Europe, May 2011, Nureberg, Germany, pp. 413-419.

[9] S. Du, J. Liu, 'A Brief Comparison of Series Connected Modular Topology in STACOM Applications, 2013 IEEE ECCE Asia Downunder (ECCE Asia), 2013, pp. 456-460.

[10] E. Behrouzian, M. Bongiorno, M. , H. De La Parra Zelaya, 'Investigation of negative sequence injection capability in H-bridge multilevel STATCOM', 2014 16th European Conference on Power Electronics and Applications (EPE'14-ECCE Europe), 2014.

[11] D. Basic, M. Geske, S. Schroeder, 'Limitations of the H-Bridge Multilevel STATCOMs in Compensation of Current Imbalance', 2015 17th European Conference on Power Electronics and Applications (EPE'15-ECCE-Europe), 2015, pp. 1-10.

[12] D. Basic, M. Geske, S. Schroeder, J. Janning, 'Injection Capability of Delta Connected STATCOM in Imbalanced Situations', 2016 18th European Conference on Power Electronics and Applications (EPE'16-ECCE-Europe), 2016, pp. 1-10.

[13] D. Basic, 'Control of Balance of Arm Capacitor Voltages in STATCOMs based on Chain Links of H Bridge Modules', US 9,590,483 B1, 7 March 2017.

Asynchronized electromechanical converter in the electrical supply system of powerful energy consumers

Aleksey G. Vorontsov
SAINT PETERSBURG ELECTROTECHNICAL UNIVERSITY "LETI"
ul. Professora Popova 5, 197376
Saint-Petersburg, Russia
Tel.: +7 (921) – 7607534
E-Mail: ag.voroncov@gmail.ru

Mikhail V. Pronin
SAINT PETERSBURG ELECTROTECHNICAL UNIVERSITY "LETI"
ul. Professora Popova 5, 197376
Saint-Petersburg, Russia
Tel.: +7 (921) – 3056783
E-Mail: mpronin1@rambler.ru

Anastasiia D. Stotckaia
SAINT PETERSBURG ELECTROTECHNICAL UNIVERSITY "LETI"
ul. Professora Popova 5, 197376
Saint-Petersburg, Russia
Tel.: +7 (960) – 2420123
E-Mail: adstotskaya@etu.ru

Vasiliy V. Glushakov
PUBLIC JOINT STOCK COMPANY "POWER MACHINES"
3A Vatutina st., 195009,
Saint-Petersburg, Russia
Tel.: +7 (921) – 5589170
E-Mail: glushvas@yandex.ru

Pavel V. Sokur
JOINT STOCK COMPANY "SCIENTIFIC AND TECHNICAL CENTER OF FGC UES"
22/3Kashirskoe sh., 115201,
Moscow, Russia
Tel.: +7 (499) – 6135722
E-Mail: sokur_pv@ntc-power.ru

Keywords

«Power supply», «Asynchronous motor», «Simulation», «Industrial application», «Energy storage»

Abstract

In power supply systems of large energy consumers, it is possible to use electric machine converters with asynchronized generator-engines. These devices provide uninterrupted power supply to consumers as well as guarantee such advantages as galvanic isolation, separate parameters stabilization, limitation of short-circuit currents, accumulation and conversion of kinetic energy with its subsequent conversion into electrical energy. Active frequency converters in the excitation systems of electrical machines allow the start-up of units and provide other modes of operation, including using the accumulated kinetic energy. The structure and operating modes of these converters are discussed.

Introduction

Metropolises, large plants, mines, according to the criterion of reliability of power supply, belong to consumers of the 1st category [1]. Their power supply is carried out from two mutually independent sources. In case of accidents, a power interruption is allowed only during the switching of sources. In some cases, the power interruption is excluded. Energy transfer to metropolises is usually carried out from power plants via transformers, switching devices and high-voltage power lines with a voltage of 110 kV, 220 kV, 330 kV etc. When the load changes, the voltages in the power lines and transformers also change, and these changes must be compensated. Difficulties in combining sections of high-voltage lines also arise in branched electric networks due to differences in voltage levels and phases as well as due to an increase in short-circuit currents. There are problems with the connection of new consumers of electricity. Transformers are made with taps to ensure the required operating modes of the power supply networks and normalize the parameters of electricity. phase-shifting transformers, current-limiting and shunt reactors, unregulated and adjustable capacitor banks, static compensators, energy storage devices and absorbers, and other devices [2] are also used.

In the power supply systems of metropolises, there are also difficulties in placing of new equipment, laying cable routes in the conditions of dense development of territories.

Asynchronized electromechanical converter

A relatively new technical solution for large consumers power supply is the use of electric machine converters (AEMFC) with asynchronized generator-engines (AGE) [3]. AGEs have well-known advantages and are used in pumped storage power plants [4,5, 6, 7, 13, 17], in asynchronized compensators and kinetic energy storage devices [8, 9, 11, 12, 14], etc. [16].

AEMFCs allow uninterrupted power supply to consumers from two sources when using the kinetic energy of rotating masses during switching of equipment (in case of accidents). AEMFCs also allow to stabilize the voltage of electric power systems (EPS) sections at specified levels, adjust the phase voltage when combining power lines switching devices, suppress voltage fluctuations in transient and emergency modes of operation of the EPS, limit short-circuit currents, participate in the distribution of loads between sources and consumers, compensate reactive power to reduce power line loads. When using the AEMFCs for the power supply of a metropolises an underground placement of an electric machine converter can be used.

The following option of power supply to the metropolis Salaryevo (district of Moscow, Russia) is considered according to the scheme in Fig. 1 due to abovementioned advantages of the AEMFC. Two mutually independent electric power grid EPG_1 and EPG_2 with a voltage of 220 kV are energy sources. Consumers are connected to two galvanically isolated power grids EPG_3 and EPG_4 with voltages of 15,75 kV. EPG_1 is connected with EPG_3 via transformer Tr_3, EPG_2 is connected with EPG_4 via transformer Tr_4. Energy consumers are connected to EPG_3 and EPG_4. The mutual connection of EPG_3 and EPG_4 is carried out through an electromechanical unit containing AGE_1 and AGE_2. Excitation of AGEs is provided by three active frequency converters FC_1, FC_2 and FC_3, each of which is designed to perform all the functions (one frequency converter is backup). Each FC contains an active rectifier AR and an autonomous voltage inverter AVI. Each FC can be connected by corresponding switching devices to EPG_1 (via Tr_1) or to EPG_2 (via Tr_2). Each FC can be connected by switching devices to the windings of the stator or rotor AGE_1 or AGE_2. At the same time energy is transferred from EPG_3 to EPG_4 or in the opposite direction through the mechanical connection of two AGEs. Uninterrupted power supply when one of the sources of electricity is disconnected is also provided including the use of energy stored in the rotating masses of the AGE in critical cases.

Thyristor protection rectifiers PR are also provided in the system. These are thyristor bridges, closed to small active resistances. In normal system operation, thyristors in PR are locked. In the event of short circuits in EPG_1 or EPG_2, the currents in the winding of the AGE stator are many times increased and transformed into a rotor. In order for these currents not to damage the transistor FCs, the rotor winding is practically inertialessly closed to a low resistance through an PR [17].

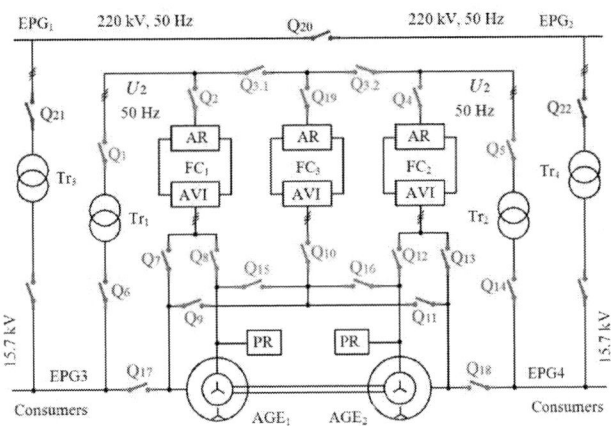

Fig. 1: Incorporation of electromechanical converter into a power grid

A design sketch of a dual-machine converter with 200 MW AGE and a rotational speed of 300 rpm is shown in Fig. 2. Unit height is equal to 16 m, unit weight is equal to 1390 t. At a power of each AGD of 200 MW, three-level FCs are used with parallel connection of less powerful converters, with the distribution of currents between parallel-connected branches using reactors and regulators. The scheme of one active FC is also presented in Fig. 2. In FC, the alternating voltage at the input and output is 1400 V, the rectified voltage is 2200 V.

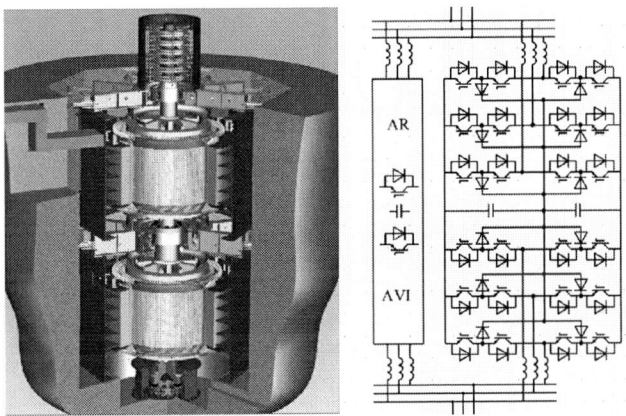

Fig. 2: Unit design with AGE and FC circuit

Operating modes are provided due to the excellent controllability of AGE. The regulation of the active powers of the stator, rotor and on the AGE shaft is provided by a relatively low-power FC in the rotor circuit. In this case, the active powers of AGE elements are distributed in accordance with Fig. 3, taking into account the fact that the FC active power P_r is equal to the product of the stator winding active power P_s and slip s (losses are neglected). The power on the shaft P_J is determined based on the law of conservation of energy.

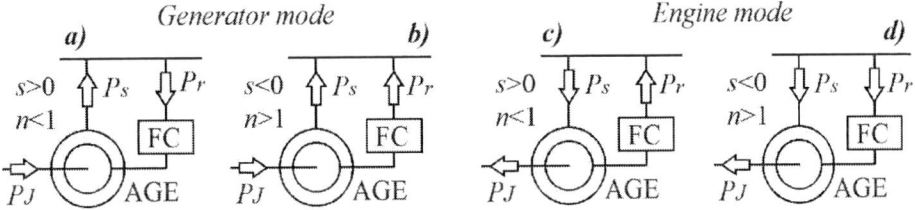

Fig. 3: AGE active power directions in various modes

Various tasks are required to be solved during the development of the installation: unit start-up, unit synchronization with EPG_3 or EPG_4, AGE speed control, regulation of the active power of the unit, adjustment of EPG_3 or EPG_4 voltages, accumulation and use of kinetic energy of the rotating masses of the unit, etc. This paper deals with the task of an asynchronized electromechanical converter start-up, some characteristics of the unit in steady state operating mode, the possibility of using the unit as a storage of kinetic energy. These tasks are solved using mathematical description and computer simulation of the installation with AGE.

AGE mathematical description

The mathematical description of the AGE was performed by dividing it into subcircuits interconnected through dependent sources of voltage and current (see Fig. 4) [13-16].

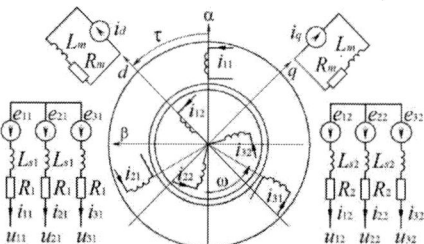

Fig. 4: AGE subsystems

The stator winding is described in the phase coordinates, it is characterized by phase voltages u_{11}, u_{21}, u_{31} and currents i_{11}, i_{21}, i_{31}. Active resistance R_1 and leakage inductance L_{s1} are taken into account in the model. The connections of the stator subcircuit with other circuits are taken into account by dependent voltage sources e_{11}, e_{21}, e_{31}.

The winding of the rotor is described in rotor phase coordinates. It is characterized by phase voltages u_{12}, u_{22}, u_{32} and currents i_{12}, i_{22}, i_{32}. Active resistance R_2 and leakage inductance L_{s2} are also taken into account in the model of rotor winding. The connections of the rotor subcircuit with other circuits are taken into account by dependent voltage sources e_{12}, e_{22}, e_{32}.

Voltages u_{12}, u_{22}, u_{32} are created by external devices and are determined in accordance with their mathematical description. The magnetization circuits are described in the dq axes.

Magnetization inductances L_m, depending on the saturation of the steel (in the calculations, this parameter is specified at each step according to the characteristic of idling AGD), and active resistances R_m are taken into account. The connections of these subcircuits with other circuits are taken into account by dependent current sources i_{ad} and i_{aq}.

During the transformation from a fixed coordinate system with axes $n=1$, 2, 3 to a rotor rotating coordinate system dq, and during reverse transitions, transformations of variables to fixed axes $\alpha\beta$ [13-17] are used.

The directions of the axes, the direction of rotor rotation with a frequency ω, the angle of rotation τ of the rotor axis d relative to the axis α are shown in the fig. 4.

In the start-up mode the moment of resistance on the shaft of a dual-machine aggregate M_c is the sum of the moment of mechanical losses of two AGEs and the moment due to losses in the stator steel of one AGE. The mechanical losses are proportional to the square of the relative rotational speed of the AGD rotor, the losses in the stator steel are proportional to the square of the product of relative flux linkage in the air gap and the relative frequency of rotation.

Model of unit with AGE

The model of the installation with a dual-machine unit with AGE is developed for start-up processes calculating. The model is built to reduce the cost of computer time for calculations. The model allows to calculate the long processes of acceleration of the unit by calculating a number of steady-state operating modes at fixed rotor speeds, followed by processing the results, taking into account the inertia of the rotor and the obtained AGE torques. A simplified diagram of the model is presented in Fig. 5.

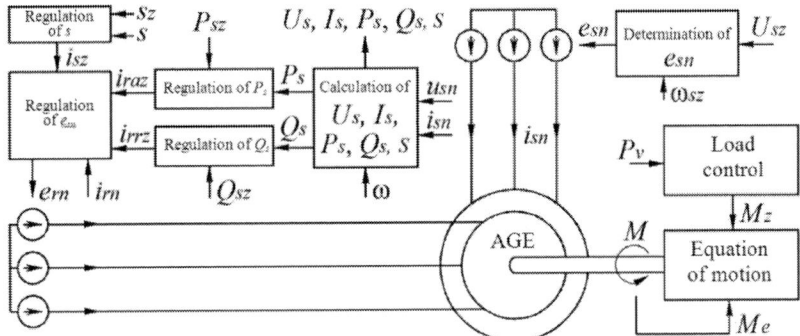

Fig. 5: The structure of the model of unit with AGE

The AGE stator winding is connected to a three-phase e_{sn} emf system. The current voltage of the stator winding U_{sz} is set. The frequency of the stator voltage differs from the synchronous one (corresponding to the given rotor speed) by the value of the given slip s_z during AGE start-up mode. The rotor winding is connected to a three-phase symmetric EMF system e_{rn} ($n=1, 2, 3$). The instantaneous values of these EMFs are formed by the CS. The signals of voltages u_{sn} and currents i_{sn} of the stator phases, currents i_{rn} of the rotor phases, given active P_{sz} and reactive Q_{sz} power of the stator, given slip s_z, the actual rotor speed ω, are accumulated by CS. The effective voltage U_s and stator current I_s, actual active P_s and reactive stator power Q_s, and actual slip s are calculated in the control system. The stator active power controller generates a task for the active component of the i_{raz} rotor current, and the stator winding reactive power controller generates a task for the i_{rrz} stator reactive power. The instantaneous values of the EMF of the phases of the rotor e_{rn} are formed using the specified settings for the current components. The model also contains blocks in which the load is regulated - the moment of resistance on the AGE shaft M_z, the shaft rotation frequency ω, the power on the shaft P_v are determined.

The described AEMFC model is implemented in the C ++ programming language and is used for calculations on personal computers of average productivity.

Results of calculation of AGE start up mode

Using the described AGE model, the steady-state operation modes of the system were calculated at fixed rotor rotation frequencies from $0,1\omega_{nom}$ to $1,1\omega_{nom}$ with a step of $0,1\omega_{nom}$.

The following nominal system parameters are used: EPG voltage – 15,750 kV, EPG frequency – 50 Hz, AGE active power – 200 MW, the stator voltage – 15,750 kV, the stator voltage frequency – 50 Hz, the power factor – 0,85, the rotor inertia – 50 tm^2 (the AGE is reduced to a two-pole design), the efficiency factor – 0,985, the coefficient of reduction of the rotor parameters to the stator – 4,902, stator phase inductance – 0,1 p.u., rotor phase inductance – 0,14 p.u., magnetization inductance – 2,52 p.u., active resistance of a stator phase – 0,0021 Ohms, active resistance of a rotor phase – 0,034 Ohms.

In the nominal mode, the sum of mechanical losses for two machines and losses in the stator steel is 2 MW, the part of mechanical losses in the specified value is 0,6. Mechanical losses are proportional to the square of the rotor speed, losses in the stator steel are proportional to the squares of the magnetic flux and speed. The calculations were performed for FC that allow an effective AVI phase current up to 4 kA and a rectified voltage up to 2,2 kV.

The AGE is started at a reduced stator voltage and low FC currents. The acceleration rate is limited by mechanical losses and inertia of the AGE rotating masses. The calculations were performed during acceleration of the unit from a stationary state to 110 % of the nominal speed. AGE operates in the "run-down" mode after FC disconnecting from the stator winding (at the end of acceleration), AGE's rotation frequency decreases by the influence of energy losses. In this case, the AGE is excited to the voltage of the EPG, then the AGE and EPG are synchronized. The calculation results of the process of unit accelerating are presented in Table I and Table II.

Table I: Voltage, current and other variables during the start-up of the unit with AGE

t	n	U_s	I_s	U_r	I_r	Ψ_m	M_J	s
s	%	V	A	V	A	%	%	%
14	10	1263	1426	1285	791	101	18	19,5
31	20	1413	2003	1415	542	56	14	19,4
57	30	1410	1998	1417	482	37	9,5	19,3
92	40	1404	1999	1417	458	28	7,1	19,1
136	50	1397	2000	1418	446	22	5,5	18,9
191	60	1393	1999	1413	439	18	4,5	18,7
258	70	1391	2000	1410	435	16	3,7	18,5
336	80	1396	2003	1409	434	14	3,2	18,1
430	90	1392	1998	1406	431	12	2,6	17,7
545	100	1378	1998	1393	429	11	2,2	17,3
687	110	1363	1997	1377	428	10	1,7	16,9

Table II: Power and other variables during the start-up of the unit with AGE

t	n	ΔP_1	ΔP_2	ΔP_{st}	ΔP_{mh}	$\Sigma\Delta P$	P_1	P_2	P_J
s	%	kW	kW	kW	kW	kW	kW	kW	kW
14	10	13	53	1	1	67	3077	650	3666
31	20	25	30	1	10	66	4861	972	5773
57	30	25	24	2	33	84	4857	965	5741
92	40	25	21	3	77	126	4850	937	5663
136	50	25	20	4	152	201	4832	927	5559
191	60	25	20	4	262	311	4804	910	5405
258	70	25	19	5	418	467	4792	894	5200
336	80	25	19	6	624	674	4831	891	5010
430	90	25	19	6	895	945	4814	861	4731
545	100	25	19	7	1218	1269	4765	832	4329
687	110	25	19	8	1621	1673	4707	801	3838

Dependences of the AGE parameters on time t during acceleration are shown in the Table I and Table II. The rotor speed n, the effective voltage U_s and the effective current I_s of the stator winding, the effective voltage U_r and the effective current I_r of the rotor winding, the flux linkage ψ_m in the AGE gap, the energy loss power in the stator winding ΔP_1 and in the rotor winding ΔP_2, the power loss in stator steel ΔP_{st}, power of mechanical losses ΔP_{mh}, total power of energy losses $\Sigma\Delta P$, active powers of stator winding P_1 and rotor winding P_2, power on shaft P_J, AGE efficiency factor η, torque on AGE shaft M_J, AGE slip s are indicated. Some of these dependencies are graphically described in Fig.6.

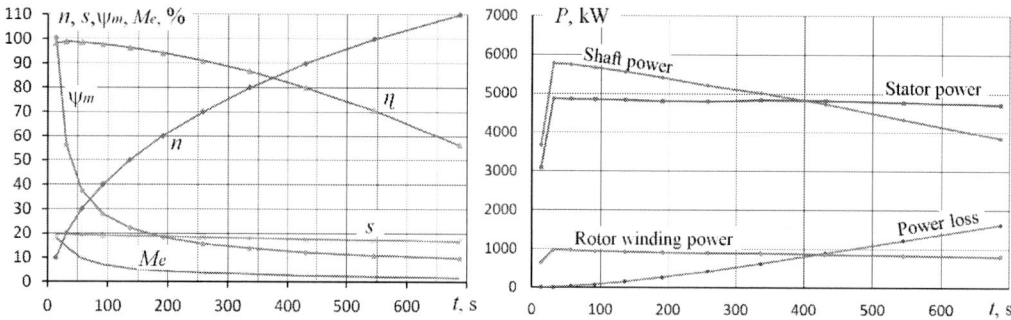

Fig. 6: AGE characteristics during the start-up of a dual-machine unit

Voltages of the stator and rotor windings were obtained close to the nominal FC voltage of 1400 V during the calculation. The AGE slip s is given 17-20 %. With increased slip, energy is transmitted to the shaft not only from the FC in the stator power circuit, but also from the FC connected to the rotor winding (see Fig. 3 d). From the calculation results, it can be noted that: the duration t of the AGE acceleration is 687 s. During AGE acceleration with a reduced stator winding voltage (about 1400 V), the flux linkage in AGE air gap decreases from 100 % to 10 %, the stator winding current is maintained at the value of 2000 A, and the AGD torque decreases from 18,2 % to 1,74 %, AGD efficiency decreases from 98 % to 56 %.

It should also be noted from Table I that with a reduced voltage of the AGE stator winding and with reduced currents, the energy losses in the windings and in the steel of the stator and rotor are relatively small and the total losses are close to the mechanical losses of the dual-machine unit.

The following result of the calculation is also significant: to start the AGE, one FC is used in the stator winding power supply circuit and another FC is used in the rotor winding power supply circuit. If one FC is used for an active power of up to 4707 kW in the stator circuit, then another FC is used for an active power of less than 1000 kW in the rotor circuit. The limitation of the FC active power in the rotor circuit is due to the limitation of the FC voltage in the stator circuit. If a more efficient acceleration of the AGE is required, then an increase in the nominal FC voltage, as well as increased insulation of the winding of the AGE rotor, is necessary.

Calculation results of the AGE steady state operating modes

Another objective of the development of an asynchronized electromechanical converter and FC is to evaluate the parameters of the system in steady-state operating modes. To calculate these modes, the computer model described above was used. The calculations were performed during the operation of the AGE in the generator and engine modes with a reactive power Q equal to 0 and an active power P_1 equal to 0 and ± 200 MW. The calculation results are presented in Fig. 7 and Fig. 8 and in Table III.

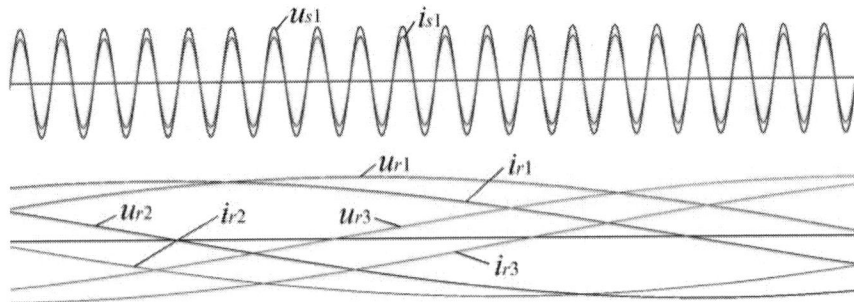

Fig. 7: AGE voltages and currents (P_1=200 MW)

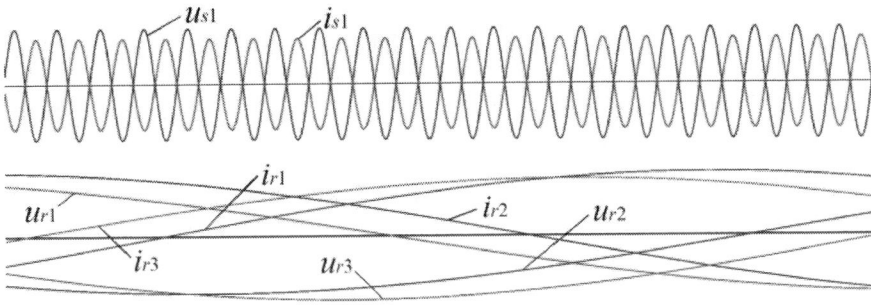

Fig. 8: AGE voltages and currents (P_1= -200 MW)

Additional advantages of an asynchronized electromechanical converter are provided when it is performed with the functions of a kinetic energy storage device. In the case under consideration, at a moment of inertia of a dual-machine unit of 50 t·m² (when the unit reaches the angular frequency of 314,15 rad/s), the kinetic energy store in rotating masses is 2567,1 MJ at a speed of 102 % and 2373,4 MJ at a speed of 98 %. With a decrease in the frequency of rotation of the AGE from 102 % to

98 %, an energy of 193,7 MJ is released. When converting this kinetic energy into electrical energy, a power of about 200 MW for 1 s (or 20 MW for 10 s, or others) can be generated in an EPG.

The indicated durations of the AGE output power to the EPG can be increased by expanding the range of variation of AGE slip and the corresponding increasing of AGE rotor winding voltage.

Table III: AGE parameters in steady-state operating modes

	Calculation #1	Calculation #2	Calculation #3	Calculation #4
Rotor speed, %	101,999	101,999	101,999	101,999
Active stator winding power, P_1, MW	199,998	-199,998	4e-06	-1e-06
Reactive stator winding power, Q_1, MVA	0,00133	0,00035	200	-200
Active rotor winding power, P_2, MW	4,193	-3,805	0,0172	0,407
Reactive rotor winding power, Q_2, MVA	-2,777	-2,877	2,006	-7,871
Current offset relative to stator voltage, φ_1, °	0,000382	179,999	89,999	-90
Current offset relative to rotor voltage, φ_2, °	-33,518	-142,901	89,506	-87,039
Effective stator phase voltage, U_{ph1}, V	9034,943	9035,986	9093,493	9093,496
Effective stator phase current, I_1, A	7379,079	7378,215	7331,248	7331,246
Effective rotor phase voltage, U_{ph2}, V	978,795	927,355	754,975	1130,093
Effective rotor phase current, I_2, A	1712,785	1714,939	885,743	2324,781
Magnetic flux in the AGE air gap, ψ_m, %	104,809	105,232	96,500	114,135
Power losses in the stator winding, ΔP_1, kW	343,023	342,943	338,624	338,624
Power losses in the rotor winding, ΔP_2, kW	299,230	299,983	80,023	551,270
Power losses in stator steel, ΔP_{st}, kW	900,878	908,165	763,695	1068,330
Mechanical power loss, ΔP_{mh}, kW	1292,736	1292,734	1292,739	1292,739
AGE shaft power, P_J, MW	201,403	-206,601	-2,458	-2,751
AGE efficiency factor, η_{AGE}, %	98,591	98,623		
AGE Torque, M_{AGE}, %	98,233	-100,768	-1,199	-1,341
Power imbalance (error), $\sum P$, kW	-47,072	-46,858	0,944	-92,617

Table III shows the parameter $\sum P$, which is the sum of all the AGE capacities in the steady-state periodic operation mode (power at the input, output, and loss) [16]. In the absence of calculation errors, the power balance ($\sum P$) should be equal to zero. However, for numerical solutions of systems of equations, there are some errors in the calculations. In the calculations under consideration, the power balance differs from zero, and this difference indicates the level of error in the calculations: $100 \cdot 92.6$ kW / 200000 kVA = 0.0463% (Table III, calculation 4).

Use of AEMFC as an energy storage

Additional advantages of an asynchronized electrical machine converter are provided when it is executed with the functions of a kinetic energy storage device. In the case under consideration, when the moment of inertia of the two-machine unit is 50 tm^2 (when the unit is brought to an angular frequency of 314.15 rad/s), the kinetic energy reserve in rotating masses is 2567.1 MJ at rotation rate of 102% and 2373.4 MJ at rotation rate of 98%. With a decrease in the frequency of rotation of the AGE from 102 to 98%, an energy of 193.7 MJ is released. When converting this kinetic energy into electric energy, a power of about 200 MW for 1 s (or 20 MW for 10 s, or others) can be generated in an EPG.

The indicated durations of the output of the AGE power in emergency conditions can be increased with an increase in the range of variation of the AGE slip and with a corresponding increase in the voltage of the rotor winding. The implementation of these changes can also reduce the duration of the start-up of the unit. However, in this case, it is necessary to adjust the design and parameters of the electric machine unit, as well as the FC.

Conclusion

The structure of redundant power supply of a metropolis using an electromechanical converter with asynchronized generators-engines and transistor exciters, also used to start an electric machine unit, is proposed.

An algorithm of an asynchronized generator-engine start-up is proposed. The stator and rotor windings of the machine receive electricity from two different frequency converters. The frequency converter in the power circuit of the rotor provides excitation of the machine, as well as transfers additional energy to the shaft.

It is proved that electromechanical converter with asynchronized electrical machines in a metropolis power supply system can be used as a storage of kinetic energy for converting it into electricity in case of emergency.

Reducing the start-up time of the AEMFC and a more complete use of the kinetic energy of the unit when active power is supplied to the EPG in emergency conditions is possible with an increase in the range of permissible changes in the AGE slip.

References

[1] Rules for the organization of electrical installations. Seventh Edition. March 15, 2019.

[2] M. Artemiev and others The concept for the development of electric networks in a metropolis based on the construction of 330 kV deep-in substations and the expanded use of 35 kV voltage class in distribution networks// Transmission and Distribution World. 2014 . №6 (27), pp. 88-92 (Rus).

[3] Dementyev, Y., Shakarian, Y., Sokur, P., Pinchuk, N., Novozhilov, V., Tretyakov, V., Dyachkov, V., Kucherov, Y., Yarosh, D., Mayorov, A., Shabash, A. Improvement of mode controllability and short-circuit currents limitation in metropolises power grid by means of electromechanical AC links as an alternative to DC links (2016) CIGRE Session 46, 2016-August

[4] Bocquel A., Janning J. 4*300 MW Variable Speed Drive for Pump-Storage Plant Applications // EPE 2003 Toulouse, France, Sept. 2003 P. 1–10.

[5] Bocquel A., Janning J. Analysis of a 300 MW Variable Speed Drive for Pump-Storage Plant Applications // EPE 05 Dresden, Germ Sept. 2005 P. 1–10.

[6] Koutnik J. Frades II -variable speed pumped storage project and its benefit to the electrical grid / Renewable Energy Conference, At Orlando, January 2012.

[7] Koutnik J. Bruns M., Hildinger T. Ein Sprung nach vorn – drehzahlvariables Pumpspeicherkraftwerk Frades II / In Wasserwirtschaft 105(5): 27-32. May 2015.

[8] Volodarskii L.G., Dovganyuk I.Ya., Mnev R.D., Plotnikova T.V., Sokur P.V., Tuzov P.Yu. The results of tests of ASK-100-4UKHL4 type asynchronized compensators at the Beskudnikovo 500 kV substation. Power Technology and Engineering, Vol. 47, No. 5, January, 2014, p.376-385.

[9] Shakaryan Y.G., Sokur P.V., Pinchuk N.D. et al. Asynchronized machines for the electric power industry / Single grid energy. 2018. № 4 (39). pp. 60-70.

[10] Dementiev Y.A., Sokur P.V., Shakaryan Y.G. Electromechanical ac converter for regime control and short – circuit current limitation in the metropolitan energi systems / Single grid energy. 2017, №5. pp. 18-27.

[11] Dovganyuk I.Y., Mnev R.D., Sokur P.V., Tuzov P.Y. An electromechanical energy-storage unit based on an asynchronized compensator // Russian Electrical Engineering. 2014. Vol. 85. № 1. pp. 53-58.

[12] Drobkin B.Z., Pronin M.V, Krutyakov E.A., Vorontsov A.G. Start up of asynchronized compensator ASK-100 // Power Technology and Engineering, 2010, № 7.

[13] Pronin M. V. et al. Electric and hydraulic process connection in hydraulic and pumped storage hydroelectric power stations //IECON 2013-39th Annual Conference of the IEEE Industrial Electronics Society. – IEEE, 2013. – C. 2033-2038.

[14] Pronin M.V., Vorontsov A.G., Kuzin M.E. Simulation and investigation of a kinetic energy storage as part of EPS // EPE'14 ECCE Europe, Finland, Lappeenranta.

[15] Pronin M.V., Shonin O.B., Vorontsov A.G., Gogolev G.A. Features of a Drive System for Pump-Storage Plant Applications based on the use of Double-Fed Induction Machine with a Multistage-Multilevel Frequency Converter // EPE-PEMC 2012, Novi Sad, Serbia.

[16] Pronin M.V., Vorontsov A.G. Electromechanotronic complexes and their modeling according to interconnected subsystems / St. Petersburg, Ladoga Publishing House, 2017 –220 p. (Rus).

[17] Pronin M. et al. A pumped storage power plant with double-fed induction machine and cascaded frequency converter //Proceedings of the 2011 14th European Conference on Power Electronics and Applications. – IEEE, 2011. – C. 1-9.

Symmetric and asymmetric operating modes of hybrid cascade frequency converters

Aleksey G. Vorontsov
SAINT PETERSBURG ELECTROTECHNICAL UNIVERSITY "LETI"
ul. Professora Popova 5, 197376
Saint-Petersburg, Russia
Tel.: +7 (921) – 7607534
E-Mail: ag.voroncov@gmail.ru

Vasiliy V. Glushakov
PUBLIC JOINT STOCK COMPANY "POWER MACHINES"
3A Vatutina st., 195009,
Saint-Petersburg, Russia
Tel.: +7 (921) – 5589170
E-Mail: glushvas@yandex.ru

Mikhail V. Pronin
SAINT PETERSBURG ELECTROTECHNICAL UNIVERSITY "LETI"
ul. Professora Popova 5, 197376
Saint-Petersburg, Russia
Tel.: +7 (921) – 3056783
E-Mail: mpronin1@rambler.ru

Anastasiia D. Stotckaia
SAINT PETERSBURG ELECTROTECHNICAL UNIVERSITY "LETI"
ul. Professora Popova 5, 197376
Saint-Petersburg, Russia
Tel.: +7 (960) – 2420123
E-Mail: adstotskaya@etu.ru

Keywords

«Multilevel converters», «Converter control», «Simulation», «Power quality»

Abstract

The hybrid cascade frequency converter contains a diode rectifier connected to the network and a three-phase autonomous voltage inverter, as well as a transformer connected in parallel to the network, supplying units with active rectifiers and single-phase voltage inverters. On the load side, three-phase and single-phase inverters are connected in series. The structure of the converter allows to use the transformer for a partial load power, increase the voltage at the output, provide the required quality of voltages and currents at the input and output of the converter, power the load when some of the blocks fail. The model of the converter and investigation results are described. Control algorithms that provide power distribution between the blocks, the required quality of voltages and currents at the input and output, as well as the efficient operation of the converter when disconnecting part of the power blocks are proposed.

Introduction

High-voltage semiconductor frequency converters (FC) are used in the energy sector, in the mining industry, in oil and gas transportation industry, etc. In wind power installations and power transmission, modular multilevel converters (MMFC) are used [1-3]. MMFC can be performed without transformers, which improves the efficiency of the system. They are able to provide a practically sinusoidal form of voltage and phase currents. However, MMFCs have a drawback - their performance decreases with a decrease in the frequency of phase currents. This makes it difficult to use MMFC in electric drives and in some other installations [3]. Three-level frequency converters (TLFC) [4-7] are also used in many cases without transformers.

Cascade frequency converters (CFC) based on low-voltage blocks [9-12] are widely used at voltages of 3-11 kV and higher. These converters provide high quality of input and output voltages and currents. CFCs have high survivability due to the fact that failed power units can be turned off, and the units remaining in operation can provide the required symmetrical load operation modes with some decrease in power [10, 13]. When the CFC is active (when the design is complicated), it is possible to recover energy from the load to the power supply network. But CFCs contain transformers at full load power, which reduce the efficiency of the installation as a whole by 1-1.5% and worsen the overall dimensions of the equipment. Transformers are multi-phase, which complicates their design, leads to the need for their placement next to semiconductor blocks.

In high voltage power grids TLFC are used [4-7, 12] in the following options:
- Option 1 - TLFC is performed with a three-phase reactor from the network side, a diode rectifier, an LC rectified voltage filter, a transistor 3-level autonomous voltage inverter (AVN).
- Option 2 - a multi-winding transformer is used, the secondary windings of which feed diode bridges and these bridges form a multi-pulse rectifier that feeds a three-level AVN.
- Option 3 - TLFC is performed with a three-phase reactor from the network side, an active rectifier (AR) and AVN.
- Other TLFC options are also possible.

If the mains voltage is 6 kV, and its possible deviations ±10%, then when the TLFC is loaded in accordance to option 1, the rectified voltage is about 6.87 kV when the switching angle of the diode bridge is 30°. With a margin of regulation of the AVN is 10%, the output TLFC is less than 4.2 kV, which requires the development of a non-standard engine.

When performing a TLFC in according to option 2 (with a transformer), the efficiency is lower and the mass of the system is greater.

If the TLFC is performed with AR, then the rectified voltage can be increased, which will allow to supply a load with a voltage of 6 kV. But with deviations of the mains voltage of ± 10% and the margin of AR for regulation of 10%, the rectified voltage of the TLFC must be at least 9.8 kV. The most high-voltage IGBT modules of the CM1000HG-130XA type are designed for a voltage of 6.5 kV, which in this case (for a three-level drive) is not enough. If transistors of the type Press-Pack are used in a TLFC, then they have another drawback - the low permissible PWM frequency (300-600 Hz). Therefore, the use of transistors of the type Press-Pack requires the use of additional AC filters at the input and output of the TLFC.

Due to the shortcomings of these inverters for the implementation of variable frequency drives, for example for 6 kV power networks, an important task is the selection and justification of the inverter structure. One possible solution is discussed in this paper.

HCFC block diagram

The considered technical solution is intended to reduce the power of the transformer in the inverter with a voltage of up to 6 kV. A hybrid cascade frequency converter (HCFC) is proposed, in which the main part of the electric power is transferred from the network to the load without a transformer, and the smaller part - through a transformer. This increases the efficiency of the installation and improves its overall dimensions. The implementation of the HCFC is possible according to the block diagram of Fig. 1. HCFC contains a high voltage FC with a diode rectifier DR and a three-level voltage inverter AVI4, made on the basis of IGBT-modules with a voltage of 6,5 kV. This unit is connected to the power grid work through the reactive coil RC. With an increase in the load power, the rectified voltage of the diode rectifier decreases and AVI4 cannot create a voltage close to the voltage of the power grid at the output. To increase the load voltage to 6 kV, single-phase three-level autonomous inverters AVI1- AVI3 are included in series in the load phases.

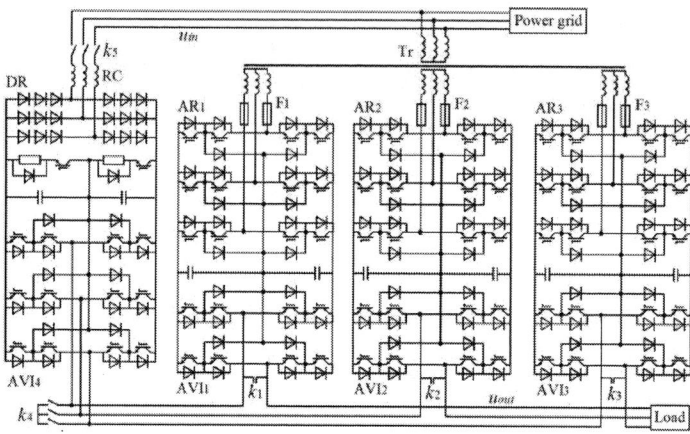

Fig. 1: HCFC diagram

The power supply of these AVIs is carried out from a power grid of 6 kV, 50 Hz through active rectifiers AR1-AR3 and a transformer Tr. AR and AVI, in addition to transferring power from the power grid to the load, filter the voltages and currents at the input and output of the HCFC.

The advantage of the FC of this type is the ability to power the load in case of failure of part of the frequency conversion blocks [13]. The shutdown of faulty units is carried out by switching devices k1-k5 and fuses F1-F3, as shown in Fig.1. It is possible to turn off the undervoltage blocks (UVB1- UVB3) using fuses and devices k1-k3, and to turn off the overvoltage block (OVB4) using devices k4-k5. When a part of the UVB are turned off HCFC characteristics are changed. These modes of operation and characteristics of the converter are discussed in this paper.

HCFC modeling based on the principle of interconnected subsystems

An analysis of the operating modes and characteristics of the HCFC is performed on a computer model constructed according to the diagram of Fig. 1. In the model, the supply power grid is represented by a three-phase regulated voltage source containing sinusoidal EMFs e_{s1}, e_{s2} and e_{s3} and inductance in phases L_s. Phase voltages being u_{s1}, u_{s2} and u_{s3}, currents being i_{s1}, i_{s2} and i_{s3}. The three-phase load is represented by the active resistances of the phases R and the inductances L. Load phase voltage being u_1, u_2 and u_3, load currents being i_1, i_2 and i_3.

The switching device at the input of the diode rectifier k_5 and fuses in the phases of the active rectifiers F_1-F_3 in the HCFC model are represented by resistors (R_{k4} and R_{k1}-R_{k3} in Fig. 2). Their resistances are equal to zero when the frequency conversion units are included in the operation, and equal to a large value (corresponding to an open circuit) when the units are turned off. On the load side, the disconnection of faulty units is carried out by acting on the corresponding control pulses of the AVI transistors.

Modeling of the installation is carried out according to the methodology for calculating systems by interconnected subsystems [12]. At the initial stage of model development the power part of the installation is separated from the control system (CS) and they are connected to each other by signals of sensors and control actions. The power part is divided into subcircuits connected by dependent sources of voltage and current. Subcircuits and their connections are mathematically described. The resulting equations are combined in a common calculation algorithm [12]. The algorithm contains an internal iterative calculation loop. Parameters of interconnections of subcircuits at each moment of time are determined. After performing of a given number of iterations, the integration of differential equations is carried out, i.e. there is a transition to the next point in time. At the next moment of time an iterative cycle is repeated, and then the integration procedure is repeated too.

One of the operations of the circuit separation into interconnected subcircuits is replacing the phases of the power grid with dependent voltage sources

$$u_{sn} = e_{sn} - L_s \frac{di_{sn}}{dt}, \quad n = 1, 2, 3. \tag{1}$$

The voltage sources u_{sn} are transferred to the phases of the DR and Tr. Phase currents and their derivatives, defined in the subsystem with DR i_{dn} and in the subsystem with Tr i_{kn}, form dependent current sources i_{sn} that are used in the subcircuit of the supply power grid

$$\frac{di_{sn}}{dt} = \frac{di_{dn}}{dt} + \frac{di_{kn}}{dt}, \qquad i_{sn} = i_{dn} + i_{kn}, \ \ n = 1, 2, 3. \tag{2}$$

Transformation (1) - (2) allows to represent the original circuit in the form of subcircuits connected by dependent sources of voltage and current, as shown in Fig. 2.

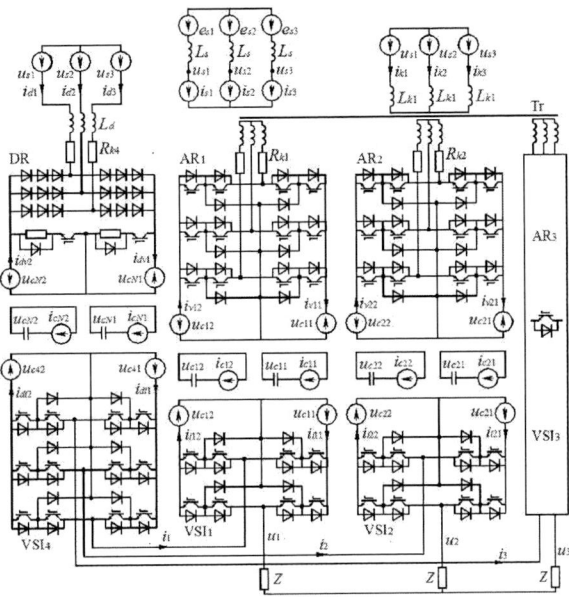

Fig. 2: HCFC model separation into subcircuits

Another circuit transformation is to replace capacitors with dependent voltage sources.

$$\left. \begin{aligned} u_{c4n} &= \frac{1}{C_4} \int i_{c4n} dt, \qquad u_{cn1} = \frac{1}{C_a} \int i_{cn1} dt, \\ u_{cn2} &= \frac{1}{C_a} \int i_{cn2} dt, \qquad n = 1, 2, \end{aligned} \right\} \tag{3}$$

where C_4 being capacitance of FC high-voltage block capacitors, C_a being capacitance of FC low voltage blocks, i_{c4n}, i_{cn1}, i_{cn2} being currents in capacitors.

The voltage sources u_{c41} and u_{cn1} are transferred to the branches of the circuit, converging at the positive poles of the rectified voltage circuits of the UVB. Dependent voltage sources u_{c42}, u_{cn2} are transferred to the branches of the circuit, converging in the negative poles of the rectified voltage circuits of the UVB. As a result, in each UVB in the rectified voltage circuit, the rectifier is connected to the AVI with one point through which current does not flow. This allows you to separate the rectifiers and AVI, as indicated in Fig. 2.

The rectified currents in the positive and negative poles of the rectifiers i_{dv1}, i_{vn1}, i_{dv2}, i_{vn2} and inverters i_{di1}, i_{in1}, i_{di2}, i_{in2} form capacitor currents that are connected to the corresponding poles of the UVB

$$\left. \begin{aligned} i_{c41} &= i_{dv1} - i_{di1}, \qquad i_{cn1} = i_{vn1} - i_{in1}, \\ i_{c42} &= i_{di2} - i_{dv2}, \qquad i_{cn2} = i_{in2} - i_{vn2}, \\ n &= 1, 2, 3. \end{aligned} \right\} \tag{4}$$

The dependent current sources obtained in expressions (4) are used in the subcircuits with capacitors shown in Fig. 2. With a further description of the power circuit, the most complex subcircuits are also divided into parts (methods for converting circuits are described in [12]).

HCFC control system features

It is assumed that the HCFC operates as part of an electric drive, the control system of which generates predetermined phase currents at the output of the HCFC. Load resistances, frequency and the set current are determined on the basis of tasks of external devices in the HCFC model. HCFC transistors are controlled by its own CS, which is included in the installation model according to Fig. 1.

HCFC CS, in particular AR CS and AVI CS perform the main and additional functions.

The main function of all AVIs is to maintain a given effective load current. To solve this task the instantaneous currents of the load phases are measured in the control system. The PI voltage controller receives signals for the given and actual load currents, the amplitudes of the main components of the control voltages of the three-phase and single-phase AVI are formed at the output. Phases of control voltages are formed as a result of integration of a given frequency of load currents. The amplitudes of the control voltages are distributed between high-voltage and low-voltage AVIs, taking into account the ratio of the rectified voltages of the FC units. The instantaneous values of the main components of the phase control voltages of all AVIs are formed based on the given voltage amplitudes and the given phases.

The main functions of all ARs are maintaining the specified rectified voltage of the FC blocks, providing a sinusoidal shape of the phase currents, forming them with a given shift relative to the phase voltages (in the ideal case, the shift is 0).

Additional functions of the ARs and single-phase AVIs are to suppress minor components in the currents of the phases at the input and output of the HCFC, as well as to ensure the operation of the HCFC from the power grid side with a power factor close to 1. These tasks are solved using numerical current filters. As a result of filtering, three-phase symmetrical components of the direct sequence of the main frequency are distinguished in the currents of the supply power grid and in the load currents. They are subtracted from the actual currents and thus minor components are determined. Minor components of the currents are regulated by a P-controller in each phase. In this case, the specified minor components of the currents are 0. Additional components of the control voltage of ARs and single-phase AVIs are formed at the output of the P-controller.

The main and additional components of the control voltage of the AR and single-phase AVI are respectively added. The obtained control voltages are used for comparison with the corresponding sawtooth reference voltages to form transistor control pulses (in the three-phase AVI, the main components of the control voltages are used).

To determine the symmetrical components of the direct sequence from a three-phase system of distorted phase currents, a well-known three-phase filter is used, in which the following calculations are performed [12]:

$$\left.\begin{aligned}
&\tau+ = \omega\Delta t, \quad e_1 = \cos(\tau), \quad e_2 = \cos(\tau - 2\pi/3), \\
&e_3 = \cos(\tau - 4\pi/3), \quad A = (e_1 i_1 + e_2 i_2 + e_3 i_3)2/3, \\
&B = \left(i_1^2 + i_2^2 + i_3^2\right)/3, \quad C+ = (B-C)\Delta t/T, \\
&I_m = \sqrt{2C}, \quad if \quad I_m \neq 0, \quad mo \quad D = A/I_m, \\
&\omega_1+ = DK_{\omega i}\Delta t, \quad \omega = \omega_1 + DK_{\omega o},
\end{aligned}\right\} \tag{4}$$

where Δt is the period of operation of AR CS or AVI_{1-3}; e_n are single sinusoidal EMF; τ is the phase of a single EMF; A, B, C, D are intermediate variables; $K_{\omega i}$, $K_{\omega o}$ are PI controller coefficients; ω is the frequency.

The operation of the three-phase filter can be explained using Fig. 3. In accordance with the figure, a 3-phase signal system i_n (instantaneous values of the currents of the phases of the network or load) is supplied to the filter inputs. These signals are multiplied phase by cosine signals e_n. The products are added and the signal A is formed. It is divided by the amplitude of the input signals I_m and a signal D is formed, the value of which is in the range from -1 to $+1$. The signal D is fed to the input of the PI controller, which changes the frequency ω of single cosine waves so that the signal D is equal to 0. The obtained frequency is integrated, and the phase τ of cosines is determined. It is used to calculate cosine values. If the parameter D is close to 0, then the parameters τ, ω, I_m are found.

EPE'20 ECCE Europe

Assigned jointly to the European Power Electronics and Drives Association & the Institute of Electrical and Electronics Engineers (IEEE)

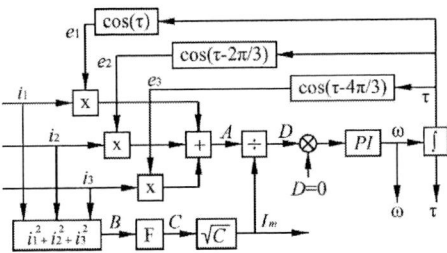

Fig. 3: Three-phase current filter circuit of HCFC

Finding the desired parameters from expressions (5) allows to determine the instantaneous values of the main components of the phase currents:

$$i_{bn} = I_m \sin\left[\tau - 2\pi(n-1)/3\right], \quad n = 1,2,3. \tag{6}$$

By the values of phase currents and the values of the main components corrective additives to control voltages are determined by:

$$\Delta u_n = K_o\left(i_n - i_{bn}\right), \quad n = 1,2,3, \tag{7}$$

where K_o is the coefficient of P-regulators in phases.

HCFC models

The HCFC simulation model is made with the programming language C++ for calculations in your own environment ComSim1_5 [14].

In this model, mathematical descriptions of subschemes and their connections in the power circuit, as well as descriptions of the operation of control devices, are combined in a single iterative calculation algorithm taking into account:

- discreteness of measurements of variables;
- synchronizing them with reference voltages;
- duration of signal processing in digital control systems and the formation of control actions [12].

Power grid currents and loads are measured several times on each PWM period and are averaged over the corresponding periods to satisfy a real installation. The discreteness of operation of the control units is determined by the duration of the PWM periods separately in UVBs.

ComSim1_5 provides the ability to display calculation results, the ability to record results and analyze them. In the model, in addition to describing the operation of power devices, the work of the control system is taken into account. It is possible to calculate the processes in the installation during an emergency shutdown of one UVB.

The approach used to simulate electrical systems by interconnected subsystems can significantly reduce the cost of computer time for calculating electromagnetic processes on computers [3, 12, 15].

HCFC simulation and calculations

Calculations of the operating modes of the converter are performed on a computer using the HCFC simulation model. The following installation parameters are used:

- The voltage of the supply power grid – 6 kV, the frequency of the supply power grid – 50 Hz, the phase inductance – 0,5 mH.
- Load power – 3000 kW, load voltage frequency – 50 Hz, load power factor – 0,95.
- Inductance of the reactive coil at the DR input – 1,07 Ohm.
- Transformer power – 1500 kVA, transformation ratio – 5,1; short circuit voltage – 5 %.
- The set rectified voltage AR – 2000 V.
- The frequency of PWM AVI₄ – 600 Hz, the frequency of PWM AR and single-phase AVIs – 4000 Hz.

The calculations were performed on a time interval of 0...2 s.

Up to a time of 1,0 s, all ARs and AVIs work properly and perform basic functions - they feed the load with a given current. In this case, the power grid and load currents are not filtered. At time 1 s, the filtering functions of the power grid currents and loads are included in the AR and single-phase AVIs.

At time 1,5 s AR₁ and AVI₁ are turned off. The reference for the load current is reduced by the shutdown signal of the indicated blocks. The calculation results are presented in Fig. 4 and 5.

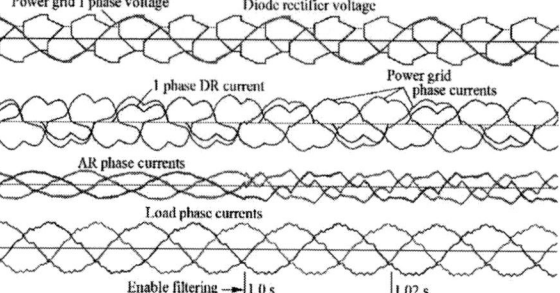

Fig. 4. The nominal operating mode of HCFC. Enabling filtering of power grid currents and load

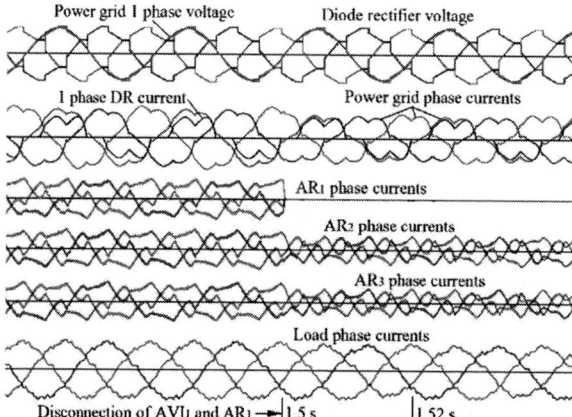

Fig. 5. The nominal operating mode of HCFC. Disconnection of single-phase AVI and one AR

The results of a harmonic analysis of voltages, currents, and power of the HCFC are presented in Table I. The analysis was performed for the cases: before turning on the filtering function of power grid and load currents; after enabling the filtering function of power grid and load currents; after one FC unit disconnecting, that puts the HCFC in asymmetric operation mode with partial load.

For given operating voltages and network load of 6000 V RMS voltage phase must be equal to 3464 W. In Table I the calculated effective voltage of the mains phases and the load on the main components accurately correspond to the tasks, both before and after filtering is turned on. With a relatively small error, the active load power also corresponds to the task.

About 70 % of the power is transferred to the load through a diode rectifier and a three-phase AVI. In particular, the total power at the input of the diode rectifier is 2079 kVA, the active power is 2024 kW. About 30 % of the active power is transferred to the load through a transformer and active blocks. However, the transformer is also loaded with higher harmonics, therefore its overall power is more than 30 %.

DR consumes currents from the power grid in which the content of higher harmonics is more than 27 %. If the ARs do not filter the power grid currents, then they work with phase currents whose shape is close to a sinusoidal (see Fig. 4) and harmonic factor is 7,76 %. This ensures a decrease in the proportion of minority components in the power grid phase currents - the harmonic coefficient of the power grid currents is 19,58 %. After turning on the filtering algorithm in the power grid currents, the proportion of minority components decreases to 10,88%. In this case, the phase currents of the active rectifiers are distorted (see Fig. 5), the content of higher harmonics increases to 27,69 %.

Electricity is consumed from a power grid with a power factor of 0,974. With the simultaneous operation of DR and AR, the power factor of the supply power grid increases to 0,985.

The rectified voltage of the DR is 7757 V, which corresponds to the voltage of the transistor module AVI₄ that is equal to 3878 V. Taking into account overvoltages and overloads, the calculated voltage and current of semiconductor elements allow to build AVI₄ based on IGBT-modules of the type

CM1000HG-130XA (1000 A, 6500 V). However, AIN4 can be performed on Press-Pack type IGBTs having a higher voltage.

Table I: Key HCFC parameters calculated on the model

	Before filtering	After filtering	After unit disconnecting
power grid 1st phase voltage, V	3456,4	3456,0	3457,1
harmonic factor, %	1,807	1,399	1,442
power grid 1st phase current, A	297,7	301,7	225,2
harmonic factor, %	19,58	10,88	19,55
power grid power factor	0,9852	0,9852	0,9784
DR power factor	0,9737	0,9737	0,9749
DR 1st phase voltage, V	3427,2	3425,4	3432,5
harmonic factor, %	12,92	12,71	1,185
DR 1st phase current, A	211,9	214,9	186,4
harmonic factor, %	27,30	27,69	28,62
Rectified voltage AVI_4, V	7752,3	7756,7	7800,5
Rectified voltage AVI_1, V	1876,7	1951,7	0
1st phase current AR_1, A	153,1	173,9	0
harmonic factor, %	7,764	27,69	0
load voltage 1st phase, V	3580,2	3641,2	3014,3
harmonic factor, %	29,73	31,71	35,61
load current 1st phase, A	300,4	303,4	247,8
harmonic factor, %	6,202	5,801	7,846
HCFC load power, kW	2933,2	2993,5	2194,8
PWM frequency of AVI_4, Hz	600		
PWM frequency of AR and $AVI_1 - AVI_3$, Hz	4000		

The rectified voltage of the AR is 1952 V, which corresponds to the voltage of one transistor module in a three-level bridge of 976 V. The calculated voltage and current of the semiconductor elements and allow to construct AR_{1-3} and AVI_{1-3} on less powerful IGBT, admitting an increased switching frequency, for example, 4000 Hz, which allows to effectively filter the phase currents.

On the load side, the phase voltages are significantly distorted - the content of higher harmonics in the voltages is 29.73% before filtering is turned on and 31.71% after filtering is turned on. However, the shape of the load currents, close to sinusoidal, in this case is provided mainly by significant phase inductances. The content of higher harmonics in phase currents is 6.2% before filtering and 5.8% after filtering is turned on.

When disconnecting a frequency conversion unit (see. Fig. 5) the current task on load current is reduced to avoid overmodulation in AVI. At the same time, the actual current decreases from 303 A to 248 A. In this case, the power decreases to 2195 kW. The share of higher harmonics in the load currents increases to 7,85 %, and in the power grid currents increases to 19,55 %. However, the symmetry of three-phase current systems is guaranteed.

It should also be noted that in the HCFC the energy recovery of the load to the supply network is provided, since active frequency conversion blocks made at partial power are used. With the converter parameters considered, recovery of 30% of the load power is possible.

Conclusion

HCFC with voltage up to 6 kV in comparison with cascade converters can reduce the weight, dimensions and cost of transformer-reactor equipment by approximately 2-3 times with satisfactory quality of currents at the input and output of the converter.

In HCFC part of the electric power is transmitted to the load through a transformer, the other part is transferred to the load, bypassing the transformer. This implies the operation of the HCFC with a higher efficiency in comparison with cascade frequency converters - approximately by 1%.

In the event of failure and disconnection of one conversion unit, the operation of serviceable units is ensured during the formation of symmetric three-phase current systems at the input and output of the converter with some reduction in load. Recovery of part of the load power to the supply network is also provided.

References

[1] Freytes J., Rault P., Gruson F., Colas F., Guillaud X. Dynamic impact of MMC controllers on DC voltage droop controlled MTDC grids / EPE'16 ECCE Europe, pp.1-10.

[2] Himmelmann P., Hiller M., Krug D., Beuermann M. A new Modular Multilevel Converter for Medium Voltage High Power Oil & Gas Motor Drive Applications / EPE'16 ECCE Europe. P.1-11.

[3] Pronin M. V., Grigoryan A. S., Glushakov V. V., Vorontsov A. G. High-speed models of systems with AC generators and modular multilevel converters // IECON-2017, Beijing, China.

[4] Damiano A., Marongiu I., Porru M., Serpi A. A Suitable PWM for DC-link Voltage Equalization of Three-Level Neutral-Point Clamped Converters // IECON-2013. Nov. 10–13, Vienna, Austria.

[5] Sintamarean C., Blaabjerg F., Wang H. Comprehensive Evaluation on Efficiency and Thermal Loading of Associated Si and SiC based PV Inverter Applications // IECON-2013. Nov. 10–13, Vienna, Austria.

[6] Vahedi H., Salem Rahmani, Al-Haddad K. Pinned Mid-Points Multilevel Inverter (PMP): Three-Phase Topology with High Voltage Levels and One Bidirectional Switch // IECON-2013. Nov. 10–13, Vienna, Austria.

[7] Narimani M., Yaramasu V., Wu B. et al. A Simple Method for Capacitor Voltages Balancing of Diode-Clamped Multilevel Converters Using Space Vector Modulation // IECON-2013. Nov. 10–13, Vienna, Austria.

[8] Cho Ja-Hwi, Ku Nam-Joon, Han Ji-Tai, et al. A Simple Control Method for Neutral-Point Voltage Oscillation Reduction of Three-level Neutral-Point-Clamped Inverter // IECON-2013. Vienna, Austria.

[9] Boonmee C., Kumsuwan Y. Control of Single-Phase Cascaded H-Bridge Multilevel Inverter with Modified MPPT for Grid-Connected Photovoltaic Systems / IECON-2013. Vienna, Austria. –C.564–569.

[10] Carnielutti F., Pinheiro H. New Modulation Strategy for Asymmetrical Cascaded Multilevel Converters Under Fault Conditions / IECON-2013. Nov. 10–13, Vienna, Austria. –C.1072–1077.

[11] Rodríguez J., Pontt J., Musalem R., Hammond P. Operation of a medium-voltage drive under faulty conditions / IEEE Transactions on Industrial Electronics. September, 2005. –C.1–5.

[12] Pronin M. V., Vorontsov A. G. Electromechanotronic complexes and their modeling according to interconnected subsystems / St. Petersburg, Ladoga Publishing House, 2017 –220 p. (Rus).

[13] Vorontsov A.G., Glushakov V.V., Pronin M.V., Sychev Yu.A. Cascade frequency converters and their control features / Notes of the Mining Institute, 2020, No. 2. (Rus).

[14] Vorontsov A.G. Computer program ComSim1_5 / Certificate number 2019661058. Date of registration with the Federal Service for Intellectual Property of the Russian Federation 19.08.2019.

[15] Drobkin B. Z., Vorontsov A. G., Pronin M. V., Krutyakov Y. A., Pavlov P. A. Debugging of microprocessor-based control systems of electric drives using mathematical models // EPE 2003, Toulouse, Fr. –pp. 1-11.

System Frequency Dynamic Response of a Novel, Self-Synchronizing Inverter in a High Renewable Penetration Grid

Christian Perenyi[1], Moath Alqatamin[2], Thibaut Harzig[1], Michael McIntyre[2] and
Brandon M. Grainger[1]

Swanson School of Engineering[1]
University of Pittsburgh
Pittsburgh, PA USA
Email:chp118@pitt.edu, thh39@pitt.edu, bmg10@pitt.edu

J. B. Speed School of Engineering[2]
University of Louisville
Louisville, KY USA
Email: moath.alqatamin@louisville.edu, michael.mcintyre@louisville.edu

Keywords

«Distributed Generation», «Frequency Dynamics», «Low-Inertia Grid», «Phase-Locked Loop», «Self-Synchronizing Inverter»

Abstract

This paper presents a self-synchronizing controller achieving a current tracking objective without knowledge of the grid parameters. An estimated rotating reference frame ($\gamma\delta$ -frame) is utilized. Within the control scheme, adaptive compensation terms facilitate the current tracking objective and, simultaneously, accounts for the unavailable grid voltage magnitude, grid frequency, and grid phase, hence eliminating the need for an additional measurement and feedback system for synchronization, such as a Phase-Locked Loop (PLL) which are sensitive to harmonics, to disturbances and to large frequency deviations causing unnecessary energy losses.

A system frequency behavior study is developed through this article by comparing a PQ-PLL controller to the novel self-synchronizing design under the conditions of a variable inertia and renewable penetration ratio. Using the novel controller in a high-renewable (low-inertia) grid indicates that the monitored performance metrics are significantly improved when compared to PLL-controlled, inverter-dominated grid. When benchmarked to the PQ-PLL controller results, increasing the penetration ratio will have a positive impact on the self-synchronizing controller due to its adaptive nature which depends only on the current control errors signals. Lowering the grid's inertia shows that the self-synchronizing inverter reaches steady-state significantly faster than the PQ-PLL one due to the internal dynamics of the adaptive control by disrupting the need for a cascaded control scheme.

I. Introduction

In today's economy, energy transition is one of Europe's top priorities. A main component of this challenge is to install a cleaner portfolio of renewable energy resources such as photovoltaic systems and wind turbines. That is why, the penetration ratio, Pr, of the renewable energy over the total energy generation (fossil + renewable) is significantly increasing; based on the German Energiewende, Germany will have a Pr of 80% by 2050 compared to 31.6% in 2016 [1].

To optimally integrate these renewable resources, it is necessary to rethink the electric power grid itself by adjusting the centralized generation model and progressively transitioning to a distributed generation (DG) based power grid architecture through power electronics and their control. Droop control, [2] a widely adopted method to integrate variable renewable sources (VRSs), is a decentralized and communication-less control, contributing to the overall frequency (and voltage) control by emulating virtual inertia (and a virtual impedance) [3]. However, the expansion of renewable energy inevitably drops the mechanical inertia of the whole power system because these generation resources cannot store kinetic energy as they do not have a rotating mass [4]. Hence, a mismatch between generation and consumption cannot always be mechanically compensated for which can cause large frequency swings. For system operation, loads and renewable sources should not be tripped [5].

To perform an appropriate DG interconnection, the phase and frequency need to be determined by a PLL within inverters. Synchronous reference frame PLLs are widely applied in three-phase DG systems. This synchronization method is the most widely proposed solution because the technique provides excellent results for balanced grid conditions but becomes quite sensitive to unbalances and harmonic disturbances in grid voltage [6], [7]. One reason for this drawback is that modern power converters require fast detection and accurate knowledge of the grid angle [6]. For faster angle detection, increasing the bandwidth of a PLL has been considered, [8], but is limited by the presence of other converters operating nearby [9], weaker grid conditions tied to short circuit ratios [9], [7], and large penetration levels of DG units which can correspond to low-frequency power oscillations and system instabilities [7]. Many of the attempts to design a better PLL to handle such conditions have not considered the coupling effects and interactions between the PLL and system impedance network, which impacts PLL tuning [8]. This coupling effect has the potential to lead to important instability issues when multiple inverters are connected together [7].

Additional hardware like voltage sensors required for PLL-type systems are expensive and introduce electrical noise and dc offsets which require compensation. A PLL system is a self-contained feedback loop that is outside the primary current control scheme, hence creating a cascaded control architecture [10]. For optimal stability and accuracy, the current control scheme should be knowledgeable of the error dynamics of the PLL system.

Most commercial inverters operate as grid-following (GFL) sources that regulate their output power by measuring the angle of the grid voltage using a PLL. These units simply follow the grid frequency and do not actively control their frequency output. In contrast, a grid-forming source (GFM) controls its frequency and voltage output. However, as GFM sources retire and are replaced with renewable based generation, GFLs begin to dominate the electric grids leading to the common problems initiated by the PLL unit. For one cause for concern with increased GFLs, consider the state of California in the United States. In one reported event in 2016 [11], California had experienced significant loss of generation, ~700MW, because the inverter PLL detected frequencies less than 57 Hz and initiated an instantaneous trip. However, the lowest measured frequency only dropped to 59.87 Hz. As [12] points out, engineers need to develop solutions for the reliable operation of inverter-dominated power systems.

In this article, a unique approach is provided for self-synchronizing, GFL inverters requiring no PLL, one three-phase current measurement, and can be modified to behave as a GFM inverter. The contribution of this work is to present results, in a highly-renewable and low-inertia grid, indicating that the discussed self-synchronizing control scheme can improve DG grid frequency behavior compared to traditional PQ reference PLL (PQ-PLL) controlled inverters.

II. Grid Model for Comparing Inverter Control Schemes

A two-source, reduced-order modelling approach is used in this work. The two-source grid model is presented in Fig. 1. On one side, the inverter voltage source aims to represent the inverter-based renewable sources where the renewable generation source itself is modeled as a DC source. This strong assumption, which implies that the dynamics linked to the source are ignored, can be justified since emphasis is put on the inverter dynamics. On the other side of the figure, the generator is modeled as a synchronous steam generator often used to represent fossil-based generators. The three phase load is composed of an inductor and resistor in series. Line impedance is negligible compared to the load impedance.

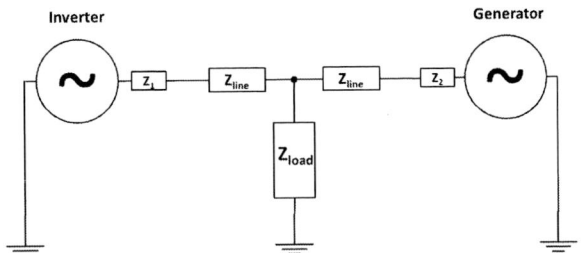

Fig. 1: Two-Source Grid Model

Generator Frequency Dynamics Modelling

The generator model accounts for the turbine, a governor and an exciter. An approximated first order linearized model of the speed-governor around the operating point, based on [13], is considered for this work. The model is an approximation because stator transients are neglected. The output of the governor is relative position of the steam valve $\Delta \hat{x}_E$ (1) that regulates the steam flowing through the turbine:

$$\Delta \hat{x}_E = \frac{K_G}{1 + T_G s}\left(P_{ref} - P_{out\_gen}\right) \tag{1}$$

where K_G and T_G are the governor's transfer function gain and time constant, P_{ref} is the generator's power reference and P_{out_gen} is the generator's output power.

The turbine is modeled as a first order transfer function that accounts for the time delay for when the steam strikes all the blades. The time constant and the gain can be taken from datasheets or be set based on experimental data.

The inertia, linked to the kinetic energy of the electric grid, helps to lower the frequency deviation when there is a mismatch between generation and consumption. In other words, the change of the grid load directly impacts the electrical power output which is proportional to the output torque, T_e, of the generator causing a mismatch between the output torque and the mechanical torque, T_m. Since the torque is proportional to the electrical power through the rotor angular speed, ω_r, H being the inertia and ω_s the synchronous speed, we can directly present the swing equation in this form:

$$\frac{2H}{\omega_s}\frac{d^2\delta}{dt^2} = P_m - P_e \tag{2}$$

where (3) and (4) define δ and θ_r as:

$$\delta = \theta_r - \omega_s t \tag{3}$$

$$\frac{d\theta_r}{dt} = \omega_r \tag{4}$$

In addition to the inertia, the damping effect, D, needs to be considered as shown in the machine inertia block in Fig. 2. The electrical grid is composed of both frequency-dependent loads (5) like motors and non-frequency dependent loads like resistive loads. That is why, the overall delta of the electrical power, ΔP_e, depends on both types of loads as shown:

$$\Delta P_e = \Delta P_l + \Delta P_L = \Delta P_l + D\Delta\omega_r \tag{5}$$

Here, $D = \frac{\Delta P_L}{\Delta\omega_r}$ is the load damping constant expressed as a percentage, ΔP_l is the non-dependent frequency load change and ΔP_L is the frequency-dependent load change.

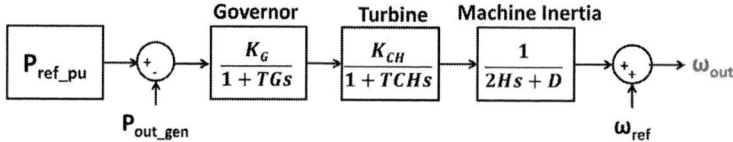

Fig. 2: Overall Generator Frequency Dynamics

Excitation System Model

The excitation model is extracted from IEEE standards [14]. A direct current commutator exciter has been selected for this work because it is widely used because of its simplicity; specifically we will be using the - IEEE Type 1 DC1A model as presented in Fig.3. For our study, the saturation model has been removed to avoid complexity in the simulation since saturation has no considerable effect on the system for this application.

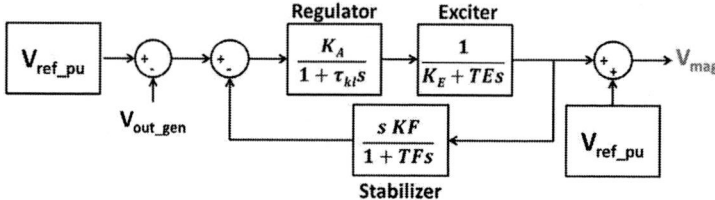

Fig. 3: Exciter Model

III. Inverter System Model: Benchmark #1 - PQ Inverter Control with a PLL (PQ-PLL)

PQ-PLL control is a commonly used control routine, which tracks real and reactive power references. The power references are than altered and used as current references in the *dq*-frame. Here, the PLL output phase angle serves a critical role in obtaining these time-independent references. Fig. 4 presents a traditional grid-connected inverter system (DC side, inverter with filter, AC side) with a PQ-PLL control.

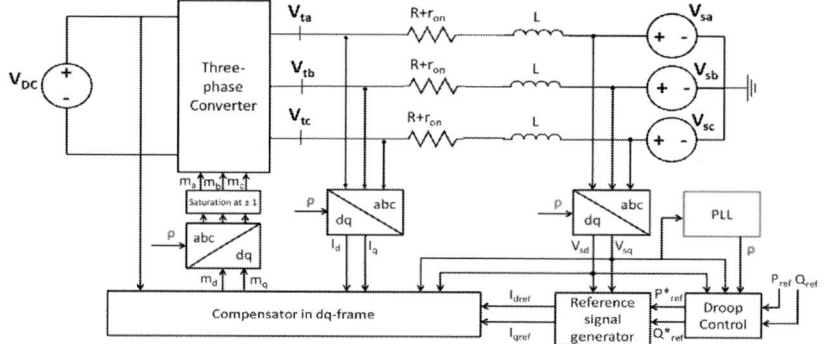

Fig. 4: Traditional Three-Phase PLL Controlled Inverter [15]

For a traditional, three-phase, balanced inverter as seen in Fig. 4, and by applying the averaging operator and the *dq*-transform, we can define the following dynamic equations:

$$\begin{cases} L\dfrac{di_d}{dt} - L\omega I_q + Ri_d = m_d \dfrac{V_{DC}}{2} - V_{sd} & (6) \\[3mm] L\dfrac{di_q}{dt} + L\omega I_d + Ri_q = m_q \dfrac{V_{DC}}{2} - V_{sq} & (7) \end{cases}$$

where $i_{d,q}$ is the current going through the filter, L the filter inductor, R the filter resistor, V_{DC} the voltage of the DC source, $V_{sd,sq}$ the voltage magnitude of the main power grid, $m_{d,q}$ the modulation index and ω the grid's frequency. Based on (6) and (7), we can express $m_{d,q}$:

$$\begin{cases} m_d = \dfrac{2}{V_{DC}}(k_d e_d - L\omega I_q + V_{sd}) & (8) \\[3mm] m_q = \dfrac{2}{V_{DC}}(k_d e_q + L\omega I_d + V_{sq}) & (9) \end{cases}$$

where $e_{d,q}$ is the current error and $k_{d,q}(s)$ the gain of the PI compensator:

$$e_{d,q} = i_{d,qref} - i_{d,q} \qquad (10)$$

$$k_{d,q}(s) = \frac{k_p s + k_i}{s} \qquad (11)$$

The detailed control is presented in Fig. 5. Both d and q-axis compensators set the system dynamics (6) and (7), while the feed forward filters, G_{ff}, (Fig. 4) prevent a peak current at startup.

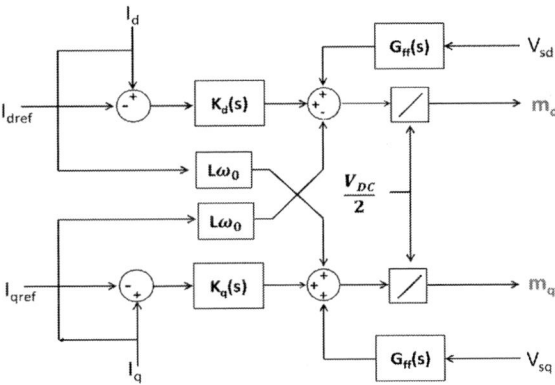

Fig. 5: Traditional Current Control Block Diagram [15]

Phase-Locked Loop (PLL)

The objective of a phase-locked loop (PLL) is to set the tuning parameter $\varepsilon(t)$ equal to $\omega t + \theta_0$. To achieve that performance, the PLL extracts the frequency from the grid. That frequency will be directly injected into the dq-transformation in order to synchronize at the same speed as the signal. The frequency measurement is based on the fact that for a balanced system the q-component of the voltage V_{sq} must be equal to 0. That way, $\varepsilon(t)$, has to be set in such way that $V_{sq} = 0$; meaning that the frame is rotating at the same speed as the signal. However, the expression of V_{sq} is a sinusoidal function $V_{sq} = V_s \sin(\omega t + \theta_0 - \varepsilon(t))$, and cannot be used directly as a feedback signal to regulate $\varepsilon(t)$. If $\varepsilon(t)$ is initially close enough to $\omega t + \theta_0$ than it is possible to state that $V_{sq} \approx (\omega t + \theta_0 - \varepsilon(t))$. A possible way to do this is to set $\dfrac{d\varepsilon}{dt}(0) = \omega_\varepsilon(0) = \omega_0$ where ω_0 is the grid's nominal frequency and always keeping ω_ε close in value to ω_0 by constraining ω_ε such as:

$$\omega_{\varepsilon min} \le \omega_\varepsilon \le \omega_{\varepsilon max}. \tag{12}$$

The phase and frequency are both estimated within a single loop as seen in Fig. 6. The PLL control loop is composed of a PI compensator which aims to set the error $V_{sq0} - V_{sq}$ equal to 0. The output of the PI compensator is the frequency adjustment. A voltage-controlled oscillator (VCO) is used as a resettable integrator in order to obtain the tuning parameter, $\varepsilon(t)$.

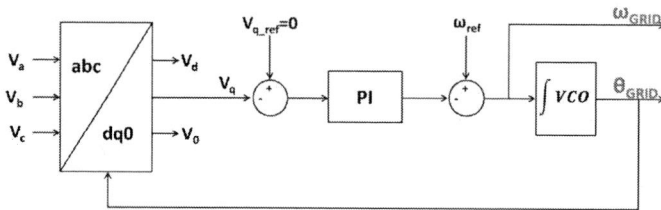

Fig. 6: Phase-Locked Loop Block Diagram

IV. Inverter System Model: Benchmark #2 - Self-Synchronizing Inverter Control

A suitable dynamic system model of a three-phase grid connected inverter as seen in Fig. 7 can be modeled in the natural abc-frame as shown in (13) where $I_a(t)$, $I_b(t)$, $I_c(t)$ are the three phase currents, V_{dc}, L, and R are the DC-link voltage, filter inductance and resistance, respectively. The control signals are the three-phase duty cycles $D_a(t)$, $D_b(t)$, $D_c(t)$, and the grid parameters are voltage magnitude $V_g(t)$ and grid phase $\theta(t)$. The grid frequency is naturally related to the phase by the following formula: $\omega = \dot{\theta}(t)$. In this paper, the grid phase is unknown so the standard dq-transformation cannot be utilized for (13). Instead an estimated $\gamma\delta$-frame is utilized by performing the dq-transformation using $\hat{\theta}(t)$ in place of $\theta(t)$, where $\hat{\theta}(t)$ is an observed grid-phase [16] to be designed subsequently. After transformation about $\hat{\theta}(t)$, (13) can be written in the $\gamma\delta$-frame as shown in (14) where the currents and the control signals have been transformed to the $\gamma\delta$-frame, and the grid-phase estimation error is defined and given in (16). The duty cycle in the the $\gamma\delta$-frame is defined in (15). Equations (17), (18) and (19), describe the full self-synchronizing controller model.

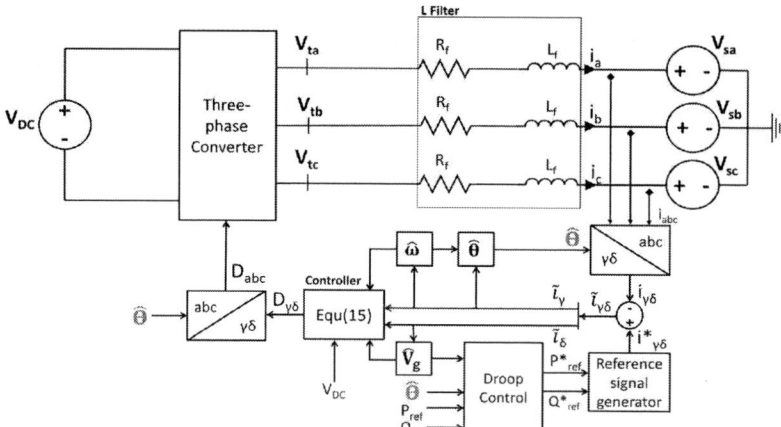

Fig 7: Full Model of the Self-Synchronizing Inverter

The validity of the current tracking performance is provided by the PLECS simulation results in Fig. 8. The stability of the controller is proved in [17] by a Lyapunov stability and convergence analysis.

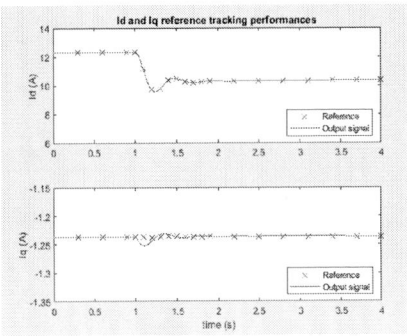

Fig. 8: Current Tracking Performance of the Self-Synchronizing Inverter

$$L \begin{bmatrix} \dot{I}_a \\ \dot{I}_b \\ \dot{I}_c \end{bmatrix} = V_{dc} \begin{bmatrix} D_a \\ D_b \\ D_c \end{bmatrix} - R \begin{bmatrix} I_a \\ I_b \\ I_c \end{bmatrix} - V_g \begin{bmatrix} \cos(\theta) \\ \cos\left(\theta - \dfrac{2\pi}{3}\right) \\ \cos\left(\theta + \dfrac{2\pi}{3}\right) \end{bmatrix} \quad (13)$$

$$L \begin{bmatrix} \dot{I}_\gamma \\ \dot{I}_\delta \end{bmatrix} = V_{dc} \begin{bmatrix} D_\gamma \\ D_\delta \end{bmatrix} - \begin{bmatrix} R & -\dot{\theta}L \\ \dot{\theta}L & R \end{bmatrix} \begin{bmatrix} I_\gamma \\ I_\delta \end{bmatrix} - V_g \begin{bmatrix} \cos\tilde{\theta} \\ \sin\tilde{\theta} \end{bmatrix} \quad (14)$$

$$\begin{bmatrix} D_\gamma \\ D_\delta \end{bmatrix} \triangleq \frac{1}{V_{dc}} \left(\begin{bmatrix} R & -\dot{\theta}L \\ \dot{\theta}L & R \end{bmatrix} \begin{bmatrix} I_\gamma \\ I_\delta \end{bmatrix} + \begin{bmatrix} \hat{V}_g \\ 0 \end{bmatrix} + k_1 \begin{bmatrix} \tilde{I}_\gamma \\ \tilde{I}_\delta \end{bmatrix} \right) \quad (15)$$

$$\tilde{\theta} \triangleq \theta - \hat{\theta} \quad (16)$$

$$\hat{\theta} \triangleq L\tilde{I}_\delta + \int_{t_0}^{t} [\hat{\omega} + (k_1 + 1)\tilde{I}_\delta]d\sigma \quad (17)$$

$$\hat{\omega} \triangleq k_\omega \left(L\tilde{I}_\delta + k_1 \int_{t_0}^{t} \tilde{I}_\delta \, d\sigma \right) \quad (18)$$

$$\dot{\hat{V}}_g \triangleq k_v \tilde{I}_\gamma \quad (19)$$

Droop Control Purpose and Implementation

In order to have a consistent comparison, a PQ reference has been set on the self-synchronizing inverter control. In order to calculate I_{d_ref} (20) and I_{q_ref} (21), the PLL controlled inverter will operate with the measured grid voltage while the self-synchronizing inverter control scheme will make use of the estimated-voltage; thus, the self-synchronizing inverter requires less sensing hardware (i.e. no voltage sensors). Furthermore, droop control has been added to ensure time-variant references for the real and reactive power so the inverter is responsible for adjusting its generation in accordance with the main grid consumption. The frequency is linked to the real power through the droop coefficient R_{droop_P} and the voltage is correlated to the reactive power through R_{droop_Q} as shown in Fig.9.

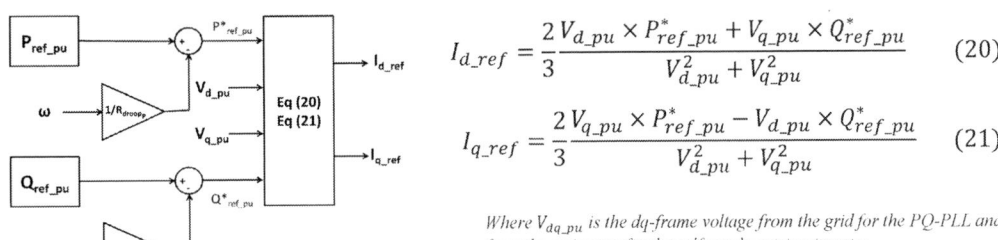

$$I_{d\_ref} = \frac{2}{3} \frac{V_{d\_pu} \times P_{ref\_pu}^* + V_{q\_pu} \times Q_{ref\_pu}^*}{V_{d\_pu}^2 + V_{q\_pu}^2} \quad (20)$$

$$I_{q\_ref} = \frac{2}{3} \frac{V_{q\_pu} \times P_{ref\_pu}^* - V_{d\_pu} \times Q_{ref\_pu}^*}{V_{d\_pu}^2 + V_{q\_pu}^2} \quad (21)$$

Where V_{dq_pu} is the dq-frame voltage from the grid for the PQ-PLL and from the estimator for the self-synchronizing inverter

Fig. 9: Droop Control Implementation droop coefficient

V. Simulation Results and Analysis

Simulations, which consist in dropping the load by 10% at time t_0 in order to study the grid's frequency dynamic behavior, have been carried out in PLECS and normalized for interpretation. Equation (22) is the frequency normalization that was applied, where f_0 is 60Hz and $f_{steady-state}$ is system frequency after a load change.

$$f(i)_{norm} = \frac{f(i) - f_0}{f_{steady-state} - f_0} \quad (22)$$

Simulation Parameters

Based on [13],[15], and [17] the following grid parameters - Table I, and machine parameters - Table II, have been chosen for the simulation:

Table I: Grid Parameters

Grid parameters			
$V_{grid-pp}$	3.3 kV	R_{Line}	≈ 0
f_{grid}	60 Hz	L_{Line}	≈ 0
R_{droop_P}	5%	**Load Drop**	10%
R_{droop_Q}	5%	**Switching Frequency, f_{sw}**	25 kHz

Table II: Generator parameters

Generator parameters			
T_g	0.1s	H	$2 - 6$ s
K_{CH}, K_G	1	D	0.01
T_{ch}	1s	K_a	1
V_{ref_pu}	1	τ_{kl}	0.02 s
f_{ref}	60 Hz	K_E	0.05
K_f	0	T_E	0.46

Impact of Mechanical Inertia

Mechanical inertia has a similar impact on both the PQ-PLL and self-synchronizing controlled inverter grid's frequency overshoot. Increasing the inertia lowers the overshoot. For example, as seen in Fig. 10 and Table III, for the Pr=20% case, a 41.1% overshoot for the self-synchronized inverter with H=2 - case (a), is observed and only 10.1% with H=6 - case (c). Another interesting observation is the hunting effect that is observed for the lower inertia, high penetration PLL, case (b), compared to the self-synchronizing routine. In addition, despite a higher overshoot for case (a), the self-synchronized inverter has a second undershoot swing that is 30% lower (7.3% against 10.1%) compared to the PQ-PLL inverter. In line with these results, it can be concluded that both inverters' overshoot is directly linked to inertia. However, the self-synchronizing controlled inverter has significantly faster dynamics and a steady-state that is always reached in less than a second while the PLL controlled inverters cannot reach steady-state within 3 seconds for low inertia, high penetration grids - case (b).

Impact of Renewable Source Penetration Ratio

The renewable source penetration ratio has a direct impact on the grid's frequency behavior for each of the inverter scenarios evaluated. Increasing the penetration ratio for the PLL based control inverter increases the overshoot. For example, for H=2, a +2.1% overall overshoot is observed when increasing Pr from 20% - case (a), to 80% - case (b). This result was already observed in [5]. On the other hand, for the self-synchronizing inverter, increasing Pr tends to reduce the maximum overshoot. For example, for H=6 - case (b), a -8.2% overall overshoot is observed by increasing Pr from 20% - case (c), to 80% - case (d); the overshoot seen at Fig. 10d is reduced by half compared to the overshoot of Fig. 10c. This key result can be explained by the self-synchronizing control itself. Elevating the penetration ratio, Pr, strengthens the role of the inverter control; hence, the inverter's adaptive control depends only the current error signals and works to minimize these current errors. In contrast, the classic PQ-PLL controlled inverter depends on the transient response of its PLL reaching a new steady-state value in the face of frequency disturbances. Then the current control scheme within the classic scheme makes adjustments to achieve its control objective. The new approach removes the slow-response time of the classic cascaded control scheme and achieves current control in the presence of uncertain grid parameters.

Table III: Overshoot, Undershoot and Steady-State Ripple for PLL (Classic - Blue) and Self-Synchronizing (Adaptive - Red) Controls for *Pr*=20% and 80% and *H*=2 and 6 extremities

	Max overshoot (%)		Max undershoot (%)		Steady-state ripple p-p (%)	
	PLL	*Self-Synch*	*PLL*	*Self-Synch*	*PLL*	*Self-Synch*
Case (a): *Pr*=20%, *H*=2	**38.0**	41.1	10.1	7.3	1.2	<0.1
Case (b): *Pr*=80%, *H*=2	40.1	**36.2**	10.6	**4.5**	4.2	<0.1
Case (c): *Pr*=20%, *H*=6	14.3	**10.1**	2.5	1.5	<0.1	<0.1
Case (d): *Pr*=80%, *H*=6	14.5	6.3	4.5	2.1	1.5	<0.1

VI. Conclusion

A self-synchronizing inverter control has been introduced in this article. After describing the grid-model, the grid dynamic frequency response to a 10% load change was examined for a system including PQ-PLL inverters and a new self-synchronizing inverter. First, increasing the mechanical inertia lowers the frequency overshoot. The self-synchronizing inverter reaches steady-state significantly faster due to the internal dynamics of the adaptive control by disrupting the need for a cascaded control scheme. Secondly, increasing the renewable source penetration ratio tends to increase the overshoot for PLL based inverters while considerably lowering the overshoot for the self-synchronizing inverters because the adaptive control only depends on the current control errors signals and not a separate system response. Based upon this study, the self-synchronizing inverter can improve the performance of traditional inverter topologies in high renewable penetration (low inertia) environments.

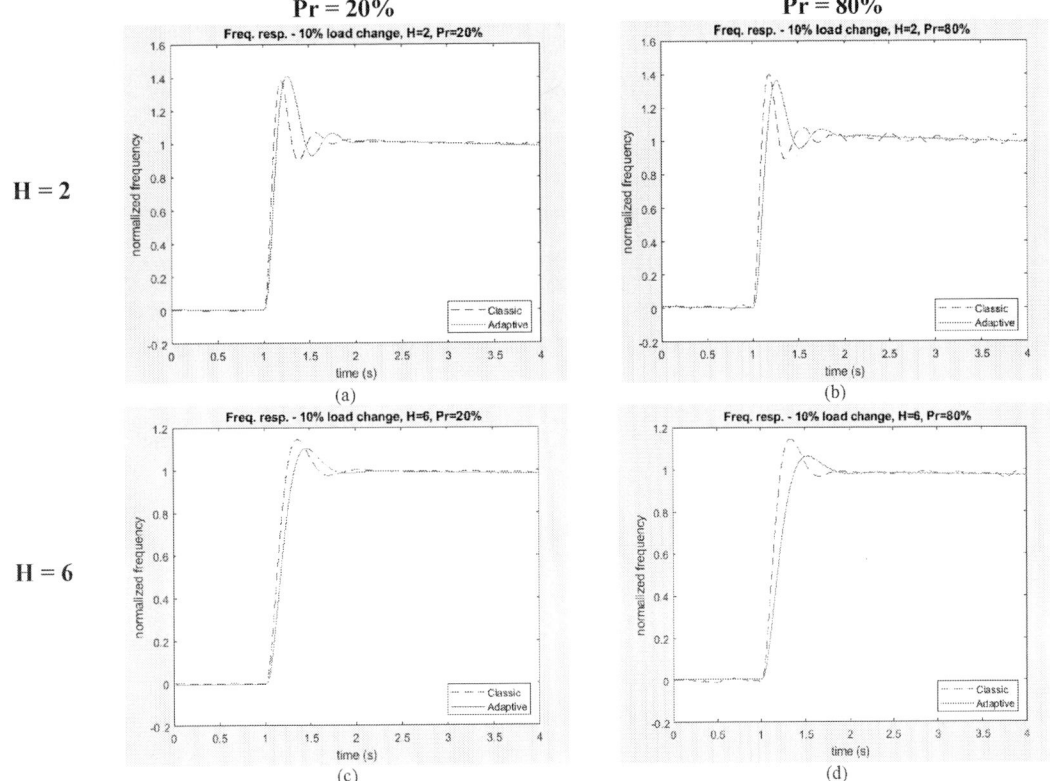

Fig. 10: PLL (Classic) and Self-Synchronizing (Adaptive) Frequency Response for *Pr*=20% and 80% and *H*=2 and 6.

References

[1] P. R. D. Federal Ministry for Economic Affairs and Energy, "The Energy of the Future – Sixth 'Energy Transition' Monitoring Report , Reporting Year 2016 – Summary," 2018.

[2] A. Kwasinski, W. Weaver, and R. S. Balog, *Microgrids and other Local Area Power and Energy Systems*. Cambridge: Cambridge University Press, 2016.

[3] J. M. Guerrero, M. Chandorkar, T. Lee, and P. C. Loh, "Advanced Control Architectures for Intelligent Microgrids—Part I: Decentralized and Hierarchical Control," *IEEE Trans. Ind. Electron.*, vol. 60, no. 4, pp. 1254–1262, 2013, doi: 10.1109/TIE.2012.2194969.

[4] B. Kroposki *et al.*, "Achieving a 100% Renewable Grid: Operating Electric Power Systems with Extremely High Levels of Variable Renewable Energy," *IEEE Power Energy Mag.*, vol. 15, no. 2, pp. 61–73, 2017, doi: 10.1109/MPE.2016.2637122.

[5] D. Pattabiraman, R. H. Lasseter., and T. M. Jahns, "Comparison of Grid Following and Grid Forming Control for a High Inverter Penetration Power System," in *2018 IEEE Power & Energy Society General Meeting (PESGM)*, 2018, pp. 1–5, doi: 10.1109/PESGM.2018.8586162.

[6] R. Rosso, G. Buticchi, M. Liserre, Z. Zou, and S. Engelken, "Stability analysis of synchronization of parallel power converters," in *IECON 2017 - 43rd Annual Conference of the IEEE Industrial Electronics Society*, 2017, pp. 440–445, doi: 10.1109/IECON.2017.8216078.

[7] D. Dong, B. Wen, D. Boroyevich, P. Mattavelli, and Y. Xue, "Analysis of Phase-Locked Loop Low-Frequency Stability in Three-Phase Grid-Connected Power Converters Considering Impedance Interactions," *IEEE Trans. Ind. Electron.*, vol. 62, no. 1, pp. 310–321, Jan. 2015, doi: 10.1109/TIE.2014.2334665.

[8] Y. Sun *et al.*, "The Impact of PLL Dynamics on the Low Inertia Power Grid: A Case Study of Bonaire Island Power System," *Energies*, vol. 12, no. 7, p. 1259, Apr. 2019, doi: 10.3390/en12071259.

[9] R. Rosso, M. Andresen, S. Engelken, and M. Liserre, "Analysis of the Interaction Among Power Converters Through Their Synchronization Mechanism," *IEEE Trans. Power Electron.*, pp. 1–1, 2019, doi: 10.1109/TPEL.2019.2905355.

[10] Qing-Chang Zhong, Phi-Long Nguyen, Zhenyu Ma, and Wanxing Sheng, "Self-Synchronized Synchronverters: Inverters Without a Dedicated Synchronization Unit," *IEEE Trans. Power Electron.*, vol. 29, no. 2, pp. 617–630, Feb. 2014, doi: 10.1109/TPEL.2013.2258684.

[11] NERC, "1,200 MW Fault Induced Solar Photovoltaic Resource Interruption Disturbance Report - Southern California 8/16/2016 Event," 2017.

[12] D. Pattabiraman, R. H. Lasseter., and T. M. Jahns, "Comparison of Grid Following and Grid Forming Control for a High Inverter Penetration Power System," in *2018 IEEE Power & Energy Society General Meeting (PESGM)*, 2018, pp. 1–5, doi: 10.1109/PESGM.2018.8586162.

[13] A. Bergen and V. Vitall, *Power Systems Analysis*, 2nd Editio. Pearson, 2000.

[14] "IEEE Recommended Practice for Excitation System Models for Power System Stability Studies," *IEEE Std 421.5-2005 (Revision of IEEE Std 421.5-1992)*. pp. 1–93, 2006, doi: 10.1109/IEEESTD.2006.99499.

[15] A. Yazdani and R. Iravani, "8.4 Current-Mode Control of Real-/Reactive-Power Controller," in *Voltage-Sourced Converters in Power Systems : Modeling, Control, and Applications*, 1 edition., Wiley-IEEE Press, 2010.

[16] G. C. Konstantopoulos, Q. Zhong, and W. Ming, "PLL-Less Nonlinear Current-Limiting Controller for Single-Phase Grid-Tied Inverters: Design, Stability Analysis, and Operation Under Grid Faults," *IEEE Trans. Ind. Electron.*, vol. 63, no. 9, pp. 5582–5591, 2016, doi: 10.1109/TIE.2016.2564340.

[17] J. Latham, M. Alqatamin, Z. T. Smith, B. M. Grainger, and and M. McIntyre, "Self-Synchronizing Current Control of a Three-Phase GridConnected Inverter in the Presence of Unknown Grid Parameters," *IEEE Appl. Power Electron. Conf. Expo.*, 2020.

Rotor Position Estimation with Hall-Effect Sensors in Bearingless Drives

Patricio Peralta, Jacopo Leo, Yves Perriard
Laboratory of Integrated Actuators, École polytechnique fédérale de Lausanne
Neuchâtel, Switzerland
Email: {juan.peraltafierro, jacopo.leo, yves.perriard}@epfl.ch
URL: https://lai.epfl.ch/

Keywords

≪Magnetic bearings≫,≪Sensor≫,≪Permanent magnet motor≫,≪Motion Control≫

Abstract

The estimation of rotor position is fundamental for the commissioning of magnetically-levitated drives. The sensors required for this measurement must be close to the rotor, often requiring constructive measures that may limit the application potential of a drive. This publication presents the estimation of rotor angle and position, based on Hall-effect sensors installed on a slotless bearingless drive. This scheme enables sensor mounting away from the rotor, thus simplifying its encapsulation.

Introduction

Contemporary research trends strive towards the miniaturization of electrical drives, which has lead to increased investigation efforts towards high-speed permanent magnet motors [1]. Nonetheless, the classical ball-bearing approach to this venture is defiant [2], as rotordynamical effects [3] and mechanical losses [4] stand in the way of reliable, high-speed operation. Contactless magnetic bearings enable getting past these issues. They are reliable and have successfully reached speeds over 10 krpm in different sizes and topologies [5, 6, 7].

Successful levitation needs information about the rotor angle in order to generate bearing forces in the desired direction [8]. Rotor angle is classically estimated by measuring the magnetic fields with Hall-effect sensors; yet these fields also contain information about the rotor position in the airgap that usually goes unexploited [9].

The information about rotor position is usually estimated by additional sensors. To measure it, capacitive, optical, and eddy current sensors are among the preferred sensor approaches [8]. They usually have to be placed inside the airgap, often requiring special constructive means that may damper the application range of the drive [10, 11]. This paper validates how only Hall-effect sensors, distributed as in Fig. 1a, enable the estimation of rotor position *and* angle.

Indeed, bearingless motors with a slotted stator have already been comissioned only by measuring the magnetic induction hall-effect sensors [12], but its specifics are not disclosed. In this paper, the principles for measuring the $x - y$ position, angle, and *additionally* axial rotor position of a one pole-pair rotor with a *slotless* stator are presented. A mathematical model is inspired by computational Finite Element (FE) method simulations, and its validity is experimentally tested. Exploiting the mathematical model of the magnetic fields, a model for rotor position estimation is developed.

The proposed sensor array can lay outside the airgap, enabling the construction of *simple* encapsulation concepts. Conveniently enough, stray fields *outside* the airgap show low intensities, compatible to *off-the-shelf* ratiometric Hall-effect sensors.

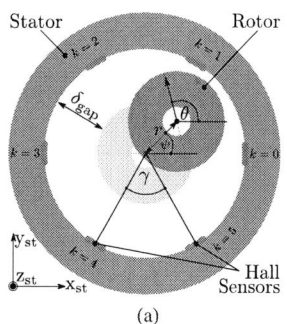
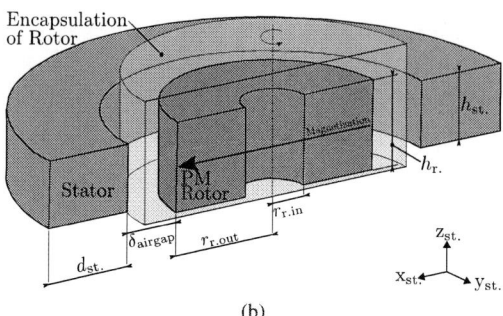

(a)   (b)

Fig. 1: (a) Top view of a drive with an exaggeratedly uncentered rotor and its symmetrical positioning of 6 Hall-effect sensors at the stator. Its radial position and magnetic orientation are determined by coordinates (r, ψ) and θ, respectively. (b) Cross section of a slotless bearingless disc drive, studied in this paper, with an encapsulated rotor and its relevant dimensions.

The mathematical derivation of angle and position estimation are general for any one pole-pair magnetically levitated (*mag-lev*) drives. Nonetheless, this paper focuses on its implementation and validation in a *disc drive* (or *slice motor*) [13], as shown in Fig. 1b. Among magnetic bearing topologies, their geometric properties (namely the large diameter-to-length ratio) offer axial and tilting passive stability thus simplifying the overall drive system and its control.

Mathematical Modelling

Field in the Airgap

A one pole-pair, diametrically-magnetized permanent magnet (PM) rotor inside a slotless stator is studied. Figure 2a depicts its cross section, with a centric and eccentric rotor along its FE-simulated radial magnetic induction. Rotor eccentricity impacts the magnetic field distribution [14].

The radial magnetic induction is sensed at two positions (0 and $\pi/2$ rad) for this example. Each sensor has a local coordinate system, with normal and tangential directions denominated with (n, τ) respectively.

Figure 2b depicts the magnetic induction of the rotating rotor measured by both sensors. The magnetic induction of increasing eccentricity r/δ_{gap} is depicted in darkening colors.

When the rotor moves parallel to a normal sensor direction, i.e. parallel to n_0, the amplitude of the magnetic induction measured by the sensor increases with rotor eccentricity. On the other hand, a movement parallel to the tangential sensor direction $\tau_{\pi/2}$ changes the phase of the measured magnetic induction.

For the a non-turning rotor, eccentricity is measured as a local change in measured magnetic induction, as shown in Fig. 2c . For small displacements, this alterations in the radial magnetic induction changes are practically linear to rotor displacement, as plotted in the Figure. These insights are implemented in the modelling of the magnetic induction in the airgap and its dependency with rotor displacement.

The motor magnetic fields are now studied by an array of six Hall-effect sensors $k = 0, 1, ..., 5$. They are symmetrically distributed on the interior of the stator, with separation $\gamma = \pi/3$, distributed as in Fig. 1a. Thus, each sensor k measures a phase-shifted magnetic flux density B_k as in

$$B_k(\theta, r, \psi, k) = \hat{B}(r, \psi, k) \cdot \cos(\theta - k \cdot \gamma + \tau_k),$$
$$\text{with } k = 0, 1, ..., 5, \text{ and } \gamma = \pi/3, \tag{1}$$

with τ_k modelled as the product of rotor displacement projected onto the tangential axis of the coordinate system of the kth-sensor $r_{\tau,k}$ and s_τ the phase-shift to displacement sensibility as in $\tau_k = s_\tau \cdot r_{\tau,k}$. This proportional phase-shift is hereafter *neglected*, as it is non-linear and theoretically distorts measurements only at large rotor deflections.

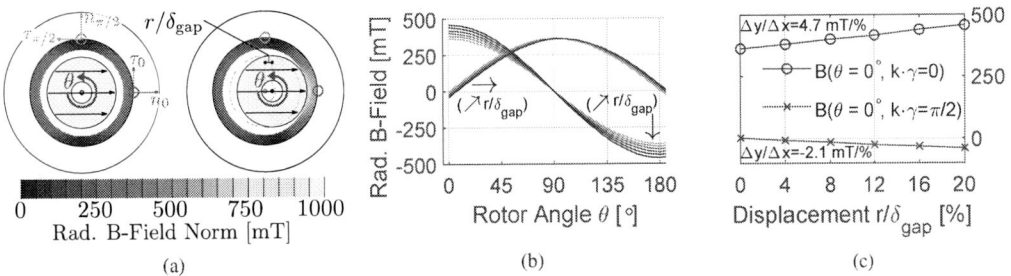

Fig. 2: (a) The one-pole pair, radial magnetic induction distribution is distorted when the rotor is eccentric. (b) In the eccentricity direction (at purple $\bigcirc$), this increases the amplitude of the radial magnetic induction. Perpendicular to the eccentricity (at green $\bigcirc$), it affects the phase of the radial magnetic induction. Darker colors represent a larger r/δ_{gap} eccentricity. (c) The latter effects have a linear impact upon the radial magnetic induction sensed by the sensors.

For feeble rotor displacements r, the peak magnetic flux density $\hat{B}(r,\psi,k)$ of Eq. (1) varies as in [9]

$$
\begin{aligned}
\hat{B}(r,\psi,k) = \hat{B}_0 \cdot \frac{1}{1 - r/\delta_{\text{gap}} \cdot \cos(\psi - k \cdot \gamma)} &\approx \hat{B}_0 \cdot [1 + r/\delta_{\text{gap}} \cdot \cos(\psi - k \cdot \gamma)], \\
&= \hat{B}_0 \cdot [1 + r \cdot s_r \cdot \cos(\psi - k \cdot \gamma)),
\end{aligned}
\tag{2}
$$

using the Taylor series approximation of $1/(1-x) \approx (1+x)$, for $|x| < 1$. The normal flux path is initially assumed to be determined by δ_{gap}, which appears as gain $1/\delta_{\text{gap}}$ of radial displacement r. Yet, this flux path is not always straight nor evident, specially in disc drives. Therefore, the equation is further generalized with a radial displacement gain s_r that depends on *where* the sensor is placed.

Hall-effect sensors also sense the axial displacements (z-direction) of the rotor. If they are *not* placed at the same axial plane as the rotor, they can sense if this moves away or approaches. This might prove useful in applications where axial vibrations need to be suppressed [15]. Provided a small displacement, the change in peak magnetic flux density $\hat{B}$ can also modelled in a linear manner as in Eq. 2.

Considering Eqs. (1) - (2) and the possibility of axial displacement, the magnetic field density that each Hall-effect sensor k measures is

$$
B_k^{\pm}(\theta,r,\psi,k,z) = \hat{B}_0 \cdot \underbrace{[1 + r \cdot s_r \cdot \cos(\psi - k \cdot \gamma)]}_{\text{polar rotor displacement}} \cdot \underbrace{(1 \pm z \cdot s_z)}_{\text{axial rotor displacement}} \cdot \underbrace{\cos(\theta - k \cdot \gamma)}_{\text{rotor rotation}},
\tag{3}
$$

with the $+$ or $-$ sign for axial rotor displacement if the Hall-effect sensor lies on over or under the rotor still stand position, respectively.

A symmetric array of 12 Hall-effect sensors (6 top and 6 bottom) is employed, so as to measure rotor angle, its radial position, and its axial position.

Model for the Estimation of Rotor Angle

The objective of rotor angle θ estimation is to generate a signal pair $(\cos\theta, \sin\theta)$ out of Hall-effect sensor measurements. For this, the effects of radial and axial displacement contained in each Hall-effect sensor measurement in Eq. (3) are to be eliminated.

The effect of axial displacement are eliminated by *adding* the respective top and bottom sensors; that is, for each top-bottom Hall-effect sensor pair $k = 0, 1, ..., 5$,

$$
\begin{aligned}
B_k^{\text{sum}}(\theta,r,\psi,k) &= B_k^+(\theta,r,\psi,k,z) + B_k^-(\theta,r,\psi,k,z) \\
&= 2 \cdot \hat{B}_0 \cdot [1 + r \cdot s_r \cdot \cos(\psi - k \cdot \gamma)] \cdot \cos(\theta - k \cdot \gamma).
\end{aligned}
\tag{4}
$$

To cancel the effect of the displaced rotor in $r \cdot s_r$, all opposing Hall-sensor pairs signals B_k^{sum} for $k =$

$0, 1, 2$ are subtracted, resulting in

$$
\begin{aligned}
B_k^{\text{sum}} &(\theta, r, \psi, k) - B_{k+3}^{\text{sum}} (\theta, r, \psi, k+3) \\
&= \left\{ 2 \cdot \hat{B}_0 \cdot [1 + r \cdot s_r \cdot \cos(\psi - k \cdot \gamma)] \cdot \cos(\theta - k \cdot \gamma) \right\} \\
&\quad - \left\{ 2 \cdot \hat{B}_0 \cdot [1 + r \cdot s_r \cdot \cos(\psi - k \cdot \gamma - 3 \cdot \gamma)] \cdot \cos(\theta - k \cdot \gamma - 3 \cdot \gamma) \right\} \\
&= 4 \cdot \hat{B}_0 \cdot \cos(\theta - k \cdot \gamma),
\end{aligned}
\tag{5}
$$

since $3 \cdot \gamma = \pi$. These signals are combined and $(\cos\theta, \sin\theta)$ is obtained as

$$
\begin{aligned}
S_{\cos} &= \left\{ B_0^{\text{sum}} (\theta, r, \psi, 0) - B_3^{\text{sum}} (\theta, r, \psi, 3) \right\} + 1/2 \cdot \left\{ B_1^{\text{sum}} (\theta, r, \psi, 1) - B_4^{\text{sum}} (\theta, r, \psi, 4) \right\} \\
&\quad - 1/2 \cdot \left\{ B_2^{\text{sum}} (\theta, r, \psi, 2) - B_5^{\text{sum}} (\theta, r, \psi, 5) \right\} \\
&= 6 \cdot \hat{B}_0 \cdot \cos\theta, \\
S_{\sin} &= \sqrt{3}/2 \cdot \left\{ B_1^{\text{sum}} (\theta, r, \psi, 1) - B_4^{\text{sum}} (\theta, r, \psi, 4) \right\} + \sqrt{3}/2 \cdot \left\{ B_2^{\text{sum}} (\theta, r, \psi, 2) - B_5^{\text{sum}} (\theta, r, \psi, 5) \right\} \\
&= 6 \cdot \hat{B}_0 \cdot \sin\theta.
\end{aligned}
\tag{6}
$$

Ultimately, the peak magnetic induction $\hat{B}_0$ can also be estimated from these signals as in

$$
\hat{B}_0 = 1/6 \cdot \sqrt{S_{\cos}^2 + S_{\sin}^2}
\tag{7}
$$

Model for the Estimation of Rotor Radial Position

Rotor position information on the $x - y$ plane is calculated analogous to Eq. (5) by *adding* opposing sensors for $k = 0, 1, 2$ as

$$
B_k^{\text{sum}} (\theta, r, \psi, k) + B_{k+3}^{\text{sum}} (\theta, r, \psi, k+3) = 4 \cdot \hat{B}_0 \cdot [r \cdot s_r \cdot \cos(\psi - k \cdot \gamma)] \cdot \cos(\theta - k \cdot \gamma),
\tag{8}
$$

thus obtaining signals which are *proportional* to polar displacement r. Once again, analogous to Eq. (6), two perpendicular signals can be obtained by combining sensor information as in

$$
\begin{aligned}
S_{X_{\text{rot}}} &= \left\{ B_0^{\text{sum}} (\theta, r, \psi, 0) + B_3^{\text{sum}} (\theta, r, \psi, 3) \right\} - 1/2 \cdot \left\{ B_1^{\text{sum}} (\theta, r, \psi, 1) + B_4^{\text{sum}} (\theta, r, \psi, 4) \right\} \\
&\quad - 1/2 \cdot \left\{ B_2^{\text{sum}} (\theta, r, \psi, 2) + B_5^{\text{sum}} (\theta, r, \psi, 5) \right\} \\
&= 3 \cdot \hat{B}_0 \cdot (r \cdot s_r) \cdot \cos(\theta + \psi), \\
S_{Y_{\text{rot}}} &= \sqrt{3}/2 \cdot \left\{ B_1^{\text{sum}} (\theta, r, \psi, 1) + B_4^{\text{sum}} (\theta, r, \psi, 4) \right\} - \sqrt{3}/2 \cdot \left\{ B_2^{\text{sum}} (\theta, r, \psi, 2) + B_5^{\text{sum}} (\theta, r, \psi, 5) \right\} \\
&= 3 \cdot \hat{B}_0 \cdot (r \cdot s_r) \cdot \sin(\theta + \psi),
\end{aligned}
\tag{9}
$$

which are the x and y coordinates of the rotor in the *rotating* coordinate system of the rotor, i.e. it depends on θ. To control the levitation, the position on the fixed coordinate system, i.e. not θ dependent, are essential [8]. This is easily obtained by taking advantage from the results of Eq. (6) and rotating $(S_{X_{\text{rot}}}, S_{Y_{\text{rot}}})$ by computing

$$
\begin{bmatrix} S_{X_{\text{st}}} \\ S_{Y_{\text{st}}} \end{bmatrix} = \frac{2}{S_{\cos}^2 + S_{\sin}^2} \cdot \begin{bmatrix} S_{\cos} & S_{\sin} \\ -S_{\sin} & S_{\cos} \end{bmatrix} \cdot \begin{bmatrix} S_{X_{\text{rot}}} \\ S_{Y_{\text{rot}}} \end{bmatrix} = (r \cdot s_r) \cdot \begin{bmatrix} \cos\psi \\ \sin\psi \end{bmatrix},
\tag{10}
$$

hence obtaining two signals projected on the fixed coordinate system. They are directly proportional to the displacement r and independent from $\hat{B}_0$.

Position Error of Estimation due to Phase-Shift

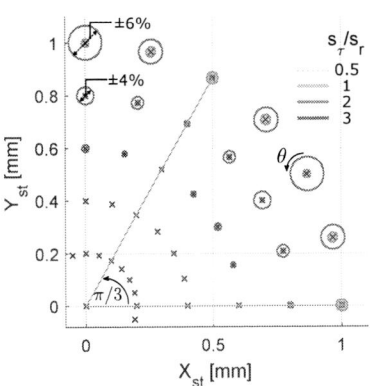

Fig. 3: Actual vs. estimated rotor position, in $\times$ and $\bigcirc$ respectively. The other position quadrants are symmetric.

The position estimation scheme is developed by neglecting the phase-shift τ_k of Eq. (1), generated by rotor displacement. This allows for a linear model and a more intuitive derivation.

Phase-shift τ_k has yet an influence upon the fidelity of position estimation. This becomes apparent for increasing rotor displacements r and phase-shift to displacement sensibility s_τ. Figure 3 depicts the true vs. model-estimated position of the rotor for increasing sensibilities s_τ. As the rotor turns, the estimated position *circles* around the true position. This behaviour is not foreseen by the estimator.

Yet, the error radius is minimized when the rotor is displaced towards a sensors. This would speak for setups with more sensors around the airgap for high-precision position estimation. Nevertheless, the depicted s_τ/s_r values are already very large with respect to values obtained by FE simulations. According to them, rather small sensitivity values $s_\tau/s_r < 1$ are expected. Therefore, position estimation error should be rather small.

Model for the Estimation of Rotor Axial Position

If only top or bottom sensors were available, signals $(S_{\cos}, S_{\sin})$ and $(S_{X_{st}}, S_{Y_{st}})$ would remain modulated by factor $(1 \pm z \cdot s_z)$. Axial position information *could* be estimated out of the angle signals that would constantly carry an offset. Yet having top and bottom sensors enables finding a tidier expression for axial displacement z, by calculating the *difference* of $(\cos\theta, \sin\theta)$ signals of the upper and lower sensors,

$$
\begin{aligned}
S_{\cos}^{\pm} = & \quad \left\{ B_0^{\pm}(\theta, r, \psi, 0) - B_3^{\pm}(\theta, r, \psi, 3) \right\} + \tfrac{1}{2} \cdot \left\{ B_1^{\pm}(\theta, r, \psi, 1) - B_4^{\pm}(\theta, r, \psi, 4) \right\} \\
& - \tfrac{1}{2} \cdot \left\{ B_2^{\pm}(\theta, r, \psi, 2) - B_5^{\pm}(\theta, r, \psi, 5) \right\} = 6 \cdot \hat{B}_0 \cdot \cos\theta \cdot (1 \pm z \cdot s_z) \\
\rightarrow S_{z-\cos} = & \, S_{\cos}^{+} - S_{\cos}^{-} = 6 \cdot \hat{B}_0 \cdot \cos\theta \cdot z \cdot s_z, \\
S_{\sin}^{\pm} = & \quad \sqrt{3}/2 \cdot \left\{ B_1^{\pm}(\theta, r, \psi, 1) - B_4^{\pm}(\theta, r, \psi, 4) \right\} + \sqrt{3}/2 \cdot \left\{ B_2^{\pm}(\theta, r, \psi, 2) - B_5^{\pm}(\theta, r, \psi, 5) \right\} \\
= & \, 6 \cdot \hat{B}_0 \cdot \sin\theta \cdot (1 \pm z \cdot s_z) \\
\rightarrow S_{z-\sin} = & \, S_{\sin}^{+} - S_{\sin}^{-} = 6 \cdot \hat{B}_0 \cdot \sin\theta \cdot z \cdot s_z,
\end{aligned}
\tag{11}
$$

which means that a signal proportional to z and then normalized by $1/\left(S_{\cos}^2 + S_{\sin}^2\right) = 1/\left(36 \cdot \hat{B}_0^2\right)$ can be obtained as in

$$
D_z = \left(S_{z-\cos} \cdot S_{\cos} + S_{z-\sin} \cdot S_{\sin}\right) / \left(S_{\cos}^2 + S_{\sin}^2\right) = z \cdot s_z.
\tag{12}
$$

Test Bench and Hall-Effect Sensor Mounting

The presented mathematical model is validated on the experimental test bench of Fig. 4. The disc drive of Fig. 1b is fabricated with the materials and dimensions of Table I and mounted.

Commercially-available, ratiometric Hall-effect sensors from *Texas Instruments* DRV5055 series [16] are contemplated, given their simple 3-pad implementation. They are available in TO-92 and SOT-23 packaging, suitable for radial and axial PCB mounting, respectively.

These Hall-effect sensors are mounted onto the fabricated PCBs of Fig. 5a. Two PCBs, one below and above the stator, help to spatially distribute the sensors as in Fig. 1a. The PCBs enable the two different DRV5055 mounting options shown in Fig. 5b.

Table I: Materials and dimensions of mounted motor.

Rotor Mat.	NdFeB42H [17]
Stator Mat.	Metglas 2605SA1 [18]
$d_{st.}$	8 mm
$\delta_{airgap.}$	5 mm
$r_{r.out}$	10 mm
$r_{r.in}$	3 mm
h_r	8 mm
$h_{st.}$	6 mm

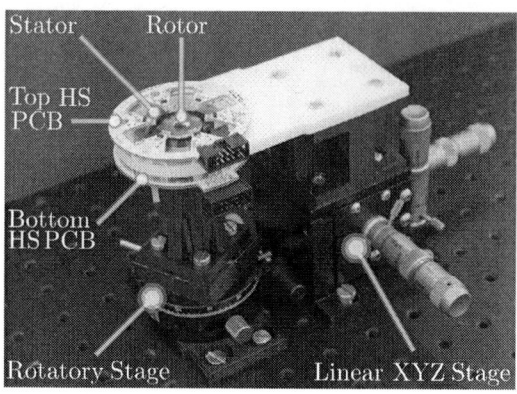

Fig. 4: Model validation test bench. The PM rotor is fixed to a rotatory stage for θ, whereas the ferromagnetic stator and and PCBs are fixed on a XYZ stage.

Model Validation

The outputs of the 12 hall-effect sensors are installed onto 12-bit ADCs. A displacement is manually imposed onto the screws of the linear or rotational stage, and the data is then logged in.

For comparison purposes, the measurements are obtained once using radial-field measuring sensors, and then axial-field measuring sensors, mounted as in Fig. 5b. For the rest of the paper, these are referred according to the orientation of the magnetic induction they measure, as radial and axial respectively. The results and their performance are then compared.

Hall-Effect Sensor Values

Figure 6a depicts the mean values of the bottom and top sensors when the rotor is turned in the center of the airgap. Given the proposed installation, axial sensors measure larger sensor amplitudes. This enhances signal-to-noise ratio, rendering axial orientation preferable to the radial for angle estimation.

The signals are combined to generate $(\cos\theta, \sin\theta)$. Despite its lower amplitude of the radially-oriented sensors, both orientation draw a linear relationship with the angle imposed on the test bench.

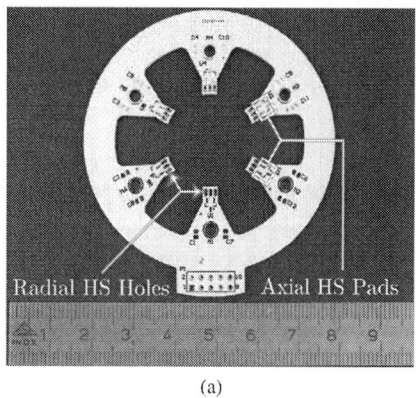

(a)

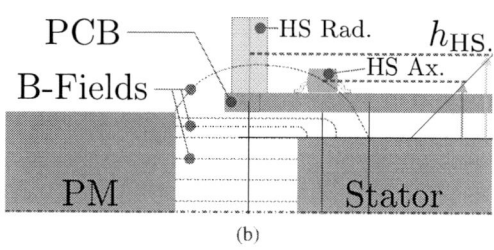

(b)

Fig. 5: (a) PCB for the mounting of 6 radial or axial Hall-effect sensors. (b) Cross-section of the hall-effect sensor placement on the PCB, with vertical axial symmetry. The radial and axial sensors are fixed at radii $(r_{HS.}, h_{HS.}) = (13, 8.3)$ mm and $(16, 7.3)$ mm, are depicted in yellow and orange, respectively.

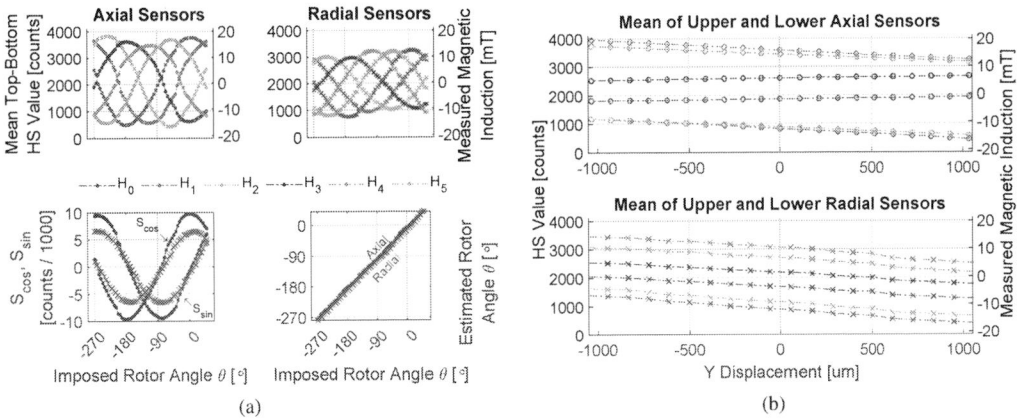

(a)　　　　　　　　　　　　　　(b)

Fig. 6: Sensor values and angle estimation when (a) rotor is turned as $\theta = [-270 : 5 : 30]^\circ$ at the center of the airgap. (b) Sensor values with rotor displaced as in $y = [-1000 : 100 : 1000]\ \mu m$ at fixed $\theta = 270^\circ$.

Validation of Rotor Radial Position Estimation

Figure 6b displays how the mean values of the bottom and top sensors change when the north pole of the rotor is between H_4 and H_5 and is displaced perpendicular to them, i.e. from H_3 and H_0.

In this case, two different behaviours are observed. The measured field of radial sensors H_0 and H_3 (perpendicular to the rotor poles) are affected by rotor displacement, and this is not foreseen by the model. In fact, the field gradients perpendiculars to the poles are as large as those next to the rotor poles, i.e. all sensors measure the same slope. Thus, parameter τ_k from Eq. (1) *should not* be neglected, so the proposed model estimates rotor position with lower accuracy.

Yet this is not the case of axial sensors. An quasi-null gradient is present in sensors H_0 and H_3, so τ_k can be neglected and the model holds valid. Axial sensors therefore enable a more accurate rotor position estimation.

With the north pole facing between H_4 and H_5, the position of the rotor is displaced in x, y and $x - y$ direction. The hall-effect sensor values are recorded and the position on stator coordinates $(S_{X_{st}}, S_{Y_{st}})$ is

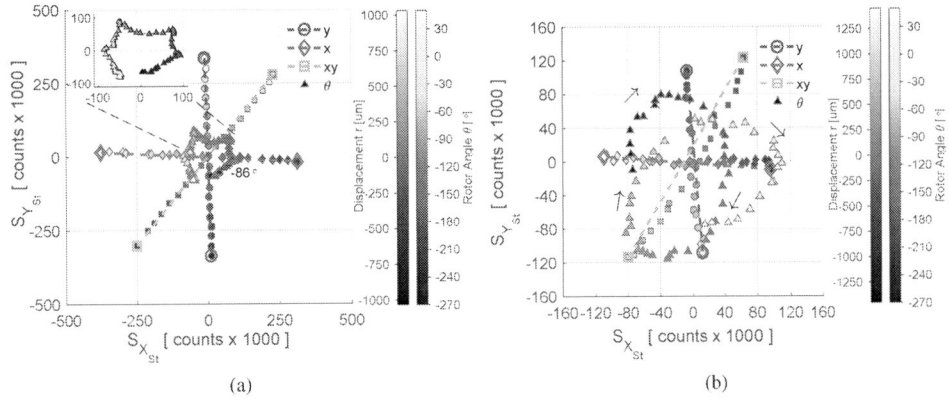

(a)　　　　　　　　　　　　　　(b)

Fig. 7: Estimated rotor position $(S_{X_{st}}, S_{Y_{st}})$ for a displacement (marked on the gray color bar) along the x, y, $x - y$ axes and half a turn of a centric rotor. The estimated position of the rotor turning in the middle of the airgap (see Fig. 6a) is marked with a $\triangle$, with rotor angle displayed on the red color bar. The position cloud for (a) axial and (b) radial Hall-effect sensors is shown.

computed and displayed on Fig. 7. The estimated position of the rotor during the rotation of Fig. 6a is also plotted.

Figure 7a shows the estimated $(S_{X_{st}}, S_{Y_{st}})$ using the axial sensors. The angle between the estimated positions is also depicted. Their non-perfect perpendicularity may be due to the a non-stiff enough test bench, so that the rotor deflects in x direction, i.e. the direction of magnetization of the rotor *during* the test, due to the stark radial stiffness of the drive configuration.

The rotor is turned for $270°$ and it is estimated that the rotor *slightly* budges. This hints once again that the test bench could slightly bend due to the radial strong stiffness of the drive configuration.

The experience is replicated using the radial sensors. The estimated positions of Fig. 7b on x and y are once again (almost) perpendicular, yet the signal amplitude is significantly reduced in comparison to the previous case.

Moreover, a second harmonic appears in the estimated position when the rotor is just turned in the middle of the airgap. The amplitude of the ripple is comparable to the amplitudes of the displacements as far away as 1 mm, making this approach not suitable for accurate position estimation. Rotor eccentricity impacts the airgap flux distribution, and axial fields are probably less sensible to this dependency.

The signal amplitude $r \cdot s_r$ of Eq. (10) and its linearized sensitivity s_r are plotted on Fig. 8 for axial and radial sensors. For the axial sensors, sensitivities s_r in the x, y and $x - y$ direction, and their differences could be due to non-perfect centering of the PCBs onto the drive, imperfect placement of the DRV5055s or test bench bending. They present only slight differences that can ultimately be corrected via software.

The sensitivity of the radially-placed sensors is approximately 3 times lower than that of the axial sensors. Moreover, it varies significantly depending on the direction of rotor displacement. This makes the radial configuration and its chosen measurement locus unsuitable for rotor position estimation purposes.

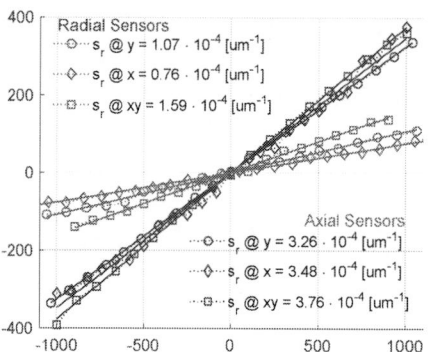

Fig. 8: Calculated radial rotor position sensitivities s_r for axial and radial sensor mounting.

Validation of Rotor Axial Position Estimation

The validity of the axial rotor position estimation is tested in Fig. 9. As expected, the amplitude of the bottom sensors decreases, whereas that from the top sensors increases. The axial sensors respond in a more linear fashion to the z displacement than its radial counterparts, which tend to be more parabolic towards -2000 μm and 2000 μm.

Signal D_z and its corresponding sensitivity s_z are linear and constant (respectively) throughout this z range. Coincidently enough, both approaches render extremely similar D_z trends and sensitivities s_z, with the axial approach being 0.6 % more sensitive.

Conclusion

This paper analyses how the rotor (x, y, z) position and its angular orientation θ can be obtained by measuring stray fields *outside* the airgap. This eases the encapsulation of a rotor, thus widening the ap-

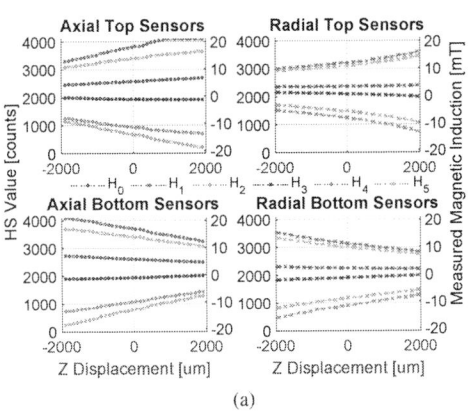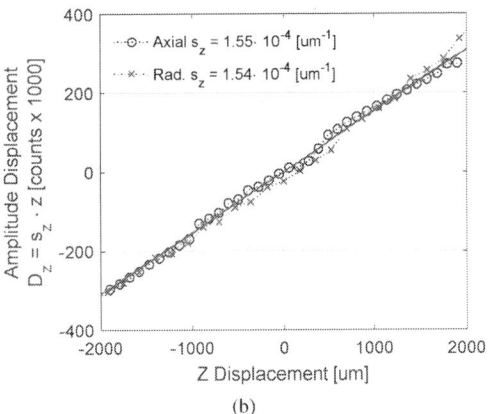

Fig. 9: (a) Measured hall-effect sensor values when rotor is displaced from $z = [-2000 : 100 : 2000]\ \mu$m with $(r, \theta) = (0\ \text{mm}, 60°)$ for axial and radial sensors and their (b) estimated axial rotor position signal D_z and sensitivity s_z.

plication spectrum of bearingless disc drive. Rotor position estimation is thus enabled by off-the-market Hall-effect sensors which are simply mounted on a PCB onto the motor. With the chosen measurement placement, axial-field sensors render better results as their radial-field mounted versions.

A simplified model correlates the displacements of the rotor and the measured fields, and it is tested on a bench test. It enables measuring the angular and axial position of the rotor seamlessly, by either measuring the radial or axial field of the rotor. Nevertheless, the proposed model fits axial rotor magnetic fields better. Therefore, the radial rotor position was only accurately estimated by using axial sensors.

Radial magnetic sensing measured additional harmonics that do not correspond to the actual rotor position at their chosen measurement locus. A more complex modelling of the fields is necessary to clarify this behaviour and propose a better sensing scheme. Thus, this publication can still be enriched by rendering the mathematical model more complex to account for higher order harmonics and phase-shifts in the airgap and the estimation errors that they might cause.

Stray flux models could be applied so that the here-studied Hall-effect sensor position can be theoretically optimized on the design phase of the drive. Undesired harmonics or effects, such as those measured by the radial Hall-effect sensors, could be therefore minimized.

Further publications might also research the estimation of rotor position for a generalized number of Hall-effect sensors and any number of pole-pairs. Similarly, the model can be tested for slotted drives. More exhaustive measurements with a stiffer test bench must be taken into considerations. Test-bench imperfections could be clarified by measuring the real rotor displacement vs. its imposed displacement on the test bench.

Furthermore, the effect of motor excitation should be scrutinized. Drive or bearing currents on motor coils generate magnetic fields, which are superposed to those from the rotor and are thus measured by the sensors. The nature of this disturbance upon rotor position estimation, and eventual current compensation scheme can be further studied.

References

[1] N. Bianchi, S. Bolognani, and F. Luise, "Potentials and limits of high speed PM motors," *38th IAS Annual Meeting on Conference Record of the Industry Applications Conference, 2003.*, vol. 3.

[2] A. Borisavljevic, *Limits, Modeling and Design of High-Speed Permanent Magnet Machines.* Springer Theses, Berlin, Heidelberg: Springer Berlin Heidelberg, 2013.

[3] R. Larsonneur, *Design and control of active magnetic bearing systems for high speed rotation.* PhD thesis, ETH Zürich, 1990.

[4] I. Takahashi, T. Koganezawa, G. Su, and K. Ohyama, "A super high speed PM motor drive system by a quasi-current source inverter," *IEEE Transactions on Industry Applications*, vol. 30, pp. 683–690, May 1994.

[5] M. H. Kimman, *Design of a Micro Milling Setup with an Active Magnetic Bearing Spindle.* PhD thesis, TU Delft, 2010.

[6] B. Klammer, A. Zochbauer, H. Mitterhofer, and W. Gruber, "Topology Comparison and Design of a Slotted Bearingless High-Speed Permanent Magnetic Synchronous Machine," in *2019 IEEE International Electric Machines Drives Conference (IEMDC)*, pp. 1–8, May 2019.

[7] J. Asama, D. Suzuki, T. Oiwa, and A. Chiba, "Development of a Homo-Polar Bearingless Motor with Concentrated Winding for High Speed Applications," in *2018 International Power Electronics Conference (IPEC-Niigata 2018 -ECCE Asia)*, pp. 157–160, May 2018.

[8] Schweitzer, Gerhard and E. H. Maslen, *Magnetic Bearings.* 2009.

[9] K. Li, G. Cheng, X. Sun, D. Zhao, and Z. Yang, "Direct Torque and Suspension Force Control for Bearingless Induction Motors Based on Active Disturbance Rejection Control Scheme," *IEEE Access*, vol. 7, pp. 86989–87001, 2019.

[10] M. Schuck, A. D. S. Fernandes, D. Steinert, and J. W. Kolar, "A high speed millimeter-scale slotless bearlngless slice motor," in *2017 IEEE International Electric Machines and Drives Conference.*

[11] M. Noh, W. Gruber, and D. L. Trumper, "Low-cost Eddy-current Position Sensing for Bearingless Motor Suspension Control," in *2017 IEEE International Electric Machines and Drives Conference.*

[12] B. Warberger, T. Reichert, T. Nussbaumer, and J. W. Kolar, "Design considerations of a bearingless motor for high-purity mixing applications," in *SPEEDAM 2010*, pp. 1454–1459, 2010.

[13] H. Bleuler, H. Kawakatsu, W. Tang, W. Hsieh, D. K. Miu, Y. C. Tai, F. Moesner, and M. Rohner, "Micromachined Active Magnetic Bearings," *4th International Symposium on Magnetic Bearings*, no. August, pp. 349–352, 1994.

[14] V. Kluyskens, C. Dumont, and B. Dehez, "Description of an Electrodynamic Self-Bearing Permanent Magnet Machine," *IEEE Transactions on Magnetics*, vol. 53, no. 1, pp. 1–9, 2017.

[15] M. Miyoshi, H. Sugimoto, and A. Chiba, "Axial vibration suppression by field flux regulation in two-axis actively positioned permanent magnet bearingless motors with axial position estimation," *Proceedings - 2016 22nd International Conference on Electrical Machines, ICEM 2016*, vol. 54.

[16] T. Instruments, "DRV5055 High accuracy 3.3 V or 5 V ratiometric bipolar Hall-effect sensor."

[17] E. Magnetics, "Rare Earth Neodymium NdFeB Magnets Datasheet," 2014.

[18] Metglas, "PowerLite C-Cores Technical Bulletin Alloy 2605SA1."

Non-unit ROCOV scheme for protection of multi-terminal HVDC systems

María José Pérez-Molina, Pablo Eguia, Marene Larruskain, Garikoitz Buigues, Esther Torres
UNIVERSITY OF THE BASQUE COUNTRY UPV/EHU
Plaza Ingeniero Torres Quevedo, 1
Bilbao, Spain
E-Mail: mariajose.perez@ehu.eus
URL: www.ehu.eus

Acknowledgements

This research was funded by the Spanish Ministry of Economy, Industry and Competitiveness (project ENE2016-79145-R AEI/FEDER, UE), the Basque Government (GISEL research group IT1191-19) and the UPV/EHU (GISEL research group 181/18).

Keywords

«Multiterminal HVDC», «Voltage Source Converter (VSC)», «Faults», «DC cable», «Bus bar».

Abstract

Nowadays, high voltage direct current transmission systems are being selected over the traditional alternating current transmission for very long transmission distances. However, there are still some unresolved challenges, most of them in terms of protection of the system. If using traditional protection systems used in alternating current grids, the characteristics of a DC fault make it complicated to clear and dangerous to power electronic components. Hence, there is a need of very fast fault detection and clearance. This paper proposes a full-selective protection system based on a rate of change of voltage. The process of selection of the threshold needed to discriminate fault conditions is described. In addition, the performance of the protection algorithm is analysed. The influence of parameters such as the fault location and fault resistance is evaluated. A comparison with similar algorithms found on the literature is presented at the end.

Introduction

Protection of high voltage direct current (HVDC) grids is still a challenging issue, which is slowing down the developing of meshed multi-terminal grids [1]. The very short range of time available to clear a fault condition, within 10 ms [2]-[5], before the damage of the power electronic devices of the Voltage Source Converters (VDC) is one of the most important limitations [5]-[9]. Another limitation is the capability of the HVDC Circuit Breakers (CBs) to interrupt the very high fault-induced overcurrents [10]. Moreover, CBs have to create a zero crossing in order to interrupt the current [7], [11], [12] and they have to dissipate the great amount of energy stored in the system inductance during a fault condition [11]-[13]. This energy is greater if there are large limiting inductors placed in the system [14].

Limiting inductors are very important if the algorithm used to protect the system is based on local measurements. This type of algorithms are very fast since they do not have to exchange information with the other end of the protection zone [15], [16] but lack selectivity [17], [18]. The limiting inductors are employed to delimit the borders of the protection zone since they damp external signals. In addition, they limit the rate of rise of the current during fault conditions giving more time to the HVDC CBs to clear the fault. However, the inductor size affects not just the energy dissipation capability of the HVDC CBs but it also could affect the stability of the system [19].

The fault clearing strategy is another aspect to take into account when designing the protection system of a grid [7], [12], [13], [16]. Existing Point-to-Point (P2P) links are usually protected by CBs placed at the Alternating Current (AC) side of the system. This way, in case of fault, the entire HVDC system is de-energised. However, this is not appropriate for a multi-terminal system. The most similar fault clearing strategy to the traditional AC approach is the so-called full-selective strategy. The aim of this strategy is to minimize the effects of a fault condition. This way, the system is divided into several protection zones, each one of them protecting a component. HVDC CBs in series with limiting inductors are located at both end of the protection zones when they are protecting a link. When DC faults appear, only the protection zone affected is isolated and de-energised while the healthy parts of the system remain operative.

In this paper, a local-measurement-based protection algorithm based on Rate Of Change Of Voltage (ROCOV) is proposed. Firstly, a description of the characteristics of this algorithm is presented. Then, the ROCOV threshold is selected by analysing the behaviour of the ROCOV values during different fault conditions. Afterwards, the performance of the ROCOV algorithm is evaluated for different fault types, fault locations and fault resistances on a four-terminal meshed HVDC grid modelled in PSCAD software. Subsequently, the proposed algorithm is compared to similar ones found on the literature. Finally, some conclusions are elaborated.

Rate of change of voltage algorithm

ROCOV algorithm is a local-measurement-based protection method. The DC voltage is measured and its derivative is calculated as in (1), where V_1 is the voltage magnitude at time t_1 and V_2 is the voltage magnitude at time t_2, being time t_2 greater than time t_1.

$$\text{ROCOV} = \frac{\Delta V}{\Delta t} = \frac{V_2 - V_1}{t_2 - t_1} \tag{1}$$

During normal operation conditions, the derivative value is zero since the DC voltage is constant. However, a sharp DC voltage drop occurs during fault conditions. Consequently, at the first instant of a fault, the derivative value quickly increases. Therefore, the ROCOV value presents good features to be a fault marker [4].

Nevertheless, the DC voltage derivate is not zero in actual systems during normal operation conditions due to fluctuations and disturbances. Then, the value of the derivative is compared to an accurate threshold value in order to discriminate between an internal fault condition and an external fault condition or normal operation changes [16].

The calculated ROCOV value is a negative value since the voltage collapses during a fault condition. Moreover, it increases sharply since the voltage drops almost instantaneously [15]. Hence, this algorithm presents great detection speed but it lacks selectivity, which is improved by using a precise threshold value and limiting inductors placed at both ends of the protection zone.

Study case

The application of a ROCOV algorithm on a multi-terminal grid is studied in this section. Firstly, the threshold is selected in order to enable fault discrimination between internal and external faults. The selection of the threshold is achieved by simulating several low and high resistance fault conditions on different points of the grid. Then, the influence of the fault location and fault resistance on the ROCOV performance is analysed.

Multi-terminal grid

The ROCOV algorithm is implemented on the four-terminal meshed grid that is presented in Fig. 1. It is modelled in PSCAD software (available in [20]). This grid interconnects two offshore wind power plants to two onshore AC grids through five links. Links 12 and 34 have a length of 100 km, link 24 is 150 km long while the length of links 13 and 14 is 200 km. The converters are Modular Multilevel Converters (MMC) with a half-bridge arrangement. The rated power of converters MMC 1 to 3 is 900 MVA while converter MMC 4 presents a rated power of 1200 MVA. The configuration of the HVDC system is symmetric monopole.

A full-selective fault clearing strategy is adopted. This way, an HVDC CB in series with a 0.1 H limiting inductor is placed at both ends of each link. With this strategy, there are five protection zones, each one of them covering a cable. The operating time of the HVDC CBs is assumed to be 2 ms, in the order of a hybrid CB.

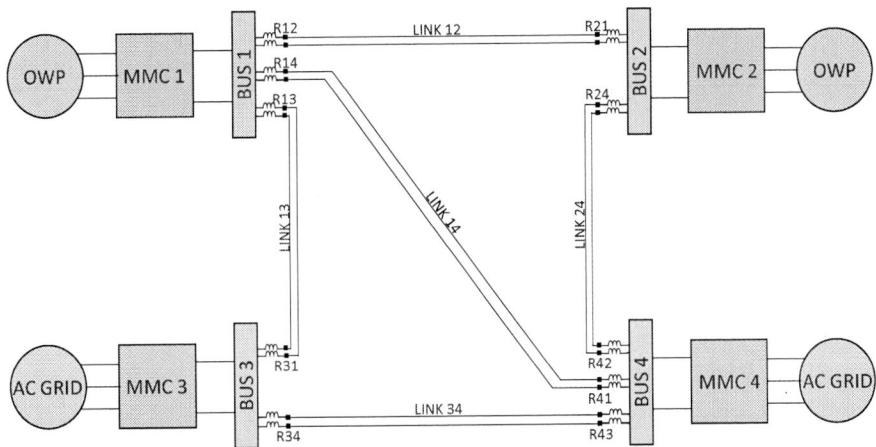

Fig. 1: Scheme of the multi-terminal MMC-based system.

Threshold selection

The selection of the threshold is essential for the performance of a local-measurement-based algorithm. The threshold value affects the selectivity and sensitivity of the algorithm. A higher value improves the selectivity but worsens the sensitivity. Consequently, a trade-off value is needed. The value is selected according to the grid conditions and algorithm characteristics through a series of low and high resistance fault simulations. With that objective, the ROCOV values of each relay of the system are calculated and compared in order to choose the most appropriate threshold value. Since the ROCOV value is negative during a fault condition, as it can be seen in Fig. 2, the critical ROCOV values which established the threshold, are the maximum ROCOV value in the case of a faulty link (represented as a red circle in Fig. 2-a) and the minimum ROCOV value in the case of a healthy link (represented as a red circle in Fig. 2-b).

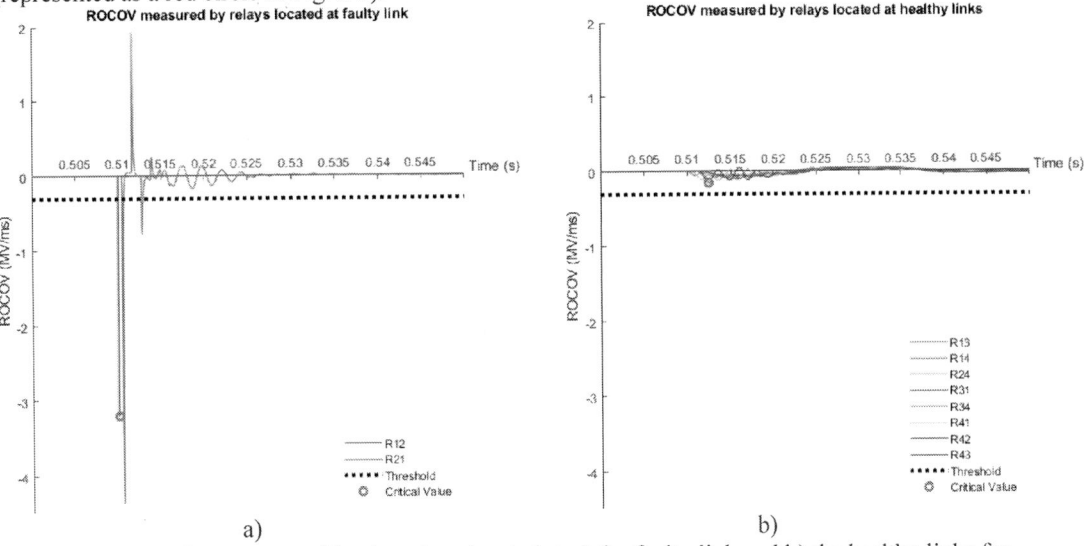

Fig. 2: ROCOV values measured by the relays located at a) the faulty link and b) the healthy links for a 0.01 Ω PtG fault condition located in link 12 (0 km).

Low resistance fault conditions

Three different 0.01 Ω fault conditions are simulated in each link. Two faults are located at the ends of the link and one in the middle of it. Both Pole-to-Ground (PtG) and Pole-to-Pole (PtP) fault conditions were simulated. Fault inception is at 0.51 s. A data window between 0.5-0.55 s was analysed to extract the critical ROCOV values.

Pole-to-Ground fault conditions

A summary of the simulations of the PtG fault conditions is shown in Table I. The critical ROCOV values are −2974083 kV/s and −170656 kV/s in the cases of faulty link and healthy link, respectively.

Table I: ROCOV values in kV/s for each relay during 0.01 Ω PtG fault conditions.

Link	Fault Location (km)	R12	R13	R14	R21	R24	R31	R34	R41	R42	R43
L12	0	-3202786	-81127	-81465	-4375052	-117938	-79819	-101916	-84424	-148165	-103876
	50	-5245025	-106035	-112658	-5245439	-102730	-102759	-111699	-105330	-133824	-95031
	100	-4379052	-117510	-132295	-3205749	-87829	-105871	-102221	-135253	-99648	-92163
L13	0	-98913	-3207581	-84077	-100884	-95000	-2988619	-126958	-87502	-100307	-139053
	100	-101535	-4393536	-140933	-118933	-82972	-4393234	-108957	-137767	-95782	-156264
	200	-106313	-2974083	-83665	-125863	-92837	-3191838	-99810	-101225	-89156	-108344
L14	0	-79929	-89660	-3212713	-82798	-105890	-92742	-101924	-2993515	-106618	-95565
	100	-82542	-141413	-4405786	-119593	-118511	-133731	-113265	-4405387	-127705	-78639
	200	-107220	-89688	-2985535	-112208	-92142	-102583	-78193	-3203913	-95758	-75213
L24	0	-109305	-80568	-105115	-100437	-3212435	-87225	-102664	-103507	-3636367	-95266
	75	-162848	-92727	-120980	-113309	-4804342	-100455	-118888	-122894	-4804098	-87412
	150	-148216	-94311	-90555	-123944	-3623259	-88002	-101257	-84861	-3200545	-96558
L34	0	-99727	-94303	-134906	-102736	-102424	-114565	-3196147	-132762	-102083	-4365833
	50	-107458	-122705	-109040	-99053	-103478	-138838	-5237680	-118776	-87413	-5237270
	100	-103846	-162909	-86368	-97085	-83789	-170656	-4375320	-86782	-82075	-3202965

Pole-to-Pole fault conditions

Table II summarises the results of the simulation of every PtP fault condition. The maximum ROCOV value for a faulty link is −2977494 kV/s and the minimum ROCOV value for a healthy link is −259031 kV/s in the case of low resistance PtP faults.

Table II: ROCOV values in kV/s for each relay during 0.01 Ω PtP fault conditions.

Link	Fault Location (km)	R12	R13	R14	R21	R24	R31	R34	R41	R42	R43
L12	0	-3206855	-145027	-145698	-4381163	-117800	-143518	-142804	-148986	-115424	-139776
	50	-5252584	-149062	-153149	-5253187	-112930	-142377	-106650	-140626	-109897	-99366
	100	-4384814	-131645	-125254	-3209811	-124573	-125255	-117819	-98824	-134068	-102154
L13	0	-142117	-3211561	-144581	-158569	-157582	-2992135	-136433	-149446	-137849	-130471
	100	-134658	-4399348	-115642	-132278	-110191	-4399828	-196850	-112800	-112371	-219794
	200	-125340	-2977494	-119029	-112416	-142263	-3196051	-150439	-120621	-141195	-142074
L14	0	-140944	-144821	-3216666	-162636	-107654	-149428	-72508	-2997005	-117186	-80603
	100	-102804	-142246	-4411250	-115644	-151392	-125668	-101121	-4411635	-170732	-111301
	200	-155792	-129221	-2988547	-119240	-200002	-110210	-168288	-3207631	-211973	-190506
L24	0	-145652	-125403	-103926	-130082	-3216248	-105652	-141517	-111189	-3640691	-134691
	75	-111295	-128308	-87475	-108997	-4810159	-106663	-152952	-141713	-4810775	-147886
	150	-119987	-111206	-167448	-128945	-3627136	-130077	-202505	-158843	-3204238	-184730
L34	0	-142191	-107290	-120176	-133278	-183900	-114438	-3200110	-164172	-189540	-4371832
	50	-160043	-103149	-112442	-162729	-154340	-122599	-5244681	-150422	-153662	-5245205
	100	-259031	-98442	-115967	-240838	-181083	-99626	-4380952	-114958	-172127	-3206920

High resistance fault conditions

As in the low impedance faults subsection, three different fault conditions are simulated in each link varying the location in the links and the fault type and the same data window is analysed. A high fault resistance value of 200 Ω is selected for the simulation in this subsection.

Pole-to-Ground fault conditions

Results of the simulation of the 200 Ω PtG fault conditions are summarised in Table III. The critical ROCOV value for a faulty link is −367769 kV/s while it is −23050 kV/s for a healthy link in the case of high resistance PtG faults.

Table III: ROCOV values in kV/s for each relay during 200 Ω PtG fault conditions.

Link	Fault Location (km)	R12	R13	R14	R21	R24	R31	R34	R41	R42	R43
L12	0	-484183	-12662	-12733	-658554	-16781	-12145	-19001	-14402	-22661	-19266
	50	-440685	-11887	-12185	-440694	-16633	-12532	-19429	-14095	-19957	-19048
	100	-659154	-13103	-14518	-484646	-18848	-14992	-18928	-17316	-18454	-18653
L13	0	-14218	-484928	-16240	-16359	-15421	-448554	-18744	-17710	-17140	-21222
	100	-15603	-367785	-14358	-15481	-16246	-367769	-18036	-15430	-16636	-17186
	200	-13480	-446389	-17400	-15083	-16383	-482561	-15239	-15894	-15590	-16405
L14	0	-13116	-16224	-485780	-13908	-17261	-17513	-15383	-449434	-16963	-14455
	100	-14611	-13637	-368773	-15034	-14620	-14855	-13688	-368662	-14521	-14411
	200	-15783	-17632	-448252	-17247	-16585	-16143	-14107	-484333	-14209	-13544
L24	0	-20543	-14868	-16832	-21402	-485814	-16448	-17869	-17205	-546630	-18288
	75	-22112	-15781	-15179	-21157	-402486	-15135	-18446	-14333	-402556	-18040
	150	-23050	-16851	-16752	-21082	-544776	-15473	-19993	-14342	-483934	-17365
L34	0	-18422	-16638	-15787	-18540	-15747	-14886	-483136	-14722	-17437	-656992
	50	-18982	-19024	-12734	-18048	-14661	-14965	-440141	-12167	-16411	-440074
	100	-18680	-21259	-13836	-18471	-14450	-16586	-658779	-13451	-14497	-484462

Pole-to-Pole fault conditions

Table IV shows the values of the ROCOV for each relay during all the simulated 200 Ω PtP fault conditions. The maximum ROCOV for a faulty link is −679356 kV/s while the minimum ROCOV for a healthy link is −28627 kV/s.

Table IV: ROCOV values in kV/s for each relay during 200 Ω PtP fault conditions

Link	Fault Location (km)	R12	R13	R14	R21	R24	R31	R34	R41	R42	R43
L12	0	-841860	-15914	-16010	-1145848	-23538	-16719	-14378	-17901	-25780	-14602
	50	-813789	-11925	-12052	-813916	-18917	-14347	-14836	-15285	-20031	-14522
	100	-1146871	-15903	-16086	-842650	-22745	-18756	-15820	-19998	-22590	-16134
L13	0	-18628	-843134	-16918	-16019	-10062	-780675	-21922	-14920	-10380	-23622
	100	-20345	-679418	-15871	-19063	-10442	-679356	-21967	-16926	-11225	-19693
	200	-19719	-776926	-14341	-18785	-8982	-838996	-19930	-17157	-10043	-19258
L14	0	-18653	-16910	-844548	-16043	-17455	-15706	-17627	-782119	-17716	-19875
	100	-20429	-14825	-681205	-20092	-18900	-16866	-19219	-681193	-16781	-19053
	200	-20972	-14186	-779992	-19902	-16769	-17121	-14947	-842122	-16315	-17078
L24	0	-28128	-8548	-16528	-24393	-844539	-7493	-19223	-14553	-951144	-20788
	75	-24270	-8305	-14523	-24010	-743390	-7833	-16953	-13595	-743409	-18445
	150	-25810	-8556	-14573	-28627	-947822	-8170	-16145	-14157	-841333	-17717
L34	0	-13948	-18402	-18097	-14784	-18847	-17150	-840062	-15152	-14655	-1143262
	50	-12866	-18442	-13056	-13778	-14015	-15054	-812758	-10571	-12934	-812636
	100	-15028	-23552	-17261	-14714	-18080	-19196	-1146076	-15457	-16399	-842106

Selection

The critical ROCOV values extracted from all the previous simulations are summarised in Table V. Low resistance fault conditions present the lowest values of ROCOV, then, they are more significant in order to select the threshold value. This threshold value should be greater than the faulty link ROCOV value and lower than the healthy link ROCOV value to enable fault discrimination. Hence, a threshold of −0.3 MV/ms (−300000 kV/s) is selected for this study case.

Table V: Critical ROCOV values in kV/s for PtG and PtP fault conditions.

Fault resistance (Ω)	Pole-to-Ground Fault		Pole-to-Pole Fault	
	Faulty Link	Healthy Link	Faulty Link	Healthy Link
0.01	-2974083	-170656	-2977494	-259031
200	-367769	-23050	-679356	-28627

Performance analysis

Both PtG and PtP faults are simulated varying their fault location and fault resistance in order to evaluate the performance of ROCOV algorithm. Four faults were simulated in each link. Two faults are located at both ends of the link while the other two faults are placed at different point of the link. Firstly, low resistance fault conditions are simulated varying their fault location. Then, the same fault distances are used for different high resistance faults.

Influence of the fault location

Several 0.01 Ω fault conditions are simulated varying the fault location on the system. Both PtG and PtP fault types were analysed. This way, the influence of the fault distance on the ROCOV values is evaluated.

Pole-to-Ground fault conditions

The results of the simulation of the PtG fault conditions are shown in Table VI. The ROCOV algorithm presents fast speed since it can detect the closest faults in 10 μs and the most remote ones in around 1 ms. On the other hand, the HVDC CBs have to deal with currents between 2.85 and 5.59 kA.

Table VI: Performance of the ROCOV algorithm against 0.01 Ω PtG fault conditions.

Link	Fault Location (km)	R_{AB}			R_{BA}		
		Detection time (ms)	Operation time (ms)	Interrupted Current (kA)	Detection time (ms)	Operation time (ms)	Interrupted Current (kA)
L12	0	0.01	2.01	4.236	0.56	2.56	3.549
	36	0.21	2.21	4.376	0.36	2.36	4.267
	84	0.47	2.47	3.996	0.10	2.10	3.909
	100	0.56	2.56	3.975	0.01	2.01	3.852
L13	0	0.01	2.01	5.219	1.11	3.11	3.299
	67	0.38	2.38	5.474	0.74	2.74	2.931
	142	0.79	2.79	5.448	0.33	2.33	3.298
	200	1.11	3.11	5.588	0.01	2.01	2.852
L14	0	0.01	2.01	4.990	1.11	3.11	4.383
	9	0.06	2.06	4.967	1.06	3.06	4.453
	153	0.85	2.85	5.368	0.27	2.27	3.787
	200	1.11	3.11	5.359	0.01	2.01	3.992
L24	0	0.01	2.01	4.456	0.83	2.83	4.002
	65	0.37	2.37	4.864	0.48	2.48	3.393
	121	0.67	2.67	4.334	0.17	2.17	3.563
	150	0.83	2.83	4.740	0.01	2.01	3.659
L34	0	0.01	2.01	3.170	0.56	2.56	4.648
	22	0.13	2.13	3.145	0.44	2.44	4.851
	70	0.39	2.39	3.432	0.17	2.17	4.820
	100	0.56	2.56	2.876	0.01	2.01	4.953

Pole-to-Pole fault conditions

Table VII shows the results of the performance analysis of the ROCOV algorithm during low impedance PtP fault conditions. Fault detection time varies between 10 μs and around 1 ms for close and remote faults, respectively, as in the case of the PtG fault conditions. Meanwhile, the HVDC CBs have to interrupt higher currents, between 3.99 and 6.16 kA.

Table VII: Performance of the ROCOV algorithm against 0.01 Ω PtP fault conditions.

Link	Fault Location (km)	R_{AB}			R_{BA}		
		Detection time (ms)	Operation time (ms)	Interrupted Current (kA)	Detection time (ms)	Operation time (ms)	Interrupted Current (kA)
L12	0	0.01	2.01	4.811	0.56	2.56	4.917
	17	0.10	2.10	4.746	0.46	2.46	4.908
	65	0.37	2.37	4.844	0.20	2.20	5.166
	100	0.56	2.56	4.483	0.01	2.01	5.131

Link	Fault Location (km)	R_AB			R_BA		
		Detection time (ms)	Operation time (ms)	Interrupted Current (kA)	Detection time (ms)	Operation time (ms)	Interrupted Current (kA)
L13	0	0.01	2.01	5.808	1.11	3.11	4.584
	79	0.44	2.44	5.495	0.67	2.67	4.143
	135	0.75	2.75	5.913	0.37	2.37	4.347
	200	1.11	3.11	6.157	0.01	2.01	4.136
L14	0	0.01	2.01	5.570	1.11	3.11	4.977
	92	0.51	2.51	5.113	0.60	2.60	4.340
	140	0.78	2.78	5.756	0.34	2.34	4.641
	200	1.11	3.11	5.929	0.01	2.01	4.633
L24	0	0.01	2.01	5.736	0.83	2.83	4.614
	44	0.25	2.25	5.360	0.59	2.59	3.986
	91	0.51	2.51	5.422	0.33	2.33	4.298
	150	0.83	2.83	6.124	0.01	2.01	4.293
L34	0	0.01	2.01	4.452	0.56	2.56	5.177
	47	0.27	2.27	4.133	0.30	2.30	5.362
	81	0.45	2.45	4.275	0.11	2.11	5.528
	100	0.56	2.56	4.231	0.01	2.01	5.591

Influence of the fault resistance

The fault resistance of several PtG and PtP fault conditions are varied to evaluate its influence on the ROCOV algorithm. The fault locations of these faults are the same of the previous subsection. The fault resistances evaluated are up to 200 Ω since 99.9% of the fault conditions, occurring in an actual grid, present a resistance within 200 Ω [2].

Pole-to-Ground fault conditions

The ROCOV algorithm detected all the simulated high resistance fault conditions. Moreover, it can be extracted from the comparison of Tables VI and VIII that the ROCOV detection time is hardly greater during high resistance faults than during low resistance faults, in the order of a few tens of microseconds. However, the HVDC CBs work with lower currents.

Table VIII: Performance of the ROCOV algorithm against high resistance PtG fault conditions.

Link	Fault Location (km)	Fault Resistance (Ω)	R_AB			R_BA		
			Detection time (ms)	Operation time (ms)	Interrupted Current (kA)	Detection time (ms)	Operation time (ms)	Interrupted Current (kA)
L12	0	182	0.01	2.01	0.724	0.57	2.57	1.068
	36	19	0.21	2.21	2.996	0.36	2.36	2.948
	84	193	0.49	2.49	0.767	0.1	2.1	1.012
	100	97	0.56	2.56	1.392	0.01	2.01	1.530
L13	0	25	0.01	2.01	3.350	1.11	3.11	1.595
	67	55	0.38	2.38	2.571	0.75	2.75	1.530
	142	31	0.79	2.79	3.354	0.33	2.33	1.329
	200	160	1.13	3.13	1.647	0.01	2.01	0.090
L14	0	183	0.01	2.01	1.224	1.13	3.13	0.277
	9	109	0.06	2.06	1.584	1.09	3.09	0.689
	153	195	0.93	2.93	1.268	0.27	2.27	0.171
	200	28	1.11	3.11	3.187	0.01	2.01	1.976
L24	0	127	0.01	2.01	1.756	0.84	2.84	0.384
	65	192	0.38	2.38	1.544	0.5	2.5	0.016
	121	15	0.68	2.68	3.607	0.17	2.17	2.554
	150	84	0.84	2.84	2.134	0.01	2.01	0.621
L34	0	7	0.01	2.01	2.799	0.56	2.56	4.298
	22	187	0.13	2.13	0.354	0.45	2.45	1.470
	70	170	0.4	2.4	0.452	0.18	2.18	1.483
	100	136	0.56	2.56	0.594	0.01	2.01	1.650

Pole-to-Pole fault conditions

Once again, the ROCOV algorithm was able to detect all the simulated fault conditions regardless of the fault resistance and its detection time is barely affected by this parameter, as it can be seen comparing Tables VII and IX. Meanwhile, the HVDC CBs interrupt higher currents than during high resistance PtG fault conditions but lower than during low resistance PtP fault conditions.

Table IX: Performance of the ROCOV algorithm against high resistance PtP fault conditions.

Link	Fault Location (km)	Fault Resistance (Ω)	R_{AB}			R_{BA}		
			Detection time (ms)	Operation time (ms)	Interrupted Current (kA)	Detection time (ms)	Operation time (ms)	Interrupted Current (kA)
L12	0	87	0.01	2.01	2.429	0.56	2.56	2.853
	17	35	0.1	2.1	3.475	0.47	2.47	3.832
	65	194	0.37	2.37	1.535	0.2	2.2	1.871
	100	7	0.56	2.56	4.246	0.01	2.01	4.816
L13	0	67	0.01	2.01	3.284	1.11	3.11	1.921
	79	15	0.44	2.44	4.875	0.67	2.67	3.490
	135	105	0.76	2.76	2.934	0.37	2.37	1.251
	200	176	1.12	3.12	2.256	0.01	2.01	0.610
L14	0	35	0.01	2.01	3.840	1.11	3.11	3.094
	92	145	0.52	2.52	2.217	0.61	2.61	1.252
	140	29	0.78	2.78	4.343	0.34	2.34	3.107
	200	44	1.11	3.11	3.771	0.01	2.01	2.587
L24	0	82	0.01	2.01	3.236	0.84	2.84	1.836
	44	171	0.25	2.25	2.454	0.6	2.6	0.915
	91	25	0.51	2.51	4.582	0.33	2.33	3.149
	150	99	0.84	2.84	3.151	0.01	2.01	1.399
L34	0	129	0.01	2.01	1.648	0.56	2.56	2.661
	47	52	0.27	2.27	2.728	0.3	2.3	3.794
	81	94	0.46	2.46	2.106	0.12	2.12	3.070
	100	39	0.56	2.56	3.045	0.01	2.01	4.132

Comparison

From the simulations, the ROCOV algorithm presents a detection speed of 0.01 ms for faults located near the relay position and of 1.11 ms for remote fault conditions. In the case of high resistance fault conditions (up to 200 Ω), the detection time is slightly increased (in the range of tens of microseconds) in comparison with the case of low resistance faults.

This algorithm is widely found in the literature. The ROCOV algorithm is analysed in reference [21] in a bipole nine-terminal system, using 0.1 H limiting inductors and HVDC CBs in each end of every link. The ROCOV threshold varies between relays in the range of -0.2 and -4 MV/ms and the ROCOV algorithm is used to detect faults up to 10 Ω (an overcurrent threshold is used for higher fault resistances). This way, a fault located at 120 km from the relay is detected in approximately 1 ms.

Similarly, the ROCOV algorithm is implemented in a bipole three-terminal system in reference [22]. Once again, 0.1 H limiting inductors are installed at both ends of each link and the ROCOV thresholds are -0.8, -2 and -5 MV/ms for the relays located at the 500, 1500 and 100 km links, respectively. The ROCOV can detect a fault located 150 km away from the relay in around 1 ms but its performance for high resistance faults is not analysed.

Likewise, the ROCOV algorithm is analysed in the same system in reference [23] but varying the size of the limiting inductors in each link. Limiting inductors of 0.1 and 0.125 H are employed in the 100 and 500 km link, respectively, while 0.05 and 0.1 H inductors are installed in the longest link of 1500 km. The ROCOV thresholds are in the range of -0.45 and 2.65 MV/ms for each relay and the algorithm performance was tested for fault resistances up to 200 Ω. The ROCOV algorithm presented difficulties in detecting faults with a resistance higher than 60 Ω, hence a directional features was added. This algorithm could detect a fault condition located at 120 km in 0.05 ms in the case of a solid fault and in 0.2 ms in the case of a 50 Ω fault resistance.

Another analysis of the ROCOV algorithm is presented in [24] carried out in a bipole three-terminal grid with four links. In this case, limiting inductors with a size of 0.04 H are selected while the ROCOV thresholds are in the range of -0.15 and -1.73 MV/ms. Its performance is tested against fault resistances up to 100 Ω and presents a detection time of 0.2 ms.

Lastly, the ROCOV algorithm proposed in reference [25] is implemented in the same grid presented in this paper, a symmetric monopole four-terminal system, with a ROCOV threshold of -0.2 MV/ms. Its performance was analysed for both PtP and PtG, low and high resistance fault conditions. It was able to detect PtP faults with a resistance up to 750 Ω and PtG faults with up to 400 Ω, in the range of 0.005 and 1.1 ms.

The proposed ROCOV algorithm in the presented paper shows a relatively faster fault detection than other ROCOV algorithms presented in the literature due to its lower threshold. In addition, common 0.1 H limiting inductors are used and the proper performance of this algorithm against high resistance fault conditions is analysed, being accurate and hardly slower than for low resistance faults (in the range of tens of microseconds).

Conclusions

A ROCOV based protection system following a full-selective fault clearing strategy is proposed in this paper. The ROCOV threshold value is selected after evaluating the behaviour of the DC voltage derivative with different fault conditions, varying the fault type, the fault location and the fault resistance. An appropriate threshold value of -0.3 MV/ms is chosen through simulations.

Then, the influence of the fault type, fault location and fault resistance on the performance of the ROCOV algorithm is analysed in a 4-terminal half-bridge MMC-based HVDC system modelled in PSCAD software. It can be extracted from the results of the simulations that the ROCOV algorithm presents great detection speed, from 10 µs for the closest faults to around 1 ms for remote fault.

In addition, its detection time is not affected by the fault type. However, the current reaches higher level during PtP fault conditions. On the other hand, the detection time is hardly affected by the fault resistance, being higher during high resistance fault conditions only for a few tens of microseconds. Conversely, the HVDC CBs have to interrupt higher currents during low resistance fault conditions. Moreover, the ROCOV algorithm shows great performance against high resistance faults since it can detect faults up 200 Ω, due to the fault-induced sharp voltage drop. In addition, a brief comparison with other ROCOV algorithms found in the literature is presented, from which is concluded that the proposed ROCOV presents a faster operation speed, due to the selection of a lower threshold value, and a better performance against high resistance faults.

References

[1] Q. Yang, S. Le Blond, R. Aggarwal, Y. Wang and J. Li, "New ANN method for multi-terminal HVDC protection relaying", Electric Power Systems Research, vol. 148, pp. 192–201, 2017.

[2] J. Descloux, P. Rault, S. Nguefeu, J. Curis, X. Guillaud, F. Colas and B. Raison, "HVDC meshed grid: Control and protection of a multi-terminal HVDC system", in CIGRÉ Session Paris, Paris, France, 26-31 August 2012, pp. 10.

[3] M. Heidemann, D. Eichhoff, C. Petino, M. Stumpe, E. Spahic and F. Schettler, "A systematic study on fault currents in multiterminal HVDC grids", in [Cigré - Lund Symposium, 27.05.2015-28.05.2015, Lund, Sweden], May. Available: http://publications.rwth-aachen.de/record/478751.

[4] R. Li and L. Xu, "Review of DC fault protection for HVDC grids", WIREs Energy and Environment, vol. 7, (2), 2018.

[5] S. Le Blond, R. Bertho, D. V. Coury and J. C. M. Vieira, "Design of protection schemes for multi-terminal HVDC systems", Renewable and Sustainable Energy Reviews, vol. 56, pp. 965-974, 2016.

[6] S. Azazi, M. Sanaye-Pasand, M. Abedini and A. Hasani, "A Traveling-Wave-Based Methodology for Wide-Area Fault Location in Multiterminal DC Systems", IEEE Transactions on Power Delivery, vol. 29, (6), pp. 2552-2560, 2014.

[7] W. Leterme and D. Van Hertem, "Classification of fault clearing strategies for HVDC grids", in CIGRÉ Lund Symposium, Lund, Sweden, 27-28 May 2015.

[8] CIGRÉ WG B4.52, "HVDC Grid Feasibility Study", Cigré, 2013. Available: https://e-cigre.org/publication/533-hvdc-grid-feasibility-study (accessed on 19 December 2019).

[9] M. E. Baran and N. R. Mahajan, "Overcurrent Protection on Voltage-Source-Converter-Based Multiterminal DC Distribution Systems", IEEE Transactions on Power Delivery, vol. 22, (1), pp. 406-412, 2007.

[10] J. Descloux, B. Raison and J. Curis, "Protection strategy for undersea MTDC grids", in 2013 IEEE Grenoble Conference, Grenoble, France, 16-20 June 2013.

[11] M. K. Bucher and C. M. Franck, "Fault Current Interruption in Multiterminal HVDC Networks", IEEE Transactions on Power Delivery, vol. 31, (1), pp. 87-95, 2016.

[12] W. Leterme, I. Jahn, P. Ruffing, K. Sharifabadi and D. Van Hertem, "Designing for High-Voltage dc Grid Protection: Fault Clearing Strategies and Protection Algorithms", IEEE Power and Energy Magazine, vol. 17, (3), pp. 73-81, 2019.

[13] E. Spahic, D. Ergin, F. Schettler, J. Dorn and C. Petino, "A closer look at protection concepts for DC systems", in Cigré 2016, Paris, France, 21-26 August 2016.

[14] D. Jovcic, G. Tang and H. Pang, "Adopting Circuit Breakers for High-Voltage dc Networks: Appropriating the Vast Advantages of dc Transmission Grids", IEEE Power and Energy Magazine, vol. 17, (3), pp. 82-93, 2019.

[15] V. Psaras, A. Emhemed, G. Adam and G. M. Burt, "Review and evaluation of the state of the art of DC fault detection for HVDC grids", in 2018 53rd International Universities Power Engineering Conference (UPEC), Glasgow, Scotland, 4-7 September 2018.

[16] I. Jahn, N. Johannesson and S. Norrga, "Survey of methods for selective DC fault detection in MTDC grids", in 13th IET International Conference on AC and DC Power Transmission (ACDC 2017), Manchester, UK, 14-16 February 2017, pp. 1-7.

[17] G. Buigues, V. Valverde, I. Zamora, D. M. Larruskain, O. Abarrategui and A. Iturregi, "DC fault detection in VSC-based HVDC grids used for the integration of renewable energies", in 2015 International Conference on Clean Electrical Power (ICCEP), Taormina, Italy, 16-18 June 2015, pp. 666-673.

[18] R. E. Torres-Olguin and H. K. Høidalen, "Inverse time overcurrent protection scheme for fault location in multi-terminal HVDC", in 2015 IEEE Eindhoven PowerTech, Eindhoven, Netherlands, 29 June-2 July 2015.

[19] J. Häfner and B. Jacobson, "Proactive hybrid HVDC breakers - A key innovation for reliable HVDC grids", in The Electric Power System of the Future - Integrating Supergrids and Microgrids International Symposium, Bologna, Italy, 13-15 September 2011.

[20] W. Leterme, N. Ahmed, J. Beerten, L. Ängquist, D. V. Hertem and S. Norrga, "A new HVDC grid test system for HVDC grid dynamics and protection studies in EMT-type software", in 11th IET International Conference on AC and DC Power Transmission, Birmingham, UK, 10-12 Feb. 2015.

[21] J. Sneath and A. D. Rajapakse, "DC fault protection of a nine-terminal MMC HVDC grid", in 11th IET International Conference on AC and DC Power Transmission, Birmingham, UK, 10-12 February 2015, pp. 1-8.

[22] J. Sneath and A. D. Rajapakse, "Fault Detection and Interruption in an Earthed HVDC Grid Using ROCOV and Hybrid DC Breakers", IEEE Transactions on Power Delivery, vol. 31, (3), pp. 973-981, 2016.

[23] N. M. Haleem and A. D. Rajapakse, "Application of new directional logic to improve DC side fault discrimination for high resistance faults in HVDC grids", Journal of Modern Power Systems and Clean Energy, vol. 5, (4), pp. 560-573, 2017.

[24] N. M. Haleem and A. D. Rajapakse, "Local measurement based ultra-fast directional ROCOV scheme for protecting Bi-pole HVDC grids with a metallic return conductor", International Journal of Electrical Power & Energy Systems, vol. 98, pp. 323–330, 2018.

[25] M. J. Pérez Molina, D. M. Larruskain, P. Eguía López and A. Etxegarai, "Analysis of Local Measurement-Based Algorithms for Fault Detection in a Multi-Terminal HVDC Grid", Energies, vol. 12, (24), 2019.

Modelling of converter systems paralleled via interphase transformers in cyclic cascade topology and optimization of PWM carrier shifts

D. Basic, H. Baërd, S. Siala

General Electric, Power Conversion, 18 Av. de Québec, Villebon-sur-Yvette, 91140, France

e-mail: duro.basic@ge.com , henri.baerd@ge.com , sami.siala@ge.com

URL: https://www.gepowerconversion.com/

Keywords

Electrical drive, Parallel operation, Interleaved converters, Multilevel converters.

Abstract

High power converters can be constructed by paralleling N interleaved converters via Interphase Transformers (ITRs) to produce multilevel output voltages while minimizing circulating currents among converters. With the ITRs connected in cyclical cascade topology, the ITR voltage stress is highly dependent on number of parallel converters and phase shifts between Pulse Width Modulation (PWM) triangular carriers. To analyse the ITR system performance a model of cross-coupling produced by ITRs is needed to establish links between the converter voltages and ITR voltages and cross currents circulating among converters. The main goal of this paper is to give a comprehensive overview of a model of the system based on the ITR cyclical cascade topology which can be used for estimation of its performance. Based on this model, opportunities for minimisation of the ITR voltages and core losses, via optimization of the converter PWM carrier shifts, are identified and discussed. Application of the theoretical results is illustrated in an example case with $N=5$ converters.

Introduction

In high power applications a number N of identical 2-level or 3-level Neutral Point Clamped (NPC) or Neutral Point Piloted (NPP) 3-phase ac/dc voltage source converters can be connected in parallel to increase power rating of the overall aggregated converter, beyond rating of individual converters [1]. Paralleling of the converters requires utilization of paralleling inductors to limit high frequency currents flowing among converters which are produced by staggered PWM switching. To increase the effective switching frequency and control bandwidth, and to reduce output voltage switching harmonics, the PWM triangular carriers of individual converters are normally interleaved [2]. With the interleaved PWM, large paralleling inductors are required. However, they produce voltage drops which can become prohibitively high, particularly if a transformer is used for the converter system coupling with a motor or ac grid. In order to reduce size of the paralleling inductors and related voltage drop it is advantageous to use ITRs [3] which provide large (magnetizing) inductance for the currents flowing among converters (cross-currents) while introducing negligible (leakage) inductance for the currents flowing from converters to the machine or ac grid (cumulative currents). The ITRs can be designed as single-phase or 3-phase units. The 3-phase, 3-core ITR design is more cost effective but it does not provide high inductance for the common mode cross currents. Hence it is applicable only when the converter dc buses, or ac terminals are mutually isolated [4]. When the converter dc buses are mutually paralleled a 4 or 5 core 3 phase ITR or single phase ITR designs must be used. The ITR electrical/magnetic stress is highly dependent on the ITR topology [3] and the converter voltage (supply) system. The theoretical framework for analysis of the ITR systems based on Fortesque's transformation and the system decomposition into the symmetrical components has been presented in [5] and applied for optimization of the supply voltage system [6].

In this paper a phase domain model of the ITR system connected in the cyclical cascade topology is presented, together with a comprehensive overview of the related voltage/current matrix transformations. By using the eigenvalue/vector analysis of the mapping matrices, effects of the PWM carriers phase shifts on the ITR stress are investigated. A procedure for minimization the ITR core losses via optimization of the PWM carrier phase shifts is presented. The theoretical results are illustrated by an example case with N=5 converters.

ITR voltages mapping and cross voltages

In this section mapping between the converter phase $V_{conv\ u,v,w\ i}$, and ITR $V_{ITR\ u,v,w\ i}$ (i=1..N) voltages is derived for a general case with N converters paralleled via cyclical cascaded connection of ITRs (Fig. 1). The ITRs used in this topology are 3-phase transformers with the transformation ratio 1:1, which magnetizing inductance L_m. is adjusted by appropriate airgaps and has a relatively high value ($\geq$ 1p.u.). The relatively small transformer leakage inductance (L_σ) can be neglected in the ITR winding voltage calculations. Typically, the converter voltages are generated by identical modulating references using the PWM triangular carriers which are mutually phase shifted (interleaved). Phase shifts of the triangular PWM carriers can be arbitrary set (among the converters and/or phases). However, certain combinations of the PWM carrier phase shifts are advantageous when minimizations of the ITR voltages, core losses or peak flux densities are of interest. The converter dc buses can be mutually isolated. Then common mode voltages of the supply system are not supported by ITRs. In the case with common dc bus (doted lines in Fig. 1) the ITRs must provide high common mode impedances to limit common mode currents. For that the 4-core (shown in Fig. 1 with dotted lines), 5-core 3-phase ITR or single phase ITRs must be used. From Fig. 1, assuming system symmetry, a set of N voltage balance equations (identical for each phase, u,v,w) can be written:

$$v_{\Sigma u,v,w} = v_{Conv\ u,v,w1} - v_{ITR\ u,v,w1} + v_{ITR\ u,v,w2}$$

$$v_{\Sigma u,v,w} = v_{Conv\ u,v,w2} - v_{ITR\ u,v,w2} + v_{ITR\ u,v,w3}$$

$$\dots\dots$$

$$v_{\Sigma u,v,w} = v_{Conv\ u,v,wN} - v_{ITR\ u,v,wN} + v_{ITR\ u,v,w1} \tag{1}$$

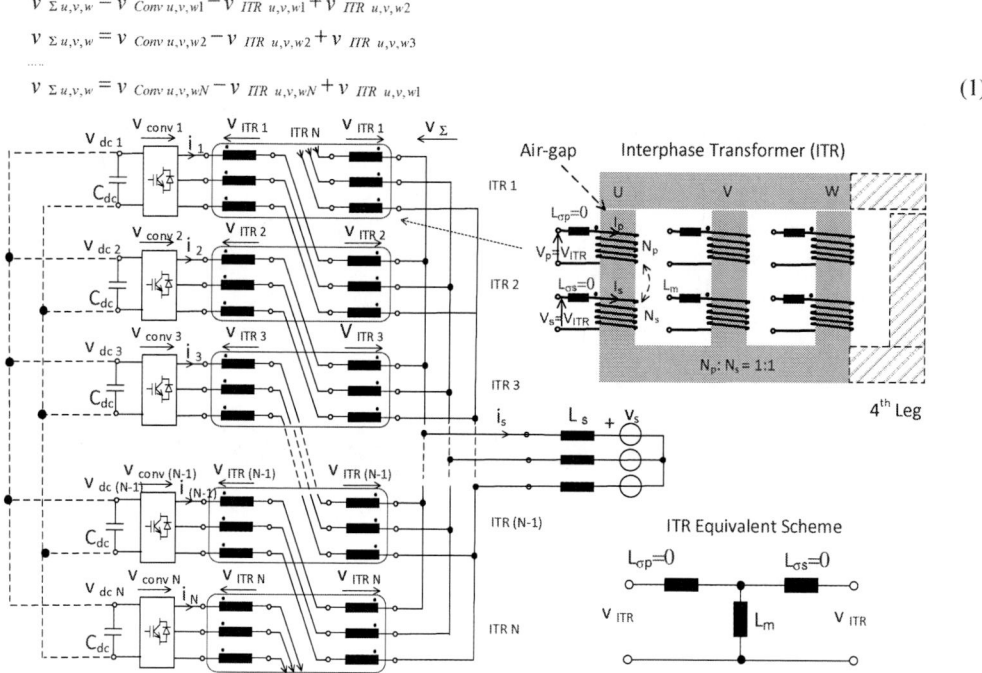

Fig. 1 : Aggregated ac/ac converter based on N converters paralleled via ITRs in cyclical cascade topology.

By summing the N voltage balance equations (1), the cumulative voltage component of all N parallel converters (indexed by Σ) can be found as an average of all converter voltages:

$$v_{\Sigma u,v,w} = \sum_{i=1}^{N} v_{Conv\ u,v,wi} \bigg/ N \tag{2}$$

The residual converter voltage components, driving the cross currents among the converters, are:

$$\Delta v_{Conv\ u,v,wi} = v_{Conv\ u,v,wi} - v_{\Sigma u,v,w} \tag{3}$$

In the further text indexes u,v,w (due to symmetry) will be omitted. In the matrix form, the voltage balance equations can be expressed in the following way:

$$\left[\Delta v_{Conv\ 1} \quad \Delta v_{Conv\ 2} \quad \cdots \quad \Delta v_{Conv\ (N-1)} \quad \Delta v_{Conv\ N} \right]^T = [C]\left[v_{ITR\ 1} \quad v_{ITR\ 2} \quad \cdots \quad v_{ITR\ (N-1)} \quad v_{ITR\ N} \right]^T \tag{4}$$

Here the ITR topology related inter-connection matrix [C] is introduced:

$$[C] = \begin{bmatrix} 1 & -1 & 0 & \cdots & 0 & 0 \\ 0 & 1 & -1 & \cdots & 0 & 0 \\ 0 & 0 & 1 & \cdots & 0 & 0 \\ \vdots & \vdots & \vdots & \vdots\vdots\vdots & \vdots & \vdots \\ 0 & 0 & 0 & \cdots & 1 & -1 \\ -1 & 0 & 0 & \cdots & 0 & 1 \end{bmatrix} \tag{5}$$

This interconnection matrix [C] defines how the ITR voltages are combined and inserted in series with the converter voltages when paralleling the converters. The inverse mapping is described in (6) by the matrix [T].

$$[\Delta v_{conv}] = [C][v_{ITR}] \qquad\qquad [v_{ITR}] = [T][\Delta v_{conv}] = [T][\Delta v_{conv}] \tag{6}$$

It is important to emphasize that the matrices $[C]$ and $[T]$ (defining the direct and inverse voltage mapping) are not directly invertible. The inverse mapping defined in the Appendix as $[T] \equiv [C]^{Inv}$ can be found via reduction of the system dimension. It should be noticed that the matrix $[T]$ rejects the cumulative components is the same way as the difference matrix $[C]$ from which it is constructed. Thus, the ITR voltages can be related directly the converter voltages in (6) without explicit removal of the cumulative component v_Σ.

ITR inductance matrix and cross currents

For magnetization of magnetic cores of ITRs, in order to produce voltages V_{ITRi}, supporting instantaneous differences between the converter voltages, certain magnetizing currents in each ITR core are needed. The required magnetizing currents can be deduced from Fig. 2 and expressed by (7):

$$[v_{ITR}] = L_m \frac{d}{dt}[i_{m1} \quad i_{m2} \quad \cdots \quad i_{m(N-1)} \quad i_{mN}]^T = L_m \frac{d}{dt}[i_1 - i_N \quad i_2 - i_1 \quad \cdots \quad i_{(N-1)} - i_{(N-2)} \quad i_N - i_{(N-1)}]^T \tag{7}$$

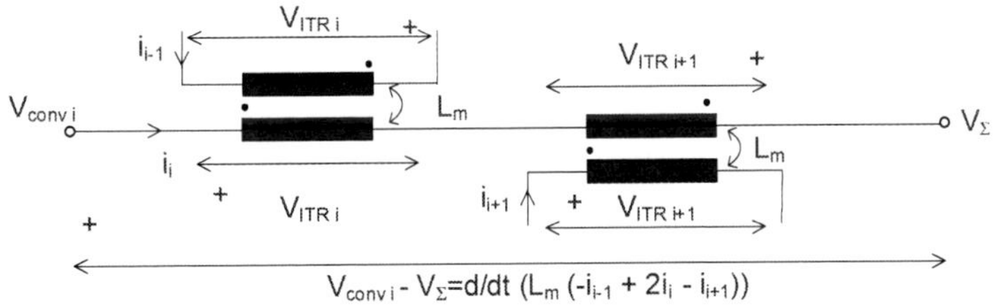

Fig. 2 : Expanded view of interconnection of one converter phase to the common paralleling point.

Similarly, as with the converter voltages, it is convenient to split the converter currents into the cumulative (i_Σ) and cross (Δi_i) current components:

$$i_\Sigma = \sum_{i=1}^{N} i_i / N \qquad\qquad \Delta i_i = i_i - i_\Sigma \quad i=1\ldots N \tag{8}$$

Due to the ITR topology (Fig. 2), only the cross currents produce magnetization the ITR cores, net effect of magnetizations by the cumulative components is zero. This is expressed by the following equation:

$$\begin{bmatrix} v_{ITR\ 1} \\ v_{ITR\ 2} \\ v_{ITR3} \\ \vdots \\ v_{ITR\ (N-1)} \\ v_{ITR\ N} \end{bmatrix} = L_m \begin{bmatrix} 1 & 0 & 0 & \cdots & 0 & -1 \\ -1 & 1 & 0 & \cdots & 0 & 0 \\ 0 & -1 & 1 & \cdots & 0 & 0 \\ \vdots & \vdots & \vdots & \vdots\vdots\vdots & \vdots & \vdots \\ 0 & 0 & 0 & \cdots & 1 & 0 \\ 0 & 0 & 0 & \cdots & -1 & 1 \end{bmatrix} \frac{d}{dt} \begin{bmatrix} \Delta i_1 \\ \Delta i_2 \\ \Delta i_3 \\ \vdots \\ \Delta i_{(N-1)} \\ \Delta i_N \end{bmatrix} = L_m [C]^T \frac{d}{dt} \begin{bmatrix} \Delta i_1 \\ \Delta i_2 \\ \Delta i_3 \\ \vdots \\ \Delta i_{(N-1)} \\ \Delta i_N \end{bmatrix} \tag{9}$$

A direct link between the converter cross voltages and currents can be also derived (10). By inspection of the matrices $[C]^T$ in (9) and $[C][C]^T$ in (10) it can be seen that the cumulative components in the input vectors have no effect on the outputs of these matrix transformation. Thus $[\Delta i]$ can be replaced by $[i]$ in (9) and (10) without explicit removal of the cumulative component i_Σ.

$$
\begin{bmatrix} \Delta v_{conv\,1} \\ \Delta v_{conv\,2} \\ \Delta v_{conv\,3} \\ \vdots \\ \Delta v_{conv\,(N-1)} \\ \Delta v_{conv\,N} \end{bmatrix} = L_m [C][C]^T \frac{d}{dt} \begin{bmatrix} \Delta i_1 \\ \Delta i_2 \\ \Delta i_3 \\ \vdots \\ \Delta i_{(N-1)} \\ \Delta i_N \end{bmatrix} = [\Lambda] \frac{d}{dt} \begin{bmatrix} \Delta i_1 \\ \Delta i_2 \\ \Delta i_3 \\ \vdots \\ \Delta i_{(N-1)} \\ \Delta i_N \end{bmatrix} \quad \text{where } [C][C]^T = \begin{bmatrix} 2 & -1 & 0 & \cdots & 0 & -1 \\ -1 & 2 & -1 & \cdots & 0 & 0 \\ 0 & -1 & 2 & \cdots & 0 & 0 \\ \vdots & \vdots & \vdots & \cdots & \vdots & \vdots \\ 0 & 0 & 0 & \cdots & 2 & -1 \\ -1 & 0 & 0 & \cdots & -1 & 2 \end{bmatrix} \quad (10)
$$

The effective inductance matrix, seen by the converter cross-voltage components, is defined by the ITR inductance matrix $[\Lambda]$. This inductance matrix links the ITR magnetizing voltages with the cross or total converter currents:

$$
[\Lambda] = L_m [C][C]^T \qquad \left[\Delta v_{conv} \right]^{\cdot\cdot} = [\Lambda] \frac{d}{dt} [\Delta i] = [\Lambda] \frac{d}{dt} [i] \qquad (11)
$$

Summary of the ITR system model

The ITR system model and related matrix transformations in the phase domain are shown in Fig. 3.

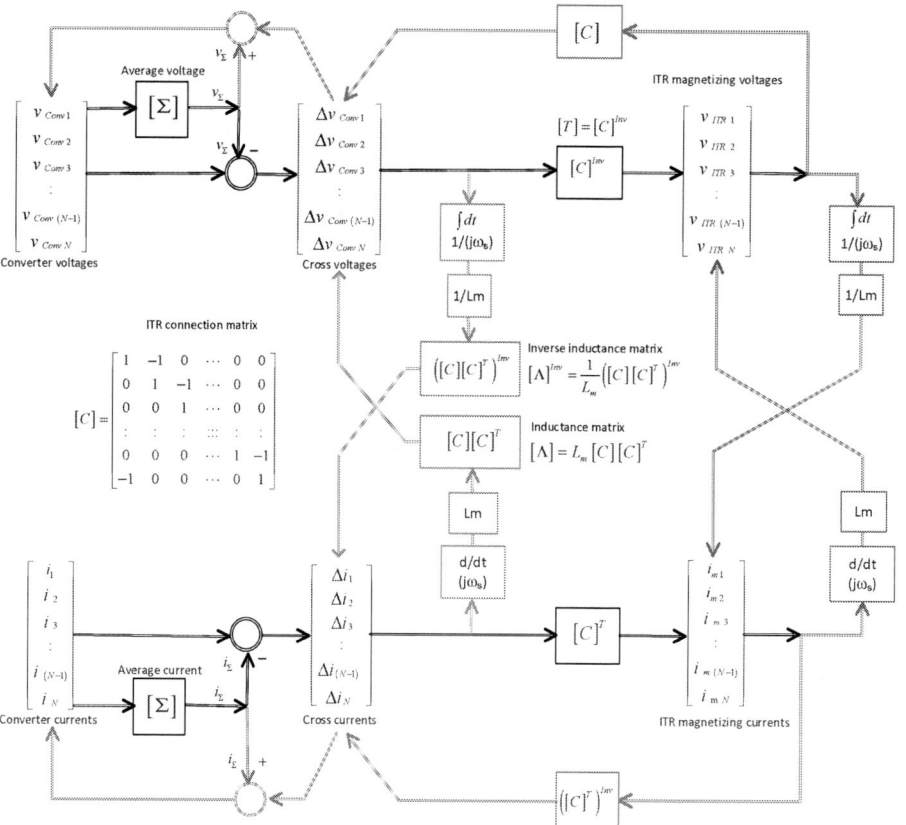

Fig. 3: Overview of the ITR system model and related matrix transformations
(ITRs connected in cyclical cascade topology).

Eigenvalues of voltage mapping and ITR inductance matrices

Eigenvalues of the voltage mapping matrix $[T]$ ($[\lambda_T] = eigen([T])$) for the drive systems with $N=2\ldots10$ converters/ITRs are given in Table I. These eigenvalues define scaling of the mapping of the voltage converter harmonics into the ITR magnetizing voltages along characteristic eigen-vector directions (which constitute base for the system decomposition into the symmetrical components using the Fortesque transform [6]). In other words, they define how the converter voltage components, located in different switching bands, project into the ITR voltage components. The first eigenvalue in Table I, associated to the cumulative components, is zero ($\lambda_{T1}=0$). The second eigenvalue λ_{T2}, associated to the mapping of the first harmonic switching band (assuming regular PWM carrier shifts by $2\pi/N$), indicates that magnitude of the ITR harmonic voltages can have values between 50% to 161.8% of the converter harmonic voltages (depending on N). In the best case, with $N=2$ and $2\pi/N=\pi$ (case with the harmonic vectors, produced by two neighbourhood converters, in counterphase), the ITR voltage components are only 50% of the converter voltages. When the phase shifts between the harmonic voltage vectors in subsequent converter voltage spectra are significantly lower than π, the ITR voltages are relatively increased. With the phase shifts lower than $\pi/6$, the ITR harmonic voltages become higher than the converter voltages. For example, with $N=10$, phase shifts between the subsequent harmonic voltage vectors are only $2\pi/N=\pi/5$. As the result, the ITR voltage harmonics increase to 161.8 % of the converter harmonic voltages.

Table I: Eigenvalues of the voltage mapping matrices $[T]$ for N=2 to N=10.

System N	Gains along eigen-vector directions									
	λ_{T1}	λ_{T2}	λ_{T3}	λ_{T4}	λ_{T5}	λ_{T6}	λ_{T7}	λ_{T8}	λ_{T9}	λ_{T10}
2	0	0.5								
3	0	0.5774	0.5774							
4	0	0.7071	0.7071	0.5						
5	0	0.8506	0.8506	0.5257	0.5257					
6	0	1	1	0.5774	0.5774	0.5				
7	0	1.1523	1.1523	0.6394	0.6394	0.5129	0.5129			
8	0	1.3066	1.3066	0.7071	0.7071	0.5412	0.5412	0.5		
9	0	1.462	1.462	0.7779	0.7779	0.5774	0.5774	0.5077	0.5077	
10	0	1.618	1.618	0.8507	0.8507	0.618	0.618	0.52576	0.52576	0.5

The eigenvalues of the inductance matrix $[\lambda] = eigen([\Lambda])$ for the systems with $N=2$ to $N=10$ converters/ITRs have been also calculated and they are summarized in Table II.

Table II: Eigenvalues of the ITR inductance matrices for N=2 to N=10.

System N	Characteristic Magnetizing Inductances									
	λ_1	λ_2	λ_3	λ_4	λ_5	λ_6	λ_7	λ_8	λ_9	λ_{10}
2	0	$4L_m$								
3	0	$3L_m$	$3L_m$							
4	0	$2L_m$	$2L_m$	$4L_m$						
5	0	$1.382L_m$	$1.382L_m$	$3.618L_m$	$3.618L_m$					
6	0	L_m	L_m	$3L_m$	$3L_m$	$4L_m$				
7	0	$0.753L_m$	$0.753L_m$	$2.445L_m$	$2.4450L_m$	$3.802L_m$	$3.802L_m$			
8	0	$0.5858L_m$	$0.5858L_m$	$2L_m$	$2L_m$	$3.4142L_m$	$3.4142L_m$	$4L_m$		
9	0	$0.4679L_m$	$0.4679L_m$	$1.6527L_m$	$1.6527L_m$	$3L_m$	$3L_m$	$3.8974L_m$	$3.8794L_m$	
10	0	$0.3820L_m$	$0.3820L_m$	$1.3820L_m$	$1.3820L_m$	$2.6180L_m$	$2.6180L_m$	$3.6180L_m$	$3.6180L_m$	$4L_m$

The eigenvalues in Table II basically indicate the effective ITR inductances seen by different symmetrical subsystems [6]. With the conventional phase shift of the PWM carriers by $2\pi/N$, the highest value of the effective impedance seen by the voltage system of the first switching band, with the phase shift of the PWM triangular carriers of π ($2\pi/N$, $N=2$), is as high as $\lambda_2=4L_m$. In the system with larger number of drives, the conventional PWM carrier shift by $2\pi/N$ results in a significantly reduction of the effective magnetizing inductance seen by the first switching side band system. For example, with $N=10$, the effective inductance is only 38.2% of L_m. From these results we can conclude that it could be beneficial (for minimization of the ITR magnetic core flux and losses or/and cross-current ripple) to re-arrange the PWM triangular carrier shifts in this way that phase shifts of the dominant PWM switching harmonic voltage vectors in the subsequent converter voltage spectra are phase shifted by the highest possible angle to achieve highest magnetizing inductance and lowest cross-current harmonics. However, this may not be possible to achieve in a symmetrical way (in all ITRs) with an arbitrary number N.

Estimation of ITR core losses

Normally, the ITR core losses have the biggest impact on the ITR design (cost) i.e. selection of maximal magnetic flux densities in the ITR cores. To reduce power losses, the ITRs are operated with low flux densities in magnetic cores which are well below saturation point of the ITR core material. Only in case when the converters are switched at low switching frequencies, it may be required to check peak values of fluxes in the ITR cores to avoid magnetic saturation. Thus, for the ITR design optimization, it is important to have an estimate of losses produced in the ITR cores by the converter PWM voltages. In this section a simplified procedure for the ITR loss estimation is presented. We start with the specific power loss in magnetic cores which is defined by the Steinmetz's equation [7]:

$$p_{Fe} = k_c \, B_m^{\,\beta} f^{\,\alpha} \; [\text{W/m}^3] \qquad\qquad 1 < \alpha < 3 \quad 2 < \beta < 3 \qquad\qquad (12)$$

where parameters kc, α and β depend on the core material and design (for example thickness of laminations). Values of the parameters α and β should be provided by the ITR manufacturer. If they are not known (during initial phase of creating ITR specification), we can adopt the following typical values for these parameters:

$$\alpha = 1.6 \qquad \beta = 2 \qquad\qquad (13)$$

The Steinmetz's equation is derived for single frequency (f) sinusoidal excitation. In presence of complex excitation (composed of multiple flux harmonics) this equation is extended to give an approximate estimate of the ITR core losses produced by multiple harmonics:

$$p_{Fe} = k_c \sum_h B_{m\,h}^{\,\beta} f_h^{\,\alpha} \qquad\qquad (14)$$

For characterisation of the core losses produced with selected PWM strategy/interleaving it is convenient to deal with harmonic voltages ($V_{m\,h}$) rather than with the flux densities which are unknown before the ITR core cross-section area (S_{Fe}) and number of turns in its windings (N_{turns}) are defined:

$$p_{Fe} = k_c \sum_h \left(\frac{1}{N_{turns} S_{Fe}} \frac{V_{m\,h}}{2\pi f_h} \right)^{\beta} f_h^{\,\alpha} = k_c' \sum_h V_{m\,h}^{\,\beta} f_h^{\,\alpha-\beta} \text{ where } k_c' = \frac{k_c}{\left(2\pi N_{turns} S_{Fe}\right)^{\beta}} \qquad\qquad (15)$$

The harmonic spectrum of the voltages across ITR can be readily obtained using the Fast Fourier Transform (FFT) of the ITR voltages:

$$\sum_h \mathbf{V}_{ITR\,h} = FFT\left(v_{ITR}\right) \qquad\qquad (16)$$

The core losses are preferably characterized using the rms values of harmonic voltages instead of peak values (using $k_c'' = k_c' \, 2^{\beta/2}$):

$$p_{Fe} = k_c'' \sum_h V_{ITR\,rms\,h}^{\,\beta} f_h^{\,\alpha-\beta} \qquad\qquad (17)$$

According to this simplified model, the core losses can be characterised by a sum of $V_{rms\,h}^{\,\beta} f_h^{\,\alpha-2}$ terms considering all significant ITR voltage harmonics:

$$p_{Fe} \propto \Gamma_{Fe} = \sum_h V_{ITR\,rms\,h}^{\,\beta} f_h^{\,\alpha-\beta} \qquad\qquad (18)$$

This sum can be readily calculated by the converter system designer even before the ITR design is completed. Thus, it can be provided in advance as a design input to the ITR manufacturer for the initial ITR design. The characteristic loss figure produced by selected PWM strategy estimated using the typical values of the parameters α=1.6 and β=2 is:

$$\Gamma_{Fe} = \sum_h V_{ITR\,rms\,h}^{2} f_h^{-0.4} \qquad\qquad (19)$$

During the detailed design, the ITR manufacturer may provide more accurate values for the α and β parameters. It may be required to work closely with the ITR manufacturer and to make several iterations to come up to more precise estimates of the core losses considering the actual ITR core characteristics.

Calculations of the characteristic core losses in ITRs can be performed in the following steps:

Step 1: Select the PWM method and PWM triangular carriers shifts. Calculate the complex voltage spectra (with the phase information preserved) for each converter phase voltage (defined with respect to dc link midpoint).

Step 2: Calculate the ITR rms voltage spectra from the converter voltage spectra using the transformation matrix [T] (number of the converters N):

$$\frac{1}{\sqrt{2}} FFT\left\{\left[\mathbf{v}_{ITR}\right]\right\} = \sum_{h}\left[\mathbf{V}_{ITR\,rms\,h}\right] = \frac{1}{\sqrt{2}} FFT\left\{\left[T\right]\left[\mathbf{v}_{Conv}\right]\right\} = \sum_{h}\left[T\right]\left[\mathbf{V}_{Conv\,rms\,h}\right] \tag{20}$$

Note: The transformation matrix [T] eliminates part of the spectra associated with the cumulative components. The calculated ITR voltage spectra directly relate to the cross-voltage (common and differential mode) components.

Step 3: Calculate specific core losses in ITR phase limbs for each ITRi ($i=1$ -N) and for cores of each phase (u, v and w). In cases when the converters share same dc bus, the ITRs voltages must support the common mode voltages, thus in the loss calculation, the complete spectra of the ITR voltages are used:

$$\Gamma_{Fe\,(u,v,w)\,i} = \sum_{h}\left|\mathbf{V}_{ITR(u,v,w)i\,rms\,h}\right|^{2} f_{h}^{-0.4} \qquad i=1...N \tag{21}$$

With isolated dc busses, the ITRs voltages must support only the differential mode cross voltage components produced by the converters, the common mode voltages are transparent for the ITRs and should be excluded from the core loss calculations:

$$\Gamma_{Fe\,(u,v,w)\,i} = \sum_{h}\left|\mathbf{V}_{ITR(u,v,w)i\,rms\,h} - 1/3\left(\mathbf{V}_{ITR\,ui\,rms\,h} + \mathbf{V}_{ITR\,vi\,rms\,h} + \mathbf{V}_{ITR\,wi\,rms\,h}\right)\right|^{2} f_{h}^{-0.4} \quad i=1...N \tag{22}$$

Step 4: Calculate characteristic core losses in the 4th ITR phase limb for each ITR (in case that the converters share common dc bus):

$$\Gamma_{Fe\,0\,i} = \sum_{h}\left|\mathbf{V}_{ITR\,ui\,rms\,h} + \mathbf{V}_{ITR\,vi\,rms\,h} + \mathbf{V}_{ITR\,wi\,rms\,h}\right|^{2} f_{h}^{-0.4} \qquad i=1...N \tag{23}$$

Effects of PWM carrier shifts-example case for N=5

In this section utilization of the transformations matrix transformations shown in Fig. 3 for calculation of the ITR voltages is illustrated in an example with $N=5$ parallel 3-level NPC converter system with isolated dc bus voltages of 5kV. The 3 Level bus Clamped (3LC) PWM is used based on the level-shifted positive and negative triangular carriers at 900Hz. The relevant voltage transformation $[T]$ and inverse inductance matrices $[\Lambda]^{Inv}$ are ($[C]$ and $[\Lambda]$ have simple forms which can be deduced from (5) and (10), thus they are not shown here):

$$[T] = [C]^{Inv} = \frac{1}{5}\begin{bmatrix} 2 & 1 & 0 & -1 & -2 \\ -2 & 2 & 1 & 0 & -1 \\ -1 & -2 & 2 & 1 & 0 \\ 0 & -1 & -2 & 2 & 1 \\ 1 & 0 & -1 & -2 & 2 \end{bmatrix} \qquad [\Lambda]^{Inv} = \frac{1}{5L_m}\begin{bmatrix} 2 & 0 & -1 & -1 & 0 \\ 0 & 2 & 0 & -1 & -1 \\ -1 & 0 & 2 & 0 & -1 \\ -1 & -1 & 0 & 2 & 0 \\ 0 & -1 & -1 & 0 & 2 \end{bmatrix} \tag{24}$$

When the PWM carriers are regularly shifted by angle $2\pi/5$, harmonics located in the 1st switching band constitute the following balanced positive sequence supply system:

$$\left[\mathbf{V}_{Conv}\right]^{T} = \begin{bmatrix} 1 & \mathbf{a} & \mathbf{a}^{2} & \mathbf{a}^{3} & \mathbf{a}^{4} \end{bmatrix} V \qquad \text{where operator } \mathbf{a} = e^{-j\frac{2\pi}{5}} \tag{25}$$

For balanced voltage excitation, it is straightforward to calculate the ITR voltages (per unit converter voltages) and effective inductance by combining Eqs. (24) and (25).

The ITR voltages created by the voltage system (25) are equal to 0.8506 of the voltages produced by the converters:

$$v_{ITR1} = \frac{1}{5}\left(2 + \mathbf{a} - \mathbf{a}^3 - 2\mathbf{a}^4\right)V = j0.8506\,\mathbf{a}^2 V \tag{26}$$

The effective inductance in this case is $1.382 L_m$:

$$j\omega\Delta\mathbf{i}_{ITR1} = \frac{1}{5L_m}\left(2 - \mathbf{a}^2 - \mathbf{a}^3\right)V = \frac{V}{1.382 L_m} \tag{27}$$

These calculations can be repeated for any other voltage supply system. Of particular interest is the voltage supply system which maximises phase shifts between the voltage vectors produced by the neighbourhood converters. For $N=5$ it is achieved if the consecutive converter voltage phasors are shifted by $\mathbf{a}^2$):

$$\left[v_{Conv}\right] = \left[1 \quad \mathbf{a}^2 \quad \mathbf{a}^4 \quad \mathbf{a} \quad \mathbf{a}^3\right]V \tag{28}$$

With this voltage excitation system, the ITR voltages are reduced to 0.5257 of that produced by the converters:

$$v_{ITR1} = \frac{1}{5}\left(2 + \mathbf{a} - \mathbf{a}^3 - 2\mathbf{a}^4\right)V = -j0.5257\mathbf{a}^4 V \tag{29}$$

In the same the effective inductance is increased to $3.618\,L_m$ ($1.382\times(0.8506/05257)^2=3.168$):

$$j\omega\Delta\mathbf{i}_{ITR1} = \frac{1}{5L_m}\left(2 - \mathbf{a}^4 - \mathbf{a}\right)V = \frac{V}{3.618 L_m} \tag{30}$$

Because different converter voltage systems result in different ratios between the ITR and converter voltages and different effective inductances, it is advantageous to arrange the PWM carriers in such a way that the dominant switching voltage harmonics create voltage systems which minimise the ITR voltages and possibly the overall core losses. Thus, if the harmonics in the 1st switching band are dominant, it could be advantageous to shift the PWM carriers in steps of $4\pi/5$ instead of using the conventional phase shifts of the PWM carriers by one angular step of $2\pi/5$ (Table III).

Table III: Phase shifts of converter PWM triangular carriers (in steps of $2\pi/5$).

Converter	1	2	3	4	5
Regular Shift	0	1	2	3	4
Optimized Shift	0	2	4	1	3

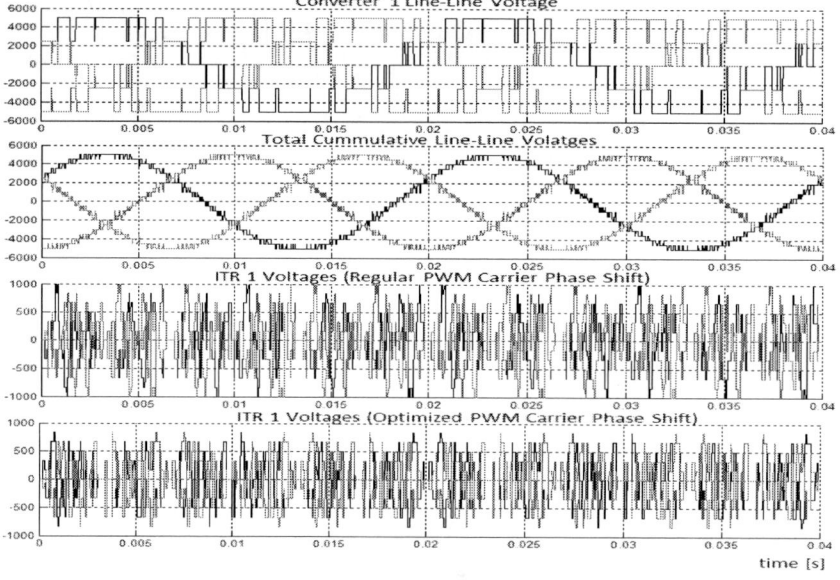

Fig. 4 : Converter 1 PWM voltage, total output voltage of $N=5$ paralleled converters and the ITR 1 voltages (with regular and optimized PWM carrier shifts).

Effect of two different arrangements of the PWM carrier shifts defined in Table III is illustrated by the simulation result shown in Fig. 4 by waveforms of the converter voltages, total average voltage produced by 5 paralleled converters (not affected by changes of PWM carrier shifts), and ITR voltages obtained with the regular and optimized PWM carrier phase shifts. A significant reduction (for about 20%) in the peak values of the ITR voltages with the optimized PWM carrier shift can be observed in the bottom plot of Fig. 4. The frequency spectra of the converter and ITR voltages (Fig. 5) show that that the dominant spectral components of the ITR voltage located in the first switching band (around carrier frequency of 900Hz±100Hz) are equal to 279.9/332.9=0.8408 (with regular PWM carrier shift) and 177.4/332.9=0.5329 (with optimized PWM carrier shift) of the converter voltage. It is important to notice that this significant reduction, consistent with the theoretical results summarized in Table I, is achieved only by reshuffling of the PWM carriers among the converters.

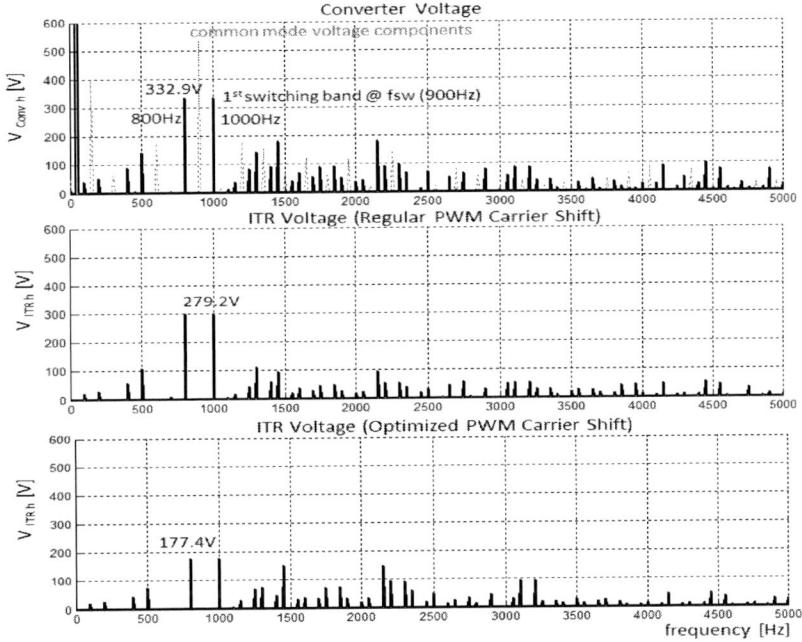

Fig. 5 :   Voltage spectra of the converter and ITR voltages (with regular and optimized PWM carrier shifts).

The PWM carrier shift increases the ITR voltages associated with the PWM harmonics located in the second switching sideband (@1800Hz in Fig. 5). Thus, effect of the PWM carrier shifts on the overall ITR core losses must be checked using the procedure outlined in the previous section. The characteristic core losses calculated in function of the PWM modulation depth M ($M=V^{ref}/ (V_{dc}/2) \times 2/\mathrm{sqrt}(3)$), assuming the PWM carrier frequency of 900Hz and output frequency of f_s=50Hz, are shown in Fig. 6. This result shows that the ITRs core losses with the optimized PWM carrier phase shifts are reduced for $M>0.3$ for approximately 30%.

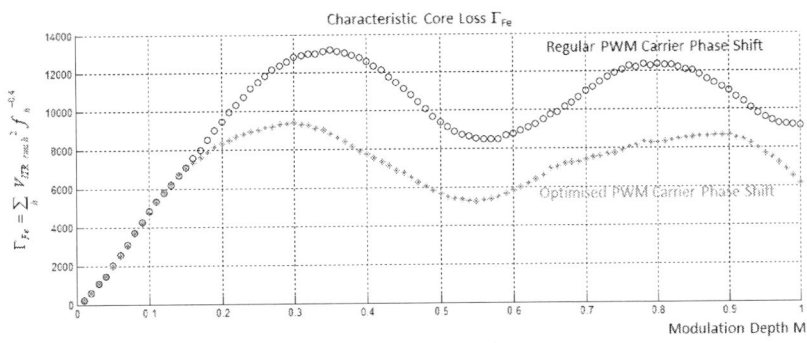

Fig. 6 :   Characteristic core loss in function of mod. depth M with the regular and optimized PWM carrier shifts.

Conclusion

A comprehensive overview of modelling of the converter systems paralleled via ITRs connected in cyclical cascade topology and a procedure for estimation of the ITR core losses has been presented. Several matrix transformations have been introduced which define the effective inductances and voltage stresses of ITRs produced by the converter PWM voltage harmonics. The presented model can serve as a base for the system simulations, ITR sizing & core loss estimation and optimization of the converter PWM carrier phase shifts aimed to minimize the ITR voltage stress and core losses. Application of the theoretical results has been illustrated using an example system with $N=5$ converters.

Appendix

In this Appendix the inverse mapping between ITR and converter voltages is defined ($[C]^{Inv}$):

$$[C][v_{ITR}] = [\Delta v_{Conv}] \qquad\qquad [v_{ITR}] = [C]^{Inv}[\Delta v_{Conv}] \qquad (31)$$

The matrix $[C]$ describing the cyclical cascaded interconnection of ITRs is not invertible (it has one eigenvalue with zero value). To make the inverse mapping to $[C]$ it is needed to reduce its dimension from N to N-1. For reduction of the system dimension the following equation is used:

$$\sum_{i=1}^{N} \Delta v_{ITR\,i} = 0 \qquad (32)$$

In the matrix form this reduction of the system dimension can be expressed using the system dimension reduction matrix $[R]$:

$$
\begin{bmatrix}
\Delta v_{ITR\,1} \\
\Delta v_{ITR\,2} \\
\Delta v_{ITR\,3} \\
\vdots \\
\Delta v_{ITR\,(N-1)} \\
\Delta v_{ITR\,N}
\end{bmatrix}_{N\times 1}
= [R]_{N\times(N-1)}
\begin{bmatrix}
\Delta v_{ITR\,1} \\
\Delta v_{ITR\,2} \\
\Delta v_{ITR\,3} \\
\vdots \\
\Delta v_{ITR\,(N-2)} \\
\Delta v_{ITR\,(N-1)}
\end{bmatrix}_{(N-1)\times 1}
\quad \text{where} \quad
[R]_{N\times(N-1)} =
\begin{bmatrix}
1 & 0 & 0 & \cdots & 0 & 0 \\
0 & 1 & 0 & \cdots & 0 & 0 \\
0 & 0 & 1 & \cdots & 0 & 0 \\
0 & 0 & 0 & \cdots & 0 & 0 \\
0 & 0 & 0 & \cdots & 1 & 0 \\
-1 & -1 & -1 & \cdots & -1 & -1
\end{bmatrix}_{N\times(N-1)}
\qquad (33)
$$

Then the interconnection matrix of the reduced system $[C]_{(N-1)\times(N-1)}$ is invertible. Consequently, the mapping inverse to that defined by the matrix $[C]$ can be found from the following matrix equation:

$$[T] = [C]^{Inv} \equiv [R]\left([R]^T[C][R]\right)^{-1}[R]^T \qquad (34)$$

where superscript T denotes transposition. Similarly, the inverse mapping $[\Lambda]^{Inv}$ is defined by (35):

$$[\Lambda]^{Inv} \equiv [R]\left([R]^T[\Lambda][R]\right)^{-1}[R]^T \qquad (35)$$

References

[1] M. Morati, D. Girod, F. Terrien, V. Peron, P. Poure, S. Saadate, "Industrial 100-MVA EAF Voltage Flicker Mitigation Using VSC-Based STATCOM With Improved Performance", IEEE Transactions on Power Delivery, Vol. 31, No. 6, Dec. 2016-, pp. 2494-2501.

[2] G.R. Walker, "Digitally-Implemented Naturally Sampled PWM Suitable for Multilevel Converter Control", IEEE Transactions on Power Electronics, Vol. 18, Issue 6, Nov. 2003, 1322-1329.

[3] I. G. Park, S. I. Kim, "Modeling and Analysis of Multi-Interphase Transformers for Connecting Power Converters in Parallel", Record 28th Annual IEEE Power Electronics Specialists Conference. PESC97, 1997, pp. 1164-1170.

[4] J. S. Siva Prasad, G. Narayanan, "Minimization of Grid Current Distortion in Parallel-Connected Converters Through Carrier Interleaving", IEEE Transactions on Industrial Electronics, Jan. 2014, Vol. 61, Issue 1, pp. 76-91.

[5] E. Laboure, A. Cuniere, T. A. Meynard, F. Forest, E. Sarraute, "A Theoretical Approach to InterCell Transformers, Application to Interleaved Converters", IEEE Transactions on Power Electronics, Vol. 23, Issue 1, pp. 464-474, Jan. 2008.

[6] F. Forest, T. A. Meynard, E. Labouré, V. Costan, E. Sarraute, A. Cunière, T. Martiré, "Optimization of the Supply Voltage System in Interleaved Converters Using Intercell Transformers", IEEE Transactions on Power Electronics, Vol. 22, No. 3, May 2007, pp. 934-942.

[7] J. Reinert, A. Brockmeyer, R.W.A.A. De Doncker," Calculation of losses in ferro- and ferrimagnetic materials based on the modified Steinmetz equation", IEEE Transactions on Industry Applications, Vol. 37, Issue 4, Jul/Aug 2001, pp. . 1055 - 1061

Measurement and Calculation Method of Wireless Power Transfer Coil Equivalent Series Resistance under the Vehicle

Norihito Kimura*, Hiroaki Yuasa**
*SOKEN, INC.
500-200, Minamiyama, Komenoki-cho, Nisshin, Aichi, Japan
**TOYOTA MOTOR CORPORATION
1, Toyota-cho, Toyota, Aichi, Japan
Tel.: +81 (561) 57–0436
Fax: +81 (561) 57–0694
E-Mail: norihito.kimura.j5g@soken-labs.co.jp
URL: https://www.soken-labs.com

Keywords

«Wireless power transmission», «Contactless Power Supply», «Hybrid Electric Vehicle (HEV)», «Electric vehicle», «Measurement», «Estimation technique»

Abstract

In order to design a highly efficient wireless power transfer (WPT) for plug-in hybrid vehicles (PHVs), a novel methodology is proposed to measure the efficiency of power transfer between transmitting and receiving coils and to calculate the equivalent series resistance (ESR) of coils under actual vehicle mounting. The proposed measurement method can accurately measure the ESR of coils at the rated current that cannot be measured by low-current general-purpose instruments such as LCR meters. The proposed calculation method can accurately calculate the ESR of WPT coils under the vehicle such as the steel body and the muffler. Experimental results under the vehicle verify the accuracy of their proposed methods.

I. Introduction

In recent years, research on wireless power transfer (WPT) has been advanced as a means of charging environmentally friendly vehicles such as plug-in hybrid vehicles (PHVs) and electric vehicles (EVs) [1]-[4]. Dynamic wireless power transfer (DWPT) is also expected to expand the driving range of EVs [5]-[7]. In case of vehicles, an 85 kHz magnetic resonance coupling is used in WPT. For highly efficient power transmission, it is important to improve the efficiency of power transfer between transmitting and receiving coils. Referring to the fact of using magnetic field in this method, it is necessary to enhance the WPT efficiency between coils in consideration of the influence of the steel body and the muffler, which counted as a magnetic substance near the receiving coil, when mounted on a real vehicle (Figure 1).

In previous research, LCR meter has been used to measure the equivalent series resistance (ESR) of the coils required for the calculation of WPT efficiency between coils [8]. However, due to the minute current measurement, the influence of the non-linear characteristic of the magnetic substance cannot be taken into consideration. Therefore the construction of a measurement method at the time of rated output is an issue. A measuring method for improving the issue has been recently studied. For example, a method based on a frequency domain approach using a parallel resonance circuit [9]. In [9], although only the coil itself has been verified, the verification with the measured efficiency between coils under the vehicle has not been confirmed. In addition, although the experimental evaluation of the bottom metal in the vehicle body has been performed [10], there is a gap between the measured value and calculated value using the finite element method (FEM). Therefore the improvement of calculation accuracy is also an issue.

In this paper, a new measurement method of the efficiency between coils is proposed. Moreover, a calculation method of the ESR is presented to determine the efficiency between coils under actual vehicle mounting at rated output, and its effectiveness is verified.

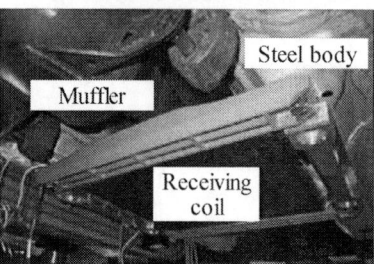

Fig.1 Under the vehicle installation example of the receiving coil

II. Measurement Method of Efficiency between Coils under the Vehicle

A. Derivation method of the efficiency between coils

Figure 2 shows the equivalent circuit representing the coils of WPT. The WPT efficiency between coils η_c is expressed by equation (1).

$$\eta_c = \frac{P_2}{P_2 + R_1 I_1{}^2 + R_2 I_2{}^2} \tag{1}$$

P_2 is the output power. I_1 and I_2 represent the currents of the primary side coil and the secondary side coil. R_1 and R_2 represent the ESR of the primary side coil and the secondary side coil respectively. The values of R_1 and R_2 include not only coil resistance components but also core loss of the magnetic core used for WPT coils, aluminum shield loss, and body loss. If these ESR can be measured accurately, the WPT efficiency between coils can be determined indirectly using equation (1). Furthermore, if these ESR of coils can be calculated accurately by the FEM, the WPT efficiency between coils can be easily designed by circuit simulation.

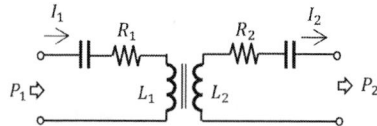

Fig.2 Equivalent circuit

Conventionally, the measurement using an LCR meter as shown in Figure 3 has been applied to the measurement of the ESR R_1 and R_2 of the WPT coil. However, since the power factor is low when using a general-purpose LCR meter, the measurement accuracy is low at the power transmission frequency of 85 kHz. For example, for a coil with a Q-factor of 300 and an inductance of about 200µ H, the measurement error of the ESR is up to about 15% [11]. Furthermore, because of the measurement of a very small current of about 20 mA, it is not possible to take into account the non-linearity of the magnetization characteristics of the magnetic material, and an error occurs in the efficiency near the rated value, which is the area of actual usage.

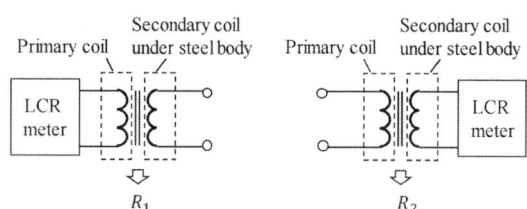

Fig.3 Conventional method for measuring ESR of coils

B. Proposed method for measuring ESR of coils

Figure 4 shows the proposed measurement method of the ESR. The proposed method includes a power supply unit, a power meter, a resonant capacitor, and a coil to be measured. By setting the power factor to 1 using offsetting the reactance component of the coil with a capacitor, only the resistance component is left. As the resonant capacitor, a capacitor having an ESR sufficiently smaller than the coil to be measured is selected. In this state, the rated current is supplied to the coil by adjusting the output of the power amplifier. Then, by measuring the effective powers P_1 and P_2 corresponding to the coil loss by a power meter, it is possible to measure the ESR of coils when the rated current is applied from equations (2) and (3).

$$R_1 = \frac{P_1}{I_1{}^2} \tag{2}$$

$$R_2 = \frac{P_2}{I_2{}^2} \tag{3}$$

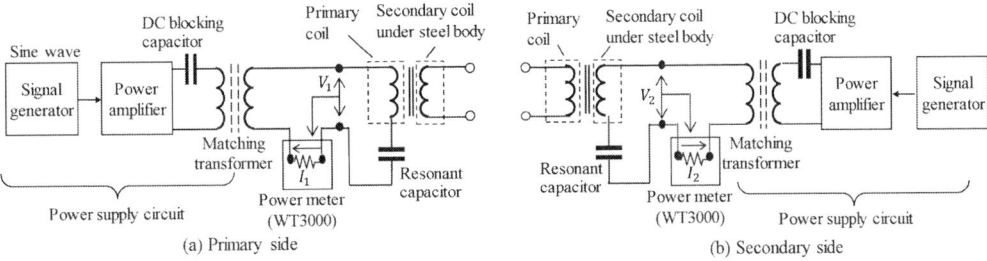

Fig. 4 Proposed method for measuring ESR of coils

The power supply unit includes a signal generator, a power amplifier, DC blocking capacitor, and a matching transformer. The DC blocking capacitor is inserted to suppress the coil magnetization due to the DC component of the amplifier output. The power amplifier cannot supply sufficient power to a load resistance different from the rated load. Therefore, impedance conversion is performed using a matching transformer, and sufficient power can be supplied.

From the above, it is possible to measure the ESR of coils at the time of high power factor and rated current condition by the proposed method. Since the measurement is performed in a state of high power factor, the measurement accuracy can be improved compared to measurement using a LCR meter in a state where the power factor of only the coil is low. In addition, the high frequency power amplifier for power supply only needs to supply the power that consumes the resistance, so that there is an advantage that the measurement can be performed with a facility simpler than the inverter for power transmission.

C. Verification result

Figure 5 shows a direct measurement system of the efficiency between coils under the actual power supply operation state. The measurement accuracy is improved by using a shunt resistor instead of a clamp-type current sensor that causes a phase measurement error as a current sensor. Figure 6 shows the configuration of the coil to be measured and table 1 shows the measurement conditions.

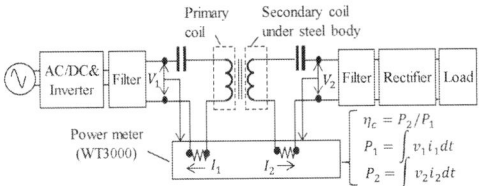

Fig.5 Coil efficiency measurement system

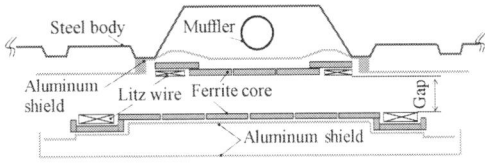

Fig.6 Coil configuration (Cross section)

Figure 7 shows the ESR measurement results of the primary and secondary coils. The current characteristics of the ESR can be measured by the proposed method. The current dependence of the ESR seems to be due to the non-linear characteristics of the magnetic properties of magnetic materials. The value measured by the conventional LCR meter is about 28% lower on the primary side and about 20% lower on the secondary side than the ESR at the rated current measured by the proposed method.

Table I: Measurement condition

Output power		3kW
Frequency		85kHz
Gap		117mm
Core size	Primary coil	600 × 400 mm
	Secondary coil	□280 mm
Body size		□1600 mm

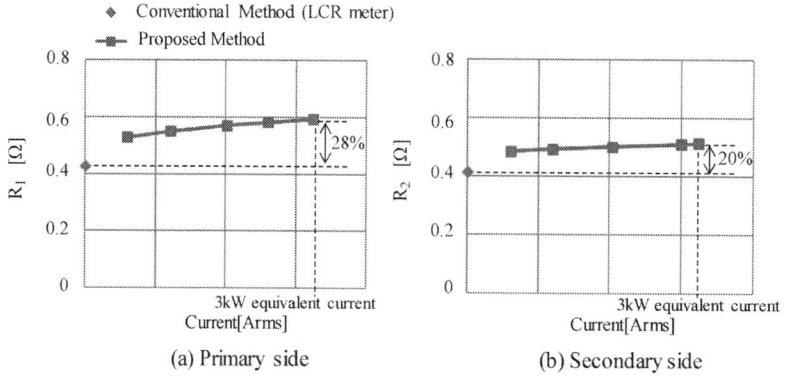

(a) Primary side (b) Secondary side

Fig.7 ESR measurement result

Figure 8 shows the directly measured value of the efficiency between coils at the rated output, and also shows the indirectly measured values calculated from equation (1) using the measured the ESR. The measurement error of the efficiency between coils using the proposed ESR measurement method is within 1%. The measurement error of the efficiency between coils using the conventional ESR measurement method by the LCR meter is 4.5%. From the above, the effectiveness of the proposed ESR measurement method is confirmed.

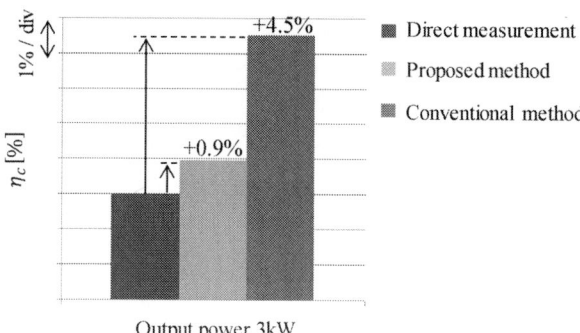

Output power 3kW

Fig.8 Coil efficiency measurement result

III. Calculation Method of Coils ESR under the Vehicle

A. Iron loss evaluation of the steel body and the muffler

In order to determine how to calculate the loss of the steel body and the muffler, a measurement sample is made to clarify the iron loss breakdown at 85 kHz. The iron loss P_c is expressed by equation (4).

$$P_c = P_h + P_e = K_h \cdot f + K_h \cdot f^2 \qquad (4)$$
$$\frac{P_c}{f} = K_h + K_h \cdot f \qquad (5)$$

P_h is the hysteresis loss, P_e is the eddy current loss, K_h and Ke are coefficients in the hysteresis loss and the eddy current loss respectively, f is the frequency. Equation (5) is obtained by dividing equation (4) by the frequency.

Therefore, by measuring the frequency characteristics of the iron loss and dividing the result by the frequency, it is possible to calculate the breakdown of the hysteresis loss and the eddy current loss.

Figure 9 shows the configuration for the iron loss measurement. A sample photograph for the iron loss measurement of the steel body is shown in Figure 10. The plate material is cut into 39mm in outer diameter and 30 mm in inner diameter by wire-cut electric discharge machining, and a primary winding and a secondary winding are applied. The board thickness is 1mm.

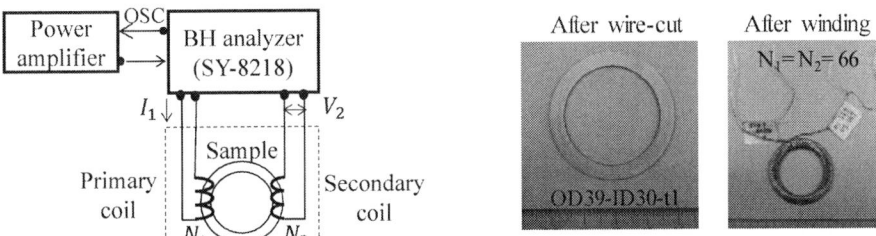

Fig.9 Measurement configuration   Fig.10 Measurement sample

Figure 11 shows the results of the iron loss measurement at 10 mT in magnetic flux density and 2 kHz to 100 kHz in frequency by a BH analyzer (SY-8218). The breakdown of the iron loss is calculated by dividing the result by the frequency. The results are shown in Figure 12. At 85 kHz, 97% in the iron loss of the steel body and 93% in the iron loss of the muffler are eddy current losses. From the above, it had found that in order to calculate the ESR of the coils under the vehicle, it is necessary to calculate the eddy current loss with high accuracy.

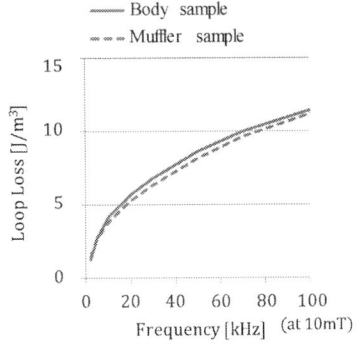
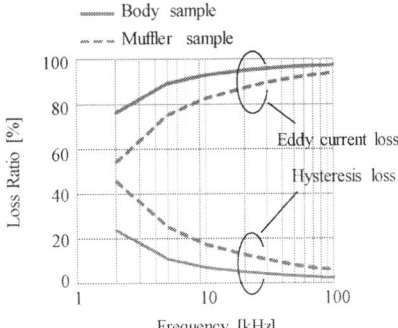

Fig.11 Loop loss measurement result   Fig.12 Frequency characteristics of the loss ratio

B. DC magnetization curve and volume resistivity evaluation of the body and muffler

The basic equation for magnetic field analysis when an eddy current flows through a conductor is expressed by equation (6) to (8) based on Maxwell's electromagnetic equation. A is a magnetic vector potential, v is a magnetic resistivity, J_0 is a forced current density, J_e is an eddy current density, σ is a conductivity, and ϕ is an electric scalar potential. The effect of an eddy current caused by the change in the magnetic field is included in equations (6) to (8).

$$rot(\upsilon rot\mathbf{A}) = \mathbf{J_0} + \mathbf{J_e} \tag{6}$$

$$\mathbf{J_e} = -\sigma\left(\frac{\partial A}{\partial t} + \text{grad}\phi\right) \tag{7}$$

$$\text{div}\mathbf{J_e} = 0 \tag{8}$$

Therefore, accurate DC magnetization characteristics and volume resistivity are required for the eddy current analysis of body materials and muffler materials.

Figure 13 shows the configuration for measuring the DC magnetization characteristics. Figure 14 shows a sample photograph of the steel body for measurement. Five piece with a thickness of 1mm, which are cut into 39mm in outer diameter and 30mm in inner diameter by wire-cut electric discharge machining, then stacking and winding are applied.

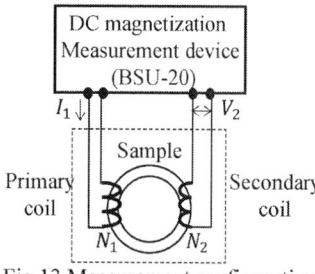

Fig.13 Measurement configuration

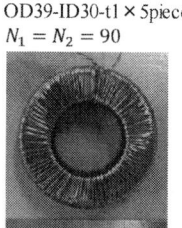

OD39-ID30-t1 × 5piece
$N_1 = N_2 = 90$

Fig.14 Measurement sample

Figure 15 shows the measurement results of DC magnetization characteristics up to 4000 A/m at a low frequency of 0.01 Hz that can be regarded as DC, together with the characteristics of SPCC and SUS430, which are default materials of the magnetic field analysis software JMAG. It can be seen that the steel body has a greater difference in the magnetization characteristics in the low magnetic field region than the SPCC.

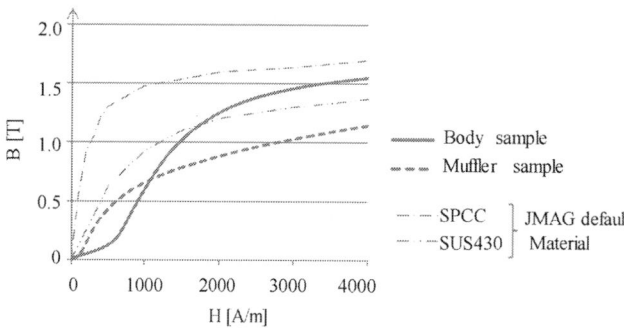

Fig.15 Measurement result of DC magnetization

Figure 16 shows the configuration for measuring the volume resistivity. A sample photograph for the volume resistivity measurement of the steel body is shown in Figure 17. The plate material is cut into 80mm in length, 3mm in width and 1mm in thickness by wire-cut electric discharge machining.

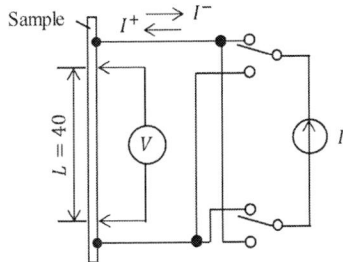

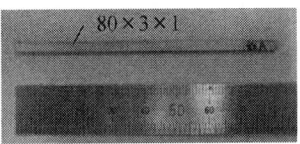

Fig.17 Measurement sample

Fig.16 Measurement configuration

A constant current is applied to the sample, and the voltage drop between the voltage measurement terminals is measured by a digital voltmeter. The volume resistivity is calculated from Equation (9).

$$\rho = \frac{(\Delta V^+ - \Delta V^-)}{2I}\frac{XY}{L} \tag{9}$$

ΔV^+ is the voltage drop when current flows in the + direction, ΔV^- is the voltage drop when current flows in the - direction, I is the DC current value, L is the length between the voltage measurement terminals, X and Y are the width and thickness of the sample. In order to remove the effect of the thermal electromotive force generated between the voltage measurement terminal and the sample made of different materials, the voltage drop is measured by changing the direction of the measurement current, and half of the difference is corrected as the voltage drop.

Figure 18 shows the measurement results of the volume resistivity. It can be seen that steel body has a resistivity of about 1.7 times that of the SPCC.

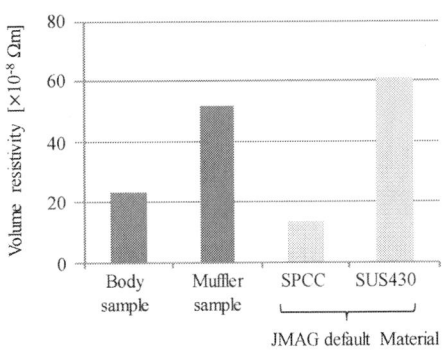

Fig.18 Measurement result of volume resistivity

C. Calculation method of coils ESR under the vehicle

The ESR of primary and secondary coils are calculated from equations (10) and (11).

$$R_1 = \frac{P_F + P_L + P_B + P_M + P_A}{I_1^{\;2}} \tag{10}$$

$$R_2 = \frac{P_c + P_L + P_B + P_M + P_A}{I_2^{\;2}} \tag{11}$$

P_F is the ferrite core loss, P_L is the litz wire loss, P_B is the steel body loss, P_M is the muffler loss, P_A is the aluminum shield loss. The ESR is calculated by dividing their total loss by the square of the applied current.

Table 2 shows the calculation method of losses for each part under the vehicle. The FEM is used to calculate losses of all parts when a constant current is applied. The losses of the steel body and the muffler are calculated only eddy current loss using the measured DC magnetization characteristics and the volume resistivity. The loss of the ferrite core is calculated using the characteristics of relative permeability and core loss to the magnetic flux density measured in advance. Since the calculation method for the alternating current (AC) resistance of a litz wire used for the coil is not considered in the loss calculation, the AC resistance value measured in the state wound around the bobbin is applied.

Table II: Calculation condition

Parts	Analysis conditions (with JMAG-Designer)	
Steel body and Muffler	Resistivity	Apply measured value
	Magnetization characteristics	Apply measured value
	Iron loss calculation method	Eddy current direct calculation (no consideration of hysteresis loss)
Ferrite Core	Magnetization characteristics	Apply measured value
	Iron loss calculation method	Iron loss map reference method
Aluminum shield	Resistivity	Apply measured value
	Loss calculation method	Eddy current direct calculation
Litz wire	AC resistance	Apply measured value

D. Verification Result

Figure 19 shows the configuration of the coil to be measured, and Table 3 shows the measurement condition. Figure 20 shows the finite element analysis model that reproduces the experimental configuration. Verification of the calculation of the ESR is performed at a gap of 65 mm between coils, which is the coil position where the influence of the steel body is large.

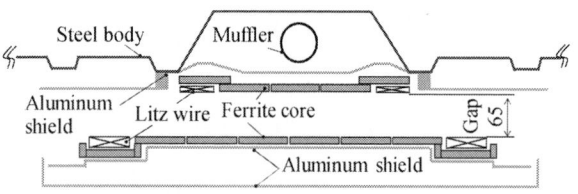

Fig.19 Coil configuration (Cross section)

Table III: Measurement condition

Coil Current		~ 3kW equivalent
Frequency		85kHz
Gap		65mm
Core size	Primary coil	600 × 400 mm
	Secondary coil	□280 mm
Body size		□1600 mm

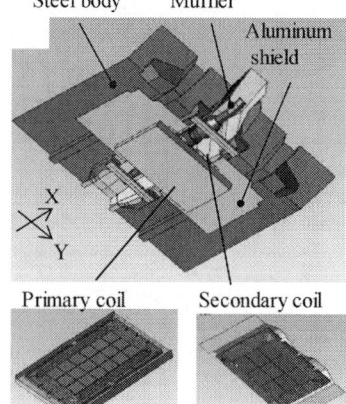

Fig.20 Finite element analysis model

Figure 21 shows the verification results of the calculation of the ESR on the primary side and the secondary side. In addition, the calculation results by using JMAG default material are also shown. The calculation error at the 3kW rated current is 1 % on the primary side and 5% on the secondary side. From the above, it was found that the ESR of the coils can be calculated with good accuracy by the proposed calculation method. The reason behind the large calculation error at low current on the primary side is interpreted as poor measurement accuracy of the DC magnetization curve in the low magnetic field region. Since the magnetic field strength of the steel body at low current is as low as 30 A/m approximately, it is considered that the calculation error is affected by an error of ± 20 A/m of the DC magnetization curve measuring apparatus used. In addition, the calculation error by using JMAG default material (SPCC and SUS430) is 26 % on the primary side and 17 % on the secondary side. Therefore, it is important to measure the proper material properties of the steel body and the muffler of the actual vehicle.

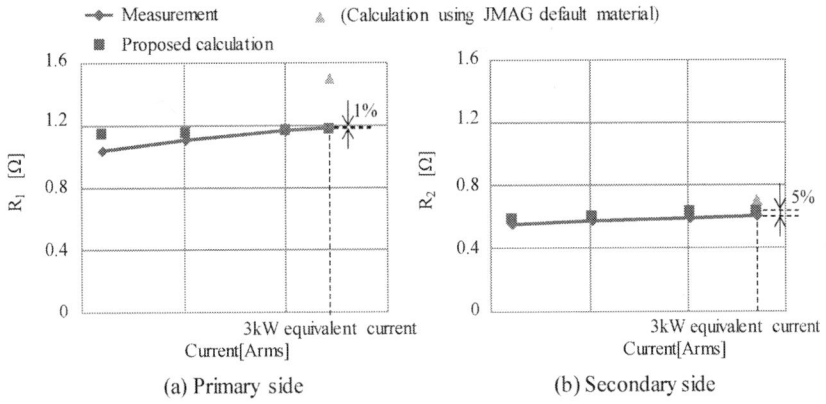

(a) Primary side   (b) Secondary side

Fig.21 ESR calculation accuracy verification result

Moreover, it is possible to analyze the breakdown of the ESR for each part by using the constructed accurate ESR calculation method. Figure 22 shows the breakdown of the ESR on the primary and secondary coils at 3kW equivalent current. The sum of the ESR of the steel body and the muffler is 68 % on the primary side and 45 % on the secondary side, which indicates that the influence on the primary side is large. This is because the primary coil size is larger than the secondary coil size, and the amount of magnetic flux from the primary side to the body increases when the distance between the coils is short.

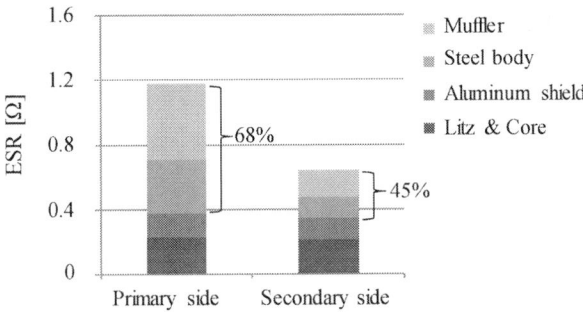

Fig.22 Breakdown of the ESR at 3kW equivalent current

E. Conclusion

The capacitor compensation method can accurately measure the ESR of the coils when the rated current is applied. The calculation method of the ESR of the WPT coils under actual vehicle mounting can be accurately calculated by measuring the DC magnetization curve and volume resistivity of the steel body and the muffler and performing loss calculation considering only the eddy current loss. By using the constructed accurate ESR calculation method, an analysis example of the ESR breakdown on each part under the vehicle has been shown. From the results, it is necessary to reduce the effects of the steel body and muffler for the sake of transmitting high-efficiency power under the vehicle. It could be argued that the proposed ESR measurement method and calculation method are useful for such studies.

References

[1] Y. Kaneko, N. Ehara, T Iwata, S. Abe, T. Yasuda and K. Ida, "Comparison of Transformer Winding Methods for Contactless Power Transfer Systems of Electric Vehicle", IEEJ Transactions on Industry, Vol.130, No.6, pp734-741, 2010

[2] T. Yamanaka, I. Fujita, Y. Kaneko, S. Abe and T. Yasuda, "Colling Structure for Large Capacity H-shaped Core Contactless Power Transformers of Electric Vehicles", IEEJ Transactions on Industry, Vol.134, No.3, pp370-375, 2014

[3] Y. Sugiyama, "Analysis of Robustness of Wireless Power System for EVs/PHVs", EVS 31& EVTeC 2018, No.20189253, 2018

[4] M. Pathmanathan, S. Nie, N. Yakop and P. Lehn, "Efficiency improvement of a wireless power transfer system using a receiver side voltage doubling rectifier", EPE'19 ECCE Europe, No.188, 2019

[5] K. Hata, T. Imura, H. Fujimoto, Y. Hori and D. Gunji, "Charging Infrastructure Design for In-motion WPT Based on Sensorless Vehicle Detection System", IEEE PLES Workshop on Emerging Technologies, Wireless Power Transfer (WoW), pp205-208, 2019

[6] S. Cui, X. Gao, B. Song and S. Dong, "A Three-Phase LCC to Single-Phase S Compensation Topology for DWPT-TS System", EPE'19 ECCE Europe, No.500 2019

[7] D. Haddad, T. Konstantinou, A. Prasad, Z. Hua, D. Aliprantis, K. Gkritza and S. Pekarek, "Data-Driven Design and Assessment of Dynamic Wireless Charging Systems", IEEE PLES Workshop on Emerging Technologies, Wireless Power Transfer (WoW), pp59-64, 2019

[8] T. Tohi, Y. Kaneko and S. Abe, "Maximum Efficiency of Contactless Power Transfer Systems using k and Q", IEEJ Transactions on Industry, Vol.132, No.1, pp123-124, 2012

[9] Gaurav R. Kalra, Matthew G. S. Pearce, Seho Kim, Duleepa J. Thrimawithana, and Grant A. Covic, "Measuring the Q-factor of IPT Magnetic Couplers", IEEE PLES Workshop on Emerging Technologies, Wireless Power Transfer (WoW), pp34-38, 2019

[10] K. Hanajiri, O. Shimizu and H. Fujimoto "Evaluation of effect of body metal on dynamic wireless power transfer to electric vehicles", SPC, IEE Japan 2018(165-168,170), pp19-23, 2018

[11] Keysight, "E4980A Precision LCR Meter User's Guide", E4980-97210

Design of a circumscribing polygon wide bandgap based integrated modular motor drive topology with thermally decoupled windings and power converters

Abdalla Hussein Mohamed
Electromechanical, Systems and Metal Engineering
Ghent University
Ghent, Belgium
Flanders Make@UGent - core lab EEDT-MP
EPE
Cairo University
Cairo, Egypt
a.hussien.rashad@gmail.com

Hendrik Vansompel
Electromechanical, Systems and Metal Engineering
Ghent University
Ghent, Belgium
Flanders Make@UGent - core lab EEDT-MP
Hendrik.Vansompel@UGent.be

Peter Sergeant
Electromechanical, Systems and Metal Engineering
Ghent University
Ghent, Belgium
Flanders Make@UGent - core lab EEDT-MP
Peter.Sergeant@UGent.be

Keywords

≪Integrated motor drives≫, ≪Wide Bandgap devices≫, ≪GaN≫, ≪DC link capacitor≫, ≪CFD≫

Abstract

In this paper, the design of an integrated modular motor drive topology based on the circumscribing polygon of the outer surface of the conventional cylindrical housing is introduced from the mechanical and the thermal point of view. The design of the shared machine and converter cooling system is optimized from the thermal point of view using computational fluid dynamics (CFD) simulations. A wide bandgap, specifically Gallium Nitride (GaN), based half-bridge converter module is designed and implemented for integration. For a case study of a machine of outer radius 75 mm, axial length 80 mm and six stator modules, the resulted surface area for each converter module is $80*87$ mm^2. The size of the converter module was reduced so as to exactly match the available surface around the machine. A method for the calculation of the maximum power per module is introduced resulting in 1032W per module for the case study considered in the paper. A method for the DC-link capacitor design is introduced and the influence of the stator phases connections on the DC-link current stress is explained. Experimental measurements are done on one segment of the proposed integration topology.

Introduction

Integrated modular motor drives (IMMDs) combine the segmented machine stator with the driving converter of each segment into the same housing [1]. This physical integration results in the elimination of the separate heatsink and enclosure needed for the driving converter. Moreover, the long cables connecting the converter output to the machine winding can be also eliminated. On the one hand, the elimination of such components reduces the total weight, volume and cost of the whole drive [2]. On the other hand, the elimination of the cables improves the electromagnetic compatibility (EMC) of the whole system [3]. Due to the existence of the converter modules near to the heat generation sources of the machine (winding, core), challenges regarding the stable mechanical mounting and the sufficient cooling of both the machine and the converter modules should be handled [4]. The modularity of the stator structure and the driving converter provides many advantages over the conventional drives. As the total drive power is divided on all the modules, the electrical and the thermal stresses of the semiconductor devices reduces, which in turn improves the reliability of the whole drive. By proper connection of the modules, the fault tolerance of the whole drive can be greatly enhanced [5].

In IMMD, the ambient temperature of the power converter switches is expected to be high due to the proximity of the power semiconductor devices to the winding and/or the core of the machine. This high ambient temperature reduces the amount of temperature rise allowed for the semiconductor junction until reaching the rated junction temperature. The maximum amount of junction temperature rise allowed depends on the power dissipation in the switch, the ambient temperature, the rated junction temperature and the thermal resistance from the junction of the semiconductor device to the ambient. Among the three commercially available semiconductor technologies (Si, GaN, SiC), the wide bandgap (WBG) devices (GaN, SiC) represent an excellent candidate for the IMMD applications for their low conduction and switching losses, high rated junction temperature and high thermal conductivity [6]-[7]-[8].

IMMD can be classified according to the location of the power converter module with respect to the machine into: axially stator iron mounted (ASM), radially stator iron mounted (RSM), axially housing mounted (AHM) and radially housing mounted (RHM) [2]. The RSM and ASM provide a more compact integrated drive with higher cooling challenges [9]

In this paper, a RSM modular topology is proposed, designed and one module is implemented to prove the concept. The paper is organized as follows: the first section introduces the idea and provides mechanical design details, section two elaborates the design of the power converter module, section three explains the thermal modelling and design of the proposed integration topology, section four provides an algorithm for the calculation of the maximum power per module and the last section is dedicated for the experimental results.

Mechanical construction of the proposed circumscribing polygon topology

In IMMD, every module consists of a stator tooth along with its driving converter module and a shared cooling system for evacuating heat from both the stator tooth and the converter module. The shared cooling system should be capable of: thermally decoupling the power converter module and the machine tooth, providing a mechanically stable surface for mounting the converter printed circuit board (PCB) and creating a good thermal contact between the thermal pad of the transistors and the cooling surface. Fig. 1 (left) shows the proposed integration topology for a six stator coils machine. The proposed shared cooling of the machine and the power converter has a polygon shaped radial cross sectional area with number of sides equal to the number of the stator coils. The heat generated by the machine and the power converter is evacuated by the cooling fluid pumped in the cooling channel. Fig. 1 (right) shows the heat flux path from the stator and the power converter to the ambient.

Power converter module design

In this section, the calculation of the mechanical dimensions and the thermal design of the power converter module are explained in addition to the DC-link capacitor design.

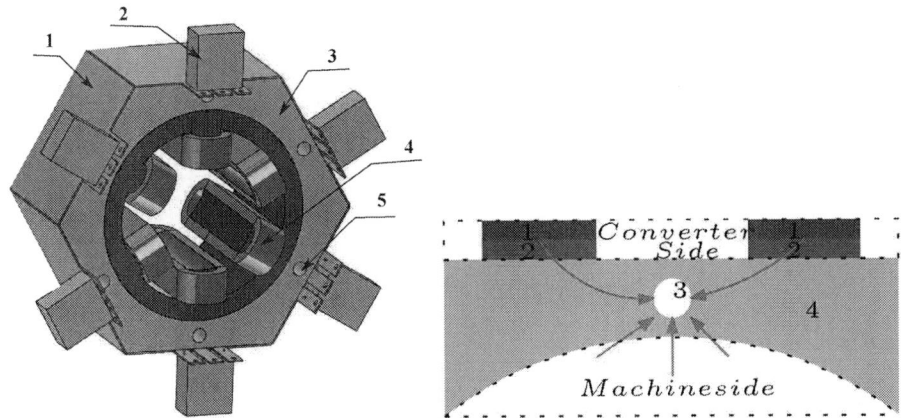

Fig. 1: (left) The proposed IMMD topology for a six teeth concentrated winding machine: (1) Converter PCB, (2) DC link capacitor, (3) Shared cooling structure, (4) One machine tooth, (5) Cooling channel. (Right) The heat flow paths (1) The junction of the transistors, (2) The PCB, (3) The cooling channel, (4) The shared cooling structure.

Mechanical dimensions of the power converter module

For the design of the power converter module PCB, the converter module dimensions should be calculated first. Since the original axial length of the machine will be maintained, the dimension of the PCB in the axial direction will be the same as the axial length of the machine (L_y). The value of the other dimension (L_x) depends on the radius of the inscribed circle which is the outer housing radius (R_{hout}) and the number of stator modules (n). From Fig. 2, the value of L_x can be calculated from (1).

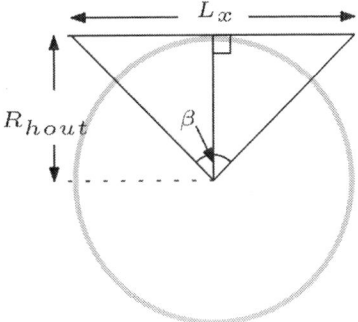

Fig. 2: Calculation of the converter dimension (L_x)

$$\begin{cases} \beta &= \frac{360}{n} \\ L_x &= 2R_{hout}\tan\frac{\beta}{2} \end{cases} \tag{1}$$

A case study with six stator modules machine and 80 mm axial length will be considered for the converter design. From (1) and the axial length of the machine, the converter module size shouldn't exceed 80*87 mm². For optimal size and efficiency, Gallium nitride (GaN) technology has been selected for the implementation of the converter. The *(GS66508B)* GaN switch has been selected for the implementation of the half bridge module, the key parameters of the *(GS66508B)* device are included in Table I. This device has been selected for two reasons, the first one is : its low losses indicated by its small figure of merit ($FOM = R_g * Q_g$) and the zero reverse recovery charge (Q_{rr}) indicating zero switching losses in the reverse conduction path. The second reason is its small package size (7.1*8.5 mm²)

Table I: *GS66508B* key specifications

Property	Value
Rated voltage (V)	650
Rated current (A)	30
R_{ds} (mΩ)	50
θ_{JC} ($^\circ C/W$)	0.5
Q_g (nC)	5.8
Q_{rr} (nC)	0

Power converter module thermal design

From Fig. 1(right), the heat transfer path from the junction of the devices to the coolant includes the PCB layers and the thermal pad electrically insulating the transistors from the shared cooling structure. Minimization of the thermal resistance from the junction to the coolant can be done by distributing a number of thermal vias beneath the thermal pad of the transistor and beyond [10]. In this design, a four-layer board is chosen for the implementation of the converter. Fig. 3 (left) shows the package layout of the (GS66508B) transistor, pin (2) is the power source pin that receives most of the heat generated in the junction. Fig. 3 (right) depicts the copper area beneath the thermal pad. Zone (1) is the inner zone just beneath the thermal pad of the transistor and zone (2) is added for further reduction of the thermal resistance of the PCB from top layer to the bottom layer.

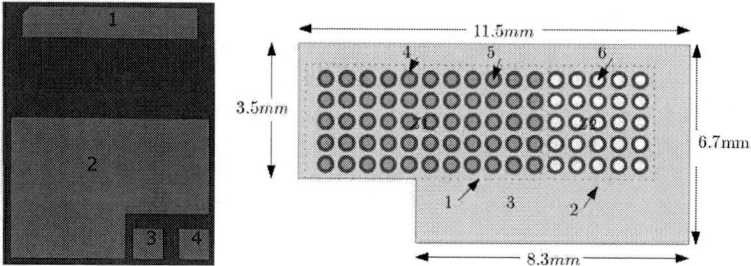

Fig. 3: (left) The (GS66508B) transistor package (1) Drain, (2) Source (thermal pad), (3) Gate, (4) Source sense. (right) Thermal vias pattern: (1) The thermal vias underneath the switch thermal pad (Zone 1), (2) The thermal vias pattern outside the thermal pad (Zone 2), (3) The copper area, (4) The via plating copper, (5) The via filling solder, (6) Air.

For the calculation of the thermal resistance with and without the thermal vias, a thermal model is built. A 1W loss is imposed on the thermal pad of zone (1) (see Fig. 3) while the bottom layer temperature is kept at 20°C. Fig. 4 (left) is the temperature distribution over the PCB without adopting the thermal vias. The maximum thermal resistance from the thermal pad to the bottom layer is 66.99 °C/W. In Fig. 4 (right), the maximum thermal resistance is 2.18 °C/W. The thermal via pattern used resulted in 98.7 % reduction in the thermal resistance from the top to the bottom layer.

Design of DC link capacitor

DC link capacitors are supplying the DC link current pulses resulting from the switching action of the inverter switches [11]. Since, the case study machine has six-stator coils, it can be configured to work as a three-phase or six-phase machine. In case of six-phase machine, the phases can be either connected to have one neutral or it can be divided into two sets of three phase coils (see Fig. 5). For the latter case, interleaving of PWM carriers can be utilized to reduce the DC-link capacitor current stress by proper selection of the interleaving angle (k), see Fig. 6.

Fig. 7 shows the variation of the capacitor rms current with the interleaving angle. The optimal interleaving angle is 90°.

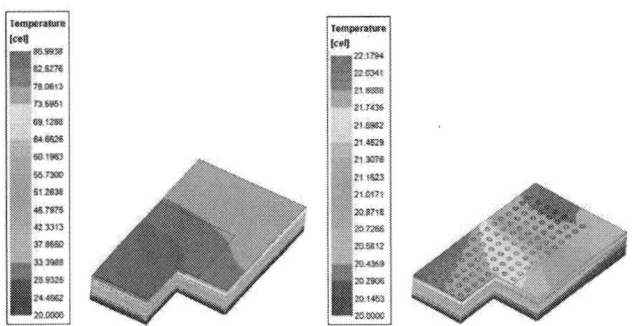

Fig. 4: The temperature distribution under the transistor thermal pad, (left) without thermal vias, (right) with thermal vias.

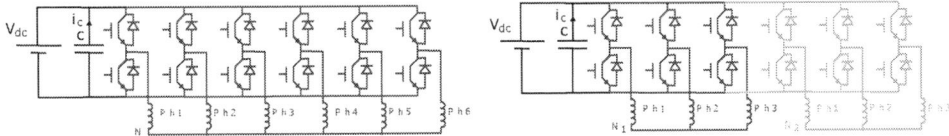

Fig. 5: Configurations of a six phase machine

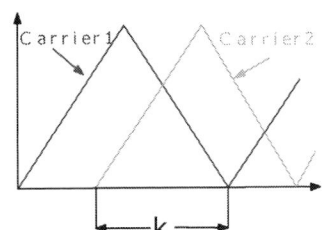

Fig. 6: Interleaved carriers

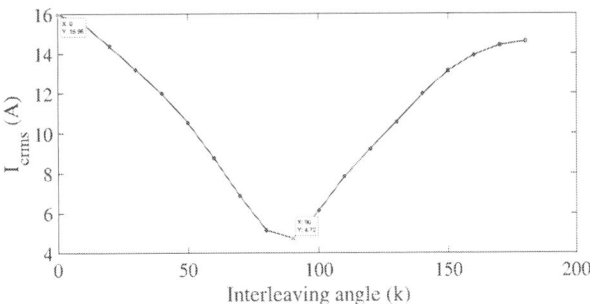

Fig. 7: Capacitor rms current versus the interleaving angle

The DC link capacitance is calculated using (2).

$$\begin{cases} C & = \frac{\Delta Q}{\Delta V} \\ \Delta Q & = \int i_c(t) \,|_{ch/disch} \, dt \end{cases} \tag{2}$$

where, ΔQ is the amount of charge that the capacitor gains or loses every switching cycle, ΔV is the peak to peak capacitor voltage ripple, $i_c(t)$ is the capacitor current during charging or discharging.

The capacitor is sized to have $\Delta V = 1\% V_{dc}$.

Table II lists the specifications of the capacitor needed for each phase configuration. It can be seen that the six phase configuration with optimal interleaving angle results in the smallest DC link capacitance which can enhance the power density of the whole drive.

Table II: DC link capacitance specifications

Configuration	Capacitance (μF)	rms ripple current (A)	Rated voltage (V)
$3 - \phi$	10.6	8	300
$6 - \phi$, 1-neutral	42.5	16	150
$6 - \phi$, 2-neutral	6	4.8	150

Thermal design of the circumscribing polygon topology

The heat transfer rate from both the machine side and the converter side depends on the radius of the cooling channel ($R_{cooling}$) and the flow rate. The value of the cooling radius should be optimized as increasing the cooling radius increases the area through which the heat transfer takes place which enhances the heat transfer rate but at the same time the speed of the coolant at a given flow rate will reduce at the same time. A CFD model has been developed for one module for the optimization of the cooling channel radius. The simulations are done for the channel radii from 2 mm to 7 mm considering 5W losses in each transistor of the half bridge and 150W losses in one machine tooth. Fig. 8 (left) shows the variation of the convection coefficient of the cooling channel for different channel radii at different flow rates. Fig. 8 (right) depicts the maximum temperature of the transistor thermal pad. From Fig. 8, the optimal channel radius in the chosen range is 2 mm providing the highest convection coefficient and the lowest transistors temperature.

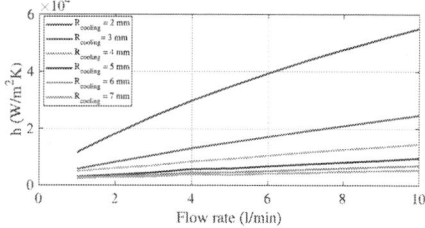 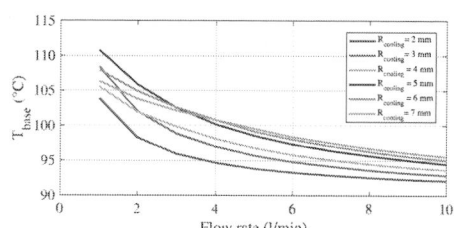

Fig. 8: (left) The convection coefficient versus the flow rate, (right) The transistor thermal-pad temperature versus the flow rate.

Fig. 9 shows the temperature distribution for one integrated tooth at 1 l/min flow rate, 87W copper losses and 5W losses in each transistor. All other surfaces are exposed to natural air convection. From Fig. 9, the peak transistor base plate temperature is 64.3°C, from the datasheet, the junction to case thermal resistance is 0.5°C which makes the junction temperature 66.8°C. The maximum coil temperature is 88.4°C.

The thermal properties of the materials of one module are given in Table III. The thermal conductivity of the non-homogeneous parts is given in the three directions (x, y, z) respectively [12]. Here, C_p is the specific heat capacity and ρ is the mass density.

Maximum output power per module

For calculation of maximum output power per module for the proposed integrated topology, the following steps are used:
- From the CFD model, calculate the transistors power losses that result in a junction temperature of 125°C, a 25°C safety margin are left.
- From a developed total transistor power losses (conduction and switching) versus temperature model (see (3) and (4)), at certain operating conditions (fundamental frequency, dc link voltage,

switching frequency, power factor (PF)), calculate the peak line current that results in the power losses calculated in (1) at 125°C junction temperature.

- Calculate the module power output from ($P_{module} = V * I * PF$), where V is the rms phase voltage, I is the rms phase current. At 1 l/min flow rate, the transistors power losses that results in 125°C junction temperature are 12W. At the operating conditions in Table IV, the peak line current is calculated to be 18A, the maximum module output power is 1032.7W. The total drive power can be calculated by multiplying the module power by the number of modules.

The power losses in the switches can be calculated from (3) and (4) respectively.

$$
\begin{cases}
P_{cons} &= I_{srms}^2 R_{ds}(T) \\
P_{son} &= V_{dc} f_o \sum_{n=1}^{n_{son}} (t_{on}(n,T) i_{on}(n) + Q_{rr}(n)) \\
P_{soff} &= V_{dc} f_o \sum_{n=1}^{n_{soff}} (t_{off}(n,T) i_{off}(n)) \\
P_{st} &= P_{cons} + P_{son} + P_{soff}
\end{cases}
\tag{3}
$$

$$
\begin{cases}
P_{cond} &= f_o \int_t^{t+T_o} i_d(t) v_d(t,T) dt \\
P_{sd} &= 0.25 V_{dc} f_o \sum_{n=1}^{n_{doff}} Q_{rr}(n) \\
P_{dt} &= P_{cond} + P_{sd}
\end{cases}
\tag{4}
$$

where, $P_{cons}, P_{son}, P_{soff}, P_{st}$ are the switch conduction, turn on, turn off and total losses respectively, V_{dc} is the DC link voltage, fo is the fundamental frequency of the switch current waveform, n_{son} and n_{soff} are the number of turn-on and turn-off transitions of the switch in one fundamental cycle respectively, $t_{on}(n,T)$ and $t_{off}(n,T)$ are the turn-on and the turn-off time of the switch at the transition number n and temperature T, $i_{on}(n)$ and $i_{off}(n)$ are the switch current at the turn on and off instants respectively, $Q_{rr}(n)$ is the reverse recovery charge of the diode, $T_o = \frac{1}{f_o}$, $i_d(t)$ and $v_d(t,T)$ are the instantaneous current and voltage drop of the diode at temperature T, P_{cond}, P_{sd}, P_{dt} are the diode conduction, switching and total losses respectively.

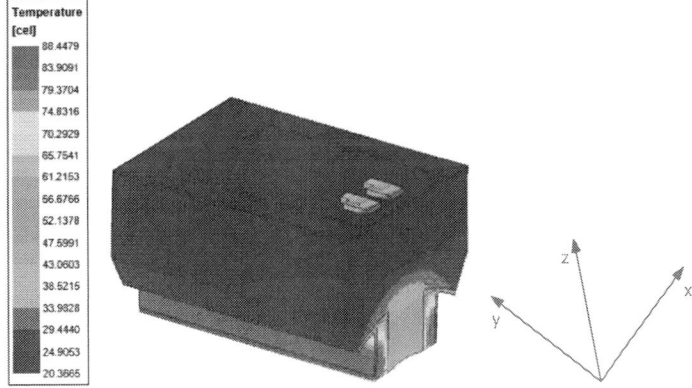

Fig. 9: The temperature distribution for one module at 1 l/min flow rate, 5W losses in transistors and 87W in windings.

Table III: Thermal properties of the materials with the conductivity given in (x,y,z) direction

Part	Conductivity (W/m.k)	$C_p(J/kg.k)$	$\rho(kg/m^3)$
Winding	(327.6, 2.58, 2.58)	393.2	7799.3
Core	(4.9, 22.2, 22.2)	440	7650
PCB	(5.4, 6.4, 1.36)	1127.2	2523.3
Cooling structure	237.5	951	2689
epoxy	0.21	1180	440

Table IV: The case study operating conditions

Variable	Value
DC link voltage (V)	300
Fundamental frequency(Hz)	333
PF	0.85
PWM technique	Sinusoidal PWM
Switching frequency (kHz)	50
Modulation index	0.9

 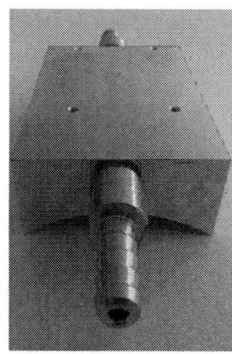 

Fig. 10: One integrated tooth setup: (left) One machine tooth, (middle) Shared cooling structure, (right) One assembled integrated tooth

Experimental results

For validation of the CFD expected transistors temperature based on which the module maximum power was evaluated, a setup consisting of one integrated tooth is built (see Fig. 10).

5W losses are generated in each transistor by connecting them in series with a dc power supply and adjusting the dc voltage until getting a 10W total supply output power. 87W losses are generated in the tooth coil by the same way. Fig. 11 shows the temperature distribution for the half-bridge converter at 1 l/min. The resulted transistors maximum temperature is 67°C indicating good correspondence with the CFD model (See Fig. 9). Note that the low power components around the transistors remain at low temperature which ensures safe operation for the converter.

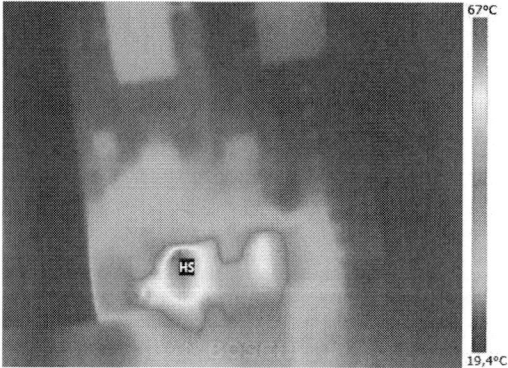

Fig. 11: The measured transistors temperature at 5W transistor losses, 87W coil losses and 1 l/min flow rate.

Conclusion

An integrated modular motor drive topology is proposed, designed and analyzed from the mechanical and the thermal point of view. The power converter module is designed and implemented using GaN technology. The designed converter GaN module of size $80*87$ mm^2 is capable of providing up to 1032W without violating the thermal or the electrical limits of the components. The DC-link capacitors are designed for every possible winding connections with the smallest DC-link capacitance value and stresses resulting for the six phase windings splitted into two three phase sets at the optimal carrier interleaving angle. A CFD model is built for one integrated module to calculate the maximum power that can be supplied by the proposed topology. A one tooth setup is built to asses the validity of the proposed integrated topology. The CFD model results are in a good agreement with the measurements.

References

[1] R. Abebe, G. Vakil, G. L. Calzo, T. Cox, S. Lambert, M. Johnson, C. Gerada, and B. Mecrow: Integrated motor drives: State of the art and future trends, IET Electric Power Applications Vol. 10, no. 8, pp. 757771, 2016

[2] Lee, S. Li, D. Han, B. Sarlioglu, T. A. Minav, and M. Pietola: A Review of Integrated Motor Drive andWide-Bandgap Power Electronics for High-Performance Electro-Hydrostatic Actuators, IEEE Transactions on Transportation Electrification, vol. 4, no. 3, pp. 684693, 2018.

[3] M. Maerz, E. Schimanek, and M. Billmann: Towards an Integrated Drive for Hybrid Traction, pp. 37, 2008.

[4] A. Tenconi, F. Profumo, S. E. Bauer, and M. D. Hennen: Temperatures evaluation in an integrated motor drive for traction applications, IEEE Transactions on Industrial Electronics, vol. 55, no. 10, pp. 36193626, 2008.

[5] B. Ioana, R. Mircea, and S. Lorand: Modular Electrical Machines A Survey, no. 15, pp. 611, 2015.

[6] S. Hazra, S. Madhusoodhanan, S. Bhattacharya, G. K. Moghaddam, and K. Hatua: Design considerations and performance evaluation of 1200 V, 100 a SiC MOSFET based converter for high power density application, 2013 IEEE Energy Conversion Congress and Exposition, ECCE 2013, pp. 42784285, 2013.

[7] M. Trivedi and K. Shenai: Performance evaluation of high-power wide band-gap semiconductor rectifiers, Journal of Applied Physics, vol. 85, no. 9, pp. 68896897, 1999.

[8] K. Shirabe, M. Swamy, J. K. Kang, M. Hisatsune, Y. Wu, D. Kebort, and J. Honea: Advantages of high frequency PWM in AC motor drive applications, 2012 IEEE Energy Conversion Congress and Exposition, ECCE 2012, pp. 29772984, 2012.

[9] M. Maerz, M. Poech, E. Schimanek, and A. Schletz: Mechatronic Integration into the Hybrid PowertrainThe Thermal Challenge, Proc. 1th International Conference on Automotive Power Electronics (APE), no. June, pp. 27, 2006. [Online]. Available: http://www.ecpe.org/download/publications/Paper Maerz APE2006.pdf

[10] Y. Shen, H.Wang, F. Blaabjerg, H. Zhao, and T. Long: Thermal modeling and design optimization of PCB vias and pads, IEEE Transactions on Power Electronics, vol. 35, no. 1, pp. 882900, 2020.

[11] X. Pei, W. Zhou, and Y. Kang: Analysis and Calculation of DC-Link Current and Voltage Ripples for Three-Phase InverterWith Unbalanced Load, IEEE Transactions on Power Electronics, vol. 30, no. 10, pp. 54015412, 2015.

[12] P. Romanazzi, M. Bruna, and D. A. Howey: Thermal homogenization of electrical machine windings applying the multiple-scales method, J. Heat Transfer, vol. 139, no. 1, pp. 18, 2017.

Limits of enhanced desaturation detection method with adaptive blanking for GaN HEMTs

Jan Schmitz, Markus Meißner, Steffen Bernet
CHAIR OF POWER ELECTRONICS, DRESDEN UNIVERSITY OF TECHNOLOGY
Helmholtzstraße 9
01069 Dresden
Tel.: +49 / (351) – 463 – 35665.
Fax: +49 / (351) – 463 – 42138.
E-Mail: jan.schmitz@tu-dresden.de
URL: https://tu-dresden.de/ing/elektrotechnik/eti/le

Acknowledgements

This project has been supported in the frame of the ECPE Joint Research Programme.

Keywords

«Wide bandgap devices», «Gallium Nitride (GaN)», «Safety», «Measurement», «Faults»

Abstract

A fast short circuit detection method was presented. This method is based on the desaturation detection with gate-voltage monitoring for adaptive blanking. Based on simulations in LTspice, first measurements have shown a proper working of the enhanced desaturation detection method. Detailed investigations have shown false detections in some normal operating points. The most relevant influences for false detections in normal operating mode besides EMI noise are common source couplings, drain currents above the nominal value and the temperature of the GaN device. This paper presents the limits of the proposed short circuit detection method for proper working in detail.

Introduction

The short circuit detection and protection of fast switching wide bandgap power semiconductors is challenging due to a fast destruction of the device under fault conditions. Gallium Nitride High Electron Mobility Transistors (GaN HEMTs) fail under short circuit condition at high dc-link voltages in several hundred nanoseconds [1], [2]. Different short circuit detection methods e.g. via shunt measurement or di/dt measurement on the stray inductance have been established over years [3], [4]. A shunt resistor or an additional stray inductance enlarges the commutation loop inductance and therefore increases overvoltage at fast switching. To enable the full performance of the GaN HEMT in a hard switching application, a fast short circuit detection based on a desaturation detection method was developed [5]. The proposed method uses a gate-voltage monitoring for an adaptive blanking and fast reaction. Measurements have shown that the adjustment of a fast and reliable detection versus false detections is difficult. After eliminating most EMI issues concerning the measurement equipment, false detections occurred in some normal operating points. This paper presents the limits of the proposed short circuit detection method concerning the switching speed, the drain current, the drain-source voltage and the case temperature at normal commutation in detail.

Principle of the enhanced desaturation detection method

In a low inductive hard switching fault (HSF) the drain-source voltage v_{ds} stays high and the dv/dt is missing. In consequence, there is no miller plateau. The gate-source voltage v_{gs} rises directly to the driver voltage VCC of +6 V. When the gate-source voltage v_{gs} exceeds the reference voltage of the maximum miller plateau voltage, the comparator output v_{mf} changes to high state and finally the output

v_{scd} – *short circuit detect* – of the logic AND gate changes to high state. Figure 1 presents the schematic of the short circuit detection method developed in [5]. The first part is based on a common desaturation detection method and includes a voltage source $V_{desat\ drive}$ depicted in Figure 1 a). To take into account the fast transient of the drain-source voltage at the input of the comparator, an RC frequency compensated voltage divider is used. Figure 1 b) presents the second part containing the gate-source voltage monitoring to enable an adaptive blanking. The reference voltage $V_{ref,mf}$ of the comparator represents the maximum miller plateau voltage $V_{reference}$. A high drain-source voltage v_{ds} and a high gate-source voltage v_{gs} is detected as short circuit. A fast logic AND gate combines the comparator outputs and generates the *short circuit detect* signal v_{scd}. Figure 2 presents the gate-source voltage v_{gs}, the comparator output signals v_{dt} and v_{mf} and the logic AND gate output signal v_{scd} under low inductive hard switching fault condition.

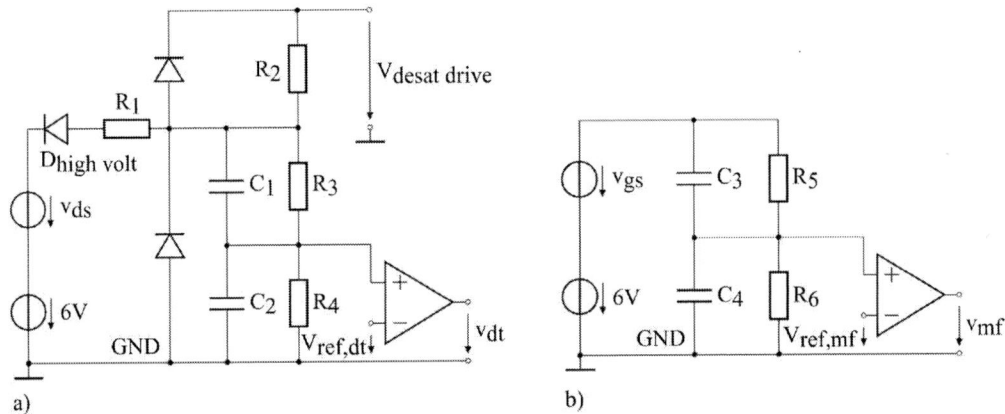

Fig. 1: Schematic of the fast short circuit detection method [5]. a) Drain-source voltage v_{ds} monitoring based on desaturation detection. b) Gate-source voltage v_{gs} monitoring for adaptive blanking.

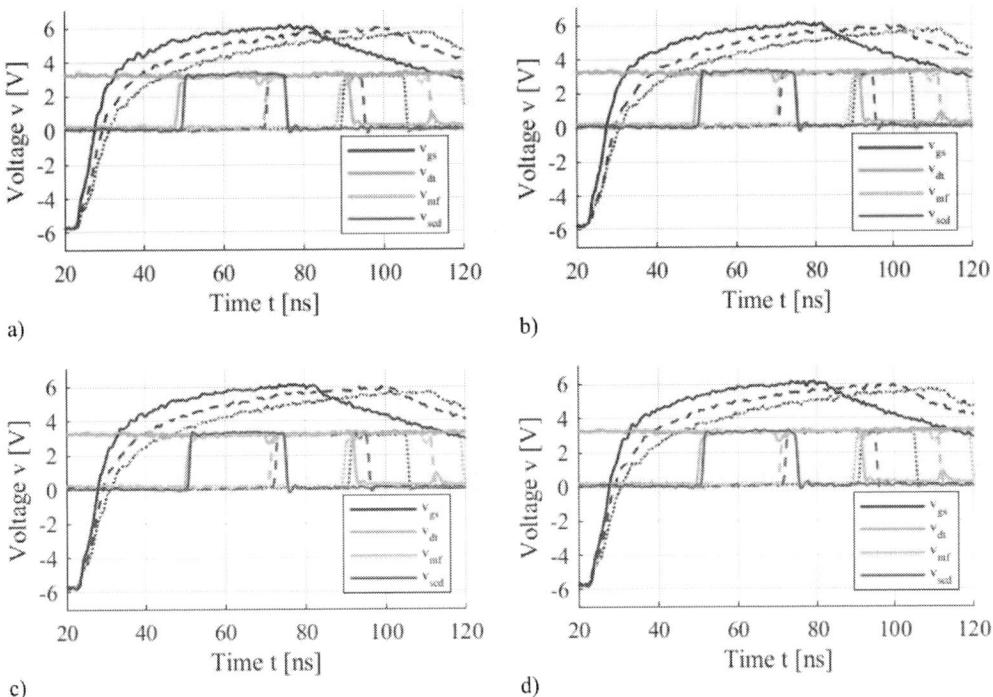

Fig. 2: Measurement of the gate-source voltage v_{gs}, the comparator output signals v_{dt} and v_{mf} and the logic AND gate output signal v_{scd} at HSF fault condition. a) V_{dc} of 100 V, b) V_{dc} of 200 V, c) V_{dc} of 300 V and d) V_{dc} of 400 V. Dotted: R_g of 15 Ω. Dashed: R_g of 10 Ω. Solid: R_g of 5 Ω. T_c of 25 °C.

Measurements have shown short circuit detection times of 30 ns in HSF condition at dc-link voltage V_{dc} of 400 V and gate resistance R_g of 5 Ω as well as 70 ns at dc-link voltage V_{dc} of 400 V and gate resistance R_g of 15 Ω at case temperature T_c of 25 °C.

A half-bridge configuration with two GaN eHEMTs GS66516T from GaN Systems is used for the measurements. All signals are measured differentially. 20 dB attenuators are used for the gate-source voltage v_{gs} measurement to protect the 50 Ω inputs of the oscilloscope against high voltage in case of breakdown. 50 Ω terminators are used for the measurement of the digital voltages. Figure 3 presents the short circuit detection times Δt_{scd}. The detection time Δt_{scd} increases with increasing gate resistance R_g. The dc-link voltage V_{dc} has almost no influence. The detection time is measured from the beginning of charging the gate of the GaN eHEMT to the rise of the logic AND gate output v_{scd}. This paper presents a detailed analysis on the relation of the switching behavior of the GaN eHEMT and the logic circuit.

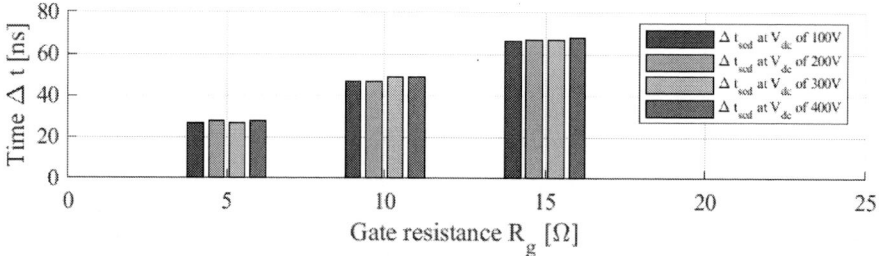

Fig. 3: Measurement of the short circuit detection times Δt_{scd} at gate resistance R_g of 5 Ω, 10 Ω and 15 Ω. Blue: V_{dc} of 100 V. Green: V_{dc} of 200 V. Cyan: V_{dc} of 300 V. Red: V_{dc} of 400 V. T_c of 25 °C.

Test setup for detailed investigations in different operating points

A double pulse setup is used for a detailed analysis on false detections under normal operating points. Figure 4 presents the schematic and the setup in the test-bench. The driver stage, the logic circuits and the half-bridge including the dc-link capacitors are located on a printed circuit board (PCB). The load inductor is applied via screw terminals. The differential measurements are done with coaxial cables and the 50 Ω inputs of the oscilloscope. Ferrite beads are used on each pair of coaxial cable to suppress EMI. Two oscilloscopes, MSO5204B and DPO5204B, from Tektronix are used and grounded close to the source pad of the low side GaN switch (device under test, DUT). Although a shunt increases the commutation loop inductance, the drain current is measured with a frequency compensated SMD shunt [6] to verify the operating point.

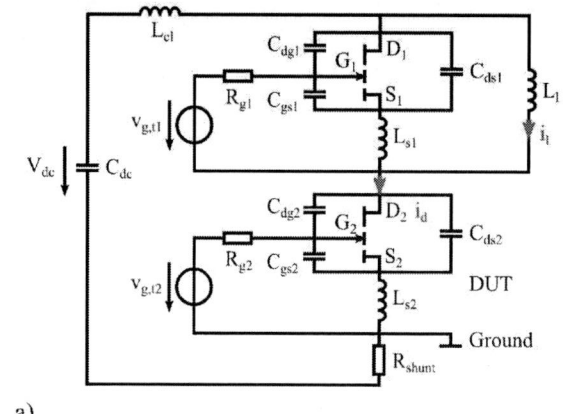
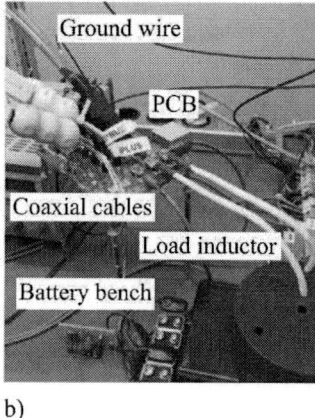

a)   b)

Fig. 4: Schematic and setup of the test-bench for measurements in double pulse configuration. a) Schematic with commutation loop, GaN eHEMT half-bridge and load inductor. b) Setup in test-bench with applied coaxial cables and ground wire to oscilloscopes. External battery bench to generate higher desaturation voltages $V_{desaturation}$ without overloading the driver stage.

The half-bridge consists of two enhancement mode GaN eHEMTs. The upper GaN eHEMT is used as freewheeling path and the lower GaN switch is used as DUT with the desaturation detection method attached. Figure 5 presents the PCB layout of the half-bridge. The commutation loop is designed on top and 1st mid layer of the four-layer PCB to minimize the commutation loop inductance L_{cl}. The gate loop is designed with additional polygons on all four layers to minimize the gate loop inductance.

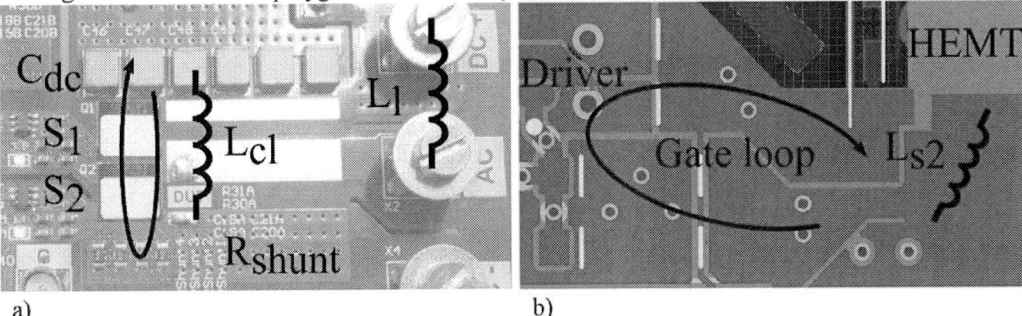

a)   b)

Fig. 5: Picture of the half-bridge configuration and the gate loop. a) Commutation loop with GaN eHEMTs S_1, S_2, commutation loop capacitor bench C_{dc}, parasitic commutation loop inductance L_{cl} and shunt. b) Gate loop with polygon arrangement and parasitic common source inductance L_{s2}.

Further, the gate signal source connection is designed as close as possible to the solder pad with a separated polygon to minimize the common source inductance L_{s2}. For the measurements at different case temperatures, a heat plate is attached on the top side cooling pads of the two GaN eHEMTs. The 650 V GaN eHEMTs GS66516T from GaN Systems with nominal drain current I_d of 60 A are used for all measurements. Table I presents the evaluated operating points of the influencing parameters.

Table I: Evaluated operating points of the influencing parameters

Parameter	Value
R_g	5 Ω, 10 Ω, 15 Ω
I_d	20 A, 40 A, 60 A, 80 / 83 A, 110 / 115 A, 130 / 135 / 140 A, 140 / 150 A
V_{dc}	100 V, 200 V, 300 V, 400 V, 450 V
T_c	25 °C, 40 °C, 50 °C, 60 °C, 70 °C, 80 °C, 90 °C, 100 °C, 110 °C, 120 °C, 125 °C

Influences of switching speed

By increasing the switching speed the duration of the miller plateau phase is reduced due to the faster charging of the miller capacitance. As a result, the time span $\Delta t_{dt,mf}$ between desaturation monitoring signal getting low and gate monitoring signal getting high is reduced as well. False detections of fault condition can occur, if the time span $\Delta t_{dt,mf}$ is too low and in the same order as the propagation delay of the logic AND gate generating the *short circuit detect* signal. Figure 6 presents the influence of the switching speed adjusted by R_g. The time span $\Delta t_{dt,mf}$ between the signal change v_{dt} and v_{mf} is measured considering the switching speed and the dc-link voltage V_{dc}.

Figure 7 presents the time span $\Delta t_{dt,mf}$ as function of the switching speed. As depicted the time span is reduced at faster switching and lower dc-link voltage V_{dc}. As mentioned, the miller plateau phase is shorter due to a faster charging of the miller capacitance. The *desat true* comparator output changes earlier to low state due to the faster decrease of the drain-source voltage v_{ds}. The *miller finished* comparator output changes much earlier to high state due to the fast charging miller capacitance. The combination of both effects overall reduces the time span $\Delta t_{dt,mf}$, which can lead to non-functioning blanking and false short circuit detections. The TLV3501 comparators with propagation delay of 4.5 ns and SN74LVC1G11 logic AND gate with propagation delay of 4.1 ns from Texas Instruments are used. A minimal time span $\Delta t_{dt,mf}$ of 12.3 ns is measured at R_g of 5 Ω and V_{dc} of 100 V. A voltage swing in the output of the logic AND gate v_{scd} can be observed by using a gate resistance R_g of 15 Ω. The voltage swing is present at all dc-link voltages V_{dc} but does not affect the proper working of the logic AND gate itself and is probably caused by the measurement equipment.

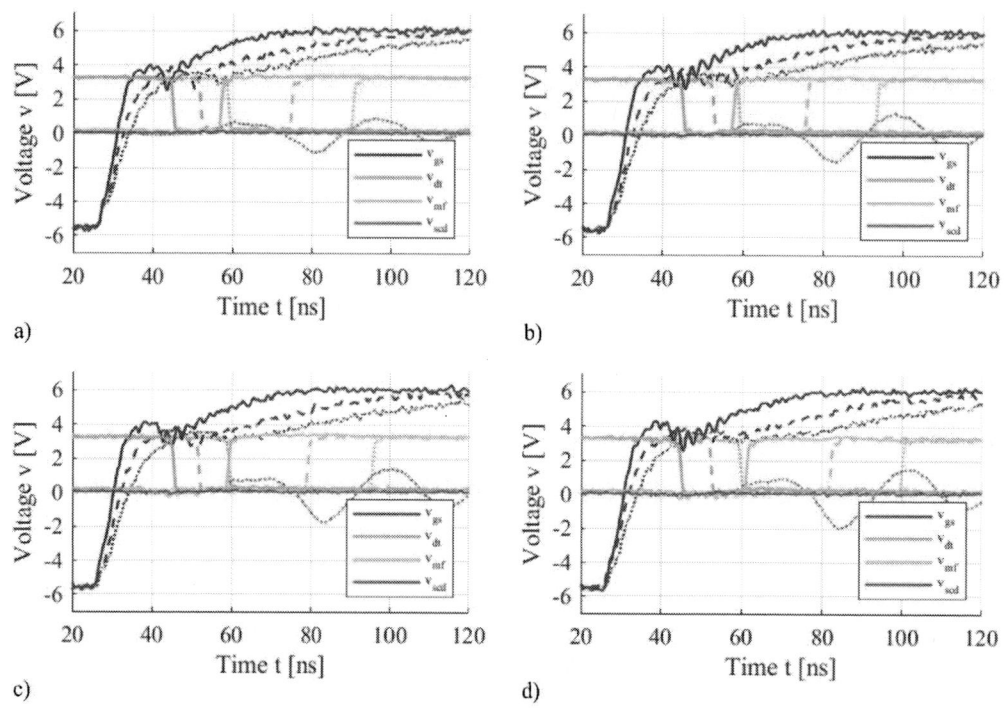

a)

b)

c)

d)

Fig. 6: Measurement of the gate-source voltage v_{gs}, the comparator output signals v_{dt} and v_{mf} and the logic AND gate output signal v_{scd} at normal commutation at drain current I_d of 60 A. a) V_{dc} of 100 V, b) V_{dc} of 200 V, c) V_{dc} of 300 V and d) V_{dc} of 400 V. Dotted: R_g of 15 Ω. Dashed: R_g of 10 Ω. Solid: R_g of 5 Ω. T_c of 25 °C.

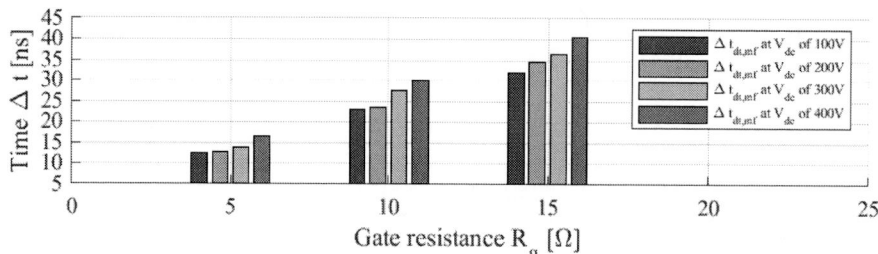

Fig. 7: Measurement of the time span $\Delta t_{dt,mf}$ between the signal change v_{dt} and v_{mf} considering the switching speed adjusted by R_g at different dc-link voltages V_{dc} and drain current I_d of 60 A. Blue: V_{dc} of 100 V. Green: V_{dc} of 200 V. Cyan: V_{dc} of 300 V. Red: V_{dc} of 400 V. T_c of 25 °C.

Influences of drain current

The height of the gate-source voltage v_{gs} at the beginning of the miller plateau depends on the rise of the drain current i_d due to the common source inductance L_s in the gate loop and the transfer characteristic of the transistor. Higher switching speed and higher drain currents therefore reduce the voltage to the defined reference voltage $V_{reference}$ of the highest miller plateau during normal operation. This voltage must not be exceeded by the gate voltage before the miller plateau phase is finished during regular operation. Figure 8 presents the behavior of the gate-source voltage v_{gs} at different drain currents I_d and different gate resistances R_g.

The gate-source voltage peak $v_{gs,peak}$ at normal commutation is enlarged by higher drain currents I_d and higher switching speed adjusted by R_g. The higher drain current I_d enlarges also the miller plateau voltage, whereas the higher switching speed does only affect the gate-source voltage peak $v_{gs,peak}$. By defining the reference voltage $V_{reference}$ the voltage peak $v_{gs,peak}$ has to be taken into account considering the desired operating condition. A reference voltage $V_{reference}$ of 4.78 V is used in the measurements. The reference voltage is crossed by the voltage peak in the operating points $\{V_{dc}, I_d, R_g\}$ of $\{300\ V, 130\ A, 5\ \Omega\}$, $\{300\ V, 140\ A, 10\ \Omega\}$ and $\{300\ V, 140\ A, 5\ \Omega\}$. Figure 9 presents the height of the voltage peak $v_{gs,peak}$ at the beginning of the miller plateau phase considering the operating points depicted in figure 8. By crossing the reference voltage, the *miller finished* comparator is triggered and the output changes to high state. This leads to a false failure detection. Figure 10 presents the output signal v_{mf} of the *miller finished* comparator concerning the operation points presented in figure 8. False triggering occurs also at $\{300\ V, 110\ A, 5\ \Omega\}$, $\{300\ V, 130\ A, 15\ \Omega\}$, $\{300\ V, 130\ A, 10\ \Omega\}$, $\{300\ V, 140\ A, 15\ \Omega\}$ and $\{300\ V, 140\ A, 10\ \Omega\}$.

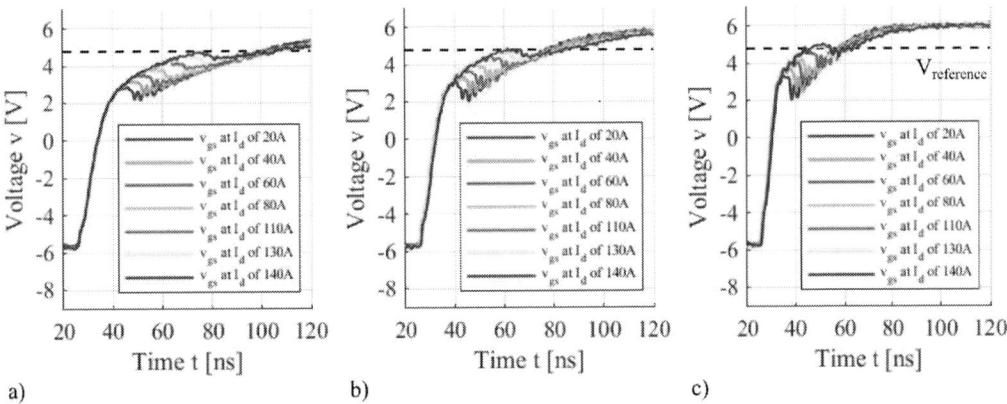

a)   b)   c)

Fig. 8: Measurement of the gate-source voltage v_{gs} at normal commutation with dc-link voltage V_{dc} of 300 V and different drain currents I_d of 20 A to 140 A. a) R_g of 15 Ω, b) R_g of 10 Ω, c) R_g of 5 Ω. T_c of 25 °C. Dashed: Defined reference voltage $V_{reference}$.

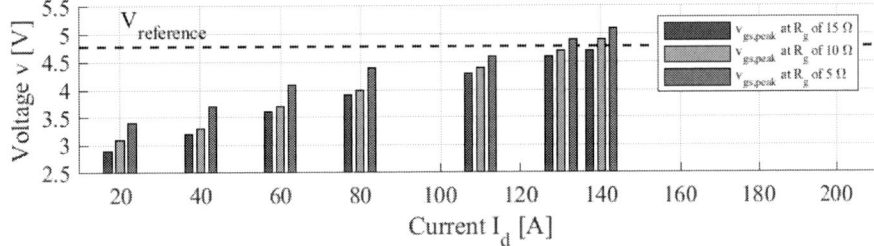

Fig. 9: Measurement of the gate-source voltage peak $v_{gs,peak}$ at the beginning of the miller plateau phase on normal commutation with dc-link voltage V_{dc} of 300 V, drain currents I_d of 20 A, 40 A, 60 A, 80 A, 110 A, 130 A, 140 A and gate resistances R_g of 15 Ω, 10 Ω, 5 Ω.

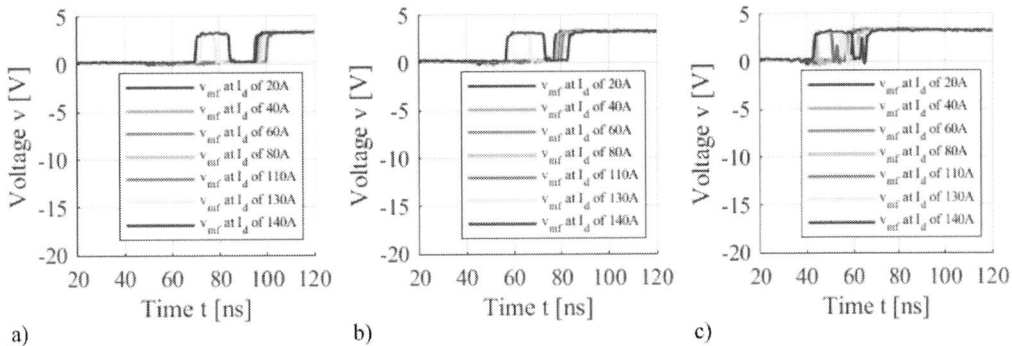

a)   b)   c)

Fig. 10: Measurement of the *miller finished* comparator output signal v_{mf} at normal commutation with dc-link voltage V_{dc} of 300 V and different drain currents I_d of 20 A to 140 A. a) R_g of 15 Ω, b) R_g of 10 Ω, c) R_g of 5 Ω. T_c of 25 °C.

Figure 11 presents the measurement of the *short circuit detect* logic AND gate output signal v_{scd} considering the operating points presented in figure 10. High drain currents near double rated current of 120 A lead to false detections, if the reference voltage $V_{reference}$ is set too low. As mentioned before, the voltage swings in the output voltage at R_g of 15 Ω have no impact on the proper working of the logic AND gate.

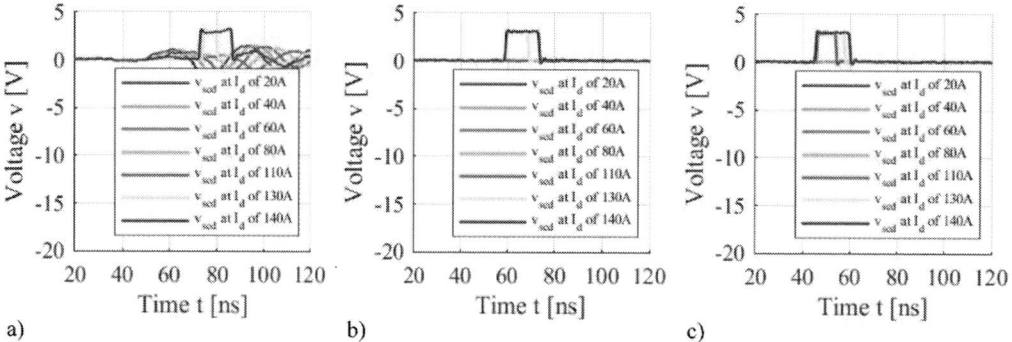

a)   b)   c)

Fig. 11: Measurement of the logic AND gate output signal v_{scd} at normal commutation with dc-link voltage V_{dc} of 300 V and different drain currents I_d of 20 A to 140 A. a) R_g of 15 Ω, b) R_g of 10 Ω, c) R_g of 5 Ω. T_c of 25 °C.

Influences of drain-source voltage

The dynamic of the drain-source voltage v_{ds} has also an impact on the behavior of the enhanced desaturation detection method, depending on the load current I_d and the gate resistor R_g. Figure 12 presents the behavior of the drain-source voltage v_{ds} at different drain currents I_d and different gate resistances R_g.

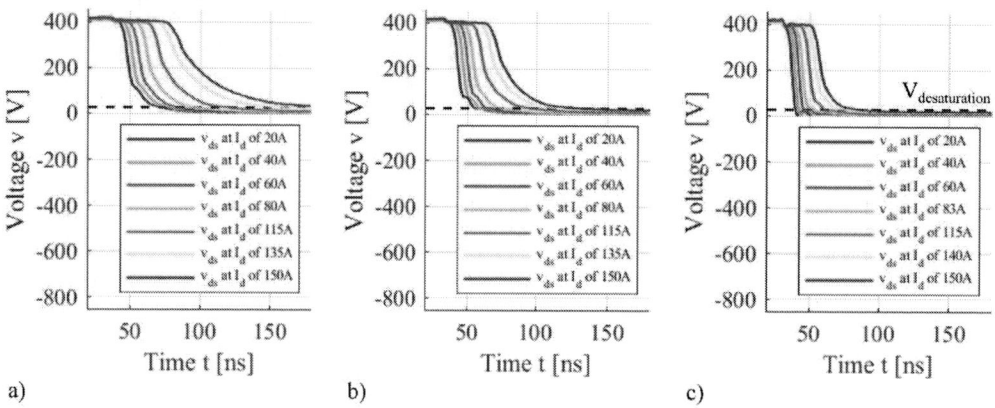

a)   b)   c)

Fig. 12: Measurement of the drain-source voltage v_{ds} at normal commutation with dc-link voltage V_{dc} of 400 V and different drain currents I_d of 20 A to 150 A. a) R_g of 15 Ω, b) R_g of 10 Ω, c) R_g of 5 Ω. T_c of 25 °C. Dashed: Defined desaturation voltage $V_{desaturation}$ of 25 V.

The crossing of the defined desaturation voltage $V_{desaturation}$ is delayed by higher drain currents I_d and higher gate resistances R_g. A desaturation voltage $V_{desaturation}$ of 25 V is used for the measurement. Figure 13 presents the behavior of the *desat true* comparator output signal v_{dt} considering the operating points depicted in figure 12. Figure 14 presents the measurement of the logic AND gate output signal v_{scd}.

The comparator reacts faster than the drain-source voltage does cross the defined desaturation voltage $V_{desaturation}$. The reason therefore is the parasitic capacitance C_{par} of the high voltage diode of the desaturation detection circuit. At high dv/dt of the drain-source voltage v_{ds} the capacitance of the diode is dominating and influences the RC voltage divider. This leads to a fast reaction of the *desat true* comparator. The transient of the drain-source voltage v_{ds} can be divided in a dynamic phase and a static phase. Figure 15 presents the measurement of the drain-source voltage v_{ds} and the *desat true* comparator input voltage $v_{dt,in}$. The IsoVu TIVH05 from Tektronix with $10\ M\Omega$ termination is used for the measurement of the comparator input voltage. The RFU02VSM8S from Rohm with maximum coupling capacitance of 4 pF is used as high voltage diode.

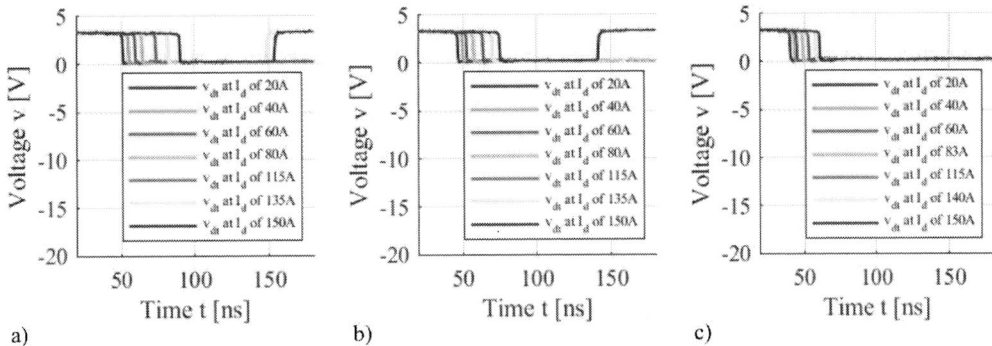

a)　　　　　　　　　b)　　　　　　　　　c)

Fig. 13: Measurement of the *desat true* comparator output signal v_{dt} at normal commutation with dc-link voltage V_{dc} of 400 V and different drain currents I_d of 20 A to 150 A. a) R_g of 15 Ω, b) R_g of 10 Ω, c) R_g of 5 Ω. T_c of 25 °C.

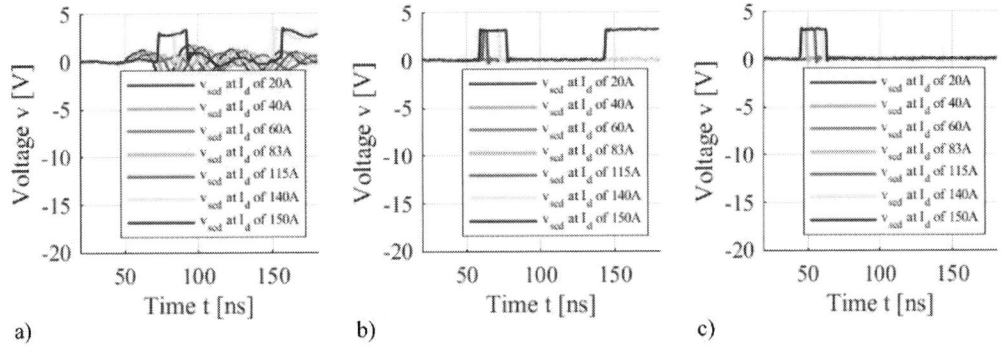

a)　　　　　　　　　b)　　　　　　　　　c)

Fig. 14: Measurement of the logic AND gate output signal v_{scd} at normal commutation with dc-link voltage V_{dc} of 400 V and different drain currents I_d of 20 A to 150 A. a) R_g of 15 Ω, b) R_g of 10 Ω, c) R_g of 5 Ω. T_c of 25 °C.

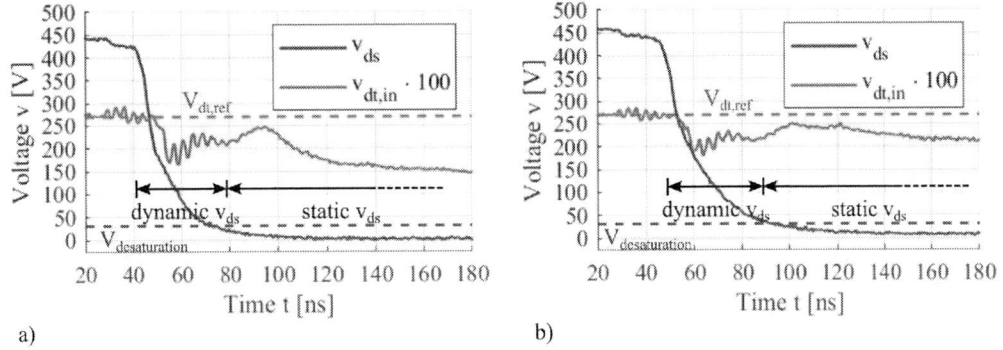

a)　　　　　　　　　　　　　　　b)

Fig. 15: Measurement of the drain-source voltage v_{ds} and the *desat true* comparator input voltage $v_{dt,in}$ at dc-link voltage V_{dc} of 450 V and drain current I_d of a) 60A and b) 80A.

The dv/dt induced current through the high voltage diode pulls down the comparator input voltage $v_{dt,in}$ during the dynamic phase. At the transition to the static v_{ds} phase, the comparator input voltage rises again considering the value of the RC voltage divider. A desaturation voltage $V_{desaturation}$ of 31 V is used for the measurements depicted in figure 15. At high drain-source voltage v_{ds} in the static phase, the comparator is triggered again and the output changes to high state. False triggering concerning high drain-source voltage v_{ds} in the static phase occurs at operating points $\{V_{dc}, I_d, R_g\}$ of $\{400\ V, 115\ A, 15\ \Omega\}$, $\{400\ V, 150\ A, 15\ \Omega\}$, $\{400\ V, 150\ A, 10\ \Omega\}$. This leads to a false failure detection.

Influences of case temperature

The temperature has an impact on the dynamic of the drain-source voltage v_{ds}. Figure 16 presents the behavior of the drain-source voltage v_{ds} at different case temperatures T_c. The dv/dt of the drain-source voltage v_{ds} is slower for higher case temperatures T_c. A higher drain current further reduces the fall time of the drain-source voltage v_{ds}. False triggering of the *desat true* comparator occurs at drain current I_d of 80 A and temperatures above 40 °C. Figure 17 presents the *desat true* comparator output v_{dt} belonging to the drain-source voltage v_{ds} depicted in figure 16.

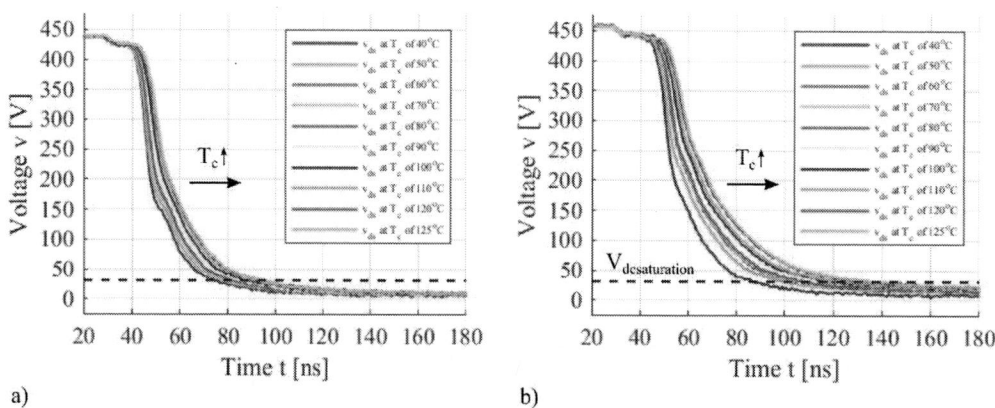

Fig. 16: Measurement of the drain-source voltage v_{ds} at normal commutation with dc-link voltage V_{dc} of 450 V and drain currents I_d of a) 60 A and b) 80 A at case temperatures T_c of 40 °C to 125 °C. R_g of 10 Ω. Dashed: Defined desaturation voltage $V_{desaturation}$ of 31 V.

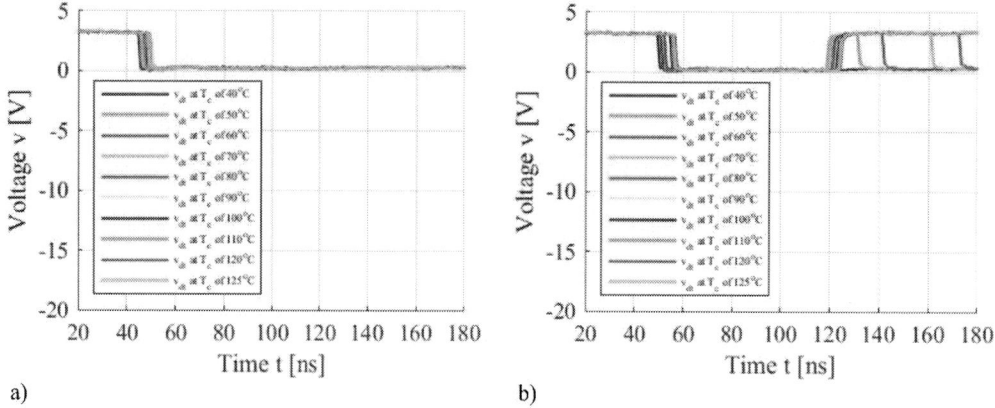

Fig. 17: Measurement of the *desat true* comparator output signal v_{dt} at normal commutation with dc-link voltage V_{dc} of 450 V and drain currents I_d of a) 60 A and b) 80 A at case temperatures T_c of 40 °C to 125 °C.

The false triggering of the *desat true* comparator in combination with the high state output of the *miller finished* comparator leads to false short circuit detections. Further, high temperatures have also an impact on the gate-source voltage v_{gs}. Figure 18 presents the behavior of the gate-source voltage v_{gs} at different case temperatures T_c. The voltage peak of the gate-source voltage $v_{gs,peak}$ at the beginning of the miller plateau phase is increased by higher case temperatures T_c. In combination with higher drain currents and higher di/dt, this can lead to false failure detections. The rise of v_{gs} to the driver on state voltage VCC of +6 V is decreased at higher case temperatures. This leads to a delayed change of the *miller finished* comparator output and results in a nearly constant blanking time regarding the time delay of the *desat true* comparator output change to low state depicted in figure 17.

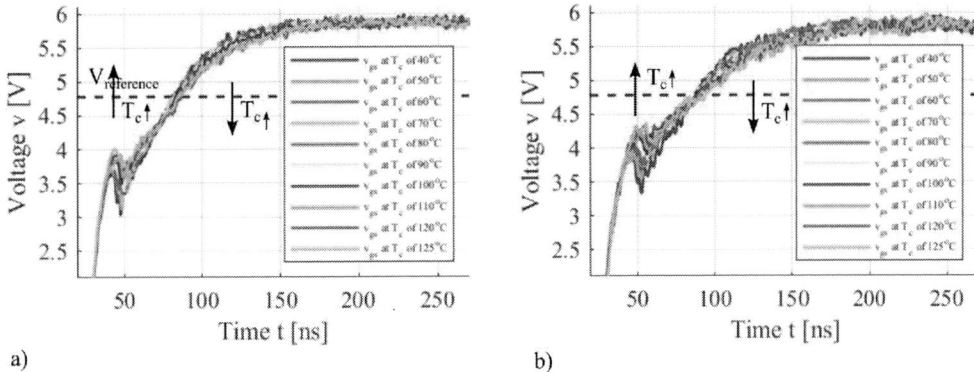

a)
b)

Fig. 18: Measurement of the gate-source voltage v_{gs} at normal commutation with dc-link voltage V_{dc} of 450 V and drain currents I_d of a) 60 A and b) 80 A at case temperatures T_c of 40 °C to 125 °C. Dashed: Defined reference voltage $V_{reference}$ of 4.78 V.

Conclusion

The enhanced short circuit detection method with adaptive blanking [5] offers fast reacting under hard switching fault condition. Further investigations have shown false detection under normal operating condition. Two reasons for false triggering can be observed. The first one is a too large voltage peak $v_{gs,peak}$ at the beginning of the miller plateau phase. The second one is the slow decrease of the drain source voltage v_{ds} while the gate-source voltage is already high. Using the GaN eHEMT GS66516T from GaN Systems with gate-source voltage of +/-6 V, the reference voltage $V_{reference}$ for the gate monitoring should be higher than +5 V to deal with double rated current and common source coupling as well as high temperature. Further, the desaturation voltage $V_{desaturation}$ should be higher than 100 V to deal with the slow decreasing drain-source voltage at high drain currents and high temperature. For distinguishing a double rated current from a hard switching fault short circuit condition by using a gate resistance higher than 10 Ω, the desaturation voltage should be even higher than 100 V.

References

[1] X. Huang, D. Y. Lee, V. Bondarenko, A. Baker, D. C. Sheridan, et al., "Experimental study of 650V AlGaN/GaN HEMT shortcircuit safe operating area (SCSOA)," in 2014 IEEE 26th International Symposium on Power Semiconductor Devices, IEEE, 2014.

[2] H. Li, X. Li, X. Wang, J. Wang, Y. Alsmadi, et al., "E-mode GaN HEMT short circuit robustness and degradation," in 2017 IEEE Energy Conversion Congress and Exposition (ECCE), IEEE, 2017.

[3] M. M. Bakran and S. Hain, "Integrating the New 2D Short circuit detection method into a power module with a power supply fed by the gate voltage," in 2016 IEEE 2nd Annual Southern Power Electronics Conference (SPEC), 2016, pp. 1-6.

[4] P. Hofstetter, S. Hain, and M. Bakran, "Applying the 2D-Short Circuit Detection Method to SiC MOSFETs including an advanced Soft Turn Off," in PCIM Europe 2018; International Exhibition and Conference for Power Electronics, Intelligent Motion, Renewable Energy and Energy Management, 2018, pp. 1-7.

[5] J. Schmitz, M. Meissner, F. Weiss and S. Bernet, "New fast short circuit detection method for SiC and GaN HEMT power semiconductors," PCIM Europe 2019; International Exhibition and Conference for Power Electronics, Intelligent Motion, Renewable Energy and Energy Management, Nuremberg, Germany, 2019, pp. 1-8.

[6] M. Meissner, J. Schmitz, F. Weiss and S. Bernet, "Current measurement of GaN power devices using a frequency compensated SMD shunt," PCIM Europe 2019; International Exhibition and Conference for Power Electronics, Intelligent Motion, Renewable Energy and Energy Management, Nuremberg, Germany, 2019, pp. 1-8.

Current Control of a Grid-Connected Single-Phase Voltage-Source Inverter with LCL Filter

Alfonso Parreño Torres*, Fco. Javier López-Alcolea**, Pedro Roncero-Sánchez**, Javier Vázquez**, Emilio J. Molina-Martínez** and Felix García-Torres***
*Institute of Industrial Development, Castilla-La Mancha Science and Technology Park
Avenida de la Investigación, 1
02006, Albacete, Spain
Tel.: +34 967555307
**School of Industrial Engineering, University of Castilla-La Mancha
Campus Universitario, S/N
13071, Ciudad Real, Spain
Tel.: +34 926295300 / Fax: +34 926295361
***Application Unit, Centro Nacional del Hidrogeno
Prolongación Fernando el Sato, S/N
13500, Puertollano, Spain
Tel.: +34 926420682
E-Mail: Alfonso.Parreno@pctclm.com, Fjavier.Lopez@uclm.es, Pedro.Roncero@uclm.es,
Javier.Vazquez@uclm.es, EmilioJose.Molina@uclm.es, felix.garcia@cnh2.es.
URL: http://www.uclm.es, http://www.cnh2.es

Acknowledgements

This work was supported by the European Regional Development Fund (ERDF) under the program Interreg SUDOE SOE3/P3/E0901 (Project IMPROVEMENT).

Keywords

Active Damping, Converter Control, Voltage Source Converter (VSC), Single Phase System

Abstract

The control in power electronic systems play an important role in the connection of renewable energy sources to the electrical grid. This paper presents the design of a current control for a grid-connected single-phase voltage source inverter (VSI). The VSI is connected to the grid through an LCL filter in order to attenuate the switching harmonics in the output signal. The proposed control provides an active damping at the resonance frequency and tracks the current reference with a fast dynamic response. The control scheme is based on a resonant regulator implemented in a structure with two nested controllers. The proposed control ensures a direct pole-placement of the closed-loop transfer function without the need for state observers and only measuring the grid current. Simulation results assess the validity of the control method.

Introduction

The increasing impact of renewable energy sources and the way in which they are connected to the electrical grid have attracted much attention for researchers during the last years. Voltage source inverters (VSI) with passive filters are typically employed to connect these distributed generation sources to the grid with the aim of maximizing the power injected to the grid. The passive filter is designed to limit the switching current harmonics generated by the VSI in order to accomplish with the different power quality standards [1]. A simple inductor can be used for the connection. However, a high inductance value must be used considering their high cost and great size. A solution to overcome these disadvantages is the use of an LCL filter, which is widely used and provides a better accomplishment of the power quality requirements as it offers an attenuation of 60 dB per decade above its cutoff frequency [2]. On the other hand, its natural resonance frequency and its high order must be carefully treated in order to obtain a good response for the current control method.

The solutions for the suppression of the LCL filter resonance can be divided into two groups: passive damping methods that are based on the physical introduction of a series resistor with the capacitor of the filter [3], and active damping methods that are focused on the compensation of the resonance by means of control systems. Owing to the superior behavior in relation with the losses and the performance, the active damping methods are preferred to the passive damping methods [4]. The active methods can also be divided into two groups. The first group is based on the concept of virtual resistance [5] [6], in which the main drawback is the need to measure the current or voltage of the capacitor, and an extra sensor is, therefore, necessary. The second group is based on adding a digital filter in cascade with a proportional-integral (PI) or proportional-resonant (PR) controller in the closed-loop system to obtain the attenuation of the high frequencies [7] [8]. Here, the filter is tuned to ensure that all the poles of the closed-loop transfer function are located inside the unit circle on the Z plane when the design is carried out in the discrete domain. Usually, low-pass or notch filters are used. The effectiveness of these schemes depends on the adjustment in the tuning of the filter parameters that allow to eliminate the resonance frequency.

A control solution with a feedback loop and, only, a PR controller is presented in [9]. However, the LCL filter design is conditioned in this approach, as the resonance frequency must be low. Another control method with only a feedback loop for the current can be found in [10], nevertheless the number of design parameters needed to design the control is very high.

This paper presents a new current control method for a single-phase VSI connected to the grid by means of an LCL filter. The control scheme is based on a structure with two nested controllers that allows to define the location of all the poles of the closed-loop transfer function. Although this could also be achieved with a state-feedback control, the proposed scheme allows to locate the poles without the need of observers reducing, thus, the number of variables to be measured, as it is only necessary a feedback loop for the grid current.

The paper is organized as follows. The transfer function of the power system is depicted in "System Modeling" Section, while the "Proposed Control Method" Section presents the design procedure for the controllers. Simulation results are shown to validate the proposed control method and evaluate its performance in Section "Simulation Results". Finally, the main conclusions are drawn in the last Section.

System Modeling

The electrical scheme of a VSI connected to the grid (or point of common coupling, PCC) through an LCL filter is shown in Fig. 1. The model of the system is defined by the LCL filter, supposing the switching frequency of the VSI to be sufficiently high. In such case, the open-loop transfer function of the system that relates the voltage generated for the VSI, $U(s)$, with the grid current, $I_2(s)$, can be expressed in the continuous domain by means of the following third-order equation:

$$P_1(s) = \frac{I_2(s)}{U(s)} = \frac{s\beta_4 + \beta_3}{s^3 + s^2\beta_2 + s\beta_1 + \beta_0} \tag{1}$$

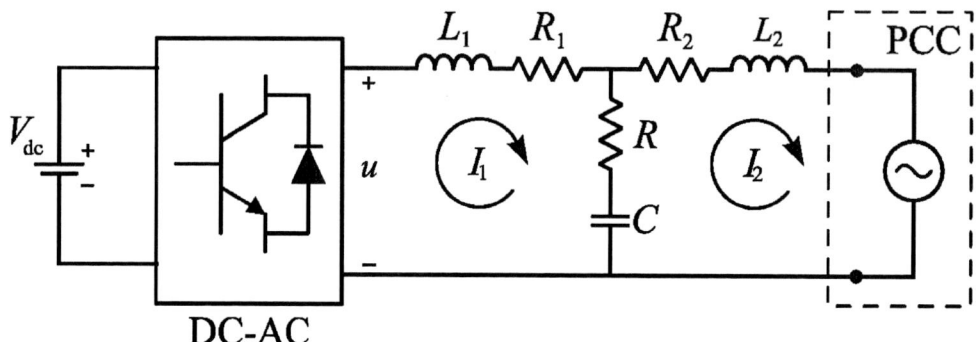

Fig. 1: Power system of the grid-connected single-phase voltage source inverter with an LCL filter.

being:

$$\beta_4 = \frac{R}{L_1 L_2}; \qquad \beta_3 = \frac{1}{L_1 L_2 C}; \qquad \beta_2 = \frac{R+R_1}{L_1} + \frac{R+R_2}{L_2}$$

$$\beta_1 = \frac{1}{L_1 C} + \frac{1}{L_2 C} + \frac{R_1 R_2 + R(R_1 + R_2)}{L_1 L_2}; \qquad \beta_0 = \frac{R_1 + R_2}{L_1 L_2 C}$$

(2)

where C is the capacitor and L_1 and L_2 are the self-inductances of the VSI side and the grid side of the LCL filter, respectively. R is the equivalent series resistance of the capacitor and R_1 and R_2 correspond to the equivalent series resistances of the inductances.

Since the control system will be implemented on a digital platform, it is convenient to include the time delay caused by the calculations needed for control purposes. Consequently, $P_1(s)$ can be rewritten as:

$$P_2(s) = \frac{e^{-\tau s}(s\beta_4 + \beta_3)}{s^3 + s^2 \beta_2 + s\beta_1 + \beta_0}$$

(3)

where τ is the time delay owing to the operations in the digital platform that can be considered equal to one sampling period.

The LCL filter is designed following the steps detailed in [2] and considering a switching frequency, f_{sw}, of 10 kHz and a nominal voltage of, U_{nom}, of 400 V. According to this procedure, the values obtained for the LCL filter are: L_1=0.9 mH, R_1=0.257 Ω, L_2=1.8 mH, R_2=0.115 Ω, C=25 µF and R=0.05 Ω.

Proposed Control Method

In this section, the proposed control method that permits a direct pole-placement of the closed-loop transfer function is introduced. Usually, for high order systems, as is the case for a VSI connected to the electrical grid via an LCL filter, the performance of a PR controller can be very poor since it cannot define the location of all the poles of the closed-loop transfer function. With the aim of solving this problem, the control scheme shown in Fig. 2 is used. The proposed control is based on a nested configuration with two controllers that results in a two degrees of freedom structure. As a result, the closed-loop transfer function can be defined as follows:

$$H(s) = \frac{P_2(s)R_1(s)}{1 + P_2(s)(R_1(s) + R_2(s))}$$

(4)

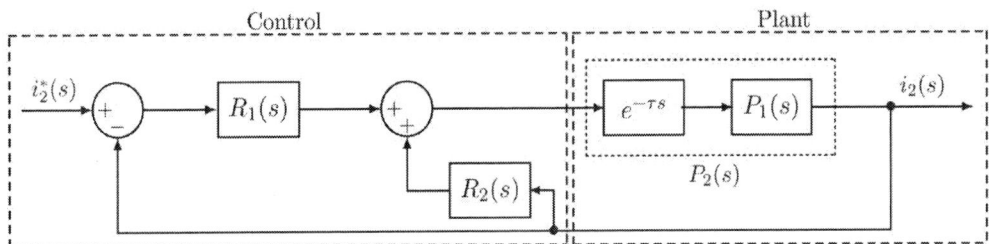

Fig. 2: Proposed control scheme.

The control methodology is focused in the definition of the poles of the closed-loop transfer function, whereas the zeros cannot be defined. A proper selection of the transfer functions $R_1(s)$ and $R_2(s)$ derives in a closed-loop transfer function in which the number of design parameters is equal to the number of poles. In order to minimize the number of zeros in the closed-loop transfer function shown in equation (1), it is useful that the numerator of $R_1(s)$ contains the smallest number of zeros as possible (minimum number of design parameters). On the contrary, the numerator of $R_2(s)$ should contain the maximum number of zeros (maximum number of design parameters). Moreover, the denominator of $R_1(s)$ should include a resonant term in order to achieve zero tracking error for a sinusoidal reference at the fundamental frequency of the grid voltage, ω_1. Considering those three conditions, along with the need for both controllers to be causal, the transfer functions for both regulators can be written as:

$$R_1(s) = \frac{s^2 K_1 + s K_0}{\left(s^2 + \omega_1^2\right)\left(s^2 + s\rho_1 + \rho_0\right)}; \qquad R_2(s) = \frac{s^2 K_4 + s K_3 + K_2}{\left(s^2 + s\rho_1 + \rho_0\right)} \qquad (5)$$

where K_0, K_1, K_2, K_3 and K_4 are the design parameters in the numerator of $R_1(s)$ and $R_2(s)$, ρ_0 and ρ_1 are the design parameters in the denominator of $R_1(s)$ and $R_2(s)$.

Operating with equations (3) and (4), the characteristic polynomial of the closed-loop transfer function, $F(s)$, can be obtained as:

$$F(s) = \left(s^3 + s^2\beta_2 + s\beta_1 + \beta_0\right)\left(s^2 + \omega_1^2\right)\left(s^2 + s\rho_1 + \rho_0\right) + \left(s^2 + \omega_1^2\right)\left(s^2 K_4 + s K_3 + K_2\right)\left(s\beta_4 + \beta_3\right)e^{-\tau s} +$$
$$+ \left(s K_1 + K_0\right)\left(s\beta_4 + \beta_3\right)e^{-\tau s} \qquad (6)$$

$F(s)$ is a seventh order polynomial which depends on the plant and the design parameters. It is important to remark that the number of the closed-loop poles is equal to the number of design parameters.

The parameters of controllers $R_1(s)$ and $R_2(s)$ can be directly calculated by substituting the desired locations of the seven poles of the closed-loop transfer function in the characteristic polynomial $F(s)$ that must be equal to zero ($F(p_i=0)$), being $p_i : i \in \{1,2,\dots,7\}$. As a result, there is a system of seven linear equations that can be written in a matrix form as:

$$\mathbf{A}x = \mathbf{B} \qquad (7)$$

where x is the matrix composed by the design parameters ($x = [\,\rho_0 \; \rho_1 \; K_0 \; K_1 \; K_2 \; K_3 \; K_4]^{\mathrm{T}}$), $\mathbf{A} \in \Re^{7\times1}$ is the matrix formed by the elements that depend of the design parameters, and $\mathbf{B} \in \Re^{7\times1}$ is the matrix composed by the elements which not depend of the design parameters. The elements of matrices $\mathbf{A}$ and $\mathbf{B}$ are defined as follows:

$$a_{i,1} = p_i^6 + p_i^5 \beta_2 + p_i^4 \left(\beta_1 + \omega_1^2 \right) + p_i^3 \left(\beta_0 + \beta_2 \omega_1^2 \right) + p_i^2 \beta_1 \omega_1^2 + p_i \beta_0 \omega_1^2$$

$$a_{i,2} = p_i^5 + p_i^4 \beta_2 + p_i^3 \left(\beta_1 + \omega_1^2 \right) + p_i^2 \left(\beta_0 + \beta_2 \omega_1^2 \right) + p_i \beta_1 \omega_1^2 + \beta_0 \omega_1^2$$

$$a_{i,3} = p_i^5 \beta_4 e^{-\tau p_i} + p_i^4 \beta_3 e^{-\tau p_i} + p_i^3 \beta_4 \omega_1^2 e^{-\tau p_i} + p_i^2 \beta_3 \omega_1^2 e^{-\tau p_i}$$

$$a_{i,4} = p_i^4 \beta_4 e^{-\tau p_i} + p_i^3 \beta_3 e^{-\tau p_i} + p_i^2 \beta_4 \omega_1^2 e^{-\tau p_i} + p_i \beta_3 \omega_1^2 e^{-\tau p_i} \qquad (8)$$

$$a_{i,5} = p_i^3 \beta_4 e^{-\tau p_i} + p_i^2 \beta_3 e^{-\tau p_i} + p_i \beta_4 \omega_1^2 e^{-\tau p_i} + \beta_3 \omega_1^2 e^{-\tau p_i}$$

$$a_{i,6} = p_i^2 \beta_4 e^{-\tau p_i} + p_i \beta_3 e^{-\tau p_i}$$

$$a_{i,7} = p_i \beta_4 e^{-\tau p_i} + \beta_3 e^{-\tau p_i}$$

$$b_i = -p_i^7 - p_i^6 \beta_2 - p_i^5 \left(\beta_1 + \omega_1^2 \right) - p_i^4 \left(\beta_0 + \beta_2 \omega_1^2 \right) + p_i^3 \beta_1 \omega_1^2 + p_i^2 \beta_0 \omega_1^2$$

The design parameters for the controllers $R_1(s)$ and $R_2(s)$ that achieve the desired closed-loop poles can be calculated by inverting matrix **A** and solving: $x = A^{-1} \cdot B$. Note that for A^{-1} to exist, matrix **A** must be square and full rank [11].

The full rank requirement for matrix **A** is fulfilled when all the desired closed-loop poles are different, that is, the multiplicity of every pole is equal to one ($m_{p_i}=1$). However, even considering that the multiplicity of any pole can be $m_{p_i}>1$, it is possible to obtain a full rank matrix composed by independent equations, as the derivative of the characteristic polynomial $F(p_i)$ is also equal to zero. Thus, a generalized equation for the derivatives of F(s) that must be calculated for a pole with a multiplicity m_{p_i} can be derived:

$$\frac{d^{m_{p_i}-1} H(p_i)}{dp_i^{m_{p_i}-1}} = 0 \qquad (9)$$

Design Example

The methodology described in the previous section has been implemented to design the current control of the system depicted in Fig. 1. The dynamical behavior of the system is defined with a proper selection of the poles of the closed-loop transfer function. It is important to remark that, as the delay is an exponential function, a fourth order Padé approximation is used to represent the final locations of all the poles [12].

The chosen location for the poles are: p_1=-1000 rad/s and p_2=-4000 rad/s with a multiplicity of 6 (m_{p_2}=6). Fig. 3 shows the root locus of the design system with the dominant poles. However, Fig. 3 does not show the 4 poles introduced by the Padé approximation for the delay. The final location of all the poles is detailed in Table I. As can be seen, the Padé approximation poles ($p_{8,9}$ and $p_{10,11}$) are not dominant and their influence in the dynamical behavior can be neglected.

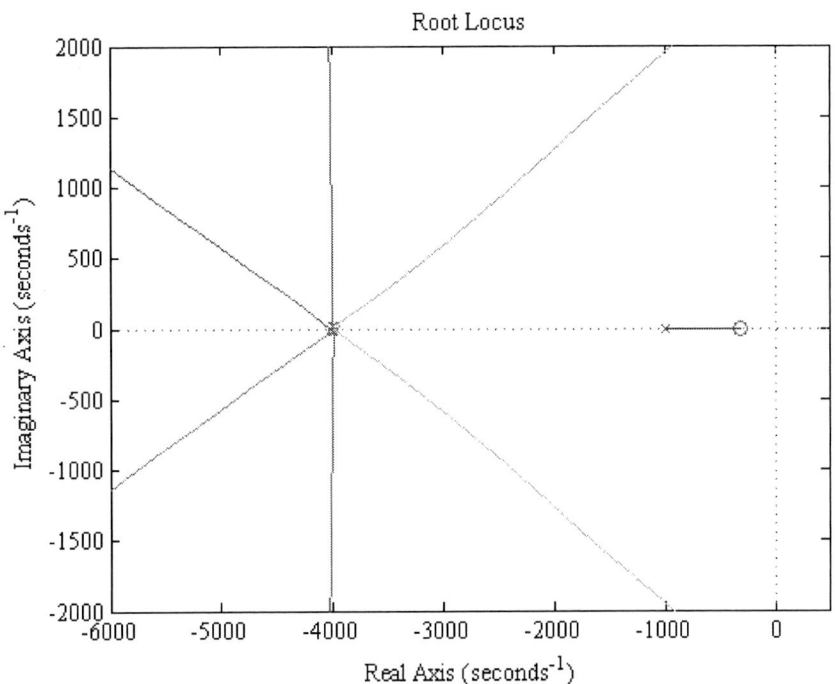

Fig. 3: Root locus plot for the proposed control designed.

Table I: Location of the closed-loop poles with the delay included

p_i	Location (rad/s)
p_1	-1000 rad/s
p_2	-4000 rad/s
p_3	-4000 rad/s
p_4	-4000 rad/s
p_5	-4000 rad/s
p_6	-4000 rad/s
p_7	-4000 rad/s
$p_{8,9}$	$-43392 \pm j53144$ rad/s
$p_{10,11}$	$-52693 \pm j4912$ rad/s

The Bode plot for the proposed control designed in the continuous domain and the one obtained after the system discretization with a sampling frequency of 10 kHz are shown in Fig. 4.

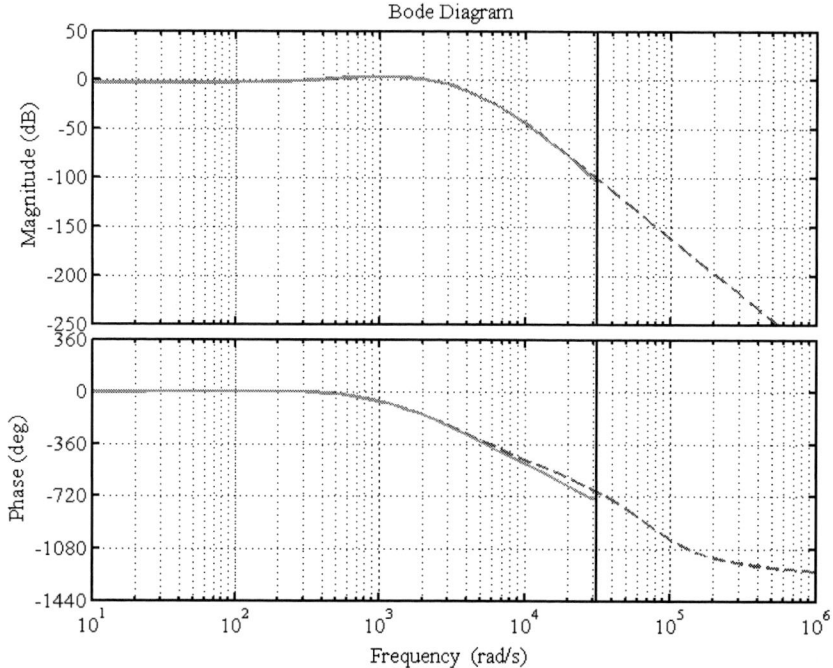

Fig. 4: Bode plot for the proposed control designed in the continuous domain (blue dashed line) and the system obtained after discretization at sampling frequency (green line).

Simulation Results

The performance of the current control presented has been tested using PSCAD/EMTDC. The power system implemented for simulations is similar to that shown in Fig.1 with a DC bus voltage of 670 V.

The simulation comprises the following steps: 1) the reference signal for the current is generated at $t=0$ s with a rms value of 8 A. 2) at $t=0.1$ s, the reference signal change to 4 A. 3) the simulation finishes at $t=0.2$ s.

Fig. 5 shows the reference signal and the grid current. As can be seen, the grid current is practically sinusoidal and the transient response is very fast (5 ms, approximately). The current in the VSI side is shown in Fig. 6. As it was expected, this current has a harmonic content higher than the current in the grid side.

Conclusion

This paper presents a current control scheme for a single-phase VSI with an LCL filter. The scheme is based on a resonant controller implemented in a structure with two nested regulators. This configuration allows to choose the location of all the poles of the closed-loop transfer function system. Unlike other approaches, it is not necessary to use state-observers and the number of variables to be measured is reduced. The simulation results corroborate the good performance of the current control.

Current Control of a Grid-Connected Single-Phase Voltage-Source Inverter with LCL Filter   PARREÑO TORRES Alfonso

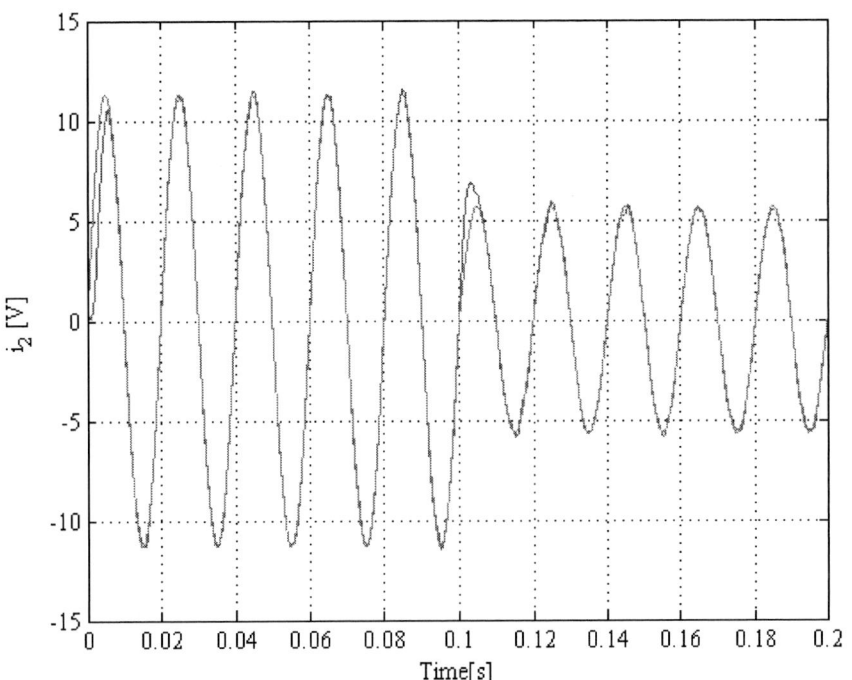

Fig. 5: Control response for the grid current (blue) and reference signal (red).

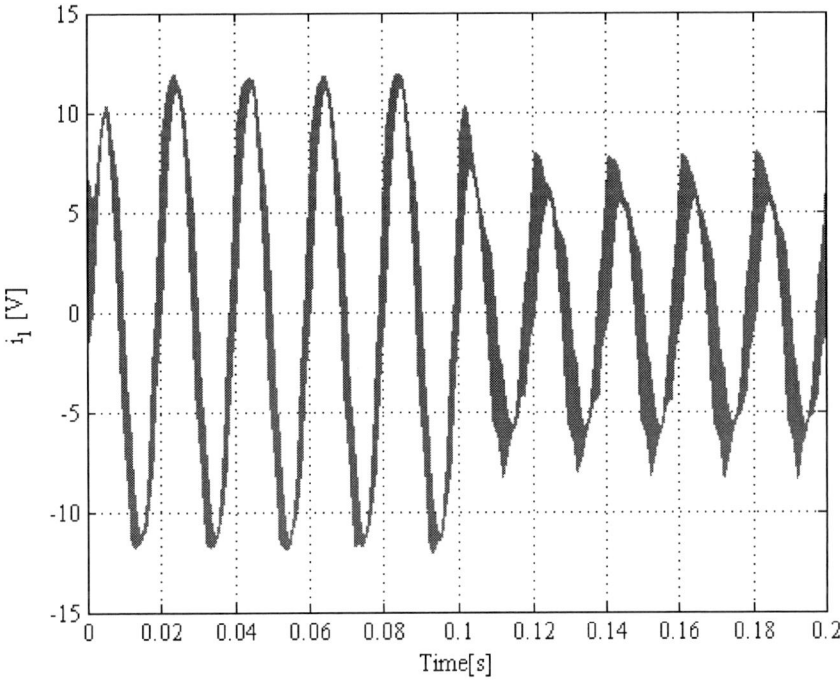

Fig. 6: Control response for the current of the voltage source inverter side.

References

[1] IEEEStd. 1547, IEEE Std. 1547-2003. *Standard for interconnecting distributed resources with electric power systems*. June 2003.

[2] M. Liserre, F. Blaabjerg, S. Hansen. "Design and control of an LCL filter-based three-phase active rectifier." *IEEE Transactions on Industry Applications,* vol. 41, no. 5, pp. 1281-1291, September 2005.

[3] R. N. Beres, X. Wang, M. Liserre, F. Blaabjerg, C. L. Bak. "A review of passive power filters for three-phase grid-connected voltage-source converters," *IEEE Journal of Emerging and Selected Topics in Power Electronics,* vol. 4, no. 1, pp. 54-69, March 2016.

[4] R. Guzman, L. G. de Vicuña, J. Morales, M. Castilla, J. Miret. "Model-based active damping control for three-phase voltage source inverters with LCL filter," *IEEE Transactions on Power Electronics,* vol. 32, no. 7, pp. 5637-5650, July 2017.

[5] X. Wang, F. Blaabjerg, P. C. Loh. "Grid-current-feedback active damping for LCL resonance in grid-connected voltage-source converters," *IEEE Transactions on Power Electronics,* vol. 31, no. 1, pp. 213-223, January 2016.

[6] Y. Chen, Z. Xie, L. Zhou, Z. Wang, X. Zhou, W. Wu, L. Yang, A. Luo. "Optimized design method for grid-current-feedback active damping to improve dynamic characteristic of LCL-type grid-connected inverter," *Electrical Power and Energy Systems,* vol. 100, pp. 19-28, September 2018.

[7] J. Dannehl, M. Liserre, F. W. Fuchs. "Filter-based active damping of voltage source converters with LCL filter," *IEEE Transactions on Industrial Electronics,* vol. 58, no. 8, pp. 3623-3633, August 2011.

[8] R. Peña Alzola, M. Liserre, F. Blaabjerg, M. Ordonez, T. Kerekes. "A self-commissioning notch filter for active damping in a three-phase LCL-filter-based grid-tie converter," *IEEE Transactions on Power Electronics,* vol. 29, no. 12, pp. 6754-6761, December 2014.

[9] R. A. Fantino, C. A. Busada, J. A. Solsona. "Optimum PR control applied to LCL filters with low resonance frequency," *IEEE Transactions on Power Electronics,* vol. 33, no. 1, pp. 793-801, January 2018.

[10] C. A. Busada, S. G. Jorge, J. A. Solsona. "Full-state feedback equivalent controller for active damping in LCL-filtered grid-connected inverters using a reduced number of sensors," *IEEE Transactions on Industrial Electronics,* vol. 62, no. 10, pp. 5993-6002, October 2015.

[11] A. Parreño Torres, P. Roncero-Sánchez, V. Feliu. "A Two Degrees of Freedom Resonant Control Scheme for Voltage-Sag Compensation in Dynamic Voltage Restorers," *IEEE Transactions on Power Electronics,* vol. 33, no. 6, pp. 4852-4867, June 2018.

[12] J. R. Partington. "Some frequency-domain approach to the model reduction of delay systems," *Annual Reviews in Control,* vol. 28, no. 1, pp. 65-73, January 2004.

Four Switch Buck/Boost Converter for DC Microgrid Applications

Matthias Schulz**, Nico Schleippmann**, Kilian Gosses**, Bernd Wunder[†], Martin März[†]
**Fraunhofer Institute for Integrated Systems and Device Technology IISB
[†]Chair of Power Electronics, Friedrich-Alexander University Erlangen-Nuremberg
Schottkystraße 10
Erlangen, Germany
Tel.: +49 / (0) – 9131 761 582
Fax: +49 / (0) – 9131 761 390
E-mail: matthias.schulz@iisb.fraunhofer.de
URL: www.iisb.fraunhofer.de

Acknowledgment

The authors acknowledge the financial support by the Federal Ministry for Economic Affairs and Energy of Germany in the framework of the project ETIBLOGG (project number 01MD18006F).

Keywords

«DC Microgrid», «Converter Circuit», «Pulse Width Modulation», «Converter Control», «Batteries»

Abstract

This paper describes a bidirectional, non-inverting, and non-isolated DC-DC converter, which connects different electronic devices up to a maximum operational voltage of 60 V. Special considerations about the transition between step-down and step-up are made and have been implemented in a fully functional prototype. The converter can operate in a stand-alone mode or can work in parallel with a DC microgrid. Therefore an algorithm was implemented to cooperate with a droop-controlled bus voltage. The paper issues an implementation of a novel modulation scheme and gives a comparison to existing schemes. A steady-state analysis shows the way of charging the bootstrap capacitor. Some effects on the length of the bootstrap interval are evaluated. Additionally the advantage of using a burst mode implementation is shown. Another important aspect in this paper is the control and design of the power electronic circuit of the converter itself. A prototype with 97.5% peak efficiency and 300 W nominal power was built. The theoretical analysis is verified by electrical measurements of the converter.

Introduction

A lot of electronic devices in the human's daily life works with direct current (DC). Often used safety extra low (SELV) voltages are between 5 V and 60 V [1]. Therefore, in modern buildings a DC Microgrid installation is becoming more and more an interesting alternative compared to the traditional alternating current (AC) grid [2]. In Fig. 1 an AC grid connected (1) DC Microgrid is depicted schematically. Installed battery storages (2) can be charged in times of cheaper electricity rate or excess regenerative energy generation [3]. For the electrical distribution inside the building a lower voltage DC (LVDC) level (3) is used because of the lower conduction losses or thinner cable diameters. Due to the lack of an international standard, different voltage levels are common for example 350 V,

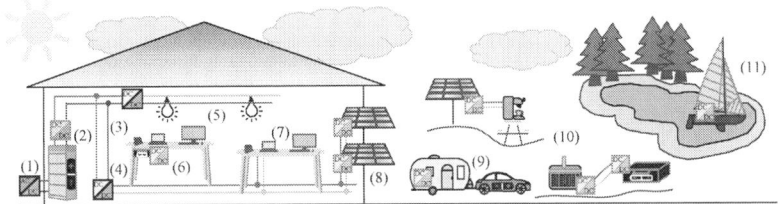

Fig. 1: Schematic of different use cases for the non-isolated and bidirectional DC-DC converter (drawn in yellow), which is the main focus of this paper.

380 V or 400 V [4]. A bidirectional and galvanically isolated DC-DC converter (4) connects these building distribution buses to a room distribution bus in one of the mentioned SELV ranges. 48 V is used for example as nominal voltage of the LED lighting installation (5), whereby the voltage range is between 36 V and 52 V [5], [6]. Mobile battery systems (6), which are utilized to decouple office desks from the room infrastructure, for example have a nominal voltage of 24 V with a voltage range from 20 V to 29 V [7]. Telecommunication or office equipment (7) works with many different voltages, in the example application 12 V, 19 V and 5 V are used with a USB-C plug. A façade photovoltaic (PV) panel (8) has a nominal voltage of 24 V and the open circuit voltage $V_{PV,oc}$ is 47.5 V [8]. The devices in the steady growing market for camping accessories (9), (10) and sailing boats (11) mostly use DC from batteries. For such outdoor activities, a nominal voltage of 12 V is common, whereby the range is defined between 9 V and 20 V [9]. In AC grids the standard converter for every single device is a galvanically isolated AC-DC converter, which is known as switched mode power supply (SMPS). In case of a DC Microgrid and to do justice to the basic ideas of a highly efficient and resource-oriented implementation an universal DC-DC converter for all the mentioned devices on the SELV level is still missing here.

1. Topology and Characteristics

For all the electronic devices as shown in Fig. 1 a topology is needed, which can operate from grid voltages above, below or equal to the desired output voltage. For charging/discharging a battery, to control the power flow inside a DC Microgrid or to connect devices independent of input and output side a bidirectional characteristic is additional necessary. Analyzation of the power demand offers a nominal power P_{nom} of 300 W and a nominal current I_{nom} of 10 A on both sides. The switching frequency f_{sw} is set to constant 200 kHz by specification. In [7] the authors did a literature research and an investigation of five eligible topologies. As a result, the H-bridge topology of the four-switch buck/boost (FSBB) converter, depicted in Fig. 2, is the most suitable one. Additional circuits to achieve soft switching can be added [10] or different modulation schemes can be implemented [11]-[13] to increase the efficiency.

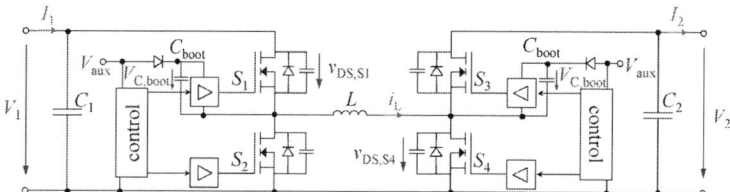

Fig. 2: Schematic of the H-bridge topology of the four-switch buck-boost DC-DC converter including the bootstrap drive circuit and the bootstrap capacitor C_{boot}.

2. Modulation scheme with triangular shaped inductor current

An overview of the different modulation schemes of the FSBB topology in step-down operation is given in Fig. 3. Next to established schemes (a)-(d) the respective novel one is illustrated (e). v_L shows the voltage across the inductor, i_L is the current through the inductor and the voltages $v_{DS,S1}$ and $v_{DS,S4}$ are the drain-to-source voltages of the transistors S_1 and S_4. The black i_L line illustrates the steady-state inductor current, the grey line the inductor current for an altering output power. In [7] the modulation scheme is shown for step-down and step-up, but there the focus is on the whole system. For a detailed analyzation of switching and the components the step-down mode is chosen in the sections below. This mode is defined by V_1 being higher than V_2, which is drawn in Fig. 2.

2.1 Operating principle of the CCM-mASC scheme in step-down

To design the DC-DC converter the high-side gate driver supply is realized via bootstrapping. Instead of additional circuitry, a modulation scheme with switched low-side transistors should be used to charge the bootstrap capacitor. Modulation schemes shown in Fig. 3 (a) and (b) enable only the use of the intrinsic diodes of S_1 and S_3 [14]. Also for (b), the switching frequency f_{sw} has to be variable.

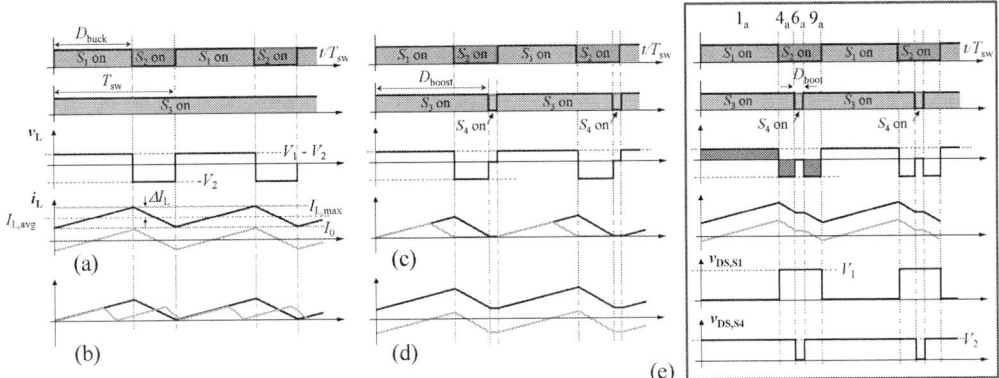

Fig. 3: Key waveforms with triangular shaped inductor current of the FSBB converter in step-down mode: (a) continuous conduction mode (CCM); (b) boundary conduction mode (BCM); (c) discontinuous conduction mode (DCM); (d) continuous conduction mode with active short-circuit (CCM-ASC); (e) novel continuous conduction mode with middle active short-circuit (CCM-mASC).

Thus, the possible modulation schemes [15], which can be implemented in the converter are at last (c), (d) or (e). The established modulation schemes (c) and (d) have fewer intervals. The novel scheme CCM-mASC has a total of ten intervals of operation. Four of them are depicted in Fig. 3 (e). All ten modes are shown in following Fig. 4 in detail.

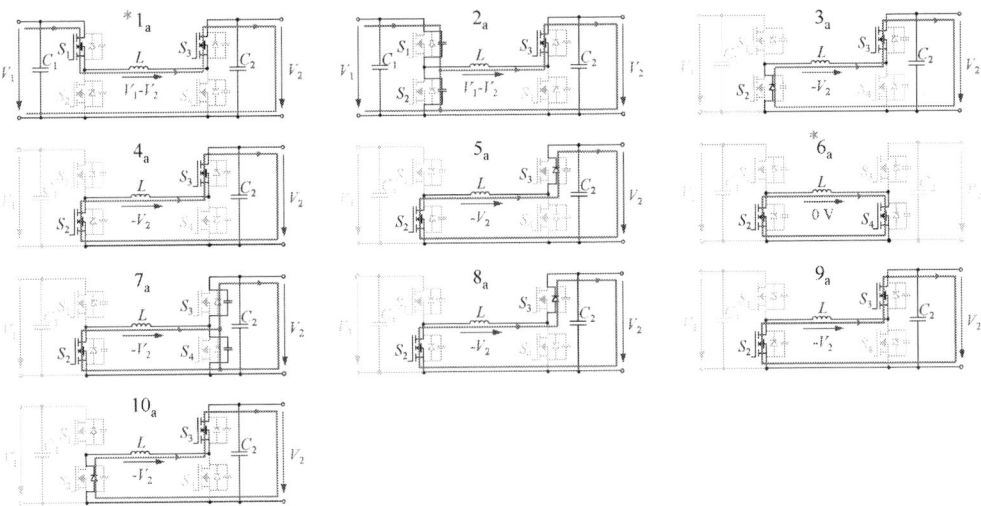

Fig. 4: Modes of operation in step-down direction for CCM-mASC scheme: The main intervals are 1_a, in which the inductor current i_{Lc} increases, intervals 4_a as well as 9_a, in which the inductor current decreases and interval 6_a. in which the inductor is shorted and the bootstrap capacitor is charged.

Mode 1_a: With the converter being in step-down operation the load voltage and current is controlled by the transistors S_1 and S_2. Starting in this interval means S_1 is turned on and the positive voltage v_L across the inductor L increases i_L. The non-switching half bridge (in this case S_3 and S_4) is supposed to conduct the whole inductor current to the load. Therefore the upper transistor S_3 is turned on to lower conduction losses caused by the voltage drop of the intrinsic diode.

Mode 2_a: In this interval the channel of the transistor S_1 is turned off and the associated parasitic drain-to-source capacitance C_{DS} is charged. No switching losses occur if the channel is turned off fast enough.

Mode 3_a: As soon as the voltage across S_2 falls below zero, i_L commutates to its intrinsic diode.

Mode 4_a: The channel of S_2 can now be turned on under zero voltage switching (ZVS) condition.

Mode 5_a: The gate charge for the upper transistor S_3 is supplied by the bootstrap capacitor. As there is a small leakage current caused by the capacitor and transistor a constant turn on is not feasible. By controlling the upper transistor S_3 as synchronous rectifier there are no switching losses for S_3 as the current i_L commutates to the intrinsic diode of transistor S_3 for this very short time instant 5_a.

Mode 6_a: In this interval the channel of S_4 is turned on hard and the bootstrap capacitor will be charged.

Mode 7_a: However for the lower transistor there are no turn off losses because the current through the inductor is split between the parasitic drain-to-source capacitance of S_4 and S_3.

Mode 8_a: Because of the direction of i_L the intrinsic diode of transistor S_3 conducts.

Mode 9_a: As soon as interval 8_a is achieved the channel of S_3 can be turned on at ZVS.

Mode 10_a: Before S_1 is turned on in the next interval the channel of S_2 is turned off and for a short time instant the intrinsic diode conducts. After this mode a new switching cycle begins. In steady-state operation the average inductor current $I_{L,avg}$ is not changed.

Overall by using CCM-mASC in two of ten modes switching losses occurs, both intervals are marked with a green starlet in Fig. 4. In the next section the waveform of the inductor current will be analyzed because at the moment the discontinuous conduction mode (DCM), illustrated in Fig. 3 (c), is another possibility for implementation as well.

2.2 Ripple current analysis

To calculate the ripple current ΔI_L the normalized voltage $|V_2|$ and the normalized current $|I_2|$ are defined as:

$$|V_2| = \frac{V_2}{V_1} = D_{buck} \tag{1}$$

$$|I_2| = \frac{L f_{sw}}{V_1} I_2. \tag{2}$$

ΔI_L is depicted in Fig. 3 (a). For Fig. 5 the maximum voltage $V_{1,max}$ of 60 V is selected to design the converter for the worst case condition. The ripple current as function of the duty cycle in step-down mode, D_{buck}, can be calculated with the following equation:

$$\Delta I_L = \frac{I_2}{2|I_2|} D_{buck}(1 - D_{buck}) \tag{3}$$

whereby D_{buck} is defined in equation (1) and valid only in CCM.

In Fig. 5 the solid lines show the ripple current ΔI_L for different inductors. The area below marks the DCM and above the CCM. Based on the example of 5 µH the trend of the duty cycle in both operation modes is illustrated. The dashed line represents half of the nominal current I_{nom}, which decreases linear due to the maximum power P_{nom} for $V_2 > V_1/2$. For an overall operation in the DCM a small value for the inductor has to be selected. However, for an inductor with 3 µH the maximum peak current $I_{L,max}$ is calculated to 25 A for I_{nom} and V_2 of 30 V. At 10 µH and higher the converter is in CCM at the nominal operating point. The peak current decreases farther, so that for example $I_{L,max}$ is equal to 11.5 A at 25 µH.

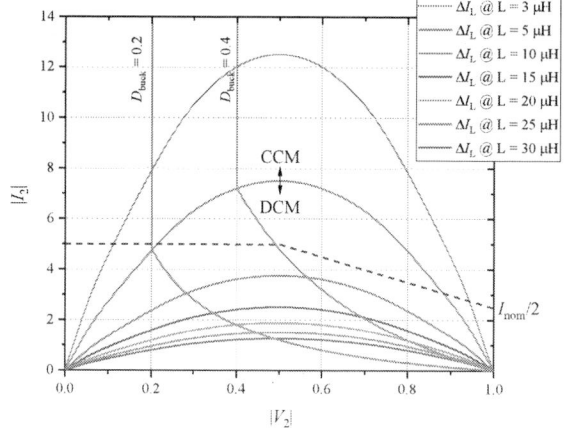

Fig. 5: Evaluation of the ripple current ΔI_L for different inductors as function of the lower voltage normalized to 60 V on the higher voltage side.

The root-mean-square (RMS) of the current through the inductor depends on the waveform and can be calculated with the equations in following Table I.

Table I: RMS equations for the inductor current i_L.

	DCM	boundary	CCM
RMS of i_L	$\dfrac{I_{L,max}}{\sqrt{3}}$	$\dfrac{2}{\sqrt{3}} I_{load}$	$\sqrt{I_{nom}^2 + \dfrac{(2\Delta I_L)^2}{12}}$

For an overall operation in the DCM a small inductor value is necessary and for an operation in the boundary mode additional the switching frequency f_{sw} depends on the momentary output power. However, the nominal current results in a higher RMS value because of the peak current. Additional the capacitors C_1 and C_2 have to compensate a higher ripple current. Therefore an operation in CCM is preferable.

The difference between using modulation scheme (d) or (e) is the level of the inductor current. During the inductor freewheeling phase there is a constant current with $I_{L,avg}$ in scheme (e) whereas the DC current is lower by ΔI_L in scheme (d). The inductor value L is calculated by:

$$L = \frac{1}{2 \cdot \Delta I_L} \frac{V_{1,max}}{4} \frac{1}{f_{sw}} \tag{4}$$

1 to 2% of the averaged value $I_{L,avg}$ at the nominal operation point is a common range of values for ΔI_L. The authors use a 22 µH inductor with PQ core (MnZn) and a DC resistance of 10.65 mΩ (WE-HCF 7443632200). For the following Fast Fourier Transformation (FFT) comparison between (a) and (e) the 22 µH inductor was used, V_1 was set to 48 V and V_2 to 28 V. ΔI_L is approximately 1.5 A and depicted in Fig. 6.

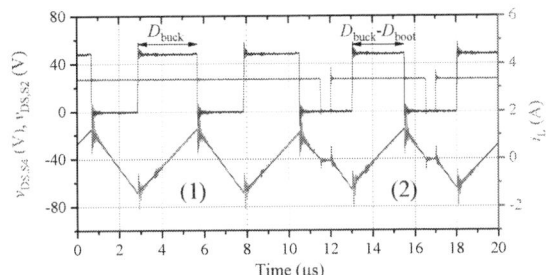

Fig. 6: Measured voltage and current waveforms for the FFT analysis.

For both (1) and (2) in Fig. 7 the output power is set to 10% of the nominal power P_{nom}. The Total Harmonic Distortion (THD) of (1) is measured to 16% and 19.1% for (2). By increasing the output power the DC offset getting higher. A low THD corresponds to a good quality and lower losses. In consequence the bootstrap capacitor should be only charged if it is absolutely necessary.

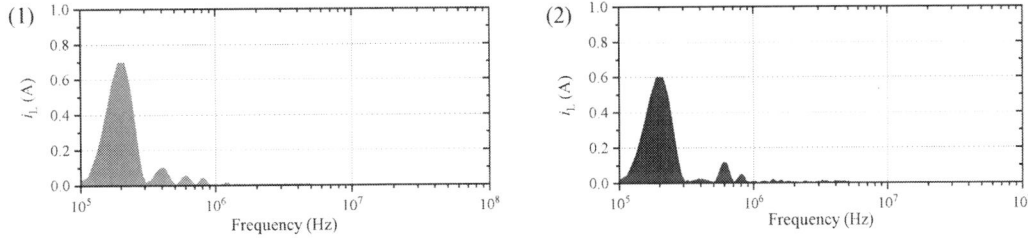

Fig. 7: FFT analysis of the continuous conduction mode (1) of Fig. 3 (a) and the CCM-mASC modulation scheme (2) at V_1 equals 48 V and V_2 of 28 V, whereby the power at higher voltage side is 30 W.

2.3 Charging the bootstrap capacitor

To evaluate the switching behavior two different silicon (Si) MOSFETs are analyzed. For the transistors the authors compared FDMT800100DC, which has one of the lowest on-resistance $R_{DS,on}$ (2.3 mΩ), but high output capacitance C_{oss} (1160 pF), input capacitance C_{iss} (5595 pF) and gate charge Q_g (79 nC) [16] and the transistor BSC040N10NS5 ($R_{DS,on}$ is equal to 4.0 mΩ, C_{oss} of 630 pF, C_{iss} of 4100 pF and Q_g of 58 nC) [17]. All values are nominal values. First, the influence of the parasitic capacitances is shown in Fig. 8. Hereby V_{grid} is set to 30 V, V_{load} equals 24 V and i_L is 1 A.

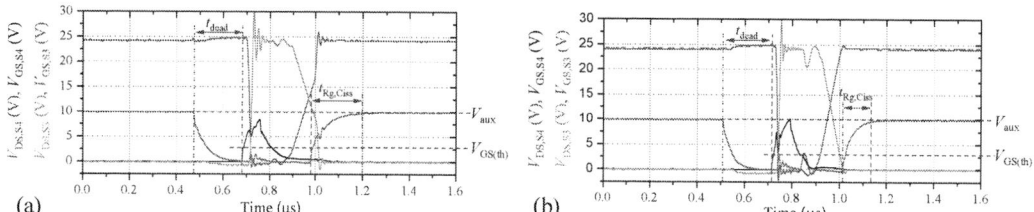

(a)   (b)

Fig. 8: Gate-to-source and drain-to-source voltages of S_3 and S_4 for two different transistors: (a) FDMT800100DC; (b) BSC040N10NS5. The input capacitance C_{iss} has an effect on the minimum duty cycle for charging the bootstrap capacitor.

200 ns are implemented as dead time t_{dead} for the half-bridge switches as shown in Fig. 8 (a) and Fig. 8 (b). Additionally the length of the on pulse is 100 ns. Subsequently the duty cycle D_{boot} is implemented to 0.06. The gate resistor R_g is selected to 5 Ω for both measurements. $t_{Rg,Ciss}$ is defined as the time for the gate-to-source voltage V_{GS} to reach 99.3% of the auxiliary supply voltage V_{aux}. By multiplication of the gate resistor R_g, the input capacitance C_{iss} of the transistor and the constant factor five (equals $5\tau_{RC}$), the time interval can be calculated. According to Fig. 8 (a) and Fig. 8 (b) a smaller on time of $V_{GS,S4}$ can be achieved, if a transistor with lower parasitic capacitances is used. Otherwise the transistor is switched hard as in Fig. 8 (a), because the capacitive charge is not fully transferred. In Fig. 8 (b) the drain-to-source voltage of S_3 is zero at the moment the channel is actively switched on. To achieve a higher efficiency, therefore the transistor BSC040N10NS5 is used in the prototype.

As section 2.1.1 and Fig. 8 show, switching losses due to bootstrapping during the turn on of the transistor S_4 occur. In order to reduce those losses, this operation can be skipped for several switching cycles. In the next section a method is shown to find the overall lowest semiconductor losses.

2.4 Bootstrap charging technique to reduce switching losses

To avoid switching losses caused by recharge pulses in every switching cycle a burst mode operation is proposed. In the linear region of the output characteristics of the transistor the drain-to-source on resistance decreases approximately exponentially with the gate-to-source voltage V_{GS}. By fitting the $R_{DS,on}$ sampling points of the datasheet [17] the $R_{DS,on}$ can be mapped to the time

$$R_{DS,on} = 3.35 \text{ m}\Omega + 127.645 e^{-0.835251 \cdot \frac{V_{GS}}{V}} \text{ m}\Omega \tag{5}$$

with
$$V_{GS} = 11.5 \text{ V} - 677.2 \cdot t. \tag{6}$$

As conduction losses decrease but switching losses increase with the burst frequency, there is an optimum for total transistor losses. The burst frequency is defined as 15 bootstrap recharge switching cycles followed by an adaptive amount of switching cycles without recharging the capacitor. The switching losses depend on both the output voltage of the non-switching half bridge and the inductor current whereas the conduction losses are just dependent on the inductor current. In Fig. 9 the total losses are mapped to V_{GS} and the output voltage of the non-switching half bridge. At the V_{GS} range of minimum losses there is only a slight change for different output voltages V_2 of the non-switching half bridge. In the present example minimum losses are estimated for V_{GS} from 6 to 9 Volts and burst frequencies f_{burst} from 120 Hz to 270 Hz. Additionally the dependency of the losses on the inductor current is depicted in Fig. 10 in which the losses are normalized to the inductor current. With the conduction losses as quad-

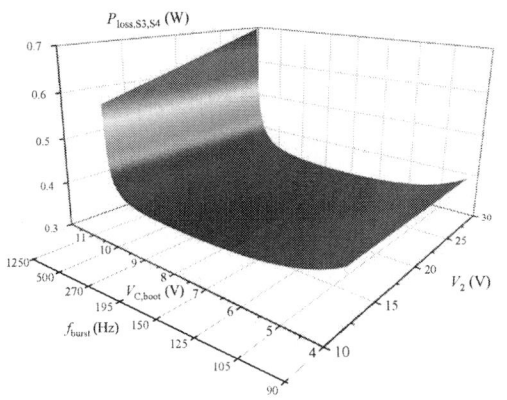 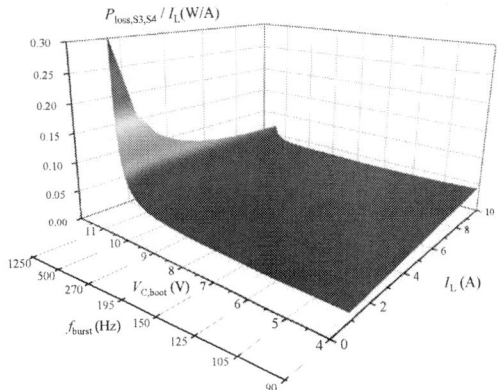

Fig. 9: Total losses of synced half bridge as function of bootstrap capacitor voltage and output voltage of the synced half bridge with I_L of 10 A.

Fig. 10: Total losses normalized to inductor current of synced half bridge as function of bootstrap capacitor voltage and inductor current with V_2 of 29 V.

ratic function of the inductor current the normalized conduction losses still rise linearly with i_L. Therefore the normalized total losses $P_{loss,S3,S4}$ in Fig. 10 increase linearly with the inductor current for very low burst frequencies. For very high burst frequencies and small inductor currents the switching losses represent a huge share of the normalized total losses as these rise significantly with lower currents.

In Fig. 11 based on the gate-to-source voltages V_{GS} of transistor S_3 and S_4 the implemented modes are shown. The implemented burst mode, according to Fig. 9 and Fig. 10, with a period T_{burst} of 3.7 ms is depicted in Fig. 11 (a). The voltage drop of the gate-to-source voltage $\Delta V_{GS,S3}$ is measured to 2.4 V. The auxiliary supply drops down from 9.9 V. Fig. 11 (b) shows the voltages for the transition mode, whereby between the switching intervals the intrinsic diode of S_3 is used for rectification. In Fig. 11 (c) both transistors will be switched with 200 kHz because in this figure the converter is in step-up mode.

Fig. 12 shows an additional measurement result of an efficiency comparison between continuous switching of transistor S_4 and burst mode implementation. The measurement conditions are V_1 equals 30 V and V_2 is 24 V. The measured benefit in efficiency is 1.1% by implementing the burst mode, according to Fig. 9 and Fig. 10 and depicted in Fig. 11.

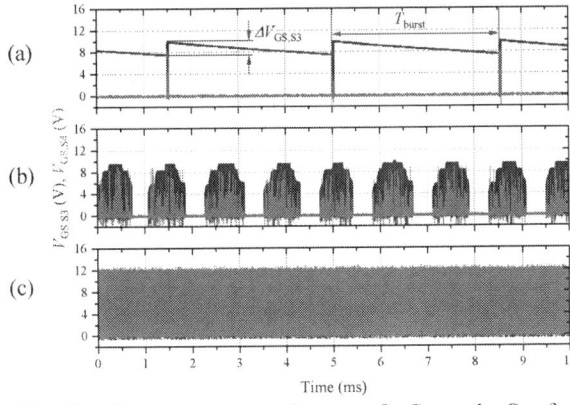 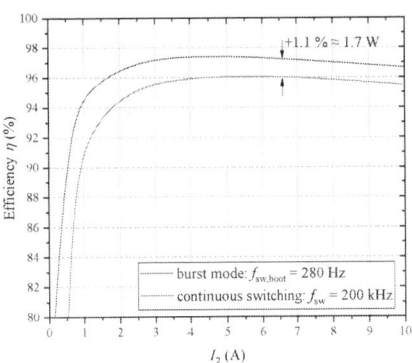

Fig. 11: Gate-to-source voltages of S_3 and S_4 for (a) $V_1 = 30$ V, $V_2 = 24$ V; (b) $V_1 = 24$ V, $V_2 = 24$ V; (c) $V_1 = 15$ V, $V_2 = 24$ V.

Fig. 12: Benefit of efficiency between burst mode and continuous switching of S_4.

3. Mode Transition Technique and Digital Control

The control scheme of the proposed DC-DC converter is depicted in following Fig. 13. A cascaded control structure, consisting of two proportional integral (PI) controllers, regulates the current I_2, the voltage V_2, the current I_1 or the voltage V_1. In the inner loop the inductor current i_L is still used. The control parameters of the outer loop can be static for stand-alone operation or altered via a communication bus from a graphical user interface (GUI).

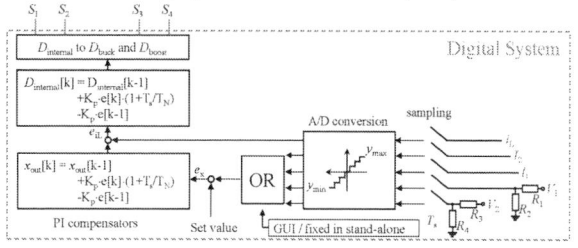

Fig. 13: Control scheme of the proposed DC-DC converter.

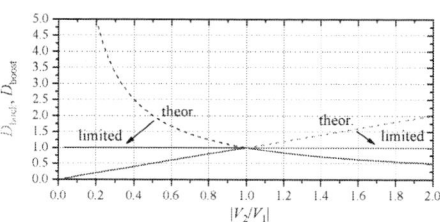

Fig. 14: Duty cycles for step-down and step-up mode, the limits as well as the resultant one.

The block "$D_{buck,internal}$ to D_{buck} and D_{boost}" limits the duty cycles as follows:

$$D_{buck,internal} \leq 1.0: \qquad \begin{aligned} D_{buck} &= D_{buck,internal} \\ D_{boost} &= 1.0 \end{aligned} \qquad (7)$$

$$D_{buck,internal} > 1.0: \qquad \begin{aligned} D_{boost} &= 2.0 - D_{buck,internal} \\ D_{buck} &= 1.0 \end{aligned} \qquad (8)$$

So for the transition between step-down ($V_1 > V_2$) and step-up ($V_1 < V_2$) mode, a fixed duty cycle limit can be used. Due to the definition of D_{buck} and D_{boost} for S_1 and S_3 respectively at the transition both duty cycles are equal. By means of clamping one duty cycle the system changes from a step-down to a step-up converter and vice versa. This illustrates Fig. 14, in which the theoretically calculated duty cycles [19] in green and blue, the limit lines as well as the resulting one in red are shown. The calculated $D_{buck,intern}$ can take values between zero and two, whereas $D_{boost,intern}$ can theoretically become infinite.

4. Prototype and experimental verification

In this section some measurement results are presented, which are performed with the converter prototype and the mobile battery system shown in Fig. 15. The parameters of the used semiconductors, the inductor as well as voltage and power ratings were mentioned before. The dimension of the prototype is 150 x 60 x 15 mm (*length* x *width* x *height*). The characteristics of the mobile battery system are listed in Table II.

Fig. 15: Photography of the prototype besides the mobile battery system.

Table II: Characteristics of the mobile battery system.

Parameters	Symbol	Values
Energy	E_{bat}	960 Wh
Nominal voltage	V_2	24 V
End-of-discharge voltage	$V_{2,min}$	20 V
End-of-charge voltage	$V_{2,max}$	29 V
Temp. range charging	T_{charge}	0°C – 45°C
Temp. range discharging	$T_{discharge}$	-20°C – 60°C
Dimension	l x w x h	240 x 190 x 95 mm
Weight	m	6 kg

In the following results the mobile battery system is connected on V_2 side, so in case of this example step-down means $V_2 > V_1$ and step-up $V_2 < V_1$.

4.1 Waveform under steady-state condition for CCM-mASC

Fig. 16 shows the voltage across the inductor v_L and the choke current i_L in case of charging the battery from V_1 side. Referring to Fig. 3, the drain-to-source voltages of transistors S_1 and S_4 are also measured.

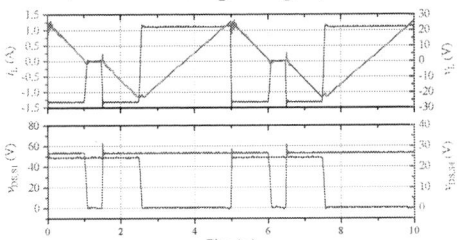

Fig. 16: Modulation scheme CCM-mASC for V_1 equal 48 V and V_2 equal 24 V; red: inductor current i_L measured with an AC current probe, green: voltage across the inductor v_L, blue: drain-to-source voltage of S_1, yellow: drain-to-source voltage of S_4.

4.2 Soft-start implementation, load transient response and hold-up time

One example of the start-up implementation is depicted in Fig. 17 (a). In this case V_1 is generated from the battery on V_2 side. The transition between step-down ($V_2 < V_1$) and step-up ($V_2 > V_1$) mode is marked and it is shown that only a little effect on the voltage v_1 is discernible. Fig. 17 (b) also shows that the converter can handle an operating point near the transition mode. The measured settling time $\Delta t_{settling}$ is 5 ms, the voltage undershoot ΔV_{under} is equal to 500 mV and the rising time Δt_{rise} of i_L is 1.9 ms. One load dump result is depicted in Fig. 17 (c) for V_1 of 48 V at approximately 288 W. At this example a cable with 0.5 m length is connected and the load current I_1 ramps down with 8 A/μs. The voltage overshoot $V_{1,dump}$ is measured to 8.42 V with an AC probe.

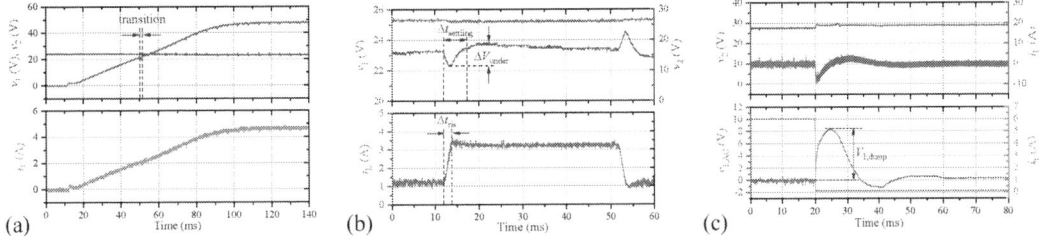

Fig. 17: (a) start-up at $V_2 = 24$ V for constant V_1 voltage (CV) to 48 V and 10 Ω grid load; (b) load step at $V_2 = 25$ V for CV ($V_1 = 23$ V), whereby I_1 is changed from 1 A to 3 A; (c) Load dump at V_1 equals 48 V and I_1 of 6 A, V_2 is set to 29 V; blue: battery voltage v_2, green: voltage v_1, red: inductor current i_L, orange: current i_1.

4.3 Transient response, EMI behavior and pyrometry

First, as shown in Fig. 18 (a), the mobile battery system is charging. From beginning of the discharging signal 500 ms passes until the battery is discharged with the same power. The measured electromagnetic interference (EMI) spectrum is shown in Fig. 18 (b). Thereby the scheme CCM-mASC without burst mode was applied and the converter works in CV on V_1 side. The mobile battery system supplies a 12 V LED module with a nominal power of 60 W. Last Fig. 18 (c) shows a pyrometer measurement of the printed circuit board (PCB) mounted on a heatsink. The converter operates in step-down mode with implemented burst mode. This measurement confirms the theoretical analyzation in Fig. 4. Transistor S_1 has the highest temperature because of the hard switching interval 1_a.

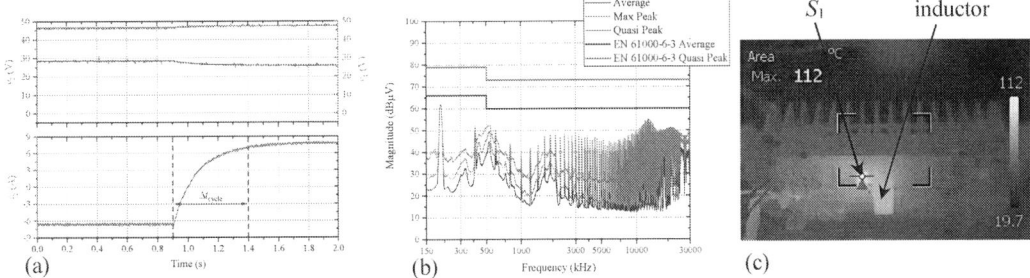

Fig. 18: (a) from charging to discharging Δt_{cycle} in 500 ms with each 200 W; (b) EMI noise spectrum; (c) pyrometer measurement of the converter at $V_{grid} = 60$ V, $V_{load} = 24$ V and $I_{load} = 9$ A after 30 min operation.

The last Fig. 19 shows the efficiency curves of the prototype. The peak efficiency was measured to be 97.5% including the auxiliary supply and is achieved at the transition mode.

Conclusion

The continuous conduction modulation scheme with middle active short-circuit is used to charge the bootstrap capacitor of the four-switch buck/boost converter in the H-bridge topology. The bidirectional DC-DC converter connects different electronic devices between 5 V and 60 V with nominal power of 300 W. In this application the continuous operation is preferable compared to discontinuous operation. This paper also shows that lower parasitic capacitances of the transistors are beneficial for the

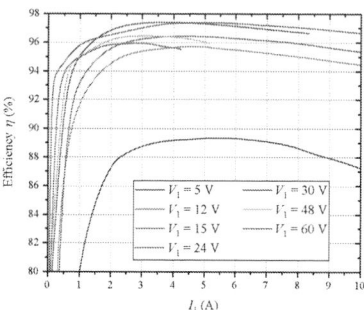

Fig. 19: Efficiency curves over the full grid voltage at a battery voltage V_2 of 24 V. The peak efficiency is 97.5%.

bootstrap interval. The analyzation of the single operation intervals depicts two hard switched instants. Using burst mode increases the efficiency over the full power range. Also the FFT analysis of the inductor current waveform shows a higher THD compared to the continuous triangular waveform, this confirmed that burst mode should be preferred. The highest efficiency of 97.5% was measured at the transition between step-down and step-up, whereby a smooth transition is implemented with the novel modulation scheme as well.

References

[1] F.W.Y. Saputra, Aripriharta, I. Fadlika, N. Mufti, K.H. Wibowo, G.J. Jong, "Efficiency Comparison between DC and AC Grid Toward Green Energy In Indonesia", in *IEEE International Conference on Automatic Control and Intelligent Systems (I2CACIS 2019)*, Selangor, Malaysia, 29 June 2019, pp. 129-134

[2] S. Peyghami, H. Mokhtari, P.C. Loh, P. Davari, F. Blaabjerg, "Distributed Primary and Secondary Power Sharingin a Droop-Controlled LVDC Microgrid WithMerged AC and DC Characteristics", *IEEE Transactions on Smart Grid*, Vol. 9, No. 3, May 2018, pp. 2284-2294

[3] C. Phurailatpam, R. Sangral, S.N. Singh, F.G. Longatt, "Design and Analysis of a DC Microgrid with Centralized Battery Energy Storage System", in *Annual IEEE India Conference (INDICON)*, New Delhi, India, 17-20 December 2015, pp. 1-6

[4] F. Zhang, C. Meng, Y. Yang, C. Sun, C. Ji, Y. Chen, W. Wei, "Advantages and Challenges of DC microgrid for Commercial Building A Case Study from Xiamen University DC Microgrid", in *IEEE First International Conference on DC Microgrids (ICDCM)*, Atlanta, Georgia, USA, 7-10 June 2015, pp. 355-358

[5] A.S. Morais, L. A.C. Lopes, "Interlink Converters in DC nanogrids and its effect in power sharing using distributed control", in *IEEE 7th International Symposium on Power Electronics for Distributed Generation Systems (PEDG)*, Vancouver, British Columbia, Canada, 27-30 June 2016, pp. 1-7

[6] "48-Volt Electrical Systems - A Key Technology Paving to the Road to Electric Mobility", *1st ed, ZVEI - German Electrical and Electronic Manufacturers' Association Electronic Components and Systems and PCB and Electronic Systems Divisions*, Frankfurt am Main, Germany, April 2016, pp. 1-40

[7] M. Schulz, N. Schleippmann, K. Gosses, B. Wunder, "Four Switch Buck/Boost Converter to Handle Bidirectional Power Flow in DC Subgrids", in *International Exhibition and Conference for Power Electronics, Intelligent Motion, Renewable Energy and Energy Management (PCIM Europe)*, Nuremberg, Germany, 7-8 July 2020, pp. 1-8

[8] Phaesun GmbH, "Solar Modules / SolarmoduleSun Peak SPR (with Standard4)", datasheet: 310293, 310294, Memmingen, Germany, June 2017, pp. 1-2

[9] "Spannungsklassen in der Elektromobilität", *ZVEI - Zentralverband Elektrotechnik- und Elektronikindustrie e. V. Kompetenzzentrum Elektromobilität*, Frankfurt am Main, Germany, December 2013, pp. 1-44

[10] H. Yun, M. Dong, Y. Jian, J. Wan, M. Shen, Y. Wang, "Application of Soft-switching Technology in Four Switch Buck-Boost Circuit", in *12th IEEE Conference on Industrial Electronics and Applications (ICIEA)*, Siem Reap, Cambodia, 18-20 June 2017, pp. 1675-1679

[11] Z. Zhou, H. Li, X. Wu, "A Constant Frequency ZVS Control System for the Four-Switch Buck–Boost DC–DC Converter With Reduced Inductor Current", *IEEE Transactions on Power Electronics*, Vol. 34, No. 7, July 2019, pp. 5996 – 6003

[12] S. Waffler, J. W. Kolar, "A Novel Low-Loss Modulation Strategy for High-Power Bidirectional Buck + Boost Converters", *IEEE Transactions on Power Electronics*, Vol. 24, No. 6, June 2009, pp. 1589-1599

[13] K. Xia, Z. Li, Y. Qin, Y. Yuan, Q. Yuan, "Minimising peak current in boundary conduction mode for the four-switch buck-boost DC/DC converter with soft switching", *IET Power Electronics*, Vol. 12, No. 4, October 2019, pp. 944-954

[14] N. Su, D. Xu, M. Chen, J. Tao, "Study of Bi-Directional Buck-Boost Converter with Different Control Methods", in *IEEE Vehicle Power and Propulsion Conference*, Harbin, China, 3-5 September 2008, pp. 1-5

[15] S. Waffler, "Hochkompakter bidirektionaler DC-DC-Wandler für Hybridfahrzeuge," Ph.D. dissertation, Power Electronic Systems Laboratory (PES), ETH Zürich, Zürich, Switzerland, 2013.

[16] ON Semiconductor, FDMT800100DC N-Channel Dual CoolTM 88 PowerTrench®MOSFET, Aurora, Colorado, USA, July 2015, Rev. 1.1

[17] Infineon Technologies AG, OptiMOS™ 5 Power-Transistor, 100 V, Munich, Germany, Revision: 2016-09-23, Rev. 2.2

[18] F. Li, R. Hao, H. Lei, X. You, C. Ke, J. Wang, "Non-inverting Three-level Buck-Boost Converter for Wide Voltage Range Application", in *IEEE Energy Conversion Congress and Exposition (ECCE)*, Portland, Oregon, USA, 23-27 September 2018, pp. 4870-4875

[19] R. W. Erickson, D. Maksimovic, "Fundamentals of Power Electronics", 2nd ed. New York, USA, Springer, 2001

Stability Investigation of Three-Phase Grid-Tied PV Inverters with Impedance-Based Method

Zhiqing Yang, Wanchao Gou, Xian Luo, Chirag Shah, Nurhan Rizqy Averous and Rik W. De Doncker
Institute for Power Generation and Storage Systems
E.ON Energy Research Center
RWTH Aachen University
Mathieustr. 10
52074 Aachen, Germany
Email: post_pgs@eonerc.rwth-aachen.de

Acknowledgments

Funded by the German Federal Ministry for Economic Affairs and Energy (BMWi, FKZ0324211D), PV-Kraftwerk2025.

Keywords

≪Impedance model≫, ≪Voltage-source converter≫, ≪Power quality≫.

Abstract

This paper presents stability investigations of three-phase grid-tied photovoltaic inverter systems using the impedance-based method. Impedance models (IMs) are established considering different control loops, and passive elements. IMs with a current control in both synchronous and stationary frames are established and compared. Impacts of different control loops, filter parameters, operation point as well as the number of inverters in the system are discussed. It is possible to identify the reason of an instability with the IMs.

Introduction

With continuous expansion of renewable energy generations, different control strategies are required in power converters to support the grid performance in the distribution level [1]. However, interactions between control loops and the grid impedance can cause instabilities. It is reported that a negative incremental resistance exists within the bandwidth of a phase-locked loop (PLL) [2], which may cause a severe instability issue. Non-optimum design of the direct-voltage control (DVC) may lead to stability problems [3]. Hence, it is necessary to pay attention to the stability issue on systems with overlapping impacts of different control loops such as photovoltaic (PV) inverter systems.

Two methods are commonly used to analyze the small-signal stability of an inverter system. The first method is the eigenvalue-based stability analysis, in which detailed system informations are required to build a state-space model (SSM) [4]. The eigenvalues (EVs) of a SSM are calculated to identify the oscillation modes and corresponding damping. However, large matrices are required to represent the high-order dynamics or to form a large-scale system. The second method is the impedance-based stability analysis (IBSA), where the system is split into a source and a load subsystem. The stability is assessed by applying the Nyquist criterion on the source-load impedance ratio [5]. The stability of a multi-inverter system can be easily analyzed by considering an equivalent total impedance [6]. The Impedance model (IM) and IBSA have been investigated in literature for generalized grid-tied inverters with L filter [7], offshore wind farms [8] and railway applications [9]. However, several points can still be improved:

1. Only few research focuses on stability issues of three-phase grid-tied inverters with *LCL* filters and multiple control loops including both PLL and DVC, e.g. [8]. It is necessary to explore the overlap influence of the PLL, DVC and other factors in a wide frequency range.

2. Most research focuses only on synchronous frame based proportional-integral (PI) alternating-current control (ACC). However, stationary frame based proportional-resonant (PR) ACC is also widely used in grid-tied applications, which needs to be analyzed and compared to PI ACC.

3. The accuracy of an IM needs to be validated especially near stability boundaries, to prove the effectiveness of the IBSA. Different analysis methods need to be summarized and compared.

This paper contributes to the modeling and stability analysis of PV inverter systems. IMs are established considering both PI and PR ACC. Different analysis methods are summarized and compared. The accuracy of the developed IM is validated by comparing it to a detailed switching model. The impedance shaping influenced by different control loops, filter parameters, maximum power point tracking (MPPT) as well as the inverter number are analyzed. The stability with PI and PR ACC are compared with a focus of the influence of the DVC and PLL.

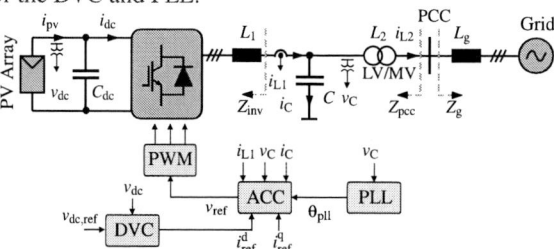

Fig. 1: System configuration of a three-phase PV inverter

Impedance Model of a Three-Phase Grid-Tied PV Inverter

The PV inverter shown in Fig. 1 is taken for the investigation. The inverter is equipped with an *LCL* filter for harmonic suppression, where L_2 is realized by the leakage inductance of a step-up transformer. IBSA requires IMs for both sides at an investigation point, i.e. inverter-side impedance at the point of common coupling (PCC) Z_{pcc} and the grid impedance Z_g. To calculate the inverter-side impedance, the dynamics of different control loops are analyzed. IMs of Z_{inv} with PI ACC have been derived in previous works [2, 8, 10]. The influence of the ACC and PLL is analyzed in [2] without DVC. While the influence of the DVC is considered in [10] without PLL. However, both models are established with only L filters and the impact of the voltage feedforward filter (VFF) is ignored. A detailed model can be found in [8] considering the ACC, PLL and DVC. However, the damping of the *LCL* filter is ignored and the dynamics of v_{dc} in the modulator is simplified. The PR ACC is also popular for grid-tied inverters, which have been seldom discussed in the literature. An IM with PR ACC is introduced in [7] based on L filter regulating the positive sequence component. This paper develops detailed IMs including complete impacts of ACC, PLL, DVC and delay introduced by the digital control. The damping of an *LCL* filter and the VFF are modeled. An extension of the IM from PI to PR ACC is also discussed. For clarity, complex vectors are represented with $\underline{x}^{dq} = x^d + jx^q$, real vectors with $x^{dq} = [x^d, x^q]^T$. Small-signal variables are noted with $\hat{x}$ and steady-state values are marked with X.

Alternating-Current Control

An ACC can be implemented either in the synchronous reference frame (SRF) with PI control or in the stationary frame with PR control as illustrated in Fig. 2. For a PI ACC, the control function G_{PI} is given in (1). G_{del} described the system delay caused by digital control and zero-order hold effect of the modulator with Pade approximation as given in (2). v_{AD} is the compensation voltage for implementation of active damping. All measured currents and voltages need to be converted from the stationary frame to a SFR defined by PLL, while the modulated voltage v_{inv} is converted back to the stationary frame in the modulator. For a PI ACC, it is possible to implement the VFF in a format of 1st-order low-pass filter as given in (3), where ω_{VFF} is the cut-off frequency. The system stability benefits from an optimum design of ω_{VFF} [11].

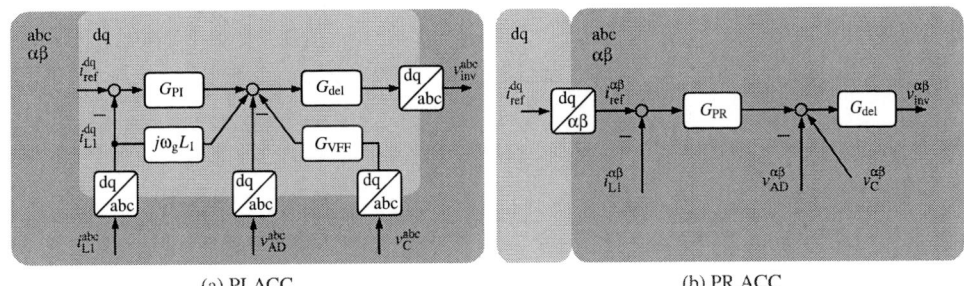

(a) PI ACC   (b) PR ACC

Fig. 2: Block diagram of ACC

$$G_{PI}^{dq} = K_p^{ACC} + \frac{K_i^{ACC}}{s} \tag{1}$$

$$G_{del} = \frac{2 - sT_{del}}{2 + sT_{del}} \tag{2}$$

$$G_{VFF} = \frac{\omega_{VFF}}{s + \omega_{VFF}} \tag{3}$$

If a PR ACC is implemented, the measured currents and voltages can be fed in the control structure with Clarke transformation, which avoids the influence of the PLL. However, the current reference command needs to be converted to the stationary frame. To establish the IM with a PR ACC, the control transfer function G_{PR} is used as given in (4). Since both the PLL and DVC are established in the SRF, it is necessary to convert (4) to the SRF as given in (5) by substituting $s^{\alpha\beta} = s^{dq} + j\omega_g$, where ω_g is the grid angular frequency.

$$G_{PR}^{\alpha\beta} = K_p^{ACC} + \frac{K_r^{ACC}s}{s^2 + \omega_g^2} = K_p^{ACC} + \frac{K_r^{ACC}}{2}\left(\frac{1}{s - j\omega_g} + \frac{1}{s + j\omega_g}\right) \tag{4}$$

$$G_{PR}^{dq} = K_p^{ACC} + \frac{K_r^{ACC}}{2s} + \frac{K_r^{ACC}s}{2(s^2 + 4\omega_g^2)} - j\frac{K_r^{ACC}\omega_g}{s^2 + 4\omega_g^2} \tag{5}$$

Phase-Locked Loop

As shown in Fig. 3, a SRF based PLL is used for grid synchronization. By regulating v_C^q to zero, a synchronization angle θ is tracked by each inverter individually. H_{PLL} represents the small-signal relation between v_C^q and θ as given in (6), where G_{PLL} is a PI controller as given in (7). Due to the nonlinearity, H_{PLL} depends on the steady-state value V_C^d. To precisely describe the impact of the PLL, an inverter is modeled with two SRFs, which are the control frame dq^c and system frame dq^s as illustrated in Fig. 4. In transients, the two SRFs are different due to the dynamics in the PLL. The reference frame DQ is a common frame defined by the grid voltage at the PCC. The relationship from the system to control frame is given in (8) for any variable x, where X^d and X^q are the steady-state values in the SRF and Θ is the steady-state phase angle between v_C and v_{pcc}. Variables in control and system frames are marked with a superscript c and s, respectively. Variables belong to the ith inverter are noted with a subscript i.

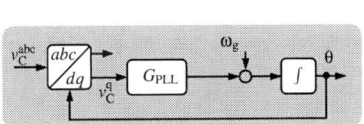

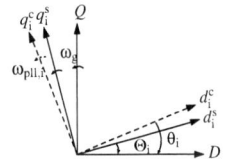

Fig. 3: Block diagram of PLL   Fig. 4: Reference frames

$$H_{PLL} = \frac{\hat{\theta}}{\hat{v}_C^q} = \frac{G_{PLL}}{s + V_C^{d,s}G_{PLL}} \tag{6}$$

$$G_{\mathrm{PLL}} = K_{\mathrm{p}}^{\mathrm{PLL}} + \frac{K_{\mathrm{i}}^{\mathrm{PLL}}}{s} \tag{7}$$

$$\begin{bmatrix} \hat{x}^{\mathrm{d,c}} \\ \hat{x}^{\mathrm{q,c}} \end{bmatrix} = \begin{bmatrix} \hat{x}^{\mathrm{d,s}} \\ \hat{x}^{\mathrm{q,s}} \end{bmatrix} + \begin{bmatrix} 0 & X^{\mathrm{q,s}}H_{\mathrm{PLL}} \\ 0 & -X^{\mathrm{d,s}}H_{\mathrm{PLL}} \end{bmatrix} \begin{bmatrix} \hat{v}_{\mathrm{C}}^{\mathrm{d,s}} \\ \hat{v}_{\mathrm{C}}^{\mathrm{q,s}} \end{bmatrix} \tag{8}$$

Direct-Voltage Control

As depicted in Fig. 5, a DVC is implemented in the cascaded outer control loop to set the d-axis reference current value as described in (9), where G_{DVC} is a PI controller as given in (10). The dynamics of v_{dc} is shown in Fig. 6, where $Y_{\mathrm{dc}} = sC_{\mathrm{dc}}$ is the dc-link impedance and p_{ac} is the transferred power calculated with (11). After linearization, the small-signal variable $\hat{v}_{\mathrm{dc}}$ is influenced by both modulation index m^{dq} and inductor current $i_{\mathrm{L1}}^{\mathrm{dq}}$ as given in (12), where M^{dq} and $I_{\mathrm{L1}}^{\mathrm{dq}}$ are the steady-state values.

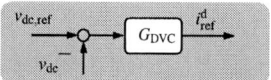
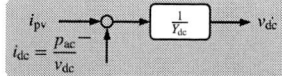

Fig. 5: Block diagram of DVC   Fig. 6: Dynamics in the dc-link capacitor

$$\begin{bmatrix} i_{\mathrm{ref}}^{\mathrm{d,c}} \\ i_{\mathrm{ref}}^{\mathrm{q,c}} \end{bmatrix} = \begin{bmatrix} G_{\mathrm{DVC}} \\ 0 \end{bmatrix} (v_{\mathrm{dc,ref}} - v_{\mathrm{dc}}) \tag{9}$$

$$G_{\mathrm{DVC}} = K_{\mathrm{p}}^{\mathrm{DVC}} + \frac{K_{\mathrm{i}}^{\mathrm{DVC}}}{s} \tag{10}$$

$$p_{\mathrm{ac}} = \frac{3}{2} \left(v_{\mathrm{inv}}^{\mathrm{d}} i_{\mathrm{L1}}^{\mathrm{d}} + v_{\mathrm{inv}}^{\mathrm{q}} i_{\mathrm{L1}}^{\mathrm{q}} \right) \tag{11}$$

$$\hat{v}_{\mathrm{dc}} = \frac{3}{4} \begin{bmatrix} \frac{M^{\mathrm{d,s}}}{Y_{\mathrm{dc}}} & \frac{M^{\mathrm{q,s}}}{Y_{\mathrm{dc}}} \\ 0 & 0 \end{bmatrix} \begin{bmatrix} \hat{i}_{\mathrm{L1}}^{\mathrm{d,s}} \\ \hat{i}_{\mathrm{L1}}^{\mathrm{q,s}} \end{bmatrix} + \frac{3}{4} \begin{bmatrix} \frac{I_{\mathrm{L1}}^{\mathrm{d,s}}}{Y_{\mathrm{dc}}} & \frac{I_{\mathrm{L1}}^{\mathrm{q,s}}}{Y_{\mathrm{dc}}} \\ 0 & 0 \end{bmatrix} \begin{bmatrix} \hat{m}^{\mathrm{d,s}} \\ \hat{m}^{\mathrm{q,s}} \end{bmatrix} \tag{12}$$

Combining all the influence of all the control loops together, the inverter impedance can be calculated based on the small-signal relation of $\hat{v}_{\mathrm{inv}}$ and $\hat{i}_{\mathrm{L1}}$ as represented in (13). The inverter-side impedance at PCC can be acquired by including also the admittance of the filter capacitor Y_{C} and the impedance of the leakage inductor in the step-up transformer Z_{L2} as given in (14).

$$Z_{\mathrm{inv}} = \frac{\hat{v}_{\mathrm{inv}}}{\hat{i}_{\mathrm{L1}}} \tag{13}$$

$$Z_{\mathrm{pcc}} = (Z_{\mathrm{inv}}^{-1} + Y_{\mathrm{C}})^{-1} + Z_{\mathrm{L2}} \tag{14}$$

Impedance Based Stability Analysis

Analysis Methods

There are three commonly used methods to analyze the stability based on the IMs, which are Bode plot [5], Nyquist plot [2] and frequency response [7]. The impedance Bode plot describes the frequency response of Z_{pcc} and Z_{g} with two-dimensional matrices. The d-axis diagonal impedances $Z_{\mathrm{pcc}}^{\mathrm{dd}}$ and $Z_{\mathrm{g}}^{\mathrm{dd}}$ are depicted in Fig. 7(a) as an example. The intersection point of two the impedances predicts the potential resonance, while the phase difference at the corresponding frequency determines the stability. The system is stable if the phase between Z_{pcc} and Z_{g} at this intersection is smaller than $180°$. This is only a necessary but not sufficient condition similar to the concept of the phase margin, as a system may lose stability under a large disturbance even with a small positive phase margin. It is possible to identify the reason of an unstable resonance with the Bode plots, as the impedance features are observed in dq frame separately. The second method to assess the system stability is the Nyquist criterion based on the EVs $\lambda_{1,2}$ derived from the impedance ratio matrix $Z_{\mathrm{pcc}}^{-1} \cdot Z_{\mathrm{g}}$. The system remains stable if no EV

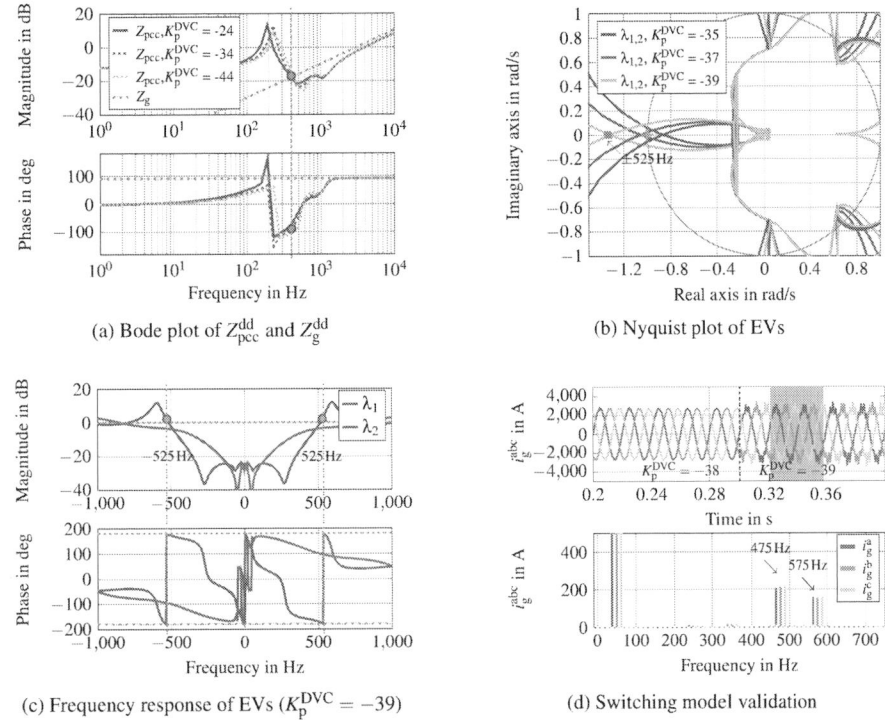

(a) Bode plot of $Z_{\mathrm{pcc}}^{\mathrm{dd}}$ and $Z_{\mathrm{g}}^{\mathrm{dd}}$

(b) Nyquist plot of EVs

(c) Frequency response of EVs ($K_{\mathrm{p}}^{\mathrm{DVC}} = -39$)

(d) Switching model validation

Fig. 7: Different stability analysis methods

encircles the point $(-1, 0)$ as illustrated in Fig. 7(b). However, frequency information is missing in the Nyquist plot. Apart from that, the stability can be investigated according to the frequency response of the EVs. A system is stable if the magnitude in dB is below zero for all the corresponding frequency points with a phase jump, which indicates that the Nyquist plot of all EVs does not enclose $(-1, 0)$. The frequency response of an unstable case is shown in Fig. 7(c). Note that both the Nyquist plot and frequency response are mirror symmetric in positive and negative frequency ranges.

IMs depend on operation points and neglect the switching behaviour. Thus, deviations between IMs and switching models are inevitable. Considering the proportional gain in DVC, the stability boundary predicted by the impedance bode plot is $K_{\mathrm{p}}^{\mathrm{DVC}} = -34$, while for the Nyquist plot and frequency response is -37. The boundary validated with a switching model in PLECS is -39 as shown in Fig. 7(d), which is slightly conservative compared to the IM. The stability assessed by Nyquist plot or frequency response is more accurate compared to the impedance plot. Because the Nyquist criterion of EVs is a sufficient and necessary condition for stability analysis [12]. For an unstable operation with $K_{\mathrm{p}}^{\mathrm{DVC}} = -39$, resonances at 475 Hz and 575 Hz are observed in the abc frame in the switching model, which are also predicted by the Nyquist plot and frequency response at ± 525 Hz in the dq frame. The 50 Hz frequency shift between abc and dq frame is explained by the coupling effect between the positive and negative frequencies [7]. The stability criterion and features of different analysis methods are summarized in Table I.

Table I: Comparison of different stability investigation methods

Analysis methods	Stability criterion (stable)	Features (+: pros; -: cons)
Bode plot of impedance	$\lvert \angle Z_{\mathrm{pcc}}(s) - \angle Z_{\mathrm{g}}(s) \rvert < 180°$ while $\lvert Z_{\mathrm{pcc}}(s) \rvert = \lvert Z_{\mathrm{g}}(s) \rvert$	+ intuitive variation, instability identification - inaccurate boundary
Nyquist plot of EVs	characteristic loci of λ encircle $(-1, 0)$	+ intuitive and accurate boundary - indirect frequency information
Frequency response of EVs	$\lvert \lambda \rvert < 0$ while $\angle \lambda = \pm 180°$	+ frequency information, accurate boundary - unintuitive boundary

Impact of Control Loops

Control loops shape an IM within the control bandwidths, which influences the system stability. Thus, the impact of different control loops on inverter impedance Z_{inv} is observed. As the stability is usually determined by the diagonal elements of a impedance matrix [2], only Z_{inv}^{dd} and Z_{inv}^{qq} are considered. The impact of the ACC is firstly observed in Fig. 8, which has a symmetric influence on both d and q axis. With an increasing bandwidth f_{bw}^{ACC}, the curves of both magnitude and phase shift towards high-frequency range. According to the stability criterion, the ACC may introduce an unstable resonance around hundreds of Hz. Potentially unstable ranges induced by the ACC are marked with shaded areas. The implemented parameters are highlighted with markers.

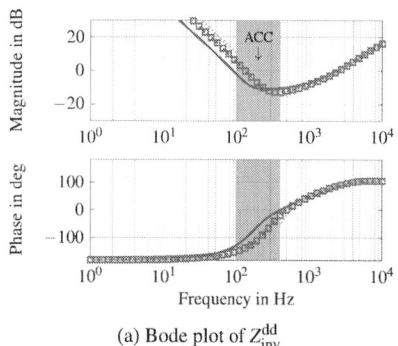
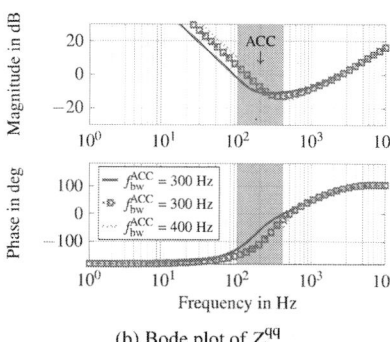

(a) Bode plot of Z_{inv}^{dd} (b) Bode plot of Z_{inv}^{qq}

Fig. 8: Bode plot of Z_{inv} influenced by ACC (implemented parameters highlighted with markers)

If a PLL is considered, a constant magnitude occurs in Z_{inv}^{qq} within the bandwidth f_{bw}^{PLL}, while the phase angle remains $-180°$ as shown in Fig. 9. The phase difference between Z_{inv} and Z_g can excess $90°$, which can lead to an instability. This indicates a negative incremental resistance caused by the PLL [7, 8]. Potentially unstable range exists until 100 Hz due to the PLL as marked with the shaded areas. Since the PLL regulates the q-axis voltage for grid synchronization, it has no impact on Z_{inv}^{dd}.

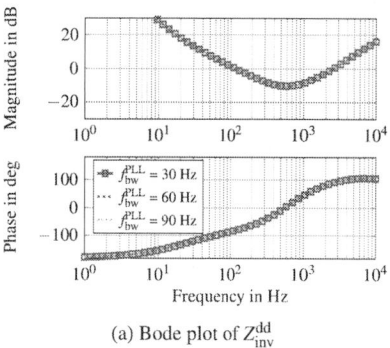
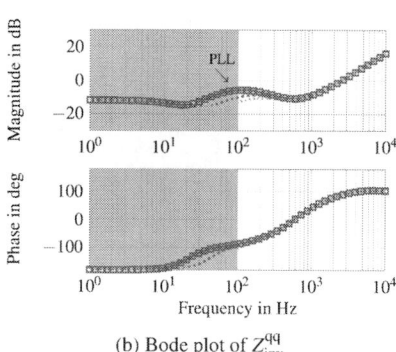

(a) Bode plot of Z_{inv}^{dd} (b) Bode plot of Z_{inv}^{qq}

Fig. 9: Bode plot of Z_{inv} influenced by PLL (implemented parameters highlighted with markers)

If a DVC is further considered, a phase jump around 100 Hz is observed in Z_{inv}^{dd}, which leads a dramatic sink in the phase curve. A resonance peak occurs in the low-frequency range if the bandwidth f_{bw}^{DVC} is not high enough, in which a phase sink occurs. Potentially unstable range with potential resonances are highlighted with the shaded areas. As the DVC provides only a d-axis reference value for the ACC, it has no impact on Z_{inv}^{qq}.

Impact of Filters

The sensitivity of filter parameters on the IM is observed for the *LCL* filter and the dc-link capacitor in Fig. 11. Only the d-axis impedance is observed as symmetric filters are assumed for simplicity. Assume that the value of all filter parameters may have a $\pm20\%$ variation to the nominal value due to the manufacture tolerance or the temperature change during the operation. Several features are observed and

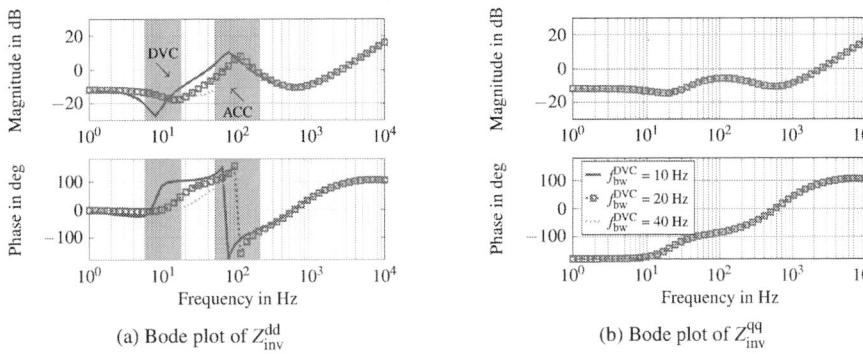

(a) Bode plot of $Z_{\text{inv}}^{\text{dd}}$ (b) Bode plot of $Z_{\text{inv}}^{\text{qq}}$

Fig. 10: Bode plot of Z_{inv} influenced by DVC (implemented parameters highlighted with markers) summarized as follows: a) L_1 and C mainly influence the frequency range from $100 - 1000\,\text{Hz}$, which are also influenced by the ACC according to Fig. 8. Because the control parameters in the ACC are highly dependent on those filter parameters. b) L_2 mainly affects the frequency range above $1\,\text{kHz}$. c) Different value of C_{dc} has almost no influence on the IM calculated based on the ac-side small-signal relationships. d) Generally, the IM is not sensitive to a variation of filter parameters. Filters have almost no influence on the frequency ranges below $100\,\text{Hz}$, which is dominantly influenced by the PLL and DVC.

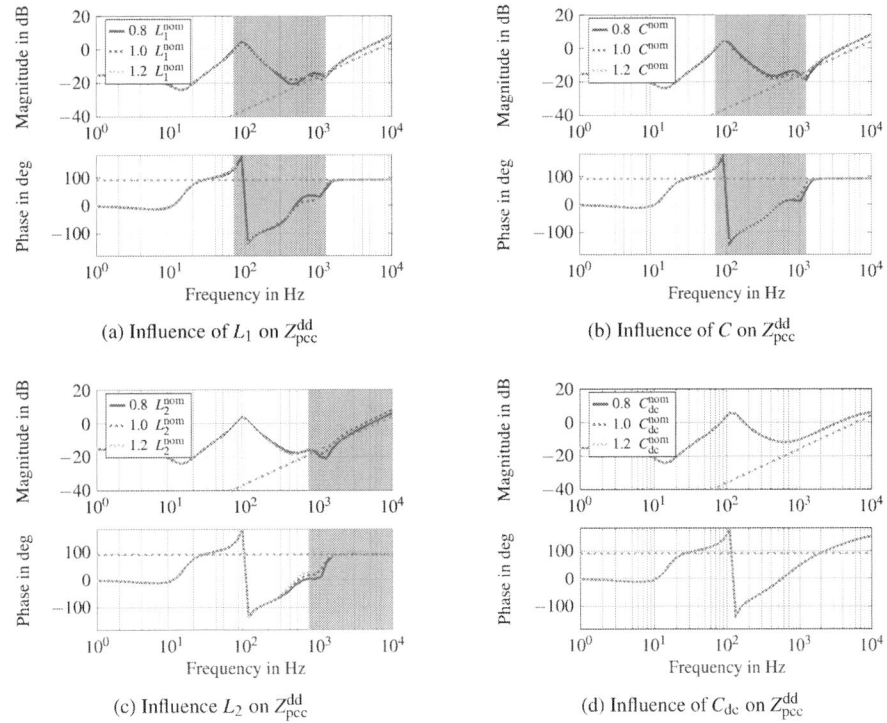

(a) Influence of L_1 on $Z_{\text{pcc}}^{\text{dd}}$ (b) Influence of C on $Z_{\text{pcc}}^{\text{dd}}$

(c) Influence L_2 on $Z_{\text{pcc}}^{\text{dd}}$ (d) Influence of C_{dc} on $Z_{\text{pcc}}^{\text{dd}}$

Fig. 11: Bode plot of Z_{pcc} and Z_{g} with different filter parameters

Impact of MPPT

Operation points affect the frequency-domain features of an IM. Both the PV current I_{pv} and the dc-link voltage V_{dc} vary according to MPPT. The influences of different I_{pv} and V_{dc} on $Z_{\text{pcc}}^{\text{dd}}$ and $Z_{\text{pcc}}^{\text{qq}}$ are presented in Fig. 12. It is observed that I_{pv} has more influence due to a wider variation range. A higher value of I_{pv} reduces the magnitude for both d- and q-axis impedances in the frequency range until hundreds of Hz. However, a resonance sink appears only in the $Z_{\text{pcc}}^{\text{dd}}$ together with a phase reduction, because the I_{pv} mainly influence the dynamics of the DVC, which dominates the d-axis impedance. Potential unstable range and dominant control loops are also marked with shaded areas.

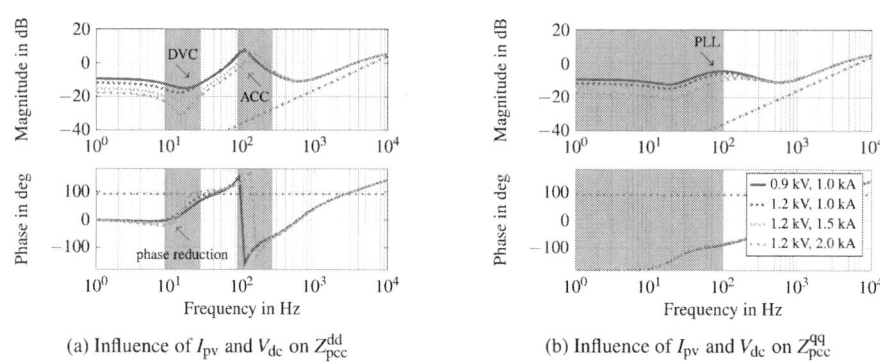

(a) Influence of I_{pv} and V_{dc} on Z_{pcc}^{dd}

(b) Influence of I_{pv} and V_{dc} on Z_{pcc}^{qq}

Fig. 12: Bode plot of Z_{pcc} and Z_g with different operation points

Impact of Inverter Number

One of the advantages using the IBSA is that the stability of a multi-inverter system can be easily investigated by the impedance aggregation as described in (15). According to Fig. 4, each inverter has an individual SRFs dq^c and dq^s which are related to its own PLL dynamics. To aggregate the dynamics of all the inverters together, all the IMs should be converted to the common frame DQ as marked in Fig. 4. The conversion can be implemented with (16), where $T_R(\Theta_i)$ is a rotation matrix defined in (17) and Θ_i is the steady-state phase angle between v_C and v_{pcc} of the ith inverter.

$$Z_{pcc,total} = \sum_{i=1}^{N} Z_{pcc,i} \tag{15}$$

$$Z_{pcc,i}^{DQ} = T_R(\Theta_i) \cdot Z_{pcc,i}^{dq} \cdot T_R^{-1}(\Theta_i) \tag{16}$$

$$T_R(\Theta_i) = \begin{bmatrix} \cos(\Theta_i) & -\sin(\Theta_i) \\ \sin(\Theta_i) & \cos(\Theta_i) \end{bmatrix} \tag{17}$$

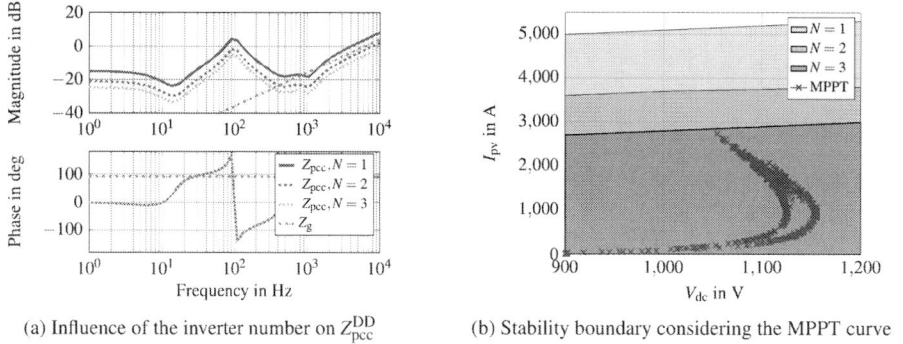

(a) Influence of the inverter number on Z_{pcc}^{DD}

(b) Stability boundary considering the MPPT curve

Fig. 13: Impact of the inverter number

The equivalent impedance of a multi-inverter system is depicted in Fig. 13(a) with different inverter numbers N. With an increasing value of N, the phase angle remains unchanged, while the magnitude of the total inverter impedance decreases, which shifts the potential resonance towards low-frequency range and deteriorates the system stability, as the phase difference increases at the intersection point. The stability boundary decreases with an increasing number of inverters N ($N = 3$ is within areas of $N = 1, 2$) as shown in Fig. 13(b). The MPPT curve can be integrated to investigate whether the operate point may cause instability for a multi-inverter system. However, this is only a rough estimation with the assumption that all the inverters are operated in the same condition. Note that the MPPT operations are close to the stability boundary with a large value of I_{pv}, which occurs usually at noon. A dc-coupled storage system is recommended to balance the load demand as well as to increase the stability during a high injection in the future.

Comparison Between PI and PR ACC

PI and PR ACC are compared with a focus of the influence of DVC and PLL. The Nyquist plots of EVs for both control structures are depicted in Fig. 14. The stability boundary of inverters with a PR ACC is more sensitive to DVC. The reason for that is the d-axis reference value of a PR ACC is influenced by both the DVC and PLL. However, inverters with a PR ACC are less influenced by PLL, as feedback of measurements are not required to convert into the dq frame. Consequently, inverters with a PR ACC are able to be operated against weaker grid conditions.

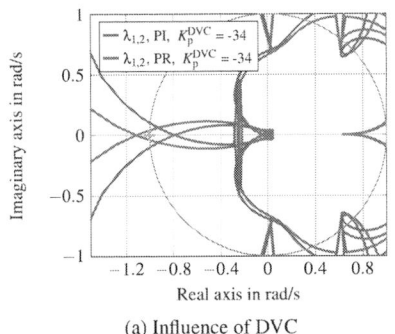 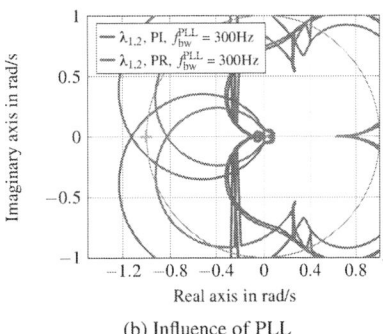

(a) Influence of DVC

(b) Influence of PLL

Fig. 14: Comparison of PI and PR ACC with Nyquist plots

The comparison between the PI and PR ACC are validated in switching models. Both controllers works properly with $f_{bw}^{PLL} = 30\,\text{Hz}$. With a step change of f_{bw}^{PLL} to $300\,\text{Hz}$, a resonance starts in the inverter with a PI ACC, while the PR ACC still remain stable as shown in Fig. 15. Both controllers lose the stability with a higher value of the PLL bandwidth.

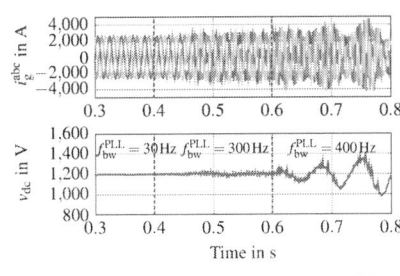 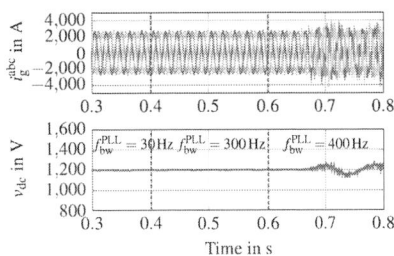

(a) Waveforms of an inverter with PI ACC

(b) Waveforms of an inverter with PR ACC

Fig. 15: Current and voltage waveforms under step changes of f_{bw}^{PLL}

Conclusion

In this paper, the stability of a three-phase grid-tied PV inverter system is analyzed with the IBSA based on the developed IM. Different analysis methods are summarized and compared. The Nyquist plot can predict an accurate stability boundary, while the Bode plot gives clear overview of the influence of individual parameters over a wide frequency range. Based on the Bode plots, the impact of different control loops, filter parameters, operation points and the number of inverters on the system stability can be thoroughly investigated. It is observed that an inverter with a PR control is less influenced by the PLL, which is more robust to against weak grid conditions.

References

[1] Lammert, G., He, T., Schmidt, M., Schegner, P., and Braun, M. (2015). Dynamic grid support in low voltage grids - Fault ride-through and reactive power/voltage support during grid disturbances. Power Systems Computation Conference.

[2] Wen, B. , Boroyevich, D. , Burgos, R. , Mattavelli, P. , and Shen, Z. . (2015). Analysis of d-q small-signal impedance of grid-tied inverters. IEEE Transactions on Power Electronics, 31(1), 675-687.

[3] D. Wang, L. Liang, L. Shi, J. Hu, and Y. Hou, Analysis of modal resonance between pll and dc-link voltage control in weak-grid tied vscs, IEEE Transactions on Power Systems, vol. 34, no. 2, pp. 11271138, 2018.

[4] Z. Yang, Q. Wang, J. Warmuz, and R. W. De Doncker, Stability assessment of a three-phase grid-tied pv inverter with eigenvalue-based methods, in 2019 10th IEEE International Symposium on Power Electronics for Distributed Generation Systems (PEDG). IEEE, 2019, p. 722-727.

[5] Sun, and Jian. (2011). Impedance-based stability criterion for grid-connected inverters. IEEE Transactions on Power Electronics, 26(11), 3075-3078.

[6] Yoon, C. , Bai, H. , Beres, R. , Wang, X. , Bak, C. , and Blaabjerg, F. . (2016). Harmonic stability assessment for multi-paralleled, grid-connected inverters. IEEE Transactions on Sustainable Energy, 1-1.

[7] Wang, X. , Harnefors, L. , and Blaabjerg, F. . (2018). Unified impedance model of grid-connected voltage-source converters. IEEE Transactions on Power Electronics, 33(2), 1775-1787.

[8] Amin, M. , and Molinas, M. . (2016). Understanding the origin of oscillatory phenomena observed between wind farms and hvdc systems. IEEE Journal of Emerging and Selected Topics in Power Electronics, 1-1.

[9] H. Tao, H. Hu, X. Zhu, Y. Zhou and Z. He, Harmonic Instability Analysis and Suppression Method Based on $\alpha\beta$ - Frame Impedance for Trains and Network Interaction System, in IEEE Transactions on Energy Conversion, vol. 34, no. 2, pp. 1124-1134, June 2019.

[10] Lu, D., Wang, X., Blaabjerg, F. (2018). Impedance-based analysis of DC-link voltage dynamics in voltage-source converters. IEEE Transactions on Power Electronics, 34(4), 3973-3985.

[11] Harnefors, L. , Zhang, L. , and Bongiorno, M. . (2008). Frequency-domain passivity-based current controller design. IET POWER ELECTRONICS, 1(4), 1.

[12] Lunze, J. (1996). Regelungstechnik 2 (Vol. 6). Berlin: Springer.

Stability Investigation of Large-Scale PV Parks with Eigenvalue-Based Method

Zhiqing Yang, Christian Bendfeld, Jin Qiang, Benedict Mortimer and Rik W. De Doncker
Institute for Power Generation and Storage Systems
E.ON Energy Research Center
RWTH Aachen University
Mathieustr. 10
52074 Aachen, Germany
Email: post_pgs@eonerc.rwth-aachen.de

Acknowledgments

Funded by the German Federal Ministry for Economic Affairs and Energy (BMWi, FKZ0324211D), PV-Kraftwerk2025.

Keywords

≪State-space model≫, ≪Voltage-source converter≫, ≪Multi-machine system≫.

Abstract

This paper introduces a state-space model (SSM) based framework to analyze the small-signal stability of large-scale photovoltaic (PV) parks. An improved SSM of PV inverters is proposed considering an accurate impact of the grid synchronization. Different methods are implemented to investigate the system stability, including eigenvalue distribution, participation factor and a proposed damping-resonance analysis. Based on that, the reason of an unstable resonance can be identified and the operation and control design can be optimized following a systematic approach.

Introduction

With increasing penetration of renewable energy generations, different control strategies are required in power converters to satisfy grid codes in the distribution level [1]. However, interactions between control loops and grid impedance can cause instabilities [2]. Unstable resonances have been observed in large-scale photovoltaic (PV) parks that involve a wide range of frequencies from tens to thousands of Hertz [3]. It is reported that negative incremental resistance exists within the bandwidth of a phase-locked loop (PLL) [2], which may cause severe instability issues. Non-optimal design of the direct-voltage control (DVC) may also lead to stability problems [4]. Hence, a systematic framework is required to analytically investigate the system stability and identify the cause of instabilities in inverter-based power systems.

To analyze the system stability, two methods are commonly used. The first method is impedance-based stability analysis, in which the system is split into a source and a load subsystem and the stability is assessed based on the source-load impedance ratio [5]. The second method is eigenvalue-based stability analysis (EBSA) [6], which requires detailed system information in order to develop a state-space model (SSM). Component connection method (CCM) is introduced in [6, 7] to develop the SSM in a systematic approach. Investigations have been made to summarize and compare these two different methods in [8]. Since SSMs contain detailed information about control structures, it is possible to identify the resonance modes and corresponding damping features with EBSA. In addition, the impact of different control loops on system stability can be evaluated through participation factor (PF) analysis [9], which facilitate to understand the resonance mechanisms. SSMs have been investigated in literature for micro-grid [6], high-voltage direct current [10] and PV [11] applications. However, several aspects can be still improved:
    1. The model of the PLL is simplified, which reduces the accuracy of the analysis.

2. Only few research focuses on the SSM of multi-inverter systems with more than two inverters e.g. [12], which requires modeling of large-scale systems in a more systematic manner.

3. A SSM can be used not only to assess the stability, but also to facilitate the control design and to identify the cause of an unstable resonance. Different applications need to be further explored.

This paper proposes a SSM-based framework for stability analysis of large-scale PV parks. SSMs of PV inverters and the grid are established. An approach to accurately model the small-signal influence of PLL is proposed. CCM is considered to model PV parks from inverter to system level. Based on established SSMs of PV parks, different methodologies for stability investigation are studied i.e. eigenvalue (EV) distributions, PF analysis and a proposed damping-resonance analysis. With the proposed framework, it is possible to analyze the system stability, to optimize the system operation and control parameters, and to identify the reason of an unstable resonance in a PV park.

PV Park Model

A PV park usually consists of several grid-tied PV inverters which are parallel connected at the point of common coupling (PCC) as shown in Fig. 1. Each inverter is equipped with an *LCL* filter for harmonic suppression, where L_2 is realized by the leakage inductance of a step-up transformer. In order to analyze the stability of the total system, SSMs are established including the control loops from a single inverter to a PV park. For clarity, complex vectors are represented with $\underline{x}^{dq} = x^d + jx^q$, while real vectors with $x^{dq} = [x_d, x_q]^T$. Steady-state values are marked with capitals i.e. X.

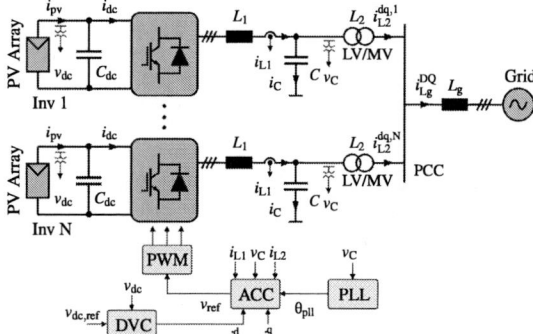

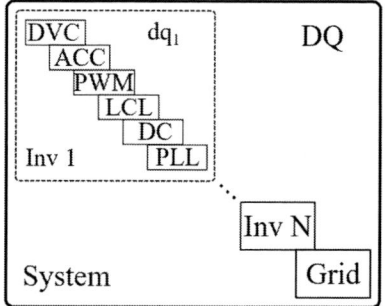

Fig. 1: System configuration of a PV park   Fig. 2: System matrix structure

Three-Phase Grid-Tied Inverter

Direct-Voltage Control

DVC is implemented to realize maximum power point tracking in the cascaded outer loop control. The state and output equations are given in (1) and (2), where $v_{dc,ref}$ is the reference dc-link voltage, v_{dc} is the measured dc-link voltage, K_p^{DVC} and K_i^{DVC} are gains in a proportional-integral (PI) controller. ψ_{dc} represents the integration of the feedback error in the controller, i_{ref}^d is the d-axis reference current.

$$\frac{d\psi_{dc}}{dt} = v_{dc,ref} - v_{dc} \tag{1}$$

$$i_{ref}^d = K_p^{DVC}(v_{dc,ref} - v_{dc}) + K_i^{DVC}\psi_{dc} \tag{2}$$

Alternating-Current Control

The synchronous reference frame (SRF) based ACC is implemented as the inner loop with the decoupling and feedforward terms. The state and output equations are given in (3) and (4). K_p^{ACC} and K_i^{ACC} are PI gains, γ^{dq} represents integrations of the feedback error in the controller, i_{ref}^{dq} is the current reference obtained from outer-loop controls and ω_g is the angular grid frequency. A voltage feedforward filter (VFF) in format of 1st-order low-pass filter is considered to conditioning the measured capacitor voltage

v_{C}^{dq} as given in (5), where $v_{\text{Cf}}^{\text{dq}}$ are filtered capacitor voltages, ω_{VFF} is cut-off frequency. Active damping (AD) is also implemented with capacitor currents feedback (6), where K_{AD} is the damping factor.

$$\frac{d\underline{\gamma}^{\text{dq}}}{dt} = \underline{i}_{\text{ref}}^{\text{dq}} - \underline{i}_{\text{L1}}^{\text{dq}} \tag{3}$$

$$\underline{v}_{\text{ref}}^{\text{dq}} = K_{\text{p}}^{\text{ACC}} \left(\underline{i}_{\text{ref}}^{\text{dq}} - \underline{i}_{\text{L1}}^{\text{dq}} \right) + K_{\text{i}}^{\text{ACC}} \underline{\gamma}^{\text{dq}} + j\omega_g L_1 \underline{i}_{\text{L1}}^{\text{dq}} + \underline{v}_{\text{Cf}}^{\text{dq}} - \underline{v}_{\text{AD}}^{\text{dq}} \tag{4}$$

$$\frac{d\underline{v}_{\text{Cf}}^{\text{dq}}}{dt} = -\omega_{\text{VFF}} \underline{v}_{\text{Cf}}^{\text{dq}} + \omega_{\text{VFF}} \underline{v}_{\text{C}}^{\text{dq}} \tag{5}$$

$$\underline{v}_{\text{AD}}^{\text{dq}} = K_{\text{AD}} \underline{i}_{\text{C}}^{\text{dq}} = K_{\text{AD}} (\underline{i}_{\text{L1}}^{\text{dq}} - \underline{i}_{\text{L2}}^{\text{dq}}) \tag{6}$$

PLL

The SRF-based PLL synchronizes the phase angle by regulating v_{C}^{q} to zero. The state equations of PLL are given in (7) and (8), where ε is the integration of v_{C}^{q} and $K_{\text{p}}^{\text{PLL}}$ and $K_{\text{i}}^{\text{PLL}}$ are the PI gains.

$$\frac{d\varepsilon}{dt} = v_{\text{C}}^{\text{q}} \tag{7}$$

$$\frac{d\theta}{dt} = \omega = K_{\text{p}}^{\text{PLL}} v_{\text{C}}^{\text{q}} + K_{\text{i}}^{\text{PLL}} \varepsilon \tag{8}$$

Modulator

As a SSM averages the dynamics over one switching period, the pulse width modulation (PWM) can be simplified as 1st-order system delay in format of Pade approximation as given in (9), where $\underline{v}_{\text{ref}}^{\text{dq}}$ and $\underline{v}_{\text{inv}}^{\text{dq}}$ are the inverter reference and output voltages. $T_{\text{del}} = 1.5\,T_{\text{s}}$ represents the total delay including computational delay T_{s} in a digital control and $0.5\,T_{\text{s}}$ due to the latching process in PWM, where T_{s} is the sampling time. It is possible to rewrite (9) in state-space format with auxiliary variables $\underline{x}_{\text{del}}^{\text{dq}}$ [7].

$$\underline{v}_{\text{inv}}^{\text{dq}} = \frac{2 - sT_{\text{del}}}{2 + sT_{\text{del}}} \cdot \underline{v}_{\text{ref}}^{\text{dq}} \quad \Rightarrow \quad \frac{\underline{v}_{\text{inv}}^{\text{dq}}}{\underline{x}_{\text{del}}^{\text{dq}}} \cdot \frac{\underline{x}_{\text{del}}^{\text{dq}}}{\underline{v}_{\text{ref}}^{\text{dq}}} = \frac{2 - sT_{\text{del}}}{2 + sT_{\text{del}}} \tag{9}$$

LCL Filter

The dynamics of the *LCL* filter are expressed with differential equations of the currents and voltages. L_1 and R_1 are the filter inductance and parasitic resistance, C is the filter capacitance, v_{pcc} is the voltage at PCC, L_2 and R_2 are the equivalent leakage inductance and resistance of the transformer.

$$\frac{d\underline{i}_{\text{L1}}^{\text{dq}}}{dt} = \frac{1}{L_1} \underline{v}_{\text{inv}}^{\text{dq}} - \frac{1}{L_1} \underline{v}_{\text{C}}^{\text{dq}} - \frac{R_1}{L_1} \underline{i}_{\text{L1}}^{\text{dq}} - j\omega_g \underline{i}_{\text{L1}}^{\text{dq}} \tag{10}$$

$$\frac{d\underline{v}_{\text{C}}^{\text{dq}}}{dt} = \frac{1}{C} \underline{i}_{\text{L1}}^{\text{dq}} - \frac{1}{C} \underline{i}_{\text{L2}}^{\text{dq}} - j\omega_g \underline{v}_{\text{C}}^{\text{dq}} \tag{11}$$

$$\frac{d\underline{i}_{\text{L2}}^{\text{dq}}}{dt} = \frac{1}{L_2} \underline{v}_{\text{C}}^{\text{dq}} - \frac{1}{L_2} \underline{v}_{\text{pcc}}^{\text{dq}} - \frac{R_2}{L_2} \underline{i}_{\text{L2}}^{\text{dq}} - j\omega_g \underline{i}_{\text{L2}}^{\text{dq}} \tag{12}$$

DC-Link Capacitor

Ignoring the inverter power losses, the dynamic performance of the dc-link capacitor can be modeled based on the power balance given in (13). The factor $3/2$ is introduced by the amplitude-invariant Park transformation.

$$\frac{dv_{\text{dc}}}{dt} = \frac{I_{\text{pv}}}{C_{\text{dc}}} - \frac{3}{2} \frac{v_{\text{inv}}^{\text{d}} i_{\text{L1}}^{\text{d}} + v_{\text{inv}}^{\text{q}} i_{\text{L1}}^{\text{q}}}{C_{\text{dc}} V_{\text{dc}}} \tag{13}$$

Reference frames

Inverter

In transients, the SRF estimated by the PLL does not match the SRF in the physical plant due to a non-zero value of v_C^q. The PLL has impact on variables with Park (or inverse Park) transformation, i.e. voltages and currents that are fed back in the controllers or fed into the modulator. The currents and voltages in an *LCL* filter are not directly influenced by the dynamics of the PLL. To precisely describe the impact of the PLL, each inverter is modeled with two SRFs, a control frame dq_i^c and a system frame dq_i^s for the ith inverter as illustrated in Fig. 3. Variables measured from the physical plant are converted to dq_i^c as the feedbacks in the control loops, while voltage references set by the ACC are converted to dq_i^s for modulations. The improvement of the proposed PLL model is validated by comparing three different SSMs against a reference switching model (SM) in PLECS simulation as summarized in Table I. Three different scenarios under a weak grid condition with a low short-circuit ratio (SCR) are selected, in which the system remains stable and the a dominant resonance is clear to be observed. Taking the waveform of i_g as an example, the proposed SSM with separate control and system frames predicts the closest dominant resonance frequency compared to the Fourier analysis results acquired by PLECS simulations. The transient waveform of the synchronization angle θ under a step response of $v_{dc,ref}$ acquired by different models are also compared in Fig. 4, which indicates that the proposed SSM enables more accurate stability investigations.

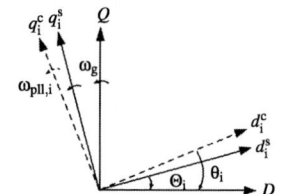

Fig. 3: Reference frames

Table I: Dominant resonance of i_{Lg} in different models

SCR	I_{pv}	SM PLECS	SSM1 [10]	SSM2 [6]	SSM3 Proposed
2	500 A	140 Hz	155 Hz	150 Hz	141 Hz
2.5	1000 A	150 Hz	176 Hz	167 Hz	155 Hz
5	1500 A	200 Hz	237 Hz	227 Hz	209 Hz

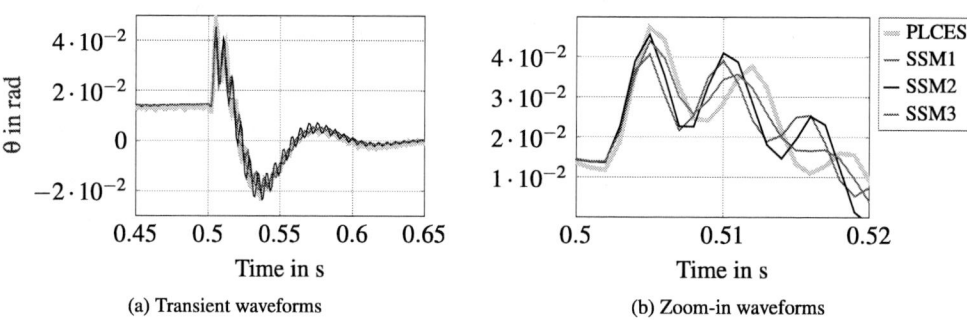

(a) Transient waveforms

(b) Zoom-in waveforms

Fig. 4: Comparison of different SSMs with the SM in PLECS

Conversions between the control and system frame for any variable x_i in the ith inverter are described in (14) with a small-signal representation, where X_d and X_q are the steady-state values in the SRF and θ_i is the phase angle between v_C and v_{pcc} in the ith inverter.

$$\Delta x_i^{dq,s} = \Delta x_i^{dq,c} + \begin{bmatrix} -X^q \\ X^d \end{bmatrix} \cdot \Delta\theta_i \tag{14}$$

Point of Common Coupling

Inverters are connected at the PCC to inject power into the grid. Since operation points and filter parameters could be different, each inverter has individual SRF. To calculate the total power at the PCC, a common frame DQ is required to characterize variables from individual SRFs dq_i^s as illustrated in Fig. 3. Currents in dq_i^s of each inverter are converted to the DQ frame, while PCC voltages estimated in the grid SSM in DQ are converted to different dq_i^s for each inverter. Transformation from individual system frame of each inverter to the common SRF defined by the PCC is given in (15) with the corresponding rotation matrix defined in (16). The Θ_i is the steady-state phase angle between v_C and v_{pcc} of the ith inverter. Based on that, the PCC voltage converted to individual SRF of each inverter is calculated by (17) for totally N inverters in the park. Note that v_{pcc}^{dq} is treated as inputs to estimate i_{L2}^{dq} according to (12). Since v_{pcc} is not a capacitor voltage, it cannot be directly determined through any state-space equation. To ensure the voltages are well defined, a virtual resistor R_v is assumed between the PCC and ground as suggested in [6].

$$x_i^{DQ} = T(\Theta_i) \cdot x_i^{dq,s} \tag{15}$$

$$T(\Theta_i) = \begin{bmatrix} \cos(\Theta_i) & -\sin(\Theta_i) \\ \sin(\Theta_i) & \cos(\Theta_i) \end{bmatrix} \tag{16}$$

$$v_{pcc,i}^{dq} = T^{-1}(\Theta_i) \cdot R_v (\sum_{i=1}^{N} T(\Theta_i) \cdot i_{L2,i}^{dq} - i_g^{DQ}) \tag{17}$$

Grid

The grid is modelled in the DQ frame with an equivalent impedance determined by the SCR. The overall inverter current is given in the grid model to estimate voltages at the PCC.

$$\frac{di_g^{DQ}}{dt} = \frac{1}{L_g} v_{pcc}^{DQ} - \frac{1}{L_g} v_g^{DQ} - \frac{R_g}{L_g} i_g^{DQ} - j\omega_g i_g^{DQ} \tag{18}$$

Multi-Inverter System

Combining (1) to (18), the SSM of a PV park can be acquired. To facilitate modeling of large-scale PV parks, CCM is adopted to rearrange SSMs in a diagonal format from component to system levels, as shown in Fig. 2. The SSM of a PV park comprised of N inverters has an order of $16N + 2$. For a single-inverter with grid, state variables are listed in (19). The nonlinearity of the model prevents direct application of classic linear analysis techniques. Therefore, differential equations shall be linearized at a given operation point.

$$x = [\psi_{dc} \quad \gamma_d \quad \gamma_q \quad v_{Cf}^d \quad v_{Cf}^q \quad x_{del}^d \quad x_{del}^q \quad i_{L1}^d \quad i_{L1}^q \quad v_C^d \quad v_C^q \quad i_{L2}^d \quad i_{L2}^q \quad \varepsilon \quad \theta \quad v_{dc} \quad i_g^D \quad i_g^Q]^T \tag{19}$$

Based on proposed modeling framework, the SSM of a PV park containing 5 inverters is established. To validate the SSM, waveforms of v_{dc} and i_g^D of each inverter are compared with a detailed SM as shown in Fig. 5. The inverters are assumed to operate at different operation points with V_{dc} range from 1140 V to 1160 V and I_{pv} range from 750 A to 1250 A. A step change of $v_{dc,ref}$ of the Inv. 1 occurs from 0.3 s to 0.6 s while a step change of i_{pv} of Inv. 2 occurs from 0.4 s to 0.7 s. Waveforms of the SSM match the SM accurately in both steady state and transient for all the state variables, which proves the model accuracy.

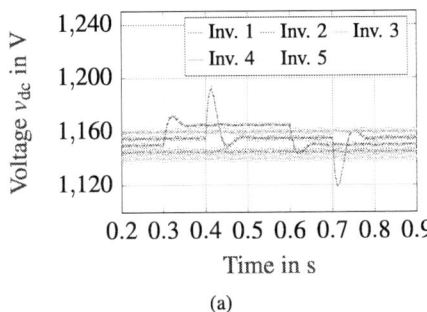
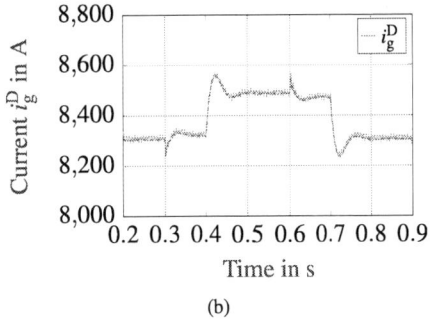

Fig. 5: Waveforms in a PV park (SM: –, SSM: -.-)

Eigenvalue Analysis

Eigenvalue Distribution

The stability can be assessed by the EVs of a system. EVs are in format of complex numbers i.e. $\lambda_i = \sigma_i + j\omega_i$ represents the ith EV in a system. The real part of EVs represents the damping features, while the imaginary part gives the information of potential resonances. A system is defined as stable if all EVs have a negative real part (positive damping). EVs of a single-inverter system are calculated and depicted in Fig. 6(a). There are in total 18 EVs, which contains maximum nine pairs of complex conjugate EVs representing different resonant modes. To quantitatively analyze system features, the damping ratio (DR) $\zeta_i = -\sigma_i/\sqrt{\sigma_i^2+\omega_i^2} = \cos(\varphi_i)$ and resonance frequency $f_i = \omega_i/2\pi$ of each EV are calculated and listed in Table II. The DR of an EV can be also represented by a cosine value e.g. for λ_1 as marked in Fig. 6(a). Two pair of conjugate EVs with a high attenuation $\lambda_{15,16} = -6.5 \cdot 10^7 \pm j314.16$ and $\lambda_{17,18} = -1.96 \cdot 10^4 \pm j36.57$ have almost no impact on the stability, which are dominantly influenced by VFF and Grid. Thus, they are regarded as auxiliary EVs and will not be shown in the following analysis. The EV distribution of a multi-inverter system is depicted in Fig. 6(b) for a five-inverter system. Inverters are designed with same control parameters but at different operation points as depicted in Fig. 5. Different marks represent the change of EVs with an increasing number of inverters from 1 to 5. With an increasing number of inverters in a park, the DRs of λ_{1-4} are increased, while the DRs of λ_{5-8} are slightly decreased. Distributions of λ_{9-14} remain unchanged.

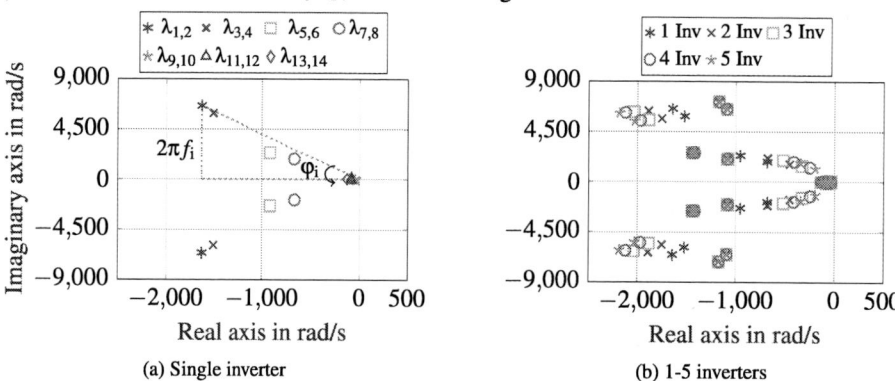

(a) Single inverter   (b) 1-5 inverters

Fig. 6: Eigenvalue distributions

Participation Analysis

To quantitatively analyze the influence of state variables to EVs, PF analysis is used. The Sensitivity of a diagonal element a_{kk} in a state matrix to the ith EV λ_i can be represented by a PF, which is equal to the product of the right-column eigenvector u_{ki} and the left-column eigenvector v_{ki} of the state matrix as $p_{ki} = \partial\lambda_i/\partial a_{kk} = u_{ki}^T \cdot v_{ki}$. A high value of the PF represents a strong impact. Based on that, dominant states of EV pairs are identified and listed in Table II.

Instability Identification

Combining the system-level SSM and PF analysis, it is possible to identify the critical inverter and the cause of an instability in a system. Take the five-inverter PV park as an example, two different unstable cases are analyzed regarding to DVC-related instability and ACC-related instability, respectively.

Table III: PFs of critical EVs in case 1 ($\lambda_{crit} = 0.41 \pm j68$)

State	PF	Influence
Inv.1-4.ψ_{dc}	< 0.01	DVC
Inv.1-4.v_{dc}	< 0.01	DC
Inv. 5.ψ_{dc}	1.00	DVC
Inv. 5.v_{dc}	0.87	DC
Grid.i_g^{DQ}	0.02 / 0.01	Grid

Table IV: PFs of critical EVs in case 2 ($\lambda_{crit} = -136 \pm j2712$)

State	PF	Influence
Inv.1.γ^{dq}	0.64 / 0.48	ACC
Inv.1.$i_{L1}^{dq}, v_C^{dq}, i_{L2}^{dq}$	0.51 − 0.97	Filter
Inv.2-5.γ^{dq}	< 0.01	ACC
Inv.2-5.$i_{L1}^{dq}, v_C^{dq}, i_{L2}^{dq}$	0.02 − 0.09	Filter
Grid.i_g^{DQ}	0.04 / 0.07	Grid

Table II: EVs and dominant states

Eigenvalue	Resonance	Damping	Dominant States	Influence
$\lambda_{1,2}$	1053 Hz	0.24	$i_{L1}^{dq}, x_{del}^{dq}, v_C^{dq}$	Filter, Delay
$\lambda_{3,4}$	944 Hz	0.25	$i_{L1}^{dq}, x_{del}^{dq}, v_C^{dq}$	Filter, Delay
$\lambda_{5,6}$	380 Hz	0.36	$i_{L1}^{dq}, x_{del}^{dq}, i_{L2}^{dq}$	ACC, Delay
$\lambda_{7,8}$	296 Hz	0.33	$i_{L1}^{dq}, x_{del}^{dq}, i_{L2}^{dq}$	ACC, Delay
$\lambda_{9,10}$	13 Hz	0.46	Ψ_{dc}, v_{dc}	DVC
$\lambda_{11,12}$	17 Hz	0.61	ε, θ	PLL
$\lambda_{13,14}$	0 Hz	1	γ^{dq}	ACC
$\lambda_{15,16}$	6 Hz	1	v_{Cf}^{dq}	VFF
$\lambda_{17,18}$	50 Hz	1	i_g^{dq}	Grid

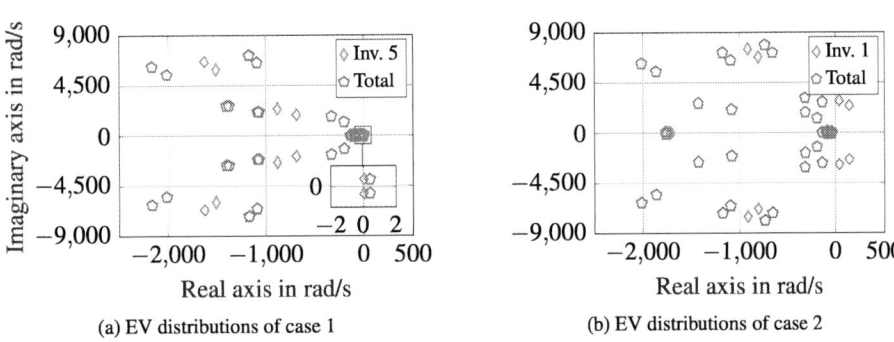

(a) EV distributions of case 1   (b) EV distributions of case 2

Fig. 7: Eigenvalue distributions of unstable cases

Case 1: DVC-Related Instability

In the first case, the instability is caused by DVC. The PV current of the Inv. 5 is increased to 1985 A, while its DVC bandwidth is reduced to 15 Hz. Parameters and operation points of other inverters remain unchanged. In this case, Inv. 5 is still stable when operating alone. Whereas, the total system becomes unstable if five inverters operate together. The EV distributions of Inv. 5 and total system are shown in Fig 7(a). PFs are analyzed for the critical EVs of the total system as listed in Table III. The PFs related to DVC and dc-link capacitor of Inv. 5 feature highest values, which indicates that the instability is dominantly induced by DVC of Inv. 5. The analysis based on EVs of the SSM is also validated with a SM as presented in Fig. 8. Inv. 5 is stable when operating alone. However, a small ripple with 10 Hz resonance is observed, which is caused by the critical EVs $0.03 \pm j68$ at the stability boundary. Inv. 5 becomes unstable when operating with other inverters together. An unstable resonance at 10 Hz is observed by v_{dc} in Fig. 8(b), which corresponds to the unstable critical EVs $0.41 \pm j68$.

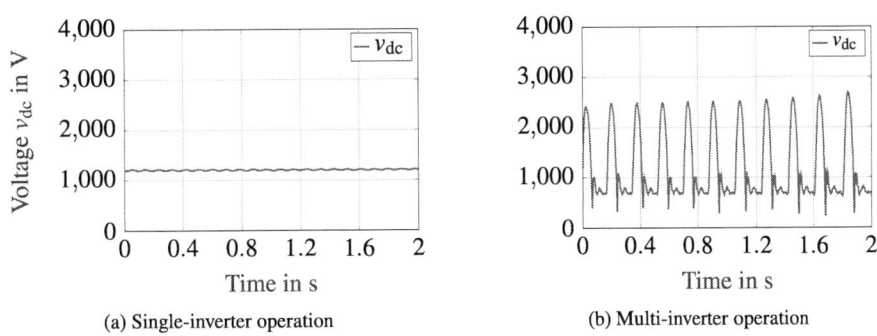

(a) Single-inverter operation   (b) Multi-inverter operation

Fig. 8: Waveforms of v_{dc} of Inv. 5 under different scenarios

Case 2: ACC-Related Instability

In the second case, the instability is caused by ACC. The ACC bandwidth of Inv. 1 is increased to 600 Hz, as an increasing number of inverters provide more damping at the PCC. Parameters and operation points of other inverters remain unchanged. In this case, Inv. 1 is unstable when operating alone. However, the total system remains stable if five inverters operate together. The EVs of Inv. 1 and total system are shown in Fig 7(b). PFs are analyzed for EVs with the lowest DR in the total PV park and listed in Table IV. The PF of state variables in the ACC and *LCL* filter of Inv. 1 are obviously higher than those in other inverters, which indicates that the ACC of Inv. 1 is the weakest part of the system, even if the total system is still stable. The analysis is also validated by comparing to a SM as shown in Fig. 9.

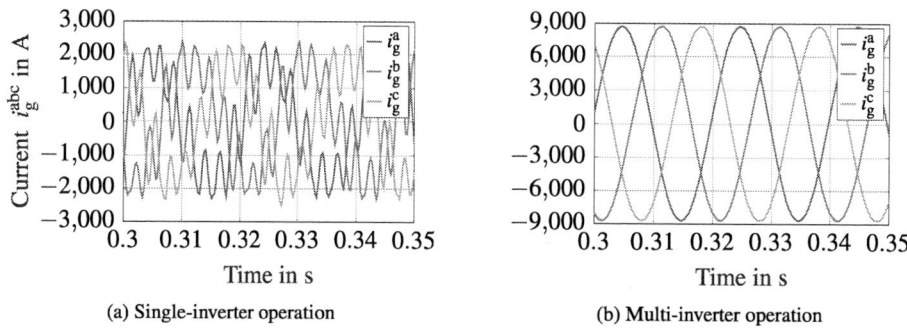

(a) Single-inverter operation   (b) Multi-inverter operation

Fig. 9: Waveforms of i_g^{abc} under different scenarios

Damping-Resonance Analysis

EVs of a system contain two important information that influence dynamics, which are the DR $\zeta_i = -\sigma_i/\sqrt{\sigma_i^2 + \omega_i^2} = \cos(\varphi_i)$ and resonance frequency $f_i = \omega_i/2\pi$. The DR of an EV determines the attenuation speed of corresponding resonance during a transient. A system remains stable only if all the resonances can be properly damped, which indicates that the minimum damping ratio (MDR) shall always be a positive value. The corresponding resonance frequency of the EVs with MDR is defined as dominant resonance frequency (DRF). To investigate the stability in a systematic approach, the damping-resonance analysis is proposed. Figures of MDR and corresponding DRF are depicted for a defined operation range. The other two dimensions can be combinations of operation conditions, control parameters or number of inverters, which enables to extend the analysis to multi-inverter structures. Based on that, the systems can be designed more robust by optimizing operation ranges as well as control loops.

Stability-Oriented Operation

The system stability depends on the operation points. Damping-resonance figures regarding to different I_{pv} and different inverter numbers are observed in Fig. 10. For simplicity, all the inverters are assumed to be identical. MDRs are shown in Fig. 10(a), where only scenarios with a positive MDR are observed to clearly show the stable range. With either an increasing number of inverters or an increasing I_{pv}, the MDR is decreased. DRFs of stable operations are shown in Fig. 10(b). DRFs jump non-continuously, which indicates a instant change of EV pairs with the MDR. The value of DRFs qualitatively implies the reason of an instability. A DRF near 20 Hz is related to the DVC and the dynamics of the dc-link capacitor, as the resonance is close to the DVC bandwidth $f_{bw}^{DVC} = 20$ Hz. A DRF around hundreds of Hz is most likely related to the ACC, as the implemented bandwidth is $f_{bw}^{ACC} = 300$ Hz. A DRF near 1000 Hz is mainly influenced by the *LCL* filter, as the resonance frequency of the filter is $f_{res}^{LCL} = 919$ Hz. To further analyze domain factors of instabilities quantitatively, PF analysis should be applied. For operations with a high value of DRF (light-color area), the system stability is determined by the physical components rather than control loops. Thus, MDRs are higher than those in DVC- or ACC-related cases, which are more preferred. Note that the DVC-related cases happen only if $I_{pv} > 1400$ A. To extend stable operation range with as more inverters as possible, it is recommended to clip I_{pv} below 1500 A.

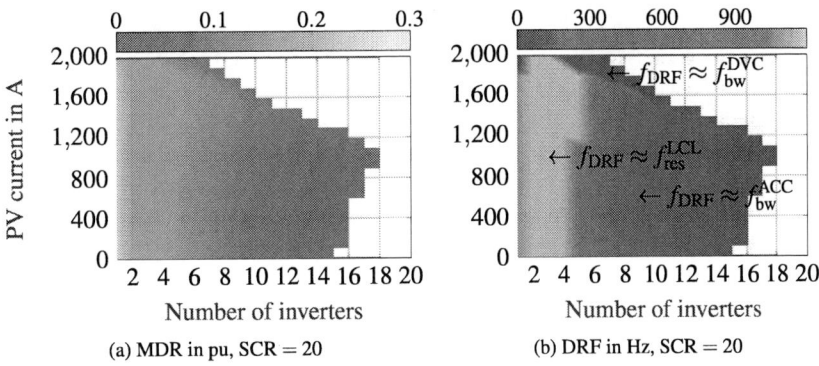

(a) MDR in pu, SCR $= 20$ (b) DRF in Hz, SCR $= 20$

Fig. 10: MDR and DRF with different inverter numbers and PV currents

Stability-Oriented Design

Damping-resonance analysis can also be used for the control bandwidths optimization. MDRs concerning different number of inverters and ACC bandwidths are depicted in Fig. 11(a)-(d) for different PV currents and SCR. To keep the system stable with as more inverters as possible, it is recommended to set the ACC bandwidth at 340 Hz, as the maximum value of the MDR is achieved. This optimum value is independent of the number of inverters and is valid under different I_{pv} and SCR. Usually, $1/10$ of switching frequency $f_{sw} = 3\,\mathrm{kHz}$ is considered as the ACC bandwidth according to the engineering experience [13]. However, the optimum ACC bandwidth may be different to the value according to the rule of thumb from the system stability point of view.

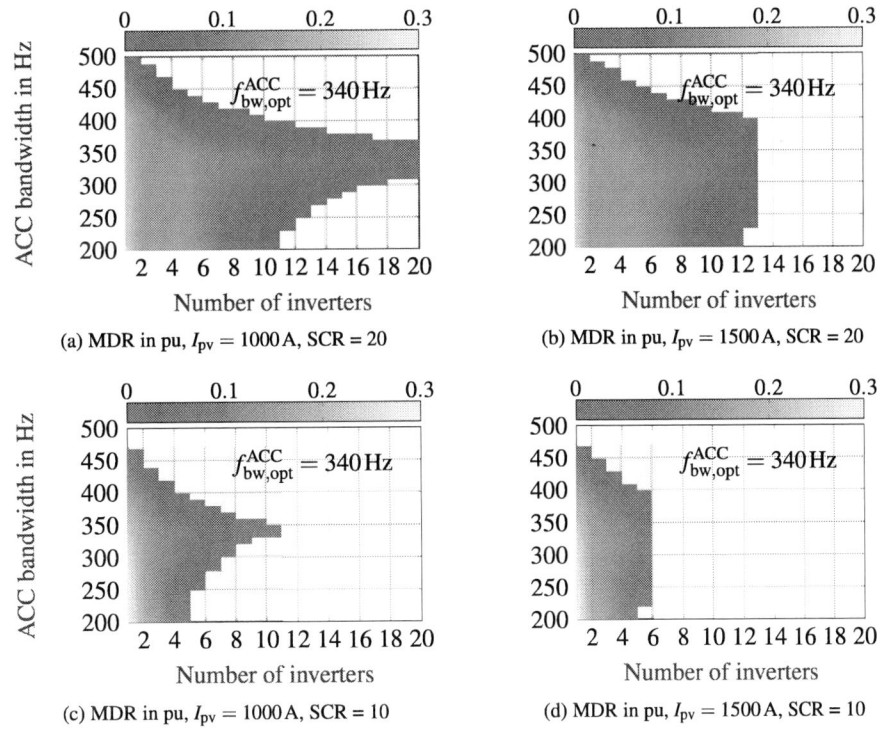

(a) MDR in pu, $I_{pv} = 1000\,\mathrm{A}$, SCR $= 20$ (b) MDR in pu, $I_{pv} = 1500\,\mathrm{A}$, SCR $= 20$

(c) MDR in pu, $I_{pv} = 1000\,\mathrm{A}$, SCR $= 10$ (d) MDR in pu, $I_{pv} = 1500\,\mathrm{A}$, SCR $= 10$

Fig. 11: Damping-resonance analysis with different ACC bandwidths

Conclusion

In this work, a SSM-based framework is established to analyze the stability of multi-inverter grid-tied PV systems from a systematic approach. Accurate SSMs of PV inverters are established from control loops to the park level using CCM. Multiple SRFs are used to model different inverters and the grid. Methodologies for stability assessment are presented including EV analysis, PF analysis and the proposed damping-resonance analysis. Potential resonances can be evaluated by EVs. PF analysis is implemented to quantitatively identify the dominant impact of state variables on each EV. Based on that, reasons of instabilities can be accurately identified in a PV park. The damping-resonance analysis is proposed to quickly investigate multi-inverter systems from the stability point of view. Based on the MDR and DRF, operation ranges and control loops can be optimized from a systematic perspective. To improve the system stability with different output powers, a dc-coupled battery storage system can be considered in the future.

References

[1] Lammert, G., Hess, T., Schmidt, M., Schegner, P., and Braun, M. (2015). Dynamic grid support in low voltage grids - Fault ride-through and reactive power/voltage support during grid disturbances. Power Systems Computation Conference.

[2] Wen, B. , Boroyevich, D. , Burgos, R. , Mattavelli, P. , and Shen, Z. . (2015). Analysis of d-q small-signal impedance of grid-tied inverters. IEEE Transactions on Power Electronics, 31(1), 675-687.

[3] Li, and Chun. (2017). Unstable operation of photovoltaic inverter from field experiences. IEEE Transactions on Power Delivery, 1-1.

[4] D. Wang, L. Liang, L. Shi, J. Hu, and Y. Hou, Analysis of modal resonance between pll and dc-link voltage control in weak-grid tied vscs, IEEE Transactions on Power Systems, vol. 34, no. 2, pp. 1127-1138, 2018.

[5] Sun, J. (2011). Impedance-based stability criterion for grid-connected inverters. IEEE Transactions on Power Electronics, 26(11), 3075-3078.

[6] Pogaku, N., Prodanovic, M., and Green, T. C. (2007). Modeling, analysis and testing of autonomous operation of an inverter-based microgrid. IEEE Transactions on power electronics, 22(2), 613-625.

[7] Wang, Y., Wang, X., Blaabjerg, F., and Chen, Z. (2016). Harmonic instability assessment using state-space modeling and participation analysis in inverter-fed power systems. IEEE Transactions on Industrial Electronics, 64(1), 806-816.

[8] Amin, M., and Molinas, M. (2017). Small-signal stability assessment of power electronics based power systems: A discussion of impedance-and eigenvalue-based methods. IEEE Transactions on Industry Applications, 53(5), 5014-5030.

[9] Verghese, G. C., Perez-Arriaga, I. J., and Schweppe, F. C. (1982). Selective modal analysis with applications to electric power systems, Part II: The dynamic stability problem. IEEE Transactions on Power Apparatus and Systems, (9), 3126-3134.

[10] Amin, M., Suul, J. A., D'Arco, S., Tedeschi, E., and Molinas, M. (2015). Impact of state-space modelling fidelity on the small-signal dynamics of VSC-HVDC systems.

[11] Z. Yang, Q. Wang, J. Warmuz, and R. W. De Doncker, Stability assessment of a three-phase grid-tied pv inverter with eigenvalue-based methods, in 2019 10th IEEE International Symposium on Power Electronics for Distributed Generation Systems (PEDG). IEEE, 2019, p. 722-727.

[12] Bakhshizadeh, M. K., Yoon, C., Hjerrild, J., Bak, C. L., Kocewiak, . H., Blaabjerg, F., and Hesselbk, B. (2017). The application of vector fitting to eigenvalue-based harmonic stability analysis. IEEE Journal of Emerging and Selected Topics in Power Electronics, 5(4), 1487-1498.

[13] Harnefors, L., Zhang, L., and Bongiorno, M. (2008). Frequency-domain passivity-based current controller design. IET Power Electronics, 1(4), 455-465.

Compact Core Loss Model Based on an Effective Frequency for Arbitrary Core Excitations Including DC-Bias

Erika Stenglein, Manfred Albach and Thomas Dürbaum
Electromagnetic Fields, University of Erlangen-Nürnberg
Cauerstr. 7, 91058 Erlangen
Erlangen, Germany
Phone: +49/ (0) – 9131 85 28947
Fax: +49/ (0) – 9131 85 27787
Email: erika.stenglein@fau.de
URL: http://www.emf.eei.fau.de

Keywords

≪Passive component≫, ≪Magnetic device≫, ≪Modelling≫, ≪Measurement≫

Abstract

To predict core losses for arbitrary excitations, the quasi-static energy losses depending on the swing and the DC-bias of the magnetic flux density are multiplied by an effective frequency. Measurement results for various shapes of the magnetic flux density (e.g. sinusoidal and triangular waveforms) with and without DC-bias verify this compact model.

Introduction

For a proper design of SMPS (switched-mode power supplies), hardware designers rely on adequate methods for core loss prediction. The approaches proposed in literature essentially fall into two categories. Either a mathematical representation of the hysteresis loop is used to determine the dissipated energy in the magnetic core during one switching cycle (e.g. Preisach [1], [2] or Jiles-Atherton [3] model) or a direct calculation of the core losses is carried out without an intermediate step of a hysteresis loop description (e.g. Steinmetz equation [4] and its extensions [5]–[12]; loss separation method [13]–[19]). Regardless of the chosen approach, developing accurate models for calculating core losses remains a challenging task as the excitation of magnetic materials varies depending on the converter topology and its operating mode. The waveform of the magnetic flux density $B(t)$ including its frequency $f = 1/T$, the magnetic flux density swing ΔB and DC-bias B_{DC} as well as core temperature ϑ affect core losses. In addition to a reasonable accuracy, developers of magnetic components require compact models that allow a simple integration into existing optimization routines and also guarantee high simulation/calculation speed. Moreover, methods with a small number of parameters are preferred that are frequency-independent, valid for arbitrary shapes of $B(t)$ as well as easily extractable from a limited number of measurements or from the data provided by manufacturers. This paper develops such an easy-to-handle model. As reliable experimental data form the basis for every model derivation, an extensive core loss study on the commercially available Mn-Zn ferrites N87, N49, 3F3 and 3C90 has been performed for different core excitations $B(t)$ with and without DC-bias B_{DC} covering a wide frequency range extending from 100 Hz to 500 kHz. The next sections present the investigated waveforms of $B(t)$ as well as the applied test rigs and procedures. Subsequently, this paper discusses the obtained measurement results. Based on the data of over 10000 experimental runs, this paper derives a compact core loss model. The proposed approach separates the core losses into specific core losses P_{spec} and the classical eddy-current losses P_{eddy}. The standard expression for the classical eddy-current losses [14],

[20] in combination with a Fourier transformation approximates P_{eddy}. Alternatively, the improved methods given in [21]–[23] can be applied. The specific core losses P_{spec} can be derived by multiplying an effective frequency f_{eff} with the quasi-static (or hysteresis) energy losses E_{hyst} occurring at low frequencies. This quasi-static component E_{hyst} comprises the influence of ΔB and B_{DC} on core loss. The effective frequency f_{eff} represents the dynamic effects and depends only on the second derivative of the magnetic flux density $B(t)$ as well as on two frequency-independent material parameters c and γ. This approach enables core loss predictions for arbitrary shapes of $B(t)$. In addition, core loss data for sinusoidal excitations without DC-bias, as usually provided by core manufacturers, suffice for the extraction of c and γ. The last section of this paper validates the proposed method and demonstrates its reasonable accuracy of 15 % for the broad range of investigated ferrites and excitations.

Investigated core excitations

The voltage waveforms across magnetic components in SMPS are mostly rectangular or a composite of rectangular shapes. Accordingly, the magnetic flux density in cores with dominating air gap exhibits a piecewise linear form. Fig. 1 shows the magnetic flux density $B(t)$ that excites the gapped core of the standard topologies buck, boost and buck-boost while operating in CCM (**c**ontinuous **c**onduction **m**ode) [5], [24], [25]. The difference between the maximum B_{max} and the minimum B_{min} of the magnetic flux density $B(t)$ equals the magnetic flux density swing ΔB, while

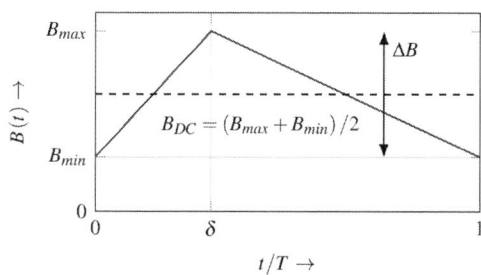

Fig. 1: CCM operation: Magnetic flux density waveform of the standard topologies buck, boost and buck-boost characterized by the duty cycle δ, the frequency $f = 1/T$, the magnetic flux density swing ΔB and the DC-bias B_{DC}.

$B_{DC} = (B_{max} + B_{min})/2$ applies for DC-bias. During material characterization, this triangular waveform $B(t)$ excites the cores under test (CUTs) for different values of the duty cycle δ, the frequency f, ΔB and B_{DC}. Furthermore, this paper obtains core loss data for sinusoidal waveforms of $B(t)$ without DC-bias ($B_{DC} = 0$), as generally done by core manufacturers. To cover a wide range of excitations, the performed core loss study includes further non-sinusoidal waveforms of the magnetic flux density $B(t)$ without DC-bias ($B_{DC} = 0$). Fig. 2 illustrates for these additional excitations the shape of $B(t)$ normalized to $2/\Delta B$ as well as the corresponding shape of the first derivative of $B(t)$ normalized to $T/(2\Delta B)$. In addition, Fig. 2 shows the triangular waveform of $B(t)$ with $\delta = 0.5$ and $B_{DC} = 0$ as well as the sinusoidal waveform. As can be seen, the selected core excitations differ significantly but have the same frequency f and the same magnetic flux density swing ΔB. The normalized waveforms of $dB(t)/dt$ and thus the imposed voltages on the core samples are of rectangular, sinusoidal and triangular shape with a point-symmetry to $t = T/2$. The fourth waveform of $dB(t)/dt$ equals the triangular waveform of $dB(t)/dt$ raised to the power of 3 and normalized to result in the same ΔB.

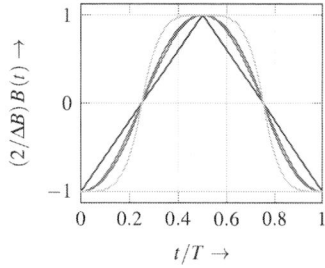

(a) Normalized waveforms of $B(t)$.

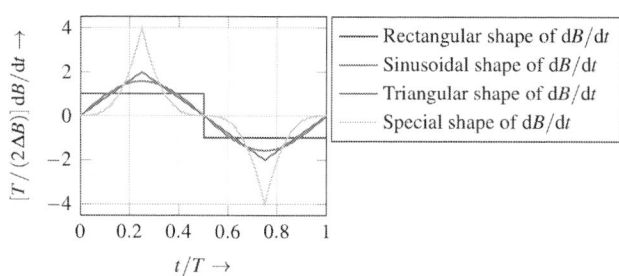

(b) Normalized waveforms of $dB(t)/dt$.

Fig. 2: Investigated symmetric core excitations: This figure shows the normalized waveforms of $B(t)$ and the corresponding normalized waveforms of $dB(t)/dt$.

Measurement set-ups for different excitations of the magnetic flux density

To investigate core losses for the wide frequency range $100\,\mathrm{Hz} \le f \le 500\,\mathrm{kHz}$, this paper applies two measurement set-ups based on the classic two-winding method [11], [26]–[30]. All experiments are conducted on closed, thin toroids with bifilar, evenly distributed windings to achieve a virtually homogeneous field inside the CUT. The primary winding with N_p turns excites the core sample. A shunt resistor R_{shunt} in series connection to the primary winding measures the corresponding primary current i_p. The secondary winding with $N_s = N_p$ turns senses the induced voltage v_s. Processing $i_p(t)$ and $v_s(t)$ yields the magnetic field strength $H(t)$, the magnetic flux density $B(t)$ as well as the core losses dissipated in the CUT. As the selected toroids of the size 16 x 9.6 x 6.3 (outer diameter x inner diameter x height) have a relatively small cross-sectional area, bulk eddy-currents and the associated losses P_{eddy} can be neglected. Consequently, the measured losses only comprise in good approximation the specific core losses P_{spec}. Note that a temperature stabilized oil bath sets the temperature ϑ of the CUTs to guarantee well-defined conditions during material characterization.

Measurement set-up for low frequencies

At sufficiently low frequencies f, the only significant contribution to the total core losses arises from the quasi-static component $P_{hyst} = E_{hyst} f$ corresponding to the specific core losses P_{spec}. The conference papers [31]–[33] describe the test rig that obtains the quasi-static energy losses E_{hyst}. Due to its high flexibility, the presented set-up is capable of imposing nearly any desired waveform of the magnetic flux density $B(t)$ including B_{DC} on the core samples. As long as the investigated ferrites show quasi-static behavior, the traversed hysteresis loops and thus E_{hyst} in steady-state only depend on ΔB as well as B_{DC}.

Measurement set-up for higher frequencies

To measure core losses at higher frequencies up to $500\,\mathrm{kHz}$, the measurement set-up thoroughly explained in [34], [35] is applied. Fig. 3 illustrates the proposed GaN half-bridge configuration to study core losses for triangular waveforms of $B(t)$ including DC-bias (see Fig. 1). Two programmable DC power supplies provide the DC voltages V_{DC}^{+} and V_{DC}^{-} across the electrolytic capacitors C_+ and C_-, respectively. The ground reference is the center point between C_+ and C_-. This allows the application of a ground referenced shunt resistor R_{shunt} for an accurate measurement of the primary current $i_p = v_{shunt}/R_{shunt}$ through the CUT. The induced voltage v_s on the secondary side of the core sample is applied to a resistive voltage divider referenced to ground with the transfer function G (not shown in Fig. 3). This resistive voltage divider adapts v_s to the input voltage range of the applied digitizer NI PXI-5122 and terminates the secondary winding with the high impedance R_{vd}. The digitizer simultaneously samples the voltages

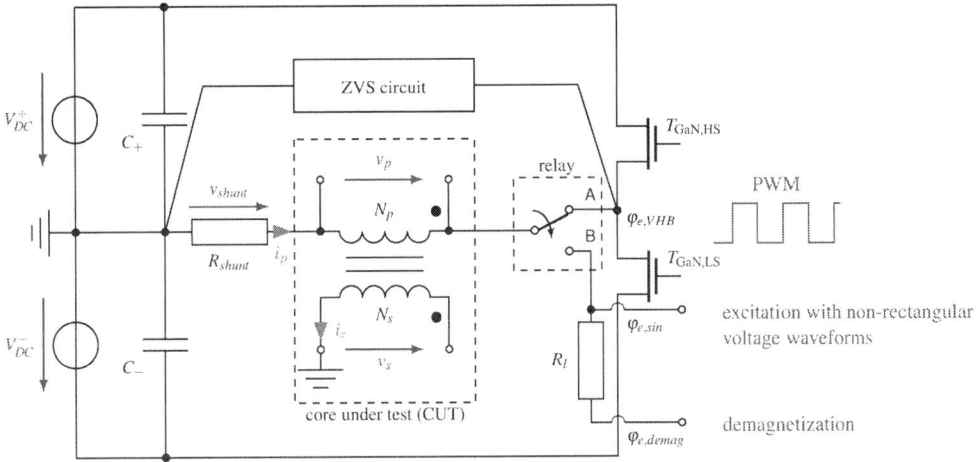

Fig. 3: GaN half-bridge configuration for the measurement of core losses: In relay position A, the half-bridge imposes a rectangular voltage on the core under test (CUT). In relay position B, the CUT can be either electrically demagnetized or excited with non-rectangular voltage waveforms.

$v_{meas} = Gv_s$ and v_{shunt} across R_{shunt}. Due to the high impedance R_{vd}, the voltage across the frequency-dependent resistance of the secondary winding is negligible. However, the secondary current i_s should be considered. Assuming the CUT has the effective magnetic path length l_e and the effective cross-sectional area A_e, the magnetic field strength $H(t)$ and the magnetic flux density $B(t)$ are given by

$$H(t) = \frac{N_p}{l_e} \left[\frac{v_{shunt}(t)}{R_{shunt}} - \frac{N_s}{N_p} \frac{v_s(t)}{R_{vd}} \right] \quad \text{and} \quad B(t) = \frac{1}{N_s A_e} \int_0^t v_s(\tau) \, d\tau + B_0, \tag{1}$$

respectively. These equations assume a high coupling between the primary and secondary side guaranteed by the bifilar winding. Plotting $B(t)$ versus $H(t)$ yields the traversed hysteresis curve.

Before the operation of the GaN half-bridge begins, the CUT is electrically demagnetized to determine the integration constant in (1) to $B_0 = 0$. In relay position B, the arbitrary waveform generator Agilent 33522A connected to a power amplifier (Rohrer: PFL 720 - U 240: amplification A) provides the required AC excitation $\varphi_{e,demag} = v_{demag}$. The amplified voltage Av_{demag} drops across the series connection of the load resistor R_l, the primary winding of the CUT and R_{shunt}. Due to the high resistance of R_l, the current $i_p \approx Av_{demag}/R_l$ flows through the primary winding. First, i_p drives the CUT into saturation. Afterwards, the voltage Av_{demag} and thus i_p slowly diminish while reducing the residual magnetization of the CUT to zero. After demagnetization, the relay position changes to A and the material under test is excited by the voltage $v_{VHB} = \varphi_{e,VHB}$. During the on-state of the GaN-HEMT $T_{GaN,LS}$ in the interval $0 < t \leq \delta T$, the voltage $V_{DC}^- = v_p + v_{shunt}$ drops across the series connection of R_{shunt} and the primary winding of the CUT with N_p turns. As $V_{DC}^- \approx v_p$ applies, the magnetic flux density in the CUT linearly increases in good approximation. During the on-state of $T_{GaN,HS}$ in the interval $\delta T < t \leq T$, the voltage v_p approximately equals $-V_{DC}^+$ causing an almost linear decrease of $B(t)$ in the CUT. To obtain a desired swing ΔB within one time period T, the two DC power supplies need to provide the corresponding voltages V_{DC}^+ and V_{DC}^-. An estimation for $V_{DC}^- = N_p \Delta B A_e f / \delta$ and $V_{DC}^+ = N_p \Delta B A_e f / (1 - \delta)$ follows from the assumption that the voltage $v_{VHB} = \varphi_{e,VHB}$ completely drops across the primary winding of the CUT. In a first step, the DC voltages V_{DC}^+ and V_{DC}^- across the capacitors equal these estimated values. Note that the two independent DC power supplies enable different duty cycles.

For the experiments conducted in this paper, the first PWM pulse of the GaN half-bridge sets the desired B_{DC} by turning on $T_{GaN,LS}$ for the time span $N_p (B_{DC} + \Delta B / 2) A_e / V_{DC}^-$.[1] The upper and lower half-bridge switch are then alternately switched on and off for several periods at the selected frequency f and duty cycle δ. Thus, the half-bridge imposes a nearly rectangular voltage on the CUT leading to an almost triangular waveform of $B(t)$ with the desired parameters B_{DC}, ΔB, δ and f. The arbitrary waveform generator Agilent 33522A provides the corresponding PWM signals to control the GaN half-bridge. Note that the application of GaN-HEMTs allows PCB designs with low parasitic inductances and thus reduces oscillations. Moreover, the ZVS circuit, comprising a series connection of C_{ZVS}, R_{ZVS} and L_{ZVS}, ensures soft-switching and thus further reduces ringing. As v_{shunt} and Gv_s are sampled during the entire excitation, $H(t)$ and $B(t)$ can be calculated by (1) with $B_0 = 0$. In the first iteration, the volt-second balance on the CUT does not equal zero due to the voltage drop across R_{shunt} and the winding resistance of the CUT as well as due to the finite rise and fall times of $v_{VHB} = \varphi_{e,VHB}$. Therefore, a stable hysteresis curve is not achieved. However, $B(t)$ can be used to calculate the deviation in the volt-second balance. Adopting V_{DC}^+ and V_{DC}^- accordingly and performing another measurement sequence repeating the entire procedure of the first step results in a stable hysteresis curve after a few iterations. The steady-state hysteresis loop is traversed after exciting the CUT for several periods with the desired triangular excitation $B(t)$. The specific core loss per unit volume $P_{v,spec}$ equals the enclosed area of this B-H-loop multiplied by the switching frequency f. The multiplication of $P_{v,spec}$ by the effective core volume $V_e = A_e l_e$ yields P_{spec}.

To study core losses for the non-rectangular waveforms of $dB(t)/dt$ shown in Fig. 2, the measurement set-up given in Fig. 3 can also be applied. The voltage $v_{sin} = \varphi_{e,sin} \approx v_p$ is imposed on the primary winding of the previously demagnetized CUT. Again, the arbitrary waveform generator Agilent 33522A

---

[1] The operation of the GaN half-bridge always starts with the turn-on of $T_{GaN,LS}$ due to the bootstrap circuit integrated in the applied half-bridge gate driver.

connected to the power amplifier provides the AC excitation v_{sin} leading to the desired waveform of $B(t)$ in the CUT. Note that the cut-off frequency of 700 kHz of the applied amplifier Rohrer: PFL 720 - U 240 represents a limitation. Therefore, only the GaN half-brigde is able to impose rectangular voltages on the CUT and thus to investigate core losses for triangular excitation of $B(t)$.

Complex test procedures to set a desired DC-bias of the magnetic flux density

In contrast to the standard approach, both measurement set-ups adopt the classic two-winding method to set a desired B_{DC} in the CUT. Core loss measurements usually impose a DC current and thus DC-bias of the magnetic field strength H_{DC} on ungapped toroids to study the influence of DC-bias [11], [36]. Yet, hardware designers require the AC and DC excitation of the magnetic flux density $B(t)$ for core loss prediction. Hence, this approach implies the derivation of B_{DC} from H_{DC}. Typically, a simple functional description, e.g. the initial magnetization curve [11], [37], the anhysteretic curve [38] or the amplitude permeability [12], [37], approximates the relationship $B_{DC}(H_{DC})$. However, experimental results presented in [32], [34], [35] demonstrate that this method is insufficient due to the hysteretic properties of magnetic materials. Magnetic history as well as the waveform of $B(t)$ including ΔB affect $B_{DC}(H_{DC})$.

Fig. 4(a) displays the measured steady-state hysteresis loops for a triangular waveform of $B(t)$ (rectangular dB/dt) characterized by ΔB, $B_{DC} = 200$ mT, $\delta = 0.5$ and $f = 100$ kHz. Note that the corresponding magnetic field strength $H(t)$ exhibits a step when the abrupt changes of dB/dt occur. The PhD thesis [12] as well as the core loss studies conducted in Dartmouth [39]–[43] also report this phenomenon. The dashed lines in Fig. 4(a) represent the corresponding H_{DC} of the hysteresis loop plotted in the same color. As can be seen, H_{DC} varies with ΔB. To further illustrate the complex relationship $B_{DC}(H_{DC})$, Fig. 4(b) shows the measured steady-state hysteresis loops for a triangular waveform of $B(t)$ with $\Delta B = 200$ mT, $B_{DC} = 0$ mT, δ and $f = 100$ kHz. The duty cycle δ affects the traversed hysteresis loops and thus H_{DC}, even though $B_{DC} = 0$ mT holds true for all investigated waveforms of $B(t)$. Hence, a simple functional approach does not suffice to model the relationship $B_{DC}(H_{DC})$. Therefore, core losses are directly measured as a function of B_{DC} to avoid an error-prone calculation of B_{DC} from H_{DC}. This can only be achieved by a rather complex measurement procedure as well as by sequencing measurements. For a detailed description of the applied test rigs and procedures, please refer to [31]–[35].

Measurement results and derivation of compact core loss model

The summation of specific core losses P_{spec} and classical eddy-current losses P_{eddy} yield the total losses in the core. Assuming that the magnetic flux density $B(t) = B_n \cos(2\pi f_n t)$ excites the CUT, the standard expression for the classical losses in a homogeneous material [14], [20]

$$P_{eddy,n} = \frac{\pi}{4} \kappa(f_n) A_e^2 l_e B_n^2 f_n^2 \tag{2}$$

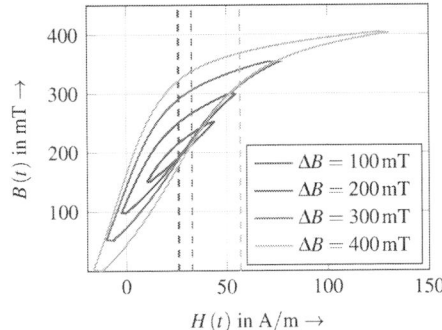

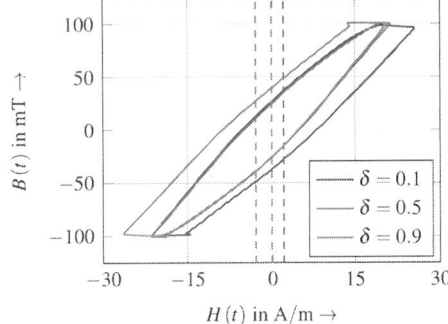

(a) Hysteresis loops for $B_{DC} = 200$ mT, $\delta = 0.5$ and different values of ΔB: The dashed lines represent H_{DC} of the corresponding hysteresis curve plotted in the same color.

(b) Hysteresis loops for $B_{DC} = 0$ mT, $\Delta B = 200$ mT and different values of δ: The dashed lines represent H_{DC} of the corresponding hysteresis curve plotted in the same color.

Fig. 4: Measured steady-state hysteresis loops for triangular waveforms of $B(t)$ with $f = 100$ kHz (N87, $\vartheta = 25\,°C$)

can be applied to estimate the corresponding eddy-current losses. To obtain the core losses for an arbitrary shape of $B(t)$, the Fourier transformation decomposes $B(t)$ into its constituent frequencies f_n. Evaluating (2) for each constituent frequency yields the corresponding eddy-current losses $P_{eddy,n}$. The frequency-dependent macroscopic conductivity $\kappa(f_n)$ can be estimated by using the expression given in [23]. The total eddy-current losses P_{eddy} result from the summation of the eddy-current losses $P_{eddy,n}$ caused by each constituent frequency f_n. Note that (2) has been derived for a homogeneous magnetic flux density $B_n \cos(2\pi f_n t)$ exciting a core of cylindrical shape with the cross-sectional area A_e and the length l_e. Thus, equation (2) is only a rough approximation for other core types. The method given in [21] provides more sophisticated approximations for cores with cylindrical and rectangular cross-sectional areas, whereas the approaches presented in [22], [23] improve the prediction of bulk eddy-current losses P_{eddy} by modeling the heterogeneous micro-structure of sintered ferrites.

To develop a compact core loss model for the specific core losses P_{spec}, an extensive core loss study has been performed. Fig. 5 displays the specific power loss density $P_{v,spec} = P_{spec}/V_e$ that occurs in the toroid samples made of N87, N49, 3F3 and 3C90 under a triangular core excitation $B(t)$ (rectangular dB/dt) shown in Fig. 1 with $\delta = 0.5$. The experiments cover different values of ΔB, B_{DC} and f. As can be seen, the dependency of core losses on B_{DC} is strongly pronounced for all frequencies. $P_{v,spec}$ significantly rises with growing values of B_{DC}, especially when the traversed hysteresis loop approaches the saturation region in the B-H-plane. Note that $P_{v,spec}/f$ for $f = 1\,\text{kHz}$ equals the quasi-static energy loss density $E_{v,hyst}$ at low frequencies. The measurement data (N87) presented in Fig. 6(a) for different duty cycles δ also indicate that B_{DC} considerably affects core losses. Further assessment of the experimental results for triangular shapes of $B(t)$ reveals that the ratio $P_{v,spec}(\Delta B, B_{DC}, f, \delta)/P_{v,spec}(\Delta B, B_{DC} = 0, f, \delta)$ is approximately independent from f and δ.

Fig. 6(b) illustrates $P_{v,spec}$ for triangular excitations $B(t)$ with different values of δ and ΔB (N87, $B_{DC} = 0$, $f = 100\,\text{kHz}$). As can be seen, complementary duty cycles cause the same specific power losses in the core. For comparison, Fig. 6(b) also shows $P_{v,spec}$ for sinusoidal waveforms of $B(t)$ characterized by ΔB, $B_{DC} = 0$ and $f = 100\,\text{kHz}$. The measurement results demonstrate that a symmetric triangular core excitation $B(t)$ with $\delta = 0.5$ (equal rise and fall time of $B(t)$) causes more than 10% lower specific core losses than the comparable sinusoidal shape with the same ΔB and f. However, if the rise or fall time of $B(t)$ increases, the related specific power loss density $P_{v,spec}$ also rises and finally exceeds the corresponding value $P_{v,spec}$ of the comparable sinusoidal excitation. The crossover point approximately occurs at $\delta \approx 0.2$ and $\delta \approx 0.8$ for all ΔB. Interestingly, this applies for all investigated Mn-Zn ferrites.

Introduction of an effective frequency

Comparing the measured values of $P_{v,spec}$ for the various core excitations (see Fig. 1 and Fig. 2) to the quasi-static energy loss density $E_{v,hyst}$ occuring at low frequencies yields the following relationship:

$$P_{v,spec} = P_{spec}/V_e = E_{v,hyst}(\Delta B, B_{DC})\, f_{eff}. \qquad (3)$$

The effective frequency f_{eff} depending on f and δ accounts for the dynamic effects related to the shape of $B(t)$. The variables ΔB and B_{DC} only affect the quasi-static energy loss density $E_{v,hyst}(\Delta B, B_{DC})$. Further evaluations of the core loss data yield the following expression:

$$f_{eff} = f\left[1 + c\left(\frac{1}{\Delta B}\int_0^T \left|\frac{d^2 B}{dt^2}\right| dt\, \frac{1}{1\,\text{Hz}}\right)^\gamma\right] \qquad (4)$$

depending only on the second derivative of the magnetic flux density $B(t)$, its period of $T = 1/f$ and two frequency-independent material parameters c and γ. Table I gives the effective frequency f_{eff} for the investi-

Table I: Effective frequency f_{eff} of the investigated core excitations shown in Fig. 1 and Fig. 2.

Shape of dB/dt	Shown in	f_{eff}
Rectangular	Fig. 1	$f\left[1 + c\left(\frac{2}{\delta(1-\delta)}\frac{f}{1\,\text{Hz}}\right)^\gamma\right]$
Rectangular with $\delta = 0.5$	Fig. 2	$f\left[1 + c\left(8\frac{f}{1\,\text{Hz}}\right)^\gamma\right]$
Sinusoidal	Fig. 2	$f\left[1 + c\left(4\pi\frac{f}{1\,\text{Hz}}\right)^\gamma\right]$
Triangular	Fig. 2	$f\left[1 + c\left(16\frac{f}{1\,\text{Hz}}\right)^\gamma\right]$
Special shape	Fig. 2	$f\left[1 + c\left(32\frac{f}{1\,\text{Hz}}\right)^\gamma\right]$

gated core excitations shown in Fig. 1 and Fig. 2. For all waveforms of $B(t)$, the dynamic effects represented by (4) are independent from the corresponding swing ΔB and DC-bias B_{DC}. Note that for the triangular core excitation $B(t)$ (rectangular dB/dt) the second derivative of $B(t)$ is zero except for dirac pulses occuring at the same time as the extrema of $B(t)$ that equal points of infinite curvature of $B(t)$.

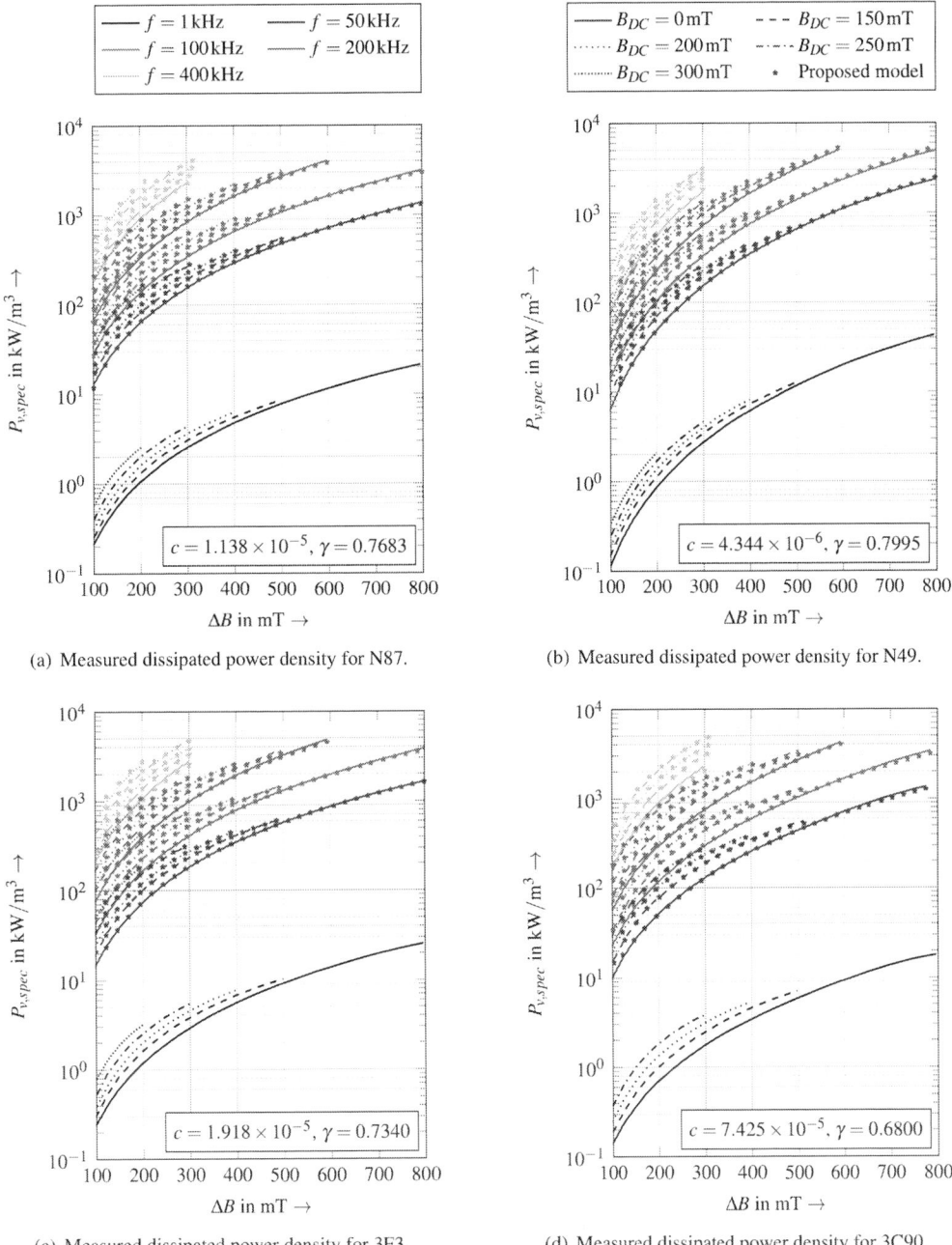

(a) Measured dissipated power density for N87.

(b) Measured dissipated power density for N49.

(c) Measured dissipated power density for 3F3.

(d) Measured dissipated power density for 3C90.

Fig. 5: Measured specific power loss density $P_{v,spec}$ for triangular core excitations $B(t)$ (rectangular dB/dt) with $\delta = 0.5$ as a function of ΔB: The parameters f and B_{DC} vary ($\vartheta = 25\,°C$). Note that $P_{v,spec}/f$ for $f = 1\,kHz$ equals the measured quasi-static energy loss density $E_{v,hyst}$. Moreover, this figures gives the two frequency-independent material parameters c and γ for all investigated materials. If $E_{v,hyst}$ is known, only these parameters are necessary to derive $P_{v,spec}$. The markers "$\star$" show the values predicted with the proposed core loss model.

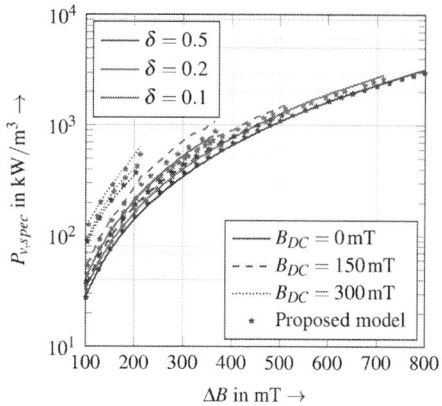 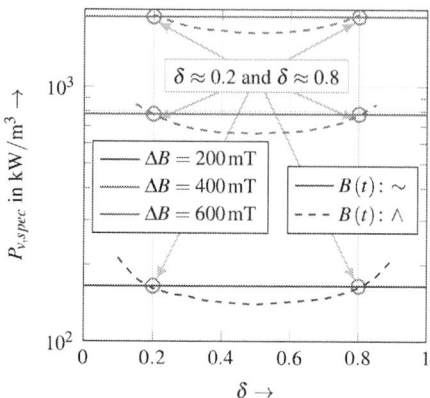

(a) $P_{v,spec}$ as a function of ΔB for triangular core excitations $B(t)$: The parameters B_{DC} and δ vary. The markers "$\star$" show the values predicted with the proposed core loss model.

(b) $P_{v,spec}$ as a function of δ for triangular core excitations $B(t)$ ($B(t)$: $\wedge$) with $B_{DC} = 0$: For comparison, this figure also shows $P_{v,spec}$ for sinusoidal core excitations ($B(t)$: $\sim$) with $f = 100\,\text{kHz}$, $B_{DC} = 0$ and the given values of ΔB.

Fig. 6: Measured dissipated power density $P_{v,spec}$ for triangular core excitations $B(t)$ (rectangular dB/dt) with $f = 100\,\text{kHz}$: The parameters ΔB, B_{DC} and δ vary (N87, $\vartheta = 25\,°\text{C}$).

Extraction of model parameters

Equations (3) and (4) enable core loss predictions for arbitrary shapes of $B(t)$ showing one maximum and minimum within one cycle, a common case in power electronics. To extract the parameters of the proposed model, a loss map needs to be built up experimentally for the magnetic materials of interest. A two-dimensional lookup table provides the quasi-static energy loss density $E_{v,hyst}$ in steady-state for different values of ΔB and B_{DC}. Interpolation or fitted curves/surfaces with mathematical expressions, e.g. the polynomial approach presented in [32], can be utilized to determine $E_{v,hyst}$ for given values of ΔB and B_{DC}. Core manufacturers could rather easily include such a lookup table in material data-sheets.

If the quasi-static energy loss density $E_{v,hyst}$ is known, equation (4) needs to be evaluated for only one waveform of the magnetic flux density $B(t)$ and one set of data (i.e. fixed ΔB and B_{DC}) to determine the frequency-independent parameters c and γ. Thus, a limited number of measurements suffices for the model parameterization. In addition, the core loss data for sinusoidal excitation of $B(t)$ without DC-bias, as usually provided by core manufacturers, can be used for the extraction of c and γ.

Core losses for complex core excitations

In some cases, the magnetic flux density waveform of $B(t)$ exhibits several minima and maxima within one cycle. This results in a hysteresis curve consisting of a major and one or more minor loops. To deal with such complex core excitations, the authors of [5] suggest a separation of the traversed hysteresis curve into subloops. The corresponding specific power losses are estimated separately for each individual subloop. The summation of all these contributions leads to the total power loss in the core. The authors of [7] and [44] have adopted this strategy to improve specific core loss prediction. According to [7] and [44], the estimated values match the experimental data well. To extend the proposed core loss model to such waveforms of $B(t)$ showing several minima and maxima, this paper also suggests to split the magnetic flux density trajectory into subloops. By means of the equations (3) and (4), the specific power loss densities are calculated for each subloop considering the corresponding section of the waveform $B(t)$. As a result, the proposed core loss model enables core loss predictions for arbitrary shapes of $B(t)$.

Verification of the core loss model

To evaluate the derived core loss model, the dissipated power density has been calculated for all investigated core excitations by using (3) and (4). The markers "$\star$" in Fig. 5 and Fig. 6(a) illustrate the predicted values for the triangular core excitation $B(t)$ (rectangular dB/dt). As can be seen, prediction and measurement agree well. Moreover, Fig. 6(b) demonstrates that a triangular core excitation $B(t)$ with

$\delta \approx 0.2$ and $\delta \approx 0.8$ causes the same specific core losses than a sinusoidal core excitation $B(t)$ with the same ΔB and f. This applies for all investigated materials. Evaluating (3), (4) and Table I confirms that this observation holds true for $\delta = 0.5 \left(1 \pm \sqrt{1 - 2/\pi} \right) \approx 0.5 \pm 0.3$ independently from the quasi-static energy loss density $E_{v,hyst} (\Delta B, B_{DC})$, the frequency f as well as from the material parameters c and γ.

Fig. 7 displays for the symmetric core excitations given in Fig. 2 the measured specific power loss density $P_{v,spec}$ as a function of ΔB (N87, $B_{DC} = 0$, $f = 100\,\text{kHz}$). Fig. 7 also includes the values calculated with (3) and (4). Again, experiment and prediction match well. Further evaluations show that the presented approach offers an accuracy of $15\,\%$ for all investigated materials and excitations.

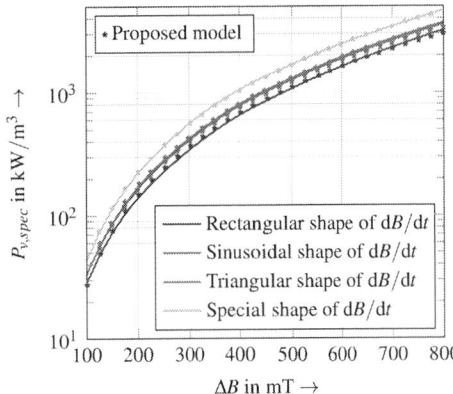

Fig. 7: Measured specific power loss density $P_{v,spec}$ for the symmetric core excitations shown in Fig. 2 as a function of ΔB for $f = 100\,\text{kHz}$ (N87, $\vartheta = 25\,°\text{C}$): The markers "$\star$" show the values derived with the proposed model.

Conclusion

This paper presents the results of an extensive core loss study comprising core excitations with, among others, sinusoidal shapes and triangular shapes of the magnetic flux density with and without DC-bias. It includes a description of the applied measurement set-ups. Based on the obtained core loss data, a compact core loss model has been derived that is valid for arbitrary core excitations including DC-bias of the magnetic flux density. While the Steinmetz equation and its extensions suffer from limited accuracy due to the frequency-dependent Steinmetz coefficients, the proposed model only requires the quasi-static core losses as well as two frequency-independent parameters for loss prediction. These parameters can by easily extracted from a limited number of measurements or from the data provided by manufacturers. The verification of the proposed approach demonstrates an adequate accuracy of $15\,\%$ for the investigated Mn-Zn ferrites and the broad range of excitations covering frequencies ranging from $100\,\text{Hz}$ to $500\,\text{kHz}$.

References

[1] F. Preisach, "Über die magnetische Nachwirkung", *Zeitschrift für Physik*, vol. 94, no. 5-6, pp. 277–302, 1935.

[2] I. D. Mayergoyz, "Hysteresis Models From the Mathematical and Control Theory Points of View", *Journal of Applied Physics*, vol. 57, no. 8, pp. 3803–3805, 1985.

[3] D. C. Jiles and D. L. Atherton, "Theory of Ferromagnetic Hysteresis", *Journal of Magnetism and Magnetic Materials*, vol. 61, no. 1-2, pp. 48–60, 1986.

[4] C. Steinmetz, "On the Law of Hysteresis", *Proceedings of IEEE*, vol. 72, no. 2, pp. 197–221, 1984.

[5] T. Duerbaum and M. Albach, "Core Losses in Transformers with an Arbitrary Shape of the Magnetizing Current", in *Proceedings of European Conference on Power Electronics and Applications (EPE)*, vol. 1, 1995, pp. 1.171–1.176.

[6] J. Li, T. Abdallah, and C. R. Sullivan, "Improved Calculation of Core Loss with Nonsinusoidal Waveforms", in *Conference Record of the IEEE Industry Applications Conference*, vol. 4, 2001, pp. 2203–2210.

[7] K. Venkatachalam, C. R. Sullivan, T. Abdallah, and H. Tacca, "Accurate Prediction of Ferrite Core Loss with Nonsinusoidal Waveforms Using only Steinmetz Parameters", in *Proceedings of IEEE Workshop on Computers in Power Electronics*, 2002, pp. 36–41.

[8] A. P. Van den Bossche, V. C. Valchev, and G. B. Georgiev, "Measurement and Loss Model of Ferrites with Non-Sinusoidal Waveforms", in *Proceedings of IEEE Power Electronics Specialists Conference (PESC)*, vol. 6, 2004, pp. 4814–4818.

[9] A. P. Van den Bossche, D. M. Van de Sype, and V. C. Valchev, "Ferrite Loss Measurement and Models in Half Bridge and Full Bridge Waveforms", in *Proceedings of IEEE Power Electronics Specialists Conference (PESC)*, 2005, pp. 1535–1539.

[10] J. Mühlethaler, J. Biela, J. W. Kolar, and A. Ecklebe, "Improved Core-Loss Calculation for Magnetic Components Employed in Power Electronic Systems", *IEEE Transactions on Power Electronics*, vol. 27, no. 2, pp. 964–973, 2012.

[11] ——, "Core Losses Under the DC Bias Condition Based on Steinmetz Parameters", *IEEE Transactions on Power Electronics*, vol. 27, no. 2, pp. 953–963, 2012.

[12] M. Mu, "High Frequency Magnetic Core Loss Study", Ph.D. dissertation, Virginia Polytechnic Institute and State University, 2013.

[13] G. Bertotti, "General Properties of Power Losses in Soft Ferromagnetic Materials", *IEEE Transactions on Magnetics*, vol. 24, no. 1, pp. 621–630, 1988.

[14] W. Roshen, "Ferrite Core Loss for Power Magnetic Components Design", *IEEE Transactions on Magnetics*, vol. 27, no. 6, pp. 4407–4415, 1991.

[15] F. Fiorillo, C. Beatrice, O. Bottauscio, A. Manzin, and M. Chiampi, "Approach to Magnetic Losses and Their Frequency Dependence in Mn-Zn Ferrites", *Applied Physics Letters*, vol. 89, no. 12, 122513 (3 pages), 2006.

[16] F. Fiorillo, M. Coïsson, C. Beatrice, and M. Pasquale, "Permeability and Losses in Ferrites from DC to the Microwave Regime", *Journal of Applied Physics*, vol. 105, no. 7, 07A517 (3 pages), 2009.

[17] F. Fiorillo and C. Beatrice, "Energy Losses in Soft Magnets from DC to Radiofrequencies: Theory and Experiment", *Journal of Superconductivity and Novel Magnetism*, vol. 24, no. 1-2, pp. 559–566, 2011.

[18] F. Fiorillo, E. Ferrara, M. Coïsson, C. Beatrice, and N. Banu, "Magnetic Properties of Soft Ferrites and Amorphous Ribbons up to Radiofrequencies", *Journal of Magnetism and Magnetic Materials*, vol. 322, no. 9-12, pp. 1497–1504, 2010.

[19] H. Zhao, C. Ragusa, C. Appino, O. de la Barrière, Y. Wang, and F. Fiorillo, "Energy Losses in Soft Magnetic Materials Under Symmetric and Asymmetric Induction Waveforms", *IEEE Transactions on Power Electronics*, vol. 34, no. 3, pp. 2655–2665, 2019.

[20] T. Kawano, A. Fujita, and S. Gotoh, "Analysis of Power Loss at High Frequency for MnZn Ferrites", *Journal of Applied Physics*, vol. 87, no. 9, pp. 6214–6216, 2000.

[21] M. Albach, *Induktivitäten in der Leistungselektronik*. Springer Vieweg, 2017.

[22] F. Fiorillo, C. Beatrice, O. Bottauscio, and E. Carmi, "Eddy-Current Losses in Mn-Zn Ferrites", *IEEE Transactions on Magnetics*, vol. 50, no. 1, 6300109 (9 pages), 2014.

[23] E. Stenglein, H. Rossmanith, and M. Albach, "Macroscopic Modeling of MnZn Ferrites for the Calculation of Eddy-Current Losses in the Frequency- and Time-Domain", in *Proceedings of IEEE Workshop on Control and Modeling for Power Electronics (COMPEL)*, 2018, pp. 1–7.

[24] C. W. T. McLyman, *Transformer and Inductor Design Handbook*, 3rd ed. Marcel Dekker, 2004.

[25] C. A. Baguley, B. Carsten, and U. K. Madawala, "The Effect of DC Bias Conditions on Ferrite Core Losses", *IEEE Transactions on Magnetics*, vol. 44, no. 2, pp. 246–252, 2008.

[26] V. J. Thottuvelil, T. G. Wilson, and H. A. Owen, Jr., "High-Frequency Measurement Techniques for Magnetic Cores", *IEEE Transactions on Power Electronics*, vol. 5, no. 1, pp. 41–53, 1990.

[27] B. Carsten, "Why the Magnetics Designer Should Measure Core Loss; With a Survey of Loss Measurement Techniques and a Low Cost, High Accuracy Alternative", in *Proceedings on Power Conversion and Intelligent Motion (PCIM)*, 1995, pp. 163–180.

[28] A. J. Batista, J. C. S. Fagundes, and P. Viarouge, "An Automated Measurement System for Core Loss Characterization", *IEEE Transactions on Instrumentation and Measurement*, vol. 48, no. 2, pp. 663–667, 1999.

[29] B. Tellini, R. Giannetti, and S. Lizón-Martínez, "Sensorless Measurement Technique for Characterization of Magnetic Material Under Nonperiodic Conditions", *IEEE Transactions on Instrumentation and Measurement*, vol. 57, no. 7, pp. 1465–1469, 2008.

[30] M. Mu, Q. Li, D. J. Gilham, F. C. Lee, and K. D. T. Ngo, "New Core Loss Measurement Method for High-Frequency Magnetic Materials", *IEEE Transactions on Power Electronics*, vol. 29, no. 8, pp. 4374–4381, 2014.

[31] E. Stenglein, D. Kuebrich, M. Albach, and T. Duerbaum, "Guideline for Hysteresis Curve Measurements with Arbitrary Excitation: Pitfalls to Avoid and Practices to Follow", in *Proceedings of PCIM Europe Conference*, 2018, pp. 1328–1335.

[32] E. Stenglein, D. Kübrich, M. Albach, and T. Dürbaum, "Novel Fit Formula for the Calculation of Hysteresis Losses Including DC-Premagnetization", in *Proceedings of PCIM Europe Conference*, 2019, pp. 1328–1335.

[33] ——, "Influence of Magnetic History and Accommodation on Hysteresis Loss for Arbitrary Core Excitations", in *Proceedings of European Conference on Power Electronics and Applications (EPE)*, 2019, P.1–P.10.

[34] E. Stenglein, B. Kohlhepp, D. Kübrich, M. Albach, and T. Dürbaum, "Novel GaN Half-Bridge Configuration for the Measurement of Core Losses Under Rectangular Voltages and DC-bias", in *Proceedings of PCIM Europe Conference*, 2020.

[35] ——, "GaN-Half-Bridge for Core Loss Measurements Under Rectangular AC-Voltage and DC-Bias of the Magnetic Flux Density", in *Transactions on Instrumentation and Measurement, Early Access (DOI: 10.1109/TIM.2020.2972140)*, 2020.

[36] A. Brockmeyer, "Experimental Evaluation of the Influence of DC-Premagnetization on the Properties of Power Electronic Ferrites", in *Proceedings of Applied Power Electronics Conference and Exposition (APEC)*, vol. 1, 1996, pp. 454–460.

[37] G. Niedermeier and M. Esguerra, "Measurement of Power Losses with DC-Bias - The Displacement Factor", in *Proceedings of International Conference on Power Conversion and Intelligent Motion (PCIM)*, 2000, pp. 169–174.

[38] A. P. Van den Bossche, V. C. Valchev, D. M. Van de Sype, and L. P. Vandenbossche, "Ferrite Losses of Cores with Square Wave Voltage and DC Bias", *Journal of Applied Physics*, vol. 99, no. 8, 08M908 (3 pages), 2006.

[39] C. R. Sullivan, J. H. Harris, and E. Herbert, "Testing Core Loss for Rectangular Waveforms", Thayer School of Engineering at Dartmouth, 2010.

[40] C. R. Sullivan and J. H. Harris, "Testing Core Loss for Rectangular Waveforms. Phase II Final Report", Thayer School of Engineering at Dartmouth, 2011.

[41] E. Herbert, "Magnetic Core Losses. The PSMA-Dartmouth Studies. Pilot Project. Phase II Project. A Supplemental Report", PSMA Magnetics Committee, 2012.

[42] ——, "Magnetic Core Losses. The PSMA-Dartmouth Studies. Phase III Project. Phase III Supplemental Report: The String of Beads Experiments", PSMA Magnetics Committee, 2013.

[43] J. H. Harris, "Magnetic Core Losses. The PSMA-Dartmouth Studies. Phase III Project. PSMA Magnetics Committee Core Loss. Phase III Final Report", Thayer School of Engineering at Dartmouth, 2013.

[44] E. Stenglein, M. Albach, and T. Dürbaum, "Separation of Magnetic Flux Density Trajectories into Subloops for the Prediction of Hysteresis Loss", in *Proceedings of European Conference on Power Electronics and Applications (EPE)*, 2020.

Assessment of Aging and Performance Degradation of Supercapacitors Integrated into a Modular Multilevel Converter

F. Errigo[1,2], L. Chédot[1],
F. Morel[1]

P. Venet[1,2], A. Sari[1,2],
A. Hijazi[1,2], R. A. Peña[1,2]

[1]SuperGrid Institute
23 Rue de Cyprian
69611 Villeurbanne, France
florian.errigo@supergrid-institute.com
https://www.supergrid-institute.com

[2]Univ. Lyon, Université Claude Bernard
Lyon 1, INSA Lyon, Ecole Centrale de Lyon,
CNRS, Ampère,
F-69622, Villeurbanne, France
http://www.ampere-lab.fr

Acknowledgements

This work was supported by a grant overseen by the French National Research Agency (ANR) as part of the "Investissements d'Avenir" Program ANE-ITE-002-01.

Keywords

«HVDC», «Multilevel Converter», «Energy storage», «Supercapacitor»

Abstract

The interest for modular multilevel converter (MMC) with energy storage systems (ESSs) has increased quickly over the last decades. Since an MMC has several hundreds of sub-modules (SMs) and ESS performances usually deteriorate quickly, the ability of the converter to provide the expected service during its lifetime must be investigated. In this work, supercapacitors are used as ESS and the reduction of their capacity is first calculated through lifetime simulation of the ESS using aging models. Then, simulations are used to analyze the available energy of the proposed converter considering dispersions for the initial parameters of ESSs and for the aging rate. Finally, the results are compared with different levels of redundancy and maintenance intervals to discuss the viability of the solution. For the depicted case study, results show that oversizing the ESSs without carrying any maintenance operations during the lifetime of the converter minimizes the total number of cells required.

Introduction

In the upcoming decades, most of the fossil-fuel power plants are planned to be replaced by renewable energy sources connected to the grid through power electronics. This development is particularly critical for the stability of power systems because the transmission grids were not developed to integrate resources with uncontrolled production, located far from the center of consumptions. One effect is the expected decrease of system inertia, which traditionally depends on the kinetic energy stored in the synchronous rotating machines. This implies that can result in dangerously low system frequencies which could lead further to a blackout [1]. This transition must inevitably be accompanied by strengthening the electrical grid.

Grid-scale Energy Storage Systems (ESSs) can be a promising solution to provide fast operating reserves. Because the required overall energy for these new short ancillary services is low compared to traditional bulk storage applications, storage technologies as supercapacitors have recently gained interests to meet these specifications [2]. Usually, utility-scale ESSs are considered to be connected to the grid through dedicated converters. Recent works have studied the opportunity to integrate ESS within converter used for power transfer capabilities such as the Modular Multilevel Converter (MMC) for HVDC (High-Voltage Direct Current) transmission in order to provide ancillary services and, in this way, to extend the range of services they can offer [3]-[5].

Even if the energy storage units can be partially distributed among Sub-Modules (SMs) [6], in this work, each SM of an MMC arm has its capacitor connected to a string of supercapacitors by means of a DC-DC converter to make an "energy storage sub-module" (ES-SM) (cf. Figure 1). Energy storage technologies are usually based on low-voltage elements while the voltage of a SM capacitor is of the order of one or few thousands of volts. Consequently, the number of components must be inherently high to sustain voltage as well as fulfill the minimum energy requirement.

In addition, the life expectancy of HVDC converters is in the range of decades. As it is often stated a 40-year converter operation is usually expected [7]. This time span is large with regard to the lifetime of electrochemical energy storage technologies. Thus, it is highly challenging for them to maintain high levels of availability throughout this period. In addition, an MMC must remain fully operational between two scheduled maintenance operations to avoid important penalties. Because a 1-GW MMC can have approximately 2400 SMs [8], with a supercapacitor stack, the availability over time of such a solution is a key performance indicator to be considered. Furthermore, the Failure In Time (FIT), 1 FIT equaling 1 failure per billion hours, of the main components in a SM (IGBTs, capacitors, etc.) varies from hundred to few hundreds [7]. Thus, it can be assumed that their failure probability is low during the lifetime of a supercapacitor cell and they are not considered in the following study.

Most of the time, a capacitance decrease or an increase of the equivalent series resistance (ESR) [9], caused by the influence of temperature and voltage, is the main root of supercapacitors performance degradation leading to noncompliance with the expected service. Wear out with age can be a good representation of the outage of supercapacitors. Moreover, the characteristics of these components show natural dispersions. Such dispersion leads to unbalanced aging and uneven energy distribution among cells that can make the system prematurely unable to provide the expected service.

This paper proposes a straightforward method that has never been done before for the assessment of the ability of an MMC with embedded supercapacitors to provide a given service over time. Thanks to physical models and accurate simulation, the lifetime of the storage unit of an ES-SM under a real mission profile is obtained. The inherent probabilistic properties of classical reliability model are avoided. Afterwards, Monte Carlo simulation is done to cope with the uncertainty of the manufacturing process of supercapacitors and the high number of cells to be considered due to the high modularity of an MMC. At the end, the aim of the proposed method is to evaluate the effectiveness of different maintenance intervals and redundancy schemes over the lifetime expectancy of the converter.

The content of the paper is organized as follows. The first section presents the main design parameters of an MMC with ESS and the case study. Secondly, the method to assess the available energy is described. Finally, the last section compares the obtained results, such as the number of supercapacitors required according to different maintenance policies during the lifetime of an HVDC project, and draws the conclusion of this work.

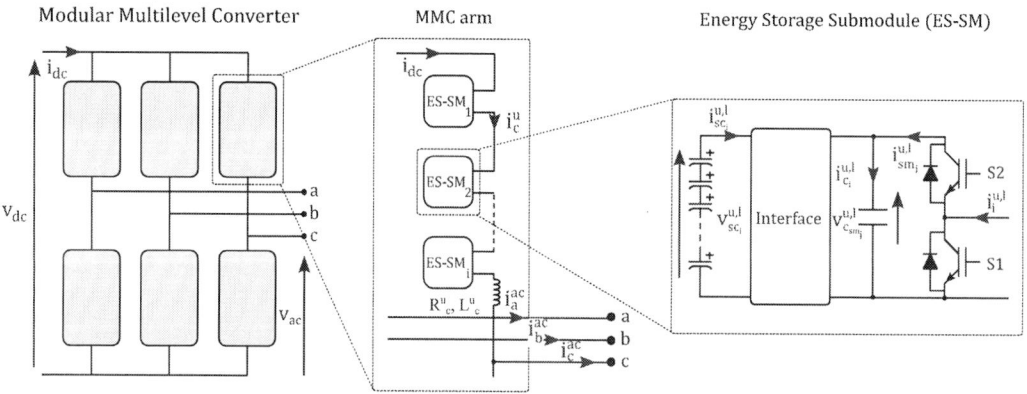

Figure 1 Topology of a modular multilevel converter with integrated energy storage systems

Modular multilevel converter with energy storage system for fast frequency response

The purpose of this part is to present the case study and the parameters for the multi-physical modelling and simulation related to the aging of the ESS. Facing future challenges of lacking inertia have introduced the need for short-term dynamic ancillary services from hundreds of milliseconds to tens of seconds, such as fast frequency support. It has been shown that ESS integrated into MMCs are well suited to bring this contribution since they can act quickly and since these service provisions require only a small part of the bulk power flow of a conventional unit, usually less than 50 MW [10].

Recently, some Transmission System Operators (TSO) have launched their own market or products because they required these new fast services from a grid point of view [11]-[14]. By definition, they are used for the purpose to mitigate uncommon system frequency fluctuation in case of disturbance and one can ask of the interest to estimate the availability of such system which are rarely used under those circumstances. However, some specific products deal with the purpose to dynamically manage the normal second-by-second frequency within a tolerance band such as in the United Kingdom (UK) where the National Grid Electricity Transmission (NGET) recently introduced the "enhanced frequency response" (EFR) [11] or in the Nordic system with the standard Normal Frequency Containment Reserves (FCR-N) [15]. The meaning of these services is to provide a one-second real-time proportional response to deviations in the grid frequency to maintain the frequency in the normal range or to provide additional power during a short time until primary reserves reach their maximal set-point or secondary reserves start in case of severe disturbances.

According to the call for tender and public grid frequency data in the UK [11] [16], a one-year load profile mission for the ESS is developed. It is assumed, that the ESS has a maximum power of 50 MW with a storage capacity of 900 MWh. It has to provide a proportional response, in either direction, to a change in system frequency with a maximum support duration of 30 seconds when a range, set at 50 Hz plus or minus 0.05 Hz, is exceed. The output of the ESS can be varied from $\pm$ 9% within the frequency deadband to manage its State of Charge (SoC). Figure 2 shows these key features.

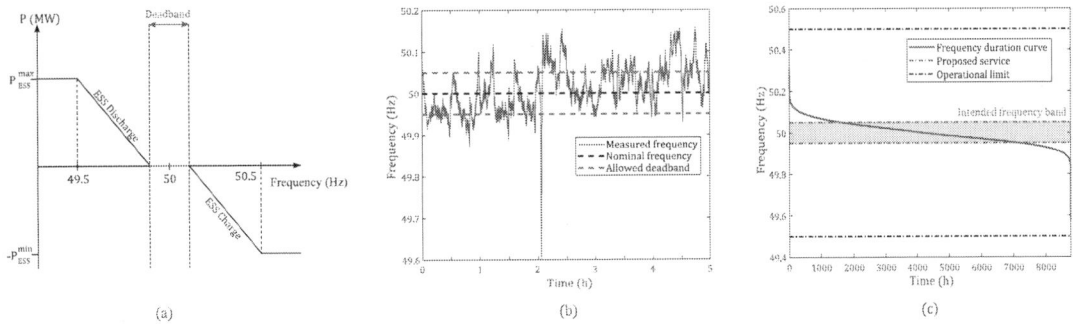

Figure 2 (a) ESS power response to grid frequency deviation (b) 5 hours extract from the one year second-by-second frequency load profile from the UK system (c) Frequency duration curve of the associated mission profile

Then, a simulation of an MMC ES-SM with a storage function coupled with a multi-physical modelling of the supercapacitor stack has been implemented under MATLAB/Simulink [17][18]. It is carried out throughout the lifetime of the ESS with the previous load profile, which is repeated for every year. The current that will flow through the ESS can be deduced from the simulation to feed the aging model and track ESS lifetime. The obtained data is used to assess the performance of the proposed solution. The main parameters of an ESS per ES-SM and the converter specifications are given in Table I. Note hereinafter in this paper, an energy of 1 p.u. is the energy available in an arm of an MMC for the expected service. 460 cells per SM are needed to provide this energy at the beginning of the lifetime of the station. Due to aging, this level of energy cannot be sustained during all the operating time of the converter and a goal of this work is to determine the number of cells, N_{cell}, to always ensure this energy content.

Table I: Grid parameters and MMC specifications for the simulation

Power rating, P_{mmc}	1 GW
DC bus voltage, v_{dc}	640 kV
AC line to ground voltage, v_{ac}	192 kV
Number of ES-SM per arm, N	400
Sub-module nominal voltage, $v_{C_{sm}}$	1600 V
Energy needed per ES-SM for the expected service, E_{ess}	375 kJ
Maximum ESS power per ES-SM, P_{ess}	20.83 kW
Minimum number of supercapacitors, N_{cell}, per ES-SM	460
Supercapacitor cell reference	Maxwell Standard Series BCAP0310
Cell capacitance, C	310 F
Cell Equivalent Series Resistance, ESR	2.2 mΩ
Ambient temperature	55°C

Energy storage remaining lifetime

Proposed method

In reliability studies, the bathtub curve (cf. Figure 3) is widely used to estimate failure rate and thus the lack of ability of an equipment to perform its function [19]. By neglecting, early failures, due to manufacturing process, the main root of failures can be divided into two parts. Random failures which occurs throughout the useful lifetime of a product, with approximately a constant failure rate λ_0, and wear out failures that increase with age.

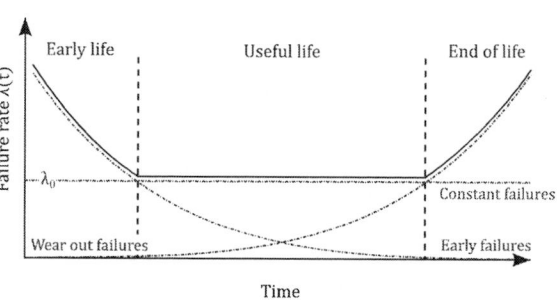

Figure 3 Bathtub curve

However, supercapacitors are electronic components whose FIT is negligible when the failure rate is constant (equal to λ_0). Only wear out "failures" can be considered. In fact, applied to supercapacitors, the main failure criteria that are generally considered are a 20% capacitance loss or a 100% increase of its equivalent series resistance [9]. It can be clearly seen that these criteria influence primarily the usable energy and does not prevent the ESS to keep working but in a degraded manner. In that case, it does not conduct to an instantaneous loss of ability of the component. Meanwhile, reliability predictions aim at calculating the rate at which a component will fail, based on an estimation of λ_0 from reliability standard, using probabilistic models for failure rate such as exponential or Weibull laws. One of the goals of this paper is to provide an estimation of the degradation of supercapacitor stack's performances integrated within an ES-SM of an MMC through damage analyses. Therefore, aging speed is first studied through electro-thermal simulation to determine ESS lifetime and then statistical analysis are carried out.

System modelling

In this purpose, the proposed method models the real behavior of components, by applying a specific mission profile and thus the real constraints, to observe the evolution of health indicators such as capacitance. Similarly, in this paper, the assumption that all the supercapacitors within a stack of one ES-SM has the same characteristics is made. Thus, they can be modelled as one equivalent capacitance. Eyring's law can be used to evaluate supercapacitor lifetime [17]. It is considered the main End of Life (EoL) criteria corresponds to a doubling of the initial equivalent series resistance or if the capacitance evolves until 80% of its initial value. However, this model is valid when a supercapacitor undergoes the same constraints all along its lifetime. This is the reason why, in this work, the modelling of a supercapacitor stack of one ES-SM is based on a dynamic lifetime model from [17]. This latter takes into account that constraints are variable over time and lead to a non-constant degradation rate of the component and so the estimation of the lifetime can change. It considers not only voltage and temperature as the main parameters of influence but also current. The dynamic lifetime of a device τ_d is estimated according to the cell voltage, its temperature and the RMS current as shown in equation (1):

$$\tau_d(t) = \frac{\tau_o}{\frac{1}{T}\int_{t-T}^{t} \exp\left(\frac{v(t)}{v_o} + \frac{\theta(t)}{\theta_o} + \frac{i_{rms}}{i_{rms_o}}\right) dt} \qquad (1)$$

where $v(t)$ and $\theta(t)$ are respectively the instantaneous value of the cell voltage (V) and temperature (°C) while i_{rms} is the value of the RMS current (A) during the time window considered. τ_o is the theoretical lifetime for a cell at 0 V and 0 °C while v_0, θ_0 and i_{rms_0} are constant whose any decrease of 10 °C, 0.2 V or 30 A double the lifetime of the component. These parameters are given as follows [17]:

$$\tau_0 = 439\,500\ years \qquad \theta_0 = \frac{10}{\ln(2)} \qquad v_0 = \frac{0.2}{\ln(2)} \qquad i_{rms_0} = \frac{30}{\ln(2)}$$

During the simulation, the electrical parameters are provided by the electrical model of the ESS whereas the thermal effects are supplied through a thermal model that analyzes heat transfers and thermal behavior of a supercapacitor cell with the ambient temperature [18]. As reminder, in the literature, the lifetime of a supercapacitor is defined as the time required for a 20% drop of the capacitance. It does not imply a failure of the component. In this work, this is only used to obtain a model of the evolution of the capacitance over time and we do not refrain from using them beyond this duration.

Simulation results

In Figure 4, the obtained results are presented. It shows the evolution of the dynamic lifetime τ_d over time. At the beginning of the simulation (t=0), when the ESS experiments a calendar aging, without current constraints, the lifetime is evaluated at 8.45 years. Then, it is clearly observed that estimation is almost constant, with only a slight decrease, when the mission profile is applied. This implies that the degradation rate of the component is not affected by the variation of the thermal and electrical parameters. Under those circumstances, supercapacitors undergo a calendar aging. Consequently, τ_d can be assimilated as a static lifetime and thus the time at the equivalent capacitance of the supercapacitor stack of one ES-SM decreases by 20%, or its ESR doubled, can be determined. It is when the dynamic lifetime estimation equals the simulation time ($\sim$ 8.36 years) as described in Figure 4.

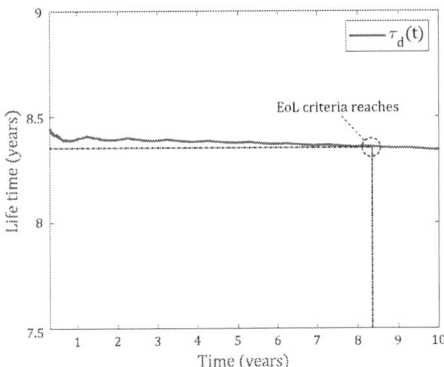

Figure 4 Evolution of the dynamic lifetime τ_d

In addition, during the simulation, the temperature of the energy storage device remains the same regardless the variation in power. This is because the grid frequency excursions are most of the time within the intended deadband during one year as seen in Figure 2.(c). It means that the power that has to be supplied is relatively low (cf. Figure 2.(a)) and as a consequence the current that flows through the ES-SM is small with regard to the nominal current that the supercapacitor cell can supply (or absorb) at full power without increasing the device temperature beyond a given value, in that case 40°C [9].

This hypothesis is confirmed during the simulation as shown in Figure 5.(b) where the current is always under the nominal current (1 p.u.). In the same manner, the voltage of the ES-SM is always kept close to the voltage reference. This voltage is calculated to make use of 75% of the maximal stored energy and handle the same amount of available energy for charging as discharging as seen in Figure 5.(a). Throughout this paper, the energy ratio is defined as the available energy from an ES-SM for the expected service and 1 p.u. is the energy content for an arm according to the initial sizing (cf. Table I).

Based on the previous results, it can be assumed in that situation an ES-SM undergoes calendar aging. Thus, the evolution of the capacitance of an ES-SM over time can be supposed to follow a square root dependency law characteristic of a floating aging that traduced the formation of a solid layer at the interface electrode/electrolyte [20], as shown in equation (5) with A_c a proportional coefficient.

$$C_{sc}(t) = C_{sc0} - A_c\sqrt{t}$$ (5)

Finally, according to the knowledge of the time at which the capacitance drops by 20% thanks to the simulation, it is easy to extrapolate the value of this coefficient with equation (5).

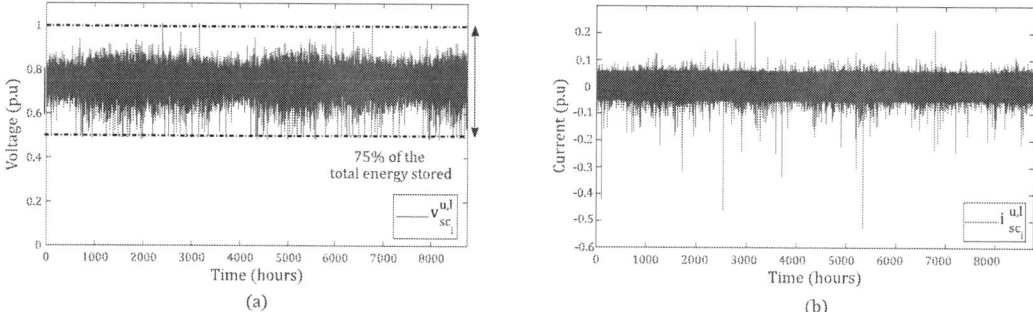

Figure 5 Simulation results for the 1st year with: (a) Voltage at the terminal of the supercapacitor stack of one ES-SM (b) Current that flows through it

Monte Carlo simulation for reliability evaluation

Probabilistic method

Generally, reliability modelling uses statistical techniques based on reliability standards such as US Military Handbook 217 F. Thus, probabilistic analysis is made assuming a constant failure rate or by introducing a time varying failure rate to define the expected lifetime of a device [7, 21]. These techniques are often criticized because of the differences between field results and predictions [19]. Moreover, these mathematical models can quickly become complex when it models periodic maintenance and the provision of redundancy, where SMs are at different levels of aging. In this paper, the wear out of components under a load profile is used to calculate performance degradation before carrying out well-known Monte Carlo methods to an MMC.

The above-mentioned simulations focus only on one ES-SM. Therefore, it is necessary to evolve from component level to converter level since an MMC contains six arms comprising hundreds of ES-SMs. as shown in Figure 1. It is assumed that all the ES-SMs are assembled identically to describe an arm with redundancy, which means they all share load during normal operation. It implies an active redundant mode. Similarly, an ESS in a SM is just a series connection of supercapacitors. Otherwise, the total stored energy of an ES-SM can be easily estimated by observing the degradation rate of a stack and subsequently the available energy in an arm to assess its lack of ability to complete its intended function. Because in average, the six arms of an MMC behave identically, it is possible to restrict the study to only one converter arm.

Note that supercapacitors have natural dispersion in capacitance or ESR due to their manufacturing but also in reason of aging. Similarly, the kinetic of degradation of the capacity also expect discrepancies. In addition, a stack of supercapacitors of one ES-SM can be replaced during maintenance operations. As a consequence performance degradation of ES-SMs within an MMC is not uniform over time because they experience different levels of aging. To define the availability of a storage function within an MMC according to these dispersions, a sequential Monte Carlo simulation has been implemented. This flexible approach allows to consider the time progress and take into account repairable systems. By random sampling of ES-SM characteristics with a given probability distribution, a high number of independent lifetime scenarios can be generated and played. From these simulations, performance indices are measured and compared. The method is as follows.

A normal probability distribution is used to randomly select the initial capacitance assuming 95% of production samples have a value around ±20% the typical capacitance value of an ES-SM as shown in Figure 6.(a). The minimum number of supercapacitors, N_{cell}, per ES-SM to provide the expected service is defined at the initialization stage. As shown, in Table I, one row of 460 cells of 310 F is necessary. It leads to an equivalent supercapacitors of 0.67 F. The same process is applied to the degradation speed of the capacitance (ie. slope deviation) around the average value of the coefficient A_c of (5), found with the simulation described in the previous paragraph, at the beginning of each scenario (cf. Figure 6.(b)).

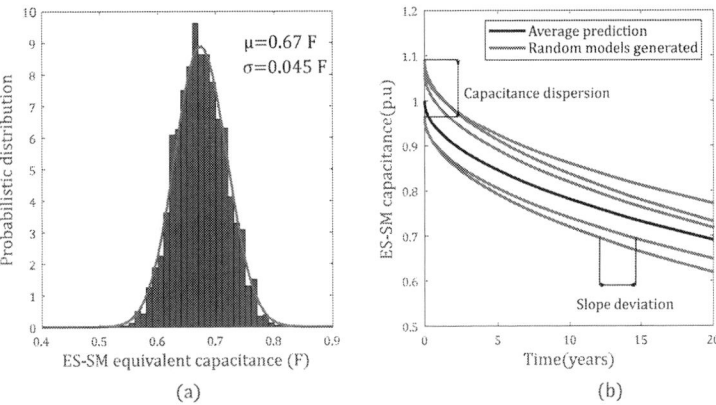

(a)   (b)

Figure 6 (a) ES-SM capacitance distribution (b) Example of a random selection of five ES-SMs with slope and initial capacitance dispersion

Then, the capacity of all ES-SMs is observed during all the life expectancy of the converter. So, the evolution of the total stored energy from the storage function can be estimated over time and thus the energy available for the service. Afterward, the availability function of the ESS of the MMC can be drawn depending on the desired energy threshold. As a reminder, in this study, the availability corresponds to the ability to provide a minimal level of energy in case of disturbance to sustain the grid. However, because aging was not considered during the design phase, it is expected that the energy requirement will not be ensured due to performance degradation. To make sure that the available energy remains above the desired threshold, redundant cells are placed in series within an ES-SM. Thus, the scenario is restarted as long as the minimum number of cells to be inserted, allowing the desired availability of the storage function, is not found. It is considered that increasing the number of cells in series allows to increase the total stored energy per stack by a factor N_{cell} if they are all charged at its nominal voltage. Note that the topology of the interface converter and the cell configurations are not looked at for these redundant designs.

Scheduled events, such as different preventive maintenance policies, are also included in the model. As an example, at each maintenance interval T_m (years), all the stacks of supercapacitors of the ES-SMs that reach a threshold capacitance value C_{min}, which corresponds to a certain decrease in the rated capacitance, are replaced by a new one while the others are still operational but with a lower capacity than when they were commissioned. Note that if an ES-SM overcomes the threshold capacitance before a maintenance interval, it is still operational even if its energy content is highly reduced.

The combination of which will provide an indication of the effects from oversizing the energy storage systems within a SM and varying the maintenance interval to maintain proper operation of the energy storage function throughout all the HVDC project lifetime. To conclude, the Monte Carlo estimation is reproduced many times for each scenario and the results analyzed for the average availability of the case study.

Simulation results

The results generated from the Monte Carlo simulations are presented in this part. The estimation was reproduced 100 times for each case. Firstly, the influence of oversizing the ES-SMs to sustain different energy levels at the end of the lifetime of the converter without any maintenance operations is described. Hence, Figure 7.(a) depicts the evolution of the energy available in one arm from the ESSs over time while Table II shows that as the energy threshold increases, the number of cells per ES-SM.

Table II: Minimum number of cells per ES-SMs to ensure different energy requirements for the expected service at the end of the lifetime of the converter without maintenance

	Energy requirements at 40 years			
Energy (p.u.)	Initial sizing	0.8	0.9	1
Number of cells N_{cells}	460	655	736	818

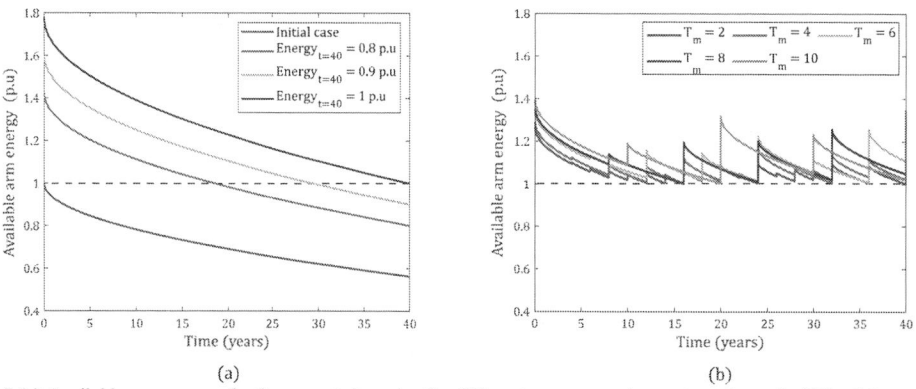

(a)   (b)

Figure 7 (a) Available arm energy for the expected service for different energy requirements at the end of life of the converter without maintenance (b) Available arm energy for the expected service for different maintenance intervals with a full availability

Obviously, the greater the initial energy available, the more redundant cells needed. It can be observed the initial sizing, that allows 1 p.u. of available energy at the commissioning, leads to less than 0.6 p.u. at the end of the lifetime of the converter. Otherwise, 818 cells per ES-SM are needed to ensure an energy level greater than 1 p.u., which implies to oversize the ESS by 80%. Finally, it should be noted in that case, each increment of 0.1 p.u. at 40 years requires a constant number of extra cells. However, this solution can face some difficulties. In fact, it is assumed supercapacitors are still used even if what is called "EoL criteria" in the literature is reached, which stands for a 20% drop in their initial capacitance. As observed in Figure 7.(a), most of the cells would lose approximately 45% of their rated value during the time span of the project which is much more than what is considered in most aging studies in the literature. Therefore, it should be clarified experimentally whether or not supercapacitors can be used with such a low capacity without leading to important concerns from a reliability point of view. In fact, the aging of supercapacitors is accompanied by a loss of performance but also with gases development. The main risk is an overpressure inside the cell that can guide to a cracking of the component or in the worst case to a blast.

In a second step, the same process is reproduced considering a maintenance interval T_m. This latter is varied from 2 to 10 years with a step time of 2 years. Indeed, an HVDC converter station scheduled outage is regularly planned for maintenance activities. All the ES-SMs, for which the capacity fade under the threshold capacitance value C_{min} at each maintenance operation, are replaced. In Figure 7.(b), that shows the evolution of the energy available from the energy storage units over time, it was decided to replace it when they reached a milestone of 70% of the initial capacitance. Similarly, an arm must keep a target availability for the expected service of 1 p.u. during the lifetime of the converter, whatever the maintenance interval, that correspond to the minimum energy needed as explained previously. Figure 7.(b) does not follows a similar shape as in Figure 7.(a) because an increase of energy is brought at each maintenance intervals. Thus, the availability is improved. Furthermore, the number of cells per ES-SM to keep an energy level for the expected service greater than 1 p.u. is reduced compared to the same case without maintenance activities.

Clearly in that case, the available energy is a combination of the total number of supercapacitors per ES-SMs and the number of time this stack has to be changed, which implies an optimum configuration. For this purpose, the relationship between the number of supercapacitors in a stack per ES-SM and the capacitance for which they are changed according to different maintenance intervals is presented in Figure 8.(a). Figure 8.(b) illustrates in the same manner the number of stacks that should be changed while Figure 8.(c) describes the total thousands of supercapacitors required during all the lifecycle of the converter.

Less frequent maintenance needs more cells per ES-SM but less stacks of supercapacitors to be replaced. This latter is explained by the fact that each stack in all ES-SMs are oversized with regard to the capacitance threshold. Similarly, decreasing the capacitance threshold allows less replacement. These choices lead to an important gain with less supercapacitors needed during the lifetime of the converter (cf. Figure 8.(c)), even if it is clearly necessary to oversize the system (cf. Figure 8.(a)). At the same time, a regular change of the stack of an ES-SM is not necessarily the best choice. Moreover, such maintenance activities are usually expensive particularly for offshore converters. In order to highlight this assessment and compare the different strategies, the total number of cells required to have at least 1 p.u. of available energy all the time, and so a full availability, with any maintenance is plotted in dashed line on Figure 8.(c). Undoubtedly, this strategy to regularly have maintenance suffers of the quick decrease of capacitance at the beginning of the lifetime of a supercapacitor, due to the square root dependency, that leads to often change cells. As shown in Figure 7.(a) with no maintenance operations, a 20% decrease in capacity is expected during the first 8.4 years but only another loss of 25% is noticed during the last 31.6 years. There is an obvious trade-off that concerns the capacitance threshold for which supercapacitors have to be changed against the risk of failures due to significant aging. Nonetheless, it seems that oversizing the energy storage system without carrying any maintenance operations or the bare minimum seems a pertinent solution.

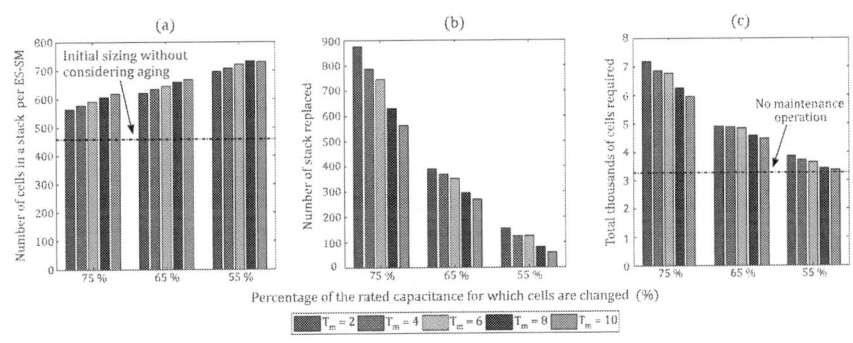

N.B : - Figures are provided for one MMC arm.
- At each maintenance step, all the cells of a stack in one ES-SM are changed (cf. Figure (a))

Figure 8 (a) Number of supercapacitors in a stack per ES-SM against capacitance threshold according to different maintenance intervals (b) Number of stacks of supercapacitors replaced against capacitance threshold according to different maintenance intervals (c) Total thousands number of cells capacitance threshold according to different maintenance intervals

Conclusion

This paper proposes a general methodology with a practical application that considers the physics of aging mechanisms to analyze performance degradation of ESSs integrated into an MMC during the lifetime of an HVDC project. A multi-physical simulation was conducted to estimate the expected lifetime of one supercapacitor stack under a high resolution mission profile. Then, a lifetime model of a supercapacitor stack to track the evolution of the capacitance as a function of time was defined. Afterwards, Monte Carlo simulations were performed to evaluate the availability of a storage service provided by an MMC under different redundancy and maintenance interval schemes.

Results show that increasing the number of planned maintenance lead to an important increase of the number of supercapacitor stacks required during the project. It was also found that oversizing the ESS at the beginning of the lifetime of the converter per ES-SM allows to minimize maintenance activities, which are expensive tasks, and to keep a good level of energy to fulfill requirements. However, it is done at a cost of using a supercapacitor beyond its normal utilization range. It will be necessary to evaluate if it does not lead to a high probability of failure. In future works also, cost assessment indicators should be equally proposed to evaluate the cost effectiveness of both strategies to define an optimum with regard to the total cost associated. Finally, the proposed methodology can potentially be extended to other energy storage technologies such as batteries.

References

[1] P. Tielens, D. Van Hertem, "The relevance of inertia in power systems", Renewable and Sustainable Energy Reviews, 2016, vol.55, pp. 999–1009.

[2] R. Alvarez, M. Pieschel, H. Gambach, and E. Spahic, "Modular multilevel converter with short-time power intensive electrical energy storage capability," in 2015 IEEE Electrical Power and Energy Conference (EPEC), Oct 2015, pp. 131–137

[3] P. Judge and T. Green, "Modular multilevel converter with partially rated energy storage with intended applications in frequency support and ancillary service provision," IEEE Transactions on Power Delivery, pp. 1–1, 2018.

[4] I. Trintis, S. Munk-Nielsen, and R. Teodorescu, "A new modular multilevel converter with integrated energy storage," in IECON 2011-37th Annual Conference on IEEE Industrial Electronics Society. IEEE, 2011, pp. 1075–1080.

[5] W. Zeng, R. Li, and X. Cai, "A New Hybrid Modular Multilevel Converter with Integrated Energy Storage," IEEE Access, pp. 1–1, 2019.

[6] F. Errigo, L. Chédot, P. Venet, A. Sari, P. Dworakowski and F. Morel, "Assessment of the Impact of Split Storage within Modular Multilevel Converter," IECON 2019 - 45th Annual Conference of the IEEE Industrial Electronics Society, Lisbon, Portugal, 2019, pp. 4785-4792.

[7] J. Wylie, M. C. Merlin, and T. C. Green, "Analysis of the effects from constant random and wear-out failures of sub-modules within a modular multi-level converter with varying maintenance periods," in 2017 19th European Conference on Power Electronics and Applications (EPE'17 ECCE Europe, 2017)

[8] J. Peralta, H. Saad, S. Dennetière, J. Mahseredjian, and S. Nguefeu, "Detailed and averaged models for a 401-level MMC–HVDC system," IEEE Transactions on Power Delivery, vol. 27, no. 3, pp. 1501–1508, 2012

[9] Maxwell Technologies, Inc., Application Note

[10] C. Mosca et al., "Mitigation of frequency stability issues in low inertia power systems using synchronous compensators and battery energy storage systems," in IET Generation, Transmission & Distribution, vol. 13, no. 17, pp. 3951-3959, 3 9 2019.

[11] National Grid. Enhanced frequency response; 2016.

[12] Australia Energy Market Operator, "Fast frequency response in the NEM," Tech.Rep., 2017.

[13] ENTSO-E, "Technical Requirements for Fast Frequency Reserves Provision in the Nordic Synchronous Area," Tech. Rep., 2020.

[14] EirGrid, "DS3 System Services Implementation Project - Regulated Arrangements," Tech. Rep., 2019.

[15] ENTSO-E, "Nordic Balancing Philosophy" Tech. Rep., 2016.

[16] National Grid. Available: https://www.nationalgrideso.com/balancing-services/frequency-response-services/historic-frequency-data

[17] P. Kreczanik, P. Venet, A. Hijazi, and G. Clerc, "Study of Supercapacitor Aging and Lifetime Estimation According to Voltage, Temperature, and RMS Current," IEEE Trans. Ind. Electron., vol. 61, no. 9, pp.4895–4902, Sep. 2014.

[18] A. Hijazi, P. Kreczanik, E. Bideaux, P. Venet, G. Clerc, and M. Di Loreto, "Thermal Network Model of Supercapacitors Stack," IEEE Trans. Ind. Electron., vol. 59, no. 2, pp. 979–987, Feb. 2012

[19] P. Venet, "Amélioration de la sûreté de fonctionnement des dispositifs de stockage d'énergie", Université Claude Bernard, Lyon, HDR 2007

[20] R. German, A. Sari, P. Venet, Y. Zitouni, O. Briat and J. Vinassa, "Ageing law for supercapacitors floating ageing," 2014 IEEE 23rd International Symposium on Industrial Electronics (ISIE), Istanbul, 2014, pp. 1773-1777.

[21] B. Wang, X. Wang, Z. Bie, P. D. Judge, X. Wang and T. C. Green, "Reliability Model of MMC Considering Periodic Preventive Maintenance," in IEEE Transactions on Power Delivery, vol. 32, no. 3, pp. 1535-1544, June 2017.

Separation of Magnetic Flux Density Trajectories into Subloops for the Prediction of Hysteresis Loss

Erika Stenglein, Manfred Albach and Thomas Dürbaum
Electromagnetic Fields, University of Erlangen-Nürnberg
Cauerstr. 7, 91058 Erlangen
Erlangen, Germany
Phone: +49/ (0) – 9131 85 28947
Fax: +49/ (0) – 9131 85 27787
Email: erika.stenglein@fau.de
URL: http://www.emf.eei.fau.de

Keywords

≪Passive component≫, ≪Magnetic device≫, ≪Modelling≫, ≪Measurement≫

Abstract

Splitting the magnetic flux density trajectory into subloops has been proposed in literature to extend core loss prediction to arbitrary waveforms. This paper evaluates the approach and demonstrates that the accommodation phenomenon as well as magnetic history hamper the estimation of hysteresis loss in the presence of minor loops.

Introduction

In the field of power electronics, optimizing inductive components achieves objectives such as increased efficiency, cost savings and miniaturization. Proper designs require the accurate calculation of core losses. However, core excitation varies depending on the selected topology and the operating mode of switched-mode power supplies. For instance, the magnetic flux density in the inductor core of the ideal boost, buck and buck-boost exhibits a triangular shape (with an idle time at the end of each switching cycle, if operated in discontinuous conduction mode) [1]. In case of the PFC boost converter, the magnetic flux density shows several maxima and minima within one mains period. If this converter operates in the boundary conduction mode, a frequently used operating mode in the low power range up to a few hundred watts, oscillations due to valley skipping occur within each high frequency period [2]. As a consequence, the magnetic material traverses one major loop and several minor loops. Thus, the calculation of core losses for arbitrary excitations is quite complex. DC-bias, the waveform of the magnetic flux density, the presence of minor loops as well as the magnetic history of the core material affect the dissipated losses in the core [3]–[8] and thus pose a great challenge for loss prediction.

In literature, a physically motivated approach has been presented that separates the power losses dissipated in the core into a static (or hysteresis) component P_{hyst} and a dynamic component P_{dyn}, which have to be investigated separately [9]–[12]. P_{hyst} equals the summation of the areas enclosed by the major and minor loops multiplied by the effective core volume and the repetition frequency f of the magnetic flux density. Thus, P_{hyst} depends linearly on frequency. The dynamic component P_{dyn} can be split up into classical eddy current loss and excess (or residual or anomalous) loss. Both effects lead to substantially increased losses at higher frequencies. At sufficiently low frequencies, the only significant contribution to the total core losses arises from the static component P_{hyst}.

In principle, a separation of the total core losses in different, physically sound components may simplify core loss prediction. However, this only applies if each component can be accurately calculated for arbitrary excitations. To deal with magnetic flux density trajectories exhibiting local extrema within one time-period, the authors of [4] suggest a separation of the traversed hysteresis loop into subloops consisting of one major and several minor loops. The corresponding power losses are estimated separately for each individual subloop. The summation of all these contributions leads to the total power loss in the core. The authors of [5] have adopted this strategy to improve specific core loss prediction. Therefore, this paper investigates whether a reliable prediction of the hysteresis losses P_{hyst} for complex waveforms can be achieved by this loop splitting method. In accordance with [4], [6]–[8], [13], this paper uses the magnetic flux density $B(t)$ to characterize core excitation. Accordingly, the minimum flux density B_{min} and maximum flux density B_{max} of each subloop have to be determined. The corresponding peak-to-peak flux density equals $\Delta B = B_{max} - B_{min}$. DC-bias of each loop is described by $B_{DC} = (B_{max} + B_{min})/2$. Note that the designation "DC" is rather misleading as the magnetic flux density averaged over time is generally not equal to B_{DC} as defined above. Nevertheless, this paper uses the index "DC" in accordance with the standard literature on DC-bias. Due to the quasi-static material behavior at low frequencies, the time derivative dB/dt does neither influence the hysteresis curve nor power dissipation. Thus, the parameters ΔB and B_{DC} fully describe the magnetic flux density waveform of each subloop.

To investigate the loop splitting method, quasi-static measurements characterize the behavior of the Mn-Zn ferrite N87 under complex excitations of $B(t)$ comprising one major and several minor loops. The next section briefly describes the applied measurement set-up. Based on the experimental data, this paper presents a physically sound strategy for separating the magnetic flux density trajectory into subloops and assesses its accuracy regarding energy loss prediction. In this context, the influence of accommodation and magnetic history, especially in the presence of minor loops, is also addressed.

Measurement Set-up for Arbitrary Excitations at Low Frequencies

The volt-amperometric method (see Fig. 1) [7], [8], [14]–[17] has been used to measure quasi-static hysteresis curves of N87 consisting of major and minor loops. The ungapped, toroidal core-under-test (CUT) carries two windings. Excitation is exerted on the core sample through the primary winding with N_p turns. An arbitrary waveform generator in combination with a power amplifier provide the source voltage $v_{src}(t)$. The high load resistance R_{load} imposes the primary current

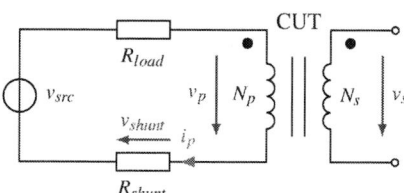

Fig. 1: Volt-amperometric method with shunt resistor.

$i_p(t) \approx v_{src}(t)/R_{load}$, which is measured by using a current-sensing resistor R_{shunt}. The secondary winding with N_s turns senses the induced voltage $v_s(t)$. Processing $v_s(t)$ and $i_p(t)$ yields the magnetic field strength $H(t)$, the magnetic flux density $B(t)$ and the dissipated energy density $E_{v,hyst}$ in the CUT. The articles [7], [8], [17] describe the measurement set-up and procedure used in this paper in great detail. This highly flexible method is capable of imposing nearly any desired waveform of the magnetic flux density $B(t)$ or the magnetic field strength $H(t)$ on the core samples. Therefore, $E_{v,hyst}$ for complex core excitations comprising major and minor loops as well as the influence of magnetic history can be studied.

Formation of Major and Minor Loops in the *B-H*-plane

To analyze the formation of subloops, an example waveform of $B(t)$ comprising a series of sine waves with a total period of $T = 30\,\text{ms}$ has been chosen. The three sine waves have a frequency of $100\,\text{Hz}$ as well as an amplitude of $300\,\text{mT}$, $100\,\text{mT}$ and $100\,\text{mT}$, respectively. Fig. 2 shows on the top left the resulting $B(t)$. After demagnetizing the CUT, this magnetic flux density $B(t)$ excites the magnetic material for several periods leading to the steady-state hysteresis curve illustrated on the top right of Fig. 2. The same color highlights a subloop and the corresponding section of $B(t)$. By inverting the second and third sine wave, the second example waveform can be obtained (see Fig. 2 bottom left). Fig. 2 also depicts on the bottom right the traversed hysteresis curve for this excitation. Note that the first sine wave with an amplitude of $300\,\text{mT}$ does not wholly belong to the major hysteresis curve. Comparing the hysteresis curves and the corresponding magnetic flux density trajectories yields the following conclusions:

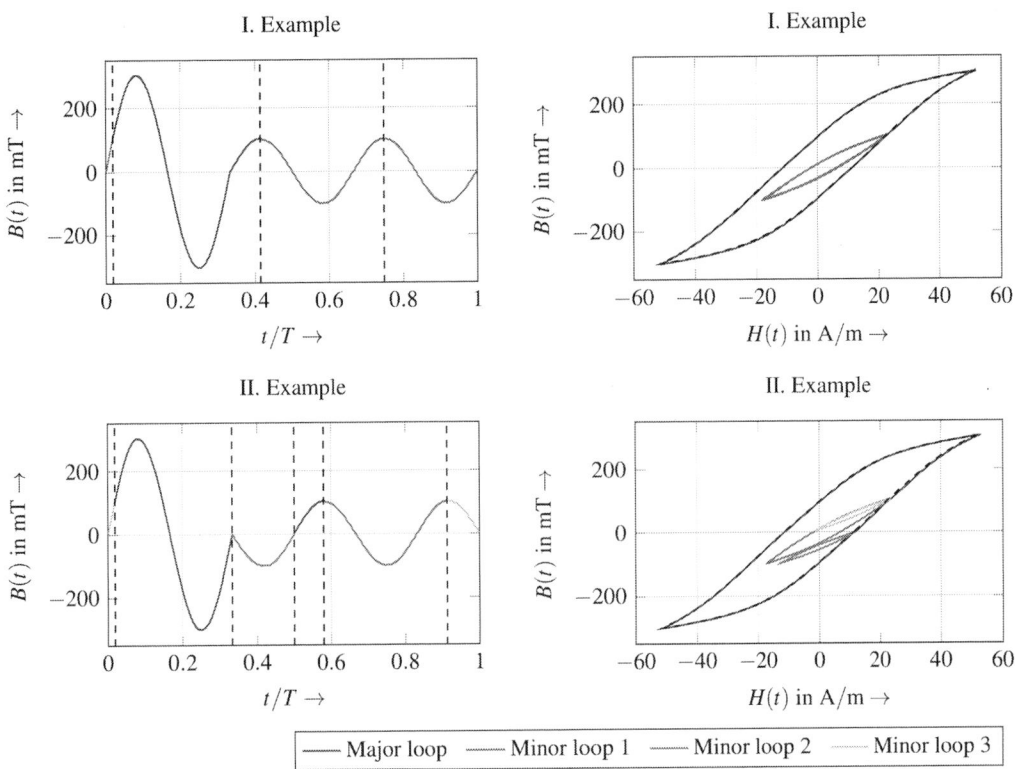

Fig. 2: Measured hysteresis curves of the material N87 for different waveforms of the magnetic flux density $B(t)$ at a temperature of 25 °C: After demagnetization, the magnetic flux densities shown on the left excite the magnetic material. The resulting hysteresis curves in the 20th period are depicted on the right. The same color highlights a subloop and the corresponding section of $B(t)$. For comparison, the dashed, black curves indicate the traversed hysteresis loop in the 20th period with a peak-to-peak value of 600 mT in the absence of minor loops.

- If N_{max} maxima and N_{min} minima occur in the magnetic flux density waveform within $0 \leq t \leq T$, the number of hysteresis loops N_{loops} within one period equals $\min(N_{max}, N_{min})$ considering the special case, where an extremum occurs both at the end and at the beginning of the cycle (see II. Example in Fig. 2). For the examples given, three subloops (one major and two minor) and four subloops (one major and three minor) can be observed, respectively.

- The major loop contains the global minimum and the global maximum of the magnetic flux density waveform. Considering the periodic nature of $B(t)$, the section of $B(t)$ between the global maximum and the global minimum belongs wholly or partly to the falling branch of the major loop, the section between the global minimum and the global maximum corresponds wholly or partly to the rising branch of the major loop. To isolate sections belonging to the major loop from the magnetic flux density trajectory, the parts of $B(t)$ forming minor loops have to be identified and removed.

- A minor loop with or without nested subloops (i. e. subloops within minor loops) between the global maximum and the global minimum begins if the time derivative of $B(t)$ changes from negative to positive (local minimum). The minor loop ends when $B(t)$ returns to the initial value of the minor loop for the first time. This also corresponds, within measurement accuracy, to the closing point of the minor loop. The entire partition of $B(t)$ between the start and the end of a minor loop belongs to the minor loop and, if present, to its nested subloops. The analogous principle applies to a minor loop between the global minimum and the global maximum. In the presence of nested subloops, two or more maxima/minima occur between the start and end of the minor loop if the minor loop begins with a minimum/maximum. Separating minor loops from nested subloops can be performed analogously to the separation of major and minor loops.

In both examples (see Fig. 2), the major loop is characterized by a peak-to-peak value of $\Delta B = 600\,\text{mT}$ and a DC-bias of $B_{DC} = 0\,\text{mT}$. The minor loops of the first example waveform exhibit a peak-to-peak value of $\Delta B = 200\,\text{mT}$ and a DC-bias of $B_{DC} = 0\,\text{mT}$. The second example waveform includes one minor loop with $\Delta B = 200\,\text{mT}$ and $B_{DC} = 0\,\text{mT}$ as well as two minor loops with $\Delta B = 100\,\text{mT}$ and a DC-bias of $B_{DC} = -50\,\text{mT}$ and $B_{DC} = 50\,\text{mT}$, respectively.

For comparison, Fig. 2 also includes the steady-state hysteresis loop with $\Delta B = 600\,\text{mT}$ and $B_{DC} = 0\,\text{mT}$ in the absence of minor loops (dashed, black curves). For both examples, only a minor deviation between the dashed, black loop and the major loop (shown in blue) occurs. Thus, an essential criterion is satisfied for calculating the core losses of the individual subloops independently. Note that the position of the minor loops in the B-H-plane with respect to the abscissa as well as their shape vary depending on the amplitude of the major loop. Accordingly, the influence of previous core excitations, i.e. magnetic history, on the minor loops and thus on hysteresis loss has to be studied. Only if this effect is negligible, a separate consideration of each subloop is possible. This issue will be addressed in a later section.

To further illustrate the formation of subloops, the waveform given in [5], [18] excites the CUT:

$$B(t) = \hat{B}(1-c)\sin(\omega t + \alpha) + \hat{B}c\sin(3\omega t + \Phi + \alpha). \tag{1}$$

The magnetic flux density waveform exhibits a fundamental frequency of $f = 1/T = \omega/(2\pi)$. Moreover, the parameters c and Φ describe the amplitude ratio and the phase shift between the first and the third harmonic, respectively. The value for α is arbitrarily chosen in such a way that $B(0) = 0$ always applies. In addition, $B(t)$ starts at $t = 0$ with the rising branch of the major loop. Fig. 3 shows the measured magnetic flux density and the traversed hysteresis curve in the 20th period for $\hat{B} = 300\,\text{mT}$. The parameter c is set to either 0.4 or 0.7. The phase angle Φ corresponds to either 0 or $\pi/4$. For the given parameters, three hysteresis loops exist in each case. The measurement results for $\Phi = \pi/4$ confirm the loop formation described above. For $\Phi = 0$, several global minima and global maxima occur within one period. Looking closely at the traversed hysteresis curve yields the following conclusion. Minor loop 1 and 2 begin and end at the global maxima and minima, respectively. The strategy for splitting the magnetic flux density trajectory into subloops derived above agrees with the recursive algorithm proposed in [5]. For the evaluation performed in the next section, the authors applied a similar but non-recursive algorithm.

Evaluation Regarding Energy Loss Prediction

For a given waveform of the magnetic flux density, magnetic history affects the position of the corresponding hysteresis curve in the B-H-plane with respect to the abscissa, and thus the shape of the major and minor loops. A separate consideration of subloops is only feasible if this effect exerts an insignificant influence on core loss.

Energy Loss Prediction in the Absence of Minor Loops

The measurement results (N87, 25 °C) presented in [7], [8] for core excitations without minor loops illustrate that magnetic history has a strong effect on the position of a steady-state hysteresis loop in the B-H-plane. Yet, the dependence of hysteresis loss on magnetic history can be neglected in steady-state for a given excitation $B(t)$. Therefore, [7] provides the polynomial approach

$$E_{v,hyst} = F_0 + F_1 \left|\frac{B_{DC}}{B_r}\right| F_2 + \left|\frac{B_{DC}}{B_r}\right|^2 + F_3 \left|\frac{B_{DC}}{B_r}\right|^3 \quad \text{with}$$

$$F_0 = a_1\left(\frac{\Delta B}{B_r}\right) + a_2\left(\frac{\Delta B}{B_r}\right)^2 + a_3\left(\frac{\Delta B}{B_r}\right)^3, \quad F_1 = b_1\left(\frac{\Delta B}{B_r}\right) + b_2\left(\frac{\Delta B}{B_r}\right)^2 + b_3\left(\frac{\Delta B}{B_r}\right)^3, \tag{2}$$

$$F_2 = c_1\left(\frac{\Delta B}{B_r}\right) + c_2\left(\frac{\Delta B}{B_r}\right)^2 + c_3\left(\frac{\Delta B}{B_r}\right)^3, \quad F_3 = d_1\left(\frac{\Delta B}{B_r}\right) + d_2\left(\frac{\Delta B}{B_r}\right)^2 + d_3\left(\frac{\Delta B}{B_r}\right)^3$$

to predict the energy loss density $E_{v,hyst}$ of a hysteresis loop characterized by ΔB and B_{DC} disregarding magnetic history. The coefficients a_n, b_n, c_n and d_n ($n = 1, 2, 3$) given in [7] and listed in Table I have

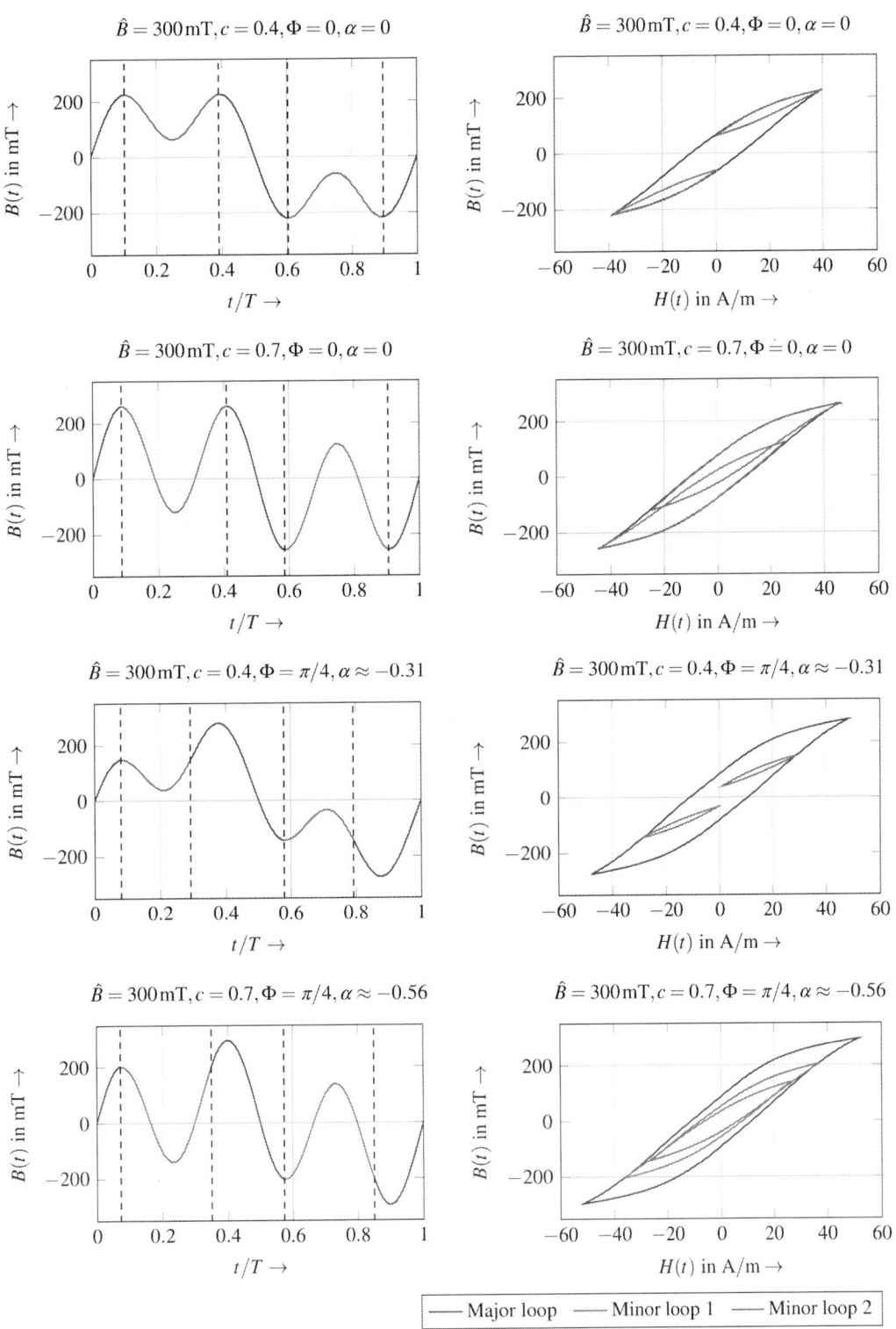

Fig. 3: Measured hysteresis curves of the material N87 for different waveforms of the magnetic flux density $B(t)$ given by (1) at a temperature of 25 °C: After demagnetization, the magnetic flux densities shown on the left excite the magnetic material. The resulting hysteresis curves in the 20th period are depicted on the right. The same color highlights a subloop and the corresponding section of $B(t)$.

been derived for a temperature of 25 °C by applying the least-squares-method twice. First the coefficients a_n have been obtained by using the measurement data for $B_{DC} = 0$. Performing the least-squares-method for the second time on the measurement data for $B_{DC} \neq 0$ yields the remaining coefficients. $B_r = 140\,\text{mT}$ refers to the remanence flux density of a hysteresis loop reaching the saturation region. Fig. 4 illustrates the dissipated energy density $E_{v,hyst}$ predicted with this fit formula as a function of ΔB for different values of B_{DC}. For comparison, the measured values of $E_{v,hyst}$ in steady-state are also displayed. For $\Delta B > 100\,\text{mT}$, the relative error between approximation and measurement is less than 10%.

Table I: Polynomial approach (2) for energy loss prediction: Coefficients valid in steady-state.

a_1	$-0.1565\,\text{J/m}^3$	b_1	$0.0953\,\text{J/m}^3$	c_1	$-0.2655\,\text{J/m}^3$	d_1	$0.0806\,\text{J/m}^3$
a_2	$0.6144\,\text{J/m}^3$	b_2	$-0.1196\,\text{J/m}^3$	c_2	$0.3693\,\text{J/m}^3$	d_2	$-0.0175\,\text{J/m}^3$
a_3	$0.0134\,\text{J/m}^3$	b_3	$0.0157\,\text{J/m}^3$	c_3	$-0.0555\,\text{J/m}^3$	d_3	$-0.0038\,\text{J/m}^3$

Note that the parameters given Table I have been determined by using the complete set of measurement data illustrated in Fig. 4. To reduce the number of experimental runs during material characterization, this paper presents the following approximation as an alternative to (2):

$$E_{v,hyst} = \underbrace{\left[a_1 \left(\frac{\Delta B}{B_r} \right) + a_2 \left(\frac{\Delta B}{B_r} \right)^2 + a_3 \left(\frac{\Delta B}{B_r} \right)^3 \right]}_{\text{dissipated energy density for } B_{DC}=0} \underbrace{\left[1 + b_1 \left| \frac{B_{DC}}{B_r} \right|^\beta \left(1 - b_2 \frac{\Delta B}{B_r} \right) \right]}_{\text{factor accounting for } B_{DC}} . \tag{3}$$

This formula also allows a fast and accurate prediction of $E_{v,hyst}$ in steady-state. The third-degree polynomial in the first square bracket approximates the dissipated energy density $E_{v,hyst}$ as a function of ΔB for $B_{DC} = 0$. The expression in the second square bracket includes the influence of DC-bias B_{DC} on $E_{v,hyst}$. Compared to (2), this approach benefits from a limited number of measurements required to determine the coefficients a_1, a_2, a_3, b_1, b_2 and β. Based on the obtained measurement results (N87, $\vartheta = 25\,°\text{C}$) for $B_{DC} = 0$, the least-squares-method yields again the parameters a_1, a_2 and a_3 given in Table I. To account for B_{DC}, the energy loss density $E_{v,hyst}$ for small values of ΔB, e.g. $\Delta B = 100\,\text{mT}$, is measured as a function of B_{DC}. In addition, the experimental runs cover larger values of ΔB, e.g.

$\Delta B = 2\,(400\,\text{mT} - B_{DC})$, that drive the material samples close to the saturation region. Based on the selected measurement data, performing the least-squares-method for the second time yields the remaining parameters $b_1 = 0.1604$, $b_2 = 0.1700$ and $\beta = 3.33$ (N87, $\vartheta = 25\,°\text{C}$). Fig. 4 shows $E_{v,hyst}$ estimated with (3). As can be seen, prediction matches experimental data well. For peak-to-peak values $\Delta B > 100\,\text{mT}$, the relative error between approximation and measurement is less than 10%. Note that (3) shows a somewhat lower accuracy than formula (2).

Disregarding the term $1 - b_2 \Delta B / B_r$ in (3) further reduces the number of experimental runs required during material characterization. In this case, the factor $1 + b_1 \left| \frac{B_{DC}}{B_r} \right|^\beta$, as proposed in [19], accounts for the influence of B_{DC}. To extract the corresponding parameters $b_1 = 0.1400$ and $\beta = 3.37$ ($b_2 = 0$, N87, $\vartheta = 25\,°\text{C}$), measurements of $E_{v,hyst}$ for small

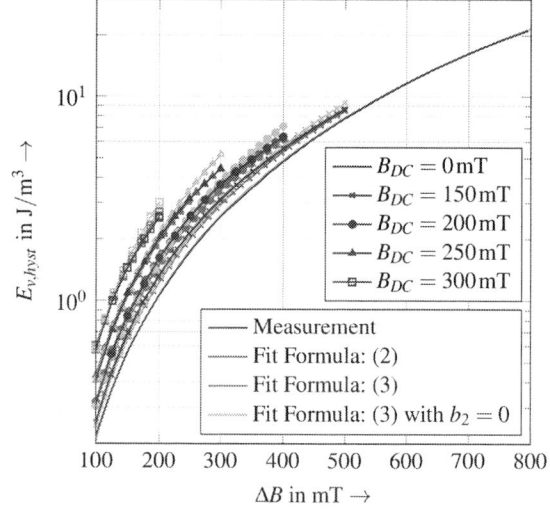

Fig. 4: Dissipated energy density $E_{v,hyst}$ predicted with the formulae (2), (3) and (3) with $b_2 = 0$: For comparison, this figure also shows the experimental data of $E_{v,hyst}$ measured in steady-state (N87, magnetic history 1, 25 °C).

values of ΔB, e.g. $\Delta B = 100\,\text{mT}$, and various values of B_{DC} suffice. Fig. 4 displays $E_{v,hyst}$ predicted with this simplified formula. In contrast to (2) and (3), however, the relative error between approximation and measurement exceeds 15 %. Depending on the desired accuracy of loss prediction and the intended effort for material characterization, this paper therefore recommends formulae (2) and (3).

Energy Loss Prediction in the Presence of Minor Loops

To assess the accuracy of the loop splitting method, this paper determines from the experimental data the energy loss density $E_{v,hyst}$ as well as ΔB and B_{DC} for each subloop of the two example waveforms depicted in Fig. 2. This evaluation uses the waveforms of $B(t)$

Table II: Relative error in the energy loss density for the examples given in Fig. 2: Comparison between measurement and formula (2).

	Major loop	Minor loop 1	Minor loop 2	Minor loop 3
I. Example	2.3 %	−10.5 %	−2.0 %	-
II. Example	1.5 %	−11.1 %	−23.0 %	−19.5 %

obtained in steady-state. Subsequently, the application of (2) using the coefficients listed in Table I estimates the dissipated energy density $E_{v,hyst}$ for each subloop. A comparison between the calculated and measured values of $E_{v,hyst}$ demonstrates the accuracy of the loop splitting method. Table II gives the relative error in the energy loss density $E_{v,hyst}$ for each subloop. Starting with the major loop, the predicted values for the energy loss density slightly overestimate the measured values for $E_{v,hyst}$. In both examples, calculation and measurement correspond well. This is, however, no longer the case for the minor loops. Underestimations up to roughly 20 % occur. Strikingly, the energy loss prediction for minor loop 2 of the first example is much more accurate than for minor loop 1, even though $\Delta B \approx 200\,\text{mT}$ and $B_{DC} \approx 0\,\text{mT}$ differ only marginally in both cases. To interpret these results correctly, the next section addresses the influence of accommodation and magnetic history.

Accommodation and Previous Magnetic History

For the investigations presented in [7], the triangular waveform of the magnetic flux density $B_\wedge(t)$ given in Fig. 5 has been imposed on the CUT made of N87 for different values of ΔB and B_{DC}. To derive coefficients valid in steady-state (see Table I), the 20th period of this triangular excitation waveform has been evaluated. Note that the performed experimental runs also differ in the magnetic history of the material:

- Magnetic history 1: A slowly decreasing and oscillating magnetic field demagnetizes the magnetic material which then traverses the virgin curve until $B = B_{DC} - \Delta B/2$. Afterwards, the triangular shape of the magnetic flux density $B_\wedge(t)$ (see Fig. 5) is imposed.
- Magnetic history 2: After demagnetization, the magnetic material is driven into the positive saturation region ($B = +B_{sat}$, I. Quadrant of the B-H-plane). The excitation level is then reduced to $B = B_{DC} - \Delta B/2$ and the imposition of the triangular waveform $B_\wedge(t)$ (see Fig. 5) begins.

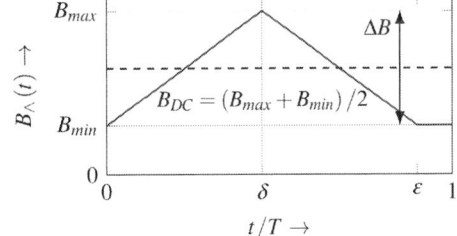

Fig. 5: Triangular waveform of the magnetic flux density $B_\wedge(t)$: As long as the magnetic material shows quasi-static behavior, the parameters δ, ε and T can be chosen freely.

- Magnetic history 3: After demagnetization, the magnetic material is driven into positive saturation followed by an excitation into the negative saturation region ($B = -B_{sat}$, III. Quadrant of the B-H-plane). Subsequently, the excitation level is increased to $B = B_{DC} - \Delta B/2$. Afterwards, the triangular waveform $B_\wedge(t)$ (see Fig. 5) is imposed.

For a given waveform of $B(t)$ in the core, magnetic history affects the position of the corresponding hysteresis curve in the B-H-plane with respect to the abscissa. Fig. 6(a) illustrates the traversed trajectories ($\Delta B = 200\,\text{mT}$, $B_{DC} = 0\,\text{mT}$) for magnetic history 1, 2 and 3 from the starting point $B = B_{DC} - \Delta B/2$ at $t = 0$ to the end of the 20th period at $t = 20\,T$. As can be seen, the hysteresis curves of magnetic history

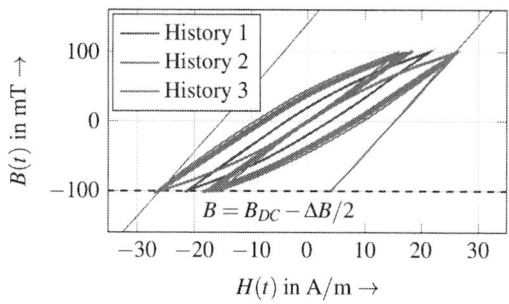
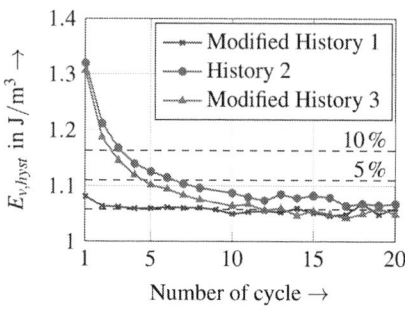

(a) The measured hysteresis curves are displayed from the starting point $B = B_{DC} - \Delta B/2$ at $t = 0$ to the end of the 20th period at $t = 20T$. The magnetic field at the starting point $B = B_{DC} - \Delta B/2$ and thus the traversed hysteresis curves depend on previous magnetic history. The black dotted lines show the saturation curve for orientation in the B-H-plane.

(b) The dissipated energy densities of the first 20 closed hysteresis loops are displayed for magnetic history 1 (modified), 2 and 3 (modified). The black dashed lines indicate the average value of $E_{v,hyst}$ for magnetic history 1 (modified) as well as the energy loss density at a deviation of 5 % and 10 % from this average value.

Fig. 6: Measurement data of N87 for $\Delta B = 200\,\text{mT}$, $B_{DC} = 0\,\text{mT}$ at a temperature of 25 °C.

1, 2 and 3 noticeably differ from each other. Moreover, differences between the trajectories of successive periods can be observed due to accommodation [8], [16], [20], [21]. In this special case ($\Delta B = 200\,\text{mT}$, $B_{DC} = 0\,\text{mT}$), the effect is much more pronounced for magnetic history 2 and 3 than for history 1.

To investigate the influence of accommodation on the energy loss density $E_{v,hyst}$, a closed hysteresis curve has to form in the first cycle of a given magnetic flux density excitation, as it is the case for minor loops. Therefore, the experiment has been adapted with regard to magnetic history 1 and 3. Instead of exciting the CUT with $B_\wedge(t)$ (see Fig. 5) starting at its minimum $B = B_{DC} - \Delta B/2$, the triangular excitation begins at its maximum $B = B_{DC} + \Delta B/2$ with the falling branch of $B(t)$. Thereby, a closed hysteresis curve already occurs in the first cycle of the excitation:

- Magnetic history 1 (modified): A slowly decreasing and oscillating magnetic field demagnetizes the magnetic material which then traverses the virgin curve until $B = B_{DC} + \Delta B/2$. Afterwards, the triangular waveform of the magnetic flux density $B_\wedge(t + \tau)$ (see Fig. 5) with $\tau = \delta T$ is imposed.
- Magnetic history 3 (modified): After demagnetization, the magnetic material is driven into the positive saturation region followed by an excitation into the negative saturation region ($B = -B_{sat}$, III. Quadrant of the B-H-plane). The excitation level is then increased to $B = B_{DC} + \Delta B/2$. Afterwards, the triangular waveform $B_\wedge(t + \tau)$ (see Fig. 5) with $\tau = \delta T$ is imposed.

The further evaluations presented in this paper are based on this modified experimental procedure. Fig. 6(b) shows the dissipated energy loss per cycle $E_{v,hyst}$ for the first 20 cycles of the triangular excitation. In case of magnetic history 1, $E_{v,hyst}$ does not exhibit a significant change over time within measurement accuracy for the given parameters ΔB and B_{DC}. The black dashed line illustrates the corresponding average value of $E_{v,hyst}$ for magnetic history 1. The other two black dashed lines indicate the dissipated energy density at a deviation of 5 % and 10 % from this average value. Moreover, the values for $E_{v,hyst}$ in case of magnetic history 1, 2 and 3 differ only slightly in the last few cycles. Accordingly, the dependence on magnetic history can be neglected in those cycles. In addition, it is assumed that the core material reaches steady-state in the 20th cycle. In comparison to steady-state, $E_{v,hyst}$ is considerably higher for magnetic history 2 and 3 in the first cycle. Due to accommodation, $E_{v,hyst}$ decreases with each cycle and thus over time. A significant drop in $E_{v,hyst}$ can already be observed in the second cycle.

The coefficients listed in Table I do not account for the increased energy losses at the beginning of the core excitation owing to accommodation. This explains the underestimation of $E_{v,hyst}$ occurring for the minor loops of the two example waveforms shown in Fig. 2. While the major loop reaches steady-state, this apparently does not hold true for the minor loops, especially for those traversed only once during the period T. Those subloops rather correspond to a hysteresis curve traversed in the first cycle of a given core excitation with the same parameters ΔB and B_{DC}. Accordingly, the application of the fit formula (2) using the coefficients valid in steady-state yields a lower value of $E_{v,hyst}$ than measured.

To improve energy loss prediction, a rather straight-forward approach offers another set of coefficients that fit $E_{v,hyst,first\ cycle}$ dissipated in the first cycle and account for the accommodation phenomenon. Applying (2) with these coefficients yields $E_{v,hyst,first\ cycle}$. When the same hysteresis curve characterized by ΔB and B_{DC} is traversed several times in succession, just a simple exponential function models the decrease in energy loss with each passing cycle m. Thus, $E_{v,hyst}(m)$ in the m-th cycle is given by

$$E_{v,hyst}(m) = E_{v,hyst,steadystate} + \left(E_{v,hyst,first\ cycle} - E_{v,hyst,steadystate}\right) e^{\left(-\frac{m-1}{M}\right)}. \tag{4}$$

$E_{v,hyst,steadystate}$ can be obtained by using (2) with the coefficients listed in Table I. Unfortunately, the dissipated energy density $E_{v,hyst,first\ cycle}$ of the first cycle strongly depends on previous magnetic history (see Fig. 6(b)). Including magnetic history in the energy loss prediction, however, poses a great challenge and requires extensive measurements for a thorough material characterization. As magnetic history is usually unknown to hardware developers, this paper proposes a simple approach intended for an easy application in practice. Based on experimental data for all three magnetic histories (see [8]), the least-squares-method derives one set of coefficients (see Table III) valid in the first cycle for all magnetic histories. Accurately fitting the first cycle is recommendable as the first cycle is always traversed in the presence of minor loops and deviates most from steady-state. The material parameter M is assumed to be independent of magnetic history, ΔB and B_{DC}. Applying the least-squares method yields $M = 2.8$. Note that a limited accuracy of the approach is to be expected as (2) does not incorporate magnetic history despite its influence.

Table III: Polynomial approach (2) for energy loss prediction: Coefficients valid in the first cycle.

a_1	$-0.0944\,\text{J/m}^3$	b_1	$0.2547\,\text{J/m}^3$	c_1	$-0.4282\,\text{J/m}^3$	d_1	$0.1307\,\text{J/m}^3$
a_2	$0.6591\,\text{J/m}^3$	b_2	$-0.2103\,\text{J/m}^3$	c_2	$0.5230\,\text{J/m}^3$	d_2	$-0.0612\,\text{J/m}^3$
a_3	$0.0039\,\text{J/m}^3$	b_3	$0.0297\,\text{J/m}^3$	c_3	$-0.0790\,\text{J/m}^3$	d_3	$-0.0013\,\text{J/m}^3$

Loss Prediction Including Accommodation: Verification

This section considers again the example waveforms shown in Fig. 2. Energy loss prediction can be significantly improved by using (4) in combination with (2) for the minor loops (coefficients: Table IV). In the first example, minor loop 2 is traversed directly after minor loop 1 having the same parameters ΔB and B_{DC}. Thus, the variable m equals 1 and 2 for minor loops 1 and 2, respectively. As ΔB and B_{DC} differ for all minor loops in the second example, $m = 1$ applies for minor loops 1, 2 and 3. For the major loops, the coefficients valid in steady-state (see Table I) in combination with (2) are used for energy loss prediction. As can be seen in Table IV, the suggested approach now slightly overestimates $E_{v,hyst}$ of the minor loops meeting the preference of hardware developers.

Table IV: Relative error in the energy loss density for the examples given in Fig. 2: $E_{v,hyst}$ is predicted by using equations (4) and (2).

	Major loop	Minor loop 1	Minor loop 2	Minor loop 3
I. Example	2.3 %	2.1 %	7.6 %	-
II. Example	1.5 %	1.8 %	8.7 %	13.5 %

To further assess the validity of the loop splitting method, the example waveform given in (1) excites the core material for different values of c ($0.2 \le c \le 0.9$) and Φ ($\Phi = 0$ and $\Phi = \pi/4$). The parameter $\hat{B}$ equals $300\,\text{mT}$ for all evaluations. Furthermore, the value for α is chosen in such a way that $B(0) = 0$ always applies. In addition, $B(t)$ starts at $t = 0$ with the rising branch of the major loop. The relative error between approximation and measurement is less than 10 % for the investigated parameter combinations. Only for excitations with $\Delta B < 100\,\text{mT}$, deviations of over 10 % occur. Altogether, equations (2) and (4) offer an acceptable degree of accuracy for major and minor loops.

Conclusion

Splitting the magnetic flux density trajectory into subloops can extent hysteresis loss prediction to arbitrary waveforms. In extension to the standard literature, the experimental data provided in this paper illustrate that magnetic history and accommodation influence core loss. As magnetic history is usually unknown to hardware developers, this paper models the decrease in energy loss due to accommodation by means of an exponential function neglecting magnetic history despite its influence. Nevertheless, the proposed approach offers an acceptable degree of accuracy for major and minor loops. As switched-mode power supplies usually operate at higher frequencies, future work will have to evaluate the loop splitting method when the dynamic component P_{dyn} contributes noticeably to core loss.

References

[1] R. W. Erickson and D. Maksimovic, *Fundamentals of Power Electronics*, 2nd ed. Kluwer Academic Publishers, 2001.

[2] M. Doebroenti, M. Schmid, and T. Duerbaum, "Matlab Based Fast and Accurate Simulation of a Power Factor Corrector Switch-Mode Power Supply in Boundary Conduction Mode with Valley Skipping," in *Proceedings of the 12th IEEE Workshop on Control and Modeling for Power Electronics (COMPEL)*, 2010, pp. 1–8.

[3] D. Y. Chen, "Comparisons of High Frequency Magnetic Core Losses Under Two Different Driving Conditions: A Sinusoidal Voltage and a Square-Wave Voltage," in *Proceedings of the 9th IEEE Power Electronics Specialists Conference (PESC)*, 1978, pp. 237–241.

[4] T. Duerbaum and M. Albach, "Core Losses in Transformers with an Arbitrary Shape of the Magnetizing Current," in *Proceedings of the 6th European Conference on Power Electronics and Applications (EPE)*, vol. 1, 1995, pp. 1.171–1.176.

[5] K. Venkatachalam, C. R. Sullivan, T. Abdallah, and H. Tacca, "Accurate Prediction of Ferrite Core Loss with Nonsinusoidal Waveforms Using only Steinmetz Parameters," in *Proceedings of the 8th IEEE Workshop on Computers in Power Electronics*, 2002, pp. 36–41.

[6] C. A. Baguley, B. Carsten, and U. K. Madawala, "The Effect of DC Bias Conditions on Ferrite Core Losses," *IEEE Transactions on Magnetics*, vol. 44, no. 2, pp. 246–252, 2008.

[7] E. Stenglein, D. Kübrich, M. Albach, and T. Dürbaum, "Novel Fit Formula for the Calculation of Hysteresis Losses Including DC-Premagnetization," in *Proceedings of the International Exhibition and Conference for Power Electronics, Intelligent Motion, Renewable Energy and Energy Management (PCIM Europe)*, 2019, pp. 1328–1335.

[8] ——, "Influence of Magnetic History and Accommodation on Hysteresis Loss for Arbitrary Core Excitations," in *Proceedings of the 21st European Conference on Power Electronics and Applications (EPE)*, 2019, P.1–P.10.

[9] G. Bertotti, "General Properties of Power Losses in Soft Ferromagnetic Materials," *IEEE Transactions on Magnetics*, vol. 24, no. 1, pp. 621–630, 1988.

[10] F. Fiorillo and A. Novikov, "An Improved Approach to Power Losses in Magnetic Laminations Under Nonsinusoidal Induction Waveform," *IEEE Transactions on Magnetics*, vol. 26, no. 5, pp. 2904–2910, 1990.

[11] W. Roshen, "Ferrite Core Loss for Power Magnetic Components Design," *IEEE Transactions on Magnetics*, vol. 27, no. 6, pp. 4407–4415, 1991.

[12] F. Fiorillo, C. Beatrice, O. Bottauscio, and E. Carmi, "Eddy-Current Losses in Mn-Zn Ferrites," *IEEE Transactions on Magnetics*, vol. 50, no. 1, 6300109 (9 pages), 2014.

[13] C. W. T. McLyman, *Transformer and Inductor Design Handbook*, 3rd ed. Marcel Dekker, 2004.

[14] V. J. Thottuvelil, T. G. Wilson, and H. A. Owen, Jr., "High-Frequency Measurement Techniques for Magnetic Cores," *IEEE Transactions on Power Electronics*, vol. 5, no. 1, pp. 41–53, 1990.

[15] B. Carsten, "Why the Magnetics Designer Should Measure Core Loss; With a Survey of Loss Measurement Techniques and a Low Cost, High Accuracy Alternative," in *Proceedings on Power Conversion and Intelligent Motion (PCIM)*, 1995, pp. 163–180.

[16] B. Tellini, R. Giannetti, and S. Lizón-Martínez, "Sensorless Measurement Technique for Characterization of Magnetic Material Under Nonperiodic Conditions," *IEEE Transactions on Instrumentation and Measurement*, vol. 57, no. 7, pp. 1465–1469, 2008.

[17] E. Stenglein, D. Kuebrich, M. Albach, and T. Duerbaum, "Guideline for Hysteresis Curve Measurements with Arbitrary Excitation: Pitfalls to Avoid and Practices to Follow," in *Proceedings of the International Exhibition and Conference for Power Electronics, Intelligent Motion, Renewable Energy and Energy Management (PCIM Europe)*, 2018, pp. 1328–1335.

[18] J. Li, T. Abdallah, and C. R. Sullivan, "Improved Calculation of Core Loss with Nonsinusoidal Waveforms," in *Conference Record of the IEEE Industry Applications Conference, 36th IAS Annual Meeting*, vol. 4, 2001, pp. 2203–2210.

[19] A. Stadler, *Simulation and Optimization of Coupled Inductors for Power Electronics Applications*, Seminarunterlagen STS Spezialseminar 2019 in Ludwigshafen-Bodman, 2019.

[20] L. J. Swartzendruber, L. H. Bennett, F. Vajda, and E. Della Torre, "Relationship Between the Measurement of Accommodation and After-Effect," *Physica B: Condensed Matter*, vol. 233, no. 4, pp. 324–329, 1997.

[21] C. Heck, *Magnetic Materials and Their Applications*. Butterworth & Co. (Publishers) Ltd, 1974.

Influence of Generalized Discontinuous Pulse Width Modulation (GDPWM) on the DC-link Current and Voltage Ripple in Battery-fed PWM Inverter Systems

Panagiotis Mantzanas[1], Alexander Bucher[2], Daniel Kuebrich[1], Alexander Pawellek[2], Christian Hasenohr[2], Harald Hofmann[2] and Thomas Duerbaum[1]

[1]ELECTROMAGNETIC FIELDS, FRIEDRICH-ALEXANDER UNIVERSITY ERLANGEN-NÜRNBERG (FAU) Cauerstrasse 7, 91058 Erlangen Erlangen, Germany

[2]VALEO SIEMENS eAUTOMOTIVE GERMANY GMBH Frauenauracher Str. 85, 91056 Erlangen Erlangen, Germany

Tel.: +49(0) – 9131 8527171
E-Mail: panagiotis.mantzanas@fau.de, alexander.bucher.jv@valeo.com, daniel.kuebrich@fau.de, alexander.pawellek.jv@valeo.com, christian.hasenohr.jv@valeo.com, harald.hofmann1.jv@valeo.com and thomas.duerbaum@fau.de
URL: [1]http://www.emf.eei.fau.de, [2]https://valeo-siemens.com

Keywords

«Pulse Width Modulation (PWM)», «Voltage Source Inverters (VSI)», «Passive component», «Simulation», «Adjustable speed drive»

Abstract

In battery-fed pulse width modulated inverters, the DC-link capacitor represents a limiting factor in terms of power density. The size of the DC-link capacitor strongly depends on its current and voltage ripple. This paper investigates the modulation method "Generalized Discontinuous Pulse Width Modulation" regarding DC-link current and voltage ripple in battery-fed pulse width modulated two-level three-phase voltage source-source inverters. "Generalized Discontinuous Pulse Width Modulation" already proved to be effective in reducing the inverter switching losses. Parameter studies performed in this paper reveal that this modulation method also enables a reduction of the DC-link voltage ripple in comparison to the widely used modulation method "centered space vector modulation".

Introduction

In battery-fed pulse width modulated (PWM) inverter systems (Fig. 1), the DC-link capacitor is a key component.

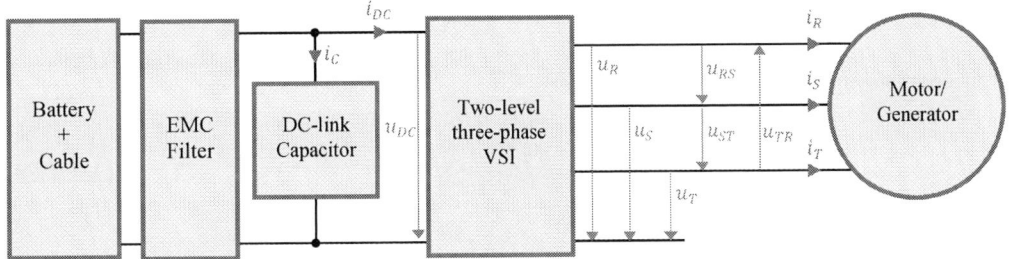

Fig. 1: Schematic representation of a battery-fed PWM two-level three-phase VSI system

First of all, it serves as a filter element preventing that the high frequency current caused by the switching actions of the inverter flows through the supply cable into the battery. Furthermore, it limits the DC-link voltage ripple providing an almost constant voltage to the voltage-source inverter (VSI) and thus reducing interference levels for other components in parallel. Unfortunately, DC-link capacitors are bulky, heavy and expensive. Therefore, they can become a limiting factor in many applications.

The DC-link current and voltage ripple have a substantial impact on the size of the DC-link capacitor. An optimization with respect to current and voltage ripple is crucial for minimizing the DC-link capacitor size. The DC-link current and voltage ripple does not only depend on the operating point of the electrical machine but also on the applied modulation method. Hence, several papers study the influence of the modulation method on the DC-link current and voltage ripple. For instance, references [1] and [2] compare different modulation methods regarding root mean square (rms) value of the DC-link voltage ripple. The articles [3] and [4] investigate the influence of various modulation methods on the rms as well as the peak-to-peak value of the DC-link voltage ripple. Additionally, reference [4] also studies the impact of the modulation method on the spectrum of the DC-link current. Other papers propose new modulation methods for reducing the rms value of the DC-link current [5], [6], [7]. Nevertheless, the high-performance modulation method "Generalized Discontinuous Pulse Width Modulation" (GDPWM) presented in [8] has not been investigated yet with respect to the DC-link current and voltage ripple. Thus, this paper studies the influence of this modulation method on the rms value of the DC-link current as well as the rms and the peak-to-peak value of the DC-link voltage ripple in battery-fed PWM two-level three-phase VSIs. Moreover, a comparison with the widely used modulation method "conventional/centered space vector modulation" (SVM) [9] is drawn.

Generalized discontinuous pulse width modulation (GDPWM)

Discontinuous pulse width modulation methods proved to be effective in reducing the inverter switching losses since each half-bridge of the inverter does perform switching actions only during two-third of the electrical motor period. The number, position and duration of the intervals in which the half-bridges do not perform switching actions (clamping intervals) depend on the modulation method. DPWMMIN, DPWMMAX as well as DPWM1-DPWM4 represent the most common discontinuous pulse width modulation methods [9], [10]. Those modulation methods have fixed clamping intervals. The article [8], however, presents a generalized discontinuous pulse width modulation method (GDPWM) which enables an adaptation of the position of the clamping intervals (within certain limits) to the operating point of the electrical machine. This chapter summarizes the implementation of this modulation method by using carrier-based PWM [9], [10].

Carrier-based PWM compares in each phase a reference signal with a high frequency carrier signal (usually a symmetrical triangular signal [9] with the carrier frequency f_{PWM}). As long as the reference signal is higher than the carrier signal, the high-side switch of the corresponding half-bridge is turned on and the low-side switch is turned off. If the carrier signal becomes higher than the reference signal, the high-side switch is turned off and the low-side switch is turned on. The reference signal $z_{ref,X}(t)$ in phase X (X represents either phase R, S or T) consists of a sinusoidal term $z_{sin,X}(t)$ as well as an injection signal $z_{ref,0}(t)$:

$$z_{ref,X}(t) = \underbrace{m \sin(2\pi f_{AC}t - \varphi_X)}_{z_{sin,X}(t)} + z_{ref,0}(t) \tag{1}$$

with $\varphi_R = 0$, $\varphi_S = 2\pi/3$ and $\varphi_T = 4\pi/3$. The frequency f_{AC} represents the desired electrical motor frequency. The amplitude of the sinusoidal term m denotes the modulation index, which regulates the amplitude of the spectral component of the inverter output voltages at the electrical motor frequency $f = f_{AC}$ [3]. The injection signal $z_{ref,0}(t)$ is identical in all three phases and depends on the applied modulation method. For GDPWM, the injection signal has to be chosen as follows [8], [10]:

$$z_{ref,0}(t) = -1 + 2\mu(t) - \mu(t)z_{sin,max}(t) - [1 - \mu(t)]z_{sin,min}(t) \tag{2}$$

with

$$z_{sin,min}(t) = \min\left(z_{sin,R}(t), z_{sin,S}(t), z_{sin,T}(t)\right), \tag{3}$$

$$z_{sin,max}(t) = \max\left(z_{sin,R}(t), z_{sin,S}(t), z_{sin,T}(t)\right) \tag{4}$$

and $\mu(t)$ a periodical square wave signal alternating between 0 and 1 with a frequency of $3f_{AC}$ and a duty cycle of 50 %. Fig. 2 illustrates the sinusoidal term $z_{sin,R}(t)$, the signal $\mu(t)$, the injection signal $z_{ref,0}(t)$ as well as the reference signal $z_{ref,R}(t)$ of the modulation method GDPWM in the phase R for an exemplary modulation index m, electrical motor frequency f_{AC} and GDPWM phase angle φ_G.

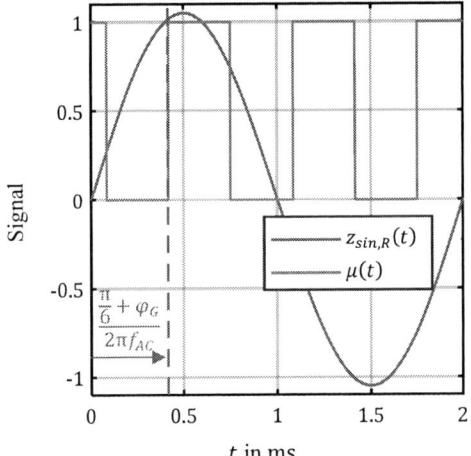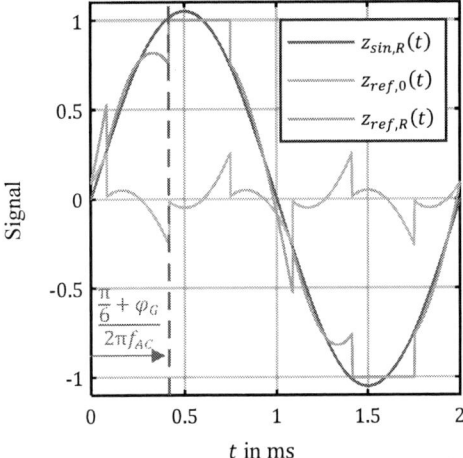

Fig. 2: Sinusoidal term $z_{sin,R}(t)$, signal $\mu(t)$, injection signal $z_{ref,0}(t)$ as well as reference signal $z_{ref,R}(t)$ for $m = 1.05$, $f_{AC} = 500$ Hz and $\varphi_G = \pi/4$ in case of GDPWM

The GDPWM phase angle φ_G (see Fig. 2) influences the relative phase shift between the sinusoidal term $z_{sin,R}(t)$, $z_{sin,S}(t)$ or $z_{sin,T}(t)$ and the signal $\mu(t)$. Hence, this phase angle has an impact on the position of the clamping intervals (intervals in which the reference signals equal -1 or 1). Note that φ_G is confined to the interval $0 \leq \varphi_G \leq \pi/3$ [8]. To achieve minimum inverter switching losses, φ_G has to be adapted to the power factor angle φ_{AC} of the electrical machine ([3]) as follows [8]:

$$
\varphi_{G,opt,P_S} = \begin{cases} \varphi_{AC} + \dfrac{\pi}{6}, & |\varphi_{AC}| \leq \dfrac{\pi}{6} \\[2mm] \dfrac{\pi}{3}, & \dfrac{\pi}{6} \leq \varphi_{AC} \leq \dfrac{\pi}{2} \\[2mm] 0, & -\dfrac{\pi}{2} \leq \varphi_{AC} \leq -\dfrac{\pi}{6}. \end{cases}
\tag{5}
$$

Simulation model

To study the impact of the modulation method GDPWM on the DC-link current and voltage ripple, a simulation model of the battery-fed PWM two-level three-phase VSI system is required. The fast analytical simulation model presented in [3] is ideally suited for this investigation since it achieves very low computational efforts and allows therefore extensive parameter variations. Fig. 3 shows the equivalent circuit of this simulation model. An ideal constant current source replaces the battery, the cable as well as the input electromagnetic compatibility (EMC) filter. This represents a suitable approximation if the resonance frequency formed by the inductance in the battery, cable or EMC filter and the DC-link capacitance is much lower

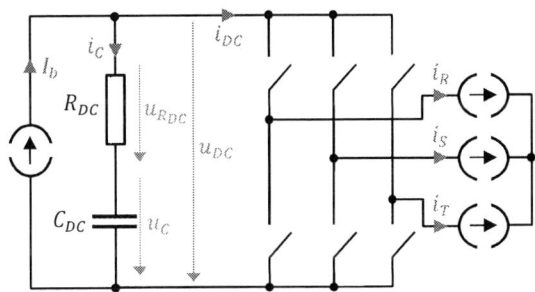

Fig. 3: Equivalent circuit of the investigated battery-fed PWM two-level three-phase VSI system

than the carrier frequency. Additionally, the ESL of the DC-link capacitor is neglected since this paper investigates the pulsating component (charging and discharging of the DC-link capacitance) resulting from the PWM and not the voltage transients caused during the switching actions of the inverter switches. Thus, a capacitance in series to a resistance models the DC-link capacitor. Furthermore, the inverter switches are assumed to be ideal (losses and switching behavior are neglected). Additionally, the motor current ripple is not taken into account. Hence, three sinusoidal current sources with the amplitude $\hat{\imath}_{AC}$ and the frequency f_{AC} approximate the motor:

$$i_X(t) = \hat{\imath}_{AC} \sin(2\pi f_{AC} t - \varphi_X - \varphi_{AC}). \tag{6}$$

Note that φ_{AC} denotes the power factor angle of the electrical machine ($\varphi_{AC} > 0$ for inductive behavior, $\varphi_{AC} < 0$ for capacitive behavior). Based on those approximations, [3] provides analytical equations for calculating the rms ($\tilde{u}_{DC,rms}$) and the peak-to-peak ($\tilde{u}_{DC,pp}$) value of the DC-link voltage ripple $u_{DC}(t) - U_{DC}$, whereas U_{DC} represents the average value of $u_{DC}(t)$. The formulae for the time dependent waveforms given in [3] also enable an analytical calculation of the rms value of the capacitor current $i_C(t)$:

$$i_{C,rms} = \sqrt{\frac{f_f}{2} \sum_{k=0}^{K} \hat{\imath}_{DCk}^2 [g_k(t_{k+1}) - g_k(t_k)] - I_b^2} \tag{7}$$

with

$$g_k(t) = t + \frac{\sin(4\pi f_{AC} t + 2\varphi_{DCk})}{4\pi f_{AC}}. \tag{8}$$

t_1, t_2, ..., t_K denote the points in time at which at least one of the half-bridges is switched, while f_f represents the fundamental frequency [3]. It should be noted that $t_0 = 0$ as well as $t_{K+1} = 1/f_f$. Equations for calculating I_b, $\hat{\imath}_{DCk}$ and φ_{DCk} can be found in [3].

GDPWM vs. SVM regarding DC-link current and voltage ripple

The simulation model described in the preceding chapter has been used to perform a comparison between the modulation method GDPWM [8] and the widely used modulation method "conventional/centered space vector modulation" (SVM) [9] regarding DC-link current and voltage ripple in a wide range for the modulation index m and the power factor angle φ_{AC}. Note that this paper focuses on the linear modulation range ($0 \leq m \leq 2/\sqrt{3}$) as well as electrical machines in the motor operating mode ($-\pi/2 \leq \varphi_{AC} \leq \pi/2$). In addition, a symmetrical triangular signal and asymmetrical regular sampling with phase delay compensation have been used as a carrier signal and PWM sampling method, respectively.

Normalization

This chapter presents further studies in a normalized form. This normalization offers the great merit that the discussed results can be easily applied to any battery

Table I: Normalized variables

Description	Normalized variable
Carrier to cut-off frequency ratio	$F_c = \dfrac{f_{PWM}}{f_c} = 2\pi f_{PWM} R_{DC} C_{DC}$
Carrier to motor frequency ratio	$F_{AC} = \dfrac{f_{PWM}}{f_{AC}}$
Normalized rms or normalized peak-to-peak (pp)	$j_{C,rms} = \dfrac{i_{C,rms}}{\hat{\imath}_{AC}/\sqrt{2}}$
	$\tilde{v}_{DC,rms/pp} = \dfrac{\tilde{u}_{DC,rms/pp} f_{PWM} C_{DC}}{\hat{\imath}_{AC}/\sqrt{2}}$

fed PWM inverter application. Table I summarizes the normalized variables. Note that f_{PWM} denotes the carrier frequency of the inverter.

Capacitor (DC-link) current ripple

Fig. 4 depicts the normalized capacitor rms current as a function of the power factor angle and the modulation index in case of SVM (Fig. 4 left) and GDPWM (Fig. 4 right) for a high carrier to motor frequency ratio ($F_{AC} = 40$). The GDPWM phase angle φ_G has been chosen according to Eq. (5). Note that the normalized capacitor rms current is independent of the carrier to cut-off frequency ratio F_c.

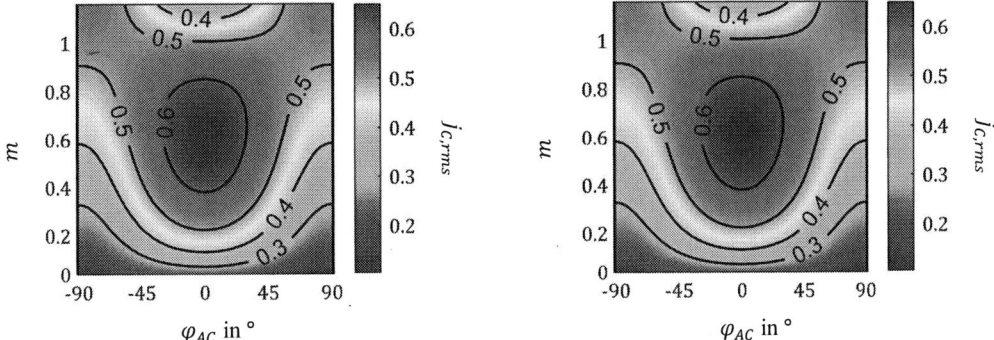

Fig. 4: Normalized capacitor rms current as a function of the power factor angle and the modulation index for $F_{AC} = 40$ (left: SVM, right: GDPWM with $\varphi_G = \varphi_{G,opt,P_S}$ according to Eq. (5))

As Fig. 4 illustrates, the capacitor rms current is basically the same for both modulation methods over the whole investigated modulation index as well as power factor angle range. This is not surprising considering the findings in [11]. This article has shown that the capacitor rms current is independent of the modulation method for high carrier to motor frequency ratios if only two adjacent active space vectors as well as one or two zero space vectors are used during one sampling period. This is the case in both SVM and GDPWM. Fig. 5 shows the relative difference in the normalized capacitor rms current between GDPWM and SVM as a function of the carrier to motor frequency ratio. Slight differences can only be observed at low carrier to motor frequency ratios.

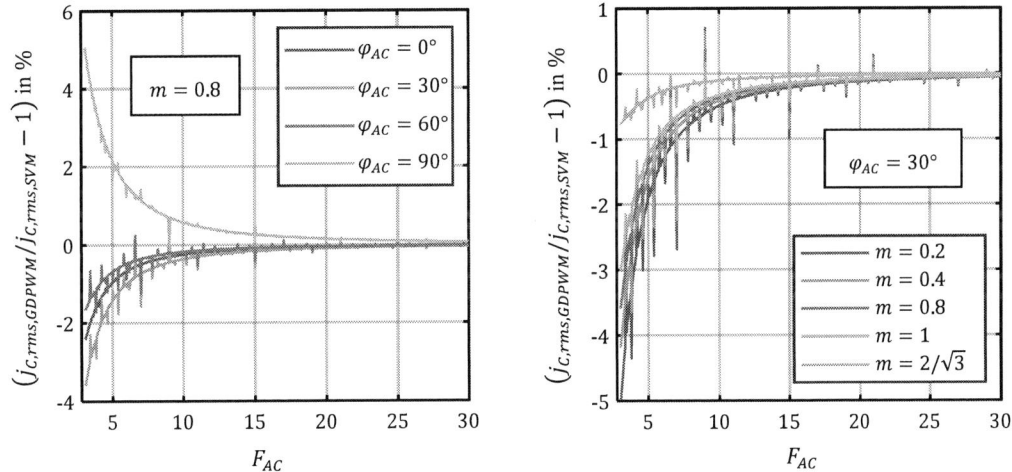

Fig. 5: Relative difference in the normalized capacitor rms current between GDPWM and SVM as a function of the carrier to motor frequency ratio for $\varphi_G = \varphi_{G,opt,P_S}$ according to Eq. (5)

DC-link voltage ripple

In contrast to the normalized capacitor rms current $j_{C,rms}$, the normalized rms $\tilde{v}_{DC,rms}$ as well as peak-to-peak value $\tilde{v}_{DC,pp}$ of the DC-link voltage ripple depends on the carrier to cut-off frequency ratio F_c. Inverters employing an electrolytic capacitor as a DC-link capacitor exhibit a high carrier to cut-off frequency ratio F_c due to a high capacitance C_{DC} as well as a high equivalent series resistance (ESR) R_{DC}. A high carrier to cut-off frequency ratio F_c leads to a considerably higher ripple in the voltage across the ESR $u_{R_{DC}}$ than in the capacitance voltage u_C (see Fig. 3). In such cases, the DC-link voltage ripple is nearly proportional to the capacitor current ripple. Nevertheless, this is no longer the case in applications with a film capacitor as a DC-link capacitor. The low capacitance and ESR value of film capacitors results in a low carrier to cut-off frequency ratio F_c. Typical battery-fed PWM two-level three-phase VSI systems employing a film capacitor exhibit a carrier to cut-off frequency ratio close to zero. This paper focuses on such applications. Therefore, the following studies will be conducted for an ideal film capacitor with $R_{DC} = 0$ ($F_c = 0$).

Fig. 6 and 7 depict the normalized rms and peak-to-peak value of the DC-link voltage ripple as a function of the power factor angle and the modulation index in case of SVM (Fig. 6 and Fig. 7 left) and GDPWM (Fig. 6 and Fig. 7 right). The GDPWM phase angle φ_G has been chosen according to Eq. (5).

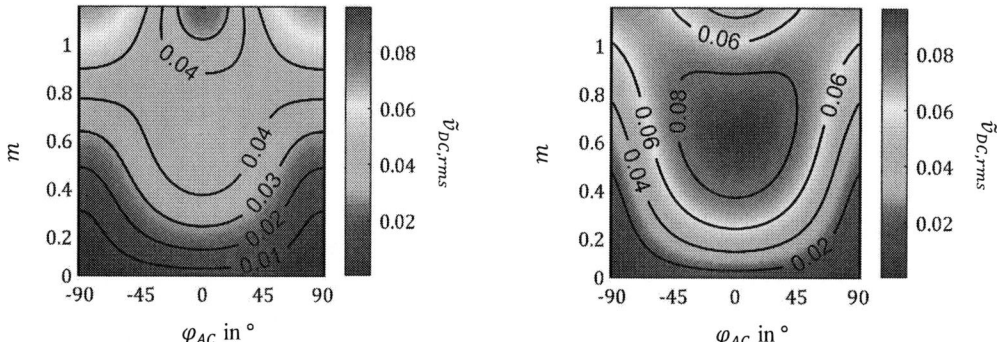

Fig. 6: Normalized rms value of the DC-link voltage ripple as a function of the power factor angle and the modulation index for $F_c = 0$ and $F_{AC} = 40$ (left: SVM, right: GDPWM with $\varphi_G = \varphi_{G,opt,P_S}$ according to Eq. (5))

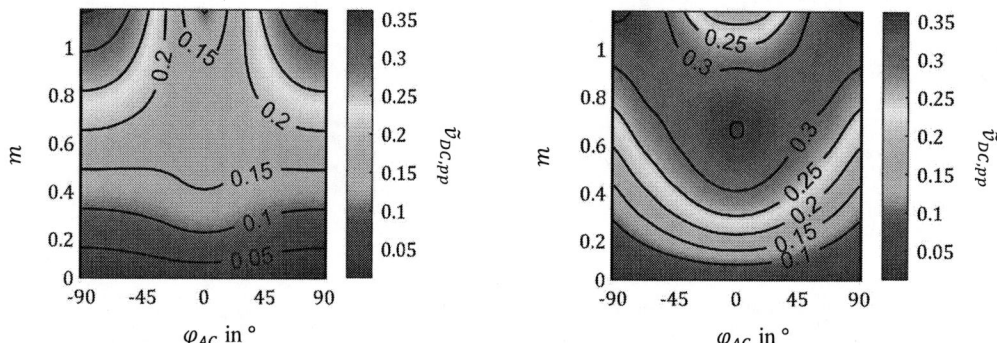

Fig. 7: Normalized peak-to-peak value of the DC-link voltage ripple as a function of the power factor angle and the modulation index for $F_c = 0$ and $F_{AC} = 40$ (left: SVM, right: GDPWM with $\varphi_G = \varphi_{G,opt,P_S}$ according to Eq. (5))

Fig. 6 and 7 reveal that the normalized rms as well as peak-to-peak value of the DC-link voltage ripple differ noticeably between GDPWM and SVM. As Table I indicates, the actual (unnormalized) rms and peak-to-peak value depend also on the capacitance value C_{DC}, the carrier frequency f_{PWM} and the peak motor current $\hat{\imath}_{AC}$. The ratio of the rms or peak-to-peak value between GDPWM and SVM at the same operating point (m, $\hat{\imath}_{AC}$, f_{AC} and φ_{AC}) as well as for the same capacitance value C_{DC} and $F_c = 0$ is given according to Table I by

$$r_{u,rms/pp} = \frac{\tilde{u}_{DC,rms/pp,GDPWM}}{\tilde{u}_{DC,rms/pp,SVM}} = \frac{f_{PWM,SVM}}{f_{PWM,GDPWM}} \frac{\tilde{v}_{DC,rms/pp,GDPWM}\big|_{F_{AC}=F_{AC,GDPWM}}}{\tilde{v}_{DC,rms/pp,SVM}\big|_{F_{AC}=F_{AC,SVM}}} \tag{9}$$

with

$$F_{AC,GDPWM} = \frac{f_{PWM,GDPWM}}{f_{PWM,SVM}} F_{AC,SVM}. \tag{10}$$

Fig. 8 shows the relative difference in the rms (Fig. 8 left) and peak-to-peak value (Fig. 8 right) of the DC-link voltage ripple between GDPWM and SVM for equal carrier frequencies.

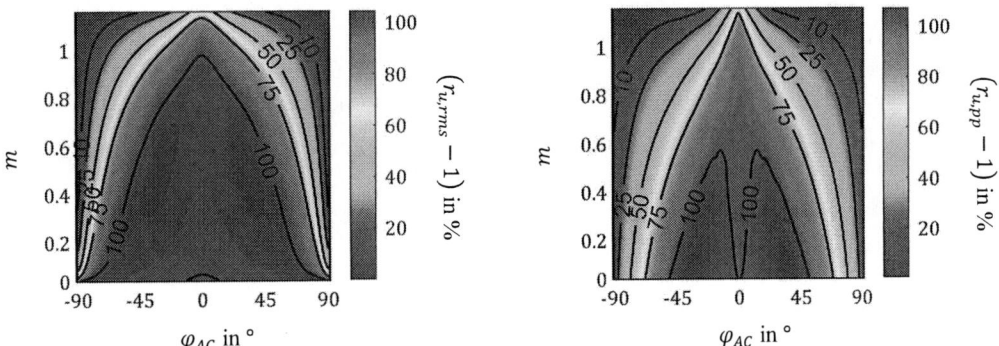

Fig. 8: Relative difference in the rms (<u>left</u>) and peak-to-peak value (<u>right</u>) of the DC-link voltage ripple between GDPWM and SVM as a function of the power factor angle and the modulation index for $F_c = 0$, $F_{AC,SVM} = F_{AC,GDPWM} = 40$, $f_{PWM,GDPWM} = f_{PWM,SVM}$ and $\varphi_G = \varphi_{G,opt,P_S}$ according to Eq. (5)

As Fig. 8 illustrates, GDPWM results in a higher rms as well as peak-to-peak value of the DC-link voltage ripple than SVM in the whole operating point range. Nevertheless, this comparison is not fair since it has been conducted under the condition that GDPWM and SVM have the same carrier frequency. The effective carrier frequency of GDPWM, however, is lower than its actual carrier frequency since each half-bridge of the inverter does perform switching actions only during two-third of the electrical motor period. Hence, GDPWM leads to lower inverter switching losses than SVM if both apply the same carrier frequency. The ratio of the GDPWM switching losses to the SVM switching losses equals

$$\frac{P_{S,GDPWM}}{P_{S,SVM}} = \frac{f_{PWM,GDPWM}}{f_{PWM,SVM}} SLF_{GDPWM}, \tag{11}$$

where SLF_{GDPWM} denotes the switching loss function of GDPWM [8]. This function represents the ratio of the GDPWM switching losses to the SVM switching losses for equal carrier frequencies ($f_{PWM,GDPWM} = f_{PWM,SVM}$). SLF_{GDPWM} strongly depends on the power factor angle φ_{AC} and the GDPWM phase angle φ_G but is approximately independent of the modulation index m. An approximate equation for determining SLF_{GDPWM} can be found in [8]. Using this equation, the relative difference in the inverter switching losses between GDPWM and SVM for equal carrier frequencies and $\varphi_G = \varphi_{G,opt,P_S}$ according to Eq. (5) have been determined (see Fig. 9 left, blue curve).

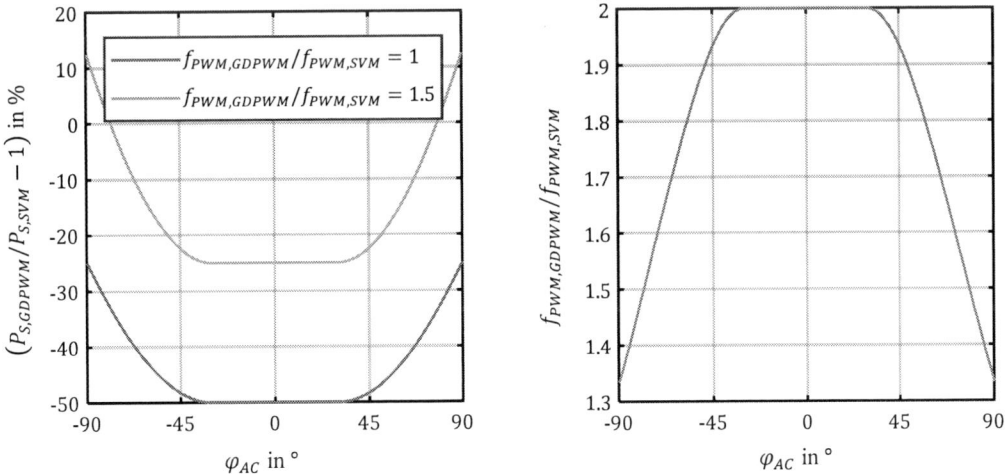

Fig. 9: Relative difference in the inverter switching losses between GDPWM and SVM (<u>left</u>) as well as ratio of the GDPWM to the SVM carrier frequency resulting in equal switching losses (<u>right</u>) as a function of the power factor angle φ_{AC} for $\varphi_G = \varphi_{G,opt,P_S}$ according to Eq. (5)

In case of equal carrier frequencies, GDPWM exhibits lower switching losses than SVM over the whole power factor angle range. Many articles claim that the effective carrier frequency of DPWM methods amounts to 2/3 of their actual carrier frequency since each half-bridge of the inverter does perform switching actions only during two-third of the electrical motor period. Thus, an increase of the DPWM carrier frequency by 50 % should lead to the same effective carrier frequency and approximately the same inverter switching losses as in case of SVM. However, this is not the case as can be seen in Fig. 9 (left, green curve). Achieving same inverter switching losses requires an adaptation of the GDPWM carrier frequency to the power factor angle according to Eq. (11) with $P_{S,GDPWM}/P_{S,SVM} = 1$. The resulting GDPWM to SVM carrier frequency ratio is shown in Fig. 9 (right). Fig. 10 shows the relative difference in the rms (Fig. 10 left) and peak-to-peak value (Fig. 10 right) of the DC-link voltage ripple between GDPWM and SVM for this carrier frequency ratio.

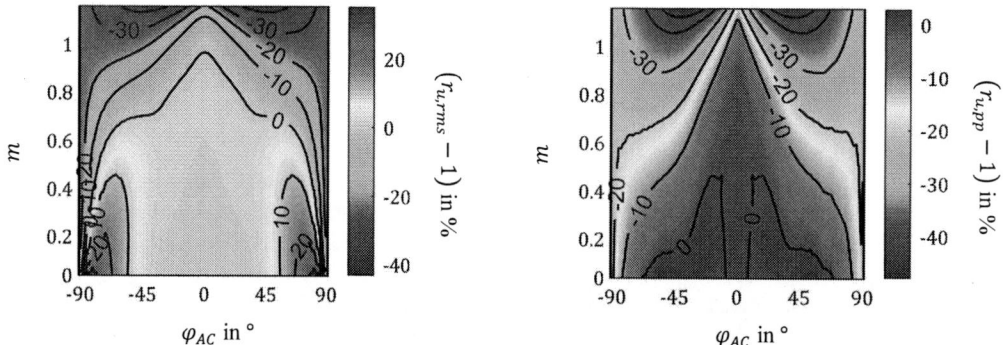

Fig. 10: Relative difference in the rms (<u>left</u>) and peak-to-peak value (<u>right</u>) of the DC-link voltage ripple between GDPWM and SVM as a function of the power factor angle and the modulation index for $F_c = 0$, $F_{AC,SVM} = 40$, $f_{PWM,GDPWM}$ according to Fig. 9 (right) and $\varphi_G = \varphi_{G,opt,P_S}$ according to Eq. (5)

Fig. 10 reveals that GDPWM with an increased carrier frequency in comparison to SVM achieves a remarkable reduction in the rms value and peak-to-peak value of the DC-link voltage ripple at operating points with a high absolute value of the power factor angle φ_{AC} (low power factor $\cos(\varphi_{AC})$) and a high modulation index m. Note that this represents the highest possible reduction under the condition that the GDPWM switching losses do not exceed the SVM switching losses.

Optimal GDPWM phase angle for minimum DC-link peak-to-peak voltage

The comparison between GDPWM and SVM in the preceding section has been performed for $\varphi_G = \varphi_{G,opt,P_S}$ according to Eq. (5), where φ_{G,opt,P_S} represents the optimal GDPWM phase angle regarding minimum inverter switching losses. Nevertheless, the question arises whether a different choice for the GDPWM phase angle φ_G is beneficial in terms of low rms or peak-to-peak value of the DC-link voltage ripple. This section focuses on determining the optimal GDPWM phase angle necessary for obtaining the lowest possible peak-to-peak value.

On the one hand, a low peak-to-peak value of the DC-link voltage ripple $\tilde{u}_{DC,pp}$ requires a small normalized peak-to-peak value $\tilde{v}_{DC,pp}$. On the other hand, the peak-to-peak value $\tilde{u}_{DC,pp}$ decreases with increasing carrier frequency. Accordingly, to enable a high increase in the GDPWM carrier frequency and therefore a low peak-to-peak value $\tilde{u}_{DC,pp}$ without exceeding the SVM switching losses, a small SLF_{GDPWM} is necessary. As a result, the GDPWM phase angle φ_G has to be adapted to φ_{AC} and m in such a way that the product $\tilde{v}_{DC,pp} \cdot SLF_{GDPWM}$ becomes minimal. Fig 11 depicts this optimal phase angle $\varphi_{G,opt,pp}$ and the corresponding SLF_{GDPWM} as a function of the power factor angle and the modulation index. To obtain equal GDPWM and SVM switching losses, an adaptation of the GDPWM carrier frequency to the modulation index as well as to the power factor angle according to Eq. (11) with $P_{S,GDPWM}/P_{S,SVM} = 1$ is necessary. Fig. 12 shows the resulting GDPWM to SVM carrier frequency ratio (Fig. 12 left) as well as the relative difference in the peak-to-peak value of the DC-link voltage ripple between GDPWM and SVM (Fig. 12 right). A comparison of Fig. 10 (right) and Fig. 12 (right) reveals that using this optimal GDPWM phase angle $\varphi_{G,opt,pp}$ with the corresponding GDPWM carrier frequency according to Fig. 12 (left) enables a further reduction of the peak-to-peak value of the DC-link voltage ripple in specific operating point ranges.

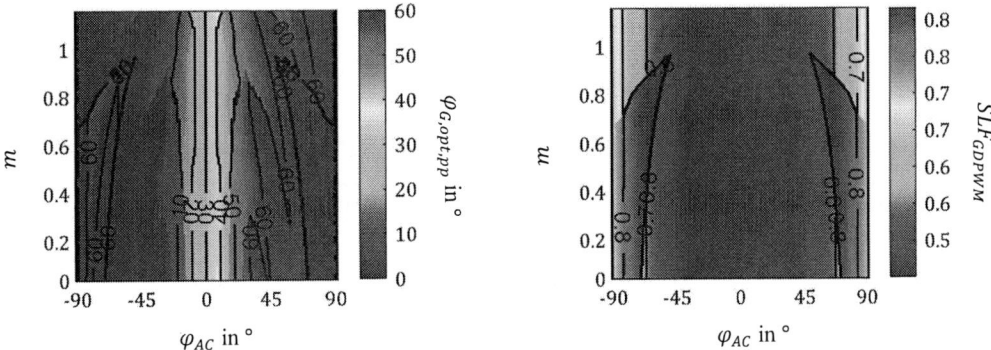

Fig. 11: Optimal phase angle $\varphi_{G,opt,pp}$ resulting in a minimal $\tilde{v}_{DC,pp} \cdot SLF_{GDPWM}$ in case of $F_c = 0$ and $F_{AC,GDPWM} = 40$ (left) and corresponding SLF_{GDPWM} (right) as a function of the power factor angle and the modulation index

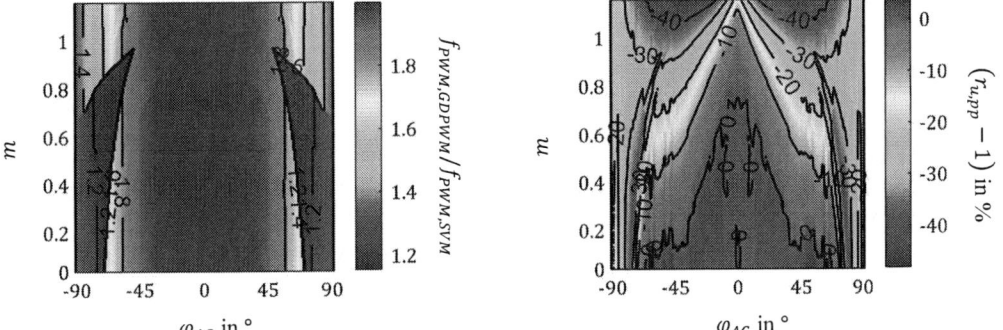

Fig. 12: Ratio of the GDPWM to the SVM carrier frequency resulting in equal switching losses in case of $\varphi_G = \varphi_{G,opt,pp}$ (left) as well as relative difference in the peak-to-peak value of the DC-link voltage ripple between GDPWM and SVM for $F_c = 0$, $F_{AC,SVM} = 40$, $f_{PWM,GDPWM}$ according to Fig. 12 (left) and $\varphi_G = \varphi_{G,opt,pp}$ (right) as a function of the power factor angle and the modulation index

Conclusion

This paper investigates the influence of the high-performance modulation method GDPWM on the DC-link current and voltage ripple in battery-fed PWM two-level three-phase VSI systems. In literature, this modulation method has not been extensively studied yet with respect to the DC-link current and voltage ripple. To allow extensive parameter variations with reasonable computational efforts, a fast analytical simulation model of the system is used for the investigation. A normalized presentation of the results offers the great merit that they can be easily applied to any battery-fed PWM inverter application. Parameter studies show that GDPWM has a negligible impact on the DC-link current ripple (capacitor rms current) in comparison to SVM. However, GDPWM with increased carrier frequency in comparison to SVM allows a reduction of the rms value and the peak-to-peak value of the DC-link voltage ripple in applications employing a film capacitor as a DC-link capacitor. At operating points with a low power factor and a high modulation index a reduction in the rms or peak-to-peak value of up to 40 % is possible under the condition that the GDPWM switching losses do not exceed the SVM switching losses.

References

[1] P. A. Dahono, Y. Sato and T. Kataoka, "An Analysis of the Ripple Components of the Input Current and Voltage of PWM Inverters" in *Proceedings of 1995 International Conference on Power Electronics and Drive Systems*, 1995.

[2] P. A. Dahono, Y. Sato and T. Kataoka, "Analysis and Minimization of Ripple Components of Input Current and Voltage of PWM Inverters" in *IEEE Transactions on Industry Applications*, 1996, vol. 32, no. 4, pp. 945-950.

[3] P. Mantzanas, A. Bucher, D. Kuebrich, A. Pawellek, C. Hasenohr, H. Hofmann, and T. Duerbaum, "A Detailed Investigation of the DC-link Voltage Ripple in Battery-fed PWM Inverter Systems" in *International Exhibition and Conference for Power Electronics, Intelligent Motion, Renewable Energy and Energy Management (PCIM Europe)*, 2019.

[4] Z. Özkan and A. M. Hava, "DC-Bus Ripple Current Characterization of Three-Phase 2/3L-VSIs Considering the Spectral Characteristics" in *9th International Conference on Power Electronics and ECCE Asia (ICPE-ECCE Asia)*, 2015, pp. 667-674.

[5] J. Hobraiche, J. P. Vilain, P. Macret, and N. Patin, "A New PWM Strategy to Reduce the Inverter Input Current Ripples" *IEEE Transactions on Power Electronics*, 2009, vol. 24, no. 1, pp. 172–180.

[6] T. D. Nguyen, N. Patin, and G. Friedrich, "Extended Double Carrier PWM Strategy Dedicated to RMS Current Reduction in DC Link Capacitors of Three-Phase Inverters" *IEEE Transactions on Power Electronics*, 2014, vol. 29, no. 1, pp. 396–406.

[7] K. Nishizawa, J. i. I. Nagaoka, A. Odaka, A. Toba, and H. Umida, "Reduction of Input Current Harmonics Based on Space Vector Modulation for Three-Phase VSI with Varied Power Factor" in *2016 IEEE Energy Conversion Congress and Exposition (ECCE)*, 2016.

[8] A. M. Hava, R. J. Kerkman, and T. A. Lipo, "A High-Performance Generalized Discontinuous PWM Algorithm" in *IEEE Transactions on Industry Applications*, 1998, vol. 34, no. 5, pp. 1059–1071.

[9] D. G. Holmes and T. A. Lipo, "Pulse Width Modulation For Power Converters" in *IEEE Press Series on POWER ENGINEERING*, 2003.

[10] E. R. C. da Silva, E. C. dos Santos and B. Jacobina, "Pulsewidth Modulation Strategies" in *IEEE Industrial Electronics Magazine*, 2011, vol. 5, no. 2, pp. 37–45.

[11] J. W. Kolar and S. D. Round, "Analytical Calculation of the RMS Current Stress on the DC-link Capacitor of Voltage-PWM Converter Systems" in *IEE Proceedings - Electric Power Applications*, 2006, vol. 153, no. 4, pp. 535–543.

Automated Design Method for Sine Wave Filters in Motor Drive Applications with SiC-Inverters

Thorben Schobre, Regine Mallwitz
TU Braunschweig, Institute for Electrical Machines, Traction and Drives
Hans-Sommer-Straße 66
Braunschweig, Germany
Phone: +49 (531) 391-3910
Email: t.schobre@tu-braunschweig.de
URL: https://www.tu-braunschweig.de/en/imab

Acknowledgments

The research leading to this publication has received funding by the German Ministry for Economic Affairs and Energy under grant number 03ET1532A-E – Ide3AL.

Keywords

≪Passive filter≫, ≪Servo-drive≫, ≪Design≫, ≪Software≫, ≪Simulation≫

Abstract

This work describes a design method for sine filters used for servo motor drive applications with a two level voltage source inverters (VSI) with 1200V SiC MOSFETs. To assess the filter effectiveness in attenuating conducted EMI emissions a worst case distortion spectrum of a PWM modulated half bridge midpoint voltage is calculated. The filter and the drive system are transformed to single phase differential (DM) and common mode (CM) equivalent circuits (EQC). The implemented algorithm enables parameter studies with various filter topologies, drive systems and values of filter components. The result of an example parameter study is analyzed and a suitable filter is built up and measured. The measured filter is then compared to the design results and the deviations are discussed.

Introduction

Drive inverters with SiC MOSFETs allow the integration of sine wave wave filters and VSI. At high operating frequencies the required filter effect can be gained with small filter elements enabling a close integration of the sine filter and inverter. For this reason SiC wide bandgap devices are perfectly suited. The higher swichting frequency allow a use of sine wave filters with a higher resonant frequency. The higher the resonant frequency is the wider is the pass band. This enables the use of sine wave filter in highly dynamic servo drive inverters with fundamental frequencies up to 600 Hz.

A sine wave filter offers numerous advantages like reduction of HF losses in machine and cable, no need of cable shielding, no reflection induced over voltage when using long cables, no bearing or shield currents and no partial discharge in the motor coil induced by high dV/dt.

The sine wave filter in this work is a combination of a switching frequency harmonic filter and an EMI filter. The design strategy presented in [1] and [2] are based on noise impedance measurements. They represents a high accuracy, however the power electronic circuit and the filter need to be developed already. Also they consider just one 2nd order filter topology which is also the case in [3]. This work focuses on the wide design space consisting of different order filter topologies, variable filter element values and variable values of DM and CM inductor couplings. Especially the potential of higher order filters already studied by [4] can contribute to the efficiency and compactness of the inverter system.

There are also several works like [5] which model the whole system with all parasitic behavior really exactly. But the more complex the model is the more computation time is needed calculate the exact behavior. In the presented method a sufficient accuracy is assumed for the area of highest disturbances by proper filter component design. The objective of this method is not to absolute exactly model the system behavior. Due to computation complexity only then it is possible to find a best suitable sine filter by calculating thousands (unlimited) of variations. When the best filter solution is narrowed down it can then be assessed by simulation with all non-linearities of e.g. core materials and parasitic elements. In this work the best solution is defined by the most attenuation at the desired frequency band with the least stored energy in the filter elements, while complying to the requirements of conducted emission [6]. The best suitable filter is then built up and analyzed with small signal vector network analyses.

Modeling of Inverter Frequency Content

A fast switching drive inverter emits a characteristic frequency spectrum which is analyzed in this section. The voltage spectrum at the midpoint of the half bridges can be determined analytical by the method presented in [7]. This spectrum is strongly dependent on the rise and fall times of the switching transitions and the duty cycle. The worst case duty cycle of 0.5, which introduces the highest magnitude in the voltage spectrum, is considered. This describes the worst case of distortion spectrum. Therefore all modulation techniques are covered. However when modulating the inverter, the energy of the distortion spectrum will be lower. The following equation (1) which is introduced by [7] describes the voltage spectrum. In this spectrum only the voltage frequency content is represented which is introduced by the switching frequency. The influence of the fundamental frequency is not relevant for the filter design and therefore not represented in this spectrum. The fundamental frequency is only required to ensure a sufficient pass band of the filter.

$$S(n) = \frac{A}{\pi n} \mid \mathrm{sinc}(\pi n \tau_{\mathrm{rise}} e^{j\pi n d}) - \mathrm{sinc}(\pi n \tau_{\mathrm{fall}} e^{j\pi n d}) \mid \tag{1}$$

Where A is the voltage amplitude respective the DC link voltage, n is the order of harmonic, d is the duty cycle and τ_{rise} and τ_{fall} are the relative rise and fall times scaled to the switching period. Based on this equation the envelope of the spectrum of the midpoint voltage of a half bridge can be calculated which is displayed in Fig. 1. It can be shown that the switching speed of the transistor significantly influences the

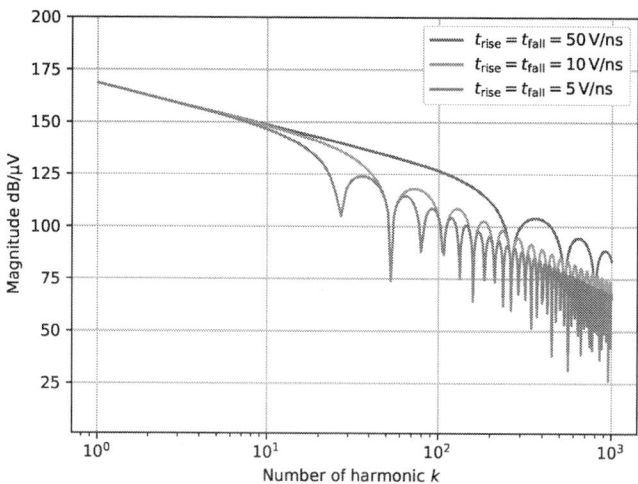

Fig. 1: Spectrum of a two Level VSI at the midpoint of one inverter leg.

magnitude of higher harmonics. When utilizing the high switching speed capabilities and the resulting low switching losses of SiC VSI the distortion energy at higher frequencies in the spectrum is higher. This obviously requires a sufficient filtering. This spectrum representation serves as an evaluation basis for the filter design. The required filter attenuation can be deduced from the distortion spectrum at different frequencies to meet EMI standards.

Modeling of Sine Wave Filter

In this work the *filter* is understood as an integration of common mode (CM) and differential mode (DM) filter effect in one filter topology. Depending on the topology the filter can have the same or different orders for the CM and DM filter effect. The remaining parts of the drive system are summarized as *system* consisting of the grid connection with rectifier, the DC link capacitor, the VSI, the motor cable and the motor.

Filter Topology

Many different filter topologies for the use as sine wave filter for a VSI application are possible. A 2nd order LC filter e.g. Fig. 2 has both a CM and DM damping effect. The CM filter effect is determined by the clamping of the filter capacitor star point in topology Fig. 2 and Fig. 3. Filter topology Fig. 4 only has a significant differential mode attenuation effect. The common mode attenuation in this configuration is only a first order filter by the CM part of the DM inductors. The different clamping options result in different equivalent circuits and capabilities of CM filtering.

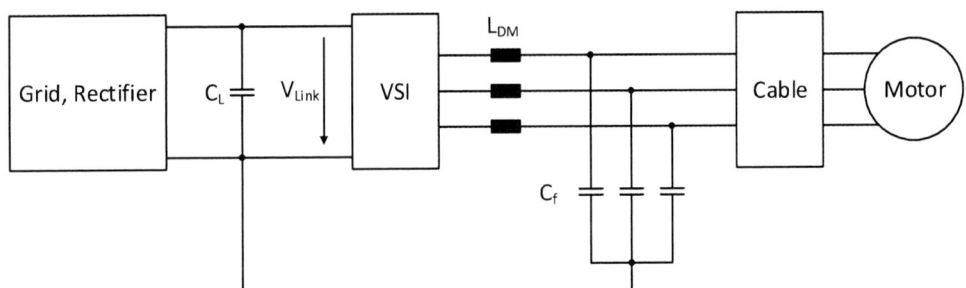

Fig. 2: 2nd Order LC filter with star point clamping to negative DC link rail.

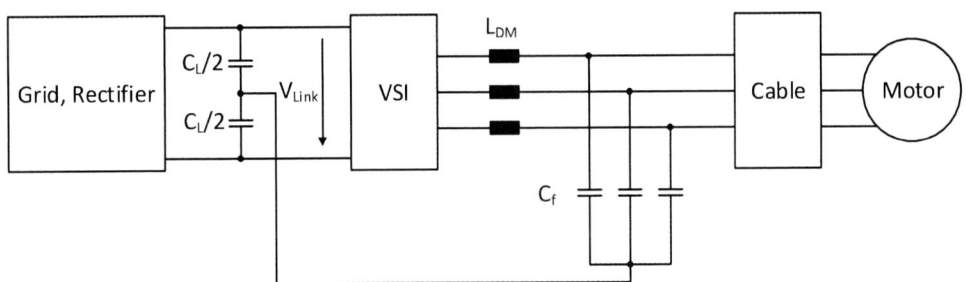

Fig. 3: 2nd Order LC filter with star point clamping to midpoint of the DC link.

The coupling of the inductors define how they work in the equivalent circuits. DM inductors (L_{DM}) can have a coupling factor in the range of -0.5 and 0. But a DM inductor with for example a coupling factor of 0 still has a CM filter effect. CM inductors with a coupling of < 1 also have DM filter effect. The location of dedicated CM inductors and DM inductors in the topology can also be varied. DM and CM inductors can be connected in series or a DM filter consisting of a DM filter capacitance and an inductance are connected to a CM filter stage consisting of CM inductors and CM capacitors. The order of the CM and DM filter can be varied.

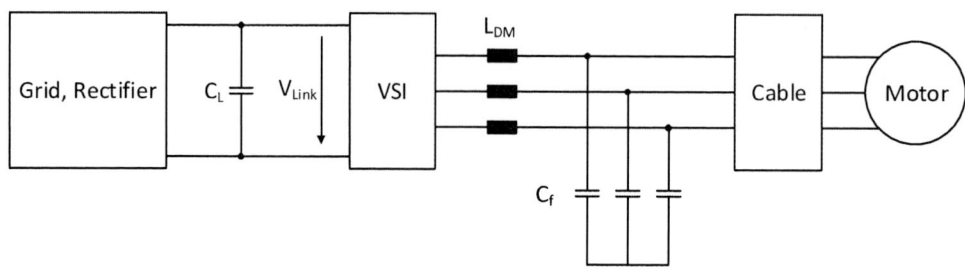

Fig. 4: 2nd Order LC filter without star point clamping.

Single Phase DM and CM Equivalent Circuits of Filter Topology

To evaluate the filter it is necessary to build a DM and CM EQC of both the drive system consisting of a grid connection, a cable and motor, and the filter. The motor can be either a synchronous machine (SM) or an induction machine (IM). The three phase system and filter is transformed to a single phase system. For all components the effective impedances for CM and DM EQC are calculated. The calculation for the inductances are more complex due to the coupling factor dependency. The computation can be derived from the inductance matrix with self and mutual inductance. A good deduction of the topic is given in [8]. The effective inductances in the DM EQC are calculated for the DM inductors in (2) and the CM inductors in (3). Depending on the coupling coefficient k, CM inductances also introduce a DM filter effect in the DM EQC. $L_{DM,DM}$ represents the DM filter effect of the DM inductance and $L_{CM,DM}$ the DM effect of the CM inductor. $L_{DM,self}$ and $L_{CM,self}$ are the self inductances of one coil of the DM and CM inductor.

$$L_{DM,DM} = L_{DM,self}(1-k) \tag{2}$$

$$L_{CM,DM} = L_{CM,self}(1-k) \tag{3}$$

The effective CM inductance $L_{CM,CM}$ of the CM inductor and $L_{DM,CM}$ of the DM inductor are presented in (4) and (5).

$$L_{CM,CM} = L_{CM,self}\frac{(1+2k)}{3} \tag{4}$$

$$L_{DM,CM} = L_{DM,self}\frac{(1+2k)}{3} \tag{5}$$

By utilizing the equations above on a system with a 2nd order LC filter consisting of a DM filter inductor and a filter capacitor clamped to the negative DC link rail, as in Fig. 2, the DM and CM EQCs can be established. The filter capacitor C_f and inductor L_{DM} are modeled with their equivalent series resistance. The DM EQC is displayed in Fig. 5 and the CM EQC in Fig. 6. Also parasitic capacitances of the inverter, the cable and the motor are considered. These DM and CM EQCs can be calculated for all different topology variants. For filter orders of maximum 4th order a number of 36 different Topologies can be generated, with and without additional CM inductors.

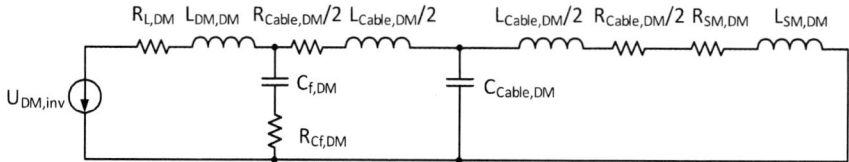

Fig. 5: Single phase DM EQC of the filter and system presented in Fig. 2.

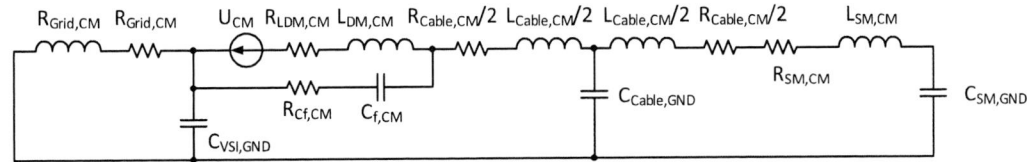

Fig. 6: Single phase CM EQC of the filter and system presented in Fig. 2.

Sine Wave Filter Design Method

To analyze all possible filter topologies and different distributions of inductances and capacitances a holistic strategy is needed which automatically calculates the CM and DM filter transfer functions. In this work an automated Spice based method is applied. The automation is realized by an algorithm implemented in Python. The Spice solver of *LTSpice* is used and controlled via batch instructions from the host program. The method is organized in different modules which are the boundary module, the variation module, the drive system module, the netlist creation module and the simulation module. The structure of the algorithm is shown in Fig. 7.

Filter Parameter Generation

The basis of the presented strategy is the determination of the design space for the filter variables. The boundaries of the design space are given by the parameters of maximum leakage current trough the capacitors in (6), the maximum voltage drop across the inductors in (7) at fundamental motor frequency and the maximum energy stored E_{st} in the components of the filter defined in (8).

$$I_{C,max} = j\omega_{fund} C U_{C,max} \tag{6}$$

$$U_{L,max} = j\omega_{fund} L I_{L,max} \tag{7}$$

$$E_{st} = \frac{1}{2}\sum_{g=0}^{b} L_g I_{g,max}^2 + \frac{1}{2}\sum_{h=0}^{c} C_h U_{h,max}^2 \tag{8}$$

Where b and c are defining the number of inductances and capacitances in the filter. For a 2nd order filter $b = 1$ and $c = 1$. Instead of the maximum stored energy it is also possible to have a boundary on the required attenuation at a certain frequency band. The boundary module calculates the maximum sum inductance and maximum sum capacitance based on the input of boundary parameters, operating conditions and number of allowed variants of inductances and capacitances. Depending on the filter type and limited by the maximum energy, parameter sets of different combinations of filter inductances and capacitances are generated. These parameter sets are then transformed into DM and CM EQCs and Spice netlists are generated. To solve the EQCs the AC-spice-solver of *LTSpice* is used. The interface of the calculation program and the AC-solver is based on netlists and automated batch instructions. The results returned as *LTSpice .raw* file are decoded with the main program and can then be analyzed and plotted.

Example for a Parameter Study

In this example a parameter study of a 2nd order filter in Fig. 2 is performed. A total of 3 different inductance values are selected which comply with the boundary of a max voltage drop of 15 volt. According to these inductances 10 capacitor variations are chosen under the boundary of the maximum capacitor current of 1 A at fundamental frequency of $f_{fund} = 600\,\text{Hz}$ and the maximum stored energy to be smaller than $E_{st} = 0.12\,\text{Ws}$. These values are set for this particular example. The total energy is a measure for size of the components. The lower the total energy, the more compact the filter will be. The DC link voltage is $V_{link} = 560\,\text{volt}$ with a rated RMS phase current of $I_{phase} = 9.5\,\text{A}$. The design switching frequency is at $f_{sw} = 128\,\text{kHz}$. The DM transfer function from the VSI to the output voltage of the filter is shown in the following figures Fig. 8 to Fig. 10 for all 30 parameter variations.

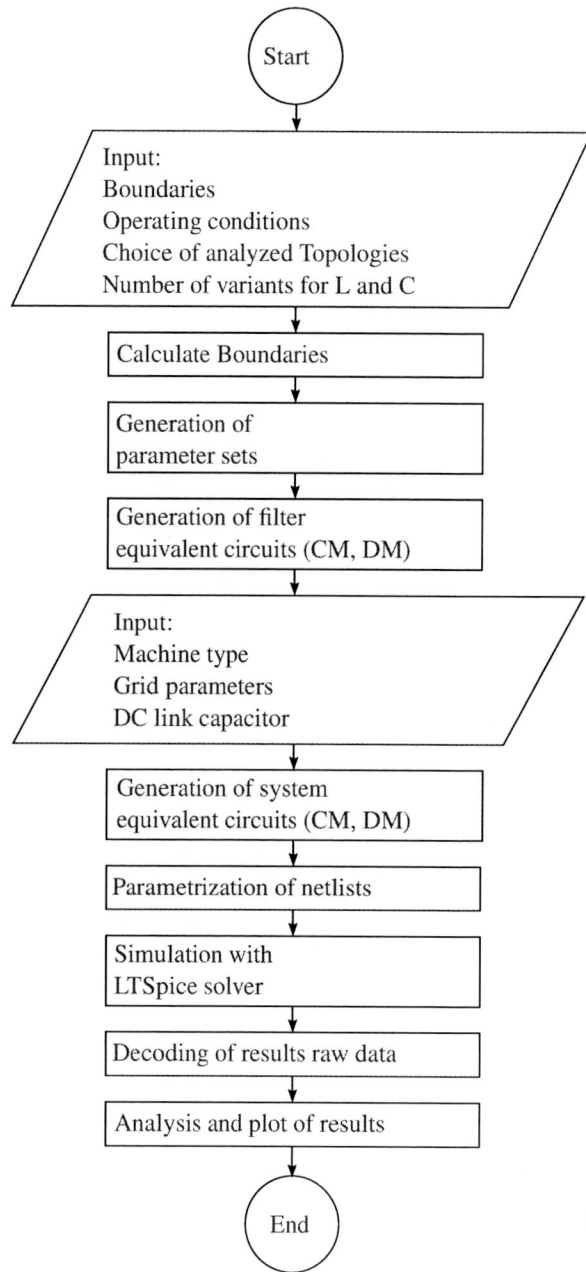

Fig. 7: Flow chart of filter design program.

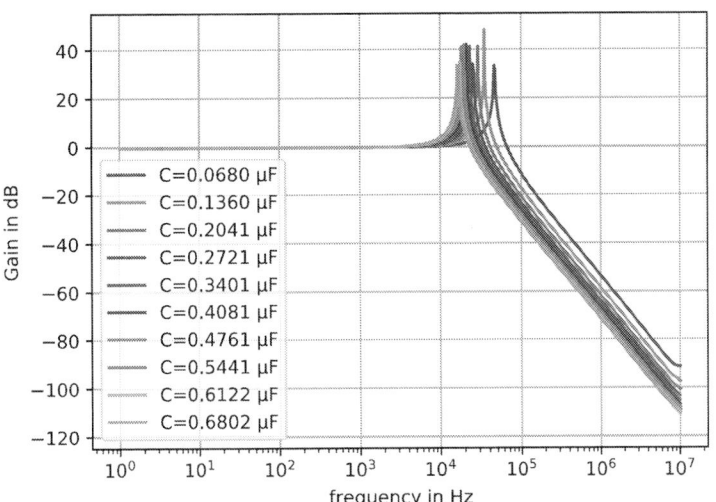

Fig. 8: Filter transfer function for a filter inductance of $139\,\mu H$ with 10 different capacitor values with a total stored energy of $E_{st} \leq 0.12\,W\,s$.

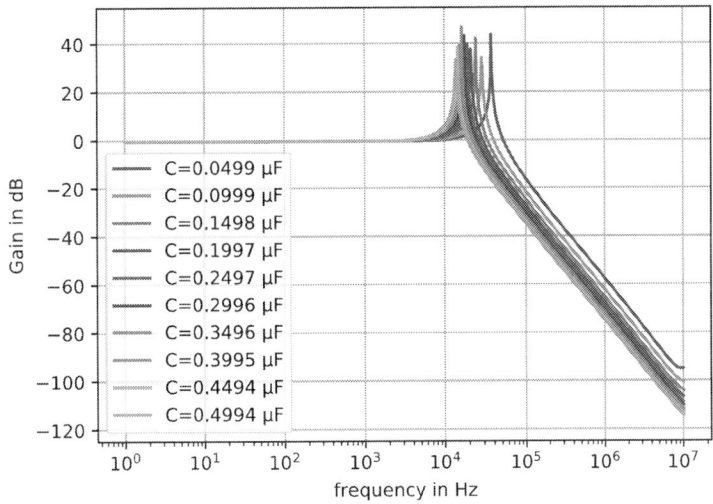

Fig. 9: Filter transfer function for a filter inductance of $279\,\mu H$ with 10 different capacitor values with a total stored energy of $E_{st} \leq 0.12\,W\,s$.

When for example an attenuation of $40\,dB$ at the switching frequency is needed, the best filter configuration can be selected from the results. Since the algorithms assumes continuity of the component values in a real filter only the nearest possible values can be chosen. The real values are discretized because of the discrete number of turns of an inductor and the commercially available filter capacitances which usually follow the E-series of preferred numbers. The closest values are chosen for the design.

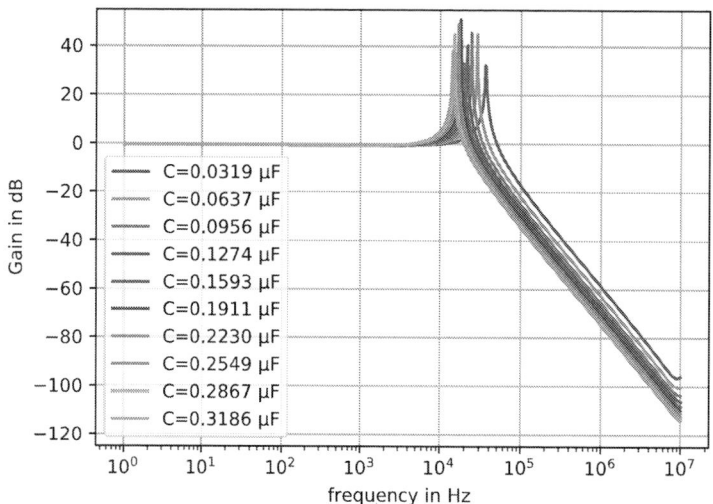

Fig. 10: Filter transfer function for a filter inductance of $418\,\mu H$ with 10 different capacitor values with a total stored energy of $E_{st} \leq 0.12\,W\,s$.

Experimental Verification

To verify the approach of finding a suitable filter a small signal voltage transfer function measurement is performed. The used measurement equipment is a *Bode100* vector network analyzer. A filter of the design procedure is build up and measured. The filter was selected to achieve an attenuation of at least $38\,dB$ at a switching frequency of $f_{sw} = 128\,Hz$. Table I shows the possible filters which enable this requirements from the parameter study in the previous section.

Table I: Selection of filter value tuples with an attenuation of at least $38\,dB$ at a frequency of $f_{sw} = 128\,Hz$

	Filter inductance (µH)	Filter capacitance (µF)
Variant 1	279	0.4494
Variant 2	279	0.4994
Variant 3	418	0.2867
Variant 4	418	0.3186

Filter variant 2 is actually built up in a test circuit on a 4 layer PCB. The capacitors are film capacitors of two $1\,\mu F$ in series and the inductors are *Sendust* toroidal inductors with a solid round wire. The measurement setup is presented in Fig. 11. The result of the voltage transfer function measurement is shown in Fig. 12.

The typical attenuation of approximately $40\,dB/decade$ can be observed almost up to $1\,MHz$. The resonant frequency at approximately $12.8\,kHz$ is not as sharp as in the more ideally simulations. This is caused by additional parasitic resistances of the components and the PCB. The attenuation at significantly higher frequencies above $2\,MHz$ is reduced, because the parasitic behavior of the inductor, capacitor and PCB come into play. Especially the inductor has an additional series capacitance caused by the inter winding capacitances. The capacitor typically adds a parasitic series inductance. These parasitic elements start to effect the filter performance for the higher frequencies. But filter requirements can still be fulfilled, because at higher frequencies, where the performance decreases, the distortion spectrum of the VSI is lower according to the calculation shown in Fig. 1. According to these measurements the sine wave filter can be used in the application with a SiC VSI drive inverter and will perform sufficiently.

Fig. 11: Setup for small signal transfer function measurements.

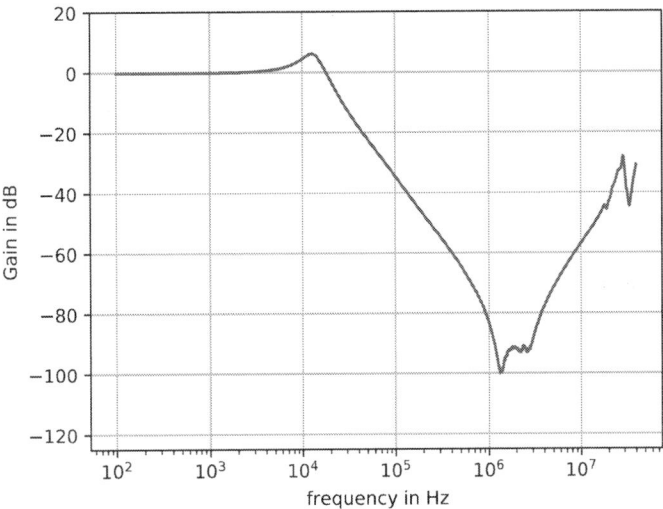

Fig. 12: Gain measurement of voltage transfer function of a 2nd order sine wave filter with $L_f = 280\,\mu H$ and $C_f = 0.5\,\mu H$.

Conclusion

The application of sine filters in drive systems have a huge potential when they can be designed small enough, enabled by high switching frequencies of SiC MOSFETs. To evaluate system behavior three phase systems are transformed to single phase DM and CM EQCs. Then different options to vary the topologies of the filter and its configurations are discussed. With the presented algorithm a suitable filter can be found either by limiting the total stored energy or by entering a desired attenuation value for a certain frequency. With the Spice based algorithm any parameter of the system or the filter can be iterated for different filter topologies and drive systems with different electric motors. Because of the modularity of the netlist representation more parasitic elements can be added to the filter representation. However the filter in the example is relatively ideal to minimize computation effort. Finally the algorithm is applied to a specific parametrization example and the results are presented accordingly. From the results of the algorithm filters which satisfy a certain requirement are selected. One filter is built up and the small signal voltage transfer function measurement is showing a good match to the design result. However in the higher frequency band parasitic effects reduce the filter effectiveness. In future work this method can be included into a multi objective optimization algorithm to find the optimal solution of the filter. Also in future work the filter design will be tested in an experimental setup with an actual VSI.

References

[1] S. Ye and Y.-F. Liu, "Emi filter design method for communication power sub-system," in Eighteenth Annual IEEE Applied Power Electronics Conference and Exposition, 2003. APEC '03, Feb 2003, pp. 483–489, vol.1

[2] S. Jiang, Y. Liu, H. Wang, G. Wang, J. Yan, and J. Peng, "Effective emi filter design method of single-phase inverter based on noise source impedance," in 2018 IEEE International Power Electronics and Application Conference and Exposition (PEAC), Nov 2018, pp. 1–6

[3] Y. Lai and P. Chen, "New emi filter design method for single phase power converter using software based noise separation method," in 2007 IEEE Industry Applications Annual Meeting, Sep. 2007, pp. 2282–2288

[4] K. Raggl, T. Nussbaumer, and J. W. Kolar, "Model based optimization of emc input filters," in 2008 11th Workshop on Control and Modeling for Power Electronics, Aug 2008, pp. 1–6

[5] X. Huang, E. Pepa, J.-S. Lai, S. Chen, and T. W. Nehl, "Three-phase inverter differential mode emi modeling and prediction in frequency domain," in 38th IAS Annual Meeting on Conference Record of the Industry Applications Conference, 2003., vol. 3, Oct 2003, pp. 2048–2055 vol.3

[6] "Adjustable speed electrical power drive systems - part 3: Emc requirements and specific test methods," 2018, iEC 61800-3:2018

[7] A. Nagel and R. W. De Doncker, "Analytical approximations of interference spectra generated by power converters," in IAS '97. Conference Record of the 1997 IEEE Industry Applications Conference Thirty-Second IAS Annual Meeting, vol. 2, Oct 1997, pp. 1564–1570 vol.2

[8] D. Kampen, "Considering mutual impacts of differential mode and common mode emissions in motor filter design for pwm inverters," in 2009 13th European Conference on Power Electronics and Applications, Sep. 2009, pp. 1–7

A Symmetrical Boost Converter with Reduced Common-Mode Leakage Currents for EV Applications

Caniggia Viana[*1], Netan Yakop[†1], Damien Frost[‡2], and Peter Lehn[§1]

[1]Department of Electrical Engineering, University of Toronto, Toronto, Canada,
[2]Brill Power, Oxford, United Kingdom

Acknowledgments

The authors acknowledge the support of the Natural Sciences and Engineering Research Council of Canada (NSERC), CRDPJ 513206 - 17.

Keywords

≪Common-Mode Current≫, ≪Leakage Current≫, ≪Transformerless≫, ≪Electric Vehicle Charger≫, ≪Symmetrical Boost Converter≫.

Abstract

An important design choice for power converters, including electric vehicles chargers, is whether or not to include a galvanic isolation transformer. While transformerless systems can be considerably lighter and cheaper, they are typically discarded as a design choice due to the potential presence of common-mode leakage currents. In this paper, a 3-switch symmetrical boost converter is proposed, analyzed, simulated, and experimentally verified. The topology is shown to be suitable for non-isolated electric vehicle charging. The converter exhibits negligible common-mode leakage current, allowing it to meet safety standards without the need for an isolation transformer.

Introduction

Charging infrastructure is a critical aspect of the current electromobility transformation [1, 2]. As electric vehicle (EV) penetration grows, so does the demand from the charging network. Such demands include power density, efficiency, and cost thereby promoting intense research and development effort from both academia and industry. Solutions introduced in literature can be divided into isolated [3, 4, 5] and non-isolated [5, 6] charging methods. The inclusion of an isolation transformer can make the prevention of leakage current much easier, as there is no leakage conduction path. As a result, several authors have recommended galvanic isolation as a necessary safety feature for grid connected EV chargers [7, 8]. However, the inclusion of a transformer adds an extra conversion step which tends to significantly impact system efficiency. Losses equivalent to 1.5% of rated power are reported by Mattson *et al.* [9]. Additionally, the isolation transformer decreases power density and increases system cost.

This work introduces a non-isolated converter, termed the symmetrical boost converter, with leakage performance comparable to isolated systems. This feature allows for more efficient, compact and cost-effective design to meet strict safety standards and regulations, such as the UL-2231 and IEC 61851 [10, 11]. Two variants of the proposed topology are shown in Fig. 1a and Fig. 1b. In both cases, the

---

[*]caniggia.castrodinizviana@mail.utoronto.ca
[†]netan.yakop@mail.utoronto.ca
[‡]damien.frost@brillpower.com
[§]lehn@ecf.utoronto.ca

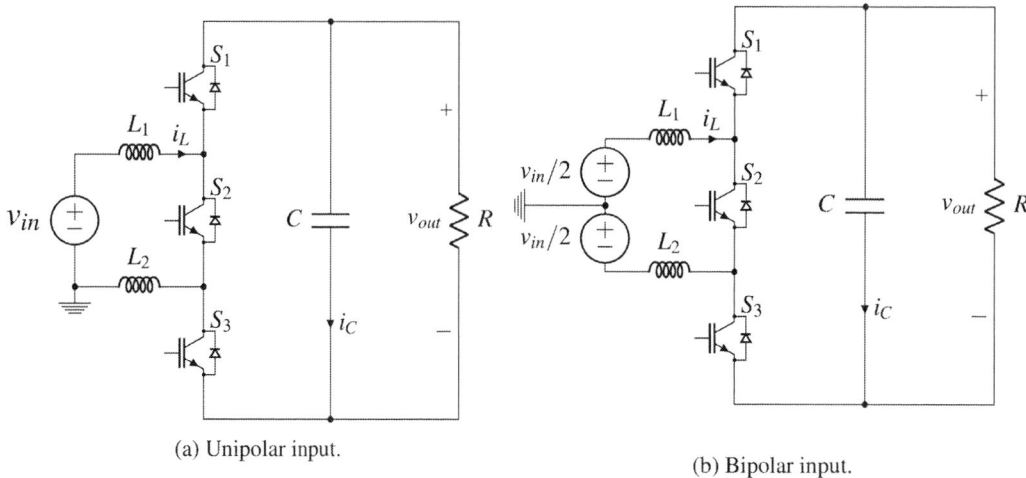

(a) Unipolar input.

(b) Bipolar input.

Fig. 1: Symmetrical boost.

converter consists of two inductors and three switches. Common to both converters is that they exhibit significantly lower leakage current when compared to the conventional boost converter.

In addition to mitigating leakage current, the variant with a midpoint grounded bipolar input voltage source, shown in Fig. 1b, meets the central grounding constraint in UL-2231, which requires the minimization of the highest potential to ground [10]. This feature is achieved without the need to establish a physical connection to ground at the midpoint of the output, which is typically not feasible if the load is an EV battery. Given the additional benefit of the variant with bipolar input voltage, and its implications regarding EV charging regulations, the circuit in Fig. 1b is the main focus of this paper.

Throughout this text, the operating principle is analyzed and the boost-like functionality is verified. The small signal model is derived and shown to be equivalent to the conventional boost converter, allowing conventional control techniques to be employed. Lastly, simulations and experimental results are included for a 6 kW prototype, where the leakage current performance is demonstrated and compared to a non-symmetrical boost converter.

Topology

The proposed topology is a boost converter variant which includes one extra switch to render the topology symmetrical, as shown in Figs. 1a and 1b. The filtering inductance is also deployed symmetrically, by use of inductors L_1 and L_2 connected to both terminals of the input voltage source, such that $L_1 + L_2 = L$, and $L_1 = L_2$.

Operating Principle

In the proposed operation, the converter is switched such that there are two switching states. The outer switches, S_1 and S_3, are switched in tandem, complementary to the middle switch, S_2. The equations describing the net inductor voltage, $v_L(t)$, and the capacitor current, $i_C(t)$, during each switching state are shown below.

$0 \leq t < DT_s \Rightarrow (S_2 \text{ on}, S_1, S_3 \text{ off})$:

$$v_L(t) = v_{in}(t) - i_L(t)(R_L + R_{on2}) \tag{1a}$$

$$i_C(t) = -\frac{v_{out}(t)}{R} \tag{1b}$$

$DT_s \leq t < T_s \Rightarrow (S_1, S_3 \text{ on}, S_2 \text{ off})$:

$$v_L(t) = v_{in}(t) - v_{out}(t) - i_L(t)(R_L + R_{on1,3}) \tag{2a}$$

$$i_C(t) = i_L(t) - \frac{v_{out}(t)}{R}, \tag{2b}$$

where

- $v_L(t) \overset{\Delta}{=} v_{L1}(t) + v_{L2}(t)$ is the induced voltage across the equivalent inductor L formed by the combination of L_1 and L_2,

- $R_L = R_{L1} + R_{L2}$ is the equivalent series resistance of the equivalent inductor L,

- R_{on2} is defined to be the on resistance of switch S_2, and

- $R_{on1,3} = R_{on1} + R_{on3}$ is the combined on resistance of switches S_1 and S_3.

Dynamic Model

In steady-state, the system conversion ratio can be described as

$$\frac{V_{out}}{V_{in}} = \frac{1}{D'} \left(1 - \frac{I_L}{V_{in}} (R_L + D R_{on2} + D' R_{on1,3}) \right), \tag{3}$$

where

- D is the average duty-cycle associated with switch S_2 and

- $D' \overset{\Delta}{=} D - 1$.

From equations (1a), (1b), (2a), and (2b), it is possible to derive the linearized dynamic model of the system. The input current dynamics are described by

$$\begin{aligned}
\hat{v}_L(t) &= L \frac{d(\hat{i}_L(t))}{dt} \\
&= \hat{v}_{in}(t) + (V_{out} + I_L(R_{on1,3} - R_{on2}))\hat{d}(t) - D'(\hat{v}_{out}(t)) - (R_L + D R_{on2} + D' R_{on1,3})\hat{i}_L(t).
\end{aligned} \tag{4}$$

The output voltage dynamics are described by

$$\begin{aligned}
\hat{i}_C(t) &= C \frac{d(\hat{v}_C(t))}{dt} \\
&= D' \hat{i}_L(t) - \hat{d}(t) I_L - \frac{\hat{v}_{out}(t)}{R}.
\end{aligned} \tag{5}$$

Combining (4) and (5), the resulting small-signal circuit model can be depicted as seen in Fig. 2, where

- $V^* \overset{\Delta}{=} (V_{out} + I_L(R_{on1,3} - R_{on2}))$ and

- $R^* \overset{\Delta}{=} (R_L + D R_{on2} + D' R_{on1,3})$.

It can be noted that the dynamic model for symmetrical boost is identical to the model of a regular boost converter. This property, allows designers to leverage control techniques developed for the conventional boost.

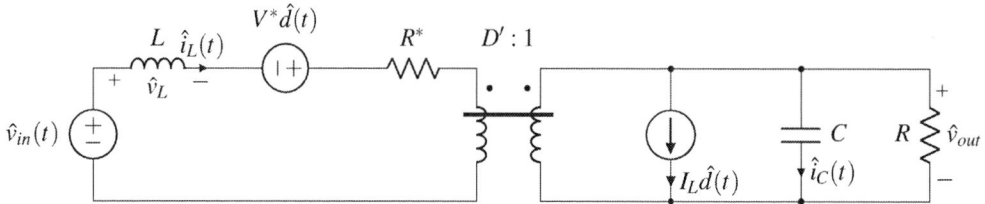

Fig. 2: Linearized small signal dynamic model for the symmetrical boost converter.

Leakage Current Reduction

Meeting strict leakage current constraints is the central challenge with non-isolated converters in general, and EV chargers in particular. In practical applications, EV chargers have Y-capacitors connected from the load (battery) terminals to the chassis which is grounded. These capacitors can be parasitic or intentionally placed and can be seen in Fig. 3. As a result, a path is created for leakage current through the ground of the grid that can trip ground fault circuit interrupter (GFCIs) or constitute a safety hazard for the user [12]. The common-mode equivalent circuit that dictates the leakage current, i_g, is derived as in [13], and is shown in Fig. 4, where g_{S2} is the gating signal for S_2.

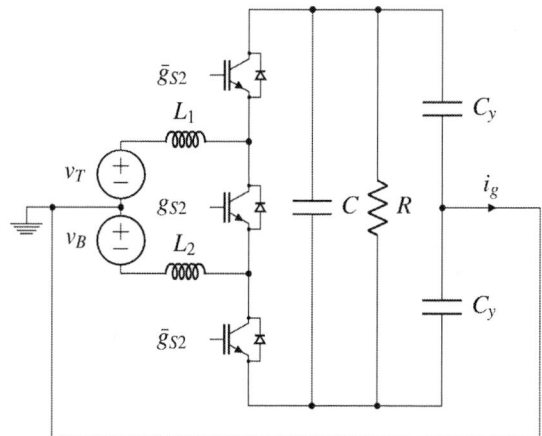

Fig. 3: Circuit with y capacitors.

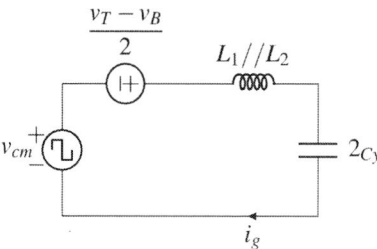

Fig. 4: Equivalent common-mode circuit.

The common-mode voltage is given by

$$v_{cm} = \left[\left(\frac{v_{in}}{2} + \frac{g_{S2} - 1}{2} v_{out} \right) \left(\frac{L_2 - L_1}{L_1 + L_2} \right) - sgn(i_g) \left(\frac{g_{S2} v_{out}}{2} \right) \right]. \tag{6}$$

Under perfect symmetry, i.e., $L_1 = L_2$,

$$v_{cm} = -sgn(i_g) \left(\frac{g_{s2} v_{out}}{2} \right), \tag{7}$$

which always opposes the flow of electrical current, since $\frac{g_{s2} v_{out}}{2}$ is strictly non-negative. Therefore, the system will generate exactly 0 leakage current.

Leakage Current Sensitivity to Nonidealities

In practice, perfect match between L_1 and L_2 is not feasible. Similarly, other asymmetries may be present in the system. A sensitivity analysis must be performed to quantify the effects of small parameter mismatch, such as inductance and parasitic capacitances. However, analytically quantifying the impact of each possible asymmetry in circuit may be impractical. Therefore, the next section employs simulations to determine common-mode ground current impacts of small asymmetry between the top and bottom voltages, the inductors and the y-capacitors.

Simulations

In this section, the circuit shown in Fig. 3 is simulated with the parameters shown in Table I under two conditions: (i) ideal symmetry, i.e., $L_1 = L_2 = 0.5\,mH$ and $v_T = v_B = V_{in}/2 = 100\,V$, (ii) non-ideal symmetry, i.e. $L_1 = 0.5\,mH$, $L_2 = 0.55\,mH$, $v_T = 105\,V$ and $v_B = 95\,V$, where both input voltages and filter inductances are mismatched by around 10%.

Parameter	Value
f_{sw}	$30\,kHz$
V_{in}	$200\,V$
V_{out}	$400\,V$
i_L	$30\,A$
P_{out}	$6\,kW$
R	$26.67\,\Omega$
L	$1\,mH$
C	$80\,\mu F$
C_y	$2.2\,\mu F$

Table I: System parameters.

The conventional, asymmetrical boost is also simulated using the parameters shown in Table I. For the conventional boost, the source impedance becomes relevant to determine the common-mode leakage current. This effect stems from the fact that the conventional boost can be seen as a special case of the symmetrical boost where $L_2 = 0\,H$. Suggesting a common-mode impedance of the conventional boost is given by $L_1 // L_2 = 0\,H$. In practice, however, L_1 and L_2 have a lower bound determined by the source impedance. The values used in this simulation reflect the parameters present in the experimental setup used in the next section, $L_{source} = 40\,\mu H$ and $R_{source} = 0.4\,\Omega$ on both the positive and the negative rails. The circuit used for the conventional boost simulation is shown in 5.

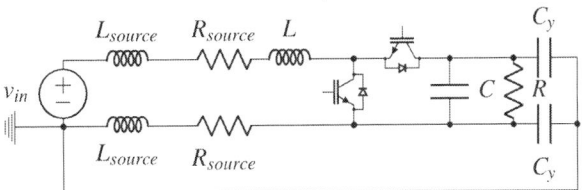

Fig. 5: Conventional boost circuit used in simulation.

The common-mode leakage performance of the three configurations is shown in Fig. 6.

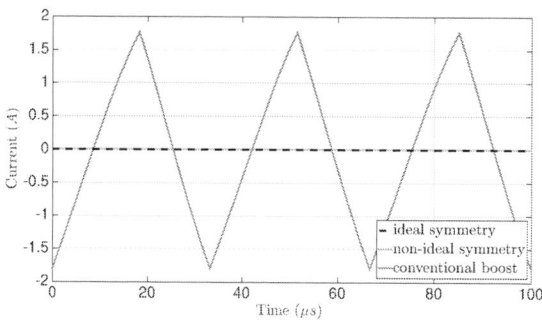

Fig. 6: Leakage current - simulation results.

The objective of this analysis is to demonstrate the efficacy of the leakage mitigation in a practical setting. To illustrate this, an example case of imperfect symmetry is studied and the leakage current waveform as well as magnitudes are presented.

In the ideal symmetry case, the common-mode current is, as expected, exactly zero as demonstrated analytically. Small mismatches in inductance create negligible currents compared to the conventional boost, with an RMS value under 0.3 mA for 10% inductor mismatch. Voltage mismatch, meanwhile, has no effect on leakage current. This can be understood by referring to the circuit shown in Fig. 4. The voltage mismatch creates a dc voltage in the common-mode equivalent circuit, which is connected in series with the equivalent capacitor, thereby generating zero current. The voltage splitting is included solely to meet the central grounding regulation, having zero impact on leakage current. Leakage current is still mitigated even if a negative grounded voltage, as shown in 1a, is used.

On the other hand, the conventional boost has a high leakage current. An RMS value of approximately 1 A, posing a severe safety hazard. The symmetrical converter reduces the leakage current by several orders of magnitude even in the presence of imperfect circuit symmetry.

Sensitivity Analysis

Once established that the proposed system is robust to small variations, it is important to determine the exact sensitivity of RMS current to circuit asymmetries. To this end, the system of Fig. 3 is simulated with the same parameters as shown in Table I and a forced inductor or capacitor mismatch. The leakage current results are recorded and shown below.

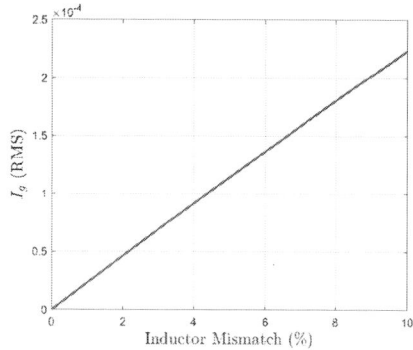

(a) Leakage current resultant from small inductor mismatch in the presence of perfect capacitor symmetry.

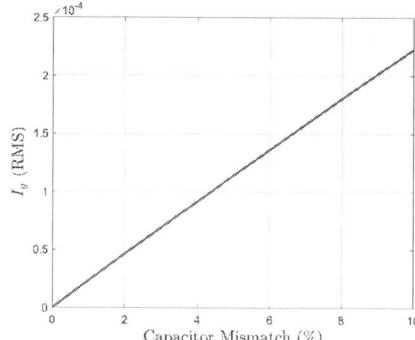

(b) Leakage current resultant from small y-capacitor mismatch in the presence of perfect inductor symmetry.

Fig. 7: Sensitivity of RMS leakage current to small circuit asymmetries.

Experimental Results

An experimental prototype is built to verify the leakage performance described in the Topology Section. The parameters shown in Table I are also used to build the experimental circuit, seen in Fig. 8.

Fig. 8: Experimental circuit.

The system under consideration is loaded to nominal operation, with an output voltage of $400\ V$ and output current of $15\ A$. In this configuration, the system outputs $6\ kW$. A discrete PI controller is used to regulate the output voltage.

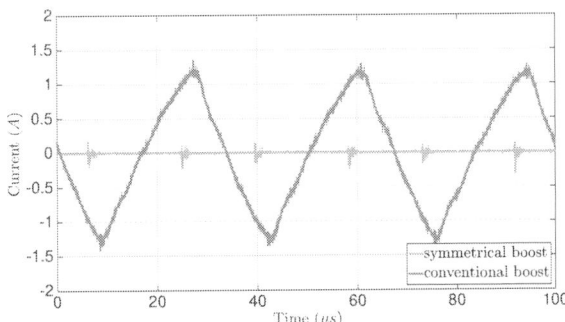

Fig. 9: Leakage current - experimental results.

The experimental results of the leakage current for the conventional and symmetrical boost converters are shown in Fig. 9. The experimental leakage current of the conventional boost is approximately 750 mA RMS, while for the symmetrical boost it is well under 20 mA RMS and approaches the noise floor of the existing current sensing, given the low signal to noise ratio. Notwithstanding the presence of sensor noise, a 40 times reduction in leakage current is verified, indicating a substantial improvement with respect to the conventional boost.

Conclusion

This work implements a symmetrical boost converter. The system differs from a conventional boost by the inclusion of an extra switch, S_3, and a split of the inductor into two inductors to mitigate common-mode current and voltage. Two variants of the topology are presented, differing from one another by the location of the ground connection. Both equally mitigate common-mode leakage current, while the inclusion of the bipolar input voltage source has the additional benefit of minimizing the highest potential from ground, thereby meeting an additional constraint of UL-2231 EV charging standard [10].

Simulations and experimental verification are developed for a $6\ kW$ system. Common-mode currents are measured on the ground path between the input and output center points. While the conventional boost converter fails to comply with regulatory requirements, the proposed converter meets EV charging standard requirements and can be used to implement a non-isolated EV charger with improved power density and lower cost, through the elimination of the isolation transformer.

References

[1] M. Yilmaz and P. T. Krein, "Review of battery charger topologies, charging power levels, and infrastructure for plug-in electric and hybrid vehicles," *IEEE Transactions on Power Electronics*, vol. 28, no. 5, pp. 2151–2169, May 2013.

[2] S. Habib, M. M. Khan, K. Hashmi, M. Ali, and H. Tang, "A comparative study of electric vehicles concerning charging infrastructure and power levels," in *2017 International Conference on Frontiers of Information Technology (FIT)*, Dec 2017, pp. 327–332.

[3] C. Viana and P. W. Lehn, "A drivetrain integrated dc fast charger with buck and boost functionality and simultaneous drive/charge capability," *IEEE Transactions on Transportation Electrification*, pp. 1–1, 2019.

[4] P. He and A. Khaligh, "Comprehensive analyses and comparison of 1 kw isolated dc–dc converters for bidirectional ev charging systems," *IEEE Transactions on Transportation Electrification*, vol. 3, no. 1, pp. 147–156, March 2017.

[5] R. Zgheib, I. Kamwa, and K. Al-Haddad, "Comparison between isolated and non-isolated dc/dc converters for bidirectional ev chargers," in *2017 IEEE International Conference on Industrial Technology (ICIT)*, March 2017, pp. 515–520.

[6] Y. Du, X. Zhou, S. Bai, S. Lukic, and A. Huang, "Review of non-isolated bi-directional dc-dc converters for plug-in hybrid electric vehicle charge station application at municipal parking decks," in *2010 Twenty-Fifth Annual IEEE Applied Power Electronics Conference and Exposition (APEC)*, Feb 2010, pp. 1145–1151.

[7] M. Yilmaz and P. T. Krein, "Review of Battery Charger Topologies , Charging Power Levels , and Infrastructure for Plug-In Electric and Hybrid Vehicles," *IEEE Transactions on Power Electronics*, vol. 28, no. 5, pp. 2151–2169, 2013.

[8] S. Y. Kim, S. Member, H.-s. Song, and K. Nam, "Idling Port Isolation Control of Three-Port Bidirectional Converter for EVs," *IEEE Transactions on Power Electronics*, vol. 27, no. 5, pp. 2495–2506, 2012.

[9] A. Mattsson, V. Vaisanen, P. Nuutinen, T. Kaipia, A. Lana, P. Peltoniemi, P. Silventoinen, and J. Partanen, "Implementation design of the converter-based galvanic isolation for low voltage DC distribution," *2014 International Power Electronics Conference, IPEC-Hiroshima - ECCE Asia 2014*, pp. 587–594, 2014.

[10] " UL 2231-1, Personnel Protection Systems for Electric Vehicle (EV) Supply Circuits: General Requirements," Underwriting Laboratories, Standard, Aug. 2016.

[11] " IEC 61851, Electric vehicle conductive charging system – Part 1: General requirements," International Electrotechnical Commission, Standard, Feb. 2017.

[12] Y. Zhang, G. Yang, X. He, M. Elshaer, W. Perdikakis, H. Li, C. Yao, J. Wang, K. Zou, Z. Xu, and C. Chen, "Leakage current issue of non-isolated integrated chargers for electric vehicles," in *2018 IEEE Energy Conversion Congress and Exposition (ECCE)*, Sep. 2018, pp. 1221–1227.

[13] B. Yang, W. Li, Y. Gu, W. Cui, and X. He, "Improved transformerless inverter with common-mode leakage current elimination for a photovoltaic grid-connected power system," *IEEE Transactions on Power Electronics*, vol. 27, no. 2, pp. 752–762, 2012.

Modeling and analysis of conducted EMI on flyback converter using power management IC with chaotic suppression EMI

Diao Jiaqi, Yang Ru, Liu Zuolian, Yang Hong, Jie Hai
Guangzhou University
230 Wai Huan Xi Road, Guangzhou Higher Education Mega Center, Guangzhou 510006,
P.R.China
+86 13342885376
Guangzhou, China
E-Mail: yangru@gzhu.edu.cn

Keywords

«EMC/EMI», «Chaotic suppression EMI», «Flyback converter», «Power management», «Chaos»

Abstract

In this paper, CM and DM transfer function of EMI models are used to simulate conducted EMI on flyback converter. The EMI model in this paper considers main parameter of the components to forecast EMI. The experimental results validate that the suggested model can roughly predict conducted EMI. Liu chaotic system, as the chosen one, is put into a power management chip because of its effective inhibition. The realization of Liu chaotic system of a hardware circuit is optimized. Using the least components to implement this chaotic circuit can reduce the power at the same time. This first chaotic power management chip on the market will be used in a flyback to elaborate its practical value.

Introduction

In electronic and electrical products, electromagnetic interference (EMI) is a major issue that must be taken into consideration to meet the electromagnetic compatibility (EMC) standards. Among these products, switched mode power supplies (SMPS) have considerable EMI due to the rapid turn-on and turn-off of switching devices. However, the harmonics produced by the converter pollutes the electromagnetic environment [1][2]. EMI filter is a common solution to reduce EMI in converters [3]. But these filters often increase the cost, weight and complexity of the circuit. After nearly a decade of development, using the chaotic signal, whose spectra is continuous, to suppress EMI has been widely accepted, and the researchers on SPWM in chaotic modulation is analyzed [4].

In [5] and [6], a model called equivalent modular-terminal-behavioral (MTB) is proposed. EMI source can be obtained based on this model. In [7]-[8], the analysis of CM interference conduction path is not accurate enough, and the high-frequency model of components is not considered, which results in unsatisfactory prediction effect. In this paper, the EMI model of flyback converter is provided in CM and DM, which can be used as transfer function to simulate conducted EMI. In order to obtain precise conducted EMI transfer function models, high-frequency parasitic parameters models of components are taken into consideration to rectify simulation results above 10MHz.

To suppress the EMI and improve the EMC of flyback converter, method of chaotic suppression is employed. The chaotic spectrum characteristics of time-frequency energy distribution based on wavelet transform was analyzed in [9], the inherent mechanism of EMI suppression from the point of energy spectrum was also addressed. An analogue chaotic carrier is designed to be usedcap in a boost converter [10]. These papers only discuss the application of chaotic spread spectrum technology in the suppression of harmonic interference, but does not analyze whether it can reduce the peak value of conducted EMI.

In 2004, Liu et al. proposed a class of third-order continuous autonomous chaotic systems with square nonlinear terms called Liu chaotic system [11]. But the realization of Liu chaotic system [12] was too complex to embed into a power management. In practical, Liu chaotic system is the chosen chaotic signal. The realization of Liu chaotic system of a hardware circuit is optimized and using fewer components to implement this chaotic circuit which can reduce the power at the same time. The results of computer simulation and circuit experiment verify the correctness of the optimized hardware chaotic circuit.

The remainder of this paper is organized as follows. The first paragraph established the EMI model of flyback converter and employing experiment to verify the accuracy of the suggested model. The optimized circuit of Liu chaotic system is analyzed and detailed experiment on a flyback converter is proposed in the second paragraph. Conclusion will be the last paragraph.

Modeling and analysis of conducted EMI on flyback converter

As shown in Fig.1, there is the typical topology of the AC/DC flyback converter with LISN. A 5W(5V/1A) flyback converter with 220VAC input voltage, 5V output voltage operating at 100kHz switching frequency is designed.

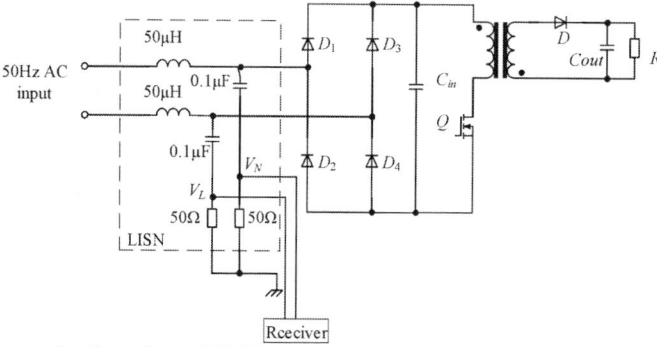

Fig. 1: Transmission path of conducted EMI

Interference loop of DM

The differential mode (DM) is called symmetric mode or Normal mode, and its noise source is connected between L line and N line. So, there is no noise current flowing through the ground plane. The EMI generated by the MOSFET flows through LISN, the rectifier bridge, primary side of transformer and MOSFET.

A simplified EMI loop in DM is shown in Fig.2, R_{cs} (0.38Ω) are high frequency parasitic resistances of the switching component provided by datasheet. The circuit elements of the transformer consist of primary inductance of 1.65mH (L_p), primary leakage inductance of 60μH ($L_{p\text{-}leak}$), secondary inductance of 8.1μH (L_s), secondary leakage inductance of 358nH ($L_{s\text{-}leak}$), inter-winding capacitance of 20pF (C_{ps}), all measured by impendence analyzer BD-100.

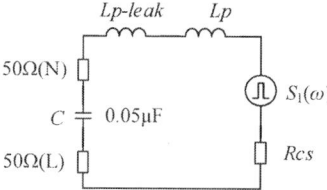

Fig. 2: EMI loop in DM

The transfer function of the loop can be written as:

$$H_{dm}(\omega) = \frac{50}{100 + \dfrac{I}{j\omega c_{lisn}} + j\omega\left(L_p + L_{p-leak}\right)} \tag{1}$$

Interference loop in CM

Common mode (CM) is called asymmetric mode or ground leakage mode. One end of the noise source is grounded and the other end is connected with L line or N line. Because the high-frequency domain of EMI is mainly affected by CM noise, the transformer should be considered. T-network model of the transformer is used to replace the conventional one.

There is only two-line input in this AC/DC converter, GND of converter is isolated to the EMI receiver which captures the conducted EMI on 50Ω. Blue line in Fig. 3 shows the real path of the CM noise. The GND is only a path for transmission

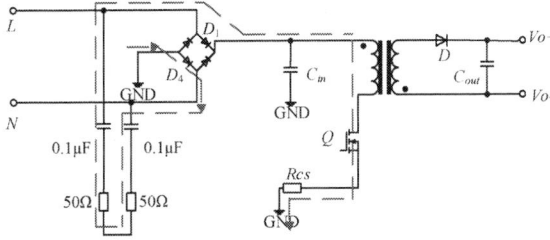

Fig. 3: EMI loop in CM of primary side

The EMI loop in CM mode is depicted in Fig. 4. C_y (1nF) is the connection Y capacitance between the ground plane and the output negative pole, therefore, C_y provides a path for EMI to returns to LISN.

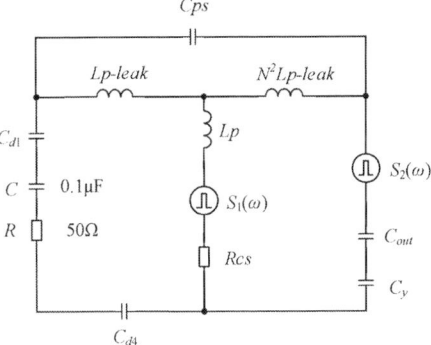

Fig. 4: EMI loop in CM

There are three paths in this loop. First, noise S_1 and S_2 flows clockwise through L_p, $N^2 L_{s-leak}$(N=15), C_{out}(470μF), C_y(1nF) and R_{cs}; secondly, noise S_1 and S_2 flows clockwise through L_p, L_{p-leak}, and then C_{ps}, C_{out}, C_y and R_{cs}; the third path goes counterclockwise through noise S1, L_p, L_{p-leak}, C_{d1}, C, R, C_{d4} and R_{cs}.

The high-frequency model of capacitor illustrates in Fig. 5, which contains R_s and L_{lead} with capacitor in series, used in the interference loop of CM mode.

$$Rs \quad L_{lead} \quad C$$

Fig. 5: High-frequency model of capacitor

Fig. 6 shows the high-frequency model of diode including stray capacitor C_j, Q_{rr} which using charge-storage effect to describe the characteristic of diode reverse recovery. The instantaneous forward characteristic curve and junction capacitance curve could be drawn based on the datasheet in Saber software, furthermore, this model can be built.

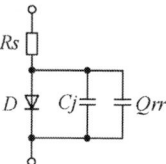

Fig. 6: High-frequency model of diode

The high-frequency equivalent model of MOSFET is plotted in Fig. 7, where C_{gd}, C_{gs} and C_{ds} are given by

$$C_{rss} = C_{gd} \tag{2}$$

$$C_{oss} = C_{gd} + C_{ds} \tag{3}$$

$$C_{iss} = C_{gd} + C_{gs} \tag{4}$$

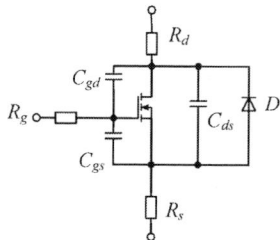

Fig. 7: High-frequency model of MOSFET

Characteristics of C_{rss}, C_{oss}, C_{iss}, R_d, R_s and R_g can be read on the datasheet also.In Fig. 8, the high-frequency models of capacitors C_{out} and C_y in the three paths mentioned above and the red dotted frame are drawn. $R_1(27m\Omega)$, $L_1(0.11uH)$ and $R_2(15m\Omega)$, L_2 (0.08uH), which can be measured by impendence analyzer.

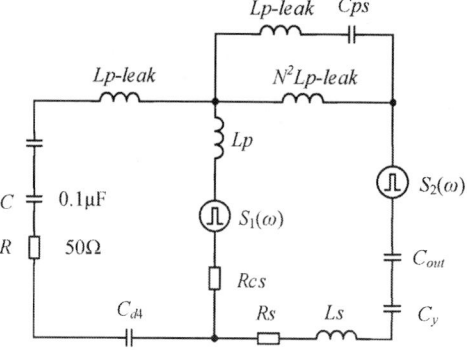

Fig. 8: Complete EMI loop in CM

In order to write the transfer function of Fig. 8, we need to simplify the expression of H_{cm}. So, we have Fig. 9. Also, the equation can be written as follow by using mesh current-method,

$$Z_1 = \left(L_p + L_{p-leak} \right) + \left(C / / Cd1 / / Cd4 + R \right),$$

$$Z_2 = \left(L_{p-leak} + C_{ps} \right) / / N^2 L_{p-leak} + \left(L_p + C_y / / C_{out} + Rs + Ls \right), \ R_s = R_1 + R_2, L_s = L_1 + L_2.$$

$$im_1 \left(Z_1 + R_{cs} \right) + \left(im_1 - im_2 \right) R_{cs} + u_s = 0 \tag{5}$$

$$im_2\left(Z_s + R_{cs}\right) + \left(im_2 - im_1\right)R_{cs} - u_S - u_D = 0 \tag{6}$$

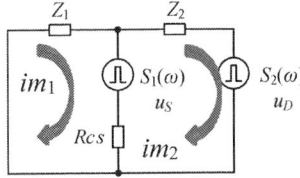

Fig. 9: Simplified EMI loop in CM

The interference voltage is captured on R by LISN, so we have (i refer to the current flows through R)

$$H_{cm}\left(\omega\right) = \frac{u_s}{i \cdot 50} \tag{7}$$

According to Fig. 10, the relationship between i and im_1 is

$$i = im_1 \cdot \frac{50}{50 + Z_1} \tag{8}$$

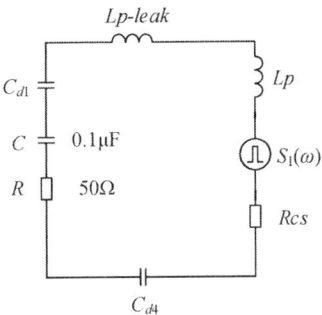

Fig. 10: Primary side of EMI loop in CM

Simulation of EMI

The EMI model has been obtained on the above, now we use Saber and MATLAB to simulate the EMI. First, we establish the complete simulation model of flyback converter with Saber software extracting the voltage data of u_S on MOSFET in frequency domain and the voltage data of u_D on the diode in frequency domain too. The model transfer functions H_{dm} and H_{cm} can be written with MATLAB. N(ω) represents the noise is captured by the EMI receiver on 50Ω resistance. Finally, the equation (13)-(16) is used to simulate the EMI. There is necessary to explain the equation (13)-(16). The EMI receiver detects the voltage on 50Ω resistance, which can be written as below (L line):

$$V_L = 50 \times \frac{1}{2} I_{cm} + 50 \times I_{dm} \tag{9}$$

Then we use N(ω) to replace the voltage, so we have:

$$N_L\left(\omega\right) = \frac{1}{2} \times N_{cm}\left(\omega\right) + N_{dm}\left(\omega\right) \tag{10}$$

$$N_{cm}\left(\omega\right) = S\left(\omega\right) \cdot H_{cm}\left(\omega\right) \tag{11}$$

$$N_{dm}(\omega) = S(\omega) \cdot H_{dm}(\omega) \tag{12}$$

Validation

To verify the EMI model established above, EMI of the AC/DC converter are measured. The EMI receiver of Rohde&Schwarz was employed in the shielding room. Under the input voltage of 220VAC and the switching frequency of 100kHz, 5W flyback prototype is adopted.The comparison shows that the simulation results of EMI model with high frequency parasitic parameters are in good agreement with the measurement results before 10MHz as illustrated in Fig. 11.

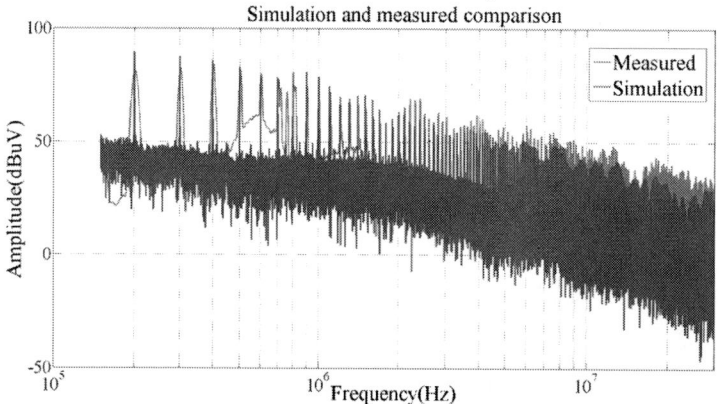

Fig. 11: Comparison of conducted EMI between simulation with high frequency parasitic parameters(blue) and measured (red)

Table I: Comparison of AV between simulation and measured EMI

Frequency/ kHz	Simulation/dBμV	Measured/dBμV	Error/ dBμV
200	88	81	-7
400	84	83	-1
600	80	80	0
800	81	73	-7
1000	79	76	-3
1200	68	70	2
1400	64	65	1
2050	63	64	1

Table I indicates that simulation EMI fits perfectly to measured EMI below 2MHz and error is within 10dBμV.

Optimize of Liu chaotic system

Liu chaotic system was proposed by Liu in 2004. The nonlinear term of Liu chaotic system is the multiply and the square. It is a three-dimensional chaotic system. The mathematical expressions of Liu chaotic system are described in equation (12)~(14).

$$\dot{x} = a(y - x) \tag{13}$$

$$\dot{y} = bx - kxz \tag{14}$$

$$\dot{z} = -ca + hx^2 \tag{15}$$

$a=10$, $b=40$, $c=2.5$, $h=4$, $k=1$.
The optimized circuit in this paper uses resistor, capacitor, operational amplifier(LM353) and multiplier(AD633). The op amp and multiplier are supplied by a power supply of ±15V.

The original 8-op amp 2 multiplier was optimized to a 3-op amp 2 multiplier without changing the original system equation. By reducing the number of op amps, power and size of the circuit are saved, while reducing the complexity of the hardware circuit. The hardware circuit designed in this way can realize the Liu Chaotic system.

Equations (13)~(15) are composed of three differential equations. An operational amplifier can realize an integral operation, so Liu chaotic system can be realized by at least three operational amplifiers. According to the operational rules of the operational amplifier, the corresponding relationship between the parameters in equations (13)~(15) and the capacitance and resistance parameters of the circuit can be obtained:

$$a = \frac{1}{R_1 C_1} \tag{16}$$

$$b = \frac{1}{R_3 C_3} \tag{17}$$

$$k = \frac{R_5}{R_3 C_3 (R_4 + R_5)} \tag{18}$$

$$c = \frac{R_9}{R_6 C_4 R_8 R_9} \tag{19}$$

$$h = \frac{1}{R_6 C_4} \tag{20}$$

$R_1 = R_2 = 100k\Omega$, $R_3 = R_4 = 25k\Omega$, $R_5 = 510\Omega$, $R_6 = R_7 = 250k\Omega$, $R_8 = 820\Omega$, $R_9 = 1k\Omega$ $R_{10} = 1.5k\Omega$, $R_{11} = 1k\Omega$, $C_1 = C_2 = C_3 = C_4 = C_5 = 10nF$.

The optimization circuit consists of a three-way differential integration amplifier circuit and two multiplier circuits. First way consists of R_1, R_2, C_1, C_2, R_{10}, R_{11} and U1A; Second way consists of R_3, R_4, C_3, R_5 and U1B; Third way consists of R_6, R_7, C_4, C_5, R_8, R_9 and U2A. The outputs of U1A, U1B, U2A are the x, y and z signals of Liu chaotic system, respectively. Fig. 12 is the simulation schematic of Liu chaotic system:

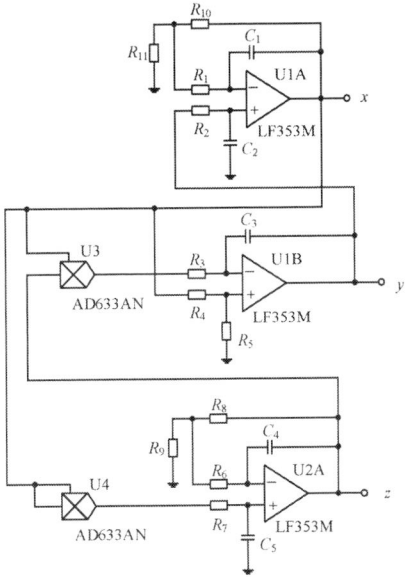

Fig. 12: Simulation schematic of Liu chaotic system

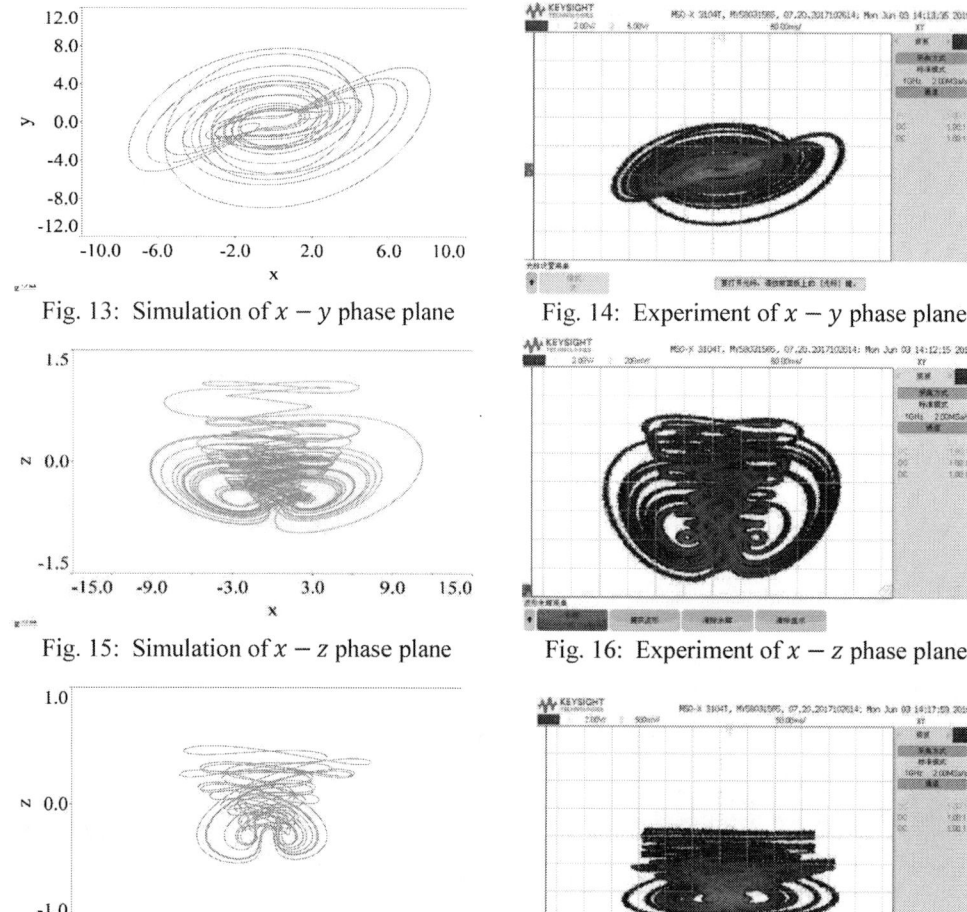

Fig. 13: Simulation of $x-y$ phase plane

Fig. 14: Experiment of $x-y$ phase plane

Fig. 15: Simulation of $x-z$ phase plane

Fig. 16: Experiment of $x-z$ phase plane

Fig. 17: Simulation of y-z phase plane

Fig. 18: Experiment of y-z phase plane

The above Figs. (13)~(18) are the phase plane comparison between the simulation and the hardware circuit experiment of Liu's chaotic system. The waveforms of three phase plane diagrams of x-y, x-z and y-z measured by oscilloscope are consistent with the numerical simulation, thus the correctness of the optimization circuit is verified.

Chaotic suppression EMI on power management IC

The prototype used in the experiment is a flyback converter designed and based on a power management chip containing chaotic spread spectrum function. The circuit is shown in Fig. 19. The operating frequency of the power management chip is set at 90kHz, the jittering frequency range is 6kHz, the input range is 80-230VAC, and the output is 5V/1A.

This power management chip only determines the frequency of PWM signal output by the external RI resistor on the 5-pin of the chip. According to the data manual of the chip, a 2nF capacitor can be connected in parallel to the RI resistor to turn on the periodic jitter function. The jitter frequency signal inside the chip is a triangular wave with fixed frequency

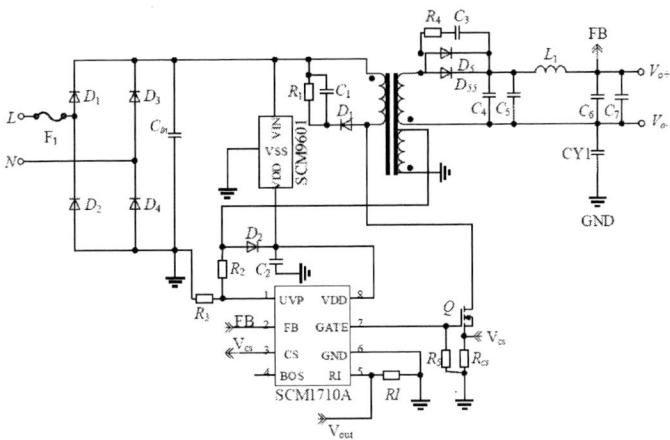

Fig. 19: Experiment circuit

The encapsulated chaotic power management chip shows in Fig. 20. The 8-pin package is the same as the common power management chip. Chaotic PWM mode is operated after the chip works. The package contains the complete optimization circuit of Liu chaotic system

Fig. 20 Power management IC for chaos

Chaotic PWM can be realized by injecting chaotic signals into the pin5 of the power management chip. In the experiment, the *x* signal of the Liu Chaotic system is used as a chaotic modulation signal. The fixed frequency, periodic spread spectrum and chaotic spread spectrum are tested respectively. As can be seen from Fig. 21, the periodic spread spectrum can suppress the conducted EMI by 5-10 dBμV. In 150kHz~1MHz, chaotic spread spectrum can suppress the conducted EMI 3~8dBμV more than periodic spread spectrum. In the frequency range of 1MHz~30MHz, chaotic spread spectrum can suppress the conducted EMI 3~5 dBμV more than periodic spread spectrum.

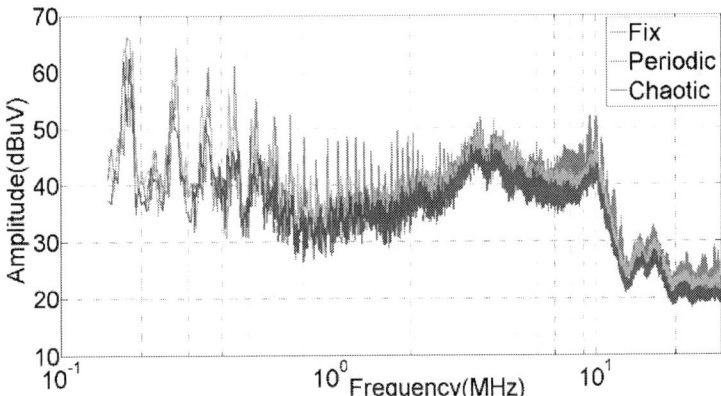

Fig. 21: Conducted EMI comparison

Conclusion

Firstly, an EMI model is established based on its transmission path and high frequency parasitic parameters of components. Simulation of conducted EMI of the low frequency domain below 10MHz basically coincides with the curve of test. The high frequency domain above 10MHz in simulation can basically meets the trend of the measured curve, but the peak value is not accurate perfectly. Additionally, this paper also optimizes the realization circuit of Liu chaotic system, simplifies the original circuit to 3-op amp, and reduces volume and the actual power consumption of the hardware circuit. Furthermore, combining the optimized circuit and power management IC, we make out the first chaotic power management IC. Through experiments, using the Liu chaotic signal as the modulation signal to suppress conducted EMI has better effect than the periodic spread spectrum.

References

[1] Yo-Chan Son and Seung-Ki Sul.: Conducted EMI in PWM inverter for household electric appliance, IEEE Transactions on Industry Applications, Vol. 38, no. 5, pp. 1370-1379, Sept.-Oct. 2002.

[2] H. Li, Z. Li, B. Zhang, F. Wang, N. Tan and W. A. Halang.: Design of Analogue Chaotic PWM for EMI Suppression, IEEE Transactions on Electromagnetic Compatibility, Vol. 52, no. 4, pp. 1001-1007, Nov. 2010.

[3] Majid, J. Saleem and K. Bertilsson.: EMI filter design for high frequency power converters, 2012 11th International Conference on Environment and Electrical Engineering, Venice, 2012, pp. 586-589.

[4] R. Yang, B. Zhang, F. Li and J. J. Jiang.: Experiment Research of Chaotic PWM Suppressing EMI in Converter, 2006 CES/IEEE 5th International Power Electronics and Motion Control Conference, Shanghai, 2006, pp. 1-5.

[5] Q. Liu, F. Wang and D. Boroyevich.: Conducted-EMI Prediction for AC Converter Systems Using an Equivalent Modular–Terminal–Behavioral (MTB) Source Model, IEEE Transactions on Industry Applications, Vol. 43, no. 5, pp. 1360-1370, Sept.-oct. 2007.

[6] Q. Liu, F. Wang and D. Boroyevich.: Modular-Terminal-Behavioral (MTB) Model for Characterizing Switching Module Conducted EMI Generation in Converter Systems, IEEE Transactions on Power Electronics, Vol. 21, no. 6, pp. 1804-1814, Nov. 2006.

[7] H. Zhang and S. Wang.: EMI noise source modeling based on network theory for power converters with mixed-mode characterization, 2018 IEEE Applied Power Electronics Conference and Exposition (APEC), San Antonio, TX, 2018, pp. 984-991.

[8] A. C. Baisden, D. Boroyevich and F. Wang.: Generalized Terminal Modeling of Electromagnetic Interference, IEEE Transactions on Industry Applications, Vol. 46, no. 5, pp. 2068-2079, Sept.-Oct. 2010.

[9] R. Yang, B. Zhang, D. Qiu and Z. Liu.: Time–Frequency and Wavelet Transforms of EMI Dynamic Spectrum in Chaotic Converter, IEEE Transactions on Power Electronics, Vol. 24, no. 4, pp. 1083-1092, April 2009.

[10] J. H. B. Deane and D. C. Hamill.: Improvement of power supply EMC by chaos, Electronics Letters, Vol. 32, no. 12, pp. 1045-, 6 June 1996.

[11] C. Liu, T. Liu, L. Liu, K. Liu.: A new chaotic attractor, Chaos Solitons & Fractals, Vol. 22, pp. 1031-1038, Jan, 2004.

[12] F. Q. Wang, C. X. Liu.: Studies on Liu chaotic system and its experimental confirmation, ACTA Physica Sinica, Vol. 55, no. 10, Oct, 2006.

High Performance Drive Inverter
for an Electric Turbo Compressor in Fuel Cell Applications

N. Langmaack, G. Tareilus, R. Mallwitz
Technische Universität Braunschweig
Institute for Electrical Machines, Traction and Drives
Hans-Sommer-Straße 66, 38106 Braunschweig
Germany

E-Mail: n.langmaack@tu-bs.de
URL: https://www.tu-braunschweig.de/imab

Acknowledgements

This work contributes to the project „Nationales Innovationsprogramm Wasserstoff- und Brennstoffzelle (NIP) - Phase II: Aufladung für Brennstoffzellensysteme durch interdisziplinär entwickelte elektrische Luftverdichter (ARIEL)" funded by the German Federal Ministry of Transport and Digital Infrastructure under the project number „03B10105D". The authors would like to thank all colleges working in the project and acknowledge the funding by the German government.

Keywords

High-speed drive, Silicon Carbide (SiC), Gallium Nitride (GaN), High power density systems, Fuel cell system

Abstract

In this paper the design of a high performance drive inverter with special requirements in terms of switching speed, power density and efficiency is presented. Solutions using different topologies, device technologies and packages are compared under the constraints given by the application. Finally an application specific optimum is found.

Introduction

Fuel cells for automotive applications need pressurized air to achieve an efficient operation and a high power density. Typically this task is performed by an electric turbo compressor system. A simple schematic of such a system is shown in figure 1. The integrated electric machine is fed by the drive inverter investigated in this paper. The 15 kW high speed (120 krpm) permanent magnet synchronous motor has a high fundamental frequency of 2 kHz and a low inductance of the stator windings. Since high current ripple will induce high additional losses in the permanent magnet rotor of the machine a high switching frequency of the inverter is needed. First estimations show that a switching frequency of at least 30 kHz for a 3-level inverter topology or 60 kHz using a 2-level topology is necessary.

The inverter is fed from a DC bus with up to 500 V making it a so called high voltage automotive component. Taking into account conventional silicon based devices as well as available wide band gap power semiconductors a very wide range of solutions can be found. They may base on different circuit topologies like 2-level, 3-level NPC or 3-level T-type, different device technologies like Si IGBT, SiC MOSFET or even GaN HEMT and also different packaging types like power modules or discrete packages.

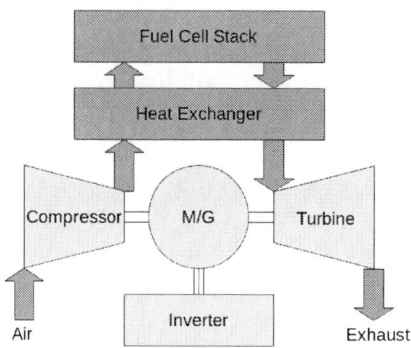

Figure 1: Electric turbo-compressor unit

In the design and optimization process, the different solutions are compared with each other using the results of a detailed loss calculation and different other criteria like predicted size, cost and overall complexity. In a second step the most promising approaches are evaluated in practical designs and application-oriented measurements. Finally a fully functional inverter prototype will be designed and build.

A comparison of different topologies and device technologies for a grid-tied application with an additional focus on the line filter has been performed in [1]. It is stated that the multilevel inverter has the better performance unless a high output fundamental frequency is needed, which is the case in this high speed drive application. A parallel branch of the ongoing research project investigates the reliability of the selectable devices for the high speed drive inverter. The results will be published separately. For a similar investigation see [2]. In terms of efficiency and power density references [5] to [8] are seen as benchmarks with different requirements.

Design and Optimization Process

To find a highly optimized solution under special application specific constraints a detailed comparison of a wide range of possible designs is performed. A novel loss calculation method is utilized which allows a simple and fast characterization of the semiconductor losses. It is based on a detailed analytical loss calculation for steady state DC operating points. Numerical integration is used to estimate the medium losses in AC inverter operation. The method supports different circuit topologies, device technologies and modulation schemes. It takes into account different effects like dead time, reverse conduction of unipolar devices and the current distribution in different parallel devices. The implemented equivalent circuit of the switching devices is able to represent any kind of power semiconductor device. Therefore it allows a fair comparison of different device technologies like silicon IGBT and diode, silicon carbide MOSFET with or without antiparallel schottky diode, gallium nitride HEMT or any kind of hybrid constellation. [3]

The performed process can be described by the following steps:
- Define design constraints and operating points
- Define possible solutions by circuit topologies and power devices
- Perform the loss calculation for all investigated solutions
- Define and weight evaluation criteria
- Select the most promising solutions for the next step: Practical evaluation

Design constraints

Table 1 summarizes the most important requirements on which the design of the inverter is based. The DC input voltage range corresponds to the HV_2b standard for electric vehicle power systems. The parameters for the AC load are defined by the electric machine. Compared to conventional industrial drives the fundamental frequency is very high. Because of the special requirements of the machine also a high switching frequency is needed to limit the distortion of the phase currents and the additional losses in the machine. Table 2 shows the typical operating points of the inverter which are important for the rating and selection of the different solutions. The operating points OP (1) and OP (2) represent typical static full load operation. OP (3) is an example for a dynamic acceleration process which is a short term overload condition.

DC voltage range	250..500 V
Load current	< 60 A_{rms}
Load power	15 kW
Load frequency	2000 Hz
Switching frequency	> 30 kHz (3-level reference)
Interfaces	CAN, resolver, HV-interlock
Ambient temperature	-25..90 °C
Coolant	water/glycol, -25..70 °C

Table 1:
Design constraints for the inverter

DC input voltage	400 V
Switching frequency	30/60/120 kHz
Load current (1)	30 A_{rms}
Modulation index (1)	0.9
Load current (2)	42 A_{rms}
Modulation index (2)	0.9
Load current (3)	60 A_{rms}
Modulation index (3)	0.5

Table 2:
Most typical operating points

Definition of Possible Designs

The choice of possible designs for the further investigations is performed according to the previously defined requirements. Due to the input voltage level, multi-level circuit topologies with more than three output voltage levels are not considered to be economical. For the topologies the investigation is therefore limited to 2-level, 3-level NPC type and 3-level T-type inverters. With 3-level designs a switching frequency of 30 kHz is found to be sufficient to limit the harmonic content of the output voltages. This can still be realized with conventional silicon IGBT technology. For the simpler 2-level topology a switching frequency in the range from 60 kHz to 120 kHz will be needed. With silicon IGBTs there will not be an efficient solution due to the massive switching losses. Therefore a strong focus of the investigation is on wide-band-gap power devices.

In the power range of 15 kW both power module and discrete devices based solutions are feasible. Especially for the wide-band-gap devices the availability of discrete devices is much higher. Another important aspect is the automotive qualification of the devices. The investigation is not limited to fully qualified devices, but these are preferred whenever possible.

The devices actually selected for the comparison are chosen based on their respective nominal ratings. A wide range of different suppliers and devices is considered. One noticeable fact is the large difference in the permissible DC link voltage for devices of the same nominal voltage class. Referring to the different suppliers, some 650 V devices can easily be used at 500 V DC due to a transient blocking voltage rating of 800 V, whereas usually 900 V or 1200 V devices would be used for this input voltage level. This is why there are silicon carbide MOSFETs with nominal ratings of 650 V, 900 V and 1200 V as well as gallium nitride devices of the 650 V class in the comparison for the same topology and voltage level.

Loss Estimation Process

The utilized loss estimation method is based on analytical loss models and numerical integration of the different loss terms over one fundamental period. Details can be found in [3]. The method is especially useful for a fair comparison of devices of different semiconductor technologies and for large parameter studies with many different devices.

The calculations result in efficiency maps for every investigated inverter configuration. Figures 2 and 3 show exemplary results for the comparison of different discrete 2-level inverter solutions using wide-band-gap power semiconductor devices. They also show the most important operating ranges for the future static and dynamic operation of the inverter.

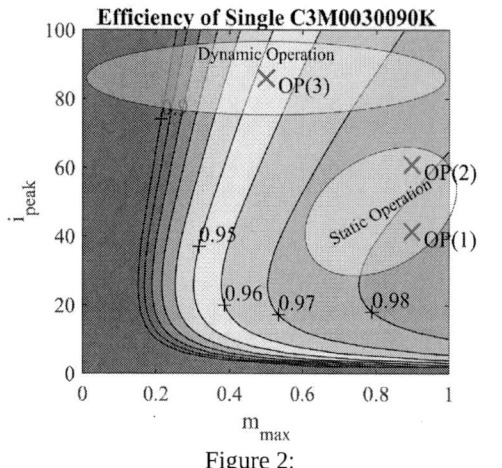

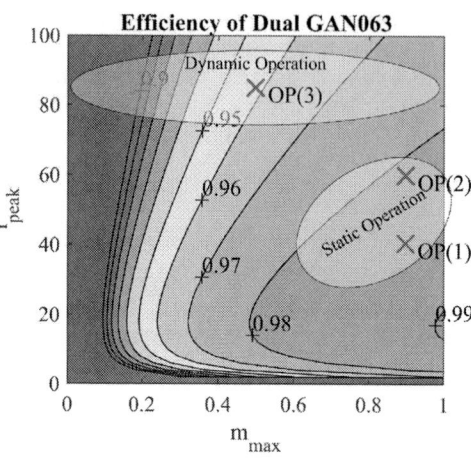

Figure 2:
Efficiency map of 2-level converter using 900V SiC MOSFETs in a discrete package [4]

Figure 3:
Efficiency map of 2-level converter using discrete 650V cascoded GaN HEMTs

Comparison and Evaluation

The selection of the most promising inverter designs and power semiconductor devices is performed using a detailed evaluation matrix with weighted criteria like

- estimated efficiency,
- estimated volume or power density,
- system complexity,
- qualification of the devices and
- assembly aspects.

A 3-level T-type inverter using silicon IGBTs and diodes in a power module package is used as the reference design. Due to the prospective increase of power density, reduction of system complexity and the good availability of AEC-Q101 qualified devices the most selected 2-level wide-band-gap inverter designs are using discrete devices, preferably in SMD packages for easy assembly. Figure 5 shows an overview of the semiconductor losses of the different inverter designs with wide-band-gap devices at a switching frequency of 60 kHz. Figure 6 shows the respective results at 120 kHz. A summary of the losses of some selected solutions at the different switching frequencies is given in figure 4 for easier comparison.

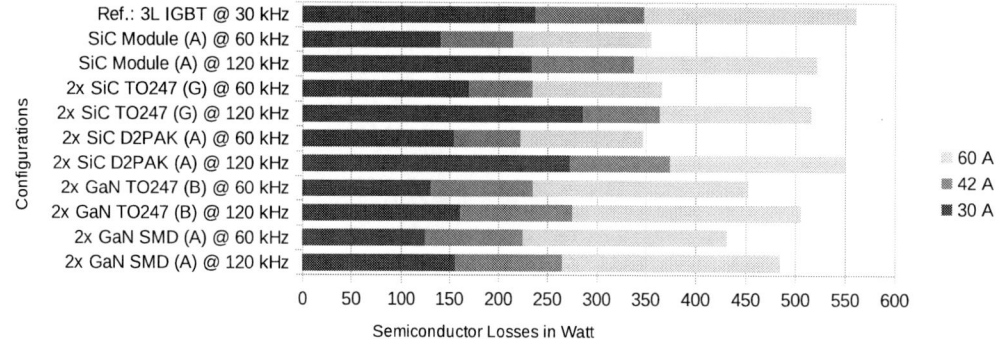

Figure 4: Summary of semiconductor losses for selected solutions

Using silicon carbide MOSFET power modules can result in inverter designs with good efficiency and high power density. The total space occupied by the SiC modules is about 1/3 to 2/3 of the area of the

T-type modules used as reference. Using power modules also allows an easy and robust thermal design with a ceramic DCB as electrical isolation. Unfortunately there is still no suitable module with AEC-Q grade availabe.

The widest range of available devices can be found for SiC-MOSFETs in TO247 packages. It has to be recognized that there are still significant differences in terms of $R_{DS,on}$ and switching losses for the different suppliers. Using two devices in parallel, most of the MOSFETs will outperform the efficiency of the reference solution at 60 kHz. At 120 kHz the switching losses of the devices in the large leaded standard package turn out to be a problem. Even with paralleled devices the total losses are high.

Using SiC MOSFETs in a standard SMD package like the D2PAK-7L seems to be a very convenient solution to the present application. With two devices in parallel ("2x SiC D2PAK (A)" or "2x SiC D2PAK (B)") the losses are fairly low in the whole range of switching frequencies from 60 kHz to 120 kHz and the achievable power density is very high. There are qualified devices available from different suppliers and the assembly and soldering process is simple. This also makes the solution very interesting in terms of cost.

Compared to SiC, the available GaN devices have higher $R_{DS,on}$ but lower switching losses. Therefore paralleling of devices is inevitable to handle the current. In return especially the solutions with two or three SMD devices (A) will result in the highest efficiency at 120 kHz by far. There is still a lack of suppliers for devices in compatible packages, but the technology will definitely be an option for high speed drives with special requirements in terms of switching frequency.

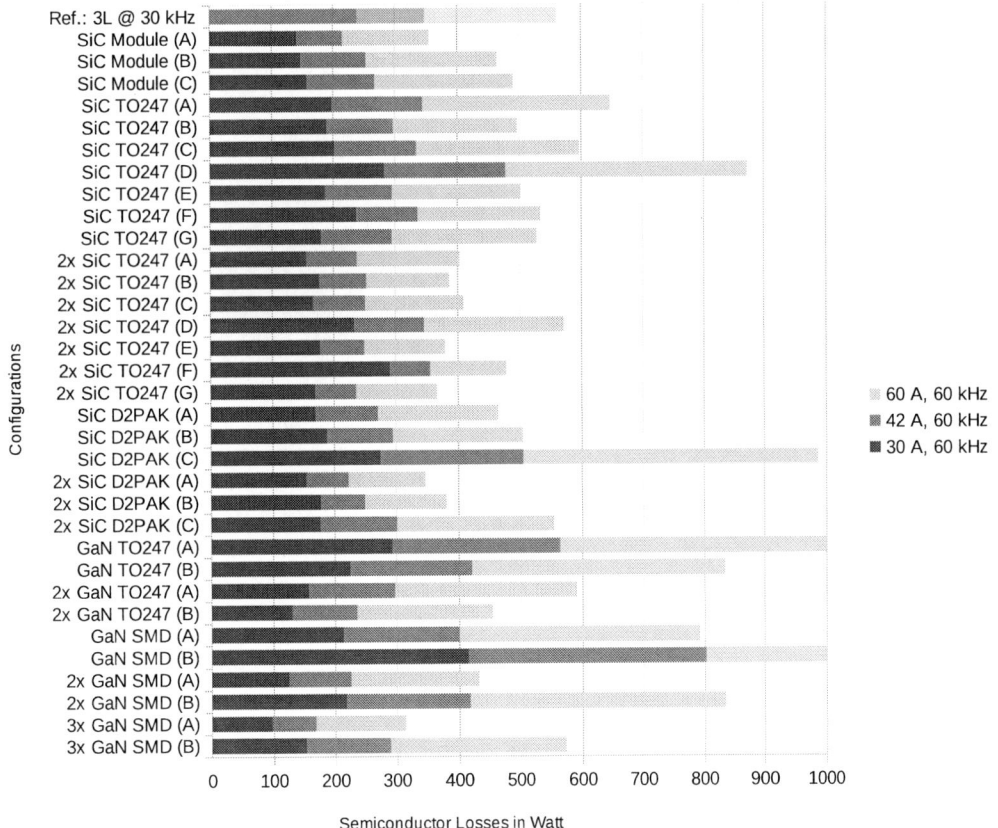

Figure 5: Losses of 2-level inverter designs in different operating points
at 60 kHz switching frequency

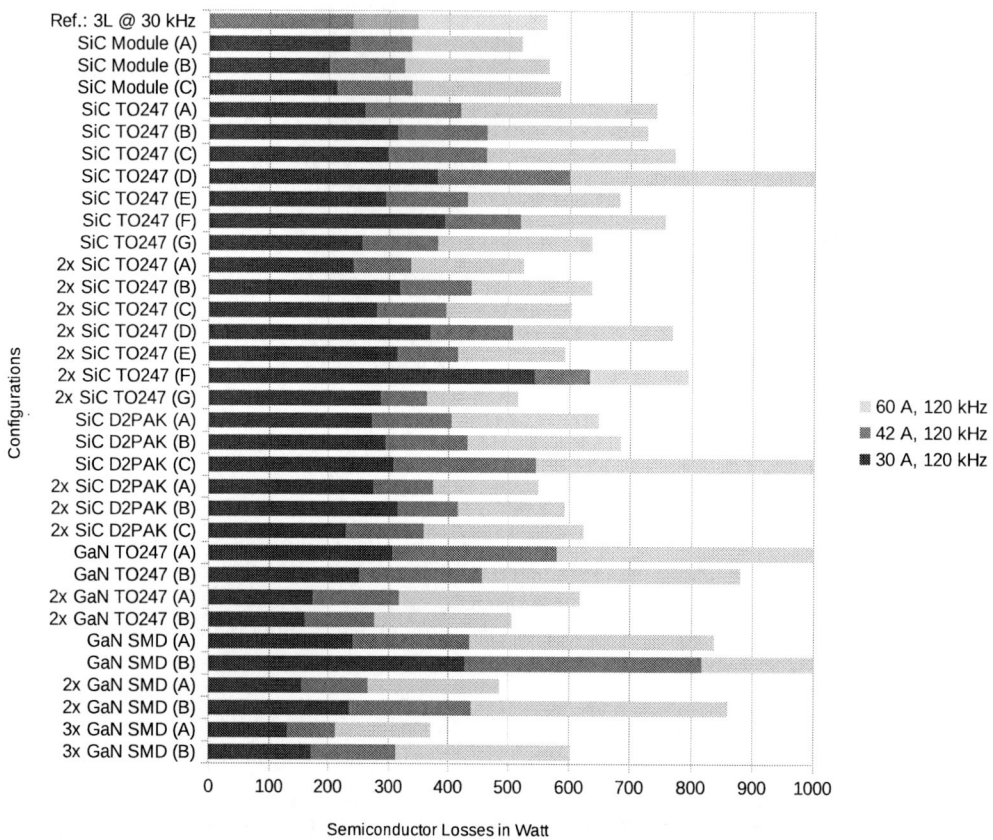

Figure 6: Losses of 2-level inverter designs in different operating points at 120 kHz

Practical Evaluation using Prototype Power Stage Designs

To evaluate the chosen power semiconductor devices, the developed loss calculation method, the implemented DC link and the gate driver designs, power stages using different power devices are designed and built. They are supposed to be operated in an inverter test bench for functional, efficiency and thermal characterization.

The evaluation process follows these steps:
- PCB design including power devices, DC link and gate drivers
- Basic functional tests
- Double-pulse tests, tuning of gate resistors, overshoot characterization, self-EMI tests
- Efficiency measurements
- Thermal characterization

PCB Designs

Although the solution using two parallel SiC-MOSFETs in D2PAK-7L is preferred according to the complete evaluation matrix, a total number of five different PCB designs is realized. These are
- SiC Module (MiniSKiiP II six-pack)
- SiC SMD (D2PAK-7L)
- SiC TO247 (TO247-4L)
- GaN SMD (CCPAK88)
- GaN TO247 (TO247-GSD)

Most of the used components like the DC link capacitors and the gate driver circuits are the same for all designs. Multiple MKP film capacitors with short pin distance are used to form a DC link with low parasitic inductance and high resonance frequency. The use of ceramic snubbers is avoided initially due to the risk of unwanted oscillations. The gate driver circuit is based on insulated half bridge driver ICs with a very high common mode dv/dt rating and an output peak current capability of 4 A. All six driver channels are supplied with individual flyback converters with bipolar output voltages to create negative gate voltages for fast and safe turn-off of the MOSFETs.

All layouts are optimized for a low parasitic inductance of the commutation loop by a suitable arrangement of the devices and the use of large area copper polygons on multi layer PCBs with 6 to 8 layers. The equal dynamic and static load of paralleled devices is supported by symmetrical layouts, equal copper trace lengths and small gate loops. Kelvin source connections to all power devices are used whenever possible.

Especially the PCB designs for the SMD devices are also optimized for their thermal performance. The surface mounted devices are cooled solely through the PCB by a liquid cold plate mounted underneath the PCB. An extensive use of filled and plated thermal vias in an analytically optimized arrangement is used to reduce the thermal resistance to a minimum. A special challenge is the thermal interface between the bottom of the PCB and the cold plate. It needs to guarantee electrical insulation, a low thermal resistance and long-term stability. Typical thermal interface materials like gap pads additionally need a defined applied mechanical pressure to ensure the rated thermal resistance.

Figures 7 and 8 show examples of the PCB layouts for the discrete SiC MOSFET devices. The DC link capacitors are placed on the left-hand side. The driver circuitry follows on the right-hand side. The load terminals are designed as small as possible to minimize the parasitic capacitance of the switching node. The figures show only two copper layers, but most of the other layers have very similar layouts.

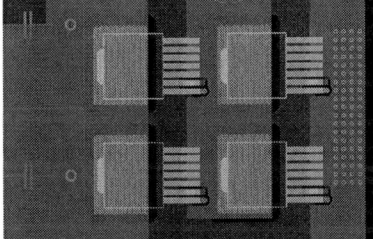

Figure 7: PCB layout of one half bridge using D2PAK-7L package for discrete SMD SiC MOSFETs

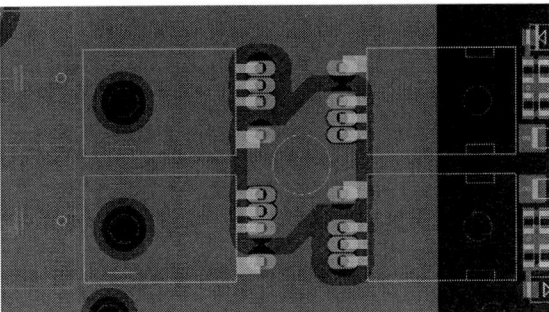

Figure 8: PCB layout of one half bridge using TO247-4L package for discrete leaded SiC MOSFETs

Basic functional and switching tests

Following the assembly all power stage prototypes are tested in a very basic way. The gate driver supply voltages are tuned and the voltage regulation and switching function of the gate driver circuits are investigated. Double-pulse tests with increasing DC link voltage are performed to investigate switching speed and voltage over-shoot performance.

Figure 9 shows a turn-off voltage transient of the utilized SiC MOSFET power module at the maximum applied DC link voltage of 820 V. This is far from the operating range specified by the application but can be handled by the 1200 V devices easily. The fastest transient recognized in this measurement is 16.4 V/ns. The maximum voltage overshoot is about 40 V at 820 V. This indicates a low parasitic inductance and an appropriate design of the DC link.

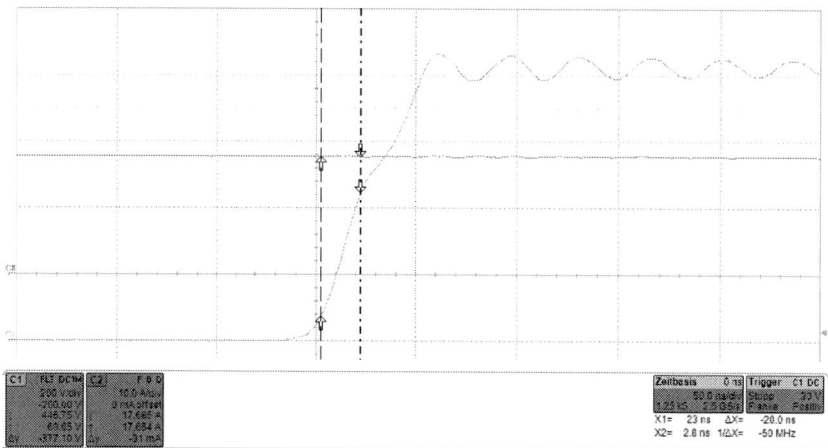

Figure 9: Turn-off gradient of SiC power module at 800V

Efficiency measurements

The efficiency measurements are performed in an inverter test bench using the components shown in figure 10. The precision power analyzer Yokogawa WT3000 is used to determine the DC and AC electric power and calculate the losses and efficiency of the device under test (DUT). The specific operating points are set by the 32 kW DC source Regatron TopCon GSS and a passive load. A liquid cooling with additional electric heater device is used to provide coolant with adjustable flow rate and temperature.

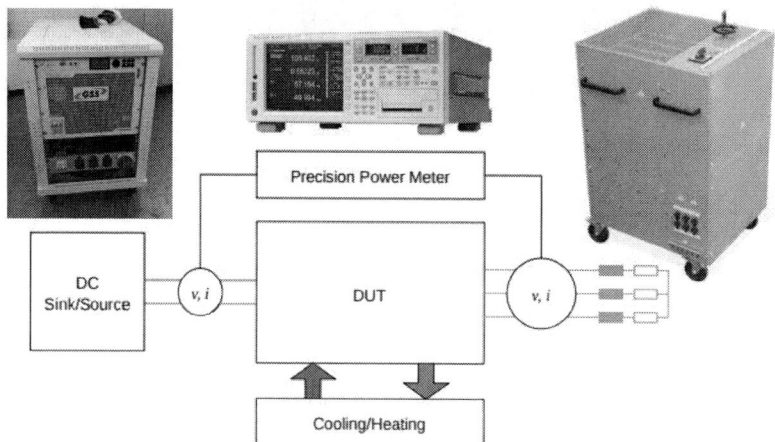

Figure 10: Drive inverter test bench setup and components

Figure 11 shows an exemplary measurement result. The efficiency map gives an overview on the operating points of the three-level silicon IGBT inverter at a switching frequency of 30 kHz. The obtained values match quite well with the calculated losses. In a typical full load operating point

(42 A, 220 V, $m = 0.9$, $\cos(\varphi) = 0.95$) the measured efficiency is about 97.1 %. The theoretical efficiency due to the calculated semiconductor losses for the utilized power module in the same operating point under optimal conditions is 97.7 %. The difference has to be explained with slightly increased switching losses and the additional losses in the DC link, the common mode filter elements and the connectors.

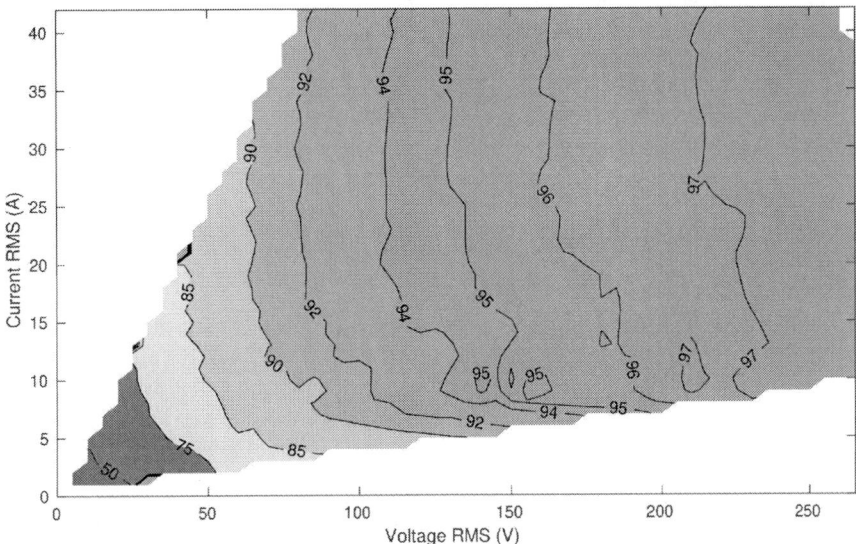

Figure 11: Efficiency map of silicon IGBT based three-level inverter

Thermal characterization

All the investigated designs are also thermally characterized. This is done in two different test setups. One is a special R_{th}/C_{th} test bench, the other is a continuous operation test in the inverter test bench. The measurement in the R_{th}/C_{th} test bench gives detailed information about the thermal path and its dynamic behaviour. A typical measurement result is shown in figure 12. The continuous operation test gives information about the static temperature of the power devices under full load operation.

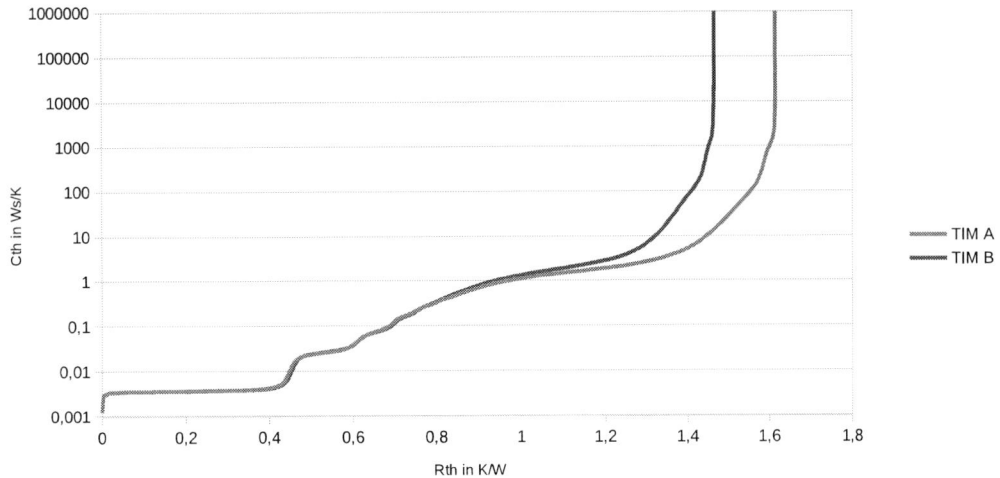

Figure 12: R_{th}/C_{th} characterization of the thermal path of silicon carbide MOSFETs in TO247 package using two different thermal interface materials (TIM)

Conclusions

An application specific optimization of a drive inverter for a fuel cell turbo compressor with special requirements in terms of frequency, efficiency and power density is performed and presented. A novel method for a fast and fair comparison of different circuit topologies and device technologies is utilized. Finally a simple topology using wide-band-gap power semiconductors is chosen as a promising solution to outperform the state-of-the-art silicon-based solution using a more complex circuit topology.

Five different PCB layouts for different device technologies are designed, optimized, built and set into operation for further evaluation. The most promising solution is based on silicon carbide power MOSFETs in a standard SMD package. The complete efficiency measurements of all designs are yet to be finished. After the detailed characterization of these different power stage prototypes an optimized complete inverter design will be realized. The collected knowledge will be ported to this full inverter design including all necessary periphery and control electronics needed for the fuel cell turbo compressor application.

In this project an inverter with a high technology readiness level (TRL), ready for real in-vehicle operation is finally designed. Therefore investigations on the reliability of the used power semiconductors are performed in a parallel branch of the project and will be published separately.

References

[1] M. Karami, R. Tallam, R. Cuzner, "Comparison of Three-Level and Two-Level Converters for AFE Application", WiPDA 2018 - 6th IEEE Workshop on Wide Bandgap Power Devices and Applications, 31 Oct.-2 Nov. 2018, Atlanta, GA, USA

[2] J. Colmenares, D.-P. Sadik, P. Hilber, H.-P. Nee, "Reliability Analysis of a High-Efficiency SiC Three-Phase Inverter for Motor Drive Applications", APEC 2016 – 31st Annual IEEE Applied Power Electronics Conference and Exposition, 20-24 Mar. 2016, Long Beach, CA, USA

[3] N. Langmaack, T. Schobre, M. Henke, "Fast and Universal Semiconductor Loss Calculation Method", PEDS 2019 - IEEE 13th International Conference on Power Electronics and Drive Systems, 09-12 Jul. 2019, Toulouse, France

[4] V. Pala, G. Wang, B. Hull, S. Allen, J. Casady, J. Palmour, "Record-low 10mΩ SiC MOSFETs in TO-247, Rated at 900V", APEC 2016 – 31st Annual IEEE Applied Power Electronics Conference and Exposition, 20-24 Mar. 2016, Long Beach, CA, USA

[5] S. Yin, K. J. Tseng, C. F. Tong, R. Simanjorang, C. J. Gajanayake, A. K. Gupta, "A 99% Efficiency SiC Three-phase Inverter Using Synchronous Rectification", APEC 2016 – 31st Annual IEEE Applied Power Electronics Conference and Exposition, 20-24 Mar. 2016, Long Beach, CA, USA

[6] J.-K. Müller, A. Mertens, "Power electronics design for a direct-driven turbo compressor used as advanced high-lift system in future aircraft", IECON 2017 - 43rd Annual Conference of the IEEE Industrial Electronics Society, 29 Oct.-1 Nov. 2017, Beijing, China

[7] N. Langmaack, G. Tareilus, M. Henke, "SiC Drive Inverter using Intelligent Gate Drivers and Embedded Current Sensing", PWTE 2019 - SIA Power Train & Electronics Conference, 12-13 Jun. 2019, Port-Marly, France

[8] K. Yamaguchi, "Design and Evaluation of SiC-Based High Power Density Inverter, 70kW/liter, 50kW/kg", APEC 2016 – 31st Annual IEEE Applied Power Electronics Conference and Exposition, 20-24 Mar. 2016, Long Beach, CA, USA

Development of an Algorithm for the Automation of the Modelling Process of Power Converters

Jon Anzola, Iosu Aizpuru and Asier Arruti
MONDRAGON UNIBERTSITATEA
Orona Ideo - Fundazioa eraikuntza; Jauregi Bailara, z.g.; 20120 Hernani (Gipuzkoa)
E-Mail: janzola@mondragon.edu

Keywords

«Power converters modelling», «GSSAM», «Switched model», «DC-DC», «DC-AC-DC».

Abstract

By the time goes by, the energetic efficiency gains more and more importance inside the strategic plans of companies that are focused on energy storage and conversion. For this reason, these same companies seek the necessity of developing algorithms capable of modelling and simulating the behaviour of power converters. Indeed, by modelling, it is possible to foresee converters' response to different input signals and comprehend better their behaviour without building a prototype. However, the process of obtaining the mathematical representation of a converter is high time consuming and, in case several topologies need to be compared, it can result in a laborious procedure. For this reason, this project has focused on the idea of developing a tool that could automatize the modelling process of a given converter. The presented algorithm is based on GSSAM and, in order to see its flexibility and precision, the results are compared to a switched model, validating the algorithm.

Introduction

Modelling a power converter means describing its behaviour by an approximated mathematical representation that is based on the physical laws that define the environment of the model [1]. In the case of power converters, these laws would be, for example, Kirchhoff's laws, Ohm's law and energy conservation laws. When it comes to comparing different modelling methods, the switched model is the less laborious and the most exact one. However, it is very complex to linearize it, which is essential for classical control implementation. On the other hand, there is the averaged model and, as its name indicates, this modelling method calculates a "sliding" average value for the desired variable during a switching period. Compared to the switched model, an averaged model is less precise, but, the implementation of classical control on it, it is much easier. Inside averaging process, there exist 2 main variants: classical state space averaging modelling (SSAM) and generalized state space averaging modelling (GSSAM). SSAM helped to describe power converters' waveforms by only considering the average value of the signal, but, this modelling method is not convenient for DC-DC converters where AC signals appear (for example, a Flyback converter) and for applications where high accuracy is required. In order to solve this problem, GSSAM is a possible solution, since it considers up to a finite number of harmonics [2], [3]. In [4], [5], GSSAM is presented for different non-isolated power converters analysis, where it is concluded that the approximation order is important for improving the model accuracy. However, the authors do only show results for first order approximation, arguing a high calculation cost. In [6], not only the GSSAM for a DAB converter is presented, but also bode plots are obtained and compared with real prototypes. There, it is concluded that, after a linearization process, GSSAM is very accurate for frequency-response analysis, which can be useful for classical control designing. In addition, [7] proposes several improvements to extent the functionalities of the modelling method presented in [6]. However, still, in [6] and [7] there is a first order approximation. In [8], authors detail a method for the automation of the calculation process of the GSSAM matrixes for a given number of harmonic. This way, the process of building the GSSAM matrixes is simplified and the accuracy of the model is increased. As it has been observed in the literature, GSSAM is a widely used method for power converter modelling and control designing. However, it can turn into a laborious mathematical process if high order approximations are required. Therefore, following the idea of [8] (where part of the modelling process is automatized), this project is motivated by the necessity of developing a tool that can automatize the hole modelling process. In other words, having the differential equations of a power converter as a starting point, it is required an algorithm that can return the GSSAM of the given converter. This way, it is expected to simplify the modelling process of any converter and shorten the working times.

GSSAM basis

Usually, when it comes to DC-DC converters analysis, SSAM is used to obtain the main voltage and current waveforms of a given converter. This averaging method neglects the ripple that could exist in the waveforms and returns a pure DC signal without taking into account a single harmonic. However, there exist different DC-DC converters that contain AC signals and that cannot be modelled with SSAM, for example: DAB and Flyback converters. In order to solve this issue, a different modelling method that could take into account AC signals and their harmonics must be used. An example of it is the GSSAM [1]. This type of method is based on the Fourier series and offers the possibility of modelling converter's signals up to the desired number of harmonic. Through this chapter, the mathematical basis that supports GSSAM will be explained.

As mentioned, the GSSAM is based on representing the waveforms of the different signals inside the converter by making use of the Fourier series (1).

$$x(\tau) = \sum_{k=-\infty}^{\infty} \langle x \rangle_k (\tau) \cdot e^{j \cdot k \cdot \omega_s \cdot t} \tag{1}$$

Where $\langle x \rangle_k(t)$ represents the complex Fourier series coefficient of the kth harmonic and $\omega_s = 2 \cdot pi \cdot f_{switching}$. Equation (1) can be also defined as:

$$\langle x \rangle_k (t) = \frac{1}{T} \cdot \int_0^T x(\tau) \cdot e^{-j \cdot k \cdot \omega_s \cdot \tau} \cdot d\tau \tag{2}$$

From equations (1) and (2) two fundamental properties can be obtained. One is the derivative of a function represented in Fourier form, shown in (3). And the second is the product of two time-variable terms, shown in (4).

$$\frac{d}{dt} \langle x \rangle_k (t) = \left\langle \frac{d}{dt} x \right\rangle_k (t) - j \cdot k \cdot \omega_s \cdot \langle x \rangle_k (t) \tag{3}$$

$$\langle fg \rangle_k = \sum_{i=-\infty}^{\infty} \langle f \rangle_{k-i} \cdot \langle g \rangle_i \tag{4}$$

Since the coefficients k and $-k$ in Fourier series are complex conjugates [8], equation (4) is rewritten in two parts: when $k = 0$ (5) and when $k \neq 0$ (6).

$$\langle fg \rangle_0 = \langle f \rangle_0 \cdot \langle g \rangle_0 + 2 \cdot \sum_{k=1}^{\infty} \langle f \rangle_{kR} \cdot \langle g \rangle_{kR} + \langle f \rangle_{kI} \cdot \langle g \rangle_{kI} \tag{5}$$

$$\langle fg \rangle_{kR} = \Re \left(\sum_{i=-\infty}^{\infty} \langle f \rangle_{k-i} \cdot \langle g \rangle_i \right) ; \ \langle fg \rangle_{kI} = \Im \left(\sum_{i=-\infty}^{\infty} \langle f \rangle_{k-i} \cdot \langle g \rangle_i \right) \tag{6}$$

Where $\Re$ and $\Im$ represent the real and imaginary parts of a complex number.

GSSAM algorithm

First of all, it must be considered that every power converter can be described by equations (7)-(9):

$$\frac{dx_i(t)}{dt} = A \cdot x_i(t) + B \cdot u_i(t) \tag{7}$$

$$x_{i+1}(t) = x_i(t) + \frac{dx_i(t)}{dt} \cdot T_s \tag{8}$$

$$y_i(t) = C \cdot x_i(t) \tag{9}$$

Where A, B and C are coefficient matrixes, T_s is the time step, $x_i(t)$ is the state variables vector and $u_i(t)$ is the input variables vector.

In order to solve the equations above, $x_i(t)$ vector is initialized (usually full of zeros) and then, an iteration process is followed. Up to this point, the modelling process seems simple. However, the most complex part is related with the way in which matrixes A, B and C are built. Indeed, the sizes of these matrixes change in function of the desired number of harmonics (n_h), the number of state variables (n_{SV}) and the number of input variables (n_{IV}). To be more precise, A is a hxh matrix, B is a hxn_{IV} matrix and C is a $n_{SV}xh$ matrix, where h is defined in (10).

$$h = \left(2 \cdot n_h + 1\right) \cdot n_{SV} \tag{10}$$

Through the next subsections, the automation of the building process of matrixes A, B and C is explained.

Building process of matrix A

As it is shown in (7), matrix A only contains the coefficients that multiply with state variables. Therefore, in order to build matrix A, the only coefficients that are required from a differential equation are the ones that multiply with a state variable. Equation (11) represents a differential equation of a power converter in Fourier's form, where only the state variables are considered.

$$\frac{d}{dt}\langle f \rangle_k = C_1 \cdot \langle h \rangle_k + C_2 \cdot \langle f \rangle_k + C_3 \cdot \langle s \rangle_k \cdot \langle j \rangle_k \tag{11}$$

Where C_1, C_2 and C_3 are constants, $\langle s \rangle_k$ is a switching function and $\langle h \rangle_k$, $\langle f \rangle_k$ and $\langle j \rangle_k$ are state variables. The first step consists of defining an identification number from zero to $n_{SV} - 1$ for each state variable:

- $n_{SV_h} = 0$. Identification number for state variable $\langle h \rangle_k$.
- $n_{SV_f} = 1$. Identification number for state variable $\langle f \rangle_k$.
- $n_{SV_j} = 2$. Identification number for state variable $\langle j \rangle_k$.

Then, analysing equation (11), 2 types of terms can be identified:

1. Single type. Consists of a constant multiplied by a state variable, for example: $C_1 \cdot \langle h \rangle_k$ and $C_2 \cdot \langle f \rangle_k$.
2. Double type. Consists of a constant and a switching function multiplied by a state variable, for example: $C_3 \cdot \langle s \rangle_k \cdot \langle j \rangle_k$.

Each term type follows a different matrix building procedure, so, the explanation will be divided in 2 parts: Single type terms and Double type terms.

Single type terms

Single type terms are those that include a constant value multiplied by a state variable. At this subchapter, the steps to place them inside matrix A are explained. For the explanation, equation (12) will be taken as an example (this term comes from (11)).

$$\frac{d}{dt}\langle f \rangle_k = C_1 \cdot \langle h \rangle_k \tag{12}$$

In first place, the values that correspond to the DC term ($k = 0$) of $\langle f \rangle_k$ are placed in matrix A (13).

$$A_{i,j} = C_1 \tag{13}$$

The values of i and j are defined in (14).

$$i = \left(2 \cdot n_h + 1\right) \cdot n_{SV_f} + 1$$
$$j = \left(2 \cdot n_h + 1\right) \cdot n_{SV_h} + 1 \tag{14}$$

Where n_{SV_f} is the identification number of $\langle f \rangle_k$ (the state variable that is in derivative form (12)), n_{SV_h} is the identification number of $\langle h \rangle_k$ (the state variable that is multiplying the constant (12)) and n_h is the number of harmonic defined by the user.

In second place, the terms of $\langle f \rangle_k$ that go from the first harmonic ($k = 1$) up to the desired one ($k = n_h$) are placed in matrix A, see (15).

$$A_{i,j} \ \& \ A_{i+1,j+1} = C_1 \tag{15}$$

In this case i and j are defined in (16).

$$\begin{cases} i = (2 \cdot n_h + 1) \cdot n_{SV_f} + 2 \cdot k \\ j = (2 \cdot n_h + 1) \cdot n_{SV_h} + 2 \cdot k \end{cases} \quad (16)$$

Where $k = 1, 2, \dots, n_h$.

Equations (13) and (15) correspond to the example shown in (12) and the same procedure should be applied for any other single type term inside the differential equation. However, there exists an exception. In case the state variable that is in derivative form and the state variable that is multiplying the constant coincide, as shown in (17), 2 new terms should be added (18). This corresponds to the negative term from equation (3).

$$\frac{d}{dt} \langle f \rangle_k = C_2 \cdot \langle f \rangle_k \quad (17)$$

$$A_{i,j+1} = \omega_s \cdot k$$
$$A_{i+1,j} = -\omega_s \cdot k \quad (18)$$

Where i and j are defined in (16).

Double type terms

Double type terms are those that include a constant value and a switching function multiplied by a state variable. Since, inside these terms exists a product of two time-dependant variables, equations (4)-(6) must be taken into account. Equation (19) is taken as an example (this term comes from (11)).

$$\frac{d}{dt} \langle f \rangle_k = C_3 \cdot \langle s \rangle_k \cdot \langle j \rangle_k \quad (19)$$

In first place, the terms that correspond to the DC term ($k = 0$) of $\langle f \rangle_k$ are placed in matrix A (20).

$$A_{i,j} = C_3 \cdot \langle s \rangle_0$$
$$A_{i,l} = 2 \cdot C_3 \cdot \langle s \rangle_{k'R}$$
$$A_{i,l+1} = 2 \cdot C_3 \cdot \langle s \rangle_{k'I} \quad (20)$$

The values of i, j and l are defined in (21).

$$\begin{cases} i = (2 \cdot n_h + 1) \cdot n_{SV_f} + 1 \\ j = (2 \cdot n_h + 1) \cdot n_{SV_j} + 1 \\ l = (2 \cdot n_h + 1) \cdot n_{SV_j} + 2 \cdot k' \end{cases} \quad (21)$$

Where $k' = 0, 1, 2, \dots, n_h$, $\langle s \rangle_0$ is the average component of the switching function, $\langle s \rangle_{kR}$ and $\langle s \rangle_{kI}$ are the real and imaginary components of the switching function for the kth harmonic and n_{SV_j} is the identification number of $\langle j \rangle_k$ (the state variable that is multiplying the constant and the switching function).

Then, the terms of $\langle f \rangle_k$ that go from the first harmonic ($k = 1$) to the desired one ($k = n_h$) are placed (22).

$$A_{i,j} \ \& \ A_{i+1,j+1} = C_3 \cdot \langle s \rangle_{k'R}$$
$$A_{i,j+1} \ \& \ A_{i+1,j} = C_3 \cdot \langle s \rangle_{k'I} \quad (22)$$

The values of i and j are defined in (23).

$$\begin{cases} i = (2 \cdot n_h + 1) \cdot n_{SV_f} + 2 \cdot k \\ j = (2 \cdot n_h + 1) \cdot n_{SV_j} + 2 \cdot k - 2 \cdot k' \end{cases} \quad (23)$$

Where $k' = 0, 1, 2, \dots, k$.

Finally, the process followed in equations (20) and (22) should be repeated for any other double type term inside the differential equation. At this point, all the terms that would correspond to matrix A have been placed.

Building process of matrix B

As it is shown in (7), matrix B only contains the coefficients that multiply with input variables. Therefore, in order to build matrix B, the only coefficients that are required from the differential equations are the ones that multiply with an input variable, see (24).

$$\frac{d}{dt}\langle f \rangle_k = C_4 \cdot \langle u_1 \rangle_k + C_5 \cdot \langle s \rangle_k \cdot \langle u_2 \rangle_k \qquad (24)$$

Where C_4 and C_5 are constants, $\langle s \rangle_k$ is a switching function and $\langle u_1 \rangle_k$ and $\langle u_2 \rangle_k$ are input variables. The first step consists of defining an identification number from 1 to n_{IV} for the input variables, for example, in case of (24):

- $n_{u_1} = 1$. Identification number for input variable $\langle u_1 \rangle_k$.
- $n_{u_2} = 2$. Identification number for input variable $\langle u_2 \rangle_k$.

Similarly to matrix A, in equation (24) there also exist Single type and Double type terms:

1. Single type. A constant is multiplied with an input variable ($C_4 \cdot u_1$).
2. Double type. A constant is multiplied with an input variable and a switching function ($C_4 \cdot u_1$).

Single type terms

As mentioned, a single type term is formed by the product of a constant and an input variable, see (25) (this term comes from (24)).

$$\frac{d}{dt}\langle f \rangle_k = C_4 \cdot \langle u_1 \rangle_k \qquad (25)$$

Since it is considered that, the dynamics of the input variables are much slower than the dynamics of the state variables, the building process of matrix B is simplified. Indeed, the first and only step consists of placing the DC term ($k = 0$) of $\langle f \rangle_k$ inside matrix B, see (26).

$$B_{i,j} = C_4 \qquad (26)$$

The values of i and j are defined in (27).

$$\begin{cases} i = \left(2 \cdot n_h + 1\right) \cdot n_{SV_f} + 1 \\ j = n_{u_1} \end{cases} \qquad (27)$$

Where n_{u_1} is the identification number of the input variable $\langle u_1 \rangle_k$.
If any other single type term would exist, equation (26) should be repeated for that specific term.

Double type terms

Double type terms are those that include a constant value and a switching function multiplied by a state variable. Since there exists a multiplication between two time dependant variables, the process differs from single type terms. For the explanation, equation (28) will be taken as an example (this term comes from (24)).

$$\frac{d}{dt}\langle f \rangle_k = C_5 \cdot \langle s \rangle_k \cdot \langle u_2 \rangle_k \qquad (28)$$

In first place, the terms that correspond to the DC term ($k = 0$) of $\langle f \rangle_k$ are placed in matrix B (29).

$$B_{i,j} = C_5 \cdot \langle s \rangle_0 \qquad (29)$$

Variables i and j are defined in (30).

$$\begin{cases} i = \left(2 \cdot n_h + 1\right) \cdot n_{SV_f} + 1 \\ j = n_{u_2} \end{cases} \qquad (30)$$

Where n_{u_2} is the identification number of input variable u_2.
Then, the terms of $\langle f \rangle_k$ that go from the first harmonic ($k = 1$) to the desired one ($k = n_h$) are placed (31).

$$B_{i,j} = C_5 \cdot \langle s \rangle_{kR}$$
$$B_{i+1,j} = C_5 \cdot \langle s \rangle_{kI} \qquad (31)$$

The values of i and j are defined in (32).

$$\begin{cases} i = (2 \cdot n_h + 1) \cdot n_{SV_f} + 2 \cdot k \\ j = n_{u_2} \end{cases} \tag{32}$$

Finally, the process followed in equations (29) and (31) should be repeated for any other double type term inside the differential equation. At this point, all the terms that would correspond to matrix B have been defined.

Building process of matrix C

As it can be observed in equation (9), matrix C is multiplied with the state variable vector and its function is to generate their estimated values by summing the harmonics of each state variable. Therefore, the size of matrix C depends on the number of state variables and the considered number of harmonics. To be more precise, the number of rows is equal to n_{SV} and the number of columns is equal to h (10). Equation (33) details the procedure to follow for placing the corresponding terms.

$$\begin{aligned} C_{i,j} &= 1 \\ C_{i,l} &= 2 \cdot \cos\left(\omega_s \cdot t \cdot k\right) \\ C_{i,l+1} &= -2 \cdot \sin\left(\omega_s \cdot t \cdot k\right) \end{aligned} \tag{33}$$

In this case, parameters i, j and l are defined in (34).

$$\begin{cases} i = n_{SV_x} + 1 \\ j = (2 \cdot n_h + 1) \cdot n_{SV_x} + 1 \\ l = (2 \cdot n_h + 1) \cdot n_{SV_x} + 2 \cdot k \end{cases} \tag{34}$$

Where $n_{SV_x} = n_{SV_h}, n_{SV_f}$ and n_{SV_j} (taking as example equation (11)) and t is the time.

At this point, all the matrixes required for solving equations (7)-(9) have been obtained.

Study case

In this section, a DAB Series Resonant Converter (DAB-SRC) is presented as example for the validation of the GSSAM algorithm, see Figure 1.

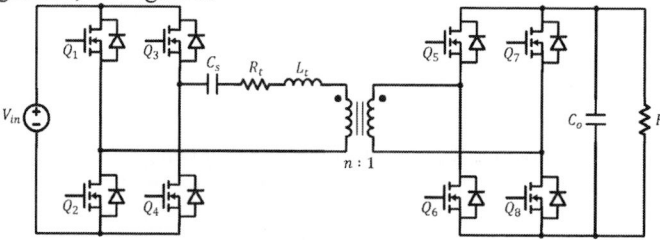

Figure 1. Simplified electrical scheme of a DAB-SRC.

Following the steps described in section **GSSAM algorithm**, the modelling of the converter starts from the differential equations that describe it. In this case, equations (35)-(37) represent the DAB-SRC from Figure 1 [9], [10]. The concerned equations are comparable to (11) and (24), where the derivative of a state variable is defined by the sum of state and input variables. Apart from that, since there exists a high frequency transformer, it must be outlined that all the variables from equations (35)-(37) have been referred to the secondary side.

$$\frac{dI_L(t)}{dt} = -\frac{R_t}{L_t} \cdot I_L(t) - \frac{1}{L_t} \cdot V_{C_S}(t) - \frac{s_2(t)}{L_t} \cdot V_{C_O}(t) + \frac{s_1(t)}{L_t} \cdot V_{in} \tag{35}$$

$$\frac{dV_{C_O}(t)}{dt} = \frac{s_2(t)}{C_O} \cdot I_L(t) - \frac{1}{R \cdot C_O} \cdot V_{C_O}(t) \tag{36}$$

$$\frac{dV_{C_S}(t)}{dt} = \frac{1}{C_S} \cdot I_L(t) \tag{37}$$

As expected, the number of equations is proportional to the number of state variables ($I_L(t)$, $V_{C_o}(t)$ and $V_{C_s}(t)$). When it comes to the switching functions, their function is to stablish the corresponding voltage polarity to the variables they are attached to and equation (38) defines their stages.

$$s_1(t) = \begin{cases} 1 & 0 < t \le T/2 \\ -1 & T/2 < t \le T \end{cases} \quad s_2(t) = \begin{cases} -1 & 0 < t \le d \cdot T/2 \\ 1 & d \cdot T/2 < t \le T/2 + d \cdot T/2 \\ -1 & T/2 + d \cdot T/2 < t \le T \end{cases} \quad (38)$$

Where d represents the delay between both signals.

Then, in order to obtain the average and the real and imaginary components of the concerned switching functions, equation (2) is applied to (38), see (39)-(41).

$$\langle s_1 \rangle_0 = \langle s_2 \rangle_0 = 0 \quad (39)$$

$$\langle s_1 \rangle_{kR} = 0 \quad \langle s_1 \rangle_{kI} = -\frac{\cos^2(\pi \cdot k) - 1}{\pi \cdot k} \quad (40)$$

$$\langle s_2 \rangle_{kR} = \frac{\sin(\pi \cdot k \cdot (1+d)) - \sin(\pi \cdot k \cdot d)}{\pi \cdot k} \quad (41)$$

$$\langle s_2 \rangle_{kI} = \frac{\cos(\pi \cdot k \cdot (1+d))}{\pi \cdot k} - \frac{\cos(2 \cdot \pi \cdot k)}{2 \cdot \pi \cdot k} - \frac{\cos(d \cdot \pi \cdot k)}{\pi \cdot k} - \frac{1}{2 \cdot \pi \cdot k}$$

Once equations (35)-(37) and (39)-(41) are obtained, matrixes A, B and C are built by following the steps described in the section called **GSSAM algorithm**. With the aim of showing the intermediates steps, the inductor current from equation (35) is used as example, see (42).

$$\frac{dI_L(t)}{dt} = -\frac{R_t}{L_t} \cdot I_L(t) \quad (42)$$

Since the term from (42) consists of a state variable and a single type term, equations (13)-(16) must be applied. In addition, as the state variable that is in derivative form and the state variable that is multiplying the constant coincide, equation (18) must also be applied. In first place, the terms that correspond to the DC term ($k = 0$) of $\langle f \rangle_k$ are placed, see (43).

$$A_{1,1} = -R_t / L_t \quad (43)$$

Then, the terms of the first harmonic ($k = 1$) are placed (44).

$$A_{2,2} \ \& \ A_{2+1,2+1} = -R_t / L_t \quad (44)$$

Next, the terms that correspond to equation (18) are placed (45).

$$A_{2,2+1} = \omega_s \cdot 1$$
$$A_{2+1,2} = -\omega_s \cdot 1 \quad (45)$$

Finally, equation (46) shows the resulting matrixes A and B for $k = 1$. If higher harmonics are considered, the size of the mentioned matrixes increases exponentially.

$$(46)$$

Results

The values of the components inside the circuit from Figure 1 are listed in Table I.

Table I. Design parameters of the DAB SRC.

Parameter	V_{in} [V]	C_s [μF]	L_t [μH]	R_t [mΩ]	C_o [μF]	R [Ω]	f_{sw} [kHz]	Delay [deg]	n
Value	100	10	26.7	50	80	21.36	25	90	0.5

Firstly, it must be mentioned that in order to develop and run both modelling models, the MatLab software has been used in a "HP ProBook 450 g5". Then, introducing the values presented in Table I inside the GSSAM algorithm and inside the switched model, Figure 2 is obtained. There, it can be observed that the GSSAM algorithm follows properly the curves obtained from the switched model. Analysing the zoomed areas, Figure 2a shows that the GSSAM output voltage curve with k=5 is slightly closer to the switched model than the GSSAM curve with k=11. However, this does not mean that it k=5 is more exact. Indeed, the shape of k=11 is more similar to the switched model. This is also confirmed in Figure 2b, where the more accurate inductor current is obtained with k=11.

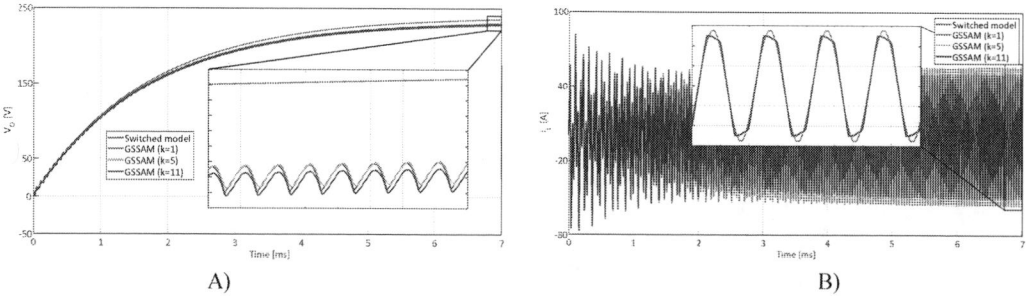

A)   B)

Figure 2. Voltage and current waveforms for the DAB SRC presented in Table I. A) Output voltage. B) Inductor current.

Finally, when it comes to the time consumed by each modelling method, the fastest solution is the Switched model with 1.4 seconds. Then, GSSAM wise, the simulation times increase exponentially as the number of harmonics increases: 2.2 seconds ($k = 1$), 16.1 seconds ($k = 5$) and 95 seconds ($k = 11$).

Conclusion

In present work, an algorithm that automatizes the modelling process of a given power converter has been presented and validated. Making use of this algorithm, the modelling of a power converter is simplified and the working time is shorten. Also, this tool can be very useful for frequency response analysis, which is essential for designing and tuning the control that will be applied on the converter. Furthermore, in other to get more accurate results, a desired number of harmonics can be easily defined. This way, the user does not have to carry out difficult mathematical operations.

References

[1]   S. Bacha, I. Munteanu, A. I. Bratcu, and others, *Power electronic converters modeling and control*, vol. 454. 2014.

[2]   S. R. Sanders, J. M. Noworolski, X. Z. Liu, and G. C. Verghese, "Generalized averaging method for power conversion circuits," *PESC Rec. - IEEE Annu. Power Electron. Spec. Conf.*, no. February, pp. 333–340, 1990.

[3]   "Sanders_Noworolski_Liu_Verghese_Pesc.Pdf." .

[4]   J. Mahdavi, A. Emaadi, M. D. Bellar, and M. Ehsani, "Analysis of power electronic converters using the generalized state-space averaging approach," *IEEE Trans. Circuits Syst. I Fundam. Theory Appl.*, vol. 44, no. 8, pp. 767–770, 1997.

[5]   V. a Caliskan, G. C. Verghese, and A. M. Stankovi, "Multifrequency Averaging of DC / DC Converters," *IEEE Trans. Power Electron.*, vol. 14, no. 1, pp. 124–133, 1996.

[6]   H. Qin and J. W. Kimball, "Generalized average modeling of dual active bridge DC-DC converter," *IEEE Trans. Power Electron.*, vol. 27, no. 4, pp. 2078–2084, 2012.

[7]   J. A. Mueller and J. W. Kimball, "An Improved Generalized Average Model of DC-DC Dual Active Bridge Converters," *IEEE Trans. Power Electron.*, vol. 33, no. 11, pp. 9975–9988, 2018.

[8]   U. Javaid and D. Dujic, "Arbitrary order generalized state space average modeling of switching

converters," *2015 IEEE Energy Convers. Congr. Expo. ECCE 2015*, pp. 6399–6406, 2015.

[9]   H. Wu, P. Wang, and Y. Li, "A control method for series resonant dual active bridge DC/DC converter," *IEEE Transp. Electrif. Conf. Expo, ITEC Asia-Pacific 2014 - Conf. Proc.*, pp. 2–6, 2014.

[10]   H. Wang, "Design and Control of Series Resonant Converters for Dc Current Power Distribution Applications," p. 211, 2018.

A Novel Fully Distributed Cost Optimal Control Method for DC Microgrid

Qingping Xia[1,2], Hua Han[1,2], Yao Liu[1,3], Zhangjie Liu[1,2], Yao Sun[1,2], Mei Su[1,2]

1. School of Automation, Central South University
2. Hunan Provinc • ial Key Laboratory of Power Electronics Equipment and Grid
Changsha, Hunan, P.R. China
3. Guangdong Power Grid Corporation Zhuhai Power Supply Bureau, China Southern Power Grid
E-Mail: qingping_xia@163.com

Acknowledgements

This work was supported in part by the National Key R&D Program of China under Grant 2018YFB0606005, in part by the National Nature Science Foundation of China under Grant 61903383, in part by the National Nature Science Foundation of China under Grant 61933011, in part by the Hunan Provincial Key Laboratory of Power Electronics Equipment and Grid under Grant 2018TP1001, and in part by the Hunan Province Strategic Emerging Industry Science and Technology Research and Major Achievement Transformation Project under Grant 2018GK4016.

Keywords

«DC microgrid», «distributed control», «economic dispatch», «ADMM algorithm», «consensus».

Abstract

This paper proposes an ADMM and observers based distributed cost optimal control method for DC microgrid. Unlike centralized control which need a central node communicate with all other nodes, the proposed fully distributed method, where DGs only change their information with their neighbors. Both global voltage recovery and total cost optimization are achieved in real time under the DGs' capacity limits. Finally, simulation results verify the anti-disturbance ability and effectiveness of the proposed method.

I. Introduction

In recent years, control schemes for DC microgrid has received more and more attention, and researches on DC microgrid have been more about the economic dispatch problem (EDP). A key issue of DC microgrid with large-scale distributed generators (DGs) is how to ensure supply-demand balance while minimize the total generation cost. Centralized methods, such as the genetic algorithm based optimal method [1-2], is adopted for the optimization of islanded microgrid operation. However, it lacks of flexibilities and cannot rapidly respond fluctuations due to complex algorithms, and the implement of centralized approach rely on high bandwidth communication links, which may easily lead to a single-point failure in the system. To avoid the communication failures, some communication free schemes, which can achieve the power dispatch according DG's local information, are proposed [3-5]. Since DGs cannot obtain the global information, there is a bus voltage drop, which may degrade the quality of the power supply. To improve the performance, many distributed methods [6-8] have been proposed to solve the EDP, where the system iterate to a global consensus using the information from their local neighbors and no central controller needed. A λ-iteration method is proposed in [9] to minimize the total cost. However, the DG's capacity limit is not considered. In [10], the load information is obtained through a global load observer, and then DGs choose the operating modes: λ-iteration or maximum output according to the look-up table related to the load. Although the output-power is considered, the optimal scheme is not flexible since the DG's look-up table is scheduled for a particular system before its operating. Moreover, all these methods are based on quadratic cost function, and not applicable to the EDP with none-quadratic functions. In [11], a leader-follower consensus algorithm, where central nodes are needed to select the leader node, which is not fully distributed, is adopted.

Motivated by the above discussion, this paper proposes a novel fully distributed control scheme to solve the EDP with general convex cost functions, which can realize global voltage recovery and economic optimization at the same time. Both the generation-demand constraint and each DG's capacity constraint are considered. The voltage observer is also constructed to observe the global average voltage, which can be recovered by each DG based on a distributed average consensus protocol. To avoid using global information or other devices to obtain the load information, a load observer is constructed through low bandwidth communication (LBC) links between DGs. Then alternating direction method of multipliers (ADMM) is introduced to optimize the output power in a fully distributed way. Finally, a test bed is built in Matlab/Simulink to validate our method. The simulation results show that the system achieves economic dispatch when DGs' capacity constraints are considered and the global average voltage can be recovered under different loads.

II. Problem formulation

DC microgrid typically consists of multiple DGs and loads. In this paper, we focus on the generation cost optimization with DGs' capacity constraints in islanded mode.

A. Formulation of the optimization problem

There are dispatchable sources such as conventional generators and nondispatchable sources including renewable sources. To reduce the total cost of power generation and increase the utilization rate of renewable energy, the nondispatchable sources are considered to operate in a decentralized way, where DGs operate in MPPT mode. Therefore, only dispatchable sources are considered in the EDP. The cost functions of common DGs, such as micro turbines and energy storage equipment, can be approximated by a none-quadratic convex function (1) [12]-[13].

$$C_i\left(P_{Gi}\right) = a_i P_{Gi}^2 + b_i P_{Gi} + c_i + d_i \exp\left(f_i P_{Gi}\right) \quad i = 1, 2, \cdots, n \tag{2}$$

where n is the number of controllable DGs and P_{Gi} is the output power of the i-th DG. Constants " a_i, b_i, c_i "and " d_i, f_i " represent, the fuel consumptions of combustion engine and the equivalent cost coefficients respectively.

Therefore, the EDP can be formulated as the following constrained optimization problem:

$$\min C\left(P_G\right) = \sum_{i=1}^{n} C_i\left(P_{Gi}\right) \tag{3}$$

$$s.t. \ P_{Gi\_min} \le P_{Gi} \le P_{Gi\_max} \quad i = 1, 2, \cdots, n \tag{4}$$

$$1_{1 \times n} P_G = P_L \tag{5}$$

where $P_G = [P_{G1}, P_{G2}, \cdots, P_{Gn}] \in \mathbb{R}^n$ and P_L denotes the total demand. P_{Gi_min} and P_{Gi_max} are, respectively, the minimum and maximum output power constraint of the i-th DG.

B. Global load observer

In this paper, the load demand is observed by DGs' local load observer, which updates its local estimate $\hat{P}_{Li}$ by processing the local power measurements and the neighbors' estimates

$$\hat{P}_{Li}\left(t\right) = n P_{Gi}\left(t\right) + \int_0^t \sum_{j \in N_i} \omega_{ij}\left(\hat{P}_{Lj}\left(\tau\right) - \hat{P}_{Li}\left(\tau\right)\right) d\tau \tag{6}$$

where N_i denotes the set of all neighbors of node i. $\omega_{ij} = 1$ if there is a communication link for data exchange between node i and node j, and $\omega_{ij} = 0$ otherwise.

All entries of the load estimates would converge to the sum of all DGs' output power [10], which also equals the total demand as expressed in (6).

$$\hat{P}_{L1} = \hat{P}_{L2} = \cdots = \hat{P}_{Ln} = \sum_{i=1}^{n} P_{Gi} = P_L \tag{7}$$

III. Proposed distributed control algorithm

A. ADMM-based distributed control scheme

Combining (2) and (4), we can get the augmented Lagrangian function as follows:

$$L_\rho\left(P_{G1}, P_{G2}, \cdots P_{Gn}, \lambda\right) = \sum_{i=1}^{n} C_i\left(P_{Gi}\right) - \lambda^T\left(\sum_{i=1}^{n} P_{Gi} - P_L\right) + \frac{\rho}{2}\left\|\sum_{i=1}^{n} P_{Gi} - P_L\right\|_2^2 \tag{8}$$

where $\lambda \in \mathbb{R}^m$ is the Lagrange multipliers or the dual variable. Constant $\rho > 0$ is incorporated as a quadratic penalty of the constraints into the Lagrangian.

To solve the EDP, an ADMM-based distributed approach is proposed. Algorithm I presents the principle of the ADMM-based distributed algorithm. It is proved that ADMM is suitable for solving the constrained optimal problem formulated in this paper [14].

Algorithm I: ADMM-based distributed algorithm

Procedure

Initialize P_G^0, λ^0, ρ with $1_{1 \times n} P_G^0 = P_L$;

For k = 0, 1, 2, ..., N

$$P_{Gi}^{k+1} = \arg\min_{P_{Gi}} C_i\left(P_{Gi}\right) + \frac{\rho}{2}\left\|\sum_{j<i} P_{Gj}^{k+1} + P_{Gi} + \sum_{j>i} P_{Gj}^{k} - P_L\right\|_2^2 \text{ with } P_{Gi\_\min} \le P_{Gi} \le P_{Gi\_\max};$$

$$\lambda^{k+1} = \lambda^k - \rho\left(\sum_{j=1}^{n} P_{Gj}^{k+1} - P_L\right);$$

If $\lim_{k\to\infty}\left\|P_{Gi}^{k+1} - P_{Gi}^{k}\right\|_2 < \varepsilon_i$

    Break;

$k = k + 1$;

End for

Return the optimal solution P_G^*;

End procedure

From Algorithm I, it is shown that when the ADMM algorithm is used to solve the EDP, the load information is needed. To ensure that the system is implemented in a distributed way, the observer constructed by (6) obtains the global load locally at DGs. Assume that the load is a constant value during each iteration, the ADMM algorithm converges to the optimal solution $P_G^* = [P_{G1}^*, P_{G2}^*, \cdots, P_{G3}^*]$ with

$$\lim_{k\to\infty}\left\|P_{Gi}^{k+1} - P_{Gi}^{k}\right\|_2 = 0 \quad i = 1, 2, \cdots, n \tag{9}$$

B. Global average voltage recovery

In an islanded microgrid, DGs are supposed to provide a reliable bus voltage for the load. Similar to the load observer, a voltage observer is established to estimate the global average voltage through a distributed communication between DGs

$$\hat{u}_i\left(t\right) = u_i\left(t\right) + \int_0^t \sum_{j \in N_i} \omega_{ij}\left(\hat{u}_j\left(\tau\right) - \hat{u}_i\left(\tau\right)\right)d\tau \tag{10}$$

where $\hat{u}_i$ is the output voltage estimate and u_i is the local measurement of the i-th DG. Similar to the load observer, estimates observed by voltage observers will iterate to the real global average voltage, namely

$$\hat{u}_1 = \hat{u}_2 = \cdots = \hat{u}_n = \frac{1}{n}\sum_{i=1}^{n} \hat{u}_i \tag{11}$$

The diagram of the proposed control strategy is modeled as a three-layer structure as shown in Fig. 1. The physical layer shows that DGs are parallel to the DC bus through a resistance. The cyber layer presents the LBC network constructed in this paper, through which DGs can communicate with their neighbors. Note that [data] denotes data is received from or sent to the DG's neighbors through the LBC network and node j denotes the neighbor of node i. The control layer include voltage recovery part and power optimization part. The power supply is guaranteed by the former part, which ensure the global average voltage keep at the reference value $u^* = 50$ v. After receiving the optimal solution P_G^* as the initial value from DG's neighbor, the DG recalculate the optimal solution according its local information and then send the new solution to the other neighbor. Since all DGs can estimate the real load demand, the P_G^* calculated by each DG will be the same. Therefore, the economic dispatch can be achieved.

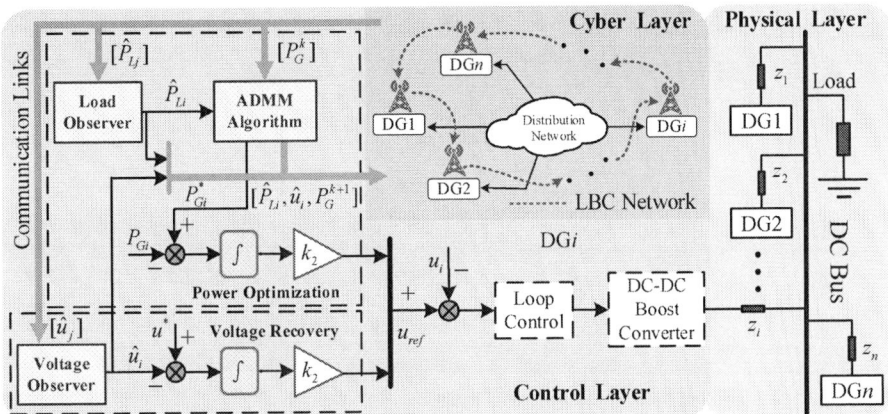

Fig. 1: Diagram of the proposed control strategy.

IV. Simulation Results

A four DGs system simulation is carried out to verify the strategy proposed in this paper. The parameters of each DG are given in Table I. Two cases are designed to test the performance when load change at the same time. The only difference between Case 1 and Case 2 is that Case 2 considers the capacity limit of DGs while Case 1 does not. Simulation results of Case 1 and Case 2 are shown in Fig. 2 and Fig.3 respectively.

Table I: Parameters of the DGs

Item	a_i	b_i	c_i	d_i	f_i	P_{Gi_min} (Kw)	P_{Gi_max} (Kw)	z_i (Ω)
DG1	0.64	0.12	50	0.2	0.001	1.0	2.0	0.1
DG2	0.74	0.1	55	0.1	0.0015	0.8	1.8	0.12
DG3	0.7	0.15	50	0.3	0.002	1.2	2.0	0.14
DG4	0.6	1.5	60	0	0	0.8	1.6	0.1

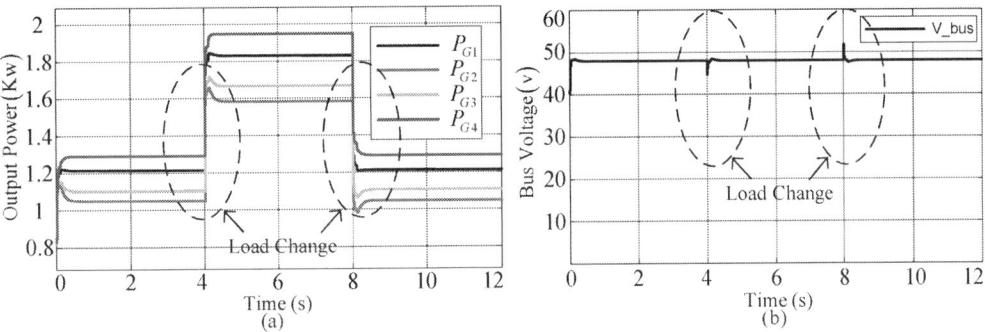

Fig. 2: Simulation results without DGs' capacity limit. (a) The output power of controllable DGs. (b) DC bus voltage.

From $t = 0$ s to $t = 4$ s, it can be found that the output power of DG3 in Fig. 2 (a) is 1.1Kw, which is less than its minimal capacity. When DGs' capacity constraints are taken into consideration, the output power of DG3 keep at the minimal value 1.2Kw as shown in Fig.3 (a). The load increase by 50% at $t = 4$ s, the system quickly enter a new balance and complete the economic dispatch. In this progress, the output power of each DG increases, and eventually DG4 exceeds its maximum power limit as shown in Fig. 2 (a), from t = 0 s to t = 4 s. The output power of DG 4 keeps at the maximal value when the DGs' capacity constraints are considered. At the time t = 8 s, the load decrease to the initial one, the equilibrium point of the system back to the former one again. As can be seen from Fig. 2 (b) and Fig. 3 (b), the power supply can be guaranteed when load changes.

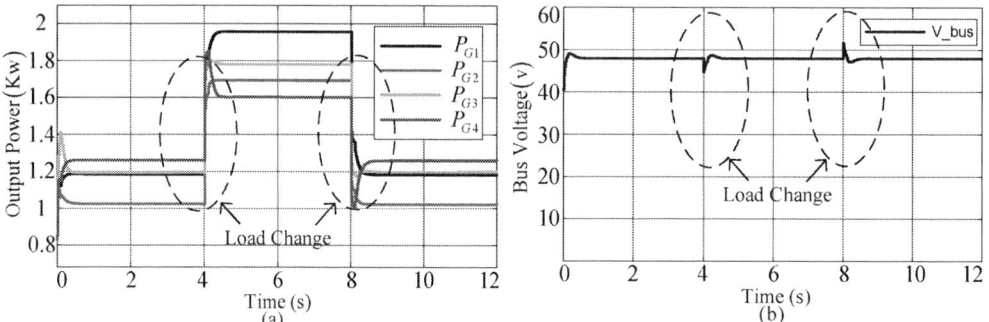

Fig. 3: Simulation results with DGs' capacity limitations. (a) The output power of controllable DGs. (b) DC bus voltage.

V. Conclusion

In this paper, a fully distributed control strategy has been proposed to solve the EDP with capacity constraints in islanded DC microgrid. The distributed algorithm realizes the total cost optimization, meanwhile the global voltage is restored, which needs only a distributed communication between DGs. Unlike most of the former studies, the control scheme proposed in this paper is suitable for the EDP with general convex functions in a fully distributed way and the system can perform economic dispatch in real time when the load changes. The effectiveness of the proposed scheme is verified in the simulations.

References

[1] A. Mazza, G. Chicco, A. Russo, Optimal multi-objective distribution system reconfiguration with multi criteria decision making-based solution ranking and enhanced genetic operators, Int. J. Electr. Power Energy Syst. 54 (Supplement C) (2014) 255–267.

[2] C. Bustos, D. Watts and H. Ren, "MicroGrid Operation and Design Optimization With Synthetic Wins and Solar Resources," in IEEE Latin America Transactions, vol. 10, no. 2, pp. 1550-1562, March 2012.

[3] Q. Xu, J. Xiao, P. Wang and C. Wen, "A Decentralized Control Strategy for Economic Operation of Autonomous AC, DC, and Hybrid AC/DC Microgrids," in IEEE Transactions on Energy Conversion, vol. 32, no. 4, pp. 1345-1355, Dec. 2017.

[4] M. Hamzeh, H. Mokhtari and H. Karimi, "A decentralized self-adjusting control strategy for reactive power management in an islanded multi-bus MV microgrid," in Canadian Journal of Electrical and Computer Engineering, vol. 36, no. 1, pp. 18-25, Winter 2013.

[5] M. Tucci, S. Riverso, J. C. Vasquez, J. M. Guerrero and G. Ferrari-Trecate, "A Decentralized Scalable Approach to Voltage Control of DC Islanded Microgrids," in IEEE Transactions on Control Systems Technology, vol. 24, no. 6, pp. 1965-1979, Nov. 2016.

[6] J. Hu, M. Z. Q. Chen, J. Cao and J. M. Guerrero, "Coordinated Active Power Dispatch for a Microgrid via Distributed Lambda Iteration," in IEEE Journal on Emerging and Selected Topics in Circuits and Systems, vol. 7, no. 2, pp. 250-261, June 2017.

[7] Y. Xu, W. Zhang, and W. Liu, ``Distributed dynamic programming-based approach for economic dispatch in smart grids," IEEE Trans. Ind. Informat., vol. 11, no. 1, pp. 166-175, Feb. 2015.

[8] Z.Yang, J. Xiang, andY. Li, "Distributed consensus based supply–demand balance algorithm for economic dispatch problem in a smart grid with switching graph," IEEE Trans. Ind. Electron., vol. 64, no. 2, pp. 1600–1610, Feb. 2017.

[9] Hua Han, Hao Wang, Yao Sun*, Jian Yang, Zhangjie Liu.A Distributed Control Scheme on Cost Optimization Under Communication Delays for DC Microgrids,IET Generation Transmission & Distribution, vol.11,no.17,pp.4193-4201 ,Nov. 2017.

[10] Z. Liu, M. Su, Y. Liu, H. Wang and H. Han, "A distributed control scheme with cost optimization and capacity constraints," IECON 2017 - 43rd Annual Conference of the IEEE Industrial Electronics Society, Beijing, 2017, pp. 521-526.

[11] Ziang Zhang, Xichun Ying and Mo-Yuen Chow, "Decentralizing the economic dispatch problem using a two-level incremental cost consensus algorithm in a smart grid environment," 2011 North American Power Symposium, Boston, MA, 2011, pp. 1-7.

[12] G. Hug, S. Kar, and C. Wu, "Consensus + innovations approach for distributed multiagent coordination in a microgrid," IEEE Trans. Smart Grid, vol. 6, no. 4, pp. 1893–1903, Jul. 2015.

[13] W. T. Elsayed and E. F. El-Saadany, ''A fully decentralized approach for solving the economic dispatch problem,'' IEEE Trans. Power Syst., vol. 30, no. 4, pp. 2179–2189, Jul. 2015.

[14] S. Boyd et al., "Distributed optimization and statistical learning via the alternating direction method of multipliers," Found. Trends Mach. Learn., vol. 3, no. 1, pp. 1–122, Jan. 2011.

Measurement of Dynamic On-State Resistance of High-Voltage GaN-HEMTs under Real Application Conditions

Benedikt Kohlhepp[1], Carsten Kuring[2], Stefan Peller[1] and Daniel Kübrich[1]

[1]Electromagnetic Fields, Friedrich-Alexander University Erlangen-Nürnberg (FAU)
Cauerstraße 7
91058 Erlangen, Germany
Tel.: +49 (0)9131 85 28951
Fax: +49 (0)9131 85 27787
E-Mail: benedikt.kohlhepp@fau.de
URL: http://emf.eei.uni-erlangen.de/

[2]Chair of Power Electronics,
Technische Universität Berlin
Einsteinufer 19
10587 Berlin, Germany
Tel.: +49 (0)30 314 23404
Fax: +49 (0)30 314 25526
E-Mail: carsten.kuring@tu-berlin.de
URL: http://www.pe.tu-berlin.de/

Keywords

«Gallium Nitride (GaN)», «Conduction losses», «Measurement».

Abstract

GaN-HEMTs gain a lot of attention to power electronics engineers. However, they could exhibit increased on-state resistance due to charge trapping. This paper focusses on measuring this effect for high voltage devices. Measurements reveal an unexpected blocking voltage dependency of on-state resistance. Therefore, a second measurement setup serves as verification.

Introduction

High efficiency or high power density commonly calls for high switching frequencies. GaN-HEMTs are said to be a key to this requirement, as they offer low on-state resistance in combination with low parasitic capacitances. Unfortunately normally-off GaN-on-Si HEMTs can exhibit increased on-state resistance due to charge trapping effects [1]. In order to make accurate loss predictions within a power electronic system, the on-state resistance present during switching operation is a crucial parameter. Therefore the literature presents different ways to determine the dynamic on-state resistance [1] [2] [3] [4].

Experimental setup

To determine the on-state resistance of GaN devices under switching conditions an appropriate test setup is vital. The half-bridge configuration is one of the most applied circuits. Therefore it is used here to study the behavior of GaN-HEMTs under clamped inductive switching (hard switching) and soft switching conditions. Compared to double pulse tests, this measurement is carried out in the target application (half-bridge circuit). This guarantees valid measurement results for power loss analysis in the targeted application. The half-bridge is driven complementarily with the conduction times $\delta \cdot T_s$ for the lower switch (DUT) and $(1 - \delta) \cdot T_s$ for the upper switch S_1 respectively. Due to the large chosen inductance value L_0, the current ripple at the applied switching frequencies is negligibly small and therefore an almost constant current flows through the DUT during on-state. Thus, the converter operates in deep CCM. This ensures approximately the same current at the on- and off-transition of the DUT. As the circuit serves as test setup to obtain the dynamic on-state resistance of GaN devices, power density, cost and size do not matter within this circuit. For this reason, using a very large inductance value, which is accompanied with a huge construction size and high cost, is reasonable [5]. In order to be able to set various currents at the converter's output, the current source in Fig. 1 (left) is realized by an electronic load, which can sink a constant but adjustable current. The right hand side of Fig. 1 depicts the circuit used for obtaining results concerning the dynamic on-state resistance under soft switching.

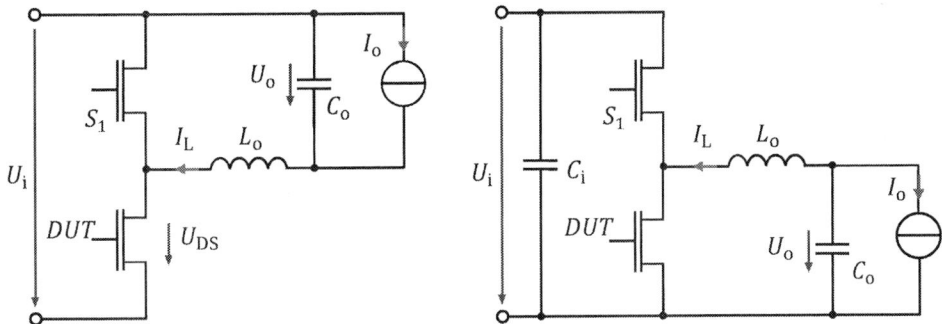

Fig. 1: Clamped inductive switching circuit for hard switching (left) and circuit for hard switching (right)

Accurately measuring the on-state voltage of power semiconductors under switching operation poses the major challenging task while dynamic on-state resistance measurement. In off-state of the device, the drain-source voltage is up to several hundreds of volts. Whereas, the voltage in the conduction phase of the DUT is only a few millivolts, resulting in a large difference (decades of magnitude) between on- and off-state of the GaN-HEMTs. Measuring the drain-source voltage directly requires a large V/div-setting in off-state of the DUT, which tremendously limits the resolution of the voltage measurement. Consequently, employing standard scopes and displaying the complete voltage waveform delivers only very noisy signals during on-state of the device. Using a V/div-setting in order to obtain more details of the voltage waveform during on-state, leads to clipping of the drain-source voltage to be measured in DUT's off-state. As a result, the scope's input amplifier will saturate and in worst case the measurement equipment can be damaged. Thus, no reliable results can be gained by these direct measurements. Typically standard scopes feature limited voltage resolution of mostly 8 bits, which therefore limits the accuracy of the measured signal because of the large voltage scale needed for measuring [6] [7]. Even if the oscilloscopes have a better resolution, there can be the problem that the input amplifier and its circuitry is often not sufficiently linear to achieve reliable measurement results in this test setup. Accordingly, using a digitizer card featuring a high resolution of 14 bit and a sampling rate of 100 MSamples/s is employed here [8]. This allows for an accurate and precise measurement of the on-state drain-source voltage of the DUT by exhibiting sufficient resolution.

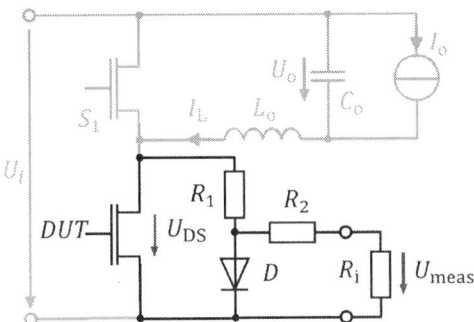

Fig. 2: Conventional clamping circuit

A challenging task during on-state resistance is the measurement of the absolute value of the drain-source voltage with high precision immediately after the falling edge of the voltage with significant amplitude. This characteristic is denoted as settling time and is a crucial but not well specified property in standard scopes [6] [7]. Measurements in [3] show that standard oscilloscopes have remarkable settling effects and as dynamic on-state resistance exhibits a similar behavior over time, only minimization of the settling effect ensures reliable measurements of dynamic on-state resistance. In order to avoid erroneously considering the settling time as a part of the dynamic on-state resistance, it is necessary to reduce its impact on the measured waveform. In addition, the measured signal could be falsified by a offset which is potentially included in the measured signal, as the dynamic on-state resistance is directly calculated from the voltage and current waveform by ohm's law. The effect of an offset is severe when testing with low currents. Besides this, the measurement result is distorted by digitalization noise because of the very low on-state voltage of the GaN semiconductor. Nevertheless, averaging of the measured signal allows for minimizing the impact of noise by using a stable triggering of the signal. Both the settling time as well as the offset and the resolution of the scope must be carefully characterized as the measurement signal during on-state of the device is at or beyond the limits of the scope.

The analysis of the scope's parameters however, is not sufficient to achieve reliable and precise measurements of the dynamic on-state resistance. Besides, the measurement task requires a clamping circuit in order to minimize the voltage at the input of the oscilloscope. Already [3] presents a very simple and accurate clamping circuit for measuring dynamic on-state resistance, which is shown in Fig. 2. A reasonable design of the clamping circuit provides voltages in the range of 60 mV during blocking of the DUT. While the DUT is conducting, the measured voltage is determined by the attenuation of the voltage divider given in (1).

$$k_\text{U} = \frac{R_1 + R_2 + R_\text{i}}{R_\text{i}} \tag{1}$$

However, this clamping circuit is only suitable for low voltage GaN-HEMTs as the power loss in R_1 is critical even for low voltages and further increases with supply voltage U_i. During blocking phase of the DUT, approximately the input voltage of the circuit U_i is applied to R_1. Therefore huge power loss occurs at this resistor. The dissipated power is in proportional to the square of the input voltage. This leads to thermal overload of the resistor at higher voltages, which in our setup allows measurements up to 80 V input voltage only. A larger package size of the resistor would enable increased power dissipation and thus higher input voltages. Nevertheless, this measure would go hand in hand with a larger construction volume and loop size, which would deteriorate accuracy by increased electromagnetic coupling. Another possible solution would be to drastically increase the resistance value of R_1 in order to be able to measure 650 V GaN devices. However, this would also increase the divider ratio k_U of the voltage divider of the clamping circuit (see (1)), so noise and limited resolution of the scope would massively impact the measurement accuracy. The aforementioned considerations show, that a useful design of this kind of clamping circuit for 650 V GaN-HEMTs is not possible.

For this reason, this paper shows a necessary improvement of the clamping circuit in order to be able to conduct measurements of dynamic on-state resistance of 650 V transistors. Fig. 3 (left) depicts the improved clamping circuit, in the following called circuit A, featuring an additional MOSFET S_aux. This auxiliary component is switched so that the input voltage is across it during most of the DUT's blocking time. In order to ensure that all transients have decayed at the time instant where the dynamic on-state resistance will be observed, switching on the auxiliary MOSFET shortly before turn-on of the DUT is mandatory. During on-state of the GaN device, the on-state $R_\text{DS,on,aux}$ of S_aux must be taken into account when calculating the voltage divider's attenuation. The DUT and the auxiliary switch are switched off simultaneously. Consequently, there is only a short period of time, slightly before switching on the GaN device, which cause losses in the voltage divider's resistance R_1. The time delay ($t_{\Delta,\text{aux}} \approx 300$ ns) between the switching instant of the auxiliary switch and the DUT during turn-on must be properly chosen in order to achieve reliable measurement results. Furthermore, the auxiliary switch is turned on only every n-th switching cycle to further reduce the losses in the voltage divider. With this modification of the clamping circuit the power loss in R_1 can be reduced due to the shorter phases where the input voltage is applied to it. Here a resistance value of 2 kΩ is used. (2) allows for calculating the power loss in the voltage divider's resistance.

$$P_\text{R1} = \frac{1}{n} \frac{\left(U_\text{i} \cdot \sqrt{t_{\Delta,\text{aux}} \cdot f_\text{s}} \right)^2}{R_1} = \frac{U_\text{i}^2}{n \cdot R_1} \cdot t_{\Delta,\text{aux}} \cdot f_\text{s} \tag{2}$$

Taking the on-state resistance of the auxiliary MOSFET into account, the voltage divider's attenuation is now given by:

$$k_{\mathrm{U}} = \frac{R_1 + R_2 + R_i + R_{\mathrm{DS,on,aux}}}{R_i} \tag{3}$$

The next section describes the selection of components suitable for applying in the proposed clamping circuit. Selecting a device for the auxiliary switch which is capable of blocking the same voltage as the *DUT* is mandatory. Besides this, the on-state resistance of it is not that important, as it is in series connection to the voltage dividers resistance R_1, which is typically much higher than it, but it should be considered within the voltage dividers attenuation in (3). Care should be taken by choosing a MOSFET with relatively small parasitic capacitances, as they must be charged and discharged during its switching transients and consequently contribute to the switching losses of the *DUT*. Here a FQT1N60C-MOSFET is used it is capable of blocking voltages up to 600 V [9]. Normally, power semiconductors are chosen in order to guarantee a safety margin, so the auxiliary MOSFET is sufficient for measurements up to 400 V with the 650 V GaN-HEMTs. For the diode within the clamping circuit a small signal Schottky diode should be chosen as they offer small forward voltage drop, which is relevant for the voltage U_{meas} during clamping (off-state of the *DUT*). Furthermore, the applied clamping diode must provide low parasitics (inductance and capacitance) in order to avoid impacting the measurement signal. As already mentioned, the power loss in the voltage divider's resistance is a crucial parameter in clamping circuit design. Therefore, Fig. 3 (right) depicts the power loss in R_1 depending on its resistance value and switching frequency. This measurement principle relies on a high impedance measurement of the drain-source voltage of >1 kΩ ensuring acceptable power losses in R_1. Besides the continuous power dissipation of the resistor R_1, care must also be taken for the pulsed power stress of it to avoid damaging the resistor. The full paper will deliver more details concerning the clamping circuit design and the selection of proper components in order to ensure valid and precise measurement results.

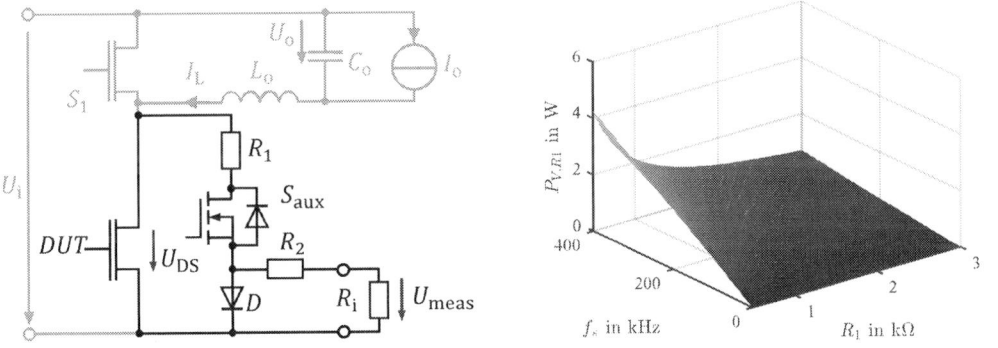

Fig. 3: Proposed clamping circuit A (left) and power loss in R_1 depending on its resistance value and switching frequency (right) for constant time delay $t_{\Delta,\mathrm{aux}} = 300$ ns

Dynamic on-state resistance measurement

As datasheets only depict the static on-state resistance of the GaN devices, measurements are the only way to gain information concerning the on-state resistance during switching operation, which can depend on various parameters like manufacturer and kind of switch, off-state blocking voltage, blocking and on-time, switching current, switching speed and temperature. As the terms switching frequency and duty cycle are common parameters of power electronic systems, these parameters are used for the following studies instead of conduction (on-state) and blocking (off-state) duration. The experiments study the impact of blocking voltage under hard switching and soft switching conditions. All parameter studies use the improved version of the clamping circuit proposed here and shown in Fig. 3 (left) with the aforementioned measurement equipment. In order to calculate the resulting on-state resistance from the measured values, the measured voltage $U_{\mathrm{meas}}(t)$ is corrected by its offset voltage and multiplied by the attenuation of the clamping circuit's voltage divider. The voltage offset can be derived during blocking phase of both *DUT* and auxiliary switch. Therefore, it is possible to accurately determine the offset only a few μs before the actual on-state resistance measurement. Dividing the obtained voltage by the drain current which is also corrected by the offset delivers the $R_{\mathrm{DS,on}}$ of the measured device.

$$R_{\mathrm{DS,on}} = \frac{k_{\mathrm{U}} \cdot (U_{\mathrm{meas}}(t) - U_{\mathrm{offset}})}{I_{\mathrm{L}} - I_{\mathrm{offset}}} \tag{4}$$

As even the static on-state resistance of power semiconductors depends on the occurring device junction temperature, the temperature dependent static on-state resistance characteristic curve is measured by putting the *DUT* into a thermo cabinet and sweeping the temperature. During dynamic on-state resistance measurement the appearing device temperature is also logged with a thermocouple mounted on the transistor package. This allows for separating the temperature dependent static on-state resistance from the increase of on-state resistance due to fast charge trapping effects (dynamic on-state resistance). In order to prevent the effects of the increase in static on-state resistance due to the rising temperature being included in the diagrams of the dynamic on-state resistance, all measurements are converted to 25°C using the static on-state resistance curve. This ensures that the increase of resistance within the diagrams are contributed to the dynamic on-state resistance and not to thermal effects which impact the static on-state resistance.

Before beginning with detailed analyses of the dynamic on-state resistance the validity of the complete measurement setup must be guaranteed. The measurement of a high voltage SiC-MOSFET serves as verification of the setup as this device shows a constant on-state resistance over time and blocking voltage independent when applying a fixed temperature. Therefore the first step is to prove that this is valid for the SiC-MOSFET (C3M0065090J) [10]. The measurements correlate with the corresponding datasheet, as it delivers the data sheet value of 65 mΩ. Despite an optimized layout, parasitic inductances and packaging parasitics cause ringing during switching of the *DUT*. Nevertheless, a constant on-state resistance of 65 mΩ is present approximately 2 μs after turn-on and is shown in Fig. 4 for soft switching.

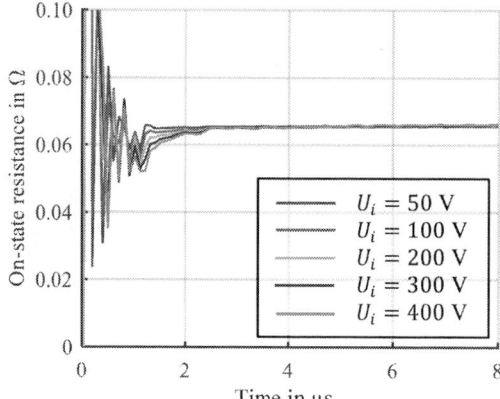

Fig. 4: Measured on-state resistance of a SiC-MOSFET depending on blocking voltage under soft switching obtained with clamping circuit A at $f_s = 50$ kHz, $|I_L| = 2$ A and $\delta = 0.5$.

The study of the dynamic on-state resistance is performed for a 650 V enhancement mode GaN-HEMT (GS66516B) which can be purchased off the shelf [11]. This device is the bottom side cooled version of a 25 mΩ 650 V GaN-HEMT. The blocking voltage dependency is the parameter to be studied. Its results for hard and soft switching are shown in Fig. 5 with a duty cycle of 50 %.

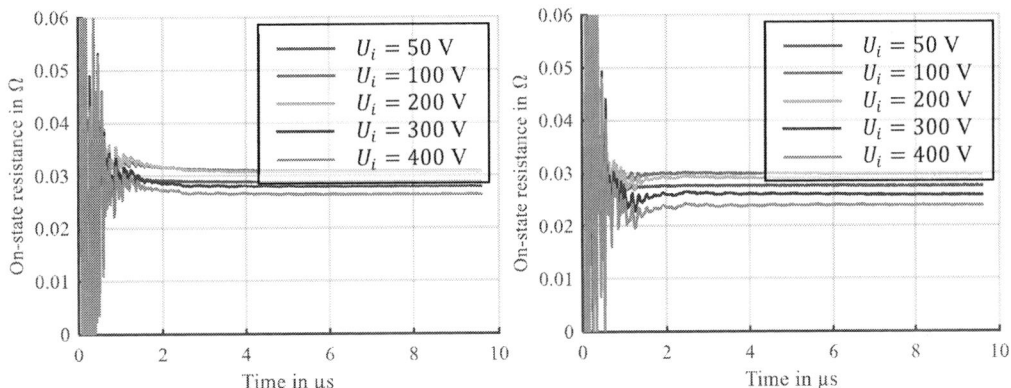

Fig. 5: Measured on-state resistance of a 650 V GaN-HEMT depending on blocking voltage under hard switching (left) and soft switching (right) obtained with clamping circuit A at $f_s = 50$ kHz, $|I_L| = 7$ A and $\delta = 0.5$.

Table I gives the device temperatures of the GaN-HEMT for the operating points of Fig. 5. As the charge trapping effects depend on the device temperature, it is useful to deliver the temperature for the operating points.

Table I: Device temperature for the operating points

Blocking voltage	Temperature hard switching	Temperature soft switching
50 V	40 °C	36 °C
100 V	43 °C	38 °C
200 V	50 °C	42 °C
300 V	59 °C	46 °C
400 V	70 °C	52 °C

During hard switching, an approximately time-constant on-state resistance can be observed for the parameters given in Fig. 5 (left) which is slightly higher than the static on-state resistance of 25 mΩ stated in the corresponding datasheet. Fig. 5 (right) depicts the dynamic on-state resistance under soft switching for the same operating point (I_L is now negative due to soft switching). The waveforms exhibit a constant on-state resistance, but the resistance is slightly lower than that under hard switching. This effect is at least partly related to the parasitic voltage drop caused by the load current affecting the internal gate-source voltage of the DUT. The variation for hard and soft switching includes the range from 50 V up to 400 V. Fig. 6 depicts the evolution of the dynamic on-state resistance over blocking voltage under hard switching with and without post-measurement temperature compensation. Concerning the temperature-compensated results, the on-state resistance achieves its maximum values at 100 V and 200 V. At voltage below 100 V as well as above 200 V, the dynamic on-state resistance is lower and reaches a minimum at a blocking voltage of 400 V.

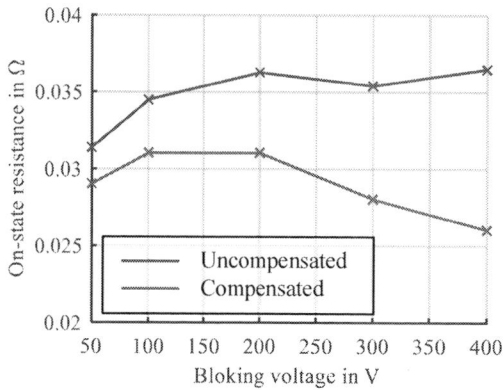

Fig. 6: Measured values of the on-state resistance of a 650 V GaN-HEMT depending on blocking voltage under hard switching obtained with clamping circuit A at $f_s = 50$ kHz, $|I_L| = 7$ A and $\delta = 0.5$.

At a first glance, this result seems astonishing, as the dynamic on-state resistance tends to increase with rising blocking voltage [3] [5]. However, similar behavior has already been observed with other GaN devices [12].

Second Test Setup for Measurement Results Validation

As the previous measurements show, there is an unexpected behavior of the dynamic on-state resistance depending on the blocking voltage. In order to be able to make reliable conclusions concerning the impact of the blocking voltage on the dynamic on-state resistance, a second test setup shall be presented, which serves to verify the results gained by the test setup with clamping circuit A.

The second implementation of the clamping circuit, hereinafter referred to as clamping circuit B (Fig. 7) has been previously used in [4] [13]. This implementation is based on a normally-off GaN transistor (GS-065-004-1-L) [14] as actively controlled clamping switch instead of a Si-MOSFET. Consequently, the on-state resistance of T_{clamp} is significantly smaller compared to the Si-MOSFET clamping switch while maintaining relatively small parasitic capacitances. Furthermore, the resistance of R_1 and R_3 are chosen significantly smaller compared to clamping circuit A. Its impedance is in the range of 50 Ω and therefore matches to the scopes input resistance R_i. Therefore, clamping circuit B relies on low impedance voltage measurement and therefore the auxiliary switch is only in on-state, when the *DUT* conducts and the on-state voltage is recorded. Otherwise, the measurement instrument will be damaged. In contrast, clamping circuit A bases on a high-impedance measurement of the drain-source voltage of the *DUT*, as $R_1 > 1$ kΩ. In this case, the auxiliary switch can be turned on slightly before *DUT*'s turn-on. When comparing both clamping circuits, it can be noticed that their circuit diagrams look similar, but the impedances differ significantly.

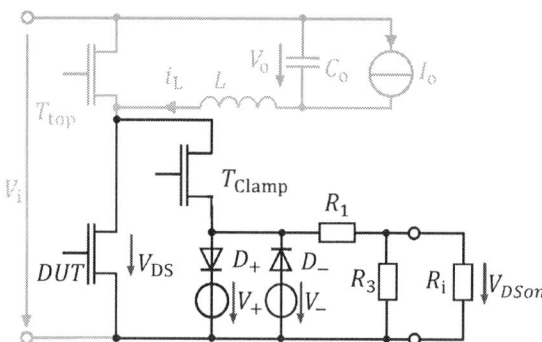

$R_1 = R_3 = R_i = 50\ \Omega$

$R_{\text{on},T_{\text{clamp}}} = 500$ mΩ

V_+ and V_- are adjusted to the expected on-state voltage of the *DUT*.

Fig. 7: Test setup with clamping circuit B

Switching transitions of the DUT causes either a rising or a falling edge of its Drain-Source voltage. In this clamping setup the clamping transistor is excited to the same voltage slew rate resulting in a displacement current through the output capacitance of T_{Clamp} and disturbing the measured on-state voltage during switching transitions. To prevent amplifier saturation or even damage of the connected measurement equipment, auxiliary clamping diodes D_+, D_- effectively suppress parasitic perturbations of the measured signal. The measuring range of the clamping circuit is tuned by adjusting the voltage sources V_+, V_- according to the specifications of the investigated DUT. Assuming an on-state voltage within the expected measuring range and consequently blocking state of the auxiliary diodes D_+, D_- the voltage attenuation of clamping circuit B is given by

$$k_{\mathrm{V}} = \frac{\left(R_1 + R_{\mathrm{on,clamp}}\right)R_3 + \left(R_1 + R_{\mathrm{on,clamp}}\right)R_{\mathrm{i}} + R_3 R_{\mathrm{i}}}{R_3 R_{\mathrm{i}}} \tag{5}$$

The clamping transistor in the second clamping circuit is a normally-off GaN-on-Si HEMT being susceptible to trapping effects just like the characterized DUT. However, the resulting measured error is sufficiently small. First, the extend of dynamic on-state resistance significantly depends on the (hard-) switched Drain-current [4] which is negligible small in case of T_{clamp}. Consequently the R_{on}-increase of the clamping transistor remains strongly limited as well. Second, the rated on-state resistance of T_{clamp} is two magnitudes smaller compared to R_1 causing a considerable small error in (5), where even an unlikely doubled on-state resistance of T_{clamp} will cause a measurement error below 0.7%. In order to verify the validity of the second clamping circuit, measurements on a normally-on GaN device (LMG3410) are carried out [15]. Since state-of-the-art normally-on devices do not exhibit charge trapping effects, dynamic on state resistance effects are not expected to occur. Fig. 8 shows the on-state resistance of the normally-on GaN transistor in a double pulse test [15]. The first pulse ends at $t = -1\ \mu s$ and the second one lasts from 0 to 1 μs. In this time ranges, the transient on-state resistance is recorded. The measured on-state resistance value is similar to the value stated in the corresponding datasheet of 75 mΩ and as expected it does not show dynamic on-state effects proving the validity of the test setup.

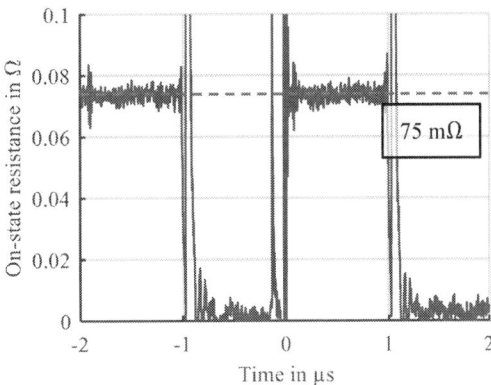

Fig. 8: Verification of clamping circuit B measuring the on-state resistance of a normally-on GaN device (LMG3410) during double pulse test at $V_i = 100$ V.

After ensuring the validity of the results obtained with clamping circuit B, measurements concerning dynamic on-state resistance of the normally-off GaN-HEMTs targeted in this paper can be carried out. In contrast to the previous measurements of the bottom side cooled device GS66516B using clamping circuit A [16], now the top side cooled counterpart GS66516T is studied. As the device characteristics in the corresponding datasheet are coherent, it is expected, that the same chip is used in both devices [11] [16]. However, small deviations between individual devices may remain due to different cooling concepts and the island technology employed by GaN Systems.

Fig. 9 shows the measurement results in continuous operation and thermal steady state depending on the blocking voltage under hard switching. The results are obtained using clamping circuit B.

This setup delivers a similar dependency of the on-state resistance on the blocking voltage compared to the results gained with clamping circuit A. The results in Fig. 9 include the measured on-state resistance as well as the value converted to 25 °C to eliminate the temperature effect of the static on-state resistance. However, the measurements of clamping circuit A and B do not exactly deliver the same results. This is related to different cooling concepts of both test setups. Circuit A is operated without any heatsink and relies on natural convection. In contrast, circuit B uses a water-cooling system and therefore the occurring device temperatures are significantly lower. The difference between the measured on-state resistance (blue) and the post-measurement temperature compensated on-state resistance (red) is also smaller with circuit B than with circuit A, since the device temperatures occurring in circuit B are significantly lower. Charge trapping and detrapping effects impacting the dynamic on-state resistance depend on the device temperature as well and thus results from both circuits are not directly comparable to each other. However, both test setups deliver the same dependency of the on-state resistance on the applied voltage. Furthermore, both measurements prove that the dynamic on-state resistance does not necessarily have the ultimate impact at the highest voltage. The studied GaN transistors exhibit their maximum on-state resistance between 100 V and 300 V. The measurements show, that the increase of on-state resistance caused by charge trapping effects is much smaller than the increase due to rising device junction temperature.

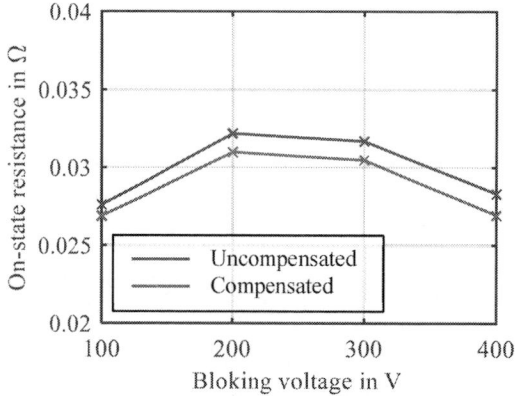

Fig. 9: Measured values of the on-state resistance of a 650 V GaN-HEMT depending on blocking voltage under soft switching obtained with clamping circuit B at $f_s = 50$ kHz, $|i_L| = 10$ A and $\delta = 0.5$.

Conclusion

In this work, an already published clamping circuit for measuring the dynamic on-state resistance of GaN-HEMTs is improved in order to extend the measuring range up to 650 V. This article discusses various problem occurring when using standard oscilloscopes for this measurement task. After a validation of the test setup by a SiC-MOSFET which complies with the on-state resistance stated in the corresponding datasheet, measurements for hard and soft switching deliver the results of the dynamic on-state resistance for a high voltage GaN-HEMT, showing its blocking voltage dependency. Due to the unexpected behavior of it depending on the blocking voltage, a second clamping circuit (B) serves to verify the results of the previous measurements. Although the circuits have a similar circuit diagram, they differ in the impedance level and thus in their function. The results demonstrate, for the same type of *DUT*, that both test setups deliver the same on-state resistance in terms of measurement accuracy. The measurements for the studied devices show, that the increase of the on-state resistance due to higher component temperatures is significantly higher than the increase caused by charge trapping effects.

References

[1] M. Meneghini, P. Vanmeerbeek, R. Silvestri, S. Dalcanale, A. Banerjee, D. Bisi, E. Zanoni, G. Meneghesso and P. Moens, *Temperature-Dependent Dynamic Ron in GaN-Based MIS-HEMTs: Role of Surface Traps and Buffer Leakage,* IEEE TRANSACTIONS ON ELECTRON DEVICES, pp. 782-787, 2015.

[2] B. J. Galapon, A. J. Hanson and D. J. Perreault, *Measuring Dynamic On Resistance in GaN Transistors at MHz Frequencies,* 2018 IEEE 19th Workshop on Control and Modeling for Power Electronics (COMPEL), 2018.

[3] B. Kohlhepp, D. Kübrich and T. Dürbaum, *Measuring and Modeling of Dynamic On-State Resistance of GaN-HEMTs,* Proceedings of 21th European Conference on Power Electronics and Applications (EPE'19 ECCE Europe), 2019.

[4] C. Kuring, M. Tannhäuser and S. Dieckerhoff, *Improvements on dynamic on-state resistance in normally-off GaN HEMTs,* Proceedings of PCIM Europe 2019, 2019.

[5] B. Kohlhepp, D. Kübrich, M. Tannhäuser, A. Hoffmann and T. Dürbaum, *Test Setup for Dynamic On-State Resistance Measurement of High- and Low-Voltage GaN-HEMTs under Hard and Soft Switching Operation,* IEEE Transactions on Instrumentation & Measurement, 2020.

[6] Datasheet: Tektronix Inc., *DPO7000,* 2018.

[7] Datasheet: Rohde & Schwarz GmbH & Co. KG, *RTO Digital Oscilloscope,* 2018.

[8] Datasheet: National Instruments, *PXIe-5122 Specifications,* 2017.

[9] Datasheet: Semiconductor Components Industries LLC (ON Semiconductor), *FQT1N60C,* 2007.

[10] Datasheet: Cree Inc., *C3M0065090J,* 2018.

[11] Datasheet: GaN Systems Corporation, *GS66516B,* 2019.

[12] C. Liu, A. Salih, B. Padmanabhan, W. Jeon, P. Moens, M. Tack and E. De Backer, *Development of 650V Cascode GaN Technology,* Proceedings of PCIM Europe 2015, pp. 994-1001, 2015.

[13] C. Kuring, N. Wieczorek, O. Hilt, M. Wolf, J. Böcker, J. Würfl and S. Dieckerhoff, *Impact of Substrate Termination on Dynamic On-State Characteristics of a Normally-off Monolithically Integrated Bidirectional GaN HEMT,* Conf. Proc. of IEEE Energy Conversion Congress and Exposition (ECCE), 2019.

[14] : *GS-065-004-1-L, Rev 200422,* 2020.

[15] Datasheet: Texas Instruments Inc., *LMG3410,* 2020.

[16] Datasheet: GaN Systems Corporation, *GS66516T,* 2019.

Analysis Of DC-side Fault Response of MMCs with Controlled Fault Blocking Capability for Different Transmission Line Types

Willem Leterme[*], Paul D. Judge[†] and Tim C. Green[‡]

[*]KU Leuven/EnergyVille, [†]The University of Edinburgh, [‡]Imperial College London

[*]Kasteelpark Arenberg 10/Thor Park 8310

[*]3000 Leuven, 3600 Genk

Email: [*]willem.leterme@esat.kuleuven.be

Keywords

≪Converter Control≫, ≪Fault Handling Strategy≫, ≪HVDC≫, ≪Multilevel Converters≫.

Abstract

MMCs with controlled fault blocking capability retain control of their currents during a dc-side fault, thereby reducing the required interruption capabilities for switchgear. To design the dc-side control to achieve this capability, it is important to take into account the interaction between the converter control and the transmission line during a short-circuit on the transmission line. This paper uses a dc-side equivalent model to assess the interactions of the converter control for two types of converters, i.e., full-bridge and hybrid, with two main types of transmission lines, i.e., cable and overhead line, during dc-side faults. In general, the dc-side voltages and currents for a full-bridge MMC connected to an overhead line show more oscillatory behavior compared to an MMC connected to a cable. Furthermore, a hybrid MMC connected to an overhead line may provide a more damped dc-side fault response due to its limited negative voltage capability.

Introduction

DC systems are increasingly used within the power system, with applications in High-, Medium- and Low-Voltage dc (HVDC, MVDC and LVDC) [1]. In these systems, converters with fault blocking capability are considered attractive given that they can eliminate the need for dc-side switching equipment with fault interruption capability, as e.g., described in [2, 3] and used in a protection strategy as discussed in [4, 5]. In HVDC, modular multilevel converters (MMCs) with controlled fault blocking capability are planned for first installation in a German project connecting the north of the country to the south [6]. The increased use of these converters calls for a unified approach towards modeling the dc-side fault behavior, as detailed converter models are computationally inefficient and control details may not be available to the users of such converters or to suppliers of other equipment connected to these converters. In this paper, we analyze the influence of controller gains as well as transmission line type on the dc-side fault behavior of converters with controlled fault blocking capability using the modeling approach introduced in [7].

Modeling for DC-side Fault Analysis

For the analysis, this paper uses the dc EMT-type model as introduced and experimentally verified in [7]. This model only models the equivalent dc-side circuit of the converter (Fig. 1b). The model assumes variations in submodule voltages to be limited. As a consequence, energy balancing controls and submodule voltage balancing controls are not considered.

The dc EMT-type model uses the circuit as shown in Fig. 1b as the electrical model for the converter and only models the dc-side component of the current control. The converter is controlled using a discrete-time current control, which includes sensor and control delays, τ_m and τ_c. A limit is applied to the

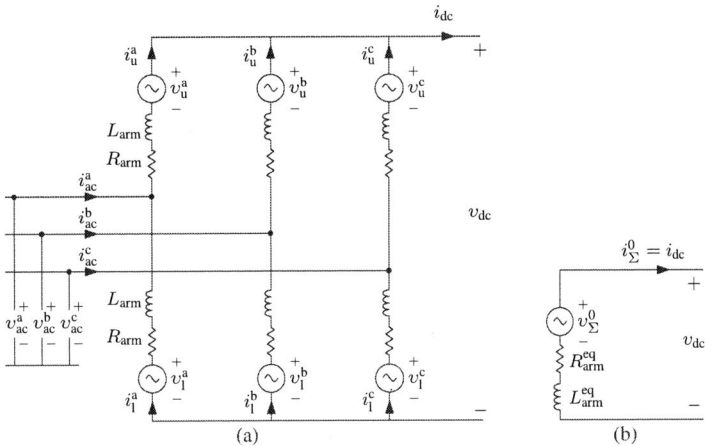

Fig. 1: MMC equivalent circuit diagram (a) and dc component (b) equivalent circuit diagrams.

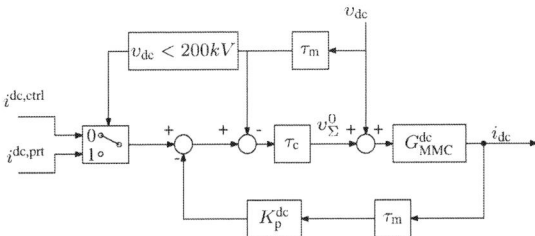

Fig. 2: Block diagram of converter dc-side current control and protection.

dc-side component of the voltage reference, to account for the limited voltage capability of the converter. A simplified version of the current control (without limit) is shown in Fig. 2. The equivalent RL-circuit shown in Fig. 1b is indicated by G_{MMC}^{dc}. The proportional constant of the feedback control is indicated by K_{p}^{dc}. The current reference is switched from the control reference ($\bar{i}^{\text{dc,ctrl}}$) to the protection reference ($\bar{i}^{\text{dc,prt}}$) based on an undervoltage criterion, but may also be based on an overcurrent criterion.

Although the dc EMT-type model only provides an approximation for the arm voltage limits, the model is used as it fits the purposes of the demonstration case studies. The purpose of the demonstration case studies are (i) to show the interactions between the converter control and transmission line type, and (ii) to demonstrate the differences between a hybrid and full-bridge MMC. The dc EMT-type model is able to demonstrate the differences qualitatively through the limit imposed on the dc-side negative arm voltage, but more accurate models, such as the three-phase EMT-type model [7] or models such as the one introduced in [8] should be used to calculate the effect of this limit more accurately.

DC-side Fault Response Under Various Transmission Line Types

Case Study

To investigate the dc-side fault response, a case study is set up as shown in Fig. 3. The case study investigates three cases, namely, an MMC connected to a stiff dc voltage (i) directly, (ii) through a cable and (iii) through an overhead line. The MMC parameters are based on the ones given in Table I and the overhead line and cable parameters are given in Table II. The cables are buried 1 m below ground and the ground resistivity is 0.5 Ω·m. To investigate the effect of close-up and remote faults, the transmission line length is varied between 25 km and 300 km. The considered sensor and actuator delays associated with the converter control are 100 μs each, taking a value close to the ones found in [3]. The converter negative voltage capability is assumed to be -650 kV in case of the full-bridge MMC and -325 kV in case of the hybrid MMC. The proportional feedback gain of the dc-side current control is varied between

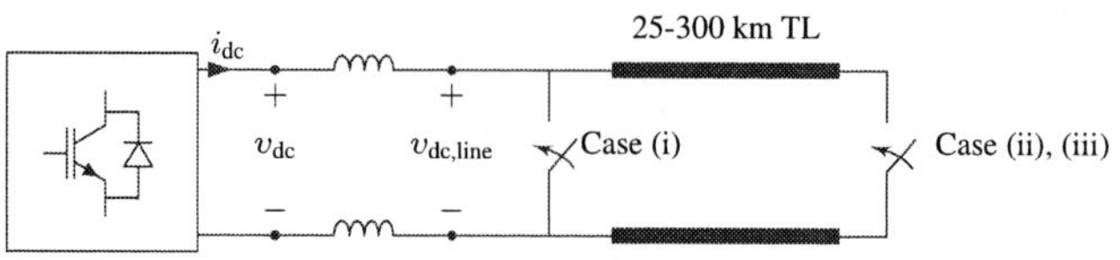

Fig. 3: Case Study Setup

Table I: Converter Parameters

Parameter Value	
Rated dc power	960 MW
dc nominal pole-to-pole voltage	640 kV
dc negative voltage capability (full-bridge)	-650 kV
dc negative voltage capability (hybrid)	-325 kV
Arm inductance L_{arm}	0.05 H
Arm resistance R_{arm}	1.6 Ω

Table II: Overhead line and cable parameters.

Parameter	Value
No. of conductors	2
Configuration	Horizontal
Conductor spacing	13.4 m
Conductor height	25 m
Conductor sag	13.9 m
Conductor outer radius	1.62 cm
Conductor DC resistance	51.9 mΩ/km
Ground resistivity	50 $\Omega \cdot$m

Conductor	Outer Radius [mm]	ρ [$\Omega \cdot m$]	μ_r	ε_r
Core	25.13	1.72e-8	1	-
Insulation	48.38	-	-	2.3
Screen	54.13	21.4e-8	1	-
Insulation	54.13	-	-	2.3
Armor	59.73	1.38e-7	-	10
Insulation	64.73	-	-	2.3

10 and 150, with intermediate steps taken as multiples of 50. The study case is modeled in EMT-type software [9], where a timestep of 10 μs is chosen for the simulation.

The transmission lines are modeled using a frequency-dependent phase model. To obtain a stable simulation, in the case of an overhead line, capacitors were put at the dc terminals of the converter, with a capacitance-to-ground of 1 μF.

Results

Effect of DC-side Proportional Control on Full-Bridge Response

The dc-side fault response for each of the three cases differs according to the influence of the converter current control on the dc-side voltage. For a stiff voltage source (case (i), no transmission line), the converter current control has no influence on the line-side voltage. The dc-side fault current shows a uniformly decaying behavior for all K_p^{dc}, except for $K_p = 150$, where the dc-side fault current undershoots before reaching zero (see Fig. 4). For the cable (case (ii)), the dc-side voltage decays more slowly due to the capacitance of the cable, and oscillates around zero before reaching steady state. In the overhead line case, the converter current control has the largest impact on the dc-side voltage. The dc-side voltage shows the most oscillatory behavior of the three cases. The peak current is the lowest of all cases, as it is limited by the line inductance, but like the dc-side voltage, it shows the most oscillatory behavior of the three cases.

In the following paragraphs, the influence of the dc-side proportional constant, K_p^{dc}, on three parameters

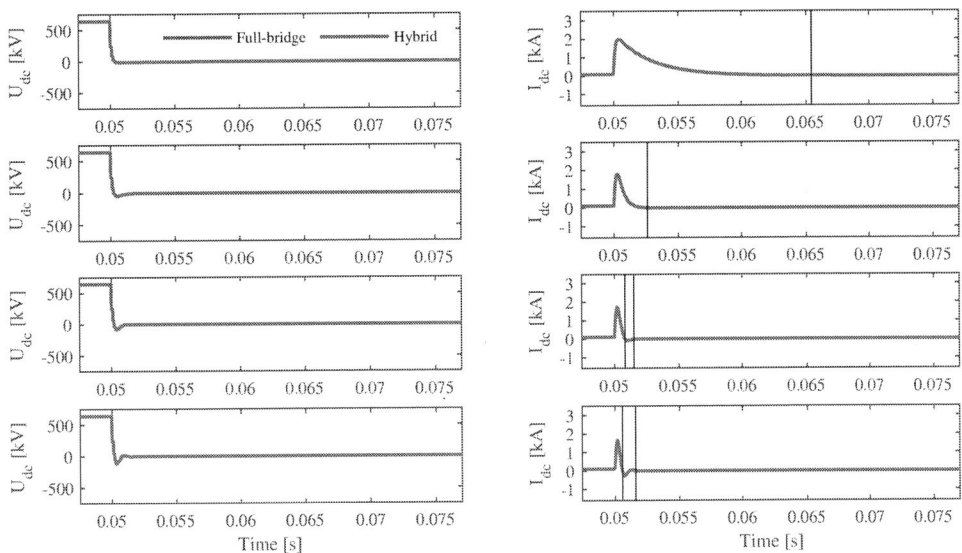

Fig. 4: Dc-side fault response for stiff dc voltage case (upper to lower plot: $K_p = 10, 50, 100, 150$). The solid black lines represent the "rise" and "settling" times, respectively.

is analyzed. A first parameter of interest is the time until the fault has reached a value at which it can be interrupted by switchgear without fault current interruption capability. To characterize this, a "rise time" and "settling time" are defined as the first instant at which the current falls below 0.01 kA and the instant at which the current permanently remains below 0.01 kA, respectively. The second parameter of interest is the peak dc-side fault current, and the third parameter of interest is the peak value of the dc-side voltage.

For cases (i) and (ii), in general, the settling time of the current decreases as K_p^{dc} increases, although this trend is only valid for $K_p^{dc} < 150$ (see Fig. 4, Fig. 5 and Fig. 6). For case (iii), the value of K_p^{dc} leading to the minimum settling time of the current depends on the length of the transmission line. For the cases of 25 km and 300 km, the settling time is the lowest for $K_p^{dc} = 50$ and 10, respectively. In general, increasing K_p^{dc} leads to an increased number of oscillations in the cases with K_p^{dc} above 50. In conclusion, for a stiff voltage source and a cable, increasing K_p^{dc} has in general a positive effect on the rise and settling times. For overhead line case, increasing K_p^{dc} leads to a lower rise time, whereas the settling time in general increases.

The peak dc-side fault current, a second parameter of interest, shows different behavior for each of the cases, and the general observed trend is that the converter control has more impact on the peak dc-side fault current whenever the slope of the dc-side voltage is less steep. In this model setup, for all cases, the peak fault current decreases with an increase in K_p^{dc}. Using the equivalent diagram of Fig. 1b, it can be seen that the dc-side current increases whenever $v_\Sigma^0 > v_{dc}$ and vice versa. For a less steep dc-side voltage, the dc-side current controller may set v_Σ^0 below v_{dc} even during the increase of the dc-side fault current. Therefore, this is more effective for remote faults in the cable case compared with the overhead line case (Fig. 6 and 8). An opposite effect can be observed for the close-by faults (Fig. 5 and 7).

The third parameter to take into account is the dc-side voltage. In the second and third case, the current controllers may influence the dc-side voltage, as can be seen in Fig. 5 and Fig. 7. For the cable case, the value of K_p^{dc} has only a minor effect on the dc-side voltage. In the overhead line case, the current controller has a considerable impact on the dc-side voltage in terms of absolute peak voltage and oscillatory behavior. For the lowest K_p^{dc}, the peak voltage and number of oscillations remain limited. For the highest K_p^{dc}, the peak voltage may increase slightly beyond the maximum voltage and oscillatory behavior is at its worst. In conclusion, for the overhead line case, increasing K_p^{dc} may not have the overall beneficial

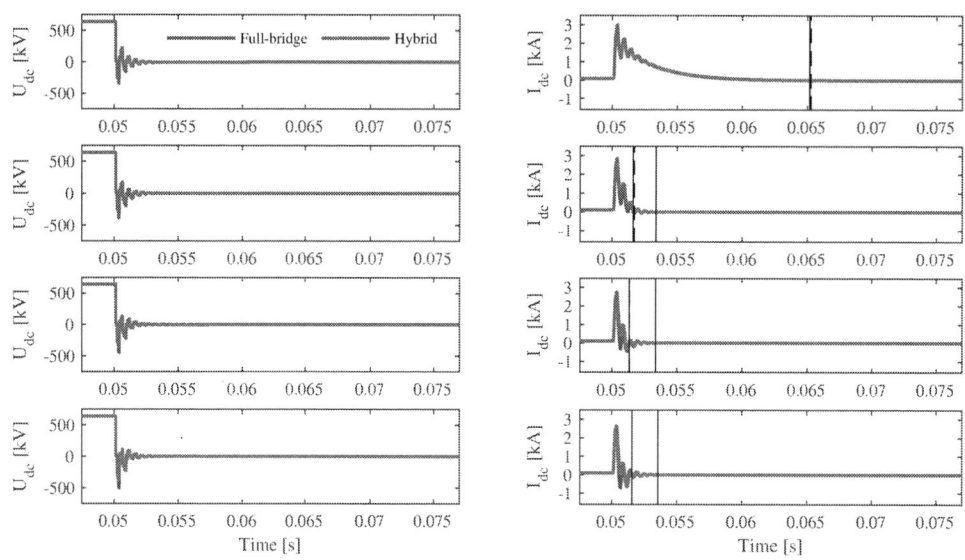

Fig. 5: Dc-side fault response for 25 km cable case (upper to lower plot: $K_p = 10, 50, 100, 150$). The solid and dashed black lines represent the "rise" and "settling" times for the full-bridge and hybrid case, respectively.

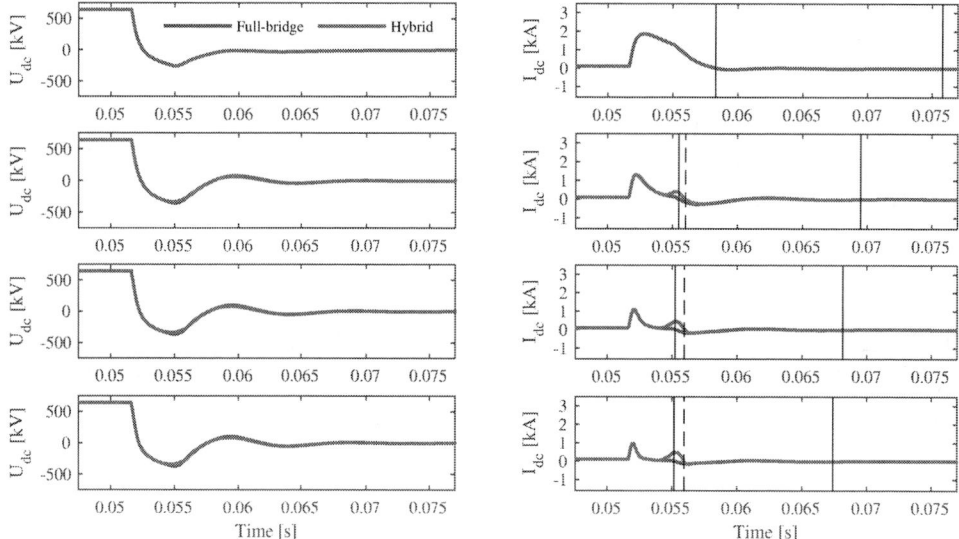

Fig. 6: Dc-side fault response for 300 km cable case (upper to lower plot: $K_p = 10, 50, 100, 150$). The solid and dashed black lines represent the "rise" and "settling" times for the full-bridge and hybrid case, respectively.

effect as seen in the case with the stiff dc-side voltage.

Difference between full-bridge and hybrid MMC

For case (i), i.e., the converter connected to a stiff dc-side voltage source, there is no difference between the responses of the full-bridge and the hybrid MMC (see Fig. 4). In this case, the negative voltage capability limit is not reached for any of the proportional control constants. As such, there is no difference in the current control response between the two cases.

In case the converter is connected to a cable, the full-bridge converter outperforms the hybrid converter

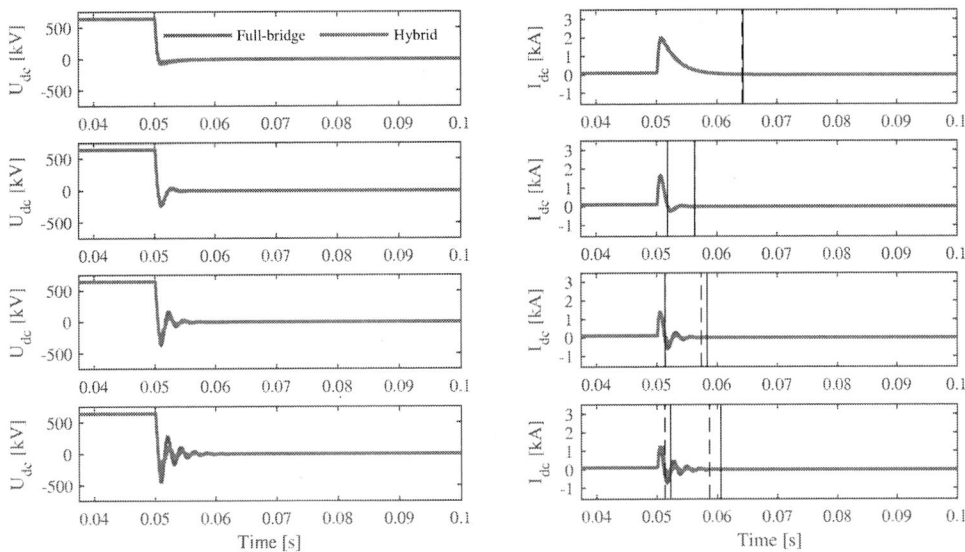

Fig. 7: Dc-side fault response for 25 km overhead line case (upper to lower plot: $K_p = 10, 50, 100, 150$). The solid and dashed black lines represent the "rise" and "settling" times for the full-bridge and hybrid case, respectively.

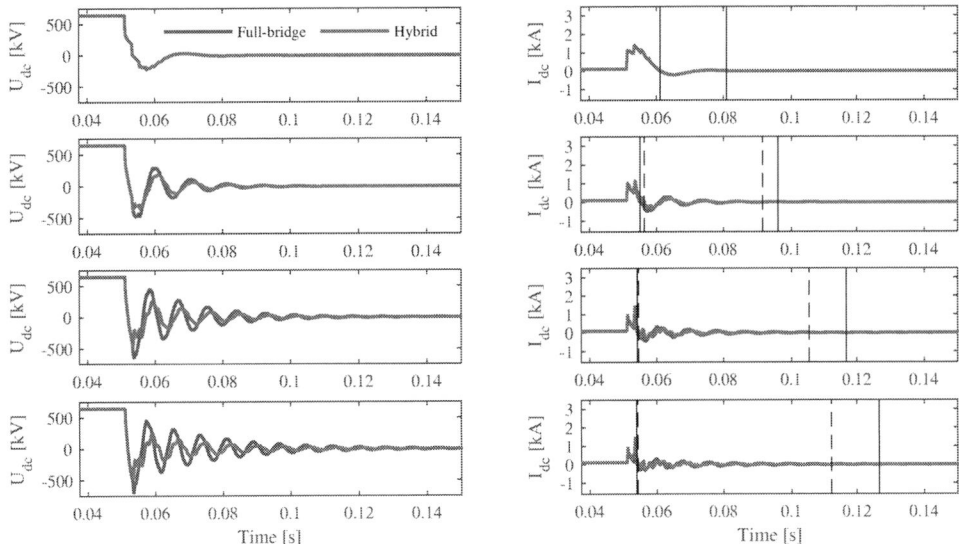

Fig. 8: Dc-side fault response for 300 km overhead line case (upper to lower plot: $K_p = 10, 50, 100, 150$). The solid and dashed black lines represent the "rise" and "settling" times for the full-bridge and hybrid case, respectively.

in controlling the dc-side current, but there is very little difference between the rise and settling times. The dc-side current of the full-bridge MMC always lies below the one of the hybrid MMC (Fig. 5 and 6). This is due to the fact that the full-bridge retains full control of its dc-side current given that the dc-side voltage never falls below its negative voltage capability. By contrast, the hybrid MMC loses control of its dc-side current whenever it cannot generate a dc-side voltage lower than the one resulting from the fault. At those instants, the dc-side current shows a temporary uncontrolled increase (e.g., see Fig. 6 for $K_p^{dc} > 50$). The difference in rise times is in the order of one millisecond and in most cases, there is no difference in settling times.

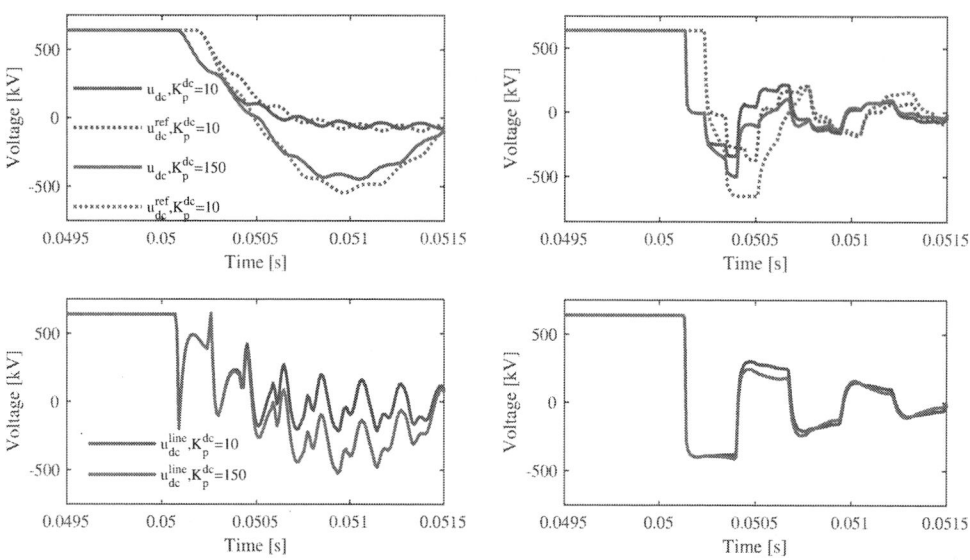

Fig. 9: DC-side fault response of full-bridge for 25 km overhead line case (left) and 25 km cable case (right).

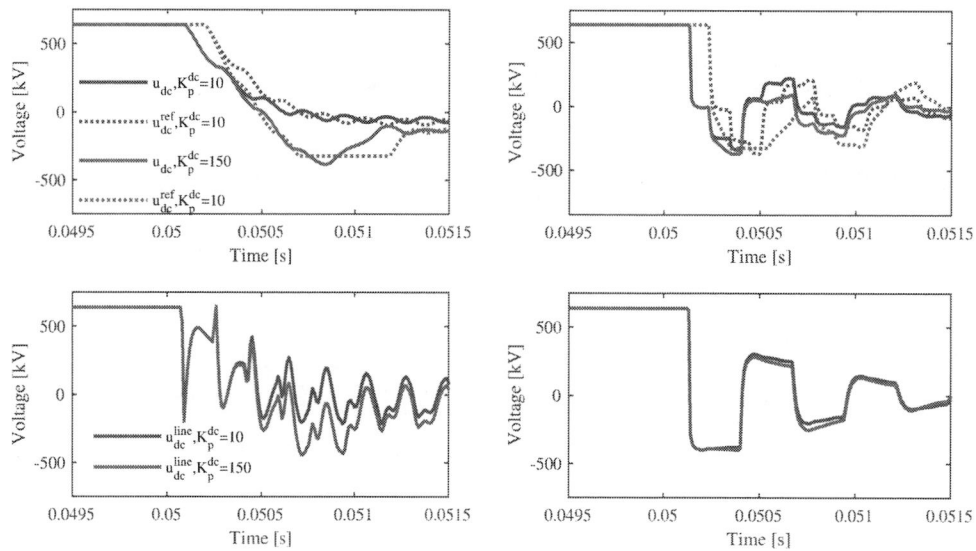

Fig. 10: DC-side fault response of hybrid MMC for 25 km overhead line case (left) and 25 km cable case (right). Voltage reference input shown in dashed lines.

For the case where the converter is connected to an overhead line (case (iii)), the converter response should be considered distinctly for the cases in which the line length is 25 km and 300 km. In case the line length is 25 km (Fig. 7), the voltages and currents as generated by the hybrid MMC are considerably less oscillatory compared with those generated by the full-bridge MMC. This reduction in oscillatory behavior, especially at high K_p^{dc}, may be attributed to the fact that the hybrid MMC is forced to retain the voltage at its negative arm voltage limit for a certain amount of time. This seems to have a damping effect on dc-side voltages and currents, causing the fault current to decay more quickly. In case the line length is 300 km, the above described effect plays a lower role and both responses show oscillatory behavior. In all cases with $K_p^{dc} > 50$, the amplitude of the voltage oscillations is higher for the full-bridge MMC in comparison with the hybrid MMC (Fig. 8).

Control response during DC-side faults

The origin of the oscillations in the DC-side fault response is different for the overhead line-case and the cable case.

For the overhead line case, the (zero-crossing) oscillations visible in Fig. 7 and Fig 8 for K_p^{dc} beyond 100 and 50, respectively, are linked to interactions between the converter control and the overhead line, but are not directly linked to reflections of traveling waves originating from the fault. In this case, the traveling wave reflections appear as small ripples superimposed to the lower frequency oscillation (Fig. 9, left plots). For the case of $K_p^{dc} = 150$, the succession of incoming waves and the delayed control response cause a cascading effect which initially pushes the dc-side voltage downwards and triggers the large oscillation. In case of a low proportional constant, and hence a less aggressive response of the controller, these oscillations do not occur. In the case of a hybrid MMC, the oscillations are damped even for high proportional constants, given that the negative voltage limit is hit in an earlier stage, resulting in a reduced magnitude of oscillations compared to the full-bridge case (see Fig. 10 and 8).

For the cable-case, the zero-crossing oscillations more closely follow the reflection pattern of the traveling waves generated by the fault (Fig. 9, right plots). The voltage reference takes a similar shape as the dc-side voltage, with a delay of 200 μs due to control and sensor delays. Given that the cable represents a more stiff voltage in comparison to the overhead line, the converter control has a relatively low influence on the shape of the DC-side voltage. In contrast with the overhead line case, the lower negative arm voltage limit of the hybrid MMC does not seem to have a beneficial effect in damping the oscillations.

Conclusion

The response of an MMC with controlled fault blocking capability depends on the transmission line type it is connected to and its negative voltage capability. This paper has used a relatively simple, but experimentally verified, dc-side equivalent circuit with proportional feedback control in a preliminary analysis of the dc-side fault response of an MMC with controlled fault blocking capability connected to an overhead line or cable. For a full-bridge MMC connected to a cable, an increase in the proportional feedback gain leads (for the range considered) to a lower settling time and a lower peak fault current. For a full-bridge MMC connected to an overhead line, an increase in the proportional feedback gain leads to a lower peak fault current, but may lead to oscillatory behavior which results in a longer settling time. Furthermore, in the latter case, the dc-side voltage may reach unacceptably high values due to the influence of the dc-side current controller on the relatively weak dc-side terminal voltage. The analysis showed that the oscillations observed in the overhead line may be attributed to interactions of the converter control with the transmission line. By contrast, the oscillations observed in the cable case may be attributed to reflections of traveling waves initiated by the fault. As a final note, a hybrid MMC may provide a more damped response compared to a full-bridge MMC when connected to an overhead line.

References

[1] M. Barnes, D. Van Hertem, S. P. Teeuwsen, and M. Callavik, "HVDC Systems in Smart Grids," *Proceedings of the IEEE*, vol. 105, no. 11, pp. 2082–2098, Nov. 2017.

[2] R. Marquardt, "Modular Multilevel Converter topologies with DC-Short circuit current limitation," in *Proc. IPEC 2011 - ECCE Asia*, Jeju, South Korea, Jun. 2011, pp. 1425–1431.

[3] M. Winkelnkemper, L. Schwager, P. Blaszczyk, M. Steurer, and D. Soto, "Short circuit output protection of MMC in voltage source control mode," in *Proc. 2016 IEEE-ECCE*, Milwaukee, WI, USA, Sep. 2016, 6 pages.

[4] P. Cairoli, R. A. Dougal, U. Ghisla, and I. Kondratiev, "Power sequencing approach to fault isolation in dc systems: Influence of system parameters," in *Proc. 2010 ECCE*, Atlanta, GA, Sep. 2010, pp. 72–78, 7 pages.

[5] P. Ruffing, C. Brantl, C. Petino, and A. Schnettler, "Fault current control methods for multi-terminal DC systems based on fault blocking converters," *The Journal of Engineering*, vol. 2018, no. 15, pp. 871–875, 2018.

[6] J. Dorn, P. La Seta, F. Schettler, J. Stankewitz, M. vor dem Berge, R. Teixeira Pinto, K. Uecker, M. Walz, K. Vennemann, B. Rusek, J. Reisbeck, and C. Butterer, "Full-bridge VSC: An essential enabler of the transition to an energy system dominated by renewable sources," in *Proc. IEEE PES GM*, Boston, MA, Jul. 2016, 5 pages.

[7] W. Leterme, P. D. Judge, J. Wylie, and T. C. Green, "Modeling of MMCs With Controlled DC-Side Fault-Blocking Capability for DC Protection Studies," *IEEE Transactions on Power Electronics*, vol. 35, no. 6, pp. 5753–5769, Jun. 2020.

[8] H. Saad, K. Jacobs, W. Lin, and D. Jovcic, "Modelling of MMC including half-bridge and Full-bridge submodules for EMT study," in *2016 Power Systems Computation Conference (PSCC)*, Jun. 2016, pp. 1–7.

[9] Manitoba Hydro International Ltd., "EMTDC TM, Transient Analysis for PSCAD Power System Simulation," 2018. [Online]. Available: https://www.pscad.com/uploads/knowledge_base/emtdc_manual_v4_6.pdf

A Hybrid Series-Parallel Microgrid and its Low-dependent Communication Control

Lang Li[1,2], Yao Sun[1,2], Hua Han[1,2], Mei Su[1,2]
1. School of Automation, Central South University
2. Hunan Provincial Key Laboratory of Power Electronics Equipment and Grid
Changsha Hunan P.R. China
E-Mail: csu_lilang@126.com

Acknowledgements

This work was supported in part by the National Key R&D Program of China under Grant 2018YFB0606005, in part by the National Natural Science Foundation of China under Grant 61933011, in part by the Major Project of Changzhutan Self-Dependent Innovation Demonstration Area under Grant 2018XK2002, in part by the Project of Innovation-driven Plan in Central South University under Grant 2019CX003, and in part by the Hunan Provincial Key Laboratory of Power Electronics Equipment and Grid under Grant 2018TP1001.

Keywords

« Distributed», « Decentralized», « Hybrid series-parallel microgrid».

Abstract

This paper proposes a new hybrid series-parallel microgrid and its low-dependent communication control scheme. The separate distributed generators (DGs) are paralleled connected as a paralleled-connected generation (PCG) module. These PCG modules are series connected to form a hybrid series-parallel microgid. Then, a low-dependent communication control scheme is proposed, in which only the local low-bandwidth communication network is needed to realize the consensus control for the PCG modules. Further, these PCG modules are synchronized autonomously without any communications. Thus, the low-dependent communication features are obtained. Finally, the feasibility and performance are verified by simulations.

Introduction

The concept of microgrid has been drawing an increased attention to integrate distributed generators (DGs) into the modern power network [1]. The inverter-based microgrid becomes an effective way due to technical advantages [2]. According to the connected way of these inverters, the microgrid can be classified into paralleled-type microgrid, cascaded-type microgird and hybrid series-parallel microgrid [3].

The paralleled-type microgrid has been widely studied due to its capabilities of extendibility and plug-and-play [4]. For the paralleled-type microgrid, the centralized control schemes depended on the complicated high bandwidth communication and centralized controller are presented [5]. Further, the distributed control schemes are proposed based on the adjacent information with the low-bandwidth communications [6]. To imitate the characteristics of the synchronous generator, the droop control schemes as a classical decentralized control are proposed [7]. The decentralized control schemes are performed only based on the local voltage and current information.

In the middle-high voltage level power network, the cascaded-type micrigrid is an alternative solution due to its convenience of boost voltage level [8]. Lots of centralized and distributed control schemes are presented in [9-10] for the cascaded-type microgrid. In order to reduce the dependency of communications, the decentralized control strategies are introduced in [11]. However, the cascaded-type microgrid is sensitive to the single-point of failure and without the capability of plug-and-play.

In order to synthesize the features of parallel-type and cascaded-type structure, the hybrid series-parallel microgrid is proposed in [12]. In this microgrid, multiple DGs are connected in series as cascaded modules, then these modules are paralleled connected to form the hybrid series-parallel microgrid. Further, a distributed control scheme is proposed. However, this structure is sensitive to the single point of failure for the cascaded modules.

To address these concerns, this paper proposed a new series-parallel microgrid and its low-dependent communication control. This structure has the capability of plug-and-play, and robust to the single point of failure. Further, a low-dependent communication control strategy is proposed, in which the pros of distributed and decentralized control are combined. Thus, the proposed scheme is a low-dependent communication solution. The contribution of this work can be summarized as follows:

➢A new hybrid series-parallel microgrid is proposed to enrich the structure of microgrids.

➢A low-dependent communication control scheme is proposed, in which the DGs are controlled with a distributed manner in a PCG modules, and the PCG modules are controlled in a decentralized manner.

Proposed hybrid series-parallel microgrid and its control

A. Structure of Hybrid Series-Parallel Microgrid

Figure 1 shows the configuration of the proposed hybrid series-parallel microgrid in the islanded mode, in which each DG unit has an independent output LC filter. The separate DG unit is connected in parallel as a parallel-connected generation (PCG) module. Then, these PCG modules are series connected to feed the AC loads through the transmission line. The system includes D PCG modules, $D \in \{1,2,3,\cdots\}$, the kth PCG module is made of n_k DG units, $n\_k \in \{1,2,3,\cdots\}$. Proposed control scheme.

In order to realize the accurate power sharing in a PCG module and proportional power sharing in different PCG modules, the proposed control scheme is introduced in this section. In a PCG module, the distributed control is adopted to realize the consistent external characteristics control. With regard to different PCG modules, the $P-\omega$ control is used to obtain the self-synchronization in a decentralized manner.

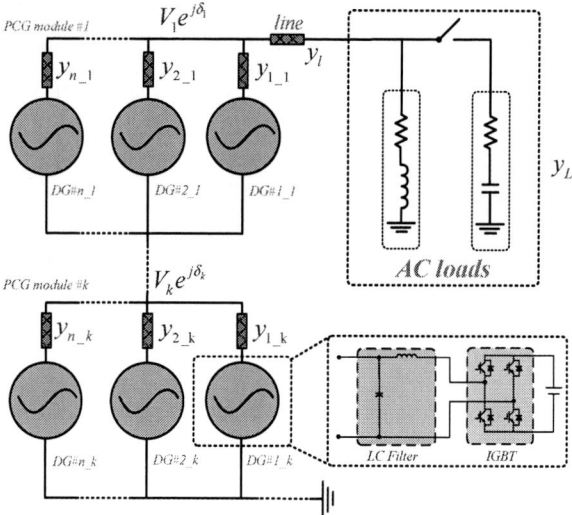

Fig. 1: Structure of hybrid series-parallel microgrid.

B. The Proposed Control Scheme

The overall schematic of the proposed scheme is depicted in Fig. 2. The proposed scheme of the i_k^{th} DG unit in a certain PCG module is given as follows,

$$\begin{cases} \omega_{i\_k} = \omega^* + m_k \, \text{sgn}\left(Q_{i\_k}\right)\hat{P}_{i\_k} + \Delta\omega_{i\_k} \\ V_{i\_k} = V^*/D \end{cases} \tag{1}$$

$$\Delta\omega_{i\_k} = \xi_\omega \sum_{j\_k \in N_i} a_{ij}\left(P_{j\_k} - P_{i\_k}\right) \tag{2}$$

where ω_{i_k}, ω^* and $\Delta\omega_{i\_k}$ are reference, nominal angular frequency and the additional correction signals, respectively. m_k is a positive coefficient of the kth module. V^* is the nominal voltage including the line drops, $\mathrm{sgn}(\cdot)$ is a signum function. $\hat{P}_{i\_k}$ is the local estimate output active power of the i_k^{th} DG unit through the average active power observer. ξ_ω is a positive coefficient. N_i denotes that the i_k^{th} DG receives information from its neighbors j_k, and $j\_k \in N_i$. $a_{ij} = 1$, if there is a communication link allowing information flow from j_k^{th} DG to i_k^{th} DG, otherwise, $a_{ij} = 0$.

Clearly, from (1)-(2), the proposed scheme among different PCG modules is performed in a decentralized manner. In a certain PCG module, only the low-bandwidth communication links are needed.

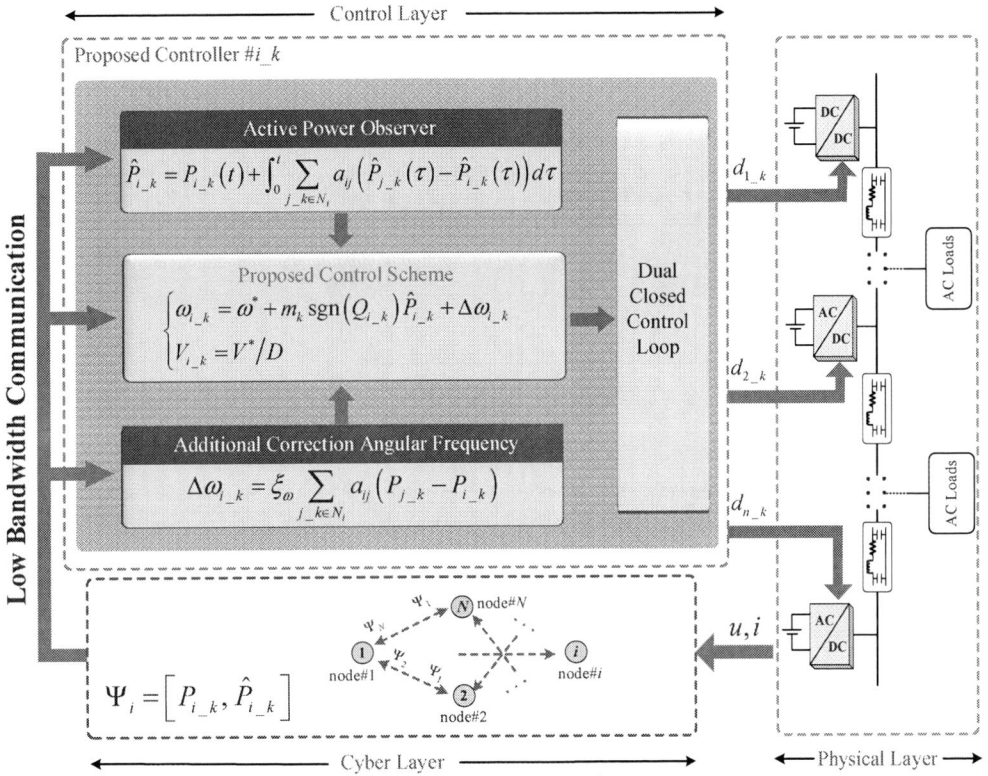

Fig. 2: Detailed configuration of the proposed scheme of the kth PCG module.

C. The Average Active Power Observer

The average output power observer is presented as

$$\hat{P}_{i\_k} = P_{i\_k}(t) + \int_0^t \sum_{j\_k \in N_i} a_{ij}\left(\hat{P}_{j\_k}(\tau) - \hat{P}_{i\_k}(\tau)\right)d\tau \tag{3}$$

where $\hat{P}_{j\_k}$ is the average estimate output power from its neighbors.

According to (3), all steady-state estimates would converge to the average output active power of the kth PCG module, namely,

$$\hat{P}_{1\_k} = \hat{P}_{2\_k} = \cdots = \hat{P}_{n\_k\_k} = \sum_{i=1}^{n\_k} P_{i\_k} \Big/ n\_k \tag{4}$$

The diagram of the average active power observer is depicted in Fig. 3.

Simulation results

TABLE I
PARAMETERS FOR SIMULATIONS

Item	Symbol	Value	Unit
Line resistance	R_{line}	0.1	Ω
Line inductance	L_{line}	1e-3	H
Line inductance	$L_{1\_1}, L_{2\_1}\ L_{3\_1}$	0.11e-3,0.12e-3,0.13e-3	H
Voltage reference	V^*	311	V
Number of PCG modules	D	2	/
Angular Frequency	ω^*	50*314	rad/s
Correction coefficient	ξ_ω	1e-3	/
Sharing coefficients	m_1, m_2	1e-4	$rad/(W \cdot s)$

To verify the effectiveness of the proposed scheme, the simulations are performed on MATLAB/Simulink platform. The considered hybrid series-parallel microgrid (See Fig.1) includes two PCG module, and each PCG module consists three DGs. The simulation parameters are lists in Table 1. In each PCG module, the bidirectional-circular communication graph is adopted shown in Fig. 4, and the corresponding in-degree matrix and adjacency matrix are given.

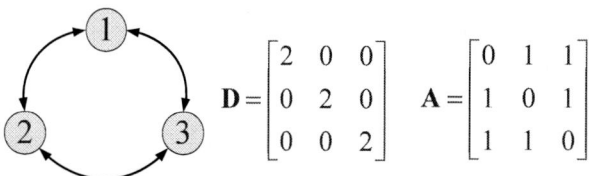

Fig. 4. Adopted communication graph.

A. Case1: Active Power Sharing Among Different PCG Modules

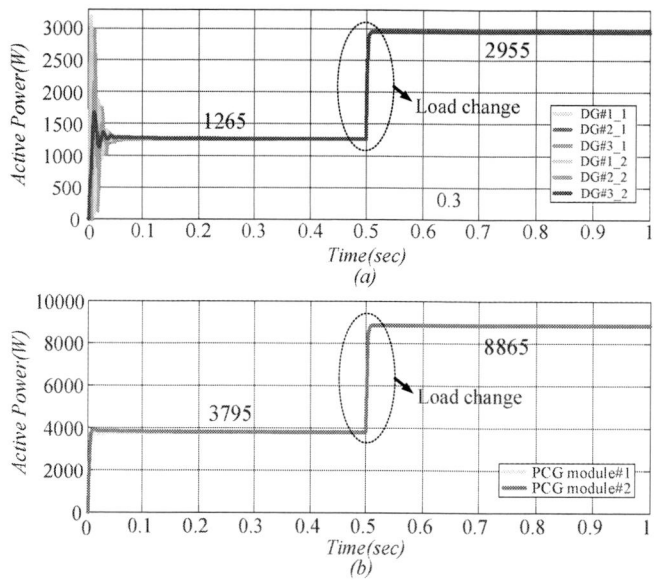

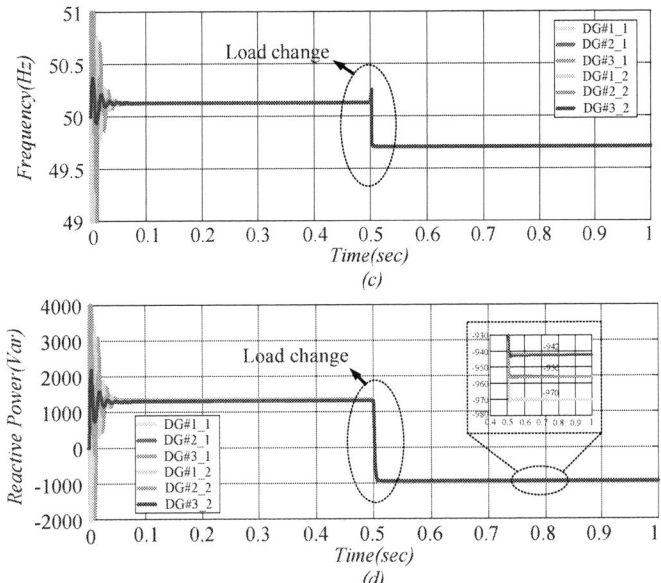

Fig. 5. Simulation results of case 1 (a) active power sharing among DGs (b) active power sharing among PCG modules (c) frequency (d) reactive power.

In this case, the accurate active power sharing among PCG module #1 and #2 with the ratio of 1:1 is verified. This simulation is performed under the resistance-inductance loads in the interval [0s, 0.5s] and resistance-capacitance loads in the interval [0.5s, 1s], respectively. In terms of the proposed scheme, the active power allocations among DGs are depicted in Fig. 5(a). In each PCG module, the accurate power sharing is obtained. From Fig. 5(b), the active power sharing among the two PCG modules is equal to the ratio of 1:1. Frequencies of DGs are shown in Fig. 5(c), in which the system can maintain the synchronization regardless of the RL loads and RC loads. The reactive power sharing results are shown in Fig. 5(d), in which the reactive power could not be shared accurately due to the effects of mismatched line impedance.

Accordingly, the proposed scheme can realize the accurate active power sharing with the ratio of 1:1 for different PCG modules of the hybrid series-parallel microgrid. Meanwhile, the frequency synchronization is obtained.

B. Case2: Plug-and-play capacity when DG#3_1 is suddenly lost

In this case, the plug-and-play capacity of the proposed scheme in the hybrid series-parallel microgrid is verified. This simulation is carried out when DG#3_1 is suddenly lost at 1s. The active power sharing waveforms are shown in Fig. 6(a), in which a new steady state operating point of the rest DGs is obtained after the DG#3_1 loss. The frequency of DGs are shown in Fig. 6(b), and the system can maintain the state operation when DG#3_1 is suddenly lost. The waveforms of reactive power allocations are presented in Fig. 6(c).

From this simulation results, it is verified that the proposed scheme holds the plug-and-play capacity in the hybrid series-parallel microgrid.

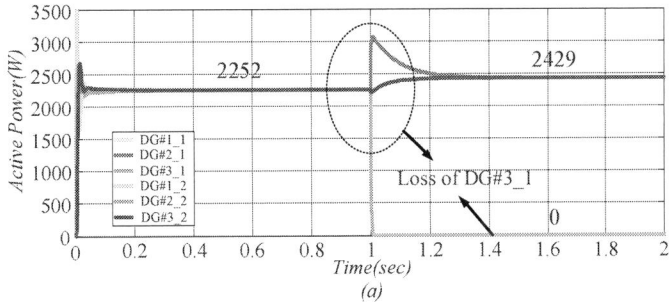

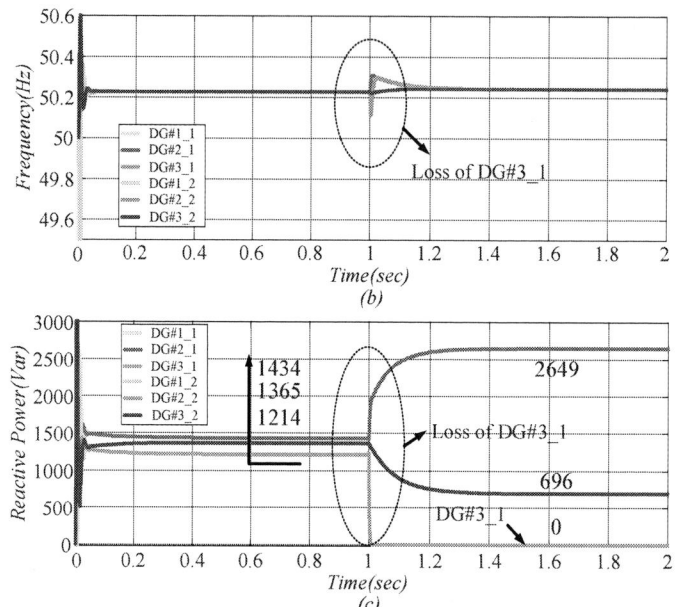

Fig. 6. Simulation results of case 2 (a) active power sharing among DGs (b) frequency (c) reactive power.

Conclusion

In this paper, a new hybrid series-parallel microgrid is proposed, which has the merits of parallel-type and cascaded-type micrigrids. This structure has the capable of plug-and-play and form a middle/high voltage power network conveniently. On the other hand, a low-dependent communication control scheme is proposed, which combines the features of distributed control and decentralized control. Moreover, the proposed scheme is robust to the communication failure and single point of failure. The effectiveness and performance of the proposed method has been verified through the simulations.

Integrating more limitations would be covered in future work, such as equivalent model of the hybrid series-parallel microgrid, steady-state analysis and more simulation results.

References

[1] R. Lasserter, A. Akhil, and C. Marnay, "The CERTS microgrid concept," in *Proc. CERTS*, 2002.

[2] Hua Han, Xiaochao Hou, Jian Yang, Jifa Wu, Mei Su, Josep M. Guerrero. Review of Power Sharing Control Strategies for Islanding Operation of AC Microgrids . *IEEE Transactions on Smart Grid*, vol.7, no.1, pp.200-215, Jan.2016

[3] Li L, Sun Y, Han H, *et al.*, "Power factor angle consistency control for decentralised power sharing in cascaded-type microgrid," in *IET Generation, Transmission & Distribution*, vol.13, no.6, pp:850-857, Mar. 2019.

[4] K. Shi, H. Ye, W. Song, and G. Zhou, "Virtual inertia control strategy in microgrid based on virtual synchronous generator technology," *IEEE Access*, vol. 6, pp. 27949–27957, 2018.

[5] Abdelaziz M M A, Shaaban, Mostafa F, Farag, Hany E, *et al.*, "A Multistage Centralized Control Scheme for Islanded Microgrids With PEVs," *IEEE Transactions on Sustainable Energy*, Vol. 5, no.3, pp: 927-937, 2014.

[6] Z. Zhang and M.-Y. Chow, "Convergence analysis of the incremental cost consensus algorithm under different communication network topologies in a smart grid," *IEEE Trans. Power Syst.*, vol. 27, no. 4, pp. 1761–1768, 2012.

[7] J. M. Guerrero, M. Chandorkar, T.-L. Lee, and P. C. Loch, "Advanced control architectures for intelligent microgrids—Part I: Decentralized and hierarchical control," *IEEE Trans. Ind. Electron.*, vol. 60, no. 4, pp.1254–1262, 2013.

[8] M. Malinowski, K. Gopakumar, J. Rodriguez, and M. A. Perez, "A survey on cascaded multilevel inverters," *IEEE Trans. Ind. Electron.*, vol. 57, no. 7, pp. 2197–2206, Jul. 2010.

[9] Y. Yu, G. Konstantinou, B. Hredzak and V. G. Agelidis, "Power Balance of Cascaded H-Bridge Multilevel Converters for Large-Scale Photovoltaic Integration," in *IEEE Transactions on Power Electronics*, vol. 31, no. 1, pp. 292-303, Jan. 2016.

[10] B. Xiao, L. Hang, J. Mei, C. Riley, L. M. Tolbert and B. Ozpineci, "Modular Cascaded H-Bridge Multilevel PV Inverter With Distributed MPPT for Grid-Connected Applications," in *IEEE Transactions on Industry Applications*, vol. 51, no. 2, pp. 1722-1731, March-April 2015.

[11] J. He, Y. Li, B. Liang and C. Wang, "Inverse Power Factor Droop Control for Decentralized Power Sharing in Series-Connected-Microconverters-Based Islanding Microgrids," in *IEEE Transactions on Industrial Electronics*, vol. 64, no. 9, pp. 7444-7454, Sept. 2017.

[12] J. He, Y. Li, C. Wang, Y. Pan, C. Zhang and X. Xing, "Hybrid Microgrid With Parallel- and Series-Connected Microconverters," in *IEEE Transactions on Power Electronics*, vol. 33, no. 6, pp. 4817-4831, June 2018.

Adaptive voltage control of islanded RES-based residential microgrid with integrated flywheel/battery hybrid energy storage system

Linda BARELLI, Gianni BIDINI, Ermanno CARDELLI, Dana-Alexandra CIUPAGEANU,
Andrea OTTAVIANO, Dario PELOSI
DEPARTMENT OF ENGINEERING, UNIVERSITY OF PERUGIA
Via G. Duranti 93, 06125
Perugia, Italy
E-Mail: linda.barelli@unipg.it, gianni.bidini@unipg.it, ermanno.cardelli@unipg.it,
dana_ciupageanu@yahoo.com, panfilo.ottaviano@unipg.it, dario.pelosi@gmail.com
URL: https://www.ing.unipg.it

Simone CASTELLINI
ERA ELECTRONIC SYSTEMS
Via G. Benucci 206, 06135
Perugia, Italy
E-Mail: simone.castellini@eraes.it
URL: https://www.eraes.it

Gheorghe LAZAROIU
POWER ENGINEERING FACULTY, UNIVERSITY POLITEHNICA OF BUCHAREST
Splaiul Independentei 313, 060042
Bucharest, Romania
E-Mail: glazaroiu@yahoo.com
URL: https://www.energ.pub.ro

Acknowledgements

This research has been performed with the funding of the Italian Ministry of Economic Development (MISE), in the framework of the TVB project (grant number CCSEB_00201). The project I.Ph.D.@UNIPG is also acknowledged.

Keywords

«Adaptive control», «Energy storage», «Fuzzy control», «Microgrid», «Photovoltaic»

Abstract

This paper aims to design an adaptive fuzzy logic voltage controller for a residential MG including PV-array and flywheel/battery hybrid energy storage implemented in a DC bus configuration. Specifically, the controller is developed for the MG islanded operation. The most relevant simulation scenario is considered the maximum load, defined based on load power profiles and correlated PV generation profiles.. Considering the DC bus voltage error and the battery state of charge as inputs, the fuzzy controller elaborates a power correction to be applied to the regular power assignments determined by the normal operation power management. Simulation results prove that the controller achieves very good performances in supplying the critical load during islanding, while maintaining adequate levels of voltage and frequency towards the critical load.

Introduction

As a consequence of accelerated technological evolution and reduced environmental impact, renewable energy sources (RES) share experienced lately an exponential growth in meeting global energy balance. Therefore, the energy transition towards RES based systems is remarkable worldwide [1], [2]. However, highly uncertain power outputs of RES plants issues great challenges in reference to power systems operation and control. Additional flexibility resources contribute to mitigate power fluctuations, enhancing power system stability and reliability features [3]. This can be achieved by implementing energy storage systems (ESS), representing a key factor in the enhancement of power generation flexibility, as they are able to satisfy production-demand unbalances caused by RES intermittency. Furthermore, connecting distributed RES with ESS in microgrid (MG) applications represents a key technology approach [4], emphasizing the ESS role to store the energy during off-peak periods and supply the load at peak-hours, while mitigating RES variability [5]. Nevertheless, specific characteristics of storage technologies confine their individual utilization. ESS hybridization brings multiple benefits at systemic level, also through the achievement of enhanced single technology operation (for instance, battery life extension) [6].

With respect to the foreseen RES generation development, it is emphasized that local energy demand could be satisfied based on MG generation and storage components exploitation. Therefore, the dependence from the power system resources and energy imports can be diminished and distributed generation widespread can be reinforced [7]. Including in the hybrid ESS (HESS) complimentary high-energy and high-power storage devices and further integrating the HESS in MG configurations brings power quality, supply reliability and system stability advantages [5]. More in detail, their matching characteristics in terms of response time and storage features enable both power and energy management optimization towards higher operational performances, while alleviating power networks transmission infrastructures burden [8], [9]. Previous research evidence that HESS (battery coupled to supercapacitor/fuel cell/superconducting magnetic energy storage system being most frequently investigated [5]) show better performances also in islanded mode, compared to single device ESS.

Depending on the connection type, two different operating regimes may occur in MGs functioning [10], [11]. If the MG is connected to the grid, in interconnected mode the unbalances between RES generation and demand in MG operation make voltage control rather difficult, particularly at the point of interface to the grid. Moreover, islanding detection techniques (implemented according to IEC 61727 and IEEE 1547 standards) may even disconnect the MG through corresponding static switches command. In islanded operation mode, the main challenge is to design a coordinative control strategy, able to consider the characteristics of both RES units (maximum power point – MPP – following) and storage devices (power and energy limitations, ramping restrictions), while keeping voltage and frequency stability [12].

The focus of this research is on the islanded operation of an interconnected MG. It is remarkable that operation of remote MGs or temporary islanded interconnected MGs is rather difficult, especially if RES penetration rate is high. Considering that RES intermittency overlaps energy demand variability, the chosen control strategy has to ensure the energy balance at each time step. Scientific literature depicts several approaches to tackle this subject, envisaging efficient exploitation of the available resources and optimal operation of each component and the overall system [13], [14]. When the MG comprises a HESS, their operational constraints (available energy, charge/discharge characteristics, thermal limits) must be always taken into account. However, as each application has particular control objectives, generality is restricted [15].

In the framework of distributed generation, centralized and distributed control methods are investigated for a wide range of MG applications [10], [16]. Because the stability of the system is strongly dependent on the droop control characteristics, which determine the adequate power sharing among MG components and the resulting voltage/frequency error. Therefore, adaptive approaches, such as artificial intelligence based techniques (fuzzy control and neural networks for instance) are generally preferred for MG regulation, being able to properly perform control in uncertain load/generation scenarios [17], [18].

This paper addresses an adaptive fuzzy controller design for islanded MG operation. In particular, a HESS comprising a Li-ion battery and a flywheel is integrated in a residential microgrid which includes a PV array. The brief time response of the flywheel enables it to cope the sudden power variations, alleviating the solicitation towards the battery [19]. When passing to islanded mode operation, the control pursues continuous critical load supply and maintaining adequate levels of power quality parameters. The control perspective addressed in this paper allows assessing the MG performances while operating in islanded mode. This contribution is complimentary to the literature available, highlighting the relevance of the research.

MG configuration description

The MG architecture under study, depicted in Figure 1, is implemented in Simulink SimPowerSystems, allowing both steady and dynamic analysis. Each subsection is indicated as it follows:

- PV array (11 kW) - yellow section,
- Li-ion battery (5 kW rated power, 9.8 kWh storage capacity) - grey section,
- flywheel (11 kW rated power, 2.1 kWh storage capacity) - blue section,
- ordinary load - green section,
- priority load - red section.

It must be mentioned that, for the purpose of this analysis, the total load (precisely 22 kW rated active power, operating at neutral power factor after reactive power compensation), is variably split in ordinary and critical loads in order to evaluate the behavior of the MG in islanded mode in different load scenarios. For what concerns the interconnected operation of the MG, the power management strategy aims maximizing its independence, as detailed in [20].

The MG is connected to the mains through a transformer; given the small value of the installed capacity within the MG, the grid is modeled as an ideal voltage source, being immune to the MG behavior. For the analysis of simulations outcomes, the equivalent configuration presented in Figure 1 is implemented in Simscape/Simpowersystems, as shown in Figure 2. The grid monitoring point is considered located at the secondary winding terminals of the transformer, while the light green section is inherent to the high hierarchical power flow control logic. The power flow management logic is customized for this particular MG, in order to maximize self-consumption and minimize the withdrawal from the mains, as previously stated. For a better understanding of the aforementioned control logic, further details are given in [19], [21].

With respect to the bus voltage level, it is remarked as the DC bus voltage (V_{DC}) is set at 650 V according to the rated voltage of commercial equipment and based on the consideration that the minimum DC voltage at the input of a three-phase inverter should be $V_{DC}^{min} = V_{AC\,ll}^{rms} \cdot \sqrt{2}$. Therefore, a $V_{AC\,ll}^{rms}$ of 400V, typical of the low voltage grid, implies a minimum DC voltage of about 566 V.

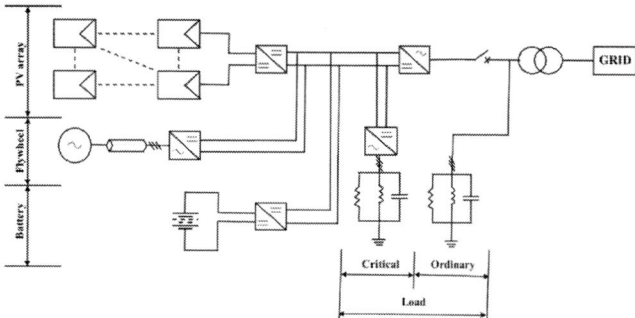

Figure 1. Equivalent representation of the DC bus MG.

Considering both ordinary and priority residential users as three-phased loads, it is necessary to implement an additional inverter in order to supply the priority load directly connected to the DC bus.

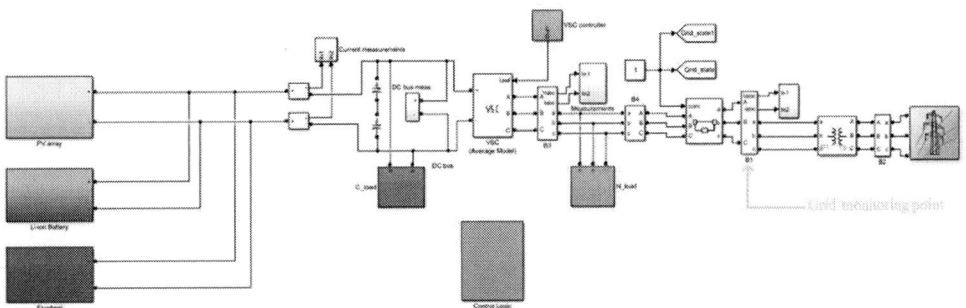

Figure 2. DC architecture implementation in Matlab/Simscape.

Adaptive control in MGs

Lacking grid support, the regulation task fulfillment is assigned to the storage devices. The PV array MPP control is based on incremental conductance technique, which is proved to be robust and effective.

Droop control technique is largely applied in microgrids, due to its reliability [22]. However, as a consequence of severe power fluctuations, traditional implementation of droop control, as given in eqs. (1) and (2), does not perform effectively in RES-based systems [23].

$$f = f_0 - m_p \cdot P \tag{1}$$

$$V = V_0 - m_q \cdot Q \tag{2}$$

where f is the actual frequency, f_0 is the frequency reference, P is the active power, V is the voltage, V_0 is the voltage reference, Q is the reactive power and m_p and m_q are the droop coefficients.

Therefore, adaptive implementations are required for RES-based MGs, particularly in islanded operation mode [17]. It is emphasized that voltage control is particularly important because of the potential harmful impact of voltage variations on equipment performances and safety.

Regularly, control performances are assessed according to the steady-state error, e.g. reference following capability during transient regimes and output wave total harmonic distortion. Droop control is a reliable control method, showing plug-n-play capability and benefiting of communication independence. Nevertheless, it has some disadvantages which restrict the applicability range, such as high dependence on the output impedance of the inverter and slow transient response, such as developed during MGs islanding.

The proposed approach aims to overcome the drawbacks of droop control by enhancing the adaptability of the control during islanding, based on the fuzzy strategies, enabling improved control performances. In this regard, the fuzzy controller intends to adjust the pre-existing power management strategy, so that the HESS is able to provide voltage support during islanding, while ensuring critical load supply.

More in detail, the fuzzy controller employs two input variables to evaluate the output variable, as depicted in Figure 3. The absolute error of the current DC bus voltage (*deltaVdc*) is the first input variable, while the battery state of charge (*SoC*) is the second one, as the battery substitutes the grid voltage support during islanding. These two inputs give insights on the voltage support requirements (*deltaVdc* being the quantification of the voltage malfunction in the MG) and the voltage support resources (*SoC* providing information about how much the battery can support the MG to reach and maintain the nominal bus voltage, as the main support). The controller elaborates the power correction factor (*deltaP*) to be applied to both HESS devices, while the islanding persists, to the power assignments determined by the main control logic already implemented. It must be mentioned that the correction *deltaP* is firstly assigned to the battery (as the main voltage support), but if the maximum power in battery charge/discharge is exceeded, the power surplus is transferred to the flywheel. It is remarkable that the proposed method operates exclusively during islanding, as an adjustment of the regular strategy, keeping a proved to be effective control strategy while normal operation [19].

The relative *deltaP* correction is multiplied by the remaining load (i.e. the critical load), obtaining the absolute power correction to be further included in the main control logic. The rule base is developed such that if the *deltaVdc* is negative and the *SoC* is high, the battery provides (in discharge) a higher additional *deltaP* to support voltage restoration. As evident in Figure 5, the *deltaP* decreases with the *SoC* diminishing and voltage error reduction (approaching zero). On the other hand, if the *deltaVdc* is positive (so the MG registers an overvoltage), the battery has to absorb power (in charge, i.e. the *deltaP* output is negative) according to the capability given by the *SoC*.

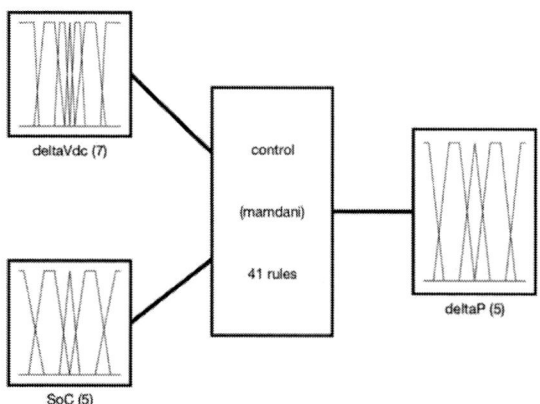

Figure 3. Fuzzy controller structure.

The linguist variables definition is given in Table I and the membership functions are depicted in Figure 4.

Table I: Linguistic variables definition

Input variables		Output variable
deltaVdc	SoC	deltaP
Large negative – LN Medium negative – MN Negative – N Zero – Z Positive – P Medium positive – MP Large positive – LP	Very low – VL Low – L Medium – M High – H Very high – VH	Large negative – LN Negative – N Zero – Z Positive – P Large positive – LP

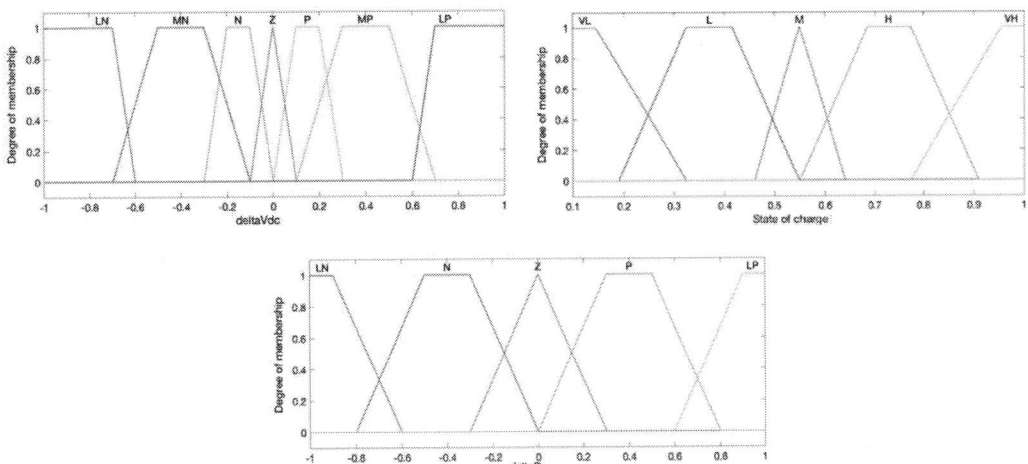

Figure 4. Membership functions.

According to the rule base defined on simulation results, the corresponding control surface is pictured in Figure 5.

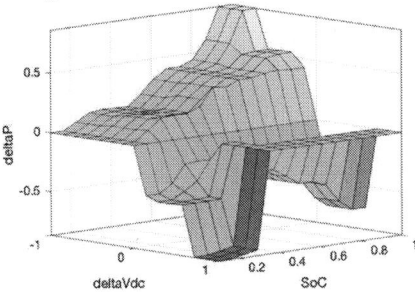

Figure 5. Control surface.

Simulations are performed in order to evaluate the MG dynamic behavior during islanding under specific operating scenarios representative of most stressful solicitations. These are selected, as different load/PV generation scenarios, according to the yearly profiles. Also the critical load percentage is varied to observe the behavior of the MG under different conditions of load connections.

Simulation results discussion

Based on the analysis of real data for the PV generation profile and load conditions, the most representative case for the islanded operation with critical load supply being considered as the maximum load. Thus, correlating the PV generation and load profiles, the simulating conditions resulted as 6.633 kW PV power and 21.502 kW load. The islanding is considered to occur between 2s and 4s. The critical load percent is set as 10% (*case 1*) and 50% (*case 2*). The voltage profiles in both simulation cases are depicted in Figure 6 at the point of interface to the grid and at the terminals of critical and ordinary loads. It can be remarked that the critical load percentage reflects mainly on the behavior of the voltage at the terminals of the critical load, while having a very reduced influence over the ordinary load and at the point of interface to the grid. This is evident also in Figure 7 which shows, in particular, the voltage behavior at the grid reconnection moment. Anyway, it is noticeable that the transition lasts for less than 2 full wave cycles (i.e. 40 ms), the voltage amplitude deviation never exceeding 10% of the nominal value.

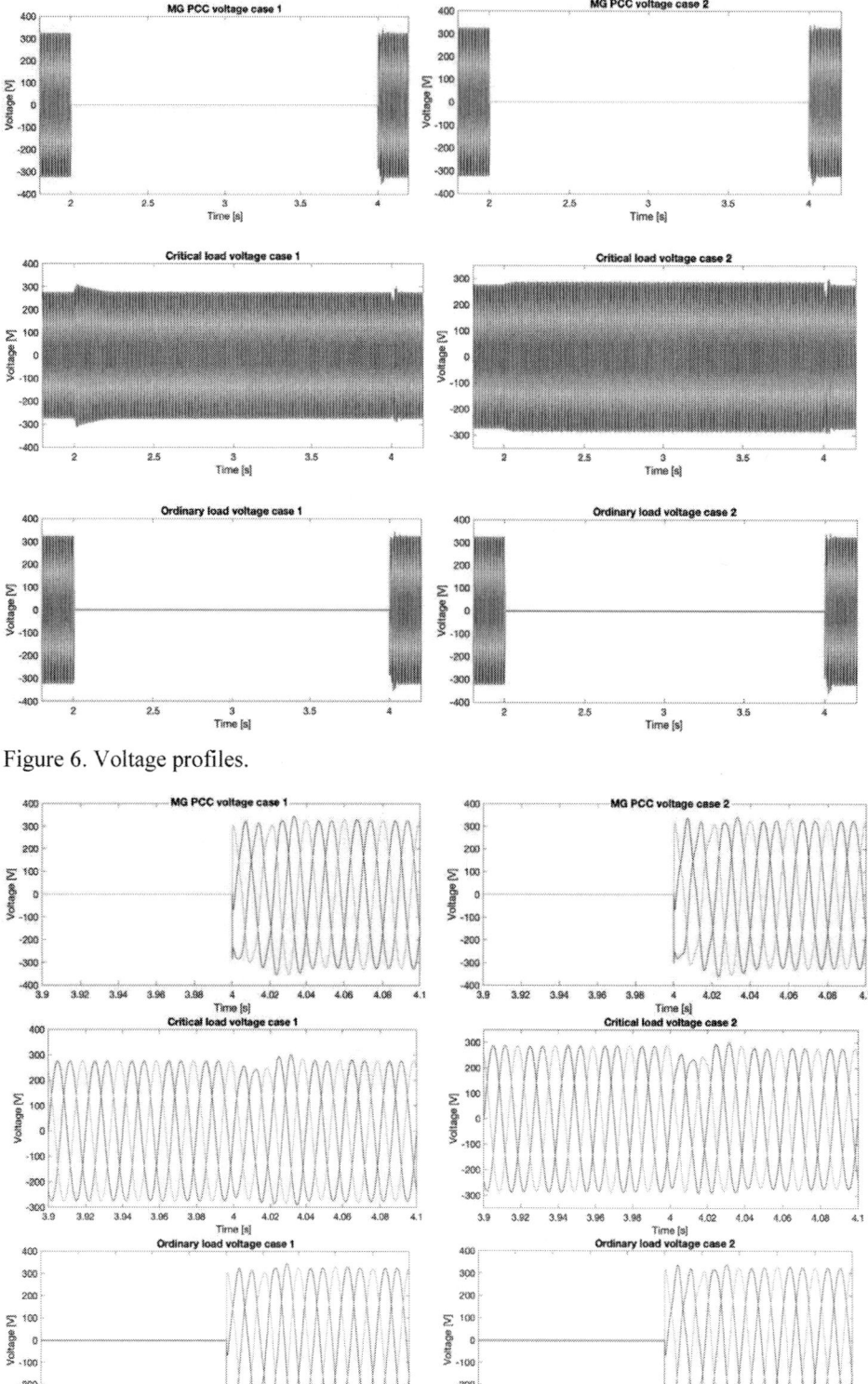

Figure 6. Voltage profiles.

Figure 7. Zoom on voltage behavior during reconnection to the grid.

The HESS components answer to the different operating conditions is depicted in Figure 8. In case 1 (i.e. 10% critical load), the flywheel maintains an almost constant speed while MG islanding emerges. The battery charges, as there is a slight overvoltage registered because of the PV generation surplus relatively to the remaining (critical) connected load. In comparison, in case 2 (i.e. 50% critical load), a PV production shortage emerges, causing an undervoltage. The power assigned to the battery (in discharge) in this case exceeds the safe operating limits, therefore in the analysis timeframe, the flywheel supplies the missing power to the critical load. Consequently, the flywheel shows an increased drop in the speed, while the battery keeps an almost constant state of charge.

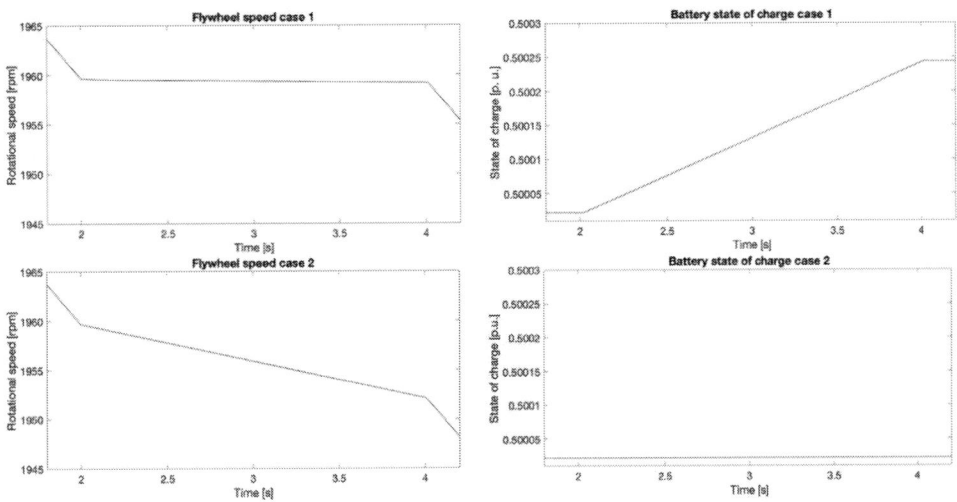

Figure 8. HESS behavior.

Figure 9 depicts the frequency evolution for the critical load and at the point of interface to the grid. It can be noticed, for the critical load, a very limited increase in the oscillation amplitude in case 2.

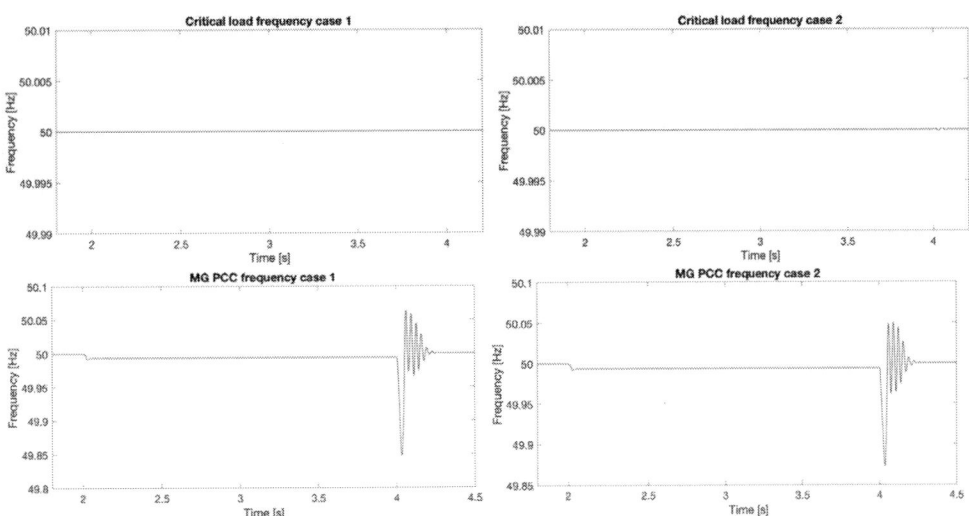

Figure 9. Frequency evolution.

Conclusion

HESS coupled with RES plants enhance systems performances, specifically in hybrid configurations, improving their reliability and overcoming the variability related issues. The control proposed in this paper is validated through simulation, proving to be robust under critical load percent variation. The proposed fuzzy logic controller enables supplying the critical load during MG islanding while maintaining adequate levels of voltage and frequency towards it. Furthermore, as it operates as an adjustment of the main control logic implemented, the reconnection to the mains is smooth, the equilibrium being reestablished in less than 40 ms (which is 2 full wave cycles). In the context of MGs concept unfolding in the wider smart grid framework, the present research contributes to improve their operation in islanding conditions, which may occur and severely impact the critical loads.

References

[1] International Renewable Energy Agency, "Electricity storage and renewables: Costs and markets to 2030," 2017.

[2] IRENA, *Global Energy Transformation: A Roadmap to 2050.* 2018.

[3] I. J. Perez-Arriaga and C. Batlle, "Impacts of Intermittent Renewables on Electricity Generation System Operation," *Econ. Energy Environ. Policy*, vol. 1, no. 2, pp. 3–18, 2012.

[4] S. Wen, S. Wang, G. Liu, and R. Liu, "Energy management and coordinated control strategy of PV/HESS AC microgrid during islanded operation," *IEEE Access*, vol. 7, pp. 4432–4441, 2019.

[5] M. Faisal, M. A. Hannan, P. J. Ker, A. Hussain, M. Bin Mansor, and F. Blaabjerg, "Review of energy storage system technologies in microgrid applications: Issues and challenges," *IEEE Access*, vol. 6, pp. 35143–35164, 2018.

[6] S. Ould Amrouche, D. Rekioua, T. Rekioua, and S. Bacha, "Overview of energy storage in renewable energy systems," *Int. J. Hydrogen Energy*, vol. 41, no. 45, pp. 20914–20927, 2016.

[7] E. Planas, J. Andreu, J. I. Gárate, I. Martínez De Alegría, and E. Ibarra, "AC and DC technology in microgrids: A review," *Renew. Sustain. Energy Rev.*, vol. 43, pp. 726–749, 2015.

[8] L. Barelli, U. Desideri, and A. Ottaviano, "Challenges in load balance due to renewable energy sources penetration: The possible role of energy storage technologies relative to the Italian case," *Energy*, vol. 93, pp. 393–405, 2015.

[9] M. C. Argyrou, P. Christodoulides, and S. A. Kalogirou, "Energy storage for electricity generation and related processes: Technologies appraisal and grid scale applications," *Renew. Sustain. Energy Rev.*, vol. 94, no. July, pp. 804–821, 2018.

[10] N. Hatziargyriou, H. Asano, R. Iravani, and C. Marnay, "Microgrids: An Overview of Ongoing Research, Development, and Demonstration Projects," *IEEE Power Energy Mag.*, vol. 5, no. 4, pp. 78–94, 2007.

[11] R. Majumder, "Some aspects of stability in microgrids," *IEEE Trans. Power Syst.*, vol. 28, no. 3, pp. 3243–3252, 2013.

[12] A. Micallef and C. S. Staines, "Voltage rise mitigation and low voltage ride through capabilities for grid-connected low voltage microgrids," *2017 19th Eur. Conf. Power Electron. Appl. EPE 2017 ECCE Eur.*, vol. 2017-Janua, pp. 1–9, 2017.

[13] I. Patrao, E. Figueres, G. Garcerá, and R. González-Medina, "Microgrid architectures for low voltage distributed generation," *Renew. Sustain. Energy Rev.*, vol. 43, pp. 415–424, 2015.

[14] B. Dong, Y. Li, Z. Zheng, and L. Xu, "Control strategies of microgrid with Hybrid DC and AC Buses," *Power Electron. Appl. (EPE 2011), Proc. 2011-14th Eur. Conf.*, no. 10, pp. 1–8, 2011.

[15] A. Bhattacharjee and H. Saha, "Design and experimental validation of a generalised electrical equivalent model of Vanadium Redox Flow Battery for interfacing with renewable energy sources," *J. Energy Storage*, vol. 13, pp. 220–232, 2017.

[16] R. H. Lasseter, "Microgrids and distributed generation," *Intell. Autom. Soft Comput.*, vol. 16, no. 2, pp. 225–234, 2010.

[17] Q. Sun, Q. Sun, and D. Qin, "Adaptive fuzzy droop control for optimized power sharing in an islanded microgrid," *Energies*, vol. 12, no. 1, 2019.

[18] T. Shu *et al.*, "Power Flow Analysis for Hybrid AC/DC Microgrid Islanding Operation Based on an Improved Adaptive Droop Control," *2nd IEEE Conf. Energy Internet Energy Syst. Integr. EI2 2018 -*

Proc., pp. 1–6, 2018.

[19] L. Barelli *et al.*, "Flywheel hybridization to improve battery life in energy storage systems coupled to RES plants," *Energy*, vol. 173, pp. 937–950, 2019.

[20] L. Barelli *et al.*, "Flywheel hybridization to improve battery life in energy storage systems coupled to RES plants," *Energy*, p. Accepted for publication, 2019.

[21] L. Barelli *et al.*, "Dynamic Analysis of a Hybrid Energy Storage System (H-ESS) Coupled to a Photovoltaic (PV) Plant," *Energies*, vol. 11, no. 2, 2018.

[22] S. Sen and V. Kumar, "Microgrid control: A comprehensive survey," *Annu. Rev. Control*, vol. 45, no. April, pp. 118–151, 2018.

[23] D. I. Makrygiorgou and A. T. Alexandridis, "Modeling and stability of autonomous dc microgrids with converter-controlled energy storage systems," *2017 IEEE 2nd Int. Conf. Direct Curr. Microgrids, ICDCM 2017*, pp. 285–291, 2017.

An Improved λ-consensus Control Method for DC Microgrids

Siqi Fu[1,2], Yao Sun[1,2], Zhangjie Liu[1,2], Hua Han[1,2], Mei Su[1,2]

1. School of Automation, Central South University
2. Hunan Provincial Key Laboratory of Power Electronics Equipment and Grid
Changsha Hunan P.R. China
E-Mail: 571330420@qq.com

Acknowledgements

This work was supported in part by the National Key R&D Program of China under Grant 2018YFB0606005, in part by the National Natural Science Foundation of China under Grant 61933011, 61903383, in part by the Major Project of Changzhutan Self-Dependent Innovation Demonstration Area under Grant 2018XK2002, in part by the Project of Innovation-driven Plan in Central South University under Grant 2019CX003, and in part by the Hunan Provincial Key Laboratory of Power Electronics Equipment and Grid under Grant 2018TP1001.

Keywords

« economical dispatch », «DC microgrids», « voltage recovery », « transmission loss »

Abstract

Economical dispatch plays an important role in the operation of DC micro-grid. This paper presents a modified λ-consensus method that can achieve the optimal economical dispatch subjected to voltage regulation constraints. Simulation results verify the feasibility of the proposed scheme.

Introduction

The economics and power quality of DC microgrids are two challenging tasks and have attracted much attention. The power losses of line impedance in electricity transmission may contribute up to 5% of the total power losses [1]. Voltage deterioration may cause abnormal operation of the load [2].

Lots of studies have been conducted to solve such problem. Among them, λ-consensus control method is an efficient way. Researchers formulate an economic dispatch problem for a DC microgrid in [3-5] and proposed a λ-consensus control method with voltage recovery. Moreover, improved methods considering the limits of output power of dc generators were proposed in [6] based on [3-4]. However, the methods they proposed ignored the transmission losses that increases the total generation cost.

The contributions of this paper are as follows: 1) A multi-objective optimization problem for reducing transmission loss and voltage regulation is illustrated in DC microgrids. 2) An efficient modified λ-consensus control method is proposed to achieve the optimal economical dispatch.

The proposed optimal power flow control for dc micrgrid

Mathematically speaking, the objective of economical dispatch problem (EDP) is to minimize the total generation cost.
In [4], the optimization problem of EDP is formulated as

$$\begin{cases} \min \sum_{i=1}^{n} C_i\left(P_i\right) \\ s.t. \sum_{i=1}^{n} P_i = P_{load} + P_{loss} \end{cases} \tag{1}$$

where P_i and $C_i(P_i)$ are the output power and cost function of the ith DG, P_{load} is the total load power. P_{loss} represents line loss of the whole system, which is a function of the output power of sources. Usually,

$C_i(P_i)$ is a convex function. The constraint in (1) represents the supply-demand balance. The main theory basis of the method proposed in [3-4] can be summarized as the follows.

Constructing the Lagrange function as

$$L(P_1, P_2, \cdots P_n, \lambda) = \sum\nolimits_{i=1}^{n} C_i(P_i) - \lambda \left(\sum\nolimits_{i=1}^{n} P_i - P_{load} - P_{loss} \right) \tag{2}$$

where λ is the Lagrange multiplier. Then, the optimal solution yields

$$\partial L / \partial P_i = \partial C_i(P_i) / \partial P_i - \lambda = 0, i = 1, 2, \cdots, n \tag{3}$$

$$\sum\nolimits_{i=1}^{n} P_i - P_{load} - P_{loss} = 0 \tag{4}$$

According to the energy conservation law, (4) always holds. That is, the global economic dispatch can be achieved as long as

$$\partial C_1(P_1) / \partial P_1 = \partial C_2(P_2) / \partial P_2 = \cdots = \partial C_n(P_n) / \partial P_n \tag{5}$$

In fact, (5) is the well-known equal incremental cost principle. Then, the main idea of [3-4] is to force the incremental cost of each DG equal by λ-consensus control. However, (3) holds if and only if

$$\partial P_{load} / \partial P_i = 0, \partial P_{loss} / \partial P_i = 0 \tag{6}$$

Clearly, (6) require that the load power and line loss does not vary with the output power of every generation. When the load is constant power load, the first equation in (6) will hold. However, the second equation does not hold in DC microgrid, because the line loss is dependent on its current which is dependent on the output power of each DG inevitably.

To illustrate the dependence of line loss on output power, we take an example of case 1 in Fig 2. The load power, total line loss power and total output power can be expressed as

$$\begin{aligned}
P_{load} &= f(i_1, i_2, i_3, i_4) \\
&\approx 50i_1^2 + 49i_2^2 + 49i_3^2 + 49i_4^2 + 99i_1 i_2 + 99i_1 i_3 \\
&\quad + 98i_1 i_4 + 98i_2 i_3 + 97i_2 i_4 + 97i_3 i_4 \\
P_{loss} &= f(i_1, i_2, i_3, i_4) \\
&\approx 7.7i_1^2 + 11i_2^2 + 9.6i_3^2 + 5.7i_4^2 - 11i_1 i_2 - 2.5i_1 i_3 \\
&\quad - 0.096i_1 i_4 - 5.7i_2 i_4 - 4.5i_3 i_4 \\
\sum\nolimits_{i=1}^{n} P_i &= f(i_1, i_2, i_3, i_4) \\
&\approx 50i_1^2 + 50i_2^2 + 51i_3^2 + 52i_4^2 + 96i_1 i_2 + 95i_1 i_3 \\
&\quad + 95i_1 i_4 + 96i_2 i_3 + 95i_2 i_4 + 96i_3 i_4
\end{aligned} \tag{7}$$

Hence, neglecting $\partial P_{load} / \partial P_i$ and $\partial P_{loss} / \partial P_i$ will make the operation point deviating from the optimal point.

Consider a DC microgrid with the known equivalent admittance matrix, which is expresses as Y. Then, the following can be obtained

$$i = Yu, i \triangleq \begin{bmatrix} i_1 & i_2 & \cdots & i_n \end{bmatrix}^T, u \triangleq \begin{bmatrix} u_1 & u_2 & \cdots & u_n \end{bmatrix}^T \tag{8}$$

where i_i and u_i are the output current and voltage of the ith DG, respectively.

Taking into account the average voltage recovery [4], the optimization problem is formulated as (9).

$$\begin{cases} \min \sum\nolimits_{i=1}^{n} C_i(P_i) \\ s.t. \quad P = [u]Yu \\ \quad u_1 + u_2 + \cdots + u_n = nu_{ref} \end{cases} \tag{9}$$

where $[u]=diag\{u\}$, y_{ij} is the corresponding entry of Y and u_{ref} is the voltage reference. The first constraint in (9) contains information about line impedance. The second constraint in (9) represents the voltage regulation.

Constructing the Lagrange function as

$$L(P,\eta,u,\lambda)=\sum_{i=1}^{n}C_i(P_i)+\eta^T([u]Yu-P)+\delta(nu_{ref}-1_n^T u) \tag{10}$$

Where $P\triangleq[P_1\ P_2\cdots P_n]^T$, $\eta\triangleq[\eta_1\ \eta_2\cdots\eta_n]^T$, $1_n\triangleq[1\ 1\cdots1]^T$, η_i and δ_i are the Lagrange multipliers. Define $k=[k_1,k_2,\cdots k_n]^T$, where $k_i=\partial C_i(P_i)/\partial P_i$. The KKT conditions for the problem in (10) are given by

$$\begin{cases}\partial L/\partial P=0\\ \partial L/\partial u=0\\ \partial L/\partial\eta=0\\ \partial L/\partial\delta=0\end{cases}\Rightarrow\begin{cases}k=\eta & \text{(11a)}\\ (diag\{\eta\}Y+Ydiag\{\eta\})u=\delta 1_n & \text{(11b)}\\ P=diag\{u\}Yu & \text{(11c)}\\ 1_n^T u-nu_{ref}=0 & \text{(11d)}\end{cases}$$

Combining (11a) and (11b), the following can be obtained

$$diag\{k\}i+Ydiag\{k\}u=\delta 1_n \tag{12}$$

Because (11c) is the physical constraint that always holds in the circuit. Thus, the optimal economical dispatch and voltage regulation can be achieved if (12) and (11d) hold. Defining $\lambda=diag\{k\}i+Ydiag\{k\}u$, then optimality condition can be achieved only if all the λ_i are equal according to (12).

Based on idea above, the proposed control method is written as

$$u_i=u_{ref}+\int a_{ij}(\lambda_j-\lambda_i)dt+\int(u_{ref}-\bar{u})dt \tag{13}$$

where $\bar{u}=\sum_{i=1}^{n}u_i/n$ and a_{ij} is the communication weight between node i and j.

The proposed λ-consensus control method structure is shown in Fig.1.

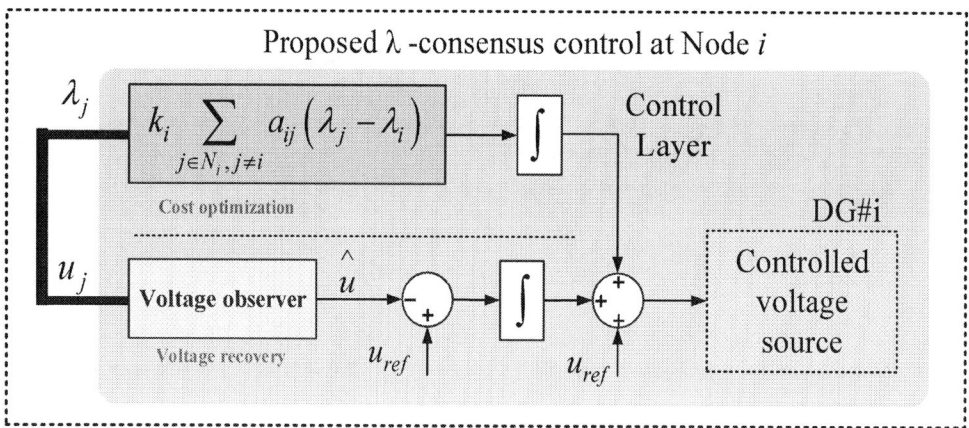

Fig 1 Proposed improved λ-consensus control structure.

CASE STUDY

To verify the correctness and effectiveness of the proposed method, four simulation cases are tested. The general operation cost function can be expressed as $C_i(P_i)=a+bP_i+cP_i^2$. A general DC microgrid with 4 energy sources and 4 resistive loads is illustrated in Fig. 2. The nominal voltage is set to 200V.

The simulation results are shown in Fig 3 and Fig 4. It can be seen in Fig.3 that all λ converge to a same value, indicating that the optimal economical dispatch of DC microgrid can be achieved. Fig. 4 shows that voltage deviation is regulated with the proposed method.

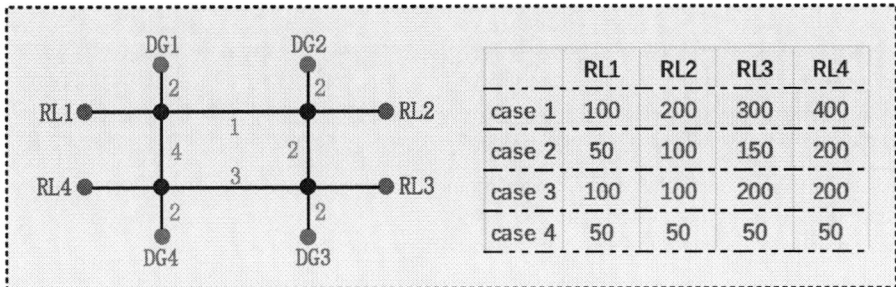

Fig 2 A general DC microgrid with 4 DGs and 4 resistive loads. The red and blue points represent DGs and loads, respectively. The black line represents the cables. The green numbers are the resistances of cables.

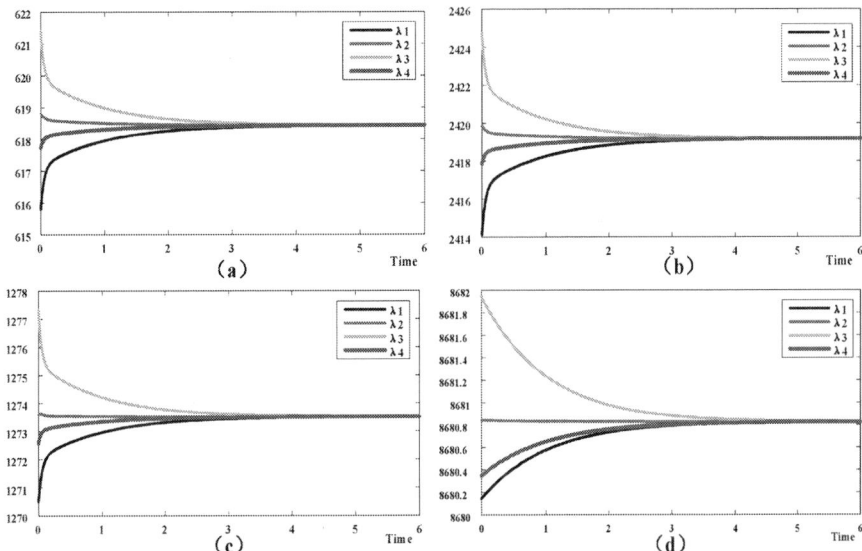

Fig 3 simulation results of λ a) case 1, b) case 2, c) case 3, d) case 4.

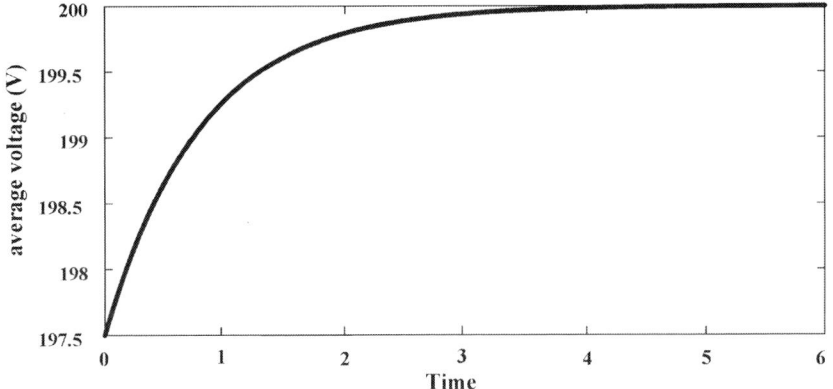

Fig 4 average output voltage of DGs

The generation cost comparison with the model in Fig.2 is shown respectively in table 1, between the method in [3-6] neglecting line power loss and the proposed method. The load power and the transmission loss comparison results of two methods are shown in table 2.

Table I: generation cost comparison

	Case 1	Case 2	Case 3	Case 4
Method in [3-6]	123762.17	484088.56	254832.83	1737838.03
Proposed method	123752.78	483960.45	254794.37	1736397.19

The results in table 1 show that the proposed method achieves lower generation cost comparatively than the method in [3-6]. In other words, the proposed method greatly reduces the power loss on the line impedance to reduce total generation cost without sacrificing the load power.

Conclusion

In this letter, we analyze deeply why the existing λ-consensus fails to realize optimization. To improve optimal economical dispatch, a modified λ-consensus method which achieves lower total generation cost by reducing transmission losses is proposed with average voltage regulation constraints. Finally, the case study results verify the correctness of the proposed method.

References

[1] Xu, Y.; Zhang, W.; Hug, G.; Kar, S.; Li, Z., "Cooperative Control of Distributed Energy Storage Systems in a Microgrid," Smart Grid, IEEE Transactions on Power System. vol.30, no 5, pp. 2780-2789. Sept. 2015.

[2] T. Sousa, H. Morais, Z. Vale, and R. Castro, "A multi-objective optimization of the active and reactive resource scheduling at a distribution level in a smart grid context," Energy, vol. 85, pp. 236–250, Jun. 2015.

[3] Han, H., Wang, H., Sun, Y., Yang, J., Liu, Z. "Distributed control scheme on cost optimisation under communication delays for DC microgrids." IET Generation, Transmission & Distribution, vol. 11, no 17, pp.4193-4201, Nov. 2017.

[4] Zaery, M., Ahmed, E. M., Orabi, M., Abdelghani, A. B. "Distributed cooperative control with lower generation cost for dc microgrid." IEEE in 2015 First Workshop on Smart Grid and Renewable Energy (SGRE),. pp.1-6, March 2015.

[5] Yu, W., Li, C., Yu, X., Wen, G., & Lü, J. "Economic power dispatch in smart grids: a framework for distributed optimization and consensus dynamics." Science China Information Sciences, vol. 61, no 1, 012204, Jan. 2018

[6] Moayedi, S., & Davoudi, A. "Unifying distributed dynamic optimization and control of islanded dc microgrids." IEEE Transactions on Power Electronics, vol.32, no 3, pp 2329-2346.

Decrease of power electronic switching losses using variable switching events

Hannes Ramm, Michael Homann, Torben A. Schulze, Faical Turki, Heiko Rabba
IAV GmbH
Rockwellstr. 16
38518 Gifhorn, GERMANY
Email: hannes.ramm@iav.de
URL: http://www.iav.de

Keywords

≪Voltage Source Inverters (VSI)≫, ≪Pulse Width Modulation (PWM)≫, ≪Switching losses≫, ≪Converter control≫, ≪Harmonics≫

Abstract

The electric range of an electric vehicle depends amongst others on the efficiency of electric components. This paper focuses on the efficiency of the inverter in one-phase operation. A comparison of three different PWM patterns is presented concerning their efforts to minimize the switching losses of the inverter and reducing the distortion of produced phase current. For one-phase systems, we show that an appropriate switching strategy will decrease the switching losses by up to 33%. Furthermore, this strategy is able to reduce the distortion factor of phase current, as a quantity of signal quality, by up to 18%.

Introduction

The electrical traction drive system in a battery electric vehicle (BEV) consists of high voltage battery, inverter and electrical machine. The electric range of the vehicle depends amongst others on the efficiency of these components. This paper focuses on the efficiency of the inverter. The main loss terms are conducting (forward power) and switching power losses, and the latter can be reduced e.g. by optimizing the gate resistance [1]. Furthermore, the switching power losses will be affected by the applied pulse-width modulation (PWM) pattern, since for a given duty cycle the switching points can be optimized in a wide range. A promising approach for the reduction of switching losses is given by the deliberate shifting of switching events, i.e. from high absolute phase current values to low ones. Fixed-frequency PWM patterns, such as the sine-triangle PWM, are commonly used, but are not feasible for a direct implementation of switching event variation. The aim of this paper is the analysis and comparison of switching losses in PWM patterns that feature variable switching frequencies. Subsequently the effort of this variable switching frequency to the signal quality of the phase current is examined.

Overview of PWM patterns (one-phase)

In this paper, three PWM patterns as shown in Fig. 1, are examined. Hysteresis control and $\Delta\Sigma$-PWM are compared to sine-triangle PWM, the latter being a fixed frequency PWM pattern. Sine-triangle PWM and hysteresis control are well known [2]. One-phase $\Delta\Sigma$-PWM as a third pattern is based on an analog class-D amplifier given in [3]. The digital implementation is known from [4]. Sine-triangle PWM and $\Delta\Sigma$-PWM are used with a cascaded current control. The set point voltage u_{sp}, as the current controller output, is shown in Fig. 1. Hysteresis control operates directly with a current set point i_{sp}, to control the phase current i_S. For hysteresis control and $\Delta\Sigma$-PWM, the parameters H and N_1 are used to define the hysteresis width in phase current and magnetic flux space, respectively. $\Delta\Sigma$-PWM operates with

$\Delta\Sigma$-Modulators ($\Delta\Sigma$-M), that can be simplified modeled as a gain $\frac{2}{U_{\max}}$. The parameter t_S represents the sampling time of $\Delta\Sigma$-PWM, which corresponds to the integrator gain of $\Delta\Sigma$-PWM. The PWM patterns drive a B2 inverter and a one-phase AC machine, realized here simply by an impedance and a induced voltage u_{emk}, shown in Fig. 2.

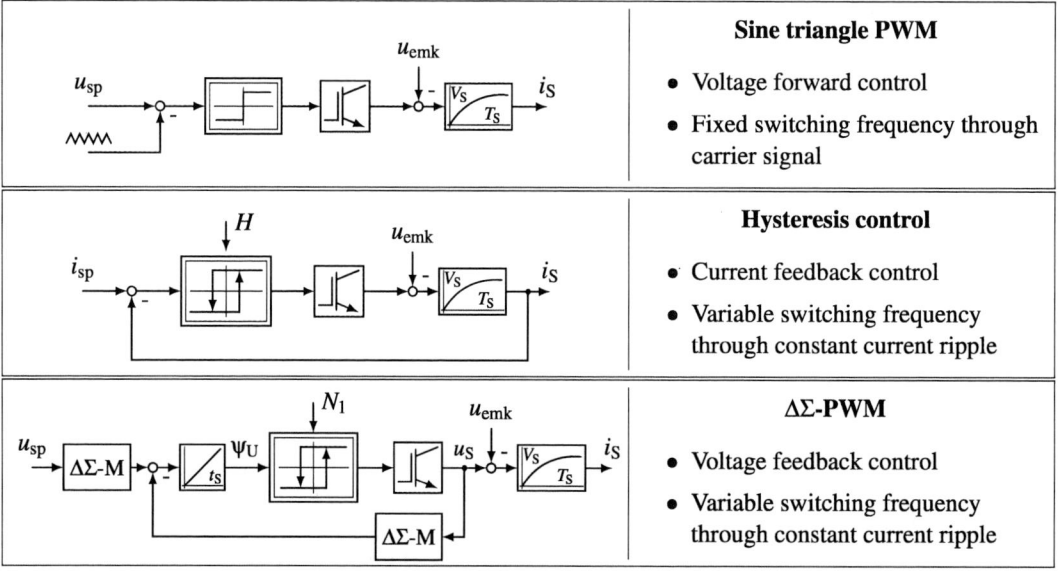

Fig. 1: One-phase PWM patterns

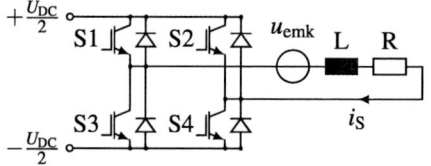

Fig. 2: B2 inverter with ohmic inductive load and induced voltage u_{emk}

Results one-phase

$\Delta\Sigma$-PWM and hysteresis control generate a variable switching frequency. The core idea, initially introduced in [5], is that this variable switching frequency will lead to differences of switching power losses. To quantify this, an instantaneous switching frequency which is changing every PWM period is derived for one-phase $\Delta\Sigma$-PWM and hysteresis control. Subsequently, the resulting switching power losses are compared with the fixed-frequency PWM pattern. As a criterion for signal quality, the phase current ripple is evaluated. It will be shown that switching power losses and phase current ripple can be analyzed with the same approach. The analytical derivation of a variable switching frequency is the basis for further evaluations of phase current ripple.

Instantaneous switching frequency Delta-Sigma-PWM

For $\Delta\Sigma$-PWM with cascaded current control, the derivation of an instantaneous switching frequency starts with an equation for the set point voltage u_{sp}, obtained by applying Kirchhoff's law in case of steady state under condition of zero resistance R, zero noise failures and by neglecting dead times:

$$u_{sp} = u_{emk} + L\frac{di_S}{dt}. \tag{1}$$

EPE'20 ECCE Europe

Assigned jointly to the European Power Electronics and Drives Association & the Institute of Electrical and Electronics Engineers (IEEE)

With equations for i_S, u_{emk} and the number of pole pairs p

$$i_S = i_{sp} = \hat{I}_S \sin(\omega_{elec}t), u_{emk} = \psi_p \omega_{mech} \sin(\omega_{elec}t) \text{ and } p = 1 \rightarrow \omega_{mech} = \omega_{elec} \qquad (2)$$

the relative set point voltage u_{sp} can be calculated as follows:

$$u_{sp} = \sqrt{(\hat{I}_S \omega_{elec}L)^2 + (\psi_p \omega_{elec})^2} \sin\left(\omega_{elec}t + \arctan\left(\frac{\hat{I}_S L}{\psi_p}\right)\right). \qquad (3)$$

The DC link voltage U_{DC}, the electrical rotational speed ω_{elec} and the peak phase current $\hat{I}_S$ are depending on the operational point. The magnetic flux ψ_p and inductance L are machine parameters. The $\Delta\Sigma$-PWM comprises a voltage control loop. The voltage error e_U including the gain of $\Delta\Sigma$-modulators is:

$$e_U = \frac{2u_{sp}}{U_{max}} - \frac{2u_S}{U_{max}}. \qquad (4)$$

The reference voltage U_{max} is given through the range of the $\Delta\Sigma$-Modulators. The voltage error e_U corresponds to the derivative of the error trajectory ψ_U (integrator of $\Delta\Sigma$-PWM, see Fig. 1). An additional equation for the derivative of the error trajectory ψ_U can be derived by means of the hysteresis behavior with the times of one switching period t_1 and t_2, see Fig. 3 (a). The calculation of voltage error and hysteresis behavior for the two operational points $U_S = -U_{DC}$ and $U_S = U_{DC}$ leads to:

$$\left.\frac{d\psi_u}{dt}\right|_{u_S=-U_{DC}} = \left(\frac{2U_{DC}}{U_{max}} + \frac{2u_{sp}}{U_{max}}\right)\frac{1}{t_S} = \frac{N_1}{t_1} \text{ and } \left.\frac{d\psi_u}{dt}\right|_{u_S=U_{DC}} = \left(\frac{-2U_{DC}}{U_{max}} + \frac{2u_{sp}}{U_{max}}\right)\frac{1}{t_S} = \frac{-N_1}{t_2}. \qquad (5)$$

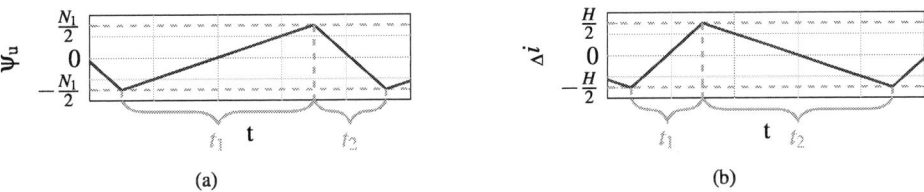

(a)   (b)

Fig. 3: Error trajectory and hysteresis behaviour for $\Delta\Sigma$-PWM (a) and hysteresis control (b) by neglecting dead times

The times t_1 and t_2 result to:

$$t_1 = \frac{N_1 t_S}{\frac{2U_{DC}}{U_{max}} + \frac{2u_{sp}}{U_{max}}} \text{ and } t_2 = \frac{N_1 t_S}{\frac{2U_{DC}}{U_{max}} - \frac{2u_{sp}}{U_{max}}}. \qquad (6)$$

With

$$f_{sw}(t) = \frac{1}{t_1(t) + t_2(t)} \qquad (7)$$

the variable switching frequency for a one-phase $\Delta\Sigma$-PWM with current control can be recast into the sum of a constant average frequency $f_{sw,av}$ and an oscillating part with amplitude $f_{sw,A}$ and angle α_{sw}:

$$f_{sw}(t) = f_{sw,av} + f_{sw,A} \sin(2\omega_{elec}t + \alpha_{sw}). \qquad (8)$$

EPE'20 ECCE Europe

Assigned jointly to the European Power Electronics and Drives Association & the Institute of Electrical and Electronics Engineers (IEEE)

where

$$f_{\text{sw,av,DS}} = \frac{U_{\text{DC}}^2 - \frac{\omega_{\text{elec}}^2}{2}(\hat{I}_S^2 L^2 + \psi_p^2)}{N_1 U_{\text{DC}} U_{\max} t_S}, \quad f_{\text{sw,A,DS}} = \omega_{\text{elec}}^2 \frac{\hat{I}_S^2 L^2 + \psi_p^2}{2N_1 U_{\text{DC}} U_{\max} t_S} \quad \text{and}$$

$$\alpha_{\text{sw,DS}} = \frac{\pi}{2} + 2\arctan\left(\frac{\hat{I}_S L}{\psi_p}\right). \tag{9}$$

Instantaneous switching frequency hysteresis control

The derivation of variable switching frequency of hysteresis control works in a comparable way. A similar derivation for a three-phase DC-chopper with neutral connection to DC midpoint was done in [6]. With Kirchhoff's law, the derivative of phase current is given through:

$$\frac{di_S}{dt} = \frac{u_S - u_{\text{emk}}}{L}. \tag{10}$$

Due to the hysteresis behavior from Fig. 3 (b), the derivative of current error e_I leads to:

$$\left.\frac{de_I}{dt}\right|_{u_S=-U_{\text{DC}}} = \frac{di_{\text{sp}}}{dt} - \left.\frac{di_S}{dt}\right|_{u_S=-U_{\text{DC}}} = \frac{H}{t_1} \quad \text{and} \quad \left.\frac{de_I}{dt}\right|_{u_S=U_{\text{DC}}} = \frac{di_{\text{sp}}}{dt} - \left.\frac{di_S}{dt}\right|_{u_S=U_{\text{DC}}} = \frac{-H}{t_2}. \tag{11}$$

The times of one switching period lead to:

$$t_1 = \frac{H}{\frac{U_{\text{DC}}+u_{\text{emk}}}{L} + \frac{di_{\text{sp}}}{dt}} \quad \text{and} \quad t_2 = \frac{H}{\frac{U_{\text{DC}}-u_{\text{emk}}}{L} - \frac{di_{\text{sp}}}{dt}}. \tag{12}$$

With (2), (7) and

$$i_{\text{sp}} = \hat{I}_S \sin(\omega_{\text{elec}} t) \tag{13}$$

the variable switching frequency for a one-phase hysteresis control can be similarly reshaped into an expression of the form given in (8), with:

$$f_{\text{sw,av,Hys}} = \frac{U_{\text{DC}}^2 - \frac{\omega_{\text{elec}}^2}{2}(\hat{I}_S^2 L^2 + \psi_p^2)}{2H U_{\text{DC}} L}, \quad f_{\text{sw,A,Hys}} = \omega_{\text{elec}}^2 \frac{\hat{I}_S^2 L^2 + \psi_p^2}{4H U_{\text{DC}} L} \quad \text{and}$$

$$\alpha_{\text{sw,Hys}} = \frac{\pi}{2} + 2\arctan\left(\frac{\hat{I}_S L}{\psi_p}\right). \tag{14}$$

(9) and (14) coincide in case of:

$$N_1 = H\frac{2L}{U_{\max} t_S}. \tag{15}$$

Hence $\Delta\Sigma$-PWM can be seen as the digital counterpart to the analogue Hysteresis control, with both techniques inducing an identical inverter behavior under the mentioned conditions. For this reason, the parameters $f_{\text{sw,av}}$, $f_{\text{sw,A}}$ and α_{sw} will be used in general for both patterns in the following.

A comparison of analytical and numerical switching frequency for one operation point is shown in Fig. 4. Differences between simulation and analysis can be explained through the mentioned conditions of the derivation of switching frequency. A comparison of phase current and instantaneous switching frequency outlines the possible advantages of switching power losses: at phase current maximum the switching frequency is minimized, at zero phase current the switching frequency is maximized. Switching events with a lower current cause lower switching power losses. The net switching power loss change is highly dependent on the angle between phase current and switching frequency as can be seen in (9) and (14).

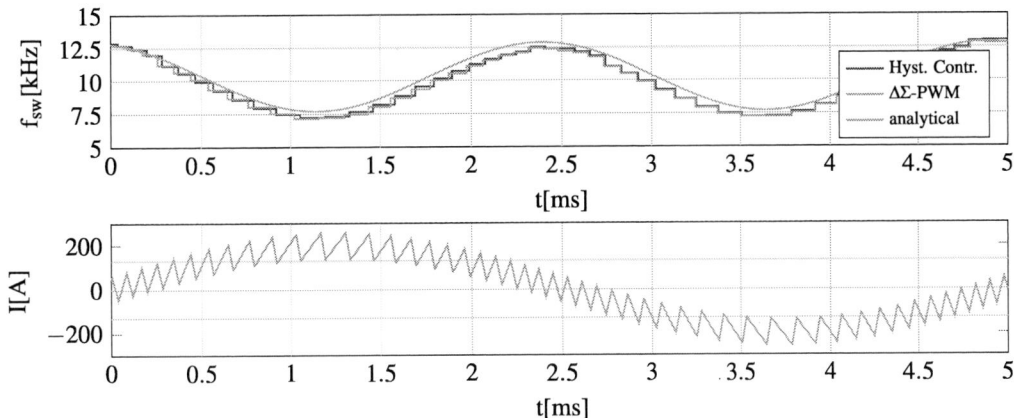

Fig. 4: Instantaneous switching frequency and typical phase current of hysteresis control and $\Delta\Sigma$-PWM

This angle α_{sw} is related to the power factor:

$$\cos(\varphi) = \cos\left(\arctan\left(\frac{\hat{I}_S L}{\psi_p}\right)\right) = \cos\left(\frac{\alpha_{sw} - \frac{\pi}{2}}{2}\right).\qquad(16)$$

Influence of dead times

In this chapter, the influence of dead times will be analyzed for hysteresis control. Thus, the derivation of the instantaneous switching frequency of hysteresis control will be expanded to dead time effects. Fig. 5 shows the behavior of hysteresis control with dead time. The error trajectory exceeds the hysteresis width during the duration of the dead time t_{dead}, which extends the time t_1. Furthermore, the dead time causes an additional term of hysteresis width H_{add}, which has to be taken into account for t_2. In the following, two different dead times are considered. First one $t_{dead,calc}$ is needed for the calculation of current control and the PWM pattern and occurs at every switching event. Second one $t_{dead,sw}$ protects the inverter against short circuits. As commonly known, the latter one only occurs at switching events with commutation from freewheeling diode to IGBT. The commutation to diode or IGBT highly depends on the current sign, thus a case definition is introduced. Three cases can be assumed: for $i_{sp} < -H$, $t_{dead,sw}$ impacts the switching event from $+U_{DC}$ to $-U_{DC}$. For $-H < i_{sp} < H$, $t_{dead,sw}$ has no impact and for $i_{sp} > H$, $t_{dead,sw}$ impacts the switching event from $-U_{DC}$ to $+U_{DC}$. The last case is shown in Fig. 5 and examined in the following.

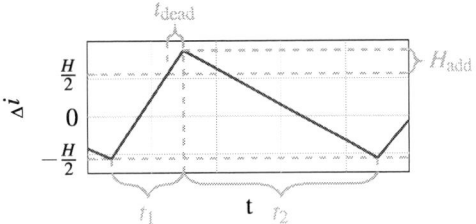

Fig. 5: Error trajectory and hysteresis behaviour for hysteresis control with influence of dead time t_{dead}

The times t_1 and t_2 from (12) are expanded to:

$$t_1 = \frac{H + H_{add,calc1}}{\frac{U_{DC}+u_{emk}}{L} + \frac{di_{sp}}{dt}} + t_{dead,sw} + t_{dead,calc} \text{ and } t_2 = \frac{-H - H_{add,calc2} - H_{add,sw2}}{\frac{-U_{DC}+u_{emk}}{L} + \frac{di_{sp}}{dt}} + t_{dead,calc}\qquad(17)$$

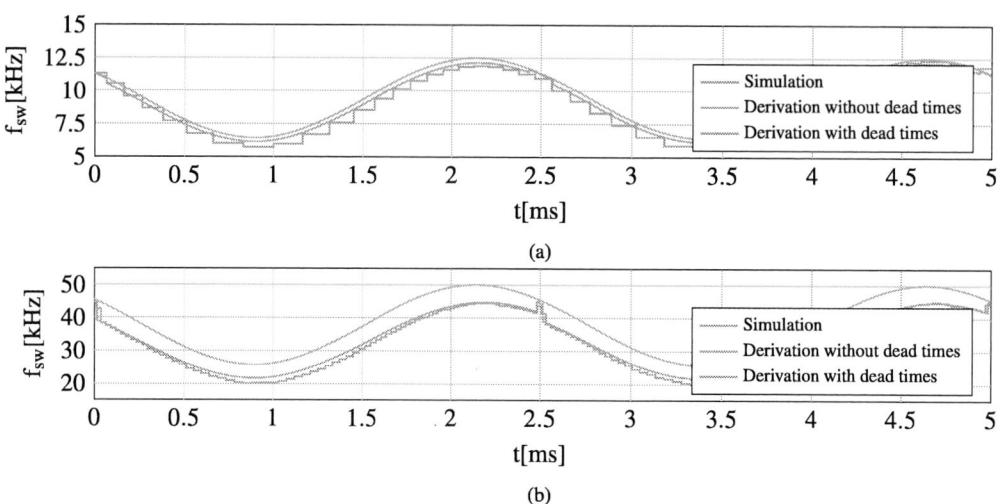

(a)

(b)

Fig. 6: Instantaneous switching frequency of hysteresis control with different hysteresis widths H

for $i_{sp} > H$. The additional hysteresis width terms can be derived with the slope of the current trajectory from (11):

$$H_{\text{add,calc1}} = -\frac{de_I}{dt}t_{\text{dead,calc}}\bigg|_{u_S=U_{\text{DC}}} , \quad H_{\text{add,calc2}} = \frac{de_I}{dt}t_{\text{dead,calc}}\bigg|_{u_S=-U_{\text{DC}}}$$

$$H_{\text{add,sw1}} = -\frac{de_I}{dt}t_{\text{dead,sw}}\bigg|_{u_S=U_{\text{DC}}} , \quad H_{\text{add,sw2}} = \frac{de_I}{dt}t_{\text{dead,sw}}\bigg|_{u_S=-U_{\text{DC}}}$$

(18)

Fig. 6 compares the derivation of instantaneous switching frequency with and without considering dead times for two different hysteresis widths and resulting different average switching frequencies. The influence of dead times is increasing with higher average switching frequencies. For $f_{\text{sw,av}} \approx 35$ kHz (Fig. 6 (b)) the derivation with consideration of dead times is significantly more accurate. At $f_{\text{sw,av}} \approx 10$ kHz (Fig. 6 (a)), which is considered in the following chapter, both variants differ only marginally due to a different ratio of dead time and PWM period time. For this reason and because of an easier definition of the instantaneous switching frequency, the dead time effect is neglected in the following.

The differences caused by the resistance R, as another main factor for inaccuracy, are increasing with higher phase current due to higher voltage drops over this resistance, which are not considered in (1). This influence is visible in Fig. 6 (b). At switching frequency minimums of the chosen operational point, the phase current passes the high or low peak, thus the voltage drop over the resistance do so as well. At these points the difference between simulation and analysis is higher in comparison to points at switching frequency maximums, where phase current and voltage drop are close to zero crossing.

Ratio of switching power losses

In this chapter, the ratio of switching power losses is derived analytically and compared to simulation results. Due to the similar switching behavior of $\Delta\Sigma$-PWM and hysteresis control, the simulation and the following derivation of switching power losses is only done for $\Delta\Sigma$-PWM. Switching power losses can be calculated according to (19) [7].

$$P_{\text{sw,IGBT}} = f_{\text{sw}}E_{\text{on+off}}\frac{\sqrt{2}}{\pi}\frac{I_{\text{out}}}{I_{\text{ref}}}\left(\frac{U_{\text{DC}}}{U_{\text{ref}}}\right)^{K_v}\left(1 + TC_{\text{Esw}}(T_j - T_{\text{ref}})\right)$$

$$P_{\text{sw,diode}} = f_{\text{sw}}E_{\text{rr}}\frac{\sqrt{2}}{\pi}\left(\frac{I_{\text{out}}}{I_{\text{ref}}}\right)^{K_i}\left(\frac{U_{\text{DC}}}{U_{\text{ref}}}\right)^{K_v}\left(1 + TC_{\text{Err}}(T_j - T_{\text{ref}})\right)$$

(19)

The curves for switching loss energy of IGBT (E_{on}, E_{off}) and diode (E_{rr}) are derived from internal data, where E_{on} and E_{off} differ only marginally. The temperature dependency is already taken into account in the curves of switching loss energy. For the calculation of relative quantities, the values for the operational point $\frac{\sqrt{2}}{\pi}$, $\frac{I_{out}}{I_{ref}}$ and $\frac{U_{DC}}{U_{ref}}$ of (19) can be neglected.

Assuming a linear linking between phase current and switching loss energy with equal switching losses of a commutation from IGBT to diode respectively from diode to IGBT, and the above mentioned assumptions, (19) can be transposed to:

$$P_{sw}(t) \sim |i_S(t)| f_{sw}(t). \tag{20}$$

A ratio of the average switching power losses of a one phase $\Delta\Sigma$-PWM (or hysteresis control instead) to fixed frequency PWM can be calculated as follows:

$$r_{Psw} = \frac{P_{sw,av,\Delta\Sigma}}{P_{sw,av,fixed}} = \frac{\frac{\omega_{elec}}{\pi} \int_0^{\frac{\pi}{\omega_{elec}}} |\hat{I}_S \sin(\omega_{elec}t)| f_{sw}(t) dt}{\frac{\omega_{elec}}{\pi} \int_0^{\frac{\pi}{\omega_{elec}}} |\hat{I}_S \sin(\omega_{elec}t)| f_{sw,c} dt}. \tag{21}$$

The variable switching frequency of $\Delta\Sigma$-PWM $f_{sw}(t)$ is mentioned in (8). The constant switching frequency of fixed frequency PWM $f_{sw,c}$ is set to the average switching frequency of $\Delta\Sigma$-PWM $f_{sw,av}$. The loss ratio can then be written as:

$$r_{Psw} = \frac{f_{sw,av} - \frac{f_{sw,A}}{3} \cos(2\varphi)}{f_{sw,av}}. \tag{22}$$

With (22) a theoretical minimal and maximal ratio of switching power losses with $f_{sw,av} = f_{sw,A}$

$$r_{Psw,min} = 66.67\% \text{ and } r_{Psw,max} = 133.33\%. \tag{23}$$

and a value for $\cos(\varphi)$ with equal switching power losses of $\Delta\Sigma$-PWM and fixed-frequency PWM can be calculated:

$$\cos(\varphi)_{neutral} = \frac{1}{\sqrt{2}} \approx 0.707. \tag{24}$$

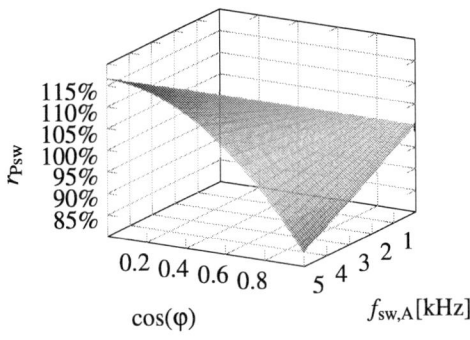

Fig. 7: Analytical switching loss ratio r_{Psw} of a one-phase $\Delta\Sigma$-PWM in comparison to fixed-frequency PWM

Fig. 7 shows a plot of (22) with $f_{sw,av} = 10\,kHz$. The simulation of ratio of switching power losses between $\Delta\Sigma$-PWM and a fixed-frequency PWM in dependence of the mentioned parameters, shown in Fig. 8, leads to a comparable result. Analytical derivation and simulation only generate marginal differences.

With a power factor close to 1, the switching power losses of a one-phase inverter with $\Delta\Sigma$-PWM are reduced up to 15% in comparison to fixed frequency PWM at the operational point $f_{A,sw} \approx \frac{f_{av,sw}}{2}$. Between hysteresis control and fixed frequency PWM a similar ratio is expected due its identical behavior concerning switching frequency variation (see (15)). The red line of Fig. 8(b) shows the 100%-line of ratio of switching loss power, corresponding to (24). In conclusion, we showed a significant enhancement regarding switching power losses of one-phase B2-inverters due to an improved PWM strategy.

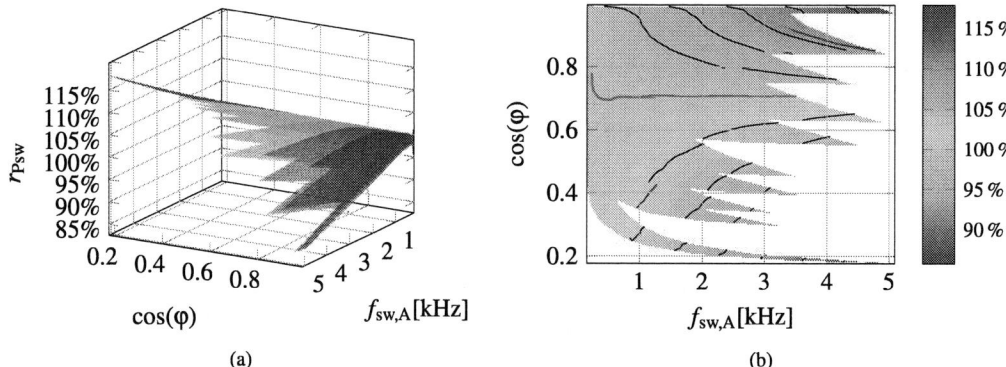

Fig. 8: Simulated switching loss ratio r_{Psw} of one-phase $\Delta\Sigma$-PWM in comparison to fixed-frequency PWM as 3D-Plot (a) and contour plot (b) in dependence of $\cos(\varphi)$ and amplitude of instantaneous switching frequency $f_{sw,A}$. The red line shows the 100%-line (see (24)). The plots were calculated under variation of the magnetic flux Ψ_p.

We note here that the same switching strategy is less beneficial for three-phase systems. Due to the constant presence of current in one of the three phases, the advantage of the deliberate switching effect reduces by a factor of 8 [5].

Distortion factor and total harmonic distortion

For the evaluation of an optimal shift strategy, switching losses alone are not sufficient and further criteria have to be evaluated. One important criterion is the signal quality of the produced phase currents, which is analyzed here using the quantities distortion factor d and total harmonic distortion THD after:

$$ d = \frac{I_{h,rms}}{I_{h,rms,six-step}} \text{ with } I_{h,rms} = \sqrt{\frac{1}{T}\int_0^T (i_S^2 - i_{S,1}^2)} \text{ and } THD = \frac{\sqrt{\sum_{n=2}^N i_{S,n}^2}}{i_{S,1}} \tag{25} $$

With both criteria the analysis of phase currents in time domain and frequency domain is covered. The quantities will be calculated from the fundamental phase current $i_{S,1}$ and its harmonics $i_{S,n}$.

Similar to the analysis of switching losses, a ratio of the distortion factors of $\Delta\Sigma$-PWM and fixed frequency PWM is derived analytically. As a first step, the phase current ripple ΔI_{max} (same as peak current error in one PWM cycle) is calculated in a similar way as the switching frequency of hysteresis control. Instead of the constant phase current ripple H and a variable switching frequency $f_{sw}(t)$, a variable phase current ripple $\Delta I_{max}(t)$ and a constant switching frequency $f_{sw,av}$ is used. The hysteresis width of hysteresis control is replaced by the peak current ripple:

$$ \Delta I_{max} = \frac{H}{2} \tag{26} $$

With Eq. (26), (11) can be written as:

$$ \frac{de_I}{dt}\bigg|_{u_S=-U_{DC}} = \frac{di_{sp}}{dt} - \frac{di_S}{dt}\bigg|_{u_S=-U_{DC}} = \frac{\Delta I_{max}}{2t_1} \text{ and } \frac{de_I}{dt}\bigg|_{u_S=U_{DC}} = \frac{di_{sp}}{dt} - \frac{di_S}{dt}\bigg|_{u_S=U_{DC}} = \frac{-\Delta I_{max}}{2t_2}. $$

$$(27)$$

With a constant switching frequency, (7) and again using the assumptions of chapter of hysteresis control, (27) lead to:

$$\Delta I_{\max}(t) = \Delta I_{\max,\mathrm{av}} + \Delta I_{\max,\mathrm{A}} \sin(2\omega_{\mathrm{elec}}t + \alpha_{\mathrm{Imax}}) \text{ with}$$

$$\Delta I_{\max,\mathrm{av}} = \frac{U_{\mathrm{DC}}^2 - \frac{\omega_{\mathrm{elec}}^2}{2}(\hat{I}_{\mathrm{S}}^2 L^2 + \psi_{\mathrm{p}}^2)}{4 f_{\mathrm{sw,av}} U_{\mathrm{DC}} L}, \ \Delta I_{\max,\mathrm{A}} = \omega_{\mathrm{elec}}^2 \frac{\hat{I}_{\mathrm{S}}^2 L^2 + \psi_{\mathrm{p}}^2}{8 f_{\mathrm{sw,av}} U_{\mathrm{DC}} L} \text{ and} \tag{28}$$

$$\alpha_{\mathrm{Imax}} = \frac{\pi}{2} + 2\arctan\left(\frac{\hat{I}_{\mathrm{S}} L}{\psi_{\mathrm{p}}}\right).$$

The striking similarity in shape between the equations for current ripple (Eq. (28)) and switching frequency (Eq. (9) and (14)) underline the reciprocal connection between switching frequency and peak current ripple.

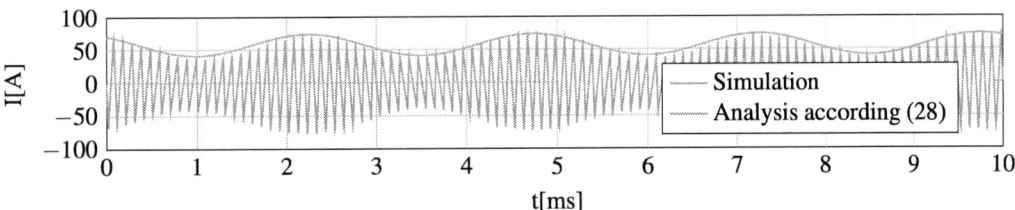

Fig. 9: Simulated current ripple and analytically derived peak current ripple of one-phase fixed-frequency PWM

Out of (9), (14), (15) and (28) follows, that an equal average switching frequency $f_{\mathrm{sw,av}}$ of $\Delta\Sigma$-PWM, hysteresis control and fixed-frequency PWM causes an equal average current ripple $\Delta I_{\max,\mathrm{av}}$. Assuming the phase current error to be uniformly distributed between $\Delta I_{\max}$ and $-\Delta I_{\max}$, only the phase current ripple $\Delta I_{\max}$ affects the distortion factor (25). The ratio of of distortion factor of $\Delta\Sigma$-PWM to fixed-frequency PWM r_{d} can be calculated with (9) as follows:

$$r_{\mathrm{d}} = \frac{I_{\mathrm{h,rms},\Delta\Sigma}}{I_{\mathrm{h,rms,ff}}} = \sqrt{\frac{\frac{1}{T}\int_0^T (\Delta I_{\max,\mathrm{av}}^2)dt}{\frac{1}{T}\int_0^T (\Delta I_{\max,\mathrm{av}} + \Delta I_{\max,\mathrm{A}} \sin(2\omega_{\mathrm{elec}}t + \alpha_{\mathrm{Imax}}))dt}} = \frac{1}{\sqrt{1 + \frac{f_{\mathrm{sw,A}}^2}{2 f_{\mathrm{sw,av}}^2}}}. \tag{29}$$

With (29) the distortion factor of $\Delta\Sigma$-PWM and hysteresis control can be minimized up to:

$$r_{\mathrm{d,min}} = 81.7\% \tag{30}$$

assuming $f_{\mathrm{sw,A}} = f_{\mathrm{sw,av}}$. This is equivalent to a margin of 18.3%. Thus, the use of our switching strategy can simultaneously reduce switching power losses and distortion. Fig. 10 shows a comparison of the analytically derived distortion factor (29) and numerical sequences. Besides the distortion factor, the THD up to 15 kHz and the THD over the full spectrum are shown. The 15 kHz limit of THD_{15} has been chosen as it approximately corresponds to the audible range, and is situated in the middle between the fundamental and first harmonic of the switching frequency. Small differences can be justified by the mentioned assumptions e.g. zero resistance and neglecting noise and dead times. The $\Delta\Sigma$-PWM produces a phase current with an average advantage of 6% for each quantity in case of an amplitude of switching frequency being half the average switching frequency. According to this result, an advantage of $\Delta\Sigma$-PWM and hysteresis control regarding to acoustic and electromagnetic compatibility (EMC) can be concluded. For a comprehensive analysis of these quantities the whole system of inverter, electric machine and their installation have to be taken into account.

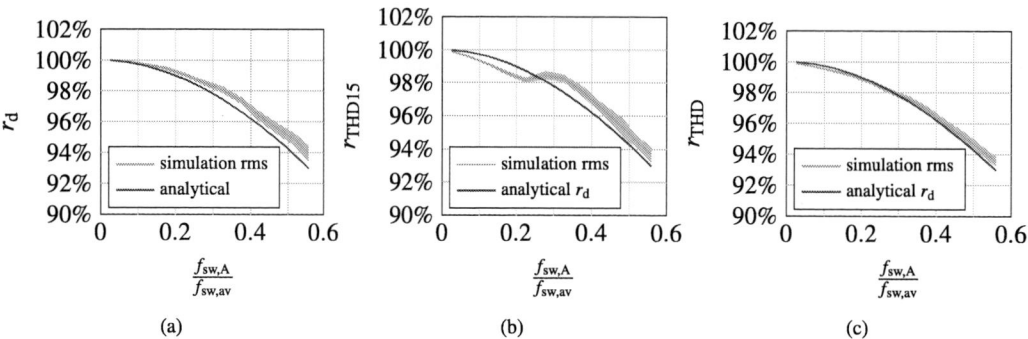

(a)   (b)   (c)

Fig. 10: Ratio of distortion factor r_d (a), total harmonic distortion up to 15 kHz r_{THD15} (b) and total harmonic distortion over full spectrum r_{THD} (c) as simulation with rms value and grey 3σ area and distortion factor as analytically derived function

Conclusion

This paper gives a comparison of switching power losses and phase current distortion in one-phase systems related to different PWM methods. First, the instantaneous switching frequency for the single-phase system is analytically derived for PWM methods with variable switching frequency, $\Delta\Sigma$-PWM and hysteresis control, and it is shown that both methods provide exactly the same results. The influence of dead times to the accuracy of this derivation is shown. Then the ratio of switching power losses of the variable frequency and a frequency fixed PWM method is analytically estimated and verified in simulation. Theoretically, a maximal reduction of 33% is possible. Depending on the power factor and operational points, simulation results show a reduction of switching power losses of up to 15%. Furthermore this advantage goes hand in hand with an decreasing of distortion factor and THD up to 6% for one-phase systems. The maximum theoretical advantage is up to 18%. This result was analytically estimated and simulated for the distortion factor. Therefore the peak current ripple of fixed-frequency PWM was derived. Summarizing the results of this paper, one-phase $\Delta\Sigma$-PWM and hysteresis control are able to reduce switching power losses and the distortion of the produced phase current at the same time. In a three-phase system, however, the described effect is reduced by at least a factor of 8 [5]. The impact of three-phase $\Delta\Sigma$-PWM to distortion factor and THD still has to be subject of further studies.

References

[1] Frank W.: Simple slew-rate control technique cuts switching losses, PCIM Europe 2019; International Exhibition and Conference for Power Electronics, Intelligent Motion, Renewable Energy and Energy Management, 2019.
[2] Holtz J.: Pulsewidth modulation-a survey, IEEE Transactions on Industrial Electronics, vol. 39, no. 5, pp. 410420, 1992.
[3] Dunlap S.K. and Fiez T.: A noise-shaped switching power supply using a delta-sigma modulator, Circuits and Systems I: Regular Papers, IEEE Transactions, vol. 51, no. 6, pp. 10511061, 2004.
[4] Homann M.: Hochdynamische Strom und Spannungsregelung von permanenterregten Synchronmaschinen auf Basis von Delta-Sigma Bitströmen. PhD thesis, TU Braunschweig, 2016.
[5] Ramm H., Homann. M., Schulze A.T., Turki F., Rabba H.: Comparison of switching power losses of fixed-frequency PWM, Hysteresis Control and Delta-Sigma PWM, accepted for publication in PCIM Europe 2020; International Exhibition and Conference for Power Electronics, Intelligent Motion, Renewable Energy and Energy Management, 2020.
[6] McMurray W.: Modulation of the chopping frequency dc choppers and pwm inverters having current-hysteresis controllers, IEEE Transactions on Industry Applications, vol. 20, no. 4, pp. 763768, 1984.
[7] Wintrich A., Nicolai U., Tursky W., and Reimann T.: Applikationshandbuch Leistungshalbleiter, 2nd edition. SEMIKRON International GmbH, 2015.

Optimization of medium-frequency transformers with large capacity and high insulation requirement

Xuan Guo, Chi Li, Zedong Zheng, Yongdong Li
Department of Electrical Engineering, Tsinghua University
NO. 30, Shuangqing Road, Haidian District
Beijing, China
E-Mail: guo-x18@mails.tsinghua.edu.cn

Acknowledgements

This work was supported by the National Natural Science Foundation of China under Grant 51777111

Keywords

«Transformer», «Design», «Insulation», «Efficiency»

Abstract

The application of traction power electronic transformers (PETs) requires high efficiency and power density, where insulation becomes a key factor in the transformer design process. This paper presents a model-based optimization design and engineering realization that considers the electromagnetic, geometric and insulating properties of the transformer. To accurately model the transformer considering the geometric asymmetry caused by high insulation requirement, a hybrid distributed inductance model is put forward. Based on the optimized pareto domain obtained by the optimization method, one can get the most suitable optimal design point of the volume and the efficiency of the transformer. A 140kW, 85kV insulation prototype has been designed and the static parameters measurement experiments and simulation results verified the theoretical design. The electricity and insulation experiments will be done in the future due to COVID-19.

Introduction

Power electronic transformers (PETs) have attracted widespread attention in academia and industry in recent years due to its advantages of power density and controllability.[1] PETs with input series output parallel (ISOP) topology can achieve high voltage ratio and become an attracting solution for medium voltage direct current (MVDC) applications like the field of power distribution and traction.[2] Medium frequency transformers, the core component of PETs, have become important factor that restrict the further improvement of the power density and efficiency of PET. The PETs of ISOP topology also put new requirements on the insulation because of the high voltage of primary winding. Insulation problem should become an important consideration of transformer design.

The area-product (AP) method, as one of the traditional transformer optimization design methods, relies on the empirical parameters, and usually has a certain margin,[3] resulting in material waste and power density reduction. On the other hand, the AP method does not consider the transformer efficiency in the design process. The efficiency characteristic need to be obtained by measurement after the transformer prototype is completed. It is unable to get the optimized design of the transformer. Therefore, the transformer needs a detailed optimization design that comprehensively considers electromagnetic, geometric and insulation characteristics.

Many literatures have proposed different transformer optimization design methods, which have achieved good results, but also have certain limitations. Zhao proposed a multi-objective optimization method for medium frequency transformer using the idea of pareto front.[4] However, the optimized free variable only contains the turns' number of the transformer's secondary side, the optimization space is small. Cao proposed a transformer optimization design method based on multi-objective genetic algorithm.[5] The optimization goal is to minimize the loss and minimize the difference between the designed distributed inductance and the given value. On the one hand, the transformer power density is not included in the optimization target, on the other hand, the optimization space is also small, the core geometry parameters are not contained in the optimization space. Leibl and Mogorovic use the pareto domain to obtain the compromise optimization design considering the transformer's efficiency and power density in the design process.[6]–[8] The geometric and electromagnetic parameters are all contained in the optimization space, but they don't take the insulation problem in-depth consideration.

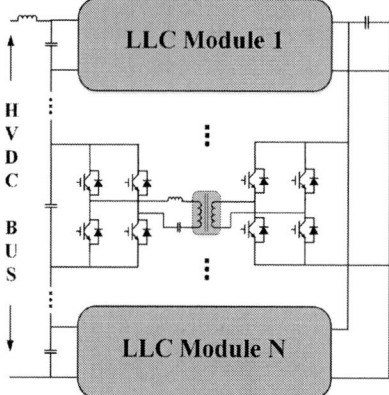

Fig. 1: The ISOP topology of PET

In this paper, for the application of the traction PET shown in Fig.1, a detailed optimization design method of medium frequency transformer considering the electromagnetic, geometric and insulating characteristics is proposed. At the same time, this paper proposes an improved distributed inductance model based on the Dowell model for the case where the height of the primary and secondary windings of medium-frequency transformers with high dielectric strength is not equal. The optimization design process proposed in this paper actually maps the multi-dimensional design points of the transformer into the two-dimensional space of the transformer's volume and efficiency, which forms the pareto domain. According to the actual needs, the designer can find the most suitable optimization design of transformer in the pareto domain.

This paper is organized as follows. The model of medium frequency transformer is proposed in part2. Part3 shows the optimization process of the transformer. Then analysis results and the simulation verification are presented in part4. The experimental results are in part5. The last part is the conclusion of this paper.

2. The model of the transformer

This paper chose EE core as the transformer's core. In the traction PET with ISOP topology shown in Fig.1, the medium frequency transformer is often placed together with the primary H-bridge in one sealed box. Because the secondary H-bridge should connect with the inverter and the motor, it is often placed outside the box. Therefore, in the design, the magnetic core and the high-voltage winding are equipotential, and the low-voltage winding is wound on the outside of the high-voltage winding with the required main insulation distance. There is also main insulation distance between the low-voltage winding with the magnetic core. The core's geometry parameters are shown as Fig. 2.

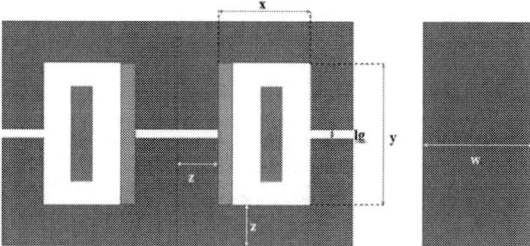

Fig. 2: The figure of core geometry

A. Coil loss model

Because of the transformer is under high-frequency operation, the influence of the proximity effect and the eddy current effect should be considered.[9] The Bessel model[10] is used to solve the copper loss of the transformer in the optimization process as shown in (1).

$$P = K_{coil} R_{dc} (FI^2 + GH^2) \tag{1}$$

Where I is the effective value of the current at the operating frequency and H is the effective value of spatial magnetic field strength amplitude. F and G are the skin effect and the proximity effect coefficient respectively consist of Bessel function as shown in (2) and (3)

$$F = \frac{\gamma}{4} \frac{ber_0(\gamma) * ber_0'(\gamma) - bei_0(\gamma) * ber_0'(\gamma)}{ber_0'(\gamma)^2 + bei_0'(\gamma)^2} \tag{2}$$

$$G = \frac{\gamma \pi^2 d_r^2}{2} \frac{ber_2(\gamma) * ber_0'(\gamma) + bei_2(\gamma) * bei_0'(\gamma)}{ber_0(\gamma)^2 + bei_0(\gamma)^2} \tag{3}$$

At the formula, γ equals $\frac{d_r}{\delta\sqrt{2}}$. d_r is the diameter of one single strand of the used litz wire and δ is the skin depth of the transformer at given operating frequency as shown in (4). f is the operating frequency, μ is permeability of the conductor and σ is the conductivity of the conductor. In addition, since the actual litz wire is not perfect litz wire (the current density is perfectly identical within each strand), it is necessary to make engineering correction to the litz wire at a certain frequency through the coefficient K_{coil}.

$$\delta = \sqrt{\frac{1}{\pi f \mu \sigma}} \tag{4}$$

B. Core loss model

There are two ways to model the transformer's core loss. The first one is solved by the improved Steinmetz model.[11] The traditional Steinmetz model is usually used to solve the core loss under sinusoidal excitation. The improved model can be applied to the solution of core loss under high frequency non-sinusoidal excitation. The Steinmetz model is shown in (5), where K, α, β are magnetic material characteristic parameters, which can be obtained by fitting the core loss characteristic curve provided by the manufacturer.

$$P_{core} = \frac{1}{T} \int_0^T k_i \left| \frac{dB(t)}{dt} \right|^\alpha |\Delta B|^{\beta - \alpha} dt \tag{5}$$

$$k_i = \frac{K}{(2\pi)^{\alpha - 1} \int_0^{2\pi} |cos\theta|^\alpha 2^{\beta - \alpha} d\theta}$$

The second one is using the modified Bertotti model to get the transformer's core loss.[12] The main idea of the Bertotti model is to divide the core loss into hysteresis loss, eddy current loss and abnormal loss. The three kinds of losses are respectively the power function of the operating frequency f and the working magnetic flux density B_{ac}. The relationship is shown in (6).

$$P_{core} = k_1 f^{a1} B^{b1} + k_2 f^{a2} B^{b2} + k_3 f^{a3} B^{b3} \tag{6}$$

The optimization process directly fits the loss curve given by the manufacturer at the operating frequency, and obtains a polynomial relationship between the core loss density and the working magnetic flux density. It is more accurate compared with obtaining K, α, β by fitting multiple sets of core loss curves under different operating frequencies and working magnetic flux densities. Therefore, the final core loss model is using a polynomial fitting function based on the modified Bertotti model.

C. Distributed inductance model

The distributed inductance value is solved by the energy method, as shown in (7) and (8), the key step is to solve the magnetic field strength distribution in the transformer window area. Most of the existing methods are based on the Dowell model,[13], [14] and the round conductor or litz wire winding is equivalent to the copper foil winding as shown in Fig. 3, under low-frequency conditions, the magnetic field strength distribution is a linear stepped distribution. At high frequencies, due to proximity effects, the magnetic field strength distribution in the winding area becomes non-linear. The Dowell model can be used to solve the actual magnetic field strength distribution, and the leakage inductance can be solved by the spatial integration of the magnetic field energy. The distributed inductance of Dowell model is shown in (9), where MLT_p, MLT_{ins}, MLT_s are the average turn length of primary winding, main insulation area, secondary winding respectively, d_{pri}, d_{ins}, d_{sec} are the thickness of the primary winding, the main insulation area, and the secondary winding, h is the window height , N is the number of turns on the primary side, m is the number of equivalent layers. For the litz wire, m = $m_l\sqrt{N_l}$, where m_l is the actual winding layer number, N_l is the strands of one conductor.

$$W_{\text{leakage}} = \frac{1}{2}\mu_0 \int H^2 dV \tag{7}$$

$$L_\sigma = \frac{2W_{leakage}}{I^2} \tag{8}$$

$$L_{Dowell} = \frac{\mu N^2}{h}\left(\frac{MLT_p * d_{pri}}{3}F_{wp} + MLT_{ins} * d_{ins} + \frac{MLT_s * d_{sec}}{3}F_{ws}\right) \tag{9}$$

$$F_w = \frac{1}{2m^2\Delta}[(4m^2-1)\varphi_1 - 2(m^2-1)\varphi_2] \tag{10}$$

$$\varphi_1 = \frac{\sinh(2\Delta) - \sin(2\Delta)}{\cosh(2\Delta) - \cos(2\Delta)}, \varphi_2 = \frac{\sinh(\Delta) - \sin(\Delta)}{\cosh(\Delta) - \cos(\Delta)}$$

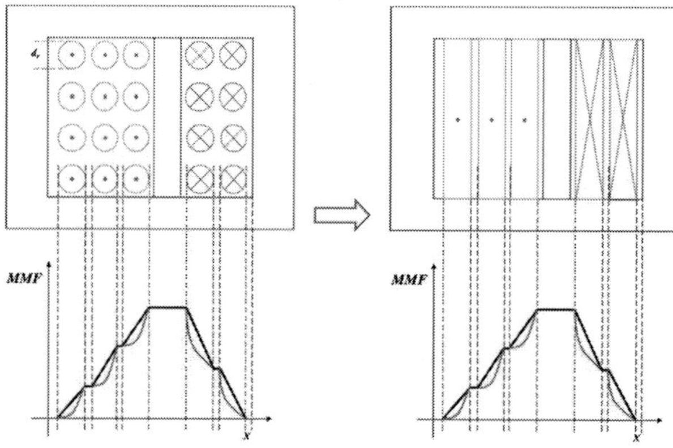

Fig. 3: The magnetic field distribution

However, the Dowell model shown in (9) does not take the irregularities of the winding geometry into account. As shown in Fig. 4, when the winding height is less than the window height, or the height of the primary and secondary windings is inconsistent, the magnetic field strength distribution is quite different from the ideal magnetic field strength distribution shown in Fig. 3. For transformer with high insulation strength requirement, the magnetic core is usually at the same potential as the high-voltage winding (or low-voltage winding), so there must be a certain insulation distance between the core and the low-voltage winding (or high-voltage winding), which causes the winding height less than the window height, and the height of the primary and secondary windings are inconsistent. In this case, a correction factor needs to be introduced to modify the Dowell model to take the geometrical factor into account.

Mogorovic introduced the Rogowski factor R_c to modify the equivalent winding height (as shown in (12) and (13)),[14] and then modified the Dowell model, which considers the geometrical factor that the winding height is less than the window height. However, it does not consider the inconsistency of the primary and secondary winding heights. Based on the model proposed by Mogorovic, this paper refers to the low frequency transformer (LFT) leakage inductance solution and introduces a correction coefficient K to further modify the Dowell model.[15] This hybrid model shown in (11) not only considers the factor that the winding height is less than the window height, but also considering the inconsistency of the primary and secondary winding heights.

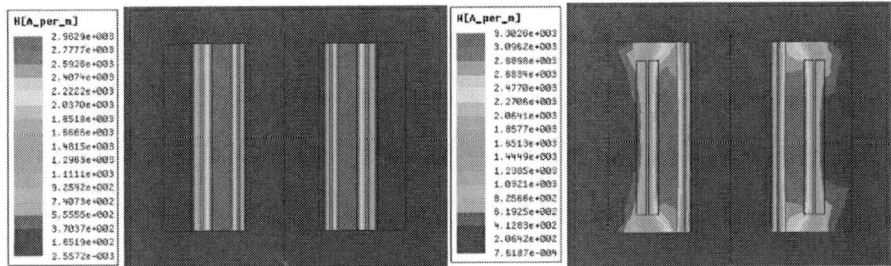

Fig. 4: The FEA simulation of magnetic field distribution (a) ideal geometry (b) actual geometry

$$L_{Dowell} = K \frac{\mu N^2}{h_{eq}} \left(\frac{MLT_p * d_{pri}}{3} F_{wp} + MLT_{ins} * d_{ins} + \frac{MLT_s * d_{sec}}{3} F_{ws} \right) \quad (11)$$

$$h_{eq} = \frac{h_w}{R_c} \quad (12)$$

$$R_c = 1 - \frac{d_{pri} + d_{ins} + d_{sec}}{h_w} \quad (13)$$

$$K = 1 + \frac{h_{w1} - h_{w2}}{2h_{w1}} \left(1 + \frac{\pi}{2} \frac{h_{w1} - h_{w2}}{d_{pri} + d_{ins} + d_{sec}} \right) \quad (14)$$

3. The optimization process of the transformer

Medium frequency transformers used in PET are usually designed under the given electrical parameters. The given electrical parameters include the rated capacity P_t, the rated voltage of the primary side V_{pri} and secondary side V_{sec} of the transformer, the insulation grade V_{ins} and the operating frequency f. The optimization goal is usually to maximize efficiency and power density, but usually the optimal design points of power density and efficiency cannot be obtained at the same time, so the problem of transformer's optimization design is a typical multi-objective optimization problem. In addition, medium frequency transformers are usually used in dual active bridge (DAB) or resonant DC-DC converters. The two topologies have certain requirements for the distributed inductance, so the value of the distributed inductance needs to be limited.

Table I: Electrical parameters of the transformer

Rated Capacity	Rated Voltage	Rated Frequency	Transformer's ratio
140kVA	3600V	5500Hz	2:1
Insulation Grade	Transformer's leakage inductance limit	Transformer's efficiency limit	
85kV	240uH	99%	

At a given operating frequency, the core material needs to be selected first, the selection principle is usually based on experience. Silicon steel cores can be selected at about 1kHz, nanocrystalline cores can be selected below 50kHz, and ferrite cores can be selected at 50kHz and above. After the core material is determined, the relevant core material parameters can be determined, including the saturation magnetic flux density and the core loss parameters involved in (5) and (6).

In this paper, nanocrystalline is selected as core material. The relevant electrical parameters are shown in Table 1. The electrical parameters and magnetic core material parameters are entered as initial values in the optimizer shown in Fig. 5 to start the optimization process. In addition to electrical parameters and core material parameters, the initial parameters also include geometrical parameters (transformer length, width and height limits $L_{lim}, W_{lim}, H_{lim}$ and the minimum insulation distance d_{ins}). The optimizer uses a brute force search algorithm, and the optimization parameters are used as loop variables. The optimization parameters include the geometric parameters of each dimension $[x, y, z, w]$ of the magnetic core shown in Fig.2, the working magnetic flux density of the transformer B_{ac}, and the working current density J. Through the value of each group of loop variables and formulas (1), (6), (11), (15), (16), the corresponding number of turns of the primary side of the transformer N, the cross-sectional area of the conductors of the primary and secondary sides of the transformer (number of litz wire strands N_l), the coil loss, the core loss and the distributed inductance can be obtained. In addition, the transformer's volume can also be calculated.

$$N = \frac{V}{K_f A_c f B_{ac}} \tag{15}$$

$$N_l = \frac{4I}{\pi d_r^2 J} \tag{16}$$

Then, each set of design points is judged in the loop body. The judgment condition is based on the geometric limit (whether the core window area is larger than the wire area, and whether the length, width and height of the transformer exceed the given limits), the electromagnetic limit (whether the transformer leakage inductance exceeds the given upper limit), the efficiency limit (whether the efficiency exceeds the given minimum requirements). Every design point that meets all of the above requirements will be retained, and the remaining design points will accumulate to form the pareto domain.

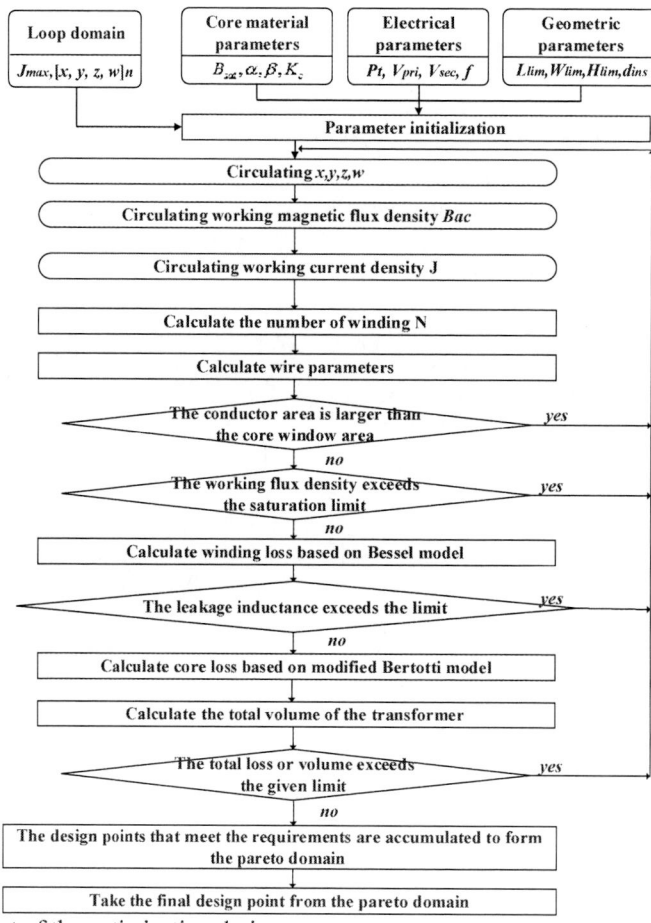

Fig. 5: The flowchart of the optimization design

4. The analysis results and the simulation verification

The pareto domain under the given electrical parameters and material parameters is shown in Fig.6. The pareto domain actually maps all feasible design points in the optimization process to the two-dimensional space of loss and volume, and each point in the domain can realize a feasible transformer design. The red star in Fig. 6 is the selected design point after considering the difficulty of transformer fabrication. Table 2 is the theoretical optimization design results of the transformer. The theoretical optimized transformer's efficiency can achieve 99.55% and its power density can achieve 13.46kW/L.

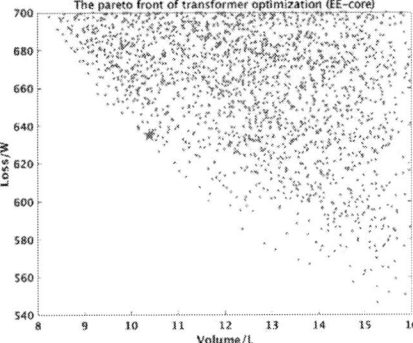

Fig. 6: The pareto domain of the transformer's optimization

Table II: The optimization design result of the transformer

Core geometry parameter x	Core geometry parameter y	Core geometry parameter z	Core geometry parameter w
6.5cm	13.5cm	2.5cm	14cm
Primary wire's cross-sectional area	The number of primary turns	The number of secondary turns	Insulation distance
5.5mm2	40	20	12mm

For the transformer with large capacity and high insulation requirement, to ensure it can work normally, it must be ensured that at the given insulation level, the electric field strength everywhere does not exceed the insulation strength of the insulating material Therefore, the insulation is verified by electromagnetic finite element simulation. The simulation results are shown in Fig.7. When the potential of the transformer core and the primary winding is 85kV and the potential of the secondary winding is 0, the maximum value of the local electric field is 13.89kV/mm, and the epoxy resin potting material to be used has an insulation strength of 25kV/mm, which meets the insulation requirements.

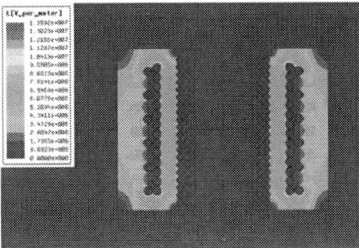

Fig. 7: The electric filed FEA simulation

5. The experimental prototype

The prototype of the unpotted transformer is shown in Fig.8(a). The results of the primary and secondary AC resistance of the transformer measuring with an impedance analyzer and the leakage inductance parameters measuring through a short-circuit experiment are shown in Table 3. The error of AC resistance and the distributed inductance between the measured value and the model predicted value is less than 10%.

In addition to the measurement error, as shown in Fig.8(a), since the secondary winding cannot be completely covered with two layers, the distance between each turn of the secondary outer layer is large, and error of distributed inductance will also be introduced in this case. Compared with the predicted value of Dowell model of 189uH, the predicted accuracy of the hybrid model mentioned in this paper is greatly improved.

Table III: The static parameters of the transformer

	The AC resistance of primary side	The AC resistance of secondary side	The distributed inductance
The measurement value	47.8mΩ	17.5mΩ	208uH
The predicted value	45.1mΩ	18.8mΩ	198uH

To ensure that under the actual loss calculated by Table 3, the transformer can work normally. it must be guaranteed that the maximum temperature rise of the transformer does not exceed the given range. The temperature FEA simulation was carried as shown in Fig. 9. The result indicates that under the given loss, the maximum temperature of the transformer is 134.31°, which meets the heat dissipation requirements.

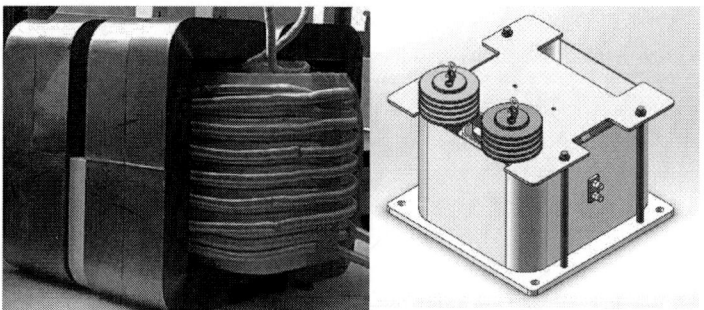

Fig. 8: The prototype (a) The unpotted transformer (b) The potting model of the transformer

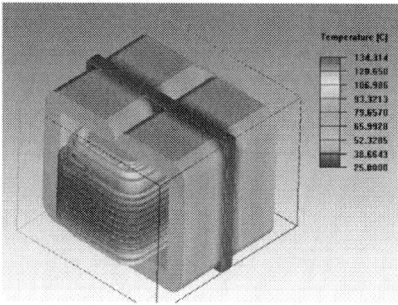

Fig. 9: The temperature FEA simulation

The potting model of the transformer is shown in Fig.8(b). The inside of the transformer is potted with epoxy resin to ensure the insulation. On the other hand, the heat of the secondary winding can be transferred to the primary side and the magnetic core through the epoxy resin, and then further dissipated by the water-cooling plate. In addition to ensuring the internal insulation distance, the terminals of the primary and secondary windings also need to consider the insulation problem. Therefore, the two terminals are led out through the insulator-shaped epoxy resin to ensure the creepage distance. The electricity experiments and insulation tests will be done in the future.

Conclusion

Power electronic transformer has become an attracting solution for MVDC applications like power distribution and traction. In this paper, an optimized design method considering electromagnetic, geometric and insulation characteristics is proposed for the key component of PET, isolated medium frequency transformer. The optimized design scheme uses the Bessel model, the modified Bertotti model and the hybrid distributed inductance model to accurately model the transformer, and then obtains the pareto domain of the transformer's power density and efficiency. From the pareto domain, this paper gets the optimized transformer design of which the efficiency can achieve 99.55% and the power density can achieve 13.46kW/L. The electromagnetic and thermal FEA simulation and the static parameters measurement experiment were carried out to verify the optimized result. The electricity and insulation experiments will be done in the future. With this optimized design method, one can get theoretical and practical guidance for magnetic component design with large capacity and high insulation requirement.

References

[1] C. Gu, Z. Zheng, L. Xu, K. Wang, and Y. Li, "Modeling and Control of a Multiport Power Electronic Transformer (PET) for Electric Traction Applications," *IEEE Trans. Power Electron.*, vol. 31, no. 2, pp. 915–927, Feb. 2016, doi: 10.1109/TPEL.2015.2416212.

[2] Y. Wang, K. Wang, C. Li, Z. Zheng, and Y. Li, "System-Level Efficiency Evaluation of Isolated DC/DC Converters in Power Electronics Transformers for Medium-Voltage DC Systems," *IEEE Access*, vol. 7, pp. 48445–48458, 2019, doi: 10.1109/ACCESS.2019.2909014.

[3] M. Kazimierczuk, *High-Frequency Magnetic Components: Kazimierczuk/High-Frequency Magnetic Components*. Chichester, UK: John Wiley & Sons, Ltd, 2013.

[4] S. Zhao, Q. Li, F. C. Lee, and B. Li, "High-Frequency Transformer Design for Modular Power Conversion From Medium-Voltage AC to 400 VDC," *IEEE Trans. Power Electron.*, vol. 33, no. 9, pp. 7545–7557, Sep. 2018, doi: 10.1109/TPEL.2017.2774440.

[5] X. CAO, W. CHEN, G. NING, G. QIAO, and C. WANG, "Optimization Design of High-power High-frequency Transformer Based on Multi-objective Genetic Algorithm," *Proc. CSEE*, vol. 38, no. 5, pp. 1348–1355, May 2018.

[6] M. Leibl, G. Ortiz, and J. W. Kolar, "Design and Experimental Analysis of a Medium-Frequency Transformer for Solid-State Transformer Applications," *IEEE J. Emerg. Sel. Top. Power Electron.*, vol. 5, no. 1, pp. 110–123, Mar. 2017, doi: 10.1109/JESTPE.2016.2623679.

[7] M. Mogorovic and D. Dujic, "100 kW, 10 kHz Medium-Frequency Transformer Design Optimization and Experimental Verification," *IEEE Trans. Power Electron.*, vol. 34, no. 2, pp. 1696–1708, Feb. 2019, doi: 10.1109/TPEL.2018.2835564.

[8] M. Mogorovic and D. Dujic, "Sensitivity Analysis of Medium-Frequency Transformer Designs for Solid-State Transformers," *IEEE Trans. Power Electron.*, vol. 34, no. 9, pp. 8356–8367, Sep. 2019, doi: 10.1109/TPEL.2018.2883390.

[9] P. L. Dowell, "Effects of eddy currents in transformer windings," *Proc. Inst. Electr. Eng.*, vol. 113, no. 8, p. 1387, 1966, doi: 10.1049/piee.1966.0236.

[10] J. A. Ferreira, "Analytical computation of AC resistance of round and rectangular litz wire windings," *IEE Proc. B Electr. Power Appl.*, vol. 139, no. 1, p. 21, 1992, doi: 10.1049/ip-b.1992.0003.

[11] K. Venkatachalam, C. R. Sullivan, T. Abdallah, and H. Tacca, "Accurate prediction of ferrite core loss with nonsinusoidal waveforms using only Steinmetz parameters," in *2002 IEEE Workshop on Computers in Power Electronics, 2002. Proceedings.*, Mayaguez, Puerto Rico, 2002, pp. 36–41, doi: 10.1109/CIPE.2002.1196712.

[12] G. Bertotti, "General properties of power losses in soft ferromagnetic materials," *IEEE Trans. Magn.*, vol. 24, no. 1, pp. 621–630, Jan. 1988, doi: 10.1109/20.43994.

[13] Z. Ouyang, J. Zhang, and W. G. Hurley, "Calculation of Leakage Inductance for High-Frequency Transformers," *IEEE Trans. Power Electron.*, vol. 30, no. 10, pp. 5769–5775, Oct. 2015, doi: 10.1109/TPEL.2014.2382175.

[14] M. Mogorovic and D. Dujic, "Medium frequency transformer leakage inductance modeling and experimental verification," in *2017 IEEE Energy Conversion Congress and Exposition (ECCE)*, Cincinnati, OH, Oct. 2017, pp. 419–424, doi: 10.1109/ECCE.2017.8095813.

[15] K. Yin, *The principals of transformer design*. China Electric Power Press, 2013.

Improved *SoC* Balancing and Active Power Sharing Control Method in

Highly Resistive Line Microgrid

Yuanhao Zhu[1,2], Hua Han[1,2], Guangze Shi[1,2], Zhangjie Liu[1,2], Yao Sun[1,2], Mei Su[1,2]
1. School of Automation, Central South University
2. Hunan Provinc • ial Key Laboratory of Power Electronics Equipment and Grid
Changsha, Hunan, P.R. China
E-Mail: 332085829@qq.com

Acknowledgements

This work was supported in part by the National Key R&D Program of China under Grant 2018YFB0606005, in part by the National Nature Science Foundation of China under Grant 61933011, in part by the major Project of Changzhutan Self-Dependent Innovation Demonstration Area under Grant 2018XK2002, in part by the Hunan Provincial Key Laboratory of Power Electronics Equipment and Grid under Grant 2018TP1001, in part by the Hunan Province Strategic Emerging Industry Science and Technology Research and Major Achievement Transformation Project under Grant 2018GK4016, and in part by the Project of Innovation-driven Plan in Central South University under Grant 2019ZZTS573.

Abstract

This paper proposes a distributed control method for AC microgrid with high resistance lines, where P-V droop control modified by average SoC obtained through the SoC average estimator to guarantee the SoC balancing and active power sharing. The results of simulations verify the effectiveness of the proposed method.

1.Introduction

As one of the most effective ways to integrate distributed power generations (DGs) [1], microgrid (MG) is getting more and more attention, and it becomes a practical approach to electrify remote areas. In the islanded mode, reasonable power sharing is required to guarantee the quality of voltage and frequency. However, the lack of inertia of converter-based DGs may lead to voltage and frequency deviation and even stability problems under disturbances. Besides, the increase of wind and solar generation units, which have intermittent and uncertain natures [2], make the above problems even more severe. Thus, the Energy Storage System (ESS) [3-6] is absolutely required to suppress the power fluctuations in MGs. To prolong the lifetime and enhance battery safety, ESSs should be utilized by balancing their *SoC*. Furthermore, the *SoC* balancing can avoid the excessive use of a specific storage module.

In the inverter-parallel MG system, the conventional *P-ω/Q-V* droop control strategy can accurately share active power without a communication network. When line impendence is highly resistive, reference [7] proposes the *P-V* droop control method for active power sharing. However, the accuracy of power sharing cannot be guaranteed. Multiple modified *P-V* droop control methods are presented in [8, 9] to improve the performance of power sharing. Nevertheless, authors of all these papers do not take into account of *SoC* balancing. Reference [10] discusses a *P-ω* droop control based *SoC* balancing strategy. However, the control strategy is not suitable for use in the condition that line impedance is highly resistive. Reference [11] proposes a control method for the parallel PWM converter for *SoC* balancing of the multiple DESS, but the centralized communication network increases the cost of the system and reduces the reliability of the system.

In this paper, a distributed *SoC* balancing and power sharing approach is proposed for islanded AC microgrids with highly resistive lines. This novel method introduces the estimate of average *SoC* of all ESSs to *P-V* droop control strategy to balance the ESSs and share the load, and only low bandwidth communication is required to obtain the *SoC* information of neighboring DESSs. Compared with centralized control methods, the proposed distributed control method is more robust, reliable, and suitable for the DESSs in MGs.

2.The Proposed Control Method

A. Droop Control Method For ESS

To deal with the stability problem led by RESs' inertia-lessness, ESSs are set in microgrids to absorb and store the volatile energy and provide reliable power. Fig. 1 shows a simplified schematics of an islanded MG. In this paper, ESS is the only power source and regarded as constant voltage source. The voltage of *i-th* Energy storage unit (ESU) can be expressed as $V_i\angle\delta_i$. And the AC bus voltage is represented by $V_b\angle\delta_b$.

Neglecting the losses of connecting impedance, the active power and reactive power of *i-th* ESU can be derived as follows:

$$P_i = \frac{1}{R_i}V_i(V_i - V_b\cos\delta_{ib}) \qquad (1)$$

$$Q_i = -\frac{1}{R_i}V_iV_b\sin\delta_{ib} \qquad (2)$$

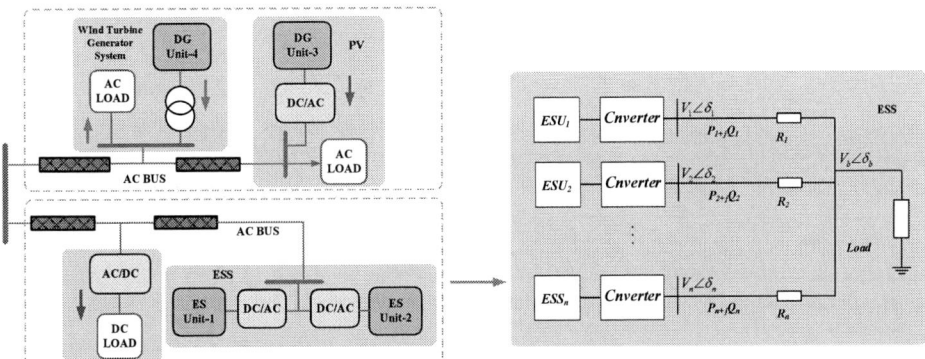

Fig. 1 Structure of an island ac microgrid

$$\delta_{ib} = \delta_i - \delta_b \qquad (3)$$

Where, P_i and Q_i are the active and reactive powers generated by *i-th* ESU. R_i is the line resistance between *i-th* ESU and Load. According to the power transmission characteristic, P_i is proportional to V_i and Q_i is proportional to δ_{ib}. The *P-V /Q-ω* droop equations shown as follows:

$$V_i = V^* - n_iP_i \qquad (4)$$

$$\omega_i = \omega^* - m_iQ_i \qquad (5)$$

Where, ω_i, V_i, n_i and m_i are the angular frequency references, voltage amplitude and the droop coefficients of the ESS, respectively. ω^* and V^* are the no-load frequency and voltage values, respectively.

The *SoC* is a significant characteristic of *i-th* ESU, which can be calculated as follows [12]:

$$SoC_i = SoC_{i0} - \frac{1}{C_i}\int i^{out\_i}dt \qquad (6)$$

Where, SoC_{i0}, C_i and i^{out_i} are the initial *SoC* value, capacity and the output current of *i-th* ESU. Supposed DC voltage of ESUs is equal $(V_{DC1} = V_{DC2} = \cdots V_{DCn} = V_{DC})$ and converter losses are neglected, the output active power can be obtained $P_{out\_i} = V_{DC}\cdot i^{out\_i}$, then the *SoC* can be written as follows:

$$SoC_i = SoC_{i0} - \frac{\int P_{out\_i}}{V_{DC}C_i} \qquad (7)$$

B. Estimated Average SoC of DESS

Each ESU has an average state estimator that uses local estimates and information from neighboring ESUs to update local average estimates $\overline{SoC}$. The distributed average consensus protocol is implemented by the state estimator implements to track dynamic signals.

For the *i-th* ESU, $\overline{SoC}_i$ is the local estimate of the average *SoC* value. N_i denotes the set of all neighbors of *i-th* ESU. a_{ij} is the communication weight. If there is a communication link allowing information flow from the *i-th* to the *j-th* ESU, $a_{ij} = 1$. The graph adjacency matrix is given by $\mathbf{A} = [a_{ij}]$. The *i-th* ESU receive average state estimates from neighbors $j \in N_{ij}$, and the $\overline{SoC}_i$ is obtained as follows:

$$\overline{SoC}_i(t) = SoC_i(t) + \int_0^t \sum_{j\in N_{ij}} a_{ij}(\overline{SoC}_j(\tau) - \overline{SoC}_i(\tau))d\tau \qquad (8)$$

Each node in the communication network has in-degree $d_i = \sum_{j=1}^N a_{ij}$ and out-degree $d_i^0 = \sum_{j=1}^N a_{ij}$. If $d_i = d_i^0$ for all nodes, the graph is balanced [13]. The overall estimator dynamic can be formulated in matrix.

$$\dot{\overline{\mathbf{SoC}}} = \dot{\mathbf{SoC}} - \mathbf{L}\overline{\mathbf{SoC}} \qquad (9)$$

$\overline{SoC} = [\overline{SoC}_1, \overline{SoC}_2, \cdots, \overline{SoC}_n]^T$ and $\mathbf{SoC} = [SoC_1, SoC_2, \cdots, SoC_n]^T$ are the vectors of local average SoC estimations and the estimated SoC. $\mathbf{D} = diag\{d_i\}$ is the in-degree matrix. And the graph Laplacian matrix is given by $\mathbf{L} = \mathbf{D} - \mathbf{A}$. $\mathbf{I}_N \in \mathbf{R}^{N \times N}$ is the identity matrix. $\overline{SoC}_i(0) = SoC_i(0)$ is known, equation (10) can be obtained.

$$\overline{\mathbf{SoC}} = s\left(s\mathbf{I}_N + \mathbf{L}\right)^{-1}\mathbf{SoC} = \mathbf{HSoC} \qquad (10)$$

Where $\mathbf{H}$ is the SoC estimator transfer function matrix. It is proofed in reference [14] that all elements of $\overline{\mathbf{SoC}}$ will finally converge to the overall average value in steady state if $\mathbf{L}$ is balanced.

$$\overline{SoC}_1^s = \overline{SoC}_2^s = \cdots = \overline{SoC}_n^s = \frac{SoC_1^s + SoC_2^s + \cdots + SoC_n^s}{N} = \overline{SoC}^s \qquad (11)$$

Where, $\overline{SoC}^s$ and SoC^s are the estimated SoC and the SoC in steady state, respectively.

C. the Proposed Control Method

SoC value of ESUs should be balanced to collaborate all ESUs. To achieve this goal, an improved droop control is proposed in this paper.

$$V_i = V^* - n_i P_i - K_i(\overline{SoC}_i - SoC_i) \qquad (12)$$

$$\omega_i = \omega^* - m_i Q_i \qquad (13)$$

The active power sharing error reduction operation can be understood with the aid of Fig. 2. As can be seen, without the correction item, the active power cannot share accurately due to the different characteristics of the ESUs. The local SoC decides the sign of $K_i(\overline{SoC}_i - SoC_i)$. If $K_i(\overline{SoC}_i - SoC_i) < 0$, the i-th ESU will output more power and the SoC will decrease faster, and vice versa. The Fig. 2 shows a span of SoC converging process that the deviation of active power decrease. The SoC of all ESUs will finally converge to $\overline{SoC}^s$ in steady state. With the SoC correct operating, the active power sharing error will converge.

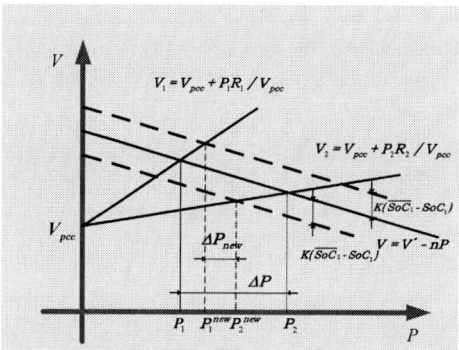

Fig. 2. Schematic diagram of the sharing error reduction operation.

3. Simulation Results

To evaluate the proposed method, an islanded MG is simulated in MATLAB/Simulink shown in Fig. 3. This MG consists of four ESUs and a load. The MG parameters have been listed in Table1.

A. Case 1: P-V Droop Control Method

In this case, the initial SoC of ESUs is set to be different. SoC_1, SoC_2, SoC_3 and SoC_4 are fixed to be 85%, 80%, 90% and 85%. The SoC and output power of ESSs are shown in Fig. 4. As can be seen, the active power cannot share completely because of different line impedances. Using P-V/Q-ω droop control makes ESUs operate stably, but the ESUs can only work independently. The initial SoC leading an unequal SoC is the drawback of the P-V/Q-ω droop control method to use in DESS.

B. Case 2: performance of the proposed control

In this case, the initial $SoCs$ are the same as case 1. Fig. 5 shows the generated reactive power of four ESUs. Compare with case 1, ESU with higher initial SoC outputs more active power at the start, and the active output power gradually declines, vice versa. Using the proposed method, the active power sharing is improved in the steady state. It can be seen that all ESUs almost generate equal active power eventually. The application of the proposed method reduces the difference of $SoCs$ of ESUs in the simulation time range from 5% to less than 1%. So the effectiveness of the proposed control method for SoC balancing is verified.

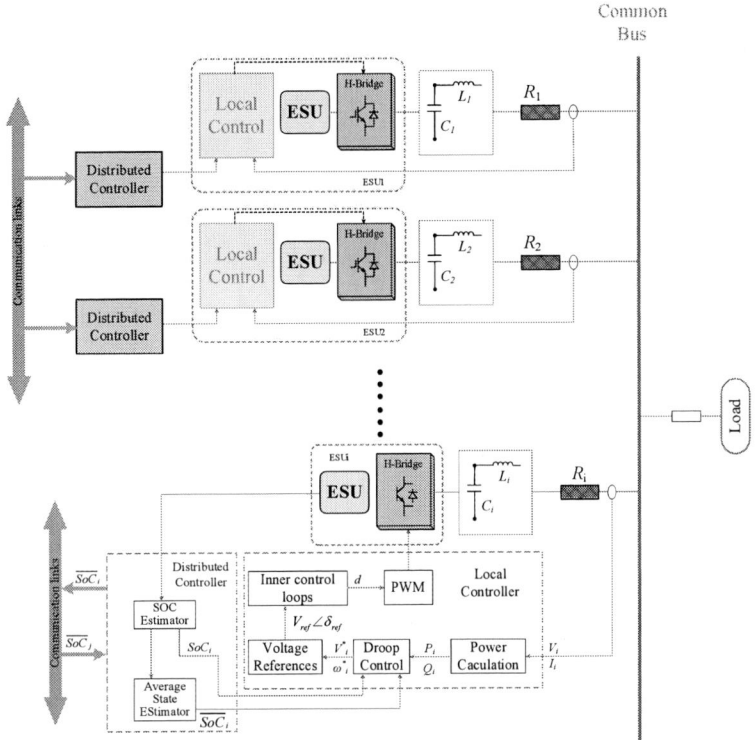

Fig. 3. Energy storage system and control system

Table 1: Parameters of simulations.

Parameter	Value
AC frequency Reference f_{ref}	50
AC voltage Reference V^*	311 V
DC voltage	400 V
ESU1 Energy Capacity C_{ESS1}	5 kW h
ESU2 Energy Capacity C_{ESS2}	5 kW h
ESU3 Energy Capacity C_{ESS3}	5 kW h
ESU4 Energy Capacity C_{ESS4}	5 kW h
n_1, n_2	$5*10^{-4}$
m_1, m_2	$5*10^{-7}$
R_1 (Line#1)	5Ω
R_2 (Line#2)	6Ω
R_3 (Line#3)	4Ω
R_3 (Line#4)	5Ω
Z_{load}	75+j0.5Ω
K	150

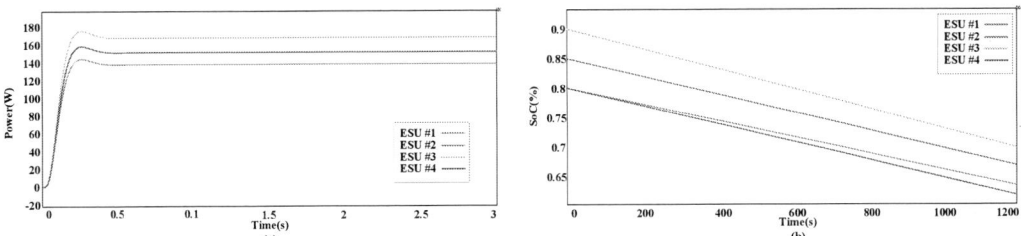

Fig. 4. Simulation results of P-V droop control (a) Power sharing. (b) SoC balancing.

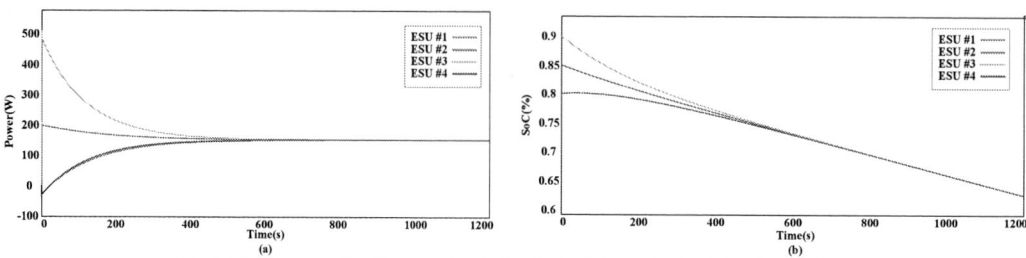

Fig. 5. Simulation results of proposed control method (a) Power sharing. (b) SoC balancing.

C. Case 3: performance of communication failure

In general, the use of distributed control reduces the risk of communication failure. Based on the mode in case2, the communication between ESU1 and ESU2 is interrupted, and the other parameters are the same as case1. The simulation results are shown in Fig. 6. The response speed of active power sharing is slower than case2. Furthermore, the accuracy of *SoC* balancing is also declined. Despite this, the active power sharing is achieved well, and the *SoC* deviation between four ESUs is 0.2% at most, which is limited within the acceptable range.

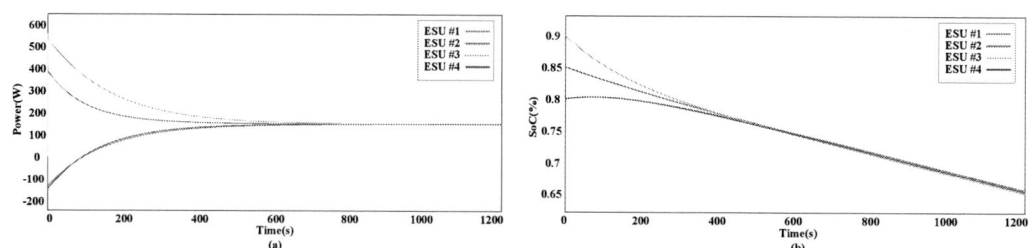

Fig. 6. Simulation results of communication failure (a) Power sharing. (b) SoC balancing.

4.Conclusion

When MGs work in the island model, the ESUs work as reliable power sources and to restrain power fluctuation. To reduce cycle count, ESUs should be utilized cooperated by balancing their *SoC*. In this paper, a new method with distributed communication is proposed for low voltage AC microgrid control. This method balances the *SoC* and can obtain power sharing at the same time. Moreover, under the condition that communication is destroyed, *SoC* balancing and power sharing can also be achieved within the acceptable range. Finally, the simulation results have proved the correctness and effectiveness of the proposed method.

References

[1] G.-C. Liao, "Solve environmental economic dispatch of Smart MicroGrid containing distributed generation system–Using chaotic quantum genetic algorithm," *International Journal of Electrical Power & Energy Systems,* vol. 43, pp. 779-787, 2012.

[2] X. Zhu, M. Xia, and H.-D. Chiang, "Coordinated sectional droop charging control for EV aggregator enhancing frequency stability of microgrid with high penetration of renewable energy sources," *Applied Energy,* vol. 210, pp. 936-943, 2018.

[3] B. Zaker, A. K. Arani, and G. Gharehpetian, "Investigating Battery Energy Storage System for Frequency Regulation in Islanded Microgrid," presented at the The 3rd Iranian Regional CIRED Conf., Tehran, Iran, May 2015.

[4] R. Hemmati, H. Saboori, and P. Siano, "Coordinated short-term scheduling and long-term expansion planning in microgrids incorporating renewable energy resources and energy storage systems," *Energy,* vol. 134, pp. 699-708, 2017.

[5] Y. Zheng, B. M. Jenkins, K. Kornbluth, and C. Træholt, "Optimization under uncertainty of a biomass-integrated renewable energy microgrid with energy storage," *Renewable Energy,* vol. 123, pp. 204-217, 2018.

[6] A. K. Arani, G. Gharehpetian, and M. Abedi, "Review on Energy Storage Systems Control Methods in Microgrids," *International Journal of Electrical Power & Energy Systems,* vol. 107, pp. 745-757, 2019.

[7] A. Engler and N. Soultanis, "Droop control in LV-grids," *2005 International Conference on Future Power Systems*, Amsterdam, 2005, pp. 6 pp.-6.

[8] C. Sao and P. Lehn, "Control and power management of converter fed microgrids," *IEEE PES General Meeting*, Providence, RI, 2010, pp. 1-1.

[9] Vandoorn, T. , Kooning, J. D. , Meersman, B. , & Vandevelde, L. . (2013). Improvement of active power sharing ratio of P/V droop controllers in low-voltage islanded microgrids. Power & Energy Society General Meeting. IEEE.

[10] Xiaonan Lu, Kai Sun, Josep Guerrero,and Lipei Huang"SoC-based dynamic power sharing method with AC-bus voltage restoration for microgrid applications," 38th Annual Conference on IEEE Industrial Electronics Society, IECON 2012, Montreal, QC, Canada,October, pp. 2012, 5677-5682.

[11] Lu, Xiaonan , et al. "SoC-based dynamic power sharing method with AC-bus voltage restoration for microgrid applications." IECON 2012 - 38th Annual Conference on IEEE Industrial Electronics Society IEEE, 2012.

[12] Lu, Xiaonan , et al. "SoC-based droop method for distributed energy storage in DC microgrid applications." (2012).

[13] V. Nasirian, Q. Shafiee, J. M. Guerrero, F. L. Lewis and A. Davoudi, "Droop-Free Distributed Control for AC Microgrids," IEEE Transactions on Power Electronics, vol. 31, no. 2, pp. 1600-1617, Feb. 2016.

[14] Morstyn, Thomas , et al. "Unified Distributed Control for DC Microgrid Operating Modes." IEEE Transactions on Power Systems 31.1(2015):1-11.

Techno-economic analysis of second-life lithium-ion batteries integration in microgrids

Camille Birou, Xavier Roboam, Hugo Radet, Fabien Lacressonnière
Université de Toulouse, LAPLACE, UMR CNRS-INP-UPS, ENSEEIHT
Toulouse, France
Tel.: +33 / (0) – 534.32.23.91
Fax: +33 / (0) – 561.63.88.75
E-Mail: fabien.lacressonniere@laplace.univ-tlse.fr

Keywords

«Microgrid», «Modelling», «Batteries», «Renewable energy systems», «Energy system management».

Abstract

Predicting ageing and performance of storage devices integrated in a global system is necessary to ensure the emergence of microgrids that promote grid services such as self-consumption. This paper deals with a techno-economic tool that allows to model a microgrid connected to the electrical grid and composed of photovoltaic solar panels, a second life lithium-ion battery and power consumers. This tool enables us to simulate the behaviour of the microgrid during a time scope of several years (life cycle based analysis). Economic indicators are also given in order to estimate the profitability of the overall system compared to the case where all the energy consumed would be purchased from the electricity grid. To achieve this, the tool has to be developed with a battery model that includes both a behavioural model (Tremblay-Dessaint) to simulate the evolution of voltage and state of charge, coupled with an ageing model to predict the capacity and power losses during cycling. The analysis highlights the strong sensitivity of storage performance and cost to economic parameters, the importance of degrading the model parameters with ageing and gives optimistic trends for the future of second-life batteries.

Introduction

Balancing electric power production vs demand becomes more and more complex due to the huge growth of alternative intermittent energies (solar and wind farms) and the increase of electric vehicles connected on the grid. In this context, grid services such as self-consumption, peak shaving and grid erasure become a prime necessity, which should lead to the emergence of smart microgrids, more performant if a storage device is used in order to guarantee a better autonomy and to limit power grid exchanges. Facing the constraints in the automotive industry, it is considered that batteries are no longer usable and must be replaced when it has reached 65% of its initial capacity. Instead of directly recycling batteries that still possess capacity, it is interesting to extend their lifespan with "a second life" on stationary applications, reducing the recycling cost for car manufacturers and lowering storage costs. The objective of this work is to design a techno-economic model of a microgrid composed of a second life battery in order to study the ageing of second life batteries for different solicitation profiles, but also to get an overview of the profitability of the storage device for the customers.

A) One case study for second life battery integration for microgrid analysis

Several applications for second-life batteries are conceivable. The aim is to extend the life of batteries from electric vehicles, which still have capacity in less restrictive applications such as in microgrids to limit the use on the grid via production-consumption balancing or to carry out grid services (peak shaving, clipping, etc.). In our case, we will focus on stationary applications in microgrids. These applications provide the advantage of being less constraining for batteries than in automotive applications because low current rates are typically required, and therefore a lower operating battery temperature, which slows down the ageing process. The installation of storage devices in a microgrid will make it possible to increase the self-consumed energy but also to mitigate the power variability on the grid due to the intermittent of renewable energies. In our case, an eco-district in Toulouse was considered, composed of twenty houses of 100 m² heated by heat pumps and equipped with roof-mounted PV system and second life batteries. This microgrid is plugged into the electricity grid. The production and consumption data are normalized from the Homer software (synthetic data): production data for 1 kWp of photovoltaic panels in Toulouse and consumption data for a normalized residential type profile for an average daily consumption of 1 kWh. On the consumption profile, a peak in the evening is observed, when the solar panels no longer produce. Hence, in order to reduce the peak energy consumption on the grid, a storage is coupled on the microgrid (Fig. 1).

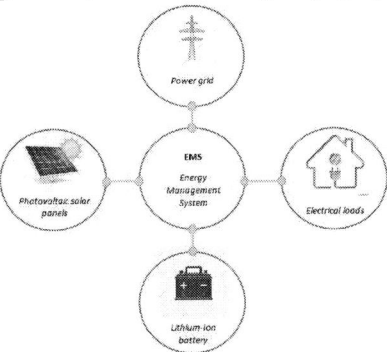

Fig. 1: Global synoptic of the microgrid

The grid service applied to the microgrid is related to the self-consumption with resale of the energy surplus as proposed in French policies for small producers (36 kWp<P_p<100 kWp). It is here possible to obtain a total 5-year photovoltaic investment bonus of 0.09 €/Wp paid annually over five years and to sell the surplus energy produced at a rate set at 0.06 €/kWh [1]. The energy purchased on the electricity grid will be invoiced at the standard Electricity of France (EDF) price of 0.147 €/kWh.

$$Cost = E_{out} \cdot €_{out} - E_{in} \cdot €_{in} \qquad (1)$$

With:

- E_{out} energy injected into the electrical grid (kWh).
- $€_{out}$ grid injection price of energy at 0.06 €/kWh.
- E_{in} energy extracted from the electrical grid (kWh).
- $€_{in}$ purchase energy price at 0.147 €/kWh.

The objective of the microgrid is firstly to satisfy the needs of energy consumers by limiting grid impact. The Energy Management Strategy (EMS) is built on simple heuristic. The issue is to use the energy produced as much as possible to avoid buying it at more expensive price or to sell it at a low price. The battery is used as soon as consumption is different from production in order to readjust the production-consumption balance. In case of underproduction, the battery provides the missing energy until it reaches its minimum of state of charge (SOC), in which case it is necessary to buy energy from the electricity grid. In a case of overproduction, firstly, the battery is charged and when it reaches the maximum SOC, energy will be injected into the grid. Given photovoltaic production data (P_{PV}) and consumption data (P_{load}) seen as the program inputs, the EMS will allocate power flows in the different devices of the microgrid while monitoring the state of the battery [i.e. both (SOC) and state of health (SOH)] to estimate the power (P_{real}) that it can actually deliver. In order to do this, it is necessary to have a battery model, which is presented in the next section.

$$P_{grid} = P_{load} - P_{PV} - P_{battery} \qquad (2)$$
$$P_{battery} = M(P_{load} - P_{PV} ; P_{real}) \qquad (3)$$

With:

- P_{load} and P_{PV} >0.
- P_{grid} and $P_{battery}$ can be positive or negative : <0 in charge et >0 in discharge.
- M=max in charge $P_{load} - P_{PV} < 0$ and M=min in discharge $P_{load} - P_{PV} > 0$.

The objective of the techno-economic model developed on Matlab from previous works [2-4] is to simulate a microgrid, including a second life lithium-ion (Li-ion) battery over a long period of time (typically the microgrid lifecycle) to get technical and economic data. In order to do that, the tool proceeds in four steps: simulation data entry, technical simulation, economic simulation and results display. Table I presents the values of parameters set during the simulations.

Table I: Parameters set

Simulation time	Step time for technical simulation	Step time for economic simulation	PV investment bonus	Purchase energy price	Redemption price of energy
30 years	10 minutes	1 year	0.09 €/Wp	0.147 €/kWh	0.06 €/kWh

B) Lithium-ion battery model

Storage elements are characterized by their energy density (Wh/kg) and power density (W/kg). Lithium-ion (Li-ion) batteries have the advantage to have a higher energy density and power density than lead or nickel-based batteries. Li-ion technology has developed strongly since the 1980s with the first commercialization in 1991 by Sony Energitech. In this part, the modelling of Li-ion battery will be discussed. Two models [5] are used (Fig.2) to design the battery as well as possible:

- A behavioural model. It simulates the electrical behaviour of battery in order to calculate its SOC and its voltage. Hence, with these values, the EMS can control the battery level before allocating the power flows in the microgrid.
- An ageing model, which allows estimating the degradation of the battery in order to get its SOH.

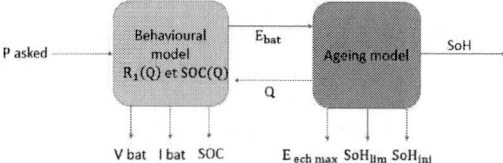

Fig. 2: The battery model.

B.1) Characteristics of the tested battery

As part of this study, we focused on second-life Carbon - Lithium Manganese Oxide (C-LMO) Li-ion batteries from electric vehicles. It is composed of a negative carbon electrode (graphite) and a positive electrode based on spinel structures: lithium manganese oxide (LMO) using mainly manganese, which has the advantage of being available in large quantities [6]. The Aging model is based on the cycle to failure curve for a temperature at 25°C and a C/2 discharge regime. It allows to know, for a given depth of discharge (DOD), the maximum number of equivalent maximum cycles. To obtain the parameters of the behaviour model, C/2 discharge characteristics (Fig. 3) was used for three SOH: 86%, 80%, 69%.

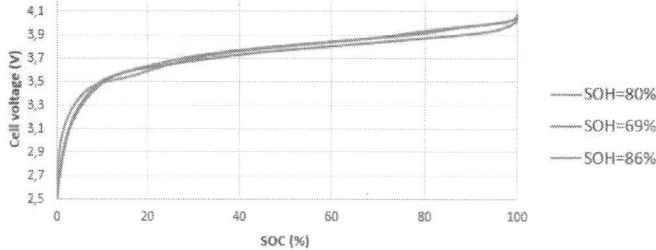

Fig. 3: Discharge curves at C/2 for different SOH

As the electric vehicle market is particularly recent, many batteries are now arriving for recycling at high SOH levels (between 80 and 90%) due to accidents or manufacturer returns. Therefore, in this study, a high initial SOH_{ini} in second life was taken at 90%. In the future, this SOH before recycling will drop to the expected 65%. Table II presents the battery's parameters set.

Table II: Battery parameters set

SOC_{min}	SOC_{max}	SOH_{ini}	SOH_{lim}	V_{min}	V_{max}	$C\text{-rate}_{max}$	Q_{nom}
10 %	90 %	90 %	45 %	2.5 V	4.1 V	C/2	50 Ah

B.2) The behaviour model

There are several types of battery models in the literature [3] [6]: mathematical models (neural networks, fuzzy logic), electrochemical models and equivalent electrical circuit models. For our application with a systemic vision of a microgrid, electrical circuit models seem to be the most suitable because they offer a good compromise between computation time and accuracy [2].

B.2.1) Description of the Tremblay-Dessaint model

The Tremblay-Dessaint [2-4] [7-8] model (Fig. 4) is a quasi-static model where voltage and current are separated. It comes from Shepherd's work in 1965 and describes the battery operation from the discharge curve (taking into account the evolution of the voltage depending on the SOC) knowing the current and the SOC.

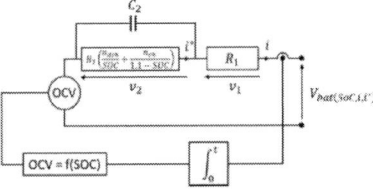

Fig. 4: Equivalent electrical circuit.

In the case of our study, the calculation time step is large enough Δt=10 min, it is possible to neglect the filtering effect modelled by C_2 [2], so we take i instead of i*. The expressions of the Tremblay-Dessaint model can be written according to the SOC where the OCV is the open circuit voltage of the cell battery and R the resistive terms [8]:

$$OCV\ (SOC) = E_0 + Ae^{-BQ_{nom}(1-SOC)} - KQ_{nom}(\frac{1}{SOC} - 1) \tag{4}$$

$$V_{bat\ in\ discharge} = OCV(SOC) - R_1\ i - R_2 i\ (\frac{1}{SOC}) \tag{5}$$

$$V_{bat\ in\ charge} = OCV(SOC) - R_1\ i - R_2 i\ (\frac{1}{1.1-SOC}) \tag{6}$$

It is necessary to identify the Tremblay-Dessaint parameters model [E_0, A, B, K, R_1, R_2] from a cell characteristics.

B.2.2) Evolution of parameters with ageing

It is important to investigate the evolution of the parameters of this model with ageing in order to modify the parameters of the Tremblay-Dessaint model. In the literature [9], OCV characteristics according to the SOC for (C-LMO) cells for different SOH are presented in Fig. 5.

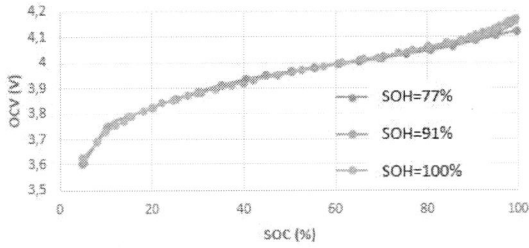

Fig. 5: OCV evolution with ageing [9]

It can be noted from Fig. 5 that, for a given SOC, the value of the OCV remains almost identical regardless of the SOH of the cell. Therefore, in the OCV equation, it is assumed that the parameters [E_0, A, B, K] remain constant despite ageing. However, ageing mainly affects two factors: a decrease in capacity and an increase of the internal resistance of the battery as shown in Fig. 6 [6]. Hence, the SOC takes into account the ageing of the battery cell with the value Q_{deg} which represents the real capacity of the cell at a given instant t_k:

$$SOC_{k+1} = SOC_k - \frac{\int_{t_k}^{t_{k+1}} i\, dt}{Q_{deg\, k}} \qquad (7)$$

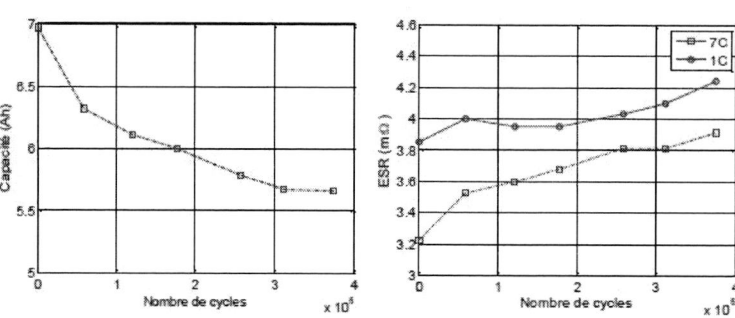

Fig. 6: Evolution of capacity and internal resistance with ageing [6]

In order to take into account the increase on internal resistance with ageing, it will then be necessary to modify the value of the internal resistance R_1 in the Tremblay-Dessaint model. From the observations of the work carried out by Akram Eddahech [6], a linear evolution of the internal resistance seems more suitable for low temperatures.

B.2.3) Method for identifying parameters

In order to identify the parameters of the Tremblay-Dessaint model [E_0, A, B, K, R_1, R_2] of the cell, an optimization algorithm is required: the non-linear least squares method was used. It consists in minimizing the error between the experimental data (y_i) and the model (f).

$$S(\theta) = \sum_{i=1}^{N}(y_i - f(x_i\,;\,\theta))^2 \qquad (8)$$

With in our case:

- N: number of cell voltage measurement points.
- x_i: inputs of the Tremblay-Dessaint model at point i, namely the current and the SOC.
- θ: unknowns of the Tremblay-Dessaint [E_0, A, B, K, R_1, R_2].

For the identification process, we decided to use the following method. First, the OCV parameters were obtained from OCV characteristic (Fig. 5). Then R1 (with an ageing law) and R2 parameters were determined from discharge curves (Fig. 3). Table III presents the accuracy performance of the model for the discharge curves studied (SOH=86%, 80% and 69%).

Table III: Errors between model and data

	SOH=86%	SOH=80%	SOH=69%
Mean-squared error	0.60 mV	0.79 mV	0.86 mV
Average relative error	0.52 %	0.45 %	0.5 %

B.3) The ageing model: "exchangeable energy model"

The ageing model allows to estimate the SOH of the battery cell. For batteries oriented energy, the SOH can be calculated as the ratio between the actual capacity of the battery and its maximal capacity. There are two types of ageing that occur in the battery operation [6]:

- Calendar ageing which corresponds to the cell performance degradation over time when the battery not operates.
- Ageing in cycling which refers to the cell performance degradation when a battery current circulate in the battery.

In the literature, there are three types of ageing models: physical, mathematical or fatigue models. In our case, with a systemic approach of a microgrid, the fatigue models seem to be the most adapted because they present a good compromise between calculation time/accuracy. Hence, the issue was simplified by considering the "exchangeable energy model" which consists in calculating the battery degradation as a function of the amount of energy (in charge and discharge) that the battery can exchange with the system throughout its life (between SOH_{ini} and SOH_{lim} [2-3]. This maximum amount of exchangeable energy is estimated [2-3] from the "cycle to failure" curve where it is possible to read the maximum number of cycles according to the chosen discharge depth:

$$E_{exch\,max} = 2. N_{cycles\,max}(DOD_{chosen}). DOD_{chosen}. E_{irated} \tag{9}$$

With:

- E_{rated}: rated energy of the battery for SOH_{ini}.
- DOD_{chosen} desired depth of discharge for the battery cycle (between 0 and 1).
- $N_{cycles\,max}$: maximum number of cycles for a DOD set to DOD_{chosen}.
- $E_{exch\,max}$: maximum amount of energy that can be exchanged by the battery during a certain period of its life (between SOH_{ini} and SOH_{lim}).

Finally, the SOH is determined at the next step from the energy exchanged during a time step according to the formula below:

$$SOH_{k+1} = SOH_{ini} + \frac{SOH_{lim} - SOH_{ini}}{E_{exch\,max}} * \sum_{i=1}^{k+1} \left| E_{exch}(i) \right| \tag{10}$$

$$\text{With } |E_{exch}| = \int_t^{t+\Delta t} |P_{bat}(t)| \, dt \tag{11}$$

C) Economic Model

In order to assess a techno economic analysis, it is necessary to calculate economic indicators to estimate the whole cost of the microgrid over its life cycle (i.e. the NPV, for Net Present Value is defined below). This latter is based on cash flows, noted F_m for the year m. Cash flows take into account the incomes due to the chosen self-consumption strategy ($cfgrid_m$) with resale of the power surplus for small producers (see part II in the case of French policies), to which the operating and maintenance costs (OPEX) must be deducted. To adjust the elements to the current value of costs for the year under consideration, it is required to multiply by $(1+e)^m$ with e the inflation rate.

$$F_m = cfgrid_m - OPEX_m \tag{12}$$
$$cfgrid_m = (E_{out} \cdot €_{out} - E_{in} \cdot €_{in}) \cdot (1 + e)^m \tag{13}$$
$$OPEX_m = (1 + e)^m \cdot OPEX \tag{14}$$

In our case, the CAPital Expenditure (CAPEX) takes into account the investment related to the photovoltaic production source $CAPEX_{PV}$ (development costs, purchase and delivery of equipment (solar panels, inverters, etc.), installation and connection) and that concerning the means of storage $CAPEX_{BAT}$. The investment is considered to be fully paid the first year (no loan). To stay in a simple economic vision, investment costs were estimated in €/kWp for the photovoltaic production source and in €/kWh for the battery system. Regarding the OPEX, there are estimated as percentage r_{OM} each year of the initial CAPEX investment. However, to be more precise, the OPEX of the battery is divided into two parts: a fixed part which will be a percentage of the CAPEX each year of the battery life and a variable part to take into account the extra cost when replacing the battery. The variable part is expressed as the cost ($CAPEX_{bat} - r_{Rep} . nb_{module}$ €) by replacing of all the batteries with r_{Rep} the cost (€) that can be saved per module compared to the initial price of the battery (recovery of certain elements, in particular electronic devices such as BMS). In a project, the net present value (NPV) is used to account for the overall cost of the microgrid over its lifetime by integrating its energy management strategy and its environment (solar irradiation, grid policies and rates). To do this, year by year, the cash flow is discounted to current value using the actual discount rate d [2].

$$NPV = -CAPEX + \sum_{year=1}^{N} \frac{F_m}{(1+d)^{year}} \tag{15}$$

Table IV: set of economic parameters

Inflation rate	Discount rate	PV CAPEX	OPEX for PV	Battery CAPEX	OPEX for Battery
e = 2 %	d = 3 %	$C_{inv\,PV} =$ 1.2 €/Wp	$r_{OM\,PV}$ = 2.5 % of CAPEX PV / year	$c_{inv\,bat} =$ 200€/kWh	**Fixed part :** $r_{OM\,bat}$ = 2% of CAPEX battery / year **Variable part :** r_{Rep} = 70€ saved / module replaced

In the economic model implemented to analyse the microgrid performance, the objective is to self-consume as much energy as possible (self-consumption case). That is why the microgrid does not allow to earn money. It is more convenient to compare the case of the microgrid with a "reference case" for which 100% of consumed energy is purchased from the electricity grid at a fixed price. The interest is to analyse if it is economically more profitable to create your microgrid (with corresponding investments and to self-consume your own energy as much possible) than to purchase energy from the grid at fixed price. In our analysis, the first key indicator will be the "relative NPV" (ΔNPV) which is the difference between the microgrid NPV (including a certain sizing of both the solar array and the storage device) and a reference system NPV_{ref}, which will obviously be negative, which would include neither storage device nor PhotoVoltaic (PV) source and which would lead to purchase the total consumed energy to the grid by considering the actual purchase rates $€_{in}$. Thus, the microgrid is profitable if:

$$\Delta NPV = NPV - NPV_{ref} > 0 \tag{16}$$

$$With\ NPV_{REF} = \sum_{year=1}^{N} \frac{-E_{consumed} \cdot €_{in}}{(1+d)^{year}} \tag{17}$$

To take account for the profitability of the microgrid in relation to the reference case, two typical economic indicators are calculated: the return on investment time (ROIT) and the profitability index (PI). The ROIT (years) represents the time from which the invested capital is recovered, i.e. when the cumulative NPV of the microgrid becomes higher than that of cumulative reference NPV_{ref}. The PI is the ratio between the financial gain at the end of the migrogrid lifetime and the CAPital Expenditure (CAPEX):

$$PI_{ref} = \frac{\Delta NPV}{CAPEX} \tag{18}$$

D) Some analysis results and discussion

With the emerging smart grids integrating intermittent renewable power sources, new pricing policies would appear to encourage renewable energy plants (solar, wind) coupled with storage to limit the use of the grid. As energy indicator, we know the "degree of autonomy" defined as the ratio between the energy produced and the energy consumed. But this latter indicator does not differ depending if the consumption takes place during production or outside production phases, thus neglecting the constraints of the grid (line congestion, etc). To quantify more conveniently the level of grid constraint, a new indicator called "degree of grid usage" (DGU) is defined as the ratio at each time step between the average grid energy (whatever if it is injected or extracted) and the energy consumed (excluding battery consumption):

$$DGU\ (in\ \%) = 100*Mean_{k=1}^{N}\left(\frac{Grid\ energy_k}{Consumed\ energy_k}\right) \tag{19}$$

with k the step time and N the number of points.

The issue about this indicator is linked with the fact that grid power must be balanced in both ways, over production as well as over consumption. A first interesting analysis aims at assessing both the peak power of the photovoltaic installation and the total stored energy for the Li-ion battery with respect to the policies and grid services with that microgrid. It is necessary to look for the best configuration in order to maximize the ΔNPV or to minimize the DGU. To do this, the mapping of figure 7 allows having a trend on the profitability and the grid usage according to the investments (sizing of devices). The following maps are obtained with E_{nom} (battery sizing) on the x-axis and P_{PV} (solar panels sizing) on the y-axis. The scale color represents the value of ΔNPV on the left and DGU on the right with in yellow the highest values and in blue the lowest values.

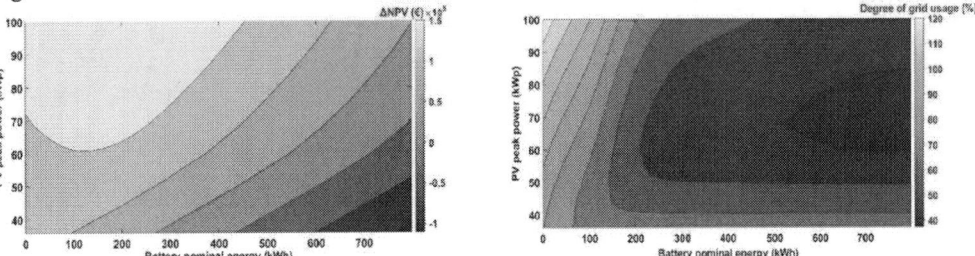

Fig. 7: Maps of the ΔNPV and the DGU versus both device sizing

It can be noted that the pairs with the best ΔNPV (yellow part) are located at high installed peak power and with a storage capacity of less than 400 kWh with assumptions made in the model. Indeed, with this strategy, the optimal point to maximize ΔNPV is: ΔNPV=199k€ for P_{PV}=100kWp and E_{nom}=198kWh. With this sizing of photovoltaic panels and based on the assumptions taken from the model, a case with storage is more profitable than a case without storage. Thus, with such grid policies and economic assessments, there is a benefit with storage by using the energy produced during the day and by postponing it to the evening to avoid buying expensive energy from the electrical grid. Even if the DGU (equal to 71.3% for the techno economic optimum point) is smaller than if the consumers would buy all the energy needs, this parameter may be improved with other battery sizing making the microgrid more and more autonomous. Looking at the Degree of Grid Usage (DGU) map (on the right), we notice that the optimal point in terms of DGU is: DGU=37%, obtained for P_{PV}=68kWp and with the maximal storage capacity E_{nom} on this map: E_{nom}=798kWh. But the ΔNPV= -4 227 € is here really negative with such a storage device: we can have here an "idea of the cost for full autonomy with stand alone microgrids". With this funding strategy, both indicators (ΔNPV and DGU) are not compatible (the yellow area on the left does not overlap the blue area on the right).

D.1) Analysis at the maximum ΔNPV point

D.1.1) Analysis of the battery performance degradation over the life cycle

Thereafter, the techno-economic analysis is carried out for the optimal sizing in terms of ΔNPV (i.e. P_{PV} = 100kWp and E_{nom}=198kWh). For this simulation, the battery model parameters change with the battery ageing (R1 and Q). Looking at the evolution on the battery State Of Health (SOH) over the full lifetime of the microgrid (in our case 30 years), it appears that it is not necessary to change the battery as the SOH reaches 59% after 30 years. From an economic point of view, the microgrid costs more money each year than it brings us (Net Present Value (NPV)= -333.3k€<0). However, analysing the relative evolution of the cumulative ΔNPV observed by comparing the net present value of the microgrid with that of the reference case (with neither PV nor storage), the obtained Return On Investment Time is less than 12 years meaning that the microgrid costs less than buying all energy from the grid. Compared to this reference, the microgrid investments will be profitable and will save 199k€ after 30 years.

D.1.2) Comparison of analysis with and without battery performance degradation

With the same optimal sizing, it is possible to compare the previous results with the ones obtained without taking into account the degradation of battery performance due to the ageing (capacity and power losses), i.e there is no loopback of SOH on behavioural model parameters (R and SOC). In this latter case, the ageing model only allows to estimate the State Of Health (SOH) in order to calculate the cashflow (taking into account the cost of battery changes when they take place). The comparative analysis between the two cases are presented in table V.

Table V: comparative analysis with and without degradation of battery performance

	ΔNPV (k€)	NPV (k€)	SOH end (%)	Qth cell (kAh)
Without degradation	208.9	-323.3	55.5	534
With degradation	198.9	-333.3	59.2	477.6
Error (%)	5	3	6	12

After 30 years, without considering the degradation of battery performance, it can be observed that more ampere hours (Qth cell) are exchanged with the storage device. Indeed, the battery without accounting the degradation (with a constant capacity) is able to maintain the exchanged energy during cycle as in its initial conditions. This will result in a lower SOH than if the degradation is taken account. The relative error between both cases (with and without accounting the degradation) is significant on technical data: 6% on SOH and 12% on Qth (Ampere-hour exchanged by the battery). However, at the economic level, the error is decreasing with only 3% on NPV and 5% on ΔNPV. Setting the simulation time at 20 years, the error on NPV is reduced by half. Let notice that taking into account ageing on the battery's behaviour improves accuracy of analysis but increases the computation time.

D.2) Discussion

The previous results show a favorable trend towards the development of renewable sources (solar PV) and the implementation of storage devices in microgrids. This trend exists not only from an economic point of view (i.e. the ΔNPV is more important than without storage at the optimal sizing point) but also from a technical point of view (limitation of grid usage with a smart energy management, etc.).

However, these results should be taken with some caution because we have placed ourselves in a very particular case study, dealing with an eco-neighbourhood with an Electricity of France (EDF) contract for the resale of the power surplus. The techno-economic model is built from hypotheses and simple models. In our case study, it is assumed that the contract for reselling the surplus with EDF remains valid for the entire life (30-years) of the microgrid (roughly corresponding with the solar panel life time), whereas it is a 20-year contract. In our model, the parameter set at the beginning of the simulation will remain the same during the 30 years, being just subject to the inflation rate (set here at 2%). However, it is obvious that certain parameters, such as the purchase price of energy, are likely to increase significantly in the next thirty years. It is thus difficult to predict economic parameters over a long period of time. For this reason, a sensitivity analysis was carried out to determine the parameters that have the greatest impact on the final ΔNPV. The sensitivity analysis revealed that the parameters which have a strong influence on ΔNPV are the inflation rate, the discount rate and the purchase price of energy from the grid. In addition, the cell temperature is considered constant around 25°C because of the low current regimes (C/2) recommendation to limit battery heating and therefore ageing. It is not necessary here to change the batteries for 30 years because the second life of the batteries was here started at high initial State Of Health ($SOH_{ini} = 90\%$). Moreover, in our model, the efficiency of power electronic devices are not taken into account and it is assumed that a perfect balancing of the cells was performed (SOH of a cell = SOH of the battery pack) thanks to the balancing system of the Battery Management System (BMS). However, in reality, there can be a significant dispersion of the capacity of the different cells constituting the battery, all the more for second life batteries (with different first lifes).

In this work, the exchangeable energy model has been retained to take into account the battery ageing during cycling by neglecting the calendar ageing. Two different approaches were discussed: the first one was to consider that the ageing model has an impact only on economic indicators via the estimation of the battery SOH without considering the degradation of battery performance. Reasoning on global systems, the objective here was to estimate the qualitative trends of ageing in order to know which parameters were the most sensitive with the greatest impact on life expectancy. The second approach has considered that the ageing model has an impact, not only on economic indicators via the SOH, but also on technical performance (by coupling ageing effects on certain battery model parameters: R1 and Q). This approach allows to have a better accuracy on analysis but the calculation time is much more important because it is necessary to carry out the technical simulation on all years of the microgrid life. On the contrary, with the first approach, it is possible to simulate the behavior of the battery over one typical year and to consider that this year is repeated on the life time. This first qualitative analysis allows to save in computing time by going from a few minutes to a few seconds.

Conclusion

This study was essentially based on the modelling of a battery within a microgrid especially in the case of second life batteries. It provides a first simplified version of a techno-economic tool for microgrid analysis which will be developed further in the future. Such tools are needed to better understand how to improve the device sizing in microgrids, the competitiveness of battery technologies and to predict their ageing in a global system. In the particular case of second life batteries, the techno economic analysis is of prime necessity. The advantage of this tool is that it is easily scalable and allows adding quickly new features or testing different case studies.

Through this first study and with the hypotheses considered, an eco-neighbourhood composed of 20 houses was analysed and showed that it was possible to take advantage of the installation of a second-life battery both from a technical point of view to reduce grid dependence and also from an economic point of view to improve profitability. In addition, the sensitivity analysis revealed that the economic parameters having a strong influence on the profitability of the microgrid ΔNPV are the inflation rate, the discount rate and the purchase price of energy from the grid. The modeling of the second life battery includes two models: a Tremblay-Dessaint behavioral model whose parameters were identified on discharge curves of second life batteries at different SOH and an exchangeable energy ageing model (taking into account the ageing in cycling and not calendar ageing). In this work, two different approaches were processed: with and without taking into account the battery degradation on model parameters. Between the two cases, it is a compromise between precision (increase of 12% for certain technical data such as Ah exchanged) and calculation time (switch from few minutes to few seconds) that needs to be made.

For the future, the challenge is to include more complex energy management strategies, but most of all to improve the accuracy of the tool by refining hypothesis for example by taking into account the losses due to power electronics, the ageing of the various components of the microgrid (production sources, power electronics, etc.), the performance losses of batteries due to the dispersion of the cells' capacity in a module over time, but also the uncertainty of the evolution of the economic parameters over the years. To improve these studies, it will be also necessary in the future to develop more complex and precise ageing models (e. g. electrochemical) to better take into account the phenomena occurring within second life batteries (impact on second-lifetime of SOH and resistance values at the end-of-first-life, related to first-life use).

References

[1] EDF ENR, Une prime pour encourager l'autoconsommation solaire, 10th december 2018, Electronic version: https://www.edfenr.com/actualites/_prime-encourage-autoconsommation/

[2] Minh, Rapport de synthèse des activités de recherche au sein du Laboratoire LAPLACE dans le cadre du lot 3.5 : Optimisation technico économique, October 2016.

[3] R. Rigo-Mariani, Méthodes de conception intégrée « dimensionnement-gestion », doctoral thesis at the University of Toulouse, 2014.

[4] Andy Varais, Modèles à échelle réduite en similitude pour l'ingénierie système et l'expérimentation simulée "temps compacté": application à un micro-réseau incluant un stockage électrochimique, doctoral thesis at the University of Toulouse, 2019.

[5] Charles Delacourt, Vieillissement des accumulateurs lithium-ion dans l'automobile, Techniques de l'ingénieur re231, July 2014.

[6] Akram Eddahech, Modélisation du vieillissement et détermination de l'état de santé de batteries lithium-ion pour application véhicule électrique et hybride, doctoral thesis at the University of Bordeaux, 2014.

[7] O. Tremblay et L.-A. Dessaint, Experimental Validation of a Battery Dynamic Model for EV Applications, World Electric Vehicle Journal, vol. 3, 2009.

[8] Javier M. Cabello, Battery dynamic model improvement with parameters estimation and experimental validation, University of Toulouse and National University of Rosario, 2015.

[9] Caiping Zhang, Jiuchun Jiang, Linjing Zhang, Sijia Liu, Leyi Wang and Poh Chaing Loh, A Generalized SOC-OCV Model for Lithium-Ion Batteries and the SOC Estimations for LNMCO Battery , November 2016.

Design, modelling, and test of a solid-state main breaker for hybrid DC circuit breaker

Jiawen Xi, Xiaoze Pei, Xianwu Zeng
University of Bath
Claverton Down, BA2 7AY,
Bath, United Kingdom.
E-Mail: j.xi@bath.ac.uk, x.pei@bath.ac.uk,
xz2478@bath.ac.uk

Liyong Niu
Beijing Jiaotong University
No.3 Shangyuancun, Haidian District
Beijing, China
E-Mail: lyniu@bjtu.edu.cn

Keywords

«DC circuit breaker», «IGBT», «Medium voltage DC-grid», «Parallel operation», «Protection device»

Abstract

Driven by the requirements of reducing air pollutant gas emissions and fuel consumption, the concepts of more electric ship and electric aircraft are attracting increasing attention. Medium voltage DC (MVDC) distribution architectures have been proposed as potential candidates to transmit and distribute energy from generators to motors in these applications. However, the low impedance in MVDC systems results in extremely fast propagation speed of fault currents. Therefore, it is necessary to interrupt the DC fault in a very short period. This paper investigates a solid-state circuit breaker with an ultrafast interruption speed as a main breaker for a hybrid DC circuit breaker. A simulation model of the hybrid circuit breaker is established using PLECS software to evaluate the performance of the main breaker. A 1 kV solid-state main breaker prototype based on series and parallel connected insulated gate bipolar transistors (IGBTs) is built. Series and parallel connection of IGBTs are implemented to increase the voltage and current level. The maximum voltage across the solid-state circuit breaker is limited to 1.8 kV during current interruption. The solid-state main breaker prototype is experimentally tested under dynamic current conditions. The solid-state main breaker prototype successfully interrupts current of 400 A within 300 microseconds and presents good voltage balancing as well as current sharing performance. The experimental results show good agreement with the simulation results.

Introduction

More electric ship and electric aircraft have attracted a lot of research interest in recent years due to their higher energy efficiency, lower environmental impact, fewer fuel burns and reduced weight. Medium voltage direct current (MVDC) system has been proposed as a potential solution to ensure a high reliability and high quality power supply for electric ship [1, 2] and turboelectric aircraft [3, 4]. However, because of the high rate of rise of the fault current and the absence of natural current zero-crossing in the DC system, the design of DC circuit breakers is more challenging than that of AC circuit breakers [5, 6]. Moreover, the fault current propagates at extremely fast speed as the DC system has low impedance. It is critical to isolate a DC fault in a very short time (within a few milliseconds). Therefore, DC circuit breakers with ultrafast interruption speed are the key enabling technology and indispensable to achieve the multi-terminal DC (MTDC) system.

There are three types of DC circuit breakers: mechanical circuit breaker (MCB), solid-state circuit breaker (SSCB), and hybrid circuit breaker (HCB) [7]. MCB has negligible on-state loss, tens-of-milliseconds interruption time is unacceptable in DC system. In contrast, SSCB can achieve ultrafast interruption speed in a few hundred microseconds. However, the on-state loss is high, which reduces the system efficiency. Xi'an Jiaotong University developed and investigated a 10 kV SSCB based on series connected press-pack insulated gate bipolar transistors (IGBTs), which successfully interrupted 5.1 kA within 1 ms [8]. Hybrid DC circuit breaker is a combination of mechanical switch and solid-state semiconductor switch, which is proposed to realize fast operation as well as low on-state loss. An HCB has lower on-state losses than SSCB and can achieve faster interruption than MCB.

A proactive hybrid DC circuit breaker concept has been proposed by ABB in 2012, as shown in Fig. 1, which is the first HCB for high voltage DC system [9]. The proactive HCB consists of a mechanical branch, a semiconductor branch and an energy absorption branch. The current normally flows through the mechanical branch, which combines a mechanical switch (MS) with a load commutation switch (LCS). The power electronic devices in the semiconductor branch act as the main breaker (MB). The energy absorption branch is made up of metal oxide varistors (MOVs) to limit the voltage across the circuit breaker during interruption and absorb any residual energy in the system inductance. An 80 kV main breaker cell has been developed by ABB based on series connection of IGBTs, which can interrupt 9 kA within 5 ms.

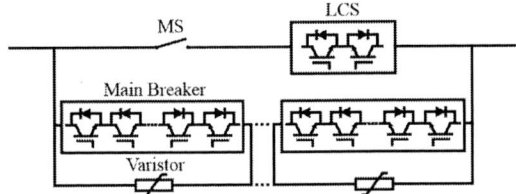

Fig. 1: Proactive hybrid circuit breaker proposed by ABB

Global Energy Interconnection Research Institute (GEIRI) has developed a $\pm$ 200 kV HCB using IGBT-based H-bridge as the main breaker for Zhoushan five-terminal HVDC project, which can break 15 kA within 3 ms [10, 11]. NR Electric Co., Ltd. has built a $\pm$ 535 kV HCB for Zhangbei four-terminal HVDC project, which is the first HVDC circuit breaker for 500 kV level. This HCB uses IGBTs in diode-based H-bridge as the main breaker and series connected IGBTs as LCS [12]. GEIRI has also developed a $\pm$ 535 kV HCB for Zhangbei project, which applies diode-based H-bridge modules as the main breaker and IGBT-based H-bridge modules as LCS [13]. Beside the design for high voltage DC system, North Carolina State University has designed a 10 kV HCB for medium voltage level. The HCB uses a 15 kV silicon carbide (SiC) emitter turn-off (ETO) thyristor in diode-based H-bridge as the main breaker. This medium voltage HCB has demonstrated a successful interruption of 100 A within 2 ms [14, 15].

Series and parallel connection of semiconductors are required to increase the voltage and current level of the solid-state main breaker. The performance of the solid-state main breaker using both series and parallel connection of IGBTs has not been discussed in detail in previous research. In this paper a 1 kV solid-state main breaker prototype based on series and parallel connected IGBTs is simulated, built and experimentally tested under fault interruption conditions. The use of series and parallel connection of IGBTs to increase the voltage and current level of the solid-state main breaker is important. The current sharing in the parallel branches and the voltage balancing in the series connection must be investigated. The operation of the solid-state main breaker in a hybrid circuit breaker is simulated in PLECS software. The voltage balancing and current sharing performance during dynamic condition are investigated through both simulation and experimental testing.

Proposed solid-state DC main breaker

Fig. 2 presents the topology of the proposed solid-state DC circuit breaker as the main breaker in the hybrid DC circuit breaker, which consists of two IGBT modules connected in parallel. Each module has two IGBTs in series and each IGBT has an anti-parallel body diode. The current sharing between parallel connected IGBTs and voltage balancing between series connected IGBTs are investigated using this topology. Balancing resistor R_s is utilized for static voltage balancing at the off state of IGBT and resistor-capacitor-diode (RCD) snubber circuit is selected for dynamic voltage balancing during current interruption [16]. Three metal-oxide varistors (MOVs) are used in the proposed solid-state main breaker to absorb the residual energy in the system. The varistor MOV1 limits the maximum voltage across the solid-state main breaker during current interruption and the varistor MOV2 and MOV3 protect individual IGBT in case the voltage is unbalanced.

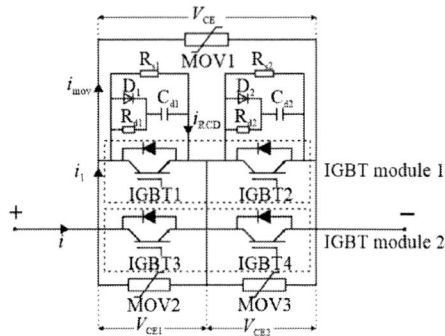

Fig. 2: Topology of the proposed solid-state main breaker

Simulation of hybrid circuit breaker

A simulation model is established in PLECS software to simulate the performance of the solid-state DC circuit breaker as the main breaker in the hybrid circuit breaker. As shown in Fig. 3, the hybrid circuit breaker is placed in a test circuit. The test circuit is an LC resonant circuit which can emulate the rising of the fault current in the DC system. The topology of the solid-state main breaker is the same as in Fig. 2. An ideal switch acts as the MS and a MOSFET is applied as the LCS in the mechanical branch. The parameters of simulation model are listed in Table II.

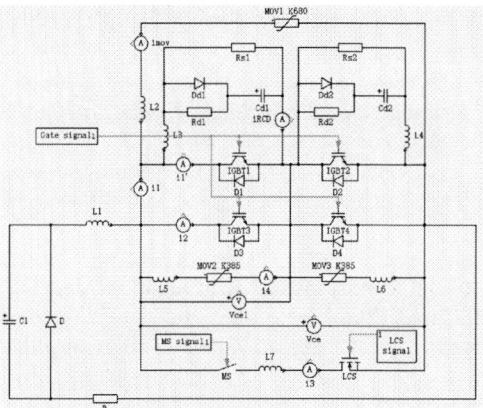

Fig. 3: Simulation of hybrid circuit breaker

Table II: Component parameters of the simulation model

Parameter	Value	Parameter	Value
DC capacitor C1	12 mF	Static balancing resistor Rs1, Rs2	20 kΩ
Line inductor L1	1 mH	Snubber resistor Rd1, Rd2	30 Ω
Line resistor R	55 mΩ	Snubber capacitor Cd1, Cd2	6 µF
Estimated stray inductance associated with MOVs L2, L5, L6	1 µH	Estimated stray inductance associated with RCD L3, L4	2 µH
Estimated stray inductance associated with mechanical branch L7	6 µH		

The interruption process of the hybrid circuit breaker is illustrated in Fig. 4. The voltage across the circuit breaker V_{CE} and voltage across one IGBT V_{CE1} are monitored. i_1', i_2 and i_3 represent the currents flowing through the IGBT module 1 branch, IGBT module 2 branch, and mechanical branch, respectively. At the beginning of the simulation, the MS and LCS are closed and the solid-state main breaker is open. The fault current i is generated by discharging the capacitor C1, and the rising rate of the fault current is limited by the line inductance L1. The fault current only flows through the mechanical branch. When the fault is detected at 2 ms, the solid-state main breaker is closed and the LCS is open. The fault current commutates from the mechanical branch (i_3) to the main breaker branch ($i_1' + i_2$). After the fault current fully transfers to the main breaker branch, the MS can be separated without arcing. In Fig. 6, the MS is simulated to be opened at 4 ms. When the current flows through the main breaker branch (from 2 ms to 5 ms), the current in two IGBT branches ($i_1' + i_2$) is twice as much as that in one IGBT branch (i_1'). Therefore, the current is shared evenly between two IGBT branches. The fault current reaches to 400 A at 5 ms, at which time four IGBTs are turned off. The voltage V_{CE} is clamped by the varistor MOV1, which is twice the value of V_{CE1}. The main breaker exhibits perfect voltage balancing during interruption.

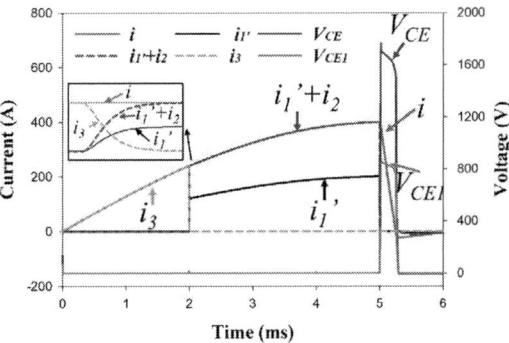

Fig. 4: Simulation of interruption process of hybrid circuit breaker of 400 A peak current value

Fig. 5 presents three currents (i, i_{MOV}, and i_{RCD}) and the voltage across the hybrid circuit breaker V_{CE} during the interruption process of the solid-state main breaker at around 5 ms. i_{MOV}, and i_{RCD} denote the currents flowing through the MOV1 branch and the RCD snubber circuit, respectively. In Fig.7(a), the IGBTs are turned off at 5 ms, the fault current is commutated to the RCD snubber circuit. The snubber capacitors are charged until V_{CE} reaches the clamping voltage of the MOV1. At the same time, the current transfers from the RCD snubber circuit to the MOV branch. And then, the fault current i gradually decreases and drops to zero at around 5.25 ms because the MOV voltage is higher than the DC system voltage. The solid-state main breaker can interrupt a 400 A DC current within 250 µs. Fig.7(b) presents interruption process from 4.98 ms to 5.05 ms in the details. It should be pointed out that there is a voltage spike of 900 V in Fig. 5(b), which is caused by the stray inductance associated with the snubber circuit and the IGBTs.

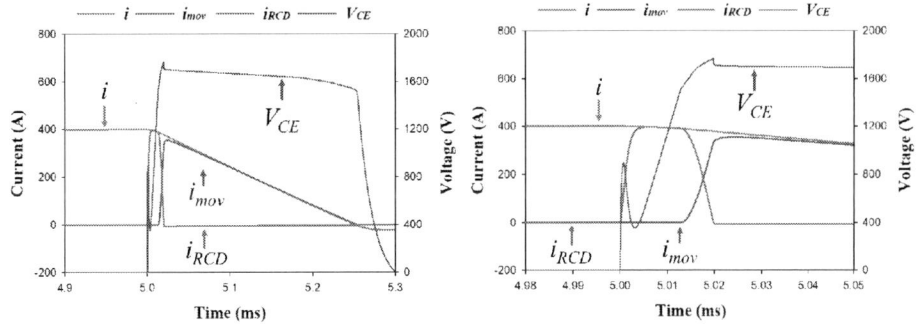

Fig. 5: Simulation result of interruption of 400 A: (a) current commutation from IGBT to RCD and MOV; (b) zoom in for (a)

Design of solid-state main breaker

Device selection

Fig. 6 illustrates an equivalent circuit of DC system with solid-state main breaker during a fault. When the solid-state main breaker is in on state, almost all the equivalent DC system voltage V_{DC} is applied on the system inductor which causes a rapid increase in fault current.

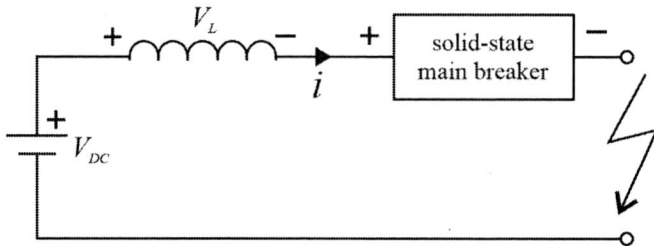

Fig. 6: Solid-state main breaker in the equivalent DC system circuit

When the fault current is detected, the solid-state main breaker starts to interrupt the current. All IGBTs are turned off simultaneously, the fault current commutates from IGBTs to the RCD snubber circuit in few microseconds and the voltage across the solid-state main breaker increases. After tens of microseconds, the voltage is clamped to a designed level, and the fault current commutates to the varistor. The voltage across the solid-state main breaker is limited by the clamping voltage ($V_{clamping}$) of the MOV1. The clamping voltage should be higher than the DC system voltage, therefore, the negative voltage applied to the inductor can reduce the fault current. All residual energy will be dissipated through the varistor. The current in the circuit can be calculated as follows:

$$i = i_0 - \frac{V_{clamping} - V_{DC}}{L} t \tag{1}$$

where i_0 is the maximum fault current flows through the varistor MOV1, and L is the total inductance in the system. The period that current flows through the varistor is:

$$T = \frac{L \times i_0}{V_{clamping} - V_{DC}} \tag{2}$$

The total energy dissipated by the varistor can be calculated by:

$$W_{total} = \int_0^T V_{clamping} \times i \, dt = \frac{1}{2} L i_0^2 \times \left(\frac{V_{DC}}{V_{clamping} - V_{DC}} + 1 \right) \tag{3}$$

High clamping voltage can reduce the interruption time and reduce the total dissipation energy of the varistor. However, the circuit breaker has to withstand the clamping voltage during interruption.

In this paper, IGBT module FF450R12KT4 from Infineon is selected for the 1 kV solid-state main breaker prototype. The IGBT module has two 1.2 kV IGBTs in series, which is more compact and convenient for connecting two IGBT modules in parallel to increase current carrying capability. The maximum voltage across the solid-state main breaker will be suppressed to 1.8 kV by varistor. In addition, varistor MOV2 and MOV3 are employed to limit the maximum voltage to 1 kV to protect IGBTs. Table I lists the devices selected for the solid-state main breaker.

Table I: Devices for the solid-state main breaker

Item	Manufacture	Product
IGBT module 1 & 2	Infineon	FF450R12KT4
MOV1	EPCOS TDK	B80K680
MOV2 & MOV3	EPCOS TDK	B80K385

Mechanical structure

Fig. 7 presents the 3D design and photo of the proposed solid-state main breaker. Two IGBT modules are assembled on a heatsink and connected in parallel using copper bars. MOVs are connected to the copper bars at both sides of the heatsink. The static and dynamic voltage balancing circuits are assembled on the top of the IGBT modules and the gate drive circuit is placed close to the IGBT modules.

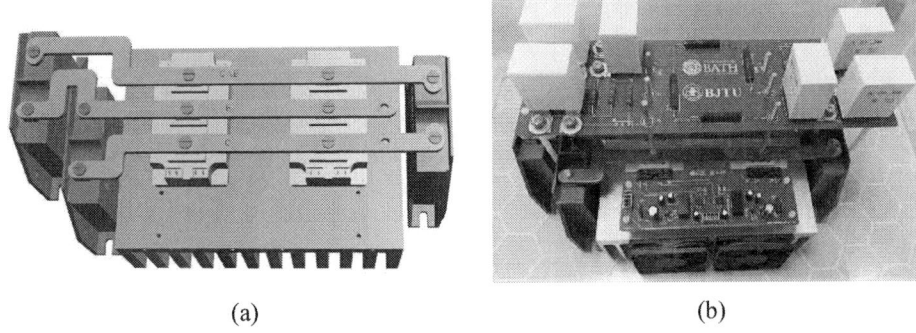

(a)   (b)

Fig. 7: Mechanical structure of the proposed topology: (a) 3D design; (b) photo of the prototype

DC fault current test

Test platform

The schematic diagram of DC fault current test circuit is presented in Fig. 8. An LC resonant circuit is used in the test circuit to emulate the rising of the fault current in the DC system. The DC fault current test circuit is designed to achieve maximum current at 5.5 milliseconds. Two 6 mF capacitors are connected in parallel and an air core inductor of 1 mH is designed. The capacitors are pre-charged by a DC power supply. The freewheeling diode is used to protect the capacitor from the reverse voltage.

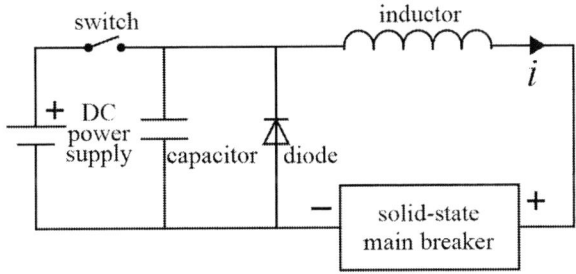

Fig. 8: Schematic diagram of DC fault test circuit

Fig. 9 shows the DC fault test circuit and the solid-state main breaker under test. A microcontroller on the control board is programmed to control the gate drive circuit to provide synchronous gate drive signals for the four IGBTs. The oscilloscope is used to display and record currents (i, i_1, i_{MOV}, and i_{RCD}) and voltages (V_{CE} and V_{CE1}) as illustrated in Fig. 2.

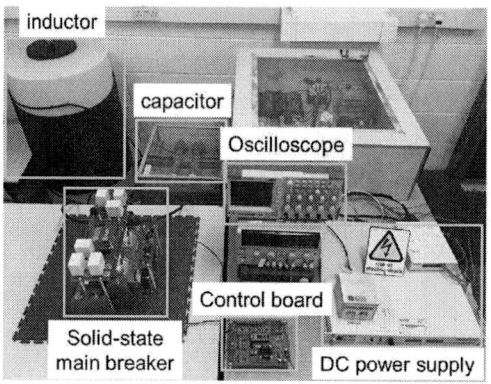

Fig. 9: Dynamic test platform

Current commutation test

In the hybrid DC circuit breaker, the current flows through the MS and LCS (mechanical branch) during normal operation. Once a fault is detected, the current is commutated to the IGBT branches of the solid-state main breaker. The test aims to investigate the current and voltage behaviors during the current commutation from the mechanical branch to solid-state main breaker. During the current commutation tests, the fault current flows through the solid-state main breaker with peak current from 25 A to 400 A. The capacitor, as shown in Fig. 8, is pre-charged to a specific level and the solid-state main breaker is turned on at 5 ms to trigger the LC resonant circuit. The total fault current i, IGBT module 1 current i_1, the total voltage drop V_{CE} and single IGBT voltage drop V_{CE1} are recorded in this test. Because almost no current flows through varistor MOV1 and snubber circuit during the current commutation test, the current i_1 represents the current flowing through IGBT module 1. Fig. 10 shows the current commutation test results when the peak current is 400A. It can be seen that the total current i is approximately twice as much as IGBT module 1 current i_1, which means that the current is evenly shared between parallel IGBT modules. Moreover, in the off state, the total voltage drop across the solid-state main breaker is about twice the voltage drop across one IGBT. Therefore, the solid-state main breaker presents a good static voltage balancing in the off state.

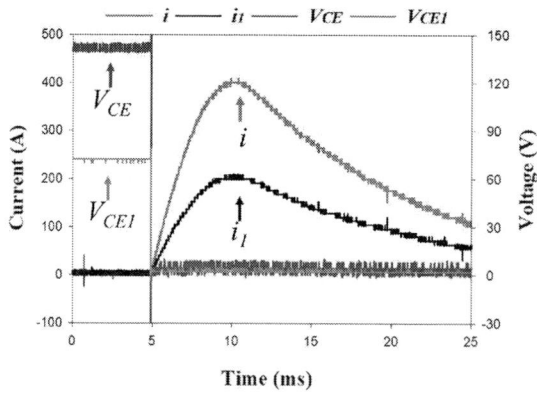

Fig. 10: Current commutation test with peak current of 400A

Current interruption test

The current interruption of the proposed solid-state circuit breaker needs to be investigated because it acts as the main breaker to interrupt the fault DC current in an HCB. The capacitor in Fig. 8 is pre-charged to a specific level, and the voltage level is increased step by step to generate different prospective fault currents. The microcontroller is programmed to turn on the solid-state main breaker to initiate the test and subsequently turn off at 5 ms, at which time the current is close to the peak fault current. Fig. 11 and Fig. 12 show the currents and voltages when the solid-state main breaker interrupts a fault current at 400 A.

Fig. 11(a) illustrates the dynamic current sharing and voltage balancing from fault ignition to current interruption. It should be noted that as shown in Fig. 2, i_1 measures the sum of the current flowing through IGBT branch 1, RCD snubber circuit, and MOV1 branch. i_1 denotes the current flowing through IGBT branch 1 before IGBTs turned off, after they turned off, i_1 represents the current in snubber circuit and MOV1. As shown in Fig. 11(a), the solid-state main breaker is turned on to initiate the test, the total fault current increases from 0 A to 400 A in 5 ms. During this period, the total current i is twice the current in IGBT1, which shows that the IGBT branches share the current evenly under dynamic current change.

Fig. 11(b) highlights the dynamic current sharing and voltage balancing during current interruption. When the IGBTs are turned off, the current commutates to RCD snubber circuit and then to MOV1, the voltage across the solid-state main breaker V_{CE} is approximately twice that of the voltage across one IGBT V_{CE1}. The solid-state main breaker presents good dynamic voltage balancing during the current interruption. When the IGBTs are turned off, the fault current i and i_1 are the same because the current flowing through IGBT branches are zero. The fault current finally reduces to zero within 300 µs.

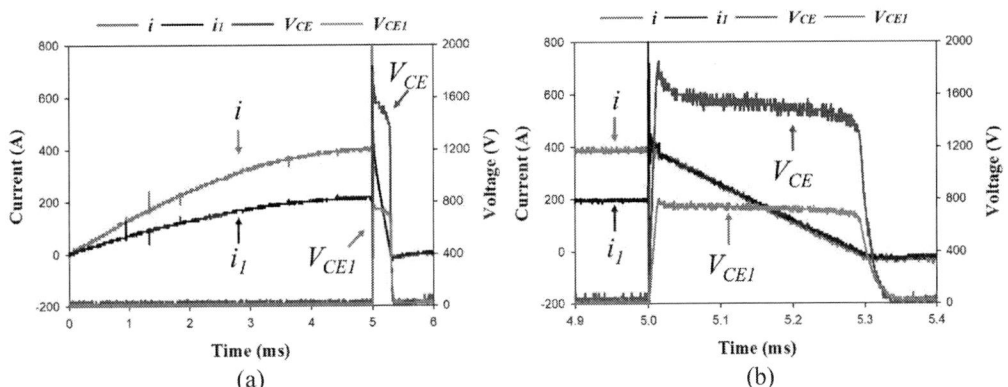

Fig. 11: Current interruption test of 400 A: (a) current share and voltage balance waveform; (b) zoom in for (a)

Fig. 12 presents the current commutation and interruption process of the solid-state main breaker. Three currents (i, i_{MOV}, and i_{RCD}) and the voltage V_{CE} are presented. The fault current i commutates to the RCD snubber circuits once the IGBTs are turned off. The current i_{RCD} flows through the diode and charges the snubber capacitors. After several tens of microseconds, the voltage V_{CE} reaches the knee point of the MOV1, the fault current i then commutates from RCD snubber circuits to the varistor MOV1 branch in several microseconds. The fault current finally reduces to zero within 300 microseconds. The experimental test results show good consistent with simulation results in Fig. 5.

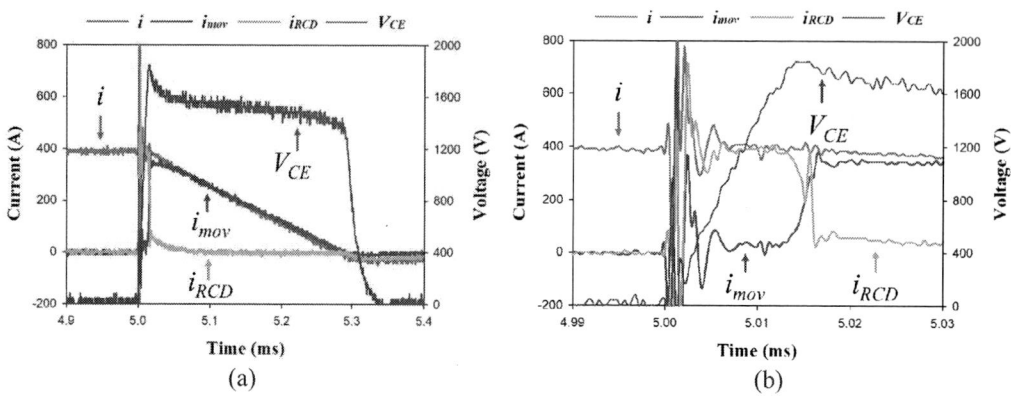

Fig. 12: Current interruption test of 400 A: (a) current commutation from IGBT to RCD and MOV; (b) zoom in for (a)

Conclusions

The performance of the solid-state main breaker in the hybrid circuit breaker is simulated in the PLECS model. A 1 kV solid-state main breaker prototype based on series and parallel connected IGBTs is designed and built. The solid-state main breaker prototype is experimentally tested under dynamic operation using the DC fault test platform. The experimental results show good consistency with the simulation results. The solid-state main breaker prototype successfully interrupts 400 A DC current within 300 µs (including the time to absorb the residual energy). The prototype demonstrates even current sharing under dynamic current change and voltage balancing under both off state and current interruption.

References

[1] Reed G.F., Grainger B.M., Sparacino A.R., and Mao Z.-H.: Ship to grid: Medium-voltage DC concepts in theory and practice, IEEE Power and Energy Magazine, 2012, 10, (6), pp. 70-79

[2] Kim K., Park K., Roh G., and Chun K.: DC-grid system for ships: a study of benefits and technical considerations, Journal of International Maritime Safety, Environmental Affairs, and Shipping, 2018, 2, (1), pp. 1-12

[3] Armstrong M.J., Blackwelder M., Bollman A., Ross C., Campbell A., Jones C., and Norman P.: Architecture, voltage, and components for a turboelectric distributed propulsion electric grid, 2015

[4] Jones C.E., Norman P.J., Galloway S.J., Armstrong M.J., and Bollman A.M.: Comparison of candidate architectures for future distributed propulsion aircraft, IEEE Transactions on Applied Superconductivity, 2016, 26, (6), pp. 1-9

[5] Franck C.M.: HVDC circuit breakers: A review identifying future research needs, IEEE Transactions on Power Delivery, 2011, 26, (2), pp. 998-1007

[6] Shen Z. J., Miao Z., and Roshandeh A. M.: Solid state circuit breakers for DC micrgrids: Current status and future trends, in 2015 IEEE First International Conference on DC Microgrids (ICDCM), 2015, pp. 228-233

[7] Pei X., Cwikowski O., Vilchis-Rodriguez D., Barnes M., Smith A., and Shuttleworth R.: A review of technologies for MVDC circuit breakers, IECON 2016-42nd Annual Conference of the IEEE, 2016, pp. 3799-3805

[8] Feng L., Gou R., Zhuo F., Yang X., and Zhang F.: Development of a 10kV solid-state DC circuit breaker based on press-pack IGBT for VSC-HVDC system, 2016 IEEE 8th International Power Electronics and Motion Control Conference (IPEMC-ECCE Asia). IEEE, 2016: 2371-2377

[9] Callavik M., Blomberg A., Häfner J., and Jacobson B.: The hybrid HVDC breaker, ABB Grid Systems Technical Paper, 2012, vol. 361, pp. 143-152

[10] Zhou W. et al.: Development and test of a 200kV full-bridge based hybrid HVDC breaker, in 2015 17th European Conference on Power Electronics and Applications (EPE'15 ECCE-Europe), 2015, pp. 1-7

[11] An T., Tang G., and Wang W.: Research and application on multi-terminal and DC grids based on VSC-HVDC technology in China, High Voltage, 2017, vol. 2, no. 1, pp. 1-10

[12] Yang B., Cao D., Shi W., Lv W., Wang W., and Liu B.: A novel commutation-based hybrid HVDC circuit breaker, in CIGRE Winnipeg Colloquium, 2017, pp. A3-15

[13] Tang G., Pang H., He Z., and Wei X.: Research on key technology and equipment for Zhangbei 500kV DC grid, in 2018 International Power Electronics Conference (IPEC-Niigata 2018-ECCE Asia), 2018, pp. 2343-2351

[14] Song X., Peng C., and Huang A.Q.: A medium-voltage hybrid DC circuit breaker, part I: solid-state main breaker based on 15 kV SiC emitter turn-OFF thyristor, IEEE Journal of Emerging and Selected Topics in Power Electronics, 2016, 5, (1), pp. 278-288

[15] Peng C., Song X., Huang A. Q., and Husain I.: A medium-voltage hybrid DC circuit breaker—Part II: Ultrafast mechanical switch, IEEE Journal of Emerging and Selected Topics in Power Electronics, 2016, vol. 5, no. 1, pp. 289-296

[16] Shammas N., Withanage R., and Chamund D.: Review of series and parallel connection of IGBTs, IEE Proceedings-Circuits, Devices and Systems, 2006, 153, (1), pp. 34-39

Model Predictive Control for Three-Phase Split-Source Inverter

Youssuf Elthokaby	Islam Mohamed	Naser Abdel-Rahim
Faculty of Engineering at Shoubra, Benha University, Egypt		Future University in Egypt
youssef.hassan@feng.bu.edu.eg	islam.ahmed@feng.bu.edu.eg	naser.abdelrahim@fue.edu.eg

Keywords

«Three-phase system», «MPC (Model predictive control) », «Modeling», «Converter circuits», «Converter control».

Abstract

This paper proposes a new topology of three-phase split-source inverter (SSI) with only one auxiliary power electronic switch inserted into the three-phase full-bridge inverter. Split-source inverter is a single stage topology, which combines the boosting stage with the DC/AC converter stage. The proposed three-phase SSI has fewer passive components than impedance source converters (like ZSI). Switching states of the conventional three-phase VSI can be used with the SSI to control the output voltage. An independent ECCE control for both DC link voltage and output AC voltage is achieved using model predictive control. The DC link voltage can be controlled via duty cycle obtained by a maximum power point tracker or by fixed duty cycle as used in this paper. Simulation results show that the proposed control achieves low THD of the AC output voltage (good performance) with both normal loading condition and sudden load change condition.

Introduction

Interest in single-stage buck/boost inverters has grown over the past few years due to the need to reduce the size as well as the cost of the overall system while increasing the efficiency of the power converter. The main application of single-stage boost inverters is in the PV-based systems. A conventional PV system consists of a DC/DC converter and a single/three phase inverter. The two-stage topology suffers from low efficiency and bulkiness, but it has simple control strategy. The single-stage buck/boost inverter topology, however, performs two advantageous functions: boosting the DC input voltage and ensuring that the output AC voltage is sinusoidal with low THD. Various circuit topologies of single-stage buck/boost topologies have been reported in the literature [1]-[3].

One of the single-stage boost inverter topologies is the split-source inverter (SSI) that has been reported in the literature [4]-[9]. SSI is a DC-AC boost inverter topology which requires a boosting inductor and a DC link capacitor, see Fig. 1. The single-phase SSI was first introduced in 2010 [4]. It consists of a single-phase H-bridge inverter with two common anode diodes connected in each leg and an inductor connected with a DC source and capacitor connected across the inverter legs, see Fig. 1.(a) [5]. The control strategy proposed is similar to that of the conventional single-phase H-bridge inverter, where, during the positive half cycle, switches Q_1 and Q_2 are turned ON/OFF and the other two switches are switched OFF. During the negative half cycle, Q_3 and Q_4 are turned ON/OFF while Q_1 and Q_2 are turned OFF. During either half cycle, the inductor is charged and discharged into the capacitor but with variable duty cycle leading to low frequency component at the DC side, which increases the conduction losses.

To overcome the problem of variable duty cycle and hence improve the overall system efficiency, an alternative SSI configuration has been introduced in [6]. The two diodes in that converter have been connected to a DC source with their common cathode configuration, see Fig. 1.(b). In that topology, the inductor was charged if one/both of the two upper switches were turned ON and discharged into the DC link capacitor when each of the upper switches were turned OFF. A modified SPWM technique has been used to achieve constant duty cycle, where the reference signals of each leg have been modified to be quasi-sinusoidal in a half cycle and constant in another half cycle. That guaranteed that one of the upper switches was turned ON/OFF constantly in each half cycle. However, both the SSIs in [5] and [6] have

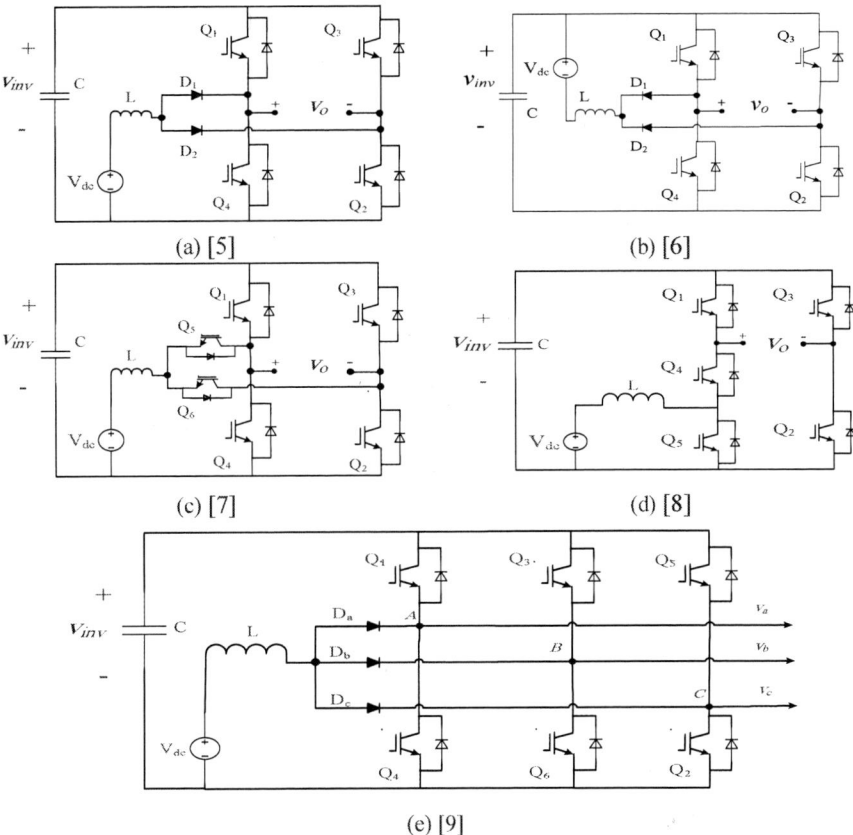

(a) [5]

(b) [6]

(c) [7]

(d) [8]

(e) [9]

Fig. 1. Existing SSI topologies in the literature

been incapable of exchanging power with the DC source, i.e. no bidirectional power flow. Also, the power diodes suffered from high frequency commutation and hence losses.

To enable bidirectional power flow and reduce the high-frequency commutation of the diodes, the two input diodes have been replaced with two MOSFETs in [7], see Fig. 1.(c). A constant duty cycle was obtained by modifying the reference signals of each leg to be hybrid quasi-sinusoidal in half cycle and constant in the other half cycle. That converter topology had the same number of switches as that of the bidirectional converter and of the conventional single-phase H-bridge inverter but with a different switching arrangement.

To cut down on the number of switching devices, a simplified SSI (S^3I) has been introduced in [8], where only one auxiliary MOSFET was connected to one of the H-bridge legs, forming a one leg with three switches parallel to the two-switched leg of the inverter, see Fig. 1.(d). That topology reduced power switches, which enhanced power efficiency, increased voltage boosting gain, and reduced output filter requirements. Sinusoidal PWM has been used to control S^3I, where two sinusoidal waveforms with phase shift and constant waveforms were compared with carrier waveforms to control the two legs of the inverter and the auxiliary switch, respectively.

A three-phase SSI was introduced in [9], where three diodes were connected to the three legs of the inverter, see Fig. 1.(e). The inductor was energized if at least one of the bottom switches was turned ON. The inductor discharged into the capacitor when all the top switches were turned ON. The three-phase SSI was controlled by modified SVPWM (MSVPWM), where the reference signals were the same as those of the SVPWM with VSI, but the lower envelopes were constant. Hence, the duty cycle was fixed. However, that topology suffered from high-frequency commutations of the input diodes, which represented additional losses. Moreover, that converter was incapable of performing bidirectional power flow with the DC source.

This paper proposes a new topology of the three-phase SSI. The new three-phase SSI replaces three diodes in [9] with only one bidirectional switch, enabling bidirectional power flow (suitable for stand-alone PV systems with batteries) and reducing the system size. The proposed topology eliminates the high-frequency commutation of the input diodes, and improves efficiency due to low turn-ON resistance of the bidirectional switch [7], [8]. However, an additional gate driver is required for the bidirectional switch. The control of the proposed three phase SSI is achieved by finite control set model predictive control (FCS-MPC). FCS-MPC is used to control the traditional three-phase SSI as in [10]-[12]. FCS-MPC is suitable for the discrete nature of the power electronics converters and can deal with multivariable systems, constraints and nonlinearities[13], [14]. An independent control for both DC-link voltage and AC output voltage can easily be achieved.

The advantages of the proposed SSI topology are: 1) continuous input current, 2) continuous DC-link voltage, 3) less passive components, 4) lower stress on the switches at higher voltage gains, 5) maintaining the same switching state of VSI, and 6) enabling bidirectional power flow. On the other hand, the disadvantages are: 1) unbalanced current distribution among SSI switches, 2) higher current stress at high voltage gains, and 3) high voltage stress at low voltage gains.

Proposed Three-Phase SSI

The proposed three-phase SSI is shown in Fig. 2. The inductor L is charged if the auxiliary switch Q_7 is ON. While inductor L is charging, the SSI can operate with eight possible switching states of the three-phase VSI as shown in Fig. 3.(a). When the auxiliary switch Q_7 is turned OFF, the inductor L is discharging into a capacitor (C). At this moment, the three-phase SSI can operate with only four possible switching states as shown in Fig. 3.(b). The switch Q_7 can be controlled with a fixed duty cycle or by a duty cycle obtained from MPPT controller of a PV system to extract the maximum available power from the PV array and hence control the DC link voltage. The DC link voltage (v_{inv}) can be obtained from (1) [8] . The output AC voltage can be controlled by the PWM dictated by the model predictive control presented later in this article. An independent control for both DC-link capacitor voltage and output AC voltage can easily be provided. However, the three-phase SSI suffers from unbalanced current sharing among the switches as shown in Table I. Yet, this is an inherent problem with all SSI topologies which can easily be resolved by paying attention to the sizing of the power electronic switches [8], [9].

$$v_{inv} = \frac{V_{dc}}{1 - D} \tag{1}$$

Model Predictive Control

The proposed three-phase SSI is controlled by finite-control set model predictive control (FCS-MPC). FCS-MPC is suitable for power electronics converters since it uses the model of the system to predict the future behavior of the controlled variables for all possible voltage vectors of the converter and uses

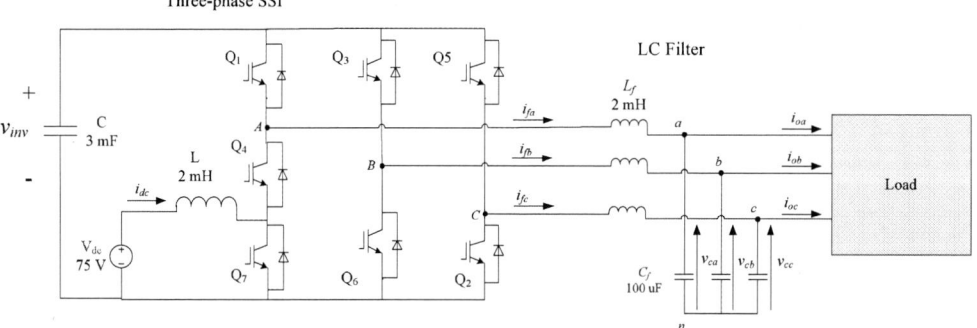

Fig. 2. The proposed three-phase SSI

Table I. Current and voltage stresses for the proposed SSI switches

Switch	Current stress	Voltage stress
Q_1: Q_6	$i_{f,\,ph}$	
Q_7	$i_{f,\,ph}+i_{dc}$	
Anti-parallel diode of Q_1	$i_{f,\,ph}+i_{dc}$	
Anti-parallel diode of Q_4	i_{dc}	v_{inv}
Anti-parallel diodes of Q_2, Q_3, Q_5, Q_6	$i_{f,\,ph}$	

the information to obtain optimal action by choosing the voltage vector that minimizes a cost function. The possible switching states and corresponding voltage vectors of the proposed three-phase SSI are shown in Fig. 3 and Fig. 4.

The Model of the System

The power circuit of a three-phase inverter with a second-order LC filter is shown in Fig. 2. The load voltage (v_c), the inductor filter current (i_f), and the load current (i_o) are given by:

$$v_c = \frac{2}{3}\left(v_{ca} + a v_{cb} + a^2 v_{cc}\right) \tag{2}$$

$$i_f = \frac{2}{3}\left(i_{fa} + a i_{fb} + a^2 i_{fc}\right) \tag{3}$$

$$i_o = \frac{2}{3}\left(i_{oa} + a i_{ob} + a^2 i_{oc}\right) \tag{4}$$

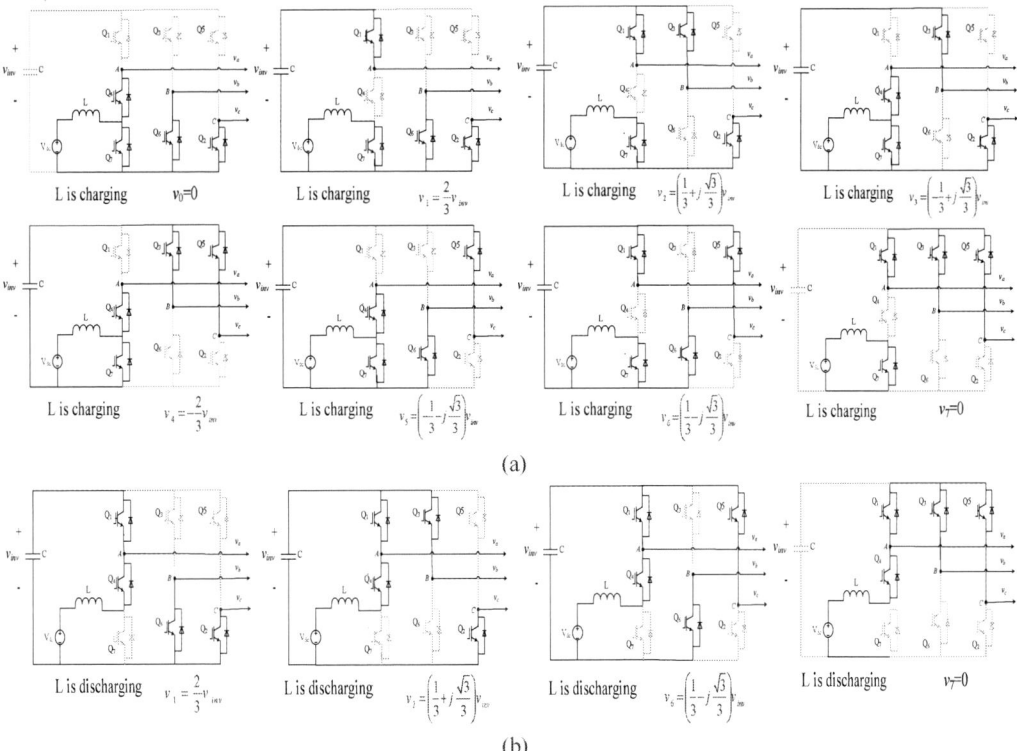

Fig. 3. Possible switching states of the proposed three-phase SSI at (a) charging L and (b) discharging L

Where $\mathbf{a} = e^{j\frac{2\pi}{3}} = -\frac{1}{2} + j\frac{\sqrt{3}}{2}$ represents the 120° phase displacement between phases. The governing differential equations of the LC filter variables are:

$$L_f \frac{di_f}{dt} = v_i - v_c \tag{5}$$

$$C_f \frac{dv_c}{dt} = i_f - i_o \tag{6}$$

Where, v_i, C_f and L_f are the SSI output voltage, filter capacitance and inductance, respectively. Equations (5) and (6) can be rewritten in a state-space variable format as:

$$\frac{dx}{dt} = Ax + Bv_i + B_d i_o \tag{7}$$

where,

$$x = \begin{bmatrix} i_f & v_c \end{bmatrix}^T \tag{8}$$

$$A = \begin{bmatrix} 0 & \frac{-1}{L_f} \\ \frac{1}{C_f} & 0 \end{bmatrix} \quad B = \begin{bmatrix} \frac{1}{L_f} \\ 0 \end{bmatrix} \quad B_d = \begin{bmatrix} 0 \\ \frac{-1}{C_f} \end{bmatrix} \tag{9}$$

The system output equation is given by:

$$v_c = \begin{bmatrix} 0 & 1 \end{bmatrix} x \tag{10}$$

The discrete time model is obtained from (7) for a sampling time T_S and expressed as:

$$x(k+1) = A_q x(k) + B_q v_i(k) + B_{dq} i_o(k) \tag{11}$$

where

$$A_q = e^{AT_S} \quad B_q = \int_0^{T_s} e^{At} B \, dt \quad B_{dq} = \int_0^{T_s} e^{At} B_d \, dt \tag{12}$$

Model Predictive Control Strategy

FCS-MPC algorithm is shown in Fig. 5. The value of the output voltage ($v_c(k)$), inductor filter current ($i_f(k)$) and the load current ($i_o(k)$) are measured at the current instance k. The condition of the switch Q_7 is observed. If Q_7 is ON, the SSI can operate with all possible voltage vectors as shown in Fig. 4. (a). If Q_7 is OFF, the SSI can only operate with four possible voltage vectors as shown in Fig. 4.(b). The output voltages, for every possible voltage vector, at next instance ($v_c(k+1)$) can be predicted using (11). Then, the cost function (13) is used to select the voltage vector that gives the lowest error between reference and output voltages at next instance. Consequently, the corresponding switching state is applied in the next instance.

$$g = \left(v^*_{c\alpha} - v_{c\alpha}(k+1)\right)^2 + \left(v^*_{c\beta} - v_{c\beta}(k+1)\right)^2 \tag{13}$$

where, $v_{c\alpha}$ and $v_{c\beta}$ are the real and imaginary parts of the predicted output voltage vector $v_c(k+1)$. The control block diagram of the system is shown in Fig. 6. In the FCS-MPC, there is no need for a PWM modulator because the output of the controller itself is the switching signal.

Simulation Results

The proposed three-phase SSI with FCS-MPC is simulated using MATLAB/Simulink with the system parameters shown in Fig. 2. The load is linear with 0.9 power factor lag and 33 Ω impedance. The DC link voltage is controlled via the switch Q_7 by a fixed duty cycle (D) 0.83 and with a switching frequency of 20 kHz. The output AC voltage is controlled via the switches Q_1 - Q_6 using the FCS-MPC with 20 kHz switching frequency. The switching frequency of Q_7 must be equal to or less than the switching frequency of Q_1-Q_6. The DC link capacitor voltage (v_{inv}) is shown in Fig. 7. (a), and equal to 429.5 V. The boosting inductor current (i_{dc}) is shown in Fig. 7. (b) and equals nearly 23 A. The three-phase output voltage waveforms (v_{ca}, v_{cb}, v_{cc}) are shown in Fig. 8. The output voltages are sinusoidal with low THD of 1.64%. The peak value of the phase voltage at steady state is 197.2 V. Fig. 9.(a) shows the three-

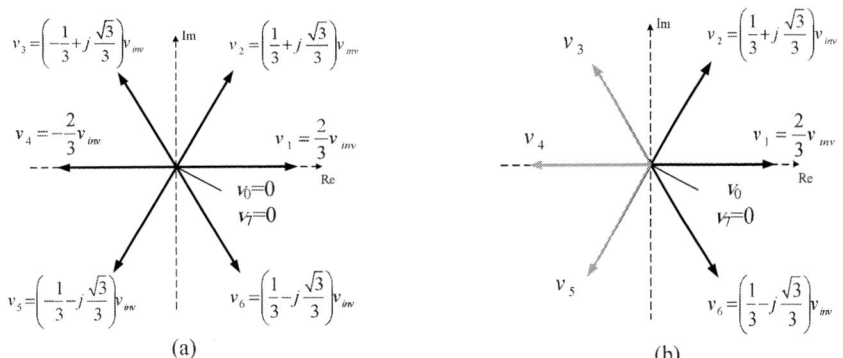

Fig. 4. Voltage vectors of the proposed three-phase SSI at (a) charging L, and (b) discharging L

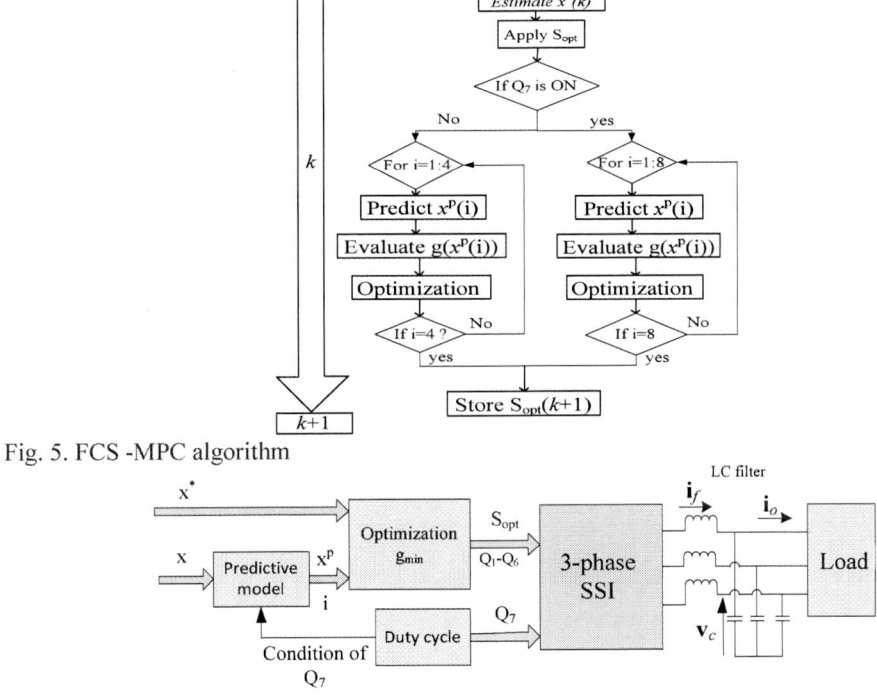

Fig. 5. FCS -MPC algorithm

Fig. 6. System control block diagram

phase load currents (i_{oa}, i_{ob}, i_{oc}), which equals 4.2 A rms with 0.55% THD. Fig. 9.(b) shows the inductor filter currents (i_{fa}, i_{fb}, i_{fc}). The figure shows that the inductor filter current is rich of harmonics with a THD of 50.11%.

Q_1 current is shown in Fig. 10, where the positive part is the current in the IGBT and the negative part of the current is current passing through the anti-parallel diode. From the figure, the current of Q_1 in the IGBT is equal to the inductor filter current while the anti-parallel diode current is equal to the inductor filter current plus the boosting inductor current. Fig. 11 shows the current of the switch Q_4 which is equal to the inductor filter current while the anti-parallel diode is equal to the boosting inductor current. Fig. 12 shows that the currents of the IGBT and the anti-parallel diode of Q_3 combined are equal to the inductor filter current. The current of the switches Q_2, Q_5, and Q_6 are similar to that of Q_3. The current of the switch Q_7 is shown in Fig. 13 to be equal to the inductor filter current plus the boosting inductor current. The anti-parallel diode of Q_7 has no current because there is no bidirectional power flow between the load and the DC source. The switches currents are shown in Table I. From the above, the

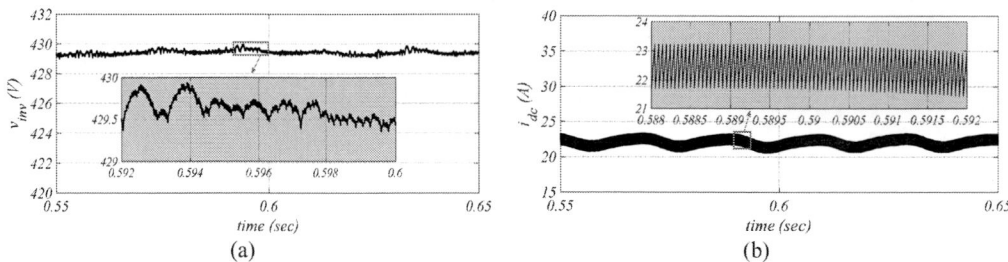

(a)   (b)

Fig. 7. (a) The DC-link capacitor voltage and, (b) the boosting inductor current

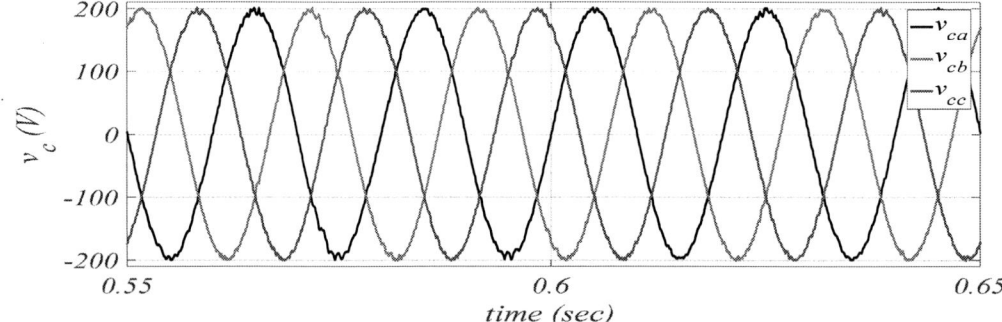

Fig. 8. Three-phase output voltage

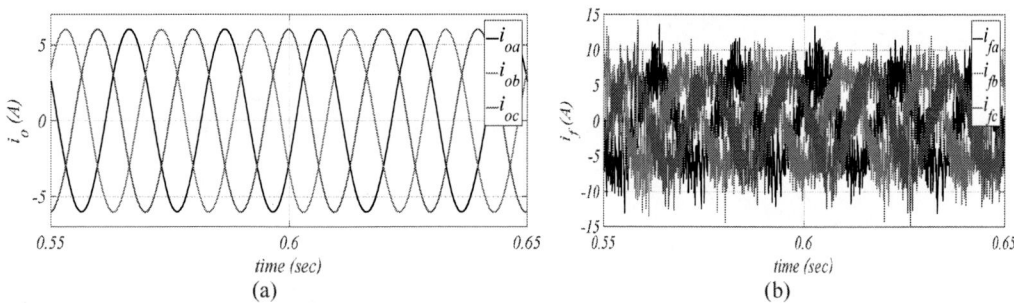

(a)   (b)

Fig. 9. Steady-state waveforms: (a) three-phase output current and, (b) three-phase inductor filter current

switches Q_1 and Q_7 should be selected to withstand the inductor filter current plus the inductor booting current. While the switch Q_4 should withstand the inductor boosting current, and the rest of the switches (Q_2, Q_3, Q_5 and Q_6) should withstand the inductor filter current.

The dynamic behaviour of the proposed inverter is investigated by applying sudden change in the load from 33 Ω at 0.9 p.f. lagging to a load of 21 Ω at 0.7 power factor lagging and is shown in Fig. 14. The load current waveforms (i_{oa}, i_{ob}, i_{oc}) before, during, and after load change are shown in Fig. 14.(a). The figure shows that the currents undergoes increase from 4.2 A to 6.7 A rms (almost 60% increase). The load currents take approximately 12.8 ms to reach its steady-state value. The three-phase output voltages (v_{ca}, v_{cb}, v_{cc}) waveforms are shown in Fig. 14.(b). The figure shows that there is a slight effect on the load voltage due to this sudden load change. The THD of the output voltages is 1.73%. The DC-link capacitor voltage (v_{inv}) is shown in Fig. 14.(c). The figure shows a slight decrease in the DC voltage from 429 V to steady-state value 426 V with an undershoot reaching 423 V (-1.3%). The boosting inductor current (i_{dc}) is shown in Fig. 14.(d), which has increased from 23 A to 28 A.

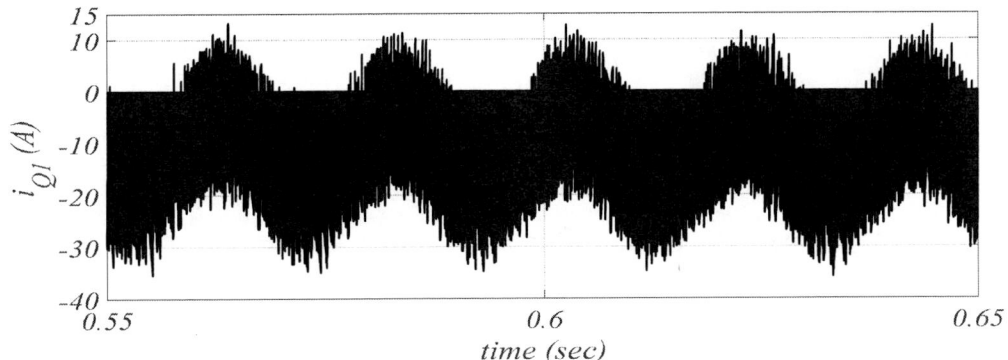

Fig. 10. The current of the switch (Q_1)

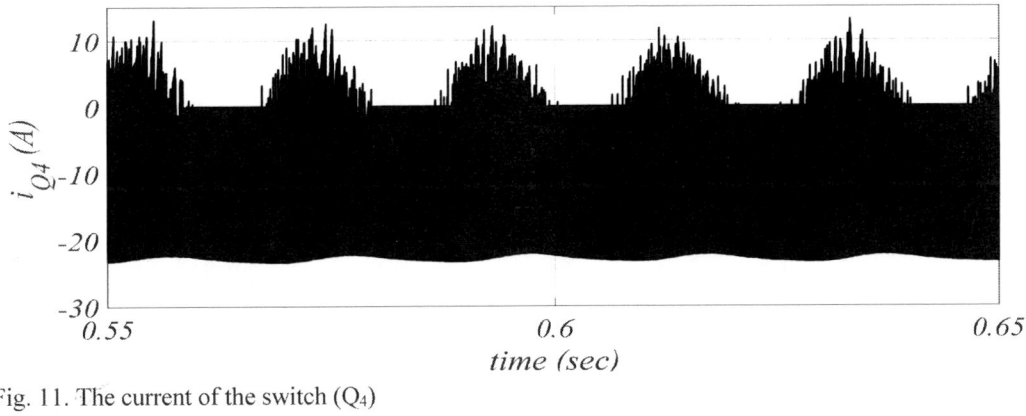

Fig. 11. The current of the switch (Q_4)

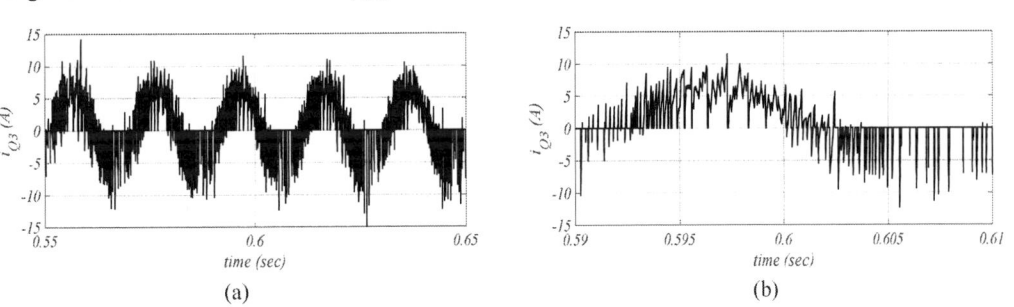

| (a) | (b) |

Fig. 12. (a) The current of the switch (Q_3), (b) Zoom view for one cycle of the switch current

Another sudden change in the load from 21 Ω at 0.7 power factor lagging to 33 Ω at 0.9 power factor lagging. The load current waveforms (i_{oa}, i_{ob}, i_{oc}) are shown in Fig. 15.(a). The figure shows that load currents have decreased from 6.7 A to 4.2 A rms. The load currents take approximately 6.32 ms to reach its steady-state condition. The three-phase output voltages (v_{ca}, v_{cb}, v_{cc}) waveforms are shown in Fig. 15.(b). The figure shows that the load voltages are slightly affected due to the sudden load change. The THD of the output voltages is 1.97%. The DC-link capacitor voltage (v_{inv}) is shown in Fig. 15.(c). The figure shows a slight increase in the DC voltage from 426 V to a steady-stae value 429 V (approximately 0.7%) with overshoot reaches to 432 V. The boosting inductor current (i_{dc}) is shown in Fig. 15.(d). The figure shows that i_{dc} has decreased from 28 A to 23 A.

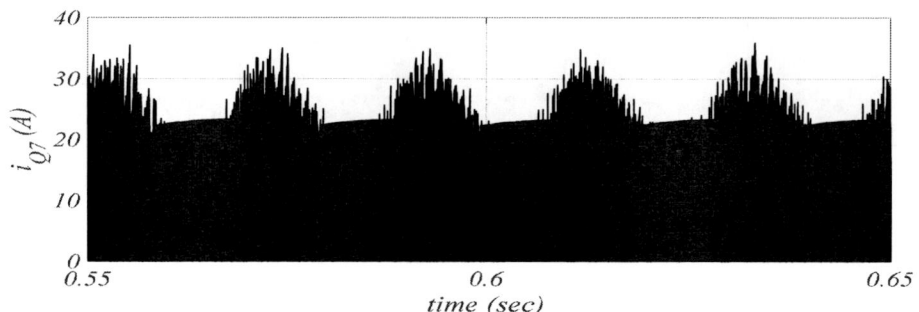

Fig. 13. The current of switch (Q7)

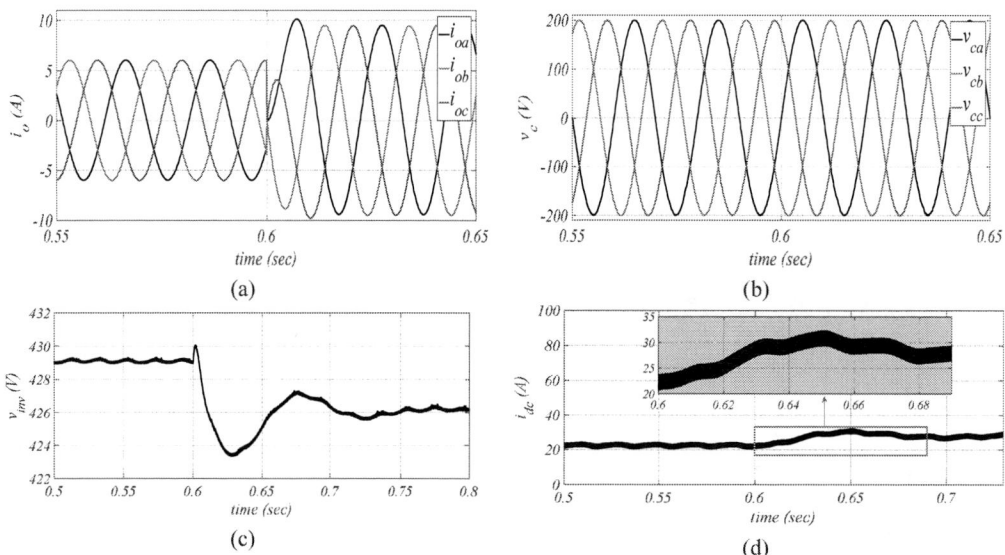

(a)

(b)

(c)

(d)

Fig. 14. Dynamic behavior of sudden increase in the load current of the proposed SSI: (a) three-phase output currents, (b) three-phase output voltages, (c) the DC-link capacitor voltage, and (d) the boosting inductor current

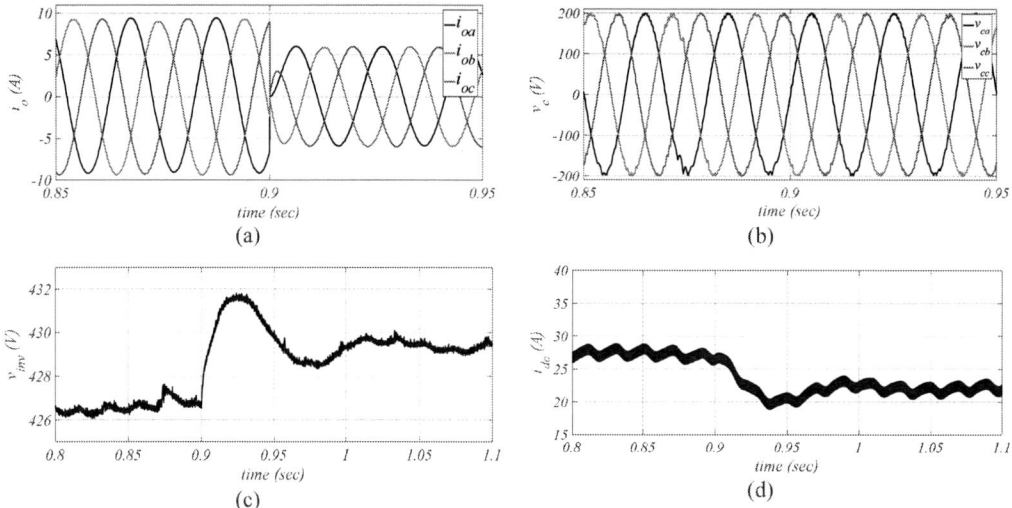

(a)

(b)

(c)

(d)

Fig. 15. Dynamic behavior of sudden decrease in the load current of the proposed SSI: (a) three-phase output current, (b) three-phase output voltage, (c) the DC-link capacitor voltage and, (d) the boosting inductor current

Conclusions

This paper proposed a new circuit topology for three-phase split-source inverter (SSI). The proposed SSI has advantage of bidirectional power flow capability, which is not provided in already existing three-phase SSI. This bidirectional power flow is specially needed for PV-based systems. The proposed SSI reduces the power semiconductor count of the three-phase SSI by replacing three diodes by only one bidirectional switch. This, in turn, reduces the high-frequency commutation of the input diodes and hence improves the overall efficiency. FCS- MPC is used as the control algorithm of the new topology. When the boosting inductor is charging, all switching states can be used, while only four switching states can be used at discharging of the boosting inductor. The proposed SSI is simulated with a linear load at both steady-state and dynamic phases. Under all operating conditions, the results show good load voltage regulation with low THD.

References

[1] Y. P. Siwakoti, F. Z. Peng, F. Blaabjerg, P. C. Loh, G. E. Town, and S. Yang, "Impedance-source networks for electric power conversion part II: Review of control and modulation techniques," *IEEE Transactions on Power Electronics*, vol. 30, no. 4, pp. 1887–1906, Apr. 2015.

[2] O. Ellabban and H. Abu-Rub, "Z-source inverter: Topology improvements review," *IEEE Industrial Electrons Magazine*, vol. 10, no. 1, pp. 6–24, Mar. 2016.

[3] J. Yuan,F. Blaabjerg, Y. Yang, A. Sangwongwanich and, Y. Shen, "An Overview of Photovoltaic Microinverters:Topology, Efficiency, and Reliability," *IEEE 13th International Conference on Compatibility, Power Electronics and Power Engineering (CPE-POWERENG)*, Sonderborg, Denmark, 23-25 April 2019.

[4] H. Ribeiro, A. Pinto, and B. Borges, "Single-stage dc-ac converter for photovoltaic systems," in *Proc. IEEE Energy Conversion Congress and Exposition,* Atlanta, GA, USA, pp. 604–610, 12-16 Sept. 2010.

[5] A. Nahavandi, M. Roostaee, and M. R. Azizi, "Single stage dc–ac boost converter," *7th Power Electronics and Drive Systems Technologies Conference (PEDSTC)*, Tehran, Iran, pp. 362–366, 16-18 Feb. 2016.

[6] A. Abdelhakim, P. Mattavelli, P. Davari, and F. Blaabjerg, "Performance evaluation of the single-phase split-source inverter using an alternative dc–ac configuration," *IEEE Transactions on industrial electronics*, vol. 65, no. 1, pp. 363–373, Jan. 2018.

[7] S. S. Lee and Y. E. Heng, "Improved single-phase split-source inverter with hybrid quasi-sinusoidal and constant PWM," *IEEE Transactions on industrial electronics*, vol. 64, no. 3, pp. 2024–2031, Mar. 2017.

[8] S. S. Lee, A. S. T. Tan, D. Ishak, and R. Mohd-Mokhtar, "Single-Phase Simplified Split-Source Inverter (S^3I) for Boost DC–AC Power Conversion," *IEEE Transactions on industrial electronics*, vol. 66, no. 10, pp. 7643-7652, Oct. 2019.

[9] A. Abdelhakim, P. Mattavelli, and G. Spiazzi, "Three-phase split-source inverter (SSI): Analysis and modulation," *IEEE Transactions on industrial electronics*, vol. 31, no. 11, pp. 7451–7461, Nov. 2016.

[10] M. A. Ismeil, "High Dynamic Performance for Split-Source Inverter based on Finite Control Set Model Predictive Control," *21st International Middle East Power Systems Conference (MEPCON)*, Cairo, Egypt, 17-19 Dec. 2019.

[11] G. M. Cocco, J. A. Borges, M. Stefanello and F. B. Grigoletto, "Finite Set Model Predictive Control of Four-Leg Split-Source Inverters," *13th IEEE International Conference on Industry Applications*, São Paulo, Brazil, 12-14 Nov. 2018.

[12] J. A. Borges, and F. B. Grigoletto, "Finite Set Model Predictive Control of Grid Connected Split-Source Inverters," Juiz de Fora, Brazil, Brazilian Power Electronics Conference (COBEP), 19-22 Nov. 2017.

[13] Y. Elthokaby, A. E. Elshafei, N. Abdel-Rahim, and E. S. Abdel-Aliem, "Finite-control set model-predictive control for single-phase voltage-source UPS inverters," *18th International Middle East Power Systems Conf. (MEPCON)*, Cairo, Egypt, pp. 261–265, 27-29 Dec. 2016.

[14] P. Cortés, G. Ortiz, J. I. Yuz, J. Rodríguez, S. Vazquez, and L. G. Franquelo, "Model Predictive Control of an Inverter With Output LC Filter for UPS Applications," *IEEE Transactions on industrial electronics,* Vol. 56, No. 6, pp. 1875–1883, June 2009.

Hardware Implementation study of Variable Speed Wind-Turbine-DFIG in Stand-alone Mode

Fayssal AMRANE [*, 1, 2] Bruno FRANCOIS [2] and Azeddine CHAIBA[1, 3],

[1] Research Automatics Laboratory of Setif (LAS) Department of Electrical Engineering,
FERHAT ABBAS SETIF-1 UNIVERSITY, SETIF, ALGERIA

[2] Laboratoire d'Electrotechnique et d'Electronique de Puissance de Lille (L2EP),
ECOLE CENTRALE DE LILLE, LILLE, FRANCE.

[3] Department of Industrial Engineering
UNIVERSITY OF KHENCHELA, ALGERIA.

[*] Corresponding author: E-mail: **amrane_fayssal@live.fr; amrane_fayssal@univ-setif.dz**;

Acknowledgements

We are very grateful to the Algerian Ministry of Higher Education and Scientific Research (**MESRS**) for funding this Research Project as part of an Algerian-French **PROFAS b+** Scholarship. We are grateful also to the **L2EP** laboratory for the financial support. Also this research work falls within the framework of the research project "**PRFU**" under the code: **A01L07UN400120190001** (Project approved from **01/01/2019**)

Keywords

«Wind Energy», «Direct Power Control», «LCL-Passive Filter», «Digital Control», «DFIG»,

Abstract

In this paper, hardware implementation study based on Direct Power Control (DPC) of Doubly Fed Induction Generator (DFIG) is proposed for Stand-alone mode in Variable Speed-Wind Energy Conversion System (VS-WECS). The proposed DPC algorithm based on robust IP (Integral-Proportional) controllers in order to control the Rotor Side Converter (RSC) by the means of the rotor current d-q axes components (I_{rd} and I_{rq}) of DFIG through AC-DC-AC converter. In order to improve the grid-side and rotor side power qualities, LCL passive filter is implemented between the DFIG's rotor and the inverter. Finally, experimental results demonstrate that the proposed control using IP provides improved dynamic responses, and decoupled power control (*with high performances: good reference tracking, short response time and low power error*) despite for sudden wind speed changement and the variation of rotor current.

1-Introduction

During the past years, the installed wind power capacity in the world has been increasing more than 30% [1-2]. DFIG-based wind turbines have many advantages over the fixed speed induction generators or variable speed synchronous generators with full-scale power converters, including variable speed operation for maximum power tracking, decoupled active and reactive power control, lower converter cost, and reduced power loss [3, 5]. With the development of nonlinear control theory, the corresponding methods can be applied for the control of doubly fed induction generator (DFIG). Among them, feedback linearization, integrator back-stepping and passivity based control are extensively applied for inductance electric machine [6]. The main DFIG's control topologies based on AC power generation for grid connected [7, 9] and stand-alone systems [10, 14] are widely available in literature. Fig.1 presents DFIG power control scheme in stand-alone mode. The popular control techniques, for control of DFIG, are stator flux based Field Oriented Control (FOC) and Direct Power Control (DPC). In [15] DPC is characterized by quick dynamic response, simple structure and low parameter

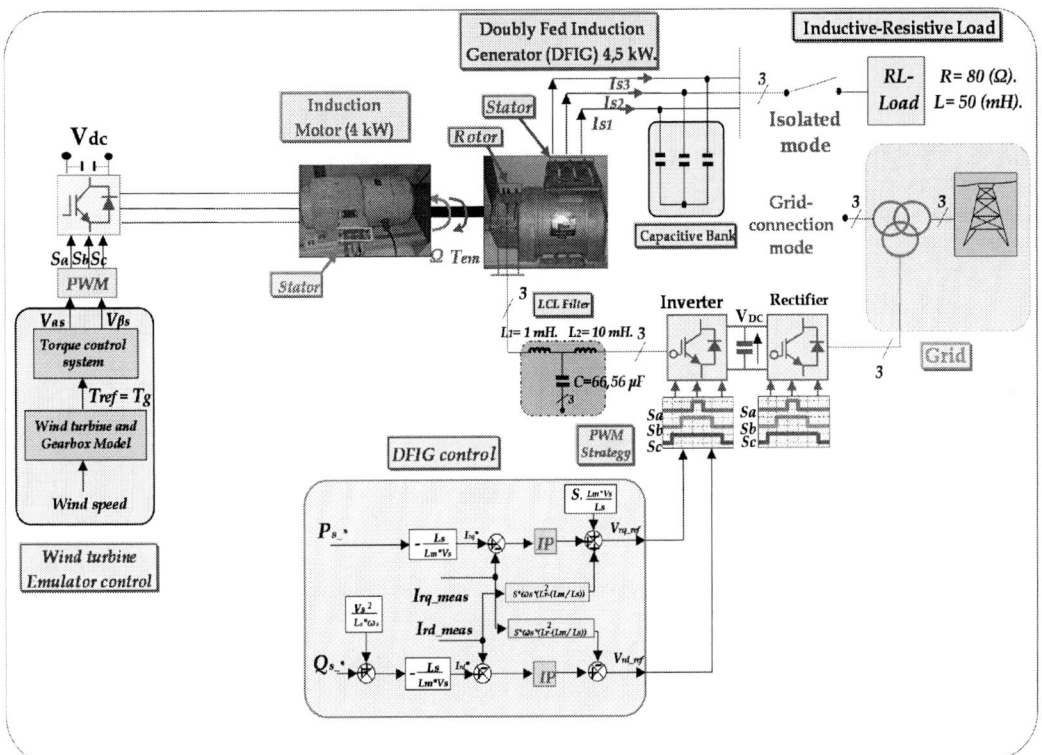

Fig.1-Proposed control scheme based on DFIG power control in stand-alone mode.

dependency. Actually, the Ps & Qs of the DFIG are controlled by regulating the amplitudes and phase angles of the stator and rotor fluxes. The application of the classical DPC control based on PI regulator generates some performance limitations' such as; power error, tracking power, over-shoot...etc. Integral-Proportional (IP) is a robust and simple controller which could be used instead the PI to avoid these drawbacks. The main contribution of this paper is the experimental validation of the DPC algorithm in the case of the stand-alone mode, under sudden wind speed variation and d-q axis rotor current changement tests, by the means the variation of the reference rotor direct and quadrature currents (I_{rd} & I_{rq}). This paper is organized as follows; firstly, the modeling of DFIG is presented in section-II. Section-III presents the proposed LCL filter which implemented between the DFIG' rotor and the inverter. The improved DPC of DFIG which based on IP controllers is illustrated in Section-IV. In section-V, experimental results are shown and discussed. Finally, the reported work is concluded.

2-Mathematical model of DFIG

The general electrical state model of the induction machine obtained using Park transformation [1-2]:
Stator and rotor voltages

$$V_{sd} = R_s.i_{sd} + \frac{d}{dt}\phi_{sd} - \omega_s.\phi_{sq} \tag{1}$$

$$V_{sq} = R_s.i_{sq} + \frac{d}{dt}\phi_{sq} - \omega_s.\phi_{sd} \tag{2}$$

$$V_{rd} = R_r.i_{rd} + \frac{d}{dt}\phi_{rd} - (\omega_s - \omega_r).\phi_{rq} \tag{3}$$

$$V_{rq} = R_r.i_{rq} + \frac{d}{dt}\phi_{rq} - (\omega_s - \omega_r).\phi_{rd} \tag{4}$$

Stator and rotor fluxes:

$$\phi_{sd} = L_s.i_{sd} + L_m.i_{rd} \tag{5}$$

$$\phi_{sq} = L_s.i_{sq} + L_m.i_{rq} \tag{6}$$

$$\phi_{rd} = L_r.i_{rd} + L_m.i_{sd} \tag{7}$$

$$\phi_{rq} = L_r.i_{rq} + L_m.i_{sq} \tag{8}$$

The electromagnetic torque is given by:

$$Tem = P.Lm.(ird.isq - irq.isd) \tag{9}$$

And its associated motion equation is:

$$Tem - Tr = J.\frac{d}{dt}\Omega + f.\Omega_{mec} \tag{10}$$

$$J = \frac{J_{turbine}}{G^2} + J_g \tag{11}$$

The DFIG model (as shown in Fig.3) can be described by the following state equations in the synchronous reference frame whose axis d is aligned with the stator flux vector as shown in Fig.2 [14].

The voltage equations and the flux equations of the stator winding can be simplified as follows [13]:

$$V_{sd} = 0 \tag{12}$$

$$V_{sq} = V_s \cong \omega_s.\emptyset_s \tag{13}$$

$$\emptyset_s = L_s.i_{sd} + L_m.i_{rd} \tag{14}$$

$$0 = L_s.i_{sq} + L_m.i_{rq} \tag{15}$$

The active and reactive powers at the stator side are defined as:

$$P_s = -V_s.\frac{L_m}{L_s}.i_{rq} \tag{16}$$

$$Q_s = \frac{V_s^2}{\omega_s * L_s} - V_s.\frac{L_m}{L_s}.i_{rd} \tag{17}$$

On the other hand the expression of the electromagnetic torque becomes:

$$T_{em} = P * \left(\frac{L_m}{L_s}\right) * \Phi_{sd} * I_{rq}. \tag{18}$$

Where; R_s, R_r, L_r, and L_s are respectively the resistances and the inductances of the stator and the rotor of the DFIG & L_m is mutual inductance, σ is leakage factor. V_{sd}, V_{sq}, V_{rd}, V_{rq}, I_{sd}, I_{sq}, I_{rd}, I_{rq}, Φ_{sd}, Φ_{sq}, Φ_{rd} & Φ_{rq} respectively represent the components along the d and q axes of the stator and rotor voltages, currents and flux. T_{em} and T_r present the electromagnetic, load, torques. J_g, $J_{turbine}$ and J are the generator, turbine and total inertia in DFIG's rotor respectively, Ω_{mec} is the mechanical speed, and G is the gain of gear box. P is number of pole pairs, ω_s is the stator pulsation, ω_r is the rotor pulsation, ω_{slip} is the slip pulsation and f is the friction coefficient. T_s and T_r are stator and rotor time-constant, and S is the slip $S = \frac{\omega_s - \omega_r}{\omega_s}$ and $\omega_s - \omega_r = S * \omega_s$. With: $T_r = \frac{L_r}{R_r}$; $T_s = \frac{L_s}{R_s}$; $\sigma = 1 - \frac{L_m^2}{L_s * L_r}$

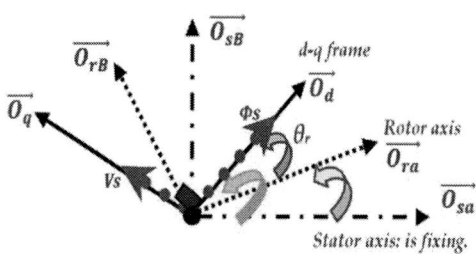

Fig.2-Stator and rotor flux vectors in the synchronous d-q frame.

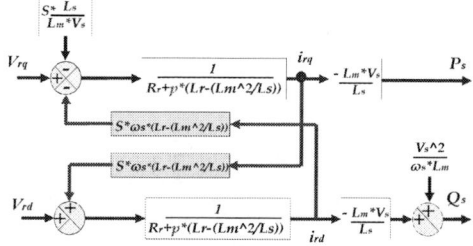

Fig.3-The doubly fed induction generator simplified model.

3-LCL Passive Filter

Transfer functions are the ratios between various input to output Laplace-transformed complex currents and voltages. The most pertinent transfer function for this LCL filter (knowing that the LCL parameters and values are described in Table.1 & Fig.4) is from the inverter voltage to the rotor current that is injected and is given by:

$$H_{LCL}(p) = \frac{I_{rotor}(p)}{V_{inv}(p)} = \frac{C.R_D.p + 1}{D_{LCL}} \tag{19}$$

The representation of equations: in state space form is:

$$\begin{pmatrix} \frac{dI_{Inv}}{dt} \\ \frac{d.I_{Rotor}}{dt} \\ \frac{dV_c}{dt} \end{pmatrix} = \begin{pmatrix} -\frac{(R_1 + R_{Inv})}{L_1} & \frac{R_D}{L_1} & -\frac{1}{L_1} \\ \frac{R_D}{L_2} & -\frac{(R_2 + R_D)}{L_2} & \frac{1}{L_2} \\ \frac{1}{C} & -\frac{1}{C} & 0 \end{pmatrix}.\begin{pmatrix} I_{Inv} \\ I_{Rotor} \\ V_c \end{pmatrix} + \begin{pmatrix} \frac{1}{L_1} & 0 \\ 0 & -\frac{1}{L_2} \\ 0 & 0 \end{pmatrix}.\begin{pmatrix} V_{Inv} \\ V_{Rotor} \end{pmatrix} \tag{20}$$

Neglecting all the three resistances, $H_{LCL}(p)$ becomes:

$$H_{LCL}(p) = \frac{1}{C.L_1.L_2.p^3 + (L_1 + L_2).p} \tag{21}$$

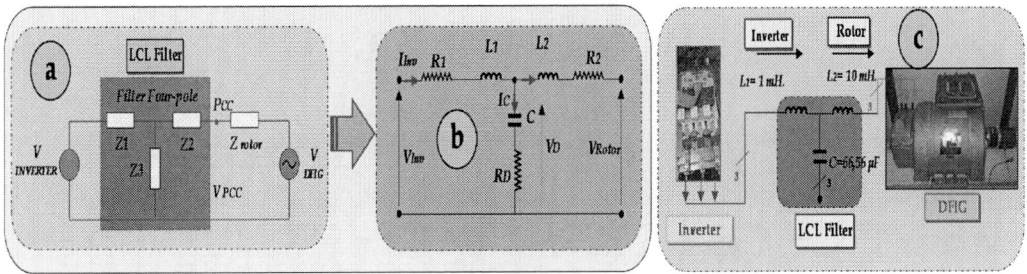

Fig. 4 "a & b": Single phase LCL-Filter schematic and "c" hardware LCL-filter implementation.

Table.1-LCL Filter's elements with their values.

LCL components:	Description and values:
V_{Inv}	Voltage inverter.
V_{Rotor}:	Rotor voltage (80 (V))
I_{Inv}:	Inverter current (A).
I_c:	Capacitance current (A).
I_{Rotor}:	Rotor current (10 (A)).
L_1:	Inverter side inductor (10 (mH)).
L_2:	Rotor side inductance (1 (mH)).
R_1:	Inverter side resistance (0.5 (Ω)).
R_2:	Rotor side resistance (0.2 (Ω)).
R_D:	Damping resistance (0.3 (Ω)).
C:	Capacitance (66.56 (µF)).

Fig.6-(with L and LCL filters) illustrate the impact of LCL filter compared to L filter in terms of stator and rotor voltage waveforms in steady states, FFT (fast fourier transform) displays the fundamental of Vs and Vr(V).

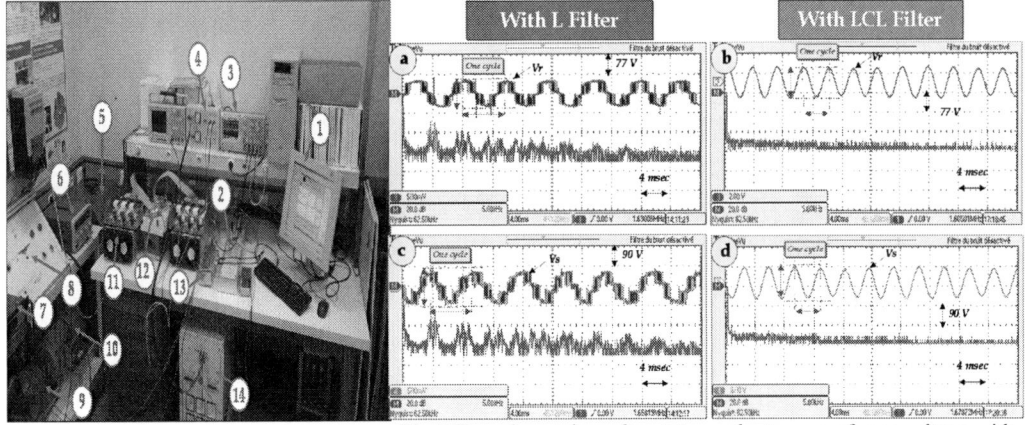

Fig.5-Experimental test bench developed in L2EP Laboratory.

Fig.6-Experimental results: stator and rotor waveforms voltage with L and LCL filter respectively.

4-Improved DPC based on Integral-Proportional (IP) Controllers

In this part IP controllers are proposed in order to control I_{rq} and I_{rd} respectively of DFIG (Fig.7-c). IP regulators are similar to PI regulators except that the proportional and integral actions are serialized unlike PI regulators, where these actions are paralleled. As described above, the system is first regulated by simplifying the system into a mono-variable model. The different devices and materials used in the tests bench are described in Fig.5. Thus the simplified model used for IP dimensioning is as flows (Figs.7-(a & b)), and the global proposed control scheme is described in Fig.7-(c). The closed loop transfer function (CLTF) with the IP controller is then written (by identification with a second order system of transfer function):

$$CLTF = \frac{Input_{ref}}{Output} = \frac{k * \omega_n^2}{p^2 + 2 * \xi * \omega_n * p + \omega_n^2} \tag{22}$$

The gains of the correctors will be expressed as a function of the parameters of the machine as follows:

$$\begin{cases} k_i = \dfrac{\omega_n^2 * (L_s * L_r - L_m^2)}{k_p * L_m * V_s} \\ k_p = \dfrac{2 * \xi * \omega_n * (L_s * L_r - L_m^2) - L_s * R_r}{L_m * V_s} \end{cases} \tag{23}$$

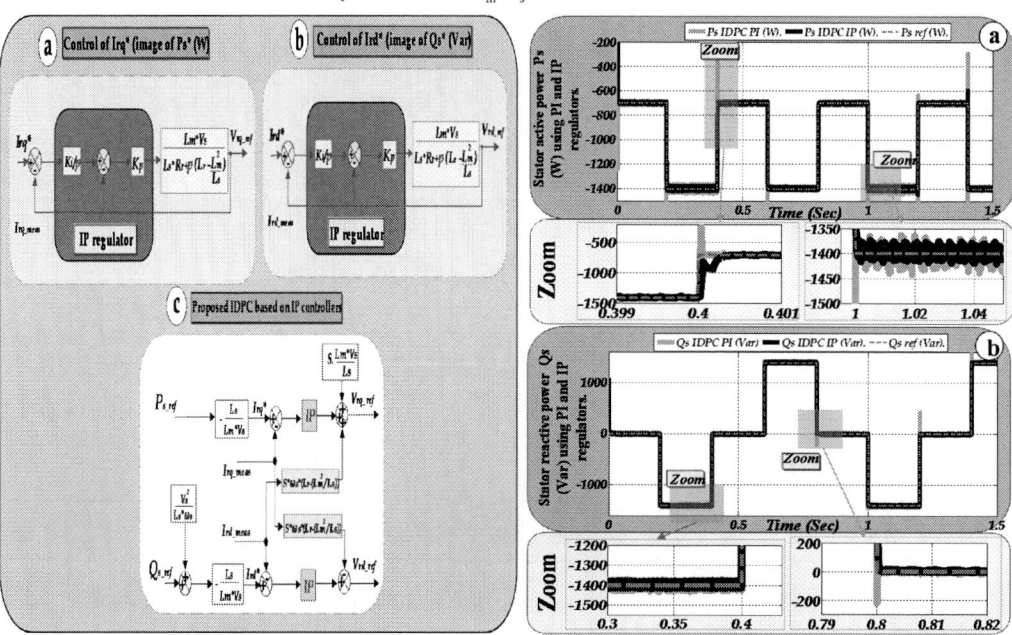

Fig. 7 Power control structure based on IP regulator (according to d and q axis).

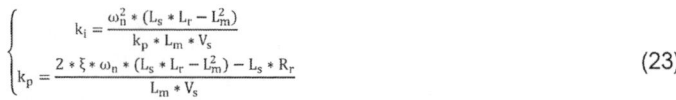

Fig. 8 "a and b": stator active and reactive power controlled by 'PI and IP' controllers respectively.

From equation: (18), it is clear that the torque can be controlled by acting on the rotor quadrature current component (I_{rq}) of the *DFIG*. In a similar manner, the rotor direct current component (I_{rd}) is used to control the generated reactive power.

$$Q_s^* = \frac{V_s^2}{\omega_s * L_s} - V_s * \frac{L_m}{L_s} * I_{rd}^* \tag{24}$$

$$I_{rq}^* = -\frac{L_s}{P. L_m. \Phi_s} * T_{em}^* \tag{25}$$

$$I_{rd}^* = \frac{\Phi_s}{L_m.} - \frac{L_s}{V_s. L_m} * Q_s^* \tag{26}$$

$$I_{rq}^* = \frac{L_s}{V_s. L_m} * P_s^* \tag{27}$$

The analysis in S-domain shows that (PI) and (I-P) controllers have the same characteristic equations, and it can be seen that the zero introduced by the (P-I) controller is absent in the case of the (I-P) controller.

Therefore the overshoot in the current for a step change in the input reference R(s), is expected to be smaller for the (I-P) controller. The IP gains values which are used in this work are presented in Table.5. The control of the DFIG through the DC-AC converter connected to its rotor must ensure the necessary torque to vary the mechanical generator speed (Ω_{mec}) in order to extract the maximum power, by imposing adequate rotor voltages to the DFIG [11]. Fig.8-(a & b) presents simulation results of the Stator active' and reactive' powers using PI and IP controllers (to control I_{rq} & I_{rd} respectively).

5-Experimental Results

Fig.5 presents the experimental test bench, the DFIG used in this real time implementation is a 4.5 kW (refer to Table.3) and the wind turbine presented by the Induction Machine is a 4.0 kW (refer to Table.4).

- **Robustness tests (d-q axis rotor currents variation):**

A- **Trapezoid and step forms:**

Figs.9-(a & b) shows experimental results of rotor direct and quadrature currents ($I_{rd}*$ and $I_{rq}*$) variation. It can be seen that the d-q axis rotor measured currents (I_{rd_meas} & I_{rq_meas}) follow exately their references respectively despite the sudden variation.

B- Sudden wind speed variation:

Figs.10-(a and b) illustrate the rotor speed below/above the synchronous speed (sub and super-synchronous operation modes), and the rotor sinusoidal current variation in transient and steady states (I_{ra_meas}) phase 'A'.

Table.2 presents brief review of power control in Stand-alone mode (which based only on experimental studies). Many criteria were been taken into account in order to show the advantage/disadvantages of each control strategy.

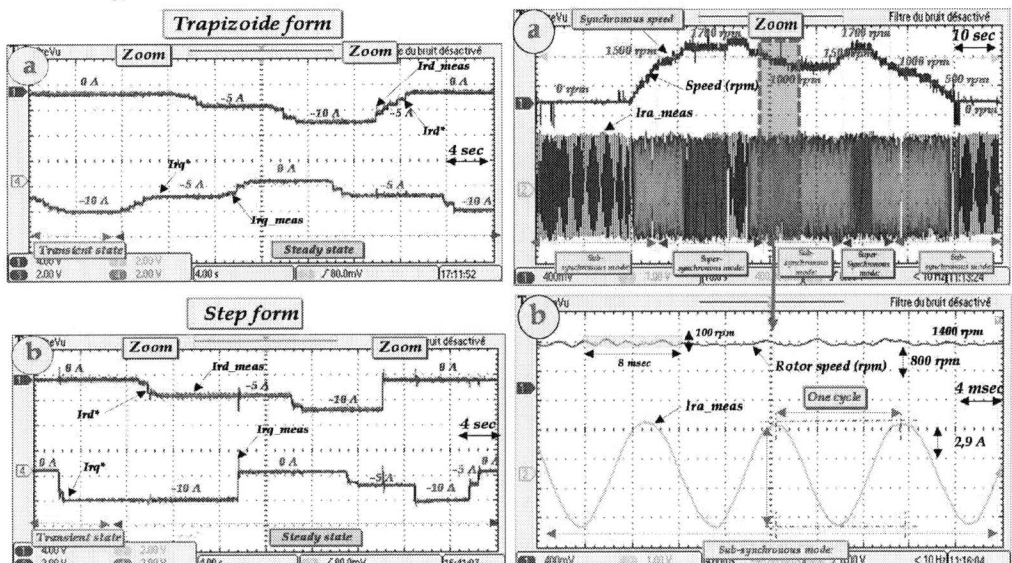

Fig.9 Robustness tests under trapezoid and step forms (rotor currents variation with RL-load).

Fig.10-A: Rotor speed variation and measured rotor current and **B:** Zoom of stable zone ≈1450 rpm.

Table.2- The review of power control in Stand-alone mode (only experimental studies).

References	Control strategy	Current/voltage controller	Load nature	Generator		Power/ current error	Robustness		Filter type:
				Type	Rated power		Tracking reference regardless power /current variation	to Wind speed variation	
[10]	VOC	PI-R	Non-linear load	DFIG	/	/	++	++	/
[11]	VOC	PR3	Nonlinear load	DFIG	2.2 KW	/	++	++	L
[12]	FOC	Hysteresis	Non-linear load	DFIG	3.5 KW	+/- 0.01 A	++	++	LC
[13]	VOC	PI	Resistive load	DFIG	7.5 KW	-	++	++	L
[16]	PDPC	PI	Nonlinear load	DFIG	20 KW	-	+++	++	/
[17]	VOC	PI	Nonlinear load	DFIG	5.5 KW	-	++	++	/
[18]	FOC	PI	Nonlinear load	DFIG	3.2 KW	-	+++	++	/
[19]	FOC	PI	Nonlinear load	DFIG	5.5 KW	-	++	++	L
[20]	FOC	PI-R	Unbalanced load	DFIG	2.2 KW	-	+++	++	L
Proposed	DPC	IP	RL load	DFIG	4.5 KW	+/- 0.1 A	+++	++	LCL

6-Conclusion

In this paper, a novel DPC of DFIG in stand-alone mode has been proposed and tested in real time. IP controllers were been proposed instead conventional ones "PI" in order to control I_{rd} and I_{rq} (images of Qs and Ps respectively). Several experimental tests have been established by rotor speed and currents variation' in order to display the Sub- and Super-synchronous operations, and on the other hand to be sure that the proposed control is able to work in different wind-speed conditions. Finally, experimental results proved high performances in terms of response time, overshoot, current/and power errors.

Appendix

Table.3-Parameters of the DFIG.

Rated Power:	4.5 KW
Stator Resistance:	Rs = 0.4 Ω
Rotor Resistance:	Rr = 0.8 Ω
Stator Inductance:	Ls = 0.082 H.
Rotor Inductance:	Lr = 0.082 H.
Mutual Inductance:	Lm = 0.081 H.
Rated Voltage Δ/Y:	Vs = 220/380 V
Number of Pole pairs:	P= 2
Rated Speed:	N=1395 rpm
Friction Coefficient:	f_{DFIG}=0.001 N.m/sec
The moment of inertia	J=0.2 kg.m^2

Table.4-Parameters of the Turbine (Squirrel cage Induction Machine).

Rated Power:	4.0 KW
Rated Voltage (Vs) Δ/Y:	Vs= 220/380 V.
Cosφ:	0.82
Gain:	G=3.9
Rated Speed:	N= 1440-2000 rpm

Table.5-IP Controller Parameters used in Experimental Study.

IP Parameters:	Gain values:
Ki:	10.
Kp:	30

References

[1] F. Amrane, A. Chaiba, B. Babes and S. Mekhilef, "Design and Implementation of high Performance Field Oriented Control for Grid-connected Doubly Fed Induction Generator via Hysteresis Rotor Current Controller", Rev. Roum. Sci. Techn.– Électrotechn. et Énerg. Vol. 61, no.4, pp.319-324, 2016.

[2] F. Amrane and A. Chaiba, "A Novel Direct Power Control for grid-connected Doubly Fed Induction Generator based on Hybrid Artificial Intelligent Control with Space Vector Modulation", Rev. Roum. Sci. Techn.– Électrotechn. et Énerg, no.3, Vol. 61, 2016.

[3] M. Jujawa," Large wind rising", *Renewable Energy World*, no. 2, Vol: 6, Mars/April 2003, pp.39-51.

[4] J. Yao, H. Li, Y. Liao, and Z. Chen, "An improved control strategy of limiting the DC-link voltage fluctuation for a doubly fed induction wind generator" *IEEE Transaction on Power Electronics*, Vol: 23, no. 3, pp. 1205-1213, 2008.

[5] J. Guo, X. Cai, and Y. Gong, "Decoupled control of active and reactive power for a grid-connected doubly-fed induction generator," Nanjing, China, pp. 2620-2625, 2008.

[6] H. Stemmler, P. Geggenbach, "Configurations of high power voltage source inverter drives", proceeding of the 5th European conference on power electronics Brighton, UK, Vol: 5, pp 7–12, 1993.

[7] D. Sun and X. Wang, "Sliding-mode DPC using SOGI for DFIG under unbalanced grid condition", IEEE Electronics Letters, vol: 53, no: 10, pp: 674–676, 2017.

[8] W. Leonhard, "Control of Electrical Dives", 3rd ed. Berlin, Germany: Springer Verlag, 2003.

[9] R. Datta and V. T. Ranganathan, "Direct power control of grid-connected wound rotor induction machine without rotor position sensors" IEEE Transactions on Power Electronics, vol: 16, no: 8, pp.: 390–399, May. 2001.

[10] V.-T. Phan, D.-T. Nguyen, Q.-N. Trinh, C.-L. Nguyen and T. Logenthiran, "Harmonics Rejection in Stand-Alone Doubly-Fed Induction Generators With Nonlinear Loads", IEEE Transactions on Energy Conversion, vol: 31, no:2, pp. 815-817, 2016.

[11] V. -T. Phan and H. -H. Lee, "Stationary frame control scheme for a stand-alone doubly fed induction generator system with effective harmonic voltages rejection", IET Electric Power Applications, vol: 5, no: 9, pp. 697-707, 2011.

[12] F. Amrane, A. Chaiba, B. Francois and B. Babes, "Experimental Design of Stand-alone Field Oriented Control for WECS in Variable Speed DFIG-based on Hysteresis Current Controller", IEEE 2017 15th International Conference on Electrical Machines, Drives and Power, Bulgaria, 2017.

[13] R. Pena, J. C. Clare, and G. M. Asher, "A doubly fed induction generator using back-to-back PWM converters supplying an isolated load from a variable speed wind turbine" IEE Proceeding. - Electronics Power Applications, vol. 143, no. 1, pp. 380–387, 1996.

[14] R. Cardenas; R. Pena; S. Alepuz and G. Asher, "Overview of Control Systems for the Operation of DFIGs in Wind Energy Applications", IEEE Transactions on Industrial Electronics, vol: 60, no: 7, pp.: 2776–2798, 2013.

[15] L. Xu and P. Cartwright, "Direct Active and Reactive Power Control of DFIG for Wind Energy Generation", IEEE Transactions on Energy Conversion, vol. 21, no: 3, 2006.

[16] Y. Zhang, J. Hu and J. Zhu, "Three-Vectors-Based Predictive Direct Power Control of the Doubly Fed Induction Generator for Wind Energy Applications", IEEE Transactions on Power Electronics, vol: 29, no: 7, pp.: 3485-3500, 2014.

[17] H. Misra, A. Gundavarapu and A. K. Jain, "Control Scheme for DC Voltage Regulation of Stand-Alone DFIG-DC System", IEEE Transactions on Industrial Electronics, vol: 64, no: 4, pp.: 2700-2708, DOI: 10.1109/TIE.2016.2632066, 2017.

[18] G. D. Marques, and M. F. Iacchetti, "Sensorless Frequency and Voltage Control in Stand-Alone DFIG-DC System", IEEE Transactions on Industrial Electronics, vol: 64, no: 3, pp.: 1949-1957, DOI: 10.1109/TIE.2016.2624262, 2017.

[19] H. Misra and A. K. Jain, "Analysis of Stand-Alone DFIG-DC System and DC Voltage Regulation With Reduced Sensors", IEEE Transactions on Industrial Electronics, vol: 64, no: 6, pp.: 4402-4412, DOI: 10.1109/TIE.2017.2669889, 2017.

[20] V. Phan and H. Lee, "Performance Enhancement of Stand-Alone DFIG Systems With Control of Rotor and Load Side Converters Using Resonant Controllers," in IEEE Transactions on Industry Applications, vol. 48, no. 1, pp. 199-210, Jan.-Feb. 2012.

Influence of Wire-Bonding Layout on Reliability in IGBT Module

Lubin Han[1], Lin Liang[1*], Wei Xin[2] and Fang Luo[3]

[1]State Key Laboratory of Advanced Electromagnetic Engineering and Technology, School of Electrical and Electronic Engineering, Huazhong University of Science and Technology, Wuhan, China.

[2]School of Optical and Electronic Information, Huazhong University of Science and Technology, Wuhan, China

[3]Department of Electrical Engineering, University of Arkansas, 72701 Fayetteville, USA

*Email: lianglin@hust.edu.cn

Acknowledgements

The authors acknowledge the financial support by National Key R&D Program of China (2018YFB0905700).

Keywords

« Power semiconductor device», «Packaging», «Reliability», «Simulation», «Power cycling»

Abstract

The wire-bonding layouts influence the profiles of current density, ohmic loss and junction temperature. The temperature profile across IGBT chip is different when the wire-bonding layouts are different. Reasonable wire-bonding layout could reduce the maximum junction temperature (T_{jmax}) of the chip and slow down the aging process of the solder and bonding wires. The uniform wire-bonding layout is proposed in this paper to improve the packaging reliability of the power module. To analyze the influence of wire-bonding layouts on reliability, the centralized wire-bonding layout and uniform wire-bonding layout of bonding wires are compared. The results of the electro-thermal simulation show that using uniform wire-bonding layout reduces the T_{jmax} of the chip by 1~3 °C. What's more, as distance between wire-bonding point and chip center increases, there is a minimum value of T_{jmax} in the uniform wire-bonding layout. The analytical model is deduced to study the phenomenon. To verify the simulation and theoretical analysis results, the IGBT modules with centralized wire-bonding layout and uniform wire-bonding layout of bonding wire are tested by power cycling tests (PCT). The experimental results show that the T_{jmax} of the module with uniform wire-bonding layout of bonding wire is 3.3 °C lower than that of the module with centralized wire-bonding layout. Correspondingly, with the help of LESIT lifetime model, the calculated lifetime of the IGBT module with uniform wire-bonding layout increases by 1/3, compared with centralized wire-bonding layout.

Introduction

The bonding wires are commonly used in commercial power devices to achieve electrical connection. The cracks and lift-off of bonding wire become the most common failure modes of IGBT module due to repetitive thermal stress [1-2]. The top side of chip is exposed in encapsulation, so it is impossible to dissipate power loss like the backside. However, the extremely small interface area between the chip and the bonding wire results in uneven current density, loss and temperature distribution. The module with lower maximum junction temperature (T_{jmax}) and junction temperature difference (ΔT_j) could achieve longer lifetime. Therefore, the layout of bonding wire should be carefully designed to maximize the thermal conduction area and reduce the T_{jmax} [3]. Multiple bonding wires (14 wires, 2.2 mm² per channel) are attached to metallization of device instead of double bond wire [4], which could achieve a lower junction temperature and more uniform temperature distribution. However, the increased number of bonding wires and copper nail head structure increase the cost of bonding process. It is found that the avalanche reliability of MOSFET devices with higher rated power is strongly dependent on wire-bonding layouts [5]. Nevertheless, it does not explain the mechanism or study the aging reliability of module with different wire-bonding layouts. The temperature distributions of power devices with different wire-bonding layouts are researched in [3], and a simple transmission line 1-D model is used to explain the non-uniform Joule heating in power devices. However, the optimum wire-bonding layout is not analyzed and the thermal impedance is not include in the model. In this paper, the reliability and lifetime of IGBT module with two types of wire-bonding layouts are compared. Furthermore, the electro-thermal models of IGBT module with the two types are established to simulate the influence of wire-bonding layouts on the junction temperature distribution. The analytical model of effective thermal conduction is used to explain the effect of wire-bonding layout and find the optimum layout. Two customized IGBT models with different wire-bonding layouts are tested by power cycling test (PCT). The experimental results verify the simulation results, and the corresponding lifetime results are predicted and compared.

Electro-thermal simulation and analysis

A. Finite element models

To accurately simulate the influence of wire-bonding layout on the junction temperature of IGBT chip, a 7-layer finite element model of IGBT module is built, as shown in Fig. 1. The material properties of each layer are shown in Table I. The conductivity of IGBT chip is calculated by on-state resistance at rated current and chip size.

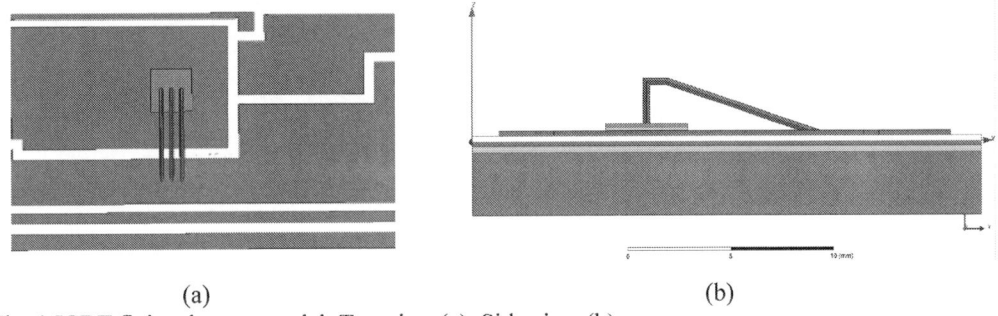

(a)   (b)

Fig. 1 IGBT finite element model. Top view (a). Side view (b).

Table I: Material parameters of IGBT module used in simulation

Material	Size (mm³)	Density (kg/m³×10³)	Conductivity (S/m)	Thermal Conductivity (W/(m·°C))
Electrode	4×4×0.005	2.7	$3.8×10^7$	238.6
IGBT chip	4×4×0.2	2.33	200	124
Chip solder	4×4×0.15	10.9	$7×10^6$	24
Upper copper	56×25×0.25	8.3	$5.8×10^7$	401
Ceramic	56×25×0.25	3.8	0	27
Lower copper	56×25×0.2	8.3	$5.8×10^7$	401
Baseplate solder	56×25×0.3	8.1	$7×10^6$	24
Baseplate	56×25×3	8.3	$5.8×10^7$	401

Generally, the diameter and number of bonding wires are determined by the rated current of chips. For simplification, each chip with three bonding wires is assumed to analyze the influence of bonding layout. The wire-bonding layout designs of higher rated current chips could follow the case in this paper. In commercial modules, the bonding wires are typically bonded along a straight line, which is called centralized wire-bonding layout, as shown in Fig. 2 (a). Centralized wire-bonding layout easily causes current crowding and temperature concentration, so the uniform distribution of wire-bonding points is proposed to avoid these problems, which is called uniform wire-bonding layout, as shown in Fig. 2 (b). In electro-thermal simulation, the distances between the bonding points in the centralized wire-bonding layout are fixed, while the distances between the bonding points and the chip center are adjusted in uniform wire-bonding layout. The five models with different layouts are shown in Fig. 3.

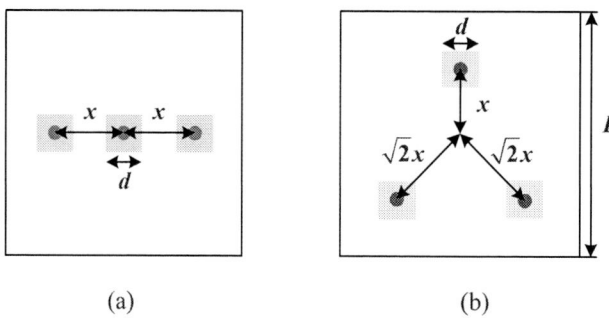

(a)   (b)

Fig. 2 Wire-bonding layout. Centralized wire-bonding layout (a). Uniform wire-bonding layout (b).

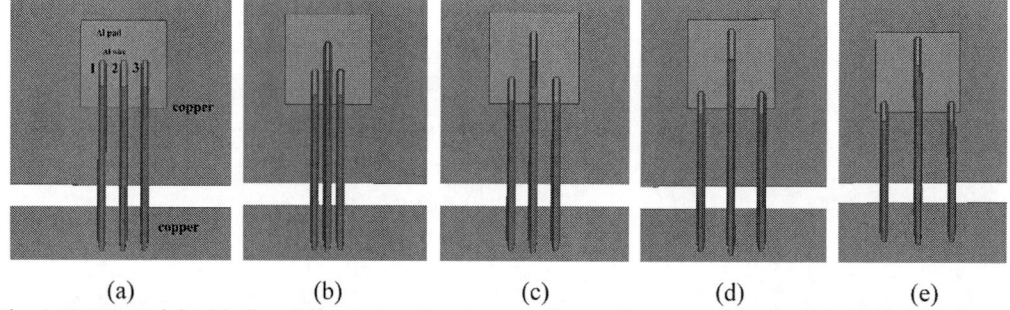

(a)   (b)   (c)   (d)   (e)

Fig. 3 IGBT model with five different bonding layouts. Centralized wire-bonding layout (a). Uniform wire-bonding layout with x=0.6mm (b), x=1mm (c), x=1.4mm (d) and x=1.6mm (e).

B. Simulation results

The current of 15A excitation flows the bonding wires and the chip, and the power loss generated by current excitation is transferred to the steady-state thermal model to simulate the temperature distribution of the module. Fig. 4 (a) shows the current density of the chip in the centralized wire-bonding layout. It can be seen that the current density decays from the bonding point. The uneven current density leads to the inhomogeneous power loss on the chip. What's more, the power density is concentrated at the bonding point, as shown in Fig. 4 (b) and Fig. 5. In addition, the electrical resistance at bonding wires also accounts for the high ohmic loss adjacent to bonding wires [6], especially at high temperature.

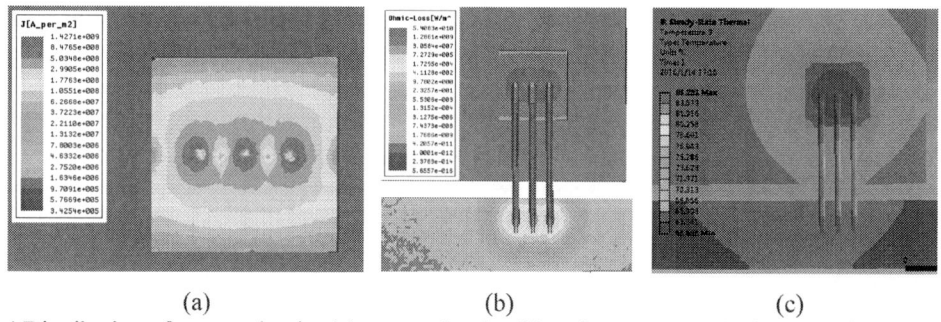

(a)   (b)   (c)

Fig. 4 Distribution of current density (a), power density (b) and temperature (c) in centralized wire-bonding layout.

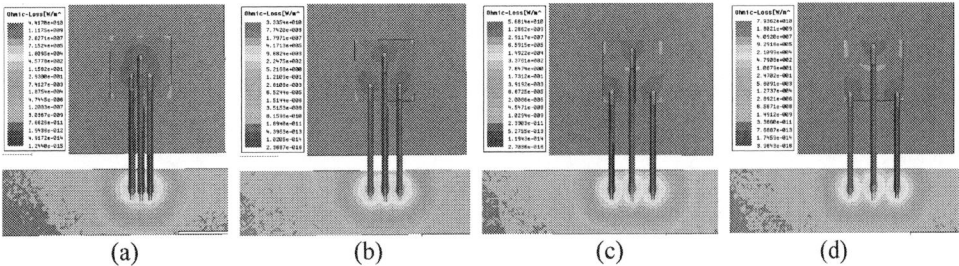

Fig. 5 Power density distribution in uniform wire-bonding layout.

In the steady-state thermal model, the convection coefficient of copper baseplate is set as 100 W/m$^2 \cdot$K [7]. The ambient temperature is set at 22 °C. The simulation results of temperature distribution are shown in Fig. 4 (c) and Fig. 6. The maximum temperature is concentrated in the chip center. However, the inhomogeneous loss caused by wire-bonding layout affects the temperature concentration of the chip and ultimately the T_{jmax}.

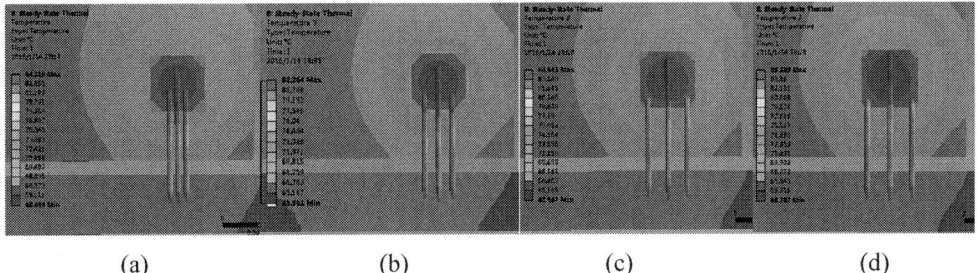

Fig. 6 Temperature distribution in uniform wire-bonding layout.

C. Analysis and discussion

The T_{jmax} is extracted from Fig. 4 (c) and Fig. 6, and shown in Fig. 7. The T_{jmax} of the centralized wire-bonding layout is higher than that of the uniform one, which indicates that the power loss dissipation capability of the centralized is poor. In uniform wire-bonding layout, with the increase of the distance between the bonding point and the chip center, the T_{jmax} decreases and then increases.

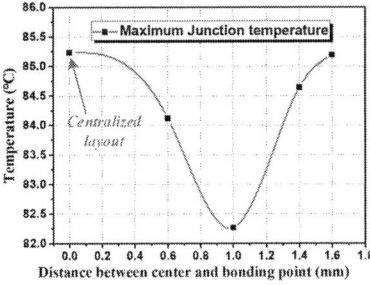

Fig. 7 T_{jmax} in different bonding layouts.

In the centralized wire-bonding layout, the power density is concentrated in the chip center, while in the uniform one, the power density is scattered on the chip, as shown in Fig. 4 (b) and Fig. 5. In addition, the effect of bonding layout on T_{jmax} can also be explained by the effective thermal conduction area (ETCA). It is assumed that the effective thermal conduction region is a square region centered on the bonding points as shown in Fig. 2. The relationship between ETCA of the two types of wire-bonding layouts ($S_{centralized}$ and $S_{uniform}$) and the distance (x) between the bonding point and the chip center is expressed as follows:

$$S_{\text{centralized}} = \begin{cases} 4dx + d^2 & 0 \le x < 0.5 \cdot d, \\ 3d^2 & 0.5 \cdot d \le x < 0.5 \cdot (L-d), \\ 2x^2 - (2L+3d)x + 0.5L^2 + d^2 + 1.5Ld & 0.5 \cdot (L-d) \le x < 0.5 \cdot L \end{cases} \tag{1}$$

$$S_{\text{uniform}} = \begin{cases} 2dx + d^2 & 0 \le x < 0.5 \cdot (L-d), \\ Ld & 0.5 \cdot (L-d) \le x < d, \\ -2dx + 2d^2 + Ld & d \le x < 0.5 \cdot L \end{cases} \tag{2}$$

Where, d is the side length of the square thermal conduction region, x is the distance between the bonding point and the chip center, and L is the chip side length. The L is fixed as 4 mm, while assuming d is 0.8 mm, 1 mm and 1.2 mm respectively. The curves of ECTA in two types of wire-bonding layouts with x under different d are shown in Fig. 8.

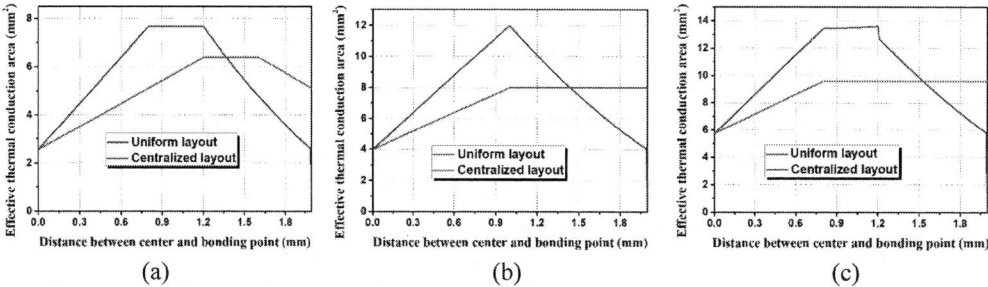

Fig. 8 The curves of ECTA in two bonding layouts with x when d=0.8mm (a), 1mm (b) and 1.2mm (c).

The ECTA relationship between uniform and centralized layouts in Fig. 8 (c) is consistent with the relationship of junction temperature between uniform and centralized layouts, which means that the practical value of d is about 1.2mm. Under different d, the uniform wire-bonding layout could achieve greater ECTA, compared with the centralized. Further, in steady state, the power loss flows through each layer of IGBT module. Moreover, it can be seen from (3) that when the ETCA increases, the temperature difference between the junction and the case decreases, so the T_{jmax} decreases as well.

$$\frac{\Delta T}{P} = \sum_{i=1}^{n} \frac{h_i}{\lambda_i \cdot S} \tag{3}$$

As analyzed above, regardless of diameter and number of bonding wires, the dispersive ohmic loss and greater ETCA in uniform wire-bonding layout account for the lower ΔT_{j}. When the diameter and number of bonding wires are different, the benefit of uniform wire-bonding layout is also varied.

Power cycling test and lifetime evaluation

According to the simulation results, two types of IGBT modules with the centralized and uniform wire-bonding layout are made. The device under test (DUT) is 650V/30A, and each IGBT chip is connected with three bonding wires. The layout of bonding wire after fabrication is shown in Fig. 9.

Fig. 9 Customized IGBT modules with centralized wire-bonding layout (a) and uniform wire-bonding layout (b).

To verify the influence of wire-bonding layout on the reliability of IGBT module, the two fabricated IGBT modules are tested by PCT [8-9]. The test conditions satisfy the PCT standard [10]. The test period is 12s with 50% duty cycle, and load current is 20A. The experimental topology of the power cycling is shown in Fig. 10 (a). The devices are cooled down by forced air-cooling through fans. The insulating black paint is evenly sprayed on the chip surface to capture the temperature of the two groups of modules. To verify and compare with the simulation results of junction temperature, the PCTs are performed until steady state. The temperature distribution of two groups of IGBT modules is shown in Fig. 10 (b) and (c). The T_{jmax} of uniform wire-bonding layout is 3.3 °C lower than that of centralized wire-bonding layout. Except for wire-bonding layouts, the test conditions, module materials and chips of two types fabricated module are the same. Therefore, the different bonding layouts induce the T_{jmax} difference between centralized and uniform wire-bonding layout. The experimental results are consistent with the simulation results.

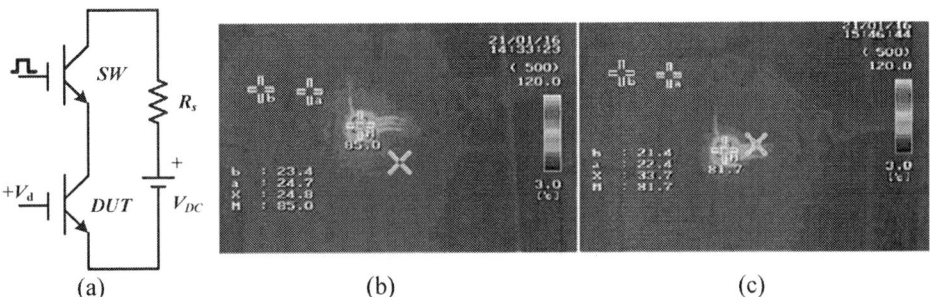

(a) (b) (c)

Fig. 10 Schematic of PCT setup (a). Temperature distribution of centralized wire-bonding layout (b) and uniform wire-bonding layout (c).

ΔT_j is the dominating factor that influences the lifetime of power module [11], compared with other factors, such as wires diameter, T_{jmax}, on-state time, current, voltage, etc. What's more, all the above factors in two layouts are the same. The contributions of these factors to lifetime are the same as well. Therefore, these factors could be considered as constant in the lifetime models. Based on above consideration, the Coffin-Manson model is used to predict the lifetime.

Since the fabricated modules are not the standard package module, the electrical and thermal conduction and failure may be different from that of commercial module. Therefore, without losing generality, the lifetime should be calculated from lifetime model of standard module to compare the influence of wire-bonding layout on reliability in IGBT module. In this paper, the parameters of lifetime model fitted by LESIT study are used to evaluate the lifetime of centralized and uniform wire-bonding layout. The lifetime model is,

$$N_f = \alpha \cdot \Delta T_j^{\beta} \tag{4}$$

Where N_f is the number of cycles when IGBT module fails. The α and β are coefficients determined by PCT. According to the test results of the standard test in the LESIT study [12], $\alpha = 2.2 \times 10^{14}$ and $\beta = -5$ are taken. ΔT_j is the difference between the T_{jmax} and the T_{jmin}. According to the PCT results, the lifetime calculation results of two groups of IGBT modules are shown in Table II. Although the ΔT_j difference between centralized and uniform wire-bonding layouts is only 2.5°C, uniform wire-bonding layout improve IGBT module lifetime by one third.

Table II: The lifetime of IGBT module with centralized wire-bonding layout and uniform wire-bonding layout

	T_{jmax} (°C)	T_{jmin} (°C)	ΔT_j (°C)	Lifetime (cycles)
Centralized wire-bonding layout	85.0	37.4	47.6	900k
Uniform wire-bonding layout	81.7	36.6	45.1	1179k

The fabricated module of Fig. 9 (b) is the optimal case where the T_{jmax} is minimum, as shown in Fig. 6, meaning that the calculated lifetime of uniform layout in TABLE II is the peak value. If the deviation x increases or decreases in Fig. 2 (b), the calculated lifetime of uniform wire-bonding layout may decrease to that of centralized wire-bonding layout due to the increasing ΔT_j. Therefore, the configuration of uniform wire-bonding layout should be carefully designed to achieve a long lifetime.

Conclusion

In this paper, the influences of the centralized and uniform wire-bonding layout on the reliability of IGBT modules are compared from electro-thermal simulation, theoretical analysis and experimental verification. The simulation results show that using uniform wire-bonding layout, the T_{jmax} can be reduced by $1 \sim 3$ °C. Adjusting the distances between bonding points, T_{jmax} could achieve a minimum value. The analytical model of effective thermal conduction area explains the effect of wire-bonding layout on junction temperature distribution. In the PCT, the T_{jmax} of IGBT module with uniform wire-bonding layout is 3.3 °C lower than that of IGBT module with centralized wire-bonding layout. Correspondingly, predicted by Coffin-Manson model, the lifetime of IGBT module with uniform wire-bonding layout is one third longer than that of IGBT module with centralized wire-bonding layout.

References

[1] K. B. Pedersen, K. Pedersen : Dynamic modeling method of electro-thermo-mechanical degradation in IGBT modules, *IEEE Transactions on Power Electronics*, vol. 31, no. 2, pp. 975-986, Feb. 2016.

[2] S. Ramminger, N. Seliger, and G. Wachutka : Reliability model for Al wire bonds subjected to heel crack failures, *Microelectronics Reliability*, vol. 40, no. 8, pp.1521-1525, Aug.- Oct. 2000.

[3] M. Ishiko, M. Usui, T. Ohuchi, and M. Shirai : Design concept for wire-bonding reliability improvement by optimizing position in power devices, *Microelectronics Journal*, vol. 37, no. 3, pp. 262-268, Mar. 2006.

[4] H. Köck, et al: Improved thermal management of low voltage power devices with optimized bond wire positions, Microelectronics Reliability, vol. 51, no. 9-11, pp. 1913-1918, Jun. 2011.

[5] H. Liu, S. Yang and W. Tang : Effects of thick Al wires bonding layout on reliability of power devices, *International Conference on Quality, Reliability, Risk, Maintenance, and Safety Engineering*, Jun 2011.

[6] Y. Chen, X. Wu, I. Fedchenia, M. Gorbounov, V. Blasko, W. Veronesi, and C. Slade : A comprehensive analytical and experimental investigation of wire bond life for IGBT modules, *IEEE Applied Power Electronics Conference and Exposition (APEC)*, Feb. 2012, pp. 2298-2304.

[7] HK. Tseng, ML. Wu : Electro-Thermal-Mechanical Modeling of Wire-bonding Failures in IGBT, *International Microsystems, Packaging, Assembly and Circuits Technology Conference*, Jan. 2013, pp. 152-157.

[8] T. Y. Hung, C. C. Wang, and K. N. Chiang : Bonding Wire Life Prediction Model of the Power Module under Power Cycling Test, *International Conference on Thermal, Mechanical and Multi-Physics Simulation and Experiments in Microelectronics and Microsystems*, Apr. 2013, pp. 1-6.

[9] U. M. Choi, S. Joergensen, and F. Blaabjerg : Advanced Accelerated Power Cycling Test for Reliability Investigation of Power Device Modules, *IEEE Transactions on Power Electronics*, vol. 31, no. 12, pp. 8371-8386, Mar. 2016.

[10] Semiconductor devices-mechanical and climatic test methods-power cycling IEC 60747-34, *IEC International Standard*, UK, 2005.

[11] R. Bayerer, T. Herrmann, T. Licht, J. Lutz, M. Feller : Model for Power Cycling lifetime of IGBT Modules-various factors influencing lifetime, *International Conference on Integrated Power Electronics Systems*, Mar. 2008.

[12] M. Held, P. Jacob, G. Nicoletti, P. Scacco, and M. H. Poech : Fast power cycling test of IGBT modules in traction application, *Second International Conference on Power Electronics and Drive Systems*, 1997, pp. 425–430.

Rail Potential Calculation Model for DC Railway Power Supply Equipped with Voltage Limiting Device

Shota KIMURA, Tsutomu MIYAUCHI, Kenji OGUMA,
Hirotaka TAKAHASHI, Keiko TERAMURA
Hitachi, Ltd.
832-2, Horiguchi, Hitachinaka, 312-0034,
Ibaraki, Japan
Tel.: +81-70-4209-4696
Fax: +81-29-353-3865
E-mail: shota.kimura.gc@hitachi.com
URL: http://www.hitachi.com

Keywords

«Simulation», «Modeling», «DC power supply»

Abstract

In DC electric railways, voltage is generated between rail and earth, i.e., rail potential. To keep the voltage below the permissible voltage limit specified by the international standard IEC 62128, a voltage limiting device (VLD) is added to railway power supply equipment. A VLD is a switch that connects the rail and the earth. The switch is closed when the rail potential rises above a permissible voltage. This lowers the voltage of the rail potential because the rail and the earth are shorted by the closed switch. We added a VLD model to a rail potential calculation model to evaluate the VLD's affection. The VLD was modeled as variable resistance that changed the resistance value depending on the open/closed status of the switch. Once the voltage between the rail and the earth exceeds the predefined limit, the status changes to close and stays closed for a predefined time before it is reopened. The results of the case study show that rail potential is reduced to almost zero by closing multiple VLDs installed in the line. However, stray current increases because it flows through the closed VLDs. The results obtained from the actual route will be compared with the simulation results in future work.

Introduction

In DC electric railways, the current travels to the trains from substations through a trolley and then returns to substations through the rail. The current that returns to the substations is called the "return current". A floating rail system is used to prevent electrolytic corrosion [1]. However, the resistance of the parts in contact with the ground, such as the sleepers and rail fastening devices, is not infinite. As a result, some of the return current leaks from the rail into the ground and returns to the substations through the ground, i.e., stray current. When stray current flows into the sleepers or the rail fastening devices, a potential difference (rail potential) is generated between the rail and the ground in proportion to the stray current and the resistance of the parts in contact with the ground. The permissible voltage of rail potential is defined in IEC 62128 [2] with the aim of preventing maintenance workers near the rail from receiving an electric shock.

A voltage limiting device (VLD) is a piece of railway power supply equipment used for preventing over-voltage of rail potential [3]. A VLD model has already been developed and analyzed [4]. The number of possible VLD installations is not limited to one; however, the affection of the rail potential and stray current from installing multiple VLDs has not yet been investigated. We developed a simulation method to calculate the rail potential of the railway with an installed VLD, and then we analyzed the behavior of multiple VLDs installed in the line.

Concept of VLD Model

The electricity flow of the return circuit with a VLD installed is shown in Fig. 1. The VLD is a short-circuit device with a switch connecting the rail and the earth. The switch is normally opened to insulate the rail from the earth so as to limit stray current. Once the voltage between the rail and the earth exceeds the predefined limit, the switch closes, and the rail and the earth are shorted to limit the voltage. The switch stays closed for a predefined time and then is reopened. While the switch is closed, the rail potential is limited, but stray current tends to flow through the VLD because the resistance of the VLD is very low. To evaluate the effect of the VLD on the rail potential and stray current, the electrical circuit must be modeled for two conditions: when the VLD switch is open and when it is closed.

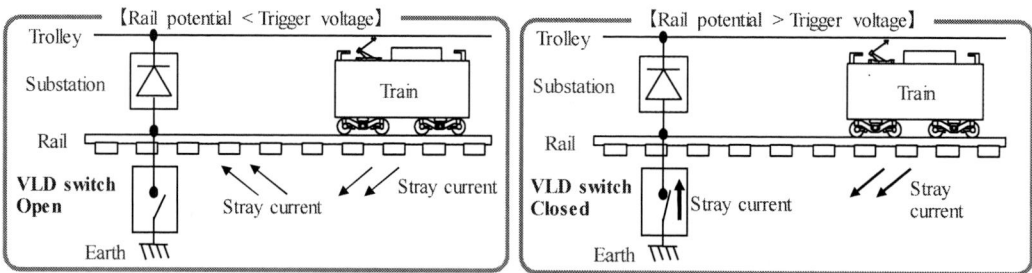

Fig. 1: Return circuit with VLD installed

Figure 2 shows the equivalent circuit of a return path with the VLD. The VLD was added to conventional return circuit models as a two-dimensional electrical circuit to calculate rail potential and stray current [1]. The VLDs were installed as circuit elements connecting the rail layer and the earth layer. When the switch is open, it can be incorporated as a high impedance resistor so that the two points are insulated; when the switch is closed, it can be incorporated as a very low impedance resistor so that the two points are short-circuited.

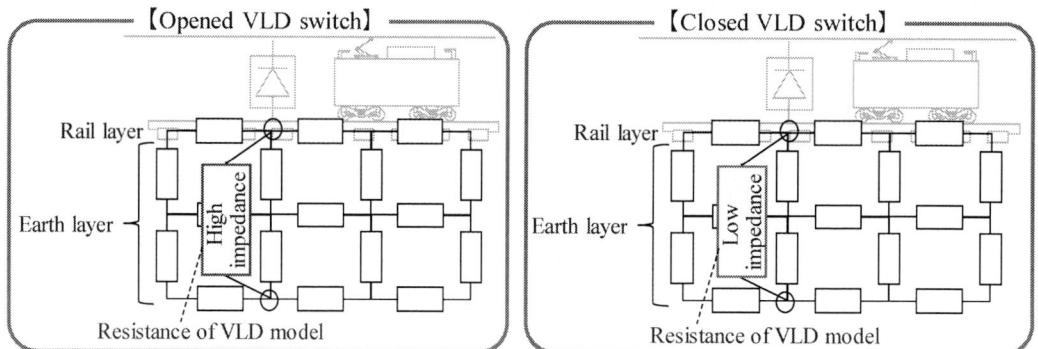

Fig. 2: Equivalent return circuit with VLD installed

Calculation sequence of rail potential using VLD model

Figure 3 shows the traditional calculation sequence for evaluating rail potential based on the train and substation behavior [5].
(Step 1) A two-dimensional equivalent circuit corresponding to the return circuit is created, including rails, track beds, civil structures, earth resistance, etc., and compressed to a one-dimensional circuit.
(Step 2) The current from trains to substations is calculated by taking into account the voltage drop in the one-dimensional equivalent return circuit.

(Step 3) The rail potential distribution is calculated by inputting the current values of the trains and substations calculated in step 2 into the two-dimensional equivalent return circuit.

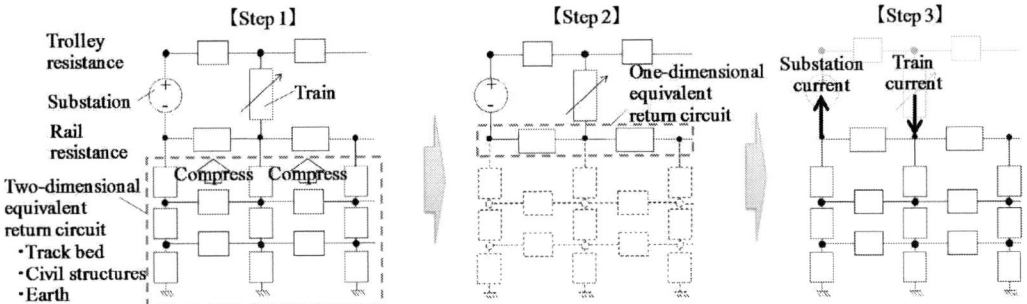

Fig. 3: Calculation sequence for evaluating rail potential based on responses from trains and substations

Figure 4 shows the implementation of the VLD model in the proposed rail potential calculation sequence. In our traditional calculation sequence, shown in Fig. 4 (a), a two-dimensional equivalent circuit is created once at the beginning of the calculations because the resistance of civil structures is constant. However, as described above, the resistance value of the VLD depends on whether the switch is open or closed; therefore, the calculation sequence was improved to reflect the status of the switch, as shown in Fig. 4 (b). That is, the open/closed status of the VLD is determined from the latest rail potential, and then the equivalent return circuit is created. The status of the VLD is updated in the rail potential feedback loop, and the calculation is repeated until all simulation periods are over.

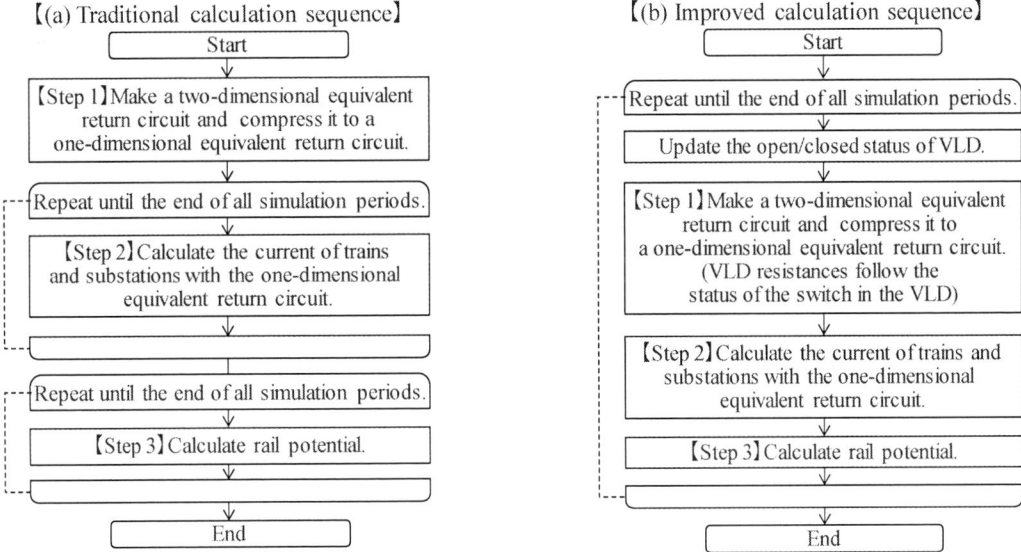

Fig. 4: Implementation of VLD model in rail potential calculation sequence

Case study of VLD model

Simulation conditions

To evaluate the calculation model with the VLD installed to limit rail potential, the rail potential is calculated for the return path with and without the VLD and then compared. Simulation conditions based on a commuter line are shown in Table I. In order to calculate the rail potential under severe conditions, it is assumed that the substation at the end of the line (substation F) fails and that its rectifier stops operating. In this situation, the rail potential increases due to the long distance between the train and the substation. Figure 6 shows a diagram of the commuter line. Trains depart from stations on both ends every 2 minutes. Figure 7 shows a schematic diagram of the return circuit model with a four-layer two-dimensional circuit model applied to simulate a viaduct return circuit; the horizontal resolution of this diagram is 200 m. In this case, the VLD connects the rail and the grounded rebar. Hereafter, rail potential is referred to as the potential difference between the rail and the rebar. Simulation parameters of the circuit elements are shown in Table II. The resistance value of each layer is set according to the common electrical characteristics of each material. The VLD trigger voltage is set to the value specified by IEC62128. The length of one simulation cycle is 1 s, which is much shorter than the amount of time the VLD is closed.

Table I: Parameters of commuter line route

Parameter		Value
Line profile	Stations and substations	Fig. 5
Substation	Feeder voltage	1575 V
	Voltage regulation	5%
	Capacity of rectifier	9000 kW
Train	Formation	6 cars
Operation	Head way	2 min (local trains only) Diagram: See Fig. 6
Return circuit		Fig. 7 and Table II

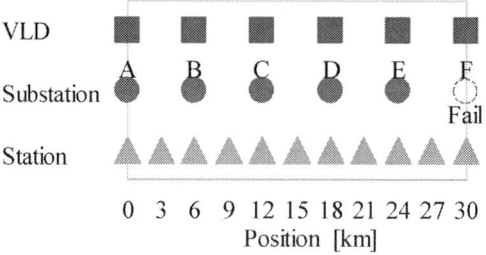

Fig. 5: Positions of VLD, substation, and station

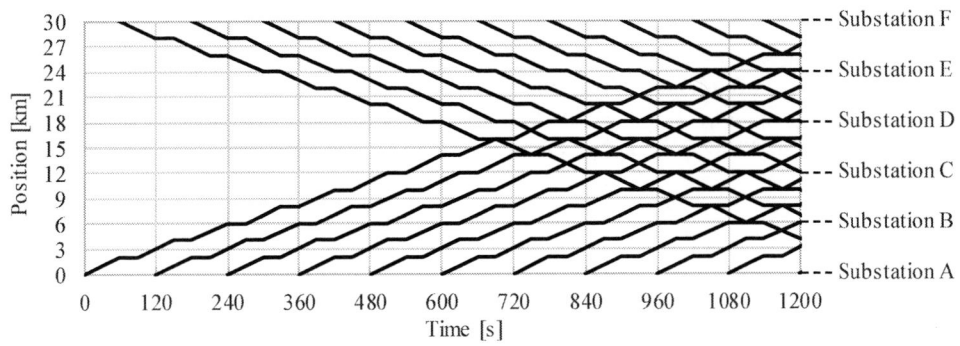

Fig. 6: Diagram of commuter line

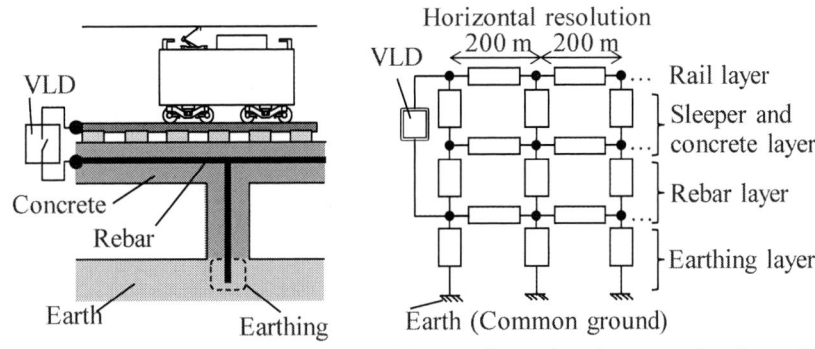

(a) Overview of civil structure   (b) Two-dimensional return circuit model

Fig. 7: Schematic view of return circuit model

Table II: Parameters of circuit elements

Parameter		Value
VLD	Resistance (Closed)	1 mΩ
	Resistance (Open)	100 MΩ
	Trigger voltage	$\pm$120 V
	Time remained closed	60 s
Rail layer	Resistance	1.6 mΩ / km
Sleeper and concrete layer	Horizontal resistance	0.38 MΩ / km
	Vertical resistance	0.77 MS / km
Rebar layer	Horizontal resistance	29.3 mΩ / km
	Vertical resistance	3.4 e11 S / km
Earthing layer	Resistance	0.01 S / km

Simulation result and evaluation

Figure 8 shows the time series waveforms of the simulated rail potential and the open/closed status of the VLD at each substation. Without the VLD, the rail potential exceeds the trigger voltage (120 V) at substation F for a long duration. On the other hand, with the VLD, the rail potential at substation F is reduced to almost zero after the VLD is closed when the rail potential exceeds the trigger voltage as expected. The VLDs at the other substations are then closed as well.

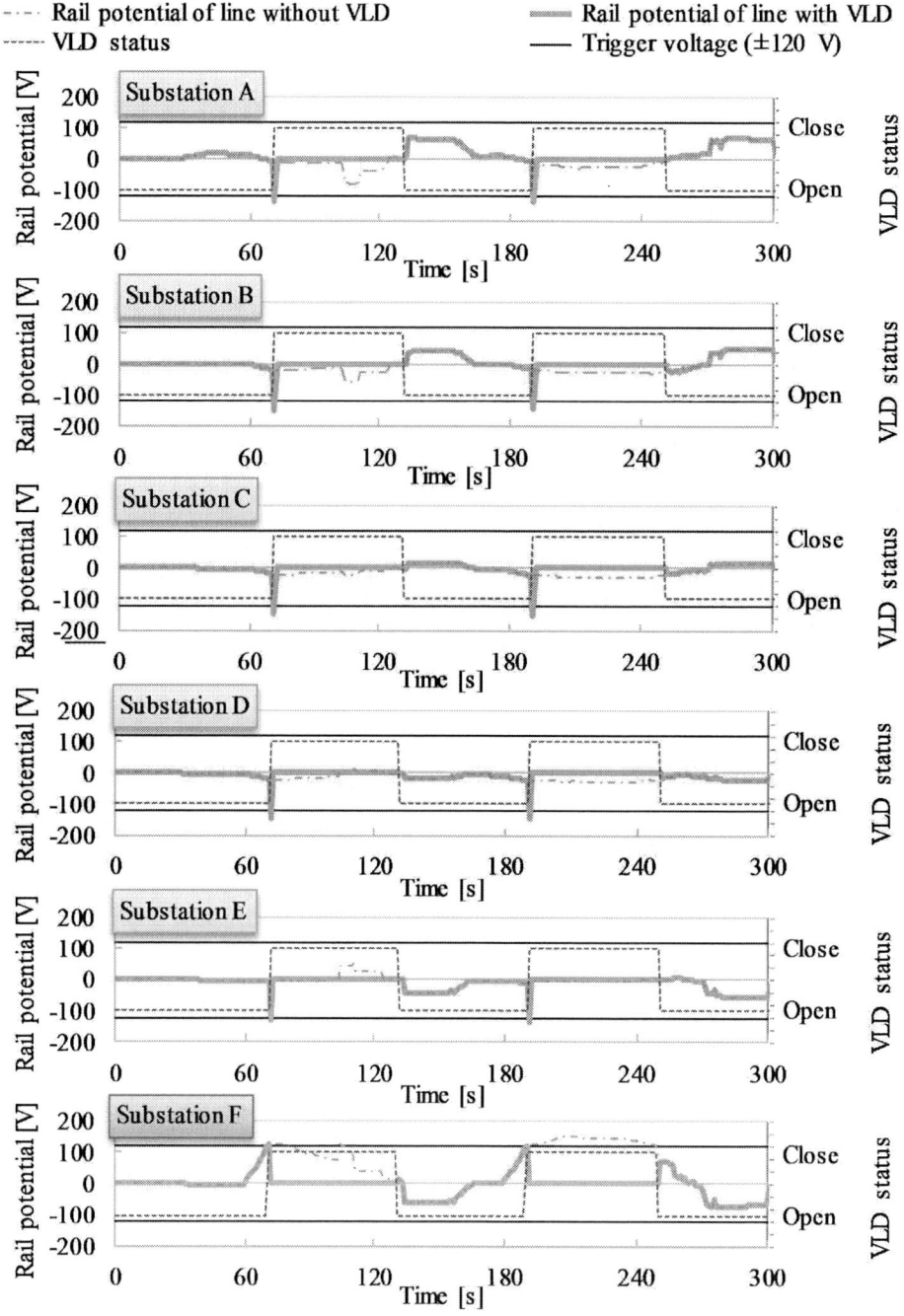

Fig. 8: Simulated time series waveform of rail potential and open/closed status of VLD at each substation

The rail potential distribution around the VLD trigger time is shown in Figure 8. Fig. 8 (a) shows the moment (190 s) when the rail potential at substation F exceeds 120 V and the VLD starts to close. Fig. 8 (b) shows the time (191 s) immediately after the VLD at substation F is closed; the rail potential at that substation is suppressed to almost zero, and the rail potential at other substations drops to -120 V or less, which is the VLD trigger voltage. Fig. 9 (c) shows the rail potential after all VLDs are closed (192 s). The rail potential at all substations is reduced to almost zero. In the 3 seconds, the current of the trains and substations did not change greatly, so the resultant rail potential is likely due to the VLDs closing.

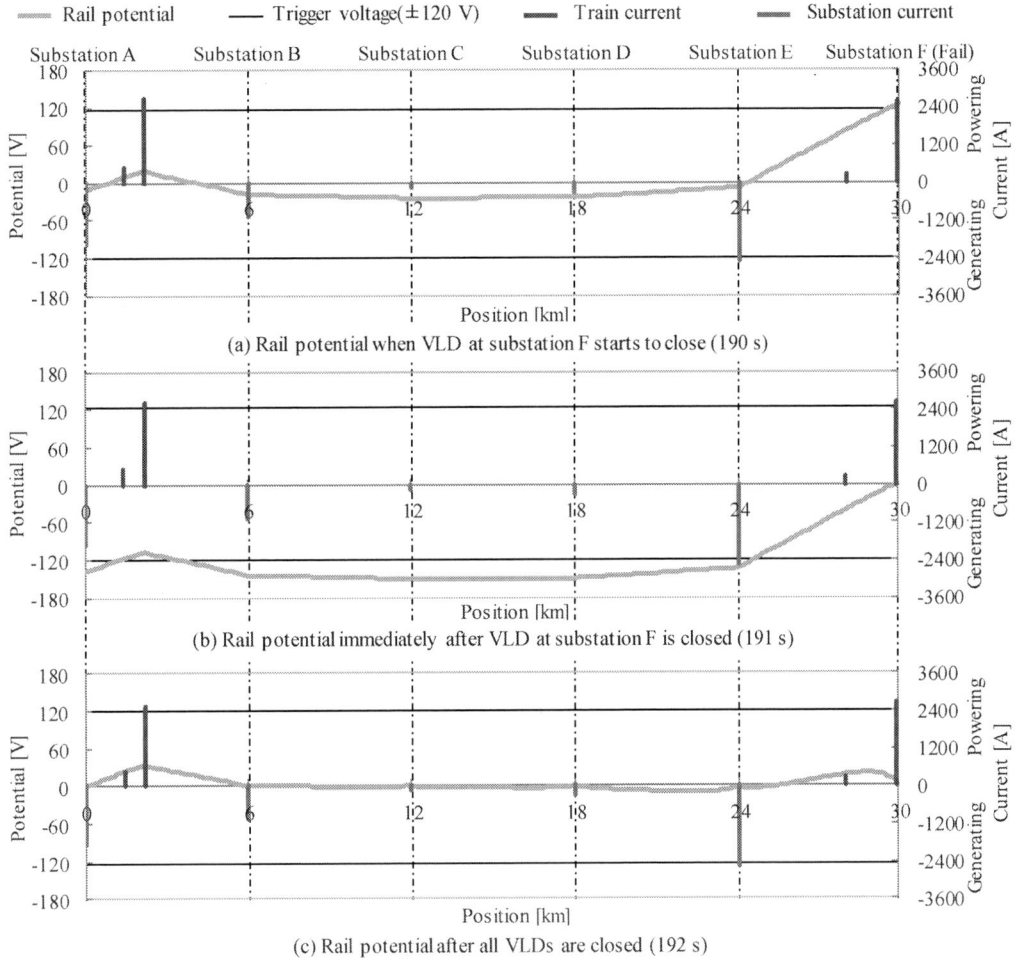

Fig. 9: Rail potential before and after the VLD closing

The distribution of the stray current is shown in Figure 10. As seen in Fig. 10 (c), the stray current increases with the addition of the VLD, compared to when there is no VLD.

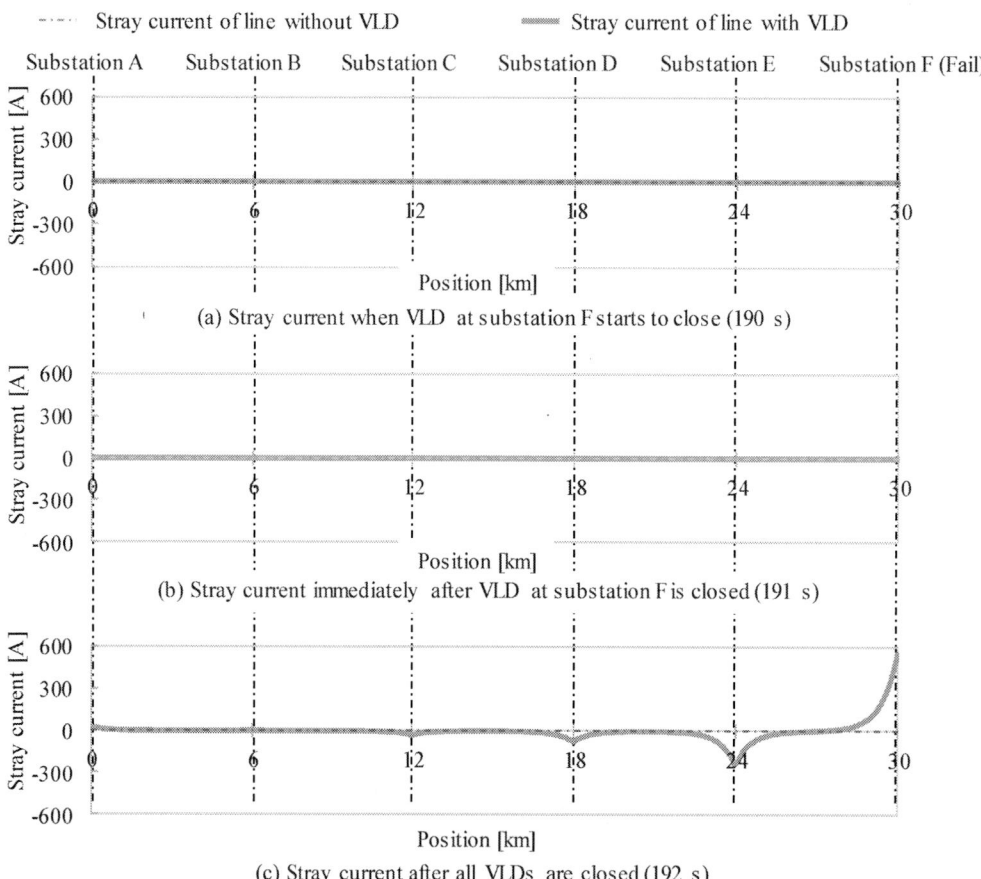

(a) Stray current when VLD at substation F starts to close (190 s)

(b) Stray current immediately after VLD at substation F is closed (191 s)

(c) Stray current after all VLDs are closed (192 s)

Fig. 10: Stray current before and after the VLD is closed

Figure 11 shows the stray current after all VLDs are closed. When several VLDs are closed, the rail and rebar are connected at all substation positions. The rebar is connected to the earth (common ground) via low impedance. Thus, the return current flowing from the rail will flow through the earth via the VLDs.

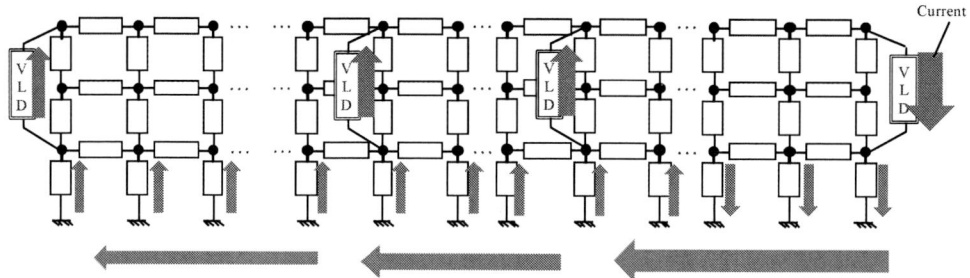

Fig. 11: Image of stray current after all VLDs are closed

As mentioned above, rail potential is limited by VLDs. At the same time, however, stray current can increase. The permissible value of stray current is also defined in IEC 62128. Our proposed rail potential calculation model can be used to determine the number of VLDs and their positions while also taking into account the increase in stray current.

Conclusion

We developed a simulation model for calculating rail potential. Our model incorporates a VLD, which is a piece of railway power supply equipment used for protection against electric shock due to over-voltage. The VLD was integrated into a conventional return circuit model as an additional circuit element. It acts as a high impedance resistor when the switch is open and as a very low impedance resistor when the switch is closed. The VLD status is updated in the rail potential feedback loop, and the calculation is repeated until the end of all simulation periods. The results of the case study verified that rail potential is reduced to almost zero by closing multiple VLDs installed in the line. However, stray current increases because it flows through the closed VLDs. The results obtained from the actual route will be compared with the simulation results in future work.

References

[1] M. Niasati and A. Gholami: Overview of Stray Current Control in DC Railway Systems, International Conference on Railway Engineering – Challenges for the Information Age, 2008-04.

[2] IEC 62128-1 Edition 2.0, Railway applications – Fixed installations – Electrical safety, earthing and the return circuit – Part 1: Protective provisions against electric shock, 2013-09.

[3] D. Achraf and R. Clement: Reverse engineering of Voltage Limiting Devices in 1500V DC Railway Lines through transient transmission line modeling, IEEE, 2019-09.

[4] M. T. Söylemez, S. Açıkbaş and A. Kaypmaz: Controlling Rail Potential of DC Supplied Rail Traction Systems, 2006-01.

[5] S. Kimura, T. Miyauchi, K. Oguma, H. Takahashi and K. Teramura : Development of simulation model for calculating rail potential during train running, ICEE 2019, ICEE19J-159.

Homogenization of Current Distribution in Parallel Connection of Interleaved Winding Layers of High-Frequency Transformers by Optimizing Distance between Winding Layers

Ryo Murata[1], Tomohide Shirakawa[1], Kazuhiro Umetani[1], Eiji Hiraki[1],
Hiroto Mizutani[2], Takaaki Takahara[3], and Osamu Mori[2]

[1]Okayama University
3-1-1, Tsushima-naka,
Kita-ku,
Okayama, Japan

[2]Advanced Technology R&D
Center
Mitsubishi Electric Corporation
8-1-1, Tsukagushi-Honmachi,
Amagasaki, Japan

[3]Living Environment Systems
Laboratory
Mitsubishi Electric Corporation
3-18-1, Oshika, Suruga-ku,
Shizuoka, Japan

Corresponding E-Mail: p9nf00b3@s.okayama-u.ac.jp

Keywords

«Transformer», «Current distribution», «Copper loss», «Extremum co-energy principle».

Abstract

The windings of high-frequency high-current transformers are required to reduce the proximity effect loss. Therefore, the Litz wire in parallel connection of interleaved winding layers is usually used as the winding of transformers. However, this can cause unbalanced AC current distribution, hindering effective reduction of the copper loss. This paper solves this problem by optimizing the distance between the winding layers based on the analytical principle called the extremum co-energy principle. The transformers used in the experiment to verify this method consists of three parallel-connected primary winding and two parallel-connected secondary winding, PQ core, thin polypropylene sheets for adjusting the distance between the winding layers. As a result, the copper loss can be recused by homogenizing the AC current distribution, and the effectiveness of this method has been clarified by experiments.

Introduction

In the isolated DC-DC power converters, the copper loss of the high-frequency transformers is one of the major causes of the power loss. For reducing the copper loss, the Litz wire is usually used to achieve the low copper loss in high frequency [1]-[4]. Furthermore, the interleaved winding layer structure is effective and commonly utilized for practical design because this structure is known to mitigate the proximity effect, which causes the eddy current inside the wire to increase the copper loss.

The interleaved winding layer structure has been commonly made with the series-connected winding layers, as the conventional low-frequency power transformers tend to have a great number of turns, which naturally requires the series-connection of the winding layers. However, recent high-frequency design can have the primary and secondary windings with a small number of turns. In this design, many transformers have parallel-connected winding layers because the multiple winding layers are commonly utilized to expand the cross-sectional area for AC current rather than to implement a large number of turns. Consequently, a number of recent transformers are designed to have a parallel connection of interleaved winding layers.

However, the AC current does not necessarily flow uniformly among the parallel-connected winding layers, unlike the DC current, which flows uniformly according to the DC resistance [5]-[11]. The reason is that the parasitic magnetic coupling among the winding layers, as well as the leakage inductance of the winding layers, [5], [8], [10]-[13] affects the AC current distribution of the parallel-connected winding layers. The unbalanced AC current distribution tends to increase in the copper loss, which hinders the effective reduction of the copper loss by the interleaved winding layer structure.

The solution to this problem may lie in the optimization of the winding layer disposition because the proximity effect is known to be significantly affected by the winding layer disposition [12], [14], [15]. The purpose of this paper is to propose a method to derive the optimum winding layer disposition to homogenize the AC current distribution in the parallel-connection of the interleaved winding layers of a high-frequency transformer. Particularly, this paper focuses on the distance between the winding layers and analytically derives the optimal distance to achieve homogeneous AC current distribution.

This paper adopts a recently proposed analytical method [16]-[19] based on the analytical principle called the extremum co-energy principle [20], [21], which is briefly reviewed in the next section. Certainly, a number of preceding studies have analyzed the proximity effect and resulting the AC current distribution by the FEM simulation [8], [14], [15]. However, the FEM simulation requires detailed geometrical data of the magnetic cores and windings,. Therefore, total recalculation is indispensable in every time the design has changed. As an alternative approach, [7], [10], [11] rather adopted a sophisticated approach utilizing the equivalent circuit model, although the model construction, as well as the circuit calculation, is somewhat complicated. The extremum co-energy principle directly utilizes the analytical solution of the magnetic field inside the transformer, which can be straightforwardly formulated by the Dowell's approximation [22]-[26], as shown later.

The extremum co-energy principle was first proposed in [16] as a simple method for analyzing the AC current distribution of the transformers with parallel-connected winding. [17] and [19] proposed a new copper loss analysis method of the transformers with parallel-connected winding by applying the extremum co-energy principle proposed in [16] to a famous copper loss analysis model called Dowell model. [17] focused on the transformers which carry only AC current, whereas [19] focused on the transformers that carry both DC and AC current. [18] proposed a method to derive the optimum winding turn allocation among winding layers to homogenize the AC current distribution in the parallel connection of the interleaved winding layers according to the extremum co-energy principle. This paper proposes a method to derive the optimum "distance between the winding layers" unlike "winding turn allocation among winding layers [18]" to homogenize the AC current distribution in the parallel-connection of the interleaved winding layers according to the extremum co-energy principle.

Review of the Extremum co-energy principle

According to the extremum co-energy principle[16], AC current in a magnetic device is distributed so that the magnetic co-energy of the device must take the extremum. This section briefly explains this reason by the transformer with parallel-connected n primary windings, P_1, P_2, ..., P_n, and parallel-connected k secondary windings, S_1, S_2, ..., S_k, as depicted in Fig. 1. For simplifying the discussion, we assume that there is no DC current flowing the windings. The extremum co-energy principle can be utilized for high-frequency AC current, for which the reactance is much larger than the resistance. Hence, this section neglects the resistance of the wire to simplify the calculation.

Let i_j and ψ_j be the AC current and AC flux linkage of winding j, respectively. Then, the total magnetic co-energy of the transformer E_{co} can be expressed as follows, if $\mathbf{i}$ and $\boldsymbol{\psi}$ are defined as $\mathbf{i} \equiv [i_{p1}, i_{p2}, ..., i_{sk}]^t$ and $\boldsymbol{\psi} \equiv [\psi_{p1}, \psi_{p2}, ..., \psi_{sk}]^t$:

$$E_{co}(\mathbf{i}) = \int_0^{\mathbf{i}} \boldsymbol{\Psi}(\mathbf{i}) \cdot d\tilde{\mathbf{i}}, \tag{1}$$

Now, we consider an arbitrary infinitesimal virtual change $\delta\mathbf{i} \equiv [\delta i_{p1}, \delta i_{p2}, ..., \delta i_{sk}]^t$ in the AC current vector $\mathbf{i}$, where $\delta i_{p1}, \delta i_{p2}, ..., \delta i_{pn}$ are the AC current change in the primary windings and $\delta i_{s1}, \delta i_{s2}, ..., \delta i_{sk}$ are that of the secondary windings. Because the total primary current is determined by the circuit operation outside the transformer, we impose the requirement that this virtual change does not affect the total primary and secondary current, i.e. $\delta i_{p1}+\delta i_{p2}+...+\delta i_{pn}=0$ and $\delta i_{s1}+\delta i_{s2}+...+\delta i_{sk}=0$. Then, the change in the magnetic co-energy δE_{co} is obtained as follows, noting that the parallel-connected windings must have the same AC flux linkage according to Faraday's law and therefore $\psi_{p1}=\psi_{p2}=...=\psi_{pn}$ and $\psi_{s1}=\psi_{s2}=...=\psi_{sk}$.

$$\delta E_{co} = \mathbf{\Psi}(\mathbf{i}) \cdot \delta \mathbf{i} = \psi_{p1}\delta i_{p1} + \psi_{p2}\delta i_{p2} + \cdots + \psi_{pn}\delta i_{pn} + \psi_{s1}\delta i_{s1} + \psi_{s2}\delta i_{s2} + \cdots + \psi_{sk}\delta i_{sk}$$
$$= \psi_{p1}\left(\delta i_{p1} + \delta i_{p2} + \cdots + \delta i_{pn}\right) + \psi_{s1}\left(\delta i_{s1} + \delta i_{s2} + \cdots + \delta i_{sk}\right) = 0. \tag{2}$$

Equation (2) indicates that the AC current of parallel-connected windings is distributed to achieve the extremum of the magnetic co-energy.

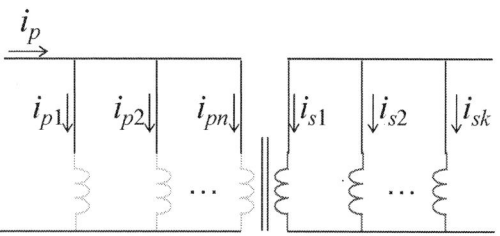

Fig. 1: Equivalent circuit of a transformer

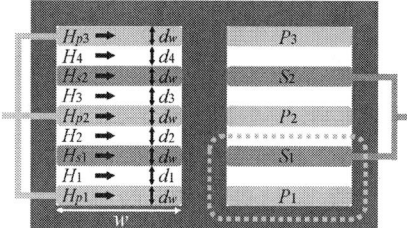

Fig. 2: Analyzed transformer structure

Homogenization of AC Current Distribution

This section analytically calculates the optimum distance between the winding layers according to the extremum co-energy principle so that the AC current distribution is homogenized. As an example, we consider the transformer structure, whose cross-sectional view is shown in Fig. 2. The windings P_1-P_3 are the parallel-connected primary winding layers, whereas the windings S_1 and S_2 are the parallel-connected secondary winding layers. The parallel-connected winding layers are assumed to have the same number of turns. In this paper, we assume that the magnetic core is lossless and has a permeability much larger than the air.

First, we formulate the magnetic co-energy of this transformer. Because the permeability of the core is large, the magnetic co-energy stored in the magnetic core is ignorable. (Note that the co-energy per unit volume is $B^2/2\mu$, where μ is the permeability and B is the magnetic flux density.) Additionally, vast open space outside the transformer provides a large cross-sectional area for the leakage flux path and therefore has a low reluctance. Hence, the open space outside the transformer has little magnetic co-energy and can be ignorable. Thus, the magnetic co-energy of the transformer is mainly contributed by the wire and the narrow space between the winding layers.

The magnetic field inside the space between the winding layers may have local inhomogeneity. However, for simplifying the calculation, we approximate that the uniform magnetic field throughout this space. This approximation is known as Dowell's approximation and widely utilized in the preceding studies [22]-[26]. According to this approximation, the magnetic field between the winding layers is easily formulated by Ampere's law. For example, the magnetic field H_2 of the space between winding layers S_1 and P_2 can be formulated as follows by applying Ampere's law along the dashed line in Fig. 2:

$$H_2 = (N_p i_{p1} + N_s i_{s1})/w, \tag{3}$$

where N_p and N_s are the numbers of turns of the primary and secondary windings, respectively, w is the width of the window space as shown in Fig. 2. Similarly, magnetic field H_1 in the space between P_1 and S_1, magnetic field H_3 in the space between P_2 and S_2, and magnetic field H_4 in the space between S_2 and P_3 are obtained as

$$H_1 = N_p i_{p1}/w, \quad H_3 = (N_p i_{p1} + N_s i_{s1} + N_p i_{p2})/w, \quad H_4 = (N_p i_{p1} + N_s i_{s1} + N_p i_{p2} + N_s i_{s2})/w. \tag{4}$$

Litz wire is not easily affected by the skin effect unlike the solid wire and the foil. Therefore, the current density inside the winding can be approximated as uniform the magnetic field inside the winding is distributed linearly as depicted in Fig. 3. Hence, when calculating the magnetic co-energy of the transformer, the magnetic field inside the winding cannot be ignored. Then, the magnetic fields are functions of x in Fig. 3, and defined as

$$H_{p1}(x)=\frac{H_1}{d_w}\cdot x, H_{s1}(x)=\frac{H_2-H_1}{d_w}\cdot x+H_1, H_{p2}(x)=\frac{H_3-H_2}{d_w}\cdot x+H_2, H_{s2}(x)=\frac{H_4-H_3}{d_w}\cdot x+H_3, H_{p3}(x)=\frac{-H_4}{d_w}\cdot x, \quad (5)$$

where d_w the width of winding layers.

For calculating the magnetic co-energy inside the winding, the integrated average magnetic fields inside the winding are defined as

$$H_{p1}=\sqrt{\frac{1}{d_w}\cdot\int_0^{d_w}\left(\frac{H_1}{d_w}\cdot x\right)^2 dx}=\sqrt{\frac{1}{3}H_1^2}, H_{s1}=\sqrt{\frac{1}{d_w}\cdot\int_0^{d_w}\left(\frac{H_2-H_1}{d_w}\cdot x+H_1\right)^2 dx}=\sqrt{\frac{1}{3}\left(H_1^2+H_1H_2+H_2^2\right)},$$

$$H_{p2}=\sqrt{\frac{1}{d_w}\cdot\int_0^{d_w}\left(\frac{H_3-H_2}{d_w}\cdot x+H_2\right)^2 dx}=\sqrt{\frac{1}{3}\left(H_2^2+H_2H_3+H_3^2\right)}, H_{s2}=\sqrt{\frac{1}{d_w}\cdot\int_0^{d_w}\left(\frac{H_4-H_3}{d_w}\cdot x+H_3\right)^2 dx}=\sqrt{\frac{1}{3}\left(H_3^2+H_3H_4+H_4^2\right)}, \quad (6)$$

$$H_{p3}=\sqrt{\frac{1}{d_w}\cdot\int_0^{d_w}\left(\frac{-H_4}{d_w}\cdot x\right)^2 dx}=\sqrt{\frac{1}{3}H_4^2}.$$

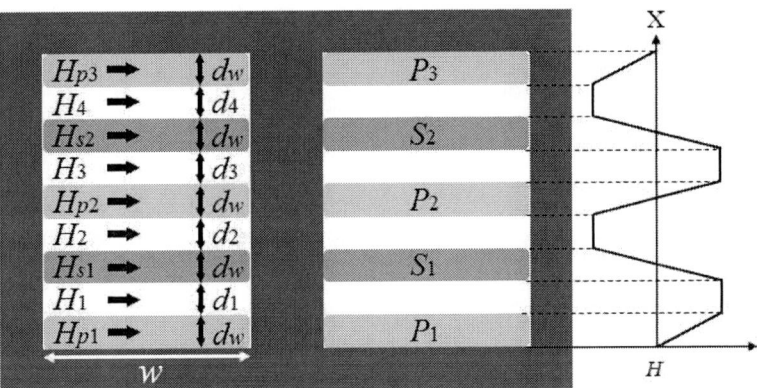

Fig. 3: Magnetic field distribution of the original transformer composed of Litz wire
$(d_1 : d_2 : d_3 : d_4 = 1 : 1 : 1 : 1)$

Because the magnetic co-energy per unit volume is $B^2/2\mu$ and $\mu H^2/2$, the total magnetic co-energy E_{co} can be obtained as

$$E_{co}(i_{p1},i_{s1},i_{p2},i_{s2})=\frac{1}{2}\mu_0\left(H_{p1}^2 Sd_w + H_1^2 Sd_1 + H_{s1}^2 Sd_w + H_2^2 Sd_2 + H_{p2}^2 Sd_w + H_3^2 Sd_3 + H_{s2}^2 Sd_w + H_4^2 Sd_4 + H_{p3}^2 Sd_w\right), \quad (7)$$

where μ_0 is the permeability of the air, and S is the top surface area of the winding layers.

Next, the AC current distribution in parallel-connected winding layers is determined according to the extremum co-energy principle. The total primary current i_p, defined as $i_p \equiv i_{p1} + i_{p2} + i_{p3}$, is given by the circuit operation outside the transformer. Besides, the total secondary current i_s, defined as $i_s \equiv i_{s1} + i_{s2}$, is required to cancel the magnetomotive force of the primary winding layers, according to the operating principle of the transformer. Hence, $N_p i_p = -N_s i_s = -N_s(i_{s1} + i_{s2})$ is imposed to the secondary current. Therefore, the extremum of the total magnetic co-energy should be searched under the constraint of $i_p = i_{p1} + i_{p2} + i_{p3}$ and $N_p i_p = -N_s(i_{s1} + i_{s2})$. For this purpose, we utilized the Lagrangian multiplier method, in which the modified function E_{co}' is introduced as

$$E_{co}' = \frac{1}{2}\mu_0\left(H_{p1}^2 Sd_1 + H_1^2 Sd_1 + H_{s1}^2 Sd_1 + H_2^2 Sd_2 + H_{p2}^2 Sd_1 + H_3^2 Sd_3 + H_{s2}^2 Sd_1 + H_4^2 Sd_4 + H_{p3}^2 Sd_1\right)$$
$$-\lambda_p\left(i_{p1} + i_{p2} + i_{p3} - i_p\right) - \lambda_s\left(N_p i_p + N_s i_{s1} + N_s i_{s2}\right),$$

(8)

where λ_p and λ_s are the Lagrangian multipliers.

The solution of the AC current in the winding layers can be obtained by solving $\partial E_{co}'/\partial i_{p1} = 0$, $\partial E_{co}'/\partial i_{p2} = 0$, $\partial E_{co}'/\partial i_{p3} = 0$, $\partial E_{co}'/\partial i_{s1} = 0$, $\partial E_{co}'/\partial i_{s2} = 0$, $\partial E_{co}'/\partial \lambda_p = 0$ and $\partial E_{co}'/\partial \lambda_s = 0$. Consequently, we have

$$i_{p1} = \frac{38d_w^3 + \left[45d_2 + 60(d_3+d_4)\right]d_w^2 + \left[(72d_2+90d_3)d_4 + 72d_2d_3\right]d_w + 108d_2d_3d_4}{171d_w^3 + \left[192(d_1+d_4)+165(d_2+d_3)\right]d_w^2 + \left\{\left[180(d_1+d_3)+144d_2\right]d_4 + 144(d_1+d_2)d_3 + 180d_1d_2\right\}d_w + 108\left\{\left[(d_1+d_2)d_3+d_1d_2\right]d_4 + d_1d_2d_3\right\}}i_p,$$

$$i_{p2} = \frac{95d_w^3 + \left[132(d_1+d_4)+60(d_3+d_4)\right]d_w^2 + \left[(180d_1+72d_2+90d_3)d_4 + 72d_1d_3 + 90d_1d_2\right]d_w + 108(d_2+d_3)d_1d_4}{171d_w^3 + \left[192(d_1+d_4)+165(d_2+d_3)\right]d_w^2 + \left\{\left[180(d_1+d_3)+144d_2\right]d_4 + 144(d_1+d_2)d_3 + 180d_1d_2\right\}d_w + 108\left\{\left[(d_1+d_2)d_3+d_1d_2\right]d_4 + d_1d_2d_3\right\}}i_p,$$

$$i_{p3} = \frac{38d_w^3 + \left[60(d_1+d_2)+45d_3\right]d_w^2 + \left[72(d_1+d_2)d_3 + 90d_1d_2\right]d_w + 108d_1d_2d_3}{171d_w^3 + \left[192(d_1+d_4)+165(d_2+d_3)\right]d_w^2 + \left\{\left[180(d_1+d_3)+144d_2\right]d_4 + 144(d_1+d_2)d_3 + 180d_1d_2\right\}d_w + 108\left\{\left[(d_1+d_2)d_3+d_1d_2\right]d_4 + d_1d_2d_3\right\}}i_p,$$

$$i_{s1} = -\frac{N_p}{N_s}\frac{57d_w^3 + (40d_1+30d_3+80d_3+88d_4)d_w^2 + \left[(60d_1+48d_3+120d_2)d_4 + 48(d_1+d_2)d_3\right]d_w + 72(d_1+d_2)d_3d_4}{114d_w^3 + \left[128(d_1+d_4)+110(d_2+d_3)\right]d_w^2 + \left\{\left[120(d_1+d_3)+96d_2\right]d_4 + 96(d_1+d_2)d_3 + 120d_1d_2\right\}d_w + 72\left\{\left[(d_1+d_2)d_3+d_1d_2\right]d_4 + d_1d_2d_3\right\}}i_p,$$

$$i_{s2} = -\frac{N_p}{N_s}\frac{57d_w^3 + (88d_1+80d_2+30d_3+40d_4)d_w^2 + \left[120d_1d_2+48(d_1+d_2)d_3 + (60d_1+48d_2)d_4\right]d_w + 72(d_1+d_4)d_1d_2}{114d_w^3 + \left[128(d_1+d_4)+110(d_2+d_3)\right]d_w^2 + \left\{\left[120(d_1+d_3)+96d_2\right]d_4 + 96(d_1+d_2)d_3 + 120d_1d_2\right\}d_w + 72\left\{\left[(d_1+d_2)d_3+d_1d_2\right]d_4 + d_1d_2d_3\right\}}i_p.$$

(9)

As an example, in the case in which all the distance between the winding layers are almost the same, i.e. $d_1 = d_2 = d_3 = d_4$ as shown in Fig. 3, the AC current distribution is not uniform; and therefore, the copper loss tends to increase compared to the uniform AC current distribution. In fact, the AC current distribution of the original transformer, i.e. $d_1 = d_2 = d_3 = d_4$ is calculated according to equation (9) as

$$i_{p1} : i_{p2} : i_{p3} = 3:7:3, \ i_{s1}:i_{s2} = 1:1.$$

(10)

Equation (9) can also determine the distance between the winding layers that makes the homogeneous AC current distribution, i.e. $i_{p1} = i_{p2} = i_{p3}$ and $i_{s1} = i_{s2}$. The necessary ratio of the distance was calculated as

$$d_1 : d_2 : d_3 : d_4 = d_1 : 2d_1 + d_w : 2d_1 + d_w : d_1.$$

(11)

As can be seen in equation (9), the AC current distribution is not dependent on the load impedance. Therefore, the necessary ratio of the distance for the uniform AC current distribution, i.e. equation (11), is effective regardless of the load condition.

This method is effective when the current density inside the Litz wire can effectively be homogenized. However, the copper loss of the proposed transformer structure ($d_1 : d_2 : d_3 : d_4 = d_1 : 2d_1 + d_w : 2d_1 + d_w : d_1$) may be larger than the copper loss of the original transformer structure ($d_1 : d_2 : d_3 : d_4 = 1 : 1 : 1 : 1$) at extremely high frequencies, which the proximity effect loss is dominant. The proximity effect loss is proportional to the square of the magnetic field [1]. As shown the proposed transformer structure ($d_1 : d_2 : d_3 : d_4 = d_1 : 2d_1 + d_w : 2d_1 + d_w : d_1$) in Fig. 4, this structure has a large magnetic field strength in the winding layer. Therefore, the proximity effect loss of the proposed transformer structure is 153.8 % larger than the proximity effect loss of the original transformer structure according to the calculation of the square of the magnetic field. On the other hand, the total loss of the DC resistance and the skin effect loss (i.e. the copper loss excluding the proximity effect loss) of the proposed transformer structure is 7.0 % smaller than the original transformer structure one. Consequently, when the proposed transformer structure is be used at extremely high frequencies, it should be designed using sufficiently high quality Litz wire.

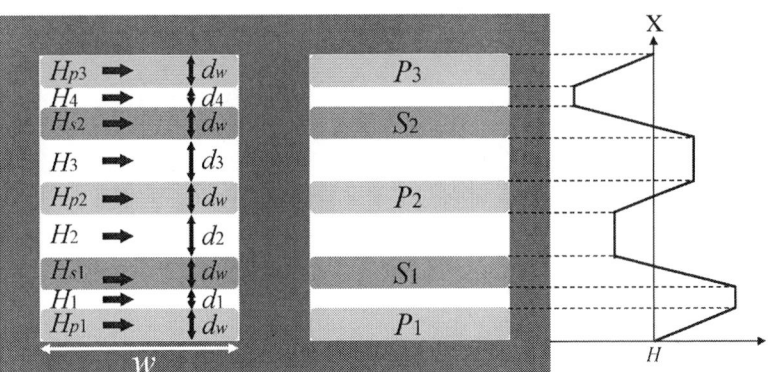

Fig. 4: Magnetic field distribution of the adjusted transformer composed of Litz wire when the AC current distribution is homogenized. (d_1: d_2: d_3: $d_4 = d_1$: $2d_1 + d_w$: $2d_1 + d_w$: d_1)

Also, the proposed method can be applied to a transformer composed of the solid wire or foil as well as the Litz wire. In the case of the solid wire or foil, note the skin effect that appears remarkably compared to the Litz wire, the skin depth in high frequency is extremely small. Therefore, the magnetic co-energy inside the winding cannot be ignored because the magnetic field inside the solid wire or foil is small due to the skin effect. Hence, the total magnetic co-energy E_{co} of a transformer composed of the solid wire or foil is generated only between the winding layers, the total magnetic co-energy E_{co} can be obtained as

$$E_{co}(i_{p1}, i_{s1}, i_{p2}, i_{s2}) = \frac{1}{2}\mu_0 \left(H_1^2 Sd_1 + H_2^2 Sd_2 + H_3^2 Sd_3 + H_4^2 Sd_4 \right). \tag{12}$$

Consequently, in the case of a transformer composed of the solid wire or foil, the proposed method can be applied in the same as the Litz wire.

This method for the homogenization of AC current distribution is based on the premise that the Dowell model can apply to the transformers. This Dowell model can apply to transformers, in which the magnetic field between the winding layers is almost uniform. Therefore, a shell-type transformer used in this paper as shown in Fig.2 can be applied the Dowell model and this proposed method, because the magnetic path through the air is short, and the magnetic field between the winding layers is almost uniform. On the other hand, the magnetic path through the air of a core-type transformer is longer than a shell-type one. However, it may be applied the Dowell model. Because if a core-type transformer in which both the primary and secondary winding wrap around the one core can almost cancel the magnetic field around each winding, and the magnetic field between the winding layers is almost uniform. And conversely, a core-type transformer in which each primary and secondary winding should wrap around another core for the insulation of the primary and secondary winding cannot be applied to this proposed method because it cannot be applied the Dowell model.

Experiment

The experiment was carried out to validate the effect of optimizing the distance between the winding layers. For this purpose, we constructed the experimental transformer by the PQ core and the parallel connection of the interleaved winding layers, as shown in Fig. 5. The experimental transformer has three parallel winding layers in the primary-side and has two parallel winding layers in the secondary-side, respectively. The total of five winding layers is arranged as shown in Fig. 2. Each primary and secondary winding layer is formed by the same alpha winding coil made of Litz wire. Fig. 6 shows the photograph of the winding layer in this experimental transformer. The designed power rating of the transformer in this paper is 1.5 kW. The specifications of the experimental transformer are shown in Table I.

Table I: Specifications of the experimental transformer

Core	Ferroxcube, 3C97
Primary and secondary winding	Litz wire 84/φ0.1
Number of turns	16 T
Height of alpha winding coil	3 mm

We measured AC current of the primary winding layers, i.e. i_{p1}, i_{p2}, and i_{p3}, under the constant total primary AC current (100 kHz, 1 Arms) by the experimental set-up as shown in Fig. 7. Then, we compared the estimated copper loss among 9 conditions of d_2 and d_3 under the premise of $d_1 = d_4 = 1$ mm, $d_2 = d_3$. These tested 9 conditions were $d_2 = 1$ mm to 9 mm. The space between the winding layers was adjusted by changing the number of thin polypropylene sheets inserted between the winding layers.

Primary copper loss P_p and secondary copper loss P_s can be expressed as

$$P_p = R(i_{p1}^2 + i_{p2}^2 + i_{p3}^2), \quad P_s = R(i_{s1}^2 + i_{s2}^2), \tag{13}$$

where R is the resistance of a winding layer. According to equation (9), P_p is expected to be dependent on these 9 conditions of the distance ratio, whereas P_s is expected to be constant regardless of these 9 conditions. Therefore, we compared the factor $i_{p1}^2 + i_{p2}^2 + i_{p3}^2$ among the 9 conditions of the distance ratio.

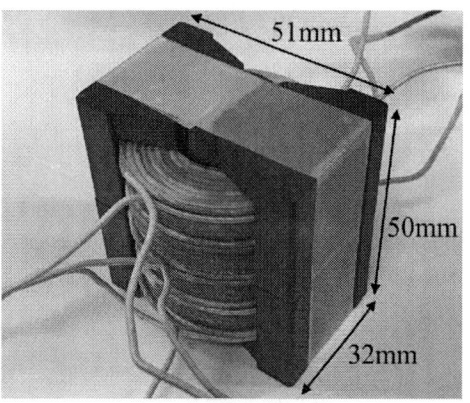

Fig. 5: Experimental transformer

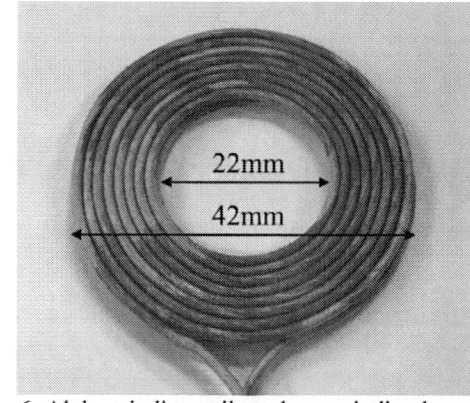

Fig. 6: Alpha winding coil used as a winding layer

Fig. 8 shows the comparison results of these 9 conditions of the distance ratio. The orange dots show the experimental results and the blue dots show the analytically calculated results according to equation (9). The horizontal axis shows d_2. (d_1 and d_3 and d_4 are omitted because of $d_1 = d_4 = 1$ mm and $d_2 = d_3$ for all conditions of the distance.)

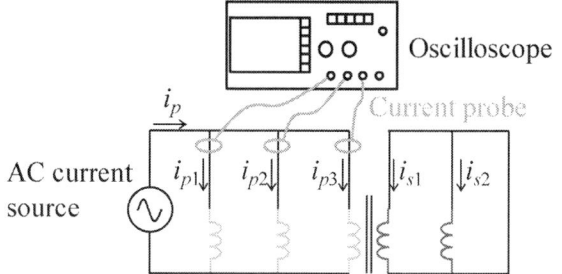

Fig. 7: Measurement system for the AC current distribution

As can be seen in Fig. 8, the experimental results agreed well with the calculated results within 2.6 %, indicating the effectiveness of the extremum co-energy principle for predicting the AC current distribution in the parallel-connected winding layers. The experimental results showed that the estimated copper loss is minimum at the condition of $d_2 = 5$ mm ($d_1 : d_2 : d_3 : d_4 = 1 : 5 : 5 : 1$), which was predicted to be an ideal structure by the theory. The experimental result suggests that the distance ratio can reduce the primary copper loss by 13.7 % compared to the original condition, i.e. $d_2 = 1$ mm ($d_1 : d_2 : d_3 : d_4 = 1 : 1 : 1 : 1$).

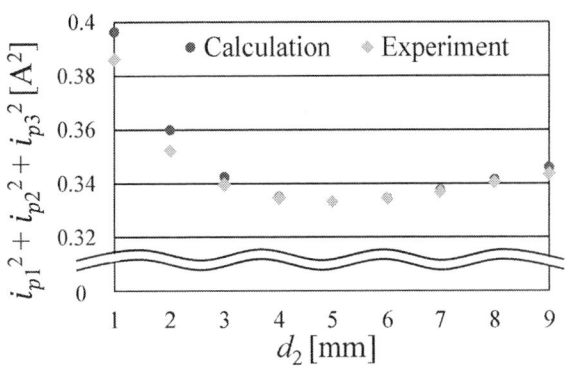

Fig. 8: Experimental and calculation results of factor $i_{p1}^2 + i_{p2}^2 + i_{p3}^2$ compared among 9 conditions of the distance

Let to focus on $d_2 = 1$ mm, which is the original transformer condition, and $d_2 = 5$ mm, in which AC current distribution is homogenized. The respective AC current distributions are shown in Table II. According to the results, the primary current distribution can be made homogenize by adjusting to the optimum distance between the winding layers. It is implied that the proposed method can homogenize the AC current distribution in the parallel connection of the interleaved winding layers. On the other hand, it was confirmed that the secondary current distribution kept homogenized as expected even if the distance between the winding layers was changed.

Table II: Measurement result of the AC current distribution

d_2[mm]	i_{p1}[Arms]	i_{p2}[Arms]	i_{p3}[Arms]	i_{s1}[Arms]	i_{s2}[Arms]
1	0.241	0.521	0.238	0.501	0.499
5	0.336	0.332	0.332	0.502	0.498

However, in the case of the estimation of the copper losses by measuring the AC current distribution, the skin effect loss and the proximity effect loss cannot be included in the copper loss. Therefore, the primary conversion AC resistances of respective conditions are measured with the secondary side short-circuited condition. The AC resistances were measured by LCR meter (ZM2376; NF co. ltd) from 10 kHz to 1 MHz. Focus on the original transformer structure ($d_2 = 1$ mm), and the adjusted transformer structure ($d_2 = 5$ mm). Fig. 9 shows the respective measurement results from 10 kHz to 100 kHz. Also, Fig. 10 shows the respective measurement results from 100 kHz to 1 MHz.

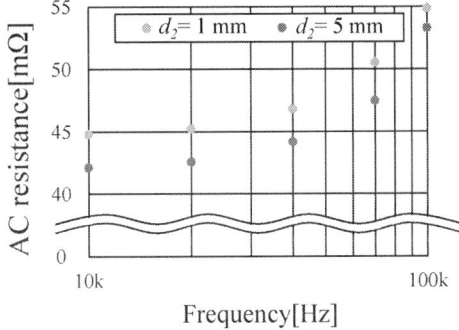

Fig. 9: AC resistance ($f = 10$ kHz ~ 100 kHz)

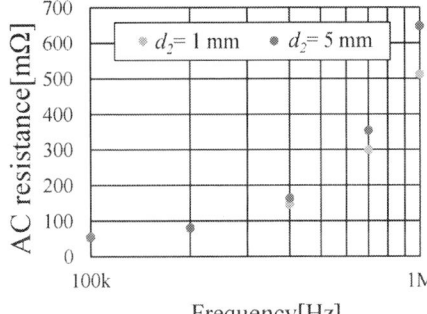

Fig. 10: AC resistance ($f = 100$ kHz ~ 1 MHz)

In the case of estimation of the AC resistance by the AC current distribution without the skin effect and proximity effect, the AC resistance of the adjusted transformer structure ($d_2 = 5$ mm) is theoretically expected as 7.0 % lower than the AC resistance of the original transformer structure ($d_2 = 1$ mm). At 100 kHz as shown in Fig. 9, the adjusted transformer structure ($d_2 = 5$ mm) has a smaller AC resistance 2.9% compared to the original transformer structure ($d_2 = 1$ mm). Because the effect of reducing the copper loss by the uniform current distribution and the effect of increasing the copper loss by the proximity effect cancel each other, the copper loss at 100 kHz cannot be decreased theoretically. At a frequency range from 10 kHz to 70 kHz, the adjusted transformer structure ($d_2 = 5$ mm) has a smaller AC resistance 5.6% to 6.1 % compared to the original transformer structure ($d_2 = 1$ mm). In this frequency range where the proximity effect does not occur so much, the copper loss can be suppressed almost as theoretically, and thus the validity of the proposed method was successfully revealed. However, at high frequency range from 200 kHz to 1 MHz, the adjusted transformer structure ($d_2 = 5$ mm) conversely has a larger AC resistance 2.7 % to 26.3 % compared to the original transformer structure ($d_2 = 1$ mm). Because the increase of the AC resistance by the proximity effect at high frequency is more than the reduction of the AC resistance by homogenizing the AC current distribution.

Therefore, the AC current distribution can be homogenized by adjusting the distance between winding layers, however, the magnetic field distribution also changes, and the proximity effect loss may increase at high frequency. Consequently, the proposed method in this paper can reduce the copper loss of the transformer by controlling the AC current distribution and magnetic field distribution according to the switching frequency.

Conclusion

The Parallel-connection of the interleaved winding layers is recently becoming used in high-frequency power transformer. However, the parallel connection of the winding layers can easily result in the unbalanced AC current distribution, which hinders the effective reduction of the copper loss. For achieving the homogeneous AC current distribution in the parallel-connected winding layers, this paper proposed optimization of the distance between the winding layers. This paper presented the technique to derive the optimum distance between the winding layers according to the recently proposed analysis method called as the extremum co-energy principle. As a result of the experiment, the AC current distribution can be homogenized by adjusting to the optimum distance between the winding layers. However, the magnetic field distribution also changes, and the proximity effect loss may increase at high frequency compared to the original transformer. Therefore, the proposed method in this paper can reduce the copper loss of the transformer by controlling the AC current distribution and magnetic field distribution according to the switching frequency. Consequently, the optimization of the distance between the winding layers for reducing the copper loss is successfully revealed effectiveness.

References

[1] M. Noah, T. Shirakawa, K. Umetani, J. Imaoka, M. Yamamoto, and E. Hiraki.: Effects of secondary leakage inductance on the LLC resonant converter-Part II: Frequency control bandwidth with respect to load variation, Proc. IEEE Appl. Power Electron. Conf. (APEC2019), Anaheim, CA, USA, Mar. 2019, pp. 1408-1414.

[2] B. A. Reese and C. R. Sullivan.: Litz wire in the MHz range: modeling and improved designs, Proc. IEEE Workshop Control Modeling Power Electron. (COMPEL2018), pp. 1-8, Jul. 2017.

[3] C. R. Sullivan.: Optimal choice for number of strands in a litz-wire transformer winding, IEEE Trans. on Pow. Electr., vol. 14, no. 2, pp.283-291, 1999.

[4] C. R. Sullivan and R. Y. Zhang.: Simplified design method for litz wire, Proc. IEEE Appl. Power Electron. Conf. Expo. (APEC2014), pp. 2667-2674, Mar. 2014.

[5] Prieto R., Cobos J. A., Garcia O., Alou P., Uceda J.: Using parallel windings in planar magnetic components, Proc. IEEE Power Electron. Specialist Conf. (PESC2001) Vol. 4, pp. 2055-2060

[6] Hu Y., Guan J., Bai X., Chen W.: Problems of paralleling windings for planar transformers and solutions, Proc. IEEE Power Electron. Specialist Conf. (PESC2002), Vol. 2, pp. 597-601

[7] Chen W., Yan Y., Hu Y., Lu Q.: Model and design of PCB parallel winding for planar transformer, IEEE Trans. Magn. Vol. 39 no. 5, pp. 3202-3204, 2003

[8] Fu D., Lee F. C., Wang S.: Investigation on transformer design of high frequency high efficiency DC-DC

converters, Proc. IEEE Appl. Power Electron. Conf. Expo. (APEC2010), pp. 940-947

[9] Margueron X., Besri A., Lembeye Y., Keradec J.-P.: Current sharing between parallel turns of a planar transformer: prediction and improvement using a circuit simulation software, IEEE Trans. Ind. Vol. 46 no. 3, pp. 1064-1070, 2010

[10] Prieto R., Asensi R., Cobos J.A.: Selection of the appropriate winding setup in planar inductors with parallel windings, Proc. IEEE Energy Conversion Congr. Expo. (ECCE2010), pp. 4599-4604

[11] Chen M., Araghchini M., Afridi K. K., Lang J. H., Sullivan C. R., Perreault D. J.: A systematic approach to modeling impedances and current distribution in planar magnetics, IEEE Trans. Power Electron. Vol. 31, no.1, pp. 560-580, 2016

[12] van Wyk Jr. J. D., Cronje W. A., van Wyk J. D., Campbell C. K., Wolmarans P. J.: Power electronic interconnects: skin- and proximity effect-based frequency selective multipath propagation, IEEE Trans. Power Electron. Vol. 20 no. 3, pp. 600–610, 2005

[13] Asensi R., Prieto R., Cobos J. A.: Automatized connection of the layers of planar transformers with parallel windings to improve the component behavior, Proc. IEEE Applied Power Electron. Conf. Expo. (APEC2012), pp. 1778–1782

[14] Suzuki Y, Hasegawa I., Sakabe S., Yamada T.: Effective electromagnetic field analysis using finite element method for high frequency transformers with Litz-wire, Proc. IEEE Intl. Conf. Elect. Mach. Syst. (ICEMS2008), pp. 4388–4393

[15] Nabaei V., Mousavi S. A., Miralikhani K., Mohseni H.: Balancing current distribution in parallel windings of furnace transformers using the generic algorithm, IEEE Trans. Magn. Vol. 46 no. 2, pp. 626–629, 2010

[16] Shirakawa T., Yamasaki G., Umetani K., Hiraki E.: Extremum co-energy principle for analyzing AC current distribution in parallel-connected wires of high frequency power inductors, Proc. IEEE Intl. Conf. Elect. Mach. Syst. (ICEMS2016), pp. 1–6

[17] Shirakawa T., Yamasaki G., Umetani K., Hiraki E.: Copper loss analysis based on extremum co-energy priciple for high frequency forward transformers with parallel-connected windings, Proc. Annu. Conf. IEEE Ind. Electron. Soc. (IECON2016), pp. 1099–1105

[18] Shirakawa T., Umetani K., Hiraki E.: Application of Extremum Co-Energy Principle for Homogenizing Current Distribution in Parallel-Connected Windings in Transformers: Design Optimization of Winding Turn Allocation among Winding Layers, 19th European Conf. on Power Electron. and Appl. (EPE'17), pp. 1-10, Sep. 11-14, 2017.

[19] T. Shirakawa, K. Umetani, E. Hiraki, Y. Itoh, and T. Hyodo.: Optimal winding layer allocation for minimizing copper loss of secondary-side center-tapped forward transformer with parallel-connected secondary windings, Proc. IEEE Energy Conversion Conf. Expo. (ECCE2019), Oct. 2019, Baltimore, ML, USA, pp. 6206-6213.

[20] Krishnan R.: Switched reluctance motor drives, CRC Press, 2000, pp. 3–7

[21] Miller T. J. E.: Electronic control of switched reluctance machines, Newns, 2001, pp. 43–45

[22] Vandelac J.-P., Ziogas P. D.: A novel approach for minimizing high-frequency transformer copper loss, IEEE Trans. Power Electron. Vol. 3 no. 3, pp. 266–277, 1988

[23] Ferreira J. A.: Appropriate modelling of conductive losses in the design of magnetic components, Proc. IEEE Power Electron. Specialist Conf. (PESC1990), pp. 780–785

[24] Robert F., Mathys P.: Ohmic losses calculation in SMPS transformers: numerical study of Dowell's approach accuracy, IEEE Trans. Magn. Vol. 34 no. 4, pp. 1255–1257, 1998

[25] Hurley W. G., Gath E., Breslin J. G.: Optimizing the AC resistance of multilayer transformer windings with arbitrary current waveforms, IEEE Power Electron. Vol. 15 no. 2, pp. 369–376, 2000

[26] Strydom J. T., J. D. van Wyk: Improved loss determination for planar integrated power passive modules, Proc. IEEE Appl. Power Electron. Conf. Expo. (APEC2002) Vol. 1, pp. 332–338

Real-time Parameters Identification of Lithium-ion Batteries Model to Improve the Hierarchical Model Predictive Control of Building MicroGrids

Daniela Yassuda Yamashita[a,b], Ionel Vechiu[a] and Jean-Paul Gaubert[b]

[a]ESTIA INSTITUTE OF TECHNOLOGY
Technopole Izarbel – 92, Allée Théodore Monod – Bidart, France
[b]LABORATOIRE D'INFORMATIQUE ET D'AUTOMATIQUE POUR LES SYSTEMES
(LIAS), UNIVERSITE DE POITIERS
4, Rue Pierre Brousse – Poitiers, France
d.yamashita@estia.fr, i.vechiu@estia.fr, jean.paul.gaubert@univ-poitiers.fr

Acknowledgements

The authors would like to thank the New Aquitaine Region for their financial support.

Keywords

Battery Management Systems (BMS), Control methods for electrical systems, MPC (Model-based Predictive Control), Device modelling, Microgrid

Abstract

Energy storage systems are key elements for enabling the design of MicroGrids in buildings, specially to deal with stochastic renewable energy resources and to promote peak shifting. However, inaccuracies in the batteries' mathematical models due to temperature and ageing effects can reduce the performance of a MicroGrid system. To tackle these uncertainties, this article presents a two-level hierarchical model predictive controller empowered with a data-driven algorithm for real-time model identification of Lithium-ion batteries. The objective is to enhance their state of charge estimation and to make their maximum use without damaging them. The results demonstrate that it improves up to three times the accuracy of state-of-charge estimation and increases about 3% the annual building MicroGrid self-consumption rate. Furthermore, the division of the building MicroGrid energy management system into two hierarchical levels soften the drawbacks arise from the inaccuracies of day-ahead data prediction while reducing the computational cost. The proposed architecture guarantees higher energetic autonomy indexes than a conventional rule-based controller in all scenarios under study.

Introduction

Building MicroGrids (BMGs) are an attractive alternative to foster the integration of renewable energy sources (RESs) into the electrical grid [1]. However, their wide implementation is restrained by the difficulty of designing a generic Energy Management System (EMS) capable of operating under stochastic variations in the power plant. Unpredictability in power generation and power consumption, along with inaccuracy of mathematical models of BMG components lead to the under exploitation of BMG resources or equipment damaging [2].

In the literature, there are various strategies to implement a reliable and efficient EMS [3]. Among the existent strategies, the hierarchical control structure enables to embed algorithms with different complexities at the same time, thanks to the parallel coordination of multiple control layers deployed at different sample time [4]. Concerning the EMS algorithms, those based on model predictive control (MPC) have proved their robustness against environmental disturbances, even with simplified plant model [5]. Nonetheless, there is a lack in evaluating the MPC performance under environmental changes, such as temperature, electric devices ageing and batteries model parameters inaccuracy.

In order to face these uncertainties, there are several techniques to estimate better the intrinsic parameters of batteries, such as the Arrhenius equation [6], or models devised from technical specifications [2]. However, they require beforehand model calibration, which can lead to uncertainties throughout the batteries' life. In this context, strategies based on data analysis are increasingly implemented. The most pertinent algorithms are incremental analysis of the voltage and capacity to estimate the batteries' state of health [7], Kalman filter estimator [8] and other machine learning methods [9].

Therefore, this article proposes a two-level Hierarchical MPC (HMPC) with a Real-Time Model Identification (RTMI) module to deal with batteries' parameter inaccuracy. Equipped with photovoltaic (PV) panels and Li-ion batteries, the purpose of the proposed HMPC is to maximise the MicroGrid self-consumption rate (τ_a) [10]. Relying on continuous data measurement and forecast data, the strategy is to optimise its internal power flow by promoting the use of renewables and reducing the energy dependency on the external grid. The results show that the RTMI enhances the State-of-Charge (SoC) estimation, leveraging τ_a and leading HMPC to outperform the traditional rule-based strategy.

The remaining of this paper is organised as follows. In the second section, the overview of the underlying hierarchical control structure associated with the proposed algorithm for identifying the battery parameters is described. The third section details the new model developed for the estimation of the SoC and the RTMI algorithm. The fourth section presents the performance of the proposed control strategy, by comparing the HMPC with RTMI module with both the conventional MPC under hierarchical and non-hierarchical architecture and a traditional rule-based controller. In the last section, the conclusions on the advantages and disadvantages of the proposed approach are summarized.

Overview of the hierarchical energy management system

The designed hierarchical EMS is a centralised controller for optimising the power flow of a grid-connected BMG that interacts with the external grid throughout a community aggregator, as illustrated in Fig. 1. Composed mainly by three control units, the objective of the proposed control architecture is to assure the building internal power balance with minimum dependency on the main grid. The following two subsections describe the hierarchical MPC and the RTMI module, respectively.

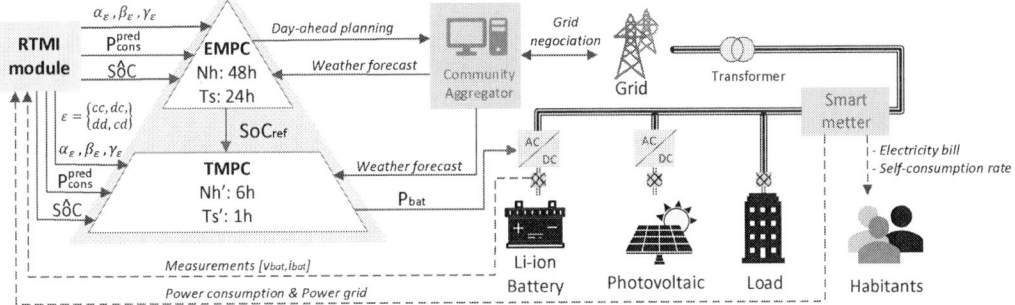

Fig. 1: Two-level hierarchical model predictive control architecture empowered with the batteries' real-time model identification module for energy management of a grid-connected building MicroGrid.

Hierarchical model predictive controller

The proposed HMPC disposes of two control levels, namely Economic MPC (EMPC) and Tracking MPC (TMPC). The upper control level performs the economic power dispatch through an EMPC and determines both the day-ahead electricity trading planning to be sent toward the community aggregator and the batteries' State-of-Charge references (SoC_{ref}) to be forward to the TMPC. Simultaneously, the lower level determines the batteries' power references (P_{bat}) based on the updated prediction data and measurements, by performing a TMPC to follow SoC_{ref}.

Considering the fluctuations of the data prediction of both PV power generation and building power consumption (P_{cons}^{pred}), the EMPC determines SoC_{ref} that, within a horizon of $N_h^{EMPC} = 48h$, minimises the external grid energy dependency. Therefore, the EMPC maximises τ_a defined by (1) and minimises additional energy imports to reduce the cost of purchasing electricity. For this, the cost function defined by (2) is optimised daily ($T_s^{EMPC} = 24h$), in which the total power injected ($P_{injected}$) and the total power purchased ($P_{purchased}$) are minimised. It is noteworthy that the horizon (N_h^{EMPC}) twice bigger than the updating time (T_s^{EMPC}) is to guarantee the most suitable SoC at the end of the day. Doubling the horizon avoids unnecessary completely discharge of batteries at the end of the day since the last half of control variables determined by the EMPC is not implemented.

$$\tau_a = \frac{Total\ energy\ produced\ from\ RES\ that\ is\ consummed\ locally}{Total\ energy\ produced\ from\ RES\ locally} = 1 - \frac{\sum_k P_{injected,k}}{\sum_k P_{PV,k}} \tag{1}$$

$$SoC_{ref} = \arg\left(\min_{SoC_{ref}} \sum_{k=1}^{N_h^{EMPC}=48} P_{injected,k} + P_{purchased,k}\right) \tag{2}$$

Parallelly to the EMPC, the TMPC is triggered at each $T_s^{TMPC} = 1h$ to follow the SoC_{ref} as much as possible, considering the updated data prediction and the $\widehat{SoC}$ estimated from the last measurements. To achieve this, the TMPC minimises hourly the cost function defined by (3). The purpose of the TMPC is to reduce the stochastic variation of the day-ahead power imbalance and battery model inaccuracies, by using a shorten horizon, simplifying the optimisation process.

$$P_{bat} = \arg\left(\min_{P_{bat}} \sum_{k=1}^{N_h^{TMPC}=6} \left(SoC_{ref,k} - \widehat{SoC}_k\right)^2\right) \tag{3}$$

Meanwhile, in the background, the EMPC supervises the performance of the TMPC. Every hour, the EMPC compares the accuracy of day-ahead power planning sent to the aggregator (P_{grid}^{ref}) and SoC_{ref} with real measurements. As soon as the absolute difference between them is higher than a predefined threshold – named Δ_{SoC}^{thr} or $\Delta_{P_{grid}}^{thr}$ at time t_{reOpt} – the EMPC determines new SoC_{ref} using the updated prediction data, but with a reduced horizon, as illustrated in Fig. 2. The reduced horizon comprehends the data prediction of the remaining period of the first EMPC optimisation and is equal to $N_h - t_{reOpt}$.

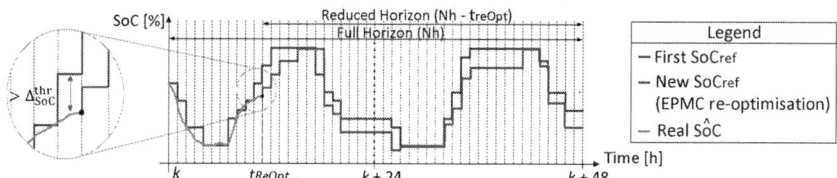

Fig. 2: Re-optimization process of EMPC with reduced horizon.

The main constraints embedded in both EMPC and TMPC are those to respect the recommendations of the French Energy Regulation Commission (ERC) for grid-connected BMGs with PV capacity over 100 kWc [11]. To prevent energy speculation with the local energy storage systems (ESSs) when trading electricity, the ERC allows charging the ESSs only from renewable power generated locally. Additionally, it imposes that $\tau_a \geq 50\%$ at the end of the year [12]. As a result, batteries can be charged (P_{bat}^{ch}) as soon as there is an internal energy surplus, while they can be discharged (P_{bat}^{dis}) when there is an energy deficit, as detailed in (4). Moreover, aiming to extend the life of batteries, the maximum power rate (P_{bat}^{max}) and the SoC limits defined by their manufacturer must be supervised [2], [13]. Consequently, it is important to well estimate future SoC (SoC_{k+1}) and to limit them. Similarly, to maximize τ_a beyond the MPC horizon, the SoC at $k = 48h$ is forced to be higher than 40%.

$$-\left|\max\left(P_{bat}^{MAX}, P_{surplus}\right)\right| \leq P_{bat,k}^{ch} \leq 0 \; ; \; 0 \leq P_{bat,k}^{dis} \leq \left|\max\left(P_{bat}^{MAX}, P_{deficit}\right)\right| \tag{4}$$

Real-time model identification algorithm

Most of the scientific studies [5] estimate SoC_{k+1} through a model composed of time-invariant parameters derived from batteries' technical specification, as formulated in (5). However, based on more realistic models of Li-ion batteries [13], the efficiency during its charge (η_{ch}) or discharge (η_{dis}), the nominal capacity (Q_{nom}) and the nominal voltage (v_{nom}) change according to the intensity of current, battery age and cell temperature (T_{cell}). According to [7], [9] the voltage variation can be around 10% of the nominal voltage when they are fully charged and discharged. Moreover, the nominal capacity can reach, at the end of their life, up to 80% of its initial value [9]. Consequently, additional uncertainties on SoC_{k+1} estimation arise, which may result in under or overuse of the batteries.

$$SoC_{k+1} = SoC_k + \frac{\eta_{ch} T_s}{v_{nom} \cdot Q_{nom}} \cdot P_{ch,k} + \frac{T_s}{v_{nom} \cdot Q_{nom} \eta_{dis}} \cdot P_{dis,k} \tag{5}$$

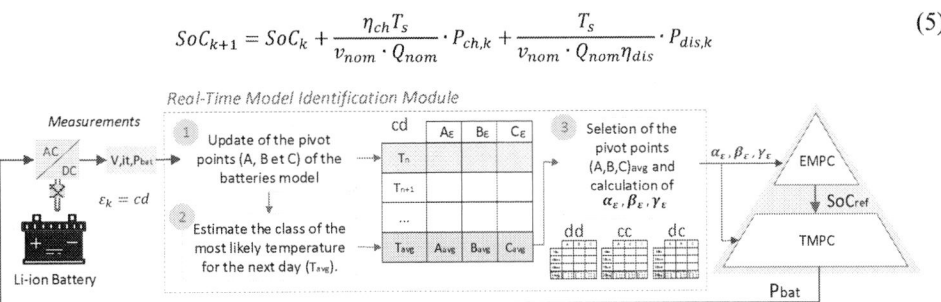

Fig. 3: Real-time parameters identification algorithm for a charge and discharge sequence ($\varepsilon_k = cd$).

To tackle this issue, the authors designed the algorithm RTMI to improve the estimation of both SoC_{k+1} and their upper (SoC_{max}^{k+1}) and lower boundaries (SoC_{min}^{k+1}). The main objective of this new algorithm is to reduce human intervention for model calibration and BMG maintenance while taking full advantages of batteries. Therefore, the RTMI module updates the parameters of SoC_{k+1}, as illustrated in Fig. 3. The limits SoC_{max}^{k+1} and SoC_{min}^{k+1} are determined to guarantee the operation of the batteries in the linear region, as shown in Fig. 4a.

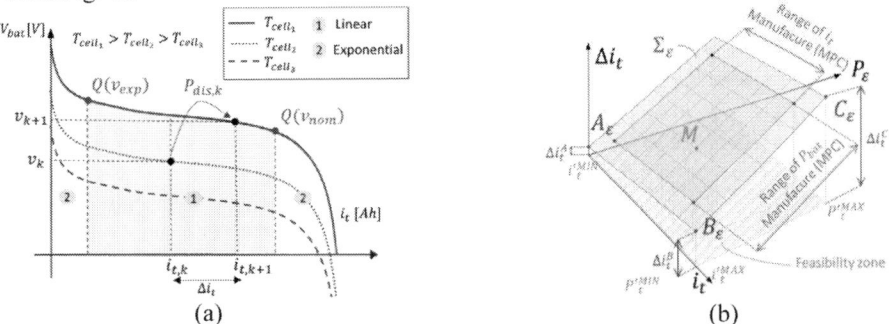

(a) (b)

Fig. 4: RTMI module operation. (a) Discharge curves of a Li-ion battery under temperature cell variations. (b) Model perspective with the three pivot points for predicting the variation of SoC (Δi_t).

The RTMI module determines the parameters α_ε, β_ε and γ_ε of the SoC model defined by (6) and (7) by implementing an unconventional linear regression based on previous data measurements. The variable $\delta_{\varepsilon,k}$ is a Boolean variable which is worth 1 when the battery is active and 0 otherwise. Remarkably, equation (7) can be interpreted as four surfaces Σ_ε of axis $xyz_\varepsilon = [P_\varepsilon, i_t, \Delta i_t]_\varepsilon$, as illustrated in Fig. 4b. Then, each surface xyz can be represented by three pivot points, namely $A_\varepsilon = [i_t^{MIN}, P_{bat}^{MIN}, \Delta i_t^A]_\varepsilon$, $B_\varepsilon = [i_t^{MAX}, P_{bat}^{MIN}, \Delta i_t^B]_\varepsilon$ and $C_\varepsilon = [i_t^{MAX}, P_{bat}^{MAX}, \Delta i_t^C]_\varepsilon$, which cover the feasibility zone which corresponds to the limits recommended by the manufacturer extended in 20%. To reduce the impact of voltage variation, the surfaces Σ_ε are classified into four main categories, named $\varepsilon = \{cd, dd, cc, dc\}$, which indicate the sequence of batteries' charge (c) and discharge (d), at time $k - 1$ and k. Consequently, the RTMI sends daily to the HMPC only the values of α_ε, β_ε and γ_ε corresponding to the most likely temperature for the next day, which is the average temperature of the previous day.

$$SoC_{k+1} = 1 - \frac{i_{t,k} + \Delta i_{t,k}}{Q_{nom_k}}; \quad P_{bat,k} = P_{bat,k}^{ch} + P_{bat,k}^{dis} = \sum_{\varepsilon = \{cd,dd,cc,dc\}} P_{\varepsilon,k} \tag{6}$$

$$\Delta i_{t,k} = \sum_{\varepsilon = \{cd,dd,cc,dc\}} \alpha_\varepsilon P_{\varepsilon,k} + \beta_\varepsilon i_{t,k-1} + \gamma_\varepsilon \delta_{\varepsilon,k} \tag{7}$$

Details of the RTMI algorithm

Synchronized with both EMPC and TMPC, the RTMI module acquires a new measurement point M, composed by the integral of the current (i_t), the reference power (P_{bat}) and the battery voltage (v) at each TMPC sampling time (i.e. T_s^{TMPC}). As detailed in the following sub-sections, the RTMI algorithm is divided into three steps, namely: classification, updating and identification of the limits for charging and discharging the batteries.

Step 1: Classification of data measurements by temperature interval

As depicted in Fig. 4a, v is directly correlated to the T_{cell}. This effect in the battery voltage impacts the batteries round-trip efficiency, reducing the accuracy of the classical model defined by (7). Higher voltage when charging the batteries or lower voltages when discharging them implies a loss of efficiency, once the storage energy variation is mainly dependent on the current flowing through the batterie cells. Therefore, aiming to improve the robustness against T_{cell} disturbance, the RTMI fits a linear model for each temperature ranges using classified measurements points. Without using any temperature sensor, the RTMI algorithm estimates the temperature interval from v measurements. The Fig. 4a and Fig. 5a show that the T_{cell} mainly involves a vertical offset in the batteries discharge curve, but almost does not affect the slope of v with respect to i_t. In this way, the maximum amplitude of v (Δv_{MAX}) due to the full charges and discharges of the batteries is almost unchanged, resulting in a quasi-constant slope along with different T_{cell}, as depicted in Fig. 5b and Fig. 5c.

Consequently, when the BMG operates for the first time, the EMS will force two full charges and discharges of the batteries to determine Δv_{MAX}, and therefore, the absolute value of the slope of the discharge curve ($|\Delta v/i_t|$). After this preliminary step, the temperature intervals are defined by identifying the lines $\vec{s}_{T_{class}}$, as shown in Fig. 5c, where T_{class} is the identification number for a specific temperature class. The lines $\vec{s}_{T_{class}}$ are outspread from a voltage interval which is chosen a priori regarding the desired precision, the computing resources and the RTMI convergence time. Typically, this voltage offset can be set as 1% of v_{nom}. As a result, a measurement M belongs to T_{class}, if and only if, the point M is between the lines $\vec{s}_{T_{class}}$ and $\vec{s}_{T_{class}+1}$.

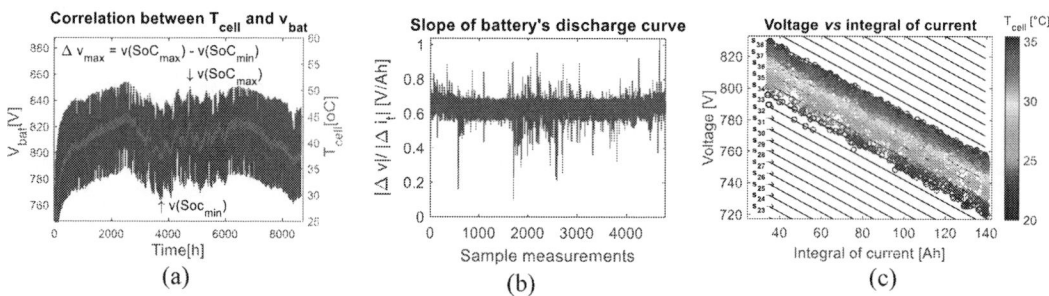

Fig. 5: One-year simulation of the BMG with SoC$_{max}$ = 80% and SoC$_{min}$ = 20%. (a) v and T_{cell} correlation. (b) Slope of battery discharge curve. (c) Correlation between voltage measurements, current integral and T_{cell}.

Step 2: Updating the values of α_ε, β_ε and γ_ε

Once determined which class of temperature the measurement M belongs to, it is possible to adjust the values of $\alpha_\varepsilon, \beta_\varepsilon$ and γ_ε of these respective T_{class}. The determination of $\alpha_\varepsilon, \beta_\varepsilon$ and γ_ε is implemented based on an iterative and real-time process. As soon as a new measurement M is acquired, the parameters of $\alpha_\varepsilon, \beta_\varepsilon$ and γ_ε are updated. Remarkably, $\alpha_\varepsilon, \beta_\varepsilon$ and γ_ε are vectors of size equals to the number of T_{class}, where each element of these vectors is the fitted values of the model defined by (8) over all measurements M. The upmost advantage of the RTMI algorithm is the non-dependency on storage of the past measurements. The knowledge of the previous measures is stored in dynamic confidence weights, named $\omega_{A,k}$, $\omega_{B,k}$ and $\omega_{C,k}$.

Starting from minimal values – typically between 0.1 and 1.0 – these confidence weights grow with the acquisition of new measurements. Since the pivot points possess fixed values in the plan $xy = [P_\varepsilon, i_t]$, the fitting process only consists of calculating the values of z-axis (i.e. $\Delta i_{t,k}^A$, $\Delta i_{t,k}^B$ and $\Delta i_{t,k}^C$). For the first iteration, the pivot points are initialized according to (8), where it is used the parameters given by the technical specification of the batteries and using the classical model defined in (7).

$$\Delta i_{t,k=0}^A = \Delta i_{t,k=0}^B = \frac{\eta_{ch} T_s}{Q_{nom} v_{nom}} \cdot P_{bat}^{MIN} ; \Delta i_{t,k=0}^C = \frac{T_s}{Q_{nom} v_{nom} \eta_{dis}} \cdot P_{bat}^{MAX} \tag{8}$$

Thereafter, the values of $\Delta i_{t,k}^A$, $\Delta i_{t,k}^B$ and $\Delta i_{t,k}^C$ are updated so that to approach the surface Σ_ε to the new measurement M. To achieve this, an intermediate surface ϕ_ε defined by the intermediary pivot points A'_ε, B'_ε and C'_ε, with z-axis values equal to z_A, z_B and z_C, are calculated through the optimisation of the cost function defined by (9). The surface ϕ_ε contains the point M and is determined to make $\Delta i_{t,k+1}^{A,B,C}$ of the next iteration as close as possible of $\Delta i_{t,k}^{A,B,C}$ of the previous period.

$$z_A, z_B, z_C = \arg\left(\min_{z_A, z_B, z_C} n_{A,k}\left(\Delta i_{t,k}^A - z_A\right)^2 + n_{B,k}\left(\Delta i_{t,k}^B - z_B\right)^2 + n_{C,k}\left(\Delta i_{t,k}^C - z_C\right)^2\right) \tag{9}$$

Subject to:

$$M \in A'_\varepsilon B'_\varepsilon C'_\varepsilon; \ z_A, z_B, z_C \geq 0$$

Intuitively, high value of $n_{A,k}$ and low values of $n_{B,k}$ and $n_{C,k}$ lead z_A approach to $\Delta i_{t,k}^A$ faster than z_B and z_C to $\Delta i_{t,k}^B$ and $\Delta i_{t,k}^C$, whereas balanced values of $n_{A,k}$, $n_{B,k}$ and $n_{C,k}$ result in a fair variation among z_A, z_B and z_C. The weights of (9) are calculated with regard the normalized distance between the measure M and the respective intermediate pivot point projected on the plan xy (i.e. operator $\|\cdot\|_{xy}$), as defined in (10) for the pivot point A'_ε.

$$n_{A,k} = \frac{\|\overline{MA}\|_{xy}}{\|\overline{MA}\|_{xy} + \|\overline{MB}\|_{xy} + \|\overline{MC}\|_{xy}} \tag{10}$$

Once determined the intermediary plan ϕ_ε, $\Delta i_{t,k}^A$, $\Delta i_{t,k}^B$ and $\Delta i_{t,k}^C$ and the confidence weights $\omega_{A,k}$, $\omega_{B,k}$ and $\omega_{C,k}$ are updated following equations (11) – (14), in the order of compilation. For the sake of simplicity, in this paper, it will only be detailed the updating process of the pivot point A_ε, but it is important to highlight that similar equations are used for B'_ε and C'_ε.

$$\omega_{A,k+1} = \omega_{A,k} + \frac{\frac{1}{\|\overline{MA}\|_{xy}}}{\frac{1}{\|\overline{MA}\|_{xy}} + \frac{1}{\|\overline{MB}\|_{xy}} + \frac{1}{\|\overline{MC}\|_{xy}}} \tag{11}$$

$$\omega_{inertia,k}^A = \sigma\left(\omega_{A,k+1} - \omega_{A,k}\right) \leq \omega_{A,k+1} \tag{12}$$

$$\Delta i_{t,k+1}^A = \frac{\Delta i_{t,k}^A \cdot \omega_{inertia,k}^A + z_A \cdot \omega_{A,k+1}}{\omega_{inertia,k}^A + \omega_{A,k+1}} \tag{13}$$

$$\omega_{A,k+1} = \omega_{A,k+1} - \tau_{decay} \cdot \left(\omega_{A,k+1} - \omega_{min}\right) \tag{14}$$

Firstly, the confidence weight $\omega_{A,k}$ is updated using (11). The closer M is to the A_ε, the more reliable the value of z_A calculated from (9) is. Consequently, $\omega_{A,k+1}$ is inversely proportional to the distance between the pivot point A_ε and the measurement M, and it is normalised according to the distance from other pivot points (i.e. B_ε and C_ε). To improve the robustness against measurement noise, the updated $\Delta i_{t,k+1}^A$ is a weighted value between the previous $\Delta i_{t,k}^A$ and the new fitted z_A. The inertial factor handles the ponderation among these two variables $\omega_{inertia,k}^A$ and the confidence weight $\omega_{A,k+1}$ as defined in (13). The equilibrium amongst $\omega_{inertia,k}^A$ and $\omega_{A,k+1}$ control the convergence time of $\Delta i_{t,k+1}^A$ to z_A. This balance can be manually set by tuning the value of σ, which indicates the importance of the new measurement regarding the previous value. Notably, $\omega_{inertia,k}^A$ is upper limited by $\omega_{A,k+1}$, to restrict the convergence time to a scale of two. Since the charge/discharge curve of batteries changes with age, the confidence weight $\omega_{A,k+1}$ gradually decreases with a time constant τ_{decay} until a minimum value ω_{min} which is equal to the starting value, as specified in (14).

After updating the plan Σ_ε, the coefficients α_ε, β_ε and γ_ε are calculated using the principles of analytical geometry. As the last step, it is still necessary to refine the developed model, because according to (7) and Fig. 4b, the plan Σ_ε does not cross the $\vec{\imath}_t$ axis, which makes batteries model inaccurate for low powers. Therefore, the model polishing consists of forcing the plan Σ_ε cross the $\vec{\imath}_t$ axis by using (15).

$$\Delta i_{t,k} = \sum_{\varepsilon=\{cd,dd,cc,dc\}} \theta_\varepsilon P_{\varepsilon,k} = \sum_{\varepsilon=\{cd,dd,cc,dc\}} P_{\varepsilon,k}\left(\alpha_\varepsilon + \frac{1}{P_{bat}^{'MAX}}\cdot\left(\gamma_\varepsilon + \frac{(i_t^{'MAX}+i_t^{'MIN})\cdot\beta_\varepsilon}{2}\right)\right) \tag{15}$$

Step 3: Identification of the limits for charging and discharging the batteries

As mentioned before and illustrated in Fig. 6a, the batteries must operate between points $Q(v_{exp})$ and $Q(v_{nom})$. The strategy consists in adjusting $i_{t,max}$ and $i_{t,min}$ to reduce the variations of the slope $|\Delta v/\Delta i_t|$. According to Fig. 6a, while i_t is inside the linear zone (zone 1), the slope $|\Delta v/\Delta i_t|$ is quasi-constant, because v is linearly dependent on i_t. However, when operating outside this zone (zone 2), $|\Delta v/\Delta i_t|$ is not constant because v is non-linear regarding i_t.

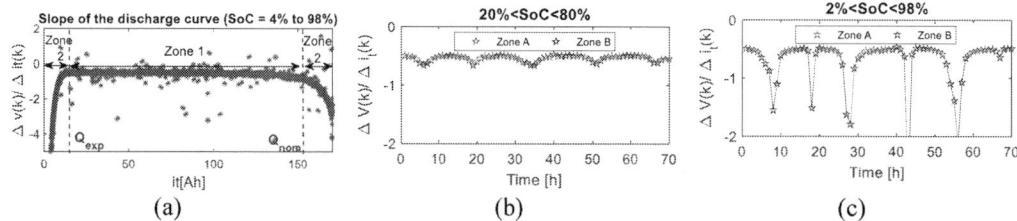

(a)   (b)   (c)

Fig. 6: Simulation of Li-ion batteries. (a) The slope of the charge/discharge curves as a function of i_t. (b) and (c) Temporal evolution of the slope when operating in linear and non-linear zones, respectively.

Based on this phenomenon, the developed algorithm for identifying the actual boundaries of i_t divides the temporal graphs into two zones, named *zone A* and *zone B*, as illustrated in Fig. 6b and Fig. 6c. The *zone A*, indicated by the red dots, represents the range where the batteries certainly operates in the linear zone. On the other hand, the *zone B*, outlined by the blue dots, is the zone for which the battery can be either in the linear or non-linear region. Therefore, *zone A* comprehends the range of 20% to 80% of the predefined limits of $i_{t,min}$ and $i_{t,max}$, whereas *zone B* is its complementary region.

Remarkably, the oscillation of the slope is more intense when zone B is in the non-linear zone (Fig. 6c) than when it is inside the linear range (Fig. 6b). As a result, to determine whether zone B corresponds to the linear or non-linear range, the average of the absolute difference between each $|\Delta v/\Delta i_t|$ measured inside *zone B* (m_{up}) and those measured inside *zone A* (m_{down}) are constant compared, through the mean deviations ϑ_{up} and ϑ_{down} calculated as detailed in Fig. 7. These deviations ϑ_{up} and ϑ_{down} are monitored by means of a proportional controller with a hysteresis at its input to maintain them within the range $\vartheta_{up,down}^{ref} \pm Tol$. The values of $Tol \cong 2\%$ and $\vartheta_{up,down}^{ref} \cong 10\%$ were manually regulated, but they can be re-adjusted to reduce the oscillations or increase the response time of $i_{t,min}$ and $i_{t,max}$. If the i_t boundaries are modified more than 10% of the previous values, the feasibility zone of Fig. 4b is readjusted accordingly and the confidence values ω_A, ω_B, ω_C are reset to ω_{min}.

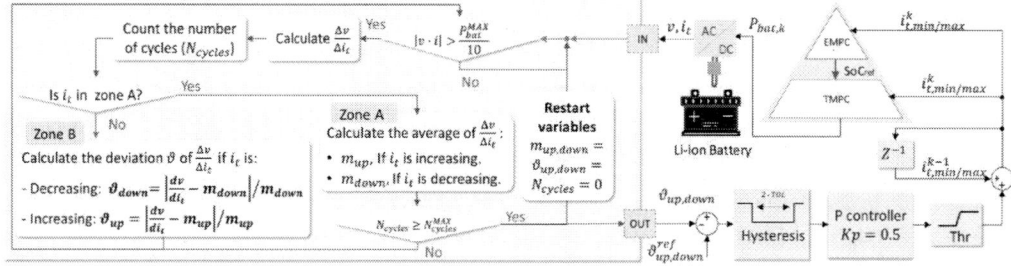

Fig. 7: Algorithm for identifying the i_t limits to guarantee the batteries' operation inside the linear zone.

Simulation results

Aiming to evaluate the performance of the proposed control architecture, a BMG equipped with PV arrays with 107kWc and Li-ion batteries with nominal voltage of 700V and nominal capacity of 167Ah was simulated for 365 days in MATLAB Simulink® under several scenarios. The simulations were carried out using real solar radiation data [14] and the estimated annual building energy consumption, resulting in 135.95 MWh energy generation and 241.85 MWh energy consumption per year. The τ_a and the total power exchanged with the main grid was used for assessing the proposed HMPC with RTMI module, a non-hierarchical MPC, and a conventional rule-based (RB) strategy with and without data prediction inaccuracy.

The RB was adapted from [15], where, using no data prediction, the batteries are charged when there is an energy surplus and discharged when there is an energy deficit. The simulation results are divided into two subsections. The first one is to validate the RTMI algorithm for SoC and batteries' capacity estimation, while the second one is for assessing the benefits of dividing the EMS hierarchically.

Performance of the algorithm for batteries' parameter identification

To show the robustness of the RTMI algorithm against inaccuracies of the parameters coming from the technical specifications, three initial values of Q were considered, namely Q_{80}, Q_{100} and Q_{120}, corresponding to 80%, 100% and 120% of the actual capacity (167 Ah), respectively. To verify the error between the day-ahead SoC_{ref} calculated by the EMPC and the real one, the TMPC in these scenarios was considered as a perfect router. Consequently, instead of optimising (3), it does implement the control variables determined by EMPC. In this manner, it is possible to decouple the effect of TMPC and highlight only the impact of the errors in EMPC state of charge estimation on the BMG performance. The graph in Fig. 8a shows that, in all study cases, the cumulative error in predicting SoC_{k+1} using the RTMI module is about 3 times lower than using the conventional model with static parameters. Furthermore, the graphs in Fig. 8b shows that the EMPC empowered by RTMI module assures τ_a about 3% higher than the EMPC without RTMI module with Q_{80} and between 2% and 4% higher than RB controller.

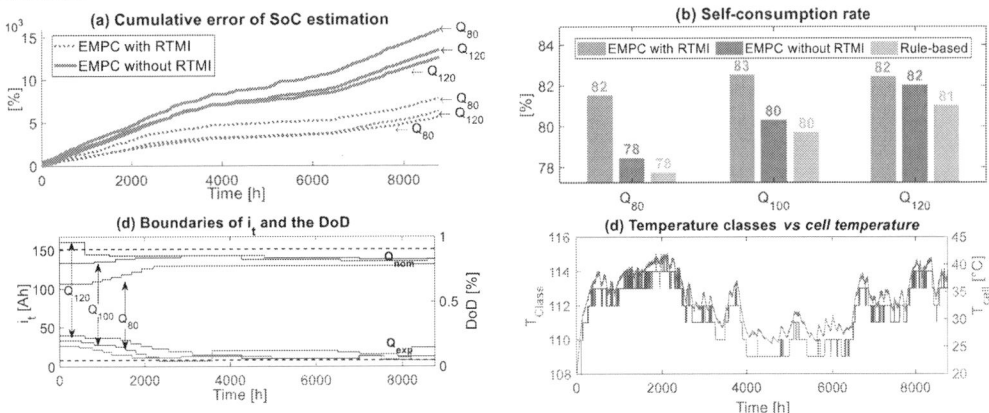

Fig. 8: Evaluation of the EMPC robustness against parameter imprecision. (a) The cumulative error of the SoC estimation. (b) Self-consumption rate comparison. (c) Maximum (in blue lines) and minimum (in red lines) boundaries of the i_t when using RTMI. (d) Estimation of T_{cell} variation through T_{class}.

The increase in τ_a is mainly due to the enlargement of the i_t boundaries closed to the real frontiers (Q_{exp} and Q_{nom}), as shown in Fig. 8c. Accurately estimating the battery capacity reinforces the potential of batteries of shifting the load toward the periods of energy surplus, resulting in an enhanced internal load matching and lower grid dependency. For this reason, the relative difference between τ_a is more remarkable when the batteries' capacity is underestimated (i.e. Q_{80}). Nonetheless, there is a tradeoff between fostering τ_a and the batteries' state of heathy. Fig. 8c shows that expanding i_t boundaries, the attained depth of discharge (DoD) is higher when employing RTMI module than not using it (at the beginning of the simulation), accelerating the batteries' degradation [7]. In addition, as shown in Fig. 8d employing the RTMI module and processing of v and i_t measurements, T_{cell} can be supervised without needing any thermal sensor, because T_{class} is an image of the temperature variation. This can provide further information to the HMPC for preserving the batteries state of healthy [2], [8].

Performance of the proposed two-level hierarchical control structure

Since the purpose of the hierarchical EMS is to soften the drawbacks provoked by stochastic variations in the internal power imbalance, the prediction data were multiplied by a random time-dependent factor (ρ), as detailed in Fig. 9a. In this manner, the error in the estimate power imbalance grows according to the horizon, attaining up to 60% at $N_h^{EMPC} = 48$, as shown in Fig. 9b and Fig. 9c. Aiming to assess the proposed HMPC, five control architectures were investigated. The first and second ones are the HMPC with and without RTMI modules. The third and fourth control structures are the non-hierarchical MPC with and without RTMI module, in which only EMPC is updated hourly with the full horizon. Finally, the fourth control disposal is the simple RB.

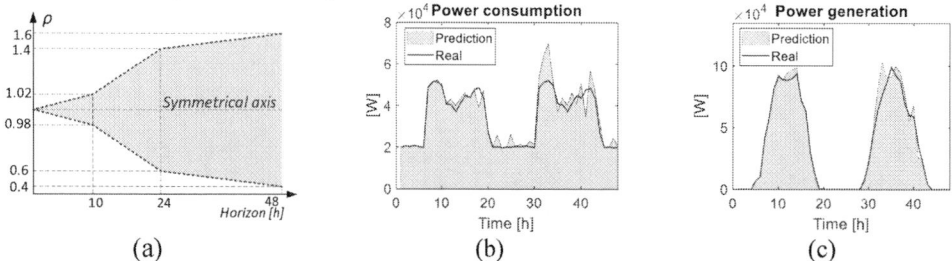

(a)   (b)   (c)

Fig. 9: Real power and its 48-hours data prediction of a summer day. (a) Time-variant factor. (b) Power consumption. (c) Power generation.

Fig. 10a and Fig. 10b show that both HMPC and the non-hierarchical MPC are robust against data prediction uncertainties, because with and without errors, the BMG imported and exported almost the same amount of energy in a year. Remarkably, when using HMPC with RTMI, it imported about 1% less energy and exported 8% less than without RTMI. Even though the non-hierarchical MPC optimises the cost function (2) around 11 times more than the HMPC, the results are very similar. Indeed, the non-hierarchical structure triggers EMPC at each hour, which means 8760 optimizations in a year, whereas only up to 631 times (number of re-optimisations plus once per day) using the proposed control structure, as detailed in Fig. 10c.

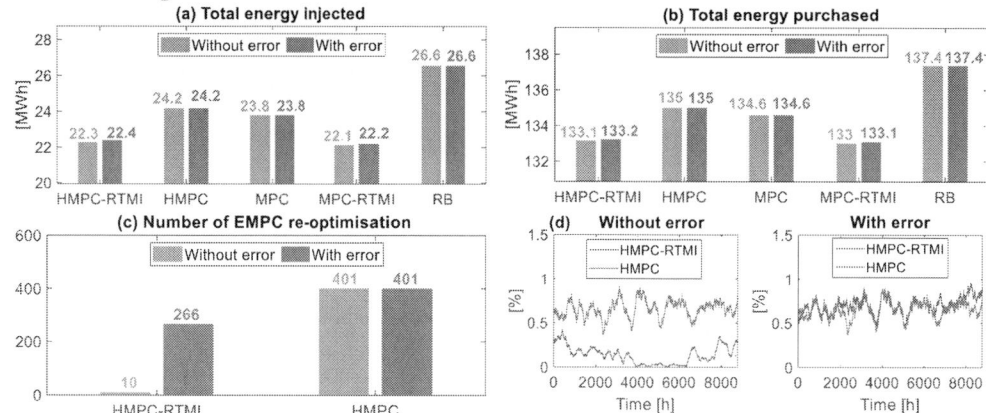

Fig. 10. Comparison between different control architectures. (a) Total power injected. (b) Total power purchased. (c) Number of EMPC re-optimization. (d) Moving average error in P_{grid}^{ref} of the last 10 days

The cooperation of two control layers – one with long and another with a short horizon – enables to handle prediction data variability without needing to optimize the laborious cost function (2) every hour, but only when the error in either SoC_{ref} or P_{grid}^{ref} is greater than $\Delta_{SoC}^{thr} = \Delta_{P_{grid}}^{thr} = 7\%$. Fig. 10d shows that errors in P_{grid}^{ref} are higher when not using the RTMI module, because of imprecisions in SoC estimation, which result in 391 more re-optimisations in the scenario without errors and 135 with errors.

Conclusions

This paper aims to develop a generic Building MicroGrid Energy Management System capable of adapting to external changes, such as Li-ion batteries modelling inaccuracy and inherent power imbalance uncertainties. The hierarchical MPC empowered with the proposed real-time parameter identification module increases the self-consumption rate regarding a well-established rule-based controller and the conventional MPC. This data-driven algorithm enables to identify the original batteries capacity and the cell temperature without any previous modelling step and without any thermal sensor, which simplify the energy management system design. Moreover, the simulation results demonstrated that the division in two control layers reduces the number of optimizations while maintaining the building less energetic dependent on the external grid even under imprecisions in data predictions.

References

[1] T. M. Lawrence *et al.*, "Ten questions concerning integrating smart buildings into the smart grid," *Build. Environ.*, vol. 108, pp. 273–283, Nov. 2016, doi: 10.1016/j.buildenv.2016.08.022.

[2] G. Cardoso, T. Brouhard, N. DeForest, D. Wang, M. Heleno, and L. Kotzur, "Battery aging in multi-energy microgrid design using mixed integer linear programming," *Appl. Energy*, vol. 231, pp. 1059–1069, Dec. 2018, doi: 10.1016/j.apenergy.2018.09.185.

[3] M. F. Zia, E. Elbouchikhi, and M. Benbouzid, "Microgrids energy management systems: A critical review on methods, solutions, and prospects," *Appl. Energy*, vol. 222, pp. 1033–1055, Jul. 2018, doi: 10.1016/j.apenergy.2018.04.103.

[4] Z. Cheng, J. Duan, and M.-Y. Chow, "To Centralize or to Distribute: That Is the Question: A Comparison of Advanced Microgrid Management Systems," *IEEE Ind. Electron. Mag.*, vol. 12, no. 1, pp. 6–24, Mar. 2018, doi: 10.1109/MIE.2018.2789926.

[5] P. R. C. Mendes, L. V. Isorna, C. Bordons, and J. E. Normey-Rico, "Energy management of an experimental microgrid coupled to a V2G system," *J. Power Sources*, vol. 327, pp. 702–713, Sep. 2016, doi: 10.1016/j.jpowsour.2016.07.076.

[6] L. Su *et al.*, "Path dependence of lithium ion cells aging under storage conditions," *J. Power Sources*, vol. 315, pp. 35–46, May 2016, doi: 10.1016/j.jpowsour.2016.03.043.

[7] X. Li, C. Yuan, X. Li, and Z. Wang, "State of health estimation for Li-Ion battery using incremental capacity analysis and Gaussian process regression," *Energy*, vol. 190, p. 116467, Jan. 2020, doi: 10.1016/j.energy.2019.116467.

[8] Z. Song *et al.*, "The sequential algorithm for combined state of charge and state of health estimation of lithium-ion battery based on active current injection," *Energy*, vol. 193, p. 116732, Feb. 2020, doi: 10.1016/j.energy.2019.116732.

[9] Y. Li *et al.*, "Data-driven health estimation and lifetime prediction of lithium-ion batteries: A review," *Renew. Sustain. Energy Rev.*, vol. 113, p. 109254, Oct. 2019, doi: 10.1016/j.rser.2019.109254.

[10] R. Luthander, J. Widén, D. Nilsson, and J. Palm, "Photovoltaic self-consumption in buildings: A review," *Appl. Energy*, vol. 142, pp. 80–94, Mar. 2015, doi: 10.1016/j.apenergy.2014.12.028.

[11] Enedis l'électricité en réseau, "Conditions de raccordement des Installations de stockage." [Online]. Available: https://www.enedis.fr/sites/default/files/Enedis-PRO-RES_78E.pdf.

[12] Commission de Régulation de l'Energie, "Cahier des charges de l'appel d'offres portant sur la réalisation et l'exploitation d'Installations de production d'électricité à partir d'énergies renouvelables en autoconsommation et situées en métropole continentale." Dec. 26, 2019.

[13] O. Tremblay and L.-A. Dessaint, "Experimental Validation of a Battery Dynamic Model for EV Applications," *World Electr. Veh. J.*, vol. 3, no. 2, pp. 289–298, Jun. 2009, doi: 10.3390/wevj3020289.

[14] "JRC Photovoltaic Geographical Information System (PVGIS) - European Commission." https://re.jrc.ec.europa.eu/pvg_tools/en/tools.html#MR (accessed Mar. 05, 2020).

[15] H. Yu, S. Niu, Y. Zhang, and L. Jian, "An integrated and reconfigurable hybrid AC/DC microgrid architecture with autonomous power flow control for nearly/net zero energy buildings," *Appl. Energy*, vol. 263, p. 114610, Apr. 2020, doi: 10.1016/j.apenergy.2020.114610.

Impact of DC fault blocking capability on the sizing of the DC-DC Modular Multilevel Converter

J. D. Paez[1], F. Morel[1], S. Bacha[1,2], Piotr Dworakowski[1], D. Frey[1,2]

[1]SuperGrid Institute
23 Rue de Cyprian
69100 Villeurbanne, France
juan.paez@supergrid-institute.com
https://www.supergrid-institute.com/

[2]G2ELab
Univ. Grenoble Alpes, CNRS, Grenoble INP*
21 Avenue des Martyrs
38000 Grenoble, France
http://www.g2elab.grenoble-inp.fr/

Acknowledgements

This work was supported by a grant overseen by the French National Research Agency (ANR) as part of the "Investissements d'Avenir" Program ANE-ITE-002-01.

Keywords

«HVDC», «Modular Multilevel Converter», «DC-DC converter», «Fault blocking».

Abstract

The capability of DC-DC converters of blocking DC faults is an important issue for the development of HVDC meshed grids. This paper analyzes the impact on the converter design of including such characteristic for the DC-DC Modular Multilevel Converter. Steady state and transient analysis are proposed and tested, showing how the converter should be oversized to include this feature. From this analysis it is concluded that including the fault blocking capability has an impact on the power losses and the investment in semiconductors but not in the investment on capacitors. The converter presents better indicators for low DC transformation ratios demonstrating the interest of the topology for these applications. However for those ratios, the impact of including fault blocking capability is the highest.

Introduction

In order to deal with the constant increase of power consumption and the integration of large scale renewable electrical sources, the actual AC system should be upgraded. A hybrid system including AC and HVDC transmission grids appears as a promising solution for the coming upgrade [1]. In order to have a stepwise evolution of HVDC grids, DC-DC converters will be necessary and are considered as a key enabling technology [2], similarly to the AC transformers on the AC system which enabled the incremental evolution of the AC system. The CIGRE studies the subject in the working group B4.76.

DC-DC converters allow the interconnection of HVDC systems that have different characteristics such as voltage levels, technology (VSC-LCC), and line architecture (monopole, bipole) [3], and could provide at the same time various functionalities like power flow control, DC voltage regulation, and fault blocking capability [4]. This characteristic can be understood as the capability of DC-DC converters to prevent the apparition of voltages or currents on one DC grid that could lead to a fault when a fault occurs on the second DC grid. To achieve this, the topology should be capable of breaking the contribution to the fault current from the healthy DC side.

In literature, it has been generally assumed that the required DC-DC converters for HVDC grids should include fault blocking capability [4], given the consequences to the transmission system of a propagation of faults between both DC grids being interconnected with the converter. Some topologies like the Front-to-Front MMC [5], and in general all the circuits that rely in a DC-AC-DC conversion chain [4], include this feature inherently but the rest of the topologies do not. These last kind of topologies are interesting comparing with those that use a complete DC-AC-DC conversion chain in regard of costs and efficiency [6], [7]. However for those circuits, modifications should be done in order to include the fault blocking capability. For example, adding bipolar sub-modules (SMs), like Full-bridge (FB) SMs, which causes an oversizing in terms of components, increasing power losses and costs. This oversize could decrease the interest on these topologies comparing to the DC-AC-DC approaches.

In this paper, the impact on the circuit design of withstanding DC faults and particularly the converter capability of breaking fault currents is studied for a promising non-isolated topology: the DC-MMC [8], [9]. The recent research on this circuit has increased given its advantages compared to other converters. Some of the works address the topology working principle, its modelling [10], [11], its control [12], the circuit sizing [13], and comparative studies [6]. However on all of these works the fault blocking capability is only mentioned by the inclusion of FB-SMs ignoring the impact on the costs and losses of including such feature. Thus, this paper aims to fill this gap. The analysis is done analytically for steady-state after the fault, as well during transients using simulations.

The Modular Multilevel DC-DC converter (DC-MMC)

The DC-DC Modular Multilevel converter (here denominated DC-MMC, but also known in literature as M2DC[12]) is a non-isolated DC-DC converter formed by the interconnection of several MMC converter legs in parallel, where the AC output ports have been connected together on the low voltage (LV) DC port through an AC filter. The AC filter can be of different types [14] and its principal function is to mitigate the propagation of AC currents into the DC grid and the magnitude of the AC circulating currents. Among the different possible filter technologies for the topology, the simplest solution is a passive filter formed by an inductor per phase as presented in Fig. 1 (a). In the figure, a three phase circuit is represented. However, the DC-MMC can be implemented with any number of phases, but given the power levels considered for the study (700 MW), a three phase DC-MMC is retained.

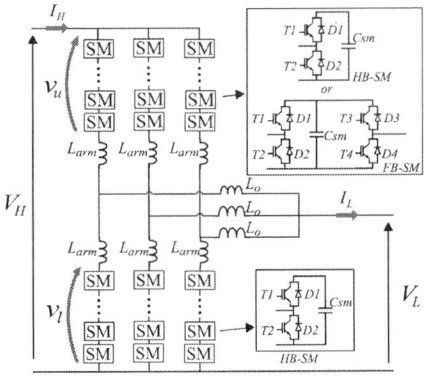

a. Three phase DC-MMC with inductive output filter L_o

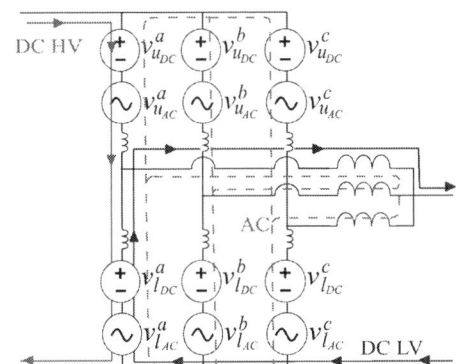

b. DC-MMC currents and voltages (for simplicity, only the DC currents for one circuit phase are represented)

Fig. 1 – DC-MMC topology with three phases.

From the analysis of one phase of the circuit (see Fig. 1), the DC quantities of upper and lower arms are obtained:

$$v_{u_{DC}} = V_H - V_L \qquad i_{u_{DC}} = \frac{I_H}{3}$$
$$v_{l_{DC}} = V_L \qquad i_{l_{DC}} = \frac{I_H - I_L}{3} \tag{1}$$

Analyzing the DC power on each arm, assuming no losses on the converter ($P_{DC} = V_H I_H = V_L I_L$), gives:

$$P_{u_{DC}} = \frac{P_{DC}}{3}\left(1 - \frac{1}{n_{DC}}\right) \qquad P_{l_{DC}} = -\frac{P_{DC}}{3}\left(1 - \frac{1}{n_{DC}}\right) \tag{2}$$

Where n_{DC} is the transformation ratio V_H/V_L. These equations it is observed that, without any AC power circulating inside the converter, the upper arms absorb energy (positive power) while the lower arms deliver the same amount (negative power). Thus, to achieve an energy balance on the arms, a circulation of AC power is needed to transfer the excess of energy from the upper arms to the lower arms.

Therefore, the converter working principle is to generate a superposition of DC and AC voltages on each converter arm by the insertion and bypass of the SM capacitors, which makes DC and AC currents to circulate. The DC currents are responsible for the DC power transfer between both DC grids while the AC currents circulate internally to balance the arm energies. Fig. 1 (b) shows the current circulation inside the circuit, highlighting the AC currents and the DC current of each DC side. A more detailed explanation of the working principle of the topology can be found in [12] where an energy based approach [15] for the topology control is proposed.

The required AC power depends of the DC voltage transformation ratio n_{DC}:

$$P_{AC_{circ}} = \left(1 - \frac{1}{n_{DC}}\right) \tag{3}$$

Consequently, the topology is particularly interesting for low transformation ratios ($n_{DC} < 2$), since less AC power is required.

In order to achieve the required AC power transfer between arms, they must generate an AC voltage defined by Eq. 5. Note that in this work it is assumed that the AC components are sinusoidal without harmonics.

$$v_{u_{AC}} = |v_{u_{AC}}|\angle\delta \quad v_{l_{AC}} = |v_{l_{AC}}|\angle 0 \tag{4}$$

In order to reduce the converter losses [12], [13], the voltage magnitudes must be equal, i.e. $|v_{u_{AC}}| = |v_{l_{AC}}| = V_{AC}$. The phase angle between both arm AC voltages is determined by the AC power and the circuit inductances, while the voltage magnitude is given by:

$$V_{AC} = \min(V_L, V_H - V_L) \tag{5}$$

Thus, the maximal voltages that each arm must generate for normal operation (no DC faults), and consequently the required number of SMs, are the sum of the DC and AC (peak) components:

$$v_{u_{max}} = V_H - V_L + \min(V_H, V_H - V_L) \tag{7}$$
$$v_{l_{max}} = V_L + \min(V_H, V_H - V_L) \tag{8}$$

Meaning that the total number of SMs required per arm under normal operation is:

$$N_{SM_{u_{normal}}} = v_{u_{max}}/V_{SM} \tag{9}$$
$$N_{SM_{l_{normal}}} = v_{l_{max}}/V_{SM} \tag{10}$$

Where V_{SM} is the nominal voltage of each SM (assuming that all SMs have the same voltage rating).

Steady state analysis of the converter after DC faults

Analyzing the behavior of the circuit when a DC fault occurs on the HV side, once the converter IGBTs are turned off, it is seen that when the DC-MMC is implemented with only HB-SMs, the fault current is propagated between both DC sides due to the direct biasing of the lower diodes on the upper arm HB-SMs, which causes a direct path for the fault current between the healthy side and the fault (Fig 2. (a)).

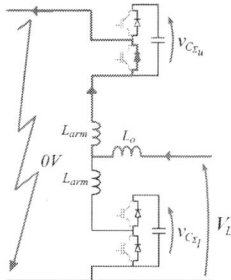

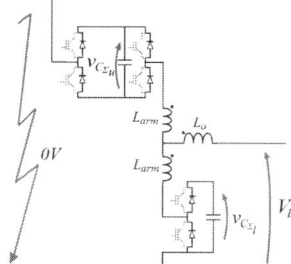

a. Upper arms with HB-SMs   b. Upper arms with FB-SMs

Fig. 2 – DC-MMC after a DC fault on HV side (only one circuit phase and only one SM representing the set of SMs on each arm are represented for simplicity)

When FB-SMs are included (Fig 2. (b)), the upper arms are capable of breaking the healthy side contribution to the fault current if the voltage on the FB-SM capacitors is enough to oppose the voltage difference between the faulty and the healthy sides. Assuming a low fault impedance, the condition on the upper arm FB-SM capacitors voltage is:

$$v_{C_{\Sigma_u}} \geq V_L \tag{11}$$

Where $v_{C_{\Sigma_u}}$ is the sum of the voltage of all the SM capacitors in the upper arm. When the condition of Eq. (11) is verified, the diodes on the FB-SMs are reverse biased and there is no current flow.

When the fault position is on the LV side (Fig. 3) the fault propagation to the HV side is stopped if the voltage on the upper arm SM capacitors (FB or HB) is enough to oppose the voltage difference between the faulty side and the healthy side given by Eq. 12, assuming low fault impedance. In that case, the diodes on the SMs are reverse biased and there is no current flow.

$$v_{C_{\Sigma_u}} \geq V_H \tag{12}$$

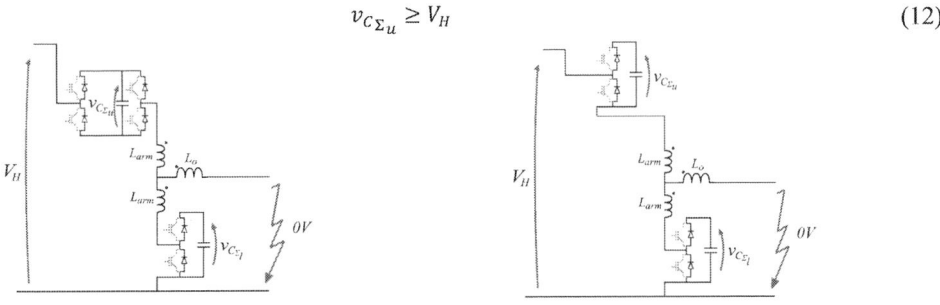

a. Upper arms with HB-SMs   b. Upper arms with FB-SMs

Fig. 3 – DC-MMC after a DC fault on LV side (only one circuit phase and only one SM representing the set of SMs on each arm are represented for simplicity)

If the upper arm SMs are not sized to withstand this voltage, the fault current will continue to circulate between HV and LV DC ports through the upper arm SM capacitors charging them and, thus, generating an overvoltage on the SMs.

From this analysis, it is demonstrated that the capability of the DC-MMC to stop the propagation of DC fault currents depends on certain conditions on the upper arm SMs. Indeed, to stop the propagation of faults when the fault is on the HV side, FB-SMs are needed. The quantity is given by:

$$N_{SM_{FB_u}} \geq V_L/V_{SM} \tag{13}$$

When the fault occurs in the LV side, the upper arm SMs should be sized to support a voltage equal to at least V_H. Therefore, additionally to the FB-SMs, a number of HB-SMs should be added:

$$N_{SM_{HB_u}} \geq V_H/V_{SM} - N_{SM_{FB_u}} \tag{14}$$

Considering the lower arms, for both DC fault positions (LV or HV side), the required number of SMs is the same that in the normal operation case (Eq. (8) and Eq. (10)).

Impact of including fault blocking capability on sizing and losses

In Fig. 4, the needed number of SMs on the upper arm to withstand DC faults and block the fault current (Eqs. (12) - (13)) are compared with the number of SMs necessary to operate the circuit in normal conditions (only HB-SMs are required according to Eqs. (7) - (10)). It is seen that for $n_{DC} < 2$ the upper arms are highly oversized since HB-SM and FB-SMs are added, while for high transformation ratios ($n_{DC} > 2$), there is no added SMs but there is still an oversizing in terms of components since FB-SMs are required instead of using only HB-SMs.

Concerning the lower arms, as explained in the previous section, there is no particular requirements in terms of number of SMs in order to withstand the voltages in case of faults. Thus, the lower arms are not oversized for the DC fault management.

The difference on the number of SMs required on the upper arms to include fault blocking capability has an impact on terms of cost and efficiency. In order to analyze this impact, the procedure proposed in [6] to compare different converters is used. Three indicators are analyzed: the utilization of semiconductors, the total energy stored in the converter SM capacitors and the power losses.

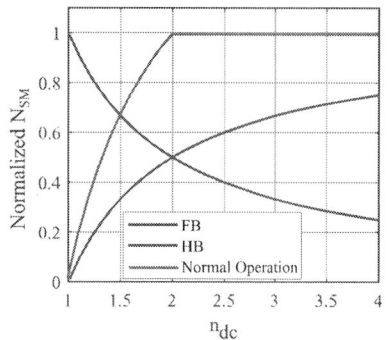

a. Installed FBs and HBs for fault management and HBs required on normal operation

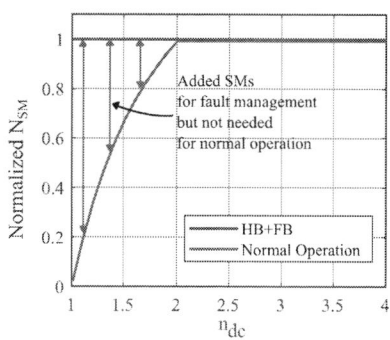

b. Total installed SMs (FBs and HBs) for fault management and HBs required on normal operation

Fig. 4 – Number of SMs on upper arms

The utilization factor is defined as the ratio between nominal power and the total rating of the semiconductor switches (Eq. (15)). It reflects the investment in semiconductors.

$$U_{SW} = \frac{P_{DC}}{\sum V_{SW_{pk}} I_{SW_{RMS}}} \tag{15}$$

If instead of representing the U_{SW} factor, its reciprocal $1/U_{SW}$ is analyzed, the interpretation could be more straightforward. In this case, the quantity $1/U_{SW}$ can be interpreted as the oversizing in terms of installed power in semiconductors to transmit a given amount of power. In such case the ideal value should be one, i.e. to transmit 1 MW, 1MVA of semiconductors should be installed.

The calculation of U_{SW}, needs the RMS current of each semiconductor device and its peak voltage value. The peak voltage is assumed to be equal to the SM voltage (V_{SM}) and the RMS currents per switch are determined by Eq. (16) [6]. Where T represents the period of the arm current, t_1 and t_2 the times where the arm current changes its sign, and $\alpha(t)$ the average duty cycle of a SM, which depends on the modulation index $m(t)$ given by Eq. (17) and the signs of current. The detailed expressions for $\alpha(t)$ can be found in [6].

$$I_{sw_{RMS}}(t) = \sqrt{\frac{1}{T} \int_{t_1}^{t_2} \left(i_{arm}(t)\right)^2 \alpha(t) dt} \tag{16}$$

$$m(t) = \frac{v_{arm}(t)}{N_{SM} V_{SM}} = \frac{v_{arm_{DC}} + v_{arm_{AC}}(t)}{N_{SM} V_{SM}} \tag{17}$$

Concerning the DC-MMC energy requirements, they can be expressed in terms of the energy factor of Eq. (18). This factor reflects the investment in SM capacitors. Lower values of E are preferable.

$$E = \frac{W_{installed}}{P_{DC}} = \frac{\sum \frac{1}{2} C_{SM} \overline{V_{SM}}^2}{P_{DC}} \tag{18}$$

To calculate the SM capacitance required to the estimation of E, Eq. (19) is used [6]. Where $V_{DC_{arm}}$ and $I_{DC_{arm}}$ represent the DC quantities on each arm (Eq. (1)), ϕ is the phase angle between arm current and voltage, N_{SM} is the number of submodules on the arm, f represents the operation frequency, and ε is the allowed capacitor voltage ripple expressed in percentage.

$$C_{SM} = \frac{V_{DC_{arm}} I_{DC_{arm}}}{2\pi f} \left(1 - \left(\frac{\cos\phi}{2}\right)^2\right)^{\frac{3}{2}} \frac{4}{\cos\phi} \frac{1}{2\varepsilon N_{SM} \overline{V_{SM}}^2} \tag{19}$$

Considering the power losses, only the semiconductor conduction losses are retained for the study. The reason of neglecting the switching losses is that they are heavily influenced by the capacitor voltage balancing mechanisms and the modulation scheme [16], which are out of the scope of the study. The conduction losses on a semiconductor switch (IGBT or diode) are approximated by:

$$P_{cond_{SW}} = V_0 I_{avg} + R_{ON} I_{RMS}^2 \tag{20}$$

Where V_0 represent the saturation voltage if the switch is an IGBT or the forward voltage if it is a diode, and R_{ON} represents the device equivalent resistance in the ON state. Both values can be obtained from the device datasheet (Infineon FZ1500R33HL3 in this study). The RMS current is determined by Eq. (16) while the average current by Eq. (21).

$$I_{sw_{AVG}}(t) = \frac{1}{T} \int_0^T i_{arm}(t)\alpha(t)dt \tag{21}$$

To analyze the impact of including fault blocking capability in the circuit using the presented indicators, the circuit parameters of Table I are used. The output filter inductance L_o has been selected sufficiently large to decrease the circulating currents to an acceptable range and the SM capacitor values have been designed to have a ripple of $\pm 10\%$ at nominal DC power.

Table I: Circuit parameters for analytical study

Circuit Parameter	Value
Nominal DC power (P_{DC})	700 MW
Nominal DC High Voltage (V_H)	640 kV
Nominal DC Low Voltage (V_L)	Determined by n_{DC}
Transformation ratio (n_{DC})	Varied between 1.1 and 4
SM Capacitors (C_{SM})	Designed to have a voltage ripple of 10% for each case
SM nominal voltage (V_{SM})	1.6 kV
Arm inductance (L_{arm})	15 mH
Filter inductance (L_o)	150 mH
AC Frequency	150 Hz

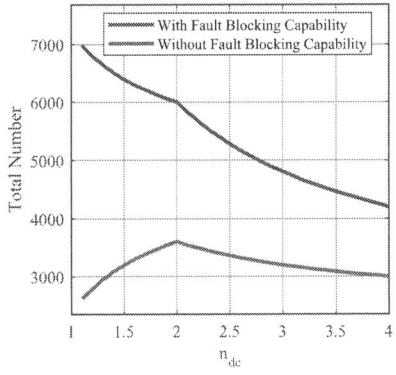
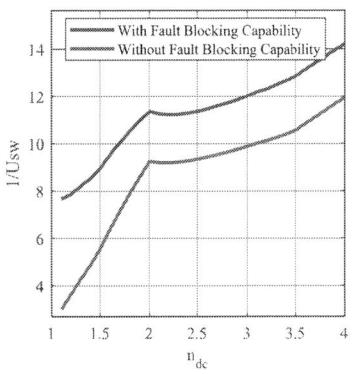

a. Total number of IGBTs   b. Oversize factor $1/U_{SW}$

Fig. 5 – Impact on semiconductor investment

Fig. 5 (a) shows the total number of IGBTs on the 3-phase DC-MMC. Is observed that for all the considered transformation ratios (n_{DC}) more IGBTs are needed to include fault blocking capability, mainly for low values of n_{DC}, which is coherent according to the number of SMs (see Fig. 4 (b)).

In Fig 5 (b), the oversize factor $1/U_{SW}$ is presented. It is observed that for low values of n_{DC}, the additional installed power needed to include fault blocking capability is significant. For example, for $n_{DC}=1.1$, 7.6 times of the nominal power is needed instead of 3 times if this feature is not implemented. As n_{DC} increases, the gap is decreased and is almost constant for $n_{DC} > 2$.

Considering the investment in capacitors, Fig. 6 presents the value of C_{SM} on the upper and lower arms in function of n_{DC}. It is observed that, as discussed in the previous section, the lower arms are not affected by including fault blocking capability, then, the same values of C_{SM} are obtained if the feature is included or not. For the upper arms, C_{SM} is influenced by the total number of SMs, which in the case of $n_{DC} < 2$ is different if the fault blocking capability is included or not (see Fig. 4 (b)). However, observing the energy factor, it is seen that for all the cases the energy requirements are the same. This comes from the fact that even if at $n_{DC} < 2$ less SMs are required when no fault blocking capability is implemented, the value of the individual SM capacitance is higher. Therefore the investment costs in capacitors are expected to be similar if the feature is included or not.

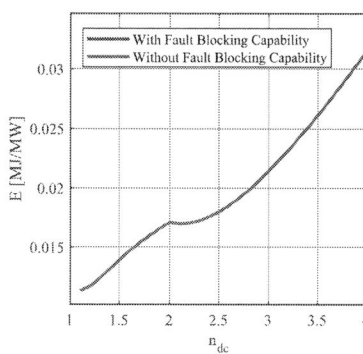

a. Submodule Capacitance b. Energy factor

Fig. 6 – Impact on investment in SM capacitors

The conduction losses at nominal power are presented in Fig. 7, as well as the ratio between the case of including fault blocking capability and the case of not including it. It is seen that in all the cases the losses are increased by including this feature, for $n_{DC} < 2$ the losses are considerably increased. For very low values of n_{DC} ($n_{DC} < 1.25$) the losses are even doubled. For $n_{DC} > 2$, the increase in the losses is kept at around 20 % independently of the value of n_{DC}.

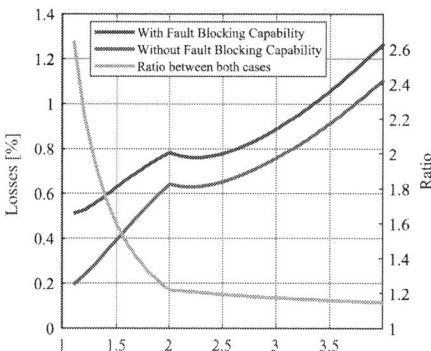

Fig. 7 – Impact on semiconductor conduction losses

From this analysis it is concluded that including the fault blocking capability on the DC-MMC has an impact on the power losses and the investment in semiconductors but not in the investment on capacitors. The DC-MMC presents better indicators for low values of n_{DC} demonstrating the interest of the topology for these applications. However for those values of n_{DC}, the impact of including the fault blocking capability is the highest.

Impact of the transient response on the fault current breaking

In the previous section, the impact of including fault blocking capability on the converter sizing, costs and power losses was analyzed considering the minimal number of SMs required for the DC fault management. In this section the transient response during fault is studied to evaluate if it impacts also the converter sizing. Simulations of the circuit for HV and LV side DC faults are done to analyze this impact. The simulations are done using an average arm model including the model during the blocked state [17]–[19] as shown in Fig. 8. The converter parameters are presented in Table II.

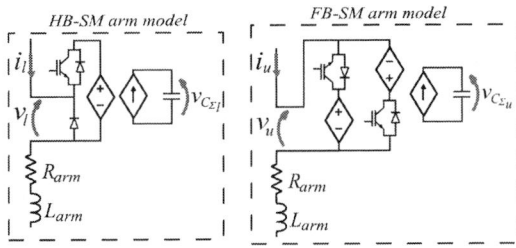

Fig. 8 – Arm average model including blocked state modelling

The simulations are done for HV and LV side faults, considering different power flow directions before the fault (from HV to LV side or from LV to HV side). Once the fault is detected, all the IGBTs in the converter are turned off. These actions are done considering a delay of 50 μs after the fault detection.

Table II: Circuit parameters for DC fault simulations

Circuit Parameter	Lower Arm	Upper Arm
Nominal DC power (P_{DC})	700 MW	
Nominal DC Voltages (V_H / V_L)	640 kV / 500 kV	
SM Capacitor (C_{SM})	1.2 mF	5.3 mF
SMs per arm (N_{SM})	400 HBs	400 FBs
SM nominal voltage (V_{SM})	1.6 kV	1.6 kV
Equivalent Arm Capacitor ($Ceq = C_{SM}/N_{SM}$)	3.1 μF	13.3 μF
Arm inductance (L_{arm})	15 mH	15 mH
Arm resistance (R_{arm})	0.5 Ω	0.5 Ω
AC voltage (RMS line-to-line voltage)	137 kV	137 kV
Filter inductance (L_o)	150 mH	
Filter series resistance (R_o)	0.08 Ω	
AC Frequency	150 Hz	

The results for the case of HV side DC fault, are presented in Fig. 10 considering a power flow from LV to HV side before the fault, the dotted line shows the moment of the converter blocking. This power flow is considered the worst case since the fault current has the same direction than the nominal power. Thus, after the fault, the current increases without passing by zero (see I_H curve on Fig. 10).

Once the converter is blocked, the current starts to decrease until all inductances in the system are totally discharged. Then, the fault current is stopped, the DC currents I_L and I_H remain at zero after the transient. However, during the process, the FB-SMs on the upper arm absorb the energy stored on the circuit inductances and also the energy coming from the LV grid. As a consequence the capacitor voltage could be highly increased (20% for phase c in Fig. 10, for this example). Thus, the converter should include more SMs to withstand the overvoltage or the capacitors should be oversized accordingly to absorb the excess of energy without a significant overvoltage. The amount depends on the system inductances (including grid inductances), and control delays.

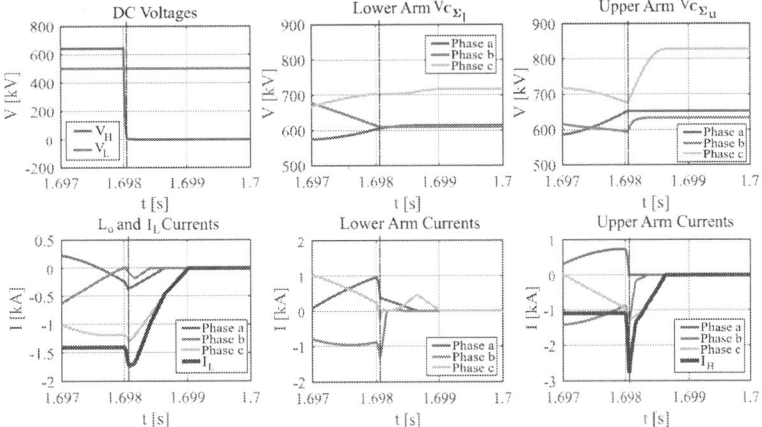

Fig. 10 – Simulation results for HV DC fault case

Fig. 11 shows the results obtained for a fault on the LV side. In this case the considered power flow before the fault is from HV side to LV side. It is observed that similarly to the previous case, the fault current is stopped but the FB-SMs on the upper arm are overcharged during the process (mainly phase b on this example). It is observed as well that fault current takes several more time to be extinguished (observe I_L curve on the figure), which in the simulation results was of 83 ms (not shown in the figure). The reason of this long discharge time is the high value of the output inductance filter L_o. In consequence, for a fault on the LV side the upper SMs must be sized to absorb the excess of energy preventing an overvoltage, and the lower SMs freewheeling diodes must be capable of withstanding the fault current during quite long time according to the value of L_o.

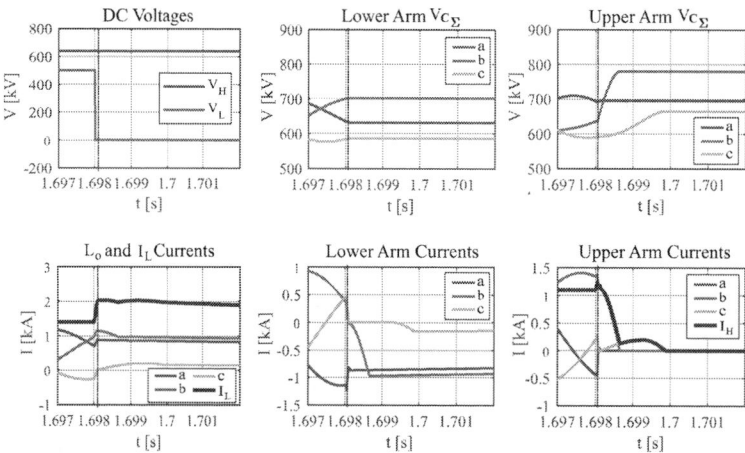

Fig. 11 – Simulation results for LV DC fault

Conclusion

The impact on the DC-MMC of DC fault management has been analyzed. It has been shown that including the fault blocking capability requires FB-SMs and/or (according to the voltage ratio) additional SMs in upper arms. This increases the converter cost and losses, mainly associated to the investment and losses in semiconductors. However the investment costs in capacitors are expected to be similar if the feature is included or not. The impact is worst for low voltage transformation ratios, for which the topology is normally more interesting. The transient response analysis showed that, in order to extinguish the contribution to the fault current, the FB-SMs on the circuit should be able to absorb the energy stored in inductances during faults which can lead to an overvoltage or an oversizing on the SM capacitance. Further works should be done to compare the DC-MMC without fault-blocking capability associated with an external DC circuit breaker, particularly for low voltage transformation ratios. In such cases the additional breaker could have an interest since the costs of including the fault blocking capability on the circuit are considerable. The paper outcomes should also be evaluated with more accurate grids models. Finally, the developed methodology can be generalized to larger set of DC-DC converters for HVDC applications.

References

[1]   N. Ahmed, S. Norrga, H.-P. Nee, A. Haider, D. Van Hertem, L. Zhang, and L. Harnefors, "HVDC SuperGrids with modular multilevel converters—The power transmission backbone of the future," in *Systems, Signals and Devices (SSD), 2012 9th International Multi-Conference on*, 2012, pp. 1–7.

[2]   D. Jovcic and B. T. Ooi, "Developing DC transmission networks using DC transformers," *IEEE Transactions on Power Delivery*, vol. 25, no. 4, pp. 2535–2543, 2010.

[3]   D. Gómez, J. D. Páez, M. Cheah-Mane, J. Maneiro, P. Dworakowski, O. Gomis-Bellmunt, and F. Morel, "Requirements for interconnection of HVDC links with DC-DC converters," in *IEEE 45th Annual Conference of the Industrial Electronics Society (IECON 2019)*, 2019.

[4] J. D. Páez, D. Frey, J. Maneiro, S. Bacha, and P. Dworakowski, "Overview of DC–DC converters dedicated to HVdc grids," *IEEE Transactions on Power Delivery*, vol. 34, no. 1, pp. 119–128, 2019.

[5] S. Kenzelmann, A. Rufer, D. Dujic, F. Canales, and Y. R. De Novaes, "Isolated DC/DC structure based on modular multilevel converter," *IEEE Transactions on Power Electronics*, vol. 30, no. 1, pp. 89–98, 2015.

[6] J. D. Páez, J. Maneiro, S. Bacha, D. Frey, and P. Dworakowski, "Influence of the operating frequency on DC-DC converters for HVDC grids," in *2019 21th European Conference on Power Electronics and Applications (EPE'19 ECCE Europe)*, 2019.

[7] A. Schön and M.-M. Bakran, "Comparison of modular multilevel converter based HV DC-DC-converters," in *Power Electronics and Applications (EPE'16 ECCE Europe), 2016 18th European Conference on*, 2016, pp. 1–10.

[8] J. A. Ferreira, "The multilevel modular DC converter," *IEEE Transactions on Power Electronics*, vol. 28, no. 10, pp. 4460–4465, 2013.

[9] S. Norrga, L. Ängquist, and A. Antonopoulos, "The polyphase cascaded-cell DC/DC converter," in *Energy Conversion Congress and Exposition (ECCE), 2013 IEEE*, 2013, pp. 4082–4088.

[10] D. Jovcic, P. Dworakowski, G. Kish, A. Jamshidi Far, A. Nami Abb, A. Darbandi, and X. Guillaud, "Case Study of Non-Isolated MMC DC-DC Converter in HVDC Grids," 2019.

[11] G. J. Kish, M. Ranjram, and P. W. Lehn, "A modular multilevel DC/DC converter with fault blocking capability for HVDC interconnects," *IEEE Transactions on Power Electronics*, vol. 30, no. 1, pp. 148–162, 2015.

[12] F. Gruson, Y. Li, P. Le Moigne, P. Delarue, F. Colas, and X. Guillaud, "Full State Regulation of the Modular Multilevel DC converter (M2DC) achieving minimization of circulating currents," *IEEE Transactions on Power Delivery*, 2019.

[13] H. Yang, J. Qin, S. Debnath, and M. Saeedifard, "Phasor domain steady-state modeling and design of the DC–DC modular multilevel converter," *IEEE Transactions on Power Delivery*, vol. 31, no. 5, pp. 2054–2063, 2016.

[14] G. J. Kish, "On the Emerging Class of Non-Isolated Modular Multilevel DC–DC Converters for DC and Hybrid AC–DC Systems," *IEEE Transactions on Smart Grid*, vol. 10, no. 2, pp. 1762–1771, Mar. 2019.

[15] A. Zama, A. Benchaib, S. Bacha, D. Frey, and S. Silvant, "High dynamics control for MMC based on exact discrete-time model with experimental validation," *IEEE Transactions on Power Delivery*, 2017.

[16] A. Zama, S. A. Mansour, D. Frey, A. Benchaib, S. Bacha, and B. Luscan, "A comparative assessment of different balancing control algorithms for modular multilevel converter (MMC)," in *2016 18th European Conference on Power Electronics and Applications (EPE'16 ECCE Europe)*, 2016, pp. 1–10.

[17] F. Xinkai, Z. Baohui, and W. Yanting, "Fast electromagnetic transient simulation models of full-bridge modular multilevel converter," in *2016 IEEE PES Asia-Pacific Power and Energy Engineering Conference (APPEEC)*, 2016, pp. 998–1002.

[18] A. Zama, S. Bacha, A. Benchaib, D. Frey, and S. Silvant, "A novel modular multilevel converter modelling technique based on semi-analytical models for HVDC application," *J. Elect. Syst.*, vol. 12, no. 4, pp. 649–659, 2016.

[19] H. Zhang, D. Jovcic, W. Lin, and A. J. Far, "Average value MMC model with accurate blocked state and cell charging/discharging dynamics," in *Environment Friendly Energies and Applications (EFEA), 2016 4th International Symposium on*, 2016, pp. 1–6.

Optimization of High Frequency Magnetic Devices with Consideration of the Effects of the Magnetic Material, the Core Geometry and the Switching Frequency

Sobhi Barg, Muhammad Farhan Alam, Kent Bertilsson
Mid Sweden University
Sobhi.barg@miun.se, FarhanMuhammad.Alam@miun.se, Kent.Bertilsson@miun.se

Keywords

« *High frequency transformer* », «*core loss*», « *winding loss* », « *optimization* », «skin and proximity effects».

Abstract

This paper presents an optimization approach of high frequency transformers based on the calculation of the optimum number of turns. The relationship between the volume, the power and the frequency is also investigated. The design approach considers the skin and proximity effects, the core geometry and the magnetic material.

Introduction

High frequency transformers are widely used in numerous dc-dc converter topologies such as the flyback, the full-bridge, the forward and the dual active bridge converters. They are the bulky components and their design is one of the most complex task in the design of power converters. Switching at high frequency, enabled by the emerging of the wide band gap devices, has indeed two major advantages which are: low volume and low cost. Another advantage which is frequently mentioned in the literature but not yet proved is the improvement of the efficiency. The prevailing opinion in the literature agrees with the possibility to increase the efficiency by increasing the frequency, however, there is no solid proof that make evidence of this opinion except of some practical design cases. The answer to this issue is one main objective of this research in addition to the development of an optimized design approach of HF transformer [1-2].

Several design approaches have been proposed in the literature to optimize the cost, the efficiency and the volume of this component. Generally, they can be classified into two main categories: the single objective design approaches in which only one objective is considered for optimization and the multi-objective design approaches, not explored in this paper, in which multiple objectives can be optimized simultaneously [3]. The first category is the most common one because of its simplicity, unlike the second one which requires advanced design techniques due to the complexity of the design problem of the transformers. The complexity results from the non-linearity of the design equations and the great interdependence between the different design variables.

Existing single design approaches mainly deal with the maximization of the efficiency. The main approaches are the area product based approaches and the core geometry coefficient approach and their derivatives [4]. The area product approach (AP) is a dimensional factor of the magnetic core (product of the core section and the winding area) which has an equivalent term function of the power, the frequency and the maximum magnetic flux density. It is not an optimized design approach and it does not also take into account, the effect of the magnetic material, the core geometry. Unlike, the AP approach, the K_{ge} design approach is based on the determination of the optimum magnetic flux density (B_{opt}) [2] and [7-8]. A core geometry coefficient is calculated for each core based on $_{Bopt}$, has an equivalent term which is mainly function of the power and the frequency. The geometrical term depends on the core section, the winding area, the mean core length, the mean turn length and β. Unfortunately, the Steinmetz parameter β is frequency dependent which makes the core geometry coefficient changing when the frequency changes. Eventually, it does not give a real view on how the volume is related to the frequency and the power. In addition to that, this approach does not account the effect of the skin and proximity effects on the AC resistance in the optimization process. In a recent works [6-7], an improved expression of Kge has relatively solved the later issue by considering the optimum AC resistance in the K_{ge} expression, however this does not reflect the real designed AC resistance. An iteration process might be a solution to determine the real AC resistance.

Petkov and Hurley design approaches, which can be considered as derivatives of the previous approaches, are also based on the determination of the optimum magnetic flux density. The main limitations of both designs are [5-6]:

• The optimum solution is not calculated for wide range of frequency and power.

• The core geometry effect is not considered in the optimization process.

• In both designs, the AC-to-DC resistance ratio is constant and independent on the real structure of the winding.

• The required core volume is not clear in Petkov approach and not accurate in Hurley approach because of some error is the model parameters.

In this paper, a novel optimization and design approach of high frequency transformer, based on the determination of the optimum volume and the optimum number of turns, is proposed to overcome the limitations of the existing design techniques. The choice of the number of turns as an optimization variable instead of the flux density is chosen for two main reasons. Firstly, it can determine the optimum flux density and eventually it minimizes the core loss. Secondly it can be used to minimize the winding loss and the core loss simultaneously with high accuracy between the theoretical design and the real structure. In other terms, it gives accurate prediction of the number of layers which can significantly affect the AC resistance due to the proximity effect. This feature is not possible with the methods based on the determination of the optimum magnetic flux density. Furthermore, the proposed approach allows to determine in second step the optimum frequency for a given input power. The full process is solved by iteration technique, however, the analytical solution for the optimum number of turns is also presented in this paper.

The main contributions of this paper are summarized as follows:

➢ To develop the relationship between the volume, the frequency and the power for high frequency transformers.

➢ To determine the minimum transformer loss with consideration of the core material, the core geometry, the skin and proximity effects.

➢ To develop the expression of the optimum number of turns.

➢ To analyze the dependency of the minimum transformer loss with respect to the optimum number of turns, the opium magnetic flux density and the frequency.

DESIGN AND OPTIMIZATION APPROACH OF HF TRANSFORMERS

Volume-Power-Frequency relationship

It is well known that the selection of the required volume is based on an accurate prediction of the transformer loss. Initially, it is quite difficult to predict the winding loss in the optimization of the magnetic core volume because it depends on several unknown such as the number of turns and the number of layer. This later depends on the winding area, the number of turns which has a great impact on the AC resistance. A good solution for this issue consists in equating the winding loss to the core loss as it was proven in the literature for an optimum design although it is not highly accurate.

The core loss can calculated using the ISE model.

$$P_c = k_f f^\alpha B^\beta V_c \tag{1}$$

Three variables need to be optimized to minimize the core loss which are the frequency, the magnetic flux density and the volume. The temperature effect is not considered in this work. Since the magnetic flux density can be expressed as a function of the frequency using Faraday's Law, then it can be eliminated. It should be noted that the Steinmetz parameters (k_f, α, β) are frequency dependent unlike the existing designs. Additionally, the volume is also a frequency dependent variable which needs to be found out.

The magnetic flux density is given as follows:

$$B = \frac{V_{in} D}{N A_c f} \tag{2}$$

Where V_{in} is the input voltage, D is the duty cycle, N_p is the primary number of turns and A_c is the core section. The number of turns are calculated as follows

$$N_p = \frac{W_{cf1} \times W_{cf1} \, A_w}{k_u \, S_p} = \frac{W_{cf1} \times W_{cf1} \, A_w}{k_u \, \pi \, \frac{d_p^2}{4}} \tag{3}$$

Where W_{cf1} is the winding configuration factor equals to 0.5 for EE type cores and equals to 1 for shell type cores. W_{cf2} is the primary portion factor from the total winding, k_u is the total filling factor, S_p is the primary conductor section, d_p is the conductor diameter and A_w is the total winding area of the bobbin. The conductor diameter (meter) is calculated by following equation:

$$d_p^2 = \frac{I_{pmax}}{7.2 * 10^6} = \frac{P_{in}}{7.2 * 10^6 \, V_{in}} \tag{4}$$

By, substituting (4) in (3), we get the expression of the number of turns:

$$N_p = \frac{28.8 * 10^6 \, W_{cf1} \times W_{cf1} \, A_w \, V_{in}}{\pi \, k_u \, P_{in}} \tag{5}$$

Substituting (5) in (2), we get the expression of B:

$$B = \frac{k_u \, P_{in} \, D}{9.16 * 10^6 \, W_{cf1} \times W_{cf1} \, A_w \, A_c \, f} \tag{6}$$

Substituting (6) in (1), we get the expression of Pc:

$$P_c = k_f f^{\alpha - \beta} \frac{k_u \, P_{in} \, D}{9.16 * 10^6 \, W_{cf1} \times W_{cf1}}^{\beta} V_c \, A_p^{-\beta} \tag{7}$$

As previously discussed, it was proven that, the minimum transformer loss are achievable when the winding-to-core loss ratio is about unit [5-6]. Hence, it can be written:

$$V_c \, A_p^{-\beta} = \frac{(1 - \eta) \, P_{in}}{k_f f^{\alpha - \beta} \left(\frac{k_u \, P_{in} \, D}{9.16 * 10^6 \, W_{cf1} \times W_{cf1}} \right)^{\beta}} \tag{8}$$

As it can be seen, the left geometrical term depends on β which in turns depends on the frequency. The required core can be chosen using the left term similarly to what is given in K_{ge} approach, however this will produce some errors because β is a frequency dependent parameter which have to be compensated. One way to achieve that is to express Ap as a function of the volume. This will enable to get a clearer view about the relationship between the volume, the frequency and the power. Additionally, it will consider the core shape effect. The parameters used to include the core shape effect are given in (9-10) and Tab.I.

$$A_p = a_1 \, V_c^{b_1} \tag{9}$$
$$A_c = a_2 \, V_c^{b_2} \tag{10}$$

Finally, we get the expression of the required optimized volume.

$$V_c = \left[\frac{(1 - \eta) \, P_{in}}{a_1^{-\beta} k_f f^{\alpha - \beta} \left(\frac{k_u \, P_{in} \, D}{1.11 \, 10^6 \, W_{cf}} \right)^{\beta}} \right]^{\frac{1}{-b_1 \beta + 1}} \tag{11}$$

The evolution of the required volume with respect to the power and the frequency for different shapes is shown in Fig.1. It can be seen that the volume is proportional and inversely proportional to the power and the frequency respectively for the investigated range of power and frequency. The UR shape is the best solution and this is because $W_{cf1} = 1$.

TAB. I. WINDING AREA AND CORE SECTION MODELS PARAMETERS

	a_1	b_1	a_2	b_2
EE	0.08401	1.378	0.2511	0.6636
UR	0.006585	1.07	0.3568	0.7312
ER	0.01135	1.233	1.201	0.7518
ETD	0.1346	1.422	0.3607	0.6985
EE planar	0.4671	1.511	0.2974	0.6372

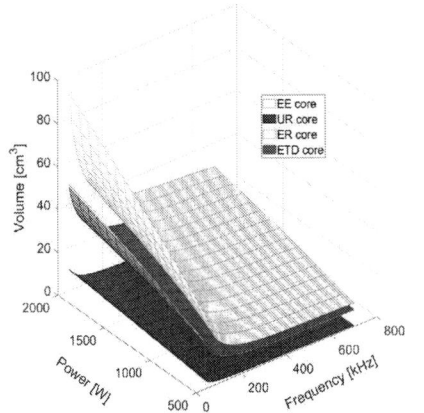

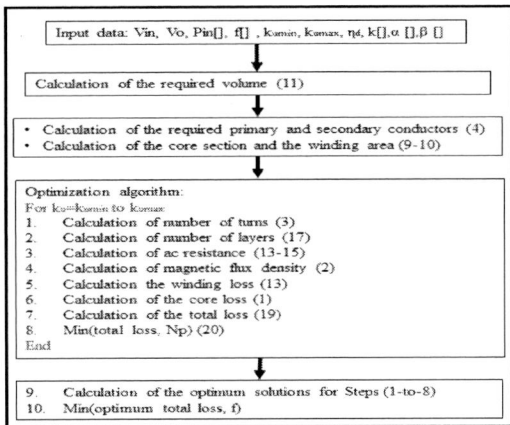

Fig. 1. Required core volume with respect to the power and the frequency for different geometries

Fig. 2. Steps of the optimization algorithm

Transformer loss optimization

The second step in the design approach is the minimization of the total loss. Once the required volume is determined for a wide range of power and frequency, the minimum transformer loss can be determined by calculating the optimum number of turns. This later has a double edge effects on the transformer loss. From one side, it decreases the core loss and from another side it increases the winding loss. Hence, an optimum solution should be calculated. This solution has to satisfy the following constraint in order to not violate the required filling factor.

$$N_{p-opt} \leq N_p \tag{12}$$

The winding loss are expressed as follows

$$P_w = R_{ac} I_{in}^2 = R_{dc} X I_{in}^2 \tag{13}$$

Where X is the AC-to-dc resistance factor which can be calculated using Dowell's equation as follows:

$$X = A \left[\left(\frac{\sinh 2A - \sinh 2A}{\cosh 2A - \cos 2A} \right) + \frac{m^2 - 1}{3} \left(\frac{\sinh A - \sinh A}{\cosh A + \cos 2} \right) \right] \tag{14}$$

$$A = d_p \sqrt{\frac{\pi \mu_0 f}{\rho}} \tag{15}$$

The DC resistance of the primary and the secondary are given by the following which is function of the number of turns.

$$R_{dcp} = \frac{\rho \, l_p}{S_p} = \frac{4 \, \rho \, l_p}{\pi \, d_p^2} = \frac{4 \, \rho \, N_p \left(d_c + \frac{d_p}{2} \right)}{\pi \, d_p^2} \tag{16a}$$

$$R_{dcs} = \frac{\rho \, l_p}{S_p} = \frac{4 \, \rho \, l_p}{\pi \, d_p^2} = \frac{4 \, \rho \, N_p \left(d_c + d_p + \frac{d_s}{2} \right)}{\pi \, d_s^2} \tag{16b}$$

Where dc is the core section diameter calculated from (10) and m is the number of layer which can be calculated by (17).

$$m = \frac{N_p \, d_p}{2 \left(\frac{A_w}{4 \, d_c} \right)} \tag{17}$$

The core loss can be also function of the number of turns.

$$P_c = k_f f^\alpha \left(\frac{V_{in} D}{N_p A_c f} \right)^\beta V_c \tag{18}$$

As formerly stated, since the core volume is fixed for a given power and a frequency using (11), then the number of turn is the only variable to optimize for the transformer loss. The total transformer loss is a non-liner function of the number of turns which requires numerical optimization technique to determine the optimum solution. The flowchart of the optimization algorithm is described in Fig.2.

$$P_{tr} = k_f f^\alpha \left(\frac{V_{in}\, D}{N_p\, A_c\, f}\right)^\beta V_c + \frac{4\,\rho\, N_p\, d_c}{d_p^{\,2}} X_p\, I_{in}^{\,2} + \frac{4\,\rho\, N_p\, d_c}{n\, d_s^{\,2}} X_s\, I_o^{\,2} \tag{19}$$

The equation to be solved is the following:

$$N_{p-opt} = \left(\frac{4\,\rho\; d_c I_{in}^{\,2}\left(\frac{X_{popt}}{d_p^{\,2}} + \frac{X_{sopt}}{n^2\, d_s^{\,2}}\right)}{\beta k_f f^\alpha \left(\frac{V_{in}\, D}{A_c\, f}\right)^\beta V_c}\right)^{\frac{1}{-\beta-1}} \tag{20a}$$

$$X_{popt} = A_p\left[X_{p1} + \frac{m_{popt}^{\,2}-1}{3}X_{p2}\right] \tag{20b}$$

$$X_{sopt} = A_s\left[X_{s1} + \frac{m_{sopt}^{\,2}-1}{3}X_{p2}\right] \tag{20c}$$

$$m_{popt} = \frac{N_{p-opt}\, d_p}{2\left(\frac{A_w}{4\, d_c}\right)} \tag{20d}$$

$$m_{sopt} = \frac{N_{s-opt}\, d_s}{2\left(\frac{A_w}{4\, d_c}\right)} = \frac{n\, N_{p-opt}\, d_s}{2\left(\frac{A_w}{4\, d_c}\right)} \tag{20e}$$

Where m_{popt} and m_{sopt} are the optimum number of layers of the primary and the secondary respectively. For ETD core, the height of the winding area is usually equal to twice the core section diameter.

DESIGN EXAMPLE: FULL BRIDGE CONVERTER

Analysis of the design results

The proposed approach is applied to design a HF transformer for full bridge converter with the characteristics given in Tab. I.

TAB. II. CONVERTER SPECIFICATIONS

Input/output voltage (V)	200/48
Input power (W)	500-5000
Magnetic material/core shape	3C90/ETD
Filling factor	0.7

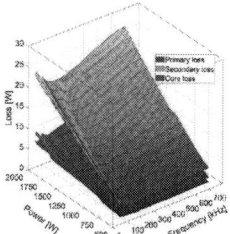

Fig. 3. Minimum transformer losses versus power and frequency

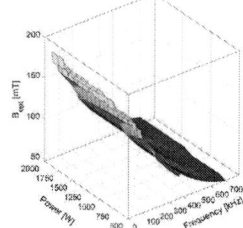

Fig. 5. Optimum magnetic flux density versus power and frequency

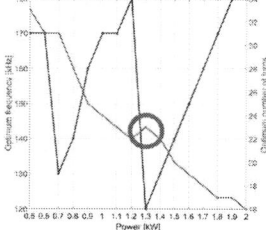

Fig. 7. Best solutions (frequency and number of turns) of the optimum design

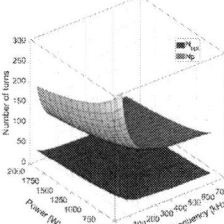

Fig. 4. N_p and N_{opt} versus power and frequency

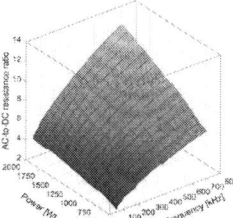

Fig.6. Optimum AC-to-DC resistance ratio

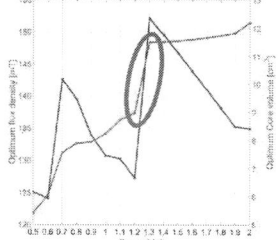

Fig. 8. Best solutions (C) of the optimum design

The evolution of the optimum transformer loss with respect the frequency and the power is depicted in Fig.3. In general, the losses increases with the power and shows a minimum for a specific frequency. The optimum number of turns has an inverse evolution with respect to the frequency (Fig.4). It has also a random variation with respect to the power, exceptionally at low frequency in the range [100-200 kHz], it is inversely proportional to the power. On the other side, the optimum magnetic flux density, which is the effect due to the variation of the number of turns and the frequency, has mostly a weak dependency with respect to the power but it shows a significant dependency on the frequency. The AC-to-DC resistance ratio of the primary shows a significant increase with the frequency due to the skin effect and a moderate increase with respect to the power resulted from the increase the required conductor diameter which in turn increases the diameter-to-skin depth ratio (15) for a given frequency. The best optimum solution are shown in Fig.7 and Fig.8 by extracting the optimum frequency with respect to the power. In general, the optimum frequency shows a random variation with the power. The optimum number of turns and the core volume have approximately a continuous decrease and increase with the power, however the rate of their variations is completely dependent on the variation of the frequency. For example, when the power changes from to 1.2 kW to 1.3 kW, the frequency decreases from 180 kHz to 120 kHz which causes the volume to increase from around 8.5 to 11.5 cm^3 and the number of turns to slightly changes the rate of variation.

Comparison with Hurley approach

The objective of this section is to compare the output results of the design case of the proposed approach and the optimized approach developed in [5] for the case of 500 W FB converter. Since in [5], the frequency is not optimized, two values are considered as shown in Tab. II. The total loss achieved using the proposed approach is about 5 times lower than the total loss obtained from Hurley approach. The main reason for that is the AC resistance is much lower resulting from the low number of layers and the low number of turns. The core size is also much lower (5470 mm^3 to 17800 mm^3 (ETD44)) which eventually reduces the cost [12]. The high core size obtained from [5] is due the inaccuracy in the volume, the core section and the area product models parameters. In general, the proposed approach is more physically realistic and can have less error with the real physical structure of the transformer. This is because the optimization variable is the number of turns, which can govern the optimum flux density and the optimum AC resistance. In contrast, in [5], the winding loss and the core loss are related through the optimum magnetic flux density which is very difficult to achieve in real design.

TAB. II. COMPARISON OF THE DESIGN OUTPUTS OF THE PROPOSED APPROACH WITH HURLEY APPROACH [4]

	Proposed Approach	Hurley Approach
Input Power (W)	500	500
f (kHz)	170 (Optimized)	170(same as the proposed solution)
ku	0.6	
Optimum flux density (T)	0.125	0.037
Selected core	ETD29	ETD44
N_p/N_s	34:8:8	54:13:13
dp/ds	0.8/1.4	
m_p/m_s	1.3/0.57	1.3/1.02
AC resistance	0.22/0.017	0.24/0.04
Core loss (W)	2.18	0.034
Winding Loss (W)	2.33	5.77
Total loss (W)	4.51	5.8

Experimental verification

In order to evaluate the proposed design approach, two transformers are built. The first one is the one obtained using the proposed approach (ETD 29/16/10 core) and the second one, formed with the ETD 44/22/15, is obtained using Hurley approach. We note that the ETD 44/22/15 volume is 3.3 times the volume of the ETD 29/16/10. The winding structures used for both transformers are the same. In fact, the primary has 1.3 layers however, the secondary has only one layer interleaved into the primary. Both transformers are tested with a full-bridge converter up to 480 W. The efficiency of the converter circuit for each transformer is shown in Fig.10. The ETD 44/22/15 transformer is tested with 170 kHz and 100 kHz.

Fig.10 shows that the converter with the ETD29/16/10 shows better efficiency than the solution of the ETD44/22/15 at the rated power 480 W and at 170 kHz. In our analysis, it can be fairly supposed that the losses of the switching devices and the output filters of both cases are same since both converters were operated at the same frequency. Therefore, it can be concluded that the efficiency of the ETD29/16/10 transformer is higher compared to the ETD 44/22/15 transformer at the rated power (480 W). This can also be synthesized from the thermal behavior of both transformers as shown in Fig.11. The maximum temperature rise measured at 480 W are 62°C and 87°C for ETD29 and ETD44 respectively.

At power lower than 440 W and 170 kHz, the converter using ETD 44/22/15 core shows higher performance which can be explain by the following. The rate of decrease of winding loss for ETD44/22/15 is higher than ETD 29/16/10 core because the winding loss is function of the current square. Since the core loss is constant, the variation of total loss depends only on the winding loss. Thus, at a given output current (around 8A), the total loss for ETD 44/22/15 becomes lower than the one of ETD29/16/10. The analysis of the variation of the theoretical calculated losses as a function of the output current is depicted in Fig.12. It gives an insightful explanation of the reasons of the efficiency variation given in Fig.10.

The efficiency of the ETD 44/22/15 solution was also tested at 100 kHz to reach 93.7% compared to 93% for the proposed solution at 480 W, which means a gain of 3.36 W. This slight increase is mostly due to the decrease in the switching devices loss and not due to the transformer because the temperature rise is still higher (71°C to 62°C) despite its higher surface (Fig.11).

Fig. 9. Pictures of the designed transformers

Fig. 10. Measured efficiency of the both solutions

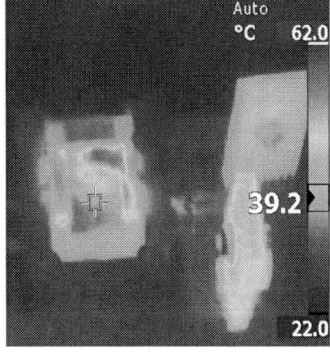
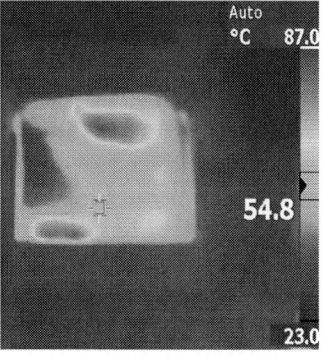
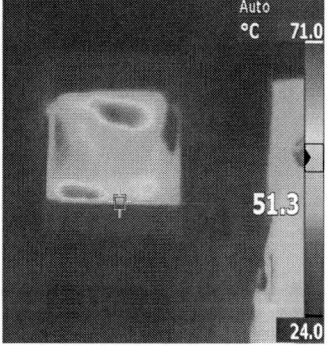

(a) ETD29, 170 kHz   (b) ETD44, 170 kHz   (c) ETD44, 100 kHz

Fig. 11. Measured temperatures of both solutions at the rated power (480 W)

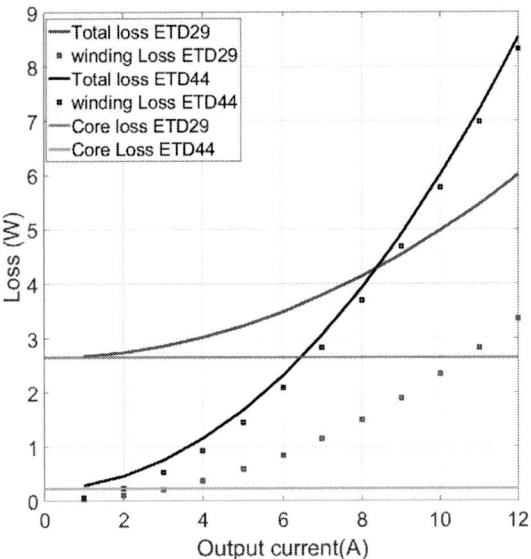

Fig. 12. Theoretical variation of the total, core and winding losses of the designed transformers with respect to the output current, Vin=260 V, f=170 kHz.

Conclusion

A novel optimization and design approach is presented in this work. Unlike previous approaches, the number of turns is chosen to be the optimization variable instead of the magnetic flux density to accurately predict the winding loss and minimize the total transformer loss. In second step, the proposed design allows to determine the optimum frequency. Comparison study and experimental verification with Hurley approach shows that the proposed approach has more advantages such as lower core volume and lower loss. The advantages are resulted from the accurate prediction of the required volume and the accuracy in the calculation of the AC resistance.

References

[1]   K.D.T. Ngo, R.P. Alley, A.J. Yerman, R.J. Charles, M.H. Kuo, "Design issues for the transformer in a low-voltage power supply with high efficiency and high power density", *IEEE Transactions on Power Electronics, Vol: 7, July 1992.*

[2]   C. T. MCL YMAN, "Transformer and inductor design handbook", 3d edition, Marcel Dekker, Inc 2004.

[3]   S. Barg, K. Bertilsson, "Multi-objective Pareto and GAs nonlinear optimization approach for flyback transformer" *Journal of Electrical Engineering, Elsevier, 25 September 2019.*

[4]   Marian K. Kazimierczuk, "High-frequency magnetic components" *2014, John Wiley & Sons, Ltd.*

[5]   W. Hurley, W. W"olfle, and 1. G. Breslin, "Optimized Transformer Design: Inclusive of High-Frequency Effects", *IEEE Transactions on Power Electronics, Vol. 13, No. 4, July 1998.*

[6]   R. Petkov,"Optimum Design of a High-Power, High-Frequency". *IEEE Transactions on Power Electronics, Vol. 11, No. 1, January 1998.*

[7]   S. Barg, K. Ammous, H. Mejbri, A. Alahdal, A. Ammous, "Optimum Design Approach of High Frequency Transformer" *2016 Asia-Pacific International Symposium on Electromagnetic Compatibility (APEMC).*

[8]   S. Barg, "Optimum Design Approach of High Frequency Transformer: Including the Effects of Eddy Currents" *2018 15th International Multi-Conference on Systems, Signals & Devices (SSD).*

[9]   C. P. Steinmetz, "On the law of hysteresis," *AIEE Trans.*, vol. 9, pp. 3–64, 1892. Reprinted under the title "A Steinmetz contribution to the ac power   evolution," Introduction by J. E. Brittain. *Proc. IEEE*,vol. 72, no. 2, pp. 196–221, 1984.

[10] P. L. Dowell, "Effects of eddy currents in transformer windings" *PROC. IEE, Vol. 113, No. 8, AUGUST 1966.*

[11] Handbook of chemistry and physics, HCP, 58th edition, p F-163.

[12] Epcos data book 2013, Ferrite and accessories, www.epcos.com.

Real Time Control Hardware in The Loop test of a novel MVDC solid-state breaker

Alessio Clerici, Riccardo Chiumeo and Chiara Gandolfi
Ricerca Sul Sistema Energetico - RSE S.p.A.
Via Rubattino 54, 20134
Milan, Italy
E-Mail: alessio.clerici@rse-web.it
URL: http://www.rse-web.it/

Acknowledgements

This work has been financed by the Research Fund for the Italian Electrical System in compliance with the Decree of April 16, 2018.

Keywords

«Real time simulation», «Current limiter», «Microcontrollers», «Programming».

Abstract

This paper assesses the development of a novel solid-state MVDC breaker and current limiter. Matlab/Simulink simulations are used as benchmark, then Real Time Control Hardware In the Loop (CHIL) simulations are performed. Same control algorithm is implemented into a microcontroller by using two different software languages: C and MicroPython. In both cases, high performances are achieved by using fast features like hardware interrupts and low level assembler code. Results are then compared in terms of simplicity, ease and speed of implementation and code readability, very important in fast prototyping activity.

Introduction

Medium Voltage Direct Current (MVDC) research is presently under great development, in particular about meshed MVDC grids, in order to combine transmission grids features in DC with traditional radial distribution in AC [1]. This approach let to face some typical transmission-like issues: in a meshed grid, fault currents must be handled to protect equipment from damage and to provide a reasonable current/time ratio to trigger protections properly [2]. As a result, in a meshed MVDC grid, advanced breaking devices become necessary [3]; mechanical DC switches are usually very expensive and, despite high speed response and robustness, they do not cover limiting functions natively: in a modern distribution grid, such features become even more crucial, in order to increase selectivity and definitely system reliability [4].

Due to above reasons, in previous works [5, 6] a novel model of integrated solid state MVDC breaker-limiter has been developed, sized and tested by means of Control Hardware In the Loop (CHIL) simulations. Results have been satisfactory by performance point of view, but development was quite complex: microcontroller algorithm was implemented in C language, and software implementation was quite long, especially due to the necessity to implement fast features like hardware interrupts, timers etc.

In this work, CHIL tests have been repeated by using another kind of programming language for microcontrollers: MicroPython.

Activity is structured as follows: first paragraph shows briefly device design and working principle, including main components sizing to achieve expected performances; second paragraph reports Matlab/Simulink simulations results, as benchmark; third paragraph presents CHIL equipment and a brief description of device control algorithm; fourth paragraph presents a small comparison between C and MicroPython philosophy and main characteristics; fifth paragraph shows CHIL simulations results, comparing C results with MicroPython results; after commenting SW and CHIL results, conclusions are provided.

Solid-state MVDC breaker-limiter concept

Many designs of solid-state breakers are under development worldwide [7, 8], as integration of existing and future DC distribution grids [9, 10].

The device here presented is quite innovative because it combines in a single unit both breaking and current limiting features; in the specific case, it is designed for 2 kV DC rated voltage (±20%).

Main target of such study is obtaining a device with following performances:

- Fully bidirectional current flow;
- Tunable protection functions (current amplitude and limiting time);
- Fast breaking time;
- Galvanic insulation in stand-by position (zero current).

A schematic diagram is shown in Fig. 1:

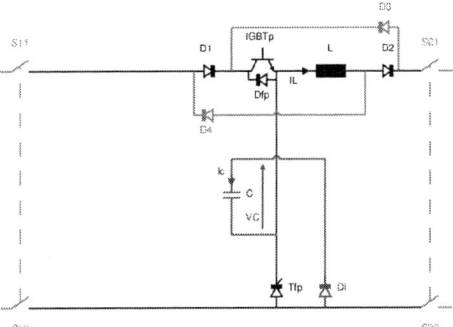

Fig. 1 – Device schematic diagram; forward current path and limit chopper (black); backward current path (red); breaking current path (blue).

Device working principle is here briefly described: during normal operation, current flows through L and $IGBTp$; system is bi-directional by means of diode couples $D1$-$D2$ (forward current) and $D3$-$D4$ (backward current), respectively (Fig. 1); in steady state, L effect is negligible. Capacitor C has been set at grid rated voltage (2 kV) by proper charging circuit; two options are available: charging directly by grid power (Fig. 2A) o by an external power supply (Fig. 2B).

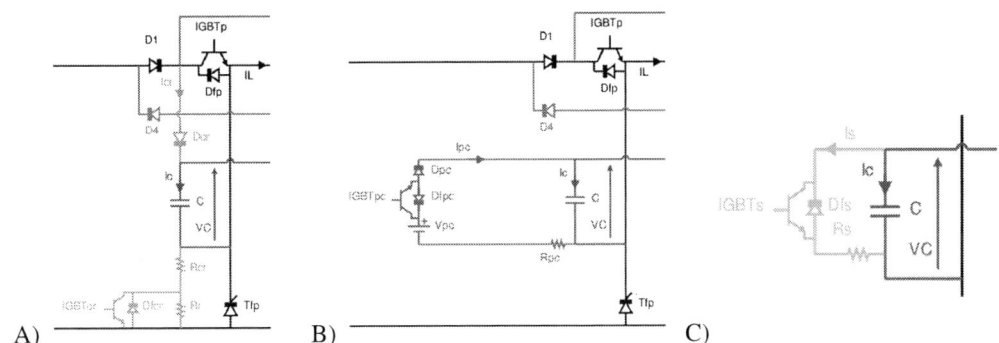

Fig. 2 – Device schematic diagram (detail). A) Capacitor charging circuit through DC grid (green); B) Capacitor charging by external power supply (purple); C) Capacitor discharge circuit (yellow)

When an overcurrent event occurs, it triggers the limitation action: a hysteresis current control operates $IGBTp$ and thyristor Tfp alternatively, like a step-down chopper, using L as smoothing reactor (Fig. 1). Current IL mean value is tunable and it is lowered at given value for a given time, also tunable.

After limitation, breaking phase is engaged: *IGBTp* is opened permanently, while *Tfp*, not fired, is not conducting; inductive current *IL* can flow only through diode *Di* to charge *C* up to current zeroing. After that, circuit is galvanically insulated by opening (off load) all disconnectors (*S11-S12-S21-S22*). Finally, *C* is set at safe voltage (e.g. 50 V) by a discharge circuit (Fig. 2C).

Due to above explanation, *L* and *C* and sizing is crucial to proper manage currents and power flows through the device. A brief sizing method and a set of expected performances is presented below.

Fig. 3 shows a time diagram of current *IL* in the most general case of:

1. device operating at nominal current (*In*);
2. overcurrent protection trip (maximum threshold *Ik*);
3. current limitation (start of limitation within given time *Tkl*, limitation for the set time *Tlim*);
4. current interruption within given time (*Ti*).

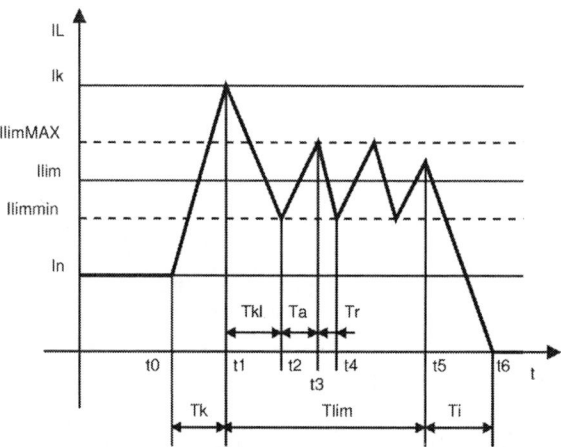

Fig. 3 – Time diagram of current *IL* in the most general case (current profile is indicative).

The graph is indicative: waveforms, timing and slopes are not to scale, they are magnified to better show the various operating phases.

It is important to point out that the main purpose of the device is the protection of the electrical system placed upstream of it, in particular the power converter (or converters) responsible for supplying the DC line [11].

According to previous considerations, following design specifications have been chosen (Table I):

Table I: Device design specifications

Quantity	Name	Value	Unit
In	Rated current	2500	A
Vn	Rated voltage	2000	V
VnMAX	Maximum voltage	2400	V
Vnmin	Minimum voltage	1400	V
VCMAX	Maximum capacitor voltage	4050	V
Ik	Fault overcurrent	5000	A
Ilim= IL mean	Limiting current (inductor current mean value)	4500	A
IlimMAX	Limiting current, maximum value	4700	A
Ilimmin	Limiting current, minimum value	4300	A
fswMAX	Maximum switching frequency for IGBT	5	kHz

Here below most important sizing equations are summarized: they result from differential equations solutions of the circuit in various phases of Fig. 3 diagram.

As to mentioned specifications, maximum fault resistance to proper trigger overcurrent protection is *RgMAX*:

$$RgMAX = \frac{Vnmin}{Ik} = 0.32 \; \Omega \tag{1}$$

Limiting current gap *ΔIl* is:

$$\Delta Il = IlimMAX - Ilimmin = 400 \; A \tag{2}$$

To proper smooth and slow fault current rise, inductance *L* is at least:

$$L \geq \frac{VnMAX \cdot IlimMAX \cdot RgMAX}{fswMAX \cdot \Delta Il(IlimMAX \cdot RgMAX + VnMAX)} = 0.5 \; mH \tag{3}$$

And capacitance *C* results:

$$C = \frac{L[Ik^2 - Ilimmin^2 + IlimMAX^2]}{[VCMAX^2 - VnMAX^2]} = 1.3 \; mF \tag{4}$$

Once main parameters have been sized, Matlab/Simulink simulations let to verify both control algorithms correctness and expected results in term of performances.

Matlab/Simulink simulation results

First set of Matlab/Simulink simulations (software only, not real time nor HW-in-the-loop) provided encouraging results by performance point of view, as shown in Fig. 4:

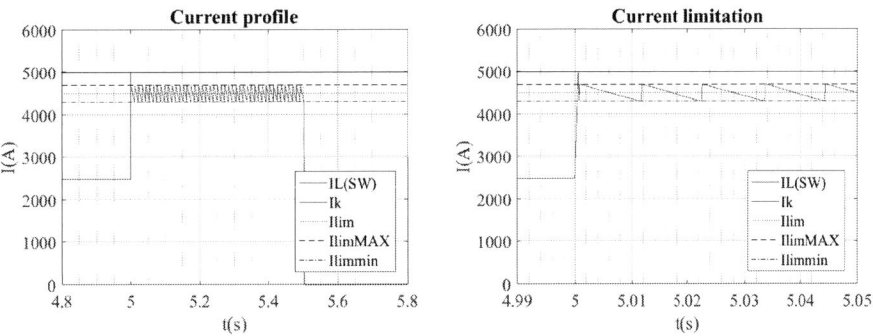

Fig. 4 – Initial SW simulation results.

With proper control logics implementation and tuning, expected current limitation and breaking profiles have been achieved: MVDC line is operating at rated current (2500 A); at t=5 s, a fault event occurs: as shown, initial peak current limit (5000 A) triggers the protection, then breaker-limiter device successfully limits the faulty overcurrent around the given value (IL mean value, 4500 A), and also upper (4700 A) and lower (4300 A) thresholds are never violated. At the end of the process, current is forced to zero by charging the capacitor, whose voltage does not reach maximum sizing level (4050 V). In practice, capacitor operates as a snubber in fault transients and also during normal operation of MVDC line [12].

Results proved design and sizing criteria goodness, but real-time tests need also a good implementation of system control and fast system responses.

In order to achieve same results, an OPAL-RT CHIL equipment has been used, as described in the following.

CHIL equipment description

In this work, main target is testing static breaker into a real HW control loop: breaker logic and measurement have been programmed into a ST-Nucleo Microcontroller board, while power electronics circuit and a sample MVDC grid have been modeled using Matlab/Simulink into FPGA module of an OPAL-RT HW-in-the-loop simulator. Communication between two systems has been performed by means of cabled digital and analog I/O, as shown in Fig. 5.

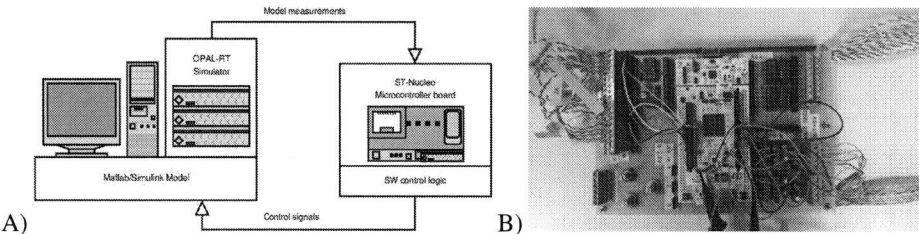

Fig. 5 – A) HW-in-the loop circuit diagram. OPAL-RT simulator interfaced to ST-nucleo microcontroller board by HW signals. B) ST-Nucleo with prototype interface board.

In ST-Nucleo board, control logic has been implemented in C code. HW interrupts have been programmed, in order to make fault reaction and limiting phase as much as possible close to ideal ones.
Control Hardware In the Loop (CHIL) simulation results are shown in the followings.

CHIL Real-Time simulations with C language

Fig. 6 shows simulation results with CHIL system of Fig. 5 and C code implemented into ST-Nucleo microcontroller board:

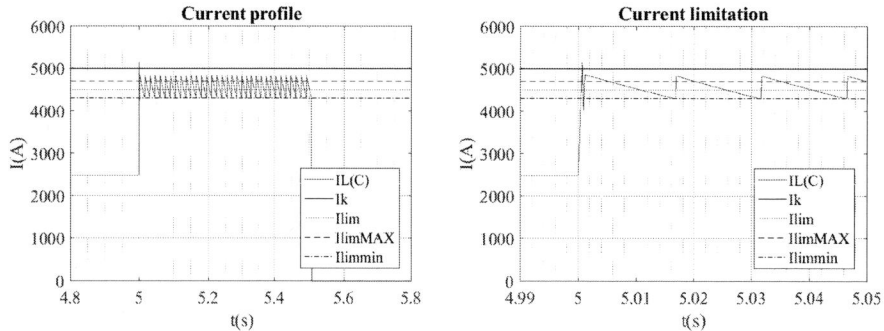

Fig. 6 – CHIL Real Time simulation results with C programmed microcontroller board.

Comparing graphics of Fig. 6 with the ones in Fig. 4, it can be noted a substantial equivalence in behavior of CHIL simulation with SW-only simulation; ST-Nucleo microcontroller board can effectively perform the expected tasks, by controlling power electronics equipment simulated into OPAL-RT program. HW interrupts in the control loop guarantee high speed response, minor differences can be appreciated in Fig. 7.

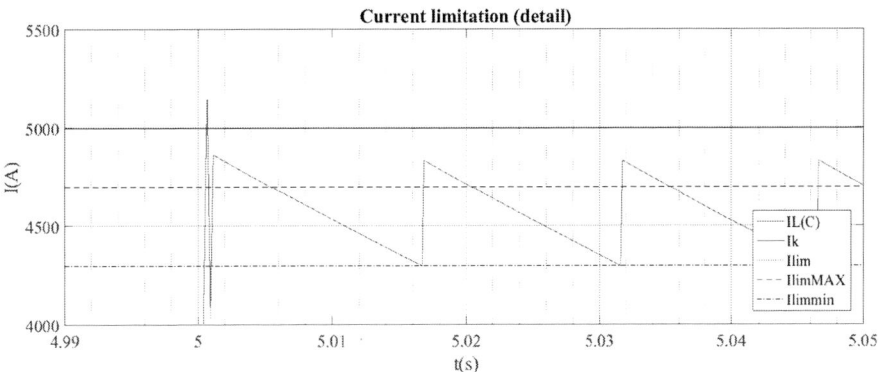

Fig. 7 – Detail of CHIL Real Time simulation detail with C programmed microcontroller board.

Due to board computational period and signals transmission delay, control has small offsets: peak overcurrent is 5147 A (+ 2.9 %) while upper and lower limits are 4835 A (+ 2.7 %) and 4299 A (-0.01 %), respectively. As a result, *IL* mean value is about 4560 A (+1.3 %).

Differences between SW simulations and CHIL ones are anyway definitively small and unavoidable due to physical performances of equipment (sampling time, transmission delays etc.).

The choice of programming language: C vs. MicroPython

C language is nowadays one of the most popular, well-known and used programming language worldwide: its success is mainly given by intrinsic robustness, reliability and speed of execution; developed in 1970s, despite it is classified as high level language, it can be used to provide low-level access to memory and to machine instructions, all with minimal runtime support. This is the main reason it is widely used in microcontrollers, as the best way to interact directly to hardware.

C language high speed execution is given by compilation: source code is processed by a program, called compiler, that translates the high-level instructions into low-level executable code. In microcontrollers, the executable is loaded into the flash memory and it is executed as soon as the board is powered on. The result is that every code modification implies re-compilation of the source code, and the reloading into the flash memory. Another issue of compilation is given by "dialects": every modification in canonical instructions makes a custom compiler necessary, with lack of compatibility; such phenomena is particularly evident in microcontroller market, where companies develop custom libraries, instructions and syntax for dedicated hardware. The result is that the code cannot be easily transferred on another board without a full translation and many dialects have to be learnt.

All above considerations make C language a very useful and efficient language for microcontrollers, but with several issues given by compilation: in particular, fast hardware prototyping is definitely difficult, because every single modification implies the full recompilation and loading of the software to be tested.

MicroPython was developed in 2013; it is a software implementation of a programming language largely compatible with Python 3, written in C, that is optimized to run on a microcontroller.

Differently from pure C language, MicroPython is an interpreted language: it means that microcontroller is loaded with a Python compiler, that is basically an interpreter. The user can interact to the board with a prompt (the REPL, see Fig. 8Fig. *8*) to execute supported commands immediately. Being based on ARM Cortex microprocessor, STM32 Nucleo F767ZI Board is MicroPython compatible [13]:

```
MicroPython v1.12-452-g801f7dca7 on 2020-05-14; NUCLEO-F767ZI with STM32F767
Type "help()" for more information.
>>>
>>> 1.265+3.44732
4.71232

>>>
>>> |
```

Fig. 8 – MicroPython command prompt (REPL) on a STM32 Nucleo F767ZI Board; as shown, supported commands are executed immediately, e.g. floating point calculations.

The interpreter converts high-level commands into low-level instructions and then the hardware can execute them. The result is that, compared to C, MicroPython has generally lower speed execution, because the compilation process has to be performed "on-the-fly". This is only partially true, because the language is optimized for microcontroller boards, then it has many low-level functions that are briefly described in the followings.

How to increase MicroPython speed

One of the most interesting features of MicroPython is the possibility to easily implement assembler code into the main Python code (inline assembler); assembler code is just a step above machine language code, so instructions are executed in very fast way, because they do not have to be processed by the interpreter. The result is a program can mix high-level Python constructs (universally recognized as much more clear and easy to read respect to C language) for the "slow" part of the code and assembler instructions for the "fast" part of the code, in practice the direct access to the hardware (input/outputs).

An example of archived performances is given in Fig. 9.

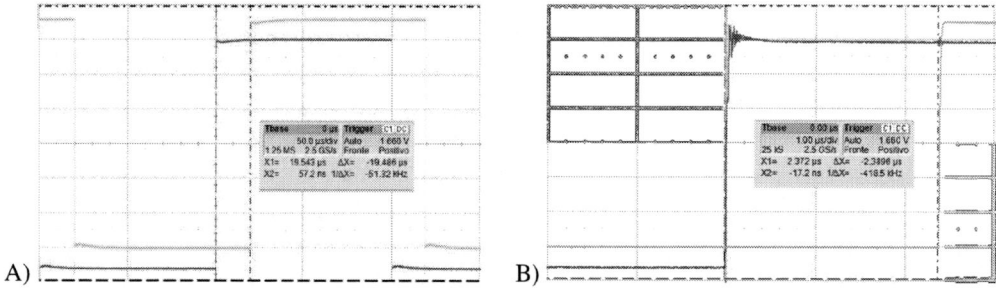

Fig. 9 – MicroPython hardware performances. A) Systematic delay in digital output changing (light green line) respect to digital input changing (blue line) by using MicroPython instructions. B) Systematic delay in digital output changing (light green line) respect to digital input changing (blue line) by using inline assembler instructions.

A simple program to change the status of a digital output (light green line) respect to the status of a digital input (blue line) has been implemented; code execution is triggered by hardware interrupt on the input, so output delay represents the execution time of the code.

In the first diagram (Fig. 9A), the output is controlled by standard MicroPython instructions, processed by the interpreter; in the second diagram (Fig. 9B), the same output is controlled by inline assembler; as shown, the execution delay between input and output passes from 20 μs of Fig. 9A to 2.4 μs of Fig. 9B (more than eight times faster).

Above considerations and examples suggest that MicroPython can be a valid alternative to C language for fast hardware prototyping, because it combines the needed speed response with the simplicity and readability of Python language; in addition, program modifications can be tested "on the fly", with no necessity of repetitive compiling. Next section shows CHIL real time simulations with a MicroPython controller board.

CHIL Real-Time simulations with MicroPython

Fig. 10 shows simulation results with CHIL system of Fig. 5 and MicroPython code implemented into ST-Nucleo microcontroller board; for direct comparison, results of Fig. 6 are shown in the same plot.

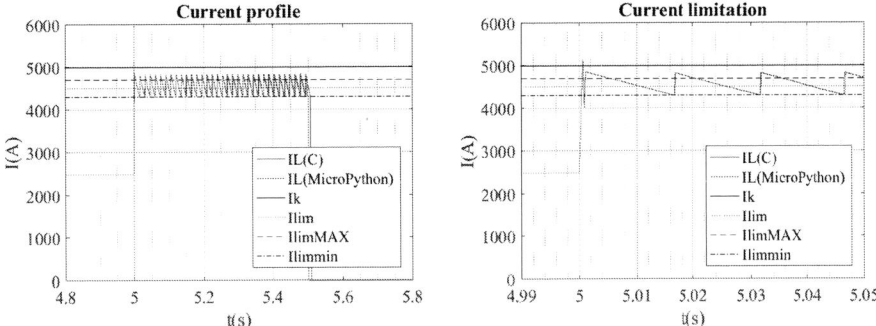

Fig. 10 – CHIL Real Time simulation results with MicroPython programmed microcontroller board (purple). C results are shown in the same plot (blue) for direct comparison.

Comparing results of MicroPython with the C language one, it can be noted a substantial equivalence of both CHIL simulations (in some cases, even better than C, even if infinitesimally); ST-Nucleo microcontroller board can effectively be programmed with MicroPython and achieve the same performances as C. HW interrupts and inline assembler in the control loop guarantee high speed response; the equivalence can be better appreciated in Fig. 11.

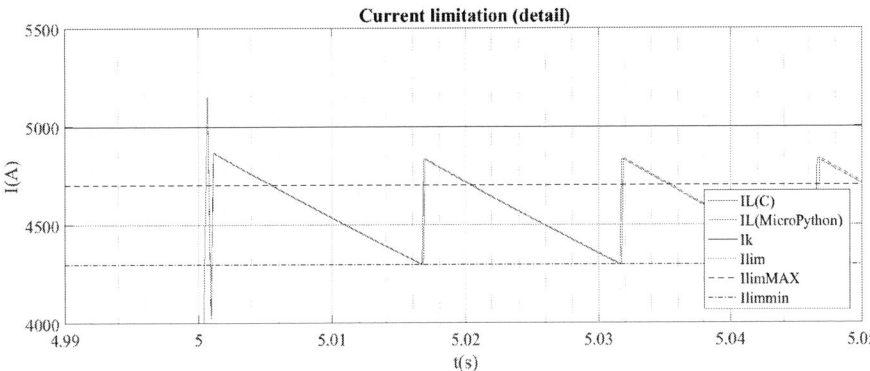

Fig. 11 – Detail of CHIL Real Time simulation with MicroPython programmed microcontroller board (purple). C results are shown in the same plot (blue) for direct comparison.

Control performances are summarized in Table I:

Table II: Simulations control performances

Quantity	SW only	Real Time CHIL with C	Real Time CHIL with MicroPython
Ik	5000 A	5147 A (+2.9 %)	5138 A (+2.7 %)
IlimMAX	4700 A	4835 A (+2.7 %)	4831 A (+2.7 %)
Ilimmin	4300 A	4299 A (-0.01 %)	4299 A (-0.01 %)
Ilim	4500 A	4560 A (+1.3 %)	4560 A (+1.3 %)

As previous, small offsets are given by Real Time system time step (20 µs), that is significantly longer than microcontroller hardware response time (less than 3 µs). The result is that mathematical model inside the simulator is updated by inputs delivered by the controller with a systematic delay.

Compared to C, MicroPython coding took about one third of the total development time to provide a full functional control scheme. It must be noted that MicroPython development was a lot easier because it did not start from scratch; anyway, time and resource gaining was still definitely relevant.

Conclusion

This work assessed the development of a novel solid-state MVDC breaker and current limiter. Matlab/Simulink simulations have been used as benchmark, then Real Time Control Hardware In the Loop (CHIL) simulations have been performed. C and MicroPython language have been adopted to implement the same control algorithm on a ST-Nucleo microcontroller board. Despite MicroPython is theoretically less performing than C in terms of speed response, high hardware performances are achieved by using fast features like interrupts and low level inline assembler.

In this project, MicroPython can be considered a valid alternative to C language for fast hardware prototyping, because it combines the needed speed response with the simplicity and readability of Python language.

Future development of the project includes the utilization of more advanced control techniques respect to "simple" hysteresis band, hopefully to increase system precision and reduce limited current ripple.

References

[1] R. Zuelli, R. G. M. Chiumeo, C. Gandolfi, A. Clerici, S. Pugliese, S. Fratti "The impact of MVDC links on distribution networks," 2018 AEIT International Annual Conference, Bari, 2018, pp. 1-5. doi: 10.23919/AEIT.2018.8577442

[2] L. Qi, A. Antoniazzi, L. Raciti, "DC Distribution Fault Analysis, Protection Solutions, and Example Implementations," IEEE Transaction on Industrial Applications, Vol. 54, Issue: 4, 2018.

[3] P. Cairoli, R. A. Dougal, "Fault detection and isolation in Medium Voltage DC microgrids: coordination between supply power converters and bus contactors", IEEE Transactions on Power Electronics, Vol. 33, No. 5, 4535 – 4546, May 2018

[4] A. Villa, F. Belloni, C. Gandolfi, R. Chiumeo, A. Clerici, "Coordination of Active Current Limiters and Hybrid Circuit Breakers for a MVDC Link Meshing MVAC Distribution Grids", 19th European Conference on Power Electronics and Applications (EPE'17 ECCE Europe), 11-14, Settembre-2017.

[5] A. Clerici, R. Chiumeo, C. Gandolfi, "MVDC solid-state breaker Control Optimization by Real Time Control Hardware in The Loop tests", 20th International Conference on Environment and Electrical Engineering EEEIC 2020, Madrid, Spain, 09-12 June 2020.

[6] A. Clerici, R. Chiumeo, C. Gandolfi, M. Cabiati "Risultati dello studio di un dispositivo integrato di limitazione e interruzione delle correnti di corto circuito in reti MVDC" (Results of the study of an integrated device for limiting and interrupting short-circuit currents in MVDC grids), RSE 20001737, 2019. Rapporto Ricerca di Sistema (Italian language).

[7] X. Pei, O. Cwikowski, D. S. Vilchis-Rodriguez, M. Barnes, A. C. Smith and R. Shuttleworth, "A review of technologies for MVDC circuit breakers," IECON 2016 - 42nd Annual Conference of the IEEE Industrial Electronics Society, Florence, 2016, pp. 3799-3805. doi: 10.1109/IECON.2016.7793492.

[8] J. Magnusson, R. Saers, and L. Liljestrand, "The commutation booster, a new concept to aid commutation in hybrid DC-breakers", CIGRE, Lund, 2015.

[9] A. Villa, F. Belloni, C. Chiappa, M. Brenna, D. Palladini "Integrazione di collegamenti in corrente continua nelle reti di distribuzione in MT: studio delle condizioni di guasto nelle reti c.a. e c.c.." RSE 16001664, 2015. Rapporto Ricerca di Sistema (Italian language)..

[10] A. Villa, A. Clerici, C. Gandolfi "Sistemi di distribuzione in Media Tensione in corrente continua multiterminali e a più livelli di tensione: simulazioni di Control Hardware in the Loop (CHIL) al simulatore RT" RSE 18000003, 2017. Rapporto Ricerca di Sistema (Italian language)..

[11] K. Corzine, A. Overstreet and P. E. T. Baragona, "Solid-state breaker protection in MVDC systems," 2017 IEEE Electric Ship Technologies Symposium (ESTS), Arlington, VA, 2017, pp. 414-418.

[12] F. Liu, W. Liu, X. Zha, H. Yang, K. Feng, "Solid-State Circuit Breaker Snubber Design for Transient Overvoltage Suppression at Bus Fault Interruption in Low-Voltage DC Microgrid" IEEE Transactions on Power Electronics 2017.

[13] UN1974 User Manual – STM32 Nucleo-144 boards, Doc ID028599, Rev 7, ST Microelectronics, 2017.

IGBT Lifetime Estimation in a Modular Multilevel Converter for bidirectional point-to-point HVDC application

Diego Velazco[1,2], Guy Clerc[1,2], Emmanuel Boutleux[1,2], François Wallart[1], Laurent Chédot[1]

[1] SUPERGRID INSTITUTE SAS
23 rue Cyprian
Villeurbanne, France
Tel.: +33 / (0) – 4 28 01 23 23
E-Mail: Diego.VELAZCO@supergrid-institute.com
URL: https://www.supergrid-institute.com/

[2] Univ. Lyon, Univ. Claude Bernard Lyon1, INSA Lyon, Ecole Centrale Lyon, AMPERE
UMR CNRS 5005

Acknowledgements

This work was supported by a grant overseen by the French National Research Agency (ANR) as part of the "Investissements d'Avenir" Program (ANE-ITE-002-01).

Keywords

«Multilevel converters», «HVDC», «Lifetime estimation», «Reliability», «Mission profile», «IGBT».

Abstract

This paper deals with the lifetime estimation of one modular multilevel converter submodule in a bidirectional point-to-point HVDC application. A mission profile of a reversible link is considered, then power losses are calculated and an electro-thermal simulation is carried out. Finally, a rain flow counting method is used for organizing thermal profiles and the life consumption is computed thanks to a new extrapolation method of aging data coming from manufacturer and the application of the Miner's rule.

I - Introduction

Modular multilevel converter (MMC) is nowadays the most promising voltage source converter (VSC) topology for distribution and transmission applications [1], [2]. MMC offers many advantages such as scalability, improved output quality thanks to a modular structure, high DC level and improved efficiency due to less switching losses than other VSC topologies [2].

As a VSC topology, MMC can control active and reactive power independently and can reverse the power flow while maintaining voltage polarity, which makes it attractive for its integration to the existing AC grids and for implementing Multi-Terminal DC (MTDC) grids [3]. In an HVDC link, power reversals can take place during converter operation, thus modifying the thermal loading of its static switches. The effects of power reversal have not yet been considered regarding reliability and lifetime of the converter. Therefore, studying the thermal loading of a point-to-point link is necessary for establishing the impact of bi-directionality on the converter.

There is extensive literature on lifetime estimation methods for converters in offshore wind farms [4], [5]. Moreover, research on the lifetime estimation of MMC for offshore applications was introduced in [6]. Those works focus on life consumption under unidirectional power flow, but one of the main characteristics of MMC is power reversibility. Thus, a detailed analysis of the impact of a reversible mission profile on a MMC submodule will be carried out throughout this paper. More particularly, thermal loading effects on the power semiconductors of a submodule will be addressed.

As stated in [6], lifetime estimation of semiconductor devices can be classified into analytical and physical modelling methods. The latter methods are not applicable to the MMC due to the large number of submodules [2]. Analytical methods are employed in this research. They estimate the number of thermal cycles that the semiconductor can withstand before failure [7], [8].

Two of the main failure mechanisms of die-attached semiconductor devices are wire bonds lift-off and solder cracking [9]. Both mechanisms are related to a mismatch of thermal expansion coefficients (TEC) of the different layers of power semiconductors. Cyclic thermal stresses are brought by the loading variations and the periodic commutations of power switching devices [10].

Section II of this paper introduces the MMC topology and a yearly mission profile of a bidirectional HVDC link is presented. Section III deals with the lifetime estimation methodology. In this section, the different steps will be developed. Section IV presents the results of this research.

II - MMC topology and reversible HVDC link mission profile

A point-to-point HVDC link allows the interaction between two AC grids. The link is composed of two MMCs connected by a cable on the DC terminals as seen in Fig. 1 (a). The MMC topology is shown in Fig. 1 (b). It consists of a 3-phase structure, where each phase is composed by an upper and lower arm. Each arm is composed of a stack of submodules (SM) and an arm inductor. The submodules can either have a full-bridge or half-bridge configuration, but in most applications the latter is employed. A half-bridge submodule contains two IGBT modules T1 and T2, as well as two antiparallel diodes D1 and D2, as seen in Fig 1 (c). The different switches will be active depending on the direction of the current and the command signals. Those switches will allow the insertion or the bypass of a storage capacitor.

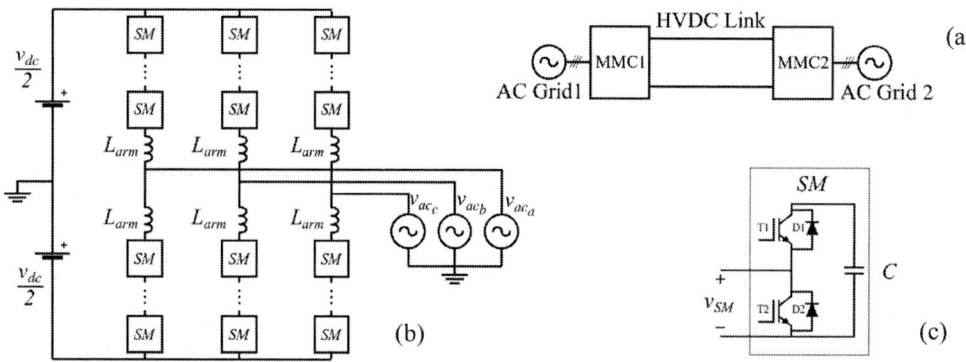

Fig. 1: (a) HVDC link (b) MMC topology, (c) Half-bridge submodule detailed structure

The lifetime estimation will focus on one of the MMC submodule semiconductors. The studied module is the ABB HiPak 5SNA 1500E330305. In order to carry out this research work, a reversible HVDC link mission profile with a time span of 1 year by step of 1h was evaluated (Fig. 2). The mission profile corresponds to the NorNed link which is based on Line Commutated Converter (LCC) technology. The data was adapted to a MMC based link in order to perform the simulations, due to the lack of public information on mission profiles for MMC based HVDC links.

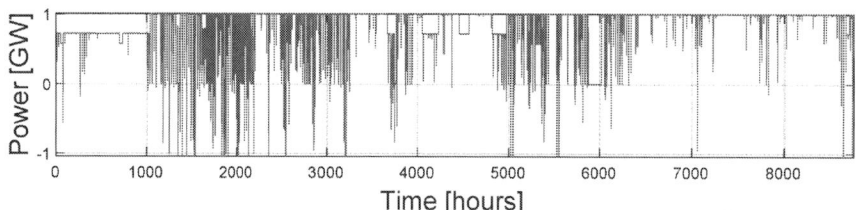

Fig. 2: Yearly mission profile of a reversible HVDC link

III - Method for lifetime estimation

The methodology for lifetime estimation is divided in 5 steps as seen in Fig. 3. The first step uses an electrical model of the MMC which will calculate the currents and voltages across the semiconductors. The next step requires a device loss model that will compute the conduction and switching losses sustained by the switches. The third step consists of a thermal model that will estimate the junction temperatures of IGBTs and diodes. Afterwards, a rain flow counting algorithm will organize the different thermal cycles experienced by the modules. Finally, the number of cycles to failure for each thermal cycle will be calculated thanks to the lifetime models provided by the manufacturer [7] and the effect of all the cycles will be added using Miner's rule. It is worth noting that the lifetime estimation will be performed under different time-scales as done in [4]. The same steps will be applied for the different time scales, but different approaches are used depending on the studied time scale. Results for the different time scales can be added thanks to the Miner's rule [11]. All these steps will be detailed hereunder in this section.

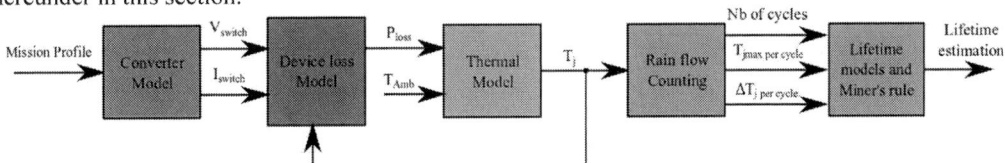

Fig. 3: Flowchart of the lifetime estimation methodology

In [4], lifetime estimation for a converter in an off-shore wind farm is carried out by analyzing 3 time scales, long-term, short-term and medium-term. In that study, long-term is dependent on environmental conditions, medium-term is dependent on the mechanical time constants and short-term is dependent on electrical behavior and its time constants.

In this research, long-term analysis is performed effortlessly, as the mission profile evaluated provides hourly operating points for the converter as previously seen in Fig. 2. Thermal time constants of semiconductors and cooling system can be disregarded for simplicity of long-term lifetime estimation.

In [4], the wind turbine system provides the dynamics for medium term analysis. In the case of a point-to-point application, there is no mechanical system whose time constants can affect the thermal profiles of the converter. Thus, another method has to be employed to perform medium-term analysis. An HVDC link can provide added capabilities to the AC grid such as improvement of transient stability, voltage and frequency support, etc. [12]. These functionalities, as well as grid fluctuations can impact the thermal loading of semiconductors. However, these conditions are hard to predict and the availability of this data is restricted to Transport System Operators (TSO). Therefore, and in order to evaluate thermal loading within a 1-hour period, hourly operation data was simulated around mean powers, ranging from -1GW to 1GW by steps of 0.1GW. The hourly profiles are generated following a normal distribution and provide the basis for the medium-term analysis.

This method was employed due to the lack of public information about converter operation, and is used to reflect possible grid fluctuations. The different profiles can be customized by modifying the value of the standard deviation employed. Results were obtained using a standard deviation of 1.33% with respect to the mean value to reflect a realistic profile. An example of an hourly profile with a mean power of 1GW and a maximum deviation of 40MW can be seen in Fig. 4. Afterwards, medium-term profiles have to be extrapolated for one-year operation using the occurrence frequency of each operating point of the long term mission profile, as seen in Fig. 5.

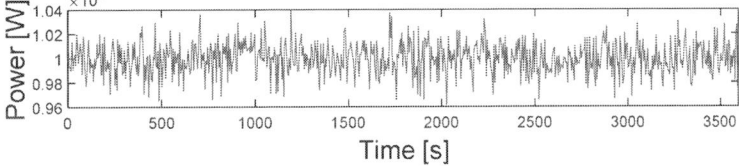

Fig. 4: Generated operating points for medium-term analysis (around 1GW average power)

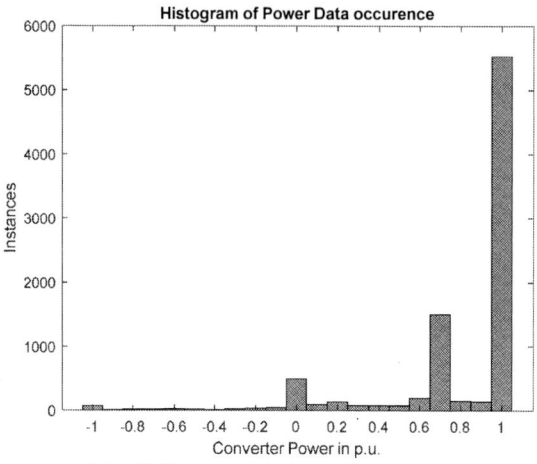

Fig. 5: Occurrence frequency of the different operating points over yearly mission profile

Medium-term analysis will take into consideration the dynamics of the power semiconductors and cooling system, as they can no longer be neglected. This analysis will be performed for time spans lower than 1-hour. Short-term analysis will not be conducted in this research, since the manufacturer models used for lifetime estimation do not consider thermal cycles lower than 2 seconds, and short term thermal cycles swing at grid frequency, typically 50 Hz.

Electrical model of the MMC

A model of the converter was conceived in order to obtain the electrical waveforms experienced by the semiconductors of a submodule. One submodule is fully detailed in order to have all currents and voltages across the devices, the others ones are considered within an average model.

It is important to highlight that the gate signal Sh of each switch, will determine whether the storage capacitor is inserted or bypassed. Upper and lower switches in Fig. 1 (c) are complementary. When Sh = 1 the upper switch is active and when Sh = 0 the lower switch is active. The direction of the current will determine which component, IGBT or diode will conduct the current. The specifications of the converter can be seen in Table I. Fig. 6 shows some waveforms of the MMC such as the arm current I_{arm}, the switching state of the submodule Sh, the capacitor voltage V_c and the capacitor current I_c.

Table I: Specifications of the converter

Parameter	Variable	Value	Unit
Rated power	P	1	GW
AC Grid voltage	Uac_{rms}	333	kV
DC-link voltage	Vdc	640	kV
AC grid frequency	f0	50	Hz
Average switching frequency	fsw	150	Hz
Arm inductor	Larm	50	mH
Number of SM per arm	Nsm	400	-
SM capacitor	Csm	10	mF
Average capacitor voltage	Vc	1.6	kV

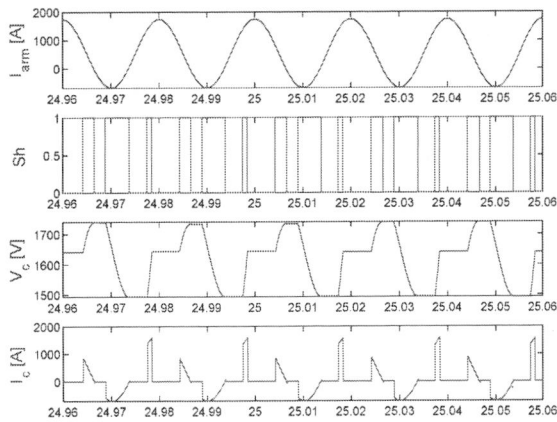

Fig. 6: Electrical waveforms of the MMC

Device loss model

A power loss model Fig. 7 (a) was built around the datasheet information of the power semiconductor, and the outputs of the electrical model described previously. In order to reduce the computation time, a 3D look-up table is used Fig. 7 (b). The latter was established to calculate the losses of the different switches for different operating points, in terms of converter power and junction temperatures T_j. This approach is employed to compute the losses in both studied time scales.

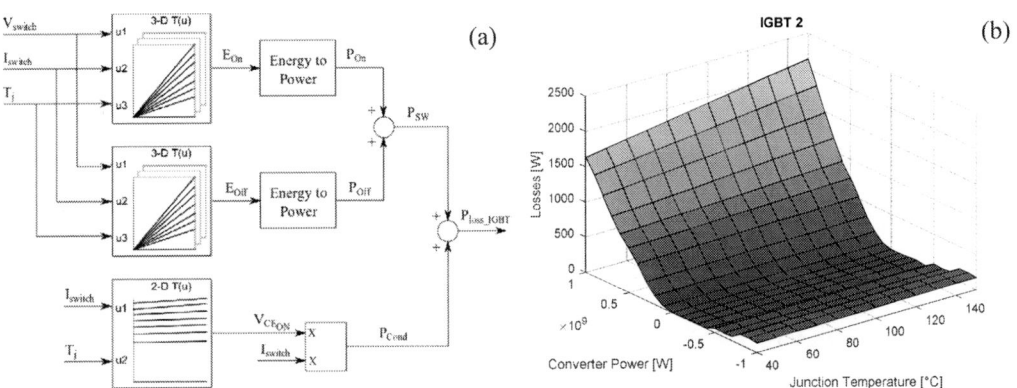

Fig. 7: (a) IGBT power loss model (b) 3D look-up table for loss calculation of IGBT2

Thermal model

As mentioned previously, in the case of long-term analysis, time constants of semiconductors and cooling system can be neglected. This means that only thermal resistances are kept in the model. The thermal model shown in Fig. 8 can be employed for this time scale.

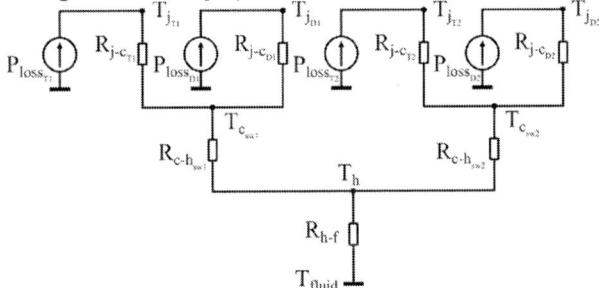

Fig. 8: Static thermal model employed for long-term lifetime evaluation

The temperature estimation of semiconductors for medium-term analysis is based on the dynamic thermal model shown in Fig. 9, which is a 4th order Cauer thermal network for each switch. The parameters of the model are obtained by calculating the Cauer equivalent of the Foster network provided by the manufacturer. The Cauer representation provides a better representation of the temperatures across the different layers of the modules. All the components have independent case-to-heatsink thermal resistances and they are all connected to the same heatsink. The network in Fig. 9 will allow the computation of the junction temperature T_j of the different switches when expressed in the form of the state space model (1).

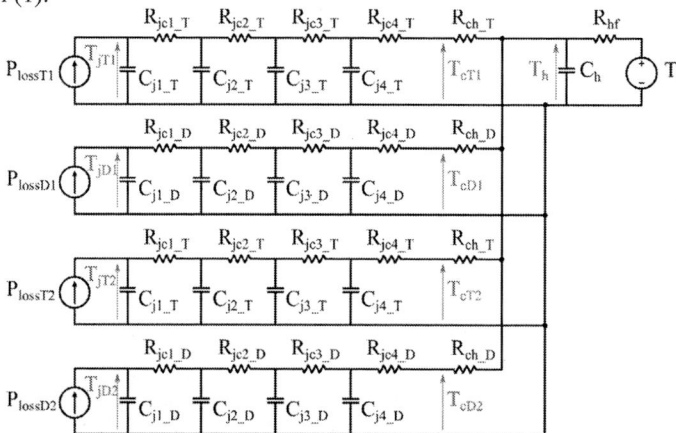

Fig. 9: Cauer electro-thermal model of a submodule

$$\frac{dx(t)}{dt} = Ax(t) + Bu(t)$$
$$y(t) = Cx(t)$$

(1)

In (1) the state vector x(t) and the output vector y(t) represent the junction temperatures across the different layers of semiconductors and the heatsink temperature. The input vector u(t) contains the losses experienced by the switches (P_{lossT1}, P_{lossD1}, P_{lossT2} and P_{lossD2}) and the thermal fluid temperature T_f. State matrix A, input matrix B and output matrix C can be obtained by inspection of the network in Fig. 9.

The long term temperature profile obtained with the static thermal model is seen in Fig. 10 and an hourly medium-term profile obtained with the dynamic model is represented in Fig. 11.

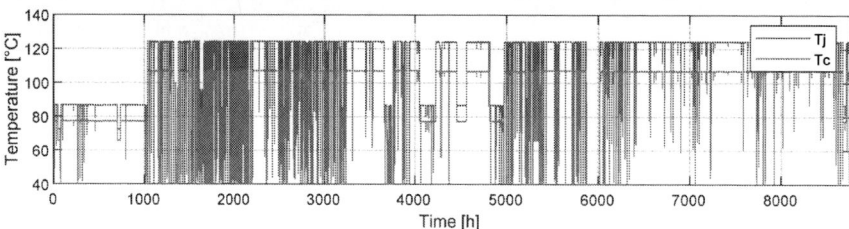

Fig. 10: Thermal profile of IGBT2 under the yearly profile

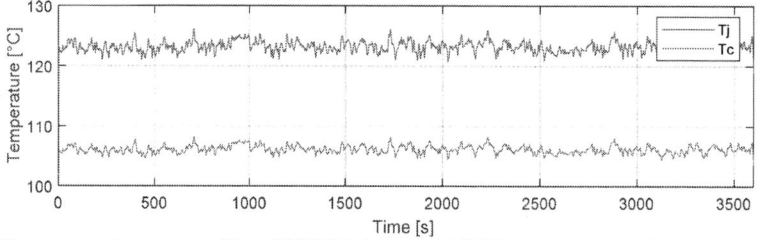

Fig. 11: Medium-term thermal profile of IGBT2 (around 1GW average power)

Rain flow counting

Semiconductors lifetime consumption depend on factors such as the range of thermal cycles ΔT_j, a temperature reference (either Tj_{max} or Tj_{min}) and the duration of the cycle t_{cycle} [4], [7]. Therefore, thermal profiles must be organized with the help of a rain flow counting algorithm [13] in order to use lifetime models provided by the manufacturer [7]. The results of the rain flow counting algorithm applied to the thermal profile of IGBT2 can be seen in Fig. 12. The rain flow counting method gives also the duration of the cycles in order to be able to use the models from the manufacturer.

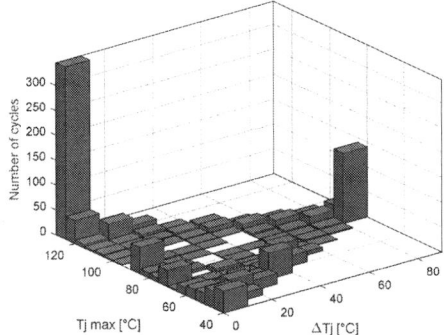

Fig. 12: ΔT_j and Tj_{max} values for IGBT2 identified by the rain flow counting algorithm

Lifetime models and Miner's Rule

The manufacturer provides its own lifetime models based on experimental results and Coffin-Manson ageing laws [7]. These models provide the B10 lifetime which represents the number of thermal cycles until a 10% failure rate (i.e. 10% of the components have failed). [7] gives aging data for the different elements of the modules: the solder joints of substrates, the solder joints of chips and the bond wires. In order to be able to apply these information, a new model-based extrapolation method was developed. The method is based on Norris-Landzberg lifetime model [14], with slight modifications to take into account the duration of the thermal cycles instead of the cycle frequency as seen in (2).

$$ N_1 = N_2 \left(\frac{\Delta T_2}{\Delta T_1} \right)^{\alpha} \left(\frac{t_2}{t_1} \right)^{\beta} \exp\left(\gamma \left(\frac{1}{T_1} - \frac{1}{T_2} \right) \right) \tag{2} $$

In (2) N_i represents the number of cycles to failure, ΔT_i the temperature swing, t_i the cycle duration and T_i the temperature reference for conditions 1 and 2 respectively. Coefficients α, β and γ can vary according to the evaluated conditions. Data from the manufacturer is put into a 3D grid where each axis represents the α, β and γ coefficients from expression (2) as shown in Fig. 13. Each point inside this grid represents the number of cycles to failure N_f found in [7]. This grid is then used for recalculating the new coefficients α, β and γ for the cycles issued from the temperature profile. The grid allows interpolation and extrapolation of the cycling data.

In case of long-term data, equation (2) becomes independent of the cycle length and a similar approach is adopted with a 2D interpolation/extrapolation. These results are mixed when the cycle length is between the longer time cycle for short-term data and the shorter one of long-term data. Results from this new method require experimental validation.

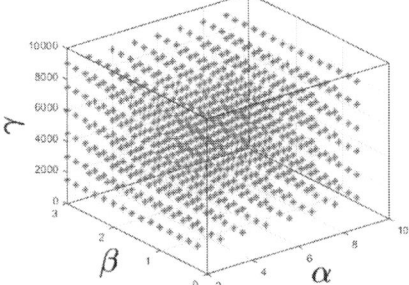

Fig. 13: 3D grid for model-based interpolation method

For long-term estimation, once the B10 lifetime is estimated for a given i[th] thermal cycle of the rain flow, the consumed lifetime (CL) for each element is calculated with the expression (3). Finally, a lifetime estimation can be performed with (4) for each element of the semiconductor.

$$CL_{i,k} = \frac{nb_i}{N_{i,k}} \cdot 100 \, (\%), k \in \{\text{chip solder, substrate solder, bond wire}\} \tag{3}$$

$$LF_k = \frac{1}{\sum_{i=1} CL_{i,k}}, k \in \{\text{chip solder, substrate solder, bond wire}\} \tag{4}$$

In expression (3) nb_i represents the number of experienced cycles from the rain flow counting method and N_i represents the number of cycles to failure for a given thermal cycle. It is worth noting that LF_k in expression (4) will give the expected lifetime in years as the evaluated period is one-year.

In the case of the medium-term lifetime estimation an additional normalization step is required. Each hourly profile must be weighted by its occurrence frequency (Fig. 5), then it has to be extrapolated to 1 year as done in [4] with the following expression (5). Factors 365 and 24 in this expression are used for the extrapolation on one year as each medium-term profile gives life consumption estimation in an hour.

$$CL_{1year-medium} = 365 \cdot 24 \cdot \left(W_{-1GW} \cdot CL_{-1GW} + W_{-0.9GW} \cdot CL_{-0.9GW} + \cdots + W_{1GW} \cdot CL_{1GW} \right) \tag{5}$$

IV - Results

Lifetime consumption in the studied time scales can be appreciated in Fig. 14. From Fig. 14 (a), it can be concluded that in the long-term time scale, for the studied application, solder joints of substrates consume more lifetime than the other elements of the semiconductor. In the case of the medium-term life consumption, Fig. 14 (b) indicates that chip solder aging is more relevant compared to the aging of the other elements.

Fig. 14 also gives relevant information about the semiconductor which suffers the most from the thermal loading profile. In both cases, long-term and medium-term, it can be seen that IGBT 2 is the element whose lifetime is more impacted by the temperature swings. This result is similar to the one found in [6] as IGBT 2 has higher temperatures than the other semiconductors.

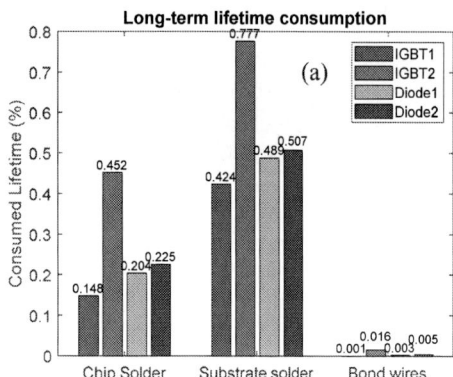 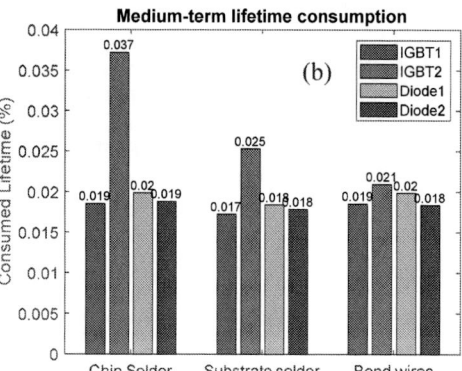

Fig. 14: Lifetime consumption per year (a) Long-term results, (b) Medium-term results

From Fig. 14 it can be concluded that long-term is more impactful in terms of life consumption than medium term. Hence, total lifetime consumption seen in Fig. 15 is dominated by long-term life consumption. This can be explained in the point-to-point application as thermal profile swings corresponding to medium-term time scale are very small when compared to long-term temperature swings. Temperature swings in medium-term range less than 10°C as seen in Fig. 11, whereas long-term temperature swings range to a few tenths of °C as seen in Fig. 10.

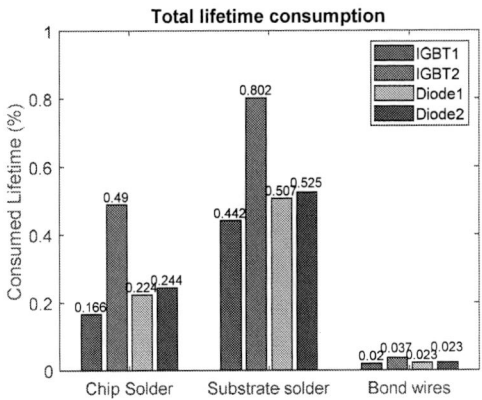

Fig. 15: Total lifetime consumption

For computing the expected lifetime of each semiconductor, only the element that has the highest life consumption is used as the input for (4). In this study the remaining lifetime of the semiconductors will be given by the life consumption of substrate solders. Results of the remaining lifetime are given in Table II. As B10 lifetime was used for obtaining these results, they can be interpreted as the time in which 90% of the components experiencing the same conditions will still be operational.

Table II: Lifetime consumption of switches in a submodule in a bidirectional application

Device	Consumed Lifetime (%) / year	Lifetime (year)
IGBT 1	0.4415	226.5
IGBT 2	0.8025	124.6
Diode 1	0.5071	197.2
Diode 2	0.5254	190.3

In [6], the element that has the highest expected lifetime is diode 2 with 653 years, while in this work, diode 2 has the second lowest expected lifetime with 190 years. Fig. 16 gives insight as to why this occurs. Fig. 16 (a) shows temperatures for a converter that delivers power to an AC grid (positive power). In this condition, IGBT 2 has the highest temperature. However, in Fig. 16 (b) power reversal has taken place and in this condition, diode 2 has the highest temperature. Therefore, reversibility of the link will strongly affect lifetime consumption for the different semiconductors.

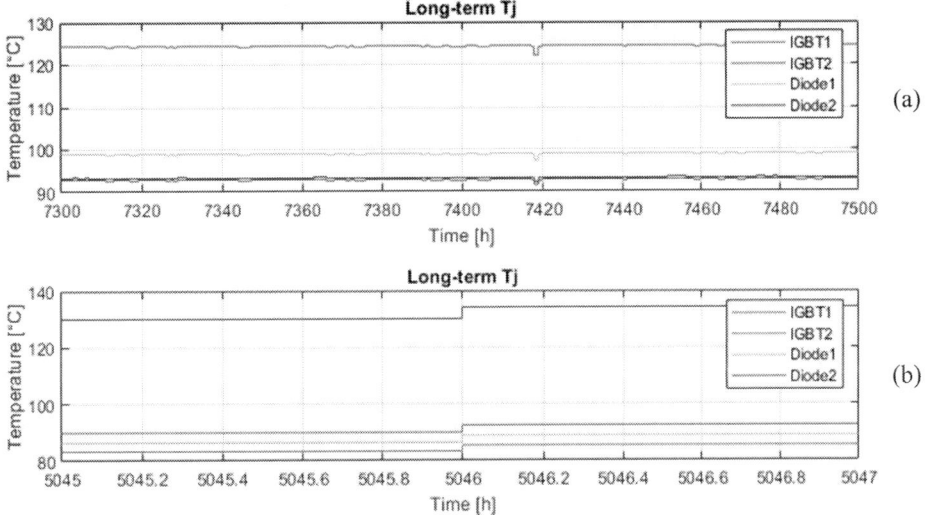

Fig. 16: Long-term temperature profiles (a) Power = 1GW, (b) Power = -1GW

The mission profile evaluated in this study is mostly unidirectional as seen in Fig. 5 and the converter on the "mostly exporting" end of the converter was chosen for the lifetime estimation. Results of this study show that, even with a few instances of power reversal, life consumption of the switches can be altered. Intuitively, if the "mostly importing" converter is analyzed, the element that will age the most will be diode 2, and will possibly have a lower expected lifetime than the one calculated for IGBT 2 in this study. Fig. 16 (b) shows that the maximal junction temperature for diode 2 is higher than the maximal temperature for IGBT 2 in Fig. 16 (a). This is explained by the thermal impedance values for IGBTs and diodes.

Overall, it can be seen that lifetime consumption for the studied mission profile is lower than lifetime consumption in an off-shore application [4], [6]. Off-shore applications have more frequent power swings when compared to bidirectional point-to-point applications and thus more frequent temperature swings for the semiconductors.

V - Conclusion

The methodology for lifetime estimation of a bidirectional MMC converter was presented. The differences for life consumption in 2 different time-scales were also detailed in this work. Additionally, a new model-based extrapolation technique was employed for lifetime estimation.

In order to improve the results of this research, real converter operating points are required for medium-term analysis. Also, a mission profile with more instances of power reversal can provide a deeper understanding on bi-directionality and its impact on the life of power semiconductors.

References

[1] R. Marquardt, A. Lesnicar, and J. Hildinger, "Modulares Stromrichterkonzept für Netzkupplungsanwendungen bei hohen Spannungen," *ETG-Conf.*, 2002.

[2] S. Allebrod, R. Hamerski, and R. Marquardt, "New transformerless, scalable Modular Multilevel Converters for HVDC-transmission," in *2008 IEEE Power Electronics Specialists Conference*, Jun. 2008, pp. 174–179, doi: 10.1109/PESC.2008.4591920.

[3] M. P. Bahrman and B. K. Johnson, "The ABCs of HVDC transmission technologies," *IEEE Power Energy Mag.*, vol. 5, no. 2, pp. 32–44, Mar. 2007, doi: 10.1109/MPAE.2007.329194.

[4] K. Ma, M. Liserre, F. Blaabjerg, and T. Kerekes, "Thermal Loading and Lifetime Estimation for Power Device Considering Mission Profiles in Wind Power Converter," *IEEE Trans. Power Electron.*, vol. 30, no. 2, pp. 590–602, Feb. 2015, doi: 10.1109/TPEL.2014.2312335.

[5] H. Wang, D. Zhou, and F. Blaabjerg, "A reliability-oriented design method for power electronic converters," in *Applied Power Electronics Conference and Exposition (APEC), 2013 Twenty-Eighth Annual IEEE*, Mar. 2013, pp. 2921–2928, doi: 10.1109/APEC.2013.6520713.

[6] H. Liu, K. Ma, Z. Qin, P. C. Loh, and F. Blaabjerg, "Lifetime Estimation of MMC for Offshore Wind Power HVDC Application," *IEEE J. Emerg. Sel. Top. Power Electron.*, vol. 4, no. 2, pp. 504–511, Jun. 2016, doi: 10.1109/JESTPE.2015.2477109.

[7] N. Kaminski, "Load-cycling capability of HiPak IGBT modules. APPLICATION NOTE 5SYA 2043-04." ABB Group, Feb. 04, 2014.

[8] Infineon Technologies AG, "PC and TC Diagrams," Infineon, Application Note AN2019-05.

[9] M. Ciappa, "Selected failure mechanisms of modern power modules," *Microelectron. Reliab.*, vol. 42, no. 4, pp. 653–667, 2002, doi: https://doi.org/10.1016/S0026-2714(02)00042-2.

[10] H. Wang *et al.*, "Transitioning to Physics-of-Failure as a Reliability Driver in Power Electronics," *IEEE J. Emerg. Sel. Top. Power Electron.*, vol. 2, no. 1, pp. 97–114, Mar. 2014, doi: 10.1109/JESTPE.2013.2290282.

[11] Miner, M.A., "Cumulative Damage in Fatigue," *J. Appl. Mech.*, no. 12, pp. A159–A164, 1945.

[12] J. C. Gonzalez-Torres, G. Damm, V. Costan, A. Benchaib, and F. Lamnabhi-Lagarrigue, "Transient stability of power systems withembedded VSC-HVDC links: Stability margins analysis and Control," *IET Gener. Transm. Distrib.*, 2020.

[13] M. Matsuichi and T. Endo, "Fatigue of metals subjected to varying stress," 1968.

[14] K. C. Norris and A. H. Landzberg, "Reliability of Controlled Collapse Interconnections," *IBM J. Res. Dev.*, vol. 13, no. 3, pp. 266–271, May 1969, doi: 10.1147/rd.133.0266.

Optimization Design for SiC Drift Step Recovery Diode (DSRD)

Xiaoxue Yan[1], Lin Liang[1*], Ziyue Wang[2], Guoqiang Tan[2]

[1]State Key Laboratory of Advanced Electromagnetic Engineering and Technology,
School of Electrical and Electronic Engineering,
Huazhong University of Science and Technology,
Wuhan, China,
[2]School of Optical and Electronic Information,
Huazhong University of Science and Technology,
Wuhan, China,
*E-Mail: lianglin@hust.edu.cn

Acknowledgements

The authors acknowledge the financial support by the National Natural Science Foundation of China (51877092).

Keywords

«Diode», «Semiconductor device», «Device simulation», «Reverse recovery», «Silicon Carbide (SiC)».

Abstract

A device structure optimization method is proposed for silicon carbide (SiC) DSRD. The bulk structure is determined by analyzing the influence of each structure parameter on the performance. For the termination structure, the 3-step etched-JTE is applied for the DSRD for the first time. The optimally designed 3-step JTE effectively reduces the peak electric field from 2.73 MV/cm to 0.95 MV/cm. In addition, the preparation process technology especially the microtrench phenomenon of SiC materials in inductively coupled plasma (ICP) etching is also studied. The samples with 1.2 kV breakdown voltage are fabricated and tested in the pulsed generation circuit.

Introduction

High-voltage pulses with fast switching speed are widely applied in many fields such as laser driving [1], power generators for electric discharge [2], particle accelerator [3], and ultra wideband radars [4]. In the early 1980s, Ioffe Physical Technical Institute proposed drift step recovery diode (DSRD) [5]. DSRDs belong to the open switches which are based on the pre-charged plasma triggering. They are often used in the inductive energy storage circuit to transfer the energy stored in the inductor to the load by their sudden opening. The pulse generator based on DSRDs can generate high voltage pulses with nanosecond rise time, MW peak power and hundreds of kHz repetition rate. DSRDs have good reliability, high repetition rate, high operating current and long lifetime. They can be applied to the electromagnetic pulse radar [6-7], ground penetrating radar [8], accelerator [9], ultra-high speed broadband beam deflection [10], ignition system of internal combustion engine [11], communication and other fields.

With the development of epitaxial growth technology, epi-Si DSRD with abrupt junctions is reported, which improves the switching performance based on high-quality epitaxial layers with the better doping concentration profile [12]. The more thorough innovation is turning to the wide bandgap material, such as silicon carbide (SiC), which has higher electric breakdown field and higher saturation velocity of carriers. SiC has been considered to fabricate DSRD. The possibility of 4H-SiC DSRD implementation is firstly demonstrated by I. V. Grekhov in [13]. In theory, the switching speed of SiC DSRD is 2-4 times of that of Si DSRD, and the breakdown voltage that SiC DSRD can bear is about 10 times of that of Si DSRD with the same thick drift region. Moreover, under the same power loss condition, the average pulse repetition frequency of SiC DSRD may be 10 times higher than that of Si DSRD [14].

SiC DSRD has been studied in Russia and Japan, etc. However, there has been no systematic study on the structure design of SiC DSRD. In this paper, the bulk structure and edge termination structure of SiC DSRD are designed and the structure parameters are optimized by the device model established in TCAD. The device sample is fabricated based on the parameter designed results, while the multiple steps are realized by the inductively coupled plasma (ICP) etching. The sample is tested in the pulse generation experiment.

Structure and principle

The structure of the SiC DSRD is shown in Fig. 1(a). The P^+PNN^+ four layers are formed by three epitaxies. The drift region is N-base. Fig. 1(b) shows the cross-sectional view with the 3-step etched-JTE termination. As an opening switch, SiC DSRD requires tens of nanoseconds of forward pumping to inject carriers into the base region and can turn off in nanoseconds.

The simulation circuit for testing the SiC DSRD is shown in Fig. 2 [15]. In the first stage the MOSFET Q_1 is turned on with the duration of ΔT, which is accompanied by the injection of non-equilibrium charge carriers of both types into the base region, so the diode is in a conductive state. The charge pumped into SiC DSRD is determined by V_{ee} and ΔT. Then Q_1 is turned off, resulting in the current through L_1 flowing through the DSRD in the reverse direction and the extraction of non-equilibrium carriers begins. After the non-equilibrium carriers injected into DSRD have been removed, the recovery process begins: majority carriers are being extracted from the base region with saturated velocity and the current is commutated to the load R_1, forming a high-voltage short pulse. Rise time of the pulse is determined by the velocity of current interruption. The circuit parameters are $L_1 = 75$ nH, $L_2 = 65$ nH, $L_3 = 45$ nH, $C_1 = 100$ pF, $C_2 = 1$ μF, $C_3 = 1$ μF, and $R_2 = 20$ Ω, and the load is $R_1 = 50$ Ω.

The behavior of DSRD and its driving circuit are simulated, by solving the coupled Possion Electron Hole equations and circuit model in TCAD. The effects of high doping and high injection levels, avalanche generation and incomplete ionization are considered in the simulation, while the switching process of SiC DSRD mainly depends on the incomplete ionization of acceptors in 4H-SiC and the narrowing of bandgap in heavily doped emission region [16].

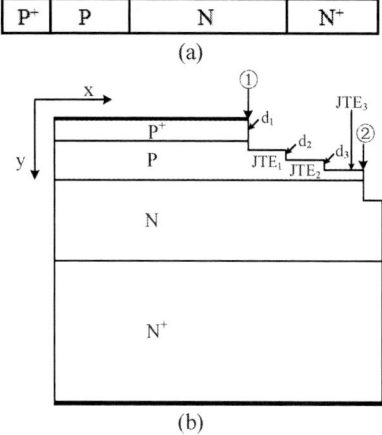

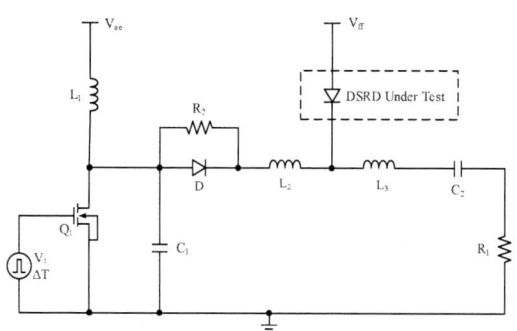

Fig. 1: (a) Structure of SiC DSRD. (b) Cross-sectional view with 3-step etched-JTE termination.

Fig. 2: Pulse generation circuit.

Optimization design for structure

Optimization design for bulk structure

The turn-off characteristics of SiC DSRD and pulse voltage on load are simulated when the doping concentration in N-base $N_{N\text{-base}}$, the width of P-base $W_{P\text{-base}}$, and the doping concentration in P-base $N_{P\text{-base}}$ change separately, shown in Fig. 3 to Fig. 5. It could be seen in Fig. 3 that the amplitude of load voltage drops sharply from 1038 V to 820 V and the rise time gets longer with the increase of $N_{N\text{-base}}$. In Fig. 4, a long P-base is usually a little better for the high voltage generation. Considering the promotion is slight while the cost rises rapidly when $W_{P\text{-base}}$ increases, a P-base layer with 2μm width is good enough. The influence of $N_{P\text{-base}}$ presented in Fig. 5 shows that lower $N_{P\text{-base}}$ has little improvement in load voltage. However, the concentration of 1E16cm^{-3} makes the breakdown voltage of the device lower than 1000V. In order to achieve a balance between dynamic characteristics and breakdown voltage, $N_{P\text{-base}}$ is determined to be 8E16cm^{-3}. Based on the simulation analysis, reasonable values are selected for these structural parameters. Fig. 6 shows the influence of the carrier lifetime in N-base on the load voltage. It is shown that the pulse voltage on load is significantly lower when the carrier lifetime is less than the forward pumping time which is 80ns here.

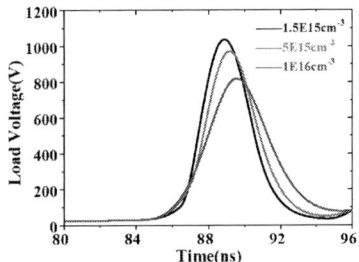

Fig. 3: Pulse voltage waveforms on load for different doping concentration in N-base.

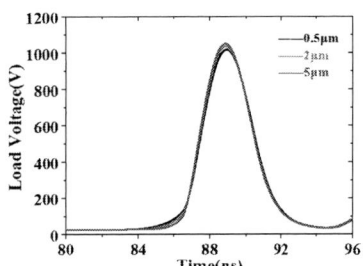

Fig. 4: Pulse voltage waveforms on load for different width of P-base.

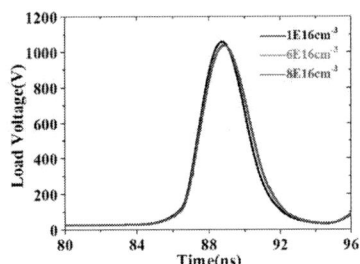

Fig. 5: Pulse voltage waveforms on load for different doping concentration in P-base.

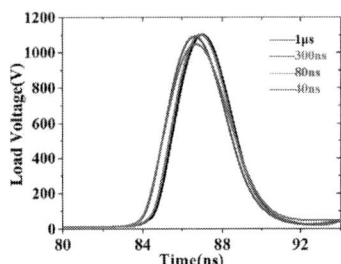

Fig. 6: Pulse voltage waveforms on load for different carrier lifetime in N-base.

As mentioned above, the space charge region (SCR) expands with saturated velocity in theory after the non-equilibrium carriers injected into DSRD have been removed. The switching time t_s can be calculated as:

$$t_s = \frac{qpWS}{\overline{i_R}} \tag{1}$$

where q is the elementary charge, p is the carrier density in the base region, W is the base thickness, S is the diode area, and $\overline{i_R}$ is the average reverse current. Due to the fact that the width of N-base is much longer than that of P-base, the switching time of DSRD is more affected by N-base, which is consistent with the simulation result. As can be seen, lower $N_{P\text{-base}}$ makes the diode switch faster. Moreover, too short carrier lifetime makes the efficiency of pumping and extraction worse and influences the t_s by the reverse current.

Fig. 7(a) shows snapshots of the hole density distribution and its time evolution inside the SiC DSRD. It could be seen that the carrier concentration in the base region reaches the peak value at 80 ns, which is the Q_1 turn-on duration. Then the plasma begins to be pulled out. The extra plasma stored in diodes is completely exhausted at 86.5ns. Therefore, the space charge region of p-n junction is formed and voltage on load rises rapidly as shown in Fig. 7(b).

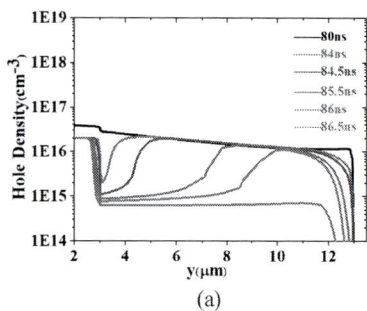 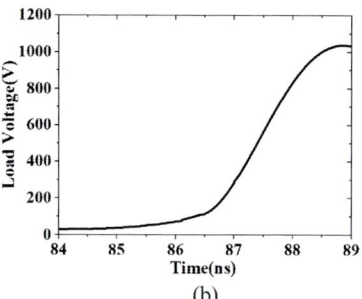

(a) (b)

Fig. 7: (a) Snapshots of the hole density distribution and its time evolution. (b) Pulse voltage waveforms on load.

Optimization design for junction termination structure

According to the structure features of SiC DSRD, the junction termination is designed to be the 3-step etched-JTE. Its parameters are also optimized. The etching depths of multiple-JTE have been proven to be the most critical, since etching depths control the charge distribution [17]. Therefore, simulation of electrical characteristics at the edge of the diode is carried out.

Fig. 8 shows that the simulated breakdown voltage of SiC DSRD with 3-step etched-JTE varies in different etching depth d_1 for different etching depths of d_2 and d_3. The result indicates that the window of etching depth d_1 gets broader with the increase of the value of d_2 and d_3 at the beginning and keeps constant when the value of d_1 exceeds 0.2 μm. The narrower the window is, the more difficult it is to manufacture the device because small deviation of the charge distribution can make the breakdown voltage miss the peak point. Therefore, the etching depths of d_2 and d_3 are determined to be 0.2 μm.

The charge distribution is controlled by d_1 when d_2 and d_3 keep constant. Fig. 9(a) and (b) show the electric fields observed vertically from points ① and ② of Fig.1(b), respectively. As it can be seen in Fig. 9(a), two obvious electric field peaks appear when the JTE layer is too thin. On the contrary, the optimally designed JTE effectively reduces the peak electric field from 2.73 MV/cm to 0.95 MV/cm at y = 2.56 μm. In Fig. 9(b), the maximum electrical field at the edge of 3-step JTE is apparently higher when the 3-step JTE is too thick.

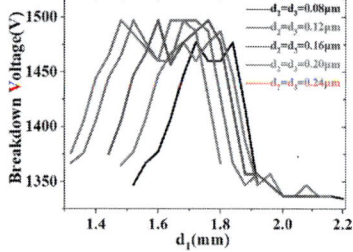 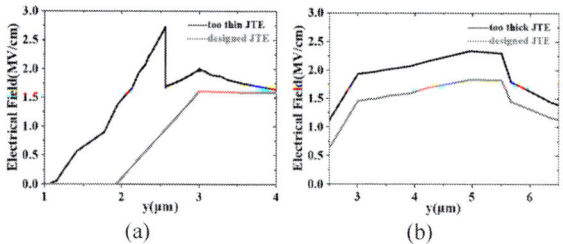

(a) (b)

Fig. 8: Simulated breakdown voltage of SiC DSRD as functions of etching depth d_1 with different d_2 and d_3.

Fig. 9: Simulated electrical field vertically along (a) the main junction edge and (b) the outmost edge of multiple-JTE.

Experimental results

Based on the simulation results, an epitaxial P⁺-P-N structure on an N⁺-substrate with the 350 μm thickness is fabricated. The thicknesses of the epilayers are 1 μm, 2 μm and 10 μm, respectively. The active area is 4mm².

The 3-step JTE termination is formed by multiple ICP etching, and the etching result directly affects the consequence of the edge termination. Microtrench at the bottom corner is a common phenomenon during the etching process due to the characteristics of the SiC material. As shown in Fig. 10(a), the apparent microtrenches appear at the bottom corner under original ICP etching recipe, which may greatly enhance the electric field at the corner under reverse blocking state. Besides, the tiny trenches are difficult to fill when depositing passivation layer. In summary, the presence of microtrenches increases the risk of premature breakdown. Optimized etching recipe is achieved by adjusting the etching parameters and SEM photos of etching steps with depths of 3.3 μm and 5.2 μm under the recipe are shown in Fig. 10(b) and (c). It can be seen that the sidewall is steep and the bottom is flat. The optimized recipe is utilized to etch the 3-step JTE. Photos of the etched-JTE under optical microscope and SEM are demonstrated in Fig. 11. Favorable trench profile with flat bottom corners can be seen in the SEM image.

Fig. 10: SEM photo of etching step under (a) original recipe, and optimized recipe with depth of (b) 3.3 μm and (c) 5.2 μm.

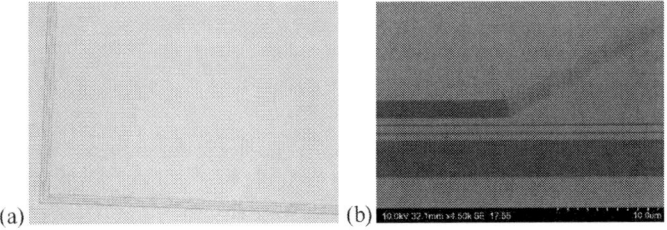

Fig. 11: (a) Optical microscope and (b) SEM Photo of Etched-JTE.

The measured reverse I-V characteristic for single chip is shown in Fig. 12. The developed sample with 3-step JTE termination has a blocking voltage of 1080 V. The measurements of I-V characteristics of the samples are carried out by a Keithley 2600 series test equipment. The sample is applied in pulse generation circuit shown in Fig. 13. The trigger duration is 80 ns through the test and the waveforms are shown in Fig. 14. The measurements are performed by an oscilloscope of Teledyne LeCroy HDO4054 with 500MHz bandwidth. V_{ee} and V_{ff} are 57 V and 45 V, respectively. The non-equilibrium charge carriers are injected into the base region from t_0 to t_1 and then pulled out until extraction at t_2 instant. The reverse current of SiC DSRD reaches its peak value with the density of 278 A/cm² at the same time. Then SiC DSRD is reversely recovered hard and quickly while the current is commutated to the load R_1, and the voltage on the load reaches the pulse peak at t_3 instant. This peak value is 508V and the full width at half-maximum(FWHM) is 6.4ns.

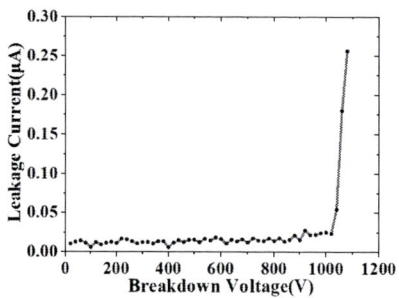

Fig. 12: Measured reverse I-V characteristics of the 4mm² DSRD sample.

Fig. 13: Photo of test circuit.

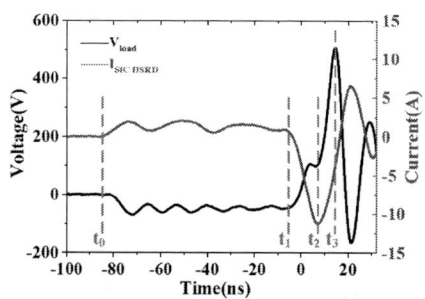

Fig. 14: Experimental pulse current waveform on SiC DSRD and voltage waveform on load.

Conclusion

The structure parameters of SiC DSRD are optimized in this paper. The influences of the bulk structural parameters for SiC DSRD, including the doping concentration in N-base $N_{\text{N-base}}$, the width of P-base $W_{\text{P-base}}$, the doping concentration in P-base $N_{\text{P-base}}$ and the carrier lifetime in N-base, on the pulse voltage on load are discussed by TCAD simulation. Parameters of edge termination are also designed optimally. The etching depth is carefully selected to ensure that the breakdown voltage of SiC DSRD reaches the peak point and the window of etching depth is broad. Based on the simulation results, the epi wafers are customized. The 3-step JTE termination is applied by ICP etching utilizing optimized ICP recipe to obtain favorable trench profile with steep sidewall and flat bottom corners. Chips with the blocking voltage of 1080V are acquired. The diode is tested before the final dicing and packaging and a pulse with amplitude of 508V is achieved when the supply voltage is 57V. Experiments under higher voltage will be studied in the future.

References

[1] S. V. Korotkov.: Semiconductor switches of laser pumping pulses of nanosecond duration, Instruments and Experimental Techniques Vol. 52, pp. 699-702

[2] A. G. Lyublinsky.: Pulse power nanosecond-range DSRD-based generators for electric discharge technologies, IEEE Trans. Plasma Sci Vol. 41 no 10, pp. 2625-2629

[3] F. Arntz.: SLIM, short-pulse technology for high gradient induction accelerators, IET European Pulsed Power Conference 2009, pp. 1-4

[4] A. Zhu.: An implementation of step recovery diode-based uwb pulse generator, IEEE International Conference on Ultra-Wideband 2010, pp. 1-4

[5] I. V. Grekhov.: Power drift step recovery diodes (DSRD), Solid-State Electronics Vol. 28 no 6, pp. 597-599

[6] Prokhorenko V P.: Electromagnetic impulse radiator. Ultrawideband and Ultrashort Impulse Signals 2004, Sevastopol, Ukraine, pp. 243-245

[7] Prokhorenko V P.: Drift step recovery devices utilization for electromagnetic pulse radiation. International Conference on Grounds Penetrating Radar 2004, pp. 195-198

[8] Prokhorenko V.: Drift step recovery diode transmitter for high-power GPR design. 8th International Conference on Ground Penetrating Radar (GPR 2000)

[9] Arntz F O.: SLIM, short-pulse technology for high gradient induction accelerators. IEEE International Pulsed Power Conference 2009

[10] Benwell A.: A 5KV, 3MHz solid-state modulator based on the DSRD switch for an ultra-fast beam kicker. IEEE International Power Modulator and High Voltage Conference 2012, pp. 328-331

[11] Tropina A.: Ignition System Based on the Nanosecond Pulsed Discharge. IEEE Transactions on Plasma Science Vol 42 no 12, pp. 3881-3885

[12] A. V. Gorbatyuk.: Theory of quasi-diode operation of reversely switched dinistors, Solid-State Electronics Vol. 31 no 10, pp. 1483-1491

[13] I. V. Grekhov.: On the Possibility of Creating a Superfast-Recovery Silicon Carbide Diode. Tech. Phys. Lett Vol. 28 no 7, pp. 544-546

[14] Kozlov V A.: New generation of drift step recovery diodes(DSRD) for subnanosecond switching and high repetition rate operation. Conference Record of the Twenty-Fifth International Power Modulator Symposium 2002

[15] L. M. Merensky.: Fast switching of drift step recovery diodes based on all epi-Si growth, Proc. IEEE Int. Conf. Microw., Commun., Antennas Electron. Syst. 2009, pp. 1-4

[16] Ilyin V A.: High-voltage ultra-fast pulse diode stack based on 4H-SiC. Materials Science Forum 2016, pp. 786-789

[17] L. Lin.: Simulation and experimental study of 3-step junction termination extension for high-voltage 4H-SiC gate turn-off thyristors, Solid-State Electronics Vol. 86, pp. 36-40

Discrete Super-Twisting Sliding Mode Current Controller for Induction Motor Drives

Tianqing Wang, Bo Wang, Yong Yu, Yangming Zhu, Dianguo Xu

Harbin Institute of Technology
No.92, Xidazhi Street, Nangang District
Harbin, China
Tel.: +86-451-86413420.
Fax: +86-451-86413420.
E-Mail: HiteeWTQ@163.com; wangbohit@hit.edu.cn; yuyong@hit.edu.cn;
zhuyangming_hitee@163.com; xudiang@hit.edu.cn

Acknowledgements

This work was supported in part by National Natural Science Foundation of China (51690182) and (51807038); in part by China Postdoctoral Science Foundation funded project (2018M630354) and (2019T120267); in part by Heilongjiang Postdoctoral Foundation funded project (LBH-Z18097); and in part by Fundamental Research Funds for the Central Universities (HIT.NSRIF.2019025).

Keywords

«Induction motor», «Sliding mode control».

Abstract

The sliding mode controller can effectively improve the robustness of induction motor (IM) current control performance, but it suffers from the drawback of severe chattering and steady-state error. To address these problems, this paper proposed a discrete super-twisting sliding mode current controller for induction motor drives. In order to improve the convergence speed of the stator current, the conventional ST-SMC is modified by a fast power-reaching-law. Then, the improved ST-SMC is discretized to realize digital control. The experiments confirm the effectiveness of the proposed method.

Introduction

The sliding mode control (SMC) has received more and more attention in induction motor (IM) drives due to its strong robustness and high applicability to nonlinear systems [1]. However, there is serious chattering problem in the conventional SMC. To address this problem, different first-order SMCs (FO-SMC) have been proposed, such as, boundary layer method, saturation function method, and power-reaching-law method [2]. But these methods sacrifice the robustness of SMC. Moreover, since these FO-SMC are the similar to K_p control, it would cause the steady-state error problem.

Compared with the FO-SMC, the second-order SMC can achieve the finite-time convergence of sliding mode surface and its first derivative [3], leading to chattering and steady-state error elimination. The super-twisting SMC (ST-SMC) is an easy-implemented second-order SMC. Since it does not require the first derivative of state variables in the sliding mode surface, the ST-SMC is suitable for first-order system [4]. The current loop of IM is a typical first-order system.

In recent years, much efforts have been put into the application of SMC in IM current control. In [5], an observer-based discrete SMC was proposed for IM current control. It can achieve the fast convergence, and provide voltage decoupling. However, the chattering and steady-state error problems still exist. In [6], a second-order SMC was applied for current control. Although the steady-state error is eliminated, its control law only contained the integral term, and thus it decreases the convergence speed of stator current. In [7], the current loop of IM is improved by a ST-SMC, leading to fast convergence speed and steady-state error elimination. However, one of the coefficients in the power reaching law was selected greater than 1. This would lead to the deterioration of transient performance.

To eliminate chattering and steady-state error, this paper proposed a discrete ST-SMC (DST-SMC) for the current controller. The main contributions of the proposed scheme can be summarized as: (1) the drawbacks of steady-state current error and chattering of FO-SMC is overcome; (2) the proposed DST-SMC can effectively improve the convergence speed of the stator current.

This paper will first introduce the IM drive system. Then, the design process and theoretical analysis of the DST-SMC are given to improve the current controller. Finally, the studied algorithm is verified on a 3.7kW industrial IM test bench.

Modeling and System

Based on the rotor-field-oriented control (RFOC), the IM mathematical model can be expressed as [9]:

$$
\begin{cases}
\dfrac{di_{sd}}{dt} = -\dfrac{R_s L_r^2 + R_r L_m^2}{\sigma L_s L_r^2} i_{sd} + \omega_e i_{sq} + \dfrac{L_m}{\sigma L_s L_r T_r} \psi_r + \dfrac{u_{sd}}{\sigma L_s} \\[4mm]
\dfrac{di_{sq}}{dt} = -\dfrac{R_s L_r^2 + R_r L_m^2}{\sigma L_s L_r^2} i_{sq} - \omega_e i_{sd} - \dfrac{L_m}{\sigma L_s L_r} \omega_r \psi_r + \dfrac{u_{sq}}{\sigma L_s}
\end{cases}
\tag{1}
$$

where i_{sd} and i_{sq} are stator current dq component; u_{sd} and u_{sq} are stator voltage dq component; R_s and R_r are stator and rotor resistances; L_s, L_r and L_m are self and mutual inductances; ψ_r is the rotor flux; ω_e and ω_r are synchronous and rotor speeds; $T_r = L_r/R_r$ is rotor time constant; and $\sigma = 1 - L_m^2 / (L_s L_r)$ is leakage coefficient.

By rewriting (1), the mathematical model of IM can be expressed in its matrix form:

$$
\frac{di}{dt} = Ai + B\psi_r + Cu
\tag{2}
$$

where $i = \begin{bmatrix} i_{sd} & i_{sq} \end{bmatrix}^T$; $A = -\dfrac{R_s L_r^2 + R_r L_m^2}{\sigma L_s L_r^2} I + \omega_e J$; $C = \dfrac{1}{\sigma L_s} I$; $B = \begin{bmatrix} \dfrac{L_m}{\sigma L_s L_r T_r} & -\dfrac{L_m}{\sigma L_s L_r} \omega_r \end{bmatrix}^T$;

$u = \begin{bmatrix} u_{sd} & u_{sq} \end{bmatrix}^T$; $I = \begin{bmatrix} 1 & 0 \\ 0 & 1 \end{bmatrix}$; $J = \begin{bmatrix} 0 & 1 \\ -1 & 0 \end{bmatrix}$.

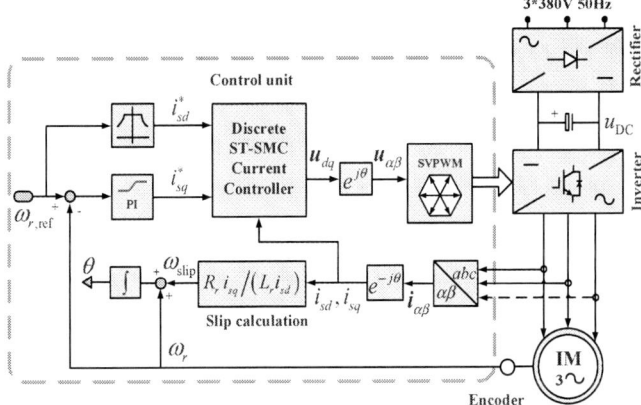

Fig. 1: Block diagram of the RFOC-based IM drive with the proposed DST-SMC

Fig.1 is the block diagram of RFOC-based IM control system. The whole system includes an IM, a rectifier, an inverter, and a control unit. The control unit adopts cascaded double-closed-loop structure. The outer loop is the speed loop with PI control. The current loop is the inner loop with the proposed DST-SMC control. The stator torque component i_{sq} reference value is determined by the output of the speed loop, and the stator current field component i_{sd} reference value is constant (within the rated speed range). An incremental encoder is applied to measure the IM rotor speed. The rotor flux position θ is obtained by integrating synchronous angular speed ω_e for coordinate transformation.

Current Controller Design

The PI controller is the most commonly used method for IM current controller. It can achieve steady-state errorless convergence of stator current by pole-zero cancellation with linear control theory [10]. However, the pole-zero cancellation PI controller design depends on the precise motor model, leading to the decreased robustness. The SMC is a strong robust control strategy, but it suffers from chattering and steady-state error problems. Thus, this section proposes a DST-SMC current controller to address the problems.

The basic theory of the conventional ST-SMC is first introduced, and then modified by introducing a fast power-reaching-law. Afterwards, to realize the proposed ST-SMC, the DST-SMC is designed in the discrete digital control system.

Basic theory of ST-SMC

Consider a system with the following mathematical model:

$$dx/dt = a(x,t) + b(x,t)u \tag{3}$$

where x is the system state variable; and u is the control input.

In the closed loop system, x follows its reference value under the control of u. Considering x_{ref} is the reference value of x, and then the sliding mode surface is designed as:

$$s = e = x_{\text{ref}} - x \tag{4a}$$
$$ds/dt = dx_{\text{ref}}/dt - a(x,t) - b(x,t)u \tag{4b}$$

The dynamic equation of ST-SMC can be expressed as [11]:

$$ds/dt = -k_1 |s|^\gamma \operatorname{sign}(s) + v \tag{5a}$$
$$dv/dt = -k_2 \operatorname{sign}(s) \tag{5b}$$

The right side of (5a) contains two parts: the power-reaching-law part $-k_1 |s|^\gamma \operatorname{sign}(s)$, and the integral part $-\int k_2 \operatorname{sign}(s)dt$. When k_2 is selected as 0, (5) can be rewritten as $ds/dt = -k_1 |s|^\gamma \operatorname{sign}(s)$, which is a typical power-reaching-law FO-SMC.

In order to force the system to converge along (5), the control input u need to be designed as:

$$
\begin{cases}
u = u_{eq} + u_n \\
u_{eq} = -b(x,t)^{-1} \left[a(x,t) - dx_{\text{ref}}/dt \right] \\
u_n = b(x,t)^{-1} \left[k_1 |s|^\gamma \operatorname{sign}(s) + w \right] \\
dw/dt = k_2 \operatorname{sign}(s)
\end{cases}
\tag{6}
$$

In (6), u_{eq} is the equivalent control term to offset $a(x,t)$ and dx_{ref}/dt, and u_n is the sliding mode control term designed according to (5). When $k_2=0$, (6) become the power-reaching-law FO-SMC.

It can be seen that (5) is based on the ideal case. However, in practical applications, the variation of system parameter and disturbance are inevitable. Considering f represents the set of parameter variation and other unmodeled disturbance, (5a) can be rewritten as:

$$ds/dt = -k_1 |s|^\gamma \operatorname{sign}(s) + v + f \tag{7}$$

Assuming $|df/dt| \le \xi$. In any parameter selection scheme [12-15], $k_2 > \xi$ is always a necessary condition for the system to achieve finite-time convergence. Thus, if $k_2 = 0$, the power-reaching-law FO-SMC can never achieve finite-time convergence, leading to the problem of steady-state error.

Fig.2 shows the system dynamic trajectory of power-reaching-law FO-SMC and ST-SMC. As the red curve in Fig.2(a), when the system is affected by disturbance, the power-reaching-law FO-SMC has the problem of steady state error. In Fig.2(b), the ST-SMC can achieve steady-state errorless convergence to $ds/dt = s = 0$ under the effect of disturbance f. At the same time, the switching function is retained in dv/dt, so ST-SMC can maintain the strong robustness of the conventional SMC. The advantages of ST-SMC can be summarized as following:

- The ST-SMC maintains the strong robustness of SMC;
- The ST-SMC realizes the finite-time convergence of the sliding surface and its derivatives;
- The control law is a continuous function, which can suppress chattering;
- The sliding mode surface requires only the information of the state variable error.

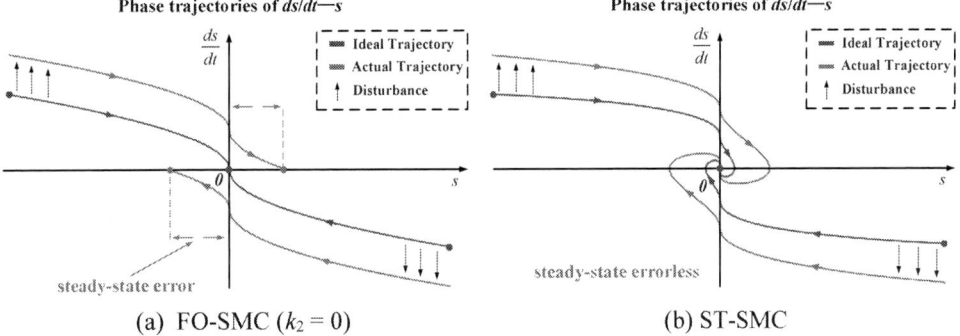

(a) FO-SMC ($k_2 = 0$)   (b) ST-SMC

Fig. 2: System dynamic trajectory of both SMC method

Based on the above analysis, the continuous ST-SMC current controller can be designed as:

$$\begin{cases} u = u_{eq} + u_n \\ u_{eq} = C^{-1}\left[-Ai - B\psi_r + (di_{ref}/dt) \right] \\ u_n = C^{-1}\left[k_1 [s]^\gamma \operatorname{sign}(s) + w \right] \\ dw/dt = k_2 \operatorname{sign}(s) \end{cases} \tag{8}$$

where, $s = i_{ref} - i = \begin{bmatrix} s_1 & s_2 \end{bmatrix}^T$, and i_{ref} is the reference value of i; $\operatorname{sign}(s) = \begin{bmatrix} \operatorname{sign}(s_1) & \operatorname{sign}(s_2) \end{bmatrix}^T$; and $[s]^\gamma$ is defined as $[s]^\gamma = \begin{bmatrix} |s_1|^\gamma & 0 \\ 0 & |s_2|^\gamma \end{bmatrix}$. By adopting reasonable control gains, the ST-SMC current controller can drive stator current converge to $ds/dt = s = 0$ in finite-time.

ST-SMC with Fast Power-Reaching-Law

In this paper, the conventional ST-SMC is slightly modified with fast power-reaching-law to increase the convergence speed. Thus, the dynamic equation is restructured:

$$\begin{cases} ds/dt = -k_1 |s|^{\gamma} \operatorname{sign}(s) - k_3 s + v \\ dv/dt = -k_2 \operatorname{sign}(s) \end{cases} \tag{9}$$

Then, the control law (8) is modified into:

$$\begin{cases} u = u_{eq} + u_n \\ u_{eq} = C^{-1} \left[-Ai - B\psi_r + \left(di_{\text{ref}}/dt \right) \right] \\ u_n = C^{-1} \left[k_1 [s]^{\gamma} \operatorname{sign}(s) + k_3 s + w \right] \\ dw/dt = k_2 \operatorname{sign}(s) \end{cases} \tag{10}$$

Comparing (8) with (10), the ST-SMC add another reaching law $k_3 s$ into the original control law. When $\|s\| < 1$, the system approaching speed is mainly determined by $k_1 [s]^{\gamma} \operatorname{sign}(s)$ .When $\|s\| > 1$, the system the system approaching speed is mainly determined by $k_3 s$, leading to faster convergence speed.

Convergence time analysis

The main idea of the proposed ST-SMC is to improve the convergence speed of conventional ST-SMC by using fast power-reaching-law. By considering the disturbance f, (9) is rewritten into:

$$\begin{cases} ds/dt = -k_1 |s|^{\gamma} \operatorname{sign}(s) - k_3 s + v + f \\ dv/dt = -k_2 \operatorname{sign}(s) \end{cases} \tag{11}$$

To ensure the system stability, we have to ensure that $k_2 > \zeta$, and k_1, k_3 are dominating with respect to k_3. According to the analysis in [7], when the value of k_2 is close to ζ, the convergence time mainly depends on the selection of the parameters of k_1 and k_3. To simplify the analysis process, we can neglect the effect of v and f on the convergence time. Thus, (11) can be rewritten as:

$$ds/dt = -k_1 |s|^{\gamma} \operatorname{sign}(s) - k_3 s \tag{12}$$

Multiply both sides of the equation by $e^{k_3 t}$, and sorting out the equation, yields:

$$\frac{d\left(e^{k_3 t} s\right)}{dt} = -k_1 \left| e^{k_3 t} s \right|^{\gamma} e^{(1-\gamma)k_3 t} \operatorname{sign}(s) \tag{13a}$$

$$\frac{d\left(e^{k_3 t} s\right)}{\left| e^{k_3 t} s \right|^{\gamma} \operatorname{sign}(s)} = -k_1 e^{(1-\gamma)k_3 t} dt \tag{13b}$$

By integrating both sides of (13b), the solution of equation (12) can be obtained. Suppose $s(0)$ is the initial value of s, and $s(0)>0$. Then the expression of $s(t)$ is:

$$s(t) = e^{-k_3 t} \left(s(0)^{1-\gamma} + \frac{k_1}{k_3} \left(1 - e^{(1-\gamma)k_3 t} \right) \right)^{\frac{1}{1-\gamma}} \tag{13c}$$

By solving the above equation, we can get T_r is the time when s converges to 0.

$$T_r = \frac{\ln\left(1 + \frac{k_3}{k_1} s(0)^{1-\gamma}\right)}{k_3(1-\gamma)} \tag{14}$$

When $k_3=0$, (12) becomes the dynamic equation of conventional ST-SMC. By neglecting the effect of v and f, the dynamic equation of conventional ST-SMC is rewritten as:

$$ds/dt = -k_1 |s|^\gamma \,\mathrm{sign}(s) \tag{15a}$$

$$s(t) = \left[s(0)^{1-\gamma} + k_1(\gamma-1)t \right]^{\frac{1}{1-\gamma}} \tag{15b}$$

The convergence time of the conventional ST-SMC T_c can be obtained by solving (15b).

$$T_c = \frac{s(0)^{1-\gamma}}{k_1(1-\gamma)} \tag{16}$$

It can be seen that $T_r < T_c$. When $s(0)<0$, the analysis process and results are the same as above. Thus, the DST-SMC with fast power-reaching-law can increase the convergence speed of state variable. Fig. 3 shows the phase trajectory of ds/dt—s, and s—t ($a > 0$). It is obvious that the system convergence speed will be improved with the increase of k_3 in a reasonable range.

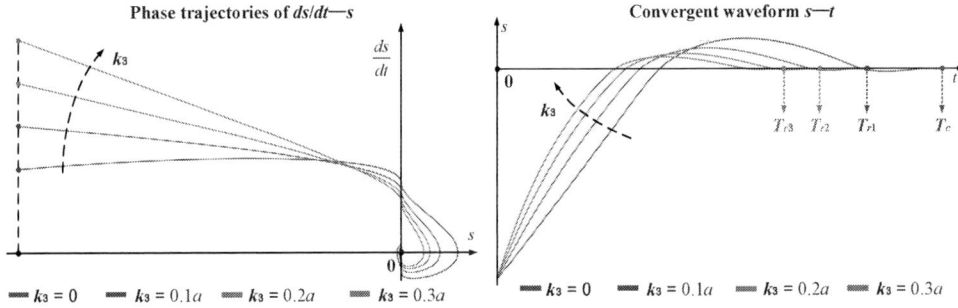

Fig. 3: System dynamic trajectory of conventional ST-SMC and the proposed ST-SMC

DST-SMC current controller

In order to realize the proposed ST-SMC in the discrete digital control system, a DST-SMC current controller is proposed in this section. (2) can be discretized by forward Euler into:

$$\boldsymbol{i}(k+1) = (\boldsymbol{I} + \boldsymbol{A}T_s)\boldsymbol{i}(k) + \boldsymbol{B}T_s\boldsymbol{\psi}_r(k) + \boldsymbol{C}T_s\boldsymbol{u}(k) \tag{17}$$

A discrete sliding mode surface is designed as: $s(k) = e(k)$, $e(k) = i_{\mathrm{ref}}(k) - i(k)$. T_s is the sampling period.

The discrete-time form of ST-SMC is given according to (9):

$$\begin{cases} s(k+1) = s(k) - k_1 T_s \left[s(k) \right]^\gamma \mathrm{sign}(s(k)) + T_s v(k) - k_3 T_s s(k) \\ v(k) = v(k-1) - k_2 T_s \,\mathrm{sign}(s(k)) \end{cases} \tag{18}$$

In order to obtain the expression of DST-SMC, we need to compute $s(k+1)$ as follow:

$$
\begin{aligned}
s(k+1) &= \mathbf{i}_{\text{ref}}(k+1) - \mathbf{i}(k+1) \\
&= \mathbf{i}_{\text{ref}}(k+1) - (\mathbf{I} + \mathbf{A}T_s)\mathbf{i}(k) - \mathbf{B}T_s\psi_r(k) - \mathbf{C}T_s\mathbf{u}(k)
\end{aligned} \tag{19}
$$

By sorting out (18) with (19), the expression of $\mathbf{u}(k)$ is designed as:

$$
\begin{cases}
\mathbf{u}(k) = \mathbf{u}_{eq}(k) + \mathbf{u}_n(k) \\
\mathbf{u}_{eq}(k) = \mathbf{C}^{-1}\left[-\mathbf{A}\mathbf{i}(k) - \mathbf{B}\psi_r(k) + \dfrac{\mathbf{i}_{ref}(k+1) - \mathbf{i}_{ref}(k)}{T_s} \right] \\
\mathbf{u}_n(k) = \mathbf{C}^{-1}\left[k_1\left[s(k) \right]^{\gamma} \operatorname{sign}(s(k)) + k_3 s(k) + \mathbf{w}(k) \right] \\
\mathbf{w}(k) = \mathbf{w}(k-1) + k_2 T_s \operatorname{sign}(s(k))
\end{cases} \tag{20}
$$

When the IM is in a stable state, $\mathbf{i}_{\text{ref}}(k+1) = \mathbf{i}_{\text{ref}}(k)$. The DST-SMC shares similar characteristics with the continuous one with small enough T_s. Thus, if the control gains k_1 and k_2 are selected according to the conditions shown in [15], and $k_3 > 0$, the stability of the DST-SMC can be ensured.

Experimental results

The proposed control method is verified on a STM32F103 ARM-based 3.7 kW IM experimental bench, as shown in Fig.4. The platform contains two motors with the same parameters: one is used as algorithm verification motor, and the other one is used as load motor in torque control mode. The IM parameters are listed in TABLE I. The reference value of i_{sd} is set to 35% of the rated current. The coefficient of speed loop PI regulator is $Kp = 2.0$, $Ki = 450$. The conventional ST-SMC gains are selected as $k_1=120$; $k_2=1700$. When $k_2=0$, it becomes the power-reaching-law FO-SMC. The DST-SMC gains are selected as $k_1=120$; $k_2=1700$; $k_3=80$. The parameters of power-reaching-law is $\gamma=0.5$.

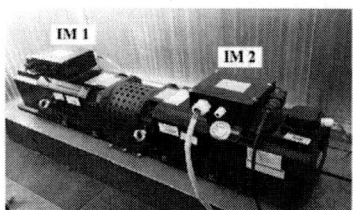

Fig.4. IM experimental bench.

TABLE I. Parameters of Induction Motor

Quantity	Value	Quantity	Value
Rated power	3.7 kW	Rotor resistance	0.825 Ω
Rated voltage	380 V	Mutual inductance	118.9 mH
Rated speed	1500 r/min	Stator, rotor inductance	124.4 mH
Rated current	8.9A	Total inertia	0.0256 kg·m²
Number of pole pairs	2	Stator resistance	1.142 Ω

Fig.5 shows system response during acceleration and deceleration with three different current controllers: the power-reaching-law FO-SMC, the conventional ST-SMC, and the proposed DST-SMC. The IM accelerates to the rated speed, and then decelerates to 600rpm. It can be seen that power-reaching-law FO-SMC has chattering and steady-state error problems in Fig.5 (a). The conventional ST-SMC can eliminate steady-state error and suppress chattering in Fig.5 (b), but it has control overshoot, leading to the decrease of current convergence speed. By contrast in Fig.5 (c), the proposed DST-SMC can eliminate steady-state error, and increase the convergence speed effectively.

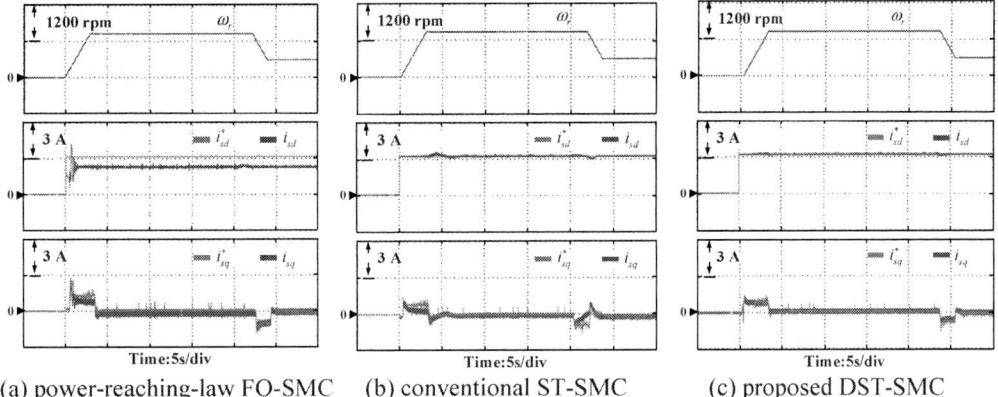

(a) power-reaching-law FO-SMC   (b) conventional ST-SMC   (c) proposed DST-SMC

Fig.5. System response during acceleration and deceleration without load. (From top to bottom: rotor speed ω, field current and its reference i_{sd}^{*} & i_{sd}, torque current and its reference i_{sq}^{*} & i_{sq})

Fig.6 are the response of two control method under sudden load change. When the IM works stably at 900 rpm, 75% of the rated load is suddenly applied. In Fig.6(a), from the tracking performance of stator current in dq axis, power-reaching-law FO-SMC has severe steady-state error. According to the rotor speed waveform, it can be seen that speed fluctuation reaches 87 rpm. In Fig.6(b), the proposed DST-SMC has improved current tracking performance and eliminates steady-state error. The speed fluctuation is reduced to 52 rpm. Thus, the proposed method can effectively improve the system capability to resist load disturbance.

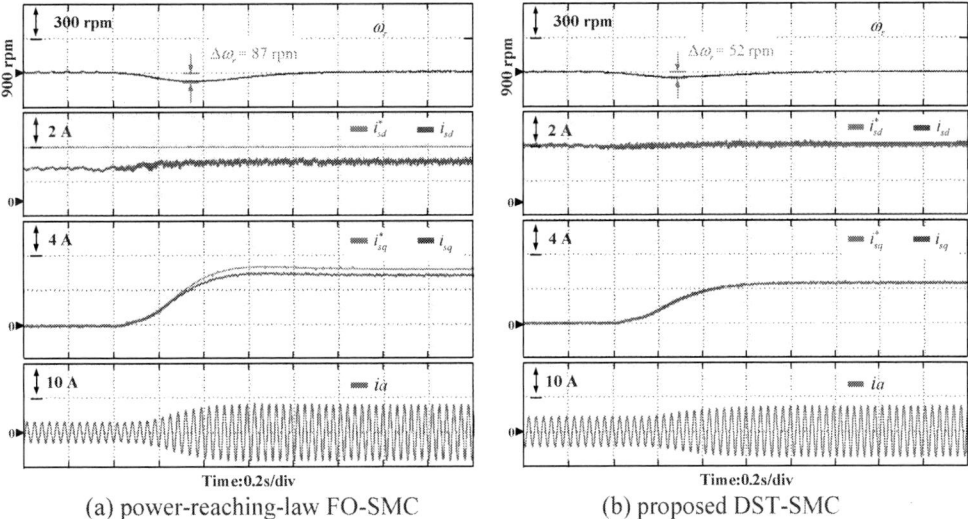

(a) power-reaching-law FO-SMC   (b) proposed DST-SMC

Fig.6. System response with load change. (From top to bottom: rotor speed ω_r, field current and its reference i_{sd}^{*} & i_{sd}, torque current and its reference i_{sq}^{*} & i_{sq}, phase current i_a)

Fig.7 presents the responses of two controllers during acceleration with rated load. Under the condition of rated load, the motor works stably at 300 rpm, and then accelerates to 1500 rpm. The waveforms of rotor speed, stator current, and phase current is recorded. The experimental results show that the proposed algorithm can effectively eliminate steady-state error and shorten the acceleration time.

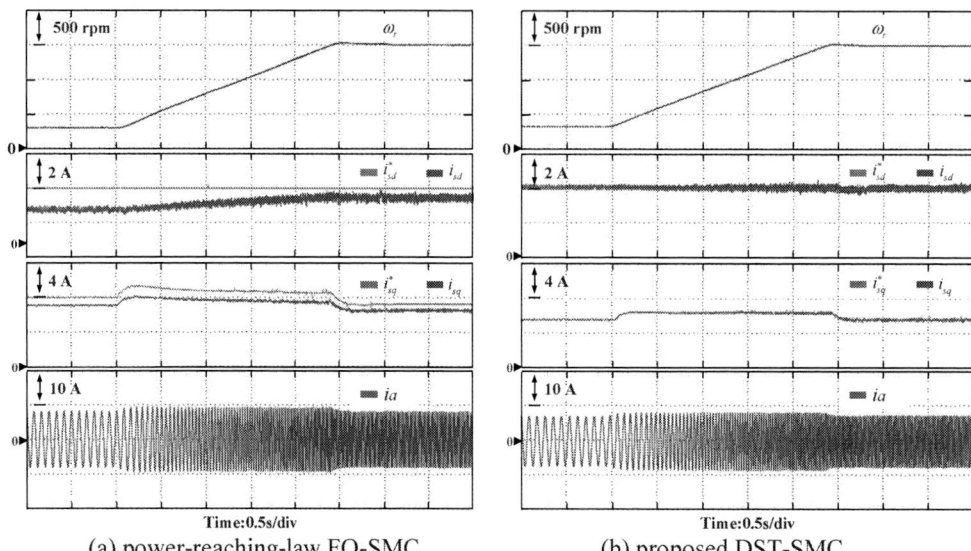

(a) power-reaching-law FO-SMC	(b) proposed DST-SMC

Fig.7. System response during acceleration with rated load. (From top to bottom: rotor speed ω_r, field current and its reference $i_{sd}^* \& i_{sd}$, torque current and its reference $i_{sq}^* \& i_{sq}$, phase current i_a)

Conclusion

This paper proposed a DST-SMC current controller for IM drives. The chattering steady-state error problems in the FO-SMC is effectively eliminated. Compared with the conventional ST-SMC, the proposed DST-SMC can increase the convergence speed of stator current by introducing the fast power reaching law. The comparative experimental results show that DST-SMC can eliminate the steady-state current error, and optimize the transient performance in comparison with the conventional ST-SMC.

References

[1] Zheng X, Feng Y, Han F, and Yu X.: Integral-Type Terminal Sliding-Mode Control for Grid-Side Converter in Wind Energy Conversion Systems. IEEE Transactions on Industrial Electronics, vol. 66, no. 5, pp. 3702-3711, 2019.

[2] Bartoszewicz A, Leśniewski P.: New Switching and Nonswitching Type Reaching Laws for SMC of Discrete Time Systems. IEEE Transactions on Control Systems Technology, Vol. 24 no. 2, pp. 670-677, 2016.

[3] Lascu C, and Blaabjerg F.: Super-twisting sliding mode direct torque control of induction machine drives. 2014 IEEE Energy Conversion Congress and Exposition (ECCE), Pittsburgh, PA, 2014, pp. 5116-5122.

[4] Wang B, Dong Z, Yu Y, Wang G, and Xu D.: Static-Errorless Deadbeat Predictive Current Control Using Second-Order Sliding-Mode Disturbance Observer for Induction Machine Drives. IEEE Transactions on Power Electronics, vol. 33, no. 3, pp. 2395-2403, 2018.

[5] Deng A, Zou J, Shao Z, Huang G, Shi L, and Yang J.: Internal and sliding mode current decoupling control of asynchronous motor. The 27th Chinese Control and Decision Conference (2015 CCDC), Qingdao, pp. 3572-3577, 2015.

[6] A. V. R. Teja, Chakraborty C, and B. C. Pal.: Disturbance Rejection Analysis and FPGA-Based Implementation of a Second-Order Sliding Mode Controller Fed Induction Motor Drive. IEEE Transactions on Energy Conversion, vol. 33, no. 3, pp. 1453-1462, 2018.

[7] Mishra J, Wang L, Zhu Y, Yu X, and Jalili M.: A Novel Mixed Cascade Finite-Time Switching Control Design for Induction Motor. IEEE Transactions on Industrial Electronics, vol. 66, no. 2, pp. 1172-1181, Feb. 2019.

[8] Salgado I, Chairez I, Bandyopadhyay B, Fridman L, and Camacho O.: Discrete-time non-linear state observer based on a super twisting-like algorithm. IET Control Theory & Applications, vol. 8, no. 10, pp. 803-812, 2014.

[9] Sala G, Mengoni M, Rizzoli G, Zarri L and Tani A.: Decoupled d − q Axes Current-Sharing Control of Multi-Three-Phase Induction Machines. IEEE Transactions on Industrial Electronics, vol. 67, no. 9, pp. 7124-7134, Sept. 2020.

[10] Lee S.: Closed-Loop Estimation of Permanent Magnet Synchronous Motor Parameters by PI Controller Gain Tuning. IEEE Transactions on Energy Conversion, vol. 21, no. 4, pp. 863-870, Dec. 2006.

[11] Moreno A J, Osorio M.: Strict Lyapunov functions for the super-twisting algorithm. IEEE Trans. Automatic Control, vol. 57, no. 4, Apr. 2012, pp. 1035-1040.

[12] Shtessel Y, Edwards C, Fridman L, and Levant A.: Sliding Mode Control and Observation. Springer, New York, 2014.

[13] Gonzalez T, Moreno A J. and Fridman L.: Variable Gain Super-Twisting Sliding Mode Control. IEEE Transactions on Automatic Control, vol. 57, no. 8, pp. 2100-2105, Aug. 2012.

[14] Lascu C, Boldea I, and Blaabjerg F.: Super-twisting sliding mode control of torque and flux in permanent magnet synchronous machine drives, IECON 2013 - 39th Annual Conference of the IEEE Industrial Electronics Society, Vienna, 2013, pp. 3171-3176.

[15] Salgado I, Chairez I, and Bandyopadhyay B.: Discrete-time non-linear state observer based on a super twisting-like algorithm. IET Control Theory Appl, vol. 8, no. 10, pp. 803–812, Jul. 2014.

New Grid-Connected Multilevel Boost Converter Topology with Inherent Capacitors Voltage Balancing Using Model Predictive Controller

[1]Rasoul Shalchi Alishah, [2]Kent Bertilsson, [3]Frede Blaabjerg, [4]Mohd. Ali Jagabar Sathik, [5]Ali Yahya Rezaee.

[1, 2, 5]Department of Industrial Design, Mid Sweden University, Sundsvall, Sweden.
[3]Department of Energy Technology, Aalborg University, Aalborg, Denmark.
[4]Department of Electrical and Electronics Engineering, SRM Institute of Science and Technology, Kattankulathur, India.
E-Mail: [1]rasoul.shalchialishah@miun.se, [2]kent.bertilsson@miun.se, [3]fbl@et.aau.dk, [4]mjsathik@ieee.org, [5]alre1802@student.miun.se.

Keywords

«Multilevel Converter», «Grid-Connected», «Model Predictive», «Injected Current».

Abstract

This paper presents a new grid-connected multilevel boost converter topology. The proposed multilevel boost converter includes several switched-capacitor units along with a developed H-bridge. The capability of voltage boosting, inherent voltage balancing of capacitors, reduction of power electronic elements, and voltage on switches are the merits of proposed topology compared to other topologies. All required mathematical analysis related to the suggested topology is presented. The proposed topology is verified by experimental results for grid-connected applications using model predictive controller.

1. Introduction

Nowadays, application of multilevel converters in the electric industry such as power system, renewable energy sources, and induction motor drives has been increased. For this aim, many researchers have been focused on multilevel converter topologies and their modulation techniques.

Flying-capacitor (FC), diode-clamped (DC) and cascaded H-bridge (CHB) multilevel converter topologies were the traditional topologies introduced in this type of power electronic converter. Each one of them has several advantages and disadvantages [1, 2]. Balancing the capacitors voltages are the main challenge in both FC and DC topologies which can be solved by complicated modulation techniques and external multi-output dc-dc converters [3]. The FC and DC multilevel converters suffer from using many capacitors and diodes, respectively. The CHB structure is a simple and suitable structure for high voltage applications. However, this circuit utilizes a large numbers of dc voltage sources which is the main restriction of the CHB topology [4, 5].

Recently, several cascaded type multilevel converters have been reported in literatures. The proposed multilevel converters in [6-11] use many dc voltage sources, gate drivers, and switches which increase the size and cost of converter. Also, these topologies cannot boost the value of input dc voltage sources. In addition, the voltage on the switches of the mentioned structures is high and rapidly increases by increasing the number of output voltage levels. Also, the number of required high voltage switches in the proposed multilevel converters in [6-8, 10] is high.

In order to control the injected power to the grid in power converters, several techniques such as model predictive controller, hysteresis controller and ramp type controller are used [12-15]. Model predictive controller is one of most popular controllers in comparison with other methods. Its performance is simple and has a fast dynamic response.

This paper presents a new grid-connected multilevel boost converter which can overcome the restrictions of suggested multilevel converters in [6-11]. In order to control the value of injected power to the grid, model predictive controller is utilized.

2. Proposed Grid-Connected Multilevel Boost Converter.

Fig. 1(a) illustrates the basic structure of presented grid-connected 7-level converter which comprises of two dc voltage sources (V_{L1}, V_R), one capacitor (C_{L1}), two power diodes (D_{L11}, D_{L21}), and nine unidirectional switches (S_{L1}, T_{L1}, N_{L1}, K_1, K_2, K_3, K_4, K_5, K_6).

Table I indicates the switching, capacitor and diodes states of proposed basic structure for producing all possible levels at output voltage waveform. According to this table, it is clear that this topology can produce seven levels at output voltage waveform. It is notable that the values of dc voltage sources for generating seven levels at output voltage waveform should be adjusted with similar amplitudes as $V_{L1}=V_R=V_{in}$. Considering the similar values for dc voltage sources, the generated levels will be 0, $\pm V_{in}$, $\pm 2V_{in}$, $\pm 3V_{in}$. The sum value of dc voltage sources in the proposed 7-level grid-connected converter is $2V_{in}$. However, the maximum value of output voltage is $3V_{in}$. It proves that the proposed converter can operate as a boost converter.

Table I. The switching, capacitor and diodes states of proposed basic structure.

Output Voltage	Switches									Diodes		Capacitor
	N_{L1}	S_{L1}	T_{L1}	K_1	K_2	K_3	K_4	K_5	K_6	D_{L11}	D_{L12}	C_{L1}
0	off	off	on	on	off	on	off	off	on	on	off	Charging
$+V_R$	off	off	on	on	off	off	off	on	on	on	off	Charging
$-V_R$	off	off	on	off	on	on	off	on	off	on	off	Charging
$+(V_{L1}+V_R)$	on	off	on	on	off	on	off	on	off	on	off	Charging
$-(V_{L1}+V_R)$	on	off	on	off	on	off	on	off	on	on	off	Charging
$+(V_{L1}+V_R+V_{CL1})$	on	on	off	on	off	on	off	on	off	off	off	Discharging
$-(V_{L1}+V_R+V_{CL1})$	on	on	off	off	on	off	on	off	on	off	off	Discharging

In order to increase the number of generated levels at output voltage waveform and voltage gain, the proposed basic structure can be extended as indicated in Fig. 1(b). In this topology the values of dc voltage sources are similar. The number of used switches (N_{switch}), dc voltage sources (N_{source}), and generated levels at output voltage waveform (N_{level}) are obtained by (1-3), respectively:

$$N_{source} = n+1 \tag{1}$$

$$N_{switch} = 3n+6 \tag{2}$$

$$N_{level} = 4n+3 \tag{3}$$

Where n represents the number of used capacitors. Also, the voltage gain of proposed converter (G) is:

$$G = \frac{V_{output}}{V_{input}} = \frac{(2n+1)V_{in}}{(n+1)V_{in}} = \frac{(2n+1)}{(n+1)} \tag{4}$$

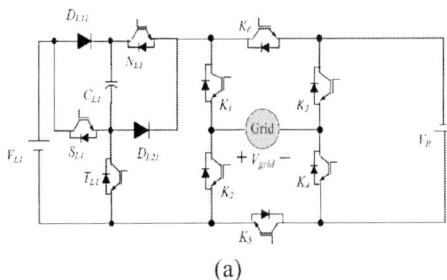

(a)

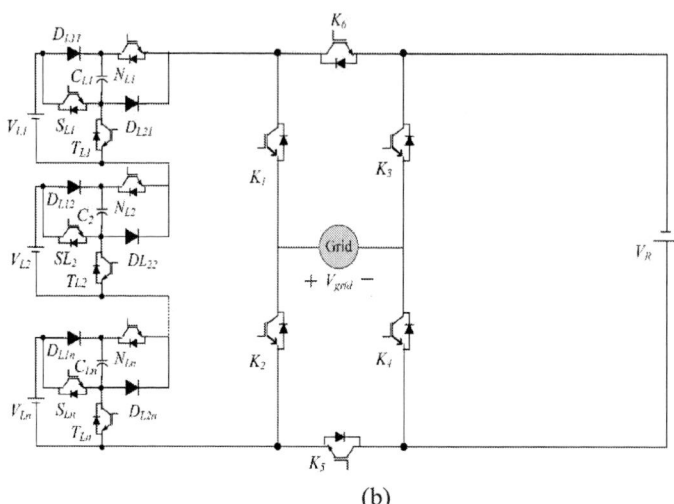

(b)

Fig. 1. (a) The basic structure of presented grid-connected 7-level converter, (b) The extended structure of proposed multilevel converter.

3. Calculation of Voltage on Switches and Power Losses.

Voltage on switches is an important criterion in multilevel converter which determines the manufacturing cost and its application. The voltage on the each switch in the proposed topology is obtained as follows:

$$V_{SLi} = V_{TLi} = V_{NLi} = V_{in} \qquad i = 1, 2, ..., n \tag{5}$$

$$V_{K3} = V_{K4} = V_{in} \tag{6}$$

$$V_{K1} = V_{K2} = 2nV_{in} \tag{7}$$

$$V_{K5} = V_{K6} = (2n+1)V_{in} \tag{8}$$

Therefore, the total voltage on switches (*TVS*) can be calculated as follows:

$$TVS = (11n+4)V_{in} \tag{9}$$

Another important parameter in multilevel converter topologies is power losses. In the proposed topology, there are three kinds of power losses which are named switching losses (P_{sw}), conduction losses (P_{cond}), and ripple losses of capacitors (P_{ripple}). Then, the total power loss (TP_{loss}) is:

$$TP_{loss} = P_{cond} + P_{sw} + P_{ripple} \tag{10}$$

When the input dc voltage sources are in parallel with the capacitor, the losses of capacitor voltage ripple is created which can be evaluated by (11):

$$P_{ripple} = \frac{1}{2T} \sum_{i=1}^{n} (C_{Li} \Delta V_{C_{Li}}) \tag{11}$$

Where ΔV_{Cj} and T are the voltage ripple across the capacitors C_{Li} and the time period, respectively.

The conduction losses of the presented multilevel boost converter structure can be calculated as follows:

$$P_{cond} = \left[\frac{x(t)}{\pi} \int_0^\pi (V_{on,diode} . I_{out}(t) + R_{diode} I_{out}^2(t)) \, dwt \right] + $$
$$\left[\frac{y(t)}{\pi} \int_0^\pi (V_{on,sw} . I_{out}(t) + R_{sw} I_{out}^{\beta+1}(t)) \, dwt \right] + \left[\frac{z(t)}{\pi} \int_0^\pi (R_{cap} I_{out}^2(t)) \, dwt \right] \tag{12}$$

Where, R_{diode}, R_{sw}, and R_{cap} represent the on-state resistances of diode, switch, and capacitor, respectively. In addition, $V_{on,diode}$ and $V_{on,sw}$ represent the voltage drop of the diode and switch, respectively. Also, β is a constant parameter which depends on the specification of the switch. Moreover, $x(t)$, $y(t)$, and $z(t)$ indicate the number of diodes, switches, and capacitors in current path, respectively.

The switching losses of switches in the proposed multilevel boost converter during the turn-on transition and the turn-off transition are obtained by (13):

$$P_{sw} = \frac{V_{sw}.I}{6T}(N_{off}.t_{off} + N_{on}.t_{on})$$

(13)

Where, N_{off} and N_{on} are illustrates the number of turning on and off switches in a period, respectively. The t_{on} and t_{off} are the turn on and turn off time of the switch, respectively.

4. Model Predictive Controller.

For the suggested grid-connected multilevel boost converter, model predictive controller method is used which anticipate the behavior of the grid current for each voltage vector produced by converter. By the use of Forward Euler approximation at the sample time T_s, the vectorial model for the injected current in a sample $m+1$ in terms of measurement in previous sample time m will be:

$$i_{out}(m+1) = \left[1 - \frac{R_f T_s}{L_f}\right] \times i_{out}(m) + \frac{T_s}{L_f} \times (V_{out}(m) - V_{grid}(m))$$

(14)

In order to select the optimal voltage vector, the below cost function (CF) should be optimized:

$$CF = \left| i_{out,ref}(m+1) - i_{out}(m+1) \right|$$

(15)

In other words, the cost function (CF) minimizes the error between its reference current $i_{out,ref}(m+1)$ and the predicted grid current $i_{out}(m+1)$. The usable model predictive algorithm for the proposed multilevel converter is indicated in Fig. 2.

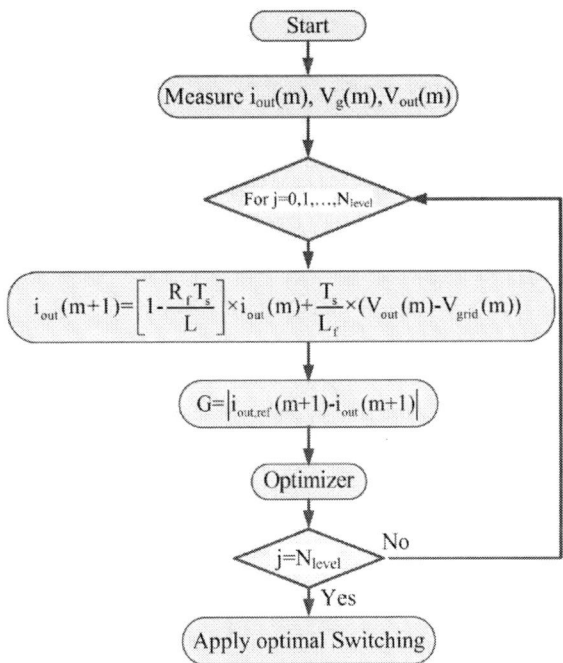

Fig. 2. Presented MPC algorithm.

5. Comparison of the Proposed Multilevel Boost Converter with Other Topologies.

In order to indicate the merits of the proposed multilevel boost converter, the comparison results are illustrated in Table II. According to this comparison, it is obvious that the proposed multilevel boost converter requires lower dc voltage sources, gate drivers, and switches compared to other structures proposed in [6-11]. Also, the number of high voltage switches in the proposed topology is lower than the proposed topologies in [6-8, 10]. These advantages cause the circuit size, manufacturing cost, and volume to be reduced. Capability of voltage boosting is another important advantage of proposed topology in comparison with other topologies.

Table II. Comparison of the proposed multilevel boost converter with other structures.

Topology	N_{source}	N_{driver}	N_{switch}	The number of high voltage switches	Capability of Voltage Boosting
Cascade H-bridge	$\frac{N_{level}-1}{2}$	$2(N_{level}-1)$	$2(N_{level}-1)$	0	No
[6]	$\frac{N_{level}-1}{2}$	$N_{level}+3$	$N_{level}+3$	4	No
[7]	$\frac{N_{level}-1}{2}$	$N_{level}+1$	$N_{level}+1$	4	No
[8]	$\frac{N_{level}-1}{2}$	$\frac{3N_{level}-3}{2}$	$\frac{3N_{level}-3}{2}$	4	No
[9]	$\frac{N_{level}-1}{6}$	N_{level}	N_{level}	0	No
[10]	$\frac{N_{level}-1}{2}$	$N_{level}+1$	$2N_{level}-2$	4	No
[11]	$\frac{N_{level}-1}{2}$	$N_{level}+1$	$N_{level}+5$	2	No
Proposed	$\frac{N_{level}+1}{4}$	$N_{level}-2$	$N_{level}-2$	2	Yes

6. Experimental Results.

In order to prove the performance of the proposed converter and model predictive controller method, the experimental results for the proposed grid-connected 7-level converter (See Fig. 3(a)) are illustrated. Fig. 3(b) indicates the photo of prototype. The value of input dc voltage sources is 127V. Fig. 3(c) indicates the output voltage waveform of the experimented topology. This figure proves that the proposed topology can generate seven levels with the maximum value of output voltage of 381V. Fig. 3(d) illustrates the voltage waveform of the switch S_{L1} which tolerate a voltage equal to 127V. According to Fig. 3(e), the maximum value of voltage on the switch N_{L1} is 127V. The total maximum value of voltage on all switches is 1905V.

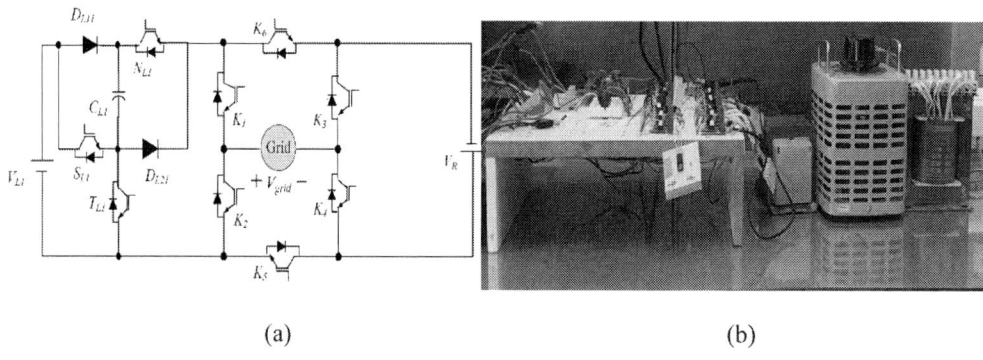

(a)   (b)

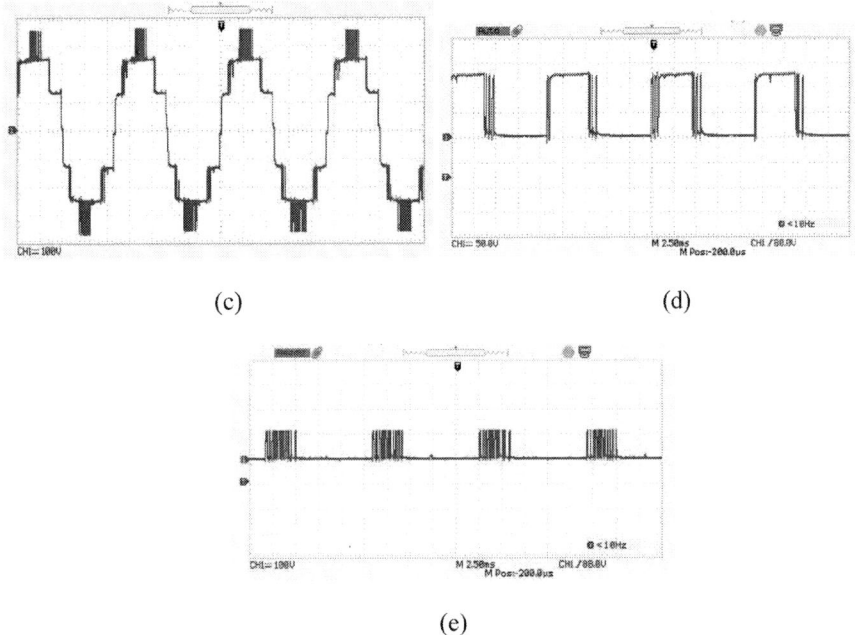

(c)   (d)

(e)

Fig. 3. (a) the studied 7-level grid-connected converter, (b) photo of prototype, (b) output voltage waveform, (c) Voltage waveform on the switch S_{L1}, (d) Voltage waveform on the switch N_{L1}.

In order to control the injected current to grid, model predictive controller is used. Fig 4(a) shows the waveforms of grid voltage and injected current to the grid. This figure proves that the experimented 7-level converter can inject the maximum power to the grid because of unity power factor. The value of injected power to the grid is 1181W. The efficiency of experimented topology is 97.21%. Fig. 4(b) indicates the waveforms of injected and reference current. This figure indicates that the injected current to the grid can track the waveform of reference current in the best possible form.

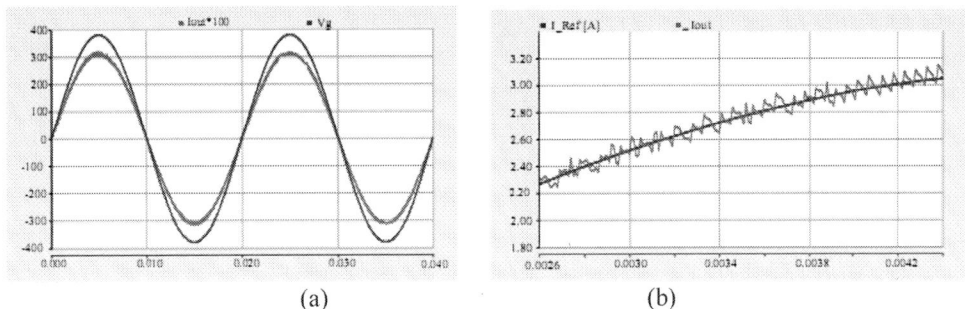

(a)   (b)

Fig. 4. (a) The grid voltage and injected current to the grid in the experimented grid-connected 7-level converter, (b) waveforms of injected current and reference current.

7. Conclusion

This paper suggested a new grid-connected multilevel boost converter which comprises of several switched-capacitor units along with a developed H-bridge. As the numbers of switched-capacitor units increase, the voltage gain and the number of generated levels increase. The capabilities of voltage boosting and inherent voltage balancing of capacitors are the main advantages of proposed topology. Moreover, the proposed multilevel boost converter requires lower dc voltage sources, gate driver circuits, and switches compared to conventional cascade H-bridge and presented topologies [6-11]. Model predictive controller was used to control the injected power to the grid. The experimental results were used to verify the performance of the suggested multilevel boost converter.

References

[1] R. Sh. Alishah, D. Nazarpour, S. H. Hosseini, and M. Sabahi: Reduction of power electronic elements in multilevel converters using a new cascade structure, IEEE Trans. Ind. Electron. 2015, vol. 62, no. 1, pp. 256–269, Jan. 2015.

[2] Ali, Jagabar Sathik Mohd, Rasoul Shalchi Alishah, and Vijayakumar Krishnasamy: A new symmetric multilevel converter topology with reduced voltage on switches and DC source, International Conference on Power, Instrumentation, Control and Computing 2018, pp. 1-6.

[3] V. Dargahi, A. K. Sadigh, M. Abarzadeh, S. Eskandari, and K. A. Corzine: A new family of modular multilevel converter based on modified flying capacitor multicell converters, IEEE Trans. Power Electron 2015, vol. 30, no. 1, pp. 138–147.

[4] Sadigh, Arash Khoshkbar, Seyed Hossein Hosseini, Mehran Sabahi, and Gevorg B. Gharehpetian. "Double flying capacitor multicell converter based on modified phase-shifted pulse-width modulation."IEEE Trans. on Power Electron. 2010, vol. 25, no. 6, pp. 1517-1526.

[5] Feng, Chunmei, Jun Liang, and Vassilios G. Agelidis. "Modified phase-shifted PWM control for flying capacitor multilevel converters." IEEE Trans. Power electron 2007, vol. 22, no. 1, pp.178-185.

[6] Babaei, E., Hosseini, S. H., "New cascaded multilevel inverter topology with minimum number of switches," Elsevier Journal of Ene. Conv. and Man., vol. 50, no. 11, pp. 2761-2767, 2009.

[7] Kangarlu, Mohammad Farhadi, and Ebrahim Babaei. "Cross-switched multilevel inverter: an innovative topology." IET Power Electronics, vol. 6, no. 4, 2013, pp. 642-651.

[8] Babaei, E., Hosseini, S. H., "New cascaded multilevel inverter topology with minimum number of switches," Elsevier Journal of Ene. Conv. and Man., vol. 50, no. 11, pp. 2761-2767, 2009

[9] Liu, Junfeng, K. W. E. Cheng, and Yuanmao Ye. "A cascaded multilevel inverter based on switched-capacitor for high-frequency AC power distribution system." IEEE Trans. Power Electron., 2014, vol. 29, no. 8, pp. 4219-4230.

[10] Alishah, R. Sh., Hosseini, S. H., Babaei, E., Sabahi, M. and Ardashir, J. F.: An Improved Symmetric H-Bridge Multilevel Converter Topology; An Attempt to Reduce Power Losses. Journal of Circuits, Systems and Computers 2018, vol. 27, no. 12, pp.1850187.

[11] Alishah, Rasoul Shalchi, Ebrahim Babaei, Seyed Hossein Hosseini, and Mehran Sabahi. "A Developed Two-Leg Ladder Multilevel Converter Structure. "Journal of Circuits, Systems and Computers 2018, vol. 27, no. 12, pp. 1850183.

[12] M. Maal Andish, T. Jalilzadeh, R. Shalchi Alishah, M. Sabahi, "Leakage Current Elimination of Grid-Connected PV Panels Using an Improved Non-Isolated DC-AC HERIC Converter," The International Conference on New Researches in Engineering Sciences-RKES 2016, University of Tehran.

[13] A. Zarrin Gharehkoushan, M. Sabahi, R. Shalchi Alishah "Hybrid High-Voltage Gain Inverter for Grid-Connected Photovoltaic System Employing Model Predictive Controller." The International Conference on New Researches in Engineering Science, 2016.

[14] Ho, C.N.M., Cheung, V.S. and Chung, H. S. H. Constant-frequency hysteresis current control of grid-connected VSI without bandwidth control. IEEE Trans. power electronics, vol. *24*, no. 11, 2009, pp. 2484-2495.

[15] Yan, Ruifeng, and Tapan Kumar Saha. "Power ramp rate control for grid connected photovoltaic system." Conference Proceedings IPEC, 2010, pp. 83-88. IEEE.

DCM Operation of Single-Switch High Step-up DC-DC Converter with Three-winding Coupled Inductor

Masataka Minami, Genki Hase
Kobe City College of Technology,
8-3, Gakuenhigashi-machi, Nishi ku,
Kobe city, Hyogo, 651-2194, JAPAN
Tel: +81 78 795 3232/Fax: +81 78 795 3314
kcct-minami@g.kobe-kosen.ac.jp

Abstract

We presented a novel high step-up DC-DC converter, which has only one active switch and a three-winding coupled inductor. And, the previous research showed the CCM operation and steady state principle of the proposed converter. This paper describes the DCM operation of the proposed converter and shows the experimental results.

Keywords

≪High voltage power converters≫, ≪Converter circuit≫

Introduction

Photovoltaic (PV) source is one of the significant energy sources in the world. However, as the output voltage of the PV arrays is relatively low with the parallel-connection, the high step-up and high efficiency DC-DC converters are required to boost the low PV voltage to a high voltage for grid-connection [1].

Ideally, a conventional boost converter is able to achieve high step-up gain with an extreme duty ratio. In practice, the step-up gain is limited by effects of the active switch, the diode, and parasitic resistor of the inductor and the capacitor. In addition, the extreme duty ratio occurs a serious reverse-recovery problem and conduction losses. Qun Zhao and Fred C. Lee [2] introduced a family of high step-up and high efficiency DC-DC converters by only adding one additional diode and a small capacitor. It is able to recycle the leakage energy and alleviate the reverse-recovery problem. Nowadays, several papers related to converter reported [3].

Masataka Minami and Kosuke Tomoeda [4] proposed a novel high step-up DC-DC converter, which has only one active switch and a three winding coupled inductor. The proposed converter based on the converter [5] combines the Cockcroft-Walton circuit [6]. While the previous papers [7, 8, 9, 10] proposed the different topologies with the three-winding coupled inductor, the proposed converter achieves higher step-up gain than the previous one [4]. However, the paper [4] showed only the CCM operation of the proposed converter. Then, this paper reveals the DCM operation of the proposed converter. In addition, the experimental results show the DCM operation.

1 DCM Operation and Steady State Principle

This section expresses the operation modes and leads the high step-up gain of the DCM operation in steady state. Fig. 1(a) shows the circuit configuration of the proposed converter. This converter consists of a power supply V_{in}, one active switch S, one three-winding coupled inductor L_1, L_2, and L_3, five diodes $D_1, \cdots, D_5$, and six capacitors $C_1, \cdots, C_5$, and C_o. To simplify the circuit analysis, the following conditions are assumed.

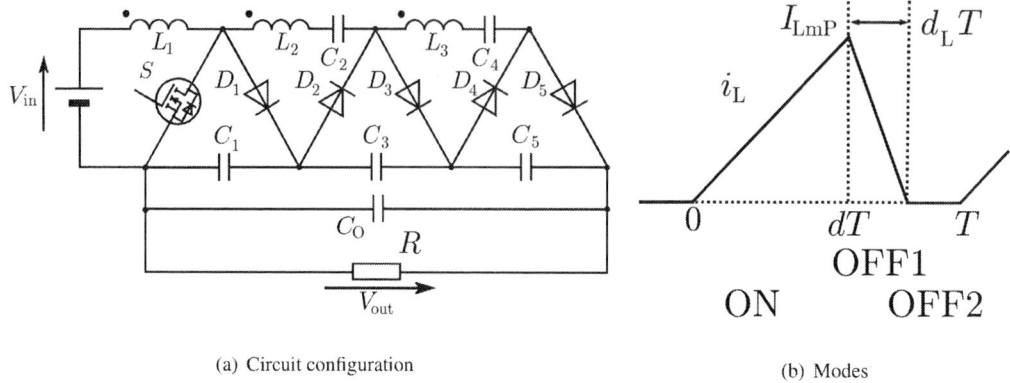

(a) Circuit configuration

(b) Modes

Fig. 1: Circuit configuration and its modes of single-switch high step-up DC-DC Converter with a three-winding coupled inductor [4].

- Capacitors $C_1, \cdots, C_5$, and C_o are large enough. Then, $V_{C1}, \cdots, V_{C5}$, and V_{out} are considered as constant voltage.

- The active switch S and the diodes $D_1, \cdots, D_5$ are treated as ideal.

- The equivalent series resistance (ESR) of the capacitors $C_1, \cdots, C_5$ and parasitic resistance of the three-winding inductor L_1, L_2, and L_3 are neglected.

- The turn ratios of the three-winding coupled inductor are $n_2 = N_2/N_1$ and $n_3 = N_3/N_2$.

In addition, in order to analyze the DCM steady-state principle simply, this paper considers the only three modes: ON, OFF1, and OFF2 modes in Fig. 1(b). In the DCM operation, the OFF period has two modes: OFF1 and OFF2 modes. OFF1 mode is the inductor current flow mode. OFF2 mode is the no inductor current mode. After the mode of ON period, the current $i_L (= i_{L1} + i_{L2} + i_{L3})$ reaches $I_{LmP} = (V_{in}/L_m)(d/f)$.

In the mode of ON period, the switch S is turned on. At the same time, the diodes D_2 and D_4 becomes conducting, the other diodes D_1, D_3, and D_5 are reverse-biased. Applying Kirchhoff's voltage law, $v_{L1}^{on} = V_{in}$. Moreover, the coupled inductor voltages v_{L2}^{on} and v_{L3}^{on} become $n_2 V_{in}$ and $n_3 V_{in}$. In addition, since the capacitors charge and discharge through the path, the equations $V_{C2} = V_{C1} + v_{L2}^{on}$ and $V_{C4} = V_{C3} + v_{L3}^{on}$ hold in this mode. In the same way,

$$v_{L1}^{off1} = -(V_{C1} - V_{in}), \tag{1}$$
$$v_{L2}^{off1} = -n_2(V_{C1} - V_{in}), \tag{2}$$
$$v_{L3}^{off1} = -n_3(V_{C1} - V_{in}), \tag{3}$$
$$V_{C3} = V_{C2} - v_{L2}^{off1}, \tag{4}$$
$$V_{C5} = V_{C4} - v_{L3}^{off1}, \tag{5}$$

are derived from OFF1 mode. Finally, OFF2 mode satisfies the follow equations;

$$v_{L1}^{off2} = v_{L2}^{off2} = v_{L3}^{off2} = 0, \tag{6}$$
$$i_{L1}^{off2} = i_{L2}^{off2} = i_{L3}^{off2} = 0, \tag{7}$$

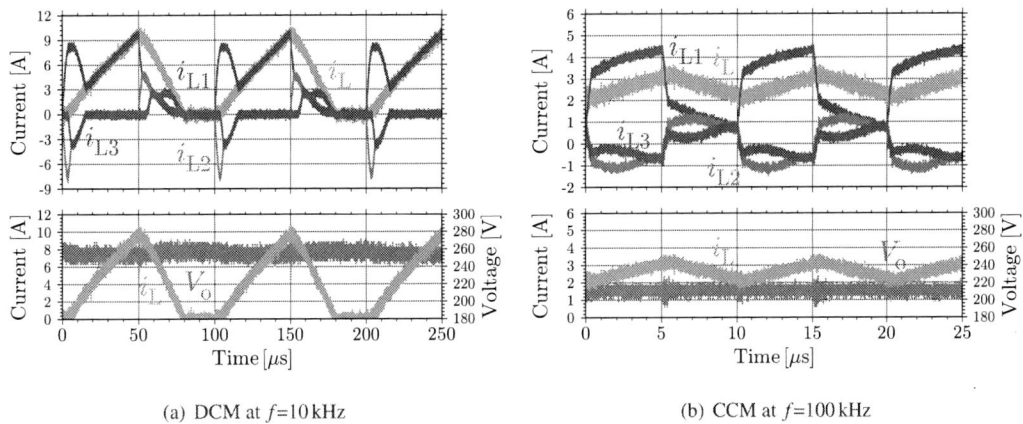

(a) DCM at $f=10\,\text{kHz}$
(b) CCM at $f=100\,\text{kHz}$

Fig. 2: Waveforms of Output voltage: V_{out} and inductor current: $i_{\text{L}1}$, $i_{\text{L}2}$, $i_{\text{L}3}$, and i_{L}. (a) DCM and (b) CCM.

$$\frac{1}{T}\left\{\int_0^{dT} v_{\text{L}1}^{\text{on}}\,\text{d}t + \int_{dT}^{dT+d_{\text{L}}T} v_{\text{L}1}^{\text{off1}}\,\text{d}t + \int_{dT+d_{\text{L}}T}^{T} v_{\text{L}1}^{\text{off2}}\,\text{d}t\right\} = 0, \tag{8}$$

$$dV_{\text{in}} - d_{\text{L}}(V_{\text{C}1} - V_{\text{in}}) = 0, \tag{9}$$

$$V_{\text{C}1} = \left(1 + \frac{d}{d_{\text{L}}}\right)V_{\text{in}}. \tag{10}$$

These equations about the voltage and current lead to the next equation: the boost ratio of DCM α_{DCM},

$$\alpha_{\text{DCM}} = \frac{3+2n_2+n_3}{2} + \sqrt{\frac{(3+2n_2+n_3)^2}{4} + \frac{d^2(3+2n_2+n_3)}{2(2+n_2+n_3)\tau_{\text{L}_{\text{m}}}}}, \tag{11}$$

where $\tau_{\text{L}_{\text{m}}}$ denotes the normalized inductor time constant as $L_{\text{m}}f/R$ [11]. Here, we omit the formula on the way for want of space. As comparison, the boost ratio of CCM: α_{CCM} becomes $(3+2n_2+n_3)/(1-d)$ from the previous paper [4].

2 Experimental Results

In order to verify the DCM operation and steady-state of the proposed converter, the prototype with the following specifications is designed in this section. 1) input DC voltage $V_{\text{in}} = 20\,\text{V}$, 2) switching frequency $f = 1/T = 10$ and $100\,\text{kHz}$, it means DCM and CCM, 3) duty ratio $d = 0.5$, 4) three-winding coupled inductor $L_1 = 87.30\,\mu\text{H}, L_2 = 87.47\,\mu\text{H}, L_3 = 88.21\,\mu\text{H}, k_{12} = 0.961, k_{23} = 0.914, k_{31} = 0.952, n_2 = 1, n_3 = 1$, measured by the impedance analyzer IM3570, 5) capacitors $C_1, \ldots, C_5 = 2.2\,\mu\text{F}$, and $C_{\text{o}} = 10\,\mu\text{F}$, 6) load resistor $R = 1\,\text{k}\Omega$, 7) switch S and diodes $D_1, \cdots, D_5$: SiC MOSFET(SCT2120AF) and SiC SBD(SCS220KG).

Figure 2 shows the measured waveforms: V_{out}, $i_{\text{L}1}$, $i_{\text{L}2}$, $i_{\text{L}3}$ and the calculated waveforms: $i_{\text{L}} = i_{\text{L}1} + i_{\text{L}2} + i_{\text{L}3}$ in the prototype proposed converter. Fig. 2(a) and Fig. 2(b) respectively illustrate DCM and CCM operations at $f = 10$ and $100\,\text{kHz}$. In Fig. 2(a), the inductor current i_{L} behaves as increase, decrease, and zero. Therefore, the proposed converter is operated as DCM in the same way from Section II. Especially, in OFF2 mode, all current waveforms: $i_{\text{L}1}$, $i_{\text{L}2}$, and $i_{\text{L}3}$ become to zero. In Fig. 2(a), the ON mode has three pattern. At first, $i_{\text{L}1}$ and $i_{\text{L}2}$ increases and decreases and $i_{\text{L}3}$ keeps zero. Secondary, $i_{\text{L}1}$ decreases and $i_{\text{L}2}$ and $i_{\text{L}3}$ become same and increase. Finally, $i_{\text{L}1}$ increases and $i_{\text{L}2}$ and $i_{\text{L}3}$ reach and keep zero.

These patterns are caused that the capacitors $C_1, \cdots, C_5$ are not large enough at 10kHz. If the capacitors are large enough, iL behaves similar as Fig.2(b) in the ON mode.

On the other hand, Fig. 2(b) shows that the proposed converter is operated as CCM. The CCM operation has already been analyzed in the previous paper [4]. Fig. 2 reveals that the operation changes due to the switching frequency.

3 Conclusion and Future works

This paper explained and demonstrated the DCM operation of the novel high step-up DC-DC converter with the single switch and the three-winding coupled inductor. As a result, the operation changes due to the switching frequency. Therefore, more work remains to analyze the boundary condition between CCM and DCM, load and frequency characteristics, efficiency and so on in the future.

References

[1] W. Li and X. He, "Review of nonisolated high-step-up DC/DC converters in photovoltaic grid-connected applications," *IEEE Transactions on Industrial Electronics*, vol. 58, no. 4, pp. 1239–1250, 2011.

[2] Q. Zhao and F. C. Lee, "High-efficiency, high step-up DC-DC converters," *IEEE Transactions on Power Electronics*, vol. 18, no. 1, pp. 65–73, 2003.

[3] M. Forouzesh, Y. P. Siwakoti, S. A. Gorji, F. Blaabjerg, and B. Lehman, "Step-up dc–dc converters: a comprehensive review of voltage-boosting techniques, topologies, and applications," *IEEE Transactions on Power Electronics*, vol. 32, no. 12, pp. 9143–9178, 2017.

[4] M. Minami and K. Tomoeda, "An analysis of operation in single-switch high step-up dc-dc converter with three-winding coupled inductor," in *2019 IEEE Applied Power Electronics Conference and Exposition (APEC)*. IEEE, 2019, pp. 2135–2137.

[5] R.-J. Wai and R.-Y. Duan, "High step-up converter with coupled-inductor," *IEEE Transactions on Power Electronics*, vol. 20, no. 5, pp. 1025–1035, 2005.

[6] J. D. Cockcroft and E. Walton, "Experiments with high velocity positive ions," *Proceedings of the Royal Society of London A*, vol. 129, no. 811, pp. 477–489, 1930.

[7] R.-J. Wai, C.-Y. Lin, R.-Y. Duan, and Y.-R. Chang, "High-efficiency DC-DC converter with high voltage gain and reduced switch stress," *IEEE Transactions on Industrial Electronics*, vol. 54, no. 1, pp. 354–364, 2007.

[8] S.-K. Changchien, T.-J. Liang, J.-F. Chen, and L.-S. Yang, "Novel high step-up DC–DC converter for fuel cell energy conversion system," *IEEE Transactions on Industrial Electronics*, vol. 57, no. 6, pp. 2007–2017, 2010.

[9] K.-C. Tseng, J.-T. Lin, and C.-C. Huang, "High step-up converter with three-winding coupled inductor for fuel cell energy source applications," *IEEE Transactions on Power Electronics*, vol. 30, no. 2, pp. 574–581, 2015.

[10] M. Khalilzadeh, M. Mahdipour, and K. Abbaszadeh, "High step-up dc-dc converter based on three-winding coupled inductor," in *Power Electronics, Drives Systems & Technologies Conference (PEDSTC), 2015 6th*. IEEE, 2015, pp. 195–200.

[11] Y.-P. Hsieh, J.-F. Chen, T.-J. P. Liang, and L.-S. Yang, "Novel high step-up DC–DC converter with coupled-inductor and switched-capacitor techniques for a sustainable energy system," *IEEE Transactions on Power Electronics*, vol. 26, no. 12, pp. 3481–3490, 2011.

Power Losses Calculation for Medium Voltage DC/DC Current-Fed Solid State Transformer for Battery Grid-Connected

E. K. Hussain[1], Mohammad Abusara[1], S. M. Sharkh[2]
[1]Exeter University, [2]University of Southampton
Penryn, UK
Tel.: +44 01326 371885.
E-Mail: M.Abusara@exeter.ac.uk
URL: http://emps.exeter.ac.uk/renewable-energy/research/

Acknowledgements

The authors wish to acknowledge the support of this work by the INTERREG Channel funding stream through the Intelligent Community Energy (ICE) project, Grant Agreement Number 5025.

Keywords

« Solid State Transformer », « Power Losses », « Battery », « Grid Connected ».

Abstract

Current fed solid-state transformers (CF-SSTs) offer substantial weight and size reduction advantages over 50/60Hz traditional transformers. CF-SSTs have low source current ripple which makes it suitable for the connection of large batteries in the medium voltage grid. However, power losses are high compare to the traditional one due to the high operating frequency range. Therefore, power losses calculations are the main key to select the main design parameters. Unlike most of the published work dealing one aspect of power losses or a specific element and only consider voltage fed Solid State Transformers topology (VF-SSTs), this paper presents a power losses calculation method for CF-SSTs considering all the power losses, the shape of the voltage waveforms, cores dimensions and manufacturer datasheets are main paper pros. ANSYS and Matlab Simulink are employed to validate the analytical equations. Due to simplicity, this method can be used for optimization.

Introduction

Battery Energy Storage System (BESS) is becoming a crucial element in the smart grid. BESS can smooth Renewable Energy Source (RES) output power and provide grid support such as frequency response and voltage control. Traditionally, BESS is connected to the medium voltage (MV) grid (1-35kV) via a bidirectional power electronic DC/DC converter, DC/AC converter and line frequency step-up transformer. These converters are responsible for ensuring grid operation codes and standards are met, and power from/to the battery is controlled. Due to limited voltage and current capabilities of the power electronic devices (up to 6.5kV/25A [1]), many power electronics topologies are proposed to meet the required voltage and current specifications.

To achieve the advantages of galvanic isolation, weight reduction and low battery voltage, Solid State Transformer (SST) can be employed [2]. SST offers many advantages over the traditional transformer such as power flow control, voltage sag compensation, fault current limitation, reduced size and weight, improved power quality, and high power density [1]. SST has mainly emerged from Dual Active Bridge converter (DAB) technology. There are two main topologies for DAB, voltage-fed DAB (VF-DAB) and current-fed DAB (CF-DAB). The voltage phase shift technique is employed to transfer the power to the load in VF-DAB. There are different control techniques to achieve the phase shift in the voltages. Single-phase shift (SPS), dual-phase shift (DPS) and triple-phase shift (TPS) [3, 4]. DPS is the most suitable control technique [5]. In the CF-DAB, a choke coil is employed to transfer the power [6]. The turn-off voltage spike is the main problem in CF-DAB due to the storge energy on the leakage inductance. RCD snubber circuits, active clamp, zero current switching and secondary modulation can be employed to overcome this problem [7-10]. VF-DAB suffers from several limitations of high input pulsating current, high circulating current through devices and magnetics [4]. Therefore CF-DAB is more suitable for grid-connected battery applications. Current fed solid-state transformer (CF-SST) topology is considered in this paper. There are three main losses in CF-SST; core losses, copper losses and power electronics losses.

Numerous publications proposed techniques to calculate power losses based on the power electronics parameters and core losses equations [11-16]. They can be divided into three categories. The first one, authors gave much detail about core losses calculation and no much details about the power electronics losses. The second one, the authors considered only power electronics losses and employed ordinary core losses equations based on sinusoidal waveforms. The last one, the authors employed a practical measuring of the core losses or power electronics losses. In this paper, a power losses calculation method is proposed based on the manufacturer datasheets. The proposed method considers all the power losses of the CF-SST, voltage waveform shape, core flux and operating frequency. Furthermore, the main dimensions of the transformer core and choke coil core are calculated to find the volume of the cores. This method can be a useful tool to find the optimum design to achieve high efficiency or smaller size due to its simplicity.

CF-SST System Configuration

A CF-SST consisting of a medium frequency transformer, choke coil and H-bridge is shown in Fig. 1. Converter modules are connected in series and/or parallel to reduce the voltage or current stress on the power electronic devices. Fig. 2 shows a simple model for CF-SST [10]. There are two modes of operation, charging mode (buck converter) and discharging mode (boost converter). In the discharging mode, the choke coil and H-Bridge 1 operate as a boost converter. In the charging mode, H-Bridge 2 operates as a buck converter and H-Bridge 1 operate as a rectifier. The choke coil is assumed high enough to smooth the battery current, i.e, ripple currents very small. The load current is also assumed as constant and ripple-free.

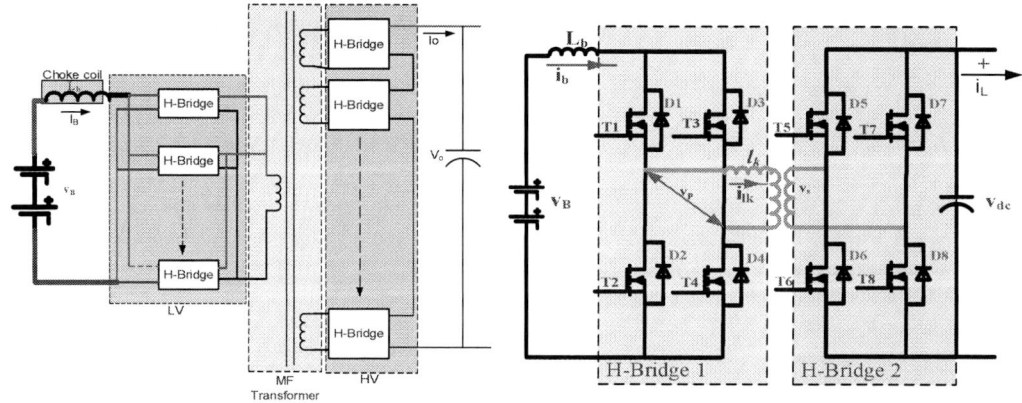

Fig. 1: SST System configuration   Fig. 2: Simplified CF-SST model

CF-SST Power Losses Calculation

The CF-SST system power losses can be divided into transformer core and copper losses, power electronics losses, choke coil core and copper losses. Details about these losses and how to be calculated are shown in the next sections.

Transformer Core and Copper Losses

Hysteresis and eddy current losses are the main iron losses in the transformer core. The hysteresis loss is related to the re-orientation of the magnetic domains while the eddy-current loss is due to the induced voltage within the magnetic core. To calculate the iron losses, the Original Steinmetz Equation (empirical equation) is normally employed [17]. This equation, however, is only applicable to a sinusoidal excitation so it is not valid for SST case where the magnetic material is excited with nonsinusoidal waveforms. Core losses under nonsinusoidal waveforms can be estimated by the Generalized Steinmetz Equation (tuned) based on the rate of change of the flux density can be estimated according to equation (1). [18, 19].

$$P_v = \frac{1}{T} k_i (\Delta B)^{\beta - \alpha} |2\Delta B|^{\alpha} (DT)^{1-\alpha}, \tag{1}$$

$$k_i = \frac{K}{(2)^{\beta-1}(\pi)^{\alpha-1}\left(1.1044 + \frac{6.8244}{\alpha + 1.354}\right)}, \tag{2}$$

The core losses in (1) are calculated per unit volume. To get the total core losses of the transformer core, the design of the transformer core is considered to calculate the core volume as in (2) [20-22]:

$$Vol_c = 2A_i\left(\sqrt{\frac{A_w}{r_w}}(r_w + 1) + \frac{d_c}{\sqrt{K_c}}\right), \tag{3}$$

where, A_i is the area product of the transformer core, A_w is the window area, r_w is the ratio of winding window height to width and d_c is the diameter of the core circumscribing. All these parameters depend on the operating frequency and core flux density. The transformer core losses can be calculated based on (1) and (2). For the copper losses, the ac resistance of the primary (R_p) and secondary voltage winding (R_s) can be calculated as in (3) [23] considering the skin effect. where ρ_c is the resistivity of the winding material, l_{pp} and l_{ps} are the total length of the primary and secondary winding respectively. δ is the skin depth. A_{cup} and A_{cus} are the cross-section of the winding conductor for the primary and secondary winding respectively. r_{op} and r_{os} are the radius of the winding conductor for the primary and secondary winding respectively.

$$R_p = \rho_c \frac{l_{pp}}{A_{cup}}\left[1 + \left(\frac{(r_{op}/\delta)^4}{48 + 0.8(r_{op}/\delta)^4}\right)\right] \tag{4}$$

$$R_s = \rho_c \frac{l_{ps}}{A_{cus}}\left[1 + \left(\frac{(r_{os}/\delta)^4}{48 + 0.8(r_{os}/\delta)^4}\right)\right] \tag{5}$$

Choke Coil Core and Copper Losses

The value of the choke coil depending on the duty cycle of the converter (D_1), the switching frequency (f), the supply voltage (V_{dc}) and the maximum allowance ripple in the DC current (ΔI_{dc}) as shown in (4) for boost operation of the SST system (battery discharging) [18]. The choke coil resistance is based on the mean length of the turn (MTL), number of turns (N_c) and cross section of the winding (A_{cuc}) as in (5).

$$L_B = \frac{(D_1 - 0.5)V_B}{f\Delta I_B} \tag{4}$$

$$R = \rho\frac{N_c \times MTL}{A_{cuc}}, \quad \text{where } MTL \cong \pi\left[\sqrt{H_c^2 + W_c^2} + W_w\right], \tag{5}$$

where H_c is the height of the core cross-section, W_c is the width of the core cross-section and W_w is the width of the core window. The winding window area (A_{wc}) equals to W_wH_w. The core cross-section area (A_c) equals to W_cH_c. The area product of the core (A_{pc} and equlas to $A_{wc}A_c$) is the main parameters to find a suitable core for the choke coil. A_{pc} can be calculated as in (6) [24].

$$A_{pc} = \frac{2En\times10^{-4}}{B_{mc}JK_u}, \tag{6}$$

where En the energy in watt-seconds, J is the current density, B_{mc} is the choke coil flux density and K_u is the choke coil window utilization factor for the core. The number of choke coil turns can be calculated as in (7) [25]. where K_{uc} is the effective window factor for the choke coil core, K_f is the filling factor, I_{B-rms} is the RMS value of the battery current and K_{cc} is the ratio of the winding are to the core cross-section area.

$$N_c = \frac{K_{uc}K_fI_{B-rms}}{J\sqrt{A_{pc}K_{cc}}} \tag{7}$$

Here, the fluctuation of the flux density of the choke is very small due to the low ripple on the choke coil current. Therefore, the core losses of the choke coil can be calculated by simple core losses equation. Based on the volume of the core, the choke coil core losses can be calculated.

Power Electronics Losses

Power electronics losses have two main losses, conduction losses and switching losses. Assuming several H-bridges are connecting in parallel at the low voltage side to reduce the current stress in the switch and several parallel H-bridges on the high voltage side to reduce the voltage across the switch as shown in Fig. 1. Based on the circuit diagram in Fig. 3 and mapping the on-state characteristic as shown in Fig. 3, the RMS and average of the switch current of the H-Bridge 1 (for switch T1) under discharge mode are [10]

$$I_{T1} = \bar{I}_B\sqrt{\frac{2 - D_1}{3}} \tag{6}$$

$$\overline{I_{T1}} = \bar{I}_B\frac{3 + D_1}{4} \tag{7}$$

The RMS and average of the switch, I_{T5}, and diode currents, I_{D5}, of the H-Bridge 2 (for switch T5) are

$$I_{T5} = \overline{I_L}\sqrt{\frac{D_2}{3}} \tag{8}$$

$$\overline{I_{T5}} = \overline{I_L}\frac{D_2}{2} \tag{9}$$

$$I_{D5} = \overline{I_L}\sqrt{\frac{5 - 4D_1 - 2D_2}{6}} \tag{10}$$

$$\overline{I_{D5}} = \overline{I_L}\frac{3 - 2D_1 - 2D_2}{6} \tag{11}$$

where D_2 is the duty cycle of the switch in H-bridge 2, and $\overline{I_L}$ is the average load current.

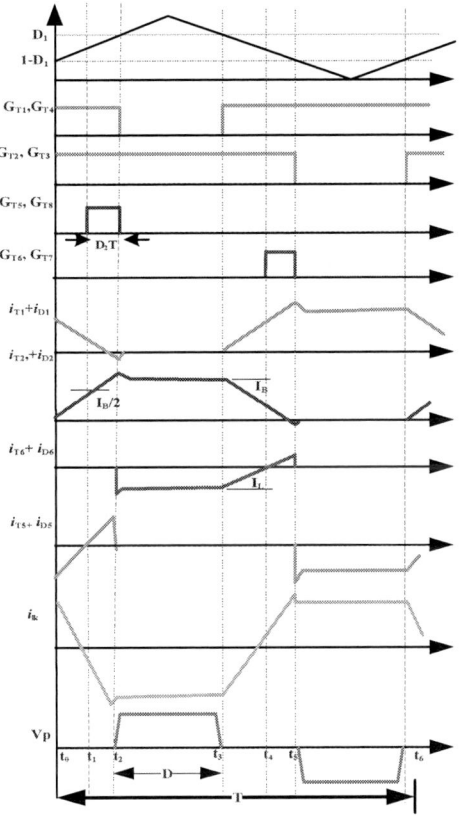

Fig. 3: Details of CF-SST operation over one cycle [10]

Based on mapping the on-state characteristic and employing equations and consider the effect of the junction temperature, the conduction losses and turn OFF power dissipation of the switches and diode can be calculated for the CF-SST configuration in Fig. 1 as [26, 27].

$$P_{cond\_T} = N_{LV}(K_1 I_{T1}^2 + K_2\overline{I_{T1}}) + N_{HV}\left(I_{T5}^2 K_1 + K_2\overline{I_{T5}}\right) \tag{12}$$

$$P_{sw\_T} = K_3 f E_{on+off} \left(N_{LV} \left[\frac{I_{T1}}{I_{ref}} \right]^{K_{is}} \left[\frac{V_B}{2V_{ref}} \right]^{K_{vs}} + N_{HV} \left[\frac{I_{T5}}{I_{ref}} \right]^{K_{is}} \left[\frac{V_o}{2N_{HV}V_{ref}} \right]^{K_{vs}} \right), \tag{13}$$

$$P_{cond\_D} = N_{HV} \left[I_{D5}^2 K_5 + \overline{I_{D5}} K_4 \right], \tag{14}$$

$$P_{sw\_D} = K_6 N_{HV} f E_{rr} \left[\frac{I_{D5}^2 K_5}{I_{ref}} \right]^{K_{id}} \left[\frac{V_o}{2N_{HV}V_{ref}} \right]^{K_{vd}}, \tag{15}$$

where

$$K_1 = V_{CE(25^0C)} + TCs_V(\Delta T_j), \tag{16}$$

$$K_2 = r_{CE(25^0C)} + TCs_r(\Delta T_j), \tag{17}$$

$$K_3 = 1 + TCs_{Esw}(\Delta T_j), \tag{18}$$

$$K_4 = V_{f(25^0C)} + TCd_V(\Delta T_j), \tag{19}$$

$$K_5 = r_{f(25^0C)} + TCd_r(\Delta T_j), \tag{20}$$

$$K_6 = 1 + TCd_{Err}(\Delta T_j), \tag{21}$$

where N_{LV} is the total number of switches at the low voltage side and N_{HV} is the total number of diodes at the high voltage side. D_2 is the duty cycle of switches of H-Bridge 2, $\overline{I_L}$ is the average load current. TCs_V and TCs_r are temperature coefficients of the on-state characteristic for the switch and TCd_V & TCd_r are the temperature coefficients of the on-state characteristic for the diode. ΔT_j is the increase in the junction temperature from the ambient temperature. The other parameters and their values are shown in TABLE I [26, 27].

TABLE I : POWER ELECTRONICS LOSSES PARAMETERS

Term	Meaning	Value
K_{is}	Exponent for the current dependency of switching losses (switch)	≈ 1
K_{vs}	Exponent for the voltage dependency of switching losses (switch)	≈ 1.3 to 1.4
K_{id}	Exponent for the current dependency of switching losses (diode)	≈ 0.6
K_{vd}	Exponent for the voltage dependency of switching losses (diode)	≈ 0.6
TCs_{Esw}	Temperature coefficient of the switching losses	≈ 0.003
TCd_{Err}	Temperature coefficients of the diode switching losses	≈ 0.006

Simulation Results

1MW DC/DC CF-SST is considered as a study model. The battery voltage is 600V and the dc-link voltage for MV is 18kV. Due to the high battery current (1.7kA), four parallel H-bridges are employed to reduce the current on the switch to 417A. For the high voltage side, 6 series H-Bridge are considered to reduce the reverse voltage. The analytical calculations are compared with the simulation based on the MATLAB Simulink. Matlab Simulink calculates the power electronics losses with the help of a 3-D lookup table from the manufactures datasheets [28]. A Simulink model for the CF-SST is built on the Matlab Simulink platform. The IGBT module 5SNG 0450X330300 [29] is considered as the power electronics switches for the CF-SST. The IGBT junction temperature is fixed at 125°C. The on-state resistance for the IGBT (r_{CE})and the diode (r_f) are not included in datasheets. To estimate these values, the typical on-state characteristic relationship between V_{CE} and I_C can be employed. Based on the V_{CE} and I_C curves, $r_{CE} = 3.2m\Omega$ and $r_f = 2m\Omega$ at $125°C$ junction temperature.

Fig. 4 shows the power electronics losses of the CF-SST by analytical equations and Matlab simulation at the different switching frequency and the full load. It seems that there is a small mismatch between the analytical and simulation results. For the core losses, finite element analysis (FEA) is employed to verify the analytical equations by ANSYS Workbench. SURA No18 is employed as a core material [30]. The 1200 V input voltage waveform is injected to the transformer primary. Under different duty cycles of low voltage H-Bridges, the core losses of the transformer calculated by the analytical equations and by ANSYS are shown in Fig. 5. There is an error between the ANSYS and analytical equations not more than 20%. The core losses increase with the decreases in the duty cycle. This is due to the increase in the duty cycle decrease the width of the injected voltage to the transformer.

Fig. 6shows transformer copper and core losses, choke coil copper and core losses and power electronics losses at different operating frequencies and a fixed flux density of 1T. At the low frequency, the copper losses are the dominant losses while power electronics losses are the main losses for frequency more than 200Hz. All these curves depend on the core materials selection and the power switching technology. The core losses of the choke coil are small due to the change of the flux density of the choke coil core is very small. The total CF-SST losses under different operating frequency and core flux density are shown in Fig. 7. The operating frequency about 600Hz seems to be the optimum value for minimum power losses.

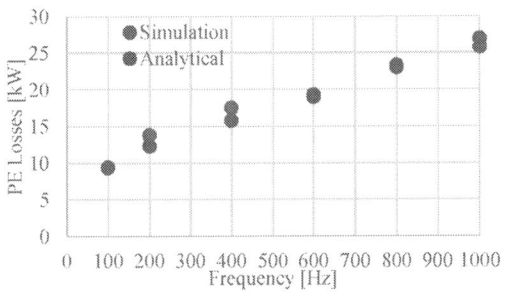

Fig. 4: Power electronics losses for the CF-SST by simulation and analytical method

Fig. 5: Transformer core losses by ANSYS and analytical method at a different duty cycle

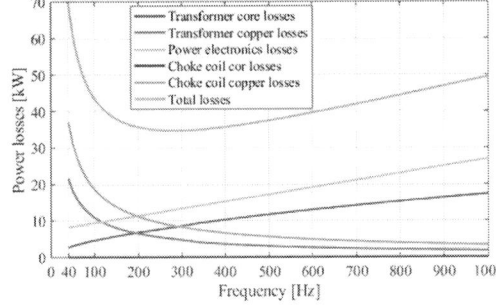

Fig. 6: CF-SST copper, core and power electronics losses

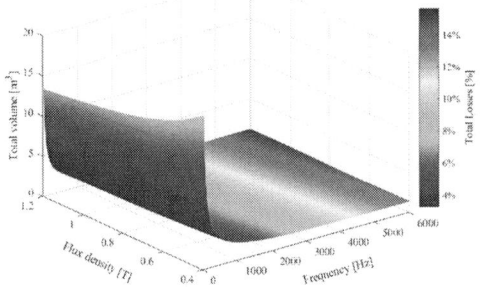

Fig. 7: Total CF-SST losses at different operating frequencies and flux density

Conclusion

This paper presents an analytical method to calculate the power losses for current fed solid-state transformer based on the manufacturer datasheets, operating frequency and core flux density. It considers the nonsinusoidal voltage waveforms to calculate the core losses and power electronic losses. ANSYS and Matlab Simulink are employed to validate the analytical results. At low frequency, the copper losses are the main losses. As the frequency increases the copper losses decreases and the power electronics and core losses increase. There are an optimum operating frequency and flux density where the total losses are minimum. This is topology can be employed to find the optimum flux density and operating frequency to gain maximum efficiency.

References

[1] X. She, A. Q. Huang, and R. Burgos, "Review of Solid-State Transformer Technologies and Their Application in Power Distribution Systems," IEEE Journal of Emerging and Selected Topics in Power Electronics, vol. 1, no. 3, pp. 186-198, 2013, doi: 10.1109/JESTPE.2013.2277917.

[2] L. Jih-Sheng, A. Maitra, A. Mansoor, and F. Goodman, "Multilevel intelligent universal transformer for medium voltage applications," in Fourtieth IAS Annual Meeting. Conference Record of the 2005 Industry Applications Conference, 2005., 2-6 Oct. 2005 2005, vol. 3, pp. 1893-1899 Vol. 3, doi: 10.1109/IAS.2005.1518705.

[3] M. N. K. R. W. G. D. M. D. E. D. Baumann, "Performance characterization of a high-power dual active bridge DC-to-DC converter," IEEE Transactions on Industry Applications, vol. 28, no. 6, Nov/Dec 1992, doi: 10.1109/28.175280. IEEE.

[4] B. Zhao, Q. Yu, and W. Sun, "Extended-Phase-Shift Control of Isolated Bidirectional DC–DC Converter for Power Distribution in Microgrid," IEEE Transactions on Power Electronics, vol. 27, no. 11, pp. 4667-4680, 2012, doi: 10.1109/TPEL.2011.2180928.

[5] B. Zhao, Q. Song, and W. Liu, "Power Characterization of Isolated Bidirectional Dual-Active-Bridge DC–DC Converter With Dual-Phase-Shift Control," IEEE Transactions on Power Electronics, vol. 27, no. 9, pp. 4172-4176, 2012, doi: 10.1109/TPEL.2012.2189586.

[6] P. Xuewei and A. K. Rathore, "Novel bidirectional snubberless soft-switching naturally clamped zero current commutated current-fed dual active bridge (CFDAB) converter for fuel cell vehicles," in 2013 IEEE Energy Conversion Congress and Exposition, 15-19 Sept. 2013 2013, pp. 1894-1901, doi: 10.1109/ECCE.2013.6646939.

[7] R. Y. Chen, R. L. Lin, T. J. Liang, J. F. Chen, and K. C. Tseng, "Current-fed full-bridge boost converter with zero current switching for high voltage applications," in Fourtieth IAS Annual Meeting. Conference Record of the 2005 Industry Applications Conference, 2005., 2-6 Oct. 2005 2005, vol. 3, pp. 2000-2006 Vol. 3, doi: 10.1109/IAS.2005.1518722.

[8] T. Wu, Y. Chen, J. Yang, and C. Kuo, "Isolated Bidirectional Full-Bridge DC–DC Converter With a Flyback Snubber," IEEE Transactions on Power Electronics, vol. 25, no. 7, pp. 1915-1922, 2010, doi: 10.1109/TPEL.2010.2043542.

[9] Y. Miura, M. Kaga, Y. Horita, and T. Ise, "Bidirectional isolated dual full-bridge dc-dc converter with active clamp for EDLC," in 2010 IEEE Energy Conversion Congress and Exposition, 12-16 Sept. 2010 2010, pp. 1136-1143, doi: 10.1109/ECCE.2010.5617843.

[10] P. Xuewei and A. K. Rathore, "Novel Bidirectional Snubberless Naturally Commutated Soft-Switching Current-Fed Full-Bridge Isolated DC/DC Converter for Fuel Cell Vehicles," IEEE Transactions on Industrial Electronics, vol. 61, no. 5, pp. 2307-2315, 2014, doi: 10.1109/TIE.2013.2271599.

[11] H. Qin and J. W. Kimball, "A comparative efficiency study of silicon-based solid state transformers," in 2010 IEEE Energy Conversion Congress and Exposition, 12-16 Sept. 2010 2010, pp. 1458-1463, doi: 10.1109/ECCE.2010.5618255.

[12] N. M. Evans, T. Lagier, and A. Pereira, "A preliminary loss comparison of solid-state transformers in a rail application employing silicon carbide (SiC) MOSFET switches," in 8th IET International Conference on Power Electronics, Machines and Drives (PEMD 2016), 19-21 April 2016 2016, pp. 1-6, doi: 10.1049/cp.2016.0196.

[13] W. Qingshan and D. Liang, "Research on loss reduction of dual active bridge converter over wide load range for solid state transformer application," in 2016 Eleventh International Conference on Ecological Vehicles and Renewable Energies (EVER), 6-8 April 2016 2016, pp. 1-9, doi: 10.1109/EVER.2016.7476340.

[14] T. Liu et al., "Frequency-domain-based complete loss model for 10 kV/1 MW solid-state transformer," The Journal of Engineering, vol. 2019, no. 16, pp. 2873-2877, 2019, doi: 10.1049/joe.2018.9145.

[15] F. Yazdani, S. Haghbin, T. Thiringer, and M. Zolghadri, "Accurate Power Loss Calculation of a Three-Phase Dual Active Bridge Converter For ZVS and Hard-Switching Operations," in 2017 IEEE Vehicle Power and Propulsion Conference (VPPC), 11-14 Dec. 2017 2017, pp. 1-5, doi: 10.1109/VPPC.2017.8330875.

[16] Y. H. Abraham, H. Wen, W. Xiao, and V. Khadkikar, "Estimating power losses in Dual Active Bridge DC-DC converter," in 2011 2nd International Conference on Electric Power and Energy Conversion Systems (EPECS), 15-17 Nov. 2011 2011, pp. 1-5, doi: 10.1109/EPECS.2011.6126790.

[17] C. P. Steinmetz, "On the law of hysteresis," Proceedings of the IEEE, vol. 72, no. 2, pp. 197-221, 1984, doi: 10.1109/PROC.1984.12842.

[18] R. J. G. Montoya, "High-Frequency Transformer Design for Solid-State Transformers in Electric Power Distribution Systems," University of Arkansas, Fayetteville, 2015.

[19] L. Jieli, T. Abdallah, and C. R. Sullivan, "Improved calculation of core loss with nonsinusoidal waveforms," in Conference Record of the 2001 IEEE Industry Applications Conference. 36th IAS Annual Meeting (Cat. No.01CH37248), 30 Sept.-4 Oct. 2001 2001, vol. 4, pp. 2203-2210 vol.4, doi: 10.1109/IAS.2001.955931.

[20] R. L. Bean, Transformers for the electric power industry. McGraw-Hill, 1959.

[21] M. G. Say, Alternating current machines. London: Pitman, 1983.

[22] B. Hochart, Power transformer handbook. London: Butterworths, 1987.

[23] W. H. W. W. G. Hurley, Transformers and Inductors for Power Electronics: Theory, Design and Applications. Wiley, 2013.

[24] W. McLyman, Transformer and inductor design handbook. New York, 1978.

[25] W. H. W. o. W. G. Hurley, TRANSFORMERS AND INDUCTORS FOR POWER ELECTRONICS. United Kingdom: John Wiley & Sons Ltd, 2013.

[26] E. I. GmbH. Application Manual Power Semiconductors, 2015.

[27] D.-I. U. N. Dr.-Ing. Arendt Wintrich, Dr. techn. Werner Tursky , Dr.-Ing. Tobias Reimann, Application Manual Power Semiconductors. SEMIKRON International GmbH, 2015.

[28] MathWorks, "Loss Calculation in a Three-Phase 3-Level Inverter," 2018.

[29] ABB. "LinPak phase leg IGBT module." https://library.e.abb.com/public/766b5fc8ed264c47ae634455dbade759/5SNG%200450X330300%205SYA% 201458-02%2012-2018.pdf (accessed 2019).

[30] S. E, "Language difficulties of international students in Australia : the effectsof prior learning e0perience," International Education Journal, vol. 6, no. 5, 2005.

Modelling and Experimental Validation of a Pole-To-Ground Protection Device in Low Voltage DC Microgrids

L. Hallemans[*†], G. Govaerts[*†], G. Van den Broeck[‡], S. Ravyts[*†],
M. M. Alam[†§], P. Van Tichelen[†§], J. Driesen[*†]

[*]KU Leuven, Dept. of Electrical Engineering, div. Electa, Kasteelpark 10, Leuven, Belgium
[‡]DCinergy, Langdorpsesteenweg 106, 3200 Aarschot, Belgium
[†]EnergyVille, Thor Park 8301, 3600 Genk, Belgium
[§]VITO, Boeretang 200, 2400 Mol, Belgium
Email: leonie.hallemans@kuleuven.be

Acknowledgements

The authors would like to thank VITO for the support in performing this research. This work has been supported by VLAIO in the Flux50 project BIDC (HBC.2018.0528).

Keywords

≪Faults≫, ≪Protection Device≫, ≪Safety≫, ≪LVDC≫

Abstract

Over the past years, the scientific interest in Low Voltage DC grids as an alternative to traditional LVAC grids has been growing steadily. This is caused by the fact that the amount of renewable energy sources and DC compatible loads in the grid has been increasing significantly. Furthermore, LVDC grids offer a higher efficiency and transmission capacity compared to their AC equivalent. However, the protection of these grids remains a major challenge for their breakthrough. This paper analyses the fault behaviour of an LVDC grid and proposes a prototype for a pole-to-ground protection device. To start with, an overview of the challenges for the protection against the different fault types in a converter-fed LVDC microgrid is presented. Subsequently, a PSCAD model of an LVDC grid for fault studies is discussed and the simulation results are analysed. Then, a test setup is developed to investigate the fault behaviour and the experimental results are presented and compared to the simulations. Finally, a prototype for pole-to-ground fault protection is proposed and its working principle is experimentally verified.

Introduction

In recent years, there has been an increasing interest in Low Voltage DC (LVDC) grids as an alternative to traditional LVAC grids. This is driven mainly by their higher efficiency and transmission capacity, along with their compatibility with the growing amount of renewable energy sources and DC loads (e.g. batteries, ICT, LED lighting) in the grid [1, 2]. However, fast and selective fault protection in LVDC grids still proves to be one of their major challenges [2, 3].

In essence, two important types of fault situations can occur in an electrical grid: pole-to-pole (P2P) or pole-to-neutral (P2N) faults on the one hand, where the main challenge is to safeguard the stability and availability of the system and protect its components, and pole-to-ground (P2G) faults on the other hand, where the focus shifts to protecting the people in the surroundings against electrocution. As this paper focusses on a TN-S grounding configuration, electrically speaking a P2G fault is comparable to a P2N fault. Physically speaking, however, a P2G fault causes a part of the installation that is not meant to be energised to become live, for instance the casing of a device, while this hazard is not present during a

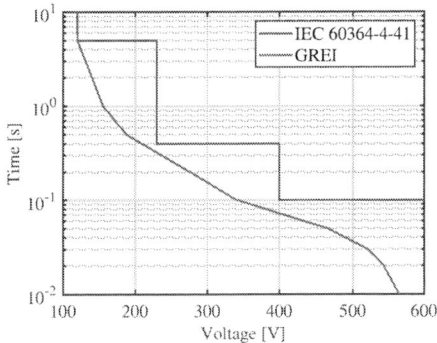

Fig. 1: Electric shock disconnection times imposed by IEC 60364-4-41 and the Belgian GREI

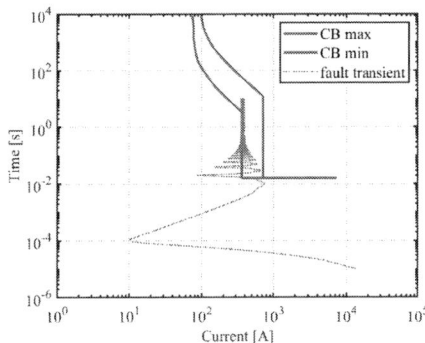

Fig. 2: Turn-off characteristic of a DC MCCB vs. a typical fault transient

P2N fault. For this reason, different boundary conditions and requirements apply for protection against these faults.

For the protection of people against a P2G fault, the required disconnection times are specified by IEC 60364-4-41 and displayed in Figure 1 [4]. In Belgium, more restrictive disconnection times are imposed by the General Regulations on Electrical Installations (GREI), as shown in Figure 1. Additionally, each electrical installation involving non-authorized personnel is required to be protected by a Residual Current Device (RCD), regardless of the grounding configuration [5]. As described in [6], finding an equivalent DC RCD is not straightforward, as no such devices are commercially available or standardized.

The challenges for P2P and P2N protection are of a more technical nature. Figure 2 shows a comparison of the turn-off characteristic of a DC Molded Case Circuit Breaker (MCCB) with a simulation of a P2N fault transient in an LVDC grid. The fault transient will be further discussed in Section II, but this figure already allows to illustrate the challenges for LVDC short circuit protection due to the specific behaviour of short circuit transients in LVDC grids. From the figure, it is clear that the high fault current peak occurs in a time range several orders of magnitude below the specified reaction time range of the protection device. As a result, it is uncertain whether and how fast the protection device will react to this current peak. On the one hand, if the device does not react to the first peak of the fault current, the DC bus voltage will collapse, which will have a negative impact on the availability of the grid. Furthermore, the steady state fault current is limited by the converter and, as a result, rather low compared to the turn-off characteristic of the CB. This poses a risk that, in some cases, a protection device might not react to a steady state short circuit current either, as shown in [7]. On the other hand, in case the protection device does react to the first fault current peak, it will be required to interrupt a very high DC current in which per definition no zero-crossing is present, and hence precautions should be taken to extinguish the arc that is formed during the breaking process [6].

Section II of this paper focusses on a simulation model of an LVDC grid for fault studies and presents simulation results of the fault types described above. Subsequently, Section III describes the test setup to perform these fault experiments in a lab environment and presents the experimental results. Finally, Section IV proposes a prototype protection device for P2G faults to protect people against electrocution by indirect contact, and shows that the device is capable to isolate the fault and bring the voltage down to a safe level within the required time limits. Section V presents the conclusion of this paper.

Fault simulations of a unipolar LVDC grid

In order to investigate the behaviour of an LVDC grid during fault conditions, the simple, unipolar LVDC grid shown in Figure 3 is modelled in PSCAD. Table I summarises the characteristics of the simulated grid. The source is modelled as an AC/DC full-bridge inverter connected to a strong AC grid. The load is modelled as a DC/DC buck converter that draws a load current. All of the converters are controlled using classic PI controllers and PWM techniques at a switching frequency of $f_s = 10\text{kHz}$. In order to control the DC bus voltage to $V_{dc} = 375\text{V}$ a cascaded voltage- and current control algorithm is applied to the AC/DC converter. The load current drawn by the DC/DC converter is controlled to $i_{dc} = 10\text{A}$ by a simple current control algorithm. The cables are modelled by their cable resistance, inductance and capacitance according to equations (1) to (3) from [8].

Figure 4 shows the fault current and DC pole-to-neutral voltage in case of a P2N fault. Figure 5 shows the fault current and the DC bus voltage for a P2G fault in an LVDC grid with a TN-S grounding configuration. When analysing the fault current, its waveform can be split into two major parts: (1) A first, very high and very fast peak caused by the discharge of the capacitance in the grid and (2) the sustained short circuit current fed by the sources in the grid (in this case the AC grid). As described in [9, 10], during the first part of the capacitor discharge, the DC bus voltage remains higher than the maximum AC voltage and the freewheeling diodes of the AC/DC converter remain in their blocked state. During the transition from the capacitor discharge phase to the sustained short circuit current phase, the DC voltage drops below the maximum AC voltage, allowing the diodes to start conducting and the AC inductors to discharge through the diodes. Finally, in the sustained short circuit current phase, the diodes of the converter work as an uncontrolled AC/DC converter and the AC grid feeds the fault.

$$R_c = \frac{\rho_{cu} \cdot l}{A_{eff}} \ [\Omega] \tag{1}$$

$$L_c = \frac{\mu_0 \cdot l}{2\pi \log(\frac{D}{GMR})} \ [\text{H}] \tag{2}$$

$$C_c = \frac{2\pi\varepsilon \cdot l}{\log(\frac{D}{r})} \ [\text{F}] \tag{3}$$

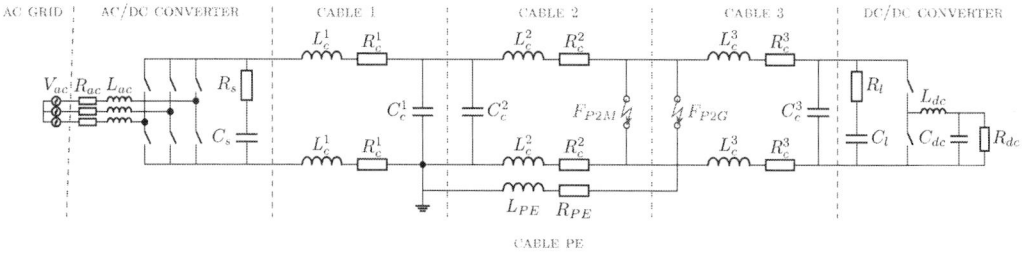

Fig. 3: LVDC grid model for fault studies

Table I: Simulation parameters of the modelled LVDC grid

AC GRID		DC LOAD	
V_{ac}^{LL}	230 V	V_{load}	100 V
L_{ac}	1.5 mH	L_{dc}	1.5 mH
R_{ac}	3 mΩ	R_{dc}	3 mΩ
CABLE 1		CABLE 3	
L_c^1	12.3 µH	L_c^3	1.36 µH
R_c^1	34.7 mΩ	R_c^3	18.8 mΩ
$C_c^{1,2}$	13.73 nF	C_c^3	0.814 nF
CABLE 2		CABLE PE	
L_c^2	13.3 µH	L_{PE}	12.8 µH
R_c^2	55.2 mΩ	R_{PE}	44.9 mΩ
CONTROL PARAMETERS		CONVERTER FRONT-ENDS	
$K_p^{i,ac}$	1.777	R_s	1 mΩ
$K_i^{i,ac}$	520.483	R_l	1 mΩ
$K_p^{v,ac}$	0.132	C_s	1 mF
$K_i^{v,ac}$	5.938	C_l	560 µF
$K_p^{i,dc}$	0.088	f_s	10 kHz
$K_i^{i,dc}$	25.767	V_{dc}	375 V

A more detailed plot of the first fault current peak has been included in the upper right corner of Figures Fig. 4a and 5a for clarity. In both fault current waveforms, a 50 Hz oscillation from the AC grid and the switching transients from the converter are can be noticed in the sustained short circuit current. Furthermore, when comparing the two fault types it seems that, although the sustained short circuit current behaves similar in both cases, the first current peak of the P2N fault is significantly faster and higher than that of the P2G fault, which in turn also causes the voltage to collapse faster during the P2N fault. This is caused by the fact that the main contributor to the first peak is the load capacitor C_l, for which the discharge return path is significantly longer for the P2G fault (PE cable – cable 2 – cable 3) than for the P2M fault (cable 3), resulting in a slower and more dampened fault current waveform. The main contributor to the sustained short circuit current, however, is the AC grid, for which the impedance to both faults is approximately equal.

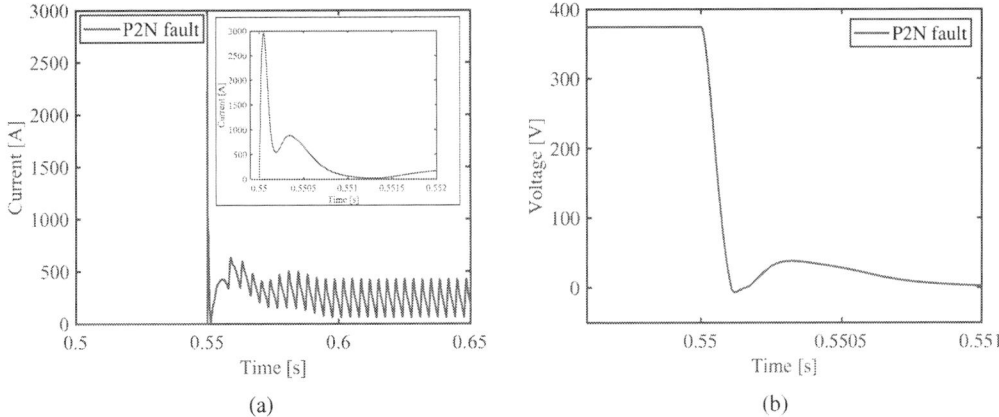

Fig. 4: Modelled pole-to-neutral fault current (a) and DC bus voltage (b)

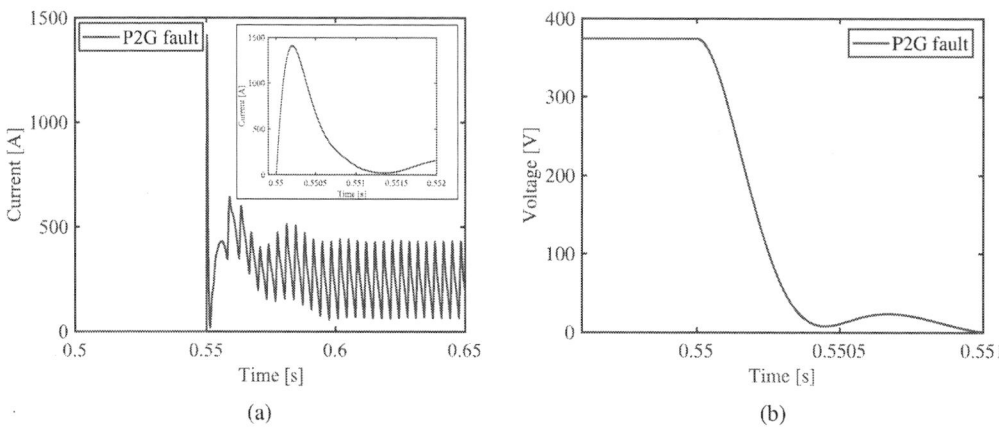

Fig. 5: Modelled pole-to-ground fault current (a) and DC bus voltage (b)

Fault experiments in a simple LVDC grid

In order to validate the simulation results presented above, the test setup shown in Figure 6 was developed to perform the fault experiments. Both for the source and the load, bidirectional Delta Elektronika power supplies were used. A fault emulator device was developed in order to perform the fault experiments safely, reliably and reproducibly. Prototypes of a pole-to-ground protection device (RCD 1 and 2) were integrated in the test setup and will be discussed in the Section IV. Two types of faults are created: a P2N fault F_{P2N} and a P2G fault F_{P2G}, with a fault resistance of approximately 40 mΩ. The test conditions are summarized in Table II. Two P2N fault experiments and two P2G fault experiments were performed, labelled experiment 1 & 2 and experiment 3 & 4 respectively. The measured fault currents are shown in Figures 7 to 9. The faults were also simulated for the specific test conditions, presented in Table II, and are shown on the figures for comparison. The main goal of the simulations is to give an estimate of behaviour of the grid in fault conditions, but the cable and converter models are not sufficiently detailed to obtain an exact prediction of the fault current waveforms that are observed in reality.

Figure 7a shows the fault current peak of a P2N fault initiated at $t = 0$ at a DC bus voltage of 375 V. The figure shows a relatively good correspondence between the measured current and the simulation, except for the very first 60 μs. The simulations show a regular, continuous capacitive discharge peak at the initiation of the fault, with a higher peak than observed in the experiment, while in reality the closing of the circuit breaker initiating the fault causes a more erratic transient as shown in Figure 7b. This may be caused by the fact that the model applies an ideal switch to initiate the fault, while in reality the mechanical switch has a measurable resistance and a closing time during which an arc could briefly be created. In other words, in reality some of the energy is dissipated in the closing process of the

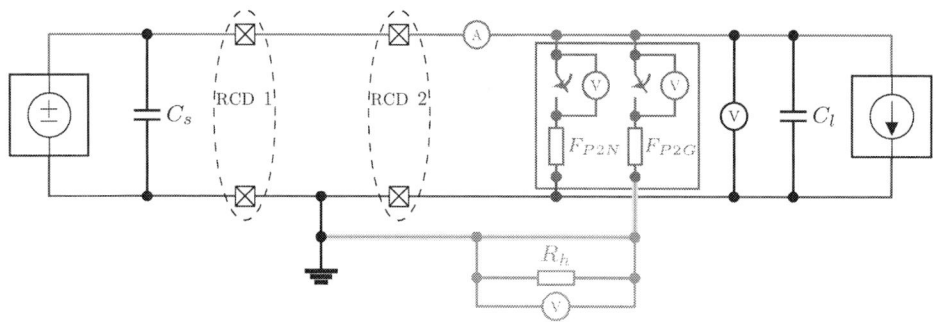

Fig. 6: Schematic of the test setup

Table II: Test conditions of the four experiments

	EXP 1	EXP 2	EXP 3	EXP 4
Fault	F_1	F_1	F_2	F_2
V_{dc}	375 V	250 V	250 V	250 V
I_l	2 A	2 A	0 A	0 A
C_s	1 mF	1 mF	1 mF	1 mF
C_l	560 μF	560 μF	560 μF	560 μF
R_h	–	–	1.2 kΩ	1.2 kΩ

switch, while in simulation all of the energy is still available when the fault is initiated, which causes a higher capacitive discharge peak, and hence an overestimation of the fault current peak, in the simulation results. For the development and testing of protection devices as presented in Section IV, however, this does not pose a problem as it adds some extra safety margin. Another phenomenon that can be observed in Figure 7a is a cut off of the fault current around 2.34 ms. This is a result of the lab grid protection system reacting to the high current peak. In order not to disturb the measurements by a tripping circuit breaker, experiments 2 to 4 were performed at a lower DC bus voltage of 250 V, as shown in Table II.

Figure 8a shows the fault current of experiment 2. It is clear that, because of the lower voltage, the current peak is lower and the measurement looks more noisy. Therefore, Figure 8b shows the same measurement filtered with a low-pass filter, which clarifies the waveform shape but also filters out some of the high frequent peaks. The same deviation between the simulated and measured current peak as in experiment 1 can be observed. Furthermore, a fast oscillation can be seen in the measurement after the first peak, which is not present in the simulation. This can be attributed to the fact that applied Delta sources probably do not behave exactly the same way as the simple AC/DC converter that was used to model the source. Finally, Figure 9 compares the fault current of the P2G experiments 3 and 4 to their simulation. There is a good similarity between the waveform shapes, although the model seems to underestimate the size of the current peak. This is likely due to an overestimation of the PE cable parameters. The experiments also show less difference between a P2N fault current and a P2G fault current than was observed in the simulations.

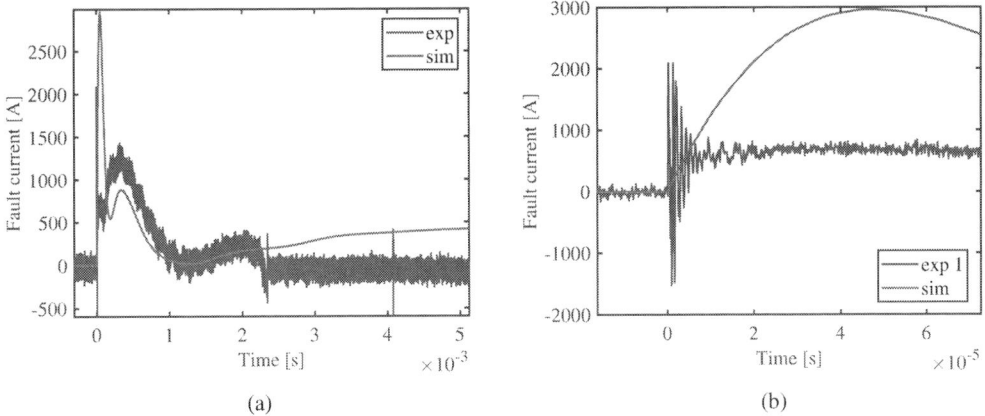

Fig. 7: The fault current in experiment 1 (a) and a detailed plot of the first 60 μs of the same waveform (b)

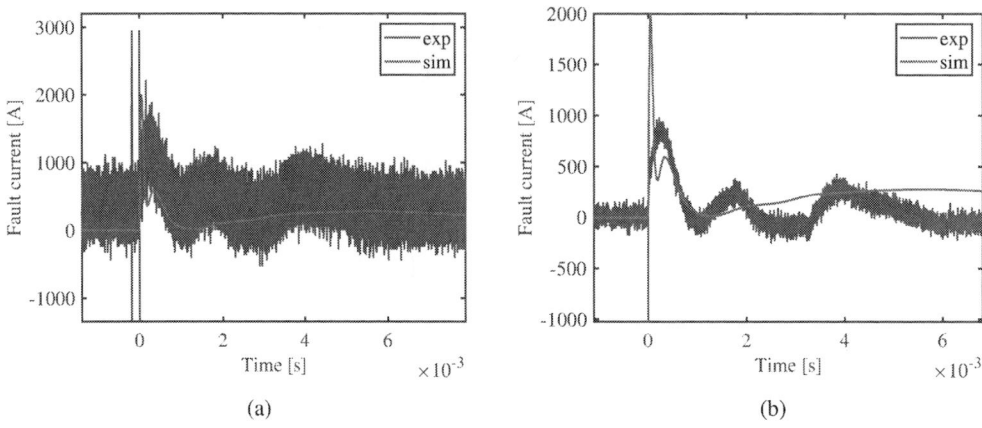

Fig. 8: The fault current in experiment 2 (a) and its filtered waveform (b)

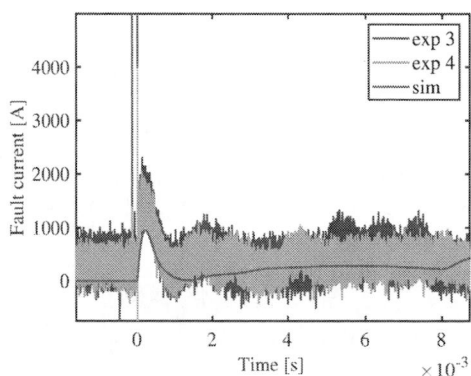

Fig. 9: The fault current in experiment 3 and 4

Prototype of a pole-to-ground protection device

The final part of this paper presents a prototype of a protection device against pole-to-ground faults in LVDC grids, based on commercially available components. The principle of the device is illustrated in Figure 10 [11]. It consists of a Programmable Logic Controller (PLC), a Bender Residual Current Monitor (RCM), which continuously monitors the sum of the current in the three wires of a bipolar grid, and an ABB DC MCCB per phase. In case the sum of the currents is not zero, the RCM flags a fault to the controller, which in turn sends a tripping signal to the DC MCCBs in each line. The setup of Figure 6 (fault F_{P2G}, experiments 3 & 4) is used to experimentally verify the functionality and speed of the prototype: A fault is initiated between the positive pole and the ground, which is connected to the neutral pole according to a TN-S grounding configuration. As the lower part of RCD 2 is bypassed by the fault current, which flows through the protective earth connection instead, its RCM is expected to measure a non-zero sum of the currents and flag a fault. In practice, the P2G fault with a low fault impedance corresponds to an indirect touch fault, i.e., a fault occurs that puts a part of the installation which is not supposed to be energised under a dangerous touch voltage, and an earthed person accidentally touches the energised part. A resistor of $1.2\,\mathrm{k\Omega}$ is placed in parallel with the fault path to emulate the human touching the faulty device. Because of practical reasons, the resistor was put in parallel with the PE conductor, which means that the earth resistance (estimated between 5–$10\,\Omega$) was neglected. As explained in the introduction, an RCD is installed to protect a person against electrocution specifically in this situation.

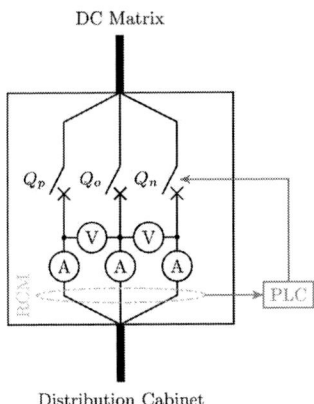

Fig. 10: Schematic diagram of the RCD prototype [11]

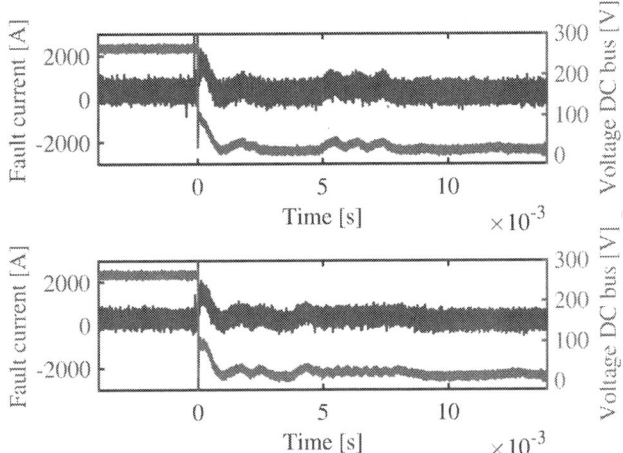

Fig. 11: The fault current and the DC bus voltage a P2G fault, experiment 3 (top) and 4 (bottom)

As described in [12], the touch voltage a person could experience during an indirect touch fault is half of the DC bus voltage, in this case 125 V. According to the IEC limits described above, this corresponds to a turn-off time of maximum 5 s. The Belgian GREI is more strict and requires a turn-off time of 4.43 s. Figure 11 shows the DC bus voltage and fault current for experiments 3 and 4. The results show that the RCD prototype proposed in this paper is able to react to the fault and bring down the DC bus voltage in approximately 10 ms. This is well within the limits defined by IEC norms and the Belgian GREI to assure people in this environment are protected against electrocution hazards by indirect touch. As discussed in Section III, the experiments conducted at the target voltage of 375 V were interrupted by the lab overcurrent protection system and, as a result, no conclusive reaction time of the RCD could be measured, as, although the RCD reacted to the fault, it was not clear which of the two protection devices reacted first. Therefore, the P2G fault experiments were repeated with a high fault impedance of 1.2 kΩ in experiment 5 and 6 (avoiding a high overcurrent), of which the test conditions are shown in Table III. In practice, a P2G fault with a high fault impedance can be compared to a direct touch fault, i.e. a person directly touches a live part of the installation, like an uninsulated conductor. The main protection measure against direct touch faults is to make live parts of the installation unreachable by barriers, casings and insulation of the components, although IEC 60364-4-41 specifies that an RCD of a low current rating may be used as an additional protection measure against this type of fault [4].

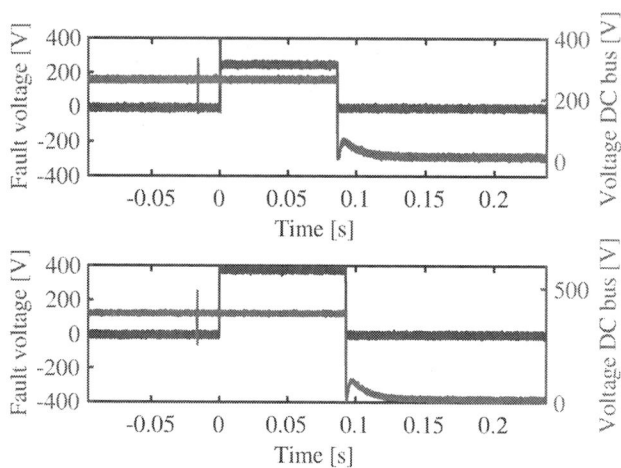

Fig. 12: The fault current and the DC bus voltage a P2G fault with high impedance, experiment 5 (top) and 6 (bottom)

Table III: Test conditions of experiment 5 and 6

	EXP 5	EXP 6
Fault	F_{P2G}	F_{P2G}
V_{dc}	250 V	375 V
I_l	0 A	0 A
C_s	1 mF	1 mF
C_l	560 μF	560 μF
R_{fault}	1.2 kΩ	1.2 kΩ

In case of a direct touch fault, the touch voltage equals the total DC bus voltage, which results into a turn-off time of 0.4 s according to the IEC norm (both for 250 V and 375 V) and 0.335 s and 0.086 s according to the GREI for 250 V and 375 V, respectively. The resulting DC bus voltage and the voltage over the fault impedance in experiment 5 and 6 are shown in Figure 12. The fault is initiated at $t = 0$ s and the RCD protection reacts after approximately 0.0866 s and 0.093 s. In case of experiment 5 at 250 V this reaction time is well within the limits of both the IEC norm and the GREI. In experiment 6, the reaction time of the protection is within the limits of the IEC norm but the device reacts just too slow according to the GREI. However, as mentioned above and shown in [12], the touch voltage in case of an indirect touch fault, which is the primary safety hazard an RCD should protect against, is only half of the DC bus voltage, resulting in a touch voltage of 187.5 V. This corresponds a turn-off time of 5 s and 0.5 s imposed by the IEC norm and the Belgian regulations respectively, which is well above the measured turn-off times in experiments 5 and 6. Furthermore, the time until turn-off is comparable for both experiments. In other words, it appears that the DC bus voltage magnitude only has a minor influence on the turn-off time of the RCD prototype. The high fault impedance, on the other hand, does seem to have a major influence on the turn-off time, as it causes the fault current to be much smaller than in experiment 3 and 4, which causes the RCD protection devices to react slower in comparison. As a result, it is expected that, in case of a low impedance P2G fault such as an indirect touch fault at a DC bus voltage of 375 V, the protection devices will be able to react in a time period of the same order of magnitude as experiment 3 and 4 and well within the time specified by international and Belgian regulations.

Conclusion

This paper studied the fault behaviour of a simple LVDC grid using simulations and experiments and proposed a prototype for a pole-to-ground protection device. Firstly, different types of faults that can occur and their specific challenges for the protection system were discussed. A distinction was made between the protection of the grid and its components against pole-to-pole or pole-to-neutral faults and the protection of people against pole-to-ground faults. Secondly, a model for the simulation of these faults in PSCAD was presented. The fault current and DC bus voltage collapse of a pole-to-neutral and a pole-to-ground fault were presented and compared. Thirdly, a test setup to perform these same faults experimentally was introduced. The experimental results of two pole-to-neutral and two pole-to-ground experiments were presented and compared to their simulations. Finally, this paper proposed a DC pole-to-ground protection device that could serve as a DC version of the traditional AC Residual Current Device. The device was built using commercially available and certified components. Its main purpose is to protect people in the surroundings of the DC grid against electrocution by indirect contact. The experiments presented in this paper show that the developed prototype is able to react to a pole-to-ground fault with low impedance, which corresponds to an indirect touch fault, and bring the DC bus voltage down to a safe level within the time limits imposed by the international and Belgian safety norms. They also show a significant difference in reaction time of the prototype to an indirect touch fault (with low impedance) and a direct touch fault (with high impedance), and, as a result, it remains uncertain whether the proposed prototype could also be used as additional protection against direct contact.

References

[1] D. Wang, A. Emhemed, G. Burt, and P. Norman, "Fault analysis of an active lvdc distribution network for utility applications," in *2016 51st International Universities Power Engineering Conference (UPEC)*, 2016, pp. 1–6.

[2] L. Mackay, T. Hailu, L. Ramirez-Elizondo, and P. Bauer, "Towards a DC distribution system - opportunities and challenges," in *2015 IEEE First International Conference on DC Microgrids (ICDCM)*, 2015, pp. 215–220.

[3] L. Zhang, N. Tai, W. Huang, J. Liu, and Y. Wang, "A review on protection of DC microgrids," *Journal of Modern Power Systems and Clean Energy*, vol. 6, no. 6, pp. 1113–1127, 2018.

[4] IEC60364-4-41, *Low-voltage electrical installations – Protection for safety – Protection against electric shock*.

[5] *AREI: Algemeen Reglement op de Elektrische Installaties.* Belgisch Staatsblad, art. 31, 86, 87.

[6] A. A. S. Emhemed, K. Fong, S. Fletcher, and G. M. Burt, "Validation of fast and selective protection scheme for an LVDC distribution network," *IEEE Transactions on Power Delivery*, vol. 32, no. 3, pp. 1432–1440, 2017.

[7] D. Wang, A. Emhemed, P. Norman, and G. Burt, "Evaluation of existing DC protection solutions on an active LVDC distribution network under different fault conditions," *CIRED - Open Access Proceedings Journal*, vol. 2017, no. 1, pp. 1112–1116, 2017.

[8] L. L. Grigsby, *Electric power generation, transmission, and distribution.* CRC press, 2016.

[9] S. Xue, C. Chen, Y. Jin, Y. Li, B. Li, and Y. Wang, "Protection for DC distribution system with distributed generator," *Journal of Applied Mathematics*, vol. 2014, 2014, Article ID 241070, 12 pages.

[10] L. Hallemans, G. Van den Broeck, S. Ravyts, M. Alam, M. Dalla Vecchia, P. Van Tichelen, and J. Driesen, "Fault identification and interruption methods in low voltage DC grids – a review," in *2019 3rd International Conference on DC Microgrids (ICDCM)*, 2019.

[11] G. V. den Broeck, "Voltage control of bipolar DC distribution sysems."

[12] S. Ravyts, M. Dalla Vecchia, G. Van den Broeck, L. Hallemans, K. Stul, and J. Driesen, "Earth fault analysis and safety recommendations for BIPV module-level converters in low-voltage DC microgrids," in *2019 3rd International Conference on DC Microgrids (ICDCM)*, 2019.

Design of a Dual Active Bridge Converter for On-Board Vehicle Chargers using GaN and into Transformer Integrated Series Inductance

K. Siebke, M. Giacomazzo, R. Mallwitz
TU Braunschweig, Institute for Electrical Machines, Traction and Drives
Hans-Sommer-Str. 66
38106 Braunschweig, Germany
Phone: +49 (0) 531-391 3910
Fax: +49 (0) 531-5767
Email: k.siebke@tu-bs.de
URL: http://www.imab.de

Keywords

Power converters for EV, Battery charger, Gallium Nitride (GaN), ZVS converters, Resonant converter.

Abstract

In this paper a DAB using GaN and an into the transformer integrated series inductance are designed for an on-board vehicle charger. The DAB design and a GaN full bridge design are described. The parasitic drain source capacitance is measured and considered in the DAB design to determine the zero voltage switching (ZVS) boundaries. Three different transformer setups are compared and evaluated in praxis. Finally a map of losses and efficiency is created using once single phase shift (SPS) modulation and once triple phase shift (TPS) modulation.

Introduction

In this paper the dc-dc converter of a bidirectional charger for electric vehicles is designed. The requirement of a bidirectional energy flow reduces the number of possible topologies. Two very common topologies are the CLLC converter (figure 1) and the DAB (figure 2). Both can be built up as a half bridge variant or as a full bridge. In this application only the full bridge converters are considered. The full bridge topologies are at 3.7 kW more efficient than the half bridge converters due to the smaller semiconductor currents and therefore smaller conduction losses [1],[2].

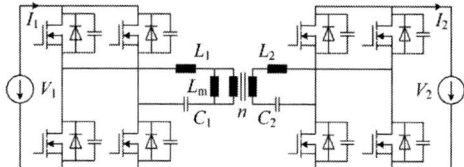
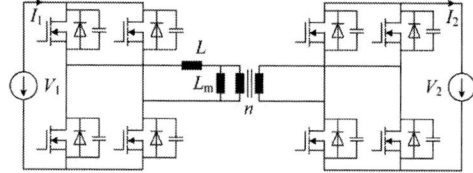

Fig. 1: Topology of CLLC converter   Fig. 2: Topology of DAB converter

The CLLC converter is mostly controlled by the switching frequency while the DAB is controlled in e. g. single phase shift (SPS) or triple phase shift (TPS). Both DAB and CLLC converter have zero voltage switching (ZVS) capabilities. While the DAB leaves the ZVS region in partial load condition the ZVS capability of the CLLC converter is even in partial load condition available. The presence of parasitic capacitance of the switches will restrict the ZVS capability in both topologies [3], [4]. In [3] two CLLC converters using GaN Systems HEMTs are compared. One CLLC converter is designed with a low resonance frequency and one with a high resonance frequency. Due to the high parasitic capacitance of the

HEMT the ZVS capability is lost in the high frequency design and in high switching frequency operating points. The low frequency design seems more suitable but the low resonance frequency leads to more voluminous passive components. The resonance inductance cannot be integrated into the transformer.

The optimum operating point of the CLLC converter is at the resonance frequency. Away from this operating point the circulating currents will increase and this will increase the conduction losses. Further in light load condition and low output voltages the switching frequency of the CLLC converter has to be increased but limited to a maximum. Due to that the output power of the CLLC cannot fall below a minimum value. To reduce switching losses of the secondary side switches a synchronous rectification should be implemented. Especially in case of GaN HEMTs which have a very high reverse conduction voltage (approx. 4 V) the synchronous rectification is necessary for full load operation [5]. But the synchronous rectification needs a lot of effort in circuit and sensors.

As a consequence of the above mentioned drawbacks of the CLLC converter in this paper a DAB is investigated. The switching frequency is selected to 500 kHz to reduce the size of the series inductance L and to allow the integration of L into the transformer as leakage inductance. In [6] a very similar DAB for vehicle charger is investigated with the focus on one transformer design. This paper will describe the DAB component design including the selection of L and GaN half bridge design. Afterwards 3 different transformer designs with integrated series inductance are proposed and analyzed in detail. The three transformers are built up and measured in praxis. Using the transformer with the lowest losses a map of losses and efficiency are measured for the DAB using once SPS modulation and once TPS modulation.

Design of DAB

The requirements of the DAB are V_1=400 V, V_2=250 V...450 V and maximum output power of P_{max}= 3.7 kW. During the design of the DAB the ZVS regions and the circulating current has to be considered. An advantage of the DAB is the soft switching capability [7]. ZVS should be ensured during the most operating points in the charging process in order to exploit a higher efficiency region. The ZVS region is effected by the parasitic capacitance C_p of the switches. As described in [4] the ZVS region is reduced significantly by the parasitic capacitance of the switches. The reduced ZVS region with consideration of C_p can be determined using the equations from [4]. To increase the ZVS region the series inductance can be increased but this causes higher circulating currents and the conduction losses will increase [6]. Therefore the value of the parasitic output capacitance C_p has a significant impact to the DAB design and this value has to be determined before starting the DAB design process. In the following subsections first the design of the GaN full bridge is described. Subsequently the parasitic output capacitance C_p is measured in the designed GaN full bridge. Finally the DAB is designed taking the parasitic capacitance C_p into account.

GaN full bridge design and parasitic output capacitance determination

For the practical realization the GaN device GS66516B (GaN Systems) is chosen. It is a 650 V, 60 A HEMT. To reduce the inductance of the commutation mesh the layout of the GaN half bridge is optimized using an eight layer PCB shown in figure 3. The eight layer PCB allows opposite current directions in adjacent layers which reduces the inductance. The blue tracks in figure 3 are the positive or negative dc link connections to the dc link (CeraLink) capacitor in the middle of the two GaN HEMTs. The red tracks are the connections between source of the high side HEMT and drain of the low side HEMT. All tracks are as flat as possible to reduce the inductance and spread the heat of the HEMTs. The heat is conducted by a high amount of vias through the PCB from the HEMT on the top side to a heat sink on the bottom side.

The whole DAB is realized using a rapid prototyping platform. Each component (GaN full bridges, transformer, dc link, output filter, controller) is realized as a separate module which can be adapted or changed by another one rapidly. In figure 4 one of the two GaN full bridges is shown. One of these modules are used for each primary and secondary side. Figure 5 shows the whole rapid prototyping environment.

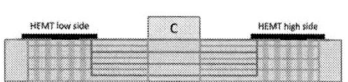

Fig. 3: Cross section of the GaN half bridge

Fig. 4: GaN full bridge module of DAB

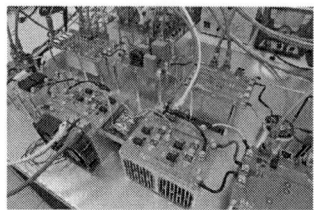

Fig. 5: Rapid prototyping environment

To measure the parasitic output capacitance C_p a measurement setup as shown in figure 6 is used. Both diagonals (M_1, M_4 and M_2, M_3) are driven by a gate signal with 50 % duty cycle and 100 kHz while the two diagonals have a phase shift of 180°. The dead time is set to 80 ns. The current and voltage waveforms are shown in figure 7. With the measured input current I_{DC} the following equation can be defined:

$$P_M = C_p V_1^2 f_s = \frac{V_1 I_{DC}}{4} \qquad (1)$$

P_M is the power loss of one HEMT, f_s is the switching frequency and V_1 is the input voltage which was changed from 0 V to 0 V. The parasitic output capacitance can be obtained rearranging equation 1:

$$C_p = \frac{I_{DC}}{4V_1 f_s} \qquad (2)$$

The results are shown in figure 8. With increasing voltage V_1 the capacitance C_p decreases. At 400 V the parasitic output capacitance is 650 pF. For comparison in figure 9 the C_{oss} is depicted as a function of the drain source voltage V_{DS} obtained from the datasheet. Both capacitance characteristics are decreasing with increasing voltage but the measured capacitance is at 400 V with 650 pF much higher than obtained from the datasheet (130 pF). These differences are due to additional parasitic capacitances caused by the 8-layer PCB with the very flat tracks of the GaN full bridge and the components of the driver circuity.

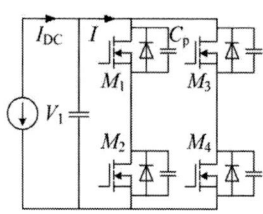

Fig. 6: Measurement of parasitic capacitance C_p

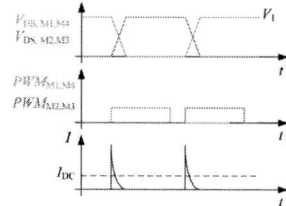

Fig. 7: Voltage and current waveforms during measurement of parasitic capacitance C_p

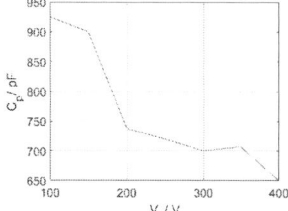

Fig. 8: In application measured capacitance characteristic

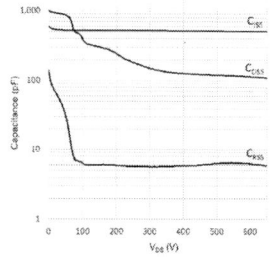

Fig. 9: Capacitance characteristic of the GS66516B

Optimization of the DAB under consideration of parasitic C_p

To minimize the current stress in the DAB using singe phase shift (SPS) in [8] an optimization method is proposed which is used in this paper. The result of the current stress optimization is a series inductance of $L=7.3\,\mu$H and a dead time of 80 ns. The design parameters of the DAB are summarized in table I. In figure 10 the ZVS regions are considered. Below the blue line the ZVS capability of the secondary side switches is lost while on the primary side the ZVS capability is lost above the red line. Due to the high value of parasitic capacitance of the GaN devices the ZVS region is very small. To increase the ZVS region, the inductance can be increased but this will lead to higher circulating currents and with increasing value of inductance it becomes more difficult to integrate the series inductance as leakage inductance in the transformer. Also in figure 10 the operating point used in the charger are depicted once as constant current (CC) and once as constant power (CP) charging line. The shown CC and CP charging lines are identical for voltages above 370 V due to the maximum power of 3.7 kW.

To increase the ZVS area and shift the ZVS boundaries to lower output currents the triple phase shift (TPS) modulation can be applied. In [9] a general analysis of TPS modulation is described and the optimum switching modes to reduce RMS current of the inductor and to achieve a wide range of ZVS or zero current switching (ZCS) are proposed. The ZVS and ZCS conditions are defined under ideal conditions with no parasitic drain source capacitance and no dead time. [10] considers the parasitic drain source capacitance and adapts the theoretical ZVS and ZCS boundaries. However the dead time is neglected in the adapted boundaries. The dead time is considered in [11] but the parasitic capacitance is neglected. Both dead time and parasitic capacitance have an effect to the DAB ZVS boundaries. In summary it is difficult to determine the ZVS boundaries considering both dead time and parasitic capacitances. In this paper the ZVS boundaries using TPS are determined in the practical experiment using the rapid prototyping platform.

Design Parameter	value
P_{max}	3.7 kW
$I_{2,max}$	10 A
V_1	400 V
V_2	250 V...450 V
L	7.3 μH
n	1
$C_{1,2}$	650 pF
f_s	500 kHz
t_{dead}	80 ns

Table I: DAB Design

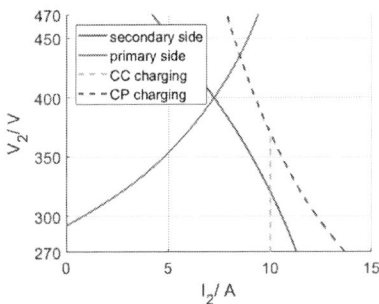

Fig. 10: ZVS boundaries of the DAB design and different charging methods

Design of transformer with integrated series inductance

The low switching losses of GaN semiconductor devices allow high switching frequencies and small sizes of passive components. To profit most from the high switching frequencies in this DAB concerning volume and weight a transformer with integrated series inductance will be designed. The series inductance will be formed by the leakage inductance of the transformer. Due to that the design of the transformer is most important. In the chapter above a series inductance of 7.3 μH is chosen which has to be integrated into the transformer. Due to the high value of series inductance it is quite challenging to integrate this inductance as leakage inductance. In this paper three different opportunities of transformer designs are investigated (figure 11). The version 1 of transformer design is shown in figure 11a which depicts the half transformer including the winding configuration. Primary (green) and secondary (blue) winding are placed with a defined distance on top of each other. A 450 strands, 0.05 mm litzwire is used. Version 2 is depicted in figure 11b with a horizontal winding configuration and interleaved configuration for the center windings. For this configuration a 2250, 0.05 mm litzwire is used. Figure 11c shows the third version which is a complete symmetric winding design. In version 3 two litzwires of 450 strands, 0.05 mm are connected in parallel. The core used in version one is a PQ40/40 core, in version 2 and 3 a PQ50/50 core is used. The core material is always 3F36 (Ferroxcube). The built up transformers are shown in figure 12.

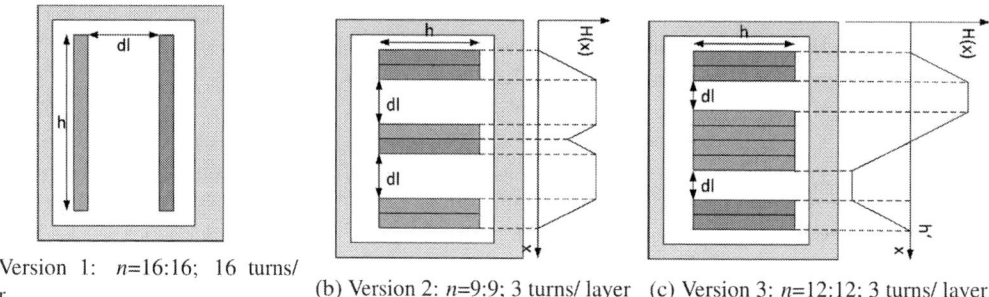

(a) Version 1: $n=16{:}16$; 16 turns/layer

(b) Version 2: $n=9{:}9$; 3 turns/ layer

(c) Version 3: $n=12{:}12$; 3 turns/ layer

Fig. 11: Realizations of different transformer versions

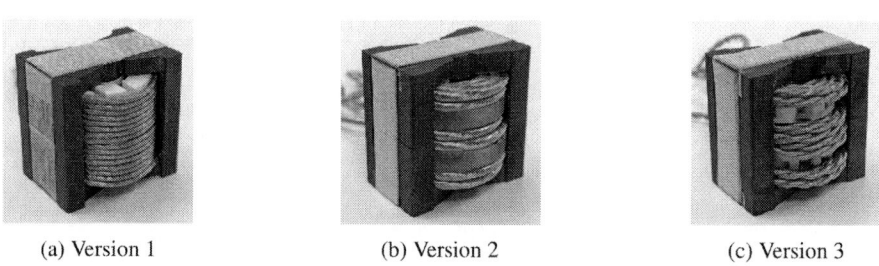

(a) Version 1

(b) Version 2

(c) Version 3

Fig. 12: Realizations of different transformer versions

The losses of each transformer version is analytically calculated as shown in equations 3 to 6. The copper losses P_{dc} in each winding are calculated with 3. The eddy current losses for each winding can be determined using equation 4 [12]. With the specific power loss of the core obtained from the core datasheet the core losses can be calculated using equation 5. Finally the overall losses are determined by equation 6.

n: Number of turns
l_w: Medium length of turns
N_s: Number of strands
A_s: Cross section of one strand
ρ: Specific copper resistance
d: Diameter of strands
T: Switching period
r_w: Radius of wire
P_v: Specific power loss of the core
A_e: Effective core Area
L_e: Effective core length
B: Constant factor depending on winding configuration: $B = 768$ for Version 1, $B = 192$ for Version 2,3

$$P_{dc} = \frac{\rho n l_w}{N_s A_s} I^2 \tag{3}$$

$$P_{eddy} = \frac{\mu_0^2 \pi n N_s l_w d^4}{B \rho r_w^2 T} \int_0^T (\frac{di}{dt})^2 dt \tag{4}$$

$$P_{core} = P_v A_e L_e \tag{5}$$

$$P_{total} = 2(P_{dc} + P_{eddy}) + P_{core} \tag{6}$$

The leakage inductance of the transformer is calculated using the volume integral of the magnetic field density in and between the layers. This energy is equal to the energy stored in the leakage inductance L

which is shown in equation 7.

$$E = \int_0^V \frac{1}{2}\mu_0 H(x)^2 dV = \frac{1}{2}\mu_0 l_{\mathrm{w}} h \int_0^{h'} H(x)^2 dx = \frac{1}{2}LI^2 \qquad (7)$$

The results of the loss calculations are summarized in table II. Also the calculated and measured values of the leakage inductance are shown. Every version reaches nearly the same series (leakage) inductance but very different losses. The total losses of version 3 are the smallest and therefore version 3 is selected as the final transformer.

	Version 1	Version 2	Version 3
L calculation	$6.7\,\mu$H	$7.2\,\mu$H	$7.2\,\mu$H
L measurement	$6.8\,\mu$H	$7.4\,\mu$H	$7\,\mu$H
L_{m} measurement	$655\,\mu$H	$292\,\mu$H	$494\,\mu$H
$2P_{\mathrm{dc}}$	$8.41\,$W	$1.21\,$W	$3.51\,$W
$2P_{\mathrm{eddy}}$	$6.93\,$W	$19.79\,$W	$22.99\,$W
P_{core}	$20.75\,$W	$40.44\,$W	$6.12\,$W
P_{total}	$36.09\,$W	$61.44\,$W	$32.62\,$W

Table II: Transformer losses

Measurement results

The DAB is built up in practice using transformer version 3. The losses and efficiency are measured in different operating points using SPS and TPS modulation. The results of the measurements are shown in figure 13. The total losses of the DAB using SPS modulation are shown in figure 13a. In the upper and lower left corner the ZVS capability of the DAB is lost and the semiconductor losses increase. In the upper right corner the maximum output power of the DAB is exceeded. Due to that no losses in the white regions were measured. The grey lines in figure 13a are the theoretical ZVS boundaries obtained from figure 10. As depicted the minimum losses occur in the operating point 7.7 A and 400 V with 40 W. Figure 13b shows the efficiency of the DAB in different operating points using SPS modulation. The highest efficiency is achieved in the operating point 7.6 A and 390 V with 98.6 %.

To increase the ZVS area and the efficiency in partial load operating points the TPS modulation is applied to the same hardware. The results are shown in the figures 13c and 13d. These figures are the results of an optimization. In TPS modulation each operating point, determined by V_2 and I_2, can be reached using different combinations of the three phase shift parameters and each combination leads to different losses for the same output voltage and current. The combination of the three phase shift parameters which lead to the minimum losses is selected and the losses and the efficiency is referenced in the figures 13c and 13d. In comparison to the SPS modulation the ZVS boundary is shifted to the left using TPS modulation. This allows the operation with lighter loads compared to the SPS modulation. The efficiency of the TPS modulation is slightly increased and the maximum is 98.74 % at 7.9 A and 400 V and the minimum losses are 40 W at 7.6 A and 400 V.

In order to evaluate the different transformer designs in praxis, the temperatures were measured in the operating point 9.25 A, 400 V, 3.7 kW. To ensure a thermally stable operation the duration of measurement is chosen to one hour for transformer version 2 and 3. All three transformers are cooled by the same and constant air flow. The temperature of version 1 reached 100 °C after 30 minutes and the measurement was finished to avoid damages. The thermal images are shown in 14. The maximum temperature of transformer version 2 is 87 °C and of version 3 66.5 °C after one hour. The maximum temperature is always located in the winding which may be due to the better cooling capability of the core compared to the winding. Further the integrated leakage inductance leads to copper losses. For transformer version

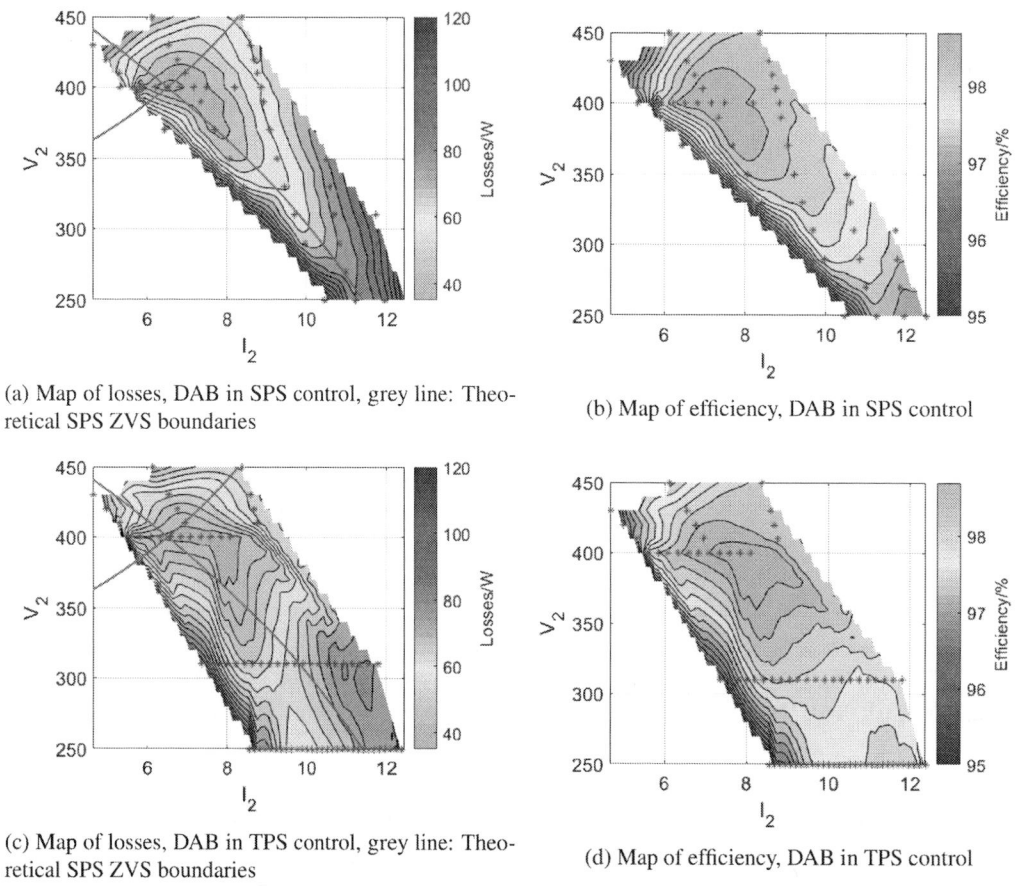

(a) Map of losses, DAB in SPS control, grey line: Theoretical SPS ZVS boundaries

(b) Map of efficiency, DAB in SPS control

(c) Map of losses, DAB in TPS control, grey line: Theoretical SPS ZVS boundaries

(d) Map of efficiency, DAB in TPS control

Fig. 13: Measurement results of DAB; Blue stars: Measurement points

2 high core losses are predicted which can be observed in the higher core temperature compared to the transformer version 3.

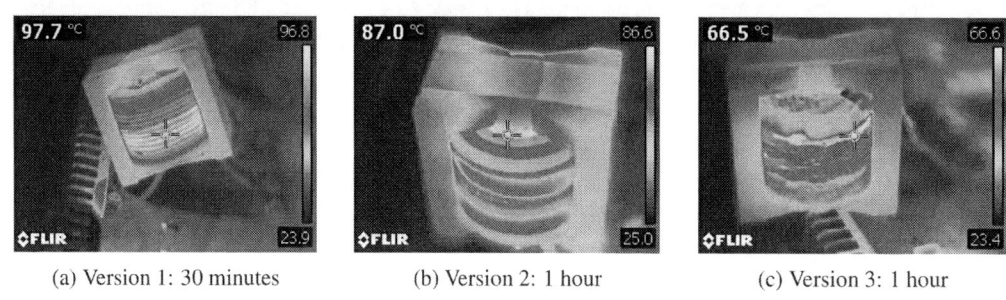

(a) Version 1: 30 minutes

(b) Version 2: 1 hour

(c) Version 3: 1 hour

Fig. 14: Temperature measurement of different transformer versions

Conclusion

In this paper a DAB for electric vehicle chargers is designed using GaN HEMTs and an into transformer integrated series inductance. The DAB topology is compared to the CLLC converter. As mentioned in [3] the parasitic capacitance of the GaN HEMT leads to the lost of ZVS capability in case of high resonant frequency (480 kHz) design and in high switching frequency operating points which are far away from the resonance frequency. To benefit most from GaN a DAB with a high switching frequency of 500 kHz is designed. The switching frequency of the DAB is limited to 500 kHz in comparison to the CLLC converter and ZVS is expected in operating points next to full load. The DAB component and hardware design is described and the parasitic drain-source capacitance is measured and considered in the DAB design. Afterwards 3 different transformer designs with integrated series inductance are proposed and analyzed to find the solution with the lowest losses and lowest temperature. The three transformers are built up and measured in praxis. The transformer version 3 offers the lowest losses and leads to the lowest temperature. The transformer version 3 is a complete symmetric winding design on a PQ50/50 core and 3F36 material. Using this transformer a map of losses and efficiency are measured for the DAB using once SPS and once TPS modulation. The peak efficiency of the DAB using TPS modulation is 98.74 % which is much better than comparable CLLC converter designs (96 %) [5].

References

[1] T. Schobre, K. Siebke, R. Mallwitz, "Design of a GaN based bidirectional CLLC converter with synchronous rectification," EPE 2019, Genoa

[2] P. He, A. Khaligh, "Comprehensive Analyses and Comparison of 1 kW Isolated DCDC Converters for Bidirectional EV Charging Systems IEEE Transactions on Transportation Electrification, vol. 3,no. 1, pp. 147-156, March 2017

[3] K. Siebke, T. Schobre, R. Mallwitz, "Comparison of GaN based CLLC converters for EV chargers operating at different switching frequency ranges," EPE 2019, Genoa

[4] R. T. Naayagi, A. J. Forsyth and R. Shuttleworth, "Performance analysis of DAB DC-DC converter under zero voltage switching," 2011 1st International Conference on Electrical Energy Systems, Newport Beach, CA, 2011, pp. 56-61

[5] T. Schobre, K. Siebke and R. Mallwitz, "Design of a GaN based CLLC converter with synchronous rectification for on-board vehicle charger," PCIM Europe 2019; International Exhibition and Conference for Power Electronics, Intelligent Motion, Renewable Energy and Energy Management, Nuremberg, Germany, 2019, pp. 1-5.

[6] M. Mu, L. Xue, D. Boroyevich, B. Hughes and P. Mattavelli, "Design of integrated transformer and inductor for high frequency dual active bridge GaN Charger for PHEV," 2015 IEEE Applied Power Electronics Conference and Exposition (APEC), Charlotte, NC, 2015, pp. 579-585

[7] A. Rodríguez, A. Vázquez, D. G. Lamar, M. M. Hernando and J. Sebastián, "Different Purpose Design Strategies and Techniques to Improve the Performance of a Dual Active Bridge With Phase-Shift Control," in IEEE Transactions on Power Electronics, vol. 30, no. 2, pp. 790-804, Feb. 2015

[8] F. Yazdani and M. Zolghadri, "Design of dual active bridge isolated bi-directional DC converter based on current stress optimization," 2017 8th Power Electronics, Drive Systems & Technologies Conference (PEDSTC), Mashhad, 2017, pp. 247-252

[9] C. Calderón, A. Barrado, A. Rodriguez, P. Alou, A. Lazaro, C. Fernandez, P. Zumel, "General Analysis of Switching Modes in a Dual Active Bridge with Triple Phase Shift Modulation," Energies, 2018, 11, 2419

[10] C. Calderón, A. Barrado, A. Rodriguez, P. Alou, A. Lazaro, C. Fernandez, P. Zumel, "General Analysis of Switching Modes in a Dual Active Bridge with Triple Phase Shift Modulation," Energies 2018, 11, 2419

[11] C. Song, A. Chen, Y. Pan, C. Du, C. Zhang, "Modeling and Optimization of Dual Active Bridge DC-DC Converter with Dead-Time Effect under Triple-Phase-Shift Control," Energies 2019, 12, 973

[12] C. R. Sullivan, "Computationally efficient winding loss calculation with multiple windings, arbitrary waveforms, and two-dimensional or three-dimensional field geometry," in IEEE Transactions on Power Electronics, vol. 16, no. 1, pp. 142-150, Jan. 2001

An Experimental Analysis of Circulating Current Control Circuit for Output Power from Vibration Generator for Vibration including the Third Harmonics

Masataka Minami, Akito Nakagaki, Genki Hase
Kobe City College of Technology,
8-3, Gakuenhigashi-machi, Nishi ku,
Kobe city, Hyogo, 651-2194, JAPAN
Tel: +81 78 795 3232/Fax: +81 78 795 3314
kcct-minami@g.kobe-kosen.ac.jp

Abstract

The vibration generators based on piezoelectric elements provide very low power. This research focuses on the method for improving in the output power. The previous research proposed a circulating current control circuit for improving in output power under only single frequency. This paper numerically and experimentally shows that the proposed circuit improves the output power for the vibration including the third harmonics.

Keywords

≪Renewable energy systems≫, ≪DC power supply≫, ≪Harmonics≫.

Introduction

Recently, the energy-harvesting system has attracted increased attention [1]. The ambient energy sources include vibration, solar, thermal energy, and so on [2]. The system is expected to become practical application [3]. This research focuses on vibration generators based on piezoelectric elements. The vibration generators convert vibration energy into electric energy [4]. As the vibration generators provide very low power around mW or μW level, the improvement of its performance is required for utilization.

The previous research proposed the improvement of the output power by LC resonant circuit [5]. This system can improve the output power in only resonance vibration frequency. Generally, the vibration includes not only the fundamental frequency but also any harmonics. Reducing the output power is undesirable because of the harmonics in the vibration. Hence, we developed a Circulating Current Control circuit (C3 circuit) for these vibrations [6]. The results [6] revealed that the C3 circuit is useful for the fundamental-frequency vibration. This paper numerically and experimentally shows that the output power of the proposed circuit grows up compared to the conventional circuit for the vibration including the third harmonics.

Equivalent circuit of piezoelectric elements

Figure 1 shows the piezoelectric elements and the experimental system in this paper. Fig. 1 (a) describes the size and structure of the piezoelectric element. The left-hand side of the piezoelectric element is clamped and vibrated by a vibration source. On the other side of the piezoelectric element, the mass is tipped for the mechanical resonance to increase the amplitude. The picture of Fig. 1 (b) is the experimental system. The piezoelectric elements are electrically connected in parallel.

In this research, the target application is the vibration of engines such as a motor vehicle [7]. When the fundamental frequency of vibration is measured, it is around 7000 rpm. This paper assumes the fundamental frequency as 120 Hz and the third harmonics as 360 Hz based on the preliminary experiment. In

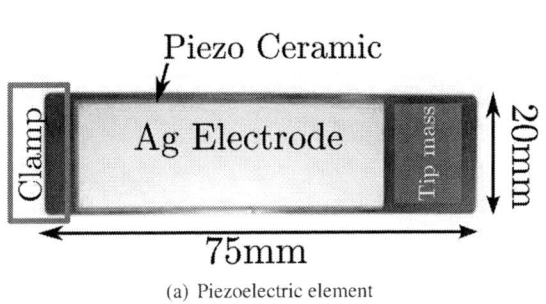

(a) Piezoelectric element

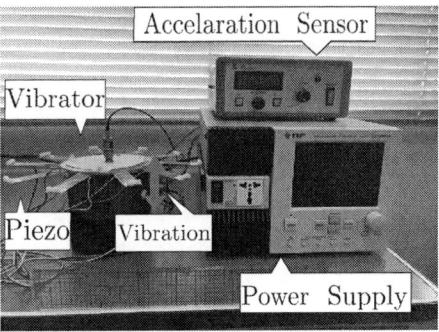

(b) Vibration generator system

Fig. 1: Piezoelectric element and experiment of system.

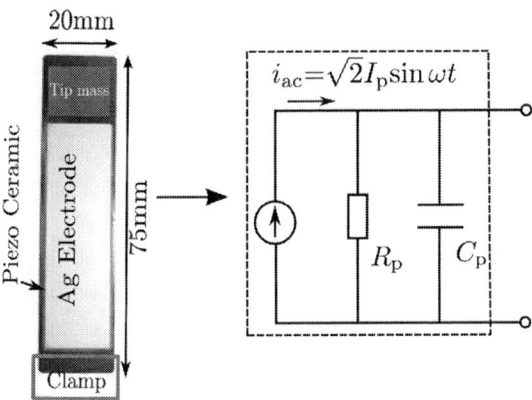

Fig. 2: Equivalent circuit of piezoelectric element [8].

addition, the ratio ε of the amplitude between the fundamental and third harmonics is decided from the preliminary experiment.

The piezoelectric elements are expressed as shown in Fig. 2 (see [8]). The vibration energy and the piezoelectric ceramics are expressed in the AC current source and internal impedance as a capacitor and a resistor [8].

Verifications of proposed circuit

Figure 3 illustrates the circuit for vibration generator. Fig. 3 (a) shows the conventional circuit of vibration generator. We call it "w/o C3 circuit". The broken line describes the equivalent circuit of the piezoelectric elements in Fig. 3 (a). The piezoelectric elements connect the diode rectifier for getting DC power.

Figure 3 (b) shows the proposed circuit (C3 circuit). The C3 circuit provides the inductor current i_L to compensate the internal capacitor current i_c for improving the output power P_o. In Fig. 3 (c), the C3 circuit of Fig. 3 (b) connects the conventional circuit of Fig. 3 (a) in parallel. The inductor current i_L is made by controlling output voltage v_{sw} from the C3 circuit of Fig. 3 (b). The controlling equation is expressed as

$$v_{sw}^* = \frac{KL}{T}\left(C_p \frac{dv_c}{dt} - i_L\right) + v_c, \tag{1}$$

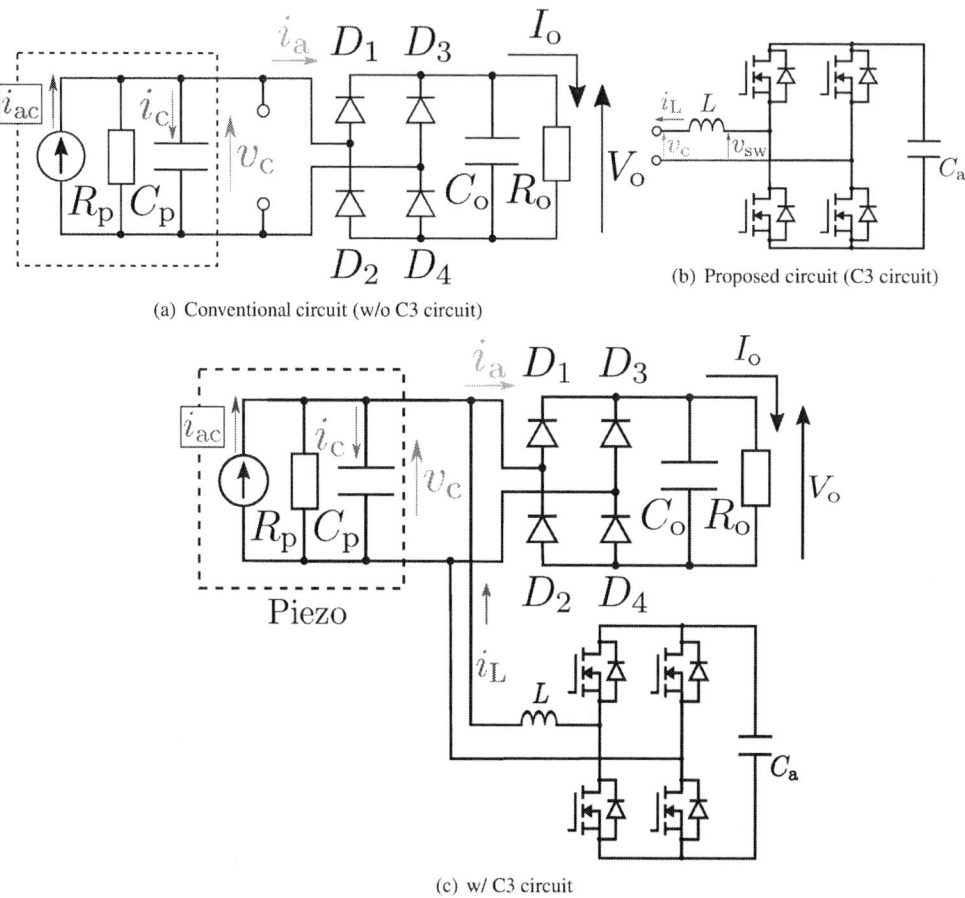

(a) Conventional circuit (w/o C3 circuit)

(b) Proposed circuit (C3 circuit)

(c) w/ C3 circuit

Fig. 3: Proposed circuit (C3 circuit) is connected to conventional circuit (w/o C3 circuit).

where the K and T denote as a stabilizing gain and a sampling time.

The vibration includes the fundamental and the small third harmonics in this paper. Thus, the AC current source is expressed as

$$i_{ac} = \sqrt{2}I_p \sin\omega t + \varepsilon\sqrt{2}I_p \sin(3\omega t + \varphi). \tag{2}$$

Setting of conditions and parameters

Table I lists the parameters of the piezoelectric elements, the rectifier, and the C3 circuit in Fig. 3. The inductor L in the C3 circuit includes the equivalent series resistor R_L. The diodes of the rectifier are numerically ideal and experimentally using schottky barrier diodes 1SS108.

Current and voltage waveforms

Figures 4 and 5 show the current and voltage waveforms of numerical and experimental results: i_{ac}, i_a, i_c, i_L, and v_c, when the load resistor R_o set to 1 kΩ. In Fig. 4 (a), the current i_a to the load is small because the internal capacitor current i_c slowly flows. In Fig. 4 (b) and Fig. 5 (b), as the inductor current i_L supports and compensates the internal capacitor current i_c, the internal capacitor current i_c can quickly

Table I: Parameters of the piezoelectric elements and the circuit

$I_\text{p} = 6.85$ mA	$C_\text{o} = 22\ \mu\text{F}$	$R_\text{L} = 0.39\ \Omega$	$L = 11.4$ mH
$\varphi = -0.5$ rad	$C_\text{a} = 10\ \mu\text{F}$	R_o=variability	$\varepsilon = 0.3$
$\omega = 2\pi \times 120$ rad/s	$C_\text{p} = 1.06\ \mu\text{F}$	$R_\text{p} = 3.6$ kΩ	$f_\text{sw} = 100$ kHz

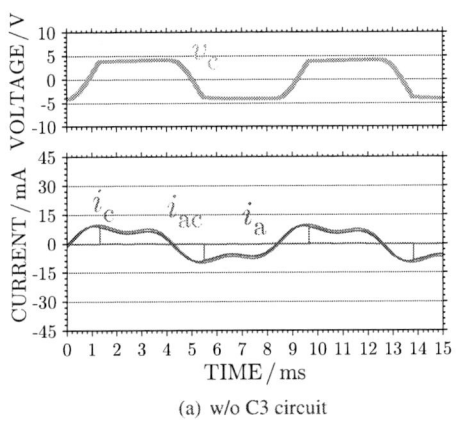
(a) w/o C3 circuit

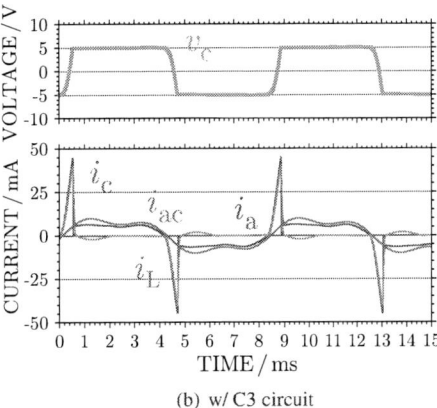
(b) w/ C3 circuit

Fig. 4: Numerical current and voltage waveforms in w/o C3 circuit and w/ C3 circuit.

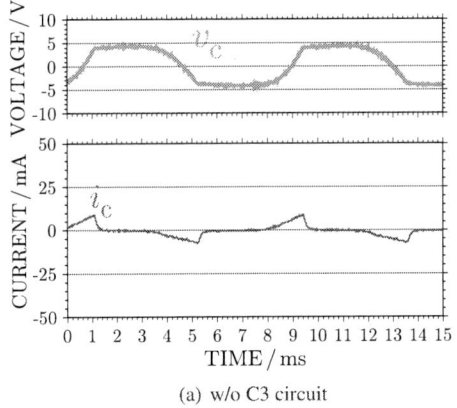
(a) w/o C3 circuit

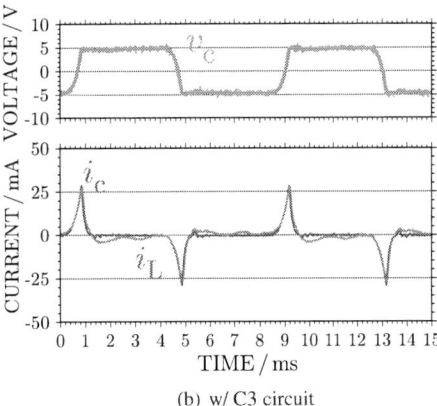
(b) w/ C3 circuit

Fig. 5: Experimental current and voltage waveforms in w/o C3 circuit and w/ C3 circuit.

flow. As a result, the current i_a to the load of Fig. 4 (a) larger than the current i_a to the load of Fig. 4 (b).

I-V and *P-V* output characteristics

Figure 6 represents the numerical and experimental *I-V* and *P-V* output characteristics. The shapes of these results are similar between the conventional and the proposed circuit. The values of the output current and power in the proposed circuit become higher and larger than the conventional one. As a result, the C3 circuit improves the output power when the vibration includes the third harmonics. In

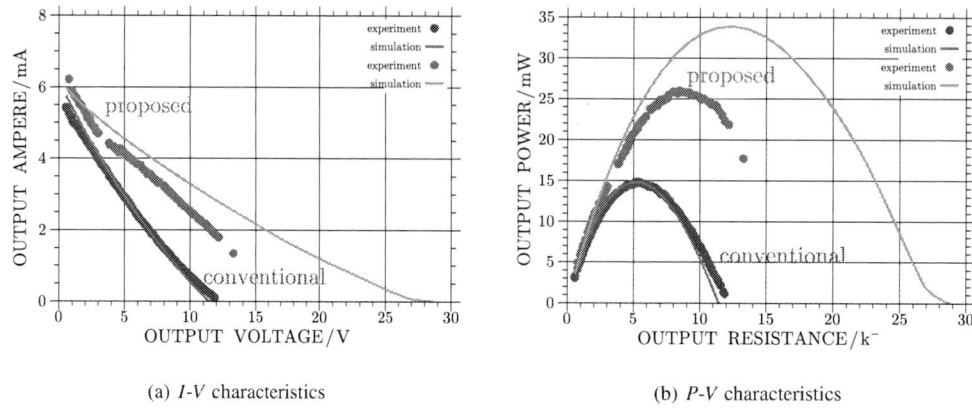

(a) *I-V* characteristics   (b) *P-V* characteristics

Fig. 6: Numerical and experimental results of *I-V* and *P-V* output characteristics in w/o C3 circuit and w/ C3 circuit.

addition, the C3 circuit can be expected to compatible for any harmonics by this result.

Conclusion

This paper proposed the C3 circuit for improvement in the output characteristics for the vibration including the third harmonics. Then, the numerical and the experimental results clarified the validity of the proposed C3 circuit in this case.

References

[1] L. Mateu, M. Echeto, and F. de Borja, "Review of energy harvesting techniques and applications for microelectronics." International Society for Optical Engineering, 2005.

[2] P. Glynne-Jones, S. P. Beeby, and N. M. White, "Towards a piezoelectric vibration-powered microgenerator," *IEE Proceedings-Science, measurement and technology*, vol. 148, no. 2, pp. 68–72, 2001.

[3] P. Spies, M. Pollak, and L. Mateu, *Handbook of Energy Harvesting Power Supplies and Applications*, 1st ed. Pan Stanford, 2015, ch. 1.

[4] A. Erturk and J. Inman, "An experimentally validated bimorph cantilever model for piezoelectric energy harvesting from base excitations," *Smart Materials and Structures*, vol. 18, no. 2, p. 025009, 2009.

[5] M. Minami, "Improvement in output characteristics using a resonator and passive rectifiers in vibration generators," *IEEE Transactions on Power Electronics*, vol. 34, no. 8, pp. 7184–7191, 2019.

[6] M. Minami, K. Kondo, S. Motegi, and M. Michihira, "Current compensation with circulating current control circuit in vibration generator (in Japanese)," *IEEJ Trans. IA*, vol. 139, no. 9, pp. 822–823, 2019.

[7] S. Hashimoto, N. Nagai, Y. Fusikura, J. Takahashi, S. Kumagai, M. Kasai, K. Suto, and H. Okada, "Multi-mode vibration-based power generation for autombiles," in *Proceedings of the Industry Applications Society Annual Meeting*, 2012.

[8] S. Roundy and P. K. Wright, "A piezoelectric vibration based generator for wireless electronics," *Institute of Physics Publishing Smart Materials Structures*, vol. 13, no. 5, pp. 1131–1142, 2004.

Implementation of Control Strategy for Step-down DC-DC Converter Based on Piezoelectric Resonator

Mustapha TOUHAMI [1], Ghislain DESPESSE [1], François COSTA [2], [3] and Benjamin POLLET.

[1] CEA, LETI, Minatec Compus, Université de Grenoble Alpes, 17 avenue des martyrs, 38054 Grenoble, France.

[2] Université Paris-Saclay, ENS Paris-Saclay, CNRS, SATIE, 91190, Gif-sur-Yvette, France,

[3] Université Paris Est Créteil, 94010, Créteil France.

E-Mail: mustapha.touhami@cea.fr, Ghislain.Despesse@cea.fr, francois.costa@satie.ens-cachan.fr, benjamin.pollet.p@gmail.com.

URL: http://www.leti-cea.com.

Keywords

«DC-DC converter», «Power conversion», «Piezoelectric Resonator», «Soft Switching», «Control Strategy».

Abstract

With the growth in demand for miniaturization in power electronics, the current solutions are starting to display their limits in dimensions, power densities and efficiency. To meet the previous demands, the new piezoelectric materials achieving high power densities and efficiency could be the solution to ensuring the requirements. The piezoelectric resonators (PRs) and the piezoelectric transformers (PTs) have been used previously. Unlike the PTs, the use of PRs in power electronics has not been fully explored, and their use has been limited to operating as switched capacitors. However, a new operating principle using PRs based on energy and electrical balance exhibits good performances in steady state. In this paper, our motivation is to investigate in the capability to control a dc-dc converter based on PRs using this operating principle. Indeed, this paper presents the control strategy of a new step-down DC-DC converter based on a piezoelectric resonator (PR), which is used as an energy storage element. The operating principle of the converter is also presented. Moreover, the control algorithm has been implemented in field programmable gate array (FPGA) to regulate the output voltage. The control principle is validated experimentally for input-output voltages 120 - 48 V, and achieving an efficiency up to 94% for large operating power range.

Introduction

In last decades, miniaturization, integration and high power densities became a serious challenge for power electronics designers. Various solutions on power electronics have been proposed for these objectives using magnetic components, switched capacitors and variable capacitors [1-3]. Magnetics components are widely used in power electronics, but their integration on silicon is difficult and the performances become limited by magnetic core at high frequency [1]. Variable capacitor converters can be integrated on silicon but need high chip area, and their efficiency is sensitive to load variations [2-3]. Concerning switched capacitor converters, their operations are limited to specific output/input voltages ratios [4].

To overcome the abovementioned limits of traditional solutions, an alternative storage technique is emerging. It is based on piezoelectric resonators (PRs) or piezoelectric transformers (PTs) which store the transient energy in a mechanical form. Indeed, the piezoelectric components can offer a high quality factor (up to 2000 inducing low losses), and it can operate at high frequency (up to 30 MHz or more inducing a high power densities). Their integration on silicon is widely achieved and continuously improved [5-6]. In addition, the low profile of piezoelectric components can be interesting for integrated converters in thin devices as smartphones, laptops and other connected things (IoT). In literature, most of the proposed converter topologies use PTs in order to replace the magnetic transformers, but these topologies need an additional inductor to increase their performances [7]. So, suppressing this additional inductor in such converters is the aim of several research works [8-13].

In this paper, we present a new conversion principle and our DC-DC converter-using PRs that operate on energy and electrical charges balancing over a mechanical resonant period of the piezoelectric material. The conversion cycle is composed of six steps, which alternate constant voltage steps (connected PRs to a voltage level), and constant electrical charge ones (isolated PRs). In fact, in steady state, the PR takes energy from the input source equal to the one restituted to the output load plus the one dissipated, maintaining mechanical oscillations of constant amplitude. The energy exchange operates during the constant voltage steps. Therefore, the isolated stages (constant charge operation) allow natural evolving of the PR's voltage up to the next constant voltage level. This enables to operate with zero voltage switching (ZVS), leading to reduce power losses and EMI level [14]. A full design of step down DC-DC converter based on a similar principle is presented in [9-10]. The concept is validated by simulation and experiments with efficiency up to 98.4 % at 160 mW for low voltage conversion [9-10]. Recently, various topologies are presented based on the similar principle [10]. The first results show good performances in terms of power densities and efficiency for low power conversion and expect to be competitive compared with traditional converters.

The main objective of this paper is to introduce a new topology of DC-DC converter dedicated to applications at low output/input voltage ratio and its related control strategy. This paper demonstrates the feasibility of controlling a DC-DC converter based on PR, which operates on a six phases conversion cycle. The aim of this control is to regulate the output voltage, ensuring the soft switching and the energy balancing. The load variation should also be compensated. This paper is organized as follows: in the first part, we present the new topology of the converter and the operating principle. In the second part, we analyze and identify the various parameters aiming to control the energy balance and to regulate the output voltage. Then, we present the regulation mechanism for each control parameter. In the experimental results part, we exhibit our prototype, and the dynamic response for $V_{in} = 120\ V$ and $V_{out} = 48\ V$ at $P_{out} = 10\ W$ is experimentally validated. Moreover, the waveforms in steady state are displayed. Then, we validate the regulation against the load variation. Finally, we conclude about this present and future works.

Topology description

The DC-DC converter is presented in Fig. 1. As shown, the converter is composed of a high quality factor piezoelectric resonator (PR), four diode D $_{\{1, 2, 3, 4\}}$ and two NMOS FET switches S_1 and S_2. The PR is modeled by a capacitance C_P in parallel with a L-C-R motional branch, the electrical equivalent circuit is displayed in Fig.2a. The material of the PR used in this converter is Lead Zirconate Titanate (PZT), this ceramic offers high quality factor and high coupling factor (see Table I). In inductive region (near the resonant frequency) [9], the motional LCR branch can be modeled by a sinusoidal current source i in series with the resistance R_p (Fig.2b). Because we consider a step-down converter, the input-output voltage gain should not exceed 0.5 ($V_{in} > 2V_{out}$). This topology allows establishing three possible voltages on the PR: $V_{in} - V_{out}$, $-V_{out}$ and $+V_{out}$. During these connections, the PR exchanges energy with the input source and the output load. Moreover, the PR can operate in open-circuit mode. In this mode, no electrical charges are exchanged with the outside and the motional branch's sinusoidal current charges/discharges the parallel capacitance C_P and makes the voltage V_P progressively change, as show in Eq.1. Thanks to this mode, the voltage V_P evolves naturally to the next constant voltage level and permit to switch-on S_1 and S_2 when the voltage across them reaches 0 V (soft switching operation), leading to reduced switching losses. Indeed, in steady state, the PR stores electrical energy when connected to $V_{in} - V_{out}$ as an increase of mechanical energy. This mechanical energy is then released as electrical energy when PR is connected to V_{out} or $-V_{out}$ (depending on the sign of current).

$$V_P(t) = -\frac{1}{C_P} \int i(t) \cdot dt \tag{1}$$

$$i(t) = I \cdot \sin(\omega \cdot t) \tag{2}$$

V_P is the PR's voltage, i is the motional branch internal current, I is its amplitude, $\omega(rad/s)$ is the pulsation of the conversion period.

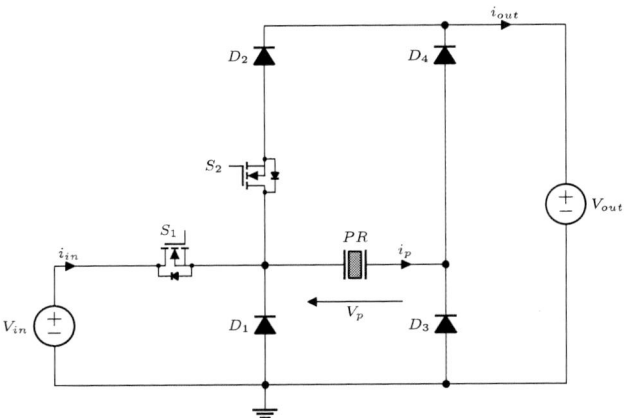

Fig. 1: Converter topology.

(a)   (b)

Fig. 2: Electrical model for PRs, (a) Circuit model [15], (b) Electrical model in inductive region.

Energy conversion cycle

The conversion cycle is presented in Fig. 3. It is composed of 6 steps, which alternates constant voltage and constant charge. Each stage is represented in Fig.3 with the cycle timeline. We separate step 6 in steps 6.a and 6.b in order to separate the part before and after the sign of the current i changes. The 6th step has for objective to let the piezoelectric voltage V_P goes from V_{out} to $V_{in} - V_{out}$, then as soon as V_P reaches $V_{in} - V_{out}$ the switch S_1 should be closed. However, the Drain-Source voltage of S_1 is not null but equals to V_{out} which induces power losses, since the diode D_3 was in conduction just before. To overcome that limitation and ensure ZVS condition, we let V_P goes up to V_{in} before going back to $V_{in} - V_{out}$ during the stage 6.b with a voltage approaching 0 across D_4 and across S_1 at $t = t_0$.

In Fig. 4, we represent the voltage V_P in function of the transfer electrical charges for a single conversion cycle. In the diagram V_P vs Q, the horizontal displacements correspond to the connected phases since V_P is constant, and the vertical displacements correspond to the isolated phases since no electrical charge is exchanged with the input source or output load. Moreover, for a displacement in positive sense, the transferred electrical charge is positive value, and for displacement in negative sense, the transferred electrical charge is negative. The area under the curve represent the electrical energy exchanged with the PR, as shown in Fig.4. The diagram V_P vs Q allow to more understanding the operating principle. Using the V_P vs Q representation enable to construct any switching sequences by alternating horizontal displacement and vertical displacement with ensured closed diagram. This representation allow to easily visualizing and understanding the energy and electrical charge transferred.

In steady state, over a full conversion cycle, the energy and electrical charges are balanced, inducing a constant oscillation amplitude. The energy transfer with the PR operates during constant voltage V_P. Therefore, the conservation of energy (CoE) during each cycle is given by Eq.3, while E_{in} is the electrical energy stored when the PR is connected to $V_{in} - V_{out}$. E_{out} is the electrical energy restored to the load, when the PR is connected to $-V_{out}$ and $+V_{out}$. Energy losses due to the PR's resistance are taken into account by Eq.3 and Eq.4.

$$E_{in} + E_{out} + E_{PR_{losses}} = 0 \tag{3}$$

$$E_{PR_{losses}} = -\frac{1}{2} R_P I^2 T \tag{4}$$

T: conversion cycle duration. (All values are algebraic)

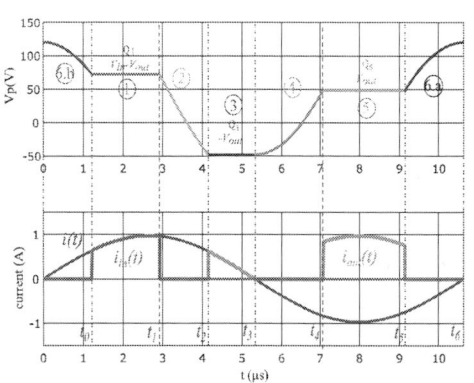

Fig. 3: Waveform of the PR's voltage and currents for 1-6b phases: $V_{in} = 120\,V$, $V_{out} = 48\,V$ and $P_{out} = 10\,W$. (Red: i_{in}, Blue: i, Green: i_{out})

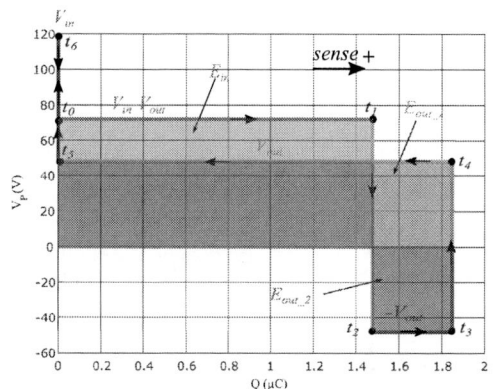

Fig.4: Voltage Vs Electrical charge of PR for 1-6b phases: $V_{in} = 120\,V$, $V_{out} = 48\,V$ and $P_{out} = 10\,W$.

The conservation of electrical charge (CoC) during a cycle of conversion is given by Eq.5:

$$Q_1 + Q_3 + Q_5 = 0 \tag{5}$$

Q_1, Q_3 and Q_5 are the algebraic electrical charges for stage 1, 3 and 5 respectively.

We can determine each transition time (t_0, t_1... t_6) of the cycle by combining the above equations in steady state and considering PR's electrical model, assuming a constant output voltage and operating in soft switching. As shown in Fig. 3, we solved the previous equations for input voltage 120 V, output voltage 48 V and output power 10 W by using the electrical parameters of the PR that are listed in Table 1. Then, we have calculated each transition time for each phase. The electrical parameters of the PR are obtained experimentally by using an impedance analyzer.

Table I: Characteristics of the electrical model of PR. (Part NO: Z0.75T25D-W (C-213). Material is Lead Zirconate Titanate (PZT), the diameter is 25 mm, and thickness is 0.75 mm, the supplier is FUJITSU)

	L	C	R_P	C_P	f_r	Quality factor Q	Coupling factor k
Value	1.1 mH	2.9 nF	0.6 Ω	8.4 nF	89 kHz	1000	52 %

Control strategy of the DC-DC converter

In this section, we propose an analysis of the control strategy in order to regulate the output voltage. As demonstrated in the previous section, in steady state, we should achieve the CoE, CoC and the operating in ZVS conditions to reduce losses. Therefore, the objective of the control is also to regulate the output voltage with high efficiency.

Synchronization with the current i

Since, the PR resonates at variable frequency, which is highly depending on the output power and on the parasitic capacitances of switches [9-10]. The resonance frequency and the phase of the internal current i should be well known in order to synchronize the drive of switches S_1 and S_2 (as shown in Fig.3). In this purpose, we propose to use the state of diode D_3 to detect the 0 crossing of the internal current i and deduce the resonant period duration Therefore, when current i is positive, diode D_3 is turned-off, and -on when negative. Hereunder, we will consider that the switches driving is synchronized with the internally current thanks to monitoring the state of D_3.

Zero Voltage Switching

In order to operate in soft switching, we have to detect the zero voltage crossing of switches S_1 and S_2. For switching-on S_2, we simply turn it-on at the half period of the conversion cycle (as shown in Fig. 3), when the voltage of the PR equals $-V_{out}$. For switching-on S_1, we have to regulate t_5 to let voltage V_P reaching the input voltage value V_{in} at the end of the period. Indeed, if t_5 is superior to the optimal value t_{5_opt} (the value to achieve ZVS for S_1, see in Fig. 3), then voltage V_P will not have time enough to reach V_{in}. Alternatively, if t_5 is inferior to the optimal value t_{5_opt}, the voltage V_P will reach V_{in} before the end of the cycle and the voltage will be clamped by the body diode of S_1, and a part of energy is restored to the input source that inducing unnecessary losses. Wherefore, a simple comparator can be used to regulate t_5 for the next conversion cycle by comparing the voltage V_{D_1} (left side of V_P) with input voltage V_{in} (see Fig. 5).

Output voltage regulation

As displayed in Fig. 3, the PR takes energy from the input source during step 1; in fact, the instant t_1 controls the amount of energy exchanged. By considering a resistor load R_{Load} in parallel with a capacitor C_{out} that filter the output voltage, we can deduce a relation between t_1 and V_{out}. Equation 6 gives the relation between t_1 and V_{out}.

$$\cos(\omega * t_1) = 1 - 4\frac{V_{out}}{V_{in}} + V_{out}\frac{C_P \omega}{I}\left(1 + 2\frac{V_{out}}{V_{in}}\right) - \pi\frac{R_P I}{V_{in}} \tag{6}$$

$\omega \ (rad/s)$ is the resonant pulsation of the PR, which is synchronized the switches driving. V_{in} is the input voltage and considered constant ($V_{in} = 120\ V$), I is the amplitude of the internal current i that is also considered constant. So, only t_1 and V_{out} are variables. By using small signals modeling, we can deduce a gain expression between t_1 and V_{out}. Then, a proportional-integral (PI) controller can be designed to regulate the output voltage by adjusting the instant t_1.

For ease of understanding, we have represented the control strategy in the block diagram depicted in Fig.5. Thus, to implement the control, we need a comparator for current sign detection to monitor the state of D_3 in order to synchronize the switching sequence with the current waveform. A second comparator compare V_{D_1} with V_{in} at the end of period to ensure ZVS operation on S_1. For output voltage regulation, the difference between the measured V_{out} and the reference is multiplied by a PI controller to calculate t_1. As shown Fig. 5, the switches S_1 is turned on at t=t_0 with ZVS. Then, V_P decrease from V_{in} to $V_{in} - V_{out}$. Ones $V_P = V_{in} - V_{out}$, the diode D_4 is turned on, and PR takes energy from input source until to turn off S_1 at t=t_1 in order to regulate V_{out}. During [t_1, t_2], V_P decrease from $V_{in} - V_{out}$ to $-V_{out}$. When $V_P = -V_{out}$, the diodes D_1 and D_4 are turned on naturally, and part of the stored energy in PR is transferred the output load until the current i becomes negative. When t=t_3, the switches S_2 is turned on, and the voltage V_P goes from $-V_{out}$ to V_{out}. Ones $V_P = V_{out}$, the diode D_2 and D_3 are turned on, and the second part of the stored energy in PR is restored to output load. This phase ends by turning off S_2 at t=t_5 in order to let the voltage V_P evolve to reach V_{in} at the end of period.

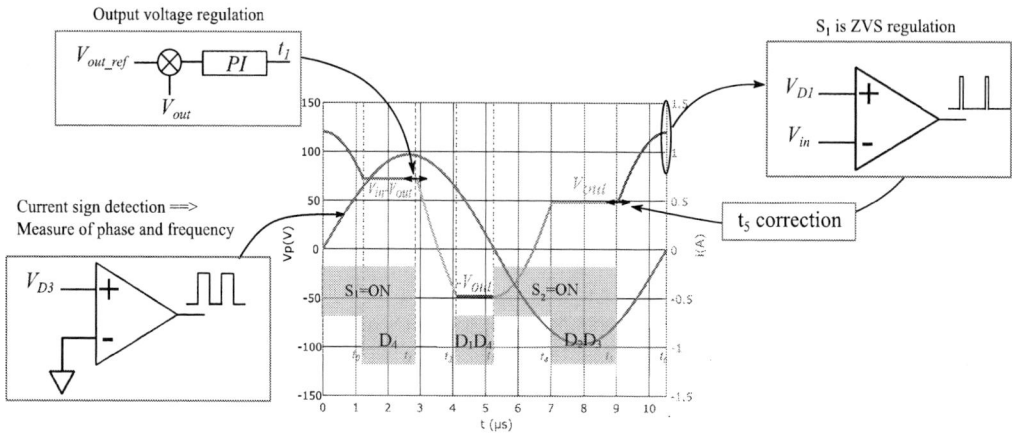

Fig. 5: Control diagram of the circuit.

Experimental Results

Setup

An actual power converter circuit is designed in the aim to test the proposed control strategy. Components references used in the prototype are listed in Table II. Figure 6 shows the prototype that is composed of two PCB circuits; power circuits and FPGA board. The control strategy is modeled in VHDL language and implemented in a FPGA SPARTAN 3 (XC3S200) design kit. The 230 Ω load resistance is calculated for 48 V / 10 W output voltage and power. The output capacitor is chosen to reduce high frequency ripple of V_{out}, the value of C_{out} is 10 µF. The output voltage is sensed by a voltage divider and then digitalized with an analog digital converter (ADC) AMC1204. The difference between the voltage reference and the measured output voltage enables to calculate the time duration t_1 by using a PI regulator, which is digitalized and implemented in the FPGA on state-machine. The FPGA receives measurements from the power circuit, and then calculates the variables t_1, t_5 and f .Ones this variables are calculate the FPGA generates the width modulated pulse (PWM) synchronized with the internal PR current phase.

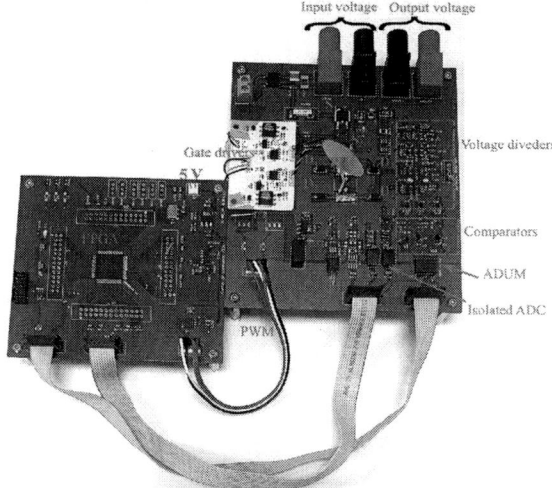

Fig. 6: Prototype circuit.

Table II: Reference of power components

	MOSFET Switch (S1, S2)	Diode (D1, D2)	Diode (D3, D4)
Reference	STD5NM50T4	US1G	VSSB310

Experimental waveforms

Dynamic response

Figure 7 shows the dynamic response for an input voltage of 120 V and an output voltage of 48 V, the output power is 10 W. As shown in Fig. 7, the converter operates firstly in open loop to increase the amplitude of oscillation, in order to detect the 0 current crosses and to measure the frequency of the PR. During this phase, the oscillation amplitude is not monitored because the switching sequences is not synchronized with the internal current. In a second step, the regulation is turned-on. As shown, the amplitude of oscillations decreases and the output voltage is regulated to the reference value. The settling time is 1.2 ms for the output voltage reach 90% of voltage reference. The conversion cycle is presented in steady state in the upper right part of Fig.7. The experimental waveform V_P is matching with the theoretical one (see Fig.3). The current i_p displayed in Fig.7 represents the internal current i measured during only the constant voltage, since no current flow during isolated phases. From the waveform of i_p, we can deduce the instantaneous current of i_{in} and i_{out}. Indeed, i_{in} equals current i_p during phase 1, and i_{out} is the current i_p during phase 1, phase 3 and $-i_p$ during phase 5. Moreover, the waveform of the internal current i can be approximated by a sinusoidal curve, as shown in Fig.7. We have measured 94.05% of efficiency at 10 W output power for 120/48 V input-output voltage considering only the power circuit. We have measured 1.1 W at 5 V power supply for the FPGA and the driving stage, which is quite high for our application, but no optimization was done for this circuit (out of the scope of this paper). However, the high efficiency and power densities of PRs-based converters could motivate the investigation in optimization of integrated digital controller (ASIC) with low cost and low power consumption.

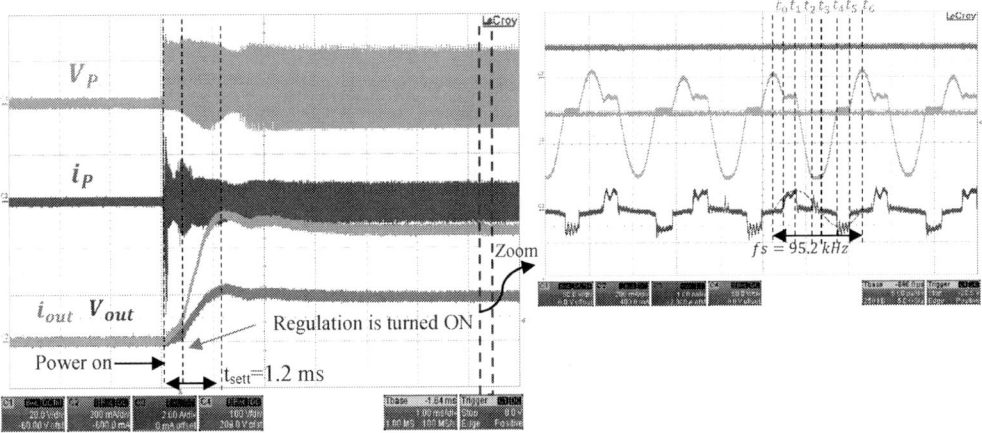

Fig. 7: Experimental dynamic response for input voltage 120 V and output voltage 48 V at 10 W.

Load variation

The load transient is examined for load variation. Fig. 8(a) shows a step change in load from 7 W to 13 W, and the output current goes from 0.145 A to 0.27 A. We can see that the output voltage V_{out} falls by 6 V before going back to the voltage reference when the load variation is applied. The output voltage is then regulated after 2.2 ms. Moreover, the resonance frequency decreases from 96.9 kHz to 94.3 kHz, as the amplitude of the oscillation increases on current i_p. It is noted that during the transient response, the switching driving still synchronized with the resonance period of the PR since no overshoot of current i_P is observed. Fig. 8(b) shows a step change in load from 13 W to 7 W. In this case, the overshoot of the output voltage is less than 2 V and the output current i_p reaches the desired 0.145 A value after 2.5ms. We can also observe an overshoot on current i_p because no ZVS of S_1 is ensured since voltage V_P did not reach V_{in} at the end of the period. However, the regulation of t_5 operates to ensure soft switching of S_1, as shown in steady state.

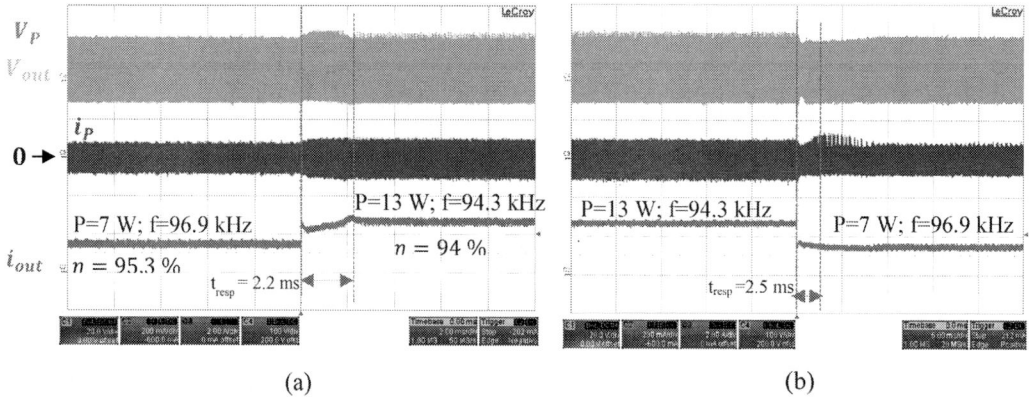

(a)   (b)

Fig. 8: Dynamic response for load variations.

Efficiency

Figure 9 shows the efficiency as a function of the output power for $V_{in} = 120\,V$ and for different gain ratios up to $\frac{V_{out}}{V_{in}} = 0.4$. Experimentally, we have noted that the maximum output power for our PR is limited by the amplitude of the PR current $I\ (\approx 1\ A)$ due to the rise of temperature. So, increasing the output voltage will increase the maximum output power allowed, as shown in Fig.9. For a fixed input voltage, the efficiency increases when the output voltage increased as the amplitude of the current decreases, which decreases the power losses by I^2. We have obtained an efficiency higher than 94% for large operating power range and a peak efficiency of 96% is obtained for $\frac{V_{out}}{V_{in}} = 0.4$ and $P_{out} = 8\,W$.

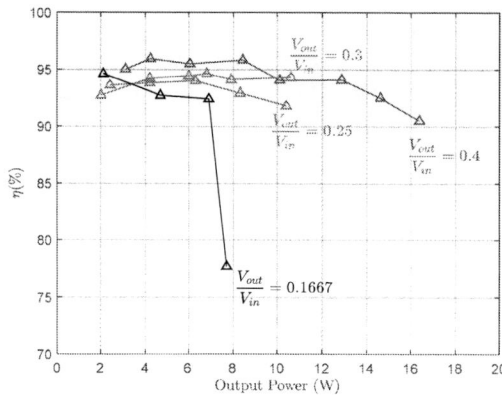

Fig. 9: Efficiency Vs Output Power for $V_{in} = 120\,V$ and for different gain ratios up to $\frac{V_{out}}{V_{in}} = 0.4$.

Conclusion and future works

The concept to control a new DC-DC step-down converter based on a piezoelectric resonator is introduced in this paper. The conversion principle is explained in details. The control strategy is modeled and implemented in a FPGA. The operating with good performances is verified. Furthermore, our specific control strategy enables to regulate efficiently the output at any voltage between $0\ V$ to $\frac{V_{in}}{2}$ in less than 1.4 ms. Experimental results validated both the dynamic response and the steady state operations. Efficiency up to 94.05 % has been measured at 10 W and at input-output voltages of 120 - 48 V.

This paper demonstrate the feasibility of controlling dc-dc converter based on PR by using commercial components, however, we can reduce consumption and increase performances by integrating the control circuit into an ASIC, which is one of the scopes of our lab. The principle using PZT PR in DC-DC converters makes it an excellent solution for very compact, high input voltage, low output voltage applications at low power and high efficiency. Many opportunities still have to be explored such as serialization and/or parallelization of PRs to increase the power. As a conclusion, this new operating principle gives new opportunities in power electronics with the aims of miniaturizing and integration. It could be a serious challenger for magnetic and electrostatic based converters.

References

[1] C. R. Sullivan, B. A. Reese, A. L. F. Stein and P. A. Kyaw, "On size and magnetics: Why small efficient power inductors are rare," 2016 International Symposium on 3D Power Electronics Integration and *Manufacturing (3D-PEIM)*, Raleigh, NC, 2016, pp. 1-23.

[2] S. Ghandour, G. Despesse and S. Basrour, "Design of a new MEMS DC/DC voltage step-down converter" *Proceedings of the 8th IEEE International NEWCAS Conference 2010*, Montreal, QC, 2010, pp. 105-108.

[3] S. V. Cheong et al., "Inductorless DC-to-DC Converter with High Power Density", IEEE Trans. Ind. Electronics, vol. 41, no.2, 1994.

[4] M. S. Makowski and D. Maksimovic, "Performance limits of switched-capacitor DC-DC converters," Proceedings of PESC '95 - Power Electronics Specialist Conference, Atlanta, GA, USA, 1995, pp. 1215-1221vol.2.

[5] R. Schulze et al., "Integration of piezoelectric polymer transducers into microsystems for sensing applications" Proceedings of ISAF-ECAPD-PFM 2012, Aveiro, 2012, pp. 1-4.

[6] S. Shanmugavel *et al.*, "Miniaturized acceleration sensors with in- plane polarized piezoelectric thin films produced by micromachining," *IEEE Transactions on Ultrasonics, Ferroelectrics, and Frequency Control*, vol. 58, no. 11, pp. 2289-2296, November 2011.

[7] M. Ekhtiari, Z. Zhang and M. A. E. Andersen, "State-of-the-art piezoelectric transformer-based switch mode power supplies," IECON 2014 - 40th Annual Conference of the IEEE Industrial Electronics Society, Dallas, TX, 2014, pp. 5072-5078.

[8] S. Moon and J.-H. Park, "High power DC–DC conversion applications of disk-type radial mode Pb(Zr,Ti)O_3 ceramic transducer," Jpn. J. Appl. Phys., vol. 50, no. 9, Sep. 2011, Art. no. 09ND20.

[9] B. Pollet, F. Costa and G. Despesse, "A new inductorless DC-DC piezoelectric flyback converter" 2018 *IEEE International Conference on Industrial Technology (ICIT)*, Lyon, 2018, pp. 585-590.

[10] B. Pollet, G. Despesse and F. Costa, "A New Non-Isolated Low-Power Inductorless Piezoelectric DC–DC Converter," in *IEEE Transactions on Power Electronics*, vol. 34, no. 11, pp. 11002-11013, Nov. 2019.

[11] J. D. Boles, J. J. Piel and D. J. Perreault, "Enumeration and Analysis of DC-DC Converter Implementations Based on Piezoelectric Resonators," *2019 20th Workshop on Control and Modeling for Power Electronics (COMPEL)*, Toronto, ON, Canada, 2019, pp. 1-8.

[12] Thenathayalan, Daniel, Chun-gu Lee and Joung-Hu Park. "Battery voltage-balancing applications of disk-type radial mode Pb(Zr · Ti)O3 ceramic resonator," 2017 Japanese Journal of Applied Physics. 56.

[13] M. Touhami, B. Pollet, G. Despesse, and F. Costa, "A new dc-dc piezoelectric converter," 9th National Days on Energy Harvesting and Storage (JNRSE), May 2019.

[14] M. M. Jovanovic, K. Liu, R. Oruganti and F. C. Y. Lee, "State-Plane Analysis of Quasi-Resonant Converters," in *IEEE Transactions on Power Electronics*, vol. PE-2, no. 1, pp. 36-44, Jan. 1987.

[15] K. S. Van Dyke, "The piezo-electric resonator and its equivalent network," Proceedings of the Institute of Radio Engineers, vol. 16, no. 6, pp. 742–764, June 1928.

Thermal impedances and temperature sensors: a combined approach for a novel thermal model of power semiconductors

Maria De Lauretis[1,3], Jonas Millinger[1], Erik Baker[1], Martin Karlsson[1], and Diane -Perle Sandik[2]

[1]Atlas Copco Industrial Technique AB, Stockholm [2]KTH Royal Institute of Technology, Stockholm

[3]Present affiliation: Luleå University of Technology, Luleå

Sweden

[1]E-Mail: {jonas.millinger, erik.baker, martin.r.karlsson}@atlascopco.com [2]E-Mail: dianes@kth.se [3]E-Mail: maria.de.lauretis@ltu.se

Acknowledgments

This work was supported by the Swedish Energy Agency (Energimyndigheten), grant number 48809-1. We would also like to thank Guillermo Bossi from Atlas Copco for his precious help and feedback.

Keywords

≪Thermal design≫, ≪Power semiconductor device≫, ≪MOSFET≫, ≪Adjustable speed drive≫, ≪Variable speed drive≫.

Abstract

Power semiconductors, or transistors, constitute the core part of adjustable speed drives, which are commonly used in motor drive applications. Overheat of the semiconductors can compromise their behavior and, eventually, shorten the lifespan of the whole system. Therefore, the thermal management of transistors is crucial both at the design stage and during operation. Commonly performed thermal simulations normally rely on transistor thermal resistance. However, this approach does not account for the thermal behavior in transient regime. Furthermore, it is inappropriate for intermittent applications, such as for drilling machines, where the motor is on and off in repetitive working cycles, and the transistors never reach the equilibrium temperature. The transient thermal behavior is described by the concept of transfer thermal impedance. The transfer junction-to-case thermal impedance is given in the datasheet and assumes a constant ambient temperature; an assumption that is, however, not true in real applications. In this paper, we overcome this main limitation by using a resistive sensor. We consider MOSFETs in a 3-phase inverter, and model their thermal behavior with a well-known algorithm that uses the junction-to-case and junction-to-ambient thermal impedances, along with application dependent parameters. The actual rise of ambient temperature of the circuit board is included in the algorithm by virtue of the resistive sensor. The method has been validated with lab measurements, for two different MOSFETs. The proposed method can be used both at the design stage and during operation of the motor drive. Future works will include refinements of the power loss formulas, of the junction-to-ambient impedance modeling, as well as aging effects of the transistors.

Introduction

Power transistors are commonly used in motor drive applications in the converter stage. The control of the motor speed and torque depends on the adopted switching strategy, which relies, ultimately, on the transistor characteristics. Failure of the power transistors can potentially cause the failure of the power electronic equipment and, eventually, of the whole system, as investigated in [1]. Overheat is a common reason for failure, which justifies the effort devoted to thermal characterization and thermal monitoring of power semiconductors both in the academia, see for example [2, 3], and in the industry, see for example [4, 5]. At the design stage, the thermal behavior can be simulated in several ways, for example by resorting to Finite Element Analysis (FEA) methods, as presented in [6]; by using algorithms that are based on datasheet diagrams, see [4, 7]; by using SPICE libraries as given by the supplier, which include internal thermal networks. In general, all models and methods must be tailored to application-dependent scenarios, and all of them will introduce approximations and errors that are difficult to predict. In fact, no model can account for different manufacturing characteristics of the same transistor, and aging phenomena are challenging to include, although work is done in that direction as well, see for example [8]. The choice of the best method to use depends on different factors. Sophisticated models or software based on FEA methods, for example, require high expertise and are limited to the design phase. SPICE models that use thermal networks are easier to understand and to use, but they cannot account for the actual rise in temperature of the board where the transistor is mounted on. So-called "pocket calculator methods" as described in [4, 7] rely on datasheet curves, are very handy to use, and provide a quick estimation of the worst-case scenario. Due to their simplicity, they are very popular among electronic designers in the industry, and are used at the design stage; however, these methods cannot provide the real temperature that the transistor will reach during operation.

In this work, we present a novel approach that combines datasheet parameters with sensors data. The method was used to characterize the temperature profile of the MOSFETs used in the motor drive of a drilling machine, whose operation should be interrupted if the temperature of the transistor goes above its maximum allowed. The printed circuit board assembly (PCBA) includes a resistive sensor placed in the nearby of the transistors. As validated in the lab, the difference between the transistors case-temperature recorded with a thermal camera and the temperature recorded by the resistor is too wide to allow using the resistor as a temperature sensor for the MOSFETs. Therefore, it was suggested to implement a thermal model that should be able to predict the temperature of the MOSFET during the operation of the tool. The available data are the datasheet of the MOSFETs, the temperature from the resistive sensor, the phase currents, and application-dependent parameters. The goal was to combine these data into a model that would be as easy to use as a "pocket calculator method", suitable for intermittent application (i.e., based on transfer thermal impedances), dependent on the temperature recorded by the resistive sensor, and able to give the temperature of the MOSFET during operation, potentially to be implemented in a real-time embedded software. As first step, we chose the "pocket calculator method" described in [4] that can be used for pulse-power operations as found in an inverter, where the transistors are subject to pulsed load due to the pulse-width-modulation (PWM) strategy commonly adopted. The method, in the following referred to as "pulse-power method", uses the junction-to-ambient and junction-to-case transfer thermal impedances. The junction-to-case transfer impedance is from the datasheet. The junction-to-ambient transfer impedance is application-dependent; the datasheet provides merely a junction-to-ambient thermal resistance for a very specific configuration and, therefore, this value should be taken as a reference for coarse-grained approximations only. Due to a lack of resources, we approximated the junction-to-ambient transfer impedance via model fitting. In particular, we considered the junction-to-ambient impedance as a scaled and translated version of the junction-to-case impedance, in alignment with the theory proposed in [5]. In the future, we will evaluate it with suitable FEA methods. As previously mentioned, the "pulse-power method" assumes a constant ambient temperature. In our model, this limitation is overcome by including the recorded temperature from the resistive sensor, which gives the offset for the simulated junction and case temperatures. Another limitation of the method [4] is that the power-loss formula used is very simple because it is intended to give a worst-case scenario. With the aim of keeping easy-to-use, albeit realistic power-loss formulas, we used power-loss average formulas commonly adopted in motor-drive applications, see [9], which include the power losses for the conduction phase, the diode re-circulation phase, the dead-time, and the gate charge. The model has been validated by comparing simulation and measurement results. In particular, we compare the measured case temperature with the simulated case temperature. For the test cases under study, the error is below 6 °C. The major advantages of the proposed model are that it can be used both at the design level, by assuming a constant base temperature, and in real applications, by combining it with a resistive sensor. The model can easily be adapted to different applications and semiconductors because it is merely based on datasheet parameters and easy-to-access application data. Further improvements of the model will involve the algorithm for the junction and case temperature computation, as well as the power losses estimation. As already mentioned, the junction-to-ambient transfer impedance will be modeled by resorting to FEA methods. For initial validation, the DC-link voltage was considered to be fixed; in the future, the model will account for possible variations of the DC-link voltage as well.

Notation and data used

Table I lists the data that are application dependent. Table II lists the data from the datasheet of the 2 MOSFETs under investigation, which are:

- BSC016N06NS from Infineon [10], in the following referred to as 60 V type; and
- BSC046N10NS3 from Infineon [11], in the following referred to as 100 V type.

The output data are listed in Table III. Their meaning should become clear in the next sections. Time-width quantities have a ⁻ superscript to avoid confusion with time values.

Table I: Data from the application

Notation	Description	Value
$\bar{t}_d$	Deadtime to avoid a short-circuit between 2 transistors in one leg	400 ns
$i_{ph}(t)$	Motor phase current	*measured*
V_{IN}	Battery input voltage	18 V
f_{sw}	Switching frequency of the PWM carrier wave signal	48 kHz
T_a	Ambient temperature measured by the built-in resistive sensor	*measured*
V_{GS}	Gate-to-source voltage	12.4 V
$\bar{t}_p$	Time-width of the power pulses related to $i_{ph}(t)$	*measured*
N_p	Number of equivalent pulses in the $i_{ph}(t)$	*measured*

Table II: Data from the MOSFET datasheet

Notation	Description
$R_{DS}(T_j)$	Drain-source on-state curve resistance, max value, as a function of the junction temperature T_j
Q_{rr}	Reverse recovery charge (reverse diode), typical value
$\bar{t}_{rr}$	Reverse recovery time, typical value
V_D	Diode forward-voltage (Rreverse diode), typical value
$\bar{t}_r$	Rise time (dynamic characteristics)
$\bar{t}_f$	Fall time (dynamic characteristics)
Z_{th-jc}	Transient thermal impedance, for single pulse (duty-factor $\delta = 0$)
$\bar{t}_p$	Loading time (pulse-width) for Z_{th-jc} or Z_{th-ja}

System under investigation and basic assumptions

The system under investigation is a circuit board with a 3-leg inverter, composed by 6 MOSFETs, feeding a PMSM (permanent magnet synchronous motor) to be used in a drilling machine. The MOSFETs have no heatsink. The circuit board has a resistive temperature sensor that does not correctly detect the temperature of the MOSFETs, which are the components that mostly heat up during operation. We assumed and measured in the lab a balanced load. As customary for balanced loads, we can consider a one-phase model, as investigated in [12]. We assume that, in each standard running cycle of the tool, all MOSFETs are equally used for the same amount of time, and that the high-side MOSFET (H)

Table III: Output data from the model

Notation	Description
$T_j(t)$	Junction temperature as a function of time
$T_{j,s}(t)$	Simulated junction temperature
$T_c(t)$	Case temperature as a function of time
$T_{c,m}(t)$	Measured case temperature as a function of time
$T_{c,s}(t)$	Simulated case temperature as a function of time
$P_{CM}(T_j)$	Conduction losses as a function T_j
P_{SW}	Switching losses, average value
P_D	Freewheeling diode losses, average value
P_G	Gate driving losses, average value
P_{dead}	Dead time losses, average value
$P(T_j)$	Total average power loss

and the low-side MOSFET (L) of the generic leg conduct in a dual manner: when H is conducting, L is not conductive, and vice-versa. The measurement observation further validated these assumptions, because the transistors warmed up in the same way during operation. Therefore, the analysis focused on 1 MOSFET of a leg, assuming that all the MOSFETs have the same temperature profile.

Test rig for temperature measurement

The test rig is depicted in Fig. 1. The measurement equipment consist of a thermal camera Flir® ETS320, a 1-pixel infrared thermometer Optris® CTlaser LT, a current probe placed on 1 phase cable, a DEWE 43A for data acquisition, and a built-in resistive sensor whose data can be read with an internal software or with the Flir® camera as well. The DEWE 43A is a data acquisition tool that reads the data from the 1-pixel camera and the current probe, allowing to analyze them in parallel with the software from Dewesoft®. The Flir® camera is a fixed-focus camera, connected via USB to the computer; the data are recorded and analyzed by using the Flir Tools+ software®. The Flir® camera detects the temperature of the whole PCBA, therefore providing a complete understanding of the temperature-critical components. The measurements are taken once the board reaches the equilibrium temperature, after having connected the voltage source.

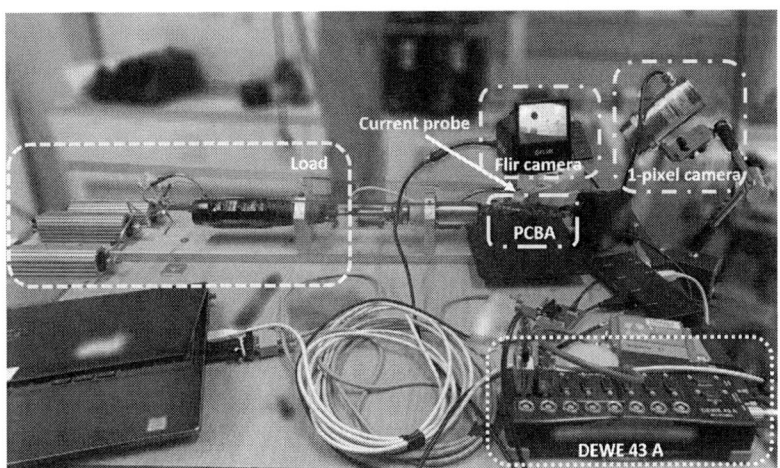

Fig. 1: Test rig used for temperature measurements

Comparison between the MOSFET and the resistive sensor temperatures for a standard run

The resistive sensor temperature and the measured MOSFET case are compared by considering a standard run with the motor on for 3 s and off for 2 s. The temperature variation for the sensor and the MOSFET are depicted in Fig. 2a with a violet and green line respectively; the black vertical line refers to the camera frame depicted in Fig. 2b, where the pointer Sp2 is placed on the MOSFET, and the pointer Sp3 on the resistive sensor. The drop in temperature between the 6th and the 7th cycle is a measurement error, because the corresponding camera frame is missing. The difference between the MOSFET temperature and the sensor temperature increases over time, as better visualized in Fig. 3 in Matlab®. This result clearly shows the low sensitivity of the built-in resistive temperature sensor in respect to the thermal behavior of the transistor.

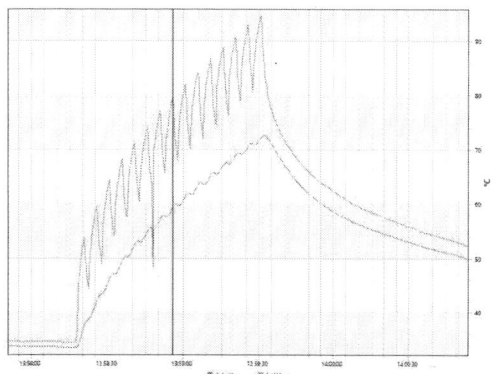

(a) Temperature for the sensor (violet curve) and the MOS-FET (green curve).

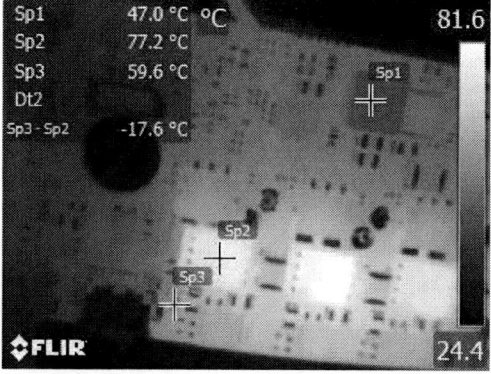

(b) Camera frame that corresponds to the time selected by the horizontal line in the plot.

Fig. 2: MOSFET vs resistive sensor temperatures.

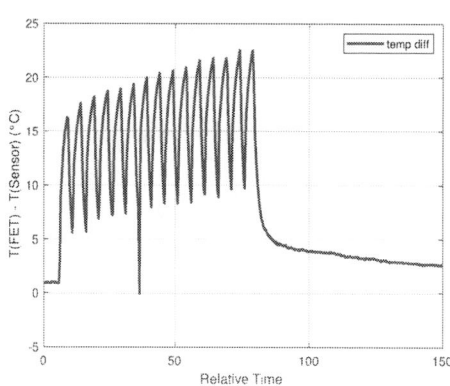

Fig. 3: Difference between the two temperatures in Fig. 2a (MOSFET and sensor temperatures)

1 Background on thermal impedances

In this section, we summarize some main concepts about thermal models for semiconductors, and the reader is referred to relevant literature for further details. The parallelism between thermal and electrical circuit allows to evaluate the junction and case temperatures by using thermal capacitances, resistances, and impedances [13, 14]. The MOSFET's temperature measured in the lab corresponds to the "case temperature" T_c; this is the temperature of the resin. The junction temperature T_j is defined as the silicon-die temperature and it is externally not accessible. The ambient temperature T_a is, in our case, given by the resistive sensor. The power P that flows from the junction to the case can be regarded as a thermal current, and the temperature difference between the junction and case can be regarded as a thermal voltage. It follows that:

$$P = \frac{T_j - T_c}{Z_{th-jc}} \qquad \text{for } T_j > T_c. \tag{1}$$

The power can equivalently be defined using the Z_{th-ja} and T_a. The equations that link the three temperatures T_c, T_j, T_a, and the power P are [13, 4]:

$$T_j = \Delta T_{ja} + T_a, \tag{2}$$
$$T_c = T_j - \Delta T_{jc}, \tag{3}$$

where

$$\Delta T_{jc} = T_j - T_c = P \cdot Z_{th-jc}(\bar{t}_p, \delta), \tag{4}$$
$$\Delta T_{ja} = T_j - T_a = P \cdot Z_{th-ja}(\bar{t}_p, \delta), \tag{5}$$

are the temperatures between junction and case (ΔT_{jc}), and between junction and ambient (ΔT_{ja}). In the equations, note the dependency of Z_{th-ja} and Z_{th-jc} on the pulse width of the power $\bar{t}_p$ and the so-called duty factor δ. In fact, in the datasheet the Z_{th-jc} reads as a family of curves based on the duty factor δ. For all curves $Z_{th-jc} \xrightarrow[\bar{t}_p \to \infty]{} R_{th-jc}$, where R_{th-jc} is the thermal resistance, junction-to-case, bottom [15]. The thermal impedance junction-to-ambient Z_{th-ja} is dependent on the board and, for this reason, the datasheet provides merely the corresponding resistive value R_{th-ja} for a specific board configuration and cooling area. Note that the thermal impedances assume a constant base temperature, i.e. T_a is fixed. However, the model proposed in this paper accounts for the variation of T_a by virtue of the resistive sensor; therefore, the final model accounts for the variation of the ambient temperature as well. In the next sections, we clarify the algorithm used to compute (2), which of the curves for Z_{th-jc} is chosen, the formula used for the power, and the approximation adopted for Z_{th-ja}.

2 The "power-pulse" method

As mentioned in the introduction, the model we developed is based on the algorithm provided in [4] that, in the following, is referred to as the "power-pulse method". In this section, we briefly summarize the main ideas and the adopted implementation; the reader is directly referred to [4] for more information. In motor-drive applications, each MOSFET is subject to burst of pulses. This type of signal can be treated as a composite waveform by superimposing several rectangular pulses (called equivalent pulses) that have a common period, with both positive and negative amplitudes. Positive pulses increase the junction temperature T_j, and negative pulses decrease it. We used this idea to combine the phase current $i_{ph}(t)$ measurement with the MOSFET's temperature. By defining a working cycle as the running time of the tool, we can use the same ideas of the algorithm in [4] for burst pulses. Accordingly to the basic hypothesis explained in the previous sections, the phase current is distributed between the two MOSFETs of the generic leg, such that the high-side MOSFET is subject to the positive part of $i_{ph}(t)$, and the low-side MOSFET is subject to the negative part. In our application, the $i_{ph}(t)$ is sinusoidal in the steady state, whereas it shows an expected different period for the starting and the run down phases, as depicted in Fig. 4. As we focus only on the high-side MOSFET, we can limit our attention to $i_{ph}(t) > 0$. In this first version of the model, we define positive rectangular equivalent power-pulses associated to $i_{ph}(t) > 0$ as follows:

Definition 2.1. Each interval $[t_i, t_{i+1}]$ of $i_{ph}(t) > 0$ associated to the generic local maxima, for $i = 1, \cdots, N_p$, with N_p number of maxima, corresponds to a rectangular pulse that has the width of the interval $[t_i, t_{i+1}]$, and the amplitude equal to the corresponding current maximum.

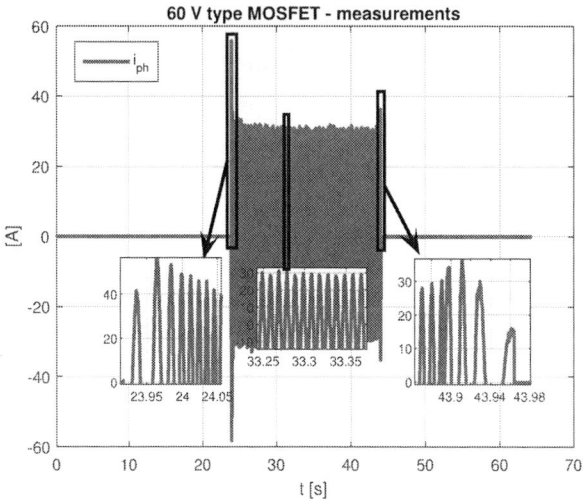

Fig. 4: Phase current $i_{ph}(t)$ for a 20 s run.

Similar definition holds for negative rectangular equivalent power-pulses associated to $i_{ph}(t) < 0$. The width of the first pulse (first time when $i_{ph}(t) > 0$) corresponds to the $\bar{t}_p$ in the Z_{th-jc} curves. The effect of the pulses is incremental, both for positive and negative pulses; in fact, to be able to add all the effects of the pulses at the generic time $t_x = t_{N_p+1}$, all pulses (negative and positive) must end at time t_x. In the following, we summarize the algorithm and we refer to [4] for a complete description. Given a total number of pulses equal to N_p, where the last pulse is assumed to end at the generic time t_x, and the time interval $[t_i, t_{i+1}]$ associated to each generic pulse i, the Algorithm 1 illustrates how to compute:

- $\bar{t}_{i,p}$, defined as the equivalent positive-pulse width at the time t_x associated to the i pulse;
- $\bar{t}_{i,n}$, defined as the equivalent negative-pulse width associated to the i pulse at the time t_x; and
- $\Delta T_{jc,t_x}$, defined as the junction-to-case temperature at the time t_x.

Algorithm 1 Pulse width and incremental case temperature computation

---

1: **procedure** EQUIVALENT PULSES AND INCREMENTAL TEMPERATURE $(\bar{t}_{i,\cdots,N_p}, P_i, Z_{th,jc})$
2:     **for** i to N_p **do**
3:         $\bar{t}_{i,p} = t_{N_p+1} - t_i$
4:         $\bar{t}_{i,n} = t_{N_p+1} - t_{i+1}$
5:     **end for**
6:     $\Delta T_{jc,t_x} = \sum_{k=1}^{N_p} P_i \cdot Z_{th-jc}(\bar{t}_{i,p}) - \sum_{k=1}^{N_p} P_i \cdot Z_{th-jc}(\bar{t}_{i,n})$
7:     **Return** $\Delta T_{jc,t_x}$
8: **end procedure**

The same procedure is used to compute ΔT_{ja}. The computed ΔT_{jc} and ΔT_{ja} are used in the final equation (2). Note that, at the design stage, the current $i_{ph}(t)$ can be considered a pure sinus function, and the total number of pulses is chosen a-priori, allowing to dramatically simplify the proposed algorithm. In real application, we do not know when a pulse is the *last* pulse. For this reason, the algorithm must be repeated for each new positive pulse that ends at t_x.

The single-pulse curve for the junction-to-case thermal impedance

The datasheet provides a family of $Z_{th,jc}$ curves based on the duty cycle. In real applications, the total duration of the burst pulses could change; moreover, we do not know the period of the burst pulses, i.e. we do not know a-priori the time between different on and off tool cycles. For this reasons, we selected the "single pulse" curve for $Z_{th,jc}$.

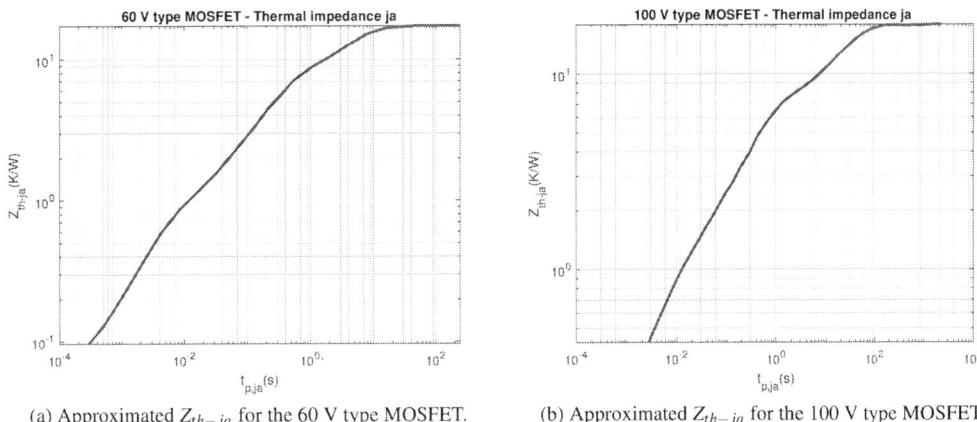

(a) Approximated Z_{th-ja} for the 60 V type MOSFET.   (b) Approximated Z_{th-ja} for the 100 V type MOSFET.

Fig. 5: Approximated transient junction-to-ambient thermal impedances Z_{th-ja}.

The junction-to-ambient thermal impedance

As previously mentioned, the junction-to-ambient thermal impedance $Z_{th,ja}$ is application dependent and it should be computed either by resorting to FEA techniques, or via suitable measurements. Due to a lack of resources, the $Z_{th,ja}$ was approximated via model fitting and it reads as a scaled and translated version of the $Z_{th,jc}$, such that:

$$Z_{th-ja} = K_{Zth} \cdot Z_{th-jc} \text{, and} \tag{6a}$$

$$t_{p,ja} = K_{time} \cdot t_{p,jc} \text{,} \tag{6b}$$

with $K_{Zth}, K_{time} \in \mathbb{R}^+$. We found K_{Zth} equal to 22 and 20, and K_{time} equal to 2.5×10^3 and 0.28×10^3, for the 100 V and 60 V type respectively. The approximated junction-to-ambient thermal impedances for the transistors under investigation are depicted in Fig. 5. With respect to the magnitude, in the datasheets of the two MOSFETs we see that the R_{th-ja} is 20 to 60 times bigger than R_{th-jc}. Via model fitting, we found $R_{th-ja} = 17.86$ for the 100 V type MOSFET, and $R_{th-ja} = 17.26$ for the 60 V type MOSFET, which are both reasonable values. With respect to the time scale, we used the results in [13] and [5]. In particular, the time scale of $Z_{th,ja}$ accounts for the rise in temperature between case and ambient, and it is in the millisecond range, as found in our approximation as well.

3   Computation of the power losses

The switching mechanism adopted for the MOSFETs, i.e. the motor control strategy used will dictate a different impact in the power losses; see for example [12], where a standard sine PWM is compared against a THIPWM (with third harmonic injection). However, in this first implementation, we followed a simplified (and rather common) approach. The proposed model uses the formula for the power losses as given in the technical reports [9, 13]. The power losses can be divided into: conduction losses P_{CM}, switching losses P_{SW}, freewheeling diode losses P_D, gate driving losses P_G, and deadtime losses P_{dead}. They are computed as:

$$P_{CM}(T_j) = R_{DS}(T_j) \cdot \frac{I_{peak}^2}{2} \text{,} \qquad P_D = Q_{rr} \cdot V_{IN} \cdot f_{sw} \text{,} \qquad P_{dead} = V_D \cdot I_{peak} \cdot (2\bar{t}_d) \cdot f_{sw} \text{.}$$

$$P_{SW} = V_{IN} \cdot I_{peak} \cdot \frac{(t_r + t_f)}{2} f_{sw} \text{,} \qquad P_G = Q_G \cdot V_{GS} \cdot f_{sw} \text{,}$$

$$\tag{7}$$

In the model, the total power (8) is computed for each equivalent pulse, where the values of I_{peak} and $R_{DS}(T_j)$ are updated at each pulse:

$$P_i(T_j) = P_{i,CM} + P_{i,SW} + P_D + P_G + P_{i,dead} \quad \text{for } i = 1, \cdots, N_p.$$

(8)

Future implementations of the model should account for the variation of V_{IN}, and the dependency of $V_D(T_j)$ and $Q_{rr}(T_j)$ on T_j as well.

4   Results

In this section, the measured case temperature $T_{c,m}$ is compared with the simulated case temperature $T_{c,s}$ as given by our model, along with the absolute error. The Algorithm 1 been implemented in Matlab®, and the peak values of the current with the corresponding widths are extracted by using the `findpeaks` function from the Signal toolbox®. The simulated junction temperatures $T_{j,s}$ are reported as well. The model was validated under different motor loads and motor speed. Due to lack of space, we present the results for a 20 s run, load condition, as depicted in Fig. 6 for the 60 V type MOSFET, and in Fig. 7 for the 100 V type MOSFET. Note that the error is higher at the start and at the stop of the run phase. In fact, the cooling phase is obtained by vertically mirroring the warming up curves for ΔT_{ja} and ΔT_{jc} and accounting for the sensor temperature by using eq. (2). Therefore, a large error in the start will result in a large error at the beginning of the cooling curve as well. We assume that these errors would be mitigated by an improved approximation for $Z_{th,ja}$, by refining the algorithm to better model the in-rush and run-down currents, and by accounting for the variation of the DC-link voltage at the start and stop of the tool. These improvements, necessary to properly model short runnings of 1 or 2 s, where the initial and final errors become crucial in the temperature modeling, will be addressed in future works.

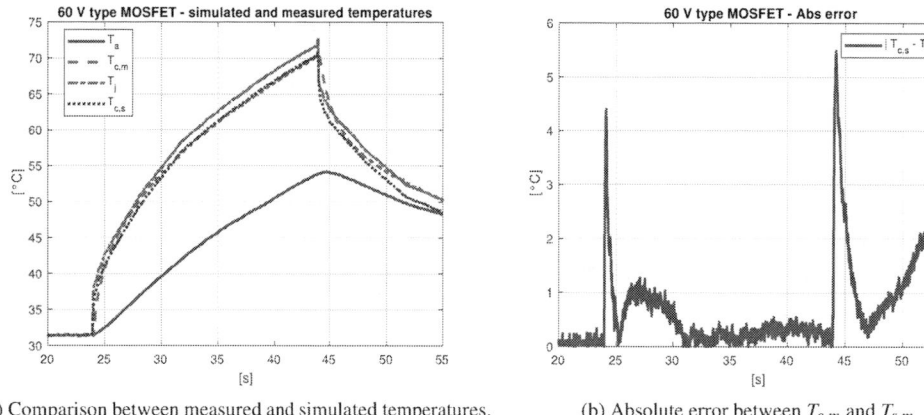

(a) Comparison between measured and simulated temperatures.

(b) Absolute error between $T_{c,m}$ and $T_{s,m}$

Fig. 6: Temperatures and error for the 20 s run test, 60 V type MOSFET.

5   Discussions and conclusions

In this paper, we propose a method for simulating the case and junction temperatures of a MOSFET as used in a 3-phase inverter configuration for a motor drive application. The method uses the transient junction-to-ambient and junction-to-case thermal impedances, this last as provided in the datasheet. The main novelty of the proposed model is that it accounts for the rise in temperature of the board by virtue of a resistive sensor placed in the vicinity of one of the MOSFETs. Therefore, it overcomes the main limitation of fixed ambient temperature commonly found in models that are based on thermal impedances. The method has been validated for 2 different MOSFETs, and proved to provide accurate results, with errors below 6 °C. The simplicity and reliability of the model allow to use it at the design as well as

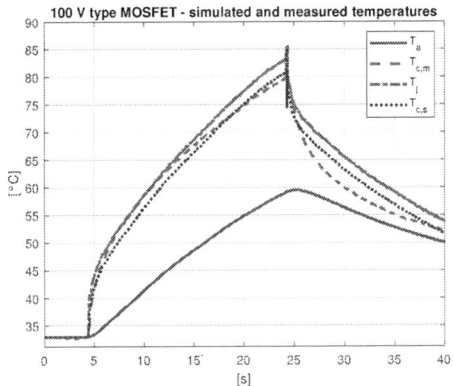
(a) Comparison between measured and simulated temperatures.

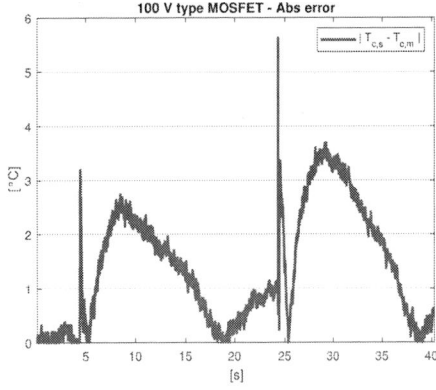
(b) Absolute error between $T_{c,m}$ and $T_{s,m}$

Fig. 7: Temperatures and error for the 20 s run test, 100 V type MOSFET.

during the operation of the tool. In the future, FEA simulations will be used to get an accurate evaluation of the junction-to-ambient thermal impedance, currently approximated via model fitting. The power loss formulas will be refined to include the temperature dependency of the reverse diode. Future versions of the model will as well include variations of the DC-link voltage and refinements of the algorithm.

References

[1] Wolfgang, E., 2007. "Examples for failures in power electronics systems. ECPE tutorial on reliability of power electronic systems", Nuremberg, Germany, pp.19-20.

[2] Nilsson, J., 2019. "Wireless High-Temperature Monitoring of Power Semiconductors: A Single-Chip Approach". Doctoral dissertation, Luleå University of Technology.

[3] Drofenik, U. and Kolar, J.W., 2005, April. "A general scheme for calculating switching-and conduction-losses of power semiconductors in numerical circuit simulations of power electronic systems". In Proc. IPEC (Vol. 5, pp. 4-8).

[4] Philips Semiconductors, 1994. "Power semiconductor applications". Chapter 7, "Thermal management".

[5] Lenz, M., Striedl, G. and Frohler, U., 2000. "Special Subject Book: SMD Packages Thermal Resistance, Theory and Practice". Infineon Technologies.

[6] Košel, V., de Filippis, S., Chen, L., Decker, S. and Irace, A., 2013. "FEM simulation approach to investigate electro-thermal behavior of power transistors in 3-D". Microelectronics Reliability, 53(3), pp.356-362.

[7] Infineon, 2017. "Dynamic thermal behavior of MOSFETs: simulation and calculation of high power pulses". Application Report No. AN_201712_PL11_001. Infineon Technologies AG.

[8] Maricau, E. and Gielen, G., 2013. "Transistor Aging Compact Modeling". In Analog IC Reliability in Nanometer CMOS (pp. 37-77). Springer, New York, NY.

[9] Mistretta, C. and Scrimizzi, F., 2018. "Low-voltage power mosfet switching behavior and performance evaluation in motor control application topologies". STMicroelectronics, application note AN5252, 2018.

[10] Infineon Datasheet BSC016N06NS, "MOSFET, OptiMOS™Power-MOSFET, 60V", Online available https://www.infineon.com/dgdl/Infineon-BSC016N06NS-DataSheet-v02_04-EN.pdf?fileId= db3a3043353fdc160135532b353c483c.

[11] Infineon Datasheet BSC046N10NS3, "MOSFET, OptiMOS™3 Power-Transistor, 100V", Online available https://www.infineon.com/dgdl/Infineon-BSC046N10NS3-DS-v02_00-en.pdf?fileId= db3a304332fc1ee7013316f966a4713c.

[12] Wang, Q., 2015. "Investigation and Implementation of MOSFETs Losses Equations in a Three-phase Inverter". Master thesis. Department of Energy and Environment, Chalmers University of Technology, Gothenburg.

[13] Melito, M. and Gaito, A. and Sorrentino, G., 2015. "Thermal effects and junction temperature evaluation of power mosfets". STMicroelectronics, application note AN4783.

[14] Infineon, 2020. "Transient thermal measurements and thermal equivalent circuit models". Application note AN2015-10.

[15] Huang, A., 2012 "Infineon OptiMOS™ power mosfet datasheet explanation". Application Note AN 2012-03, Infineon Technologies Austria AG .

A 3A Low Voltage Laser Diode Driver IC in a CMOS technology for an iToF-based 3D image sensor

Romain David[1,2], Bruno Allard[1], Xavier Branca[2], Charles Joubert[1]

[1]University of Lyon, University Claude Bernard Lyon 1, INSA Lyon, CNRS, Ampere, F-69621
Villeurbanne, France

[2]STMicroelectronics, Analog, MEMS & Sensors Group, 12 Rue Jules Horowitz, 38019
Grenoble B.P. 217, France

Email: romain.david@etu.univ-lyon1.fr

Keywords

≪High-speed drive≫, ≪Pulsed power converter≫, ≪Integrated Circuit (IC)≫, ≪Diode≫, ≪Sensor≫.

Abstract

This paper presents a Laser Diode Driver (LDD) intended for a 3D image sensor used in mobile phones and based on the indirect Time-of-Flight (iToF) measurement. The work is focused on the study and design of the Integrated Circuit (IC) considering constraints related to mobile applications (low input voltage range, strong integration). The architecture relies on a classical driving topology where a buck-boost DC/DC converter is employed for controlling the current through the laser diode with a high power efficiency while a switching element is connected in series with the diode for generating laser pulses. The novelty here concerns the feasibility of integrating the whole solution (except laser diode and off-chip passive components) on a single chip. A LDD prototype has been implemented in a 130nm CMOS technology from STMicroelectronics. It is able to provide current pulses up to 3A with a 2.5ns pulse width at a maximum 200MHz Pulse Repetition Frequency (PRF) for a 3.6V supply voltage. Under theses conditions, the prototype delivers an average output electrical power of 4.5W to the laser diode with an electrical efficiency of 63%.

Introduction

Three-dimensional (3D) image sensors are key enablers for unlocking emerging applications in consumer electronics such as facial recognition, presence detection, gesture control or Augmented Reality (AR). These sensors mostly rely on range measuring techniques such as structured-light or Time-of-Flight (ToF) principles [1]. The indirect Time-of-Flight (iToF) principle offers the advantage of a simple, reliable and low cost solution for mobile applications by using a laser emitter and an image sensor [2][3]. Its operating principle is to calculate a distance by measuring the phase shift between a modulated infrared laser signal and the optical signal received by the sensor after reflection on an object from the scene. The depth z to each point of the scene is then given by

$$z = \frac{c}{4\pi f}\varphi \tag{1}$$

where φ is the measured phase shift, f is the modulation frequency and c is the speed of light. The block diagram of a typical iToF-based 3D image sensor is depicted in Fig. 1.

Laser pulses with a duty cycle close to 50% are usually sent through the scene as modulated signal [4]. The modulation frequency as well as the peak optical power of laser pulses are key parameters in iToF applications. Indeed, a better depth accuracy is obtained by increasing the modulation frequency and the

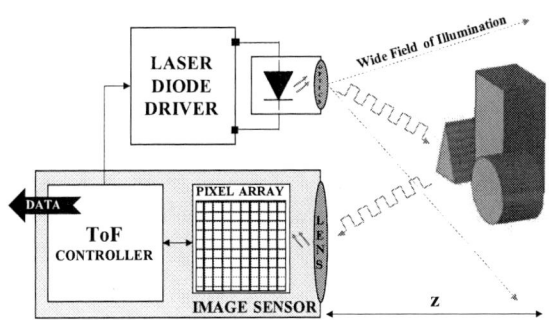

Fig. 1: Block diagram of a typical iToF-based 3D image sensor

peak optical power, which is also a direct function of the laser diode current, as defined by

$$\sigma_z \propto \frac{c}{4\sqrt{2}\pi f} \cdot \frac{1}{\sqrt{P_{OPT}}} \qquad (2)$$

where σ_z is the depth uncertainty, f is the modulation frequency, P_{OPT} is the peak optical power and c is the speed of light [5][6]. Moreover, multiple frequencies are needed to extend the maximum unambiguous range without impacting the measurement precision [7]. Depending on several parameters of the system such as the sensor efficiency, the choice of the laser diode, the object reflectivity or the maximum measurable range, a frequency from tens to hundreds of MHz with a peak current up to few Amps are required to reach a millimeter depth accuracy at few meters of distance. Nevertheless, short pulse trains are needed to avoid thermal dissipation issues and human eye damages according to laser safety standards [8].

These specifications lead to new challenges in the design of a compact, cost-effective and efficient Laser Diode Driver (LDD). Main issues concern the high level of current and the shaping of short-duration pulse at high frequency because of the parasitic elements in passive components, packages and PCB interconnections. In addition, mobile applications bring strong constraints in design choices such as accommodating the limited input voltage range of battery, from 2.5V up to 4.8V in a worst case scenario, as well as dealing with the restricted available space. Finally, a high power efficiency proves essential in order to save battery lifetime and to minimize the self-heating of the device.

Several works have been published presenting laser diode drivers for ToF applications with various current, pulse width and repetition rate specifications. A current pulse generator based on a LC-resonant topology can be used to generate current pulses by charging a capacitance at a predefined voltage and then discharging it through a switch and a load, as reported in [9][10]. Major drawbacks concern the pulse width depending on the capacitance and stray inductance values, and the relatively high voltage required to achieve a high peak current. No additional circuit for managing this high voltage has been reported. Another issue is the long time needed to charge the capacitance which limits the Pulse Repetition Frequency (PRF). An architecture developed in [11] uses a half-bridge synchronous rectifier feeding an inductance to provide a controlled current to a laser diode. An assembly of three switches are employed to shunt the current to and from the load and achieve short current pulses according to a Vernier method. Even if a monolithic LDD working at a 5V input voltage is reported, high breakdown voltage devices, which are usually area and power consuming, are required for supporting high voltage spikes during transients. Another method relies on a current source using a cascode transistor in saturation region and connected in series to a switch. It allows to generate configurable short current pulses through a laser diode at a high repetition rate, as reported in [12]. Although a full integrated LDD is stated, the main concern is the dissipated power through the load path due to the large voltage headroom required for saturating the cascode transistor. The maximum pulse width is thus restricted because of thermal dissipation issues. Furthermore, no additional circuit is mentioned to manage the biasing of the laser diode.

These circuits are mostly intended for LiDAR applications, where constraints in term of power losses are more flexible than for mobile applications. Technologies used might also be specific and incompatible with a low cost requirement. The major objective of this work is to develop a low-cost integrated driver in standard 130nm CMOS technology capable of generating short and high current pulses at a high PRF while operating at a low input voltage range (2.5V-4.8V) and maximizing the power efficiency.

This paper describes the design, implementation and characterization of an integrated LDD prototype. It relies on a current-mode buck-boost DC/DC converter for controlling the current through the laser diode according to the input voltage range. A switching element is connected in series with the laser diode and triggered by a fast modulation signal for generating short current pulses. The novelty here concerns the feasibility of integrating the whole solution (except laser diode and passive components). Current pulses are programmable up to 3A for a PRF from 50MHz up to 200MHz with a 50% duty cycle. Section 2 quickly presents the operating principle of the proposed architecture for the driver. Section 3 describes the design and implementation of the integrated prototype. Simulation results are detailed in section 4. Section 5 covers measurements performed on the prototype with a laser diode. Finally, last section concludes the paper.

Operating Principle

Fig. 2 illustrates a simplified schematic of the proposed driver architecture. A buck-boost DC/DC converter, acting as a current-controlled voltage source, provides a constant output voltage for biasing the laser diode. A switch is connected in series to generate current pulses according to a modulation signal. A current-control loop is employed in order to control the average current through the laser diode, I_{AVG}, by adjusting the output voltage according to the targeted current and the input voltage. The ON-state value of the pulse I_{ON} is controlled by the duty cycle α of the high frequency modulation, given by

$$I_{ON} = \frac{I_{AVG}}{\alpha} \tag{3}$$

The operating frequency of the converter can be sufficiently lower ($<$10MHz) than the PRF (50MHz-200MHz) for benefiting from both flexibility for implementing the feedback loop and a stable ON-state current pulse due to the output voltage ripple seen relatively constant against the pulse width. A power inductor with low value (few hundreds of nH) and an output capacitor with high value (few F) can be selected to take advantage of small passive components while still ensuring a low output voltage ripple.

Even though a buck-boost structure is required here to step up or step down the output voltage according to the input voltage and the targeted current, the converter only provides the average current through the laser diode. Assuming a 3A ON-state current through the laser diode and a 50% duty cycle, an average output current of 1.5A is required. Depending on the DC/DC converter architecture, it could be a significant benefit for reducing the voltage drop and power losses across resistive elements of the power stage, which is critical in the design of an efficient solution specifically in mobile applications.

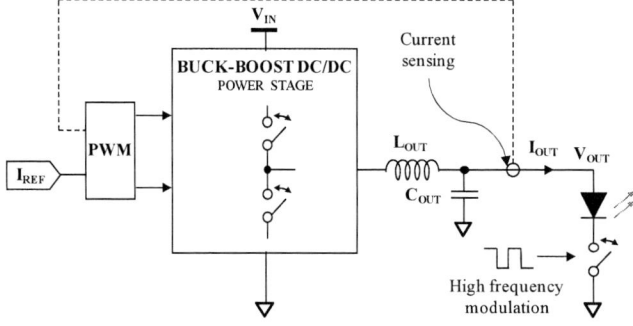

Fig. 2: Simplified schematic of the proposed driver architecture

Design of the Integrated Laser Diode Driver

This section details the design and implementation of an integrated LDD prototype based on the architecture previously described. An integrated solution proves essential in order to develop a compact, fast and low cost system. Indeed, it allows to overcome parasitic elements due to packages and interconnections thus improving transient performances, reducing losses and saving PCB footprint compared to a discrete approach. The 130nm CMOS technology employed here is low-cost and well adapted to accommodate analog, digital and power functions into a single chip.

A block diagram is depicted in Fig. 3. A current-mode buck-boost DC/DC converter IC has been reused from a previous development in order to save time and design resources. It employs a standard Pulse Width Modulation (PWM) method using a current-control loop in order to adjust the output voltage according to the targeted current value set by a current-level reference, as described in [13]. It is not detailed here. The main contribution of this work concerns the design of the high frequency switching block (HFS block for short) which aims to switch the current through the laser diode at a high frequency.

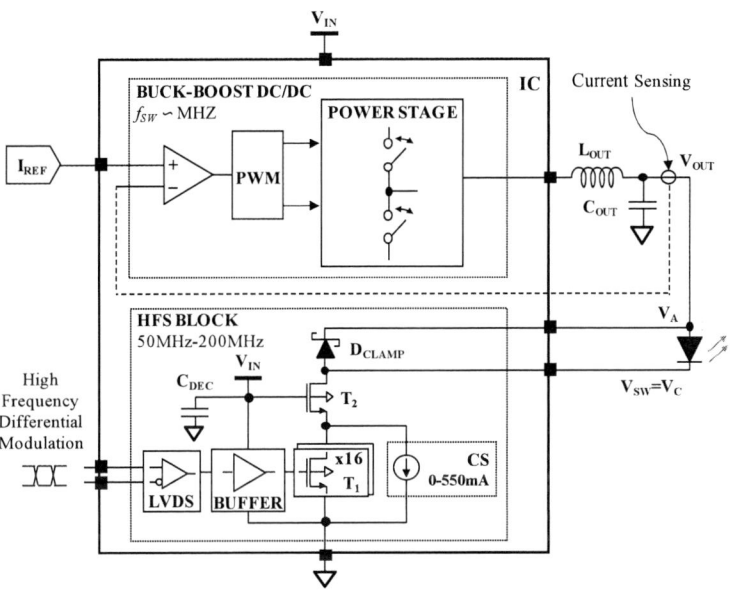

Fig. 3: Simplified block diagram of the integrated laser diode driver prototype. The current-control loop is inherent to the DC/DC converter.

Implementation of the HFS block

A thick oxide n-channel MOSFET transistor (NMOS), named T_1, ensures the current switching at high frequency. It sustains voltages up to 4.8V. Since the pulse current (up to 3A) and the PRF (50MHz-200MHz) are flexible, the gate width of T_1 has been divided into 16 equal parts for optimizing both conduction and switching losses. It is configurable here from 5.4mm to 86.4mm by step of 5.4mm. This transistor is driven by an external differential modulation signal through an appropriate circuitry. It consists of a simple tapered buffer and a Low Voltage Differential Signal (LVDS) receiver used to convert the fast differential signal into a single-ended modulation signal. A cascode transistor, named T_2, as well as a clamping diode are used to protect devices at the switching node V_{SW} from strong over voltage due to parasitic inductances occurring during fast transient currents. Transistor T_2 has been implemented using a thick oxide NMOS device with a 24mm gate width sized by considering a trade-off between its on-resistance and output capacitance. The clamping diode has been implemented using a schottky diode sized large enough to evacuate the residual energy from parasitic inductances. In addition, a programmable current source, named CS, is also employed in order to pre-bias the laser diode near its threshold current thus reducing the turn-on delay between electrical and optical pulses [14]. A current mirror technique is implemented and connected in parallel of transistor T_1. The current is configurable up to 550mA by step of 35mA.

Layout considerations

Some layout considerations have been employed such as using isolation wells and guard rings in order to isolate the bulk of power devices from the substrate. It helps to reduce substrate coupling and minimize latch-up susceptibility due to carrier injection during fast transients. A dedicated power supply path with an assigned pin for each analog, digital and power domains has been used for minimizing the impact of noise between each block. Higher levels of metal with a large number of Vertical Interconnect Access (VIA) in parallel are used for power routing. An integrated Metal-Insulator-Metal (MIM) capacitance, named C_{DEC}, has been implemented in the remaining silicon area for an effective decoupling of the buffer power supply, thus reducing the voltage ripple. For easy readability, only the physical layout of the HFS block, except the clamping diode, is depicted in Fig. 3. The overall silicon area of the chip is around 7mm^2.

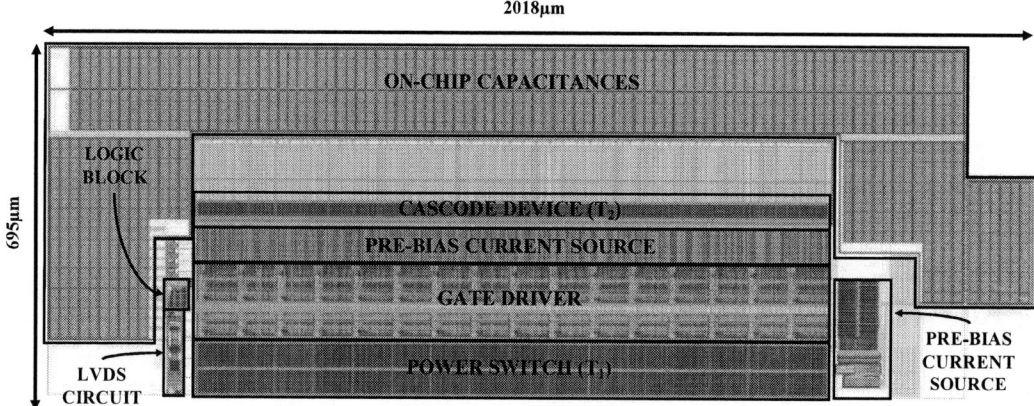

Fig. 4: Physical layout of the HFS block

Simulation Results

Post-layout simulations with parasitic models of package and PCB have been carried out with a SPICE simulator to validate the circuit implementation and to evaluate performances of the LDD prototype. An electrical SPICE- based model of a laser diode is considered for simulation, stating a 2.7V forward voltage for reaching a 3A peak current. Parasitic inductances in the laser diode path have been evaluated at 900pH. Results for a 3A ON-state current with a 2.5ns pulse width at 200MHz over various input voltages are summarized in Table I. Gate width of transistor T_1 has been properly chosen for minimizing total losses. It has not been possible to evaluate the impact of the pre-bias current source through simulations due to a lack of optical model for the laser diode. Transient waveforms of interest for a 3.6V input voltage, are illustrated in Fig. 5.

From Table I, a pulse width around 2.5ns is obtained for each case (2.5V ; 3.6V ; 4.8V). Nevertheless, the rising time is slower than the falling time causing a distortion on the laser diode current waveform, as illustrated in Table I. This slow rising time is due to the relatively large time constant for charging parasitic inductances. Indeed, the circuit when transistor T_1 is switched on can be seen at first order as a series RL circuit where L and R are the equivalent series inductance and resistance in the laser diode path respectively. The high-frequency distortion on laser diode current and voltage is due to small parasitic capacitances within the laser diode electrical model.

Table I: Post-layout simulation results for a 3A ON-state current with a 2.5ns pulse width at 200MHz. Typical conditions have been assumed (typical process, 27°C).

Parameter	Value		
Input voltage	2.5V	3.6V	4.8V
Pulse width (50%)	2.2ns	2.3ns	2.7ns
Rising time (10%-90%)	1.7ns	1.8ns	1.8ns
Falling time (90%-10%)	400ps	700ps	700ps
Peak current	2.3A	4.0A	4.2A
Average current	1.1A	1.8A	1.9A
Output voltage	3.5V	4.7V	4.8V
Voltage spike at switching node	4.8V	6.8V	7V
Average output electrical power	2.5W	5.2W	5.4W
Driver input power	3.9W	8.3W	8.9W
Dissipated power in IC	1.3W	2.9W	3.4W
Electrical efficiency	64%	63%	61%

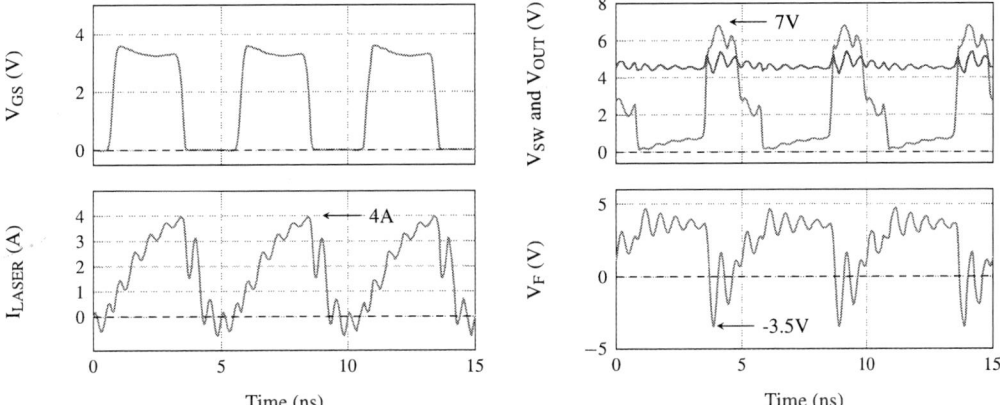

Fig. 5: Simulation transient waveforms: gate-source voltage of transistor T_1 (upper-left corner), laser diode current I_{LASER} (lower-left corner), switching node voltage V_{SW} (upper-right corner, red), output voltage V_{OUT} (upper-right corner, blue) and laser diode forward voltage V_F (lower-right corner) for a 3A ON-state current with a 2.5ns pulse width at 200MHz, under a 3.6V input voltage. Typical conditions have been assumed (typ. process, 27°C)

The average current for 3.6V and 4.8V input voltages is above the expected value of 1.5A, which is explained by the clamping diode inducing a low reverse current. The average current regulation compensates this phenomenon by reaching a 4A peak current. However, a 2.3A peak current is notified for a 2.5V input voltage. This is due to limited performances of the DC/DC converter at low input voltage causing significant voltage drops across resistive elements of power stage thus limiting the output voltage.

Voltage spikes are noticed at the switching node when transistor T_1 is turning off due to parasitics. Even if over voltages above 4.8V are reported for some cases, spikes are absorbed to some extent by the clamping diode. Furthermore, the cascode transistor T_2 still ensures a protection for transistor T_1.

Finally, an electrical efficiency above 60% is noticed for each case. Dissipated powers in IC (not including off-chip components) are indicated for a continuous emission. Practically, short pulse trains are considered for reducing overall losses and self-heating.

Experimental Setup and Measurements Results

Test bench description

A test bench as shown in Fig. 6 has been set up for performing electrical and optical measurements. The laser diode, integrated LDD prototype and off-chip passive components have been assembled on a test board designed for easy configuration. The board features as well optimized interconnections for maximizing performances.

DC voltages have been generated with a Keysight N6705B DC power analyzer. The differential signal (LVDS) as well as an envelope signal for providing pulse trains have been generated with an Agilent 81110A pulse/pattern generator and a Keysight 33600A waveform generator respectively. The laser pulse shape has been acquired using a iCHaus ic212 high-speed free-space photodetector with a 1.4GHz bandwidth. Signals have been measured with a Keysight MSOS254A 2.5GHz mixed-signal oscilloscope and Keysight N2752A 6GHz active differential probes. A Newport 819D-SL-5.3-CAL2 integrating sphere as well as a Newport 918D-SL-OD1R optical power detector have been used for optical power measurements. A short illumination pattern has been set to reduce self-heating effects as much as possible. Finally, a STM32F407 microcontroller from STMicroelectronics has been used for configuring the integrated circuit.

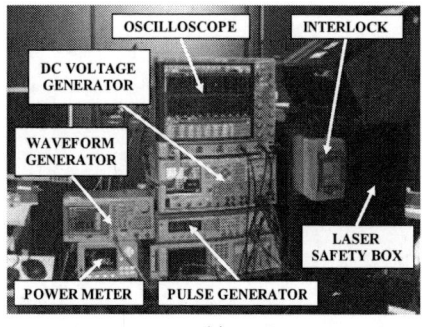

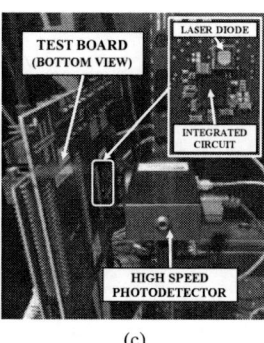

(a) (b) (c)

Fig. 6: Test bench for performing electrical and optical measurements. Due to laser safety standards, measurements have been performed in a laser safety box protected with an interlock. Description of photographs: (a) full test bench, (b) test board mounted on integrating sphere (viewed from inside the box) and (c) test board with LDD prototype coupled to photodetector (viewed from inside the box, a zoom presents the IC, laser diode and passive components)

Results and EMI issues

Primary electrical results have shown some disturbances on an analog block within the chip while the current is switching. This block is a bandgap voltage reference providing a fixed voltage of 1.2V used for the current-control loop. Transient waveform of this bandgap voltage V_{BG} when generating a pulse train for a 3A ON-state current with a 10ns pulse width at 50MHz under a 3.6V input voltage is illustrated in Fig. 7. A pulse train of 200μs has been configured. The bandgap voltage has been directly measured through a dedicated output pin.

Disturbances on bandgap voltage occurs between 150μs and 350μs corresponding to the pulse train duration. An average voltage of 1.18V is noticed when the current is switching while a 1.22V bandgap voltage is expected. More severe disturbances have been noted for higher frequency droping down to 1.15V at 200MHz, causing the current regulation to be less accurate. These disturbances are due to a coupling between the HFS block and bandgap circuit that may arise from internal and/or external interconnections through the substrate of the chip, metallization, PCB traces and ground planes. This could lead to further investigations. Despite these disturbances, the circuit can still be used in an open-loop manner by adjusting the ON-state current value.

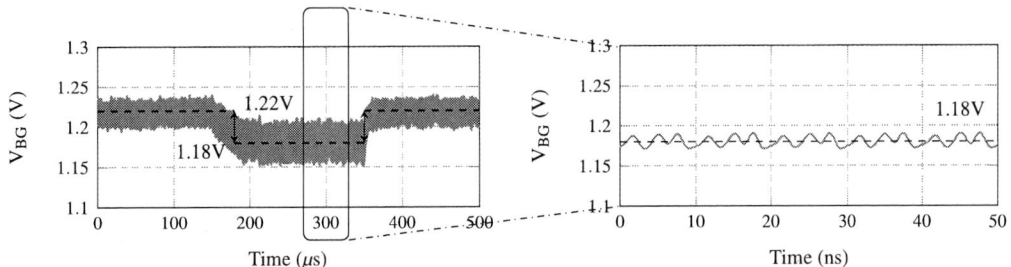

Fig. 7: Measurement transient waveforms : bandgap voltage V_{BG} for a 3A ON-state current with a 10ns pulse width at 50MHz, under a 3.6V input voltage. A zoom is presented on the right. A "peak detect" sampling mode was used for acquired the waveform during a relatively long time (500μs, left curve) while a high resolution mode (real time with averaging) was used for a short time (50ns, right curve)

LDD prototype performances have been measured under typical use case conditions. Short pulse trains have been set in order to reduce the heat generation inside the device. Evaluated performances are transient performances such as rising and falling time of the laser pulse as well as power performances such as the electrical efficiency, defined as

$$\eta = \frac{P_{OUT}}{P_{IN}} \qquad (4)$$

where P_{IN} is the driver input power and P_{OUT} is the average electrical power provided by the driver to the laser diode during a pulse train. Since the output current cannot be measured, P_{OUT} has been estimated from the optical power measurement using the laser diode power conversion efficiency (ratio of output optical power to input electrical power) which is 28% at 27°C. The peak current value has been estimated as well using the laser diode slope efficiency (ratio of output optical power to input current) which is 90% at 27°C. These parameters are only known for 27°C due to lack of date from the supplier. Thus estimations must be considered with precaution.

Performance measurements have been carried out targeting a peak optical power of 2.5W at laser diode level for a 2.5ns pulse width at 200MHz under a 3.6V input voltage. The ON-state current value as well as the pulse width of the external differential modulation signal have been finely tuned for facing a potential laser signal distortion and disturbances on the bandgap voltage reference. The gate width of transistor T_1 has been properly chosen to minimize losses. No pre-bias current has been set for this set of measurements. Transient waveforms of interest are illustrated in Fig. 8. It must be specified that optical and electrical waveforms are not in phase due to impossibility to calibrate the scope with the output signal from the photodetector. Performance results are summarized in Table II.

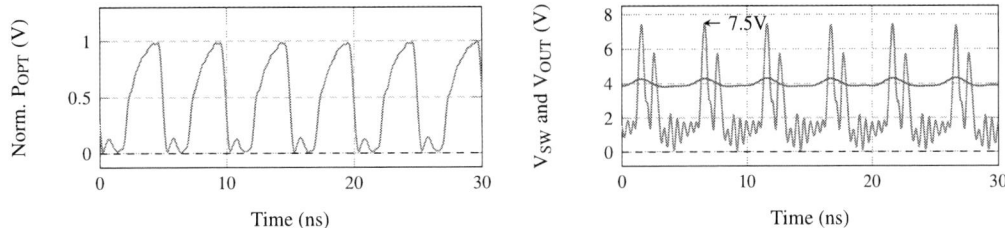

Fig. 8: Measurement transient waveforms : Normalized optical power P_{OPT} of laser pulses (left), switching node voltage V_{SW} (right, red) and output voltage V_{OUT} (right, blue) for a 3A ON-state current with a 2.5ns pulse width at 200MHz, under a 3.6V input voltage. A peak optical power of 2.5W has been targeted here. Y-axis for optical power has been normalized with respect to the peak optical power

First of all, results show that the LDD prototype is able to generate laser pulses. However, even if a pulse

Table II: Performance results obtained for a 3.6V input voltage.

Performance	Value
Peak optical power	2.44W
ON-state current (estimation)	3.01A
Output voltage	4.0V
Average output electrical power (estimation)	4.44W
Driver input power	7.06W
Electrical efficiency (estimation)	63%
Pulse width (50%)	2.5ns
Rising time (10%-90%)	1.6s
Falling time (90%-10%)	410s

width of 2.5ns is measured at 200MHz, the laser signal is slightly distorted. The rising time is slower than the falling time confirming simulation results. In addition, a slight optical bounce is noticed when the laser pulse is cutting off. It is caused by the strong ringing observed on the switching node voltage where it becomes sufficiently low during a brief moment for activating the laser diode anew. Voltage spikes around 7.5V are also noticed at the switching node suggesting that the clamping diode is less efficient than expected. Furthermore, it can be noticed that a 4V output voltage against a 3.6V input voltage is required for reaching targeted objectives, confirming that a buck-boost structure is essential for this kind of series configuration. Finally, an electrical efficiency of 63% is estimated from measurements what is consistent with simulation results.

Conclusion and Outlooks

An integrated laser diode driver prototype for ToF applications has been developed in a 130nm CMOS technology. Its architecture relies on a current-mode buck-boost DC/DC converter connected in series with a laser diode and a power switch for generating short laser pulses. Optical and electrical measurement results show that the prototype is able to provide laser pulses of a 2.5W peak optical power with a pulse width of 2.5ns at 200MHz under a 3.6V input voltage. Under these conditions, a relatively high efficiency is noticed which is quite satisfying in the context of mobile applications. Unfortunately it may not be compared to the state-of-art as efficiency values are rarely mentioned and the application is different in context. Nevertheless, an architecture based on another classical driving topology, known as shunt topology, for same requirements, has been tested. This will be discussed in another paper.

A second set of measurements (by sweeping the PRF, peak current and input voltage for instance) is essential to fully validate performances of the laser diode prototype. The measurement test bench should be consolidated as well to improve measurement accuracy. The programmable pre-bias current source, that has not been studied yet, should be deeply evaluated for confirming its benefit. In addition, more investigations are required in order to confirm the source of issues observed on the bandgap voltage reference. Similarly, issues about the effectiveness of the clamping diode should be further examined. Lastly, the prototype should be tested when operating with a ToF image sensor for validating the overall performances of the system at application level.

References

[1] R. Lineback, "O.S.D. Report - A Market Analysis and Forecast for Optoelectronics, Sensors/Actuators, and Discretes - Section 4", IC Insights Inc., pp. 85-87, 2019.
[2] C. Tubert, L. Simony, F. Roy, T. Arnaud, L. Pinzelli, and P. Magnan, "High Speed Dual Port Pinned-photodiode for Time-Of-Flight Imaging", *Proc. International Image Sensor Workshop*, 01 2009.
[3] A. Suss, V. Rochus, M. Rosmeulen, and X. Rottenberg, "Benchmarking time-of- flight based depth measurement techniques," *Smart Photonic and Optoelectronic Integrated Circuits XVIII* (S. He, E.-H. Lee, and L. A. Eldada, eds.), vol. 9751, pp. 199 217, International Society for Optics and Photonics, SPIE, 2016.

[4] C.L. Niclass, "Single-Photon Image Sensors in CMOS: Picosecond Resolution for Three-Dimensional Imaging", *PhD Thesis*, Lausanne, EPFL, 2008, pp. 6-8.

[5] R. Lange and P. Seitz, "Solid-State Time-of-Flight Range Camera", *IEEE Journal of Quantum Electronics*, vol. 37, no. 3, pp. 390-397, 2001.

[6] B. Bttgen and P. Seitz, "Robust Optical Time-of-Flight Range Imaging Based on Smart Pixel Structures", *IEEE Transactions on Circuits and Systems*, Vol. 55, No. 6, 2008, pp. 1512-1525.

[7] A. P. P. Jongenelen, D. A. Carnegie, A. D. Payne and A. A. Dorrington, "Maximizing precision over extended unambiguous range for TOF range imaging systems", *IEEE Instrumentation & Measurement Technology Conference Proceedings*, Austin, TX, 2010, pp. 1575-1580.

[8] International Electrotechnical Comission, "Safety of laser products - Part 1: Equipment classification and requirements", 2014, IEC 60825-1:2014.

[9] J. Nissinen and J. Kostamovaara, "A 1A Laser Driver in 0.35um Complementary Metal Oxide Semiconductor Technology for a Pulsed Time-of-Flight Laser Rangefinder", *Review of Scientific Instruments*, Vol. 80, 2009.

[10] J. Glaser, "High Power Nanosecond Pulse Laser Driver using a GaN FET", *PCIM Europe 2018*, 2018.

[11] E. Abramov, M. Evzelman, O. Kirshenboim, T. Urkin and M. M. Peretz, "Low Voltge Sub-ns Pulsed Current Driver IC for High-Resolution LiDAR Applications", *IEEE Applied Power Electronics Conference and Exposition*, 2018.

[12] G. Blasco, D. Drich, H. Reh, R. Burkard, E. Isern and E. Martin, "A Sub-ns Integrated CMOS Laser Driver With Configurable Laser Pulses for Time-of-Flight Applications", *IEEE Sensors Journal*, Vol. 18, No. 16, 2018, pp. 6547-6556.

[13] C. Fai Lee and P. Mok, "A Monolithic Current-Mode CMOS DC-DC Converter With On-Chip Current-Sensing Technique", *IEEE Journal of Solid-State Circuits*, vol. 39, no. 1, pp. 3-14, 2004.

[14] J. M. Senior, "Optical Fiber Comunications, Principles and Practice (third edition)", Pearson Education Limited, 2009, p686.

Comparison of Decoupling Techniques via Discrete Luenberger Style Observer for Voltage Oriented Control

Gyanendra Kumar Sah, Michael Schütt, Hans-Günter Eckel
UNIVERSITY OF ROSTOCK
Albert-Einstein-Str.2
D-18059 Rostock, Germany
Tel.: +49 / (0) 381 – 498 7116.
Fax: +49 / (0) 381 – 498 7102.
E-Mail: michael.schuett@uni-rostock.de
URL: http://www.iee.uni-rostock.de

Acknowledgments

This paper was made within the framework of the research project *Netz-Stabil* and financed by the European Social Fund (ESF/14-BM-A55-0015/16). This paper is part of the qualification program *Promotion of Young Scientists in Excellent Research Associations - Excellence Research Programme of the State of Mecklenburg-Western Pomerania*.

Keywords

«Digital Control», «Vector Control», «Converter Control», «Voltage Source Inverters (VSIs)», «Frequency Domain Analysis».

Abstract

This work focuses on the implementation, analysis, and comparison of discrete-time synchronous frame current controllers for grid-connected three-phase VSIs using *Discrete Luenberger Style Observers* (DLSO) in both synchronous (dq), and stationary (αβ) frames including the controller's computational delay compensation. Further, the unique cross-coupling effects of the voltages and currents in the discrete domain are discussed. *Pole-zero* cancellation techniques are used for the tuning of the current controllers. A discrete *Phase-Locked Loop* (PLL) with decoupled nonlinearities is proposed. For each topology, Frequency Response Function (FRF) plots and the step responses are obtained and analyzed.

Introduction

VSIs are widely used for controlling the power exchange between the electric grid and electric loads, as well as between renewable power plants and the grid. Discrete control techniques are the standard implementation methods due to the rapid advancement in digital computing devices, cost reduction, and reduced hardware complexity. Unlike in continuous-time (analog hardware) control, the microcontroller needs finite time for computing the required duty cycle for the switches; hence computational delay is introduced in the discrete-time controller. Therefore, for designing robust controller computation, the delays in the plant need to be compensated, which can be achieved by making use of discrete Luenberger style observers [1].

The main goal of this paper is to extend the work in [1]. The advantage of implementing current control in the dq-frame is that the steady-state AC states become DC in the dq-frame. However, designing a robust controller needs extra care because of the computational delay of the microcontroller and the unique cross-coupling effects present in the discrete dq-frame [1, 2]. These obstacles demand an accurate discrete-time model, which plays a vital role in designing a robust controller and discrete Luenberger style observers [1-4]. The advantage of the αβ-frame model is that the states are not cross-coupled, unlike in the dq-frame. However, controlling AC states introduces phase errors between the reference and the controlled states.

EPE'20 ECCE Europe

Assigned jointly to the European Power Electronics and Drives Association & the Institute of Electrical and Electronics Engineers (IEEE)

This problem can be solved by adding *Feed Forward* (FF) in the control, which is very parameter sensitive, though. The synchronous frame current controller can be implemented using DLSO in either the dq or the αβ-frame. Proper decoupling of the grid voltage is also significant, which defines the control plant as the filter and thus achieving higher robustness of the control in obtaining a robust discrete model [1-3]. A comparative study of different current controllers using discrete Luenberger style observers in both dq and αβ-frames with computational delay compensation and a unique method for the grid voltage decoupling is presented in this paper.

Average Circuit Model Development

Fig. 1 illustrates the basic circuit diagram of the three-phase PWM VSI connected to the grid with *L*-filter (RL load). Further, Fig. 2 shows the simplified average circuit diagram. The discrete current controller takes $i_{123}[k]$, $v_{123}[k]$, and $v_{dc}[k]$ as the input signals and gives suitable switching signals (PWMs) to the switches as the output signal. Since the neutral point of the grid (N) is floating, the zero-sequence current is absent ($i_0 = 0$). Fig. 3 depicts the average circuit model of three-phase PWM VSIs connected to the grid with *L*-filter (RL load) in the αβ-frame [3].

The voltages and currents are sampled at the extremes of the carrier waveform, which is shown in Fig. 4. Further, the duty cycle (d_{123}) or PWMs are updated at these extremes [3]. In this case, the sampling frequency ($f_S = 1/T_S = 2f_{SW}$) is equal to twice the switching frequency ($f_{SW} = 1/T_{SW}$) of the carrier waveform, where T_S and T_{SW} are the sampling time and the switching period, respectively. The controller's computational time (Δt_X) introduces one sample time delay in the abc or αβ-frame. Fig. 7 illustrates the state block diagram of a grid-side VSI with *L*-filter (*RL*-load) using complex vector representation in both αβ and dq-frames, including the computation delay [1,4]. In this figure, the *Zero-Order Hold* (ZOH) is used to sample and hold the signal until the next sampling event is triggered, representing the A/D interface and average behavior of the PWM [4]. The effect of computational time delay can be compensated by computing $(k+1)^{th}$ duty cycle ($d[k+1]$) using present (k^{th}) input samples and updating it in the next or $(k+1)^{th}$ sampling event. The delay model in the dq-frame can be obtained by transforming the delay in the αβ-frame using $z^{-1}|_{\alpha\beta} \rightarrow e^{-j\omega_e T_s} z^{-1}|_{dq}$ [3], which is shown in Fig. 7b.

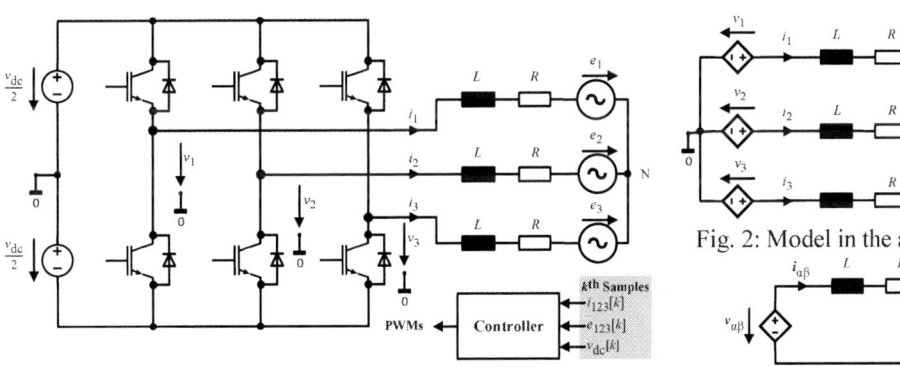

Fig. 1: Three-phase VSI connected to the grid using *L*-filter.

Fig. 2: Model in the abc-frame.

Fig. 3: Model in the αβ-frame.

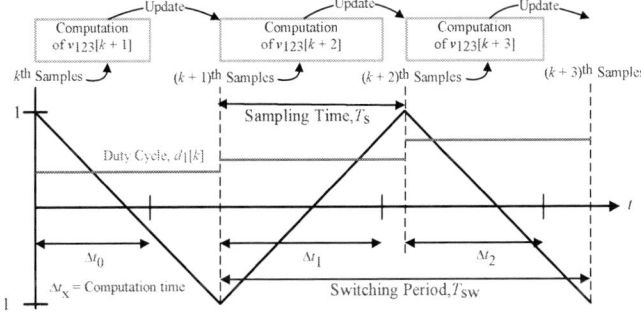

Fig. 4: PWM implementation for the first leg of VSI.

Fig. 5: Controller parameter tuning.

Discrete Phase-Locked Loop with Decoupled Nonlinearities

The proposed discrete PLL with the decoupling of nonlinearities is shown in Fig. 6. $\hat{E}_m = \sqrt{(\hat{e}_d)^2 + (\hat{e}_q)^2}$ represents the estimated phase to neutral grid voltage magnitude, where $\hat{e}_d$ and $\hat{e}_q$ are the estimated grid voltages in the d-axis and q-axis, respectively. The PLL synchronizes the grid voltage angle with the d-axis by controlling the grid voltage in the q-axis ($\hat{e}_q = 0$). For symmetrical three-phase grid voltages ($e_1 = E_m \cos(\theta_e)$, $e_2 = E_m \cos(\theta_e - 2\pi/3)$, and $e_3 = E_m \cos(\theta_e + 2\pi/3)$), the general expression for the grid voltage magnitude in the q-axis is $e_q = E_m \sin(\theta_e - \hat{\theta}_e)$, where θ_e and $\hat{\theta}_e$ are the actual and estimated grid voltage angles, respectively. Fig. 6 illustrates the decoupling of nonlinearities. The closed-loop transfer function of the PLL, which is obtained by solving (1), is given in (2). The closed-loop poles of the PLL are placed at two eigenvalues (at f_1 and f_2) by comparing the denominator of (2) with $z^2 - z\,(e^{-2\pi f_1 T_s} + e^{-2\pi f_2 T_s}) + e^{-2\pi(f_1 + f_2)T_s}$ and solving the equations. Fig. 6 shows the general expressions for computing K_P and K_I. The closed-loop eigenvalues of the proposed PLL are placed at $f_2 = 1\,\mathrm{Hz}$ and $f_1 = 10\,\mathrm{Hz}$. The estimated grid voltage angle ($\hat{\theta}_e[k+1]$) and angular frequency ($\hat{\omega}_e$) are used for the Park's and the Inverse Park's transformations.

$$\left(\theta_e(z) - \hat{\theta}_e(z)\right)\left(K_p + \frac{K_I\,T_s}{1 - z^{-1}}\right)\left(\frac{T_s\,z^{-1}}{1 - z^{-1}}\right) = \hat{\theta}_e(z) \tag{1}$$

$$\frac{\hat{\theta}_e(z)}{\theta_e(z)} = \frac{T_s\left(z\left(K_p + K_I\,T_s\right) - K_p\right)}{z^2 - z\left(2 - T_s(K_p + K_I\,T_s)\right) + \left(1 - T_s\,K_p\right)} \tag{2}$$

Direct-Discrete Model Development

The impact of the sampled input voltage ($V_{\alpha\beta}(z)$) on sampled current ($I_{\alpha\beta}(z)$) in the αβ-frame can be obtained by transforming the fundamental differential equation of the plant, obtained from Fig. 7a, into a difference equation and applying z-Transformation [1, 3], which is shown in (3), where $\tau\,(= L/R)$ represents the electric time constant. (4) represents the overall transfer function of the system in the αβ-frame, including computational delay. Since the PWM is implemented in the αβ-frame, the impact of the sampled input voltage ($V_{dq}^*(z)$) on the sampled current ($I_{dq}(z)$) in the dq-frame is obtained by transforming (4) using $z^{-1}|\alpha\beta \rightarrow e^{-j\omega_e T_s}\,z^{-1}|dq$, which is given in (5) [3], where ω_e represents the synchronous speed of the grid voltage. The discrete-time grid voltage model in the dq-frame cannot be obtained using a model in the αβ-frame because this fails to present the continuous sinusoidal nature of the grid voltage. However, a perfect grid assumption would yield a constant voltage in the dq-frame and therefore, can be represented accurately also in the discrete domain.

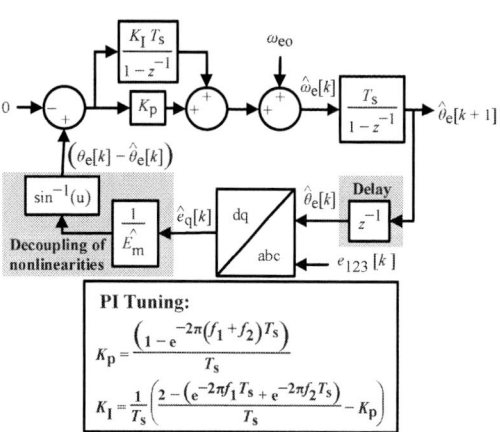

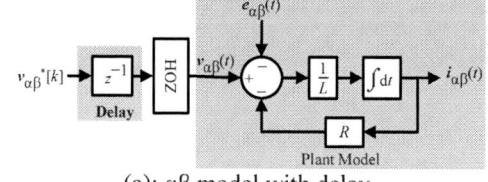

(a): αβ model with delay.

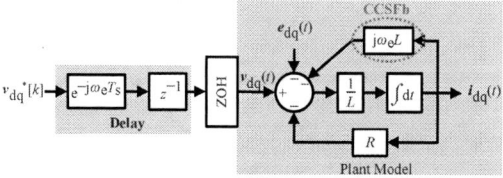

(b): dq model with delay.

Fig. 6: Proposed discrete-time Phase-Locked Loop (PLL) for grid voltage angle and angular frequency computation.

Fig. 7: Complex state block diagram of grid-side VSI with L filter (RL load) in continuous time with delay.

$$\frac{I_{\alpha\beta}(z)}{V_{\alpha\beta}(z)} = \frac{z^{-1}\left(1 - e^{-T_s/\tau}\right)}{R\left(1 - e^{-T_s/\tau}z^{-1}\right)} \tag{3}$$

$$\frac{I_{\alpha\beta}(z)}{V_{\alpha\beta}^{*}(z)} = \frac{z^{-1}\left(1 - e^{-T_s/\tau}\right)}{R\left(1 - e^{-T_s/\tau}z^{-1}\right)} z^{-1} \tag{4}$$

$$\frac{I_{dq}(z)}{V_{dq}^{*}(z)} = \frac{z^{-1}e^{-j\omega_e T_s}\left(1 - e^{-T_s/\tau}\right)}{R\left(1 - e^{-(j\omega_e + 1/\tau)T_s}z^{-1}\right)} z^{-1}e^{-j\omega_e T_s} \tag{5}$$

By following the same steps used to obtain (3), the impact of $E_{dq}(z)$ on $I_{dq}(z)$ is achieved using a model in the dq-frame first (refer Fig. 7b) [3], which is shown in (6). Then the transfer function $I_{\alpha\beta}(z)/E_{\alpha\beta}(z)$ is obtained by inversely transforming (6) $(I_{dq}(z)/E_{dq}(z))$ using $z^{-1}|^{dq} \rightarrow e^{j\omega_e T_s}z^{-1}|^{\alpha\beta}$, which is shown in (7). The accurate dq-frame discrete-time model can be obtained by superimposing (5) and (6), which is shown in (8). Further, the accurate $\alpha\beta$-frame discrete-time model can be obtained by superimposing (4) and (7), which is shown in (9). (8) and (9) describe the discrete-time complex state block diagram of the grid-side VSI with L filter (RL load), including the computation delay in the dq and $\alpha\beta$-frame and are shown in Fig. 8 (top right) and Fig. 10 (bottom right), respectively.

$$\frac{I_{dq}(z)}{E_{dq}(z)} = -\frac{z^{-1}\left(1 - e^{-(j\omega_e + 1/\tau)T_s}\right)}{(R + j\omega_e L)\left(1 - e^{-(j\omega_e + 1/\tau)T_s}z^{-1}\right)} \tag{6}$$

$$\frac{I_{\alpha\beta}(z)}{E_{\alpha\beta}(z)} = -\frac{z^{-1}e^{j\omega_e T_s}\left(1 - e^{-(j\omega_e + 1/\tau)T_s}\right)}{(R + j\omega_e L)\left(1 - e^{-T_s/\tau}z^{-1}\right)} \tag{7}$$

$$I_{dq}(z) \doteq \frac{z^{-1}\left(1 - e^{-T_s/\tau}\right)}{R\left(1 - e^{-(j\omega_e + 1/\tau)T_s}z^{-1}\right)}\left(\left(z^{-1}e^{-2j\omega_e T_s}\right)V_{dq}^{*}(z) - \frac{R\left(1 - e^{-(j\omega_e + 1/\tau)T_s}\right)}{(R + j\omega_e L)\left(1 - e^{-T_s/\tau}\right)}E_{dq}(z)\right) \tag{8}$$

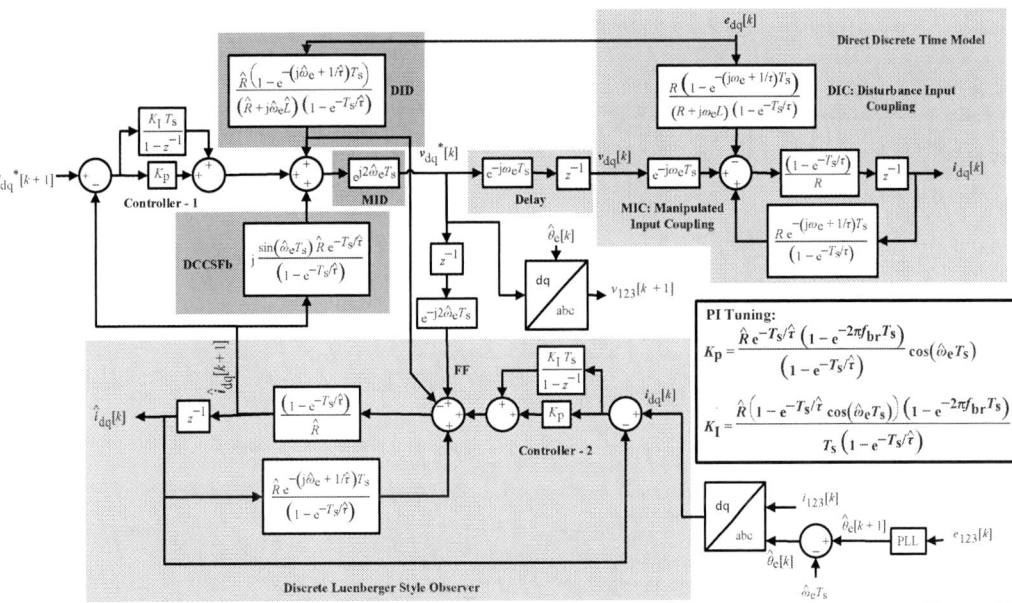

Fig. 8: Proposed discrete-time complex vector synchronous frame PI current controller with computational delay compensation using a Discrete Luenberger Style Observer (DLSO) implemented in the synchronous (dq) frame with discrete MID, DID and DCCSFb.

$$I_{\alpha\beta}(z) = \frac{z^{-1}\left(1 - e^{-T_s/\tau}\right)}{R\left(1 - e^{-T_s/\tau}\,z^{-1}\right)}\left(z^{-1}V_{\alpha\beta}{}^*(z) - \frac{R\left(e^{j\omega_e T_s} - e^{-T_s/\tau}\right)}{\left(R + j\omega_e L\right)\left(1 - e^{-T_s/\tau}\right)}E_{\alpha\beta}(z)\right) \tag{9}$$

Discrete Current Controllers with Computational Delay Compensation

For computing the future duty cycles ($d_{123}[k+1] \propto v_{123}[k+1]$) using k^{th} input samples, the future currents ($i_{123}[k+1]$) need to be controlled. But, in the present (k^{th} sample), it is not possible to sample future currents ($i_{123}[k+1]$) from the actual physical system. Hence, future currents ($\hat{i}_{123}[k+1]$) need to be estimated using present sampled currents ($i_{123}[k]$). This can be realized by using a DLSO. A DLSO can be implemented in the abc, $\alpha\beta$, or dq-frame. Fig. 8 (button left) depicts an example of the implementation of a DLSO for the estimation of future currents using present sampled currents in the dq-frame with a PI controller. The mathematical model of the plant is implemented in the DLSO. The manipulated input (v_{dq}) of the current controller is added as a *Feed-Forward* (FF), and the sampled current represents the reference signal. Since the plant model is implemented mathematically in the DLSO, it provides access to additional state information. In Fig. 8, the plant is rearranged to get future estimated current ($\hat{i}_{dq}[k+1]$) as the output signal. In this paper, four synchronous (dq) frame current controller techniques are discussed. These current controllers can be further sub-divided into two sections based on the implementation of DLSO in the dq-frame or the $\alpha\beta$-frame.

Synchronous Current Controller with DLSO implemented in the dq-Frame

A. *Direct-Discrete dq-Current Controller with DLSO Implemented in the dq-Frame using PI Regulator*

The complex vector state block diagram of the proposed discrete-time dq-current controller with discrete *Decoupling Cross-Coupling State Feedback* (DCCSFb), *Manipulated Input Decoupling* (MID), *Disturbance Input Decoupling* (DID), and the DLSO implemented in the dq-frame using PI controllers is shown in Fig. 8. Here, $\hat{L}$, $\hat{R}$, and $\hat{\tau}$ ($=\hat{L}/\hat{R}$) are the estimated inductance, resistance and electric time constant of the L-filter, respectively. It can be observed that there are two current controller loops (a) *Reference Command* ($I_{dq}{}^*[k+1]$) *Current Control Loop* (RC-CCL) and *DLSO Current Control Loop* (DLSO-CCL). The closed-loop eigenvalues of RC-CCL and DLSO-CCL can be designed independently. The DLSO-CCL has low pass filter attributes for the sampled currents. On the one hand,

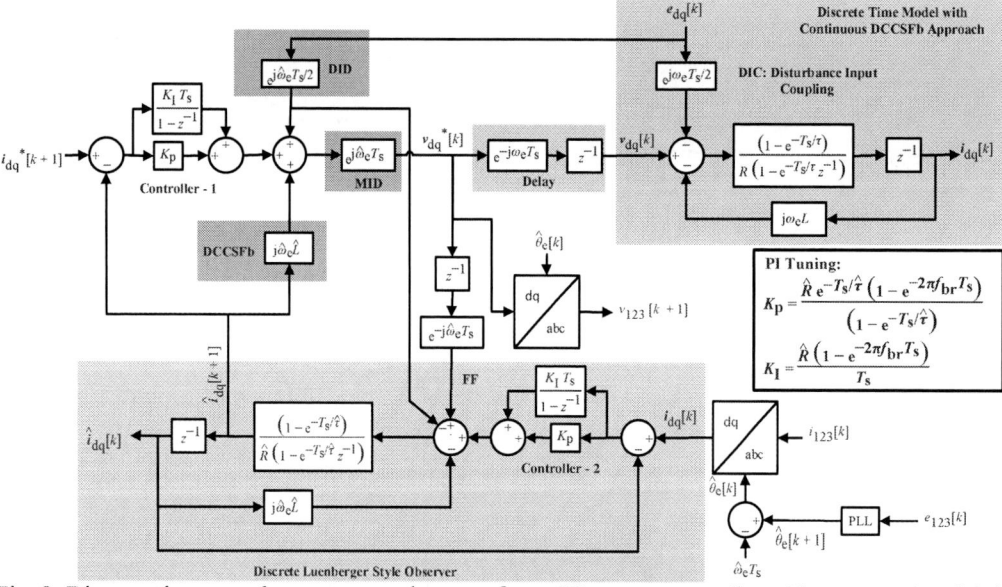

Fig. 9: Discrete-time complex vector synchronous frame PI current controller with computational delay compensation using discrete Luenberger Style Observer (DLSO) implemented in synchronous (dq) frame with discrete DID and continuous DCCSFb approach.

decreasing its bandwidth can help reducing noise in the current signal. But a lower bandwidth also lowers the overall robustness of the system. Thus, the bandwidth has to be designed appropriately regarding robustness and noise requirements.

Control Parameter Tuning

Fig. 5 shows the simplified state block diagram for the tuning of the PI controller, which is obtained by ignoring the *Cross-Coupling State Feedback* (CCSFb), *Manipulated Input Coupling* (MIC) and *Disturbance Input Coupling* (DIC) because of the implementation of DCCSFb, MID and DID. (10) provides the open-loop transfer function ($G_{dq}(z)$) of RC-CCL (see also Fig. 5). The necessary condition for pole-zero cancelation is given in (11), and the closed-loop transfer function of the RC-CCL with PZ cancelation is given in (12). The closed-loop eigenvalue of the transfer function (12) can be placed at a set break frequency f_{br} by comparing its denominator with $z - e^{-2\pi f_{br}T_s}$; the result of the comparison is shown in (13). Solving (11) and (13) yields the general expressions for K_P and K_I, which is shown in Fig. 8.

$$G_{dq}(z) = \left(\frac{(K_p + K_I T_s)\left(1 - e^{-T_s/\hat{\tau}}\right)}{\hat{R}(z-1)} \right)\left(\frac{z - \dfrac{K_p}{K_p + K_I T_s}}{z - e^{-T_s/\hat{\tau}}\cos(\hat{\omega}_e T_s)} \right) \tag{10}$$

$$\frac{K_p}{K_p + K_I T_s} = e^{-T_s/\hat{\tau}}\cos(\hat{\omega}_e T_s) \tag{11}$$

$$\frac{\hat{I}_{dq}(z)}{I_{dq}{}^*(z)} = \frac{G_{dq}(z)}{1 + G_{dq}(z)} \triangleq \left(\frac{(K_p + K_I T_s)\left(1 - e^{-T_s/\hat{\tau}}\right)}{\hat{R}\left(z - 1 + (K_p + K_I T_s)\left(1 - e^{-T_s/\hat{\tau}}\right)/\hat{R}\right)} \right) \tag{12}$$

$$1 - (K_p + K_I T_s)\left(1 - e^{-T_s/\hat{\tau}}\right)/\hat{R} = e^{-2\pi f_{br}T_s} \tag{13}$$

By following the same steps used to tune Controller-1, it can be seen that the open-loop transfer function for tuning DLSO-CCL is the same as the open-loop transfer function obtained for tuning RC-CCL, which is shown in (10). Hence, the expression for K_P and K_I, shown in Fig. 8, can also be used for tuning both RC-CCL and DLSO-CCL. The same procedure has been followed for tuning both RC-CCL and DLSO-CCL for all upcoming discrete current controllers, and the general expressions for tuning the controllers are shown in their respective state block diagrams.

B. Discrete dq-Current Controller with Continuous DCCSFb Approach and DLSO Implemented in the dq-Frame using PI Regulator

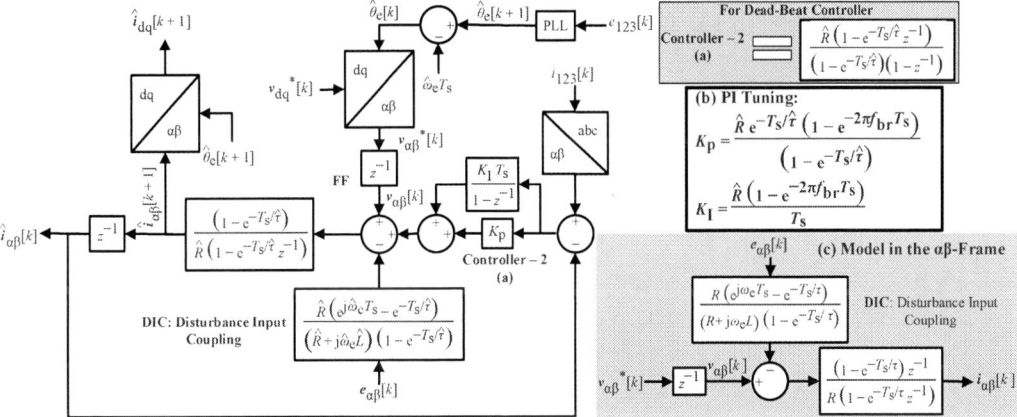

Fig. 10: DLSO implemented in the αβ-frame using (a) dead-beat Controller and (b) PI controller, and (c) direct-discrete model in the αβ-frame.

The complex vector state block diagram of the discrete-time PI current controller with continuous DCCSFb approach [1, 2, 5] and DLSO implemented in the dq-frame using a PI controller is shown in Fig. 9. Using this approach, the cross-coupling between the d and q currents (CCSFb) in continuous-time is retained in the discrete-time domain (see Fig. 7b), and the remaining state block diagram is used for obtaining the discrete plant model in the dq-frame which is shown in Fig. 9 (top right).

Synchronous Current Controller with DLSO implemented in the αβ-Frame

Since cross-coupling of the states in the αβ-model is not present, the DLSO can be implemented in the αβ-frame, and the current controller can be designed in the dq-frame to utilize the advantages of both frames. The DLSO in the αβ-frame is realized using (9), which is shown in Fig. 10. Its controller can be tuned using either a PI controller or a dead-beat controller (where $\hat{I}_{dq}(z)/I_{dq}(z) = z^{-1}$).

Direct-discrete dq-Current Controller with DLSO Implemented in the αβ-Frame using a Dead-Beat Regulator

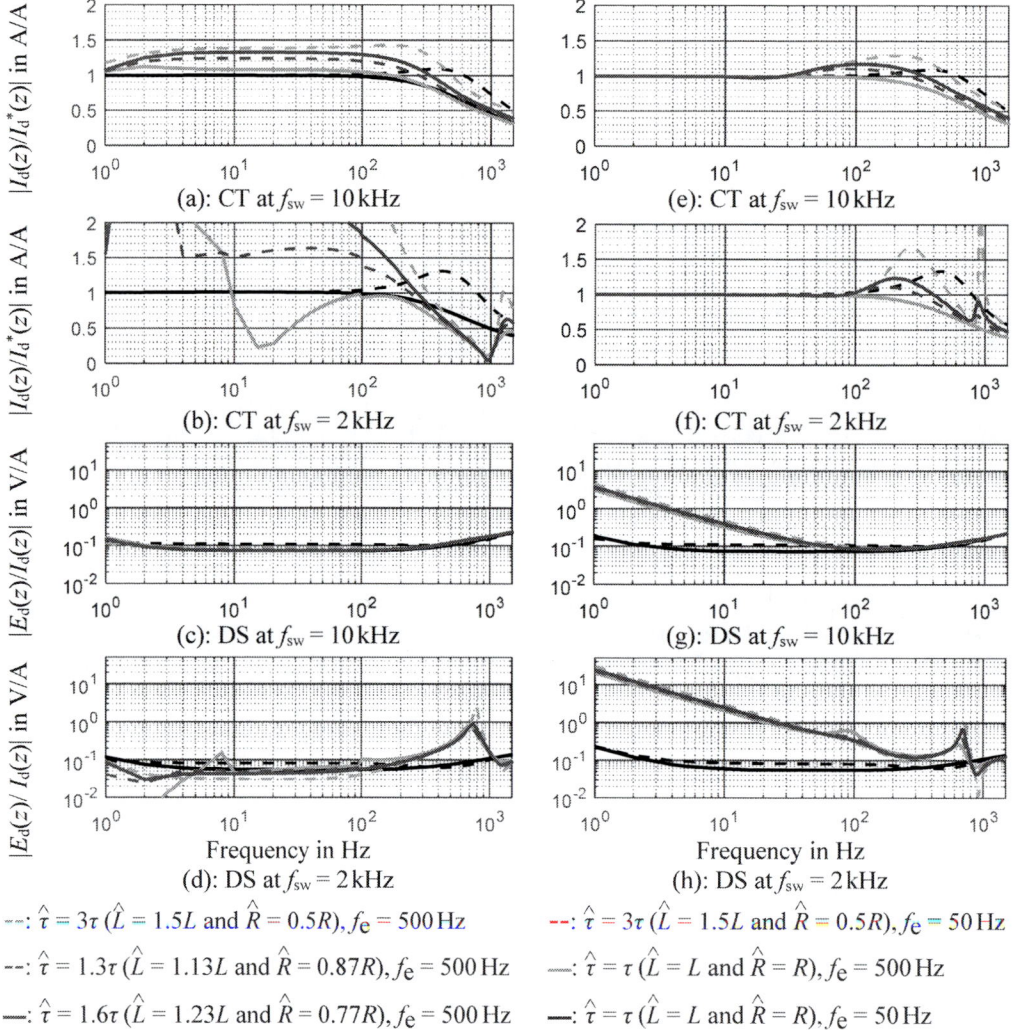

$--:\ \hat{\tau} = 3\tau\ (\hat{L} = 1.5L\ \text{and}\ \hat{R} = 0.5R), f_e = 500\,\text{Hz}$

$--:\ \hat{\tau} = 1.3\tau\ (\hat{L} = 1.13L\ \text{and}\ \hat{R} = 0.87R), f_e = 500\,\text{Hz}$

$—:\ \hat{\tau} = 1.6\tau\ (\hat{L} = 1.23L\ \text{and}\ \hat{R} = 0.77R), f_e = 500\,\text{Hz}$

$--:\ \hat{\tau} = 3\tau\ (\hat{L} = 1.5L\ \text{and}\ \hat{R} = 0.5R), f_e = 50\,\text{Hz}$

$—:\ \hat{\tau} = \tau\ (\hat{L} = L\ \text{and}\ \hat{R} = R), f_e = 500\,\text{Hz}$

$—:\ \hat{\tau} = \tau\ (\hat{L} = L\ \text{and}\ \hat{R} = R), f_e = 50\,\text{Hz}$

Fig. 11: Command tracking and dynamic stiffness frequency response functions of 500 Hz bandwidth synchronous frame PI current controllers for 10 kHz and 2 kHz switching frequencies at low and high synchronous speed w/ and w/o parameter estimation error for different control structures with DLSO in the dq frame: (a – d) Continuous DCCSFb controller with DLSO implemented in the dq-frame, (e – h) Discrete DCCSFb controller with DLSO implemented in the dq-frame for the parameters $R = 0.3\,\text{m}\Omega$, $L = 25\,\mu\text{H}$ and $E_{LL(\text{rms})} = 690\,\text{V}$.

Dead-beat controller showcases the fastest possible response in discrete time. The command is reached in one sample period. Therefore, DLSOs based on dead-beat controllers provide no noise filtering. (14) provides the expression for the dead-beat controller (Controller-2) with $\hat{I}_{dq}(z)/I_{dq}(z) = z^{-1}$ (also see Fig. 10).

$$Controller\text{-}2 \ (DB) = \left(\frac{z^{-1}}{1 - z^{-1}}\right)\frac{V_{\alpha\beta}(z)}{\hat{I}_{\alpha\beta}(z)} = \frac{\hat{R}\left(1 - e^{-T_s/\hat{\tau}}z^{-1}\right)}{\left(1 - e^{-T_s/\hat{\tau}}\right)\left(1 - z^{-1}\right)} \tag{14}$$

Simulations Results

The simulation results are obtained for the parameters $R = 0.3\,\text{m}\Omega$, $L = 25\,\mu\text{H}$ and $E_{\text{LL(rms)}} = 690\,\text{V}$. The PI controllers are all tuned to a bandwidth (f_{br}) of 500 Hz using the pole-zero cancelation technique. For different synchronous speeds, switching frequencies, and parameter estimation errors, the dynamic

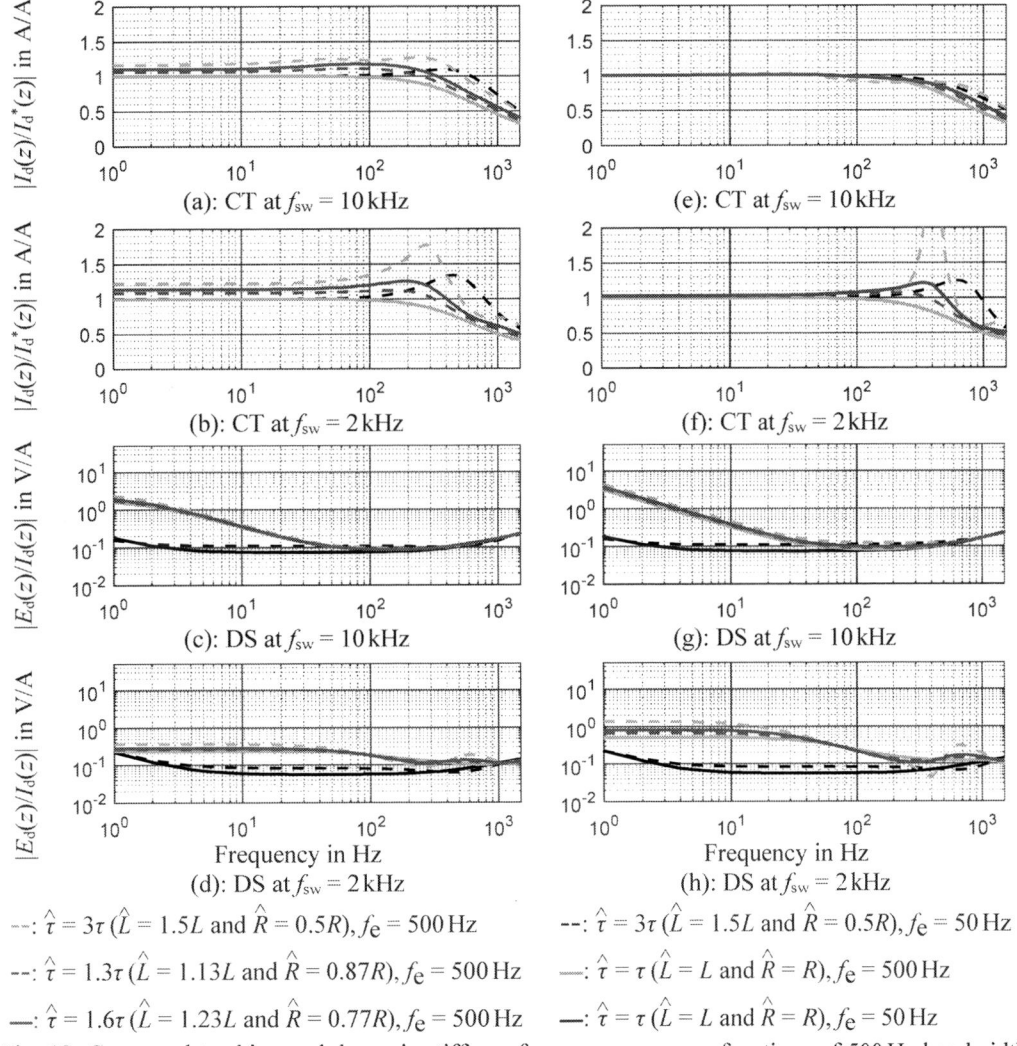

--: $\hat{\tau} = 3\tau \ (\hat{L} = 1.5L$ and $\hat{R} = 0.5R)$, $f_e = 500\,\text{Hz}$ --: $\hat{\tau} = 3\tau \ (\hat{L} = 1.5L$ and $\hat{R} = 0.5R)$, $f_e = 50\,\text{Hz}$

--: $\hat{\tau} = 1.3\tau \ (\hat{L} = 1.13L$ and $\hat{R} = 0.87R)$, $f_e = 500\,\text{Hz}$ —: $\hat{\tau} = \tau \ (\hat{L} = L$ and $\hat{R} = R)$, $f_e = 500\,\text{Hz}$

—: $\hat{\tau} = 1.6\tau \ (\hat{L} = 1.23L$ and $\hat{R} = 0.77R)$, $f_e = 500\,\text{Hz}$ —: $\hat{\tau} = \tau \ (\hat{L} = L$ and $\hat{R} = R)$, $f_e = 50\,\text{Hz}$

Fig. 12: Command tracking and dynamic stiffness frequency response functions of 500 Hz bandwidth synchronous frame PI current controllers for 10 kHz and 2 kHz switching frequencies at low and high synchronous speed w/ and w/o parameter estimation error for different control techniques with DLSO implemented in the $\alpha\beta$-frame: (a – d) 500 Hz bandwidth PI based observer controller, (e – h) Dead-Beat based observer controller for the parameters $R = 0.3\,\text{m}\Omega$, $L = 25\,\mu\text{H}$ and $E_{\text{LL(rms)}} = 690\,\text{V}$.

analysis metrics: *Command Tracking* (CT) and *Dynamic Stiffness* (DS) [1] for the studied controllers are shown in Fig. 11, and Fig. 12.

It is observed that the continuous DCCSFb approach controller becomes unstable at low switching frequencies and high synchronous speed even with accurate parameter estimation ($\hat{\tau} = \tau$). With perfect parameter estimations, all direct-discrete controllers are found to have stable dynamic behavior with one exception. Resonant properties are present in the DS plot at 800 Hz for $\omega_e = 2\pi 500$ rad/s and $f_{sw} = 2$ kHz for the direct-discrete current controller with DLSO implemented in the dq-frame. [1] shows that the eigenvalues of the DS move closer to each other with increasing synchronous speed for dq-current control techniques, which can cause resonant issues (see also Fig.11h). This issue can be solved by implementing dq-current controllers with DLSO in the αβ-frame, which have well-behaved dynamics performance at perfect parameter estimation, and the dip in the DS curve for $\omega_e = 2\pi 500$ rad/s at $f_{sw} = 2$ kHz is not present (see Fig. 12).

With poor parameter estimation ($\hat{\tau} = 3\tau$), resonant properties in CT are observed for the direct-discrete dq-current controller with DLSO in the dq-frame and dead-beat controller-based DLSO in the αβ-frame.

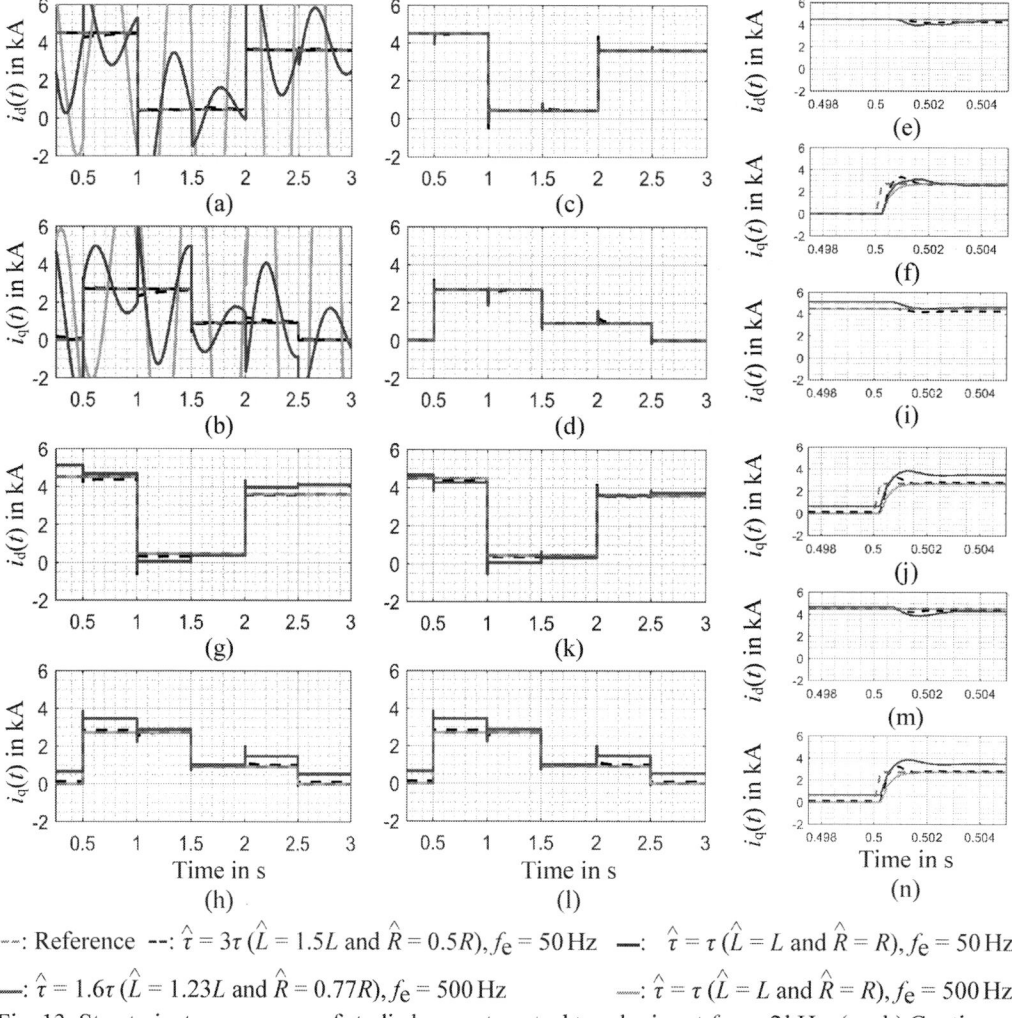

--: Reference   --: $\hat{\tau} = 3\tau$ ($\hat{L} = 1.5L$ and $\hat{R} = 0.5R$), $f_e = 50$ Hz   —: $\hat{\tau} = \tau$ ($\hat{L} = L$ and $\hat{R} = R$), $f_e = 50$ Hz

—: $\hat{\tau} = 1.6\tau$ ($\hat{L} = 1.23L$ and $\hat{R} = 0.77R$), $f_e = 500$ Hz   —: $\hat{\tau} = \tau$ ($\hat{L} = L$ and $\hat{R} = R$), $f_e = 500$ Hz

Fig. 13: Step trajectory response of studied current control topologies at $f_{sw} = 2$ kHz; (a – b) Continuous DCCSFb controller with DLSO implemented in the dq-frame, (c – f) Discrete DCCSFb controller with DLSO implemented in the dq-frame, (g – j) Discrete DCCSFb controller with DLSO implemented in the αβ -frame using PI controller, and (k – n) Discrete DCCSFb controller with DLSO implemented in the αβ-frame using Dead-Beat controller for the parameters $R = 0.3$ mΩ, $L = 25$ μH and $E_{LL(rms)} = 690$ V.

In contrast, the synchronous frame current controller using PI controller based DLSO in the αβ-frame is found to be stable for all conditions but with higher offset error and lower damping in CT compared to dead-beat controller-based DLSO in the αβ-frame and PI controller based DLSO in the dq-frame. At synchronous speed, DLSO in the αβ-frame does not track the grid current correctly compared to the DLSO in the dq-frame. Further, the current controller in the dq-frame does not make up for the magnitude and phase error introduced by DLSO in the αβ-frame due to wrong parameter estimation. Direct-discrete synchronous frame current controller with DLSO in the dq-frame and dead-beat controller-based DLSO in the αβ-frame is found to be well-damped for reasonable parameter estimations, $\hat{\tau} \leq 1.3\tau$ and $\hat{\tau} \leq 1.6\tau$, respectively, at $\omega_e = 2\pi 500\,\text{rad/s}$

Fig. 13 illustrates step responses of the studied current controllers at $f_{\text{sw}} = 2\,\text{kHz}$ for different synchronous speeds and parameter estimations. For all direct-discrete controllers, currents in the d and q axes are entirely decoupled at $\hat{\tau} = \tau$, and the currents track the reference command without any overshoot. With $\hat{\tau} = 1.6\tau$ at $\omega_e = 2\pi 500\,\text{rad/s}$, currents in the d and q axes are not decoupled entirely and current controllers with DLSO implemented in the αβ-frame show offset problems. However, the synchronous frame current controller with DLSO in the αβ-frame has the advantage of reduced coupling effects and well-behaved disturbance rejection properties. As discussed before, it yields phase and magnitude errors for wrong parameter estimations at synchronous speed, unlike with DLSO in the dq-frame where the AC signal is DC.

Conclusion

The performance analysis of different synchronous frame discrete current controllers for grid-connected three-phase VSIs with the controller's computational time delay compensation using *Discrete Luenberger Style Observer* (DLSO) is presented and discussed. The DLSO can be implemented in either the dq or the αβ-frame. Dynamic plots (command tracking, dynamic stiffness) and step responses are used to analyze and compare the current controllers. The difference in the dynamic performance for the studied decoupling and modeling techniques in the discrete domain varies with synchronous speed, switching frequency, and parameter estimation. None of the topologies are found to be simultaneously superior regarding robustness, dynamic performance, and noise filtering. The discrete controller with DLSO implemented in the αβ-frames shows improved disturbance rejection properties compared to the proposed discrete controller with DLSO implemented in the dq-frame. Synchronous frame current controllers with DLSO in the αβ-frame introduces offset errors at extreme cases of parameter estimation error, low switching frequencies, and high synchronous speed. Further resonant controllers within the DLSO could potentially solve this issue. In the case of dead-beat, the noise filtering in the αβ-DLSO is further reduces, but the dynamic performance gain is significant.

References

[1] M. Schütt and H. Eckel, "Design and Analysis of Discrete Current Controllers for VSIs," *2018 20th European Conference on Power Electronics and Applications (EPE'18 ECCE Europe)*, Riga, 2018, pp. P.1-P.10.

[2] M. Schütt and H. Eckel, "Design and analysis of complex vector current controllers for modular multilevel converters," *2017 19th European Conference on Power Electronics and Applications (EPE'17 ECCE Europe)*, Warsaw, 2017, pp. P.1-P.10.

[3] C. H. van der Broeck, R. W. De Doncker, S. A. Richter and J. von Bloh, "Discrete time modeling, implementation and design of current controllers," *2014 IEEE Energy Conversion Congress and Exposition (ECCE)*, Pittsburgh, PA, 2014, pp. 540-547.

[4] H. Kim, M. Degner, J. M. Guerrero, F. Briz and R. D. Lorenz, "Discrete-time current regulator design for AC machine drives," *2009 IEEE Energy Conversion Congress and Exposition*, San Jose, CA, 2009, pp. 1317-1324, doi: 10.1109/ECCE.2009.5316077.

[5] R. Pena, J. C. Clare and G. M. Asher, "Doubly fed induction generator using back-to-back PWM converters and its application to variable-speed wind-energy generation," in *IEE Proceedings - Electric Power Applications*, vol. 143, no. 3, pp. 231-241, May 1996, doi: 10.1049/ip-epa:19960288.

[6] A. G. Yepes, A. Vidal, O. López and J. Doval-Gandoy, "Evaluation of Techniques for Cross-Coupling Decoupling Between Orthogonal Axes in Double Synchronous Reference Frame Current Control," in IEEE Transactions on Industrial Electronics, vol. 61, no. 7, pp. 3527-3531, July 2014, doi: 10.1109/TIE.2013.2281160.

Variable Switching Point Parallel Predictive Current Control (VSP^{3}CC) for Induction Motor

Qing Chen, Ralph Kennel
Chair of Electrical Drive Systems and Power Electronics
Technical University of Munich
Arcisstr.21
Munich, Germany
Phone: +49 (0)89 289-28445
Fax: +49 (0)89 289-28336
Email: qing.chen@tum.de
URL: http://www.eal.ei.tum.de

Acknowledgments

This work is supported by EAL and China Scholarship Council (CSC). Qing Chen would like to thank CSC for the scholarship (No.201706420070).

Keywords

≪Variable Switching Point≫, ≪Predictive current control≫, ≪VSP^{3}CC≫, ≪Induction motor≫.

Abstract

This paper presents a variable switching point parallel predictive current control (VSP^{3}CC) strategy for an induction motor. VSP^{3}CC is introduced as one method of the Finite-Set Model Predictive Control (FS-MPC) methodology. VSP^{3}CC implements parallel predictive current control (PPCC) algorithm by means of a variable switching point. In this way, not only can a switching state be applied for less than a whole sample cycle which results in a lower current ripple, but weighting factor can also be eliminated. Experimental results are provided to verify the performance of VSP^{3}CC strategy.

Introduction

Electric motors are often at the spearhead of modern production systems. Initially, direct current (DC) motors drew the attention of drive developers because of its inherently simple speed and torque control. Over time, two basically alternating current (AC) motors control approaches for electrical drives have dominated high performance industrial applications: field-oriented control (FOC) [1], [2] and direct torque control (DTC) [3], [4]. The first ideas about predictive control have been published in the 1960s by Emeljanov [5] und applied to power converters started in the 1980s [6].

In the last decade, finite-set MPC (FS-MPC) [7] are widely studied for power electronics and electrical drives by many researchers in the academic field [8]-[12]. Inevitably, FS-MPC suffers mainly three major drawbacks:

- The change of switching state takes place at the beginning of a sampling cycle, which leads to high ripples on controlled variables.
- Complicated design of weighting factor with more than two controlled variables.
- The calculation effort will rise exponentially with an increase in the prediction horizon or level of the converter.

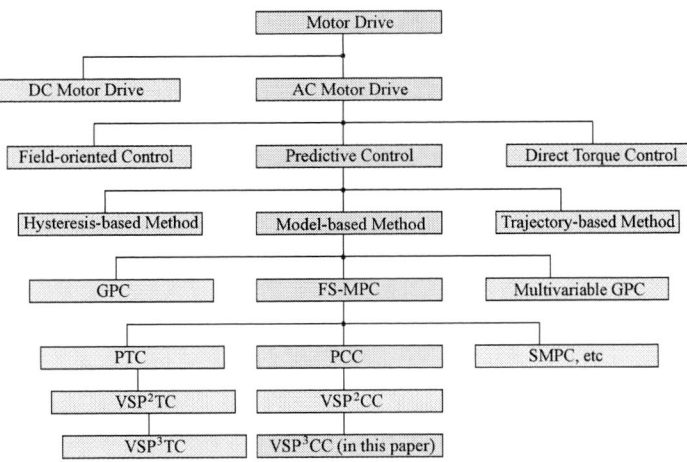

Fig. 1: Family tree of machine control

In order to solve the first problem, a method to calculate an optimal switching time point for predictive torque control (PTC) was first proposed in [8] and developed into variable switching point predictive torque control (VSP^{2}TC) in [9]. Furthermore, variable switching point parallel predictive torque control (VSP^{3}TC) was proposed recently [10]. This problem occurs also in predictive current control (PCC) [11] and was solved by a variable switching point predictive current control (VSP^{2}CC) strategy in [12].

In order to solve the second problem, a parallel predictive torque control (PPCC) algorithm was proposed in [13]. Contrary to the traditional principle of FS-MPC, PPCC uses parallel cost functions to select the optimal voltage vector, which can eliminate the weighting factor.

With regard to the third problem, VSP^{3}CC will be validated experimentally for a two-level voltage source inverter (2L-VSI) fed induction motor with one step prediction. Therefore, it will not be discussed.

In this paper, a variable switching point parallel predictive current control (VSP^{3}CC) implements the PPCC algorithm based on calculation of a variable switching point, which can decrease high ripples on controlled variables and eliminate the design of weighting factor, simultaneously. In summary, a briefly family tree of motor drive is depicted in Fig. 1.

Physical System

Two-Level Voltage Source Inverter

As shown in Fig. 2(a), a 2L-VSI is used to feed the induction motor. This 2L-VSI has eight possible switching states, which produce seven different voltage vectors depicted in Fig. 2(b).

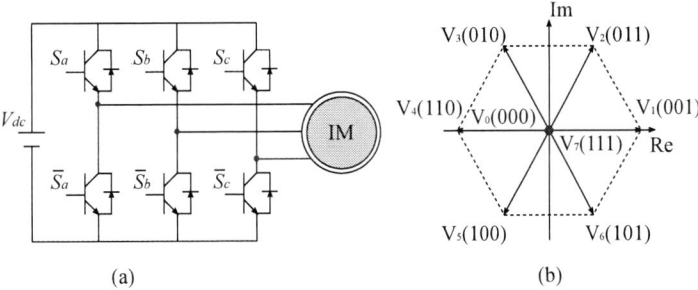

Fig. 2: (a) Topology of 2L-VSI. (b) Voltage vectors

The switching state of a 2L-VSI can be summarized in a single complex switching function:

$$S = \frac{2}{3}(S_a + a \cdot S_b + a^2 \cdot S_c) \tag{1}$$

where $a = e^{j2\pi/3}$, $S_i = 1$ means ON, $\bar{S}_i$ means OFF with $i = a,b,c$.

The voltage vector v is related to the switching state S by

$$v = V_{dc}S \tag{2}$$

where V_{dc} is the dc-link voltage.

Induction Motor

The system equations of a squirrel-cage induction motor can be described in terms of space vector quantities using a stationary reference frame.

$$v_s = R_s i_s + \frac{d\psi_s}{dt} \tag{3a}$$

$$0 = R_r i_r + \frac{d\psi_r}{dt} - j\omega_m \psi_r \tag{3b}$$

$$\psi_s = L_s i_s + L_m i_r \tag{3c}$$

$$\psi_r = L_r i_r + L_m i_s \tag{3d}$$

where v_s is the stator voltage vector; i_s, i_r are the stator and rotor currents, respectively; R_s and R_r are the equivalent stator and rotor resistances, respectively; ψ_s and ψ_r are the stator and rotor flux, respectively; L_s, L_r and L_m are the stator, rotor and mutual inductances, respectively; ω_m is the rotor angular speed.

The motor dynamics equations derived from (3) in terms of state variables ψ_r and i_s are presented as :

$$\tau_r \frac{d\psi_r}{dt} + \psi_r = L_m i_s + j\omega_m \tau_r \psi_r \tag{4a}$$

$$\tau_\sigma \frac{di_s}{dt} + i_s = \frac{1}{r_\sigma} v_s + \frac{k_r}{r_\sigma \tau_r}(1 - j\omega_m \tau_r)\psi_r \tag{4b}$$

where $\tau_r = L_r/R_r$, $\sigma = 1 - (L_m^2/L_s L_r)$, $\tau_\sigma = L_s \sigma/r_\sigma$, $k_r = L_m/L_r$ and $r_\sigma = R_s + R_r k_r^2$.

The electromagnetic torque can be expressed in terms of stator current and stator flux as:

$$T_e = \frac{3}{2}p(\psi_s \times i_s) = \frac{3}{2}p(\psi_r \times i_r) \tag{5}$$

where p is the number of pole pairs.

The equation of the mechanical system can be described as:

$$\tau_m \frac{d\omega_m}{dt} = T_e - T_L. \tag{6}$$

Variable Switching Point Parallel Predictive Current Control (VSP³CC)

In Fig. 3, the block diagram of variable switching point parallel predictive current control (VSP³CC) is shown.

Control Motivation

The control objective of the proposed strategy is not only to regulate the current to their reference values. Furthermore, the controller should aim to minimize the current ripple.

The theoretical maximum switching frequency of FS-MPC methods is equal to half of the sampling frequency. A comparison of the average switching frequency of VSP2CC and PCC in [7] is shown in

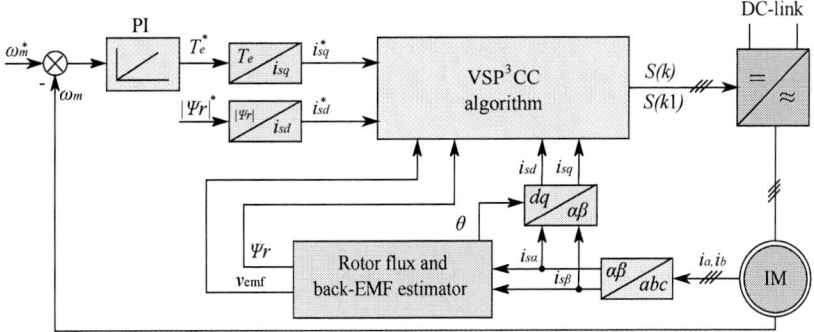

Fig. 3: Block diagram of variable switching point parallel predictive torque control (VSP³CC)

Fig. 4. Obviously, the average switching frequency was proven far lower than the theoretical maximum switching frequency. It depends on both the speed of induction motor and the control strategy. The average switching frequency of VSP²CC is higher than PCC under the same speed. The effect of increased switching frequency is the reduction of current ripple.

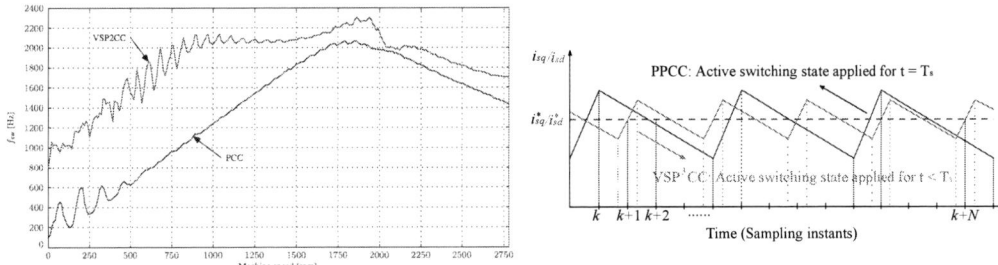

Fig. 4: Average switching frequency of VSP2CC and PCC ($T_s = 100\ \mu$s) [7]

Fig. 5: Principle of VSP³CC for current ripple reduction

As mentioned in [10], the relationship of the average switching frequency: VSP³TC > PPTC > VSP²TC > PTC. The increased average switching frequency will result in the reduction of the torque ripple. We believe that the same phenomenon can also be occurred in the current control methods.

In order to achieve a better control result, VSP³CC is proposed based on the above two facts. Fig. 5 illustrates the basic principle of VSP³TC for current ripple reduction.

Calculation of the variable switching point

The core part of VSP³CC strategy is calculation of the variable switching point. The calculation principle is illustrated in Fig. 6.

In order to keep the calculation simple, we assume that the current changes linearly if a fixed switching state is applied with the time t:

$$i_{sq}(t) \approx i_{sq}(k) + m(t-k)T_s, \tag{7}$$

where $i_{sq}(k)$ is the quadrature-axis current at step k and m is the current slope with the applied switching state S.

With regard to a 2L-VSI, the current slope m_i for all voltage vectors can be calculated to:

$$m_i = \frac{i_{sqi}(k+1) - i_{sq}(k)}{T_s}, i \in \{0, 1, \cdots, 6, 7\}. \tag{8}$$

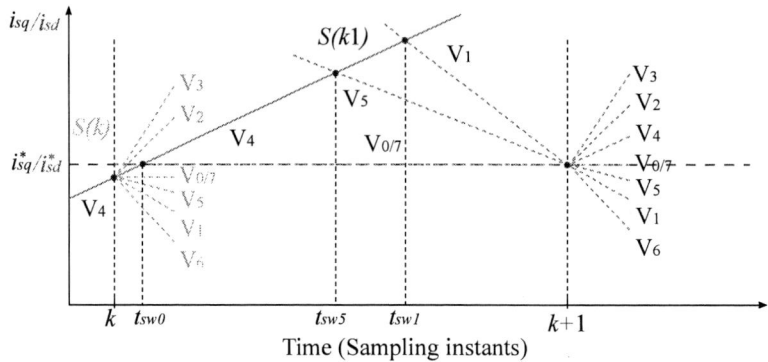

Fig. 6: Variable switching points where the current slopes of $V(k)$ and $V(k1)$ intersect are calculated.

The location of the switching points is to guarantee that the quadrature-axis current at step $k+1$ is equal to its reference:

$$i_{sqj}(k+1) = i_{sq}^*, j \in \{0,1,5,7\}. \tag{9}$$

And then the quadrature-axis current at step $k+1$ can be calculated to:

$$i_{sqj}(k+1) = i_{sq}(t_{swj}) + m_j(k+1-t_{swj})T_s. \tag{10}$$

where m_j is the current slope of the new switching state $S(k1)$.

After further calculations, the variable switching point can be calculated to:

$$t_{swj} = \frac{i_{sq}^* - i_{sq}(0) - m_j T_s}{(m_i - m_j)T_s}, \tag{11}$$

and the variable switching point in the range of $t_{swj} \in [0,1]$ should be chosen.

Basic equations

In order to implement VSP³CC by means of a digital system, the discrete-model of physical system must be first obtained. The rotor flux for the next sample can be estimated with

$$\hat{\psi}_r(k+1) = \left(\frac{\tau_r - T_s}{\tau_r} + j\omega_m T_s\right)\hat{\psi}_r(k) + \frac{L_m T_s}{\tau_r} i_s(k), \tag{12}$$

and stator current can be predicted with

$$\hat{i}_s(k+1) = \frac{\tau_\sigma - T_s}{\tau_\sigma} i_s(k) + \frac{T_s}{\tau_\sigma r_\sigma}(v(k) - v_{emf}(k)). \tag{13}$$

The back-EMF can be estimated with

$$v_{emf}(k) = -\frac{k_r}{\tau_r}\hat{\psi}_r(k) + jk_r\omega_m\hat{\psi}_r(k). \tag{14}$$

As shown in Fig. 3, the electromagnetic torque reference is generated by a speed PI controller, the rotor flux reference is set to a constant value. The corresponding reference values for the field- and torque-producing currents are given by

$$i_{sd}^* = \frac{|\Psi_r|^*}{L_m} \tag{15a}$$

$$i_{sq}^* = \frac{T_e^*}{\frac{3}{2}\frac{L_m}{L_r}|\Psi_r|^*}. \tag{15b}$$

The clarke to parke angle transform converts the $\alpha\beta$ components in a stationary frame to dq components in a rotating reference frame. For an α-phase to d-axis alignment, the transform uses the equation:

$$\begin{pmatrix} i_{sd} \\ i_{sq} \end{pmatrix} = \begin{pmatrix} \cos\theta & \sin\theta \\ -\sin\theta & \cos\theta \end{pmatrix} \cdot \begin{pmatrix} i_{s\alpha} \\ i_{s\beta} \end{pmatrix} \tag{16}$$

where the rotor flux angle is given by $\theta = \arctan\frac{\Psi_{r\beta}}{\Psi_{r\alpha}}$.

The parallel cost functions of PPCC are first designed as

$$g_d = |\, i_{sd}^* - \hat{i}_{sd}(k+1)\,| \tag{17a}$$

$$g_q = |\, i_{sq}^* - \hat{i}_{sq}(k+1)\,|. \tag{17b}$$

The parallel cost functions of PPCC with consideration of time delay compensation are designed as

$$g_d = |\, i_{sd}^* - \hat{i}_{sd}(k+2)\,| \tag{18a}$$

$$g_q = |\, i_{sq}^* - \hat{i}_{sq}(k+2)\,|. \tag{18b}$$

Finally, the parallel cost functions of VSP³CC are designed as:

$$g_d = |\, i_{sd}^* - \hat{i}_{sd}(t_{sw})\,| + |\, i_{sd}^* - \hat{i}_{sd}(k+2)\,| \tag{19a}$$

$$g_q = |\, i_{sq}^* - \hat{i}_{sq}(t_{sw})\,| + |\, i_{sq}^* - \hat{i}_{sq}(k+2)\,|. \tag{19b}$$

Experimental Results

Fig. 7 shows the experimental setup, which consists of two 2.2 kW squirrel-cage induction motors. One motor, driven by a Danfoss VLT FC-302 3.0 kW inverter, is used as load motor. The working motor is driven by a modified SERVOSTAR620 14kVA inverter which provides full control of the IGBT gates. The real-time computer system of this test bench is a 1.4 GHz Pentium M CPU. The sampling frequency is set to be 16 kHz. The parameters of the working motor are also given in Fig. 7.

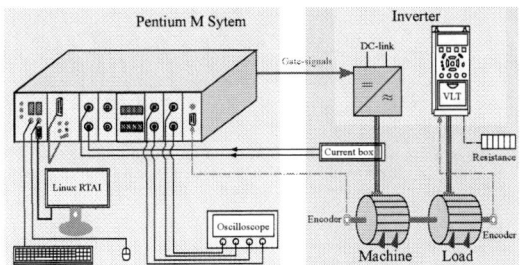

Description	Parameter	Value
Stator resistance	R_s	2.68 Ω
Rotor resistance	R_r	2.13 Ω
Mutual inductance	L_m	275.1 mH
Stator inductance	L_s	283.4 mH
Rotor inductance	L_r	283.4 mH
Pole pairs	p	1.0
Nominal speed	$\omega_{m,nom}$	2772.0 rpm

Fig. 7: Test bench description and the parameters of the working motor

Investigation of Steady-state Behavior

The steady-state behavior of the drive system was examined for both PPCC and VSP³CC. In order to investigate the machine current behavior at different speed, the reference speed is set to be 277.2 rpm (10% nominal speed), 1386(50% nominal speed), and 2772 rpm (nominal speed), respectively. Fig. 8, Fig. 9, and Fig. 10 show the experiment results.

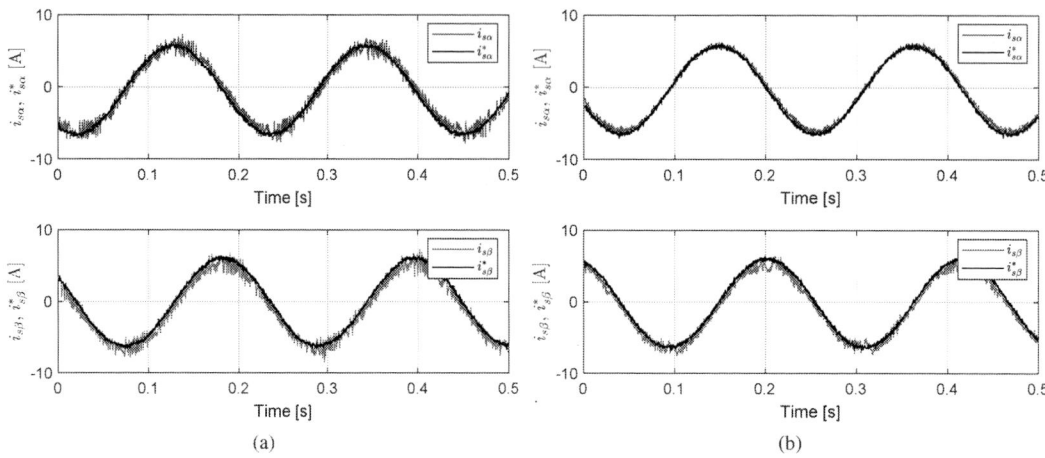

Fig. 8: Steady-state current at 277.2 rpm without load. (a) PPCC. (b) VSP^{3}CC.

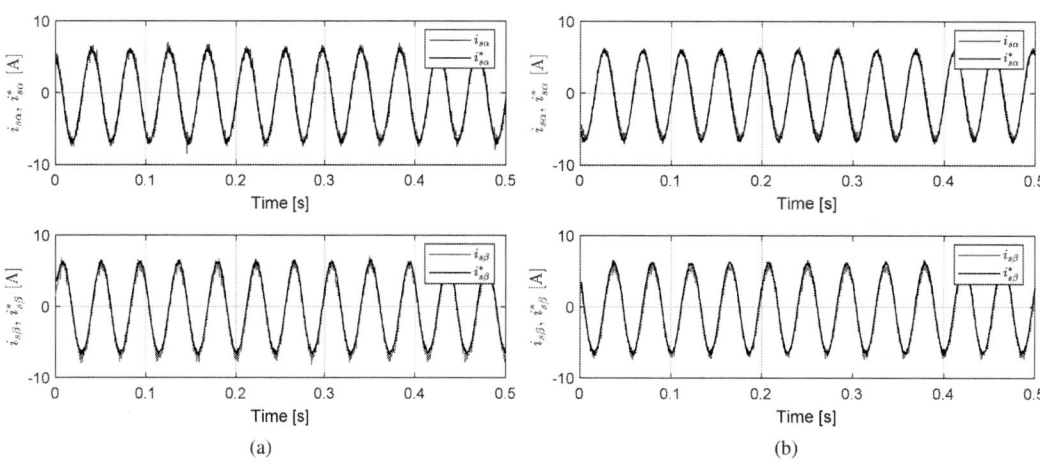

Fig. 9: Steady-state current at 1386 rpm without load. (a) PPCC. (b) VSP^{3}CC.

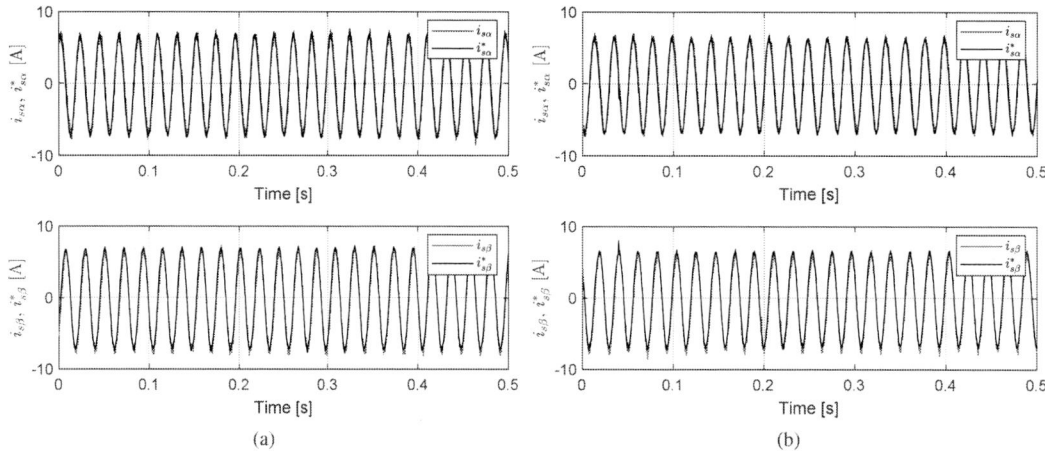

Fig. 10: Steady-state current at 2772 rpm without load. (a) PPCC. (b) VSP^{3}CC.

In comparison with PPCC, it is clearly that VSP³CC method can reduce the current ripple, especially at low speed (277.2 rpm). As already mentioned, the reduction of current ripple comes with the cost of increased frequency. A comparison of the average switching frequency of PPCC and VSP³CC will be given later.

Investigation of Torque Step Response

The performances of PPCC and VSP³CC during transients were also tested and are shown in Fig. 11. In this test, the load machine operates at a speed closed loop mode with a half nominal speed 1386 rpm, whereas the torque reference of the working machine alters from 0 to 4.0 Nm through control algorithm.

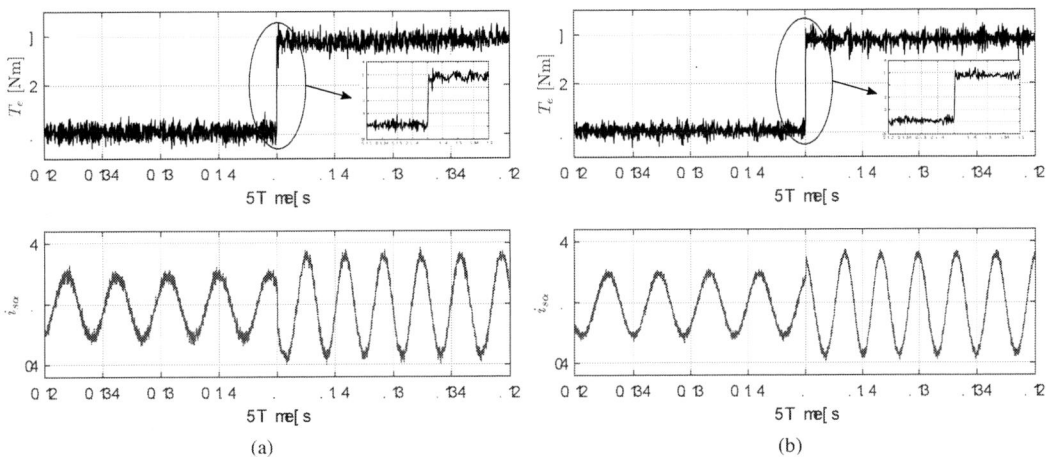

Fig. 11: Experimental results of torque step response. (a) PPCC. (b) VSP³CC.

Through the zoom part of torque step response shown in Fig. 11, it can clearly be seen that the controller of the two methods reacts very fast to the reference torque change and rejects the disturbance in less than 1 ms. Unlike VSP³TC method in [10], VSP³CC can not obviously reduce torque ripple. However, it can obviously reduce current ripple, because the current as control variable in this method.

Investigation of Speed Reversal

In order to observe the performance of VSP³CC in the range of whole speed, a speed reversal from the positive nominal speed +2772 rpm to the negative nominal speed -2772 rpm without load was performed. The result is shown in Fig. 12a, which verify the feasibility of VSP³CC in the range of whole speed.

At last, the comparison of average switching frequency of PPCC and VSP³CC is shown in Fig. 12b.

Conclusion

The average switching frequency of VSP²CC is higher than PPCC under the same condition, especially in low speed range. Since the application target of this strategy is the low voltage drives field, in which the switching losses are not mainly problem.

VSP³CC will proved to be a promising and attractive method due to its ability to reduce the current ripple and the absence of weighting factor in the low voltage drives field.

References

[1] F. Blaschke, "The principle of field orientation as applied to the new transvector closed-loop system for rotating-field machines," *Siemens review*, vol. 34, no. 3, pp. 217–220, 1972.

[2] W. Leonhard, *Control of electrical drives*. Springer Science & Business Media, 2001.

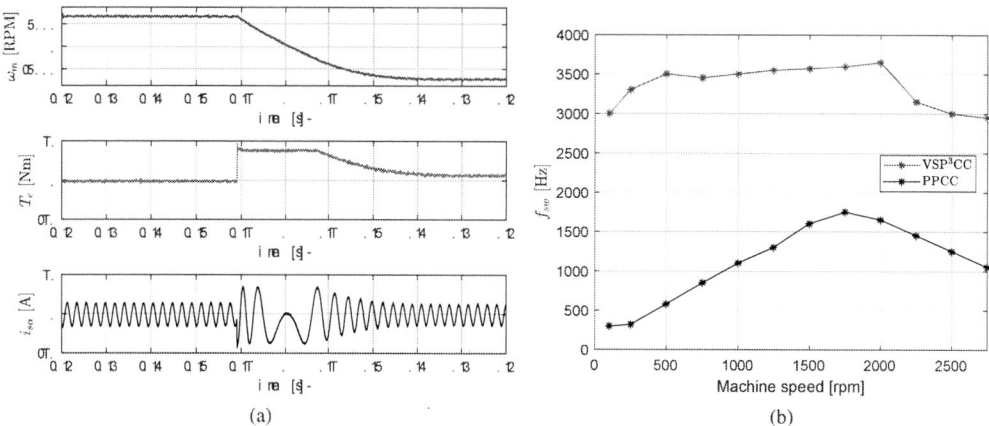

Fig. 12: (a) Experimental results of VSP³CC in the range of whole speed. (b) Comparison of average switching frequency of PPCC and VSP³CC.

[3] I. Takahashi and T. Noguchi, "A new quick-response and high-efficiency control strategy of an induction motor," *IEEE Transactions on Industry Applications*, vol. IA-22, no. 5, pp. 820–827, Sept 1986.

[4] E. Jääskeläinen and P. Pohjalainen, "Direct torque control (dtc) a motor control technique for all seasons," *ABB White Paper*, 2015.

[5] S. V. Emel'janov, *Automatische Regelsysteme mit veränderlicher Struktur*. Oldenbourg, 1969.

[6] J. Holtz, "A predictive controller for the stator current vector of ac machines fed from a switched voltage source," *Proc. of IEE of Japan IPEC-Tokyo'83*, pp. 1665–1675, 1983.

[7] P. J. Stolze, "Advanced finite-set model predictive control for power electronics and electrical drives," Ph.D. dissertation, Technische Universität München, 2014.

[8] P. Landsmann and R. Kennel, "Saliency-based sensorless predictive torque control with reduced torque ripple," *IEEE Transactions on Power Electronics*, vol. 27, no. 10, pp. 4311–4320, Oct 2012.

[9] P. Karamanakos, P. Stolze, R. M. Kennel, S. Manias, and H. du Toit Mouton, "Variable switching point predictive torque control of induction machines," *IEEE Journal of Emerging and Selected Topics in Power Electronics*, vol. 2, no. 2, pp. 285–295, June 2014.

[10] Q. Chen, H. Xie, and R. Kennel, "Variable switching point parallel predictive torque control (vsp3tc) for induction motor," in *2020 IEEE International Conference on Industrial Technology (ICIT)*, 2020, pp. 1112–1117.

[11] P. Karamanakos, T. Geyer, and S. Manias, "Direct model predictive current control strategy of dc–dc boost converters," *IEEE Journal of Emerging and Selected Topics in Power Electronics*, vol. 1, no. 4, pp. 337–346, Dec 2013.

[12] P. Stolze, P. Karamanakos, M. Tomlinson, R. Kennel, T. Mouton, and S. Manias, "Heuristic variable switching point predictive current control for the three-level neutral point clamped inverter," in *2013 IEEE International Symposium on Sensorless Control for Electrical Drives and Predictive Control of Electrical Drives and Power Electronics (SLED/PRECEDE)*, Oct 2013, pp. 1–8.

[13] F. Wang, K. Zuo, A. Xia, Z. Zhang, Y. Zhang, and J. Rodriguez, "An optimized predictive current control for induction machines with a parallel cost function," in *2019 IEEE International Symposium on Predictive Control of Electrical Drives and Power Electronics (PRECEDE)*, May 2019, pp. 1–6.

Operation of an Externally Excited Synchronous Machine with a Hybrid Multilevel Inverter

C.Terbrack, M.Sc.; J. Stöttner, M.Sc.; Prof. Dr.-Ing. C. Endisch
TECHNICAL UNIVERSITY OF APPLIED SCIENCES INGOLSTADT
Institute of Innovative Mobility (IIMo)
Esplanade 10, 85049 Ingolstadt, Germany
Tel.: +49 (0) 841 / 9348-6487; +49 (0) 841 / 9348-6506.
Fax: +49 (0) 841 / 9348-996487; +49 (0) 841 / 9348-996506.
E-Mail: Christoph.Terbrack@thi.de, Julia.Stoettner@thi.de, Christian.Endisch@thi.de
URL: https://www.thi.de/forschung/institut-fuer-innovative-mobilitaet-iimo/els

Keywords

«AC machine», «Batteries», «Converter circuit», «Electrical drive», «Electrical machine», «High voltage power converters», «Multilevel converters», «Power converters for EV», «Synchronous motor».

Abstract

An innovative hybrid multilevel inverter is proposed which is able to deliver a separate DC voltage while a three-phase system is provided. The voltages are independently, but delivered by shared battery cells. The new system operates an electrically excited synchronous machine's rotor and stator current circuit in a vehicle drivetrain.

Introduction

One of the most controversial discussed topics of today is the electrification of vehicles. Advantages such as lowering local emissions or smooth and quiet driving comfort make electric driving attractive. However, facts like limited range, charging time and the high weight of the built-in battery system as well as safety gaps comber the complete breakthrough of this technology [1-3].

New concepts are emerging with the aim to increase the efficiency of the overall powertrain and to decrease its cost and weight [1]. A promising technological solution is the multilevel inverter (MLI), which has the potential to improve state-of-the-art (SotA) converter technology [4,5].

Even more, its integration into the drive train of an electric vehicle (EV) represents a veritable revolution of the traditional traction system. In an MLI, the conventional construct of the battery system in combination with the conventional power electronics merges to one single component [6]. This includes significant advantages e.g. lowering the total harmonic distortion (THD) resulting into reduced filter requirements [4-12]. Furthermore, in contrast to conventional inverters, the MLI has a very high efficiency throughout the overall operating area, especially in the partial load range what makes it particularly interesting for EV [13,14]. Approaches can be found for example in [6,7,13,15-17] where new MLI topologies are proposed and compared to conventional two-level inverters. A closer look at the electrical machine (EM) in such drive systems reveals that permanent magnet synchronous machines (PMSM) or induction machines (IM) are most frequently used [16,18]. However, IM and PMSM have their strengths and weaknesses.

The PMSM can be built very compact and is highly efficient in the basic speed range. Nonetheless, the use of permanent magnets means that no current is required for excitation, but cost-intensive permanent magnet material. Additional, the permanent magnets cannot be switched off, this leads to drag losses in no-load operation and high currents may be required for field weakening at higher speeds. The IM convinces in the field weakening range, whereas it has only a low torque in the basic speed range. Its efficiency is lower than that of the PMSM [18,19]. In contrast, the electrically excited synchronous machine (EESM) is characterized by the fact that it combines the desired characteristics of both machines, avoiding many of their disadvantages. It achieves a high system efficiency in all operating ranges, since it offers an additional degree of freedom with an adjustable rotor flux using a directly controllable impressed rotor current. It thus enables a high torque density at low speeds like the PMSM and low losses at high speeds, comparable to the IM [20]. In fact, this variant of EM requires an energy transfer to the rotor circuit. This is mostly realized with a slip ring, which is afflicted with mechanical wear. But there are concepts that counteract this by employing a wireless inductive energy transfer system (IETS) [18,19]. Since an external excitation source is required, another aspect is the provision of an additional adjustable power supply. In order to be able to use the advantages of an EESM in an optimal way, the controllability of this power supply is essential. Helling et al. present in [17] a modular way to generate different circuits for the provision of multiple different DC or AC voltages. Therefore, they add one so-called low voltage power supply (LVPS) unit to the phase string in an MLI system. In contrast to the other battery modules in the phase string, it consists out of a capacitor instead of a battery cell. The connection to an outer auxiliary consumer or an energy supply can be made directly or by means of additional DC/DC or DC/AC converters. With this proposed concept, a bidirectional power exchange between the LVPS module and an external device is possible. However, since the LVPS module is connected in the load path and has to be charged and discharged by reconfiguration and pulsing, no direct DC output can be generated and the controllability of the voltage is limited. Furthermore, the LVPS operation must be scheduled into the load operation, which makes the operation strategy quite complex. In [21] another possible variant is shown by supplying two independent three-phase engines with one three-phase inverter system. Instead of maintaining the same battery size, they have to enlarge the entire system to ensure a sufficient supply to both consumers at nominal point.

A possible solution is demonstrated in [22], where controlled DC besides AC power lines are provided by a single battery system using shared battery cells. It is therefore possible to generate several voltages without increasing the number of cells in the battery. The publication shows the presence of an unexploited cell reserve in SotA MLI systems. Nota bene, the term 'reserve' represents the battery cells within an overall MLI system that are temporarily not needed to provide the output voltage of the inverter. The cells contribute to the supply of various electrically isolated circuits by being temporarily connected to them. For this reason, an innovative MLI concept called Hybrid MLI is presented, which makes the cell reserve accessible and generates output voltages for several separate consumers. Therefore, it achieves an effectively exploitation of all built-in battery cells in an overall battery system. In this publication, the MLI concept proposed in [22] is combined with the idea of supplying an EESM in an EV. Thus, the presented concept shows an attractive possibility for a highly efficient modern electric powertrain.

After a brief summary of the basic idea in [22], a MLI drivetrain concept of an EV, especially in combination with an EESM, is demonstrated. Therefore, a variant of the novel MLI design proposed in [22] is outlined in a three-phase fashion together with the newly integrated terminals for the control of the exciter circuit of the EESM. Simulation results, obtained with a Matlab/Simulink model of an overall power train, show the functionality of the fundamental idea and thus, the successful operation of the EESM with the hybrid MLI.

Temporarily unused battery cells in multilevel inverter

MLI are voltage converters that transform a control signal into a power signal. They are particularly standing out for their ability to directly set several different voltage levels. This is achieved through reconfiguration by switching immanent power sources into or out of the load current path. Depending on the requested output voltage, the associated control sequence of the MLI selects the combination of cell voltages that best matches the requested value. For this purpose, in a modular MLI a number of voltage sources, e.g. battery cells, are connected in series [7,23,24]. The cells are surrounded by a circuit design, which is able to connect and disconnect each cell individually to the serial string of cells. The latter is connected to the load [25,26]. The cell in combination with its switching circuitry is often referred to as module [6,7,16].

Several publications do investigations to find an advantageous circuit topology for the inverter by adjusting both the number of switches and the structure of the modules to achieve maximum functionality and efficiency [25-28]. Other researchers try to compare and choose favourable switching strategies to achieve an optimal operation [28,29]. All these investigations concentrate on the optimized utilization of active cells providing output power. However, no attention is paid to the exploitation of the battery cells, while they are disconnected from the load during operation.

In general, it is necessary that the MLI is designed in such a way that a nominal maximum output voltage of the inverter is achievable. For this purpose, a certain number of modules are connected in series. If the voltage demand of the load is less than the maximum deliverable output voltage of the built-in battery system, the inverter is forced to switch one or more cells out of the serial string. This results in a number of cells, which are temporarily galvanically isolated from the load and become inactive by bridging. Especially when the MLI provides a sinusoidal voltage curve, a periodic output voltage is generated. This results in a recurring and predictable amount of inactive cells over the operating time [22]. Moreover, the maximum amplitude is only requested twice in one period. Therefore, during most of the operation time not all battery cells are employed to reproduce the required voltage at the output of the MLI. The findings in [22] show that for a multiphase system, providing periodic phase-shifted output voltages with more than two phases, the inactive reserve of cells becomes nearly constant over time. Even more, an amount of static inactive cells of over 30% of the total battery system is present for maximum level control. Fig. 1 visualizes these outcomes, where an exemplary inverter delivers a three-phase voltage system. Each phase is composed out of eight cells. The phase voltages are displayed rectified, since in this example the same cells are used to generate both the positive and the negative output voltage. A module with the functionality of a four-quadrant controller is assumed to be used. For the voltage of each phase, the temporarily needed number of cells located below each sinusoidal curve is depictured in light colour. These cells are necessary to provide the requested sinusoidal voltage. The darker depicted cells remain deactivated.

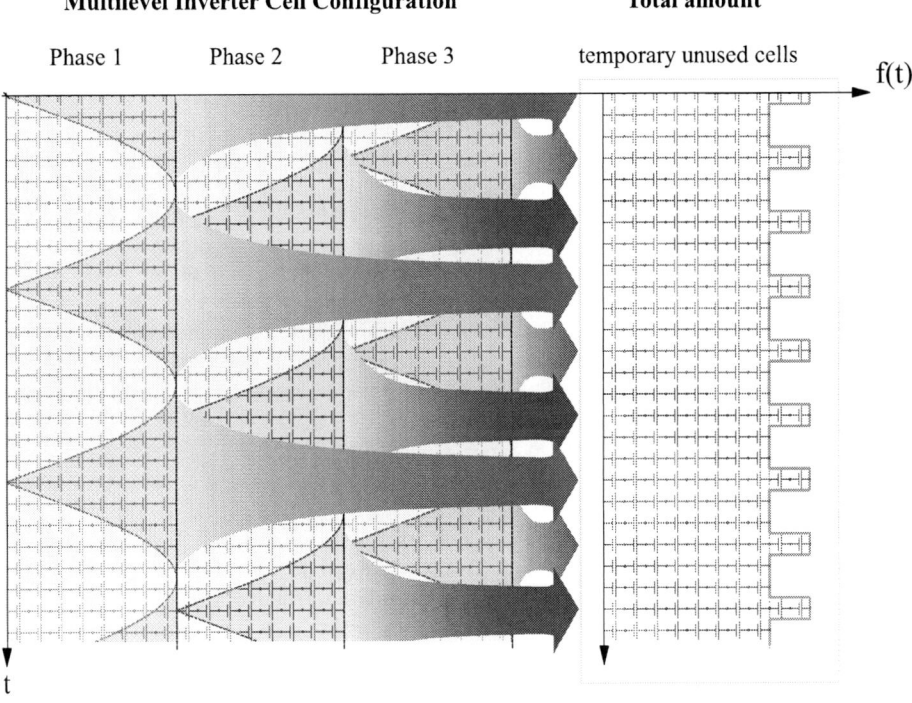

Fig. 1: Three-phase output voltage system provided by a MLI with eight cells per phase and the associated allocation of utilized (light red, light blue, light green colored) and unused (dark red, blue, green colored) battery cells over time and the sum of all temporarily inactive battery cells [22]

As pictured, an amount of at least eight cells is permanently out of operation, while the overall cell capability is 24. A maximum output level is presented, but the reserve can be even greater in the partial load range and thus at lower output levels.

The basic concept aims at an optimal usage of the entire battery system. Therefore, it is the idea to make this temporarily inactive cell reserve accessible for the supply of additional consumers.

Hybrid MLI supplying an EESM

To unlock the discovered cell reserve for further applications, an additional surrounding circuitry has to be designed, which allows access to the inactive cells. This circuitry must provide extra terminals where further loads can be tapped. Within this concept, the cells contribute to the supply of various electrically isolated circuits by being temporarily connected to them. For this reason, no rigidly connected separate circuits are required, but each individual cell can be assigned to the different circuits as required. A SotA MLI is able to activate and bypass specific cells. A circuit for supplying multiple consumers needs to provide this functionality for each sub-circuitry [17]. While a cell is used in one circuit, it must be fully electrically isolated from the others.

Fig. 2 depicts the developed MLI structure in case of a single-phase string. The generated main output voltage can be tapped at the two terminals, m_{strand} and p_{strand}, while the secondary load can be connected to terminal m_{second} and p_{second}.

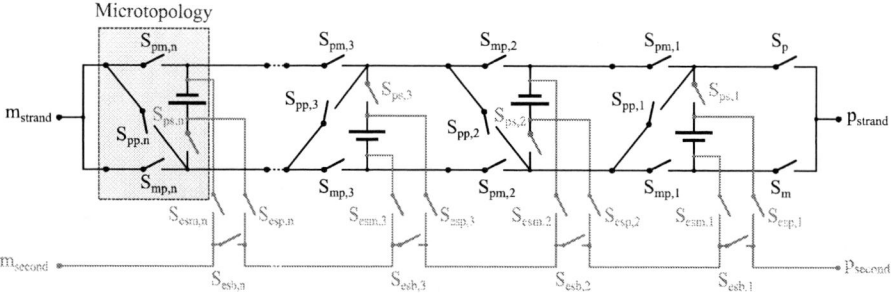

Fig. 2: Extended ECIN topology [22] (black colored) with additional switches forming a second circuitry (orange colored), which is used to supply auxiliary consumer

The proposed topology shown in Fig. 2 consists out of the so-called ECIN topology (black colored), extended by an additional second MLI (orange colored) to a new design, as it was published in [22]. Thus, a new hybrid MLI system accrues, combining two independently operable MLI structures, providing two independent output voltage systems but sharing the same energy supply units. In Fig. 3 the arrangement is enlarged to a three-phase AC system ($L1, L2, L3$) for the primary application as it is usually utilized in an EV powertrain. The secondary circuits are serially connected to one overall secondary circuit (m_{second}, p_{second}) providing the characteristic presented in Fig.1. With the here selected design, the secondary circuit is able to supply one large consumer with a DC voltage or any AC voltage.

In general, the arrangement can be extended to a system, which is able to supply several additional consumers. Therefore, the secondary circuit is divisible into smaller partial circuits that can be galvanically isolated from each other with the immanent switches. However, this increases the number of additional connections for auxiliary loads, and the advantageous characteristic of having a constant output voltage on the secondary terminals might be compromised, depending on the chosen design.

Another possibility is to set up a common DC bus as in the conventional EV on-board power system, which is fed by the constant voltage at the secondary terminals of the hybrid MLI. Further auxiliary consumers, such as on-board electronics, heating or air conditioning compressors, can be easily integrated by attaching additional DC/AC or DC/DC converters [17].

Especially, for the case of an EESM as the main traction machine in the drivetrain of an EV, the basic principle described above is applicable. With the proposed hybrid MLI both circuits of an EESM, the stator as well as the rotor circuit are supplyable. This is particularly interesting because the output voltage of the secondary inverter is fully adjustable as well. This means that, in addition to the field-oriented control in the stator circuit, the current control in the rotor circuit is used to set an optimum operating point. The combination of the hybrid MLI and EESM creates a compact and highly efficient electric drive train.

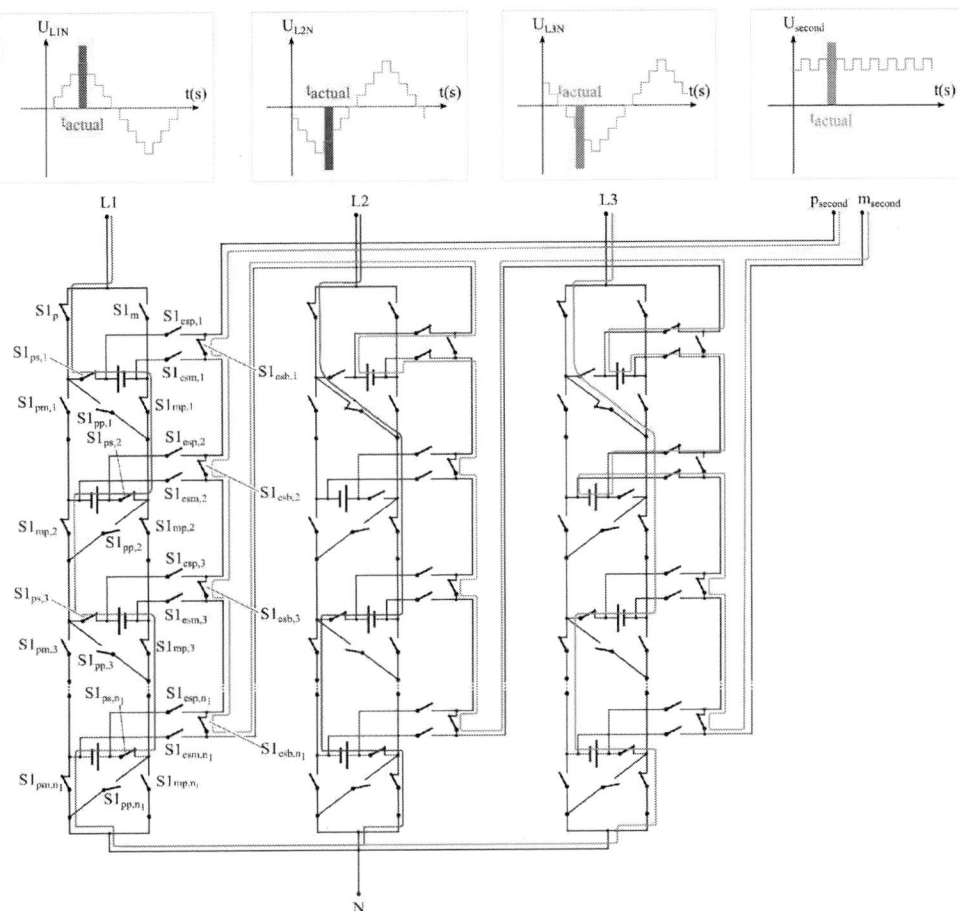

Fig. 3: Hybrid MLI composed out of a primary circuit designed as a three-phase system and a common secondary intermediate circuit

Simulations

Setup

The setup is investigated using a simulation model of an EV powertrain, in which especially a modular MLI and an EESM is modelled. The focus is to show that the proposed hybrid MLI can operate the EESM effectively by controlling rotor and stator circuit basing on the circuit configuration illustrated in Fig. 3. Simulations are carried out in Matlab/Simulink with a fixed step size solver using the Euler integration method with a simulation step time t_{step} of 0.1 μs. The Euler method is computationally efficient and for this reason fast, but comparatively inaccurate. It is chosen because the spread between the small simulation step size to reproduce the pulse width modulation pattern (PWM) of the MLI and the rather long simulation time to record the dynamic behaviour of the EESM and the battery cells is challenging. However, due to the small simulation step size, the use of this method is a justified compromise in order to be able to reproduce most effects in the components of the drive train and to obtain acceptable simulation times.

A 18650 lithium-ion cell from LG Chem was selected for the energy supply of the system, the data of which are shown in [14] and [30] and summarized in Table II. It has a nominal capacity C_N of 2800 mAh and a state of charge (SoC) dependent voltage between 2.5 V and 4.2 V. The phase peak voltage $\hat{u}_{s,max}$ of the chosen EESM is 173.2 V.The number of cells to be connected in series for the design of the MLI to meet the rated voltage is determined in the stationary state of the cells using the rated point of the electrical machine.

$$n_{ser,ph} = \lim_{t \to \infty} \frac{\hat{u}_{s,max}}{V_{cell}(t,I_N)} = \frac{\hat{u}_{s,max}}{U_{Cell,n} - \hat{i}_{s,max} \cdot \frac{\left(R_{ohm} + \Sigma_{i=1}^{3} R_{RC,i}\right)}{n_{par}}} \approx 53 \tag{3}$$

Since the maximum cell current $I_{Cell,max}$ is $2C$, a number of $n_{par} = 54$ cells is necessary, taking into account the current carrying capacity of the cells and the EESM. The cells and its load dependent dynamics are modelled via Matlab. The characteristic of each cell is represented with an individual equivalent circuit model (ECM). It consists out of an open-circuit voltage source, one ohmic resistance R_{ohm} and three series-connected RC elements composed of the respectively resistance $R_{RC,i}$ and the capacity $C_{RC,i}$ to cover the dynamic cell properties [14,30]. The underlying ECM and the associated equations correspond to the descriptions in [31].

Table I: Battery cell parameters [14,30]

Parameter		Value
Nominal voltage	$U_{Cell,n}$	3.72 V
Nominal energy	$E_{Cell,n}$	10.42 Wh
Ohmic resistance	R_{ohm}	41.53 mΩ
Resistance of RC elements	$R_{RC,1}$	5.02 mΩ
	$R_{RC,2}$	7.32 mΩ
	$R_{RC,3}$	3.23 mΩ
Capacitance of RC elements	$C_{RC,1}$	75.44 mF
	$C_{RC,2}$	339.5 mF
	$C_{RC,3}$	3.625 F

Table II: Parameters of the EESM [32]

Parameter		Value
Nominal power	P_n	87 kW
Nominal speed	n_n	4000 rpm
Number of pole pairs	p	4
Maximum phase voltage	$\hat{u}_{s,max}$	173.2 V
Maximum phase current	$\hat{i}_{s,max}$	400 A
Excitation current	$i_{E,n}$	16 A
Excitation inductance	L_E	0.4 H
Excitation resistance	R_E	5.765 Ω
Mutual inductance	L_m	0.0123 mH

For the simulation of the switching elements the IRL7472L1TRPbF power MOSFET is chosen. The related parameters can be found in the datasheet [33]. Both the switching elements and the connection resistances between the cells are modelled in a state space model according to [31]. The state space model has a switching vector as input, via which the modulation reconfigure the connections between the cells. A detailed description of the model is unfortunately beyond the scope of this paper.

To control the switching states of the inherent semiconductor devices in the simulation, the space vector control (SVC) method proposed in [34] is applied.

Control and Dynamics of the Externally Excited Synchronous Machine

For the modelling of the EESM a saturation-based model is used, which is presented in [32] and implemented with the parameters depicted in Table I.

The electromechanical torque of a synchronous machine can be calculated using the formula

$$T_{el} = \left[L_m(i_m) \cdot i_E \cdot i_q + \left(L_d(i_m) - L_q(i_m) \right) \cdot i_d \cdot i_q \right], \tag{1}$$

where i_E represents the excitation current in the rotor, n_{phases} and p the number of phases and pole pairs. L_m, L_d and L_q designate the mutual inductance between rotor and stator as well as the d- and the q-component of the stator inductance [20,35]. Thereby, the saturation-dependent values of the inductances are determined depending on the actual magnetizing current i_m and inserted into (1) [32]. Thereby, the inductances L_d and L_q consist of a main (L_{hx}) as well as a leakage component (L_σ). The main proportion of the excitation inductance is represented by the mutual inductance L_m. Its leakage component is nominated as $L_{\sigma E}$. The corresponding equations of L_m, L_d and L_q are as follows.

$$
\begin{aligned}
L_d &= L_{hd} + L_\sigma \\
L_q &= L_{hq} + L_\sigma \\
L_E &= L_m + L_{\sigma E} \\
L_{hq} &= m \cdot L_{hd}
\end{aligned}
\qquad with \qquad \psi_h = L_{hd} \cdot i_m
\tag{2}
$$

The magnetizing current is represented as a function of the currents i_d, i_q and the excitation current i_E.

$$
i_m = \sqrt{\left(i_d + \frac{i_E}{u}\right)^2 + \left(m \cdot i_q\right)^2}
\tag{3}
$$

Here m denotes the transmission ration between the main components of the q- and d-inductance, whereas u represents the transmission relation between rotor and stator reference system. The parameters $L_\sigma, L_{\sigma E}, m$ and u in equation (2) and (3) are constant and can be taken from [32]. The main component L_{hd} can be calculated with the knowledge of the main flux linkage ψ_h and its dependency of the magnetizing current i_m. The nonlinear relationship between ψ_h and i_m is investigated in [32]. In the context of the studies presented here, it is approximated using a polynomial for modelling the EESM. For this purpose, a third degree polynomial is used with the following formula.

$$
\frac{\psi_h}{Vs}(i_m) = a \cdot i_m^{\,3}\left[\frac{1}{A^3}\right] + b \cdot i_m^{\,2}\left[\frac{1}{A^2}\right] + c \cdot i_m\left[\frac{1}{A}\right] + d
\tag{4}
$$

$with \quad a = 1.1583 \cdot 10^{-9}, b = -1.4854 \cdot 10^{-6}, c = 6.6402 \cdot 10^{-4}$ and $d = -1.8571 \cdot 10^{-4}$

The resulting relationship between the main flux linkage ψ_h and the magnetizing current i_m is shown in Fig. 4a. With equation (4), reference points (red marked) are approximated by the polynomial curve (blue colored).

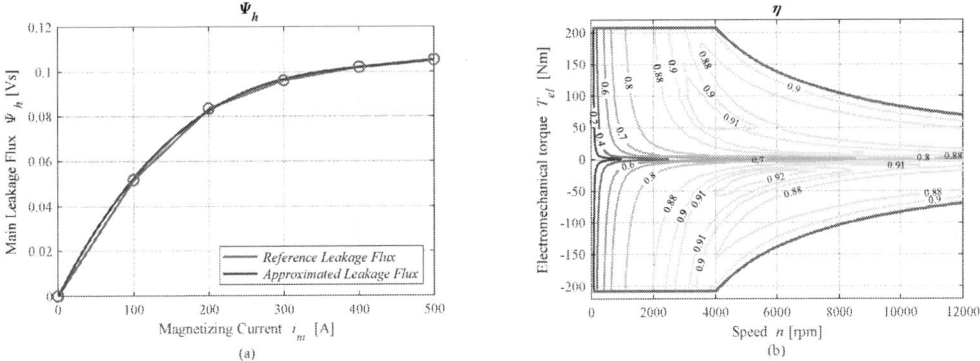

Fig. 4: (a) Magnetization characteristic curve and (b) efficiency characteristic of the EESM

Additionally, in Fig. 4b the efficiency characteristic of the chosen EESM for the entire torque-speed-range is depicted. Thus, for a given set of i_d, i_q and i_E, the actual magnetizing current i_m and all further parameters in any operating point are determined.

Control of the excitation current

Compared to other EM, the EESM offers an additional degree of freedom with the directly adjustable rotor current. This provides a setting lever for an optimal distribution of losses between stator and rotor. For the control of the EESM, a regulation must be found how the total current of the machine is distributed to the torque-forming components of the current. The aim is to find the combination of $\{i_d, i_q, i_E\}$ for each valid operating point $\{T_{el}, n\}$ in the entire operating range of the EESM such that the total losses are minimal. Therefore, an analytical target function is numerically solved gaining values of torque T_{el} and rotational speed n representing the smallest entire losses. The minimization problem to be solved can be formulated as

$$\min_{i_d, i_q, i_E} P_{V,ges}(i_d, i_q, i_E, n) = \begin{cases} T_{el}(i_d, i_q, i_E, n) & = T_{ref} \\ u_d^2 + u_q^2 & \leq u_{s,max}^2 \\ i_d^2 + i_q^2 & \leq i_{s,max}^2 \\ i_E & \leq i_{E,n} \end{cases} . \tag{5}$$

The operating range of the EESM is given by the voltage and current limits $\hat{u}_{s,max}$ and $\hat{i}_{s,max}$, which must not be exceeded. For the excitation current i_E the upper bound is set by the nominal excitation current $i_{E,n} = 16A$. The entire losses $P_{V,ges}$ are calculated by the sum of the copper, iron, friction and inverter losses ($P_{V,Cu}, P_{V,Fe}, P_{V,Fr}, P_{V,Inv}$). Another term $P_{V,Add}$ covers all other additional effects that cause power losses, such as the occurrence of stray fields or eddy current losses. [32]

$$P_{V,ges} = P_{V,Cu} + P_{V,Fe} + P_{V,Fr} + P_{V,Inv} + P_{V,Add} \tag{6}$$

The result of the optimization problem is a value of i_d, i_q and i_E for each pair of valid values T_{el} and n. In Fig. 5, the current characteristics of i_d, i_q as well as i_E are depicted.

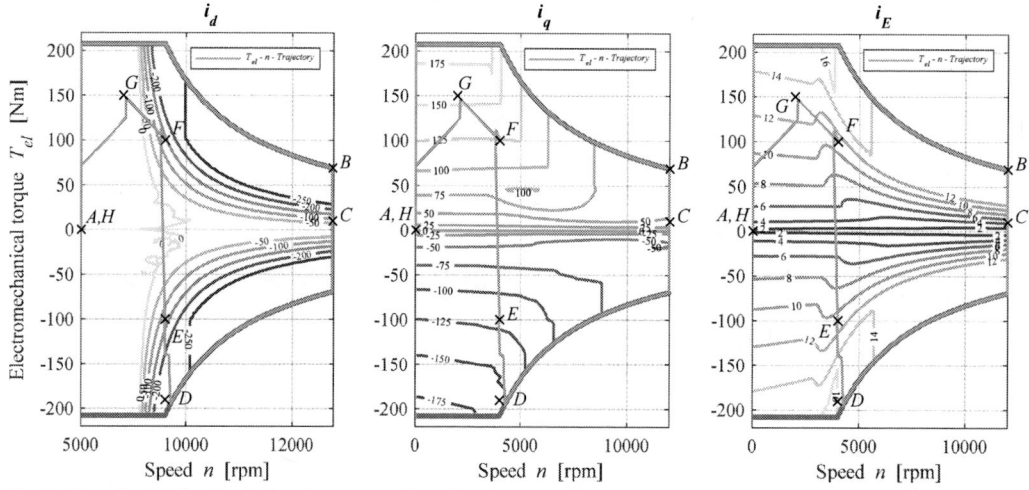

Fig.5: Results of the optimization procedure for the optimal stator and rotor currents of the EESM with exemplary torque-speed-trajectory

Fast and simple procedures are required for dynamic control tasks. Since the online estimation of the currents is computationally complex, the offline gained results are used as lookup tables (LUT) for the online control of the EESM [32,35]. The resulting simulation setup with the current controller of i_d, i_q and i_E, the MLI including the SVC modulation unit and its connection to the EESM is depictured in the following diagram.

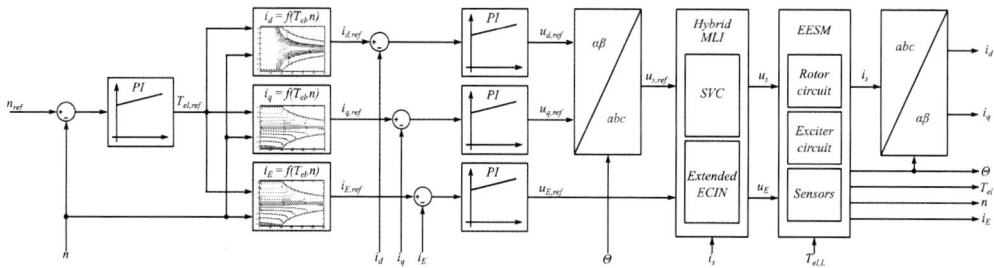

Fig. 6: Main simulation setup composed of the current control, MLI as well as the EESM

In [18] the fast demagnetization of the rotor inductance by impressing a counter voltage is discussed. By this a higher dynamic for the EESM can be achieved. However, while the operation is evaluated in the mentioned publication with a buck-boost converter and only with positive rotor voltages, the advantages of a bipolar rotor voltage are discussed in the outlook. For the presented hybrid MLI an H-Bridge could be attached to the secondary circuit as an option. By connecting an H-Bridge a bipolar voltage can be provided at the secondary terminals of the hybrid MLI, resulting in an increase of the control dynamics of the machine. This enables the inverter to be operated with maximum degrees of freedom by providing both positive as well as negative voltages. For the presented simulation an additional external H-Bridge is added.

Evaluation of operating the EESM with the hybrid MLI

Fig. 7a shows the simulated speed and torque profile of the EESM. Different load torques as well as speed specifications are approached at various time steps to show that the principle works in any operating point of the torque-speed-range. The simulated torque-speed-trajectory is explicitly drawn as a thin green line in Fig. 5 for all three currents. Additionally, special operating points of the drawn trajectory are also marked in Fig. 5 as well as in Fig. 7 (A to H).

At the beginning, a start-up with maximum acceleration is demonstrated up to $n_{max} = 12000 \, rpm$ within a time of $t_{start,up} = 2.5 \, s$. In the start-up, the electromechanical torque T_{el} first arises to its maximum of $T_{el,max} = 207 \, Nm$, before it decreases indirectly proportional to the rotational speed in the field weakening range. While the EESM accelerates to maximum speed, the hybrid MLI demonstrates the ability of the concept to utilize the full operating range of the EESM by moving the machine exactly along its system boundaries, as shown with the trajectory from A to B in Fig. 5. After reaching point B, the load torque is varied while the maximum speed remains constant. The trajectory moves along a vertical line to reference point C at maximum speed.

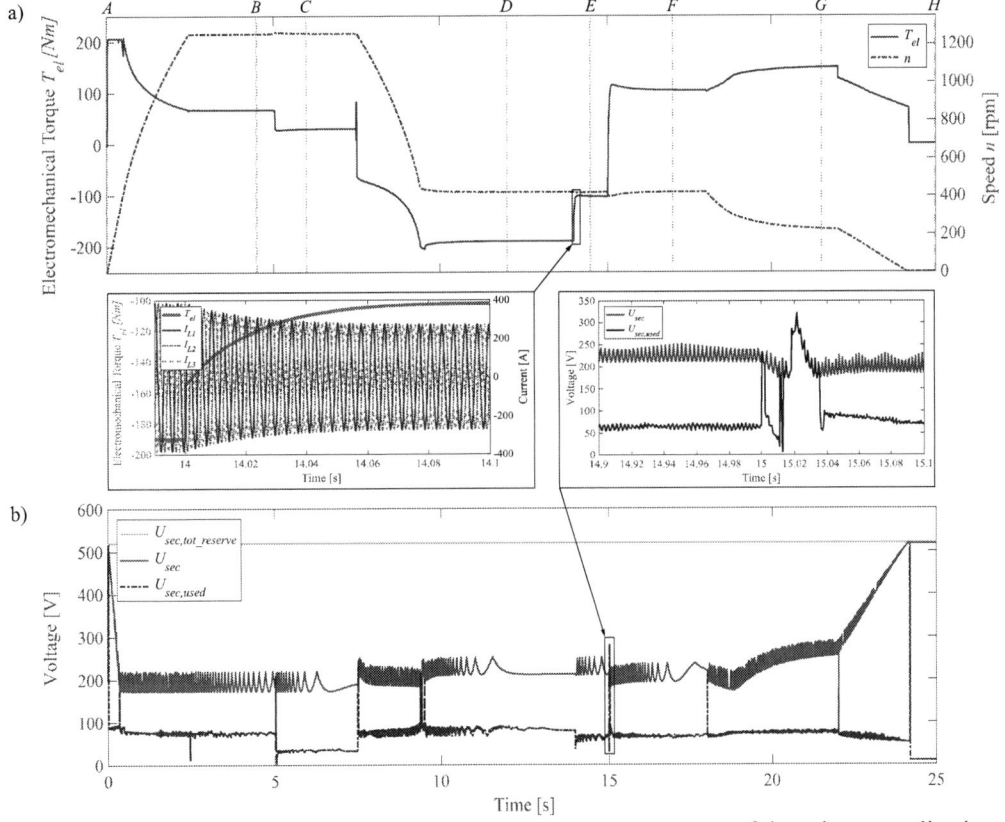

Fig. 7: Simulations results: (a) Machine torque, speed and stator currents of the primary application as well as (b) the used and total cell reserve for supplying secondary consumers

After adjusting operating point C, the speed is decreased to the nominal value n_n, whereby three different loads are requested in succession (D to F). Here, both motor and generator points are approached, which emulate traction and recuperation.

Before the system is braked to a standstill, the operating point G at a speed of $n_n/2$ and a load of $150 \, Nm$ is set.

Figure 7b shows that the remaining cell reserve U_{sec} is sufficient for each part of the simulation. In most cases the cell reserve is larger than the DC component shown in Fig. 1, even if a demanding load curve is evaluated. For partial load parts e.g. from $t_3 = 17 \, s$ to $t_4 = 23 \, s$ the cell reserve increases additionally. This provides an even greater cell reserve, which not only satisfies the request of the current controller $U_{sec,used}$ for normal operation, but also offers an additional control reserve. The latter is useful for driving the current through the relatively high inductance of the rotor circuit. This is particularly advantageous for high load changes, such as at $t_1 = 7.5 \, s$ and $t_2 = 15 \, s$, what enables a more dynamic operation.

The simulation results point out that the primary and the secondary circuit do not effect each other for the operation of the EESM. They are operated separately and are controlled in order to meet the requirements of the respective load and speed applications. The secondary voltage for the EESM rotor circuit is provided at the terminals p_{second} and m_{second} in the way as it is pictured in Fig. 3.

Conclusion

A new hybrid multilevel inverter concept operating an electrically excited synchronous machine is presented. The inverter is able to provide a secondary constant output voltage while simultaneously generating a three-phase voltage system. In comparison to a state-of-the-art multilevel inverter no additional cells or any further secondary supply units are required. The hybrid multilevel inverter only uses temporarily unexploited battery cells and a special developed circuit design to share the battery cells between the main and the secondary voltage system. This capability is utilized to supply and control a stator and rotor current of an electrically excited synchronous machine. A detailed model of a powertrain consisting of lithium-ion battery cells, controllers, power electronics and an externally excited synchronous machine is built and the concept is tested in dynamic operation as well as under full load. The simulation results show the feasibility of providing a three-phase AC voltage system to the exciter circuit of an electrically excited synchronous machine with the hybrid multilevel inverter, while the inverter simultaneously delivers a constant voltage to the rotor circuit. The operation and control of this type of electrical machine is achieved without influences between stator and rotor supply, even under extreme load changes and during the speed-up of the machine.

References

[1] A. M. Lulhe and T. N. Date, "A technology review paper for drives used in electrical vehicle (EV) & hybrid electrical vehicles (HEV)," *2015 International Conference on Control, Instrumentation, Communication and Computational Technologies (ICCICCT)*, Kumaracoil, 2015, pp. 632-636.

[2] Y. A. Alamoudi, A. Ferrah, R. Panduranga, A. Althobaiti and F. Mulolani, "State-of-the Art Electrical Machines for Modern Electric Vehicles," *2019 Advances in Science and Engineering Technology International Conferences (ASET)*, Dubai, United Arab Emirates, 2019, pp. 1-8.

[3] A. Emadi, Y. J. Lee and K. Rajashekara, "Power Electronics and Motor Drives in Electric, Hybrid Electric, and Plug-In Hybrid Electric Vehicles," in *IEEE Transactions on Industrial Electronics*, vol. 55, no. 6, pp. 2237-2245, June 2008.

[4] L. G. Franquelo, J. Rodriguez, J. I. Leon, S. Kouro, R. Portillo and M. A. M. Prats, "The age of multilevel converters arrives," in *IEEE Industrial Electronics Magazine*, vol. 2, no. 2, pp. 28-39, June 2008.

[5] J. Rodriguez, Jih-Sheng Lai and Fang Zheng Peng, "Multilevel inverters: a survey of topologies, controls, and applications," in *IEEE Transactions on Industrial Electronics*, vol. 49, no. 4, pp. 724-738, Aug. 2002.

[6] F. Helling, J. Glück, A. Singer, T. Weyh and H.-J. Pfisterer, "The AC Battery – A Novel Approach for Integrating Batteries into AC Systems," in *International Journal of Electrical Power & Energy Systems*, vol. 104, Feb. 2018.

[7] A. Singer, F. Helling, T. Weyh, J. Jungbauer and H. Pfisterer, "Modular multilevel parallel converter based split battery system (M2B) for stationary storage applications," *2017 19th European Conference on Power Electronics and Applications (EPE'17 ECCE Europe)*, Warsaw, Sep. 2017, pp. 1-10.

[8] J. Rodriguez, S. Bernet, B. Wu, J. O. Pontt and S. Kouro, "Multilevel Voltage-Source-Converter Topologies for Industrial Medium-Voltage Drives," in *IEEE Transactions on Industrial Electronics*, vol. 54, no. 6, pp. 2930-2945, Dec. 2007.

[9] S. D'Arco, L. Piegari and P. Tricoli, "Power and balancing control considerations on modular multilevel converters for battery electric vehicles," *2013 15th European Conference on Power Electronics and Applications (EPE)*, Lille, 2013, pp. 1-9.

[10] A. Hillers, M. Stojadinovic and J. Biela, "Systematic comparison of modular multilevel converter topologies for battery energy storage systems based on split batteries," *2015 17th European Conference on Power Electronics and Applications (EPE'15 ECCE-Europe)*, Geneva, 2015, pp. 1-9.

[11] S. Rohner, S. Bernet, M. Hiller and R. Sommer, "Modulation, Losses, and Semiconductor Requirements of Modular Multilevel Converters," in *IEEE Transactions on Industrial Electronics*, vol. 57, no. 8, pp. 2633-2642, Aug. 2010.

[12] L. D'Errico, A. Lidozzi, V. Serrao and L. Solero, "Multilevel converters for high fundamental frequency application," *2009 13th European Conference on Power Electronics and Applications*, Barcelona, 2009, pp. 1-14.

[13] F. Chang, O. Ilina, M. Lienkamp and L. Voss, "Improving the Overall Efficiency of Automotive Inverters Using a Multilevel Converter Composed of Low Voltage Si mosfets," in *IEEE Transactions on Power Electronics*, vol. 34, no. 4, pp. 3586-3602, April 2019.

[14] A. Kersten et al., "Inverter and Battery Drive Cycle Efficiency Comparisons of CHB and MMSP Traction Inverters for Electric Vehicles," *2019 21st European Conference on Power Electronics and Applications (EPE '19 ECCE Europe)*, Genova, 2019, pp. 1-12.

[15] M. Quraan, P. Tricoli, S. D'Arco and L. Piegari, "Efficiency Assessment of Modular Multilevel Converters for Battery Electric Vehicles," in *IEEE Transactions on Power Electronics*, vol. 32, no. 3, pp. 2041-2051, Mar. 2017.

[16] C. Korte, E. Specht, M. Hiller and S. Goetz, "Efficiency Evaluation of MMSPC/CHB topologies for automotive applications," *2017 IEEE 12th International Conference on Power Electronics and Drive Systems (PEDS)*, Honolulu, Dec. 2017, pp. 324-330.

[17] F. Helling, M. Kuder, A. Singer, S. Schmid and T. Weyh, "Low Voltage Power Supply in Modular Multilevel Converter based Split Battery Systems for Electrical Vehicles," *2018 20th European Conference on Power Electronics and Applications (EPE'18 ECCE Europe)*, Riga, 2018, pp. 1-10.

[18] S. Köhler and B. Wagner, "Control of the excitation current of an externally excited synchronous machine supplied by an inductive energy transfer system," *13th Conference on Ecological Vehicles and Renewable Energies (EVER)*, Monte-Carlo, Apr. 2018, pp.1-8.

[19] S. Köhler, B. Wagner and S. Endres, "Estimation of the Excitation Current and the Rotor Resistance of an Externally Excited Synchronous Machine With an Inductively Supplied Excitation," *PCIM Europe 2016; International Exhibition and Conference for Power Electronics, Intelligent Motion, Renewable Energy and Energy Management*, Nuremberg, May 2016, pp. 1-8.

[20] R. Mocanu and A. Onea, "Robust optimal control of Externally Excited Synchronous Machine based on passivity theory," *2017 18th International Carpathian Control Conference (ICCC)*, Sinaia, 2017, pp. 162-166.

[21] M. Kuder, A. Singer and T. Weyh, "Multi-Engine in Modular Multilevel Converter based Split Battery Systems for Electric Vehicles," *PCIM Europe 2019; International Exhibition and Conference for Power Electronics, Intelligent Motion, Renewable Energy and Energy Management*, Nuremberg, 2019, pp. 1-7.

[22] J. Stöttner, C. Terbrack and C. Endisch, "Intelligent Usage of Temporary Inactivated Battery Cells In a Multilevel Inverter Gaining an Optimized Usage of Battery System," in preparation.

[23] S. D'Arco, L. Piegari and P. Tricoli, "A modular converter with embedded battery cell balancing for electric vehicles," *2012 Electrical Systems for Aircraft, Railway and Ship Propulsion*, Bologna, 2012, pp. 1-6.

[24] S. M. Goetz, A. V. Peterchev and T. Weyh, "Modular Multilevel Converter With Series and Parallel Module Connectivity: Topology and Control," in *IEEE Transactions on Power Electronics*, vol. 30, no. 1, pp. 203-215, Jan. 2015.

[25] M. F. Kangarlu, E. Babaei and M. Sabahi, "Cascaded cross-switched multilevel inverter in symmetric and asymmetric conditions," in *IET Power Electronics*, vol. 6, no. 6, pp. 1041-1050, July 2013.

[26] M. F. Kangarlu and E. Babaei, "Cross-switched multilevel inverter: an innovative topology," in *IET Power Electronics*, vol. 6, no. 4, pp. 642-651, April 2013.

[27] E. Babaei, "A Cascade Multilevel Converter Topology With Reduced Number of Switches," in *IEEE Trans. Power Electronics*, vol. 23, no. 6, pp. 2657-2664, Nov. 2008.

[28] N. S. Hasan, N. Rosmin, D. A. A. Osman and A. H. Musta'amal@Jamal, "Reviews on multilevel converter and modulation techniques, " in *Renewable and Sustainable Energy Reviews, Elsevier*, vol. 80(C), pp. 163-174, May 2017.

[29] A. O. Arslan, M. Kurtoglu, F. Eroglu and A. M. Vural, "Comparison of phase and level shifted switching methods for a three-phase modular multilevel converter," *2018 5th International Conference on Electrical and Electronic Engineering (ICEEE)*, Istanbul, 2018, pp. 91-96.

[30] LG Chem, "Product Description: ICR18650 C2 2800mAh, " https://www.batteryspace.com/prod-specs/5702_5.pdf, (Accessed on 03/06/2020).

[31] T. Bruen and J. Marco, "Modelling and experimental evaluation of parallel connected lithium ion cells for an electric vehicle battery system," in *Journal of Power Sources*, vol. 310, pp. 91-101, Feb. 2016.

[32] R. Grune, "Verlustoptimaler Betrieb einer elektrisch erregten Synchronmaschine für den Einsatz in Elektrofahrzeugen," *Phd Thesis*, October 2012.

[33] Infineon, "Technical Information: IRL7472L1TRPbF, " https://www.mouser.de/datasheet/2/196/Infineon-IRL7472L1-DS-v02_00-EN-1732064.pdf, (Accessed on 03/06/2020).

[34] H. Zhang, Y. Meng, L. Ning, Y. Zou, X. Wang and X. Wang, "Fast and simple space vector modulation method for multilevel converters," in *IET Power Electronics*, vol. 13, no. 1, pp. 14-22, Jan. 2020.

[35] M. Märgner and W. Hackmann, "Control challenges of an externally excited synchronous machine in an automotive traction drive application," *2010 Emobility - Electrical Power Train*, Leipzig, 2010, pp. 1-6.

A facility for mixed flowing gas testing of and experimentation with power electronic components and systems

Juuso Rautio[1], Janne Jäppinen[1], Tommi J. Kärkkäinen[1], Markku Niemelä[1], Pertti Silventoinen[1], Mika Kiviniemi[2], Joonas Leppänen[2], Jonny Ingman[2]

[1]LUT University
Yliopistonkatu 34
Lappeenranta, Finland
Email: juuso.rautio@lut.fi
URL: http://www.lut.fi

[2]ABB Oy, Drives
Hiomotie 13
Helsinki, Finland

Keywords

≪Reliability≫, ≪Environment≫, ≪Test bench≫, ≪Power semiconductor device≫, ≪Corrosion testing≫.

Abstract

Corrosion phenomena have been found to inflict new types of failures on power electronics. Currently, the precise corrosion mechanisms and effects causing problems are largely unknown. A facility dedicated for environmental reliability and lifetime testing of power electronic components and systems is presented.

Introduction

The increasing use of power electronic devices such as frequency converters in solar, wind and automotive applications, has led to the diversification of the environmental conditions the devices are exposed to. The topical matters keeping up this trend are the electrification of vehicles and ambitions of increasing the energy efficiency of systems. This trend has increased the number of devices put into environments that are more and more extreme in terms of their environmental conditions. In particular, humid environments with chemically active compounds, for example water treatment plants, mining and paper industry plants, have been found to be problematic by field experience. Such environments can cause corrosion, eventually leading to the failure of the converter. Corrosion is manifested e.g. as dendritic growth in insulations (Fig. 1) which leads to a short circuit. The reliability of power electronics has been the subject of many publications. Present literature does not, however, fully discuss corrosion phenomena and related reliability issues.

Today a common climatic accelerated aging testing method for power electronic switch components is the high humidity, high temperature and high voltage reverse bias (H3TRB) test, where corrosion is induced into the devices under test by controlled humidity, temperature and bias voltage [1, 2, 3]. While this testing method does allow studying aging phenomena in humid environments, it does not include controlling the amount of impurities and ion concentrations in the exposure air. Because of this, the test does not simulate environments with corrosive gases present. With no control on the impurities, the test results may vary and even contradict each other [4].

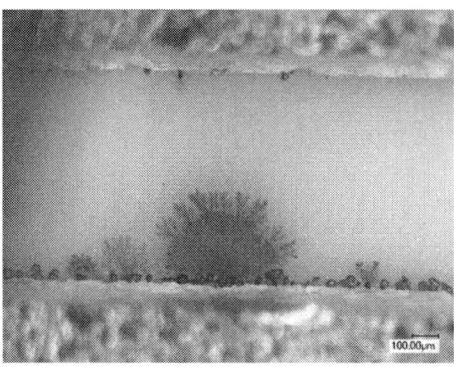

Fig. 1: Dendrite growth on an IGBT module aluminum oxide substrate. Module was recovered from a water treatment plant.

Corrosion test conditions using pollutant gases have evolved from single gas tests with high concentrations to mixed flowing gas (MFG) tests carried out with multiple corrosive gases at low concentrations [5]. While single gas tests are still used alongside MFG tests [6], MFG tests achieve high acceleration factors with lower gas concentrations because of synergistic reactions between the gases [7]. MFG tests can be run with similar humidity and electrical conditions as H3TRB, albeit at lower temperatures because of issues with exhausting and transporting moisture-laden gas to external analytical equipment without condensation [5]. The gases used in MFG tests are hydrogen sulfide H_2S, sulfur dioxide SO_2, chlorine Cl_2 and nitrogen dioxide NO_2 [8]. Some standardized test conditions such as all Battelle tests use three gases leaving out SO_2, as when developing these standards, it was found that adding SO_2 to the mix did not significantly affect copper corrosion [5]. It is also stated that SO_2 may be included to account for materials not present in the original Battelle tests. This means that SO_2 should be present when testing materials other than copper, for example power semiconductor switches.

Previously, mixed flowing gas testing has been mostly used for printed circuit boards and small components, motivated by the increasing number of failures associated with the RoHS directive taking effect in 2006 [9] [10]. Despite the long history of the method in many electronics applications, it has not made a breakthrough in component testing for power electronics. However, an increasing interest towards corrosion issues has been taken by industrial electronics manufacturers [11]. While many papers on MFG tests on small components and printed circuit boards exist, the papers do not describe the test system in detail.

Creating the appropriate test conditions requires carefully selected instruments for dosing the chemicals and analyzing the resulting gas concentrations. The test chambers are, in principle, typical environmental test chambers capable of producing a desired air temperature and humidity, that also are built to withstand the corrosive gas mixture used in the experiments. It is also necessary to monitor the gas concentrations for the purposes of occupational safety in order to avoid poisoning the persons carrying out the experiments.

Methods

To reduce the time needed to complete the experimental facility, the entire test facility is to be constructed of readily available systems and equipment. A major part of the challenge in choosing and procuring the equipment is defining the requirements for each element in the system. For producing the desired test conditions, two separate but connected problems are identified: how to dose the gases appropriately into the test chamber while also controlling the humidity and temperature, and how to verify that the resulting conditions satisfy the requirements.

The test setup is realized according to the the example system introduced in IEC 60068-2-60 [12]. The setup consists of a modified climatic chamber with an internal exposure chamber, gas dosing unit and gas

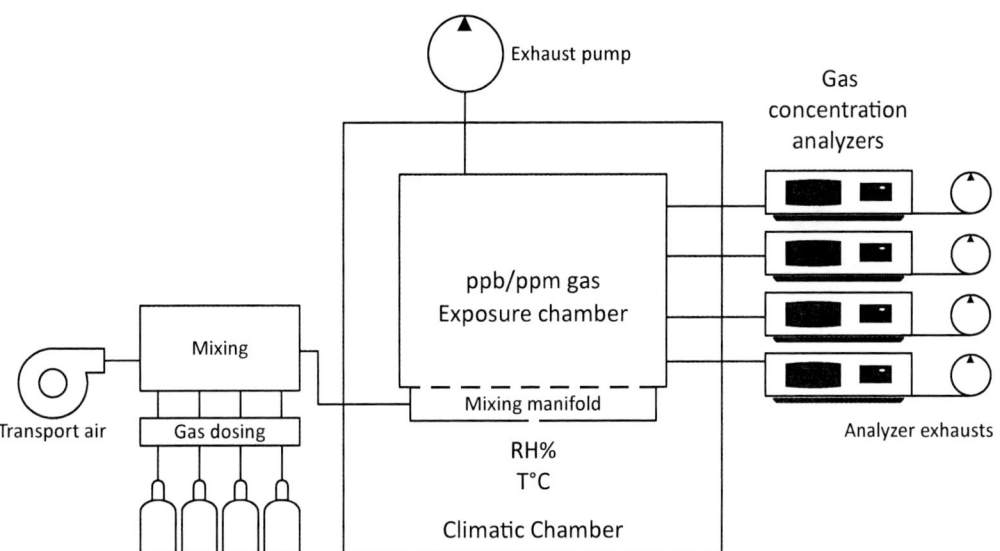

Fig. 2: Test setup with the essential components. A climatic chamber provides the desired humidity and ambient air temperature for the tests. A dosing and mixing system feeds the gases into the exposure chamber, where the devices under test are located. Gas concentration analyzers verify the test conditions.

analyzers (Fig. 2). The climatic chamber controls temperature and humidity of the system. An exposure chamber is placed inside it, into which the gases are then dosed. The exposure chamber is constructed of materials that are not harmed by the corrosive gasses. The working volume of the test system is the volume of the inner exposure chamber.

The amount of corrosive gases is controlled by mass flow controllers. The volumetric flow from the gas bottles is quite low, which means the gases would take a long time to reach the exposure chamber on their own. To mitigate this, the gases are mixed into filtered and dehumidified pressurized air called transport air. This is also the first dilution step in the gas dosing process. Transport air has a flow controller of its own.

Air flow to the exposure chamber is controlled by the exhaust system. The exhaust air pump reduces the pressure inside the exposure chamber, which allows air to diffuse into the chamber from the mixing manifold. The mixing manifold receives the mix of transport air and gases, but it also has an opening to the climatic chamber. This allows the corrosive gases to mix with temperature and humidity controlled air. This is the final dilution step. The air flowing from the mixing manifold into the exposure chamber matches the requirements in terms of temperature, humidity and gas concentrations.

The exhaust system creates a small underpressure in the exposure chamber to make sure the corrosive gases do not exit into the climatic chamber. The gas concentration analyzers sample the air from the working volume of the exposure chamber. Gas concentrations must be analyzed from the exposure chamber instead of e.g. the exhaust air because the gas concentrations decrease due to adsorption and loss by chemical reactions [5]. Measuring the exhaust air would result in smaller concentration readings than what the actual gas concentrations in the working volume are. Controlling the gas dosing based on such incorrect measurements would result in higher concentration in the working volume.

Air from the mixing manifold to the exposure chamber flows through small holes that cover the floor of the exposure chamber. The goal is to keep the flow as uniform as possible for the whole volume of the chamber. Despite this, local differences in corrosivity are possible in the chamber [12]. Local corrosivity is mapped using standard copper and silver corrosivity coupons that are spread around the exposure chamber. In addition it is also possible to use a rotating carousel structure to hang small devices from. The carousel can be used to slowly rotate the devices under test so that possible local differences in corrosivity are mitigated.

The variables affecting the gas concentrations are the exchange rate, and the flow rates or transport air, conditioned air and corrosive gases. Exhange rate – the rate at which air inside the exposure chamber is changed – is set with a valve that controls the amount of air pumped out of the exposure chamber. In standards such as the IEC 60068-2-60 the exhange rate is speficied by rate of ventilations per hour, in this case 3 to 10 ventilations per hour [12]. Exhange rate is set first as it is the only variable that is directly specified in test standards. The gas concentration analyzers exhaust their sampling air out from the exposure chamber and so their flow rate needs to be added to the total exhaust flow.

According to the system manufacturer the air in the exposure chamber contains a mix of roughly 10 % of transport air including the corrosive gases and 90 % of conditioned air (humid and temperature controlled air inside the climatic chamber). This mix is then exhausted by the exhaust air pump and analyzers. To calculate the correct settings for gas dosing a few formulas are needed. The basis for the calculations is given by the manufacturer as

$$\dot{V}_{out} = \frac{V_{EC} r_{ex}}{60\,min/h},$$ (1)

where $\dot{V}_{out}$ is the exhaust air in litres per minute, V_{EC} is the volume of the exposure chamber in litres and r_{ex} is exchanges per hour. After the $\dot{V}_{out}$ is calculated the transport air flow is simply $\dot{V}_{Trans} = 0.1 \cdot \dot{V}_{out}$ and the conditioned air is similarly $\dot{V}_{Cond} = 1.1 \cdot \dot{V}_{out}$. Then the mass-flow controller values V_{Gas} in millilitres per minute can be calculated with

$$\dot{V}_{Gas} = \dot{V}_{out} \frac{C}{M},$$ (2)

where C is the gas concentration wanted in the exposure chamber and M is the concentration in the gas cylinders. To achieve a desired humidity in the exposure chamber, the climatic chamber humidity setting has to be set higher. This is because the mixture of transport air and corrosive gases is dry, lowering the humidity of the air mass inside the exposure chamber.

To verify the conditions, accurate gas analyzers are required for each gas used in the experiments. For each test condition and each gas, the IEC 60068-2-60 and other standards specify a nominal gas concentration and a tolerance, e.g. (200 ± 25) ppb. The measurement uncertainty of the gas analyzer is not insignificant, and further reduces the allowed range of the displayed value (Fig. 3). Therefore, the higher the accuracy of the gas analyzer, the more forgiving the experimental system is for the gas dosing system.

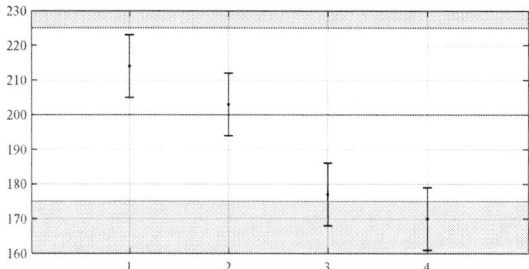

Fig. 3: Illustration of data points with uncertainty error bars reducing the range at which the displayed value can lie. In this example, the permitted range is 200 ± 25 units. The first two data points are within the specification. The third reading of the third measurement is within the specification, but conformity with the specification is unclear due to measurement uncertainty. The fourth measurement is clearly out of specification.

To compare and select gas analyzers, it was necessary to compare the measurement uncertainties. Since the instruments were not available for testing at the comparison stage, the uncertainty estimates had to rely on the manuals of the instruments. There is significant variation in how manufacturers declare the uncertainty or accuracy figures for each instrument. Some report a simple number labeled as "accuracy" or "precision", while others provide multiple sources of uncertainty associated with the instrument itself.

For each instrument, the provided sources of uncertainty were estimated for a selected, representative reading, combined into a combined uncertainty and then into an expanded uncertainty (coverage factor k=2.6, confidence level 99%) according to the JCGM Guide to the Expression of Uncertainty in Measurement [13]. Uncertainties for the different analyzers were calculated and compared at different test method levels [14]. For IEC 60068-2-60 Method 4 levels uncertainties are calculated at upper concentration limit according to tolerance (Table I). As concentration levels and tolerances for IEC 60068-2-60 Method 3 are different, the tolerances and uncertainties are calculated separately for this method (Table II). Included in tables are also analyzers that were up to consideration.

Among requirements for selecting appropriate analyzers is that they have to provide real-time measurements without interrupting the exposure test. This excludes any instrument that requires manual sample extraction or otherwise opening the chamber. The analyzers that fit the description have internal or external pumps which allow for sample air extraction from the exposure chamber. Other requirements include rack-mounting and measurement range corresponding to the gas concentrations used.

Results

The gas concentrations inside the exposure chamber are analyzed using external analyzers. For the measurement of sulfur dioxide and hydrogen sulfide ultraviolet fluorescence spectroscopy is used. Nitrogen dioxide is measured using chemiluminescence. Both measurement methods are affected by the high humidity in the test chamber, an effect which will have to be characterized before the actual exposure tests [14]. Chlorine is measured with a chemcassette-based method. Chosen analyzers are on a green background on Tables I and II with those up to consideration on white background.

Table I: Gas concentration analyzers with their associated combined and expanded (99%, k = 2.6) uncertainties calculated at IEC 60068-2-60 Method 4 levels. Analyzers on green background have been chosen for MFG gas concentration measurements and analyzers on white background were considered but were not chosen.

Gas	Analyzer	Method	Target concentration [ppb]	U [ppb]
SO_2	Ecotech Serinus 50	UV-Fluorescence Spectroscopy	200 ± 25	± 16
SO_2	Thermo Scientific 43i	UV-Fluorescence Spectroscopy	200 ± 25	± 16
H_2S	Ecotech Serinus 55	UV-Fluorescence Spectroscopy	10 ± 5	± 3
H_2S	Thermo Scientific 450i	UV-Fluorescence Spectroscopy	10 ± 5	± 4
NO_2	Ecotech Serinus 40	Chemiluminescence	200 ± 20	± 9
NO_2	Teledyne T500U	Cavity Attenuated Phase Shift	200 ± 20	± 16
Cl_2	Honeywell SPM Flex	Chemcassette (colorimetry)	10 ± 5	± 7

Honeywell SPM Flex was found to be the most suitable for measuring chlorine gas concentrations, that also fit the other criteria of real-time measurement. For a confidence level of 99 %, the extended uncertainty exceeds the tolerances given in the standards (Tables I-II). The uncertainty is calculated based on very limited information on the performance of the instrument. For this reason, the responsibility for proper Cl_2 concentrations falls more on the mass flow controller than with other gases. It is to be noted that with a confidence level of 95% the test tolerance accuracy on Table I is met.

Table II: Combined and expanded (99%, k = 2.6) uncertainties U calculated at IEC 60068-2-60 Method 3 levels for gas concentration analyzers. Analyzers on green background have been chosen for MFG gas concentration measurements and analyzers on white background were considered but were not chosen.

Gas	Analyzer	Method	Target concentration [ppb]	U [ppb]
H_2S	Ecotech Serinus 55	UV-Fluorescence Spectroscopy	100 ± 10	± 3
H_2S	Thermo Scientific 450i	UV-Fluorescence Spectroscopy	100 ± 10	± 4
NO_2	Ecotech Serinus 40	Chemiluminescence	200 ± 50	± 9
NO_2	Teledyne T500U	Cavity Attenuated Phase Shift	200 ± 50	± 16
Cl_2	Honeywell SPM Flex	Chemcassette (colorimetry)	20 ± 5	± 11

Power electronics switches, such as IGBT modules, are under high voltage bias in the tests to allow corrosion to take place. Both the exposure and the climatic chambers have inlets for analyzer hoses and electric cables. To keep the leakage through the inlets as little as possible panels covering the inlet holes can be customized for different amount and size of cables and hoses.

At first the test setup uses one 1000-liter climatic chamber with a 380-liter inner exposure chamber. The size of the chamber allows testing of components and small systems like printed circuit boards. The inner volume of the exposure chamber can't be filled too full or the flow of air may be restricted causing local differences in corrosivity. The system will be expanded with a larger 4200-liter climatic chamber with 1900-liter exposure chamber. The bigger volume is for testing assembled devices like frequency converters. The operating principles of the larger chamber are the same as with the smaller one.

Validation

The system performance was evaluated using gas concentration analyzers. First the parameters including the corrosive gas flow rates were set according to calculations by user manual. The test was conducted at 30°C temperature and 40% relative humidity. The target concentration for both NO_2 and SO_2 gases was 200 ppb.

Control parameters for transport air, conditioned air and corrosive gases were calculated from (1) and (2). With these parameters both gas concentrations became steady at around 140% of the intended value (Table III). The initial corrosive gas flow rates were corrected based on measurements and the concentrations became steady very close to the target value of 200 ppb.

Table III: Average gas concentrations inside exposure chamber and associated standard deviations per test run. Standard deviation expresses the variation in the concentrations over the observation period. Expanded (99%, $k = 2.6$) measurement uncertainty U for NO_2 analyzer Ecotech Serinus 40 and SO_2 analyzer Ecotech Serinus 50.

Measurement	Concentration [ppb]	U [ppb]
NO_2 uncorrected	291.2 ± 3.3	± 9.0
SO_2 uncorrected	283.1 ± 2.0	± 15.4
NO_2 corrected	198.1 ± 0.6	± 8.2
SO_2 corrected	194.8 ± 0.9	± 15.4

One of the reasons for higher-than-expected uncorrected values is the coarse scale of the exhange rate and conditioned air valves that affect all gas concentrations. For example the target for exhange rate valve was $1.14\,\mathrm{m}^3/\mathrm{min}$ and closest scale increments are $1\,\mathrm{m}^3/\mathrm{min}$ and $1.2\,\mathrm{m}^3/\mathrm{min}$. However this does not account for the whole error as the possible deviation is in this case less than 20%.

Other observations

From using the test system a few practical observations of possible problems in similar MFG test setups were made. The H_2S concentration measurement is based on a process where the analyzer first removes SO_2 from the sampling air, converts the remaining H_2S into SO_2 and finally measures the SO_2. The H_2S to SO_2 converter was poisoned by chlorine gas that was present in the sampled air mix. In this case 20 ppb concentration over a few days was enough to poison the converter. Poisoning reduces the conversion efficiency drastically, leading to lowered concentration readings. Readings of over 90% less than the actual concentration were observed.

Humidity in the air mix is known to affect gas concentration measurements using UV-Fluorescence spectroscopy and chemiluminescence methods [15, 16]. To keep the gas concentrations repeatable between tests with different humidity targets, they are measured before increasing humidity in the exposure chamber. The ideal relative humidity range for measurements is not speficied but it should be kept constant between tests.

When running tests with higher than ambient temperatures it is possible that the humid air condenses on the analyzer sampling tubes. One possible point where condensation occurs is metal connectors on the sampling lines. The probability of this happening can be reduced by starting the tests with low humidity and letting the sample lines warm up before increasing the humidity to desired value. If the ambient temperature is lower than the dew point of sample air, the sample lines can be heated or water condensation traps used in addition.

Conclusions

A system for Mixed Flowing Gas (MFG) tests built out of readily available instruments is described. The system can be used for research of corrosion-related failure modes of power electronic components. The tests carried out with the setup complement the more common H3TRB tests that are limited to failure modes in humid environments with no pollutants.

The requirements for equipment on a mixed flowing gas test system are hard to meet due to very small gas concentrations. This leads to fewer options especially for gas concentration analyzers. In the case of chlorine the measurements extended uncertainty (99%, $k = 2.6$) is larger than the allowed error in the example test methods from IEC 60068-2-60. The importance of gas concentration measurements is shown in the validation section: Without measurements the gas concentration values can deviate considerably from the target values. Measurements also help test repeatability by minimizing the effect of human error. In the case of this test setup two of the four parameters controlling the resulting gas concentrations are prone to such errors.

Mixed flowing gas tests require fine control of exposure parameters and managing system consisting of many devices, but they are far from impossible to conduct properly with thorough and attentive work. The practical capabilities and effects of mixed flowing gas tests on power semiconductors and assembled devices is determined in further work.

References

[1] J. Jormanainen, E. Mengotti, T. Batista Soeiro, E. Bianda, D. Baumann, T. Friedli, A. Heinemann, A. Vulli, and J. Ingman, "High humidity, high temperature and high voltage reverse bias - a relevant test for industrial applications," in *PCIM Europe 2018; International Exhibition and Conference for Power Electronics, Intelligent Motion, Renewable Energy and Energy Management*, pp. 1–7, 2018.

[2] J. Leppänen, "Humidity related failure mechanisms in power semiconductor devices," master's thesis, Aalto University, http://urn.fi/URN:NBN:fi:aalto-201706135367, 2017.

[3] C. Zorn and N. Kaminski, "Acceleration of temperature humidity bias (thb) testing on igbt modules by high bias levels," in *2015 IEEE 27th International Symposium on Power Semiconductor Devices IC's (ISPSD)*, pp. 385–388, 2015.

[4] N. L. Sbar and R. P. Kozakiewicz, "New acceleration factors for temperature, humidity, bias testing," *IEEE Transactions on Electron Devices*, vol. 26, no. 1, pp. 56–71, 1979.

[5] W. H. Abbott, "The development and performance characteristics of mixed flowing gas test environment," *IEEE Transactions on Components, Hybrids, and Manufacturing Technology*, vol. 11, no. 1, pp. 22–35, 1988.

[6] T. Wassermann, O. Schilling, K. Müller, A. Rossin, and J. Uhlig, "A new high-voltage h2s single noxious gas reliability test for power modules," *Microelectronics Reliability*, vol. 100-101, p. 113468, 2019. 30th European Symposium on Reliability of Electron Devices, Failure Physics and Analysis.

[7] W. Abbott, "Effects of industrial air pollutants on electrical contact materials," *IEEE Transactions on Parts, Hybrids, and Packaging*, vol. 10, no. 1, pp. 24–27, 1974.

[8] "Standard guide for mixed flowing gas (mfg) tests for electrical contacts," standard, American Society for Testing and Material, 100 Barr Harbor Drive, PO Box C700, West Conshohocken, PA 19428-2959, United States, Aug. 2013.

[9] Ping Zhao and M. Pecht, "Mixed flowing gas studies of creep corrosion on plastic encapsulated microcircuit packages with noble metal pre-plated leadframes," *IEEE Transactions on Device and Materials Reliability*, vol. 5, no. 2, pp. 268–276, 2005.

[10] R. Schueller, "Creep corrosion of lead-free printed circuit boards in high sulfur environments," in *SMTA International Proceedings, Orlando, FL*, vol. 21, 2007.

[11] G. K. Morris, R. A. Lukaszewski, and C. Genthe, "Environmental contamination and corrosion in electronics: The need for an industrial standard and related accelerated test method that makes sense," in *2018 Annual Reliability and Maintainability Symposium (RAMS)*, pp. 1–7, 2018.

[12] "Environmental testing – part 2-60: Tests – test ke: Flowing mixed gas corrosion test," standard, International Electrotechnical Commission, Geneva, CH, June 2015.

[13] JCGM, "Jcgm 100: Evaluation of measurement data - guide to the expression of uncertainty in measurement," tech. rep., Joint Committee for Guides in Metrology, 2008.

[14] J. Jäppinen, "Measurement of gas concentrations in an environmental mixed flowing gas test chamber," master's thesis, Lappeenranta-Lahti University of Technology LUT, http://urn.fi/URN:NBN:fi-fe2019100731571, 2019.

[15] A. B. Bluhme, J. L. Ingemar, C. Meusinger, and M. S. Johnson, "Water vapor inhibits hydrogen sulfide detection in pulsed fluorescence sulfur monitors," *Atmospheric Measurement Techniques*, vol. 9, no. 6, pp. 2669–2673, 2016.

[16] P. Visamo, "Evaluating the measurement uncertainty for nitrogen oxides in air quality measurements," Master's thesis, Lappeenranta University of Technology, Skinnarilankatu 34, 53850 Lappeenranta, Finland, 2009.

Impact of implementation of auxiliary bias-windings on controllable inductors for power electronic converters

Jonas Pfeiffer ⊕, Pierre Küster ⊕, Yeliz Erenler, Ziyad H. S. Qashlan and Peter Zacharias
UNIVERSITY OF KASSEL
Department of Electrical Power Engineering (EVS)
Wilhelmshöher Allee 71
34121 Kassel, Germany
Tel.: +49 / (0)561 - 804 6344
Fax: +49 / (0)561 - 804 6521
E-Mail: jonas.pfeiffer@uni-kassel.de
URL: http://www.uni-kassel.de/eecs/evs

Acknowledgements

Parts of the work presented in this paper have been supported by the German Federal Ministry of Education and Research (BMBF). Project funding reference number:16EMO0234. Responsibility for the contents of this publications lies with the authors.

The authors thank the *SUMIDA Components & Modules GmbH* in Obernzell, Germany, for manufacturing or machining most of the ferrite cores which were used to generate the presented measurement results.

Keywords

«Device characterization», «Device modeling», «Magnetic device», «Hardware (not only Software)», «Passive component»

Abstract

Magnetic devices are an important part of electronic converters. They can be used to control some properties of the converters. There are three different control methods using an auxiliary bias-winding. In the presented paper, the impact of the implementation of the auxiliary bias-winding was studied for mixed biasing. Seven different designs were investigated in small signal measurements and 3D-FEM-simulations. The results were proven by calculated results of equivalent diagram circuits, which have been developed. Regarding the standard crosswise current arrangement all designs show the same results. In a horizontal aux. current arrangement, the effect of the VAG technique is mitigated. Investigations of a vertical aux. current arrangement results in an unsymmetrical magnetization curve.

Introduction

An important focus of every power electronic circuit regarding controllability and loss reduction still lies on semiconductor devices. In consequence of wide band gap technologies, it becomes possible to increase switching frequency by simultaneously decreasing switching and conduction losses. Before using wide band gap semiconductor devices in an application several investigations relating to overload and thermal conditions as well as error management are essential.
In contrast, magnetic devices are characterized by a significant higher robustness and a higher reliability regarding thermal stresses or cosmic radiation. However, in power electronic applications magnetic components typically are the devices which occupy a significant portion of the overall volume and weight. Furthermore, they tend to cause the highest component costs. Especially in mobile applications, controllable magnetics can be used to afford size and cost reduction. This paper is

focused on current controlled magnetic devices. Ferrite materials are used for simulations and verification measurements.

Motivation and application approaches

The motivation of this paper is to get a deeper understanding of the usage of controllable magnetic devices. With their usage, an additional degree of freedom is gained over the course of the application's design process.

Current controlled magnetic devices can be used e.g. for size and cost reduction as it is shown in [1]. Especially in automotive applications, a significant size and weight reduction is essential. Further uses are possible, especially in resonant topologies, to improve the overall efficiency by decreasing the resonant circuit's frequency range and its reactive power losses.

Investigations and discussions of applications for controllable inductors in power electronic converters are presented in [2].

Systematization

A magnetic device gets adjustable by flux interactions of differently aligned magnetic fields. Flux interactions can be achieved by using permanent magnets. Their usage in inductors for power electronic converters is investigated and discussed in [3].

A feasible way to adjust and even control magnetic devices is to use auxiliary windings which are wound on or introduced into the core material. A current, which flows through the auxiliary winding, leads to an additional magnetic flux which saturates parts of the core and effect its overall reluctance. There are three different methods of superimposing magnetic fields. They are shown in Fig. 1. Reference [2] gives an overview of the three different methods of basic control approaches and shows possible applications.

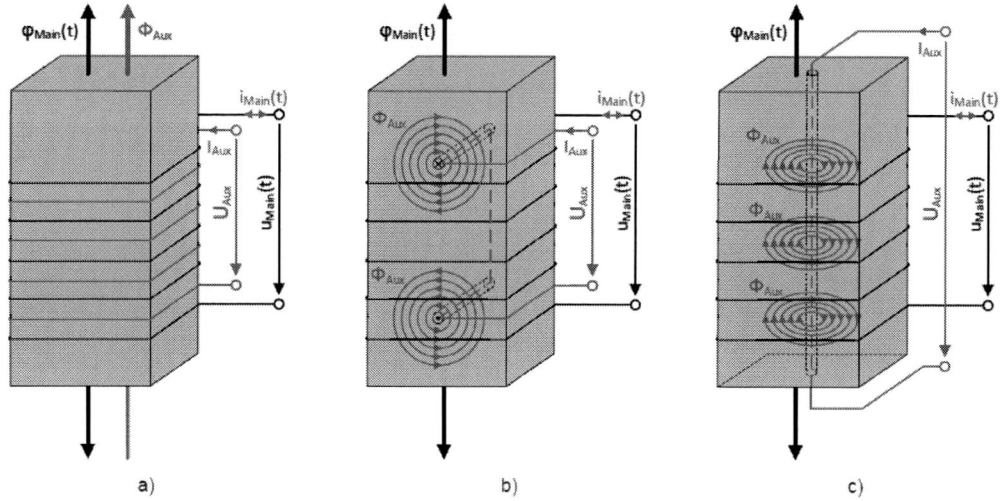

Fig. 1: Three different methods of superimposing magnetic fields by using an additional winding wound on or introduced into the core material. a): Parallel biasing; b): Mixed biasing; c): Orthogonal biasing [2]

For the following investigations, the method of mixed biasing in the form of the virtual air gap (VAG) technique is used. In this technique the auxiliary winding is introduced into the magnetic core through drill holes which are orthogonal aligned to the main flux. The auxiliary bias current occurs an auxiliary magnetic field. This field generates an additional auxiliary magnetic flux which is either parallel or orthogonal to the main flux. A detailed analysis and discussion of a virtual air gap variable reactor (VAG-VR) is presented in [4]. References [5] – [11] give additional information about the fundamental functionality of the VAG technique.

Manufacturing and machining process

Such as ceramics, ferrite is a very fragile and brittle material. If machined wrong, the magnetic core will be damaged or even breaks and becomes useless. A suitable procedure to drill the required holes into a magnetic core is the usage of an Ultrasonic/Sonic Driller/Corer (USDC). To prevent any damage to the drilling tool, it is important to comply with the manufacturing limits and tolerances of the UDSC, regarding its drilling depth in conjunction with the hole diameter.

If high quantities are necessary, e.g. in case of an industrial series production, the required holes can be integrated into the press dies.

Impact of implementation of auxiliary bias windings

Depending on the magnetic device's characteristics, high bias currents are needed in the auxiliary winding to saturate the area around the drill holes. In this situation, a suitable approach is to increase the number of turns of the auxiliary winding to decrease the needed bias current:

$$\Phi = \frac{\mathcal{F}}{R_{m,total}} = \frac{N * I}{R_{m,total}} \tag{1}$$

With:
- Φ: Magnetic flux
- $\mathcal{F}$: Magnetomotive force
- $R_{m,total}$: Total magnetic reluctance
- N: Number of turns
- I: Electric current

According to (1), the bias current can be decreased by the factor the number of turns is increased to generate a similar magnetic flux. A lower auxiliary current positively impacts the efficiency as well as the volume of the auxiliary current-source circuit.

There are various possibilities to implement an auxiliary bias-winding with multiple turns. To examine any impact of the auxiliary winding's arrangement, three different designs shown in Fig. 2 were investigated. The designs vary in the return path of the auxiliary bias-winding.

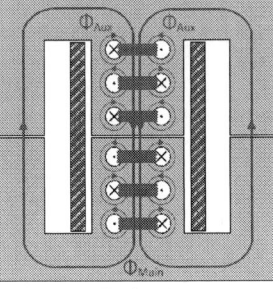 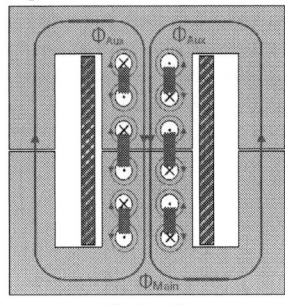

| Design A | Design B | Design C |

Fig. 2: Various designs of implementation of an auxiliary bias-windings. The main winding (red) encloses the auxiliary windings (blue) with multiple turns.

An E55/28/21 core made of "N87 material" (*EPCOS/TDK*) was used to realize the devices. The main winding consists of ten turns of litz-wire. Furthermore, an overall air gap of 0,14 mm was implemented.

For all three designs measurements started with one turn of the auxiliary bias-winding. The number of turns was increased to two turns and again up to four turns. By increasing the number of turns, the auxiliary bias-current was decreased according to (1) to keep the magnetomotive force even, whereby a comparison of the bias points is ensured.

All shown experimental results are small signal measurements by using a *Wayne Kerr Precision Magnetics Analyzer 3260B*. The analyzer measures the magnetic device's differential inductivity by increasing the main current from 1 A to 30 A in steps of 1 A. The measurement results have been numerically integrated to show the magnetic flux linkage as a function of the magnetomotive force of the main winding. Between the measurement sweeps, the auxiliary current was increased gradually from 0 A to 100 A in steps of 10 A.

In a first step, it was investigated if the increasing number of turns of the auxiliary bias-winding within a defined design has any impact on the behavior of the magnetic device.

Figure 3 a) shows the experimental results of Design B for one, two and four turns of the auxiliary bias-winding. The different graphs of the same auxiliary magnetomotive force show a comparable course. The other investigated designs A and C show similar results. Therefore, it must be concluded that, according to (1), the number of turns of the auxiliary bias-winding within a defined design has no impact on the behavior of the magnetic device.

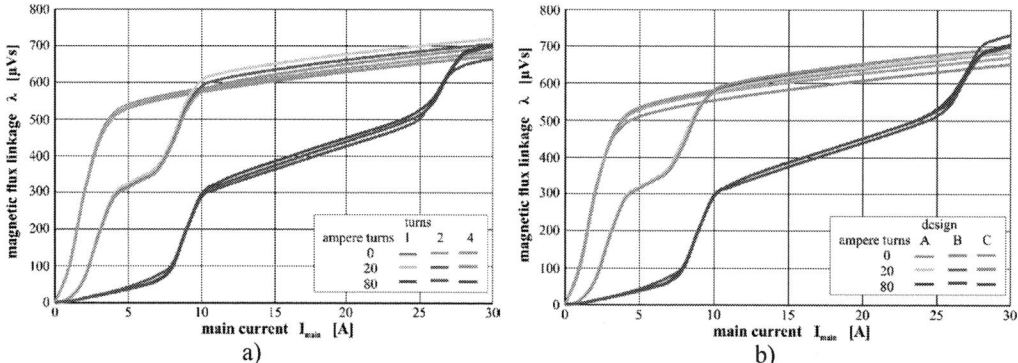

Fig. 3: Experimental results
a): Comparison of Design B with different number of turns of the auxiliary bias-winding;
b): Comparison of Design A, B and C with four turns of the auxiliary bias-winding

In a second step, it was investigated if the implementation of the auxiliary bias-winding has any impact on the behavior of the magnetic device. Therefore, the results of the three investigated designs were compared for one, two and four turns of the auxiliary bias-winding.

The comparison of the three designs for four turns is shown in Fig. 3 b). Again, the courses of the graphs with the same auxiliary magnetomotive force are comparable. The comparison of the results for one and two turns of the auxiliary bias-winding are similar. The experimental results lead to the conclusion that the arrangement of the auxiliary bias-winding has no impact on the magnetic device's behavior.

On closer inspection, multiple risings of the courses in Fig. 3 a) and b) are notable. The reasons for that effect is not completely clear. A possible explanation are flux interactions between the resulting different saturated zones around the drill holes and the main magnetic flux.

For this purpose, the effects of individual nonlinear magnetized sections were investigated using simplified equivalent circuit diagrams. Fig. 4 shows the corresponding procedure.

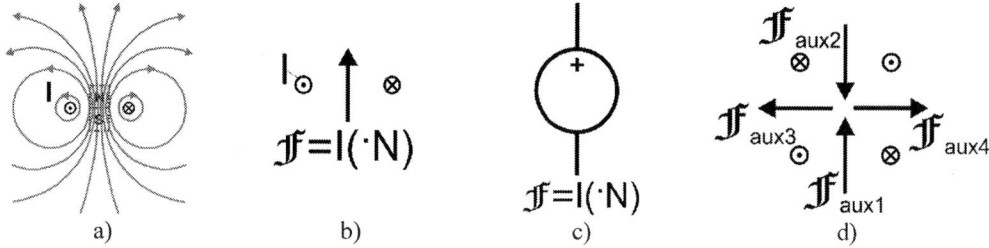

Fig. 4: Development of the equivalent circuit diagram
a): Magnetic field of a conductor pair; b): Its magnetomotive force effect;
c): Equivalent circuit diagram; d): Interaction of several conductor pairs

The magnetic flux is driven by the electromagnetic forces in a network of non-linear magnetic resistances. By superposition with the main magnetic field, the characteristic magnetization curve λ(I_main) is obtained. As a simple application example, Fig. 5 a) shows the structure of a model for design variant D. Fig. 5 b) shows the simulation results obtained, which reflect the behavior of the magnetization characteristic curve qualitatively, but also in part quantitatively.

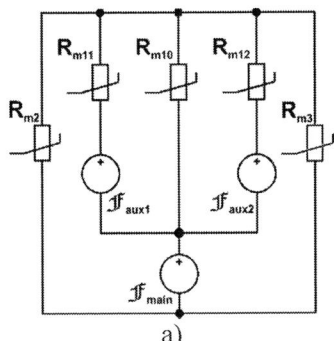
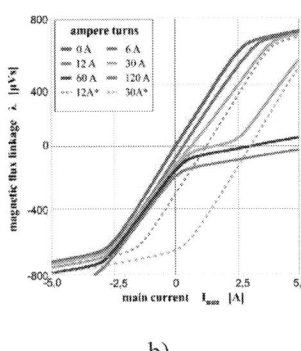

a)   b)

Fig. 5: Rough reluctance model for the winding construction according to design D with vertical auxiliary flux.
a): Construction of the model with similarly directed field of the premagnetization;
b): Characteristic curves determined by simulation based on the reluctance network, (dotted lines: same premagnetization without using the holes)

In Fig. 5 b) you can also see for comparison the effect of a simple winding under the main winding with the same premagnetization as a dotted line. One can clearly see the different effects on the symmetry of the magnetization curve. These could also be proven by measurement (see below).

Impact of arrangement of the auxiliary current

When using mixed biasing, the directions of the auxiliary current in adjacent drill holes are opposite aligned what results in a crosswise arrangement as it is shown in in Fig. 2. To get a deeper understanding of the impact of the auxiliary bias-winding on the magnetic device, the arrangement of the current direction was investigated.

In a first step, a horizontal current arrangement was applied by using a variation of the designs B and C. The directions of the auxiliary current are no longer crosswise arranged. Instead, the auxiliary current of horizontally adjacent drill holes is same aligned. The auxiliary current in vertically adjacent drill holes remains opposite aligned. Figure 6 shows the vertical current arrangement for the new designs D and E.

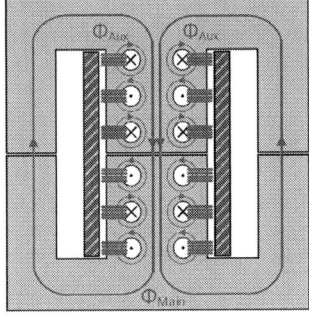
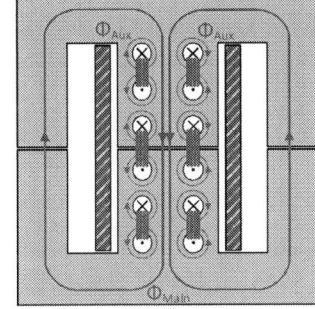

Design D   **Design E**

Fig. 4: Horizontal arrangement of the auxiliary current for two designs

The experimental results, which are shown in Fig. 7, verify the simulation results of the adjusted Design E. As in a standard current arrangement the increase of the number of turns of the auxiliary bias-winding has no impact on the magnetic device's behavior. According to simulation results, the principle effect of mixed biasing still exists in a horizontal current arrangement and is similar to a standard crosswise arrangement.

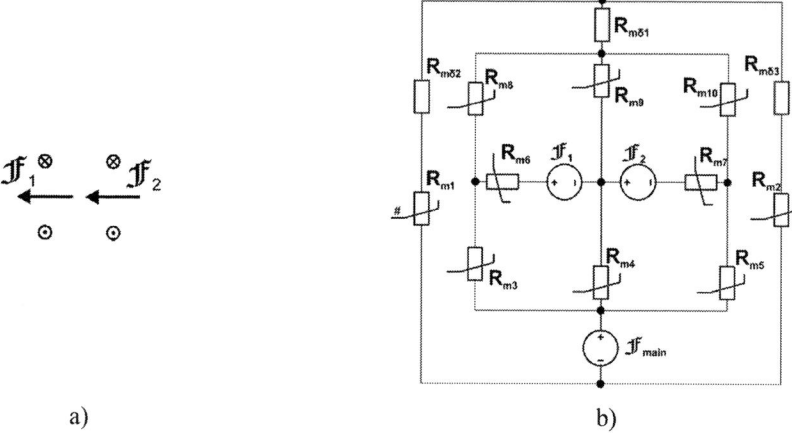

a)   b)

Fig. 7: Simplified model structure for a pre-magnetization field transverse to the main field.
a): Orientation of the MMF, b): Model structure

Figure 8 a) shows the curves of the magnetizing winding measured for design E at different numbers of turns and flooding of the premagnetization. The measurements clearly show that the higher MMF of the premagnetization leads to an increasing pinch-off of the magnetization range. This causes the first threshold in the magnetization curve to drop and the course of the flux to change. These processes cannot be represented by the model in Fig. 7b and require 3D field calculations.

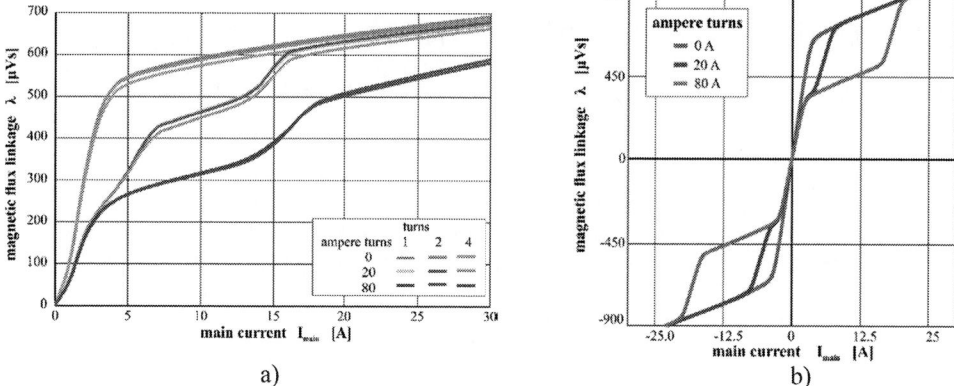

a)   b)

Fig. 5: Experimental results of Design E (a) with different numbers of turns of the auxiliary bias-winding and calculated results with the model of Fig. 7 b)

However, compared to the results of the standard arrangement in Fig. 3, the graphs of the horizontal current arrangement in Fig. 7 show different courses. At an auxiliary magnetomotive force of 20 A, the "plateau" is much more pronounced, what results in a "delayed" saturation at a higher main magnetomotive force. At an auxiliary magnetomotive force of 80 A the graphs rise immediately but their courses are less steep than in a standard arrangement, what likewise "delays" the saturation of the magnetic device.
Because of the flatter courses in Fig. 8 a), it can be said that the effect of mixed biasing gets mitigated in a horizonal current arrangement.

In a second step, a vertical current arrangement was applied by using a variation of the designs A and B. The directions of the auxiliary current of vertically adjacent drill holes is same aligned. The auxiliary current in horizontally adjacent drill holes is opposite aligned. Figure 8 shows the vertical current arrangement for the designs F and G.

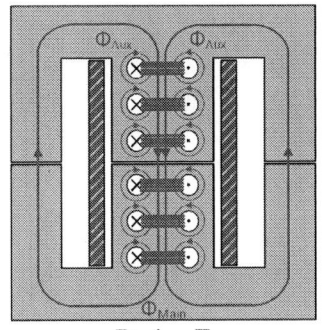

Design F

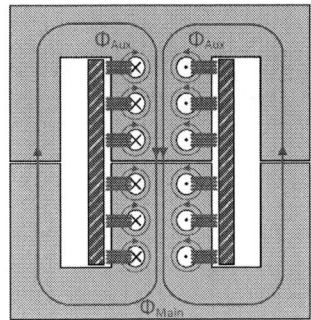

Design G

Fig. 6: Vertical arrangement of the auxiliary current for two designs

Due to the vertical arrangement the auxiliary current generates an auxiliary magnetic flux which effects the entire core geometry. Depending on the arrangement, the auxiliary magnetic flux and the main magnetic flux can be same or opposite aligned. The principle of that arrangement is similar to parallel biasing. Figure 1 a) shows the method with an auxiliary winding wound on the core instead of being introduced through drill holes. The method of parallel biasing was used in transductors (see Ref. [12]) and is still used in bias-inductors (see Ref. [13]).

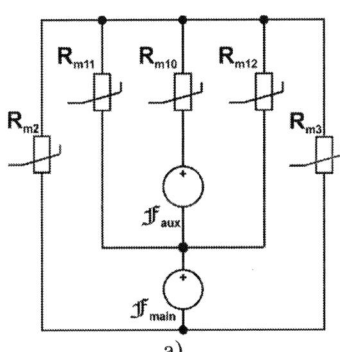

a)

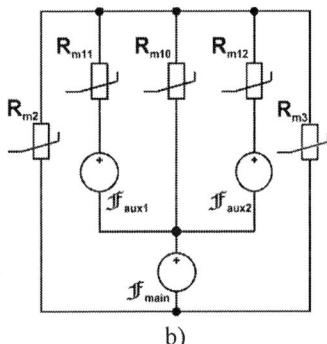

b)

Fig. 10: Simplified reluctance models to simulate the variable characteristics for the two design variants F and G. a): Design F; b): Design G

The results of the simulations with the models in Fig. 10 a) and b) are practically identical. Fig. 11 shows the comparison of practical measurements and model calculations. It should be noted that the magnetic state of space cannot be measured directly with the method used. By integrating the small-signal measurements (see above), it is possible to determine the progressions only down to an unknown constant. Therefore, the flux value for I_main = 50 A for all curves was chosen to be the same as without premagnetization. In the simulation, non-linear reluctance characteristics are calculated in the same way as with resistors. Thus, the result of the superposition of different magnetic fields is also better represented. Due to the core shape and the arrangement of the holes one has an inner and an outer magnetic circuit. At higher MMFs, there is a superposition of the main field and premagnetisation in such a way that the apparent saturation of the material shifts to lower values for I_main < 50 A.

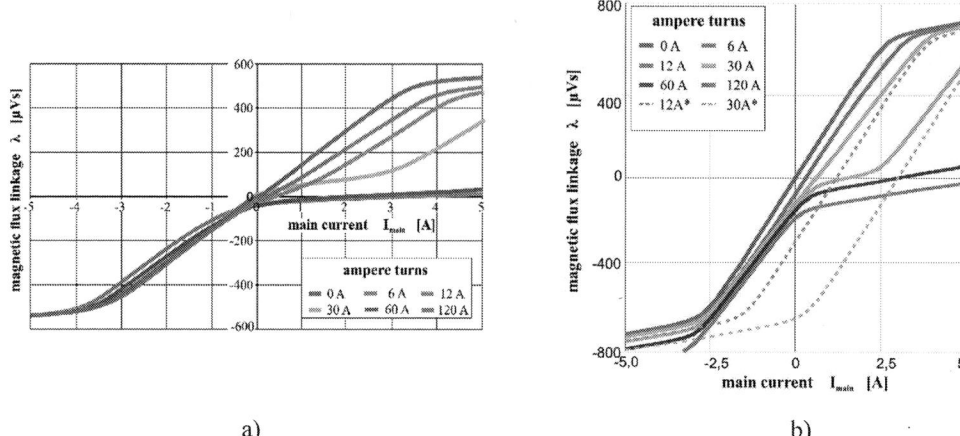

a)   b)

Fig. 11: Results for Design F. a): Experimental results; b): Calculation results

Design F was simulated using *ANSYS Maxwell*. The simulation result for a main current of 5 A and an auxiliary magnetomotive force of 2,5 A is shown in Fig. 12.
Like the standard crosswise arrangement, the center leg of the E-core is saturated while the outer legs are less stressed. However, the flux density in the outer legs is higher in the vertical current arrangement than in the crosswise arrangement.

Regarding Fig 12 a) and b), the different behavior of the ferrite core because of the interaction with the auxiliary magnetic flux can be seen in the simulation results. Especially the outer legs show a higher flux density if both, main and auxiliary magnetic fluxes are same aligned. By increasing the main and the auxiliary current that effect leads to a saturation of the whole magnetic core as it is shown in Fig. 13.

The measurement results for Design G show similar results but twisted. The outer legs will saturate if the auxiliary magnetic flux is opposite aligned. Using a same aligned auxiliary flux, the outer legs are less effected. That also results in an unsymmetrical magnetization curve like it is shown in Fig. 11, which is rotated by 180°.

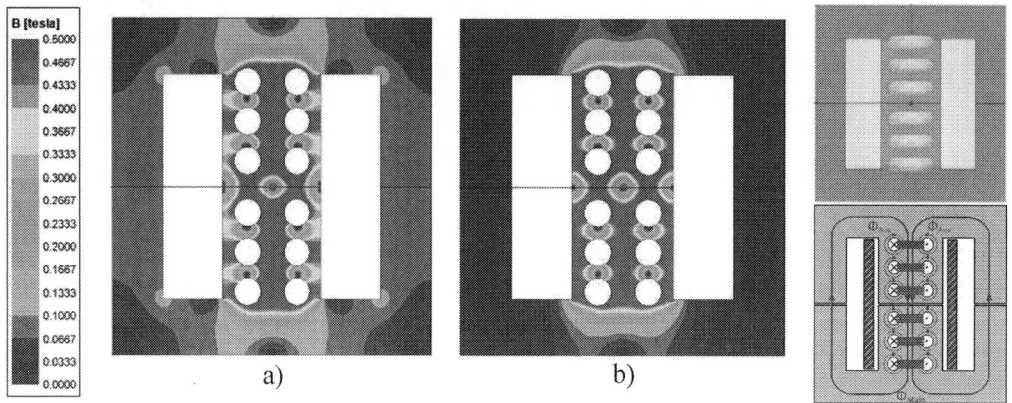

Fig. 7: Simulation results (flux density) of Design F with a main current of 5 A and an auxiliary magnetomotive force of 2,5 A. a): magnetic fluxes same aligned; b) magnetic fluxes opposite aligned

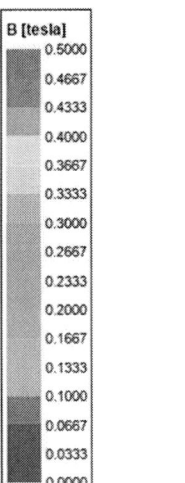 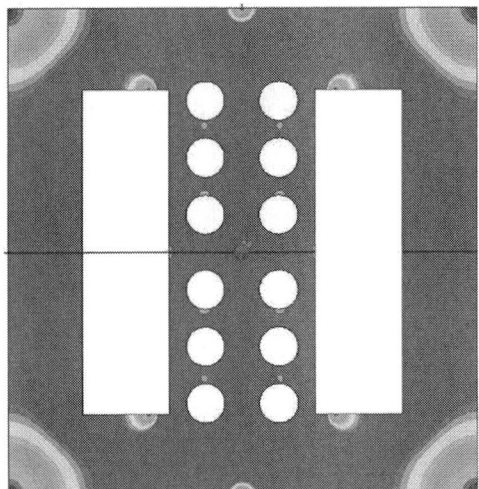 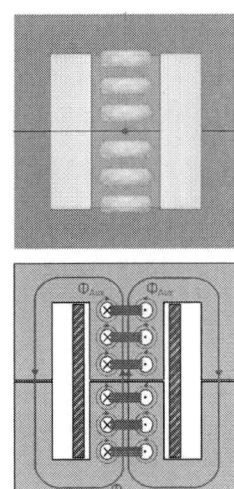

Fig. 13: Simulation results (flux density) of Design F with a main current of 25 A and an auxiliary magnetomotive force of 20 A (same aligned)

Open research questions

Due to the performed investigations there a several open research questions:
As it was shown there are multiple risings within the magnetization curves (e.g. Fig. 3). The exact reason for that courses is not completely clear. Furthermore, further research activities should focus the reduced saturation flux density in ferrite cores with drill holes.
In general, it should be investigated if it is possible to form the magnetization curve by using a special arrangement or diameter of the drill holes.

Beside the mentioned points, there are further open research questions which should be considered:
One example is the impact of the auxiliary bias-winding on the magnetic core losses. Another point is the thermal behavior of a controllable magnetic device by using an auxiliary bias-winding. A possible impact of the core geometry is also not clear. Especially for mixed and orthogonal biasing, further investigations have to be made.

Conclusion

In the presented paper the impact of the implementation of the auxiliary bias-winding was studied for mixed biasing. Seven different implementation designs were investigated in small signal measurements and 3D-FEM-simulations. The results were proven by calculated results of equivalent diagram circuits, which have been developed.

For the standard crosswise current arrangement (Designs A to C) experimental and simulation results are similar. That leads to the conclusion that there is no impact of the implementation of the auxiliary bias-winding on the magnetic device. However, a reduced saturation flux density of the ferrite core with drill holes was noted.

In further investigations the arrangement of the auxiliary current was adapted in a horizontal current arrangement (Designs D and E). According to the crosswise arrangement, no impact of the implementation of the auxiliary bias-winding on the magnetic device could be detected. However, the effect of mixed biasing is mitigated using that arrangement.
The current arrangement was changed to a vertical auxiliary current arrangement (Design F and G). Due to an auxiliary magnetic flux which effects the whole magnetic core, the magnetization curve becomes unsymmetrical, resulting in a behavior like a "magnetic diode". Simulation results for both designs are different in relation to the implementation of the auxiliary bias-winding. The

magnetization curve can be rotated by 180° if Design G is used instead of Design F without changing the directions of main and auxiliary current.

The performed investigations, especially the calculations show that the magnetization curve of a ferrite core can be calculated by integrating the small signal measurement results. The prerequisite is that measurements are performed by using the "parallel equivalent circuit" setting of the measuring instrument.

This paper shows that there new possibilities of controlling magnetic devices by using drill holes within the magnetic core material. However, further investigations have to be made to develop an explanation.

References

[1] Dennis Eichhorst, Jonas Pfeiffer and Peter Zacharias. (2019, 05). Weight reduction of DC/DC converters using controllable inductors. Presented at PCIM Europe.

[2] Peter Zacharias, Thiemo Kleeb, Florian Fenske, Jiajing Wende and Jonas Pfeiffer. (2017, 09). Controlled magnetic devices in power electronic applications. Presented at EPE ECCE Europe.

[3] Jens Friebe, "Permanentmagnetische Vormagnetisierung von Speicherdrosseln in Stromrichtern," Ph.D. dissertation, Dept. of Elect. Eng., University of Kassel, Kassel, Germany, 2014.

[4] Dale S. L. Dolan, "Modelling and performance evaluation of the virtual air gap variable reactor," Ph.D. dissertation, Grad. Dept. of Elect. and Comp. Eng., University of Toronto, Toronto, Canada, 2009.

[5] V. Molcrette, Jean-Luc Kotny, J.-P. Swan and Jean-Francois Brudny. "Reduction of inrush current in single-phase transformer using virtual air gap technique," IEEE Transactions on Magnetics., vol. 34, no. 4, pp. 1192-1194, Jul. 1998.

[6] J. Avila-Montes and Enrique Melgoza. (2012, 09). Scaling the virtual air-gap principle to high voltage large power applications. Presented at ICElMach.

[7] Enrique Melgoza, J. Avila-Montes and Manuel Madrigal. (2013, 11). Analysis of the magnetic characteristics of virtual-gap reactors. Presented at ROPEC.

[8] Adalbert Konrad and Jean-Francois Brudny. "Virtual air gap length computation with the finite-element method," IEEE Transactions on Magnetics., vol. 43, no. 4, pp. 1829-1832, Mar. 2007.

[9] Ali A. Abrishami and Hossein Heydari. (2014, 05). Improved accuracy for finite element modeling in virtual air gap lenght computation. Presented at EEEIC.

[10] Jean-Francois Brudny, Guillaume Parent and Ines Naceur. "Characterization and modeling of a virtual air gap by means of a reluctance network," IEEE Transactions on Magnetics., vol. 53, no. 7, sequence no. 8002007, Jul. 2017.

[11] S. Magdaleno and C. P. Rojas. (2010, 10). Control of the magnetizing characteristics of a toroidal core using virtual gap. Presented at CERMA.

[12] Carrol W. Lufcy, "A survey of magnetic amplifiers," Proceedings of the IRE., vol. 43, no. 4, pp. 404-413, Apr. 1955.

[13] Marina S. Perdigao, Maikel Menke, Alysson R. Seidel, Rafael A. Pinto and Jose M. Alonso. (2014, 10). A review on variable inductors and variable transformers: Applications to lighting drivers. Presented at IEEE Industry Application Society Annual Meeting.

Approximated sliding-mode control of parallel-connected grid inverters

Albrecht Gensior
TECHNISCHE UNIVERSITÄT DRESDEN
Helmholtzstr. 9, 01069
Dresden, Germany
Phone: +49 (0) 351-46334087
Email: albrecht.gensior@tu-dresden.de
URL: https://www.tu-dresden.de

Acknowledgment

This work was supported by ENERCON GmbH, Aurich, Germany. The author thanks the control electronics team of the company, especially Ingo Mackensen and Menko Bakker for their support and Sebastián Rojas and Felix Weiß from TU Dresden for their assistance regarding implementation issues.

Keywords

≪Converter control≫ ≪Sliding-mode control≫ ≪Interleaved converters≫ ≪Multilevel converters≫ ≪Non-linear control≫ ≪Parallel operation≫ ≪Three-phase system≫

Abstract

An approximated sliding-mode controller for an inverter system consisting of parallel connected voltage source converters is presented. The model-based control design considers the whole inverter system and leads to a controller carrying out the control of the grid currents and circulating currents in separate tasks. Due to using hysteresis controllers for both of them, the scheme benefits from advantages of sliding-mode implementations such as fast response to reference changes and good suppression of disturbances. Thus, even sudden changes in the configuration caused by the loss of communication to one of the converters can be mitigated almost perfectly. Furthermore, it features a variable switching frequency avoiding unnecessary switching actions. Since it leads to an interleaved operation of the converters, it offers a lower harmonic content of the grid currents than a benchmark solution with individual hysteresis controllers for each converter with comparable losses. The results are verified experimentally.

Introduction

For wind or photovoltaic applications utilizing two-level voltage-source converters, it is common practice to connect inverters in parallel in order to increase the total power of the inverter system. If they share not only the ac but also the dc side, measures have to be taken in order to control the circulating currents. Current control for these systems can roughly be split in pulse-width modulation (PWM)-controlled systems and systems using switching controllers. In order to obtain a low harmonic content in the grid currents, interleaving techniques can be used for the former as in [1, 2].

Switching controllers can usually be discussed using the sliding-mode theory and are implemented frequently as hysteresis controllers. Lacking a constant switching frequency, interleaving is not that easily possible. Instead, most references deal with single converters only and target the decoupling of the sliding functions in the switches for the purpose of obtaining a constant switching frequency [3–7] or for its prediction [8]. Interleaving techniques with sliding modes have been presented for dc-dc converters [9, 10] and single-phase applications [11] only. Thus, sliding-mode controlled three-phase inverter

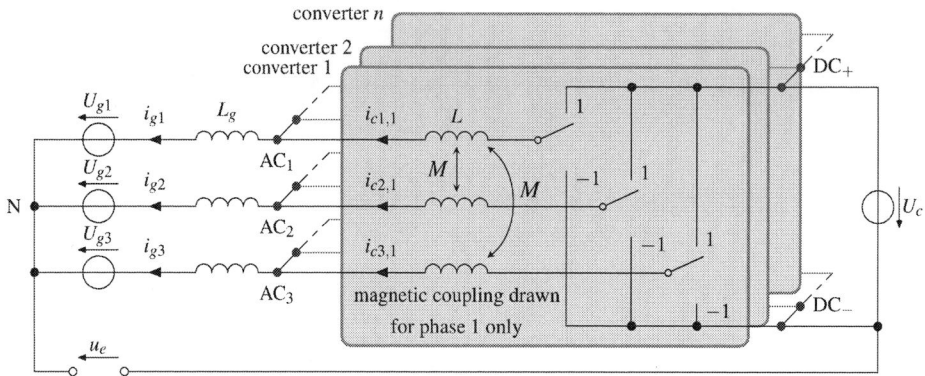

Fig. 1: Configuration of the converter system consisting of n converters. To ease the representation, only the circuit of converter 1 is drawn in detail in the foremost grey rectangle. The other converters sketched by the grey rectangles in the background have the same internal circuit. As indicated by the connections in blue, all converters share the ac connection points AC_p, $p = 1, 2, 3$ and the dc connection points DC_+ und DC_-. There is no connection between DC_- and the neutral of the grid at point N.

systems usually use individual hysteresis controllers for each converter. In this case, the quality of the grid currents depends on the choice of the tolerance bands of the individual converters.

The present paper tries to contribute here by considering the parallel connection of converters as a single converter system. Starting with the modeling of the system, a current controller is designed that allows to specify tolerance bands for the resulting grid currents. Furthermore, the control of the circulating currents and the load-sharing are enabled by hysteresis bands for the individual converters. Experimental results show the value of the results.

Modeling of the system

Consider the configuration depicted in Fig. 1. It consists of n two-level voltage source converters connected in parallel such that they share both, the ac and dc terminals. The dc supply is modeled by a voltage source U_c. On the ac side, the converters are connected to the grid via a transformer. The grid inductance, which is dominated by the transformer, is modeled by the inductance L_g and the grid voltages are denoted by U_{gp}, $p = 1, 2, 3$. Each converter contains a choke, whose inductance matrix can be approximated as

$$L_c = \begin{pmatrix} L & M & M \\ M & L & M \\ M & M & L \end{pmatrix}. \tag{1}$$

The following model equations can be extracted from the circuit diagram:

$$U_{g123} = -L_g \tfrac{d}{dt} i_{g123} - L_c \tfrac{d}{dt} i_{c123,j} + \frac{U_c}{2}(s_{123,j} + \mathbb{1}) + u_e \mathbb{1}, \quad j = 1, 2, \ldots, n \tag{2}$$

$$\sum_{j=1}^{n} i_{cp,j} = i_{gp}, \quad p = 1, 2, 3 \tag{3}$$

$$\sum_{p=1}^{3} i_{gp} = 0 \tag{4}$$

with $U_{g123} = (U_{g1}, U_{g2}, U_{g3})^{\mathrm{T}}$, $i_{g123} = (i_{g1}, i_{g2}, i_{g3})^{\mathrm{T}}$, $i_{c123,j} = (i_{c1,j}, i_{c2,j}, i_{c3,j})^{\mathrm{T}}$, and $s_{123,j} = (s_{1,j}, s_{2,j}, s_{3,j})^{\mathrm{T}}$ where $s_{p,j} \in \{-1, 1\}$, $p = 1, 2, 3$ and $j = 1, 2, \ldots, n$ are the switching positions. For compactness, $(1, 1, 1)^{\mathrm{T}} = \mathbb{1}$.

Adding the n equations (2) and dividing by n one obtains

$$U_{g123} = -\left(L_g I + \frac{L_c}{n}\right)\frac{\mathrm{d}}{\mathrm{d}t}i_{g123} + \frac{U_c}{2n}s_{\Sigma123} + \frac{U_c}{2}\mathbb{1} + u_e\mathbb{1} \tag{5}$$

with

$$\sum_{j=1}^{n}s_{123,j} = s_{\Sigma123} = (s_{\Sigma1}, s_{\Sigma2}, s_{\Sigma3})^{\mathrm{T}} \tag{6}$$

and I being the identity matrix. Transforming the three-phase quantities using

$$T_{\alpha\beta0} = \frac{1}{3}\begin{pmatrix} 2 & -1 & -1 \\ 0 & \sqrt{3} & -\sqrt{3} \\ 1 & 1 & 1 \end{pmatrix} \tag{7}$$

yields

$$U_{g\alpha\beta0} = -L_{\alpha\beta0}\frac{\mathrm{d}}{\mathrm{d}t}i_{g\alpha\beta0} + \frac{U_c}{2n}s_{\Sigma\alpha\beta0} + \left(\frac{U_c}{2} + u_e\right)(0,0,1)^{\mathrm{T}} \tag{8}$$

with $L_{\alpha\beta0} = T_{\alpha\beta0}(L_g I + \frac{L_c}{n})T_{\alpha\beta0}^{-1}$, $U_{g\alpha\beta0} = T_{\alpha\beta0}U_{g123} = (U_{g\alpha}, U_{g\beta}, U_{g0})^{\mathrm{T}}$, $i_{g\alpha\beta0} = T_{\alpha\beta0}i_{g123} = (i_{g\alpha}, i_{g\beta}, i_{g0})^{\mathrm{T}}$, and $s_{\Sigma\alpha\beta0} = T_{\alpha\beta0}s_{\Sigma123} = (s_{\Sigma\alpha}, s_{\Sigma\beta}, s_{\Sigma0})^{\mathrm{T}}$. Matrix $T_{\alpha\beta0}$ is used for the transformation and all three-phase quantities have been collected in vector notation. The vector $s_{\Sigma123}$ contains the sum of the switching positions of all converters. In these coordinates, $i_{g0} = 0$. Thanks to the property that L_c is of the form (1), the matrix $L_{\alpha\beta0} = \mathrm{diag}(L_{\alpha\beta}, L_{\alpha\beta}, L_g + (L + 2M)/n)$ with $L_{\alpha\beta} = L_g + (L - M)/n$.

Control design

Control design for the grid currents

The following control design extends the idea behind the references [3–8] to the situation with multiple converters. Since the sum of the grid currents i_{g0} is constrained to be zero, the third component $s_{\Sigma0}$ of $s_{\Sigma\alpha\beta0}$ does not have an influence on the grid currents as can be seen from (8). In order to give this input signal a meaning in the controller design, the virtual subsystem

$$L_v\frac{\mathrm{d}}{\mathrm{d}t}i_v = \frac{U_c}{2n}s_{\Sigma0} \tag{9}$$

is introduced that will be solved online. The quantity i_v can be interpreted as a virtual current if the parameter L_v has the unit of an inductance.

Dropping the third component in (8) and using (9) instead, the system reads

$$U_{g\alpha\beta v} = -L_{\alpha\beta v}\frac{\mathrm{d}}{\mathrm{d}t}i_{g\alpha\beta v} + \frac{U_c}{2n}s_{\Sigma\alpha\beta0} \tag{10}$$

with $U_{g\alpha\beta v} = (U_{g\alpha}, U_{g\beta}, 0)^{\mathrm{T}}$, $i_{g\alpha\beta v} = (i_{g\alpha}, i_{g\beta}, i_v)^{\mathrm{T}}$, and $L_{\alpha\beta v} = \mathrm{diag}(L_{\alpha\beta}, L_{\alpha\beta}, L_v)$.

Assume there exist reference trajectories for the currents in $i_{g\alpha\beta v}$ which are denoted as $i_{g\alpha\beta v}^*$, one can introduce the control error $e_{\alpha\beta v} = i_{g\alpha\beta v} - i_{g\alpha\beta v}^*$. Using this definition and (10), the error system can be written in natural coordinates as

$$\frac{\mathrm{d}}{\mathrm{d}t}e_{123} = T_{\alpha\beta0}^{-1}L_{\alpha\beta v}^{-1}T_{\alpha\beta0}\left(-(U_{g123} - U_0\mathbb{1}) + \frac{U_c}{2n}s_{\Sigma123}\right) - T_{\alpha\beta0}^{-1}\frac{\mathrm{d}}{\mathrm{d}t}i_{g\alpha\beta v}^* \tag{11}$$

with $e_{123} = T_{\alpha\beta0}^{-1}e_{\alpha\beta v} = (e_1, e_2, e_3)^{\mathrm{T}}$ which can also be written as

$$e_{123} = i_{g123} - i_{g123}^* + \mathbb{1}\left(i_v - i_v^*\right). \tag{12}$$

Since $L_{\alpha\beta\nu} = \mathrm{diag}(L_{\alpha\beta}, L_{\alpha\beta}, L_\nu)$ the choice $L_\nu = L_{\alpha\beta}$ leads to $L_{\alpha\beta\nu} = L_{\alpha\beta}I$. This decouples the equations in a sense that the switching position $s_{\Sigma p}$ only influences the corresponding error e_p, $p = 1, 2, 3$.

Due to this decoupling, the lines of (11) can be considered separately. In order to establish a sliding-mode regime, the controller must ensure $(\frac{\mathrm{d}}{\mathrm{d}t} e_p) e_p < 0$, $p = 1, 2, 3$ which boils down to the requirement

$$e_p \left(-U_{gp} + U_0 + \frac{U_c}{2n} s_{\Sigma p} - L_{\alpha\beta} \left(\tfrac{\mathrm{d}}{\mathrm{d}t} i^*_{gp} + \tfrac{\mathrm{d}}{\mathrm{d}t} i^*_\nu \right) \right) < 0, \quad p = 1, 2, 3. \tag{13}$$

This can be met by ensuring

$$\frac{U_c}{2n} s_{\Sigma p} \begin{cases} < U_{gp} - U_0 + L_{\alpha\beta} \left(\tfrac{\mathrm{d}}{\mathrm{d}t} i^*_{gp} + \tfrac{\mathrm{d}}{\mathrm{d}t} i^*_\nu \right) - U_{res} & \text{for } e_p > 0 \\ > U_{gp} - U_0 + L_{\alpha\beta} \left(\tfrac{\mathrm{d}}{\mathrm{d}t} i^*_{gp} + \tfrac{\mathrm{d}}{\mathrm{d}t} i^*_\nu \right) + U_{res} & \text{for } e_p < 0 \end{cases}, \quad p = 1, 2, 3 \tag{14}$$

where the controller parameter $U_{res} \geq 0$ has been introduced in order to guarantee that condition (14) can be fulfilled also in presence of unmodeled voltage errors. These can be due to voltage drops across the switches or errors in the measurement of the voltages.

A control law can now be derived as follows: According to (6), $s_{\Sigma p}$ may take discrete values only from the set $\{-n, -n+2, \ldots, n-2, n\} =: \mathbb{S}$. Thus, $s_{\Sigma p}$ must be picked from $\mathbb{S}$. However, for a given error e_p, multiple values from the set $\mathbb{S}$ may satisfy (14). Here, for $e_p > 0$, the largest possible value s_p^- fulfilling (14) is chosen while for $e_p < 0$ the smallest one s_p^+ is chosen which is accomplished by

$$s_{\Sigma p}^{\mp} = \pm 2 \left\lfloor \pm \frac{1}{2} \left(\left(U_{gp} - U_0 + L_{\alpha\beta} \left(\tfrac{\mathrm{d}}{\mathrm{d}t} i^*_{gp} + \tfrac{\mathrm{d}}{\mathrm{d}t} i^*_\nu \right) \mp U_{res} \right) \frac{2n}{U_c} + n \bmod 2 \right) \right\rfloor - n \bmod 2, \tag{15}$$

where $\lfloor \cdot \rfloor$ denotes the floor-function. Values exceeding the limits of $\mathbb{S}$ must be limited. In order to avoid too frequent changes of $s_{\Sigma p}$, a hysteresis is implemented such that the same switching position is used unless $|e_p| > h_\Sigma$. A time-discrete implementation is given by

$$s_{\Sigma p}[k] = \begin{cases} s_{\Sigma p}^- & \text{for } e_p[k] > h_\Sigma \\ s_{\Sigma p}^+ & \text{for } e_p[k] < -h_\Sigma \qquad p = 1, 2, 3. \\ s_{\Sigma p}[k-1] & \text{else,} \end{cases} \tag{16}$$

The implementation (15), (16) maximizes the times between switchings and leads to the smallest absolute value of the derivative $\frac{\mathrm{d}}{\mathrm{d}t} e_p$ necessary to maintain a trajectory tracking of the reference currents.

Summing up, the propsed implementation fixes $s_{\Sigma p}$ and ensures that the grid currents follow its references i^*_{gp}, $p = 1, 2, 3$ (fulfilling $\sum_{j=1}^n i^*_{gp} = 0$) and the virtual current follows its reference i^*_ν.

Control design for the internal currents

Since the sum of the switching functions of the individual converters is fixed by the controller described in the previous section, it remains to decide how to distribute this law across the converters and how to limit the internal or circulating currents. Therefore, the following scheme is introduced for each phase separately since it has been shown that they are decoupled: First, references $i^*_{cp,j}$ for the individual converter currents are fixed fulfilling (3), i.e. $\sum_{j=1}^n i^*_{cp,j} = i^*_{gp}$. This allows for a customized loading of the individual converters. Second, the errors

$$e_{p,j} = i_{cp,j} - i^*_{cp,j} \tag{17}$$

are calculated. Third, a list is generated storing the converter indices in the order they are considered to be turned on, i.e. $s_{p,j} = 1$. In other words, this is a one-to-one assignment $I_k = i$ with $k, i \in \{1, 2, \ldots, n\}$ such that for two converters with indices i and j, $I_k = i$ and $I_l = j$ with $k < l$ the following priorities apply (first takes precedence over second and so on):

1. $e_i < -h$ and $e_j > -h$, i.e. a converter is given precedence over another one if its lower tolerance band has been violated and the tolerance band of the other has not been violated or $e_i < h$ and

$e_j > h$, i.e. a converter is given precedence over another one if its upper tolerance band has not been violated and the tolerance band of the other has been violated,

2. $s_{p,i} = 1$ and $s_{p,j} = -1$, i.e. a converter is given precendence over another one if its switch is in position 1 and the switch of the other is in position -1,

3. $e_i > e_j$, i.e. a converter is given precendence over another one if its tracking error is larger than the one of the other converter.

Fourth, the switching positions are assigned as

$$s_{p,I_1} = \ldots = s_{p,I_k} = 1 \quad \text{for } s_{\Sigma p} > -n \tag{18a}$$

$$s_{p,I_{k+1}} = \ldots = s_{p,I_n} = -1 \quad \text{for } s_{\Sigma p} < n, \tag{18b}$$

with $k = \frac{s_{\Sigma p}+n}{2}$.

As a result of this scheme, the command $s_{\Sigma p}$ produced by the controller of the grid currents is always maintained and the internal currents are controlled as well. However, the control of the grid currents savors a higher priority than the control of the circulating currents.

Choice of the reference trajectories

The reference trajectories for the grid currents are calculated as

$$i_{g123}^* = T^{-1}(i_d^* \cos(\varphi), i_q^* \sin(\varphi), 0)^{\mathrm{T}} \tag{19}$$

with φ being the grid angle detected by a phase-locked loop (PLL). The quantities i_d^* and i_q^* denote the reference currents in rotating coordinates.

The reference trajectory for the virtual current i_v^* is calculated by integrating

$$\frac{\mathrm{d}}{\mathrm{d}t}i_v^* = -\frac{U_c}{4nL_v}\left(\min(\{s_{\Sigma p}^*|p = 1,2,3\}) + \max(\{s_{\Sigma p}^*|p = 1,2,3\})\right) \tag{20}$$

with

$$s_{\Sigma p}^* = \frac{2n}{U_c}\left(U_{gp} - U_0 + L_{\alpha\beta}\frac{\mathrm{d}}{\mathrm{d}t}i_{gp}^*\right). \tag{21}$$

As shown in [8], this choice exploits the available dc voltage U_c best.

Implementation

The control systems consists of a central unit and n local units, one for each converter. The former is equipped with a so-called system-on-chip (SoC) which is connected to optical transceivers in order to establish a communication with the local units containing an field programmable gate array (FPGA). The current and voltage measurements are taken locally and the values are sent via a serial protocol to the central unit. The commands for the switches are sent in a similar way to the local units. The datagrams come with a frequency of 640 kHz in both directions.

The implementation of the control law on the central unit has been split such that a part of it is running on the processing system with a rate of 10 kHz and another part is running on the FPGA with an update rate of 640 kHz. The block diagram in Fig. 2 illustrates the concept used for partitioning the algorithm: All equations depending on quantities changing slowly in time, e.g. the dc and grid voltages, are calculated on the processing system while the part of the control law depending on the currents is implemented in the FPGA inside the SoC. The control of the circulating currents resides inside the gray shaded rectangle in order to point out that it can be interpreted a a subsequent control problem: The behavior of everything inside it can be approximated by (10). Consequently, the control of the grid currents is solved by the blocks outside the shaded region.

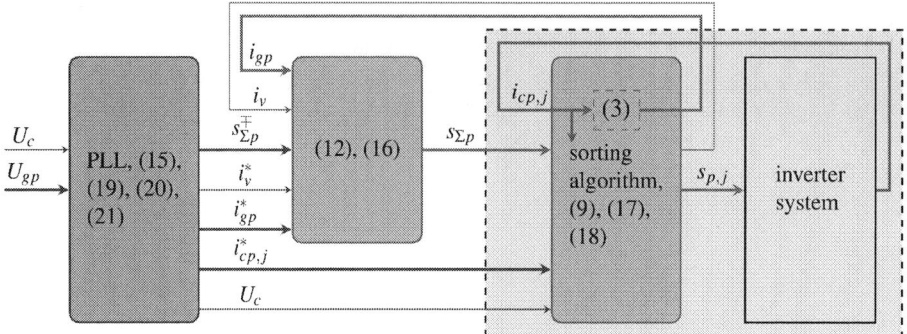

Fig. 2: Block diagram illustrating the control scheme and the update frequencies: The indices p and j are considered to be $p = 1, 2, 3$ and $j = 1, 2, \ldots, n$. The blocks and signals drawn in red are updated with a frequency of 640 kHz and are implemented in the FPGA inside the SoC and the ones drawn in blue are updated with a frequency of 10 kHz and are implemented in the processing system of the SoC. The behavior of the subsystem indicated by the dashed grayed rectangle can be approximated by (10) in natural coordinates.

Experimental results

The proposed control scheme has been tested experimentally in an industrial setup with $n = 4$ converters. Each converter is designed for a rated voltage of 630 V (line-to-line, rms) and a total power of 500 kVA. The converters are connected to a medium-voltage grid via a transformer. The dc voltage is supplied by a diode front end, fed by a variable-ratio transformer with the star-point left unconnected. This setup is parametrized as follows: the ac voltages U_p, $p = 1, 2, 3$ correspond to a three-phase system with a fundamental of 50 Hz and a peak value of $\sqrt{2/3} \, 630$ V provided by a transformer with a rated power of 1 MVA. The inductances are given by $L_g = 75 \, \mu H$, $L = 320 \, \mu H$, and $M = 35 \, \mu H$.

The measurements of the currents were taken with the control system in the following way: Since measurements are available inside the FPGA of the SoC with an update frequency of 640 kHz, average values over a sliding window consisting of 64 samples are calculated for the converter currents $i_{cp,j}$ which are sampled with the update rate of the processing sytem of 10 kHz. For the grid currents i_{gp}, the sliding window consists of 32 samples only but updated with twice this rate, i.e. 20 kHz. Thus, the sample rate for the grid currents is high enough in order to calculate the harmonics up to 9 kHz as demanded by the standard [12] which is applied here for the analysis of the harmonics. Furthermore, the control system collects information about the average switching position and the number of switching events of the semiconductors within consecutive intervals of $100 \, \mu s$ length. Together with a loss model of the semiconductors being available in the post-processing of the data, this allows for the computation of the switching and conduction loss.

A first experiment demonstrates the transient response of the system when one of the converters stops operation, e.g. due to a loss of communication. This is shown in Fig. 3 where the load current is first shared by all four converters. After the event at $t = 0$, the total current to be delivered to the grid is shared between the remaining three converters. As one can see, the grid current remains almost unaffected.

A second experiment has been performed in order to relate the results obtained with the proposed controller to an existing solution which serves as a benchmark. The latter is given by a distributed control scheme, where a controller in each converter regulates the currents of the respective converter only. For the current controllers in the benchmark solution, three independent hysteresis controllers are used. As a base of comparison, the schemes have been parametrized such that they lead to approximately the same losses in the semiconductors. For this experiment, only three converters were operating which is why $i_{cp,4} = 0$, $p = 1, 2, 3$. The dc voltage is 940 V and the setpoint is defined by $i_d^* = 312$ A, $i_q^* = 0$. The controller is parametrized by $U_{res} = 10$ V, $h_\Sigma = 12.5$ A and $h = 45$ A. The waveforms of the currents are shown in Fig. 4. The larger current ripple in the proposed solution corresponds to a larger circulating

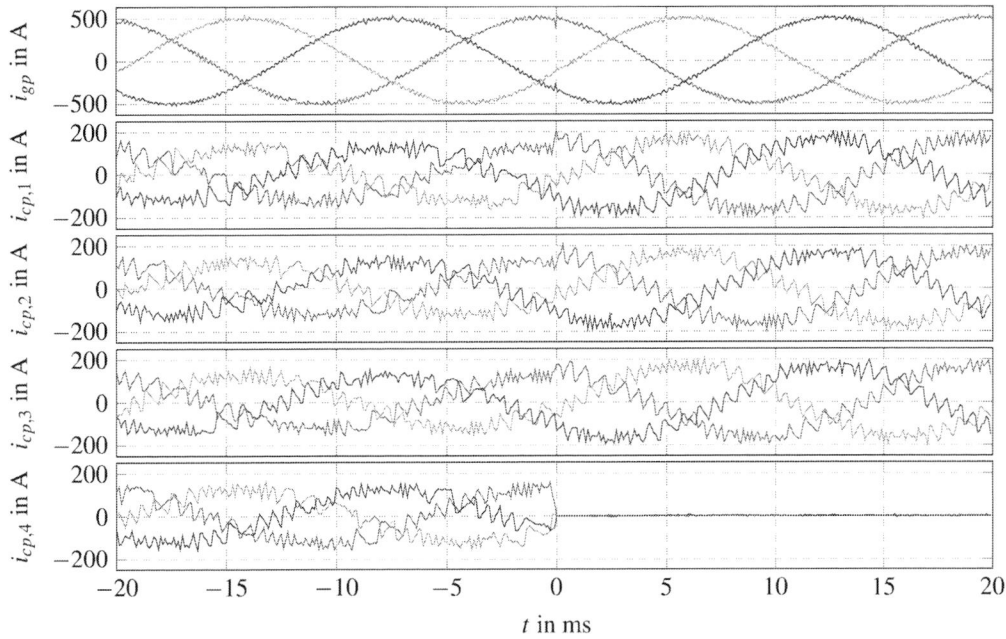

Fig. 3: Current waveforms of phases $p = 1, 2, 3$ after converter 4 suddenly stops its operation. The dc voltage is 950 V and the setpoint is defined by $i_d^* = 50\,\text{A}$ and $i_q^* = 500\,\text{A}$.

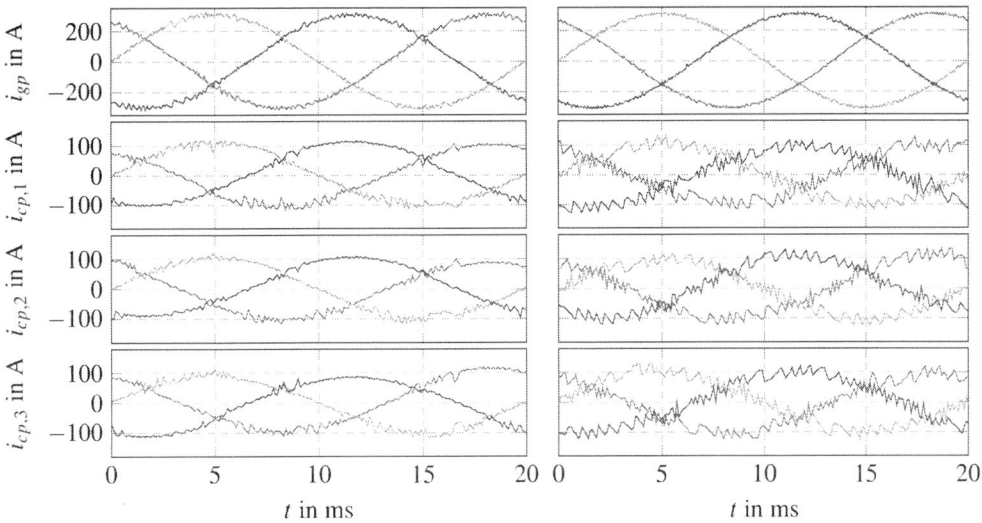

Fig. 4: Comparison of the currents for phases $p = 1, 2, 3$ between the benchmark solution (left column) and the proposed scheme (right column). The fourth converter is not used.

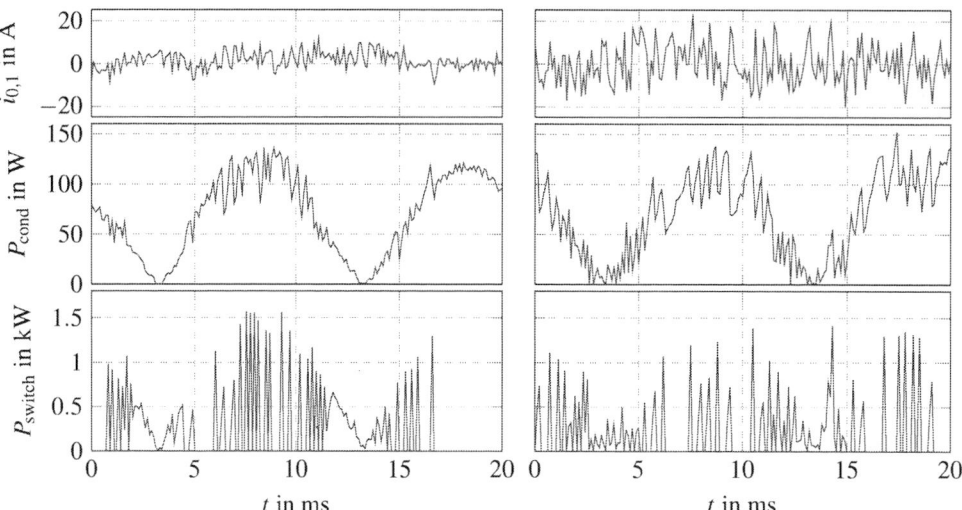

Fig. 5: Comparison of the circulating current of converter 1, i.e. $i_{0,1} = (i_{c1,1} + i_{c2,1} + i_{c3,1})/3$ and the conduction and switching loss P_{cond} and P_{switch} in phase 1 of converter 1 between the benchmark solution (left column) and the proposed scheme (right column). The measurements correspond to those shown in Fig. 4.

current that is shown in Fig. 5 for converter 1. One may assume larger losses in the magnetic components in this case but this issue has not been investigated here. The losses in the semiconductors for phase 1 of converter 1, depicted in the same figure, reveal that the benchmark solutions tends to switchings at high current with relatively short times between them whereas the proposed solution leads to switching actions throughout the whole period with larger times between them. This leads to comparable switching losses in both cases and the conduction losses are also comparable because the fundamental is the same for both cases.

Note that for the proposed solution, the grid currents themself are subject to control and their tolerance bands can be chosen independently from the ones used for controlling the internal currents. This has been made use of by choosing the individual bands wider than in the benchmark solution which can be seen by comparing the last three rows in Fig. 4 from left (benchmark) to right (proposed solution) or even the circulating current in Fig. 5. Furthermore, the dedicated control of the grid currents leads to a lower harmonic content up to 6.4 kHz as one can see in Fig. 6, although the total semiconductor losses are similar in both cases. In particular the fifth and seventh order harmonics can be reduced significantly.

Conclusion

An approximated sliding-mode controller for an inverter system consisting of parallel connected voltage source converters has been presented. Since the width of the tolerance bands can be given separately for the grid currents and for the individual currents, the proposed scheme offers a lower harmonic content then the benchmark solution. Furthermore, it shows a fast response with respect to unexpected events such as the loss of communication between the central unit and one of the locals.

References

[1] B. Cougo, et al. PD modulation scheme for three-phase parallel multilevel inverters. *IEEE Trans. Ind. Electron.*, 59(2):690–700, Feb. 2012.

[2] Ch.-Ch. Hou. A multicarrier PWM for parallel three-phase active front-end converters. *IEEE Trans. Power Electron.*, 28(6):2753–2759, Jun. 2013.

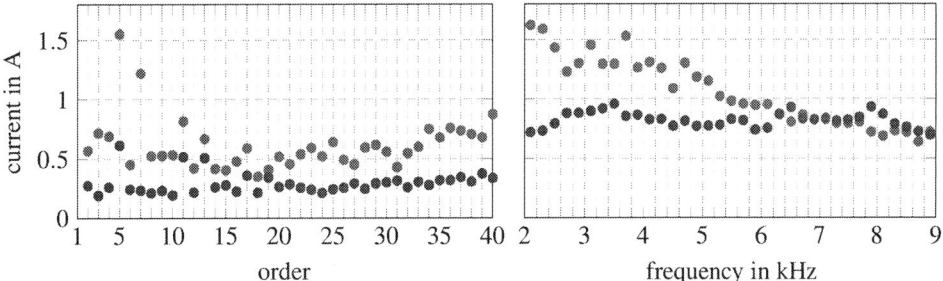

Fig. 6: Comparison of the benchmark (red) and the proposed scheme (blue) when applying the standard [12]: The left diagram shows the order of the harmonics, where the first harmonic is $(312/\sqrt{2})\,\mathrm{A}$ for both cases. The right diagram shows the harmonic content for frequencies in the range $[2,9]\mathrm{kHz}$ with bins of width $0.2\,\mathrm{kHz}$. The inter-harmonics have been omitted for compactness since they show a similar result as the harmonics.

[3] L. Malesani and P. Tenti. A novel hysteresis control method for current-controlled voltage-source pwm inverters with constant modulation frequency. *IEEE Trans. Ind. Appl.*, 26(1):88–92, Jan./Feb. 1990.

[4] L. Malesani, P. Mattavelli, and P. Tomasin. Improved constant-frequency hysteresis current control of VSI inverters with simple feedforward bandwidth prediction. *IEEE Trans. Ind. Appl.*, 33(5):1194–1202, Sep./Oct. 1997.

[5] L. Dalessandro, et al. A novel hysteresis current control for three-phase three-level PWM rectifiers. In *Applied Power Electronics Conference and Exposition, APEC*, volume 1, pages 501–507, Austin, Texas, Mar. 2005.

[6] L. A. Serpa, S. D. Round, and J. W. Kolar. A virtual-flux decoupling hysteresis current controller for mains connected inverter systems. *IEEE Trans. Power Electron.*, 22(5):1766–1777, Sep. 2007.

[7] W. Yan, et al. Sliding mode pulsewidth modulation. *IEEE Trans. Power Electron.*, 23(2):619–626, Mar. 2008.

[8] H. Fehr and A. Gensior. On trajectory planning, backstepping controller design and sliding modes in active front-ends. *IEEE Trans. Power Electron.*, 31(8):6044–6056, Aug. 2016.

[9] J. M. Schellekens, et al. Interleaved switching of parallel zvs hysteresis current controlled inverters. In *The 2010 International Power Electronics Conference - ECCE ASIA*, pages 2822–2829, Jun. 2010.

[10] V. Repecho, et al. Fixed-switching frequency interleaved sliding mode eight-phase synchronous buck converter. *IEEE Trans. Power Electron.*, 33(1):676–688, Jan. 2018.

[11] M. Meyer and A. Sonnenmoser. A hysteresis current control for parallel connected line-side converters of an inverter locomotive. In *The 5th European Conference on Power Electronics and Applications, EPE'93*, pages 102–109, Brighton, UK, Sep. 1993.

[12] Electromagnetic compatibility (EMC) - Part 4-7: Testing and measurements techniques – General guide on harmonics and interharmonics measurements and instrumentation, for power supply systems and equipment connected thereto. Standard, DIN and VDE, Frankfurt Main, Germany, Dec. 2009.

Equivalent Model and Control of a Neutral Point Supply SynRM Drive

Xiaokang Zhang, Jean-Yves Gauthier, Xuefang Lin-Shi
Laboratoire Ampère, CNRS UMR5005, Université de Lyon - INSA de Lyon
25 avenue Jean Capelle - Bât. St. Exupéry
69621 VILLEURBANNE Cedex, France
xiaokang.zhang@insa-lyon.fr, jean-yves.gauthier@insa-lyon.fr, xuefang.shi@insa-lyon.fr

Keywords

≪Converter control≫, ≪modelling≫, ≪synchronous motor≫, ≪variable speed drive≫, ≪reluctance drive≫.

Abstract

Three-phase AC motors fed by the neutral point supply (NPS) topology integrate a DC/DC converter function. In order to clearly explain the mechanism of the integrated DC/DC converter function, the average model and an equivalent circuit of the special topology are developed. Then, a new control strategy is proposed for the topology, in which the motor and the DC-link voltage can be independently controlled. Experiments are implemented on a 1.5 kW synchronous reluctance motor to verify the correctness of these theoretical analyses and the effectiveness of the proposed control strategy.

Introduction

The neutral point supply (NPS) topology is becoming a cost-effective solution for some three-phase AC motor drives with limited DC-source.The NPS integrates a DC/DC converter function by connecting the DC-source to the neutral point of a three-phase AC motor [1] [2]. Earlier applications of the NPS topology were proposed for the electric traction system of hybrid electric vehicles (HEVs), in which the system is called as multi-function converter system (MFCS) [3] [4]. The essence of the NPS topology and the MFCS is to utilize the zero-sequence inductance of AC motors and the drive inverter to indirectly construct an equivalent DC/DC converter function. The largest step-up ration of the equivalent DC/DC converter is 2 [1] [2]. The effectiveness of the equivalent DC/DC converter function is verified on induction motors (IMs) and permanent magnet synchronous motors (PMSMs) [1–6]. However, the uncoupled closed-loop controls respectively for the motor and the DC-link voltage are not achieved. The key point of the NPS topology and the MFCS is to find the independent control degree of freedoms (DOFs) respectively for the motor and the DC-link voltage. The zero-sequence voltage compensation strategy proposed in [3] is derived from a steady-state equation, which is feasible in steady-states but insufficient in transient states such as acceleration deceleration and load change. In the previous literature, only steady-state experiments were presented and the dynamic tests were absent [3] [4]. [1] and [2] proposed a variable DC-link voltage control strategy for the NPS topology by changing the mean duty-cycle of the inverter. But this strategy is still an open-loop control for the DC-link voltage. By adding an assistant chopper to provide a new control DOF for the neutral point current, an improved MFCS was presented in [6]. However, the assistant chopper increases the cost and volume of the whole system. For a three-phase AC motor driven by the NPS topology, the neutral current ripple is directly related to the value of its zero-sequence inductance [3]. Normally, a motor with larger zero-sequence inductance is more suitable for the NPS topology because smaller current ripple results in less iron losses [7].

Recently, synchronous reluctance motors (SynRMs) are becoming a contender to PMSMs and IMs due to the lack of rare-earth materials and the absence of winding losses in their rotors. These advantages make it robust, cheaper and easier to manufacture [8]. In order to provide sufficient torque, the stator inductance of SynRMs is usually large (larger than that of IMs and PMSMs for a given size) [9]. Therefore, SynRMs are more appropriate for the NPS topology.

This paper intends to study a SynRM drive fed by the NPS topology. The paper is organized as follows: Firstly, from the average model point of view, the average model of the drive are built in A-B-C and d-q-0 coordinates respectively. Further, an equivalent circuit is developed for well explaining the nature of the NPS topology. Secondly, the 0-axis duty-cycle is found to be an independent control DOF for the DC-link voltage. Then, a new control strategy is proposed, in which the DC-link voltage can be controlled in a closed-loop way without assistant choppers. Experimental validations are carried on a 1.5 kW SynRM. Conclusions are given in the end.

SynRM drive based on the NPS topology

A conventional three-phase SynRM drive is illustrated in Fig. 1(a). In addition to the SynRM, there are three main parts: a DC-source U_{in}, a DC-link capacitor C and a three-phase two-level voltage source inverter (VSI). A SynRM drive fed by the NPS topology is shown in Fig. 1(b). All the components employed in the NPS topology are same as those in the conventional topology. The only difference between the two topologies is the location of U_{in}. S_A, S_B, S_C, S_A', S_B' and S_C' are the complementary PWM signals.

Average model in A-B-C coordinate

The average modeling method is adopted in this paper and all the following variables are average values in one PWM period (T_s). Defining the inverter's duty-cycles in Fig. 1(b) respectively as α_A, α_B and α_C. Neglecting the voltage drop and dead time, the average model of the NPS-based SynRM drive can be formulated as:

$$\begin{bmatrix} U_{AO} \\ U_{BO} \\ U_{CO} \end{bmatrix} = \begin{bmatrix} \alpha_A \\ \alpha_B \\ \alpha_C \end{bmatrix} U_{DC} = \begin{bmatrix} U_{AN} \\ U_{BN} \\ U_{CN} \end{bmatrix} + \begin{bmatrix} U_{in} \\ U_{in} \\ U_{in} \end{bmatrix} \tag{1}$$

$$i_{DC} = C\frac{dU_{DC}}{dt} = -\begin{bmatrix} \alpha_A & \alpha_B & \alpha_C \end{bmatrix} \begin{bmatrix} i_A \\ i_B \\ i_C \end{bmatrix} = -\alpha_{ABC}^T \mathbf{i}_{ABC} \tag{2}$$

where the U_{AO}, U_{BO} and U_{CO} are the output voltages of the inverter; U_{AN}, U_{BN} and U_{CN} are the phase voltages of the SynRM; i_A, i_B and i_C are the phase currents; i_{in} and i_{DC} are respectively the input current and DC-link current; U_{DC} is the DC-link voltage; α_{ABC} represents the duty-cycle vector, $\mathbf{i}_{ABC}$ represents the current vector.

Moreover, the mathematical model of the SynRM can be expressed by:

$$\begin{bmatrix} U_{AN} \\ U_{BN} \\ U_{CN} \end{bmatrix} = \begin{bmatrix} R & 0 & 0 \\ 0 & R & 0 \\ 0 & 0 & R \end{bmatrix} \begin{bmatrix} i_A \\ i_B \\ i_C \end{bmatrix} + \frac{d}{dt}\left(\begin{bmatrix} L_{AA} & M_{BA} & M_{CA} \\ M_{AB} & L_{BB} & M_{CB} \\ M_{AC} & M_{BC} & L_{CC} \end{bmatrix} \begin{bmatrix} i_A \\ i_B \\ i_C \end{bmatrix} \right) \tag{3}$$

With:

$$L_{AA} = L_\sigma + L_\Sigma + L_\Delta \cos 2\theta_e$$
$$L_{BB} = L_\sigma + L_\Sigma + L_\Delta \cos 2\left(\theta_e - 2\pi/3\right)$$
$$L_{CC} = L_\sigma + L_\Sigma + L_\Delta \cos 2\left(\theta_e + 2\pi/3\right)$$
$$M_{AB} = M_{BA} = -L_\Sigma/2 + L_\Delta \cos 2\left(\theta_e + 2\pi/3\right)$$
$$M_{BC} = M_{CB} = -L_\Sigma/2 + L_\Delta \cos 2\theta_e$$
$$M_{AC} = M_{CA} = -L_\Sigma/2 + L_\Delta \cos 2\left(\theta_e - 2\pi/3\right)$$
$$L_\Sigma = \left(L_d + L_q - 2L_\sigma\right)/3$$
$$L_\Delta = \left(L_d - L_q\right)/3$$
$$L_\sigma = L_0$$

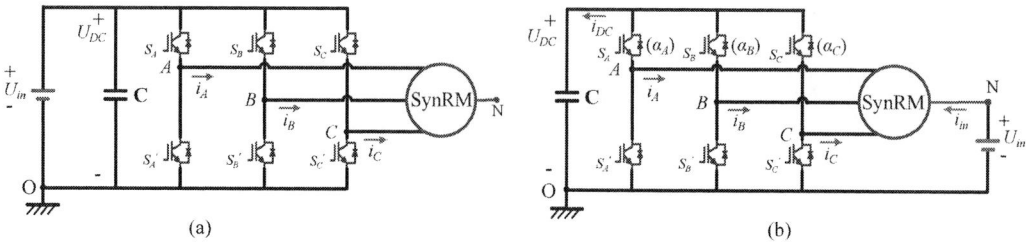

Fig. 1: Topologies of three-phase SynRM drives: (a)The conventional topology; (b)The NPS topology.

R is the stator resistance; L_{AA}, L_{BB} and L_{CC} are the self-inductance; M_{AB}, M_{BA}, M_{AC}, M_{CA}, M_{BC} and M_{CB} are the mutual inductance; L_σ, L_Σ and L_Δ are respectively the leakage inductance, the average inductance, and the inductance with respect to the rotor position. Noting that for a surface PMSM, $L_\Delta = 0$; for an interior PMSM, $L_\Delta < 0$ and for a SynRM, $L_\Delta > 0$.

Average model in d-q-0 coordinate

The Park transformation $\boldsymbol{T}_P$ (power invariant) is shown in (5)

$$
\boldsymbol{T}_p = \sqrt{\frac{2}{3}}
\begin{bmatrix}
\cos\theta_e & \cos\left(\theta_e - \frac{2\pi}{3}\right) & \cos\left(\theta_e + \frac{2\pi}{3}\right) \\
-\sin\theta_e & -\sin\left(\theta_e - \frac{2\pi}{3}\right) & -\sin\left(\theta_e + \frac{2\pi}{3}\right) \\
\sqrt{\frac{1}{2}} & \sqrt{\frac{1}{2}} & \sqrt{\frac{1}{2}}
\end{bmatrix}
\tag{4}
$$

where θ_e is the electrical position.

By using $\boldsymbol{T}_P$, (1) and (3) can be transformed into (5) and (6) respectively, as

$$
\begin{bmatrix} U_{dO} \\ U_{qO} \\ U_{0O} \end{bmatrix} =
\begin{bmatrix} \alpha_d \\ \alpha_q \\ \alpha_0 \end{bmatrix} U_{DC} =
\begin{bmatrix} U_d \\ U_q \\ U_0 \end{bmatrix} +
\begin{bmatrix} 0 \\ 0 \\ \sqrt{3} \end{bmatrix} U_{in}
\tag{5}
$$

$$
\begin{bmatrix} U_d \\ U_q \\ U_0 \end{bmatrix} =
\begin{bmatrix} R & 0 & 0 \\ 0 & R & 0 \\ 0 & 0 & R \end{bmatrix}
\begin{bmatrix} i_d \\ i_q \\ i_0 \end{bmatrix} +
\begin{bmatrix} L_d & 0 & 0 \\ 0 & L_q & 0 \\ 0 & 0 & L_0 \end{bmatrix}
\frac{d}{dt}
\begin{bmatrix} i_d \\ i_q \\ i_0 \end{bmatrix} +
\omega_e
\begin{bmatrix} -L_q i_q \\ L_d i_d \\ 0 \end{bmatrix}
\tag{6}
$$

where U_{dO}, U_{qO}, U_{0O} are the inverter output voltages in d-q-0 coordinate; α_d, α_q and α_0 are the inverter's duty-cycles converted into d-q-0 coordinate; U_d, U_q, U_0 are the motor's phase voltages in d-q-0 coordinate; i_d, i_q and i_0 are the phase currents in d-q-0 coordinate; L_d, L_q and L_0 are respectively the d-axis, q-axis and zero-sequence inductance; ω_e is the electrical speed. In the conventional topology, the 0-axis terms (U_0 and i_0) are usually omitted under the condition $(i_A + i_B + i_C) = 0$. However, they can't be neglected because the connecting of the neutral point results in $(i_A + i_B + i_C) \neq 0$ in the NPS topology.

According to (5), the inverter's output voltages (U_{dO} and U_{qO}) are respectively equal to the motor's voltages (U_d and U_q). However, the 0-axis components (U_{0O} and U_0) are unequal. Their deviation is $\sqrt{3}U_{in}$. Substituting the 0-axis term of (6) into (5), and combining with the current relation $i_0 = (i_A + i_B + i_C)/\sqrt{3}$ (according to the Park transformation), the correlation between U_{in} and U_{DC} is derived by:

$$
U_{in} = \frac{R}{3} i_{in} + \frac{L_0}{3} \frac{di_{in}}{dt} + \frac{\alpha_0}{\sqrt{3}} U_{DC}
\tag{7}
$$

Equation (7) corresponds to the average model of a single phase DC/DC converter, with $L_0/3$ as the filtering reactor, $R/3$ as the resistance and $\alpha_0/\sqrt{3}$ as the duty-cycle of a chopper. By applying the

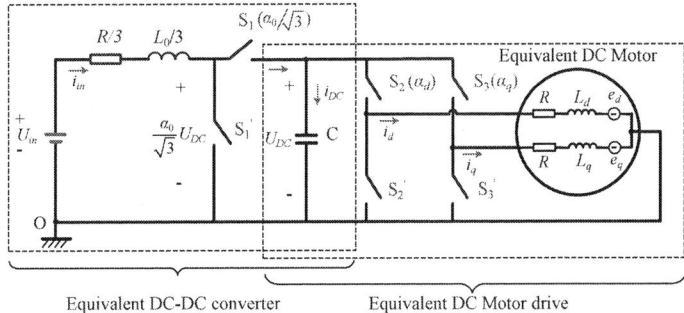

Fig. 2: The equivalent circuit of the NPS-based SynRM drive in d-q-0 coordinate

following transformations to (2), i_{DC} can be expressed by

$$
i_{DC} = C\frac{dU_{DC}}{dt} = -\boldsymbol{\alpha}_{ABC}^T \mathbf{i}_{ABC} = -\boldsymbol{\alpha}_{ABC}^T \mathbf{T}_P^{-1}\mathbf{T}_P\mathbf{i}_{ABC} = -(\mathbf{T}_p\boldsymbol{\alpha}_{ABC})^T \mathbf{i}_{dq0} = -\boldsymbol{\alpha}_{dq0}^T\mathbf{i}_{dq0}
$$
$$
= -\alpha_d i_d - \alpha_q i_q - \alpha_0 i_0 = -\alpha_d i_d - \alpha_q i_q + \frac{\alpha_0}{\sqrt{3}}i_{in}
$$

(8)

where $\boldsymbol{T}_P^{-1}$ is the inverse Park transformation.

Thanks to the technology of vector control, AC motors can be controlled in the way of separately excited DC motors in d-q coordinate. Therefore, the SynRM is equivalent to a DC motor. Combining with the 0-axis components, the NPS-based SynRM drive circuit in stationary coordinate can be converted to d-q-0 coordinate according to (5), (7) and (8). The equivalent circuit is shown in Fig. 2. Six virtual switches (S_1, S_2, S_3, $S_1^{'}$, $S_2^{'}$ and $S_3^{'}$) are assumed in the equivalent circuit. The duty-cycles of S_1, S_2 and S_3 are respectively $\alpha_0/\sqrt{3}$, α_d and α_q. e_d and e_q are respectively equal to $-\omega_e L_q i_q$ and $\omega_e L_d i_d$. Consequently, the NPS-based SynRM drive is equivalent to a DC/DC converter with a DC motor drive as its load. Based on this, the DC-link voltage U_{DC} can be controlled in a closed-loop way by $\alpha_0/\sqrt{3}$.

Control scheme

According to the proposed equivalent circuit, the performance of the SynRM is only related to the d-q components. The performance of the equivalent DC/DC converter is determined by the 0-axis components. Therefore, it is feasible to regulate the DC-link voltage U_{DC} by adjusting α_0 while the speed

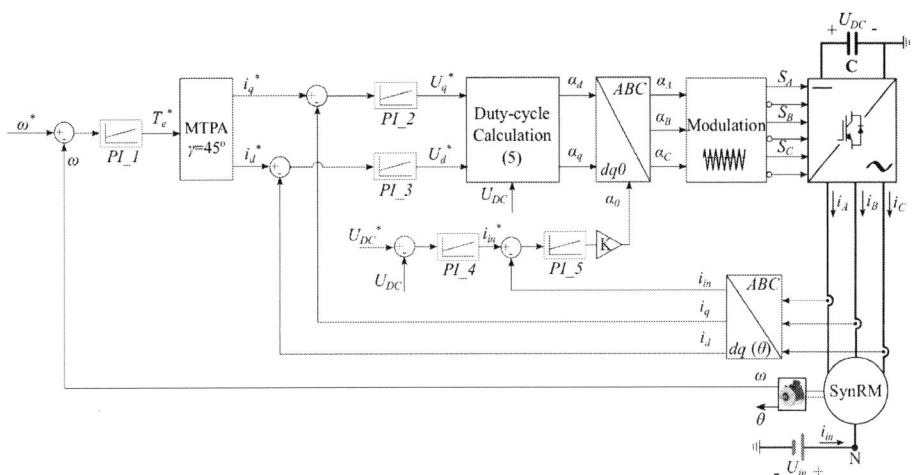

Fig. 3: The control strategy of the NPS topology

control of the SynRM can be achieved by α_d and α_q. A simple control strategy for the NPS topology is designed as shown in Fig. 3. There are five proportional integral (PI) controllers. Firstly, for a given speed reference ω^*, the speed PI regulator outputs the torque reference T_e^*. By using the maximum torque per ampere algorithm in which the current vector angle $\gamma=45°$ without considering the magnetic saturation effects, the d-q axis current references i_d^* and i_q^* are obtained. Secondly, the two current PI regulators output the d-q voltage references U_d^* and U_q^*. Substituting U_d^* and U_q^* into (5), the d-q duty-cycles α_d and α_q can be calculated. Thirdly, a common cascaded PI regulator is employed to control the DC-link voltage U_{DC}, in which the voltage PI outputs the current reference i_{in}^* and the current PI outputs the 0-axis duty-cycle α_0. The coefficient K is $1/\sqrt{3}$. After getting α_d, α_q and α_0, it is easy to calculate α_A, α_B and α_C by the inverse Park transformation. Finally, comparing the α_A, α_B and α_C with a triangular wave, the PWM signals S_A, S_B, S_C, S_A', S_B' and S_C' are generated.

Experimental results

Experiments are carried out on a 1.5 KW SynRM to verify the theoretical analyses and the proposed control strategy. The measurement method of zero-sequence inductance proposed in [10] is adopted. The parameters of the SynRM drive is shown in Table. I. If the conventional topology is adopted, the required DC-source of the drive is 400 V. Thanks to the integrated DC/DC converter function, a 200 V DC-source is sufficient for the drive when the NPS topology is used. The algorithms are implemented on a dSPACE DS1202 MicroLabBox and the test-bench is shown in Fig. 4. The parameters of PI regulators are shown in Table. II.

The experiment setup including speed regulation and sudden loading/deloading is as follows: the speed reference ω^* is 500 rpm before 2 s then it is set to 1000 rpm between 2 and 6 s. After 6 s, it is reset to 500 rpm again. Meanwhile, a 5 N·m sudden load is imposed at 10 s and removed at 15 s.

The experimental results are shown in Fig. 5. Firstly, the DC-link voltage U_{DC} is boosted at 400 V stably

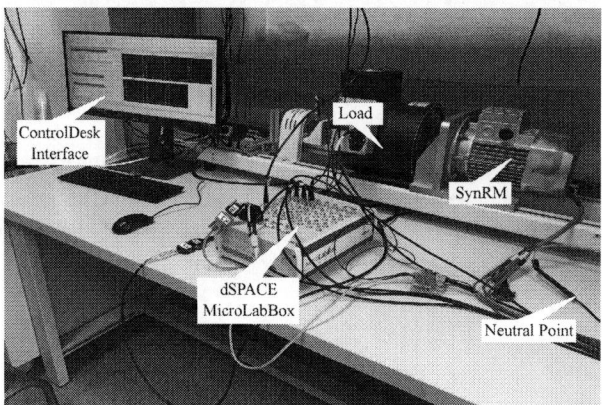

Fig. 4: Test bench

Table I: Parameters of the SynRM Drive

Parameters	Value	Parameter	Value
Rated power P_n	1.5(kW)	DC-link voltage U_{DC}	400(V)
Number of pole pairs p	2	Rated torque	9.5(N·m)
Resistance R	2.6 (Ω)	Rated current	4.1(A)
d-axis inductance L_d	0.33(H)	Rated speed	1500(rpm)
q-axis inductance L_q	0.08(H)	Switching frequency	20(kHz)
0-axis inductance L_0	0.048(H)	DC-bus capacitor C	6.6 mF

Table II: Parameters of PI regulators

Regulator	Kp	Ki	Regulator	Kp	Ki
Speed PI_1	0.30	0.23	q-axis current PI_2	84.70	2598.16
d-axis current PI_3	328.90	2600.00	DC-link voltage PI_4	0.21	1.25
Input current PI_5	0.09	23.98	Input current PI_5	0.09	23.98

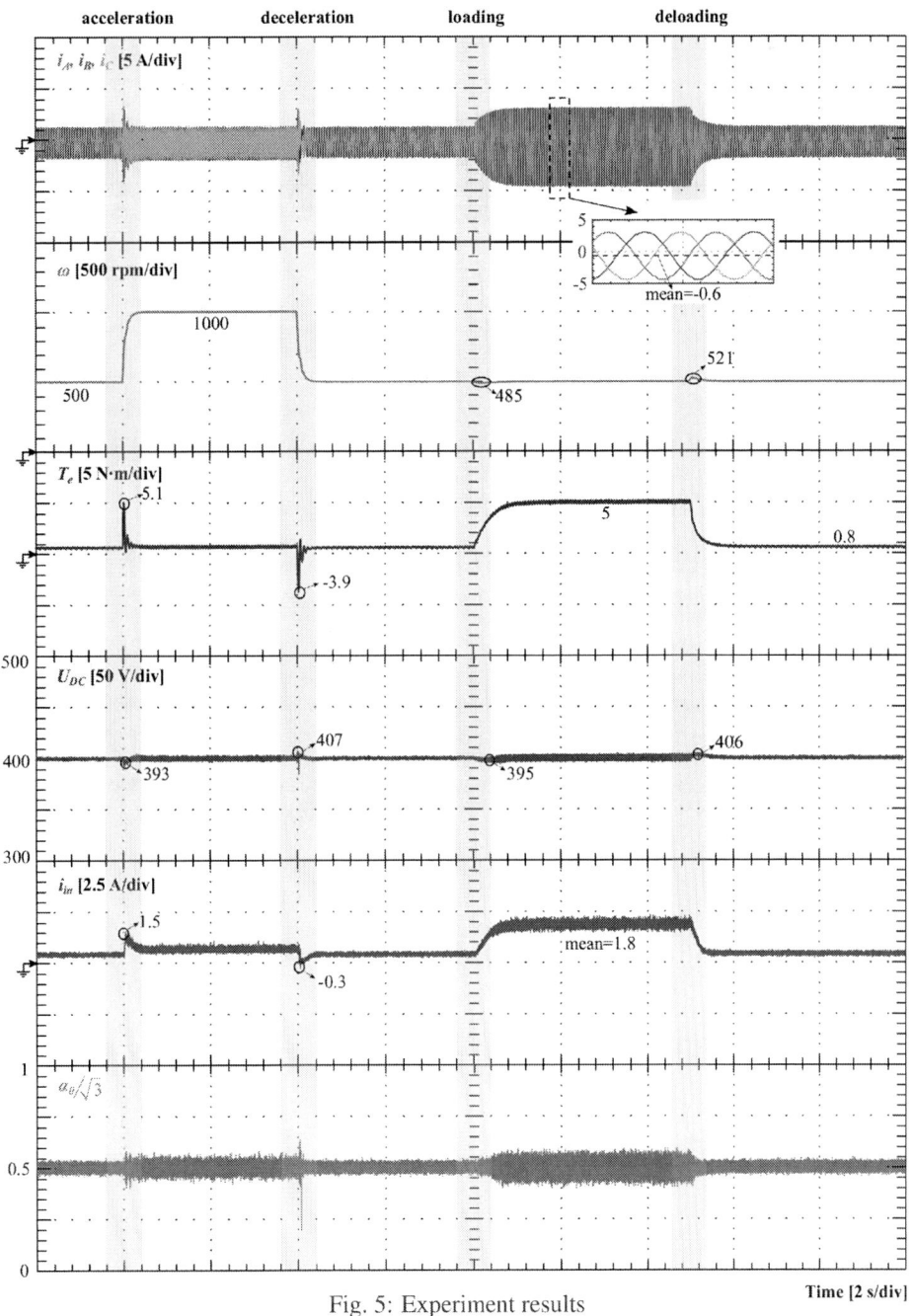

Fig. 5: Experiment results

during the test process while only a 200 V DC-source is used. The value of $\alpha_0/\sqrt{3}$ is around 0.5. This demonstrates the phenomenon of the integrated DC/DC converter function in the NPS-based SynRM drive. It also verifies the correctness of the proposed equivalent circuit for the NPS topology. Secondly, during the acceleration, deceleration, loading and deloading processes, U_{DC} shows slight perturbations with around 6 V (1.5%) but it turns to steady state quickly. Specifically, in order to oppose the voltage dip, the input current i_{in} increases to 1.5 A instantly for charging the DC-bus capacitor when the motor accelerates suddenly. On the contrary, for preventing the voltage rise, i_{in} decreases to -0.3 A instantly for bleeding the energy stored in the DC-link capacitor when the motor decelerates instantly. The good dynamic performance verifies the effectiveness of the closed-loop control strategy for the DC-link voltage by considering the 0-axis components of the motor. Thirdly, the motor speed can track the reference well. The motor shows a fast response when it accelerates and decelerates, respectively with 5.1 N·m and -3.9 N·m torque outputs. When a 5 N·m load is imposed and removed, the speed shows slight dips and rises respectively with 15 rpm (3.0%) and 21 rpm (4.2%) but it recovers to steady state rapidly. Under the condition of 500 rpm and 5 N·m, it can be seen that the phase currents (i_A, i_B and i_C) have an offset. This is caused by the zero-sequence current flowing into the motor. The mean of phase currents is about -0.6 A, which is 1/3 of the mean (1.8 A) of i_{in}. In general, the remarkable steady-state performance and dynamic performance verify the effectiveness of proposed control strategy , in which the motor and the DC-bus voltage can be controlled simultaneously.

Conclusion

This paper investigates a NPS-based SynRM drive. By adopting the average modeling method, the average model of the drive is built. Based on the average model, an equivalent circuit is developed which clearly explains the mechanism of the NPS. In order to control the motor and the DC-link voltage independently, a new control strategy is proposed by considering the 0-axis components of the motor. The remarkable experiment results including steady-state and dynamic operations verify the effectiveness of the proposed control strategy. This work contributes to the practical applications for the NPS topology and reveals that the NPS topology is more suitable for SynRMs.

References

[1] J.-Y. Gauthier and X. Lin-Shi, "Voltage boost by neutral point supply of AC machine," in *Proc. ELECTRIMACS 2019*. Springer, 2020, pp. 243–256.

[2] X. Zhang, J.-Y. Gauthier, and X. Lin-Shi, "Neutral point supply scheme for PMSM drive to boost DC voltage," in *Proc. IECON 2019 - 45th Annual Conf. of the IEEE Industrial Electronics Society*, vol. 1, 2019, pp. 3215–3220.

[3] K. Moriya, H. Nakai, Y. Inaguma, H. Ohtani, and S. Sasaki, "A novel multi-functional converter system equipped with input voltage regulation and current ripple suppression," in *Proc. Fourtieth IAS Annual Meeting. Conf. Record of the 2005 Industry Applications Conf*, vol. 3, Oct. 2005, pp. 1636–1642 Vol. 3.

[4] H. Nakai, K. Moriya, H. Ohtani, H. Fuma, and Y. Inaguma, "Overview of multi-functional converter systems," in *R&D Review of Toyota CRDL*, vol. 39, no. 3, 2004, pp. 27–32.

[5] J. Itoh and D. Ikarashi, "Investigation of a two-stage boost converter using the neutral point of a motor," *IEEE Transactions on Industry Applications*, vol. 49, no. 3, pp. 1392–1399, May 2013.

[6] G. T. Chiang and J. Itoh, "DC/DC boost converter functionality in a three-phase indirect matrix converter," *IEEE Transactions on Power Electronics*, vol. 26, no. 5, pp. 1599–1607, May 2011.

[7] T. Hackner, J. Pforr, H. Polinder, and J. A. Ferreira, "Optimization of the winding arrangement to increase the zero-sequence inductance of a synchronous machine with multifunctional converter drive," *IEEE Transactions on Industry Applications*, vol. 48, no. 6, pp. 2277–2286, Nov. 2012.

[8] F. P. Scalcon, T. S. Gabbi, R. P. Vieira, and H. A. Gründling, "Decoupled vector control based on disturbance observer applied to the synchronous reluctance motor," in *Proc. 21st European Conf. Power Electronics and Applications (EPE '19 ECCE Europe)*, 2019, pp. P.1–P.8.

[9] Z. Mynar, P. Vaclavek, and P. Blaha, "Synchronous reluctance motor parameter and state estimation using extended Kalman filter and current derivative measurement," *IEEE Transactions on Industrial Electronics*, p. 1, 2020.

[10] L. De Sousa and H. Dogan, "Method of evaluating the zero-sequence inductance ratio for electrical machines," in *Proc. 14th European Conf. Power Electronics and Applications*, 2011, pp. 1–10.

Improvements on signal-to-noise ratio in feedback measurement in DC/DC converters

Fernando Davalos Hernandez

Skolkovo Institute of Science and Technology
Bolshoi Boulevard, 30
Moscow, Russia

Facultad de Ingeniería
Universidad Panamericana
Aguascalientes,
México
F.Davalos@skoltech.ru

Federico Ibanez

Center of Energy Science and Technology

Skolkovo Institute of Science and Technology
Bolshoi Boulevard, 30
Moscow, Russia
FM.Ibanez@skoltech.ru

Sebastian Gutierrez

Facultad de Ingeniería

Universidad Panamericana

Aguascalientes,
México
jsgutierrez@up.edu.mx

Wilmar Martinez

KU Leuven
Belgium
wilmar.martinez@kuleuven.bu

Keywords

«DC/DC converter», «isolated», «power electronics», «precision», «voltage measurement»

Abstract

This paper presents different techniques for sensing the feedback signals of isolated DC/DC converters. The paper compares the traditional measuring chain with two other options that improve the signal-to-noise (SNR) significantly. A theoretical analysis and experimental tests are presented. The results indicate that, with this SNR improvement.

I. Introduction

IN DC/DC converters, there are commonly one or more feedback signals that should be well measured in order to correctly control the desired power signals [1], [2]. As in every close-loop system, there is a high impact on the quality of the power output signals due to the noise in the feedback signals (FSs)[3], [4].

Some authors proposed scheduled gain for feedback signals in order to improve the converter's performance [5], [6], others tried to improve the controller using nonlinear techniques [7], [8]. The aim of this paper is to compare different techniques for measuring the FSs. Thus, the designer can be aware of this issue and can select the feedback chain properly.

Usually, the FSs are measuring using a transducer that can be a voltage or current transducer and an acquisition system, which provides a low pass filtering to limit the noise and the proper voltage range to the analog-to-digital converter (ADC). A main diagram for an isolated DC/DC is in Fig.1.a, in which the output voltage is considered as the FS. The diagram includes the power stage, the isolated transducer, which is a hall effect sensor transducer, LEM LV-25, a low pass filter with its amplifying gain, the ADC, digital controller and isolated drivers.

Due to the limited voltage input range of the ADC, as the output voltage increase, the ADC is less sensible to ripple and low perturbations. In designs with high precision, the signal to noise ratio in measurements must be maximized.

In the process of designing any non-isolated DC/DC converter, attention is paid on the output voltage sensing circuitry, even if the controller is digital or analog, the procedure is always the same, one of the common methods is to place a voltage divider on the output of the converter and then adjust the output of the divider to match the reference voltage of the controller [2]. But in the case of isolated DC/DC controllers care must be taken to avoid unnecessary conservative design with an excessive bandwidth reduction in the controller that could make the transient response of the converter slower [9]. Some authors put the controller on the input side of the converter and acquire an output voltage sample and transmit it in an isolated fashion, this can be done through a transformer or an optocoupler [9], [10].

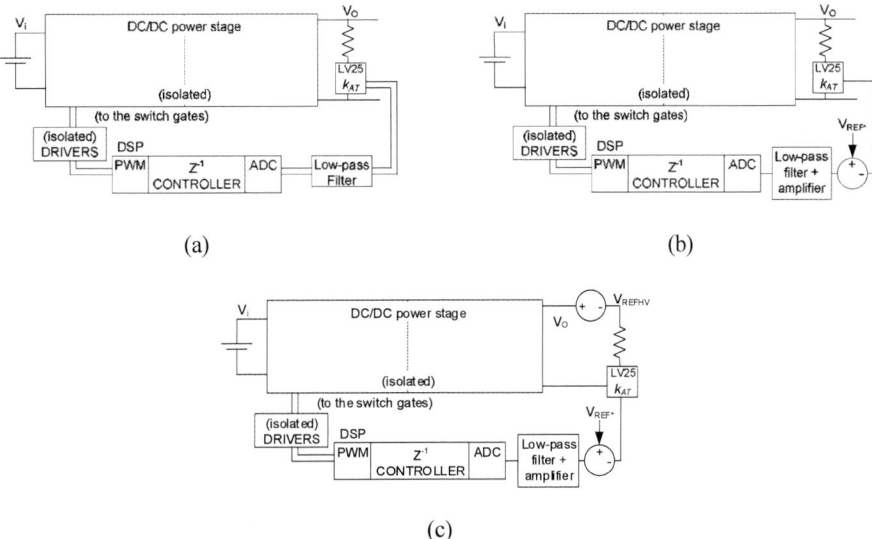

Fig. 1. DC/DC converter diagram: (a) trivial implementation, (b) first improvement and (c) second improvement

Other authors prefer to put their controller in the output side and send the isolated gate signals to the switches in the input side [7], [11]. Other methods include to use isolation amplifiers, for example the AMC1200 from Texas Instruments [12], [13]. But if isolation is desired between the controller and the input and output, then special care must be taken to select a good relationship between accuracy and bandwidth.

The main contribution of this paper is to propose different alternatives to improve the accuracy of the FSs. These proposals are evaluated theoretically and tested experimentally.

The paper is organized as follows. Section II analyses the different stages of the feedback signals for each alternative. Section III analyses the noise of the three presented methods, Section IV presents the experimental results with a comparison among alternatives, including accuracy, cost and complexity, and Section V concludes the paper.

II. Measurements of feedback signals

Three methods for measuring FBs will be detailed. The standard (traditional) method, the limited dynamic voltage range method using low voltage compensation and the limited dynamic voltage range using high voltage techniques.

A. Standard (traditional) method

In case of power isolated DC/DC converters, the standard implementation is to use an isolated sensor, in the present example LEM LV25, which reduces the output-voltage signal from values up to 1000V down to around 3V. Thus, this can be directly attached or filtered to an ADC channel as shown Fig.1a.

This technique is analyzed by considering that the output-voltage is FS and it is required to detect a

ΔV_0 around the mean output voltage value, V_0. The ADC converter should detect $k_{AT}.k_F.\Delta V_0$, which k_{AT} and k_F are the transducer attenuator factor and filter gain. The converter's controller requires several bits to detect this perturbation, so it requires N_{min} counts to satisfy the specifications. The voltage in the ADC input is, according to Fig. 1:

$$v_{ADC} = k_{AT} \cdot k_F \cdot v_0 + n_1 = k_{AT} \cdot k_F \cdot (V_0 + \Delta v_0) + n_1 \tag{1.a}$$

where $n_1 = \sqrt{(k_F \cdot n_T)^2 + n_F{}^2}$,

and the register after the conversion is: \hfill (1.b)

$$N_{(vo)} = \frac{k_{AT} \cdot k_F \cdot (V_0 + \Delta v_0) + n_1}{V_{ADCMAX}} 2^N$$

where $v_0(t)$ is the output converter's voltage, V_0 is the desired voltage and ΔV_0 is the minimum voltage that is needed to distinguish. V_{ADCMAX} is the maximum voltage of the ADC, commonly 3V, N is the number of bits for conversion (10, 12 or 16 bits) and n_T, n_F, n_l are the RMS noise of the transducer, filter, ADC input.

Two types of added noise were considered, the noise added due to the transducer and the noise due to the filter which is based in an operational amplifier.

The minimum voltage must be captured with a certain number of bits:

$$\frac{k_{AT} \cdot k_F \cdot \Delta v_0}{V_{ADCMAX}} 2^N > N_{min}. \tag{2}$$

N_{min} depends on the noise floor which is quite high in power electronics due to the switching process [14]. Thus, the importance to select the proper number of bits that are needed and to have some alternatives to the basic diagram of Fig. 1.a.

In order to calculate attenuation factor and filter gain, the maximum and minimum acquired voltage must be obtained:

$$V_{ADCMAX} = k_{AT} \cdot k_F \cdot V_{0MAX} = k_{AT} \cdot V_0(1 + g), \tag{3}$$

and $V_{ADCMIN}=0$ at $v_0(t)=0$. So, the k_{AT}. k_F is:

$$k_{AT} \cdot k_F = \frac{V_{ADCMAX}}{V_0(1 + g)}, \tag{4}$$

the g factor means that the maximum value that the ADC can convert is $V_0(1+g)$, thus the converter can regulate the output voltage. This factor can be modified, for example if g=0.15, it means that the dynamic range is from zero to 115% of the desired value. Beyond that, the sensor saturates its output. If this factor is reduced further the converter cannot detect a possible overvoltage, the dynamic range becomes too narrow.

Based in (1.a) and (4), and by setting signal-to-noise ratio to certain limit, (S/N), the minimum value of the measurement output voltage is:

$$\frac{\Delta v_0}{V_0} > \left(\frac{S}{N}\right) \frac{(1 + g) \cdot n_1}{V_{ADCMAX}} \quad (case\ A) \tag{5}$$

If $(1+g)$. n_l is too large for the desired voltage measurement, other methods must be used. Details of added noise are presented in Section III.

B. Limited dynamic voltage range method using low voltage compensation

It can be noticed that the controller does not require the entire FS range, only the variation of this signal around the desired value. Therefore, an improvement can be done if the signal is compared to a reference before the ADC input as is shown in Fig. 1.b. In this case, the input ADC voltage is:

$$v_{ADC} = k_{AT} \cdot k_{F2} \cdot (V_0 + \Delta v_0) - V_{REF} + n_2 \tag{6}$$

where $n_2 = \sqrt{(k_{F2} \cdot n_T)^2 + n_{F2}{}^2}$

where k_{F2} is the gain of the amplifier in this case. Thus, the ADC voltage range can be for example +/- 15% (g=0.15) of the desired output value. This value can be selected by adjusting the amplifier gain of the low-pass filter in Fig. 1.b. In this case, following the same procedure in Section II.A about the maximum and minimum voltages, k_{AT}. k_{F2} and V_{REF} are obtained:

$$k_{AT} \cdot k_{F2} = \frac{V_{ADCMAX}}{V_0(2g)}, \tag{7}$$

and

$$V_{REF} = k_{AT} \cdot k_{F2} \cdot V_0(1-g). \tag{8}$$

Notice that as the voltage range in the transducer does not change, the attenuator factor is the same as in case A, however, the filter gain is larger. By comparing (4) and (8), the filter gain is obtained:

$$k_{F2} = \frac{(1+g)}{2g} k_F. \tag{9}$$

Then, applying the same procedure, the voltage sensitivity in this case is:

$$\frac{\Delta v_0}{V_0} > \left(\frac{S}{N}\right) \frac{(2g) \cdot n_2}{V_{ADCMAX}} \ (case\ 2). \tag{10}$$

As 2g < (1+g), the ADC will be able to detect lower voltages if n_1 is equal to n_2.

C. Optimization of the dynamic range using high voltage compensation

Particularly, in step-up converters, an even better improvement is to subtract the DC value at the high output voltage by using a high voltage reference. Thus, the voltage sensor with an accuracy of 1% is not measuring the total voltage anymore but it is measuring only the dynamic range around the desired voltage. The measured voltage is lower, so the absolute error is lower and the accuracy of the measurement increases. This can be achieved by designing a high voltage reference source at a voltage level close to the output value. The voltage sensor has a lower attenuator factor (k_{T2}) and so the higher S/R is expected in comparison to case A and B.

Unfortunately, using this high voltage reference will increase the cost of the converter. So, this method is only used if it is really needed. In addition, this high voltage reference can also introduce a new source of noise.

In this case, the ADC input is:

$$v_{ADC} = k_{AT3} \cdot k_{F3} \cdot (V_0 + \Delta v_0) - k_{AT3} \cdot k_{F3} \cdot V_{REFHV} + n_3. \tag{11}$$

where

$$n_3 = \sqrt{(k_{AT3} \cdot k_{F3} \cdot n_{REF})^2 + (k_{F3} \cdot n_T)^2 + n_F{}^2},$$

where k_{T3} and k_{F3} are the attenuator and filter gain in this case. One more time, using the same procedure of case A, k_{AT}. k_{F2} and V_{REF} are obtained:

$$k_{AT3} \cdot k_{F3} = \frac{V_{ADCMAX}}{V_0(2g)}, \tag{12}$$

$$V_{REFHV} = V_0(1 - g).$$

in this case, the attenuation factor is reduced and the k_{F3} can be the same as in case A: $k_{F3} = k_{F1}$ so,

$$k_{AT3} = \frac{(1 - g)}{(2g)} k_{AT} \ and \ k_{F3} = k_F \tag{13}$$

The voltage sensitivity in this case is:

$$\frac{\Delta v_0}{V_0} > \left(\frac{S}{N}\right) \frac{(2g) \cdot n_3}{V_{ADCMAX}} \ (case\ 3). \tag{14}$$

In this case the sensitivity does not increase very much in comparison with the case B, however, to understand the difference the noise n_1, n_2 and n_3 must be analyzed.

III. Noise analysis

The added noise in the three methods is different. In case A, the noise is the sum of the transducer noise and the filter stage noise. The bandwidth of the system is limited to BW=50kHz, using a first order filter, which is equivalent to a noise bandwidth NBW= $(\pi/2)$ BW=78kHz [15]. The transducer datasheet does not provide any noise measurement, so a measurement was tested at 0V and the noise was 280uVRMS, in the defined bandwidth.

The noise provided by the filter stage is very low, details will be given in the experimental section, its value is lower than 90uVRMS. Therefore, $n_T^2 >> n_F^2$, so the dominant noise is the transducer, in addition, for case A,

$$n_A \approx k_F n_T. \tag{15}$$

In case B, the k_{F2} is bigger so, V_{REF} is considered noiseless:

$$n_B \approx k_{F2} \cdot n_T = \frac{1 + g}{2g} \cdot k_F n_T. \tag{16}$$

So, as the noise is amplified the same amount that the signal, no improvement is noticed. The only benefit is that the number of bits is bigger so, after a digital filter a noise can be reduced.

In case 3, the noise is:

$$n_C = \sqrt{(k_{AT3} \cdot k_{F3} \cdot n_{REF})^2 + (k_{F3} \cdot n_T)^2 + n_F^2}, \tag{17}$$

Again, the noise of the amplifier is neglected in comparison with the noise of the transducer. Thus, it is important to have low noise in the high voltage reference in comparison with the noise in the transducer. If so,

$$n_C \approx n_T \tag{18}$$

Therefore, by comparing (5), (10) and (14), the ratio between (5) and (10) is the same. Thus, the noise ratio is not improved but only the counts per volt, so using case B and a proper digital filtering (or error integrator) it is possible to distinguish a smaller voltage error value than in A.

In case C is different, the noise is reduced, in comparison with the signal. Thus, a better resolution is obtained. Experimental results will validate these assumptions.

IV. Experimental results

In order to validate the proposed model, the three methods were developed and tested. Fig. 2 shows the test bench, it allows changing k_F and adding or not the reference at low voltage or at high voltage. It consists of a DSP that acquires the analog signal, the programmable (by changing resistors) voltage sensor, the high voltage reference, a switching power supply and some loads.

In the experimental case, the voltage range of V_O is from 250-330V, the power is around 1kW, and the three measurement methods were tested.

For the standard approach, a LEM LV-25 voltage sensor was connected to the HV power source using eight resistors of 4.7kOhm (R_T). The output resistor (R_0) was fixed to 330 Ohm and then a filter stage amplifier adapts the signals to 3V to the ADC input of the DSP. The circuit can be understood from Fig. 2.a with R_T=8 x 4.7kOhm, V_{REF}=0 V and the gain according to (5). In practice the gain in the filter is not one ($k_F \neq 1$) because, for having the best accuracy, it is better to have a transducer at the maximum voltage, that is why R_0 was fixed to 330Ohm and the input current of the transducer fixed at 25mA, which is the maximum allowable current.

In the optimized dynamic range approach, the same number of resistors were used to reach the maximum voltage range. However, the range was limited using a differential amplifier for keeping the dynamic range from 250 to 330 V. For that reason, a voltage reference was added in the measurement stage and the gain of the filter stage (k_F) was increased. The circuit is the one in Fig. 2.a with V_{REF}, R_T=8 x 4.7kOhm.

In the third approach, a voltage reference at high voltage was added to reduce the measurement range in the LEM-LV-25. The voltage reference is based in a linear regulator that keeps a voltage drop constant at 240V. So, the LEM-LV-25 must measure voltages from 10V to 90V. In this case, the circuit fits Fig. 2.b.

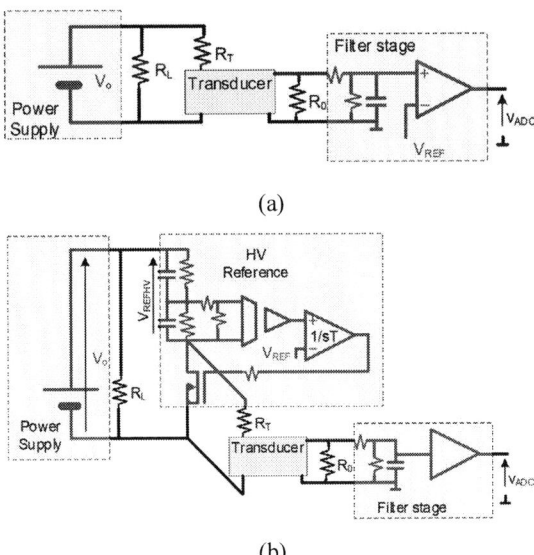

(a)

(b)

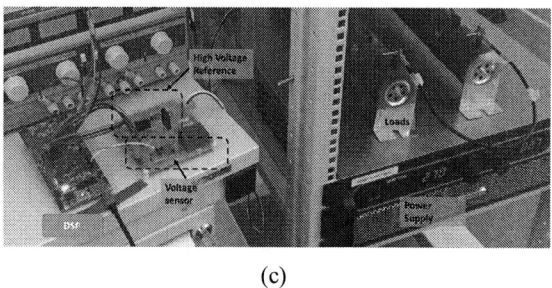

(c)

Fig. 2: Test bench: (a) basic diagram for case A and B, (b) case C and (c) prototype

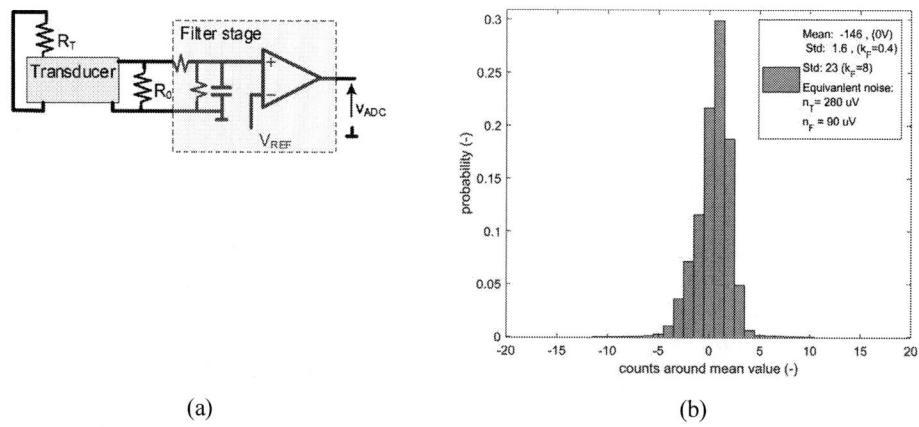

(a)   (b)

Fig. 3: Noise of the transducer at $V_0 = 0$.

The first measurement was performed to estimate the noise introduced with the transducer. For that, the LEM-LV-25 was short circuited in the input and the filter stage increase the gain up to 20 times, and the noise was measured using an oscilloscope with a filter at 100kHz. The noise was 280uVRMS. The test diagram and measurement values are presented in Fig. 3.a and 3.b.

A second measurement was to evaluate the impact of the high voltage reference in the noise. For that reason, the same power supply was used but with a lower voltage at $V_0 = 300V - V_{REFHV}$. The noise was not reduced, a standard deviation of 2.2 counts. A 16-bits ADC is used, however, only 15-bits are used due to the sign of the ADC.

The third measurements were the ones that test the different cases. A power supply which emulated the DC/DC converter at a nominal voltage of 300V was used. The results are shown in Fig. 4. Fig. 4.a shows the standard approach, Fig. 4.b shows the optimized range in the low voltage side, and Fig.4.c shows the optimized range in the high voltage side. The diagrams of each approach are in Fig.1. The measurements were plotted as histograms; thus, the standard deviation can be appreciated. In digital counts, case A and C have the lower standard deviation than case B, however, the scale is different so a better accuracy can be obtained from case C. In case A, the measured noise in the ADC is 2.2 counts which is equivalent to 21 mV, in case C the measured noise is also 5.5 counts but, as $k_{T3} > k_T$ is bigger, the noise is equivalent to 14 mV. In case B, the measured noise is 8.5 counts, almost twice the noise but as the range is smaller, it is equivalent to 23 mV. Therefore, case B does not increase the accuracy notably.

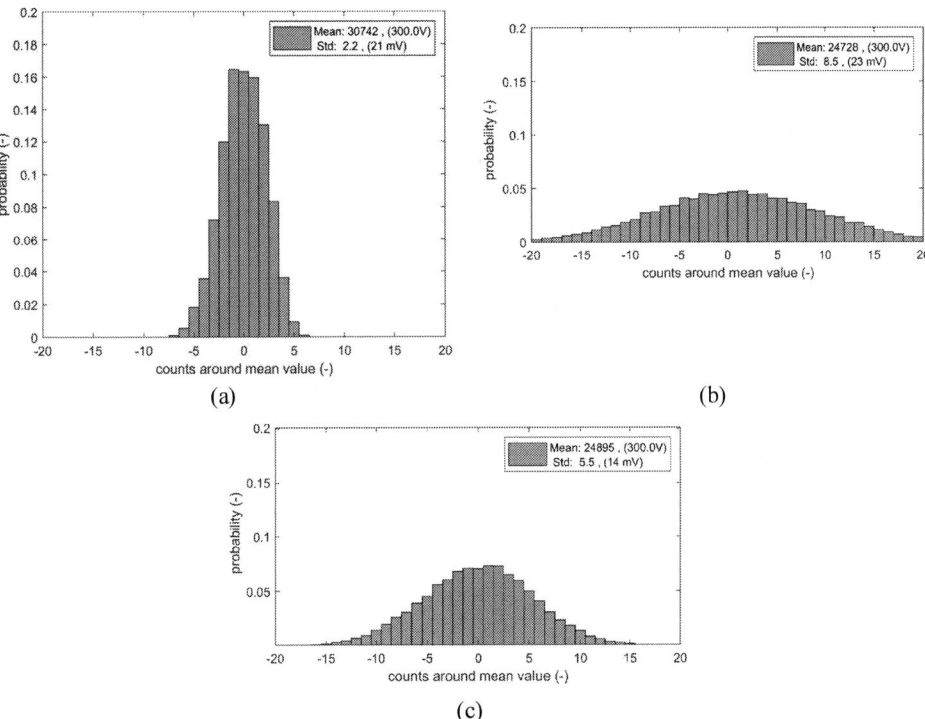

Fig. 4: Noise measurements (equivalent voltage) after the ADC acquisition.

Finally, the fourth test consisted of the exact same cases as in the third test, but a switched load was introduced at the same sampling frequency of the ADC (F_{ADC}) to simulate the PWM of a switched DC-DC converter. Fig. 5 shows the histograms for the 3 cases tests.

In comparison to the third test measurements, all the cases presented an increase in the standard deviation, this is mainly because of the noise attributed to the load switching, but the case C maintains a better accuracy in comparison to case A and B.

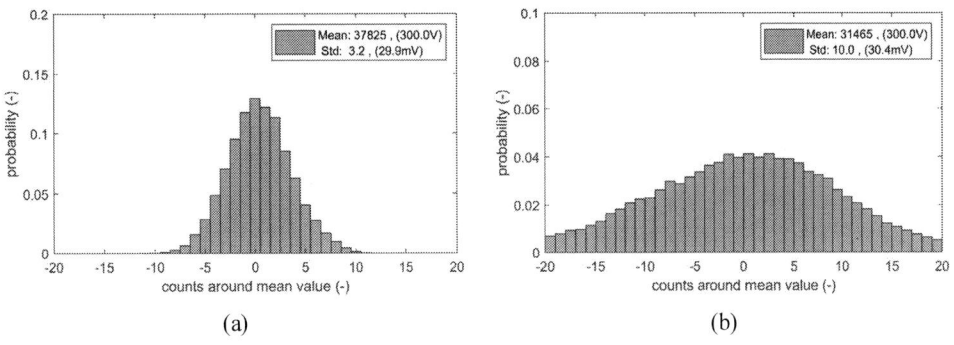

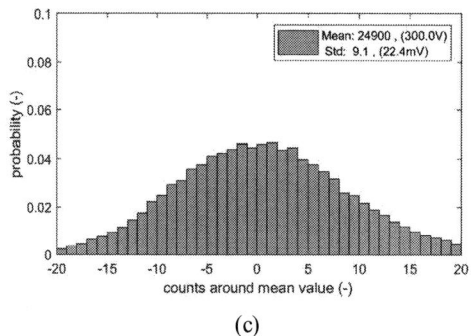

(c)

Fig. 5: Noise measurements (equivalent voltage) after the ADC acquisition with a switch at F_{ADC}.

Table 1 shows a comparison of the three methods, with the switching process at F_{ADC} and without it, it also includes cost and complexity of the circuits.

Table I: Comparison of different methods

	Case A	Case B	Case C
Complexity	Very low	Low	High
Cost	Medium	Medium	High
Accuracy	Average	Average	Improved
No switch	21mV/300V	23mV/300V	14mV/300V
Switch at F_{ADC}	29.9mV/300V	30.4mV/300V	22.4mV/300V

V. Conclusions

This paper presents three different methods for measuring feedback signals in DC/DC converters: the standard one, and the other two with a reduced dynamic range, so accuracy is improved. For that, a voltage reference was used to subtract the "offset" or bias of the error signal. Two kinds of voltage references were analyzed, one at the low voltage side and one at the high voltage side. The low voltage alternative produces an increase in the signal but also amplifies the noise, so the benefit is slightly better. The high voltage reference on the contrary increases the signal amplitude but not the noise, therefore this is the one with better results. The three approaches were tested experimentally in two conditions: when the DC/DC converter drives a constant current and when the system is triggering some switches synchronized with the acquisition. In both cases, the high voltage reference approach shows better results.

However, this circuit is the most complex, with more power losses and more expensive. In order to have a reference at high voltage with low noise and high accuracy, a linear power supply was used. This intruded high losses in the circuit.

References

[1] F. M. Ibañez, J. Vadillo, J. M. Echeverria, and L. Fontán, "100kW bidirectional DC/DC converter for a supercapacitor stack," *2013 4th IEEE/PES Innov. Smart Grid Technol. Eur. ISGT Eur. 2013*, pp. 1–5, 2013.

[2] C. Y. Wang, Y. C. Ou, C. F. Wu, and M. T. Shiue, "A voltage-mode DC-DC buck converter with digital PID controller," *2015 12th Int. Conf. Fuzzy Syst. Knowl. Discov. FSKD 2015*, pp. 2322–2326, 2016.

[3] B. Friedland, *Control System Design : An Introduction to State-Space Methods*, vol. 9, no. 1. Dover Publications, 2012.

[4] Dr. Farid Golnaraghi and D. B. C. Kuo., "Automatic Control Systems," Tenth Edit., McGraw-Hill Education, 2017.

[5] J. Liu, W. Ming, and F. Gao, "A new control strategy for improving performance of boost DC/DC converter based on input-output feedback linearization," *Proc. World Congr. Intell. Control Autom.*, no. 1, pp. 2439–2444, 2010.

[6] T. Tarczewski, L. J. Niewiara, M. Skiwski, and L. M. Grzesiak, "Gain-scheduled constrained state feedback control of DC-DC buck power converter," *IET Power Electron.*, vol. 11, no. 4, pp. 735–743, 2018.

[7] N. George, V. N. Panchalai, and E. Sebastian, "Digital feedback control of a full-bridge DC-DC converter with input voltage based gain scheduling," *Proc. - 2014 4th Int. Conf. Adv. Comput. Commun. ICACC 2014*, pp. 347–351, 2014.

[8] W. Zhenjun and W. Lifang, "converter Based on a Novel Chaos Control," no. 3, pp. 1031–1034, 2008.

[9] Y. Panov and M. Jovanović, "Small-signal analysis and control design of isolated power supplies with optocoupler feedback," *Conf. Proc. - IEEE Appl. Power Electron. Conf. Expo. - APEC*, vol. 2, no. 4, pp. 777–785, 2004.

[10] T. Uc, "APPLICATION NOTE THE UC1901 SIMPLIFIES THE PROBLEM OF ISOLATED U-94."

[11] O. Ibrahim, N. Z. Yahaya, N. Saad, and K. Y. Ahmed, "Design and simulation of phase-shifted full bridge converter for hybrid energy systems," in *International Conference on Intelligent and Advanced Systems, ICIAS 2016*, 2017, pp. 0–5.

[12] G. Gaggini, P. F. Manfredi, P. Maranesi, and G. Triulzi, "Correction to 'Isolation amplifier with combined magnetic and optical coupling,'" *IEEE Trans. Instrum. Meas.*, vol. IM–25, no. 1, pp. 89–89, Mar. 1976.

[13] H. L. Skolnik, "Design Considerations for Linear Optically Coupled Isolation Amplifiers," *IEEE J. Solid-State Circuits*, vol. 17, no. 6, pp. 1094–1101, Dec. 1982.

[14] P. Horowitz and W. Hill, "Voltage Regulation and Power Conversion," in *The Art of Electronics*, Third Edit., Cambridge University Pres, 2015, pp. 635–638.

[15] H. w. Ott, "Equivalent noise bandwidth," in *Electromagnetic Compatibility Engineering*, John Wiley & Sons, 2009, pp. 334–337.

Approach of an Active Device Protection for Drive Inverters against Short Circuit Faults in an Open Industrial DC Grid

Simon Puls[1], Urs Obernolte[2], Martin Ehlich[3], Holger Borcherding[4]

[1,2,3]Lenze SE
Breslauer Str. 3
32699 Extertal, Germany
Phone: +49 5154 82-1528
Email: simon.puls@lenze.com
URL: http://www.lenze.com

[4]OWL University of Applied Sciences and Arts
Campusallee 12
32657 Lemgo, Germany
Phone: +49 5261 702-5217
Email: holger.bocherding@th-owl.de
URL: http://www.th-owl.de

Acknowledgment

The Authors gratefully acknowledge financial support by the German Federal Ministry of Economic Affairs and Energy (BMWi) via grant Numbers 03ET7558A to N (Project acronym: DC-INDUSTRIE).

Keywords

≪DC grid≫, ≪fault handling strategy≫, ≪DC-power supply≫, ≪active protection≫, ≪drive inverters≫

Abstract

An open industrial DC grid has a lot of advantages. Also new challenges arise by coupling several DC link capacitors of inverters. This paper presents an approach and measurements of an active device protection to withstand possible faults that can occur in the DC grid. In particular, robustness in the event of faults plays a key role.

Introduction

DC grids offer a series of advantages in the industrial sector, especially for voltage source inverter-based electric drives. For this reason, proprietary DC grids in the cabinet have been state of the art for many years. Hence there is a need for extended, open DC grids in the industrial environment. The elimination of the rectifier and the direct electrical coupling makes it possible to exchange energy. The generative energy of an electric drive in braking operation can be used directly by another electric drive or consumer. Braking resistors can be completely eliminated and 20 - 30 % energy can be saved. The possible savings increase significantly during intermittent operation, for example in intralogistics for storage and retrieval machines or industrial robots with short and fast accelerations and decelerations [1].

Concept for an Open Industrial DC Grid

The concept of DC-INDUSTRIE [2] is to connect industrial devices, such as electric drives, to each other via an open DC grid. The AC/DC conversion, which is always necessary with controlled industrial drives, can be carried out at a central point to the AC grid. By eliminating the need for uncontrolled rectifiers, the overall efficiency of plants can be increased and the emitted disturbance of harmonics significantly reduced. A further advantage of DC grids is the easy coupling of storage units and decentralized energy generators, such as photovoltaic systems. If the DC grid is controlled with intelligent grid management, it can react flexibly to grid voltage changes and failures. This increases the availability of the plants. The

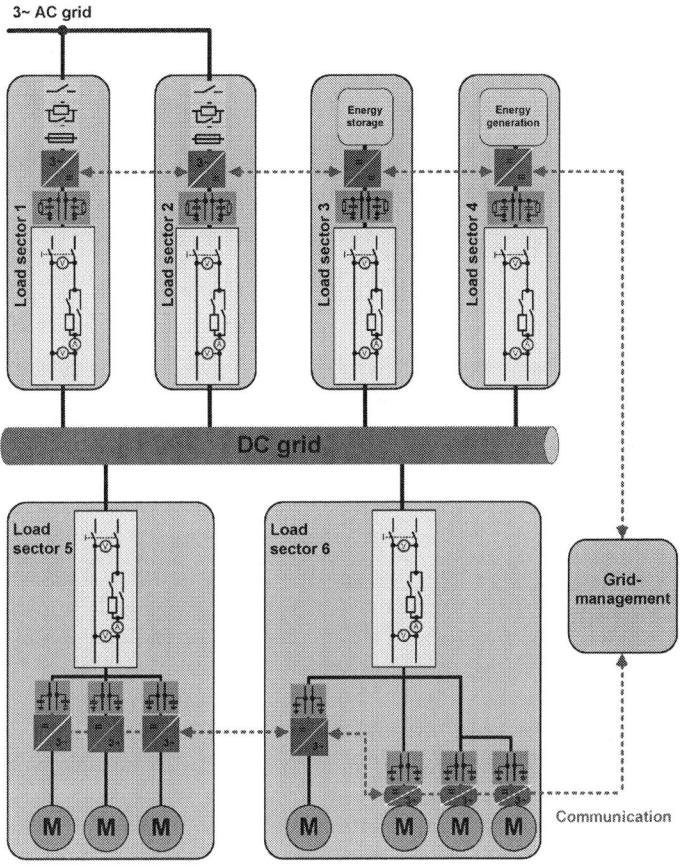

Fig. 1: Concept for an open industrial DC grid [2]

idea of DC-INDUSTRIE is to integrate devices for dedicated machine functions in load sectors. Each load sector is connected to the DC grid with a DC connection box. The task of the DC connection box is to protect all devices within the load sector from disturbances in the DC grid (overcurrent, over- and undervoltage) and to disconnect the load sector from the DC grid in case of short-circuit faults etc. in the inner circle. For this task, very fast DC switches in semiconductor- or hybrid-technology are needed. Some requirements of the DC connection box are summarized in Table I.

Table I: Some requirements of the DC connection box

1. Fast circuit breaking
2. Precharging and discharging
3. Measurement of current and voltage
4. Isolation switches for maintenance
5. Line fusing for emergency cases
6. Earth fault detection

Infeed sections are managed as single load sectors. The DC grid can be operated at a star point grounded AC transformer or designed as an isolated DC grid (typically with an Active Infeed Controller). Fig. 1 shows an overview of components that are necessary for the reliable operation of a DC grid. In addition to drive inverters (centralized or decentralized) and power supply units for coupling to the AC grid, innovative switching and protective devices as well as electromagnetic compatibility (EMC) input filters are required.

Fig. 2: Project logo "DC-INDUSTRIE"

In the project DC-INDUSTRIE there are 21 industrial partners, 4 research institutes and the ZVEI which work together on the specification of an open industrial DC grid, on the development of the necessary components and on the implementation in reference systems. The project is funded by the Federal Ministry of Economics and Energy (BMWi) within the framework of the 6th Energy Research Programme and supervised by Project Management Jülich.

Illustration of possible Short-Circuit Faults

Fig. 3 symbolically shows equivalent circuits of two DC drives with the most relevant components. These are a common-mode choke as input filter, an electrolytic DC link capacitor and the inverter section. Also shown are on the right side electric motors and on the left side a DC grid with cable connections. Various short-circuit (SC) errors can occur during operation, as shown in Fig. 3.

Short-circuit faults that occur on the output side of the motor connection are comparable with conventional AC devices (e.g. SC4). Due to the low-impedance coupling of the DC link circuits of the individual devices, the short-circuit faults SC1 in the grid and SC2 respectively SC3 are of particular importance in the DC grid. If such a fault occurs, all connected electrical storages, for example the DC link capacitors of devices, feed directly into the fault location. In the DC link of devices usually polarized electrolytic capacitors or film capacitors are used for voltage stabilization.

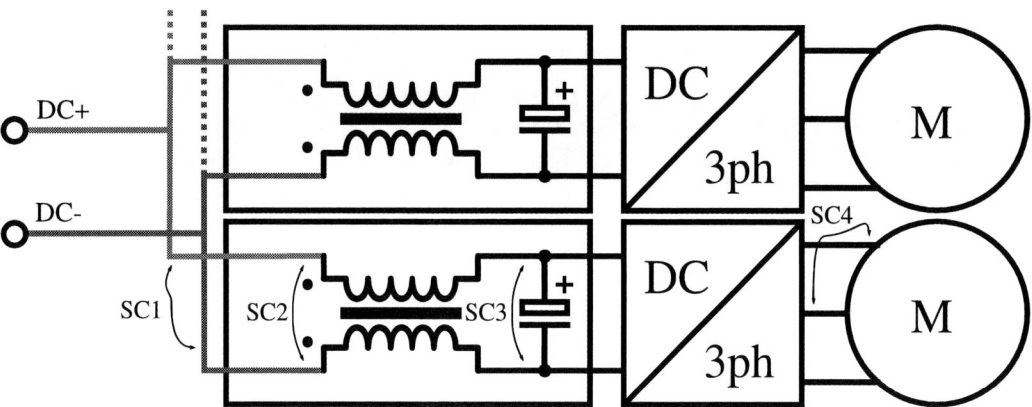

Fig. 3: Illustration of two drives coupled via a DC grid and different short-circuits faults that can occur (SC1: SC in DC Grid, SC2: SC in inverter, before filter, SC3: SC in the DC link, SC4: SC on the drive-side)

Short-Circuit Fault in the DC Grid

If a short-circuit fault occurs in the DC grid (SC1) or in a neighbouring device (SC2, SC3) while the DC link capacitors of the connected devices are fully charged, they will become discharged rapidly. Some more details of the equivalent DC drive circuit are shown in Fig. 4: The three-phase inverter module consisting of six IGBTs with antiparallel free-wheeling diodes on the right side. In the middle an electrolytic capacitor C_L as DC link. And on the left side an EMC input filter consisting of a choke L_{CM} and an unpolarized film capacitor C_X. As L_{CM} for reducing line related interference emissions is a common-mode choke, only the stray inductance and parasitic inductances effect the current profile as it is differential.

Furthermore a short-circuit fault in the DC grid is symbolized (SC1). The fault is modelled with an ohmic (R_{cab}) and an inductive (L_{cab}) part in the cable connection. In addition, the current I_{SC} located at the DC connector, the current I_{CX} located at the input filter capacitor C_X, the current I_{CL} located at the DC link capacitor C_L and the current I_m located at the DC link between the inverter module and the DC link capacitor are shown in Fig. 4.

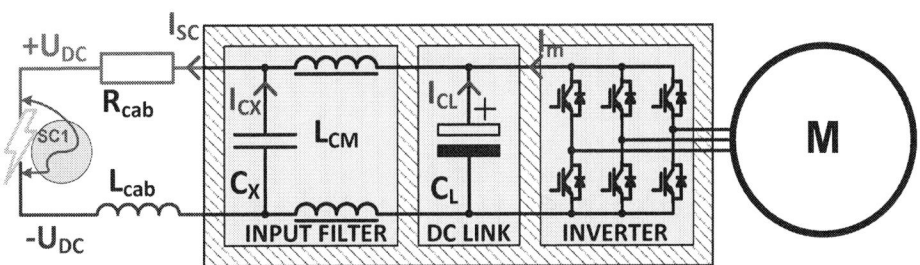

Fig. 4: Symbolized drive inverter with a short-circuit fault in the DC grid

The DC link capacitance C_L is not variable and determined by the manufacturer of the device. The voltage U_{DC} depends on the voltage supplied by the DC grid. It is usually measured by every device.

The maximum discharge current in the event of an external short-circuit like SC1 results from the voltage in the DC link and ohmic resistances in the short-circuit path, for example conductor paths and the mentioned parasitic ohmic part in components. Although inductances dampen the current rise and reduce the maximum current too, but they do not decrease the amount of energy.

Stress on DC Link Capacitors due to High Discharge Currents and due to Negative Voltage

The combination of a capacitor in the DC link and at least one serial inductance basically results in a series-resonant circuit, as shown in Fig. 4. The initial state is a charged capacitor and a nearly current-free inductance. Only the normal operating current flows. In the event of a short-circuit in the DC grid, the mesh of capacitance and inductance is closed. Ohmic resistances contained in the mesh such as those of the cable, the conductor paths in the device or the internal R_{ESR} of the capacitor dampen the resonant circuit. From a functional point of view, parasitic ohmic components are unwanted, as they cause energy losses during normal operation. Corresponding to this, they are kept as small as possible, for example through extended conductor paths.

Fig. 5 shows the voltage and the current measured in a real application at the DC terminals of a 0.37 kW-inverter at a short-circuit fault in the DC grid (SC1). The inverter has an internal DC link consisting of electrolytic capacitors with a capacity of 60 µF and a common-mode choke with a stray-inductance of $\sim$140 µH and a DC resistance of $\sim$200 mΩ. The short-circuit path is built up with a remote controlled semiconductor based switch. With this switch a short-circuit can be activated in a fast and reproducible way.

The voltage at the DC terminals rapidly drops from 900 V (maximum voltage) to zero immediately. The current rises from around −0.5 A (normal operation) to nearly 850 A in less than 40 µs. The current peak directly at the beginning results from a film capacitor in the input filter and has relatively little energy. This current can cause an increased stress to the electrolytic capacitor and the conduction paths. It should be noted that a negative current supplies the inverter and a positive current is fed by the inverter.

An effect due to parasitic inductors in the short-circuit path is that the current continues to flow after the

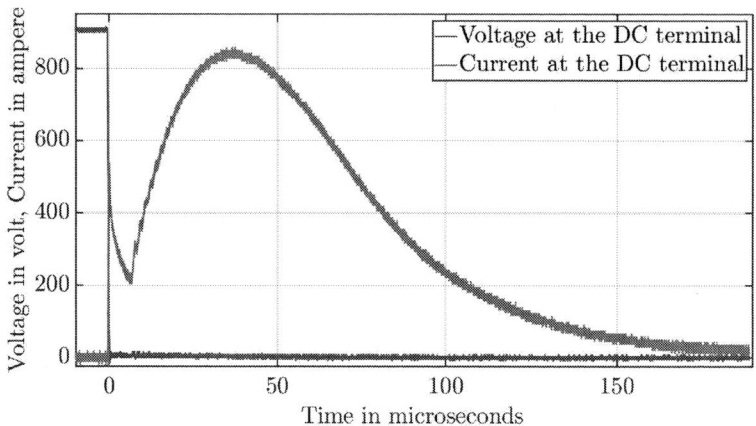

Fig. 5: Discharge-process of a device caused by a short-circuit in the DC grid: Measured voltage (blue) and current (red) at the DC terminal

DC link capacitor has been discharged to 0 V.

Fig. 6 shows a negative voltage at the device terminals after a short-circuit, which was done with approx. 15 m additional cable. The measured voltage falls down to −25 V until it reaches the 0 V potential. Because electrolytic capacitors are commonly used for this application, the negative voltage might be a problem: In contrast to film capacitors electrolytic capacitors are sensitive to negative voltage. There is a risk that the oxide layer will degrade and damage the capacitor. It is well known that even low negative voltages can cause damage to electrolytic capacitors. Negative voltage above approx. 2 V is sufficient to cause the oxide layer to degrade. The result is an addition reduction in capacity and an increase in resistance [3]. The capacitors are already subjected to intense stress during normal operation and are usually the components that limit the lifespan of devices like drive inverters [4].

It should be mentioned in addition that the free-wheeling diodes of the inverter module become conductive by a negative voltage in the DC link and absorb energy according to their forward characteristics. Both components, the capacitor and the diodes, can be overloaded and may be destroyed [5].

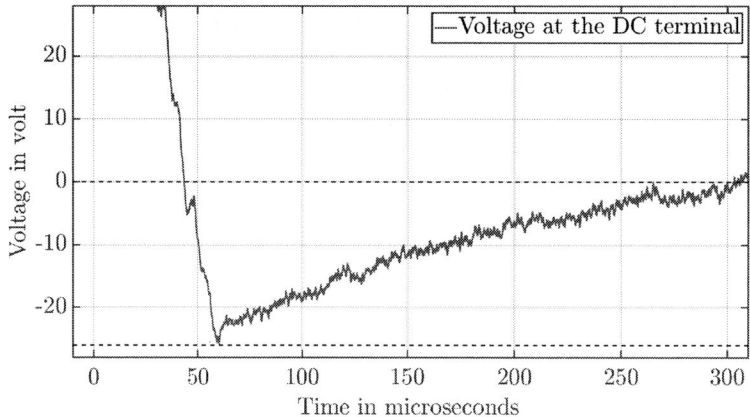

Fig. 6: Measured negative voltage at the DC terminal after a short-circuit fault

Comparison to AC

In AC devices, a line circuit breaker usually limits the energy that is fed into the device. Furthermore, in an AC system, devices arranged in parallel are decoupled from the mains by the diodes of the input rectifier so that they are not discharged. Only if a device is able to feed energy back into the grid it is possible to feed significant amounts of energy from the device into the mains. In this case a circuit breaker also works. Nevertheless, in the event of a fault, energy can flow into the fault location both from the AC grid and from the device.

In the DC grid, however, there is no decoupling. The amount of energy stored in the capacitors of devices can be very large. Without limiting the number of devices or the power of them operated in parallel this amount is unknown. In the devices, especially in drive inverters, capacitors are built in that store electrical energy. The use of capacitors in the DC link strongly smoothes the current consumption from the mains. This is necessary to provide energy for the highly volatile current demand of the inverter locally in the device. The grid is not sufficiently dynamic for such a current consumption and a heavily alternating grid voltage would be the consequence.

The amount of energy is well known and corresponds to:

$$E_{\text{device}} = \frac{1}{2} \cdot C_{\text{device}} \cdot U_{\text{grid}}^2 \tag{1}$$

The amount of the energy stored electrically for a given $0.37\,\text{kW}$ device with $60\,\mu\text{F}$ at $900\,\text{V}$ is (typical values):

$$E_{\text{device}} = \frac{1}{2} \cdot 60\,\mu\text{F} \cdot 900\,\text{V}^2 \tag{2}$$

$$\approx 24\,\text{J} \tag{3}$$

Since several devices are connected in parallel within the load sector, the energy amounts stored electrically add up to:

$$E_{\text{LZ}} = \sum_{k=1}^{n} E_{\text{device}}(\text{k}) \tag{4}$$

For the total energy stored in a load sector, this means that the amount can be a multiple of a single device. For example, if ten devices of the same $0.37\,\text{kW}$ type are operated in parallel, the value of energy stored in the load sector is $E_{\text{LZ}} \approx 240\text{J}$. Furthermore, energy can flow through the DC connection box from the superior DC grid. The energy flow is only interrupted after the current of the circuit-breaker I_{CB} has been switched off and is calculated by:

$$E_{\text{CB}} = \int_0^{T_{\text{off}}} U_{\text{grid}}(t) \cdot i_{\text{CB}}(t)\,\mathrm{d}t \tag{5}$$

A semiconductor based DC circuit-breaker can switch off within a few microseconds, for example in $150\,\mu\text{s}$. If the current raises to $1000\,\text{A}$ caused by a short-circuit fault, the load sector is approximately fed with the additional energy from the higher-level DC grid:

$$E_{\text{CB}} = \int_0^{150\,\mu\text{s}} 900\,\text{V} \cdot 6,\overline{6}\,\text{MAs}^{-1}\,\mathrm{d}t \tag{6}$$

$$= 900\,\text{V} \cdot \frac{1}{2} \cdot \frac{1000}{150}\,\text{A}/\mu\text{s} \cdot 150\,\mu\text{s} \tag{7}$$

$$\approx 68\,\text{J} \tag{8}$$

In this example, this is $\sim 30\,\%$ in addition to the energy stored electrically in the load sector. In total the sum of the electrical stored energy in the load sector and the energy fed from the higher-level DC grid is given by:

$$E_{\text{total}} = E_{\text{LZ}} + E_{\text{CB}} \tag{9}$$

In addition, mechanical energy can be stored in motors. If the motors are externally excited, this energy is also fed into the DC grid via the free-wheeling diodes of the inverter in the event of a short-circuit fault. However, the currents are much smaller and very slow compared to the current from the capacitors [5].

Approach of an Active Device Protection

An active protective circuit is supposed to disconnect the faulty device from the DC grid in time. This should ensure that a device, such as a drive inverter, does not have to absorb all the energy in the event of a short-circuit fault in the DC grid. The goal is to ensure that the device connected to the DC grid does not take any damage.

Requirements for an Active Device Protection

The active device protection circuit serves to limit the current and energy that fed by a device in the event of a low impedance error in the DC grid. Accordingly, the active protection must be able to detect the fault condition and electrically disconnect a defective device. A fault condition exists if the current at the DC terminal of the device is outside the specification. The nominal operating current of a 0.37 kW device is a maximum of about 1 A within the voltage ranges of DC-INDUSTRIE [2]. This value applies to both motor and generator operation. A current of 1 A can only be slightly exceeded in temporary overload operation and never reaches ten times the normal operating value.

For the installation of a comparable AC device, a class "B10" circuit breaker is mandatory [7], which corresponds to a continuous operating current of about 10 A (thermal trigger). That means, the device can withstand a current up to 10 A in general by the fact, an AC circuit breaker will trip accordingly. For a short period of time, a class "B10" circuit breaker can be assumed to have an even higher current (magnetic trigger). From this, initial requirements can be derived. An active device protection must be able to withstand a similar high current in the event of a fault. Also, the shutdown threshold can be set to a value just above the normal operating current, for example $I_F = 10$ A. In addition, the protective circuit must be able to function independently and must not disturb normal operation.

Design of an Active Device Protection Circuit

A protection circuit is designed that can disconnect the device from the DC grid to be protected in the event of a DC grid or a device fault. The protection circuit should not affect the normal operation. To meet the requirements functional blocks are designed:

1. Power switch circuit
2. Load current measurement
3. Switch driver and protection circuit
4. Control logic

Additional requirements to the active device protection are that it works autonomously, that it is robust and in addition that it is inexpensive. Since the switch-off must be precise, wear during switch-off is unwanted and the current rating is flow-directional, melting fuses cannot be used. Furthermore, since only one device is to be protected and the device to be protected is well known, the active protection circuit can be designed accurately. Just with a few components the circuit can be realised.

Power Switch Circuit

An IGBT is selected as a switch, which has relatively low forward losses and tolerates a high over-current for a long time, compared to SiC-FETs in the event of a short-circuit fault. A second switch in anti-serial direction would be needed for an energy flow and the ability to switch off in both directions. To safely disconnect from the mains, also a contactor is used. The switching elements shown in Fig. 7 are just idealized components.

Load Current Measurement

A current sensor is used to detect a slowly increasing over-current. In the event of a device fault, e.g. due to an increased leakage current of the DC link capacitor, such a slightly increased current can occur. This can also be used as a monitoring device for overload conditions.

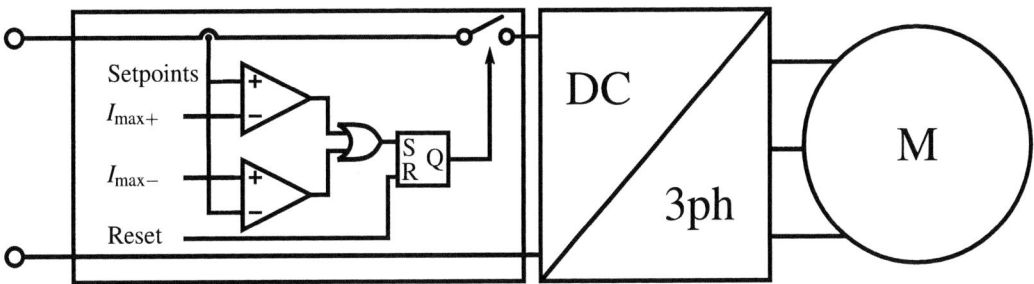

Fig. 7: Symbolized configuration of a Drive Inverter combined with an Active Protection Device with idealized components

Switch Driver and Protection Circuit

The driving circuit needs to be designed to be able to shut off the switches in overload and short-circuit conditions. To safely shut down, a two-level turn-off is being used. The saturation voltages of the IGBTs are being measured and analysed with a suitable driver [6]. This is used to detect a fast rising short-circuit current as shown in Fig. 5 above.

Control logic

The current levels, measured by the current sensor, are filtered and compared to maximum positive and negative values. After that, they are combined with the saturation signals with an OR-gate. To prevent automatic turn-on after the current reaches to zero, a R-S-flipflop is used (Fig. 7).

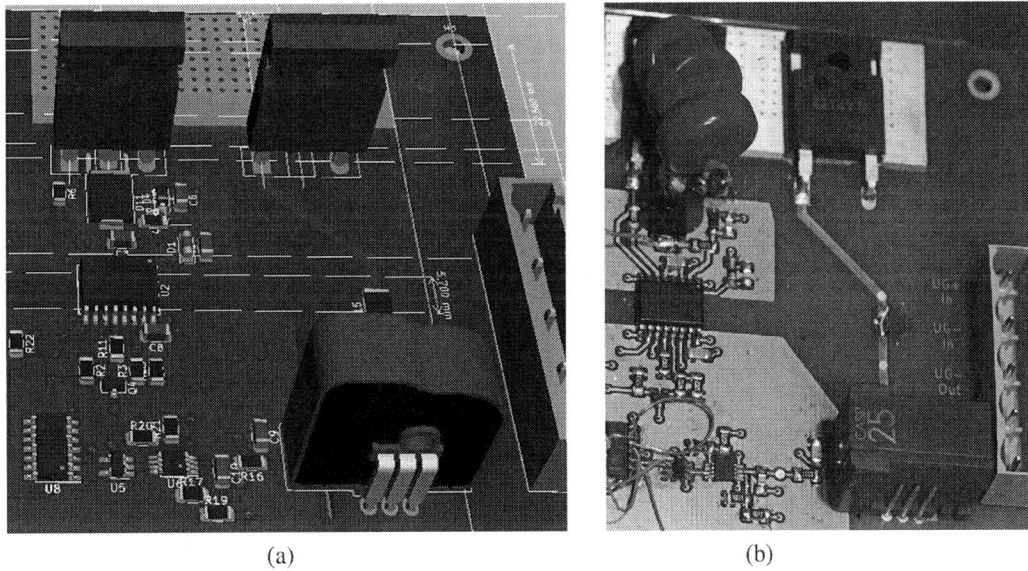

| (a) | (b) |

Fig. 8: (a) 3D model of the Active Protection Device circuit with the most important components (IGBT top left, additional antiparallel diode top right, current sensor bottom right, flip-flop bottom left), (b) Picture of realisation the Active Protection Device circuit with some modifications based on test results

Test Results

Fig. 9 shows the voltage u_{load} at the load, the voltage u_{IGBT} at the IGBT on the collector-emitter path and the short-circuit current i_{IGBT}. The current is measured directly at the protection device by a sensor. The current u_{IGBT} rises rapidly and with this, the voltage u_{IGBT} also raises. Then the switch-off process starts at about 0.5 μs due to the overcurrent. The driver detects this and reduces the control voltage. In

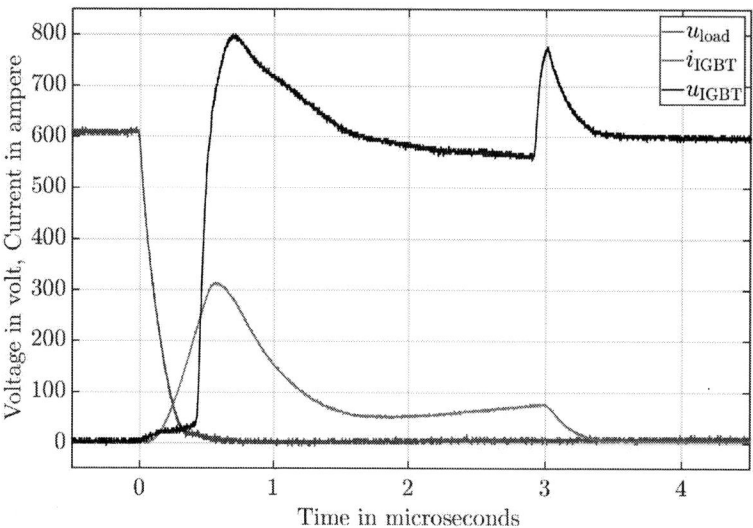

Fig. 9: Measured voltage and current curves of a switch-off process

order to avoid an incorrect tripping, it is not switched off completely immediately. If the current does not decrease again to the normal operating current, it is completely switched off after a short waiting period of approx. 2 µs. At 3 µs the driver switches off the control voltage completely. Due to this two-stage switching mode the amplitude of the current rise is significant reduced. In addition the overvoltage is reduced: At an operating voltage of 600 V, the overvoltage to 800 V corresponds to about one third, which is still far below the permissible 1200 V of semiconductors typically used in industrial drives. Three things should be mentioned here:

- The current is switched off in only 3 µs after the fault occurs
- The maximum current is significantly lower
- There is no negative voltage

Fig. 10 shows the real configuration of a drive inverter and an active protection device during operation on a DC grid. The active protection device is directly connected to the DC grid. The DC grid can be an open industrial DC grid with all its characteristics. The drive inverter in this case is a small device with low power. However, a much more powerful drive inverter can also be protected by the same principle of active protection device.

Used T&M equipment

- Oscilloscope: Tektronix MSO4054
- Current probe 1: ILA-SMZ 3000
- Current probe 2: Tektronix TCP303 with TCPA400
- Differential voltage probe: PMK Bumblebee with PS-03

Conclusion

With an active protection device, it is possible to operate drives in an open industrial DC grid without causing damage to the device in the event of faults in the grid. If a fault occurs, for example a short-circuit fault, the device will be disconnected from the DC grid within a few microseconds. This significantly reduces high short-circuit currents and dramatically reduces the load on the components used in the device. The estimated lifespan of e.g. electrolytic capacitors is still achieved and DC grid failures in particular can be reduced or even completely avoided by maintaining the DC grid voltage.

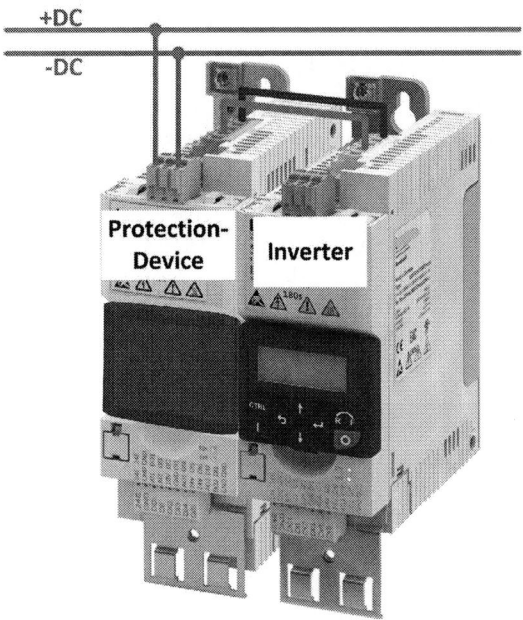

Fig. 10: Real Configuration of an Active Protection Device (left) connected to a DC grid (above) and a drive inverter (right)

References

[1] M. Pelliciari et al.: AREUS-Innovative Hardware and Software for Sustainable Industrial Robotics, IEEE International Conference on Automation Science an Engineering, Gothenburg 2015

[2] H. Borcherding et al.: Concepts for a DC Network in Industrial Production, 2017, IEEE 2nd International Conference on DC Microgrids (ICDCM), Nuremberg

[3] A. Albertsen: Mit Abstand am besten Spannungsfestigkeit von Elkos, Jianghai Europe Electronic Components GmbH, 2018

[4] S. Puls et al.: Lifetime Calculation for Capacitors in Industrial Micro DC grids, 2019, IEEE 3rd International Conference on DC Microgrids (ICDCM), Matsue

[5] S. Puls et al.: The Influence on Drive Inverters under the Effects of Short Circuits in an Open Industrial DC grid, 2019, IEEE 21st European Conference on Power Electronics and Applications (EPE), Genua

[6] J. Fuhrmann et al.: Enhancing short-circuit capability of high-performance IGBTs by gate-drive unit, 2019, IEEE International Exhibition and Conference for Power Electronics, Intelligent Motion, Renewable Energy and Energy Management (PCIM), Nurenberg

[7] Project planning manual i550 Inverter
https://download.lenze.com/TD/i55AE_Inverter%20i550%20Cabinet%200.25-132kW_v10-0_DE.pdf

A new design of an air core transformer for Electric Vehicle On-Board Charger

Valentin Rigot[1,2], Tanguy Phulpin[1], Daniel Sadarnac[1], Jihen Sakly[2]

1: University of Paris-Saclay, CentraleSupélec, CNRS, Group of electrical engineering - Paris, 91192, Gif-sur-Yvette,France
Sorbonne University, CNRS, Group of electrical engineering - Paris, 75252, Paris, France

2: VEDECOM, 23bis allée des marroniers, Versailles, France, Valentin.rigot@vadecom.fr

Keywords

«Transformer», «High Power Density Sytems», «Electric Vehicle», «Battery Charger».

Abstract

This paper compares two geometries of air core transformer, relevant for high frequency converters and high power densities: One logical toroidal, one innovative cylindrical. The manufacturing, the coupling parameter, the volume, and the flux leakage are examined in these topologies. The efficiency symbolizes a crucial parameter on power electronics, and a trade-off must ever be optimized between size and power. The proposed innovative structure permits to strongly increase the converter power density for an identical efficiency

Introduction

The autonomy of electric vehicles is a crucial challenge for the electric vehicle, and each part of the electrical power architecture must be optimized by reducing its size and weight [1]. For this purpose, active device development such as GaN and SiC allows increasing the power converter's switching frequency, which leads to an increase the power density by reducing the size of passive devices [2]. However, these passive components still present the most significant part of the converter in term of volume [3], They also limit the efficiency at high frequency because of the eddy effect and the copper losses [4]. Thus, investigate the passive components and specifically the coreless magnetics components appears as relevant for on-board battery charger development.

The paper is structured as follows: the first part presents the air-core transformer technology and its possible conductor. In a second part, a toroidal geometry with several techniques for magnetic field canalization is detailed before a new proposed transformer description. Then, the transformer model, with magnetic simulation using Finite Element Method are analyzed. To conclude, the experimental prototypes are tested, and results are compared to show the effectiveness of the proposed air-core transformer.

Air core transformer

Generally, air-core transformers are used in few applications such as inductive charges, superconductors, medical scanners, or magnetic sensors [5]. In our study, we want to use air core transformer to increase the power density of the converter. Indeed, the passive components represent usually one of the biggest places in the converter as shown in the following Fig.1 [6]-[10]. Nevertheless, a transformer without magnetic material, emits a magnetic field in its environment. To get a good coupling coefficient and avoid the CEM parasitic problems for an isolated converter, it is necessary to canalize the magnetic field. As presented by eq (1), the coupling coefficient expression in the transformer depends on the transmitted magnetic flux from its primary to its secondary. Thus, it becomes crucial for the efficiency of our converter to drive the magnetic field.

PART OF TRANSFORMER AND COOLING IN DC-DC CONVERTER

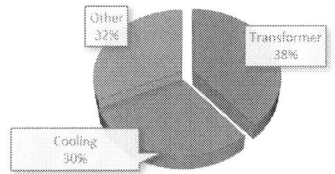

Fig. 1: Part of transformer and cooling in DC-DC converter

Moreover, as shown by eq (2), the inductance value of the transformer depends on the core magnetic permeability. A weak permeability such as air permeability induces a weak inductance value. The current crossing the inductance is increased with the voltage and decreased with the frequency as expressed by the following eq (3) and eq (4). As for power converter design, the voltage is commonly imposed, the frequency has to be the highest as possible for obtaining a high power density. As with an air-core transformer, there is no saturation of the magnetic circuit, the frequency limit is pushed further. The use of air core transformer is therefore relevant for high frequency or low voltage applications Low permeability of the air core induces however more magnetic field leakage. A solution is to minimize the magnetic field outside of the transformer by appropriate adjustment of coil's turns.

$$k = \frac{\varphi_{12}}{i_1} \tag{1}$$

Where φ_{12} and i_1 are, respectively, magnetic flux and primary current.

$$L = \frac{n^2 \mu S_e}{l_e} \tag{2}$$

Where n is the turns ratio, μ is the permeability, S_e is the equivalent section and l_e is the equivalent length.

$$u = U sin(\omega t) = L \frac{di_1}{dt} \tag{3}$$

Where u is the transformer voltage, U is the voltage amplitude and ω is the pulsation.

$$i_1 = -\frac{U}{\omega L} \cos(\omega t) \tag{4}$$

Conductor choice

Losses due to skin effect in the conductors increase with square root of the frequency following the eq (5) and eq (6), the conductor type selection is determinant for the transformer design. Three main kind of wires were considered as shown in Table I. The Litz wire which is a combination of twisted thin wires strands individually insulated to reduce the skin effect in each strand, the ribbon to get thin conductor and the copper plating.

$$e = \sqrt{\frac{\rho}{\mu_c \pi f}}$$
(5)

Where e is the thickness of the skin effect, μ_c is the copper permeability ρ is the conductivity and f is the frequency

$$P_{skin} = \frac{K}{e} I_{rms}{}^2$$
(6)

Where P_{skin} is the losses due to skin effect, I_{rms} is the root mean square current value and K is a coefficient depending on the conductor shape and the copper section.

Table I: Conductors adapted to high frequencies

Conductor	Litz wire	Ribbon	Copper plating
Section			
Example			

The ribbon is thinner, but it supposes to realize planar or solenoid geometry. Moreover, the thickness becomes higher with the frequency, increasing the section width and making the conductor more difficult to manipulate. In our case, as the conductor must be easily manipulated, the Litz wire is selected. But the insulation of each strand decreases the filling rate in the cross section of the wire [11]. Copper plating would be a good alternative, easy manipulated and with mostly full copper in the cross section but the process to mark off turns are not well developed yet. Therefore, we decide to study transformers with Litz wire conductors in air-core transformer for manufacturing arguments.

Field canalization by superposition theorem

For improving the design of an air-core transformer, it is essential to limit the magnetic field leakage. The superposition theorem could be the right solution to analyse this problematic. The result of the contribution of all coil turns is the sum of the effects of each coil turns.

With this method, the magnetic field produced by one coil turn is balanced by another coil turn, as in fig.2a. As shown by simulation with the Finite Element Method by FEMM, the magnetic field is however still spreading even with symmetry properties or with additional turns as plotted in Fig.2b. The best solution looks to add coil turns until reaching continuous turns around the asymmetric revolution axis of the transformer, meaning that the toroidal geometry is the suitable choice in this case, as shown by Fig.2c.

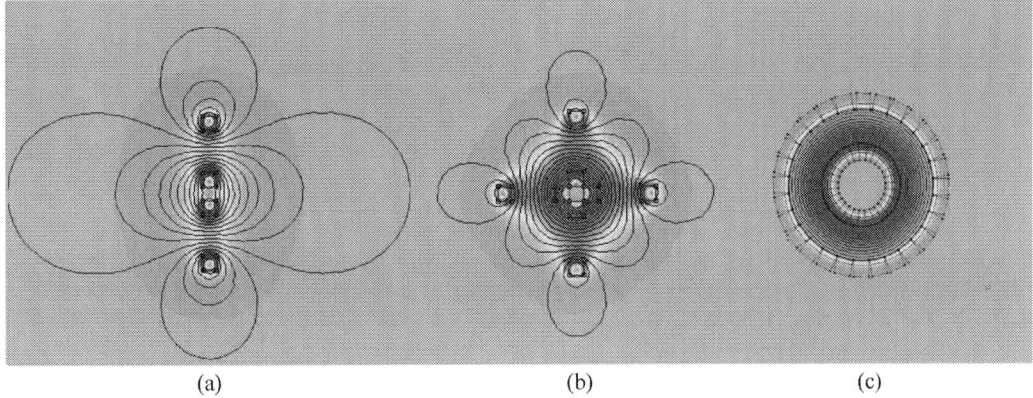

(a)   (b)   (c)

Fig. 2: Compensation of magnetic radiation by superposition method: (a) two coil turns, (b) four coil turns, (c) toroidal geometry

Unfortunately, this geometry is difficult to realize because of the difference between inside and outside radius. If the turn number in the inside radius creates a continuous conductor circle, it also creates some empty spaces in the outside circle. To double inside's turns allows us to obtain mostly continuous conductors around the transformer if considering a larger diameter twice longer than the inside diameter. The winding ensures minimal proximity loss by placing secondary winding close to each loop of the primary winding as showed in the Fig. 3 [12]. For the second winding, four layers in the inside circle are needed to keep a permanent conductor. Much wires are hence required and generate more losses, heaviness, and volume [13].

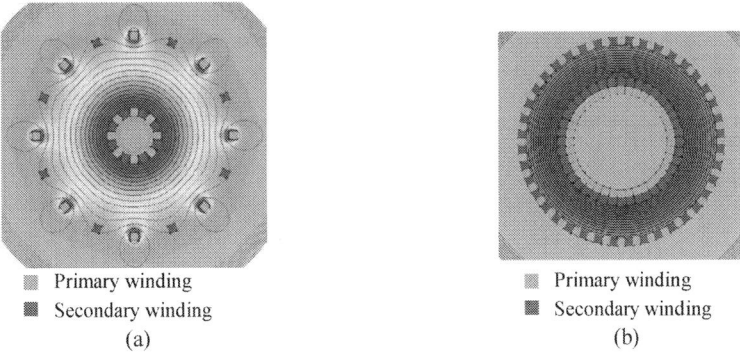

■ Primary winding   ■ Primary winding
■ Secondary winding   ■ Secondary winding
(a)   (b)

Fig. 3: Toroidal transformers: (a) common transformer (b) proposed transformer

Moreover, the realization of this kind of transformer is manual and difficult because of the few empty spaces in the outside circle. It is necessary to pull in and pull out wires like a sewing technique and requires manual interventions. The equation (7) shows the method to compute self-inductance in a toroid transformer [12].

$$L = \frac{n\,\varphi}{i} = \frac{n^2 H}{2\pi} ln\left(\frac{R_{out}}{R_{in}}\right) \tag{7}$$

For testing, we realized a prototype of a toroidal air core transformer with litz wire. We select the outside radius R_{out} twice longer than the inside radius R_{in} and the height H, the equation (7) and (8) become the equation (9)

$$\begin{cases} R_{out} = 2R_{in} \\ H = R_{in} \\ R_{in} = R \end{cases} \tag{8}$$

$$L = \frac{n^2 R}{2\pi} ln(2) \tag{9}$$

Proposed air core transformer

Because of the difficulties related to the toroidal transformer manufacturing for industry, we design a new structure of air-core transformer achievable in holding only once the wires with as main. The first goal is to keep the lowest leakage magnetic field. The second goal is to obtain a good coupling coefficient to transmit power efficiently.

In the contrary of the toroidal geometry, based on the magnetic field control for increasing the inductance value, our design takes first in consideration the inductance value in priority, then try to use the existing magnetic field. To increase the magnetic field crossing a turn Fig. 4a the best way to concentrate the magnetic flux consist in overlaying turns to make magnetic field of each turns cross all the turns as shown in Fig. 4b.

We can see that it is more interesting to increase coil turn number without extending the transformer length by increasing the number of layers as represented in Fig. 4c. This solenoid is the usual manufactured cored transformers geometry, so it is a mastered process easy to manufacture. The inductance value of the usual solenoid is showed by the eq (10).

$$L = \mu_0 \frac{N^2 S}{l} \tag{10}$$

Where S is the average section and l is the length of the solenoid.

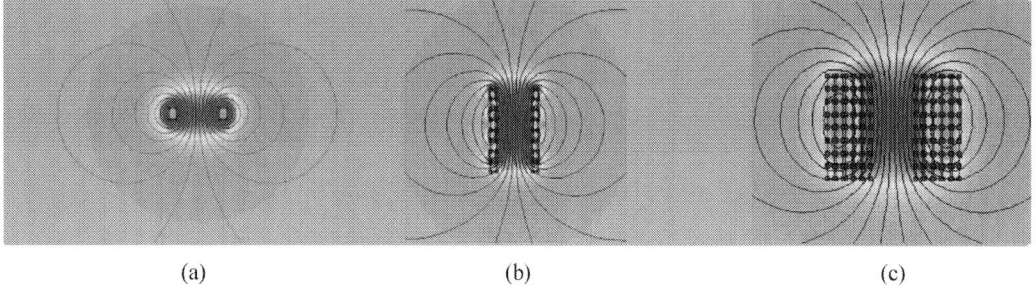

(a)   (b)   (c)

Fig. 4: Maximization of magnetic flux in each coil: (a) One coil turns, (b) Simple layer solenoid, (c) Multiple layers solenoid

Usually, they are used because of manufacturing issues, but the magnetic field is not canalized in air core transformer. Indeed, its central part spreads above and below the solenoid. That is why our idea was to complete this geometry by adding turns in the opposite direction.

As expected, it drastically reduces in simulation the leakage magnetic field, as represented in Fig 5b. It is possible to go further by adding some opposite turns at each side of the transformer, as in Fig. 5c where we simulate the magnetic field at 5cm from the center of the transformer by the Finite Element Method.

Moreover, the self-inductance of the first coil was calculated in each case by Biot & Savart law. The magnetic field looks well canalized, and the field lines appear similar to those corresponding to a magnetic core transformer. The outside magnetic field is drastically reduced by more than 300 times from the center as written in table II in the emission line.

Table II: Results and magnetic radiation for air core transformers

Transformer	Simple solenoid transformer	Solenoid transformer with turn added above and below	Transformer developed
Finite element method simulation	Fig. 5: (a)	Fig. 5: (b)	Fig. 5: (c)
Inductance	16.79 µH	16.77 µH	15.22 µH
Emission	7.12 mT	0.13 mT	0.02 mT
Volume	0.02 L	0.05 L	0.10 L

The inductance value mainly depends of the initial solenoid turn because the magnetic field inside the added turns was designed to obtain an average value near zero to limit the magnetic field. In our example, the magnetic field used to compensate the emission at the top and at the bottom create a loss of 0,1% of the total inductance and the compensation at each side create a loss of 9,2%, totalling at 9,3% less inductance than the original value. This latter was approximate with the usual solenoid's eq (10). In a second time, with a finite element method, the sides turn position, minimizing the magnetic emission is determined around the transformer.

Comparison and possible applications of two transformers

The design of the new transformer and of the toroidal transformer were realized on a 3D model software and are represented in the following Fig. 6.

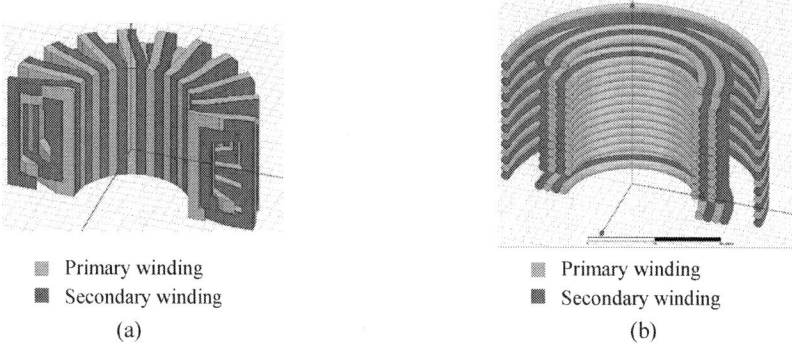

    ▨ Primary winding   ▨ Primary winding
    ■ Secondary winding   ■ Secondary winding
            (a)   (b)

Fig. 6: Model of both type of air core transformer: (a) toroidal geometry, (b) developed geometry

We have created two prototypes of air core transformers according to the best toroid selection and to the new cylindrical geometry. The toroid transformer was designed to reach 7,8µH following the eq (7), and a Finite element method computed a self-inductance of 11µH. The toroid transformer support was manufactured, and the cylindrical transformer support was 3D printed to be ventilated. The model for 3D printing is shown in Fig. 7a. Both prototypes can be viewed in Fig. 7b and Fig. 7c. Values were calculated by the finite element method and verified by experimental measurement with an RLC meter during short-circuit and open-circuit tests. The test bench can be viewed as Fig. 8.

        (a)   (b)   (c)

Fig. 7: Air-core transformer: (a) support model (b) cylindric prototype (c) toroidal prototype

Fig. 8: Test bench for air-core transformer

We compute the results in table III.

Table III: Characteristics of two air core transformers

	100kHz				500kHz				1MHz				2MHz				Volume (L)	Weight (kg)
	L (μH)	Lf (μH)	k	R ($mOhms$)	L	Lf	k	R	L	Lf	k	R	L	Lf	k	R		
Proposed toroid	4.80	1.13	0.76	23	4.89	1.09	0.78	135	4.90	1.09	0.78	449	5.04	1.05	0.79	1 410	0.48	0.88
Proposed cylindric	10.35	1.90	0.82	26	10.30	1.87	0.82	160	10.26	1.89	0.82	545	10.10	1.80	0.82	1 700	0.26	0.84

First, we note that the toroidal transformer presents worse results than the expected proper inductance and weak inductance. It is due to the winding complexity, where the exact number of turns wasn't reached and the symmetry isn't as efficient as the simulation instead of the cylindrical innovative transformer which present result almost identical than the simulation(6% lower proper inductance). Concerning the comparison, the new geometry is twice smaller, and its self-inductance L is twice more significant. That can be resumed as four times better self-inductance by volume unit. The leakage inductance Lf is more important in the cylindric transformer, but the coupling coefficient stays equivalent with respectively on average value 77% and 81%. Finally, the coupling coefficient is conserved and the self-inductance by volume unit is four time better. We can also notice that the Litz wires were not yet optimized and the resistance R logically increase over 100kHz with the frequency rising. According to the Fig. 1, the part of transformer is usually 38% of the converter volume. In theory, a converter with this transformer would be around 0.68 L for a power of 10kW achieving a power density of 14,3 kW/L and losses of 16W in the transformer while core converters reach 12kW/L [6]-[10][14] with the same losses.

Conclusion

This article proposes a study and a comparison of different geometries of coreless transformer adapted for power converter design. A methodology to produce an air-core transformer keeping an equivalent coupling coefficient such as a toroidal transformer is presented. We notice the difficulty of simulation and design realization because any analytics techniques are yet developed. A cylindrical prototype is therefore developed with a power density twice bigger than for the toroidal and the comparison indicates the benefits in terms of possible industrialization. Further optimization on the structure and on the conductors are planned to reach better performances even if we already observed some strong improvement in the power converter design. A thermal study and the integration into a power converter should validate the power density increase and the success of its new structure

References

[1]	A. Khaligh and M. Dantonio, "Global Trends in High-Power On-Board Chargers for Electric Vehicles," *IEEE Trans. Veh. Technol.*, vol. 68, no. 4, pp. 3306–3324, 2019.

[2]	Taurou E. : "Utilisation des transistors GaN dans les chargeurs de véhicule électrique" thesis, Université Paris-Saclay, France, 2018

[3]	P. Meyer and Y. Perriard, "Skin and proximity effects for coreless transformers," *2011 Int. Conf. Electr. Mach. Syst. ICEMS 2011*, vol. 1, no. 3, pp. 1–5, 2011.

[4]	R. P. Wojda and M. K. Kazimierczuk, "Winding Resistance and Power Loss of Inductors with Litz and Solid-Round Wires," *IEEE Trans. Ind. Appl.*, vol. 54, no. 4, pp. 3548–3557, 2018.

[5]	T. Kim, K Andrews, W.S., Kim, "3D printed Flexible Coreless Transformers," *2018 internationnal Flexible Electronics Technology Conference (IFETC)*, pp.1-4, 2018.

[6]	X. Jia, D. Xu, S. Du, C. Hu, M. Chen, and P. Lin, "A high power density and efficiency bi-directional DC/DC converter for electric vehicles," *9th Int. Conf. Power Electron. - ECCE Asia "Green World with Power Electron. ICPE 2015-ECCE Asia*, pp. 874–880, 2015, doi: 10.1109/ICPE.2015.7167885.

[7]	Q. Tong, H. Zhang, and D. Zhang, "Research on a High Power Density DC/DC Converter Based on Weinberg Topology," *Proc. 2018 2nd IEEE Adv. Inf. Manag. Commun. Electron. Autom. Control Conf. IMCEC 2018*, no. Imcec, pp. 2346–2350, 2018.

[8]   P. Deck and C. P. Dick, "High power density DC / DC-converter using coupled inductors Acknowledgments Keywords Volume comparison with interleaved converter (SII)," *EPE'17 ECCE Europe*, pp. 1–10, 2017.

[9]   G. Calderon-Lopez and A. J. Forsyth, "High power density DC-DC converter with SiC MOSFETs for electric vehicles," *7th IET Int. Conf. Power Electron. Mach. Drives, PEMD 2014*, pp. 1–6, 2014.

[10]   J. Biela, U. Badstuebner, and J. W. Kolar, "Impact of power density maximization on efficiency of DC-DC converter systems," *IEEE Trans. Power Electron.*, vol. 24, no. 1, pp. 288–300, 2009.

[11]   T. Guillod, J. Huber, F. Krismer, and J. W. Kolar, "Litz wire losses: Effects of twisting imperfections," *2017 IEEE 18th Work. Control Model. Power Electron. COMPEL 2017*, 2017.

[12]   D. Sadarnac, *Du composant magnétique à l'électronique de puissance - Analyse, modélisation, conception, dimensionnement des transformateurs, inductances, convertisseurs - Cours et exercices corrigés*, Éditions Ellipses. 2014

[13]   M. Nigam and C. R. Sullivan, "Multi-layer folded high-frequency toroidal inductor windings," *Conf. Proc. - IEEE Appl. Power Electron. Conf. Expo. - APEC*, pp. 682–688, 2008.

[14]   P. He and A. Khaligh, "Comprehensive Analyses and Comparison of 1 kW Isolated DC-DC Converters for Bidirectional EV Charging Systems," *IEEE Trans. Transp. Electrif.*, vol. 3, no. 1, pp. 147–156, 2017.

Enabling foil windings of medium-frequency transformers for high currents

Thomas B. Gradinger[1], Uwe Drofenik[2], and Filip Grecki[2]

[1] ABB Power Grids Switzerland Ltd.
Segelhofstrasse 1A
5405 Baden-Dättwil, Switzerland
Tel.: +41 (0)58 586 82 64
E-Mail: thomas.gradinger@ch.abb.com
URL: http://www.abb.com

[2] Corporate Research Center ABB
Segelhofstrasse 1K
5405 Baden-Dättwil, Switzerland
Emails: uwe.drofenik@ch.abb.com
filip.grecki@ch.abb.com
URL: https://new.abb.com

Keywords

«Solid-State Transformer», «Medium-Frequency Transformer», «Foil Winding», «Circulating Current».

Abstract

In foil windings of medium-frequency transformers rated for several hundred Ampères and operating at ten or several tens of kHz, parallel connection of foils is necessary to provide sufficient conductor cross-section. In this case, careful winding design is required to keep circulating currents among the foils under control. Such currents were investigated by means of an analytical model, a 2-d finite-element model, and measurements on a reduced-scale transformer for the case of two parallel connected foils. Simulations and measurements yield a consistent picture and show the potential of high extra losses in foil windings with unmitigated circulating currents. In particular, spiral windings of many turns may incur a circulating current that exceeds the useful net current by far if the inductance of the foil connections is small. Practically, it can be expected that the inductance of the foil connections leads to a noticeable reduction of the circulating current. This reduction, however, is not sufficient to bring the AC losses down to acceptable levels, such that it is recommended to transpose the foils between series-connected winding portions. Transposition largely cancels out the axial magnetic flux in the radial gaps between the parallel connected foils. The effectiveness of transposition in balancing the foil currents at frequencies up to 40 kHz is shown both theoretically and experimentally for aluminum foils of 0.2 mm thickness. With proper transposition, the extra AC losses due to circulating currents between foils can be reduced to practically negligible levels.

Introduction

Medium-frequency transformers (MFTs) are key components in Solid-State Transformers (SSTs), providing the galvanic insulation in the DC-DC conversion cells. Currently, various applications of SSTs being discussed, that are in the Megawatt range, and where a connection is provided to a medium-voltage grid [1–3]. The design of MFTs for such applications is a great challenge because of the requirements specification, that includes, in combination: (i) high currents, typically of several hundred Ampères; (ii) frequencies in the range of 10 to 20 kHz or higher; and (iii) insulation voltage in the medium-voltage range, typically of ten to several tens of kV [4]. Further requirements such as that of low stray inductance can lead to additional difficulties. Finally, high efficiency, high power density, and low cost are imperative.

Next to the selection of a suitable core material for the targeted frequency [5], a significant challenge is constituted by the design of the windings. Two popular choices for the conductor are litz wire and foils. When using litz wire, next to selecting an appropriate strand diameter [6], it is important to provide proper transposition on the level of bundles, subbundles, and strands. Failure to do so potentially results in strong circulating currents in the wire and increase of the losses [7, 8]. If several litz wires are parallel connected to reach higher currents, transposition among the wires may be necessary to avoid circulating currents among the wires [9].

Compared to litz wire, the winding design based on foils is generally more delicate. While litz wire provides subconductors that are small in both cross-sectional directions of the conductor, foils are small only in the radial direction of the winding, and extended in the axial direction. This bears a greater risk for current unbalances and necessitates careful design based on physical understanding. In a 2-d configuration of a foil winding, where the foil normal points in radial direction of the winding, one can distinguish four different effects that cause additional AC losses in the foil: (i) skin effect in radial direction; (ii) proximity effect in radial direction; (iii) edge effect, corresponding to skin and proximity effect in axial direction; and (iv) circulating currents – or current unbalance – between parallel connected foils. One can show that all these effects are mutually orthogonal such that their losses and AC resistances can be added.

AC losses in foil windings have been subject to numerous investigations. For effects (i) and (ii), 1-d skin and proximity effect, a classical model is that of Dowell [10] that provides closed-form expressions for the losses derived under controlled assumption. These assumptions are, however, not well met in many practical applications, for example due to large distance of the windings from the yoke of the core. In such cases, effect (iii), the edge effect, becomes important, which redistributes the current toward the edges of the foils, contributing to the 2-d shielding of the winding from the external H-field [11, 12]. In contrast to 1-d effects, there is no generally valid, closed-form model for 2-d effects. What the literature offers are on the one hand closed-form expressions derived under limiting assumptions [13, 14]. Typically, the expressions are complex, containing e.g. series expansions, and do not allow one to easily see functional dependencies. On the other hand, 2-d numerical models have been used to analyze the edge effect. They can provide physical understanding by studying sample cases [15], and are used as a basis for data regression to construct pseudo-empirical models [16, 17]. However, these models, while providing simple expressions, are again limited in their range of validity.

In the present paper, we specifically focus on effect (iv) – circulating currents between parallel connected foils – and the mitigation thereof. The need to parallel connect foils comes from the necessity to provide sufficient conductor cross-sections for several hundred Ampères. To limit proximity-effect losses, the foil thickness if often restricted to 0.2 mm or less. This again can lead to an unacceptably large axial winding length (corresponding to foil width) when using only a single foil. The present study complements – and in certain aspects extends – our work on parallel connected litz wire [9]. We first present an analytical model and a 2-d finite-element model to calculate circulating currents and the associated losses for the case of two parallel connected foils. We then present an experimental investigation on a reduced-scale test transformer and show how circulating currents can be successfully mitigated.

Analytical model

To provide a fundamental understanding of circulating currents in foil windings, an analytical model was developed for the case of two parallel connected foils. As will be seen, this model corresponds to a generalization of the model for parallel connected litz wires in helical windings [9]. A primary winding based on foils is illustrated in Fig. 1 on the left.

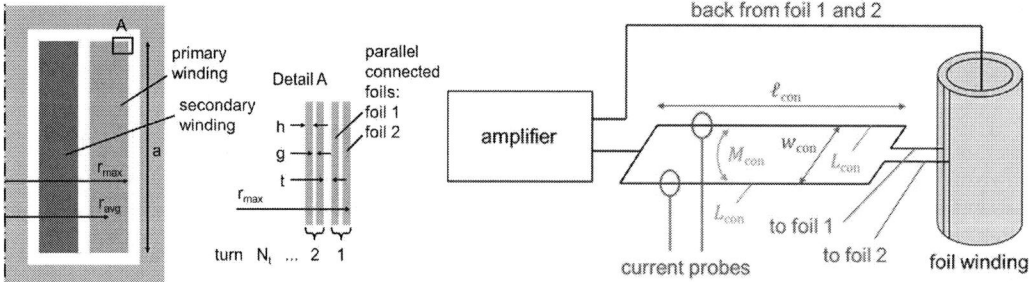

Fig. 1: Left: Core-window cross-section showing foil winding. Right: Inductance of foil connections.

Of the parallel connected foils, foil 1 is defined as the one closer to the other winding, and foil 2 is the one more distant. In Fig. 1 on the left, h designates foil thickness, g insulation thickness between parallel connected foils, t insulation thickness between turns, a axial winding length or foil width, r_{min} minimum winding radius, r mean winding radius, and N_t number of turns.

The slight difference in length between foil 1 and 2 is neglected and the DC resistance $R_{f,DC}$ of a foil is calculated as $R_{f,DC} = \rho N_t 2\pi r/(ah)$, where ρ is the foil resistivity. The total current is $i = i_1 + i_2$, and in case of DC, the current of foil k ($k = 1$ or 2) is $i_{k,DC} = i/2$. In case of AC, a deviation Δi from the DC currents is introduced:

$$i_1 = i/2 + \Delta i, \tag{1}$$
$$i_2 = i/2 - \Delta i. \tag{2}$$

The AC resistance of an individual foil is expressed using the AC-to-DC resistance ratio F_{Rf}:

$$R_f \equiv 2F_{Rf}R_{DC}, \tag{3}$$

where R_{DC} is the DC resistance of the entire winding. It is assumed that F_{Rf} is the same for both foils, which implies the assumption of equal proximity effect in both foils. For the foil-winding geometries of present interest, in particular with many turns such as 12 or 18, this assumption is reasonable. In case of AC, the voltages v_1 and v_2 over foils 1 and 2, respectively, are then:

$$v_1 = 2F_{Rf}R_{DC}(i/2 + \Delta i), \tag{4}$$
$$v_2 = 2F_{Rf}R_{DC}(i/2 - \Delta i). \tag{5}$$

In the experimental setup described further below, on the outside of the winding, the foil ends were connected to each other after a short length ℓ_{con} as shown in Fig. 1 on the right to enable mounting of the current probes. The connections have their own inductance $L_{con,1}$ and $L_{con,2}$, and are generally coupled via a mutual inductance M_{con}. This is also considered in the analytical model. In the following, it is assumed that $L_{con,1} = L_{con,2} \equiv L_{con}$, such that the reactive voltage drops $v_{ind,con,1}$ and $v_{ind,con,2}$ due to the connections can be written as

$$\begin{Bmatrix} v_{ind,con,1} \\ v_{ind,con,2} \end{Bmatrix} = \begin{bmatrix} L_{con} & M_{con} \\ M_{con} & L_{con} \end{bmatrix} \begin{Bmatrix} di_1/dt \\ di_2/dt \end{Bmatrix}. \tag{6}$$

The sum of the difference between v_1 and v_2, and between $v_{ind,con,1}$ and $v_{ind,con,2}$, must equal the induced voltage v_{ind}, i.e. $v_1 - v_2 + v_{ind,con,1} - v_{ind,con,2} = v_{ind}$, such that

$$\Delta i = \frac{v_{ind} - \left(v_{ind,con,1} - v_{ind,con,2}\right)}{4F_{Rf}R_{DC}}. \tag{7}$$

To obtain v_{ind}, Ampère's law is considered, assuming negligible field strength in the core. The turns are numbered from $n = 1$ to N_t in the direction toward the other winding, which is radially inward for the primary winding. For turn n,

$$i_2 + (n - 1)i = H_n a. \tag{8}$$

Here, H_n is the mean magnetic field strength in the core window between the mean radial position of foil 1 and 2 in turn n, i.e. between $r_{1,n}$ and $r_{2,n}$. Using Faraday's law and Eq. (8),

$$v_{ind} = \frac{d\Phi}{dt} = \frac{di_2}{dt}\underbrace{\frac{\pi\mu_0}{a}\sum_{n=1}^{N_t}\left(r_{2,n}^2 - r_{1,n}^2\right)}_{A_0} + \frac{di}{dt}\underbrace{\frac{\pi\mu_0}{a}\sum_{n=1}^{N_t}(n-1)\left(r_{2,n}^2 - r_{1,n}^2\right)}_{A_1}. \tag{9}$$

Using Eqs. (2) and (7), one gets

$$\frac{i}{2} - i_2 = \frac{1}{4F_{Rf}R_{DC}}\left\{A_0\frac{di_2}{dt} + A_1\frac{di}{dt} - (L_{con} - M_{con})\left(\frac{di_1}{dt} - \frac{di_2}{dt}\right)\right\}. \tag{10}$$

In the following, the stray inductance of the connections is defined by

$$L_{\sigma,con} \equiv L_{con} - M_{con}. \tag{11}$$

Defining the characteristic circular frequencies ω_0 and ω_1 by

$$\frac{1}{\omega_0} \equiv \frac{A_0 + 2L_{\sigma,con}}{4F_{Rf}R_{DC}} \quad \text{and} \quad \frac{1}{\omega_1} \equiv \frac{A_1 - L_{\sigma,con}}{4F_{Rf}R_{DC}}, \tag{12}$$

respectively, Eq. (10) can be rewritten as

$$i_2 = \frac{i}{2} - \frac{1}{\omega_0}\frac{di_2}{dt} - \frac{1}{\omega_1}\frac{di}{dt}. \tag{13}$$

Except for the added di/dt term and for the different definition of ω_0, this is the same expression for i_2 as for helical litz-wire windings [9]. Introducing harmonic expressions such as $i = \hat{i}e^{j\omega t}$, and referring all currents to their DC limits by defining

$$i_k^* \equiv \frac{2i_k}{i}, \tag{14}$$

yields, after some algebra,

$$\hat{i}_2^* = \frac{1 - j2\dfrac{\omega}{\omega_1}}{1 + j\dfrac{\omega}{\omega_0}}. \tag{15}$$

Defining $\omega^* \equiv \omega/\omega_0$ and

$$\kappa \equiv \frac{\omega_0}{\omega_1} = \frac{A_1}{A_0} = \frac{\left[\sum_{n=1}^{N_t}(n-1)\left(r_{2,n}^2 - r_{1,n}^2\right)\right] - L_{\sigma,\mathrm{con}}a/(\pi\mu_0)}{\left[\sum_{n=1}^{N_t}\left(r_{2,n}^2 - r_{1,n}^2\right)\right] + 2L_{\sigma,\mathrm{con}}a/(\pi\mu_0)}, \tag{16}$$

Eq. (15) can be rewritten as

$$\hat{i}_2^* = \frac{1 - j2\kappa\omega^*}{1 + j\omega^*}. \tag{17}$$

Note that, formally, for $\kappa = 0$, the expression is the same as for the helical litz-wire windings. Finally, one gets for the first foil:

$$\hat{i}_1^* = \frac{2\hat{i}_1}{\hat{i}} = 2 - \hat{i}_2^* = \frac{1 + j2(1+\kappa)\omega^*}{1 + j\omega^*}. \tag{18}$$

In the harmonic case, the time-averaged losses in foil 2 are $P_2 = R_f|\hat{i}_2|^2/2$. Noting that $|\hat{i}_2| = |\hat{i}_2^*||\hat{i}|/2$ and defining the effective total current $I \equiv |\hat{i}|/\sqrt{2}$, one obtains

$$P_2 = \frac{F_{Rf}R_{DC}}{2}|\hat{i}_2^*|^2 I^2 = P_{f,DC}F_{Rf}|\hat{i}_2^*|^2. \tag{19}$$

The losses of foil 1 are likewise calculated as

$$P_1 = \frac{F_{Rf}R_{DC}}{2}|\hat{i}_1^*|^2 I^2, \tag{20}$$

where $\hat{i}_1^*$ is determined from Eq. (18). The quantity of ultimate interest is the ratio of AC losses $P = P_1 + P_2$ to DC losses $P_{DC} = R_{DC}I^2$ of the winding. This ratio is equal to the resistance ratio F_R on the winding level. One obtains

$$F_R \equiv \frac{P}{P_{DC}} = F_{Rf}\frac{|\hat{i}_1^*|^2 + |\hat{i}_2^*|^2}{2} \equiv F_{Rf}F_{Rc}', \tag{21}$$

where $F_{Rc}' = (|\hat{i}_1^*|^2 + |\hat{i}_2^*|^2)/2$. Because of the mentioned orthogonality of the different AC effects, F_R should be written as the sum of a loss ratio on foil level and of that due to circulating currents:

$$F_R = 1 + (F_{Rf} - 1) + (F_{Rc} - 1). \tag{22}$$

Comparing Eqs. (21) and (22), we find for the loss ratio due to circulating currents:

$$F_{Rc} = 1 + F_{Rf}(F_{Rc}' - 1). \tag{23}$$

Case of a single turn

In case of only a single turn ($N_t = 1$) and $L_{\sigma,\mathrm{con}} = 0$, we have $\kappa = 0$, such that from Eq. (17),

$$\hat{i}_2^* = \frac{1}{1 + j\omega^*}, \tag{24}$$

which is the expression obtained for helical litz-wire windings [9]. Only the definition of ω_0 and hence ω^* is a bit different, since the turns are created in a different way in the litz-wire winding.

Limit of infinite frequency

It is interesting to study the limit of infinite frequency, i.e. $\omega^* \to \infty$. According to Eq. (17), $\hat{\imath}_2^* \to -2\kappa$. As an approximation, in case of $L_{\sigma,\mathrm{con}} = 0$,

$$\kappa \cong \frac{(r_2^2 - r_1^2)\sum_{n=1}^{N_t}(n-1)}{(r_2^2 - r_1^2)\sum_{n=1}^{N_t}1} = \frac{N_t - 1}{2}, \tag{25}$$

such that $\hat{\imath}_2^* \to 1 - N_t$ and $\hat{\imath}_1^* \cong 2 - \hat{\imath}_2^* = 1 + N_t$. For the example of $N_t = 18$, one gets $\hat{\imath}_1^* \cong 19$ and $\hat{\imath}_2^* \cong -17$, yielding

$$F'_{Rc} \cong \frac{1}{2}[(1 + N_t)^2 + (1 - N_t)^2] = 1 + N_t^2 \cong 325. \tag{26}$$

This shows the potential of high extra losses due to circulating currents in case of many turns. A graphical representation is provided in Fig. 2 for $N_t = 1$, 3, and 18, showing that the currents redistribute to minimize the mean H-field magnitude – and hence the magnetic energy – in the radial gap between the parallel connected foils.

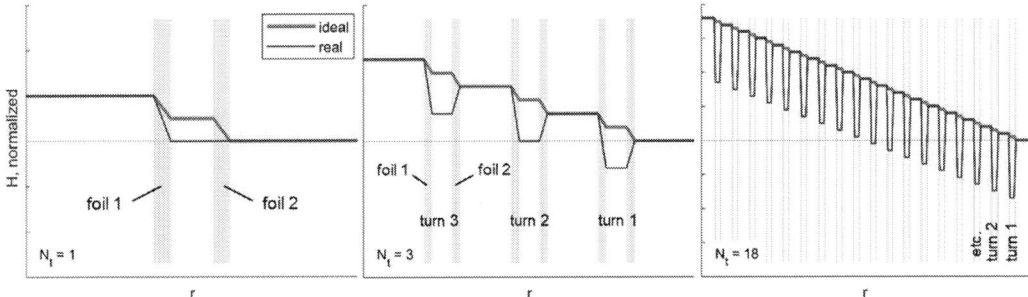

Fig. 2: Meridian section of circular winding, showing normalized axial component of H-field over radius r in winding with two parallel connected foils. For ideal current distribution ($\hat{\imath}_1^* = \hat{\imath}_2^* = 1$), and for real current distribution according to analytical model in case of $\omega^* \to \infty$, for $L_{\sigma,\mathrm{con}} = 0$.

Finite-element model

To provide yet another comparison with the experimental results presented below, a 2-d rotationally symmetric finite-element model was created of the test transformer. COMSOL Multiphysics 5.4 was used for this purpose; the geometry is shown in Fig. 3.

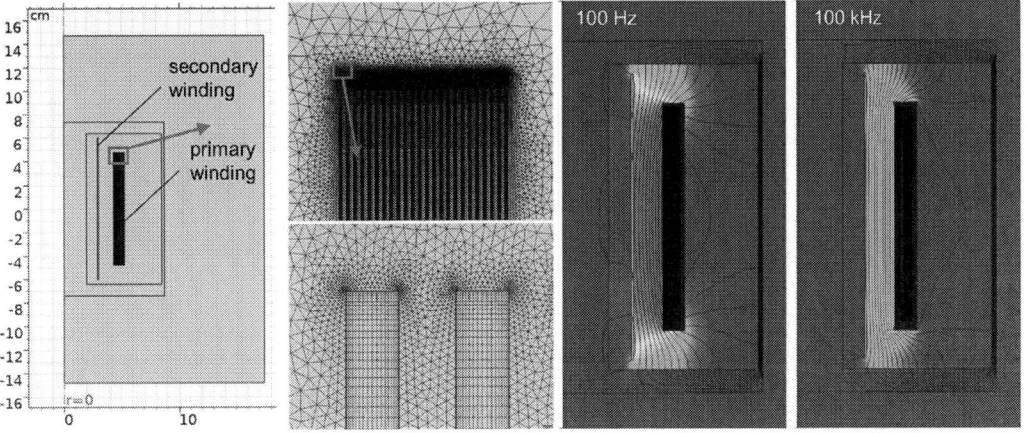

Fig. 3: 2-d finite-element model of test transformer. Left: Geometry. Middle: Mesh. Right: Magnitude and field lines of H-field.

Like in the real setup, current was imposed on the primary winding, while the secondary winding was shorted. Simulations were made up to 280 kHz, and – as shown in Fig. 3 – a fine mesh was used to ensure mesh-independent resolution of all 2-d AC effects up to this frequency.

Experimental investigation

Configurations studied

The circulating currents between parallel connected foils were experimentally studied with a transformer of reduced scale as shown in Fig. 4. Two 3-d printed plastic bobbins were mounted on a core leg with the inner bobbin carrying a secondary winding of 65 turns of litz wire from 175 copper strands of 0.1 mm diameter. The purpose of this winding was only to short the transformer and to realistically shape the H-field in the radial gap between the two windings. The outer bobbin received the foil winding under test, which was the primary winding of the transformer. In all configurations, two foils were parallel connected as shown in Fig. 1. The foils were from aluminum of thickness $h = 0.2$ mm and of width $a \cong$ 96 mm, which was somewhat less than the net height of the bobbin of 120 mm. Nomex paper of thickness 0.13 mm was used both as turn-to-turn insulation (gap t in Fig. 1), and to insulate the two parallel connected foils from each other (gap g in Fig. 1). Different setups were investigated with either one bobbin pair only, i.e. with one empty core leg as shown in Fig. 4; or with two bobbin pairs, one around each leg. In case of two bobbin pairs, each of the outer bobbins carried a primary-winding portion, and the two winding portions were series connected to form the primary winding. Likewise, the two inner bobbins carried two series-connected secondary winding portions.

The different configurations studied are summarized in Table I. Configurations 1 and 2 featured only a single bobbin pair and foil-winding portion, and differed in the connection type. Configurations 3 and 4 had two bobbin pairs and foil-winding portions. This allowed foil transposition between the winding portions, i.e. switching of the inner and outer foil to cancel the fluxes that induce the circulating current between the foils. Transposition was possible at the point where the foils were connected as shown in Fig. 5 on the right. For reasons of space and accessibility, the connection between the two winding portions was not exactly at the center, but moved a bit to the left winding portion. Hence, the flux cancellation was not perfect, but good enough as will be seen.

From measurements of the total radial thickness of the foil windings, it was concluded that the radial gap between adjacent foils layers was not equal to the Nomex paper thickness of 0.13 mm, but was larger by a factor of 1.67 due to air inclusions. In all the simulation models, the effective thickness of 1.67×0.13 mm was considered.

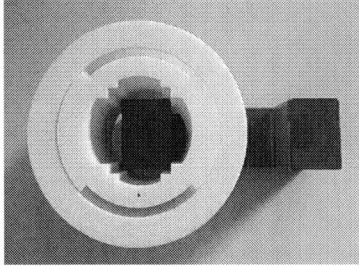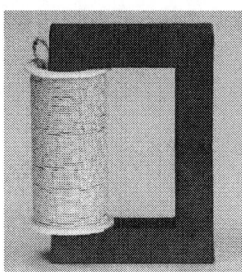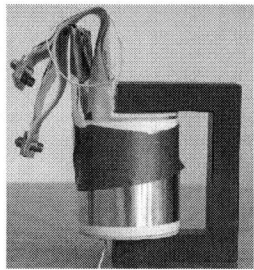

Fig. 4: Transformer for foil-winding tests. Left: View on coaxial bobbins. Middle: Core with inner bobbin, carrying secondary winding. Complete transformer with primary and secondary winding.

Table I: Winding configurations studied

No.	Description
1	One bobbin pair, $N_t = 12$. Foils connected via cables (see Fig. 5, left)
2	Like 1, but connection cables omitted and foils folded to fit through current sensors. This connection type was kept for the subsequent configurations.
3	Two bobbin pairs, foils untransposed (see Fig. 5, top right), $N_t = 12$.
4	Like 3, but foils transposed (see Fig. 5, bottom right)

Measurement setup

The measurements were conducted with small signals. The setup is shown in Fig. 6 and comprised a signal generator that created a sinusoidal output fed into a linear amplifier that then imposed the total current onto the primary side of the transformer. The foil currents were measured individually and displayed on an oscilloscope. The current amplitudes and phases were recorded and post-processed in MATLAB. The frequency range in which we could measure was limited to 40 kHz by the linear amplifier.

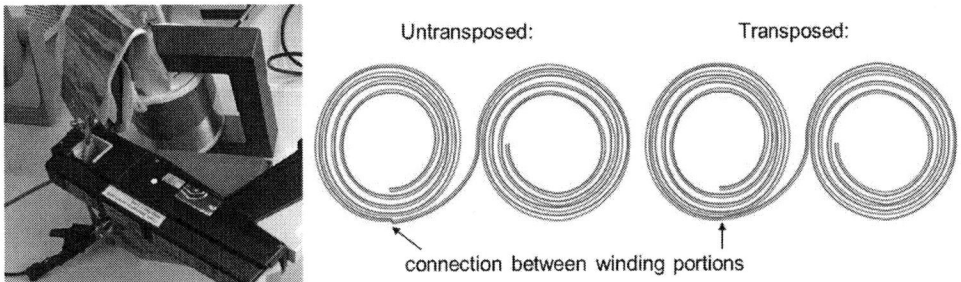

Fig. 5: Left: Foils connected via cables. Right: Two foil-winding portions without and with transposition.

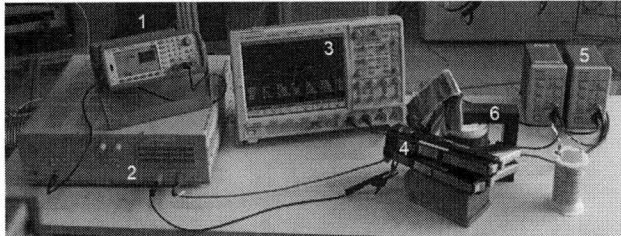

1 function/waveform generator (Agilent 33522A)
2 linear amplifier (Toellner 7621)
3 digital oscilloscope (Agilent InfiniiVision DSO7104B)
4 current probes (Tektronix TCP 303)
5 current-probe amplifiers (Tektronix TCP A300)
6 transformer

Fig. 6: Measurement setup.

Results and discussion

Single winding portion

In Fig. 7, the results of simulations and measurements are presented in terms of magnitude and phase angle of the currents in the two parallel connected foils. The analytical model was first evaluated without account of the inductance $L_{\sigma,\mathrm{con}}$ of the foil connections, see curves "analyt. $L_{\sigma,\mathrm{con}} = 0$". For a single winding portion, without transposition between the foils, there are huge circulating currents that exceed by far the net current. This is as anticipated from the above discussion and would be totally unacceptable in a real design. The 2-d FEM simulations, that also do not include $L_{\sigma,\mathrm{con}}$, are in reasonable agreement with the analytical model. The small differences are likely caused by differences in F_{Rf} between the foils that are not considered in the analytical model.

Interestingly, the first experiment conducted with a single winding portion – see red lines of "config. 1" in Fig. 7 – showed much less circulating current between the foils. The extension of the analytical model revealed that this is due to the balancing effect of the cable connections, see Fig. 1 on the right and Fig. 5 on the left. To accentuate the circulating currents, $L_{\sigma,\mathrm{con}}$ was reduced in a subsequent experiment by omitting the cables and fitting the foils directly through the current sensors by folding them. Indeed, this increased the circulating current as shown by the blue lines of "config. 2" in Fig. 7.

While the geometry of the foil winding is well defined as described above, the geometry of the foil connections is complex and not well defined. Therefore, to incorporate the connection inductance in the analytical model, a rough estimate is necessary for the inductances L_{con} and M_{con}. We used the model of two pieces of straight wire as sketched in Fig. 1 on the right. The AC self-inductance of a piece of straight wire of radius r_{con} and length ℓ_{con} in air is

$$L_{\mathrm{con}} \cong \frac{\mu_0 \ell_{\mathrm{con}}}{2\pi}\left[\ln\left(\frac{2\ell_{\mathrm{con}}}{r_{\mathrm{con}}}\right) - 0.75\right], \qquad \text{for } \ell_{\mathrm{con}} \gg r_{\mathrm{con}} \tag{27}$$

and the mutual inductance of two parallel pieces of straight wire is

$$M_{\mathrm{con}} \cong \frac{\mu_0 \ell_{\mathrm{con}}}{2\pi}\left[\ln\left(\frac{2\ell_{\mathrm{con}}}{w_{\mathrm{con}}}\right) - 1 + \frac{w_{\mathrm{con}}}{\ell_{\mathrm{con}}}\right], \qquad \text{for } \ell_{\mathrm{con}} \gg w_{\mathrm{con}}. \tag{28}$$

The parameter values used were $\ell_{\mathrm{con}} = 10$ cm, $w_{\mathrm{con}} = 5$ cm, and $r_{\mathrm{con}} = 1$ mm, yielding $L_{\sigma,\mathrm{con}} = L_{\mathrm{con}} - M_{\mathrm{con}} \cong 73$ nH. It turned out that with these values that – while being rough estimates – were not further tweaked, the analytical model was in surprising agreement with the measurement, see the dashed lines "analyt." in Fig. 7.

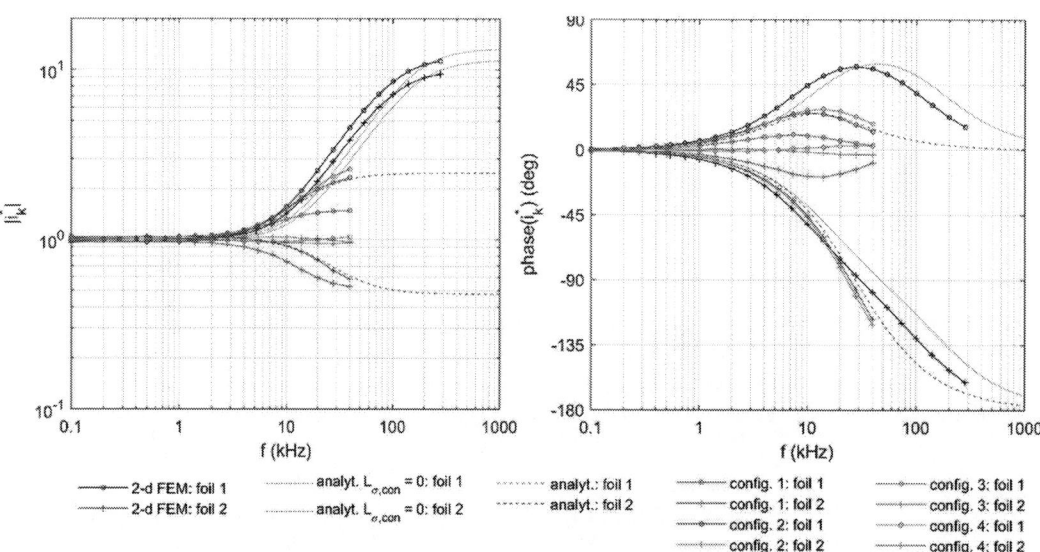

Fig. 7: Simulated (FEM and analytical model) and measured foil currents. Magnitude (left) and phase angle (right) as a function of frequency f.

Two series connected winding portions

The influence of $L_{\sigma,\mathrm{con}}$ is good news regarding the design of foil-based MFTs, since a certain connection impedance will always be present in a realistic design. However, the effect is not strong enough to render further current-balancing means unnecessary. For example, at 20 kHz, $|\hat{\imath}_1^*| \cong 2$, meaning that the AC losses will be more than doubled just due to circulating current between the foils. Therefore, further tests were made with the foil winding composed from two series connected winding portions, one fitted around each core leg, offering the possibility to transpose the foils between the winding portions. As a benchmark, in configuration 3, no transposition was used yet, which resulted in circulating currents similar as in configuration 2, see the blue lines of "config. 3" in Fig. 7.

The transposition, as illustrated in Fig. 5 on the bottom right, was implemented in configuration 4. Even though the point of transposition was not perfectly in the middle, but shifted to one winding for reasons of space and access, the transposition worked almost perfectly, suppressing the circulating current to a level at which it does not matter any more. This is shown by the beige lines of "config. 4" in Fig. 7.

Another way of displaying the foil currents is to draw the amplitudes $\hat{\imath}_1^*$ and $\hat{\imath}_2^*$ in the complex plane as shown in Fig. 8. For each foil, $\hat{\imath}_1^*$ and $\hat{\imath}_2^*$ form a semicircle with center on the real axis. The origin, $\hat{\imath}_k^*$ for $\omega^* = 0$, is at 1, and the radius grows with N_{t} and decreases with increasing $L_{\sigma,\mathrm{con}}$. The point symmetry about 1 of the two semicircles is obvious from Kirchhoff's current law, $\hat{\imath}_1^* + \hat{\imath}_2^* = 2$. For the special case of $N_{\mathrm{t}} = 1$ and $L_{\sigma,\mathrm{con}} = 0$, the radius is 0.5 and $\hat{\imath}_2^*$ approaches 0 as $\omega^* \to \infty$. For $N_{\mathrm{t}} > 1$ and $L_{\sigma,\mathrm{con}} = 0$, the current in foil 2 gets reversed as $\omega^* \to \infty$. From the plots in the complex plane, it becomes particularly apparent how $L_{\sigma,\mathrm{con}}$, and to an even greater extent the transposition, balances the foil currents.

While the current balancing works extraordinarily well in configuration 4, it needs to be mentioned that a near-perfect balancing is not necessary in practice. If the normalized currents are decomposed into a balanced part and a deviation $\Delta \hat{\imath}^*$,

$$\hat{\imath}_1^* = 1 + \Delta \hat{\imath}^*, \tag{29}$$

$$\hat{\imath}_2^* = 1 - \Delta \hat{\imath}^*, \tag{30}$$

then, F_{Rc}' only depends on the square of $|\Delta \hat{\imath}^*|$:

$$F_{Rc}' = 1 + |\Delta \hat{\imath}^*|^2. \tag{31}$$

Suppression of the circulating current to a value of $|\Delta \hat{\imath}^*| \leq 0.2$ would, for example, limit the increase of F_{Rc}' to 4 %, which would probably be acceptable.

Albeit not done here, the results found can be generalized to windings with more than two foils connected in parallel.

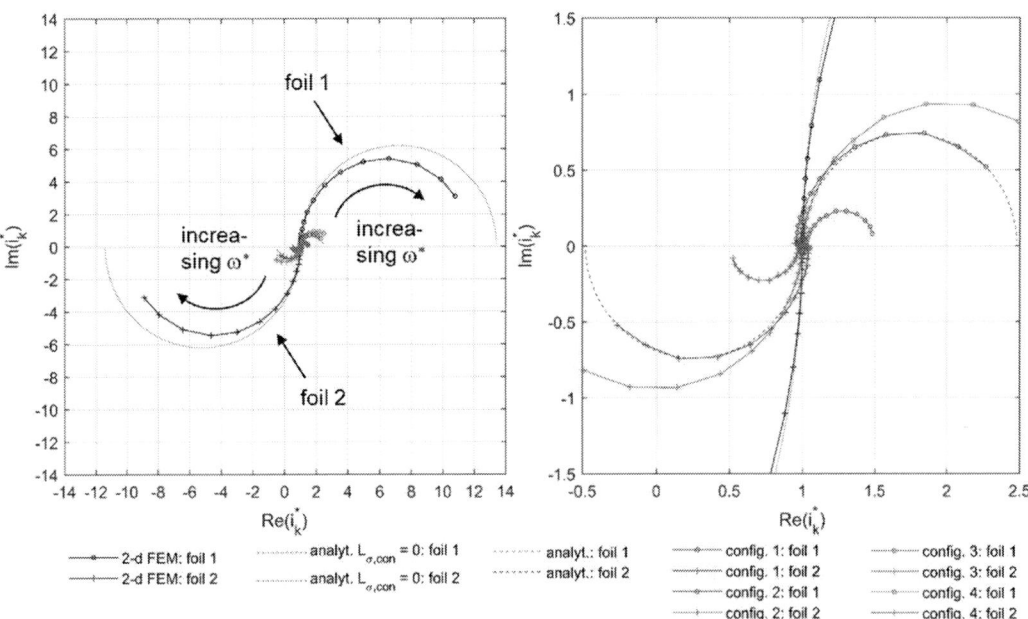

Fig. 8: Normalized complex amplitudes of simulated and measured foil currents. Zoomed view on the right.

Conclusion

For future SST applications, there is a need of MFTs in the power range of several hundred kW, operating at frequencies of 10 to several tens of kHz, and rated for currents of several hundred Ampères. The primary options for the winding design of such MFTs are litz wires and foils. A key in the design of foil windings is the ability to parallel connect foils to provide sufficient conductor cross-section. In this case, careful winding design is required to keep circulating currents among parallel connected foils under control.

Such currents were investigated by means of an analytical model, a 2-d finite-element model, and measurements on a reduced-scale transformer for the case of two parallel connected foils. Simulations and measurements yield a consistent picture and show the potential of high extra losses in foil windings with unmitigated circulating currents. In particular, spiral windings of many turns may incur a circulating current that exceeds the useful net current by far if the inductance of the foil connections is small. Practically, it can be expected that the inductance of the foil connections leads to a noticeable reduction of the circulating current. This reduction, however, is not sufficient to bring the AC losses down to acceptable levels, such that it is recommended to transpose the foils between series-connected winding portions. Transposition largely cancels out the axial magnetic flux in the radial gaps between the parallel connected foils. The effectiveness of transposition in balancing the foil currents at all frequencies of interest, i.e. up to 40 kHz in the experiment, was shown both theoretically and experimentally for aluminum foils of 0.2 mm thickness. With proper transposition, the extra AC losses due to circulating currents between foils can be reduced to practically negligible levels.

References

[1] Essakiappan S., Krishnamoorthy H. S., Enjeti P., Balog R. S., and Ahmed S.: Multilevel medium-frequency link inverter for utility scale photovoltaic integration, IEEE Trans. Power Electronics, Vol. 30 no 7, July 2015

[2] Lakshmanan P., Liang J., and Jenkins N.: Assessment of collection systems for HVDC connected offshore wind farms, Electric Power Systems Research, Vol. 129, 2015

[3] Smith P.: Edison vs Tesla: A rematch in the telecom data center, INTELEC 2015, 18-22 October, Namba, Osaka, Japan

[4] Gradinger T. B., Drofenik U., and Alvarez S.: Novel insulation concept for an MV dry-cast medium-frequency transformer, EPE 2017, 11-14 September, Warsaw, Poland

[5] Ortiz G., Biela J., and Kolar J. W.: Optimized design of medium frequency transformers with high isolation requirements, IECON 2010, 7-10 November, Glendale, USA

[6] Sullivan C. R. and Zhang R. Y.: Simplified Design Method for Litz Wire, APEC 2014, 16-20 March, Fort Worth, TX, USA

[7] Ortiz G., Leibl M., Kolar J. W., and Apeldoorn O.: Medium frequency transformers for solid-state-transformer applications – design and experimental verification, PEDS 2013, 22-25 April, Kitakyushu, Japan

[8] Guillod T., Huber J., Krismer F., and Kolar J. W.: Litz wire losses: effects of twisting imperfections, IEEE COMPEL 2017, 9-12 July, Stanford, CA, USA

[9] Gradinger T. B. and Drofenik U.: Managing high currents in litz-wire-based medium-frequency transformers, EPE 2018, 17-21 September, Riga, Latvia

[10] Dowell P.: Effects of eddy currents in transformer windings, Proc. IEE, Vol. 113, pp. 1387-1394, August 1966

[11] Pereira A., Lefebvre B., Sixdenier, F., Raulet, M. A., and Burais, N.: Comparison between numerical and analytical methods of AC resistance evaluation for medium frequency transformers: validation on a prototype, EPEC 2015, London, ON, Canada

[12] Kurita N., Hatakeyama T., Kimura M., Nakamura K., and Ichinokura O.: 500 kVA Medium-frequency core-type amorphous transformers with alternately wound sheet winding for offshore DC grid, IEEJ J. Industry Appl. Vol. 8 no 5, pp. 756-766

[13] Lotfi A. W. and Lee F. C.: Two dimensional field solutions for high frequency transformer windings, PESC 1993, 20-24 June, Seattle, WA, USA

[14] Lotfi A. W. and Lee F. C.: Two-dimensional skin effect in power foils for high-frequency applications, IEEE Trans. Magnetics, Vol. 31 no 2, March 1995

[15] Dimitrakakis G. S. and Tatakis E. C.: High-frequency copper losses in magnetic components with layered windings, IEEE Trans. Magnetics, Vol. 45 no 8, August 2009

[16] Robert F., Mathys P., Velaerts B., and Schauwers J.-P.: Two-dimensional analysis of the edge effect field and losses in high-frequency transformer foils, IEEE Trans. Magnetics, Vol. 41 no 8, August 2005

[17] Bahmani M. A., Thiringer T., and Ortega H.: An accurate pseudoempirical model of winding loss calculation in HF foil and round conductors in switchmode magnetics, IEEE Trans. Power Electronics, Vol. 29 no 8, August 2014

A High-Efficiency Wireless Power Transfer System for Unmanned Aerial Vehicle Considering Carbon Fiber Body

Kai Song[1], Peng Zhang[1], Zhengxin Chen[2], Guang Yang[1], Jinhai Jiang[1]*, Chunbo Zhu[1]

1.School of Electrical Engineering & Automation, Harbin Institute of Technology
150001, Harbin, China
2. Research Center for Science and technology Innovation, 100012, Beijing, China
Tel.: +86 (451) 8641 3621.
Fax: +86(451) 8641 3621-809.
E-Mail: kaisong@hit.edu.cn, jiangjinhai@hit.edu.cn

Acknowledgements

This work was supported by National Natural Science Foundation of China under Grant No. 51977043.

Keywords

Wireless power transfer, Eddy current loss, Carbon fiber, Unmanned aerial vehicle, Finite element analysis simulation.

Abstract

The application of wireless power transfer in the unmanned aerial vehicles (UAV) is becoming more and more extensive. In order to meet the requirements of miniaturization and lightweight, the fuselage of UAV often uses carbon fiber materials and the influence on the magnetic field distribution and energy transmission cannot be ignored. In this paper, the eddy current loss of the carbon fiber fuselage is simulated and analyzed, and the effect of the aluminum shielding method is studied. The study found that by optimizing the parameters such as the size, thickness and position of the shielded aluminum ring, the total system loss will be better improved, and the weight is light, which will not affect flight. Finally, a 500 W prototype is established to validate the method.

Introduction

Over past decades, the unmanned aerial vehicle (UAV) has been widely applied in the industry and the agriculture such as powerline inspection and wildlife protection [1], [2]. To charge the UAV more easily, wireless power transfer (WPT) technology is introduced since it does not need charging cables and is convenient to operate [3]. For wireless UAV charging, so far, some methods have been proposed. [4] studied a lightweight and integrated wireless charging system for a 10 W micro-drone. [5] concentrated on the system weight reduction and proposed a 100 W WPT system for the UAV. It is worth noting that the system considers the influence of the UAV position and designs a position alignment device. [6] investigated the multi-coil structure and corresponding wireless charging modules for a 3D Printed UAV. To improve the charging efficiency and lower the electromagnetic interference to the UAV, [7] proposed a novel magnetic coupler and proved that the system could realize a DC-DC efficiency of 90% at 624W output power.

However, in most of the presented researches, WPT systems for UAVs are equivalent to ideal coils, neglecting the effect of the UAV material. Since the plastic UAV models are used in most of the researches, and the air gaps are always small, the material of the UAV has little impact on the system. However, for a wireless charging system of an industrial UAV, especially a military UAV, the body made of carbon fiber can provide an alternate path for magnetic flux due to its relatively high permeability and produce eddy current and the reverse magnetic field due to its relatively high conductivity [8]. Thus, for loosely coupled system with large air gap, the loss produced by the UAV itself must be considered to achieve high efficiency.

In this paper, the characteristics of a WPT system for a UAV made of carbon fiber are investigated in detail. The loss caused by the UAV is simulated and validated by the experiments. Then, aluminum (Al) is applied to prevent the effects of the carbon fiber. The shape, dimensions, thickness and the location of the Al ring are optimized for higher efficiency. The experimental results show that the transmission efficiency can be improved significantly when the optimized Al ring is added compared with the system without the Al ring.

System Design

A typical WPT system is shown in Fig.1. As mentioned above, all parts of the system will produce power loss, and the magnetic coupler is the focus of this paper.

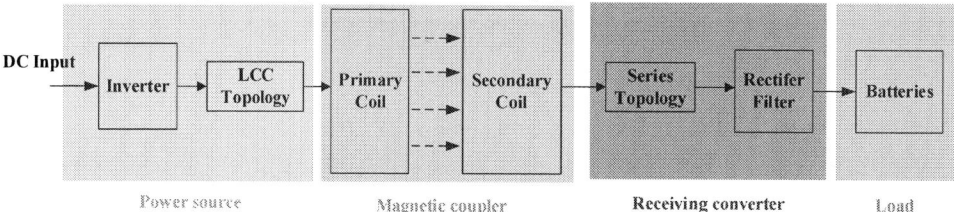

Fig. 1: Block diagram of UAV wireless power transfer system

A. Design of High Efficiency Magnetic Coupler

This paper needs to meet these requirements: i) The air gap is 300 mm. ii) The output power is not less than 500 W. iii) The transmission efficiency is higher than 90%.

Based on the lightweight principle, this paper selects the circular coil with less wire consumption and better electromagnetic compatibility (EMC). Considering the volume restraint, the diameter of the secondary coil is 400 mm, and the number of turns is set as 4. The relationship between the coupling coefficient k and the diameter of the primary coil is obtained by simulation, as shown in Fig.2. When the outer diameter is about 700 mm, k reaches the maximum value.

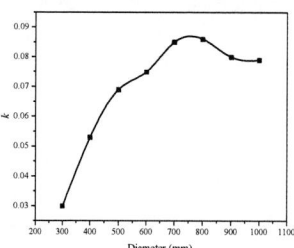

Fig. 2: The trend of k against diameter

Then, the relationship between k, mutual inductance M and the coil width d of the primary coil is simulated and analyzed, as shown in Fig.3 and Fig.4, respectively. To obtain the maximum transmission efficiency, d is selected as 120 mm.

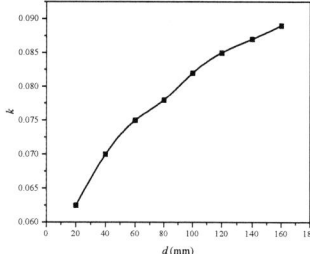

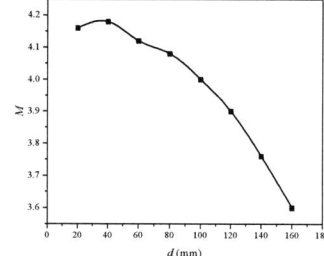

Fig. 3: The trend of k against d Fig. 4: The trend of M against d

Thus, the magnetic coupler is designed according to the parameters listed in Table I. The transmitter coil is wound with two Litz wires in parallel, single layer, the purpose is to reduce the AC internal resistance of the coil.

Table I: Parameter of the magnetic coupler

	Transmitting Coil	Receiving Coil
Coil structure	Circular	Circular
Outer diameter/mm	700 mm	400 mm
Turn Ratio	9 turns; double winding	4 turns; single winding
Self-inductance/µH	135	14
M/µH	3.9	
k	0.09	

B. Eddy Current Loss Analysis

Considering the practical UAV made of carbon fiber which has high carbon content, high strength and corrosion resistance, inevitable eddy current will be generated and cause eddy current loss.

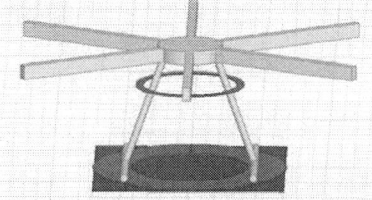

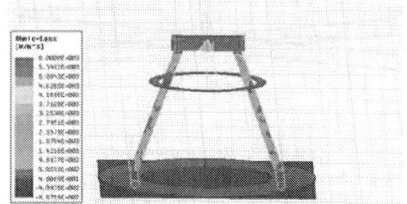

Fig. 5: Magnetic coupler with UAV   Fig. 6: Eddy current loss distribution

Fig.6 shows the distribution of eddy current loss when the UAV is added. It can be seen that eddy current loss mainly exists in the body of the UAV above the receiving coil. Due to the large eddy current loss of the body, the transmission efficiency reduces significantly, so it is crucial to consider the impact of the body.

C. Efficiency Analysis

LCC-S topology is employed to maintain the output voltage constant, as shown in Fig.7. The operating frequency is 200 kHz.

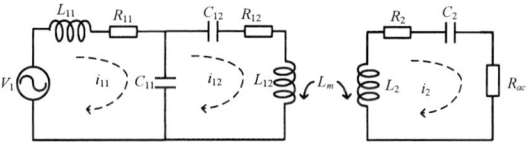

Fig. 7: LCC-S topology

Assuming that the system working in tuning condition, the primary coil current is obtained as (1)

$$i_{12} = -j\omega_0 C_{11} v_1 \tag{1}$$

The secondary side coil current can be deduced as (2)

$$i_2 = \frac{j\omega_0 L_m i_{12}}{R_2 + R_{ac}} = \frac{\omega_0^2 C_{11} L_m}{R_2 + R_{ac}} v_1 = \frac{L_m}{L_{11}(R_2 + R_{ac})} v_1 \tag{2}$$

The output power is expressed as (3)

$$P_L = |i_2|^2 R_{ac} = \frac{(\omega_0^2 C_{11} L_m)^2 R_{ac}}{(R_2 + R_{ac})^2} V_1^2 \tag{3}$$

The power consumed by the coil resistance can be obtained as (4)

$$P_C = i_{12}^2 R_{12} + i_2^2 R_2 \tag{4}$$

Therefore, the transfer efficiency of the magnetic coupler is given by (5)

$$\eta = \frac{P_L}{P_L + P_C + P_{UAV}} \tag{5}$$

where PUAV is the eddy current loss of the UAV.

The transmission efficiency with and without UAV can be shown in Fig.8. Again, it is clear that the body of the UAV makes the transmission efficiency decline significantly.

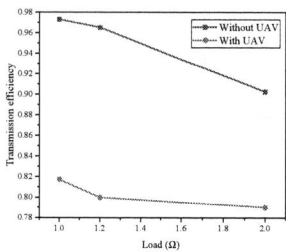

Fig. 8: Transmission efficiency with and without UAV

Proposed Method to Reduce Eddy Current Loss

To reduce the influence of eddy current loss of the UAV body, aluminum with high conductivity is applied. The different design parameters including the shape, dimensions, thickness and the location are analyzed.

A. Using Al Plate

As shown in Fig.9, the rectangular Al plate is located above the receiving coil.

Fig. 9: Al plate above the receiving coil

When the Al plate covered the coil, the shielding effect is better and the eddy current loss is reduced. But the mutual inductance of the coupling mechanism is also reduced to a large extent, which makes the output efficiency of the system significantly reduced. At the same time, too large Al plate area may affect the normal flight of UAV. Therefore, an improved shielding Al ring is proposed.

B. Using Al Ring

Based on the above conclusions, the width of the Al ring, the vertical distance between Al ring and coils, and the thickness of the Al ring are analyzed. The efficiency of Al ring width from 0 to 20 mm, and the distance from 0 to 10 mm are simulated.

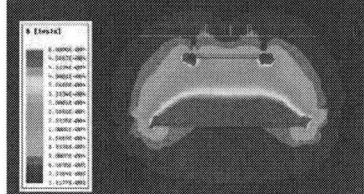
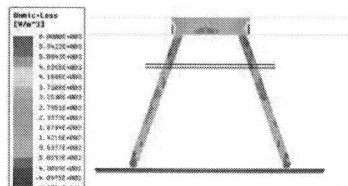

Fig. 10: Magnetic field distribution with Al ring   Fig. 11: Eddy current loss with Al ring

Through the simulation of the eddy current loss and efficiency for the system after adding the Al ring, it is found that the selection of the size and position of the Al ring will have a more obvious impact on the system performance. Therefore, it is necessary to select the appropriate size of Al ring to improve the efficiency of the system. Fig.12 shows the influence of the width and distance of the Al ring on the efficiency.

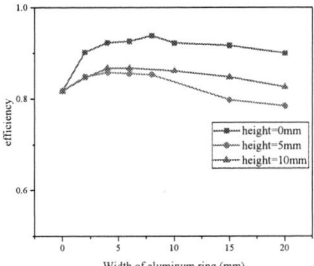 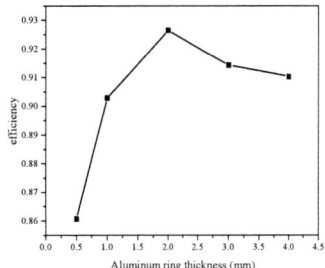

Fig. 12: Influence of the width and height of Al ring   Fig. 13: Influence of the thickness of Al ring

Fig.12 shows that the transmission efficiency is higher when the Al ring and coils are close to each other. Fig.13 shows the system transmission efficiency increases first and then decreases with the increase of Al ring thickness. When the Al ring thickness is 2mm, the system transmission efficiency is the largest.

In summary, when the Al ring is closed to the secondary coil, the width is 8 mm, and the thickness is 2 mm, the designed system has the maximum transmission efficiency. The power proportion of each part on the system is shown in Fig.14. The eddy current loss of Al ring is 4.73W, accounting for 0.89%. The eddy current loss of the UAV is 9.5W, accounting for 1.8%. The loss of the coils is 18W, accounting for 3.44%.

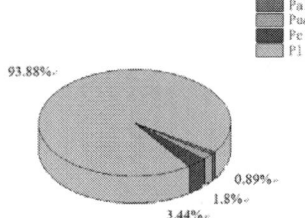

Fig. 14: Power loss distribution

Experimental Validation

A. Experimental Setup

In order to validate the proposed method, an experimental system as shown in Fig.15 is built according to the parameters listed in Table I.

Fig. 15: Experimental setup for the UAV

B. Experimental Verification

During the experiment, the UAV is stopped on the platform by operating the remote controller, then the inverter power device is started. The input, output current and transmission power are recorded by the oscilloscope and power analyzer. The results of the UAV's wireless power transfer system with the Al ring are shown in Fig.16.

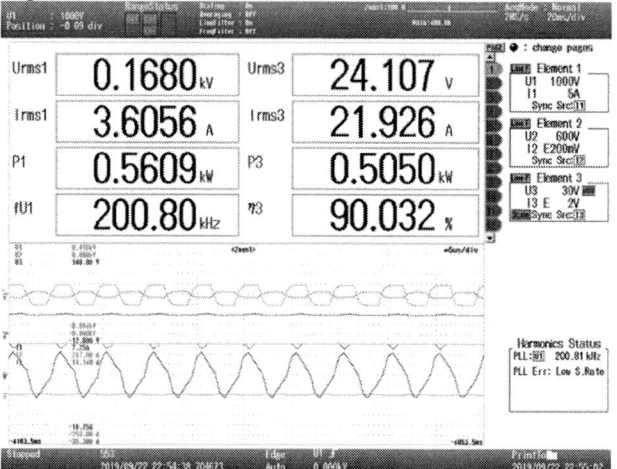

Fig. 16: System output with the UAV and Al ring

In Figure 16, the input voltage at the transmitter $Urms1$ is 168 V and the input current $Irms1$ is 3.6 A. The input power $P1$ is 560.9 W. The output voltage at the receiving end $Urms3$ is 24.10 V, the output current $Irms3$ is 21.92 A. The output power $P3$ is 505 W. The transmission efficiency $\eta3$ is 90.03%. This experiment verifies the correctness of the above simulation analysis. With the designed Al ring, the transmission efficiency is 8% higher than that without the Al ring.

Conclusion

In this paper, considering the influence of carbon fiber body, a high efficiency WPT system for the UAV is designed. Compared with conventional prototypes without UAV model or with plastic models, the practical UAV made of carbon fiber will inevitably bring much power loss and reduce the transmission efficiency. Thus, Al plate is adopted to shield and prevent the eddy current produced in the body. Aiming at high efficiency, it is found that the Al ring shows better performance compared with the Al plate. Furthermore, the design parameters, such as the shape, dimensions, thickness and the location, are optimized using FEA simulation. Experimental results show consistency with the simulations. With the designed Al ring, the transmission efficiency is over 90% at 500W output, which is 8% higher than that without the Al ring.

References

[1] M. Lu, M. Bagheri, A. P. James and T. Phung.: Wireless Charging Techniques for UAVs: A Review, Reconceptualization, and Extension, in IEEE Access, vol. 6, pp. 29865-29884, 2018.

[2] A. M. Jawad, H. M. Jawad, R. Nordin, S. K. Gharghan, N. F. Abdullah and M. J. Abu-Alshaeer.: Wireless Power Transfer With Magnetic Resonator Coupling and Sleep/Active Strategy for a Drone Charging Station in Smart Agriculture," in IEEE Access, vol. 7, pp. 139839-139851, 2019.

[3] L. Qiong, W. Huayun, W. Wenbin, M. Tianqi and W. Yongyue.: Optimal Design of Magnetic Resonance Wireless Charging Coil for Unmanned Aerial Vehicles1, 2018 3rd International Conference on Smart City and Systems Engineering (ICSCSE), Xiamen, China, 2018, pp. 469-472.

[4] S. Aldhaher, P. D. Mitcheson, J. M. Arteaga, G. Kkelis and D. C. Yates.: Light-weight wireless power transfer for mid-air charging of drones, 2017 11th European Conference on Antennas and Propagation (EUCAP), Paris, 2017, pp. 336-340.

[5] T. Campi, S. Cruciani, M. Feliziani and F. Maradei.: High efficiency and lightweight wireless charging system for drone batteries, 2017 AEIT International Annual Conference, Cagliari, 2017, pp. 1-6.

[6] J. Chen, R. Ghannam, M. Imran and H. Heidari.: Wireless Power Transfer for 3D Printed Unmanned Aerial Vehicle (UAV) Systems, 2018 IEEE Asia Pacific Conference on Postgraduate Research in Microelectronics and Electronics (PrimeAsia), Chengdu, 2018, pp. 72-76.

[7] C. Cai, S. Wu, M. Qin and Z. Yang.: A Novel Magnetic Coupler for Unmanned Aerial Vehicle Wireless Charging Systems, 2018 IEEE International Power Electronics and Application Conference and Exposition (PEAC), Shenzhen, 2018, pp. 1-5.

[8] Q. Zhu, Y. Zhang, C. Liao, Y. Guo, L. Wang and F. Li : Experimental Study on Asymmetric Wireless Power Transfer System for Electric Vehicle Considering Ferrous Chassis," in IEEE Transactions on Transportation Electrification, vol. 3, no. 2, pp. 427-433, June 2017.

Analytical computation of normal and fault-tolerant active short circuit operation of anisotropic synchronous double star machines

Michael Gleissner[1], Johannes Häring[1], Wolfgang Wondrak[2], Mark-M. Bakran[1]

[1]UNIVERSITY OF BAYREUTH
DEPARTMENT OF MECHATRONICS
CENTER OF ENERGY TECHNOLOGY
Universitaetsstrasse 30
Bayreuth, Germany
Phone: +49 (0) 921-55 7804
Email: michael.gleissner@uni-bayreuth.de
URL: http://www.mechatronik.uni-bayreuth.de

[2]MERCEDES-BENZ AG
Hanns-Klemm-Strasse 45
Boeblingen, Germany
Phone: +49 (0) 7031-4389 205
Email: wolfgang.wondrak@daimler.com
URL: www.daimler.com

Acknowledgments

This project has received funding from the Electronic Components and Systems for European Leadership Joint Undertaking under European Union's Horizon 2020 project grant agreement No 737469 (AutoDrive).

Keywords

≪Fault handling strategy≫, ≪Fault tolerance≫, ≪Optimal control≫, ≪Variable speed drive≫, ≪Electrical drive≫, ≪Permanent magnet motor≫, ≪Voltage Source Converter (VSC)≫

Abstract

The analytical computation of anisotropic synchronous machines enables a fast and simple assessment of the speed-torque operation range depending on the machine parameters. A known analytical approach considering stator resistance and mutual inductance for three-phase machines is extended to double star machines for normal and fault-tolerant mode. The decisive parameters for a successful degraded operation with an active short-circuit of one of the two winding systems are evaluated, which reduces the number of disconnection elements for short-on failures of switches and windings.

Introduction

The multiphase structure of double star machines enables fault-tolerance after machine and inverter failures for safety critical applications. They are already installed in electric steering applications [1–3], but can also be applied for traction applications with high safety requirements like autonomous vehicles. Multiphase machines reduce the total power per phase and the torque ripple [4,5]. The inverter as well as the machine are designed with more than three phases and thus enable also fault-tolerance for some machine failures [6,7]. Multi-star machines are favoured over single star machines with more than three phases, because short-on failures affect only a part of the system, when no separation elements are installed. In literature, mostly open mode failures are addressed, because they are much easier to handle and enable the operation with only one phase less [7,8]. Moreover, the cost and space for additional disconnection elements are no problem for laboratory setups. Nevertheless, for real fault-tolerance also short-on failures have to be regarded. Normally, short-on failures of semiconductors as well as machine windings require disconnection elements, e.g. in the AC phase lines (see Fig. 1). Fuses are not applicable in automotive powertrain applications, because they are very bulky for high currents, they are sensitive to vibrations and the wide temperature range of automotive application. Thus, only mechanical or semiconductor based disconnection elements are feasible. But they increase the volume as well as cost and worsen the efficiency.

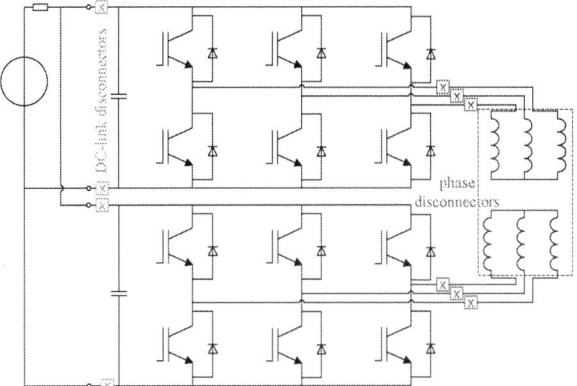

Fig. 1: Parallel inverters for double star six phase machine. Possibly required disconnection elements are marked with 'x'.

Disconnection elements at the machine can be avoided, if an active short-circuit (ASC) is allowed to be applied to the three phases of the failed system while the remaining system is still operating [3, 9, 10]. In three-phase machines, the ASC is applied to avoid induced voltages higher than the DC-link voltage when a permanent magnet machine is running in the field-weakening range. In three-phase machines, the ASC is a fail-safe measure. In multiphase machines, the ASC of one system combined with a still operable system is a fail-operational measure.

The torque behaviour and the short-circuit currents of a double star fractional slot double-layer permanent magnet machine with different winding configurations have been practically evaluated with a prototype in [11]. Experimental tests with a double star induction machine for fault-tolerant operation with different open- and short-circuited distributed winding arrangements are presented in [12]. The operability of a fault-tolerant double star flux switching permanent magnet synchronous machine with very low inductive coupling including experiments with the short-circuit braking torque and currents of a prototype are shown in [9] including the idea of compensating the short-circuit torque of a failed system with the healthy system. A similar approach of a double star permanent magnet synchronous machine with very low coupling is evaluated in [10] including thermal tests under short-circuit conditions. Simulations of concentrated and distributed winding configurations of double star permanent magnet synchronous machines are analysed in [13]. The challenges of magnetic coupling are addressed in [5] for a quadruple star permanent magnet machine.

In [9–13], important aspects and design criteria for fault-tolerant double star machines with ASC have been presented, but the feasible speed-torque operation range after a fault-tolerant reconfiguration compared to normal mode has not been addressed, which is especially relevant for traction applications. The speed-torque operation range depends on the maximum inverter voltage and current as well as machine parameters like the coupling of the winding systems. The effects on the speed-torque operation range in fault-tolerant ASC mode are shown by a measurement of a double star permanent magnet machine for electric steering in [3]. Therefore, this paper extends the known analytical computation approach for anisotropic synchronous three-phase machines, which considers stator resistance as well as mutual inductance of the same winding set [14, 15], to double star machines in normal and fault-tolerant mode. The influence of system parameters can be identified much faster with an analytical computation than with numerical simulations. Thus, the relevant machine parameters for a successful fault-tolerant reconfiguration, such as the coupling of the winding systems and the ratio of maximum inverter current to characteristic machine current, can be easily identified.

Analytical computation of anisotropic synchronous double star machines

Fundamental model

The model of an anisotropic double star permanent magnet synchronous machine in the dq-reference frame according to [16] can be expressed in matrix/vector notation like in [14, 15]

$$
\overbrace{\begin{pmatrix} u_{d1} \\ u_{q1} \\ u_{d2} \\ u_{q2} \end{pmatrix}}^{=:u_k^+} = R_s \overbrace{\begin{pmatrix} i_{d1} \\ i_{q1} \\ i_{d2} \\ i_{q2} \end{pmatrix}}^{=:i_k^+} + \omega \overbrace{\begin{bmatrix} 0 & -1 & 0 & 0 \\ 1 & 0 & 0 & 0 \\ 0 & 0 & 0 & -1 \\ 0 & 0 & 1 & 0 \end{bmatrix}}^{=:J^+} \overbrace{\left(\underbrace{\begin{bmatrix} L_d & L_m & M_d & M_m \\ L_m & L_q & M_m & M_q \\ M_d & M_m & L_d & L_m \\ M_m & M_q & L_m & L_q \end{bmatrix}}_{=:L^+} i_k^+ + \underbrace{\begin{pmatrix} \Psi_{pm} \\ 0 \\ \Psi_{pm} \\ 0 \end{pmatrix}}_{=:\psi_{kpm}^+} \right)}^{=:\psi_{ks}^+\left(i_k^+\right)} + \frac{d}{dt} \psi_{ks}^+ \left(i_k^+ \right) \quad (1)
$$

where a constant inductance matrix $L^+ \in \mathbb{R}^{4x4}$ and stator resistance R_s are assumed. Saturation effects are neglected here ($L^+ \neq f\left(i_k^+\right)$), because a general comparison between different machine configurations is intended. These effects can also be integrated in an analytic calculation, when a current-inductance map for a specific machine is known. $L_{d/q}$ are the self-inductances, L_m is the mutual inductance of the same winding set, $M_{d/q/m}$ are the mutual inductances of the other winding set, the subscript 1 refers to the first winding set and 2 to the second winding set. Both winding sets are assumed to be symmetrical. The electric frequency ω, the voltages as well as currents are time dependent. They are constant in steady state operation of the machine.

The total machine torque is:

$$
\begin{aligned}
T &= \frac{3}{2} n_p \left[i_k^{+\top} J^+ \psi_{ks}^+ \left(i_k^+ \right) \right] \\
&= \frac{3}{2} n_p \left[\Psi_{pm} \left(i_{q1} + i_{q2} \right) + \left(L_d - L_q \right) \left(i_{d1} i_{q1} + i_{d2} i_{q2} \right) + \left(M_d - M_q \right) \left(i_{d1} i_{q2} + i_{d2} i_{q1} \right) \right. \\
&\quad \left. + L_m \left(i_{q1}^2 + i_{q2}^2 - i_{d1}^2 - i_{d2}^2 \right) + 2 M_m \left(i_{q1} i_{q2} - i_{d1} i_{d2} \right) \right]
\end{aligned} \quad (2)
$$

In order to enable an analytical computation including the stator resistance and mutual coupling like in [14, 15], the degree of freedom of four currents i_{d1}, i_{q1}, i_{d2}, i_{q2} has to be reduced to a degree of freedom of two. Therefore, several scenarios are possible, which are given in the following subsections.

Normal operation with two active and equally stressed winding systems

In normal machine operation with two active winding systems, equal dq-currents in both winding systems $\left(i_d = i_{d1} = i_{d2}, i_q = i_{q1} = i_{q2} \right)$ are assumed. With this assumption, (1) is simplified to

$$
\overbrace{\begin{pmatrix} u_d \\ u_q \end{pmatrix}}^{=:u_k} = R_s \overbrace{\begin{pmatrix} i_d \\ i_q \end{pmatrix}}^{=:i_k} + \omega \overbrace{\begin{bmatrix} 0 & -1 \\ 1 & 0 \end{bmatrix}}^{=:J} \overbrace{\left(\underbrace{\begin{bmatrix} L_d + M_d & L_m + M_m \\ L_m + M_m & L_q + M_q \end{bmatrix}}_{=:L_{2A}} i_k + \underbrace{\begin{pmatrix} \Psi_{pm} \\ 0 \end{pmatrix}}_{\psi_{kpm}} \right)}^{=:\psi_{ks2A}(i_k)} + \frac{d}{dt} \psi_{ks2A} \left(i_k \right) \quad (3)
$$

where a constant inductance matrix $L_{2A} \in \mathbb{R}^{2x2}$ is assumed. In normal mode, both inverter and winding systems operate at their maximum voltage and current rating and the maximum total system power is achieved. An over-rating of the sub-systems to enable the same maximum output power in fault-tolerant mode is not considered.

Operation with only one active winding system

In machine operation with an active winding system and a deactivated second system, it is $i_d = i_{d1}, i_q = i_{q1}, i_{d2} = i_{q2} = 0$. With this assumption, (1) is simplified to

$$
\overbrace{\begin{pmatrix} u_d \\ u_q \end{pmatrix}}^{=:u_k} = R_s \overbrace{\begin{pmatrix} i_d \\ i_q \end{pmatrix}}^{=:i_k} + \omega \overbrace{\begin{bmatrix} 0 & -1 \\ 1 & 0 \end{bmatrix}}^{=:J} \overbrace{\left(\underbrace{\begin{bmatrix} L_d & L_m \\ L_m & L_q \end{bmatrix}}_{=:L_{1A}} i_k + \underbrace{\begin{pmatrix} \Psi_{pm} \\ 0 \end{pmatrix}}_{\psi_{kpm}} \right)}^{=:\psi_{ks1A}(i_k)} + \frac{d}{dt} \psi_{ks1A}(i_k) \tag{4}
$$

where a constant inductance matrix $L_{1A} \in \mathbb{R}^{2x2}$ is assumed. With only one active system, there can be restrictions in the field weakening range depending on the machine parameters due to the missing respectively reduced field weakening ability of the deactivated system as will be seen later.

One system active short-circuit mode

The currents of the winding system in single ASC mode depend on the electric frequency ω and the currents i_{d1}, i_{q1} of the active system. The equations can be derived from (1) by setting $u_{d2} = u_{q2} = 0$ [13]:

$$
i_{d2,\text{SingleASC}} = -\frac{\omega^2 L_q \Psi_{pm} + \omega^2 L_q M_d \cdot i_{d1} - \omega M_q R_s i_{q1}}{R_s^2 + \omega^2 L_d L_q}
$$

$$
i_{q2,\text{SingleASC}} = -\frac{\omega \Psi_{pm} R_s + \omega M_d R_s i_{d1} + \omega^2 L_d M_q i_{q1}}{R_s^2 + \omega^2 L_d L_q} \tag{5}
$$

Consequently, for the double star machine with one operational winding system and one system in ASC mode, there are also only two degrees of freedom. By inserting (5) in (1), the system order of four can be reduced to two $\left(i_d = i_{d1}, i_q = i_{q1} \right)$:

$$
\overbrace{\begin{pmatrix} u_d \\ u_q \end{pmatrix}}^{=:u_k} = R_s \overbrace{\begin{pmatrix} i_d \\ i_q \end{pmatrix}}^{=:i_k} + \omega \overbrace{\begin{bmatrix} 0 & -1 \\ 1 & 0 \end{bmatrix}}^{=:J} \overbrace{\left(\underbrace{\begin{bmatrix} L_d - \frac{\omega^2 L_q M_d^2 + \omega M_d M_m R_s}{R_s^2 + \omega^2 L_d L_q} & L_m + \frac{\omega M_d M_q R_s - \omega^2 L_d M_q M_m}{R_s^2 + \omega^2 L_d L_q} \\ L_m - \frac{\omega M_d M_q R_s + \omega^2 L_q M_d M_m}{R_s^2 + \omega^2 L_d L_q} & L_q - \frac{\omega^2 L_d M_q^2 - \omega M_q M_m R_s}{R_s^2 + \omega^2 L_d L_q} \end{bmatrix}}_{=:L_{1ASC}} i_k + \underbrace{\begin{pmatrix} \Psi_{pm} - \frac{\omega^2 L_q M_d \Psi_{pm} + \omega M_m R_s \Psi_{pm}}{R_s^2 + \omega^2 L_d L_q} \\ -\frac{\omega M_q R_s \Psi_{pm} + \omega^2 L_q M_m \Psi_{pm}}{R_s^2 + \omega^2 L_d L_q} \end{pmatrix}}_{\psi_{kpm1ASC}} \right)}^{=:\psi_{ks1ASC}(i_k)} + \frac{d}{dt} \psi_{ks1ASC}(i_k) \tag{6}
$$

Here, the inductance matrix $L_{1ASC} \in \mathbb{R}^{2x2}$ and the permanent flux matrix $\psi_{kpm1ASC}$ depend on the electric frequency ω due to the active short-circuit in one of the two winding systems.

The ASC in a multiphase machine is applied by disconnecting the inverter of the failed system from the voltage source, e.g. a battery or rectifier, and closing either all upper or all lower switches of the two-level inverter. Optionally, all six switches can be closed, when the semiconductor switches can withstand the short-circuit current by discharging the DC-link capacitor. This depends on the stored capacitor energy as well as semiconductor type and package and might be useful, when one semiconductor switch has failed in an undefined state. The ASC of one system can be applied after a semiconductor or machine winding short-on failure in all speed ranges. Even if only a single phase is affected by a short-circuit, the ASC of all three phases is the best choice to achieve a symmetric condition without torque pulsations [3].

Moreover, the ASC can be applied after a semiconductor or machine winding open failure, e.g. after a gate driver failure, in the field-weakening range. In the field-weakening range, the permanent magnet flux induces a voltage higher than the DC-link voltage, which leads to an uncontrolled rectifier operation via the inverter diodes, when the phase system is inactive. The ASC of one winding system generates a braking torque, while the other winding system is operable. The resulting overall torque can be analytically calculated by combination of (2) and (5).

Double system active short-circuit mode

The double ASC mode can be applied as fail-safe measure after failure of both inverters until the system comes to standstill. The currents of the equal winding systems in double ASC mode depend on the electric frequency ω and cannot be influenced by the inverter. The equations can be derived from (1) by setting $u_{d1} = u_{q1} = u_{d2} = u_{q2} = 0$:

$$
\begin{aligned}
i_{d1,\text{DoubleASC}} = i_{d2,\text{DoubleASC}} &= -\frac{\omega^2 \left(L_q + M_q\right) \Psi_{pm}}{R_s^2 + \omega^2 \left(L_d + M_d\right)\left(L_q + M_q\right)} \\
i_{q1,\text{DoubleASC}} = i_{q2,\text{DoubleASC}} &= -\frac{\omega \Psi_{pm} R_s}{R_s^2 + \omega^2 \left(L_d + M_d\right)\left(L_q + M_q\right)}
\end{aligned}
\tag{7}
$$

Calculation of optimal reference currents

In [14, 15] an analytical computation of the optimal reference currents for Maximum-Torque-per-Current (MTPC), Maximum-Torque-per-Voltage (MTPV) or Maximum-Torque-per-Flux (MTPF) operation of anisotropic single star synchronous machines considering stator resistance and mutual inductance of the same winding set has been presented. Before, either the stator resistance and/or the mutual inductance have been neglected due to the mathematical complexity or numerical approximations have been necessary for considering both. Analytical solutions are easier to implement, faster to compute and more accurate compared to numerical solutions. The optimal reference currents for a double star machine in MTPC, MTPV and MTPF operation can be analytically calculated for the three above mentioned active cases based on (3), (4) and (6). Therefore, the equations are written in implicit form, which results in quadrics. The optimal reference currents can be determined by invoking the Lagrangian formalism. These quadrics result in cone sections in the dq-plane, which represent trajectories like e.g. the maximum current circle, the maximum voltage ellipse or the maximum torque per current hyperbola (see Fig. 2). The intersection points of the quadrics can be calculated by solving quartic equations.

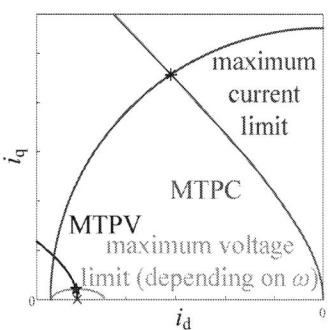

Fig. 2: Analytically calculated trajectories in the dq-current plane

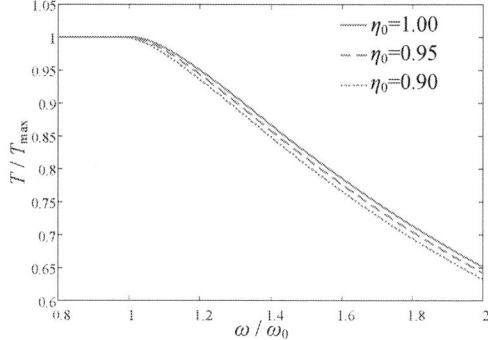

Fig. 3: Influence of stator resistance on speed-torque characteristic and efficiency. The selected stator resistance results in the respective efficiency η_0 at ω_0 and maximum torque.

The maximum speed-torque characteristic depending on the operation strategy in the following section are derived by calculating the transition frequency ω_0 between base speed and field weakening range as intersection of the MTPC hyperbola and the voltage ellipse, which depends on ω. Afterwards for infinite maximum speed machines, the maximum torque is limited by the maximum voltage and current until the intersection of maximum current circle and MTPV hyperbola. Afterwards, only the voltage is limiting and the MTPV hyperbola defines the maximum torque. For finite maximum speed machines, the maximum torque is limited by the maximum voltage and current until the maximum current circle and the voltage ellipse have no intersection. [17]

The influence of considering or neglecting the stator resistance R_s on the maximum speed-torque range is illustrated in Fig. 3. The value of the stator resistance is the only parameter to influence the efficiency of the machine, when all other losses are neglected. Thus, the efficiency of the machine is equal to one, when the stator resistance is zero. The selected stator resistance results in the respective efficiency η_0 at ω_0. The maximum torque in the base speed range is not influenced by the stator resistance. Only the transition frequency to the field weakening range depends on the stator resistance. In the field weakening range, the maximum torque at identical speed decreases with increasing stator resistance. A decreased efficiency η_0 of 5 % reduces the maximum torque at given speed in the field weakening range by approximately 1 % to 2 %. Thus, neglecting the stator resistance for calculations of high efficiency machines is tolerable, but in order to achieve the most accurate values, the analytical approach with consideration of stator resistance as well as mutual inductance of the same winding should be applied.

Comparison of feasible speed-torque operation ranges depending on machine parameters and operation mode

The speed-torque operation range of electric drives is influenced by machine parameters like permanent magnet flux, inductances, stator resistance, the maximum inverter voltage as well as current and can be analytically calculated with the procedure explained in the previous section.

The coupling of the two winding systems of a double star machine influences the mutual inductances $M_{d/q/m}$. In machines with high coupling, the values of $M_{d/q/m}$ are in the range of $L_{d/q/m}$. Typically, machines with distributed windings have a high mutual coupling depending on the winding arrangement and number of layers in a slot (see Fig. 4a, Fig. 5a). For machines with no coupling of the winding systems it is $M_d = M_q \approx 0$. Machines with concentrated single layer windings typically achieve a very low coupling close to zero (see Fig. 4b, Fig. 5b). [3, 13, 18]

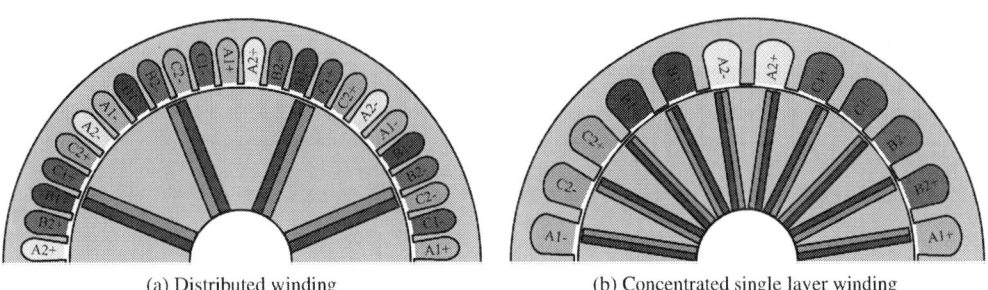

(a) Distributed winding   (b) Concentrated single layer winding

Fig. 4: Double star stator winding configurations with spoke type interior permanent magnet rotor

The stator inductances of both machine configurations depicted in Fig. 4 have been simulated with Ansys Maxwell. The stator inductances can be transformed in dq rotor reference frame [5, 19] and are shown in Fig. 5 depending on the electrical rotor position θ. The distributed winding arrangement has an average mutual coupling in q-direction in the range of 0.57, which is similar to the value obtained in [13]. The concentrated single layer winding has an average mutual coupling in q-direction in the range of 0.03. For the following analytical calculation, constant dq-inductances independent from the rotor angle are assumed. The simulation results are just depicted to illustrate, that the stator winding type significantly influences the mutual inductances.

In this paper, it is assumed for simplification that the coupling in d- and q-direction is identical:

$$k_L = \frac{M_d}{L_d} = \frac{M_q}{L_q} \tag{8}$$

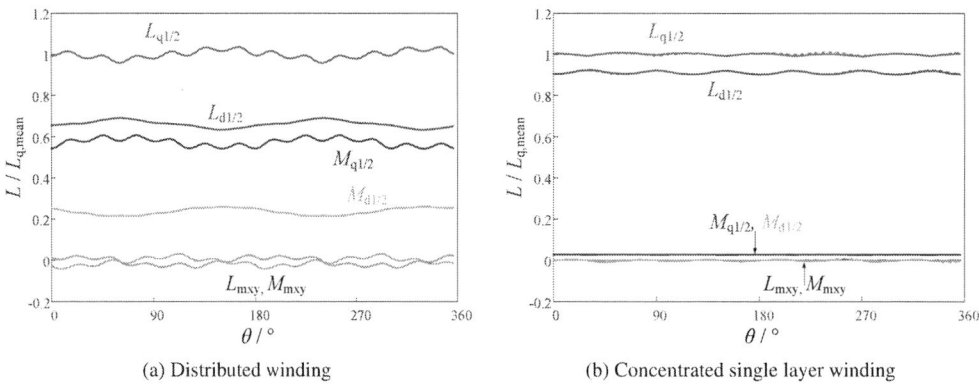

(a) Distributed winding   (b) Concentrated single layer winding

Fig. 5: Ansys Maxwell simulation of anisotropic synchronous double star machine inductances in the dq rotor reference frame depending on the electric rotor angle

Moreover, the ratio k_i of the maximum inverter/machine current i_{max} and the characteristic current i_0 for a complete compensation of the permanent flux is relevant:

$$k_i = \frac{i_0}{i_{max}} = \frac{\Psi_{PM}}{(L_d + M_d) \cdot i_{max}} \tag{9}$$

i_0 is equal to the maximum short-circuit current of the machine, when the stator resistance is neglected. Typically, k_i is in the range of 0.8 to 0.9 for traction motors. For infinite speed motors it is $k_i < 1$. For motors with $k_i > 1$ (finite maximum speed), the short-circuit current in ASC mode cannot always be handled by the inverter because the losses can get too high.

The analytically calculated speed-torque operation ranges depending on the coupling k_L and current ratio k_i each for following operation modes are illustrated in Fig. 6:

- both winding systems active (normal operation),

- one winding system active & one winding system inactive (part-load or fault-tolerant mode in base speed range for open failures),

- one winding system active & one winding system in ASC mode (fault-tolerant mode for short-on failures in all speed ranges or open failures in field weakening range),

- one winding system inactive & one winding system in ASC mode (failure mode) and

- both windings in ASC mode (failure mode).

The dashed lines in Fig. 6 indicate that either the maximum current or voltage of one winding system is exceeded and thus a steady-state operation at this speed is not feasible. For $k_i < 1$ (see Fig. 6a, Fig. 6c, Fig. 6e), the maximum current limit is never exceeded. For $k_i > 1$ (see Fig. 6b, Fig. 6d, Fig. 6f), the maximum current limit is exceeded in all operation modes with a single ASC, except for very low speeds. For no coupling it is also exceeded in double ASC mode. For these machine parameters, these operation modes are not possible without additional disconnection elements. Independent of k_i, the voltage limit is exceeded in the field weakening range, when one system is inactive. With increasing coupling k_L this limit is shifted to higher speeds and thus not depicted in Fig. 6c, Fig. 6e and Fig. 6f. With increasing coupling, the resulting flux in the deactivated system is decreased and thus higher speeds are possible. When the voltage limit is exceeded at higher speeds in the field-weakening range, there is a current flow over the diodes of the deactivated system, when no disconnection elements are installed.

EPE'20 ECCE Europe

Assigned jointly to the European Power Electronics and Drives Association & the Institute of Electrical and Electronics Engineers (IEEE)

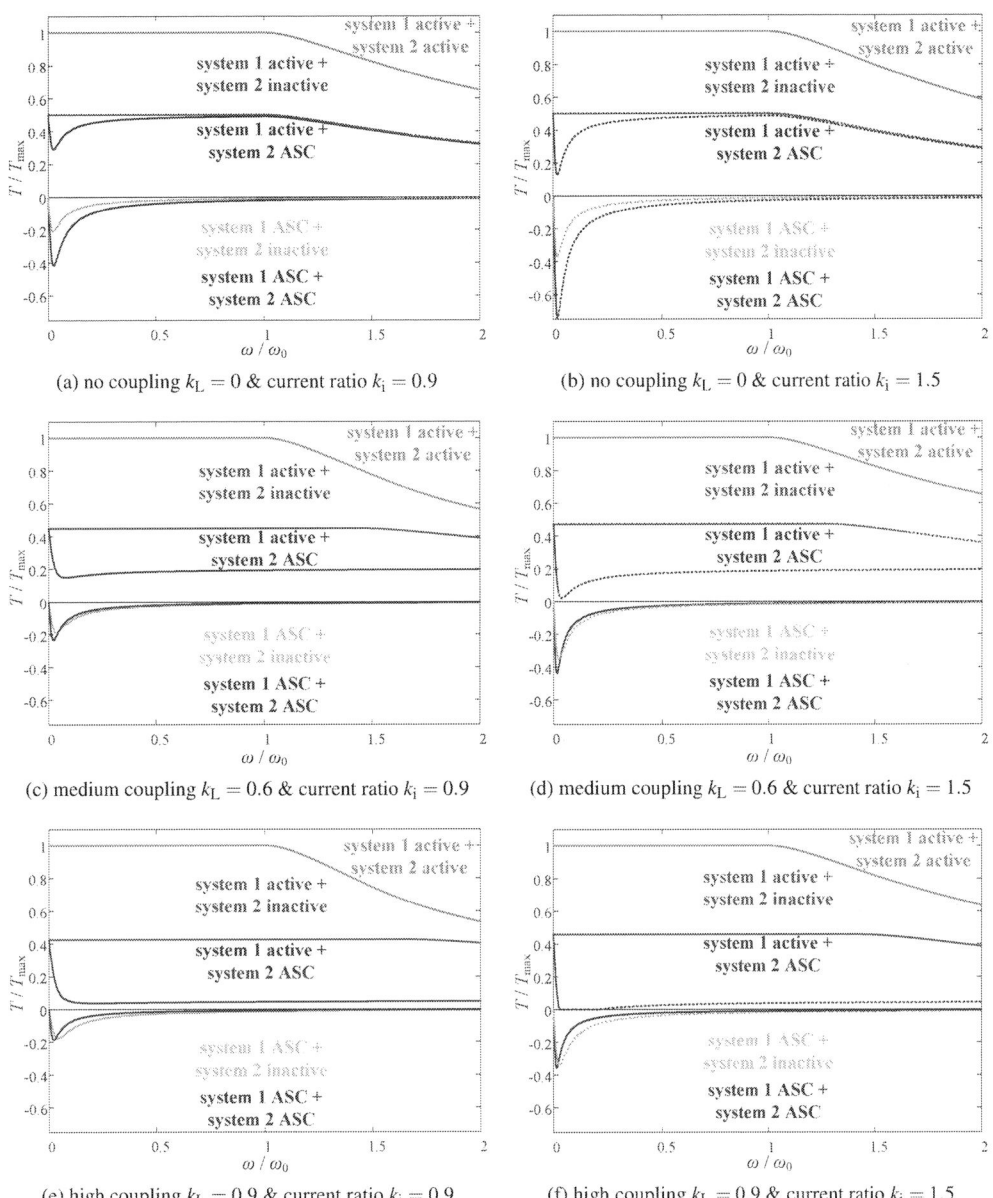

(a) no coupling $k_L = 0$ & current ratio $k_i = 0.9$

(b) no coupling $k_L = 0$ & current ratio $k_i = 1.5$

(c) medium coupling $k_L = 0.6$ & current ratio $k_i = 0.9$

(d) medium coupling $k_L = 0.6$ & current ratio $k_i = 1.5$

(e) high coupling $k_L = 0.9$ & current ratio $k_i = 0.9$

(f) high coupling $k_L = 0.9$ & current ratio $k_i = 1.5$

Fig. 6: Analytically calculated speed-torque operation ranges of anisotropic synchronous double star machines $L_q = \frac{3}{2}L_d$ depending on the mutual coupling k_L of the two winding system, the ratio of characteristic to maximum inverter current k_i and the operating mode of the two windings systems. A safe operation is only possible within areas limited by the full lines. The dashed lines indicate, that the maximum voltage or current in at least one of the two winding systems is exceeded.

The maximum torque is reduced to 50 % for $k_L = 0$ or less for $k_L > 0$ according to (2) due to the mutual reluctance torque with one active and one inactive system.

The fault-tolerant single ASC mode with one active system is only possible for machines with low coupling of both winding systems and a relatively low short-circuit current (see Fig. 6a). This proposed degraded fault-tolerant single ASC mode does not require disconnection elements in the AC-phases. Thus cost, space and losses can be reduced.

The corresponding calculated voltages and currents for maximum torque operation are depicted in Fig. 7 for the machine parameter configuration of Fig. 6a and operation with winding system 1 active and winding system 2 either inactive or in ASC mode. The voltages and currents of system 1 are identical for both operation modes for $k_L = 0$. In inactive mode, the current of system 2 is zero and the voltage of system 2 exceeds the maximum voltage limit in the field weakening range due to the missing field weakening current. In ASC mode, the voltage of system 2 is zero and the short-circuit current of system 2 is below the maximum inverter and machine current due to $k_i < 1$.

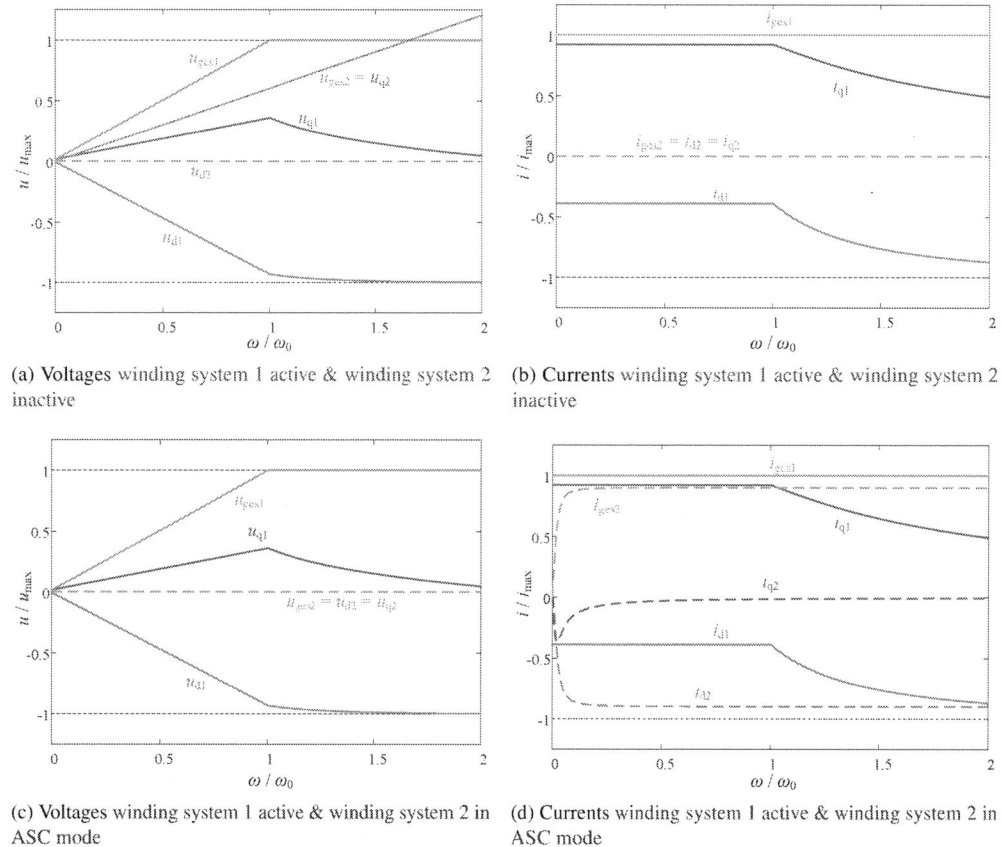

(a) **Voltages** winding system 1 active & winding system 2 inactive

(b) **Currents** winding system 1 active & winding system 2 inactive

(c) **Voltages** winding system 1 active & winding system 2 in ASC mode

(d) **Currents** winding system 1 active & winding system 2 in ASC mode

Fig. 7: Calculated voltages and currents of both winding systems for $k_L = 0$ & current ratio $k_i = 0.9$

Conclusion

In this paper, the known analytical computation approach for three-phase anisotropic synchronous machines including the stator resistance and mutual coupling has been extended to double star machines for normal and fault-tolerant operation. The degree of freedom of four currents in the dq-system of double star machines has to be reduced to only two currents in order to apply the calculation with quadrics and quartics presented in [14, 15]. The assumptions and calculations for normal operation with two active winding systems, only one active winding system and one deactivated winding system as well as one active winding system and the other system in active short-circuit mode have been presented. Thus, the influence of mutual coupling of the two windings systems as well as the ratio of the short-circuit current to the maximum inverter current can be quickly and easily assessed. The fault-tolerant operation with one winding system in active short-circuit mode is proposed in order to reduce the number of disconnection elements in the AC-phases, which normally increase cost, volume and loss. The machine requirements for this fault-tolerant operation mode are a very low inductive coupling of the winding systems and a maximum inverter current, which is higher than the characteristic current of the permanent magnet synchronous machine.

References

[1] B. Basler and T. Greiner, "Fault-Tolerant Strategies for Electronic Power Steering Systems under Functional Safety Requirements," in *Proceedings of the Global Symposium on EMC, Signal Integrity, Safety and Product Compliance Engineering*, 2014.

[2] B. Basler, T. Greiner, and P. Heidrich, "Fault-tolerant strategies for double three-phase PMSM used in Electronic Power Steering systems," in *2015 IEEE Transportation Electrification Conference and Expo (ITEC)*. IEEE, 14.06.2015 - 17.06.2015.

[3] N. Truemmel, T. Poetzl, and H.-C. Reuss, "Improvements on availability and comfort of electric drives for electric power steering application," in *8th IET International Conference on Power Electronics, Machines and Drives (PEMD 2016)*, 2016.

[4] F. Barrero and M. J. Duran, "Recent Advances in the Design, Modeling, and Control of Multiphase Machines–Part I," *IEEE Transactions on Industrial Electronics*, vol. 63, no. 1, pp. 449–458, 2016.

[5] O. Dieterle and T. Greiner, "Impact of the magnetic coupling in a quadruple-star permanent magnet synchronous machine with segmented stator windings," in *2017 IEEE International Electric Machines and Drives Conference (IEMDC)*. IEEE, 21.05.2017 - 24.05.2017.

[6] S. Calligaro, D. Frezza, R. Petrella, M. Bortolozzi, M. Mezzarobba, and A. Tessarolo, "A Fully-Integrated Fault-Tolerant Multi-Phase Electric Drive for Outboard Sailing Boat Propulsion," in *Proceedings of the 21st European Conference on Power Electronics and Applications (EPE ECCE Europe)*, 2019.

[7] M. J. Duran and F. Barrero, "Recent Advances in the Design, Modeling, and Control of Multiphase Machines–Part II," *IEEE Transactions on Industrial Electronics*, vol. 63, no. 1, pp. 459–468, 2016.

[8] H. M. Eldeeb, A. S. Abdel-Khalik, and C. M. Hackl, "Postfault Full Torque–Speed Exploitation of Dual Three-Phase IPMSM Drives," *IEEE Transactions on Industrial Electronics*, vol. 66, no. 9, pp. 6746–6756, 2019.

[9] M. Aboelhassan, T. Raminosoa, A. Goodman, L. de Lillo, and C. Gerada, "Performance Evaluation of a Vector Control Fault-Tolerant Flux-Switching Motor Drive," *IEEE Transactions on Industrial Electronics*, 2012.

[10] P. Giangrande, V. Madonna, S. Nuzzo, C. Gerada, and M. Galea, "Braking Torque Compensation Strategy and Thermal Behavior of a Dual Three-Phase Winding PMSM During Short-Circuit Fault," in *2019 IEEE International Electric Machines & Drives Conference (IEMDC)*, 2019, pp. 2245–2250.

[11] M. Barcaro, N. Bianchi, and F. Magnussen, "Faulty Operations of a PM Fractional-Slot Machine With a Dual Three-Phase Winding," *IEEE Transactions on Industrial Electronics*, vol. 58, no. 9, pp. 3825–3832, 2011.

[12] L. Alberti and N. Bianchi, "Experimental Tests of Dual Three-Phase Induction Motor Under Faulty Operating Condition," *IEEE Transactions on Industrial Electronics*, vol. 59, no. 5, pp. 2041–2048, 2012.

[13] M. Kozovsky and P. Blaha, "Double three-phase PMSM structures for fail operational control," *IFAC-PapersOnLine*, vol. 52, no. 27, 2019.

[14] C. M. Hackl, J. Kullick, H. Eldeeb, and L. Horlbeck, "Analytical computation of the optimal reference currents for MTPC/MTPA, MTPV and MTPF operation of anisotropic synchronous machines considering stator resistance and mutual inductance," in *Proceedings of the 19th European Conference on Power Electronics and Applications (EPE ECCE Europe)*, 2017.

[15] H. Eldeeb, C. M. Hackl, L. Horlbeck, and J. Kullick, "On the optimal feedforward torque control problem of anisotropic synchronous machines: Quadrics, quartics and analytical solutions," *arXiv:1611.01629 (see https://arxiv.org/pdf/1611.01629)*, 2017.

[16] J. Karttunen, S. Kallio, P. Peltoniemi, P. Silventoinen, and O. Pyrhonen, "Dual three-phase permanent magnet synchronous machine supplied by two independent voltage source inverters," in *International Symposium on Power Electronics Power Electronics, Electrical Drives, Automation and Motion*. IEEE, 20.06.2012 - 22.06.2012, pp. 741–747.

[17] W. L. Soong, "Field-weakening performance of brushless synchronous AC motor drives," *IEE Proceedings - Electric Power Applications*, vol. 141, no. 6, 1994.

[18] A. M. El-Refaie, "Fractional-Slot Concentrated-Windings Synchronous Permanent Magnet Machines: Opportunities and Challenges," *IEEE Transactions on Industrial Electronics*, vol. 57, no. 1, pp. 107–121, 2010.

[19] S. Kallio, M. Andriollo, A. Tortella, and J. Karttunen, "Decoupled d–q Model of Double-Star Interior-Permanent-Magnet Synchronous Machines," *IEEE Transactions on Industrial Electronics*, vol. 60, no. 6, pp. 2486–2494, 2013.

Full-Silicon 98.7% Efficient Three-Phase Five-Level 3-port UPS Architecture with Wide Voltage Range Battery based on Multiplexed Topology

Kepa Odriozola[1] , *Student, IEEE* — Thierry A. Meynard[2], *Fellow, IEEE* — Alain Lacarnoy[3]
Energy Management[1] — LAPLACE Laboratory[2] — *Industrial Automation*[3],
Schneider Electric IT France[1] — CNRS-INPT-UPS[2] — Schneider Electric Indus. SAS[3]
Grenoble, France[1,3] — Toulouse, France[2]
kepa.odriozola@se.com — meynard@laplace-univ.tlse.fr — alain.lacarnoy@se.com

Keywords

≪Multilevel converters≫, ≪Emerging topology≫, ≪Efficiency≫, ≪Uninterruptible Power Supply (UPS)≫, ≪Energy Storage≫, ≪Batteries≫, ≪Modulation strategy≫.

Abstract

This paper proposes a new three-phase Multilevel 3-port converter topology based on Multiplexed topology concept which is especially intended for high-power Uninterruptible Power Supplies (UPS) applications with wide battery voltage range and well adapted for high step-down voltage conversion ratio. Compared to other high-voltage conversion ratio Multi-Cell converters, the Multiplexed concept allows cascading power conversion stages without any intermediate filter and optimizing power density. In the studied case it will be used to transfer power between a 1200V DC source and 400V/480V AC load and grid. The proposed converter topology achieves very high efficiency, up to 98,7% in direct AC-to-AC power conversion mode using reduced number of silicon low blocking voltage semiconductor devices. The Hardware In the Loop (HIL) assessment has enabled the development of the modulation strategy and validated the performance of the converter. To validate the feasibility of the Multiplexed concept, both at the topological level and the modulation strategy, of the 5-Level Multiplexed 3-port topology for UPS applications a 100-kW hardware demonstrator featuring a volumetric power density of 3.68 kW/dm^3 (60.3 W/in^3) has been built.

Introduction

The growing penetration of renewable energies at worldwide scale is playing a leading role in the evolution of power electronics systems. Therefore, new high efficient power-electronic systems and associated control strategies are needed to make this transition possible. From the introduction of the first well known multi-level topologies [1, 2, 3] to their industrial maturity, many applications have taken advantage of their improved performances [4].

Regarding the already consolidated 1500V technology, several studies [5, 6, 7, 8] have shown in the past decade the economic benefits in the PV sector. Many manufacturers have already developed and installed photovoltaic inverters throughout the world and several devices (converters, contactors, combiner boxes, etc.) are already approved by global safety certification companies [9, 10]. Along with this, it should be noted that the grid integration of distributed energy sources such as Photovoltaic (PV) systems accompanied by Battery Energy Storage Systems (BESS), both at utility-scale and Behind-the-Meter (BTM) or residential-industrial level [11] is becoming more important. Proof of this are the products that are already available on the market [12, 13, 14]. This trend, today well established for system-oriented UPSs is now observed for the next generation Giant Data Centers. The installed power capacity in Data-Centers

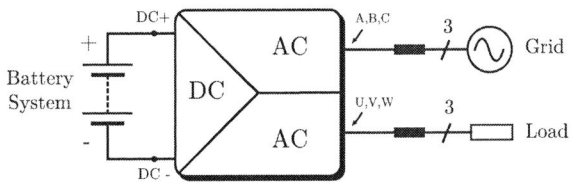

Fig. 1: General schematic of the proposed 3-port DC-AC-AC Multiplexed Power Converter. DC port side is connected to the battery pack system of the UPS. Each AC port is connected either to the grid and the load respectively.

is increasing exponentially with the never ending expansion of the global data sphere [15] and therefore battery-based UPS systems are becoming a key factor for today's Original Equipment Manufacturers (OEMs) and technology firms in this area.

Among the wide variety of industrial applications, UPS systems [16, 17] as well as other applications such as Medium-Voltage (MV) drives, energy storage systems (ESS), grid-tied systems for distributed generation sources such as PV, wind turbines and microturbines and HVDC (High-Voltage DC) transmission systems [19, 18, 20, 21, 22, 23, 24, 25] which make use of back-to-back configuration. Typical back to back configuration systems use mainly both identical topologies for inverter and AFE (Active Front End) converter for decoupling the grid from the loads. Therefore, the total cost of this type of conversion systems is double that of a single inverter stage. Therefore, obtaining high efficiency and high power density in this type of conversion architecture, whatever the application, with current semiconductor and magnetics technologies at low cost is still a big challenge.

This paper proposes a new three-phase multilevel 3-port converter topology based on multiplexed topology concept [26] and which is especially intended for high-power UPS applications with wide battery voltage range, which achieves very high efficiency, up to 98,7% in direct AC-to-AC power conversion mode using reduced number of silicon low blocking voltage semiconductor devices.

Table I: Specifications of the 5-Level three-phase 3-port DC-AC-AC Multiplexed Power Converter.

Parameter	Symbol	Value
Nominal Power	$P_{nominal}$	50kW
DC bus voltage	V_{DC}	900V-1200V
Grid voltage	$U_{(a,b,c)}$	400V-480V
Grid frequency	f_{Grid}	50-60 Hz
Load voltage	$U_{(u,v,w)}$	400V-480V
Load frequency	f_{Load}	50-60 Hz
Chopper switching frequency	f_{Chop}	8 kHz
Inverter/Rectifier switching frequency	f_{Sw}	16 kHz
Flying Capacitor	C_{F1}, C_{F2}	150 μF
RLC circuit inductance	L_{S1}, L_{S2}	200 μH
RLC circuit capacitance	C_{S1}, C_{S2}	2 μF
Grid filter inductance	$L_{f(a,b,c)}$	200 μH
Grid filter capacitance	$C_{f(a,b,c)}$	50 μF
Load filter inductance	$L_{f(u,v,w)}$	200 μH
Load filter capacitance	$C_{f(u,v,w)}$	50 μF

The structure and conversion stage topologies of this Multi- Level Multiplexed topology will be described in **Section II - 3-Port DC-AC-AC Multiplexed Converter**. The modulation strategy for the proposed

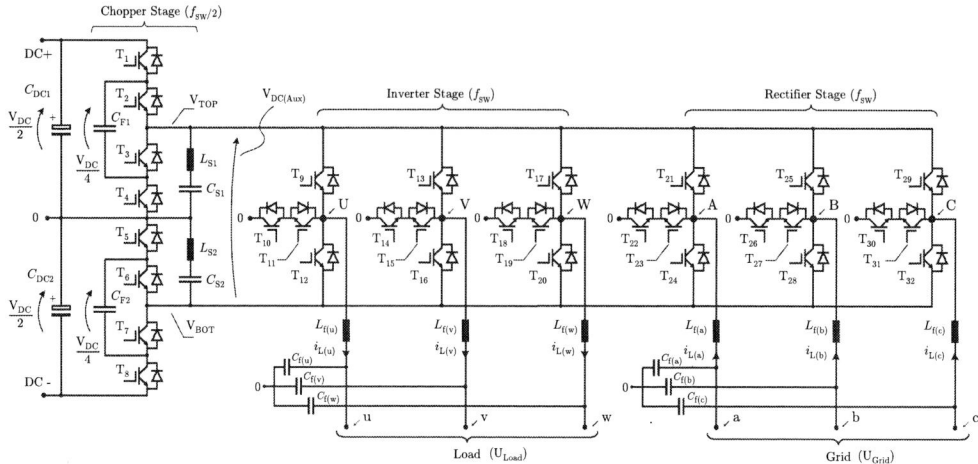

Fig. 2: Detailed schematic of the 5-Level three-phase 3-port DC-AC-AC Multiplexed Power Converter. The chopper stage consists of two 3L-Flying Capacitor converters. Both inverter and rectifier stage consists of three-phase 3L-T type NPC converters.

topology is the described in **Section III - Modulation Strategy**, starting with the basic principle and then describing more in-depth the chopper and inverter modulation. Then, control strategy is described in **Section IV - Control Strategy**. Simulation results are presented in **Section V - Simulation Results**. Hardware implementation is introduced in **Section VI - Experimental Results**. Finally, the conclusions are given in **Section VII - Conclusions**.

3-Port DC-AC-AC Multiplexed Converter

The proposed power conversion architecture (see **Fig. 1**) is based on multiplexed converter topology presented in [26, 27]. As explained above, the most typical configuration in UPS systems is back-to-back. Therefore, the most conventional would be to serialize two 5-Level Multiplexed DC-AC Converter stages that share the same DC-link.

However, the concept of control associated with multiplexed converters allows in this case, to mutualize the chopper stages. Thus, the inverter output stage and the rectifier input stage share the same chopper stage. In this way, it is possible to reduce the number of switches drastically and obtain better efficiency with respect to the back-to-back configuration.

The proposed power conversion topology is shown in **Fig. 2**. In principle, the DC-DC chopper and DC-AC inverter can be any converter with two or more voltage levels, under the condition of adapting the modulation scheme according to the selected topological variant. In the studied case, for a maximum voltage of 1200V at the DC-link, different combinations of multi-level topology can be derived. Going from 3 to n levels, where n is the number of voltage levels at the output of each converter. **Fig. 2** shows the power circuit of the three-phase 5-Level 3-port DC-AC-AC Multiplexed Power Converter with 3-level Flying Capacitor chopper and 3L-T-NPC inverter and rectifier configuration. In this paper only this configuration will be analyzed.

Semiconductor Selection

One of the main features being efficiency and low cost, the selection of semiconductors becomes essential. Only silicon components are used in order to reduce the total cost but of course wide band gap components could be used to reduce losses and improve efficiency and power density.

The first originality of this topology is that the choppers deliver square voltage waveforms with steps of one fourth of the DC-bus; assuming that an appropriate control pattern is used, the voltage applied to the inverter switches may thus be limited to 3/4 or even 1/2 the DC-bus voltage in order to ensure high immunity to cosmic rays [28] and obtain a high reliability, which of course helps maintaining a high efficiency. In our case, the inverter and rectifier stages are realized with 1200V, 75A High Speed IGBTs IKY75N120CS6 from Infineon. The typical voltage across the semiconductors of the flying capacitor chopper is 1/4 of the DC-bus voltage, and 650V, 75A Low-VCEsat IGBTs IKZ75N65ES5 from Infineon have been selected here. In order to handle the high current of the load and to reduce the conduction losses, each switch consists of four TO-247 package devices in parallel.

Voltage Balance on Flying Capacitors

In order to limit the overvoltage on the inverter and rectifier we must guarantee a correct balancing of the flying capacitors. In the offline power conversion mode and in the battery recharge mode, the power flow circulates through the choppers flying capacitors. In this way, under normal circumstances, it is possible to obtain a natural balance of the capacitors [29]. However, in the AC-AC direct energy conversion mode, the power flow circulates from the grid directly to the load via the variable DC-bus. Therefore, the current circulating through the flying capacitors is negligible even though the choppers continue to operate to generate in the variable DC-bus ($V_DC(Aux)$) the maximum and minimum part of the voltage of the two three-phase systems (grid and load). Due to imperfections of the control there will also be small unwanted current components; the inherent self-balancing property of these converters might be lost which is not acceptable because unbalanced voltages could destroy the semiconductors.

To guarantee balancing of the flying capacitors there are various techniques and we can distinguish active and passive methods. In order to avoid introducing additional distortions on the differential voltage pattern of the choppers by means of the duty cycle correction [30] the passive methods will be preferred here. Among the passive methods there are different circuits described in the literature like [31, 32]. However, these linear circuits as well as the non-linear variants are designed considering several assumptions, one of these being a constant supply voltage of the converter. A balance-booster design [33], be it internal or external, that follows the input voltage variations becomes complicated, mainly due to its limitations from the implementation point of view.

For this reason, it has been decided to use an RLC circuit [34]. This shunt passive filter essentially acts as a notch filter connected in parallel to the output of each chopper offering a very low impedance at the switching frequency of these choppers. When the voltage of the flying capacitor is unbalanced, each chopper generates at the output, voltage harmonics at the switching frequency. Since the RLC filter is tuned to the switching frequency (i.e. 8kHz) it presents very low impedance, Z_{LC} $(2 \cdot \pi \cdot f_{Chop})$, forcing the circulation of harmonics of current at the same frequency that favor the rebalancing of the capacitors. The inductance and capacitance values used in simulation for the RLC circuit are given in **Table I** and the design values in **Table II**.

Modulation Strategy

Unlike conventional three-phase topologies, the multiplexed topology does not have independent phase legs. Therefore, each conversion stage module must be controlled separately. As a basic principle of modulation, sinusoidal PWM modulation can be used to make it simple.

Similar to the modulation strategy described in [26] the main concept is to generate respectively the highest and the lowest part of the 6 voltages (3 of the grid and 3 of the load) by respectively Top and Bot choppers. Doing so, 2 of the 6 inverter legs can be saturated and stop switching. The 4 other inverter/rectifier legs need to switch to generate the appropriate grid side and load side three-phase voltage systems.

Chopper Modulation

Regarding the control of DC-DC choppers, the main objective is to generate the highest and lowest side of the both grid side and load side three-phase voltage systems. The chopper section involving capacitors has a strong requirement to ensure that the average current is zero and that the voltage of this capacitor can be regulated. Another objective will be to minimize the static voltage seen by the inverter stage in order to respect the voltage rating of the inverter switches. As a first approach, self-balancing

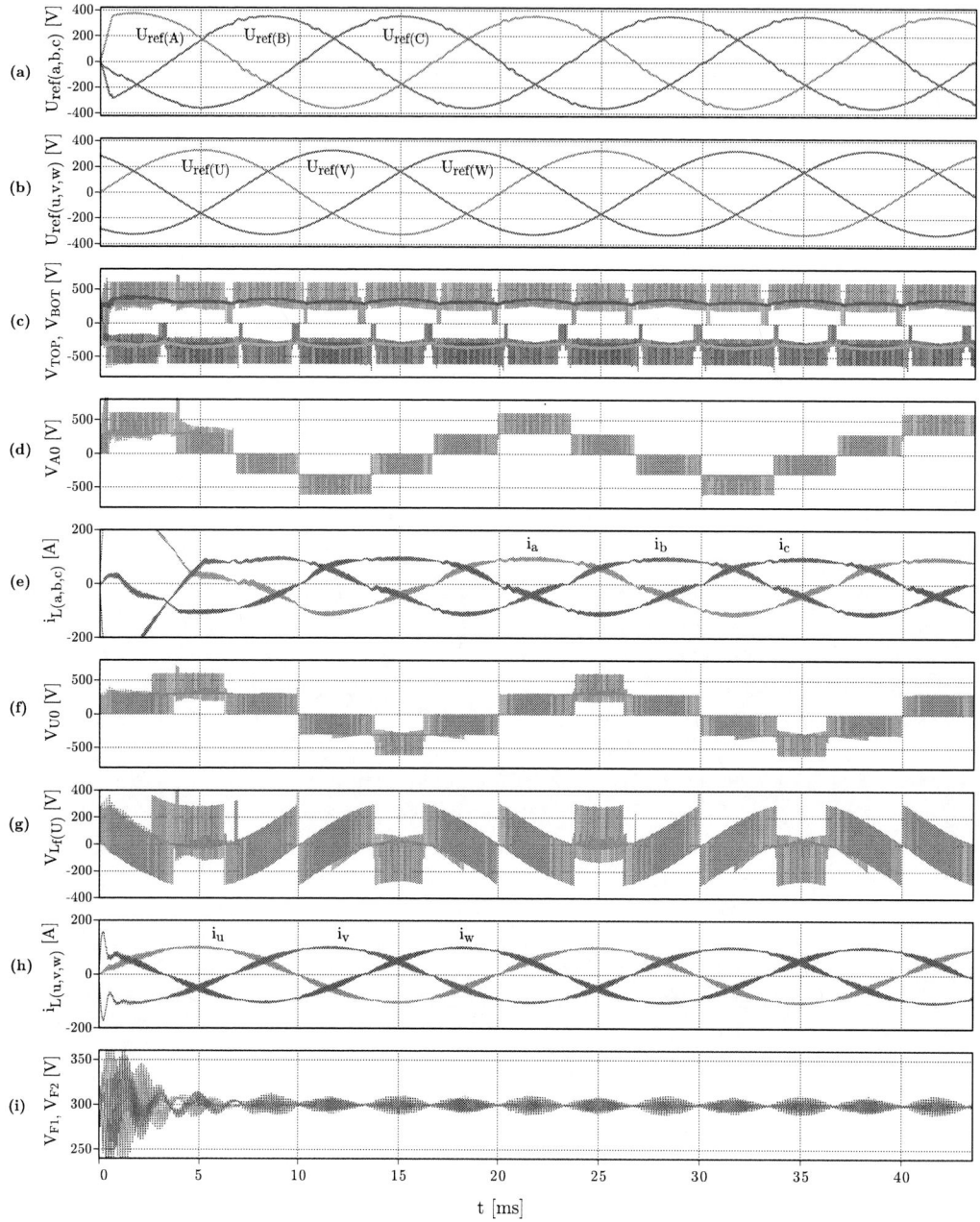

Fig. 3: Main simulation waveforms of the 5-Level three-phase 3-port DC-AC-AC Multiplexed Power Converter (see **Fig. 2**) operating with a DC-link (Battery) voltage of 1200V during direct AC-AC power conversion mode (cf., **Fig. 4a**) and a load output power of 50kW. (a): Grid reference voltages; (b): Load reference voltages; (c): Chopper output voltages, V_{TOP} in green, average value of V_{TOP} in blue, V_{BOT} in red, average value of V_{BOT} in yellow; (d): Multi-level voltage of the node A at the input of the Rectifier stage referenced to the DC-link midpoint (cf. **Fig. 2**); (e): Grid phase currents; (f): Multi-level voltage of the node U at the output of the Inverter stage; (g): Inverter filter inductor voltage of phase U; (h): Load phase currents; (i): Flying Capacitor voltages. Please note that the grid and load voltages are phase-shifted by $60°$ making the choppers generate the maximum of the 6-voltage system.

mechanism can be tried by applying a carrier based PWM modulation scheme such as Phase Shifted (PS) [35]. Another possibility is to use Phase Disposition (PD) by means of state machines like in [36] and [37], but in this kind of schemes extra commutations are needed and the balancing of flying capacitor could become complex. We therefore use PS sawtooth carriers at 8kHz switching frequency are used for having the absolute control of turn-on and turn off moments or rising-up and falling-down instants of the voltage created by each chopper.

Rising sawtooth carriers for the Top chopper and falling sawtooth carriers for the BOT chopper have been selected. In order to minimize the static voltage, $V_{DC,Aux}$ (cf., **Fig. 2**) seen by the inverter and rectifier stages over the majority operating points, there will be a relative phase shift of some switching periods between the Top and Bot choppers carrier signals. In this way when high modulation index is required, consecutive steps of voltage levels on the differential voltage of the chopper outputs can be avoided. The feature of the Flying Capacitor converter to double the apparent frequency at the output leads to a chopper switching frequency equal to half that of the inverter. Consequently, the switching losses are reduced too. This switching frequency ratio between chopper and inverter must be respected if the calculation of the inverter and rectifier duty cycle (introduced in the next section) is to be accurate.

Inverter Modulation

The voltage generated at the output of the inverter and rectifier must be synthetized by means of the voltage already switched by choppers. Hence, the duty cycle for inverter and rectifier is defined in function of choppers duty cycles and inverter and rectifier voltage reference. For that, duty cycles of each output inverter and rectifier arm are determined separately by means of two-dimensional look-up-tables using a pre-calculated PWM method. Contrary to the disadvantages of this type of techniques [38, 39], in this case, the computational cost of the algorithm is not high and duty cycles can be calculated with a high accuracy. The calculation of duty cycles is made off-line throughout two piecewise interpolations and explained more in depth in [26].

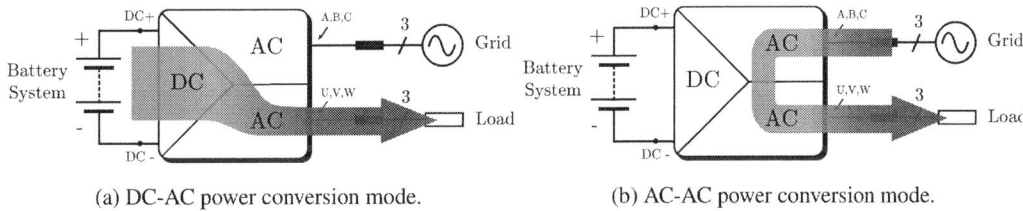

(a) DC-AC power conversion mode.   (b) AC-AC power conversion mode.

Fig. 4: Power transfer flow in the proposed three-phase 3-port Multiplexed Converter for different modes of power conversion.

The applied modulation scheme has the intrinsic property of freezing the inverter phase legs when necessary. Once the duty cycle is computed, a dedicated scheme of modulation must be applied for the inverter and rectifier output stages. Gate signals must be generated in function of selected topology. For our configuration, three-phase three-level T-type NPC has been chosen. For being compatible with the chopper modulation and mainly with the inverter duty cycle offline calculation we must choose carriers at 16kHz switching frequency such that they are in counterposition to the chopper carriers. In this case, the upper switch cell of the inverter must depend on the carriers of the Top chopper and vice versa. Therefore, having selected rising sawtooth carriers for the Top chopper then the carrier modulating the upper cell of the inverter must be of type falling sawtooth carrier. In this way, we ensure that the interpolation equations with which the duty cycle table is constructed to control the inverter stage are fair. Moreover, each switch cell will guarantee ZVS (Zero Voltage Switching) operation for inverter and rectifier arms (at least one turn-on or turn-off per switching period in function of the selected rising or falling front carrier configuration) thanks to the extra slight phase shift introduced between Top and Bot choppers and inverter stage carriers. Consequently, the desired switching pattern is obtained for the inverter and rectifier, thus only two phase legs are swiching at a time when both three-phase voltage systems have same amplitude, one on the grid-side, the other on the load-side.

Control Strategy

New regulations (i.e. IEC, IEEE 1547, etc.) impose more functionalities on electrical energy conversion devices in order to interact with the different types of current networks. These devices must be able, for example, to provide a certain amount of reactive power during a certain period of time in order to cope with eventual accidents or instabilities that may occur.

In practice, to fulfill the requirements imposed on today's new UPS equipment, different types of control laws can be applied to the proposed topology. Depending on the mode of operation, the management of the direction of the active (P_{Load}, P_{Grid}) or reactive power (Q_{Load}, Q_{Grid}) flow will be different. Thus, we can distinguish mainly the inverter mode of operation, the battery charging mode in rectifier operation (cf., **Fig. 4a**), the grid following or grid forming mode of operation, and the AC-to-AC direct energy conversion mode (cf., **Fig. 4b**) which is enabled thanks to the mutualization of the chopper stage.

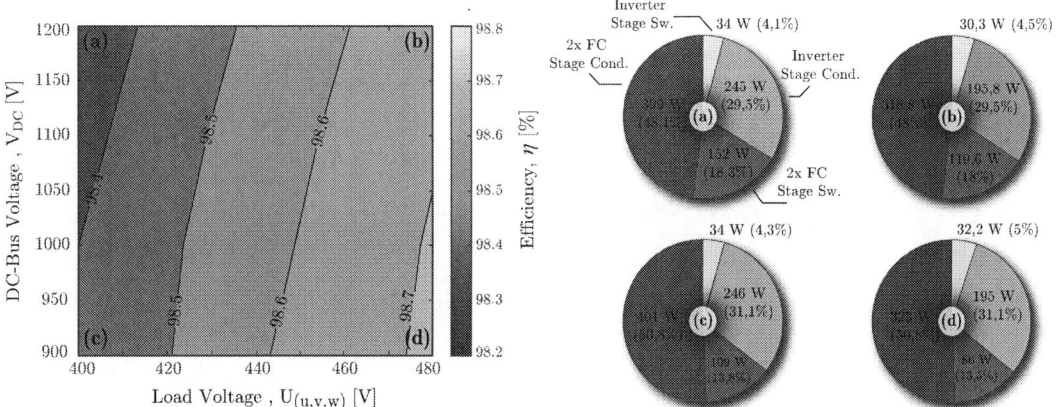

Fig. 5: Efficiency map of the 5-L three-phase 3-port DC-AC-AC Multiplexed Power Converter in single DC-AC power conversion mode (cf., **Fig. 4a**) feeding a 50 kW resistive load for different line-to-line load voltage and DC-bus voltage values

Fig. 6: Loss breakdown for 4 (corner: a,b,c,d) operating points at **Fig. 5** of the 5-L three-phase 3-port DC-AC-AC Multiplexed Power Converter in single DC-AC power conversion mode at 50kW.

In the AC-to-AC direct power conversion mode, unlike conventional back-to-back configurations in which all energy passes through the DC-link, the proposed topology partially bypasses the DC-link: power may be transferred from grid to load via only the inverter and rectifier.

It is necessary to emphasize that for all the above-mentioned operating modes, it is necessary to apply a preload strategy of DC-bus capacitors voltage as well as chopper flying capacitors, in order to guarantee a balanced distribution of the total DC link voltage [40, 41]. At any moment of the preload, it is necessary to ensure that the flying capacitor is charged to its nominal voltage. This is achieved by combining a classic preload method via a resistance that is later bypassed and a DC-bus voltage control loop that takes care of charging the DC link to the battery voltage in boost power conversion. The RLC filters connected in parallel to the output of each chopper will ensure that in case of unbalance the distribution of the voltage is compensated.

Simulation Results

The performance of the three-phase 5-Level DC-AC-AC Multiplexed Converter has been simulated in the PLECS®software using specifications described in **Table I**. All operating modes have been tested in simulation as well as the thermal performance of the converter using semiconductors described in **Subsection - Semiconductor Selection**.

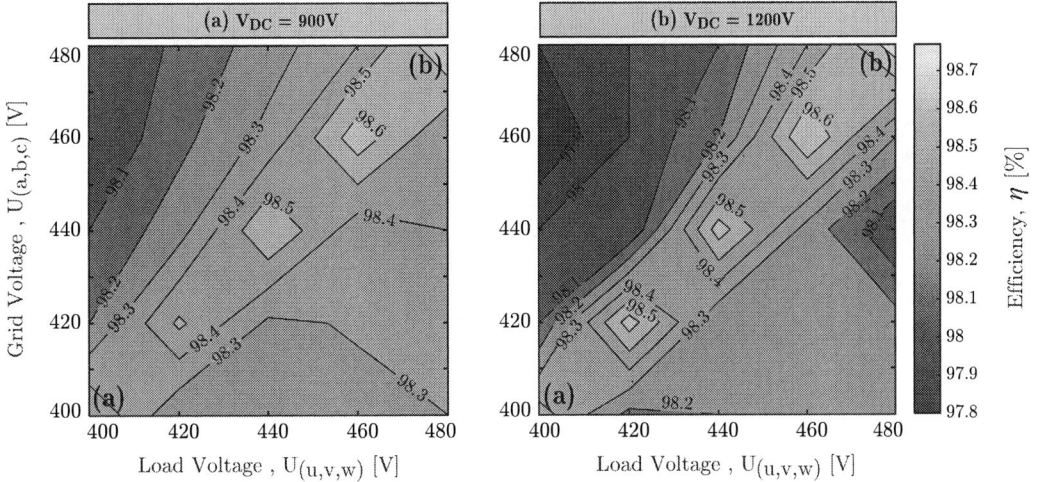

Fig. 7: Efficiency map of the 5-L three-phase 3-port DC-AC-AC Multiplexed Power Converter in direct AC-AC power conversion mode (cf., **Fig. 4b**) feeding a 50 kW resistive load for different line-to-line grid and load voltage values and identical phases and for V_{DC}= 900V and V_{DC}=1200V.

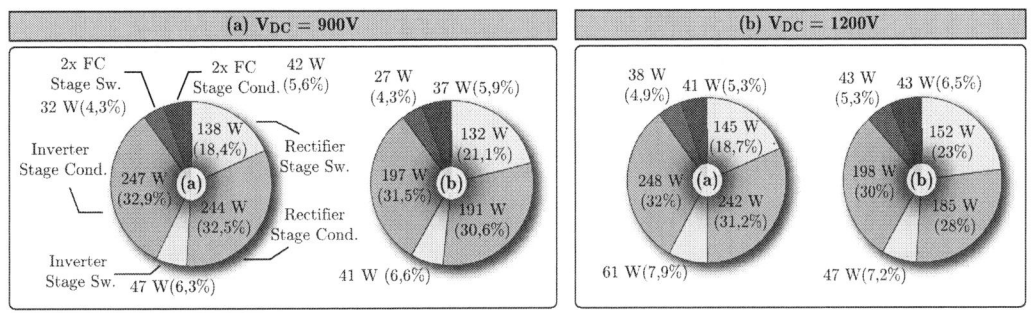

Fig. 8: Loss breakdown for 2 (corner: a,b) operating points per each value of DC-bus (battery) voltage at **Fig. 7** of the 5-L three-phase 3-port DC-AC-AC Multiplexed Power Converter in single DC-AC power conversion mode at 50kW.

Fig. 3 shows the multilevel performance of the three-phase 5-Level Multiplexed Converter (illustrated in **Fig. 2**) over two periods of fundamental output frequency. Regarding the simulation conditions for the illustrated scenario, the DC-bus (battery voltage) is 1200V, the inverter works in open loop with a reference voltage between phases equal to 400V/50Hz while the rectifier works in PFC mode with closed loop control of the grid currents. In this mode, power is transmitted directly (AC-AC power conversion mode, **Fig. 4b**) from the mains to the load without passing through the DC-bus, i.e. the battery, thus reducing overall losses and increasing efficiency. The grid voltage is 440V/50Hz with a relative phase shift of 60 with respect to the load voltage. It can be observed that at all times the choppers are responsible for synthesizing the maximum and minimum of the two three-phase systems (grid and load). It can also be seen that the voltage ripple of the flying capacitors is kept below 5% of the nominal voltage thanks to the RLC filters.

Fig. 5 shows the efficiency map of the converter in single DC-AC power conversion mode (cf., **Fig. 4a**) for different values of load DC-bus (battery) voltage. The same efficiency is expected for the converter in grid-forming mode when rectifier stage is working as inverter. Loss distribution per each conversion stage (2x 3-Level Flying Capacitor Chopper Stage and Inverter Stage) of the theoretical design is shown in **Fig. 6**.

Fig. 7 shows the efficiency map of the converter operating the system in AC-AC power conversion mode (cf., **Fig. 4b**) for different values of load and mains voltage and two constant values of battery voltage. Loss distribution per each conversion stage (2x 3-Level Flying Capacitor Chopper Stage, Inverter Stage and Rectifier Stage) of the theoretical design is shown in **Fig. 8**. We can observe this time how the losses within the Flying Capacitor choppers have been drastically reduced with respect to the inverter mode of operation. Most of the losses, around 90%, are located in the inverter and rectifier stages.

Experimental Results

To validate the proposed control strategy and the suitability of the 5-Level Multiplexed 3-port topology for UPS applications, the HIL test system and hardware implementation of the three-phase 5-Level 3-port Multiplexed topology will be presented in the following.

Hardware-In-the-Loop (HIL)

The digital implementation of the proposed modulation and control strategy and all operating modes, including start-up precharge strategy of flying capacitors and DC-bus, has been validated via HIL. The control is implemented by coding two Real-Time Imperix B-Box RCP 3.0 that interfaces with the Typhoon-HIL 602+ system. **Fig. 9** shows the HIL waveforms of the proposed converter during inverter operating mode.

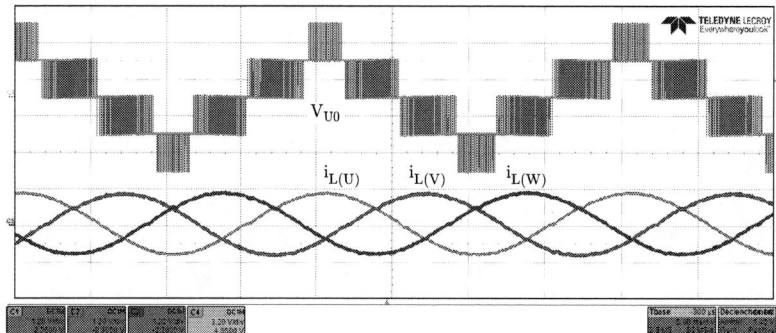

Fig. 9: Hardware In the Loop (HIL) waveforms of the five-level output voltage of the phase U of the 5-Level three-phase 3-port DC-AC-AC Multiplexed Power Converter shown in **Fig. 2** (300 V/div, referenced to the DC-link midpoint, 5 ms/div) and the three-phase (u,v,w) load currents (120 A/div) during inverter operating mode at 50 kW. The HIL real-time waveforms are in accordance with the simulation results shown in **Fig. 3**.

Hardware Implementation

To validate the feasibility of the multiplexed concept of the 5-Level Multiplexed 3-port topology for UPS applications, a 100-kW hardware demonstrator shown in **Fig. 10** has been built using components described in the **Table II**.

Experimental Waveforms

The main measured waveforms taken with a resistive load and operating the system in inverter mode are presented in **Fig. 11** for 15.7 kW operation. The five output voltage levels are shown for phase U together with the Top chopper output voltage (V_{TOP}), as well as the load phase voltage and currents. We can deduce from the shape of the 5-Level voltage of the phase U (V_{U0}) that the voltage of the flying capacitors is kept balanced thanks to the RLC filters.

Conclusions

In this paper, a 98.7% 3,68kW/dm³ three-phase 3-port multilevel converter topology based on *Multiplexed* structure was proposed. This topology has the ability to achieve a high step-down voltage con-

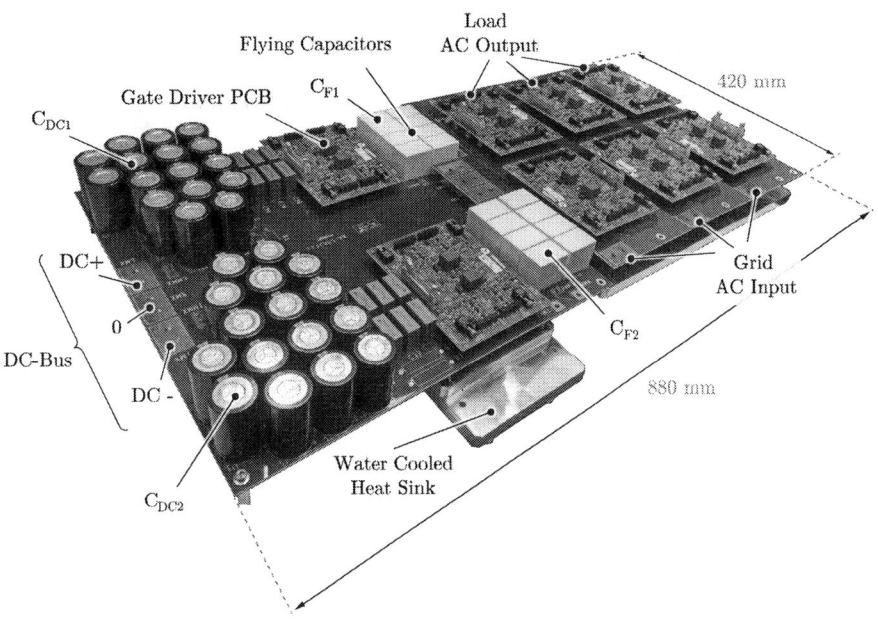

Fig. 10: Hardware prototype of the 100kW three-phase 5-Level DC-AC-AC Multiplexed Converter, measuring 420mm 880mm 110mm (16.53 in 34.64 in 4.33 in). The final volumetric power density (without considering RLC Shunt filters, output and input LC filters of each AC port and auxiliary power) is 3.68 kW/dm^3 (60.3 W/in^3).

Table II: Main components of the hardware prototype.

Component	Value	Part Number
Chopper Stage Switches		Mitsubishi 1200 400A
(3L-FC Switches)		FMF400BX-24A
Inverter/Rectifier Stage Switches		Infineon 1200V 300A
		FF300R12KE4
(3L-T-NPC Switches)		Infineon 1200V 300A
		FF300R12KE4_E (Common Emitter)
Gate Driver		Custom PCB Schneider Electric
C_{DC1}, C_{DC2}	4800 μF	Epcos TDK B43643
C_{F1}, C_{F2}	280 μF	Kemet C4AEGBW5350A3JJ
L_{S1}, L_{S2}	200 μH	75 turns, 100 strand x 200 μm Litz wire
		Core: Micrometals T300-8/90 T300
C_{S1}, C_{S2}	1,88 μF	Kemet R75UR34704040J
$L_{f(a,b,c)}$, $L_{f(u,v,w)}$	150 μH	Traftor T10125-M09
$C_{f(a,b,c)}$, $C_{f(u,v,w)}$	132 μF	Epcos TDK B32354S

version ratio which makes it suitable for new low-medium voltage applications with wide range of input voltage such as UPS systems.

Cascading inverter and choppers without intermediate filtering elements leads to a reduction of the voltage applied to the switches of the inverter and rectifier stages (in some cases ZVS operation is possible), thus reducing the switching losses and indirectly the conduction losses (since switches with a lower

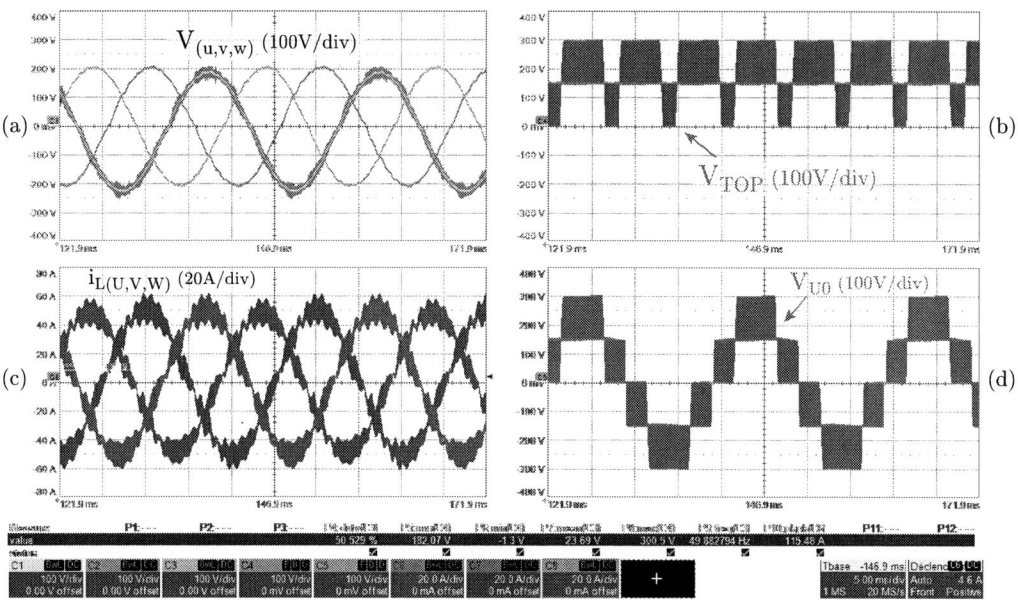

Fig. 11: Experimental waveforms of the five-level 5-Level three-phase 3-port DC-AC-AC Multiplexed Power Converter shown in **Fig. 2** operating the system in inverter mode for 15.7 kW operation (V_{DC}=600V, V_u=150V, $i_{L(U)}$=35A). (a): $V_{(u,v,w)}$ (100 V/div), Load phase voltages; (b): V_{TOP} (100 V/div), output voltage of the Top 3-Level Flying Capacitor Chopper; (c): $i_{L(U,V,W)}$ (20 A/div), output filter inductance currents; (d): V_{U0} (100 V/div), five level output voltage of the phase U referenced to the DC-link midpoint.

voltage rating can be used). The switching losses in the inverter are also reduced because the Max and Min voltage waveforms of both grid and load three-phase systems are synthesized by the choppers, so at any time two of the inverter legs are not switching. These features enables switching at relatively high frequency and helps in the end to obtain a high power density. Moreover, compared to conventional single three-phase multilevel topologies it reduces the number of modules needed because of the mutualization of the chopper stage. It should also be noted that it only uses only silicon low blocking voltage semiconductors devices reducing the overall cost of converter.

Furthermore, a higher efficiency can be obtained in AC-to-AC power conversion mode by allowing the main power transfer via the inverter stages only, with a very small portion of currents flowing via the chopper stages. In this paper an appropriate control strategy for the novel topology is presented. Simulations have been conducted on the PLECS software environment with good results especially in terms of efficiency. Real time feasibility of the control strategy has been checked using two Imperix B-Box RCP 3.0 and Typhoon HIL 602+ devices and a scale 1 prototype has been built and is now ready for testing. The first experimental results (up to 55 kW) have globally demonstrated a good behaviour of the converter, being able to validate at the same time the feasibility of the topological architecture of the converter based on the "Multiplexed" concept as well as the modulation strategy proposed for it.

We can already identify improvement margins for the final prototype. Due to the high efficiency of the topology, the size of thermal cooling system could be significantly reduced. It would also be possible to replace the power modules by paralleled 4-pin TO-247 package switches (as proposed at the theoretical study) which would greatly simplify the PCB design and reduced parasitic stray inductances between the chopper and inverter stages. The RLC Shunt filters could also be embedded on the PCB. These, improvements could lead to optimize overall power density of the converter and better fit to the standard rack-size of the Data-Center oriented UPS systems.

References

[1] Akira Nabae, Isao Takahashi, and Hirofumi Akagi. A New Neutral-Point-Clamped PWM Inverter. *IEEE Transactions on Industry Applications*, IA-17(5):518–523, 1981.

[2] T. A. Meynard and H. Foch. Multi-level conversion: High voltage choppers and voltage-source inverters. In *PESC Record - IEEE Annual Power Electronics Specialists Conference*, pages 397–403. Institute of Electrical and Electronics Engineers Inc., 1992.

[3] M. Marchesoni, M. Mazzucchelli, and S. Tenconi. Non-conventional power converter for plasma stabilization. In *PESC Record - IEEE Annual Power Electronics Specialists Conference*, pages 122–129. Publ by IEEE, 1988.

[4] Leopoldo G. Franquelo, Jose Rodriguez, Jose I. Leon, Samir Kouro, Ramon Portillo, and Maria A.M. Prats. The age of multilevel converters arrives. *IEEE Industrial Electronics Magazine*, 2(2):28–39, 2008.

[5] Yaosuo Xue, Kurthakoti C. Divya, Gerd Griepentrog, Mihalache Liviu, Sindhu Suresh, and Madhav Manjrekar. Towards next generation photovoltaic inverters. *IEEE Energy Conversion Congress and Exposition: Energy Conversion Innovation for a Clean Energy Future, ECCE 2011, Proceedings*, pages 2467–2474, 2011.

[6] Eirini Gkoutioudi, Panagiotis Bakas, and Antonios Marinopoulos. Comparison of PV systems with maximum DC voltage 1000V and 1500V. *Conference Record of the IEEE Photovoltaic Specialists Conference*, pages 2873–2878, 2013.

[7] Emanuel Serban, Martin Ordonez, and Cosmin Pondiche. DC-Bus Voltage Range Extension in 1500 v Photovoltaic Inverters. *IEEE Journal of Emerging and Selected Topics in Power Electronics*, 3(4):901–917, 2015.

[8] Zoltan Čorba, Bane Popadić, Vladimir Katić, Boris Dumnić, and Dragan Milićević. Future of high power PV plants - 1500V inverters. *19th International Symposium on Power Electronics, Ee 2017*, 2017-Decem:1–5, 2017.

[9] ABB. Contactors for DC switching - Motor protection and control — ABB.

[10] Mersen. Mersen is fully 1500VDC Ready — Mersen.

[11] Holger Hesse, Michael Schimpe, Daniel Kucevic, and Andreas Jossen. Lithium-Ion Battery Storage for the GridA Review of Stationary Battery Storage System Design Tailored for Applications in Modern Power Grids. *Energies*, 10(12):2107, dec 2017.

[12] Ingeteam. INGECON SUN STORAGE Power B Series 1500Vdc. Technical report.

[13] General Electric. GE Energy Storage Unit RSU-4000. Technical Report January, 2020.

[14] Sungrow. ST2740KWH-2500HV. Technical report, 2019.

[15] David Reinsel, John Gantz, and John Rydning. The Digitization of the World - From Edge to Core. *IDC White Paper*, (November):US44413318, 2018.

[16] Dezhi Dong, Linglin Chen, Haijin Li, Min Chen, and Dehong Xu. Design of hybrid AC-DC-AC topology for Uninterruptible Power Supply. *2015 IEEE 2nd International Future Energy Electronics Conference, IFEEC 2015*, 2015.

[17] Jinghang Lu, Mehdi Savaghebi, Yajuan Guan, Mingshen Li, and Josep Guerrero. Multi-mode operations for on-line uninterruptible power supply. *Conference Proceedings - IEEE Applied Power Electronics Conference and Exposition - APEC*, 2018-March:180–187, 2018.

[18] Samir Kouro, Mariusz Malinowski, K. Gopakumar, Josep Pou, Leopoldo G. Franquelo, Bin Wu, Jose Rodriguez, Marcelo A. Perez, and Jose I. Leon. Recent advances and industrial applications of multilevel converters. *IEEE Transactions on Industrial Electronics*, 57(8):2553–2580, aug 2010.

[19] Samir Kouro, Jose Rodriguez, Bin Wu, Steffen Bernet, and Marcelo Perez. Powering the future of industry: High-power adjustable speed drive topologies. *IEEE Industry Applications Magazine*, 18(4):26–39, jul 2012.

[20] Juan Manuel Carrasco, Leopoldo Garcia Franquelo, Jan T. Bialasiewicz, Eduardo Galván, Ramón C. Portillo Guisado, Ma Ángeles Martin Prats, José Ignacio León, and Narciso Moreno-Alfonso. Power-electronic systems for the grid integration of renewable energy sources: A survey, jun 2006.

[21] Jose Rodriguez, Steffen Bernet, Peter K. Steimer, and Ignacio E. Lizama. A survey on neutral-point-clamped inverters, jul 2010.

[22] José Rodríguez, Steffen Bernet, Bin Wu, Jorge O. Pontt, and Samir Kouro. Multilevel voltage-source-converter topologies for industrial medium-voltage drives, dec 2007.

[23] Haitham Abu-Rub, Joachim Holtz, Jose Rodriguez, and Ge Baoming. Medium-voltage multilevel converters State of the art, challenges, and requirements in Industrial applications. *IEEE Transactions on Industrial Electronics*, 57(8):2581–2596, aug 2010.

[24] Venkata Yaramasu, Bin Wu, Paresh C. Sen, Samir Kouro, and Mehdi Narimani. High-power wind energy conversion systems: State-of-the-art and emerging technologies. *Proceedings of the IEEE*, pages 740–788, may 2015.

[25] José R. Rodríguez, Juan W. Dixon, José R. Espinoza, Jorge Pontt, and Pablo Lezana. PWM regenerative rectifiers: State of the art. *IEEE Transactions on Industrial Electronics*, 52(1):5–22, feb 2005.

[26] Kepa Odriozola, Thierry A. Meynard, and Alain Lacarnoy. Multi-Level Multiplexed Power Converter Topology for 1500V Applications. *IECON Proceedings (Industrial Electronics Conference)*, 2019-Octob:4406–4410, 2019.

[27] Kepa Odriozola, Thierry A. Meynard, and Alain Lacarnoy. Multi-Level Inverter Topologies for Medium- and High-Voltage Applications, 2019.

[28] Christian Felgemacher, Samuel Vasconcelos Araujo, Peter Zacharias, Karl Nesemann, and Artjom Gruber. Cosmic radiation ruggedness of Si and SiC power semiconductors. In *Proceedings of the International Symposium on Power Semiconductor Devices and ICs*, volume 2016-July, pages 51–54. Institute of Electrical and Electronics Engineers Inc., jul 2016.

[29] Richardt H. Wilkinson, Thierry A. Meynard, and Hendrik du Toit Mouton. Natural balance of multicell converters: The two-cell case. *IEEE Transactions on Power Electronics*, 21(6):1649–1657, nov 2006.

[30] Amer M.Y.M. Ghias, Josep Pou, Mihai Ciobotaru, and Vassilios G. Agelidis. Voltage-balancing method using phase-shifted PWM for the flying capacitor multilevel converter. *IEEE Transactions on Power Electronics*, 29(9):4521–4531, 2014.

[31] Thierry Meynard. *Analysis and design of multicell DC/DC converters using vectorized models*. John Wiley & Sons, Inc., 2015.

[32] Fang Z. Peng and Donald J. Adams. Harmonic sources and filtering approaches-series/parallel, active/passive, and their combined power filters. In *Conference Record - IAS Annual Meeting (IEEE Industry Applications Society)*, volume 1, pages 448–455. IEEE, 1999.

[33] Panteleimon Papamanolis, Dominik Neumayr, and Johann W. Kolar. Behavior of the Flying Capacitor Converter Under Critical Operating Conditions. *2017 IEEE 26th International Symposium on Industrial Electronics (ISIE)*, pages 628–635, aug 2017.

[34] T. A. Meynard. Modeling of multilevel converters. *IEEE Transactions on Industrial Electronics*, 44(3):356–364, 1997.

[35] Thierry A. Meynard, Henri Foch, Philippe Thomas, Jacques Courault, Roland Jakob, and Manfred Nahrstaedt. Multicell converters: Basic concepts and industry applications. *IEEE Transactions on Industrial Electronics*, 49(5):955–964, oct 2002.

[36] Bernardo Cougo, Guillaume Gateau, Thierry Meynard, Malgorzata Bobrowska-Rafal, and Marc Cousineau. PD modulation scheme for three-phase parallel multilevel inverters. *IEEE Transactions on Industrial Electronics*, 59(2):690–700, feb 2012.

[37] Brendan Peter McGrath, Thierry Meynard, Guillaume Gateau, and Donald Grahame Holmes. Optimal modulation of flying capacitor and stacked multicell converters using a state machine decoder. *IEEE Transactions on Power Electronics*, 22(2):508–516, mar 2007.

[38] Jose I. Leon, Samir Kouro, Leopoldo G. Franquelo, Jose Rodriguez, and Bin Wu. The Essential Role and the Continuous Evolution of Modulation Techniques for Voltage-Source Inverters in the Past, Present, and Future Power Electronics. *IEEE Transactions on Industrial Electronics*, 63(5):2688–2701, may 2016.

[39] Mohamed S.A. Dahidah, Georgios Konstantinou, and Vassilios G. Agelidis. A Review of Multilevel Selective Harmonic Elimination PWM: Formulations, Solving Algorithms, Implementation and Applications. *IEEE Transactions on Power Electronics*, 30(8):4091–4106, 2015.

[40] Hossein Sepahvand, Mostafa Khazraei, Keith A. Corzine, and Mehdi Ferdowsi. Start-up procedure and switching loss reduction for a single-phase flying capacitor active rectifier. *IEEE Transactions on Industrial Electronics*, 60(9):3699–3710, 2013.

[41] Amer M.Y.M. Ghias, Josep Pou, Vassilios G. Agelidis, and Mihai Ciobotaru. Initial capacitor charging in grid-connected flying capacitor multilevel converters. *IEEE Transactions on Power Electronics*, 29(7):3245–3249, jul 2014.

On-grid/off-grid DC microgrid optimization and demand response management

Wenshuai BAI, Manuela SECHILARIU, and Fabrice LOCMENT
Sorbonne University, Université de Technologie de Compiègne, AVENUES
Rue du docteur Schweitzer, CS 60319, 60203
Compiègne, France
Tel.: +33 / (0)3. 44.23.52.98.
E-Mail: wenshuai.bai@utc.fr, manuela.sechilariu@utc.fr, fabrice.locment@utc.fr
URL: http://avenues.utc.fr

Keywords

«Microgrid», «Optimal control», «Power management», «Energy system management», «Renewable energy systems».

Abstract

Microgrid tends to be an advanced and promising technology in the future smart power supply system, which integrates renewable energy sources under power and energy management. The power management can keep the constant power balance while the energy management can achieve the optimal power flow of a microgrid, where the power prediction should be considered by using the influenced metadata to pre-schedule the power flow of a microgrid. This paper proposes a supervisory system for energy and power management in on-grid/off-grid DC microgrid, which combines sources such as: photovoltaic, storage, public grid connection, diesel generator, and supercapacitor, and supplies a dynamic load. In addition, a load shedding optimization algorithm is proposed to solve the demand response problem. Based on mixed-integer linear programming the energy management layer is used for techno-economic dispatching optimization whereas the power management layer keeps the common DC bus stable. The simulation results, based on real data, prove that: (i) the DC microgrid operates continuously while automatically switching between on-grid/off-grid modes following constraints; (ii) the 24 hours day-ahead power flow optimization achieves a power pre-schedule to decrease the operation cost of the DC microgrid.

Introduction

A microgrid consists of multiply sources and storages, and a real-time power controller achieves different users' load demand, which controls the instantaneous power balance. In reference [1] authors provide an adaptive fractional fuzzy sliding mode control strategy for a hybrid renewable energy source system to maintain the power balance, while in [2] a decentralized power management strategy in an islanded microgrid is shown and a droop control is applied to achieve the power balance. However, the power management system cannot give power trends based on the analytic data collection. Thus, energy management becomes the key method to analyze the collected data from the power monitor and users' load demand. The evident objective of the energy management is to minimize the operating cost of a microgrid. A problem formulation is the key point to solve optimization problems, and it highly depends on a study about microgrid modelling. The computation burden is based on the problem type, linear problem, non-linear problem, etc. In fact, real microgrid modelling is a non-linear and complex system. Fortunately, there are many methods to linear a non-linear constraint or to decompose a complex problem, sometimes, a suboptimal solution is better in the compromise between the computation burden and the computation time limitation than an optimal solution. In reference [3] a mathematical model is developed and integrated into an energy management system for isolated microgrids, in which the optimization problem is divided into unit commitment and optimal power flow subproblems due to the computation burden of the non-linear question, however, it only operates in the islanded model. The reference [4] proposes a mixed-integer algorithm based on the day-ahead forecasting for the optimal dispatch of a storage system; however, there is no load prediction information. In reference [5] mixed-integer quadratic programming is applied to minimize the operation cost of an isolated microgrid using DG and energy storage, the cost of the microgrid is optimized by optimal dispatch of generators and storage; however, the intermitted renewable energy source is not considered. Comparing to an AC microgrid, a DC microgrid is more efficient in energy transfer and simple in the system design and control [6]. In reference [7] a problem formulation is built considering some constraints of the reactive power, which is more complex than a problem formulation in a DC microgrid. In real application, multiple important objectives mostly need to be optimized. In reference [8] a multi-objective optimization introduced, and the Pareto optimal front of the constructed multi-objective problem is obtained using the elitist nondominated sorting genetic algorithm. In real-time controlling, the optimal controller can give a good performance considering multiple constraints, however, the more complex the constraints are, the more difficult the optimization problem. In reference [9] the regulation of AC voltage is implemented through a finite control set model predictive control based active front end rectifier, while direct power MPC is used to control the power during grid-connected operation.

In this paper, a two-layer supervisory system for a DC microgrid, operating in on-grid/off-grid modes, is proposed, which consists of photovoltaic (PV) panels, battery storage system (BS), public grid connection (PG), diesel generator (DG), and supercapacitor (SC) and DC common bus. The SC is used to compensate for power deficiency when the DG is turned on. The proposed two-layer supervisory system consists of a power management layer and an energy management layer. A rule-based method in the power management layer is used to achieve the instantaneous power balance including load shedding optimization. A day-ahead optimization is done in the energy management layer to give an optimal pre-schedule considering weather and users' load demand uncertainties. The supervisory system considers the constraints of the physical components in the microgrid and can operate continuously for 24 hours. The load is assumed to be controllable for the users and the microgrid operator, therefore, a real-time load optimization is applied. Users can turn on and turns off the load randomly. The critical load can be shed and restored through the real-time load optimization according to the load priority, load time constraints, and load power characteristics. The rest of the paper is organized as follows: microgrid modelling, energy management based on day-ahead optimization problem, power management strategy, simulation results, and conclusions.

Microgrid modelling

Fig. 1 illustrates the proposed DC microgrid including its supervisory system, which consists of an energy management layer and a power management layer. The goal of power management is to control the real-time power flow for reaching power balance. The objective of energy management is to optimize the energy cost by dispatching power sources according to the PV and load power predictions and measurement data under the constraints of every physical component.

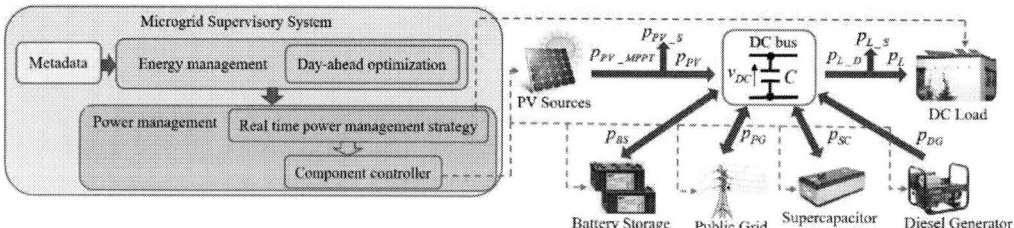

Fig. 1: DC microgrid structure.

For keeping the power balance expressed by (1) and common DC bus voltage stable, a proportional-integral (PI) controller is introduced to calculate the power to be compensated by the PG and BS.

$$\triangle p = p_{PV} - p_L - p_{PI} = p_{BS} + p_{PG} - p_{DG} + p_{SC} \tag{1}$$

where p_{PV} is the power of PV sources, p_L is the DC load power, p_{PI} is the system dynamic power for the PI controller, $\triangle p$ is the compensation power, p_{PG} is the PG power, p_{BS} is the BS power, p_{DG} is the DG supply power, and p_{SC} is the SC power.

The PV modelling comes from [10] and the maximum power point tracking (MPPT) method and the limited power control for PV come from [6]. The PV power is expressed by (2):

$$p_{PV} = p_{PV\_MPPT} - p_{PV\_S} \tag{2}$$

where p_{PV_MPPT} is the MPPT PV power and p_{PV_S} is the shedding power of PV. The PG power p_{PG} is limited as in (3):

$$-P_{G\_MAX} \leq p_{PG}(t) \leq P_{G\_MAX} \tag{3}$$

where P_{G_MAX} is the maximum power that the PG can buy, $-P_{G\_MAX}$ is the maximum power that the PG can sell. The BS can charge and discharge power to keep the microgrid power balance. The state of charge soc_{BS} is calculated according to (4), where C_{REF} is the battery capacity, v_{BS} is the BS voltage, and soc_{BS} is limited between SOC_{BS_MIN} and SOC_{BS_MAX} in (5). The BS charging and discharging powers are limited by P_{BS_MAX} and $-P_{BS\_MAX}$ respectively, as in (6).

$$soc_{BS}(t+\Delta t) = soc_{BS}(t) + \frac{100\%}{3600 C_{REF} v_{BS}} \int_{t}^{t+\Delta t} p_{BS}(t)dt \tag{4}$$

$$SOC_{BS\_MIN} \leq soc_{BS}(t) \leq SOC_{BS\_MAX} \tag{5}$$

$$-P_{BS\_MAX} \leq p_{BS}(t) \leq P_{BS\_MAX} \tag{6}$$

It is assumed that the DG works at duty cycle mode as in [11]. The DG power p_{DG} is limited by its maximal DG supply power P_{DG_MAX} as given in (7).

$$0 \leq p_{DG}(t) \leq P_{DG\_MAX} \tag{7}$$

SC is suggested to compensate for power deficiency during DG start-up stage [12]. The SC power p_{SC} is limited to its maximal SC charging power P_{SC_MAX} and maximal SC discharging power $-P_{SC\_MAX}$ as in (8).

$$-P_{SC\_MAX} \leq p_{SC}(t) \leq P_{SC\_MAX} \tag{8}$$

Due to the natural self-discharging [12, 13], SC is recharged at a certain time to keep its lowest energy for DG start-up compensation by using the soc_{SC} following limitations $SOC_{SC\_MIN\_MIN}$, $SOC_{SC\_MIN\_MAX}$, $SOC_{SC\_MAX\_MIN}$, $SOC_{SC\_MAX\_MAX}$.

The DC load is electrical appliances of a building, in which a load shedding real-time optimization [14, 15] is applied; the problem is formulated and solved by MILP in IBM CPLEX [16]. The load power p_L and the load shedding power p_{L_S} are given respectively by (9) and (10), where p_{L_OPT} is the load power after the load real-time optimization, p_{AVAIL} is the total available DC microgrid power, and p_{L_D} is the load demand power.

$$p_L = \begin{cases} p_{L\_OPT} & if \ p_{AVAIL} < p_{L\_D} \\ p_{L\_D} & if \ p_{AVAIL} \geq p_{L\_D} \end{cases} \tag{9}$$

$$p_{L\_S} = p_{L\_D} - p_L \tag{10}$$

The coefficient k_{L_CRIT} represents the percentage rate defined by the users as the minimum amount of load demand that must attend; it is defined in (11) where p_{L_CRIT} is the power of critical load that must attend.

$$k_{L\_CRIT} = p_{L\_CRIT} / p_{L\_D}, \ k_{L\_CRIT} \in [0\%, 100\%] \tag{11}$$

Energy management

The objective of energy management is to optimize the energy cost by dispatching power flow according to the power prediction and measurement data under the constraints of the modelling above. The optimization objective is to minimize the total energy cost given by (12).

$$C_{TOTAL} = C_{PV\_S} + C_{L\_S} + C_{BS} + C_{PG} + C_{DG} \tag{12}$$

where C_{PV_S} is the PV shedding energy cost, C_{L_S} is the load shedding energy cost, and C_{BS} is the BS energy cost, C_{PG} is the BS energy cost, and C_{DG} is the DG energy cost, which is consists of the fuel cost C_{DG_F}, and the DG maintains cost $C_{DG\_O\&M}$. The C_{PV_S} is calculated according to the amount of PV shedding power and its tariff in (13), the C_{L_S}, C_{BS}, C_{PG}, and C_{DG_F} is calculated as the same method in (13). The $C_{DG\_O\&M}$ is an exception, and calculated as (14).

$$C_{PV\_S} = \frac{1}{3.6 \times 10^6} \sum_{t_i = t_0}^{t_F} T_{PV\_S}(t_i) \cdot \Delta t \cdot p_{PV\_S}(t_i) \tag{13}$$

$$C_{DG\_O\&M} = \frac{1}{3.6 \times 10^3} \sum_{t_i = t_0}^{t_F} T_{DG\_O\&M}(t_i) \cdot \Delta t \cdot (p_{DG}(t_i) > 0) \tag{14}$$

This optimization is considered under the following constraints:

$$s.t. \begin{cases} p_{PV}(t_i) + p_{DG}(t_i) = p_L(t_i) + p_{BS}(t_i) + p_{PG}(t_i) \\ if \ p_{DG}(t_i) > 0 \ then \ p_{PG}(t_i) = 0 \\ if \ soc_{BS}(t_i) > SOC_{BS\_MIN} \ then \ p_{L\_S}(t_i) = 0 \\ if \ soc_{BS}(t_i) < SOC_{BS\_MAX} \ and \ p_{DG}(t_i) = 0 \ then \ p_{PV\_S}(t_i) = 0 \\ p_{DG}(t_i) = p_{DG}(t_{i-1}) \begin{cases} if \ rem(t_i / dt_{DG}) \neq 0 \\ p_{DG}(t_i) \in \{0\} \cup \left[p_{DG\_ON\_MIN} \quad p_{DG\_ON\_MAX} \right] \end{cases} \\ t_i = \{t_0, t_0 + \Delta t, t_0 + 2\Delta t, ..., t_F\} \end{cases} \tag{15}$$

where p_{DG} is zero when DG is off, p_{DG} is limited between its minimal DG output power $P_{DG\_ON\_MIN}$ and maximal DG output power $P_{DG\_ON\_MAX}$ when DG is on, t_i is the time samples, and the time interval between two samples is defined as Δt. Additional constraints are given in (16) and (17):

$$s.t.\begin{cases} if\ p_{PV\_MPPT}(t_i) \geq p_{L\_D}(t_i)\ then\ \begin{cases} p_{PG}(t_i) \geq 0 \\ p_{BS}(t_i) \geq 0 \end{cases} \\ if\ p_{PV\_MPPT}(t_i) \leq p_{L\_D}\ (t_i)\ and\ p_{DG}(t_i) = 0\ then\ \begin{cases} p_{PG}(t_i) \leq 0 \\ p_{BS}(t_i) \leq 0 \end{cases} \\ p_{L\_S}(t_i) \leq (1 - k_{L\_CRIT}) \cdot p_L(t_i) \end{cases} \tag{16}$$

$$s.t.\ p_{L\_S}(t_i) \leq (1 - k_{L\_CRIT}) \cdot p_L(t_i) \tag{17}$$

The additional constraints (16) includes the constraint for the load shedding power and the constraints that the BS and the PG cannot directly exchange power. The additional constraints (17) only constrain the load shedding power, thus the BS and the PG can exchange power under these additional constraints. The coefficients k_D and k_{DG} are given to introduce the day-ahead optimization results into the power management layer and to decouple the system operation between the power management layer and the energy management layer. They are calculated according to (18).

$$k_D = p_{BS} / (p_{BS} + p_{PG}), k_D \in [-\infty, 1]\ \text{ and }\ k_{DG} = \begin{cases} 1 & if\ p_{DG} > 0 \\ 0 & others \end{cases} \tag{18}$$

where k_D is the power distribution rate between BS and PG and k_{DG} represents whether the DG is turned on or not.

Power management

The proposed power management strategy is introduced in the flow chart structed in Fig. 2, Fig. 3 and Fig. 4, which is a rule-based method. The k_D and k_{DG} calculated in the energy management are introduced in the real time power management strategy.

In Fig.2, it is can been seen that the proposed power management strategy consists of 12 cases, the real-time load power optimization is integrated. When Δp is positive or equals to 0 and the load power is greater than its critical load or equals to its critical load, the case 1, 2, 3, 4 can happen. In case 1, PV shedding happens because the BS and the PG are limited. In case 2, 3 the BS and the PG can support Δp. In case 4, the SC is recharging by Δp because soc_{SC} reaches $SOC_{SC\_MAX\_MIN}$, the SC can stop recharging when soc_{SC} is less than $SOC_{SC\_MAX\_MAX}$ or recharging time t_{SC_CH} reaches its maximal recharging time T_{SC_MAX}. The case 5, 6, 7, 8, 9, 10 can happen when Δp is negative or the critical load is shedding. The case 11 shows the process while the DG is turned on.

In Fig.3, the detailed process of the case 5, 6, 7, 8, 9, 10 can be seen. The case 5, 6 can happen when soc_{SC} reaches $SOC_{SC\_MIN\_MAX}$, the load shedding happens in case 5 because there is no more in the BS and the PG to support Δp, the SC is recharging in case 6. There are two flow chart, sub flow-chart a and b describing in the case 7, 8, 9. The sub flow-chart a can be chosen according to the condition when k_D is positive or 0 and Δp is less or equals to $-P_{PG\_MAX}$, the sub flow-chart b can be chosen in the opposite condition. In the sub flow-chart a, the BS and the PG can supply power to the microgrid. In the sub flow-chart b, the PG can support the BS and the microgrid by selling power. The start-up signal is sent in case 10 when soc_{BS} is less or equals to SOC_{BS_MIN} and the critical load is shedding.

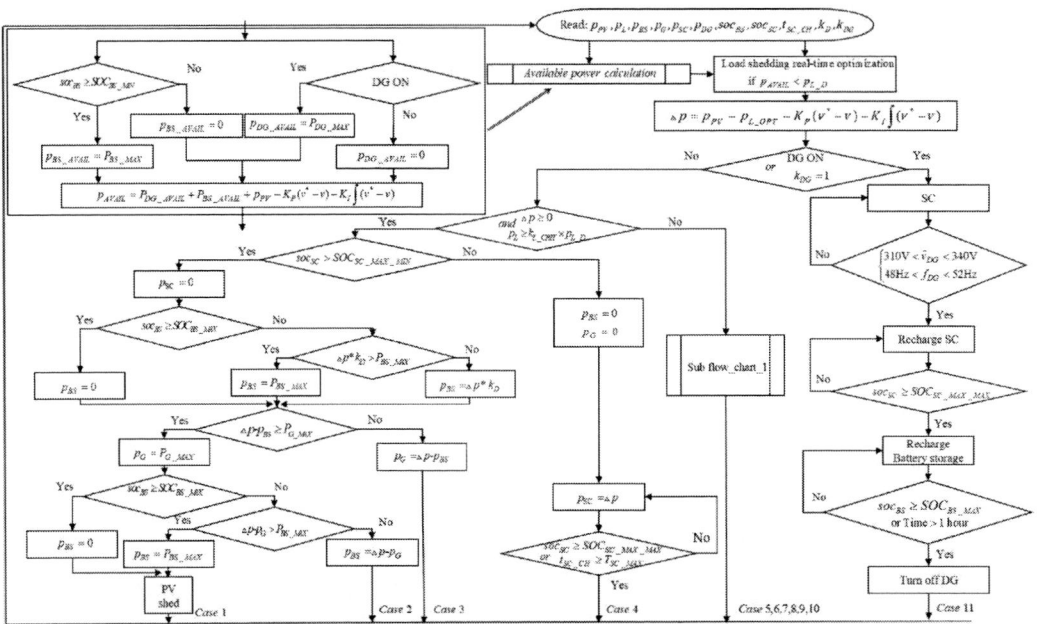

Fig. 2: Full DC microgrid power management flow chart in power management layer.

In Fig.4, the detailed flow chart of case 7, 8, 9 is shown. In case 7 of sub flow-chart a, the load shedding happens because the BS and the PG cannot support $\triangle p$, in case 8, 9 of sub flow-chart a, the BS and the PG can support $\triangle p$. In case 7 of sub flow-chart b, the PG only supply $\triangle p$, in case 8, 9 of sub flow-chart b, the PG can support $\triangle p$ and the BS.

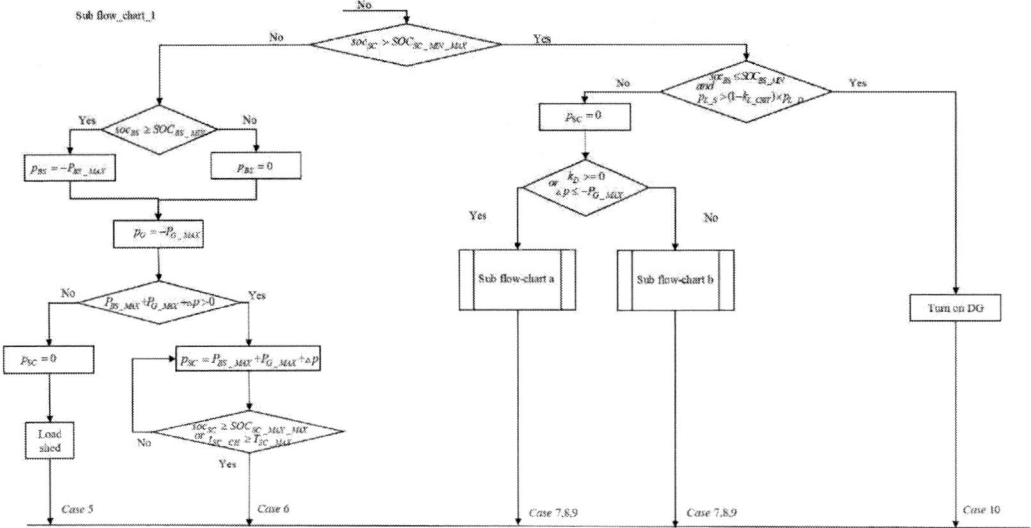

Fig. 3: Full DC microgrid power management sub flow chart 1 in power management layer.

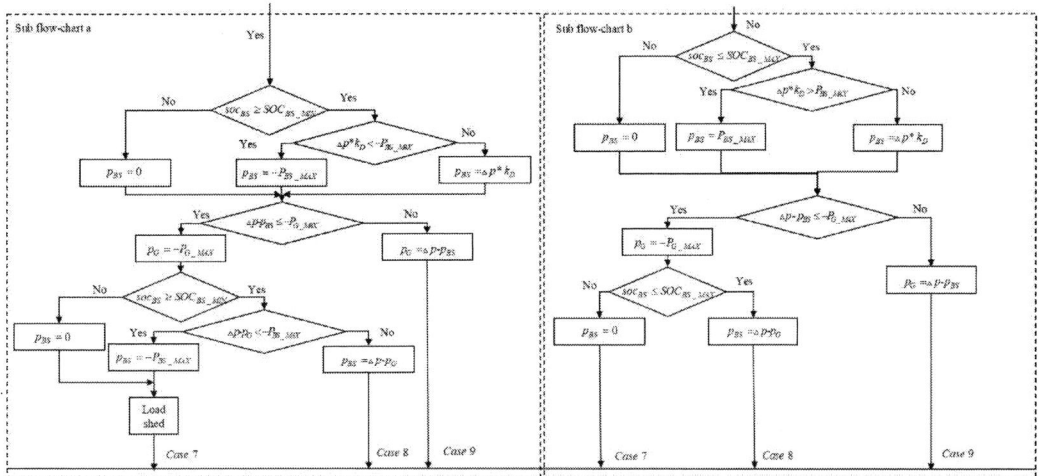

Fig. 4: Full DC microgrid power management sub flow chart a and b in power management layer.

Microgrid simulation results

The studied microgrid is a real application based on a university building in the Université deTechnologie de Compiègne. The PV power curve is calculated according to the weather data recorded on the 20th June 2018 at Compiègne. The PG is considered as a single-phase voltage power source. The load power curve is scaled according to a real daily consumption profile of the university building, which consists of 49 appliances as stated in [15]. The buy-in tariffs and the sell-off tariffs of the PG are the same, which is given according to the TOU method. The high tariff is 0.7 €/kWh in the period, 10:00-12:00 and 17:00-22:00; the middle tariff is 0.1 €/kWh in the period, 6:00-10:00 and 12:00-17:00; the low tariff is 0.01 €/kWh in the period, 0:00-6:00 and 22:00-24:00. The fixed energy tariffs are defined as [6]. The tariff of BS is set to 0.05 €/kWh. The tariff of PV shedding is set to 1.5 €/kWh. The tariff of load shedding is set to 1.8 €/kWh. The tariff of SC is set to 0.3 €/kWh. The fuel tariff of DG is set to 1.2 €/kWh. The average DG operation and maintenance (O&M) tariff is assumed to be 0.63 €/h. The simulation parameters are given in Table I, where v_{DC}^{*} is the reference voltage of the DC bus and the parameters $T_{DG\_ON\_MAX}$ and $T_{DG\_OFF\_MIN}$ are simplified to be a fixed duty cycle.

Table I: Parameters

Parameter	Values	Parameter	Values	Parameter	Values
v_{DC}^{*}	400V	P_{SC_MAX}	1500W	P_{DG_MAX}	1500W
P_{PV_STC}	1750W	T_{SC_MIN}	180s	$P_{DG\_ON\_MIN}$	50W
P_{BS_MAX}	1000W	$SOC_{SC\_MAX\_MAX}$	85%	$P_{DG\_ON\_MAX}$	1500W
SOC_{BS_MIN}	20%	$SOC_{SC\_MAX\_MIN}$	75%	$T_{DG\_ON\_MAX}$	3600s
SOC_{BS_MAX}	80%	$SOC_{SC\_MIN\_MAX}$	45%	$T_{DG\_OFF\_MIN}$	1200s
SOC_{BS_0}	50%	$SOC_{SC\_MIN\_MIN}$	35%	P_{PG_MAX}	200W, 600W
C_{REF}	6.6Ah	SOC_{SC_0}	75%	k_{L_CRIT}	80%, 100%

The simulation time horizon is 24 hours, divided into five time periods: 0:00-6:00, 6:00-8:00, 8:00-17:00, 17:00-22:00 and 22:00-24:00. In every period, the DC microgrid can operate in on-grid/off-grid modes. During 0:00-6:00 and 22:00-24:00 it is assumed that the load demand power is low respecting to the additional constraints expressed by (17), the coefficient k_{L_CRIT} is set to be 100% because all the load demand is the critical load that cannot be shedding, and the PG power is limited to 600W. During 6:00-8:00 the PV energy production is different according to the seasons and it is assumed that the additional constraints given in (16) are used, the coefficient k_{L_CRIT} is set to be 80%, and the PG power is limited to 200W. During 8:00-17:00 the load demand varies according to the energy demand of the university building and it is assumed that the power management uses the additional constraints presented in (16), the coefficient k_{L_CRIT} is set to be 80%, and the PG power is limited to 200W; during 17:00-22:00 the load demand for tertiary buildings is low but for residential buildings are very high so that the PG is highly stressed, therefore, it is assumed that the additional constraints group 1 is used, the coefficient k_{L_CRIT} is set to be 80%, and the PG power is limited to 200W.

The day-ahead optimization results are shown in Fig. 5. In the period, 0:00-6:00, the k_D is negative, representing that the PG should exchange power with the BS. In period, 6:00-24:00, the k_D is positive or zero representing the PG and BS can charge or discharge the microgrid at the same, the k_{DG} is positive in two times representing the DG should be turned on at the two time periods.

The power management results without considering the day-ahead optimization are shown in Fig. 6 and Fig. 7. DC bus voltage v_{DC} is stable to prove the power balance. The PG is charging the BS and supporting the load demand in period 0:00-6:00; the SC is recharging at 5:00. Then, the BS has a higher priority than the PG to supply the microgrid. The microgrid can automatically switch between on-grid/off-grid modes. In the period, 6:00-8:00, the DG is turned on three times due to low PV power and low PG power limitation. In the period, 8:00-17:00, the DG is on at 8:00; then the DG is turned on three times; the PV shedding happens at 12.50, 15:00 and 16:00; the SC is recharging four times. In the period, 17:00-22:00, the PV power is decreasing; the DG is turned on four times. In the period, 22:00-24:00, the PG is supporting the load demand.

The power management results by using the day-ahead optimization results are shown in Fig. 8 and Fig. 9. In the period 0:00-6:00, the PG is charging the BS and supporting the load demand; the SC is recharging at 5:00. Then, the priorities of the PG and the BS supporting the microgrid are changed according to the k_D. In period 6:00-22:00, there are 8 times that the DG is turned on; the PV shedding happens at 11:00, 15:20 and 16:00; the SC is recharging six times. In period 22:00-24:00, the PG only supports the load demand except when the DG is turned on.

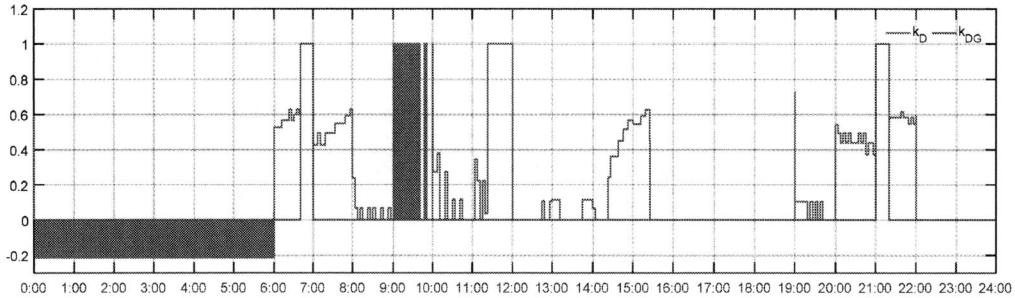

Fig. 5: The k_D and k_{DG} curves of the day-ahead optimization

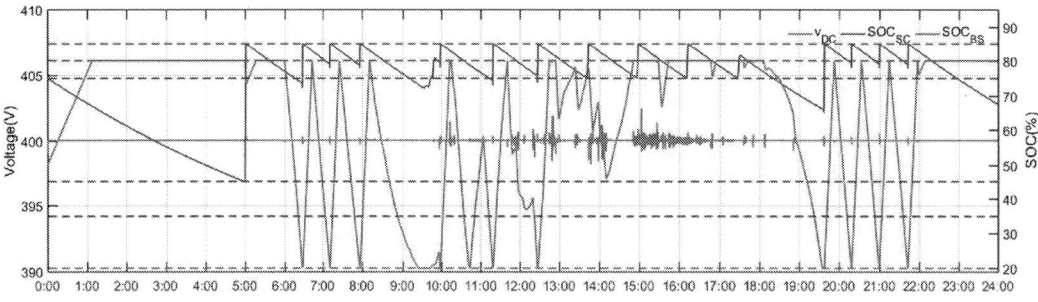

Fig. 6: The v_{DC}, soc_{BS}, and soc_{SC} curves without day-ahead optimization.

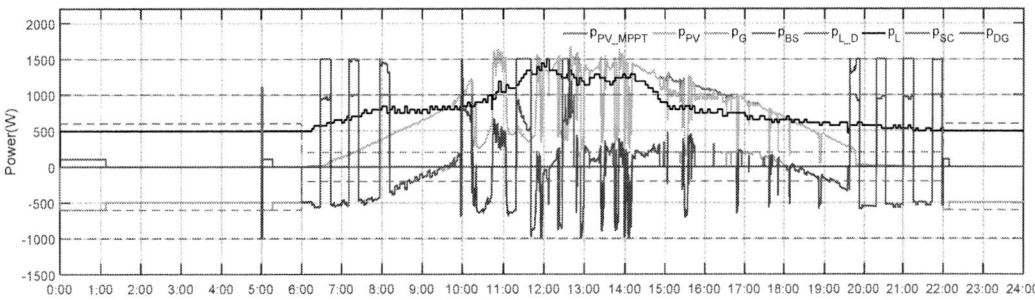

Fig. 7: The power curves without day-ahead optimization.

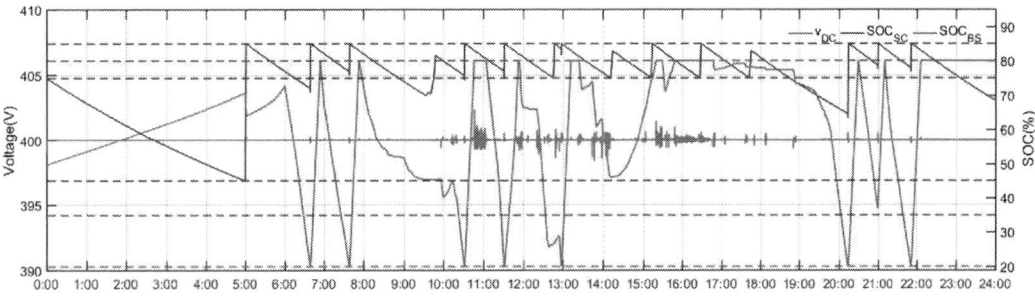

Fig. 8: The v_{DC}, soc_{BS}, and soc_{SC} curves with day-ahead optimization.

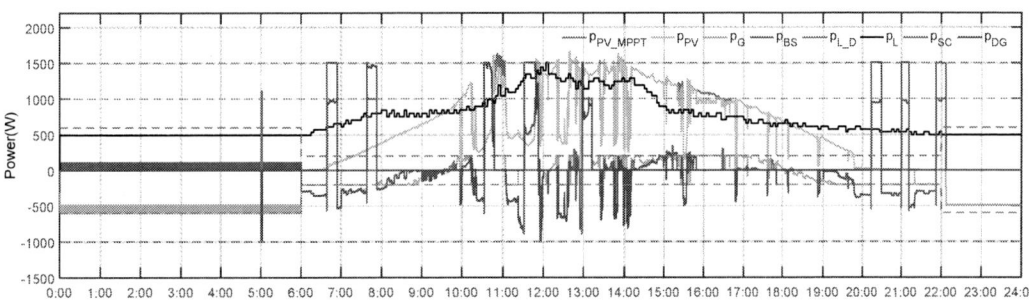

Fig. 9: The power curves with day-ahead optimization.

The cost comparations are listed in Table II. It is evident that the cost of DG is high due to its high tariff.

Table II: The cost comparations

Cost	Without day-ahead optimization (c€)	With day-ahead optimization (c€)
C_{BS}	30.496	21.294
C_G	-7.042	34.527
C_{PV_S}	28.740	22.775
C_{L_S}	0.842	0.018
C_{DG}	609.704	466.057
C_{SC}	4.438	4.380
C_{TOTAL}	667.180	549.054

The PG cost is negative representing that the cost that the PG buys power from the microgrid is higher than the cost that the PG sells power to the microgrid without day-ahead optimization. After the day-ahead optimization, the C_G is increased to positive that the cost that the PG buys power from the microgrid is less than the cost that the PG sells power to the microgrid; the C_{BS}, C_{PV_S}, C_{L_S}, C_{DG} and C_{SC} are reduced after the day-ahead optimization; especially, the 98% of C_{L_S} is reduced, it means that a better service can be given by the microgrid after the day-ahead optimization. Moreover, the total cost of the power management results with day-ahead optimization is 18% less than the total cost of the power management results without day-ahead optimization. Thus, after the day-ahead optimization, the DG start-up and the priorities of BS and PG power are more reasonably set considering the PV and load demand prediction power. The results can prove that the day-ahead optimization in the energy management layer can reduce the total cost and give a good pre-schedule for the power management.

Conclusion

This paper presents a DC microgrid that operates in on-grid/off-grid modes. This supervisory system includes power management to achieve the instantaneous power balance and energy management to give sources' pre-schedule. In the power management layer, the sluggish dynamic of the DG and the self-discharging characteristic of the SC are taken into consideration. Meanwhile, real-time load power optimization is applied. The simulation results show the feasibility and superiority of the system in decreasing the total energy cost. Based on the simulation model described in this study, future research will focus on the nonlinear optimization for DC microgrid taking converter dynamic efficiency into consideration, and analyze the impact of a new optimization algorithm on DC microgrid energy management.

References

[1] Sedaghati R., Shakarami M.R.: A novel control strategy and power management of hybrid PV/FC/SC/battery renewable power system-based grid-connected microgrid, Sust. Cities and Society 2019, Vol. 44, pp. 830-843

[2] Mahmood H., Jiang J.: Decentralized Power Management of Multiple PV, Battery, and Droop Units in an Islanded Microgrid, IEEE Transactions on Smart Grid 2019, Vol. 10 no 2, pp. 1898-1906

[3] Sauter P.S., et al.: Electric Thermal Storage System Impact on Northern Communities' Microgrids, IEEE Transactions on Smart Grid 2019, Vol. 20 no 1, pp. 852-863

[4] Conte F., et al.: Mixed-Integer Algorithm for Optimal Dispatch of Integrated PV-Storage Systems, IEEE Transactions on Industry Applications 2019, Vol. 55 no 1, pp. 238-247

[5] Sufyan M., et al.: Dynamic Economic Dispatch of Isolated Microgrid with Energy Storage Using MIQP, International Conference on Intelligent and Advanced System (ICIAS) 2018

[6] Sechilariu M., Locment F.: Urban DC Microgrid, 2016, Elsevier (Butterworth-Heinemann), pp. 35-91

[7] Zaree N. and Vahidinasab V.: An MILP formulation for centralized energy management strategy of microgrids, Smart Grids Conference (SGC) 2016

[8] Dissanayake A.M. and Ekneligoda N.C.: Multiobjective Optimization of Droop-Controlled Distributed Generators in DC Microgrids, IEEE Transactions on Industrial Informatics 2020, Vol. 16 no 4, pp. 2423-2435.

[9] Rehman S., et al.: Optimal Design and Model Predictive Control of Standalone HRES: A Real Case Study for Residential Demand Side Management, IEEE Access 2020, Vol. 8, pp. 29767-29814.

[10] Dolara A., Leva S., ManzoliniG.: Comparison of different physical models for PV power output prediction, Solar Energy 2015, Vol. 119, pp. 83-99

[11] Yin C., et al.: Power Management Strategy for an Autonomous DC Microgrid, Applied Sciences 2018, Vol. 8 no 11.

[12] Yin C., Sechilariu M., and Locment F.: Diesel generator slow start-up compensation by supercapacitor for DC microgrid power balancing, IEEE International Energy Conference (ENERGYCON) 2016

[13] Yin C., et al.: Energy management of DC microgrid based on photovoltaic combined with diesel generator and supercapacitor, Energy Conversion and Management 2017, Vol. 132, pp. 14-27

[14] Tran Dang K., et al.: Load shedding and restoration real-time optimization for DC microgrid power balancing, IEEE International Energy Conference (ENERGYCON) 2016.

[15] Trigueiro dos Santos L., Sechilariu M., Locment F.: Optimized Load Shedding Approach for Grid-Connected DC Microgrid Systems under Realistic Constraints, Buildings 2016, Vol. 6 no 4, pp. 50-64

[16] IBM ilog cplex optimizer. Available from: http://ibm.com.

Shedding and restoration algorithms for an EV charging station to maximize available power

Dian WANG, Fabrice LOCMENT, and Manuela SECHILARIU
Sorbonne University, Université de Technologie de Compiègne, AVENUES
Rue du docteur Schweitzer, CS 60319, 60203
Compiègne, France
Tel.: +33 / (0)3. 44.23.73.17.
E-Mail: dian.wang@utc.fr, fabrice.locment@utc.fr, manuela.sechilariu@utc.fr
URL: https://avenues.utc.fr

Acknowledgements

Thanks are due to ADEME France that funded this research in the context of the call for projects APRED 2017, project MOBEL_CITY grant number #1766C0006.

Keywords

«Electric vehicle», «Charging Infrastructure for EV´s», «Microgrid», «Power management», «Optimal control».

Abstract

Due to its green and environmentally friendly characteristics, the number of electric vehicles continues to increase, which leads to a greater demand for power, therefore, how to meet driver needs while maintaining electric vehicle charging station power balance is the major problem needed to be solved. This paper presents shedding and restoration optimization algorithms for an electric vehicle charging station based on a DC microgrid to maximize the utilization of available power to meet driver needs, meanwhile, taking into consideration the intermittency of the photovoltaic source, the capacity limitation of the storage, and the power limitation of the public grid. The simulation results show that compared with rule-based algorithm, the proposed EV shedding and restoration optimization algorithms respect the user's choice while reducing total charging time, increasing the full rate, and maximizing the available power utilization, which shows the feasibility and effectiveness of EV shedding and restoration optimization algorithms.

Introduction

With the sharp reduction of nonrenewable energy and the increasing awareness of environmental protection, electric vehicles (EVs) will play a vital role in the future [1], [2]. However, the large number of EVs connected to the public grid will bring a new round of rapid load growth, will cause a peak-to-valley difference in power load, and will increase the huge power supply pressure [3], [4]. In addition, as a power load, the distribution of EVs in time and space has great randomness and uncertainty, which may imply a bad impact to the public grid [5], [6]. Therefore, an issue that needs to be studied is how to control and manage the EVs connected to the public grid so that they can interact with each other to maintain the stability of the public grid under the high EVs' penetration and to maximize the available power of the electric vehicle (EV) charging station.

To reduce the consumption of fossil fuels and achieve the true meaning of low-carbon life, accompanied by sharp cost reductions for photovoltaic (PV) sources, in particular, grid-connected PV sources systems like EVs charging stations centered on a microgrid based PV sources generation are proposed [7], [8]. The management of the EV charging station should consider both the needs of users and the revenue of the charging station. An optimization model for EVs scheduling in a smart grid is proposed in [9], which considered a problem including the economic cost of energy production/acquisition from the public grid and the cost relevant to the delay in the satisfaction of the customers' demand. A multi-objective function is introduced to balance the tradeoff between maximizing the microgrid revenue and minimizing the microgrid operating cost [10]. The reference [11] proposes an optimal scheduling mathematical model for a DC microgrid consisted of PV system and EV charging station, and solves the model to acquire the Pareto solution with cost of electricity purchasing and energy circulation of storage batteries. In addition, the state management of each EV should be fully considered, especially EV shedding and EV restoration. The reference [12] presents the practical experience achieved with the operation of ten EVs and solar station during two years, taking different states of EVs into consideration: "EV is waiting", "EV is scheduled", "EV is charging", and "EV charging is finished". On the other hand, the reference [13] presents a load shedding/restoration real-time optimization for DC microgrid building-integrated. The objective of this optimization is solving the problem of selection and controlling loads in the emergency conditions and/or limitation of available power by fixing a priority coefficient to each load that needs to be controlled. However, in this case, the load types and power demand are deterministic while the EV power demand varies according to the user's need, *i.e.* charging modes and state of charge (soc_{EV}) in real-time.

Therefore, this paper presents EV shedding and restoration optimization algorithms (SROA) for an EV charging station which is based on a DC microgrid. Simulation results show the increase in available power utilization and prove the technical feasibility of the proposed optimization algorithms. The rest of the paper is organized as follows. The following sections provide details on the structural frame of an EV charging station, optimization algorithm and mathematical formulation, simulation results and analysis. Finally, conclusions are presented.

EV charging station based on a DC microgrid

A microgrid is a multi-source and multi-load system based on renewable energy sources and storage. Compared with the AC microgrid, the DC microgrid may be the main power supply structure for EV charging stations in the future due to its simplicity, economy and high efficiency.
The studied EV charging station is designed based on a DC microgrid [14]. As illustrated in Fig. 1, it is composed of PV sources, an electrochemical storage system, a public grid connection, and EV charging terminals. These components are all connected directly to the DC bus.

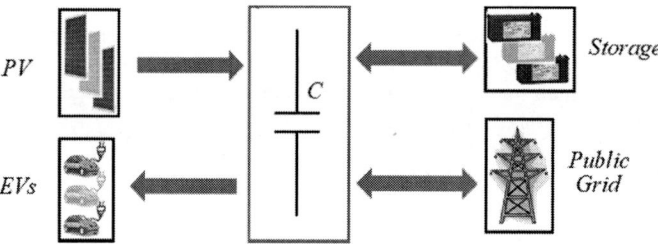

Fig. 1: DC microgrid for an EV charging station

Power balance

The power balance of the microgrid is expressed by (1):

$$p_{EVs} = p_{PV} + p_S + p_G \tag{1}$$

where p_{EVs} is the power demanded by the EVs, p_{PV} is the power of PV sources, p_S is the power of the storage, p_G is the public grid power.

PV sources

To operate a PV source within its maximum power point, whatever the solar irradiance and cell temperature variations, a maximum power point tracking (MPPT) algorithm is needed to find and maintain the peak power. Conversely, when the storage and public grid reach their upper limits, the PV production should be constrained. Therefore, in this study, PV sources work in two modes, PV-constrained production control, and PV MPPT control, and p_{PV} is expressed by (2).

$$p_{PV} = p_{PV\_MPPT} - p_{PV\_S} \tag{2}$$

where p_{PV_S} is the shedding power of PV sources, p_{PV_MPPT} is the MPPT power of PV sources.

Storage

To avoid storage damage or overcharging/over-discharging, the microgrid storage constraints are expressed by (3) for the power and by (4) for its state of charge soc.

$$-P_{S\_MAX} \leq p_S \leq P_{S\_MAX} \tag{3}$$

where $\left|P_{S\_MAX}\right|$ is limited by the maximum current I_{S_MAX} of the storage charging or discharging.

$$SOC_{MIN} \leq soc \leq SOC_{MAX} \tag{4}$$

where SOC_{MIN} is the soc lower limit and SOC_{MAX} is the soc upper limit.

Public grid

The public grid power p_G is constrained as follows:

$$-P_{G\_LIM} \leq p_G \leq P_{G\_LIM} \tag{5}$$

where the maximum output power of the public grid P_{G_LIM} is the absolute value of grid injection power limitation (positive) and grid supply power limitation (negative).

EV charging station operation

Fully considering the public grid constraints and the storage constraints, this multi-source system provides an available power, p_{EVs_lim}, namely the power limitation for the EV charging station and expressed by (6):

$$p_{EVs\_lim} = p_{PV\_MPPT} + P_{G\_LIM} + P_{S\_MAX} \quad \text{with } (soc > SOC_{MIN}) \tag{6}$$

where the storage can output power only when soc is greater than SOC_{MIN}.

When the EV arrives at the charging station, the user selects the charging mode or departures according to the available power of the charging station. After an EV is connected to the charger, the soc_{EV} is detected in real-time and the relative power and voltage are set according to soc_{EV} and the charging mode, and then charging is started. Once the users choose to charge, it means that when the power is insufficient, they can accept the downshift or waiting; also, the user can choose the departure at any time.

Shedding and restoration algorithms

This paper presents two algorithms for EV shedding and restoration: rule-based algorithm (RBA) and SROA, which are demonstrated in an EV charging station to simultaneously support five EVs.

RBA algorithm

EV shedding and restoration decisions of the RBA are obtained by comparing the value of the power demand, p_{EVs_D}, and the power limitation of the EV charging station. Flow chart of RBA is shown in Fig. 2. When the power limitation is insufficient to attend the EV power demand, the EV charging station automatically performs the EV shedding operation in the order from the 5th EV, i.e. EV5, to 1st EV, i.e. EV1. When the power limitation is sufficient to attend the EV power demand, the EV charging station automatically performs the EV restoration operation in the order from EV1 to EV5, if there are EVs on waiting connection to the charger.

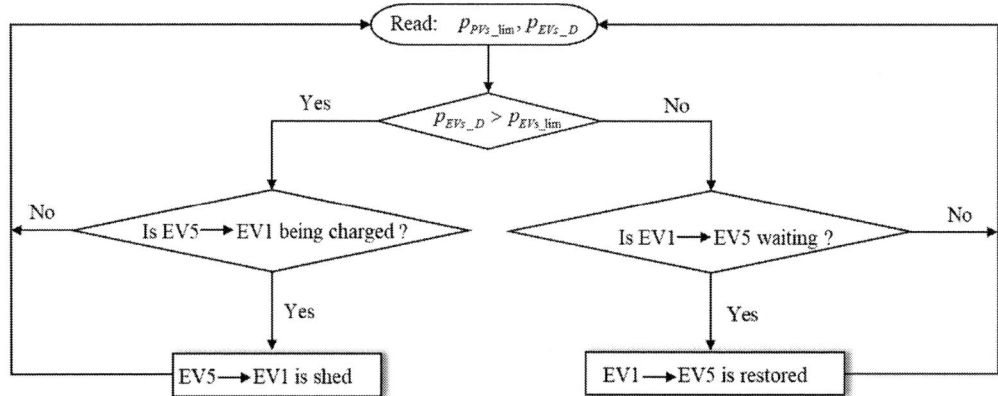

Fig. 2: Flow chart of RBA.

SROA algorithm

The goal of the SROA is to maximize the utilization of available power from the microgrid. The optimal EV shedding and restoration decisions are gained according to the power demand and the power limitation of the EV charging station. Flow chart of SROA is shown in Fig. 3.

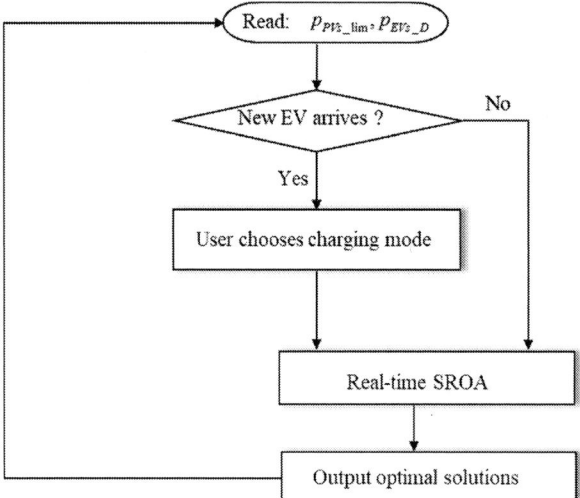

Fig. 3: Flow chart of SROA.

When the power limitation is insufficient to attend the EV power demand, the algorithm calculates the charging power that needs to be reduced, and then, the algorithm calculates the power value of which

EV downshifts or stops charging is closest to and greater than the power value needed to be reduced. In order to maximize the available power, the algorithm performs the downshift or stopping charging of the EV. When the power limitation is sufficient to attend the EV power demand, the algorithm calculates the remaining power value. If there are EVs on waiting or downshifted, the algorithm performs the EV restoration operation or upshift whose power value is closest to and less than the remaining power value.

The optimization problem, objective function and constraints of shedding and restoration for the EV charging station, are presented by following. Regarding the charging modes, it is considered that each EV has four choices corresponding to the following powers: P_{FAST_MAX} is the maximal charging power demanded by the fast mode, P_{AVER_MAX} is the maximal charging power demanded by the average mode, P_{SLOW_MAX} is the maximal charging power demanded by the slow mode, and 0 for waiting. The objective function to minimize the real-time remaining available power of the EV charging station that should be greater than or equal to zero is expressed by (7):

$$p_{\text{remaining}} = p_{EVs\_\lim} - p_{EVs\_D} \geq 0 \tag{7}$$

where $p_{\text{remaining}}$ is the real-time remaining available power of the EV charging station, p_{EVs_D} is calculated in (8), i represents the EV order and x represents the EV user possible choices as constrained in (9):

$$p_{EVs\_D} = \sum_{i=1}^{5} (P_{FAST\_MAX} \cdot x_{(i-1)\cdot 4+1} + P_{AVER\_MAX} \cdot x_{(i-1)\cdot 4+2} + P_{SLOW\_MAX} \cdot x_{(i-1)\cdot 4+3} + 0 \cdot x_{(i-1)\cdot 4+4}) \tag{8}$$

$$x_{(i-1)\cdot 4+1} + x_{(i-1)\cdot 4+2} + x_{(i-1)\cdot 4+3} + x_{(i-1)\cdot 4+4} = 1, \; i \in \{1,2,3,4,5\} \tag{9}$$

where $x_{(i-1)\cdot 4+1}, x_{(i-1)\cdot 4+2}, x_{(i-1)\cdot 4+3}, x_{(i-1)\cdot 4+4}$ are the choices, i.e. fast mode, average mode, slow mode, and waiting, for EV1, EV2, EV3, EV4, and EV5; the domains of $x_{(i-1)\cdot 4+1}, x_{(i-1)\cdot 4+2}, x_{(i-1)\cdot 4+3}, x_{(i-1)\cdot 4+4}$ are set to $\{0\}$ or $\{0,1\}$ by the users when an EV is coming.

Simulation results and analysis

The simulation for SROA is performed with MATLAB/Simulink and CPLEX that is an optimization software package for solving mixed-integer linear programming. Five chargers are considered. The simulation parameters are detailed in Table I, where P_{PV_STC} is the estimated PV power under standard test conditions and v_C^* is the reference voltage of the DC bus.

Table I: Parameters of DC microgrid

Parameters	Values	Parameters	Values
P_{FAST_MAX}	83kW	I_{S_MAX}	115A
P_{AVER_MAX}	27kW	SOC_{MIN}	20%
P_{SLOW_MAX}	7kW	SOC_{MAX}	80%
P_{PV_STC}	100kW	Number of chargers	5
P_{G_LIM}	50kW	v_C^*	400V

Fig. 4 and Fig. 5 show the DC bus voltage evolution v_C by using the RBA and SROA, respectively. The voltage deviation in Fig. 4 is 0.23%, and in Fig. 5 is 0.35%, because the calculation speed of SROA is faster than that of RBA. The standard deviation of voltage is 5.75 both in Fig. 4 and Fig. 5, so the voltage stability is the same.

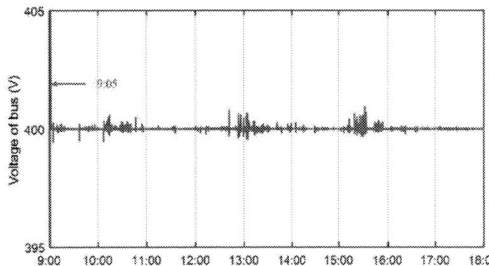 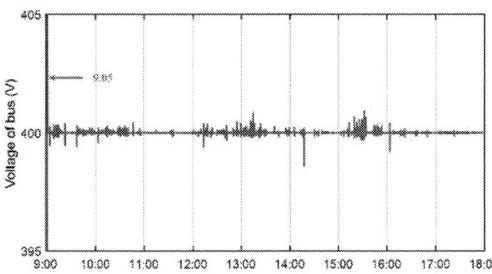

Fig. 4: DC bus voltage after RBA.   Fig. 5: DC bus voltage after SROA.

Fig. 6 and Fig. 7 show the soc_{EV} evolution of each EV by using the RBA and SROA, respectively. The waiting time of EVs in Fig. 7 is significantly less than that of Fig. 6, because when the power is insufficient, the EV in Fig. 6 directly chooses to wait, and the EV in Fig. 7 first selects the downshift. Fifteen EVs with the same battery characteristics choose to charge both in the RBA and SROA. Eleven EVs in Fig. 6 are fully charged and, four EVs are not fully charged, thus, the full rate is 73.3%. Twelve EVs in Fig. 7 are fully charged and, three EVs are not fully charged, thus, the full rate is 80%. The SROA effectiveness is therefore shown.

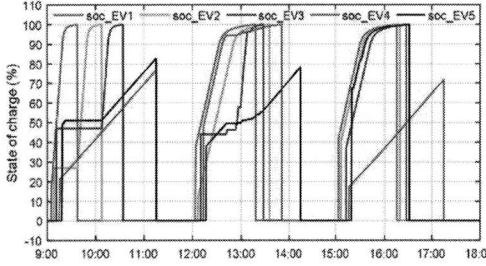 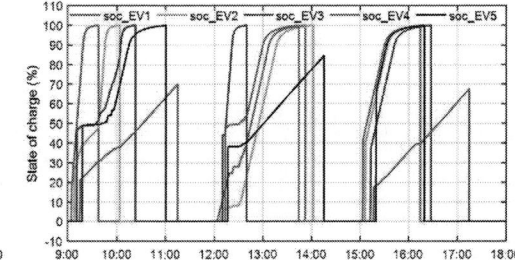

Fig. 6: SOC_{EV} evolution of each EV after RBA.   Fig. 7: SOC_{EV} evolution of each EV after SROA.

Fig. 8 and Fig. 9 show the power evolutions of PV MPPT and PV by using the RBA and SROA, respectively. The same meteorological conditions are considered for RBA and SROA. The PV energy produced remains a priority for EVs. In case of insufficient energy, system security is ensured thanks to the storage system and public grid connection.

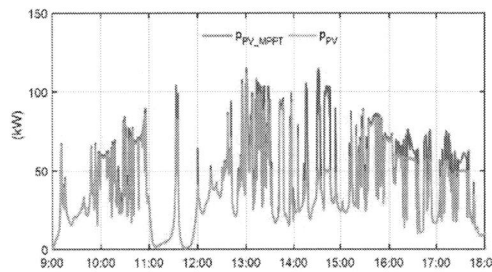 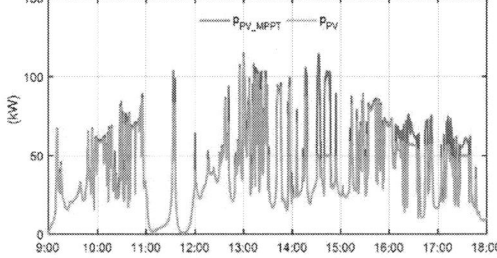

Fig. 8: PV MPPT and PV powers after RBA.   Fig. 9: PV MPPT and PV powers after SROA.

Fig. 10 and Fig. 11 show the power evolutions of power limitation and EVs' total power by using the RBA and SROA, respectively. The total power variation after SROA is smaller than that of RBA, because when the power is insufficient, downshift is considered firstly instead of waiting under SROA. The total charging time in Fig. 10 and Fig. 11 are 21.3h and 20.7h, respectively. The power utilization in Fig. 10 and Fig. 11 are 20.5% and 21.8%, respectively. So, the SROA shortens the total charging time and improves power utilization.

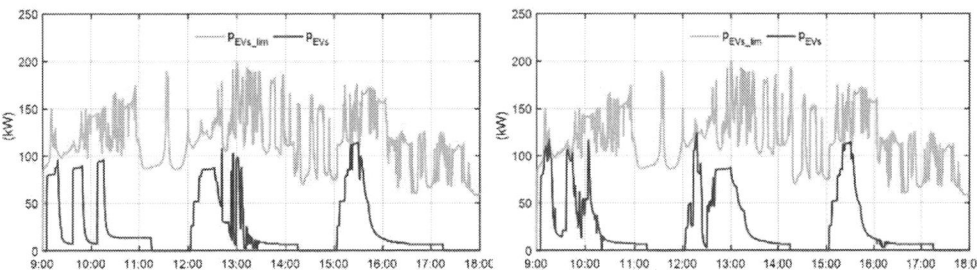

Fig. 10: EVs' power limitation and total power after RBA.

Fig. 11: EVs' power limitation and total power after SROA.

Fig. 12 and Fig. 13 show power evolutions of storage and public grid by using the RBA and SROA, respectively, where the public grid power limitation is 50 kW. The grid power is positive means that the public grid receives power; the grid power is negative means that the public grid supplies power to EVs.

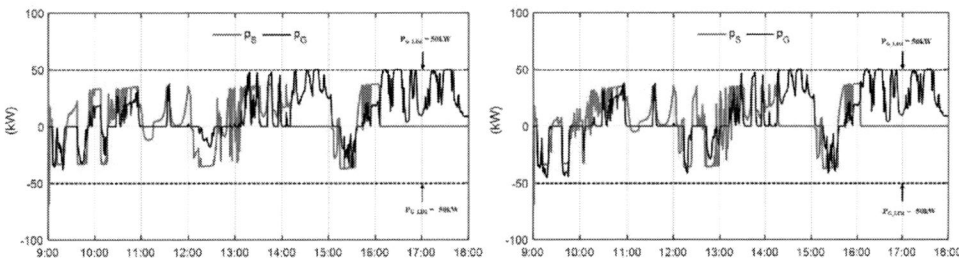

Fig. 12: Storage and public grid powers after RBA.

Fig. 13: Storage and public grid powers after SROA.

Fig. 14 and Fig. 15 show storage current by using the RBA and SROA, respectively, and present the storage current limitation for a value of 115 A.

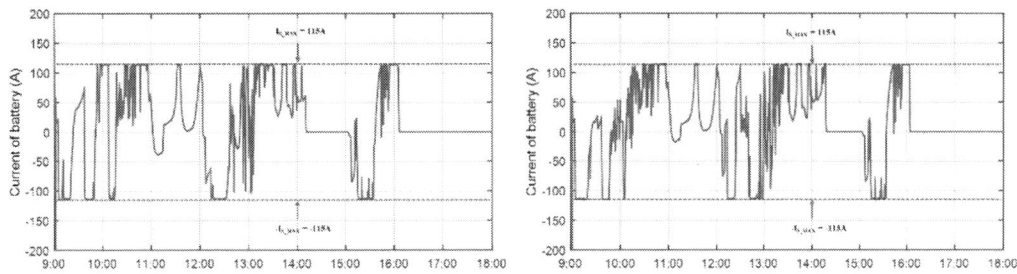

Fig. 14: Storage current after RBA.

Fig. 15: Storage current after SROA.

The comparison of results obtained by using the RBA and SROA are detailed in Table II. Due to the faster voltage regulation, the voltage deviation value by using SROA is greater than the voltage deviation value by using RBA, but the standard deviation of voltage is the same, so voltage stability has not been changed. The total charging time is shortened, the full charging rate becomes higher, and the utilization of available power becomes larger, demonstrating the availability and effectiveness of SROA.

Table II: Comparison of results by using RBA and SROA

Algorithm	Voltage deviation	Standard deviation of voltage	Total charging time	Full rate	Power utilization
RBA	0.23%	5.75	21.3h	73.3%	20.5%
SROA	0.35%	5.75	20.7h	80%	21.8%

Conclusion

As a low-noise, zero-direct emission vehicle, EVs have received unprecedented popularity. The explosive growth in the number of EVs causes to an increase in power demand, resulting in energy challenges of EV charging stations, therefore, how to maximize the available power of the EV charging station under the high EVs' penetration is an urgent problem. Compared with the RBA, the proposed SROA respects the user's choice while reducing total charging time, increasing the full rate, and maximizing the available power utilization, which provides an effective technical means for the stable operation of an EV charging station based on the microgrid. In addition, the SROA has certain reference significance for reducing the impact of EV charging on the peak load of the public grid.

References

[1] Mahmud K., Town G. E., Morsalin S., Hossain M. J.: Integration of electric vehicles and management in the internet of energy, Renew. Sustain. Energy Rev. 2018, Vol. 82 no 3, pp. 4179-4203

[2] Locment F., Sechilariu M.: DC microgrid for future electric vehicle charging station designed by Energetic Macroscopic Representation and Maximum Control Structure, IEEE International Energy Conference (ENERGYCON) 2014.

[3] Gong Q., Midlam-Mohler S., Serra E., Marano V., Rizzoni G.: PEV charging control considering transformer life and experimental validation of a 25 kVA distribution transformer, IEEE Trans. Smart Grid 2015, Vol. 6 no 2, pp. 648–656

[4] Saldaña G., San Martin J.I., Zamora I., Asensio F.J., Oñederra O.: Electric Vehicle into the Grid: Charging Methodologies Aimed at Providing Ancillary Services Considering Battery Degradation, Energies 2019, Vol. 12 no 12, pp. 2443

[5] Dang Q. Y.: Electric Vehicle (EV) Charging Management and Relieve Impacts in Grids, IEEE International Symposium on Power Electronics for Distributed Generation Systems (PEDG) 2018.

[6] Wang D., Wu H., Sechilariu M., Locment F.: Management strategy of an electric vehicle charging station under power limitation, International conference on theory and application of modelling, simulation, analysis, design optimization, identification and diagnostics in electrical power engineering (ELECTRIMACS) 2019.

[7] Locment F., Sechilariu M.: Modeling and Simulation of DC Microgrids for Electric Vehicle Charging Stations, Energy 2015, Vol. 8 no 5, pp. 4335-4356

[8] Sechilariu M., Locment F., Darene N.: Social Acceptance of Microgrids Dedicated to Electric Vehicle Charging Stations, International Conference on Renewable Energy Research and Applications (ICRERA) 2018.

[9] Ferro G., Laureri F., Minciardi R., Robba M.: An optimization model for electrical vehicles scheduling in a smart grid, Sustainable Energy, Grids and Networks 2018, Vol. 14, pp. 62-70

[10] Nguyen D. T., Le L. B.: Optimal energy trading for building microgrid with electric vehicles and renewable energy resources, Innovative Smart Grid Technologies (ISGT) 2014.

[11] Lu X., Liu N., Chen Q., Zhang J.: Multi-objective optimal scheduling of a DC micro-grid consisted of PV system and EV charging station, IEEE Innovative Smart Grid Technologies - Asia (ISGT ASIA) 2014.

[12] Ha D., Guillou H., Martin N., Cung V., Jacomino M.: Optimal scheduling for co- ordination renewable energy and electric vehicles consumption, IEEE International Conference on Smart Grid Communications (SmartGridComm) 2015.

[13] Tran Dang Khoa, Dos Santos L. T., Sechilariu M., Locment F.: Load shedding and restoration real-time optimization for DC microgrid power balancing, IEEE International Energy Conference (ENERGYCON) 2016.

[14] Wang D., Locment F., Sechilariu M.: Modelling, simulation and management strategy of an electric vehicle charging station based on a DC microgrid, Applied Sciences 2020, Vol. 10, pp. 2053-2074.

AUTHOR INDEX

Aarniovuori, Lassi 2829
Abbate, Carmine..2802
Abbosh, Amin...1006
Abdel-Rahim, Naser 352
Abdelhakim, Ahmed...2220
Abdelrahem, Mohamed 900
Abramson, Rose A...1934
Abusara, Mohammad..471
Aganza-Torres, Alejandro..1813
Aguglia, Davide..3330
Ahmad, Bilal...3348
Ahmad, Faheem..2987
Ahola, Jero..2753
Ait-Ahmed, Mourad 3289
Aizpuru, Iosu 251, 1205
Alam, M. M. 480, 1551
Alam, Muhammad Farhan 416
Alatise, Olayiwola 2241
Alawieh, Hadi..1685
Albach, Manfred 173, 193
Alexandre, Philippe 1905
Ali, Ahmed Ismail M..1417
Ali, Marwan 1118, 2039
Ali, Mohammad..2743, 2763
Ali, Waqas...871
Alisar, Ibrahim..1205
Alishah, Rasoul Shalchi..460
Alkama, Kouceila...2564
Allard, Bruno 522, 829, 1470, 1874
Almaksour, Khaled..1700
Almeida, Bruno F...3217
Alonso, C..919
Alqatamin, Moath..65
Am, Sokchea..3172
Ammann, Ulrich...3137
Amrane, Fayssal..362
Anders, Erik...944
Andrade, Fabio 1400, 1841, 1850
Andresen, Jan...2303, 2545
Anzola, Jon..251
Aoustin, Yannick...1923
Arandia, Nerea...1524
Arazi, M...2881
Arrizabalaga, Antxon...1205
Arrozy, Juris..1067
Arruti, Asier..251
Artiglia, Melissa..2791
Aríztegui, Raquel González.......................................1515

Asllani, Besar...1279
Avenas, Yvan...1685, 2564
Averbukh, Moshe..2439
Averous, Nurhan Rizqy..153
Azizian, Mohammadreza..2860
Baburajan, Silpa...2573
Bacha, S...406
Bacha, Seddik..820
Baërd, H..95
Bagaber, Bakr..2613
Bahman, Amir Sajjad..2704
Bahrani, Behrooz...787
Bai, Wenshuai..667
Baker, Erik..512
Bakran, Mark-M..............644, 686, 1252, 1533, 1831, 1885,
   2106
Bakri, R...2554
Bakri, Reda..2078
Balkowiec, Tomasz..2029
Barazi, Yazan..1057
Barelli, Linda...292
Barg, Sobhi..416
Barwig, Markus...1460
Basic, D...37, 95
Basic, Duro..2938, 2957
Bauer, Pavol 1224, 1233, 1561, 3422
Bazin, Pascal..2467
Beczkowski, Szymon Michal..2987
Beerten, J...1551
Belhaouane, Moez...1158, 1756
Bello, Guilherme...2049
Benchaib, A..745
Benchaib, Abdelkrim..820, 1215
Bender, Vitor C..3217
Bendfeld, Christian..163
Benjamin, Sébastien..3156
Benkhoris, Mohamed Fouad...3205
Bensebaa, S..1363
Bentivegna, N..2293
Benzagmout, Abdelhadi..1905
Beranger, Bruno..2467
Berkani, M...1363
Bernet, Steffen..124, 1569
Bertele, Felix...3137
Bertilsson, Kent...416, 460
Betto, Kento...3071
Betz, Robert Eric..927
Bevilacqua, Pascal...1279

Beza, Mebtu 969, 1952
Bhajana, V. V. Subrahmanya Kumar 2068
Bidini, Gianni 292
Biela, Juergen 2331, 2583, 2791
Biela, Jürgen 2230, 2409, 2446, 2475, 2513, 2524,
.... 2684, 2712, 2780, 2946
Bier, Anthony 2638
Bikinga, Wendpanga Fadel 2564
Binder, Andreas 2049
Birou, Camille 332
Bissal, Ara 871
Blaabjerg, F. 3237
Blaabjerg, Frede 1, 460, 810, 927, 2088, 2135,
.... 2220, 2393, 2573, 2888, 2898, 2928, 3119
Blanco, Marcos 1076
Blanquez, Francisco R. 1336
Blaquiere, Jean-Marc 1057
Blinov, Andrei 2996
Blume, Sebastian 2684
Böcker, Jan 2341
Böcker, Joachim 1638, 3024
Boersma, S. 745
Bohlen, Oliver 707
Bohnke, M. 1613
Boige, François 1057
Boisaubert, Emile 3172
Bolzan, Thais E. 3217
Bolzoni, A. 1306
Bombois, X. 745
Bongiorno, Massimo 969, 1952
Borcherding, Holger 608
Boulaud, Etienne 3172
Bourennane, Abdelhakim 1096
Bourguet, Salvy 2907
Bouscayrol, Alain 3330
Boutleux, Emmanuel 433
Boutry, Arthur 2366
Bozorg, Mokhtar 3247
Boškovic, N. 2812, 2820
Branca, Xavier 522
Briff, Pablo 9
Bringezu, Thilo 2780
Brockhage, Torben 1766
Brooks, Michael 2773
Bruyere, Antoine 1756
Brückner, Thomas 2938, 2957
Bründlinger, Roland 2723
Büdel, Johannes 1718
Bucher, A. 3403
Bucher, Alexander 203
Budo, Kohei 1450
Buigues, Garikoitz 85

Burgos, Rolando 2366
Burgos-Mellado, Claudio 1354
Busatto, G. 3210
Busatto, Giovanni 2802
Buttay, Cyril 1106, 2265, 2366
Cacciato, M. 909
Cai, Pei 2135
Camail, Philippe 2000, 2265
Camara, M. B. 2881
Camurca, L. 3305
Cardelli, Ermanno 292
Cárdenas, Roberto 1354
Carnielutti, Fernanda 2851
Caron, Hervé 1700
Carpita, Mauro 3247
Carpiuc, Sabin 962
Cascino, S. 2293
Castellazzi, Alberto 2210
Castellini, Simone 292
Castelltort, Arnaud 1747
Castiglia, V. 3237
Catellani, Stéphane 2638
Cavallaro, D. 2293
Chaiba, Azeddine 362
Chakraborty, Sajib 2320, 3111
Cheaito, Hassan 829
Chen, Linglin 1224, 1233
Chen, Qing 542
Chen, Yu 804
Chen, Zhengxin 637
Cheshire, Christoph 3137
Chevalier, Florian 1895
Chillón-Antón, Cristian 853, 1542
Chiumeo, Riccardo 424
Chraye, Hélène 3427
Chrin, Phok 3172, 3376
Chrzan, Piotr J. 3054
Chédot, L. 183
Chédot, Laurent 433, 1106
Ciupageanu, Dana-Alexandra 292
Clerc, Guy 433, 829
Clerici, Alessio 424
Cochelin, Anne-Sophie 3426
Coelho-Medeiros, Rafael 1479
Colak, Ilknur 871
Colas, F. 1579, 2554
Colas, Frédéric 1158
Colmenero, Manuel 1336
Connaughton, Alexander 1866
Cordier, Julien 3272
Corentin, Darbas 1803
Cornea, Octavian 2192

Costa, François 503
Coujard, Clementine 1270
Cravero, Jean-Marc 3330
Crebier, Jean-Christophe 2010
Da Cunha, Julian 2312
Dabbabi, Asma 2907
Dahmen, Christopher 843
Dai, Jing 820, 1215, 1479
Dakyo, B. 2881
Dang, Ziyue 804
Danzer, Michael A. 707
Darivianakis, Georgios 998
Davari, Pooya 726, 1006, 2573, 3119, 3295
David, Romain 522
Davidson, Colin 2366, 3425
Davoodi, Amirali 2393
De Doncker, Rik W. 153, 163, 1627
De Jaeger, Jean-Claude 1895
De Jódar, Esther 1658
De La Grandiere, Hubert 3424
De Lauretis, Maria 512
De Mora, Pablo Rodriguez 1533
De Oliveira, Eduardo F. 2535
De, Dipankar 2210
De-Preville, Guillaume 9
Defrance, Nicolas 1895
Degrenne, N. 2865
Delamea, R. 2865
Delarue, Philippe 1158, 3330
Delette, G. 1613
Delhommais, Mylène 1737
Delpech, F. 2554
Demidov, Iurii 2419
Denis, Guillaume 1270
Dennetière, S. 2967
Derbey, Alexis 2039
Derkacz, Pawel B. 3054
Despesse, Ghislain 503
Despouys, Olivier 1373
Dessante, Philippe 1858
Devos, Guillaume 1858
Di Gregorio, Francesco 1747
Dieckerhoff, Sibylle 2274, 2341, 2603
Dierks, Rebecca 3091
Dietz, Armin 1316
Dincan, Catalin 2873
Dinkel, Daniel 1297
Dinulovic, Dragan 2773
Djerioui, Ali 3205
Dong, Dong 2366
Doppelbauer, Martin 1589
Douine, Bruno 1373

Doumiati, Moustapha 2251
Drabek, Pavel 2068
Dragicevic, Tomislav 2393, 2898
Driesen, J. 480
Driesen, Johan 27
Drofenik, Uwe 627
Duarte, J. 2812, 2820
Duarte, Jorge L. 1067, 2656
Duarte, Renan R. 3217
Duerbaum, Thomas 203
Dujic, Drazen 1776, 2486
Dürbaum, Thomas 173, 193, 1460
Dworakowski, Piotr 406, 1106, 2000, 2265, 3006
Džonlaga, Bogdan 1479
Ebersberger, Janine 3340
Ebrahimi, Reza 2860
Eckel, Hans-Günter 532, 1666, 2059, 2126, 3282
Eckerle, Richard 765
Ecrabey, Jacques 2467
Egrot, Philippe 1479
Eguia, Pablo 85
Ehlich, Martin 608
El Baghdadi, Mohamed 2320
El Jihad, Hamza 1982
Elizondo, Laura Ramirez 1561
Ellul, Racquel 1025
Elsabrouty, Ibrahim 871
Elsied, Moataz 3156
Elthokaby, Youssuf 352
Endisch, C. 551
Enjeti, Prasad 1390
Erenler, Yeliz 571
Errigo, F. 183
Escobar, Gerardo 1390
Escofficr, Réne 1470
Esfetanaj, Naser Nourani 2860, 3295
Eskandari, Bahman 863
Eslamian, Morteza 3064
Eslampanah, Vahid 2860
Espina, Enrique 1354
Etoz, Burhan 2241
Fadel, Maurice 3101
Fauth, Leon 3081
Fazli, Nastaran 2126
Fehr, Hendrik 1030, 2116
Ferreira, Jan Abraham 892
Ferrieux, Jean-Paul 2039
Finkenzeller, Michael 835
Fischer, Manuel 1168, 1605, 1709
Fogsgaard, Martin Bendix 2258
Foray, Etienne 1874
Forsyth, A. J. 1306

Fort, Jiri	1086
Founier, Etienne	3101
Francois, Bruno	362, 1700, 1756
Frédèric, Poitiers	1803
Frey, D.	406
Frey, David	2695
Freytes, Julián	9
Friebe, Jens	2743, 2763, 3081
Fromme, Christopher	2603
Frost, Damien	223
Fruchier, Olivier	1905
Fu, Siqi	302
Fuchs, Simon	2409, 2513
Fürst, Markus	1885
Galeshi, Soleiman	2695
Gamatié, Abdoulaye	1747
Gandolfi, Chiara	424
Ganjavi, Amir	726, 1006
Gao, Fei	1224, 1233
Gao, Jianbo	1262
Garate, José Ignacio	1524
Garbuio, Lauric	1685
García-Torres, Felix	134
Garnier, Laurent	2467
Gaubert, Jean-Paul	396, 2284
Gauthier, Jean-Yves	590, 736
Gautier, Cyrille	1118, 1858, 2039
Gautier, Maxime	1923
Geng, Zeyang	3198
Gensior, Albrecht	581, 1030, 2116
Gentejohann, Marius	2274
Georges, Didier	820
Gerada, Chris	1944
Gerada, David	1944
Geramirad, Hadiseh	2000
Gerstner, Michael	1316
Geske, Martin	2938, 2957
Geury, Thomas	2320
Geyer, Tobias	2145
Ghamrawi, Ahmad	2284
Ghanes, Malek	3205
Gholami-Khesht, Hosein	3119
Giacomazzo, M.	490
Gierschner, Sidney	2059, 2126
Giewont, William	1016, 1972
Giotakos, Panagiotis I.	2172
Girbau-Llistuella, Francesc	1542
Gireada, Mihaita	2192
Glac, Antonín	3257
Gladen, Marcel	1148
Glasberger, Tomas	3166, 3314
Gleissner, Michael	644, 686
Glushakov, Vasiliy V.	47, 56
Gnärig, Jan Lasse	1030
Golluccio, G.	3210
Golsorkhi, Mohammad S.	2733, 2898
Gomez, Juan S.	1354
Gomis-Bellmunt, Oriol	1542, 2977
Gonzalez, Jose Ortiz	2241
Gonzalez-Torres, J-C.	745
González-Fontderubinat, Paula	1542
Gosses, Kilian	143
Gou, Wanchao	153
Govaerts, G.	480
Gradinger, Thomas B.	627
Grainger, Brandon M.	65
Grbovic, Petar J.	2723
Grecki, Filip	627
Green, Tim C.	276
Green, Tim	2977
Griepentrog, Gerd	1675
Gruson, F.	1579
Gruson, Francois	1158
Gu, Chunyang	1944
Guerrero, Josep. M	3289
Gui, Qiuye	1030
Guichon, Jean Michel	2564
Guillaud, X.	1579
Guillaud, Xavier	1158, 1756, 1952
Guo, Mingzhu	718
Guo, Xuan	317
Gutierrez, A.	919
Gutierrez, Sebastian	598
Gärtner, M.	3403
Götting, Gunther	1289
Hackl, Christoph	900
Hage-Hassan, Maya	1858
Hai, Jie	231
Hallemans, L.	480
Hamid, Muhammad	1289
Hammerer, Horst	2182
Hammes, David	2059
Han, Hua	260, 285, 302, 326
Han, Lubin	370
Han, Weiji	765
Haq, Omer Ikram Ul	3393
Häring, Johannes	644, 686
Harnefors, Lennart	1952
Hartmann, Michael	2258
Harzig, Thibaut	65
Hase, Genki	467, 498
Hasenohr, C.	3403
Hasenohr, Christian	203
Hatori, Kenji	1489

Haug, Martin	2773
He, Maojun	804
He, Yuying	1410, 1435
Hegazy, Omar	2320, 3111
Hein, Lukas	3413
Helle, Lars	2873
Heller, M.	3403
Hénaux, Carole	3101
Henninger, Stefan	3179
Heredero-Peris, Daniel	853, 1542
Herkommer, Christian	1718
Hernandez, Fernando Davalos	598
Herwig, Daniel	1766
Heucke, Sören	2341
Heydari, Rasool	2733, 2898
Hideaki, Yano	1442
Higashihata, Takeshi	1489
Hijazi, A.	183
Hiller, Marc	3366
Hillermeier, Claus	1297
Himker, Niklas	979
Hinkkanen, Marko	2495
Hiraki, Eiji	386
Hirayama, Hiroshi	2376
Hiwatari, Daichi	2376
Hofer, Matthias	2403
Hoffmann, Felix	954
Hofmann, Harald	203
Homann, Michael	307
Hong, Yang	231
Horrein, Ludovic	3330
Horvatic, Iréna	3156
Houari, Azeddine	3205, 3289
Hu, Anliang	2946
Hu, Rui	2594
Huang, Han	1128
Huang, Pin-Yu	2666
Huang, Xingxuan	1972
Huisman, Henk	1067, 1186
Hulea, Dan	2192
Hussain, E. K.	471
Iannuzzo, Francesco	2258, 2704
Ibanez, Federico	598, 3034
Idarreta, Aitor	1205
Idir, Nadir	1793, 1895, 2078
Iman-Eini, Hossein	3305
Ingman, Jonny	563
Inoue, Sadayuki	19
Iraola, Unai	1205
Isaksson, Dan	1016
Ishihara, Hiroki	19
Isobe, Takanori	3358

Itoh, Jun-Ichi	1380, 2200
Jackiewicz, Krzysztof	2029
Jaeger, Johann	3179
Jakob, Roland	2957
Jaritz, Michael	2684
Jasim, Omar	9
Jean-Christophe, Olivier	1803
Jeannin, Pierre-Olivier	3054
Jehle, Andreas	2475
Jelena, Popovic	892
Jeong, Min	2409, 2513
Ji, Shiqi	1972
Jia, Ming	1627
Jiang, Jinhai	637
Jiaqi, Diao	231
Joebges, Philipp	1627
Jonokuchi, Hideki	2376
Jorge, Tenorio	1400
Joryo, Satoshi	3071
Jotwani, Ankit	1138
Joubert, Charles	522
Judge, Paul D.	276
Jun-Ping, He	1599
Junge, Patrick	2613
Juntunen, Raimo	3227
Junyent-Ferré, Adrià	2977
Justin, Elissa Cresenta Anak	2265
Jäppinen, Janne	563
Järvisalo, Heikki	1016
Jørgensen, Asger Bjørn	2987
Kado, Yuichi	2666
Kadri, R.	2554
Kahl, Tino	2603
Kahle, Karsten	1336
Kaiser, Ingmar	1666
Kallfass, Ingmar	1915
Kaminski, Nando	954
Kampen, Dennis	2049, 2153
Kanchan, R. S.	3393
Kang, Yong	797, 804
Karaventzas, Vasilios	2583
Karlsson, Martin	512
Kaszewski, Arkadiusz	2029
Kawabata, Yoshitaka	1177, 1243
Keel, Oliver	2791
Kefer, K.	3403
Kehl, Zdenek	3166
Keller, Christian	2938, 2957
Kennel, Ralph	542, 900, 1262, 2386, 3272
Kersten, Anton	765
Kesbia, Nasreddine	1685
Kestelyn, X.	1579

Ketchedjian, Vasken	1605
Keysan, Ozan	1823
Khanzadeh, Babak	1040
Kharezy, Mohammad	3064
Kikuchi, Naoto	2200
Killeen, Peter	777
Kim, Bunthern	3172, 3376
Kimura, Norihito	105
Kimura, Shota	377
Kindl, Vladimir	1086
Kirchenberger, U.	3403
Kirowitz, Thomas	2403
Kitagawa, Wataru	1425
Kitamura, Taishi	1177
Kiviniemi, Mika	563
Kjaer, Martin Vang	2888
Kjær, Philip	2873
Klass, Stefan	3272
Klier, Samantha	707
Koch, Dominik	880, 1915
Kohlhepp, Benedikt	266, 1460
Kojabadi, Hossein Madadi	2860
Komma, Thomas	835
Komrska, Tomáš	3257
Kondo, Keiichiro	2675
Kone, Lamine	1793
Kopacz, Rafal	2457
Korhonen, Juhamatti	1016
Kosan, Tomas	3166
Kouchaki, Alireza	3015, 3129
Koutroulis, Eftychios	3146
Krall, Felix	1196
Krim, Youssef	1700
Krischan, Klaus	1866
Kroneisl, Michal	1992
Krug, Dietmar	2059
Kucka, Jakub	2020, 3091
Kuder, Manuel	765
Kuebrich, Daniel	203
Kuhlmann, Kai	1718
Kukkola, Jarno	2495
Kumar, Dinesh	726, 1006, 2573
Kuring, Carsten	266
Kusaka, Keisuke	1380, 2200
Kuwana, Kazuki	1425
Kuwata, Akiko	19
Kwasinski, Alexis	2594
Kyyrä, Jorma	3348
Kärkkäinen, Hannu	2829
Kärkkäinen, Tommi J.	563
Kübrich, Daniel	266
Küster, Pierre	571

Labiano, Daniel	1205
Labouré, Eric	1858
Lacarnoy, Alain	654
Lacressonnière, Fabien	332
Ladoux, Philippe	1106, 2265
Lafon, Frederic	1793
Lafoz, Marcos	1076
Lagier, Thomas	1106, 2000, 2265
Lana, Andrey	2419
Langbauer, Thomas	1866
Langmaack, N.	241
Langwasser, M.	3305
Lapassat, N.	37
Larruskain, Marene	85
Lautner, Frank	1831
Lazaroiu, Gheorghe	292
Le Moigne, Philippe	1158, 2078
Le Métayer, Pierre	3006
Le, Hoai Nam	2200
Lee, Seong-Yong	2486
Leedham, Rob	871
Lefebvre, Bruno	1106, 2000, 2366
Lefebvre, S.	1363
Lefebvre, Stéphane	1118
Lehmann, Franziska	944
Lehn, Peter	223
Lembeye, Yves	2010, 2695
Lemmen, Erik	2350, 2358
Leo, Jacopo	75
Leppänen, Joonas	563
Leterme, Willem	276
Letrouvé, Tony	1700
Lexow, Daniel	3282
Li, Boyang	1289
Li, Chi	317
Li, Dingrui	1972
Li, Hui	2704
Li, Jiaqi	9
Li, Jing	1944
Li, Lang	285
Li, Qi	1262
Li, Tao	3044
Li, Weilin	2135
Li, Yongdong	317, 1944
Li, Yu	2386
Li, Zheming	2106
Liang, Chaohui	3044
Liang, Lin	370, 443
Liao, Jianquan	3385
Liao, Yuefeng	1498
Licari, John	1025
Lima, Glauber De Freitas	2010

Lin, Lei 937
Lin-Shi, Xuefang 590, 736
Liserre, Marco 3305
Liu, Cuicui 2505
Liu, Fuxin 1410, 1435
Liu, Libo 1289
Liu, Xudan 804
Liu, Yao 260
Liu, Zhangjie 260, 302, 326
Liukkonen, O. 2630
Llanos, Jacqueline 1354
Llonch-Masachs, Marc 1542
Locment, Fabrice 667, 677
Loisel, Rodica 2907
Loiselay, Florent 1648
Lomonova, E. A. 2812
Lorenz, Malte 2020
Ludois, Daniel C. 777
Lunz, B. 3403
Luo, Fang 370
Luo, Xian 153
Lutz, Josef 3413
Lutze, Marcel 1675
López-Alcolea, Fco. Javier 134
Ma, Yixiao 1435
Mabe, Jon 1524
Machmoum, Mohamed 2907, 3205, 3289
Maekawa, Sari 2838, 2844
Maerz, Martin 1316
Magambo, Jean Sylvio Ngoua Teu 2078
Maharana, Manoj Kumar 2068
Maharana, Suman 2210
Mahr, Florian 3179
Maier, Robert W. 1252, 2106
Mäkelä, Juha 3348
Mallwitz, R. 241, 490
Mallwitz, Regine 213, 1515
Maneiro, Jose 1106, 3006
Mannen, Tomoyuki 3358
Mannerhagen, Felix 3198
Mantellini, Mattia 1076
Mantzanas, Panagiotis 203
Mao, Saijun 892
Marcault, E. 919
Marchesoni, Mario 3321
Marciano, D. 3210
Marciano, Daniele 2802
Margueron, Xavier 2078
Marmolejo, Narciso G. 1589
Marquardt, Rainer 843, 1297
Martin, Christian 1874
Martin, Jérémy 2638

Martinez, Wilmar 598, 3034
Martire, Thierry 1905
Martínez-Gómez, Manuel 1354
März, Martin 143
Mateos, Felix Rodriguez 2583
Mattar, Rita 1118
Mattsson, Aleksi 1016
Maussion, Pascal 3376
Mayorga, John Paul 1658
Mazuela, Mikel 1205
McIntyre, Michael 65
Mehdi, Driss 2284
Meißner, Markus 124, 1569
Mendoza-Araya, Patricio 2917
Meneses, Javiera 2917
Meng, Qingchao 2446
Mercier, Adrien 1858
Mercier, Sylvain 2467
Mertens, Axel 979, 1766, 2020, 2303, 2545, 2613,
....2743, 2763, 3091, 3340
Mesbahi, Tedjani 3205
Meynard, Thierry A. 654
Mezrag, Bachir 2564
Mezzetti, Margarita 944
Micallef, Alexander 1025
Miceli, R. 3237
Milas, Nikolaos T. 2172
Miletic, Zoran 2723
Millinger, Jonas 512
Minami, Masataka 467, 498
Mirtchev, Alex V. 2429
Mitani, Kohei 1425
Miyauchi, Tsutomu 377
Mizutani, Hiroto 386
Mohamed, Abdalla Hussein 115
Mohamed, Islam 352
Molina-Martínez, Emilio J. 134
Mollov, S. 2865
Molnar, Jan 3166
Monmasson, Eric 1118, 1823
Montero, E. Rodriguez 2098
Montesinos-Miracle, Daniel 853, 1542
Moraes, Tiago José Dos Santos 1693
Morel, Cristina 2251
Morel, F. 183, 406
Morel, Florent 2000, 2366
Morel, Hervè 1279, 1648
Mori, Osamu 386
Morici, Riccardo 1076
Morizane, Toshimitsu 1047, 3071
Mortimer, Benedict 163
Motegi, Shin-Ichi 1786

Mourouvin, Rayane 820
Mourtzis, Dimitris A. 2172
Muehlbauer, Markus 707
Muetze, Annette 1196
Müller, Jan-Kaspar 2763, 3340
Mumtaz, Muhammad Adnan 1326
Munk-Nielsen, Stig 2987
Muñoz, Fredy 3189
Muñoz, Javier 3189
Muntean, Nicolae 2192
Murata, Ryo 386
Musznicki, Piotr 3054
Nabatirad, Mohammadreza 787
Nada, Kaho 19
Nadh, Greeshma 1619
Nair, Durga S. 1619
Najera, Jorge 1076
Najjar, Mohammad 3015, 3129
Nakagaki, Akito 498
Nakamura, Keiichi 1489
Nakashima, Osamu 2376
Nakatani, Shota 1047
Nakazawa, Yosuke 2675
Narula, Anant 969, 1952
Natori, Kenji 2675
Navarro, Gustavo 1076
Navas, Alex F. 1354
Ndagijimana, Fabien 2010
Nee, Hans-Peter 2220
Neumann, Jessica 3101
Nevaranta, N. 2630
Ngoua-Teu, J-S 1613
Nguyen, Ngac Ky 1693
Nguyen, Van Sang 2638
Nicolas, Ginot 1803
Nie, Cheng 1972
Nie, Qingqing 797
Niemelä, M. 2630
Niemelä, Markku 563, 2829
Nikowitz, Mario 2403
Nisch, A. 3403
Nishida, Yasuyuki 1786
Nitzsche, Maximilian 880, 1605, 1709, 1915
Niu, Liyong 342
Norambuena, Margarita 2851
Nymand, Morten 3015, 3129
Obernolte, Urs 608
Oberschelp, Wolfgang 1727
Odriozola, Kepa 654
Oguma, Kenji 377
Ohta, Takahiro 2666
Okamori, Daichi 1047

Okazaki, Akihiro 2838
Okazaki, Yuhei 1040
Oliveira, Joao 1648
Olivier, Jean-Christophe 2251
Omori, Hideki 1047
Omrane, A. 2554
Ordoño, Ander 1524
Orfanoudakis, Georgios I. 3146
Ortega-Perez, Carmen 3330
Ota, Kenji 1489
Ottaviano, Andrea 292
Oumaziz, Amirouche 1096
Pace, Loris 1895
Páez, Juan 1106
Paez, J. D. 406
Palensky, Peter 810, 1561
Pallier, Joris 829
Palm, Herbert 707
Pan, Xuejiao 1128
Passalacqua, Massimiliano 3321
Patarroyo-Montenegro, Juan F. 1841
Patti, Davide 3210
Paulus, Sebastian 2303, 2545
Pavlicek, Vladimir 1086
Pawellek, A. 3403
Pawellek, Alexander 203
Payman, A. 2881
Pei, Xiaoze 342
Peller, Stefan 266
Pelosi, Dario 292
Peltoniemi, Pasi 3227
Peña, R. A. 183
Pendharkar, Ishan 1346
Peng, Han 797, 804
Peng, Hao 804
Peng, Tao 1498
Penin, Carolina 1905
Peralta, Patricio 75
Perenyi, Christian 65
Peretti, Luca 3393
Pérez-Molina, María José 85
Peric, V. 745
Peroutka, Zdenek 3257, 3314
Perriard, Yves 75
Petit, M. 1363
Petit, Mickael 1118, 3054
Petkovic, Marko 1776
Peyghami, Saeed 810, 2573
Pfeifer, Markus 1675
Pfeiffer, Jonas 571, 2535
Phulpin, Tanguy 618
Pidancier, Thomas 3247

Pietrzak-David, Maria..3376
Pilawa-Podgurski, Robert C. N...1934
Pinheiro, Humberto...2851
Pinomaa, Antti...2419
Pinto, Rafael A...3217
Pirsto, Ville...2495
Pitel, Ira...1390
Planson, Dominique...1279, 1648
Plaza, Jesus D. Vasquez...1841
Plissonnicr, Marc...1470
Poebl, Monika...835
Polacek, Libor...1086
Pollet, Benjamin...503
Pommier-Petit, Pascal...829
Pool-Mazun, Erick I...1390
Popuri, Madhuchandra..2068
Pouresmaeil, Edris..863
Pöyhönen, Santeri...2753
Prevost, Thibault..1270
Prieto, Dany..3101
Prieto-Araujo, Eduardo...2977
Pronin, Mikhail V...47, 56
Puls, Simon..608
Pulvirenti, M..909, 2293
Pursiainen, Jooa...3227
Putkonen, A..2630
Pyrhönen, Juha..2829
Pyrhönen, O..2630
Pyrhönen, Olli...2419
Qashlan, Ziyad H. S...571
Qiang, Jin...163
Qoria, T..1579
Qoria, Taoufik...1756
Queval, Loic..1473, 1479
Rabba, Heiko..307
Rabkowski, Jacek..2457, 2996
Radet, Hugo...332
Rahmoun, Yasser..2182
Rahul, Arun S..1619
Rajput, Shailendra..2439
Ramm, Hannes...307
Ramírez-Scarpetta, J. M..1850
Rasmussen, Tonny Wederberg...1138
Rathnayake, Hansika...726, 1006
Rault, P..2967
Raute, R...989, 1506
Rautio, Juuso...563
Ravyts, S...480
Ravyts, Simon..27
Rayati, Mohammad...3247
Razzaghi, Reza..787
Rehlaender, Philipp...1638

Reigosa, Paula Diaz...1346, 2258
Reißenweber, Lukas...3263
Rekola, Jenni..3227
Ren, Chunpin..718
Restrepo, Jose Alex...1850
Retianza, Darian V..1067
Retianza, Darian Verdy..1186
Rezaee, Ali Yahya..460
Richardeau, Frédéric..1057, 1096
Rietmann, Stefan...2712
Rigot, Valentin...618
Risch, Raffael...2230
Riu, Delphine..3156
Roa, Claudio...3189
Robet, Pierre-Philippe..1923
Roboam, Xavier..332
Robyns, Benoit...1700
Rodriguez, Jose...900, 2851
Rodriguez, José Luis..1205
Rokrok, Ebrahim...1756
Roncero-Sánchez, Pedro...134
Roose, T...1551
Röser, Tobias..3137
Roth-Stielow, Jörg.....................................880, 1168, 1605, 1709
Rouger, Nicolas..1057
Routimo, Mikko...2495
Rouzbehi, Kumars..863
Ru, Yang..231
Ruan, Xinbo...1410, 1435
Rubenbauer, Hubert..3179
Rufer, Alfred...696
Rute, Erwin..1354
Ruthardt, Johannes.....................................1168, 1605, 1709
Saad, H..2967
Saad, Yamen..3295
Saber, Christelle..1118
Sácz, Doris..1354
Sadarnac, Daniel...618
Saeedian, Meysam..863
Saggio, M..2293
Sah, Gyanendra Kumar...532
Sahin, Ilker...1823
Saim, Abdelhakim...3289
Sakai, Kazuto..1442
Sakai, Norikazu..1489
Sakai, Yasuhiro..1489
Sakaria, Omar Ahmed...3295
Sakly, Jihen..618
Sakurazawa, Yoshiki..2675
Saleh, Bassem..2648
Salem, Qusay...1289
Sallot, P..1613

Salvo, L.	909
Sánchez-Sánchez, Enric	2977
Sandelic, Monika	2088
Sandik, Diane -Perle	512
Sangwongwanich, Ariya	1, 2088
Sanseverino, A.	3210
Sanseverino, Annunziata	2802
Santos, Miguel	1076
Sari, A.	183
Sarraute, Emmanuel	1096
Sassatelli, Gilles	1747
Sathik, Mohd. Ali Jagabar	460
Saudemont, Christophe	1700
Savaghebi, Mehdi	2733, 2898
Savarit, Elise	1982
Sawada, Takashi	2623
Sayed, Mahmoud A.	1417
Scarpetta, Jose Miguel Ramirez	1400
Scelba, G.	909
Schafmeister, Frank	1638, 3024
Schanen, Jean-Luc	1685, 2039, 3054
Schiesser, Matthias	962
Schleippmann, Nico	143
Schlesinger, Richard	2524
Schlüter, Michael	2274
Schmidt, Dimitri	1168
Schmitt, Alexander	2182
Schmitt, N.	1363
Schmitt, Stefan	954
Schmitz, Jan	124, 1569
Schobre, Thorben	213, 1515
Schröder, Günter	1727
Schrödl, Manfred	2403
Schulte, Hendrik	1709
Schulz, Matthias	143
Schulz, Nicola	1346
Schulze, Torben A.	307
Schwabe, Christian	3413
Schütt, Michael	532
Sciacca, A. G.	909
Scicluna, K.	989, 1506
Sechilariu, Manuela	667, 677
Segur, R.	745
Seiler, Pascal	2331
Semail, Eric	1693
Sergeant, Peter	27, 115
Shah, Chirag	153
Sharkh, S. M.	471
Sharkh, Suleiman M.	3146
Shi, Guangze	326
Shinoda, Kosei	820, 1215
Shiozaki, Koji	2623

Shirakawa, Tomohide	386
Shousha, Mahmoud	2773
Si-Yuan, Cai	1599
Siala, S.	95
Siala, Sami	1982
Siebke, K.	490
Siemens, Ag	2603
Silventoinen, Pertti	563, 1016
Simola, Aleksi	2753
Singer, Arthur	765
Skala, Bohumil	1086, 3257
Smailus, Erik	1675
Šmídl, Václav	1992
Smit, A.	3403
Snook, Mark	871
Soeiro, Thiago Batista	1224, 1233, 1561
Sokur, Pavel V.	47
Soltau, Nils	1489
Song, Kai	637
Soumaoro, Ousmane	2251
Soupremanien, U.	1613
Sprunck, Sebastian	2535
Stadler, Alexander	3263
Staines, C. Spiteri	989, 1506
Stathis, Spyridon	2684
Staudt, Volker	1148
Stecca, Marco	1561
Steckler, Pierre-Baptiste	736
Stengl, Josef	1086
Stenglein, Erika	173, 193
Štepánek, Jan	3257
Stock, Alexander	1718
Stöckl, Johannes	2723
Stöttner, J.	551
Stotckaia, Anastasiia D.	47, 56
Stras, Andrzej	2029
Streit, Lubeš	3257
Strittmatter, Tobias	1346
Strunk, Robin	979
Su, Guoxing	1410
Su, Mei	260, 285, 302, 326, 1498
Suarez, Camilo	3034
Sugiyama, Kohei	1177
Sun, Jian	1962
Sun, Yao	260, 285, 302, 326, 1498
Svensson, Jan R.	1952
Tadano, Hiroshi	2623
Taheri, Shamsodin	863
Takahara, Takaaki	386
Takahashi, Hirotaka	377
Takahashi, Hiroyuki	1243
Takano, Tomihiro	19

Takeshita, Takaharu	1417, 1425, 1450
Takeuchi, Somi	1243
Talbert, Thierry	1905
Talla, Jakub	3257
Tan, Guoqiang	443
Tanaka, Ami	2844
Tanaka, Miwako	19
Tanaka, Nobuhiko	1489
Tang, Bojin	718
Tang, Houjun	1224, 1233
Tang, Xiaohu	1589
Tannhäuser, Marvin	2603
Tant, Jeroen	27
Tareilus, G.	241
Tarisciotti, Luca	1224, 1233
Tárraga, Sergio	1658
Tatakis, Emmanuel C.	755, 2172, 2429
Taul, Mads Graungaard	927
Tedesco, Davide	2802
Teigelkötter, Johannes	1718
Teirelbar, Ahmed	2648
Tenorio, Jorge	1850
Teramura, Keiko	377
Terbrack, C.	551
Thal, Eckhard	1489
Thiringer, Torbjörn	765, 1040, 3064, 3198
Tibola, Gabriel	2656
Tikhonov, Sergey	1638
Todd, R.	1306
Tolbert, Leon M.	1972
Torres, Alfonso Parreño	134
Torres, Esther	85
Torres, Fernando	3189
Torres, Jorge	1076
Torres, Jose Rueda	810
Touhami, Mustapha	503
Tran, Dai-Duong	2320, 3111
Traoré, Bakou	2251
Tremmel, Werner	2723
Tremouilles, D.	919
Trillaud, Frédéric	1373
Trochimiuk, Przemyslaw	2457
Tröster, Nathan	1168
Tsolaridis, Georgios	2331
Tsoumas, Ioannis	998, 2145, 2163
Turjanica, Pavel	1086
Turki, Faical	307
Twardon, N.	3403
Ufnalski, Bartlomiej	2029
Ulissi, Gabriele	2486
Umetani, Kazuhiro	386
Unruh, Roland	3024

Uwai, Shuto	1177
Vaccaro, Luis	3321
Valtee, Mikko	3227
Van Den Broeck, G.	480, 1551
Van Duivenbode, Jeroen	1186
Van Mierlo, Joeri	2320, 3111
Van Tichelen, P.	480
Vanfretti, L.	745
Vannier, Jean-Claude	1479
Vansompel, Hendrik	115
Vashishtha, Anushruti	1138
Vasquez-Plaza, Jesus D.	1850
Vázquez, Javier	134
Vecchia, Mauricio Dalla	27
Vechiu, Ionel	396
Velardi, F.	3210
Velardi, Francesco	2802
Velazco, Diego	433
Venet, P.	183
Verbelen, Florian	27
Vermeersch, Pierre	1158
Vermulst, Bas	2350, 2358
Viana, Caniggia	223
Videt, Arnaud	1793, 1895
Vienot, Stephane	1793
Vieto, Ignacio	1962
Villarejo, José	1658
Villegas, Carlos	962
Vip, Stephan	2303, 2545
Vitan, Danut	2192
Voborník, Ales	1086
Vogelsberger, M.	2098
Voigt, Matthias	944
Voldoire, Adrien	2039
Vollaire, Christian	2000
Vollmaier, Franz	1866
Vorontsov, Aleksey G.	47, 56
Votava, Martin	3314
Vu, Duc Tan	1693
Wada, Keiji	3358
Wallart, François	433, 736, 1106
Wang, Bo	450
Wang, Dian	677
Wang, Feng	2505
Wang, Fred	1972
Wang, Huai	1, 2888, 3295
Wang, Meiqi	1944
Wang, Qianggang	3385
Wang, Qiwu	1262
Wang, Tianqing	450
Wang, Xuehua	1410, 1435
Wang, Ziyue	443

Wankhede, Yugandhara H.	3081
Wasfi, Amr	2648
Watanabe, Hiroki	1380
Weicker, Martin	2049, 2153
Weimer, Julian	880, 1915
Weinert, Tristan	1727
Weiss, Sébastien	1793
Wernicke, Laurenz	2603
Weyh, Thomas	765
Wickramasinghc, Thilini	1470
Wijnands, Korneel	2350, 2358
Wilk, Andrzej	2265
Will, Frank	944
Winzer, Patrick	2182
Wolbank, T.	2098
Wondrak, W	3403
Wondrak, Wolfgang	644, 686
Wrona, Grzegorz	2457, 2996
Wu, Hailong	1693
Wu, Xiaohua	2135
Wunder, Bernd	143
Xi, Jiawen	342
Xia, Qingping	260
Xiang, Yusheng	1289
Xie, Jian	1289
Xin, Wei	370
Xiong, Weijing	1498
Xu, Chaoqun	718
Xu, Chen	937
Xu, Dianguo	450
Xu, Guo	1498
Xu, Junzhong	1224, 1233
Xu, Lie	1944
Xu, Xinwei	2656
Yakop, Netan	223
Yamada, Shota	1243
Yamashita, Daniela Yassuda	396
Yamazaki, Osamu	2675
Yamdeu, Mathias Tientcheu	3101
Yan, Xiaoxue	443
Yan, Zheng	1326
Yang, Bo	1944
Yang, Guang	637
Yang, Y	3237
Yang, Yongheng	2135, 2393, 2888
Yang, Zhiqing	153, 163
Yano, Takahiro	1177
Yao, Ran	2704
Yao, Wenli	2135
Ye, Shuaichen	2928
Ye, Zichao	1934
Yin, Tianxiang	937

Yu, Qihao	2350, 2358
Yu, Yong	450
Yu, Zhanqing	718
Yuasa, Hiroaki	105
Yüce, Firat	3366
Yuki, Kazuaki	2675
Yuratich, Michael A.	3146
Zacharias, Peter	571, 1813, 2535
Zafar, Talha	2773
Zanetti, E.	2293
Zaoskoufis, Konstantinos	755
Zare, Firuz	726, 1006
Zdanowski, Mariusz	2996
Zehelein, Matthias	880, 1915
Zeller, Valentin	1460
Zeng, Guang	3413
Zeng, Xianwu	342
Zhai, Dongling	718
Zhang, Haibo	1158, 1756
Zhang, He	1944
Zhang, Li	1128
Zhang, Peng	637
Zhang, Xiaokang	590
Zhang, Zhenbin	2386
Zhang, Ziqian	2505
Zhao, Biao	718
Zheng, Zedong	317
Zhou, Dao	2860, 2928
Zhou, Niancheng	3385
Zhu, Chunbo	637
Zhu, Q.	1306
Zhu, Yangming	450
Zhu, Yuanhao	326
Zhuo, Fang	2505
Zi-Fan, Li	1599
Ziegler, Philipp	1168, 1605, 1709
Zinchenko, Denys	2996
Zucuni, Jordan P.	2851
Zuolian, Liu	231

IEEE
445 Hoes Lane
Piscataway, NJ 08854-4141

ISBN 978-1-7281-9807-1